Elementary Algebra

Concepts and Applications

Tenth Edition

Marvin L. Bittinger
Indiana University Purdue University Indianapolis

David J. Ellenbogen
Community College of Vermont

Barbara L. Johnson
Ivy Tech Community College of Indiana

Director, Courseware Portfolio Management:	Michael Hirsch
Courseware Portfolio Manager:	Cathy Cantin
Courseware Portfolio Management Assistant:	Alison Oehman
Content Producer:	Ron Hampton
Managing Producer:	Karen Wernholm
Media Producer:	Jon Wooding
Manager, Courseware QA:	Mary Durnwald
Manager, Content Development:	Rebecca Williams
Marketing Manager:	Kyle DiGiannantonio
Field Marketing Managers:	Jennifer Crum; Lauren Schur
Marketing Assistant:	Fiona Murray
Senior Author Support/ Technology Specialist:	Joe Vetere
Manager, Rights and Permissions:	Gina Cheselka
Manufacturing Buyer:	Carol Melville, LSC Communications
Associate Director of Design:	Blair Brown
Program Design Lead:	Barbara T. Atkinson
Text Design:	Geri Davis/The Davis Group, Inc.
Editorial and Production Services:	Martha Morong/Quadrata, Inc.
Composition:	Cenveo
Illustrations:	Network Graphics
Cover Design:	Cenveo
Cover Image:	© Getty/Alex Bramwell

Library of Congress Cataloging-in-Publication Data

Bittinger, Marvin L. | Ellenbogen, David. | Johnson, Barbara L.
Elementary algebra : concepts & applications / Marvin
   L. Bittinger, Indiana University Purdue University Indianapolis, David J.
   Ellenbogen, Community College of Vermont, Barbara L. Johnson,
   Ivy Tech Community College of Indiana
Elementary algebra
10th edition. | Boston : Pearson, c2018.
LCCN 2016020676 | ISBN 9780134441375 (hardcover: student edition) | ISBN 0134441370
   (hardcover: student edition) | ISBN 9780134441634 (hardcover: AIE) |
   ISBN 013444163X (hardcover: AIE)
LCSH: Algebra—Textbooks.
LCC QA152.3 .B53 2018 | DDC 512.9—dc23
LC record available at https://lccn.loc.gov/2016020678

2   17

ISBN 13: 978-0-13-444137-5
ISBN 10: 0-13-444137-0

Contents

Preface   vii

CHAPTER 1
Introduction to Algebraic Expressions   1

1.1 **Introduction to Algebra   2**
Evaluating Algebraic Expressions • Translating to
Algebraic Expressions • Translating to Equations

1.2 **The Commutative, Associative,
and Distributive Laws   12**
The Commutative Laws • The Associative Laws •
The Distributive Law • The Distributive Law
and Factoring

1.3 **Fraction Notation   19**
Factors and Prime Factorizations • Multiplication,
Division, and Simplification of Fractions •
Addition and Subtraction of Fractions

1.4 **Positive and Negative Real Numbers   28**
Whole Numbers and Integers • Rational
Numbers • Real Numbers and Order •
Absolute Value

MID-CHAPTER REVIEW   37

1.5 **Addition of Real Numbers   38**
Adding with the Number Line • Adding Without
the Number Line • Problem Solving •
Combining Like Terms

1.6 **Subtraction of Real Numbers   44**
Opposites and Additive Inverses •
Subtraction • Problem Solving

1.7 **Multiplication and Division
of Real Numbers   51**
Multiplication • Division

CONNECTING THE CONCEPTS   57

1.8 **Exponential Notation and
Order of Operations   60**
Exponential Notation • Order of Operations •
Simplifying and the Distributive Law •
The Opposite of a Sum

TRANSLATING FOR SUCCESS   69

STUDY SUMMARY   72
REVIEW EXERCISES   75
TEST   78

CHAPTER 2
Equations, Inequalities, and
Problem Solving   79

2.1 **Solving Equations   80**
Equations and Solutions • The Addition
Principle • The Multiplication Principle •
Selecting the Correct Approach

2.2 **Using the Principles Together   88**
Applying Both Principles • Combining Like
Terms • Clearing Fractions and Decimals

2.3 **Formulas   96**
Evaluating Formulas • Solving for a Variable

MID-CHAPTER REVIEW   103

2.4 **Applications with Percent   104**
Converting Between Percent Notation and Decimal
Notation • Solving Percent Problems

2.5 **Problem Solving   112**
The Five Steps for Problem Solving • Percent
Increase and Percent Decrease

2.6 **Solving Inequalities   127**
Solutions of Inequalities • Graphs of
Inequalities • Set-Builder Notation and Interval
Notation • Solving Inequalities Using the
Addition Principle • Solving Inequalities Using
the Multiplication Principle • Using the Principles
Together

CONNECTING THE CONCEPTS   134

2.7 **Solving Applications with
Inequalities   137**
Translating to Inequalities • Solving Problems

TRANSLATING FOR SUCCESS   144

STUDY SUMMARY   146
REVIEW EXERCISES   149
TEST   151

CUMULATIVE REVIEW: CHAPTERS 1–2   152

CHAPTER 3
Introduction to Graphing   153

3.1 **Reading Graphs, Plotting Points,
and Scaling Graphs   154**
Problem Solving with Bar Graphs and Line
Graphs • Points and Ordered Pairs • Numbering
the Axes Appropriately

iii

3.2 Graphing Linear Equations   163
Solutions of Equations • Graphing Linear
Equations • Applications

3.3 Graphing and Intercepts   173
Intercepts • Using Intercepts to Graph •
Graphing Horizontal Lines or Vertical Lines

3.4 Rates   181
Rates of Change • Visualizing Rates

3.5 Slope   190
Rate and Slope • Horizontal Lines and
Vertical Lines • Applications

MID-CHAPTER REVIEW   202

3.6 Slope–Intercept Form   203
Using the y-intercept and the Slope to Graph
a Line • Equations in Slope–Intercept Form •
Graphing and Slope–Intercept Form •
Slope and Parallel Lines

3.7 Point–Slope Form   211
Writing Equations Using Point–Slope Form •
Graphing and Point–Slope Form • Estimations and
Predictions Using Two Points

CONNECTING THE CONCEPTS   216

VISUALIZING FOR SUCCESS   221

STUDY SUMMARY   223
REVIEW EXERCISES   225
TEST   227

CUMULATIVE REVIEW: CHAPTERS 1–3   228

CHAPTER 4
Polynomials   229

4.1 Exponents and Their Properties   230
Multiplying Powers with Like Bases • Dividing
Powers with Like Bases • Zero as an Exponent •
Raising a Power to a Power • Raising a Product
or a Quotient to a Power

**4.2 Negative Exponents and
Scientific Notation   237**
Negative Integers as Exponents • Scientific
Notation • Multiplying and Dividing Using
Scientific Notation

CONNECTING THE CONCEPTS   243

4.3 Polynomials   246
Terms • Types of Polynomials • Degree and
Coefficients • Combining Like Terms •
Evaluating Polynomials and Applications

**4.4 Addition and Subtraction of
Polynomials   254**
Addition of Polynomials • Opposites of
Polynomials • Subtraction of Polynomials •
Problem Solving

MID-CHAPTER REVIEW   263

4.5 Multiplication of Polynomials   264
Multiplying Monomials • Multiplying a Monomial
and a Polynomial • Multiplying Any Two
Polynomials

4.6 Special Products   271
Products of Two Binomials • Multiplying Sums
and Differences of Two Terms • Squaring
Binomials • Multiplications of Various Types

4.7 Polynomials in Several Variables   279
Evaluating Polynomials • Like Terms and
Degree • Addition and Subtraction •
Multiplication

4.8 Division of Polynomials   287
Dividing by a Monomial • Dividing by a
Binomial

VISUALIZING FOR SUCCESS   293

STUDY SUMMARY   295
REVIEW EXERCISES   298
TEST   300

CUMULATIVE REVIEW: CHAPTERS 1–4   301

CHAPTER 5
Polynomials and Factoring   303

5.1 Introduction to Factoring   304
Factoring Monomials • Factoring When Terms Have
a Common Factor • Factoring by Grouping

**5.2 Factoring Trinomials of the Type
$x^2 + bx + c$ 311**
When the Constant Term Is Positive • When
the Constant Term Is Negative • Prime
Polynomials • Factoring Completely

**5.3 Factoring Trinomials of the Type
$ax^2 + bx + c$ 319**
Factoring with FOIL • The Grouping Method

**5.4 Factoring Perfect-Square Trinomials and
Differences of Squares   328**
Recognizing Perfect-Square Trinomials • Factoring
Perfect-Square Trinomials • Recognizing
Differences of Squares • Factoring Differences of
Squares • Factoring Completely

MID-CHAPTER REVIEW   335

5.5 Factoring: A General Strategy   336
Choosing the Right Method

**5.6 Solving Quadratic Equations
by Factoring   341**
The Principle of Zero Products • Factoring
to Solve Equations

CONNECTING THE CONCEPTS   346

5.7 Solving Applications   350
Applications   •   The Pythagorean Theorem

TRANSLATING FOR SUCCESS   360

STUDY SUMMARY   363
REVIEW EXERCISES   365
TEST   367

CUMULATIVE REVIEW: CHAPTERS 1–5   368

CHAPTER 6
Rational Expressions and Equations   369

6.1 Rational Expressions   370
Restricting Replacement Values   •   Simplifying
Rational Expressions   •   Factors That Are Opposites

6.2 Multiplication and Division   377
Multiplication   •   Division

**6.3 Addition, Subtraction, and
Least Common Denominators   383**
Addition When Denominators Are the Same   •
Subtraction When Denominators Are the Same   •
Least Common Multiples and Denominators

**6.4 Addition and Subtraction with
Unlike Denominators   391**
Adding and Subtracting with LCDs   •
When Factors Are Opposites

MID-CHAPTER REVIEW   399

6.5 Complex Rational Expressions   400
Using Division to Simplify   •   Multiplying
by the LCD

6.6 Rational Equations   407
Solving Rational Equations

CONNECTING THE CONCEPTS   411

**6.7 Applications Using Rational Equations
and Proportions   414**
Problems Involving Work   •   Problems Involving
Motion   •   Problems Involving Proportions

TRANSLATING FOR SUCCESS   426

STUDY SUMMARY   428
REVIEW EXERCISES   431
TEST   433

CUMULATIVE REVIEW: CHAPTERS 1–6   434

CHAPTER 7
Systems and More Graphing   435

7.1 Systems of Equations and Graphing   436
Solutions of Systems   •   Solving Systems of
Equations by Graphing

**7.2 Systems of Equations and
Substitution   442**
The Substitution Method   •   Problem Solving

**7.3 Systems of Equations and
Elimination   449**
The Elimination Method   •   Problem Solving

CONNECTING THE CONCEPTS   454

7.4 More Applications Using Systems   457
Total-Value Problems   •   Mixture Problems

MID-CHAPTER REVIEW   466

7.5 Linear Inequalities in Two Variables   467
Graphing Linear Inequalities   •   Linear Inequalities
in One Variable

7.6 Systems of Linear Inequalities   472
Graphing Systems of Inequalities

**7.7 Direct Variation and Inverse
Variation   476**
Equations of Direct Variation   •   Problem Solving
with Direct Variation   •   Equations of Inverse
Variation   •   Problem Solving with Inverse Variation

VISUALIZING FOR SUCCESS   483

STUDY SUMMARY   485
REVIEW EXERCISES   487
TEST   489

CUMULATIVE REVIEW: CHAPTERS 1–7   490

CHAPTER 8
Radical Expressions and Equations   491

**8.1 Introduction to Square Roots and
Radical Expressions   492**
Square Roots   •   Radicands and Radical
Expressions   •   Irrational Numbers   •   Square
Roots and Absolute Value   •   Problem Solving

**8.2 Multiplying and Simplifying Radical
Expressions   498**
Multiplying   •   Simplifying and Factoring   •
Simplifying Square Roots of Powers   •
Multiplying and Simplifying

8.3 Quotients Involving Square Roots   504
Dividing Radical Expressions   •
Rationalizing Denominators with One Term

**8.4 Radical Expressions with Several
Terms   509**
Adding and Subtracting Radical Expressions   •
More with Multiplication   •   Rationalizing
Denominators with Two Terms

MID-CHAPTER REVIEW   514

8.5 Radical Equations   515
Solving Radical Equations   •   Problem Solving
and Applications

CONNECTING THE CONCEPTS   518

8.6 Applications Using Right Triangles 521
Right Triangles • Problem Solving •
The Distance Formula

8.7 Higher Roots and Rational Exponents 530
Higher Roots • Products and Quotients
Involving Higher Roots • Rational Exponents

TRANSLATING FOR SUCCESS 535

STUDY SUMMARY 537
REVIEW EXERCISES 539
TEST 541

CUMULATIVE REVIEW: CHAPTERS 1–8 542

CHAPTER 9
Quadratic Equations 543

9.1 Solving Quadratic Equations: The Principle of Square Roots 544
The Principle of Square Roots • Solving Quadratic
Equations of the Type $(x + k)^2 = p$

9.2 Solving Quadratic Equations: Completing the Square 548
Completing the Square • Solving by Completing
the Square

9.3 The Quadratic Formula and Applications 553
The Quadratic Formula • Problem Solving

MID-CHAPTER REVIEW 562

9.4 Formulas 563
Solving Formulas

9.5 Complex Numbers as Solutions of Quadratic Equations 567
The Complex-Number System • Solutions
of Equations

CONNECTING THE CONCEPTS 569

9.6 Graphs of Quadratic Equations 571
Graphing Equations of the Form $y = ax^2$ •
Graphing Equations of the Form $y = ax^2 + bx + c$

9.7 Functions 578
Identifying Functions • Function Notation •
Graphs of Functions • Recognizing Graphs
of Functions

VISUALIZING FOR SUCCESS 587

STUDY SUMMARY 589
REVIEW EXERCISES 591
TEST 593

CUMULATIVE REVIEW/FINAL EXAM: CHAPTERS 1–9 594

Appendixes 597

A. Factoring Sums or Differences of Cubes 597
Factoring a Sum or a Difference of Cubes

B. Mean, Median, and Mode 599
Mean • Median • Mode

C. Sets 602
Naming Sets • Membership • Subsets •
Intersections • Unions

Tables 605
Fraction and Decimal Equivalents 605
Squares and Square Roots with Approximations
to Three Decimal Places 605

Photo Credits 606
Answers A-1
Glossary G-1
Index I-1
Index of Applications I-9

A KEY TO THE ICONS IN THE EXERCISE SETS

⤹ Concept reinforcement exercises, indicated by blue exercise numbers, provide basic practice with the new concepts and vocabulary.

Aha! Exercises labeled Aha! indicate the first time that a new insight can greatly simplify a problem and help students be alert to using that insight on following exercises. They are not more difficult.

⌨ Calculator exercises are designed to be worked using either a scientific calculator or a graphing calculator.

📉 Graphing calculator exercises are designed to be worked using a graphing calculator and often provide practice for concepts discussed in the Technology Connections.

📝 Writing exercises are designed to be answered using one or more complete sentences.

✓ A check mark in the annotated instructor's edition indicates Synthesis exercises that the authors consider particularly beneficial for students.

Preface

Welcome to the tenth edition of *Elementary Algebra: Concepts and Applications*, one of three programs in an algebra series that also includes *Elementary and Intermediate Algebra: Concepts and Applications,* Seventh Edition, and *Intermediate Algebra: Concepts and Applications,* Tenth Edition. As always, our goal is to present the content of the course clearly yet with enough depth to allow success in future courses. You will recognize many proven features, applications, and explanations; you will also find new material developed as a result of our experience in the classroom as well as from insights from faculty and students.

Understanding and Applying Concepts

Our goal is to help today's students learn and retain mathematical concepts. To achieve this, we feel that we must prepare students in developmental mathematics for the transition from "skills-oriented" elementary algebra courses to more "concept-oriented" college-level mathematics courses. This requires the development of critical-thinking skills: to reason mathematically, to communicate mathematically, and to identify and solve mathematical problems.

Following are aspects of our approach that we use to help meet the challenges we all face when teaching developmental mathematics.

Problem Solving We use problem solving and applications to motivate the students wherever possible, and we include real-life applications and problem-solving techniques throughout the text. Problem solving encourages students to think about how mathematics can be used, and it helps to prepare them for more advanced material in future courses.

In Chapter 2, we introduce our five-step process for solving problems: (1) Familiarize, (2) Translate, (3) Carry out, (4) Check, and (5) State the answer. Repeated use of this problem-solving strategy throughout the text provides students with a starting point for any type of problem they encounter, and frees them to focus on the unique aspects of the particular problem. We often use estimation and carefully checked guesses to help with the *Familiarize* and *Check* steps (see pp. 113 and 414).

Applications Interesting, contemporary applications of mathematics, many of which make use of real data, help motivate students and instructors. In this new edition, we have updated real-world data examples and exercises to include subjects such as accessibility design (p. 200), social cost of carbon (p. 218), and video viewing (p. 441). For a complete list of applications and the page numbers on which they can be found, please refer to the Index of Applications at the back of the book.

Conceptual Understanding Growth in mathematical ability includes not only mastering skills and procedures but also deepening understanding of mathematical concepts. We are careful to explain the reasoning and the principles behind procedures and to use accurate mathematical terminology in our discussion. In addition, we provide a variety of opportunities for students to develop their understanding of mathematical concepts, including making connections between concepts, learning through active exploration, applying and extending concepts, using new vocabulary, communicating comprehension through writing, and employing research skills to extend their examination of a topic.

Guided Learning Path

To enhance the learning process and improve learner outcomes, our program provides a broad range of support for students and instructors. Each person can personalize his or her learning or teaching experience by accessing help when he or she needs it.

PREPARE: Studying the Concepts

Students can learn about each math concept by reading the textbook or eText, watching the To-the-Point Objective videos, participating in class, working in the *MyMathGuide* workbook—or using whatever combination of these course resources works best for him or her.

Enhanced! **Text** The exposition, examples, and exercises have been carefully reviewed and, as appropriate, revised or replaced. New features (see below) include more systematic review and preparation for practice, as well as stronger focus on the real-world applications for the math.

Enhanced! **MyMathLab** has been greatly expanded for this course, including adding more ways for students to personalize their learning path so they can effectively study, master, and retain the math. (See p. xii for more details.)

To-the-Point Objective Videos is a comprehensive program of objective-based, interactive videos that can be used hand-in-hand with the *MyMathGuide* workbook. Video support for Interactive Your Turn exercises in the videos prompts students to solve problems and receive instant feedback.

MyMathGuide: Notes, Practice, and Video Path is an objective-based workbook (available in print and in MyMathLab) for guided, hands-on learning. It offers vocabulary, skill, and concept review; and problem-solving practice with space for students to fill in the answers and stepped-out solutions to problems, show their work, and write notes. Students can use *MyMathGuide*—while watching the videos, listening to the instructor's lecture, or reading the textbook or eText—to reinforce and self-assess their learning.

PARTICIPATE: Making Connections through Active Exploration

Knowing that developing a solid grasp of the big picture is a key to student success, we offer many opportunities for active learning to help students practice, review, and confirm their understanding of key concepts and skills.

New! **Chapter Opener Applications with Infographics** use current data and applications to present the math in context. Each application is related to exercises in the text to help students model, visualize, learn, *and* retain the math. We also added many new spotlights on real people sharing how they use math in their careers.

Algebraic–Graphical Connections, which appear occasionally throughout the text, draw explicit connections between the algebra and the corresponding graphical visualizations. (See p. 439.)

Exploring the Concept, appearing once in nearly every chapter, encourages students to think about or visualize a key mathematical concept. (See pp. 166 and 525.) These activities lead into the **Active Learning Figure** interactive animations available in MyMathLab. Students can manipulate Active Learning Figures through guided and open-ended exploration to further solidify their understanding of these concepts.

Connecting the Concept summarizes concepts from several sections or chapters and illustrates connections between them. Appearing at least once in every chapter, this feature includes a set of mixed exercises to help students make these connections. (See pp. 243 and 518.)

Technology Connection is an optional feature in each chapter that helps students use a graphing calculator or a graphing calculator app to visualize concepts. Exercises are included with many of these features, and additional exercises in many exercise sets are marked with a graphing calculator icon to indicate more practice with this optional use of technology. (See pp. 158 and 410.)

Student Notes in the margin offer just-in-time suggestions ranging from avoiding common mistakes to how to best read new notation. Conversational in tone, these notes give students extra explanation of the mathematics appearing on that page. (See pp. 116 and 193.)

Study Skills, ranging from time management to test preparation, appear once per section throughout the text. These suggestions for successful study habits apply to any college course and any level of student. (See pp. 98 and 521.)

Chapter Resources are additional learning materials compiled at the end of each chapter, making them easy to integrate into the course at the most appropriate time. The mathematics necessary to use the resource has been presented by the end of the section indicated with each resource.

- *Translating for Success* and *Visualizing for Success.* These are matching exercises that help students learn to translate word problems to mathematical language and to graph equations and inequalities. (See pp. 144 and 483.)
- *Collaborative Activity.* Students who work in groups generally outperform those who do not, so these optional activities direct them to explore mathematics together. Additional collaborative activities and suggestions for directing collaborative learning appear in the *Instructor's Resources Manual with Tests and Mini Lectures.* (See pp. 362 and 427.)
- *Decision Making: Connection.* Although many applications throughout the text involve decision-making situations, this feature specifically applies the math of each chapter to a context in which students may be involved in decision making. (See pp. 222 and 536.)

PRACTICE: Reinforcing Understanding

As students explore the math, they have frequent opportunities to practice, self-assess, and reinforce their understanding.

Your Turn Exercises, following every example, direct students to work a similar exercise. This provides immediate reinforcement of concepts and skills. Answers to these exercises appear at the end of each exercise set. (See pp. 215 and 372.)

New! Check Your Understanding offers students the chance to reflect on the concepts just discussed before beginning the exercise set. Designed to examine or extend students' understanding of one or more essential concepts of the section, this set of questions could function as an "exit ticket" after an instructional session. (See pp. 139 and 194.)

Mid-Chapter Review offers an opportunity for active review in the middle of every chapter. A brief summary of the concepts covered in the first part of the chapter is followed by two guided solutions to help students work step-by-step through solutions and a set of mixed review exercises. (See pp. 202 and 466.)

Exercise Sets

- *Vocabulary and Reading Check* exercises begin every exercise set and are designed to encourage the student to read the section. Students who can complete these exercises should be prepared to begin the remaining exercises in the exercise set. (See pp. 159 and 217.)

- *Concept Reinforcement* exercises can be true/false, matching, and/or fill-in-the-blank and appear near the beginning of many exercise sets. They are designed to build students' confidence and comprehension. Answers to all concept reinforcement exercises appear in the answer section at the back of the book. (See pp. 139 and 196.)

- *Aha!* exercises are not more difficult than neighboring exercises; in fact, they can be solved more quickly, without lengthy computation, if the student has the proper insight. They are designed to encourage students to "look before they leap." An icon indicates the first time that a new insight applies, and then it is up to the student to determine when to use that insight on subsequent exercises. (See pp. 208 and 236.)

- *Skill Review* exercises appear in every section beginning with Section 1.2. Taken together, each chapter's Skill Review exercises review all the major concepts covered in previous chapters in the text. Often these exercises focus on a single topic, such as solving equations, from multiple perspectives. (See pp. 413 and 441.)

- *Synthesis* exercises appear in each exercise set following the Skill Review exercises. Students will often need to use skills and concepts from earlier sections to solve these problems, and this will help them develop deeper insights into the current topic. The Synthesis exercises are a real strength of the text, and in the annotated instructor's edition, the authors have placed a ✓ next to selected synthesis exercises that they suggest instructors "check out" and consider assigning. These exercises may be more accessible to students than the surrounding exercises, they may extend concepts beyond the scope of the text discussion, or they may be especially beneficial in preparing students for future topics. (See pp. 162, 285–286, and 528.)

- *Writing* exercises appear just before the Skill Review exercises, and at least two more challenging exercises appear in the Synthesis exercises. Writing exercises aid student comprehension by requiring students to use critical thinking to explain concepts in one or more complete sentences. Because correct answers may vary, the only writing exercises for which answers appear at the back of the text are those in the chapter's review exercises. (See pp. 111 and 534.)

- *Quick Quizzes* with five questions appear near the end of each exercise set beginning with the second section in each chapter. Containing questions from sections already covered in the chapter, these quizzes provide a short but consistent review of the material in the chapter and help students prepare for a chapter test. (See pp. 136 and 210.)

- *Prepare to Move On* is a short set of exercises that appears at the end of every exercise set. It reviews concepts and skills previously covered in the text that will be used in the next section of the text. (See pp. 270 and 521.)

Study Summary gives students a fast and effective review of key chapter terms and concepts at the end of each chapter. Concepts are paired with worked-out examples and practice exercises for active learning and review. (See pp. 220 and 485.)

Chapter Review and Test offers a thorough chapter review, and a practice test helps to prepare students for a test covering the concepts presented in each chapter. (See pp. 149 and 225.)

Cumulative Review appears after every chapter beginning with Chapter 2 to help students retain and apply their knowledge from previous chapters. (See pp. 301 and 368.)

Acknowledgments

An outstanding team of professionals was involved in the production of this text. Judy Henn, Laurie Hurley, Helen Medley, Tamera Drozd, and Mike Penna carefully checked the book for accuracy and offered thoughtful suggestions.

Martha Morong, of Quadrata, Inc., provided editorial and production services of the highest quality, and Geri Davis, of the Davis Group, Inc., performed superb work as designer, art editor, and photo researcher. Network Graphics provided the accurate and creative illustrations and graphs.

The team at Pearson deserves special thanks. Courseware Portfolio Manager Cathy Cantin, Content Producer Ron Hampton, and Courseware Portfolio Management Assistant Alison Oehmen provided many fine suggestions, coordinated tasks and schedules, and remained involved and accessible throughout the project. Product Marketing Manager Kyle DiGiannantonio skillfully kept in touch with the needs of faculty. Director, Courseware Portfolio Management Michael Hirsch and VP, Courseware Portfolio Manager Chris Hoag deserve credit for assembling this fine team.

We thank the following professors for their thoughtful reviews and insightful comments: Shawna Haider, *Salt Lake Community College*; Ashley Nicoloff, *Glendale Community College*; and Jane Thompson, *Waubonsee Community College*

Finally, a special thank-you to all those who so generously agreed to discuss their professional use of mathematics in our chapter openers. These dedicated people all share a desire to make math more meaningful to students. We cannot imagine a finer set of role models.

M.L.B.
D.J.E.
B.L.J.

Resources for Success

MyMathLab® Online Course

The course for *Elementary Algebra: Concepts and Applications,* 10th Edition, includes all of MyMathLab's robust features and functionality, plus these additional highlights.

New! Workspace

Workspace Assignments allow students to work through an exercise step by step, showing their mathematical reasoning. Students receive immediate feedback after they complete each step, and helpful hints and videos are available for guidance, as needed. When students access Workspace using a mobile device, handwriting-recognition software allows them to write out answers using their fingertip or a stylus.

New! Learning Catalytics

Learning Catalytics uses students' mobile devices for an engagement, assessment, and classroom intelligence system that gives instructors real-time feedback on student learning.

New! Skill Builder Adaptive Practice

When a student struggles with assigned homework, Skill Builder exercises offer just-in-time additional adaptive practice. The adaptive engine tracks student performance and delivers questions to each individual that adapt to his or her level of understanding. When the system has determined that the student has a high probability of successfully completing the assigned exercise, it suggests that the student return to the assignment. When Skill Builder is enabled for an assignment, students can choose to do the extra practice without being prompted. This new feature allows instructors to assign fewer questions for homework so that students can complete as many or as few questions as needed.

Interactive Exercises

MyMathLab's hallmark interactive exercises help build problem-solving skills and foster conceptual understanding. For this tenth edition, Guided Solutions exercises were added to Mid-Chapter Reviews to reinforce the step-by-step problem-solving process, while the *new* Drag & Drop functionality was applied to matching exercises throughout the course to better assess a student's understanding of the concepts.

Resources for Success

In addition to robust course delivery, the full eText, and many assignable exercises and media assets, MyMathLab also houses the following materials to help instructors and students use this program most effectively according to his or her needs.

Student Resources

To-the-Point Objective Videos
- Concise, interactive, and objective-based videos.
- View a whole section, choose an objective, or go straight to an example.
- Interactive *Your Turn Video Check* pauses for the student to work exercises.
- Seamlessly integrated with *MyMathGuide: Notes, Practice, and Video Path*.

Chapter Test Prep Videos
- Step-by-step solutions for every problem in the Chapter Tests.
- Also available in MyMathLab

MyMathGuide:
Notes, Practice, and Video Path
ISBN: 0-13-445332-8
- Guided, hands-on learning in a workbook format with space for students to show their work and record their notes and questions.
- Objective-based, correlates to the *To-the-Point Objective Videos* program.
- Highlights key concepts, skills, and definitions; offers quick reviews of key vocabulary terms with practice problems, examples with guided solutions, similar Your Turn exercises, and practice exercises with readiness checks.

Student's Solutions Manual
ISBN: 0-13-444168-0
- Contains step-by-step solutions for all odd-numbered text exercises (except the writing exercises), as well as Chapter Review, Chapter Test, and Connecting the Concepts exercises.

Instructor Resources

Annotated Instructor's Edition
ISBN: 0-13-444163-X
- Answers to all text exercises.
- Teaching tips and icons that identify writing and graphing calculator exercises.

Instructor's Solutions Manual
(download only)
ISBN: 0-13-444169-9
- Fully worked-out solutions to the odd-numbered text exercises.
- Brief solutions to the even-numbered text exercises.

Instructor's Resource Manual with Tests and Mini Lectures (download only)
ISBN: 0-13-444171-0
- Designed to help both new and adjunct faculty with course preparation and classroom management.
- Teaching tips correlated to the text by section.
- Multiple-choice and free-response chapter tests; multiple final exams.

PowerPoint® Lecture Slides
(download only)
- Editable slides present key concepts and definitions from the text.
- Also available for download through MyMathLab or via Pearsonhighered.com/IRC.

TestGen®
TestGen (www.pearsoned.com/testgen) enables instructors to build, edit, print, and administer tests using a computerized bank of questions developed to cover all the objectives of the text.

Chapter 1

Introduction to Algebraic Expressions

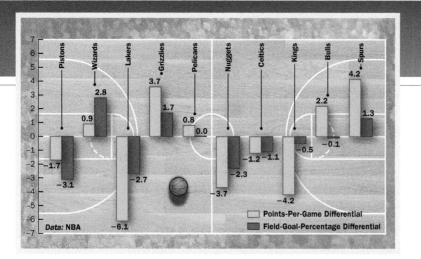

My Team's the Best!

During a season, sports teams are ranked by several statistics. One basketball statistic used is the points-per-game differential—the average of the difference of the team's score and that of its opponents and an indication of how close the games are that the team has played. Another is the field-goal-percentage differential, which measures the team's shooting accuracy. These statistics are illustrated in the graph for selected NBA teams in a recent year. How would you determine which team is "the best"? *(See Exercises 25 and 26 in Exercise Set 1.4.)*

1.1 Introduction to Algebra

1.2 The Commutative, Associative, and Distributive Laws

1.3 Fraction Notation

1.4 Positive and Negative Real Numbers

MID-CHAPTER REVIEW

1.5 Addition of Real Numbers

1.6 Subtraction of Real Numbers

1.7 Multiplication and Division of Real Numbers

CONNECTING THE CONCEPTS

1.8 Exponential Notation and Order of Operations

CHAPTER RESOURCES

Translating for Success
Collaborative Activities
Decision Making: Connection

STUDY SUMMARY

REVIEW EXERCISES

CHAPTER TEST

Using analytics, I gain insight into problems ranging from the global impact of a sponsorship to the emotional impact of attending a sporting event.

Stefanie Francis, Vice President of Analytics and Insight at Glidescope in New York, New York, uses math in her research to determine the most impactful elements of the sports world in decision making.

ALF *Active Learning Figure*   Explore the math using the Active Learning Figure in MyMathLab.

SA *Student Activity*   Do the Student Activity in MyMathLab to see math in action.

roblem solving is the focus of this text. In Chapter 1, we lay the foundation for the problem-solving approach that is developed in Chapter 2 and used in all chapters that follow. This foundation includes a review of arithmetic, a discussion of real numbers and their properties, and an examination of how real numbers are added, subtracted, multiplied, divided, and raised to powers.

1.1   Introduction to Algebra

A. Evaluating Algebraic Expressions   **B.** Translating to Algebraic Expressions   **C.** Translating to Equations

This section introduces some basic concepts and expressions used in algebra. Our focus is on the wordings and expressions that often arise in real-world problems.

A. Evaluating Algebraic Expressions

Probably the greatest difference between arithmetic and algebra is the use of *variables.* Suppose that n represents the number of tickets sold in one day for a U2 concert and that each ticket costs \$90. Then a total of 90 times n, or $90 \cdot n$, dollars will be collected for tickets.

The letter n is a **variable** because it can represent any one of a set of numbers.

The number 90 is a **constant** because it does not change.

The multiplication sign $\cdot$ is an **operation sign** because it indicates the **operation** of multiplication.

The expression $90 \cdot n$ is a **variable expression** because it contains a variable.

An **algebraic expression** consists of variables and/or numerals, often with operation signs and grouping symbols. Other examples of algebraic expressions are:

$t - 37$;   This contains the variable t, the constant 37, and the operation of subtraction.

$(s + t) \div 2$.   This contains the variables s and t, the constant 2, grouping symbols, and the operations of addition and division.

Multiplication can be written in several ways. For example, "90 times n" can be written as $90 \cdot n$, $90 \times n$, $90(n)$, $90 * n$, or simply (and usually) $90n$. Division can also be represented by a fraction bar: $\frac{9}{7}$, or $9/7$, means $9 \div 7$.

To **evaluate** an algebraic expression, we **substitute** a number for each variable in the expression and calculate the result. This result is called the **value** of the expression. The following table lists several values of the expression $90 \cdot n$.

Student Notes

Notation like "$90 \times n$" is not often used in algebra because the "$\times$" symbol can be misread as a variable.

Cost per Ticket (in dollars), 90	Number of Tickets Sold, n	Total Collected (in dollars), $90n$
90	150	13,500
90	200	18,000
90	250	22,500

Student Notes

At the end of each example in this text, you will see YOUR TURN. This directs you to try an exercise similar to the example. The answers to these exercises appear at the end of each exercise set.

1. Evaluate $m - n$ for $m = 100$ and $n = 64$.

2. Using the formula given in Example 2, find the area of a triangle when b is 30 in. (inches) and h is 10 in.

EXAMPLE 1 Evaluate each expression for the given values.

a) $x + y$ for $x = 37$ and $y = 28$

b) $5ab$ for $a = 2$ and $b = 3$

SOLUTION

a) We substitute 37 for x and 28 for y and carry out the addition:

$$x + y = (37) + (28) = 65.$$ Using parentheses when substituting is not always necessary but is never incorrect.

The value of the expression is 65.

b) We substitute 2 for a and 3 for b and multiply:

$$5ab = 5 \cdot 2 \cdot 3 = 10 \cdot 3 = 30.$$ $5ab$ means 5 times a times b.

YOUR TURN

EXAMPLE 2 The area of a triangle with a base of length b and a height of length h is given by the formula $A = \frac{1}{2}bh$. Find the area when b is 8 m (meters) and h is 6.4 m.

SOLUTION   We substitute 8 m for b and 6.4 m for h and then multiply:

$$A = \frac{1}{2}bh$$
$$= \frac{1}{2}(8\,\text{m})(6.4\,\text{m})$$
$$= \frac{1}{2}(8)(6.4)(\text{m})(\text{m})$$
$$= 4(6.4)\,\text{m}^2$$
$$= 25.6\,\text{m}^2, \text{ or } 25.6 \text{ square meters.}$$

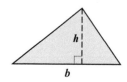

Note that we use square units for area and $(\text{m})(\text{m}) = \text{m}^2$.

YOUR TURN

B. Translating to Algebraic Expressions

Before attempting to translate problems to equations, we must be able to translate certain phrases to algebraic expressions. Any variable can be used to represent an unknown quantity; however, it is helpful to choose a descriptive letter. For example, we can use w to suggest weight and p to suggest population or price. It is important to write down what each variable represents, as well as the unit in which it is measured.

Important Words	Sample Phrase or Sentence	Variable Definition	Translation
Addition (+)			
added to	700 lb was added to the car's weight.	Let w represent the car's weight, in pounds.	$w + 700$
sum of	The sum of a number and 12	Let n represent the number.	$n + 12$
plus	53 plus some number	Let x represent "some number."	$53 + x$
more than	8000 more than Detroit's population	Let p represent Detroit's population.	$p + 8000$
increased by	Alex's original guess, increased by 4	Let n represent Alex's original guess.	$n + 4$
Subtraction (−)			
subtracted from	2 grams were subtracted from the weight.	Let w represent the weight, in grams.	$w - 2$
difference of	The difference of two scores	Let m represent the larger score and n represent the smaller score.	$m - n$
minus	A team of size s, minus 2 players	Let s represent the number of players.	$s - 2$
less than	9 less than the number of volunteers	Let v represent the number of volunteers.	$v - 9$
decreased by	The car's speed, decreased by 8 mph	Let s represent the car's speed, in miles per hour.	$s - 8$
Multiplication (·)			
multiplied by	The number of guests, multiplied by 3	Let g represent the number of guests.	$g \cdot 3$
product of	The product of two numbers	Let m and n represent the numbers.	$m \cdot n$
times	5 times the dog's weight	Let w represent the dog's weight, in pounds.	$5w$
twice	Twice the wholesale cost	Let c represent the wholesale cost.	$2c$
of	$\frac{1}{2}$ of Rita's salary	Let s represent Rita's salary.	$\frac{1}{2}s$
Division (÷)			
divided by	A 2-lb coffee cake, divided by 3	*No variables are required for translation.*	$2 \div 3$
quotient of	The quotient of 14 and 7	*No variables are required for translation.*	$14 \div 7$
divided into	4 divided into the delivery fee	Let f represent the delivery fee.	$f \div 4$
ratio of	The ratio of $500 to the price of a new car	Let p represent the price of a new car, in dollars.	$500/p$
per	The number of students per teacher.	Let s represent the number of students and t represent the number of teachers.	s/t

Student Notes

Try looking for "than" or "from" in a phrase and writing what follows it first. Then add or subtract the necessary quantity. (See Example 3a.)

3. Translate to an algebraic expression: Twenty less than the number of students registered for the course.

EXAMPLE 3 Translate each phrase to an algebraic expression.

a) Four inches less than Ava's height, in inches

b) Eighteen more than a number

c) A day's pay, in dollars, divided by eight

SOLUTION To help think through a translation, we sometimes begin with a specific number in place of a variable.

a) If the height were 60, then 4 less than 60 would mean $60 - 4$. If we use h to represent "Ava's height, in inches," the translation of "Four inches less than Ava's height, in inches" is $h - 4$.

b) If we knew the number to be 10, the translation would be $10 + 18$, or $18 + 10$. If we use t to represent "a number," the translation of "Eighteen more than a number" is $t + 18$, or $18 + t$.

c) We let d represent "a day's pay, in dollars." If the pay were $78, the translation would be $78 \div 8$, or $\frac{78}{8}$. Thus our translation of "A day's pay, in dollars, divided by eight" is $d \div 8$, or $d/8$.

YOUR TURN

Student Notes

At the end of each example in this text, you will see YOUR TURN. This directs you to try an exercise similar to the example. The answers to these exercises appear at the end of each exercise set.

1. Evaluate $m - n$ for $m = 100$ and $n = 64$.

EXAMPLE 1   Evaluate each expression for the given values.

a) $x + y$ for $x = 37$ and $y = 28$

b) $5ab$ for $a = 2$ and $b = 3$

SOLUTION

a) We substitute 37 for x and 28 for y and carry out the addition:

$$x + y = (37) + (28) = 65.$$ Using parentheses when substituting is not always necessary but is never incorrect.

The value of the expression is 65.

b) We substitute 2 for a and 3 for b and multiply:

$$5ab = 5 \cdot 2 \cdot 3 = 10 \cdot 3 = 30.$$ $5ab$ means 5 times a times b.

 YOUR TURN

EXAMPLE 2   The area of a triangle with a base of length b and a height of length h is given by the formula $A = \frac{1}{2}bh$. Find the area when b is 8 m (meters) and h is 6.4 m.

SOLUTION   We substitute 8 m for b and 6.4 m for h and then multiply:

$$A = \frac{1}{2}bh$$
$$= \frac{1}{2}(8\,\text{m})(6.4\,\text{m})$$
$$= \frac{1}{2}(8)(6.4)(\text{m})(\text{m})$$
$$= 4(6.4)\,\text{m}^2$$
$$= 25.6\,\text{m}^2, \text{ or } 25.6 \text{ square meters.}$$

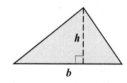

2. Using the formula given in Example 2, find the area of a triangle when b is 30 in. (inches) and h is 10 in.

Note that we use square units for area and $(\text{m})(\text{m}) = \text{m}^2$.

 YOUR TURN

B. Translating to Algebraic Expressions

Before attempting to translate problems to equations, we must be able to translate certain phrases to algebraic expressions. Any variable can be used to represent an unknown quantity; however, it is helpful to choose a descriptive letter. For example, we can use w to suggest weight and p to suggest population or price. It is important to write down what each variable represents, as well as the unit in which it is measured.

Important Words	Sample Phrase or Sentence	Variable Definition	Translation
Addition (+)			
added to	700 lb was added to the car's weight.	Let w represent the car's weight, in pounds.	$w + 700$
sum of	The sum of a number and 12	Let n represent the number.	$n + 12$
plus	53 plus some number	Let x represent "some number."	$53 + x$
more than	8000 more than Detroit's population	Let p represent Detroit's population.	$p + 8000$
increased by	Alex's original guess, increased by 4	Let n represent Alex's original guess.	$n + 4$
Subtraction (−)			
subtracted from	2 grams were subtracted from the weight.	Let w represent the weight, in grams.	$w - 2$
difference of	The difference of two scores	Let m represent the larger score and n represent the smaller score.	$m - n$
minus	A team of size s, minus 2 players	Let s represent the number of players.	$s - 2$
less than	9 less than the number of volunteers	Let v represent the number of volunteers.	$v - 9$
decreased by	The car's speed, decreased by 8 mph	Let s represent the car's speed, in miles per hour.	$s - 8$
Multiplication (·)			
multiplied by	The number of guests, multiplied by 3	Let g represent the number of guests.	$g \cdot 3$
product of	The product of two numbers	Let m and n represent the numbers.	$m \cdot n$
times	5 times the dog's weight	Let w represent the dog's weight, in pounds.	$5w$
twice	Twice the wholesale cost	Let c represent the wholesale cost.	$2c$
of	$\frac{1}{2}$ of Rita's salary	Let s represent Rita's salary.	$\frac{1}{2}s$
Division (÷)			
divided by	A 2-lb coffee cake, divided by 3	*No variables are required for translation.*	$2 \div 3$
quotient of	The quotient of 14 and 7	*No variables are required for translation.*	$14 \div 7$
divided into	4 divided into the delivery fee	Let f represent the delivery fee.	$f \div 4$
ratio of	The ratio of $500 to the price of a new car	Let p represent the price of a new car, in dollars.	$500/p$
per	The number of students per teacher.	Let s represent the number of students and t represent the number of teachers.	s/t

Student Notes

Try looking for "than" or "from" in a phrase and writing what follows it first. Then add or subtract the necessary quantity. (See Example 3a.)

3. Translate to an algebraic expression: Twenty less than the number of students registered for the course.

EXAMPLE 3 Translate each phrase to an algebraic expression.

a) Four inches less than Ava's height, in inches

b) Eighteen more than a number

c) A day's pay, in dollars, divided by eight

SOLUTION To help think through a translation, we sometimes begin with a specific number in place of a variable.

a) If the height were 60, then 4 less than 60 would mean $60 - 4$. If we use h to represent "Ava's height, in inches," the translation of "Four inches less than Ava's height, in inches" is $h - 4$.

b) If we knew the number to be 10, the translation would be $10 + 18$, or $18 + 10$. If we use t to represent "a number," the translation of "Eighteen more than a number" is $t + 18$, or $18 + t$.

c) We let d represent "a day's pay, in dollars." If the pay were $78, the translation would be $78 \div 8$, or $\frac{78}{8}$. Thus our translation of "A day's pay, in dollars, divided by eight" is $d \div 8$, or $d/8$.

 YOUR TURN

Student Notes

The standard way of translating phrases containing "the sum of," "the difference of," and so on, is to write the quantities in the order in which they are mentioned. For example, "the sum of x and y" translates to $x + y$, and "the difference of c and d" translates to $c - d$. This is important for subtraction and division, since the order in which we subtract and divide affects the answer.

4. Translate to an algebraic expression: Half of the sum of two numbers.

> *CAUTION!* The order in which we subtract and divide affects the answer! Answering $4 - h$ or $8 \div d$ in Examples 3(a) and 3(c) is incorrect.

EXAMPLE 4 Translate each phrase to an algebraic expression.

a) Half of some number

b) Seven pounds more than twice Owen's weight, in pounds

c) Six less than the product of two numbers

d) Nine times the difference of x and 10

e) Eighty-two percent of last year's enrollment

SOLUTION

Phrase	Variable(s)	Algebraic Expression
a) Half of some number	Let n represent the number.	$\frac{1}{2}n$, or $\frac{n}{2}$, or $n \div 2$
b) Seven pounds more than twice Owen's weight	Let w represent Owen's weight, in pounds.	$2w + 7$, or $7 + 2w$
c) Six less than the product of two numbers	Let m and n represent the numbers.	$mn - 6$
d) Nine times the difference of x and 10	*The variable is already given.*	$9(x - 10)$
e) Eighty-two percent of last year's enrollment	Let r represent last year's enrollment.	82% of r, or $0.82r$

↩ YOUR TURN

C. Translating to Equations

The **equals** symbol, $=$, indicates that the expressions on either side of the equals sign represent the same number. An **equation** is a number sentence with the verb $=$.

It is important to be able to distinguish between expressions and equations. Compare the descriptions given in the following table.

Expression	Equation
No $=$ sign appears.	An $=$ sign appears.
Compare to an English *phrase*, like "The interesting book."	Compare to an English *sentence*, like "The book was interesting."
May be of any length:	May be of any length:
x	$x = 7$
$3(x - 5) + 4y - 17(3 - y)$.	$3(x - 5) + 4y = 17(3 - y)$.

Although we do not study equations until Chapter 2, we can translate certain problem situations to equations now. The words "is the same as," "equal," "is," "are," "was," and "were" often translate to "$=$."

> **Words indicating equality ($=$):** "is the same as," "equal," "is," "are," "was," "were," "represents"

When translating a problem to an equation, we first translate phrases to algebraic expressions, and then the entire statement to an equation containing those expressions.

EXAMPLE 5 Translate the following problem to an equation.

What number plus 478 is 1019?

SOLUTION We let *y* represent the unknown number. The translation then comes almost directly from the English sentence.

$$
\underbrace{\text{What number}}_{y} \quad \underset{+}{\text{plus}} \quad \underset{478}{478} \quad \underset{=}{\text{is}} \quad \underset{1019}{1019?}
$$

Note that "what number plus 478" translates to "$y + 478$" and "is" translates to "=."

5. Translate to an equation: What number times 12 is 672?

↳ YOUR TURN

Sometimes it helps to reword a problem before translating.

EXAMPLE 6 Translate the following problem to an equation.

The Beipanjiang Duge Bridge in China is the world's highest bridge. At 1854 ft, it is 574 ft higher than the Baluarte Bridge in Mexico. How high is the Baluarte Bridge?

Data: www.highestbridges.com

SOLUTION Since one bridge is 574 ft *higher* than the other, its height is 574 ft *more than* the other's height. We let *h* represent the height, in feet, of the Baluarte Bridge and reword and translate as follows:

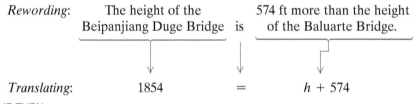

| *Rewording*: | The height of the Beipanjiang Duge Bridge | is | 574 ft more than the height of the Baluarte Bridge. |
| *Translating*: | 1854 | = | $h + 574$ |

Digital representation of the Beipanjiang Duge Bridge superimposed on a photograph of its location

6. Translate to an equation: Valley College has 13 science instructors. There are 5 more science instructors than math instructors. How many math instructors are there?

↳ YOUR TURN

When we translate a problem into mathematical language, we say that we **model** the problem. A **mathematical model** is a mathematical representation of a real-world situation. Note that the word *model* can be used as either a verb or a noun.

Information about a problem is often given as a set of numbers, called **data.** Sometimes data follow a pattern that can be modeled using an equation.

EXAMPLE 7 *Music.* The following table lists the amount charged for several purchases from an online music store. We let *a* represent the amount charged, in dollars, and *n* the number of songs. Find an equation giving *a* in terms of *n*.

Number of Songs Purchased, *n*	Amount Charged, *a*
2	$1.98
3	2.97
5	4.95
10	9.90

SOLUTION To write an equation for *a* **in terms of** *n* means that *a* will be on one side of the equals sign and an expression involving *n* will be on the other side.

↳ **Check Your**
UNDERSTANDING

A student's score on a test is the difference of 100 and *x*, where *x* is the number of points missed.

1. Translate to an algebraic expression that gives a student's score.

2. Evaluate the expression for $x = 11$.

Student Notes

The standard way of translating phrases containing "the sum of," "the difference of," and so on, is to write the quantities in the order in which they are mentioned. For example, "the sum of x and y" translates to $x + y$, and "the difference of c and d" translates to $c - d$. This is important for subtraction and division, since the order in which we subtract and divide affects the answer.

4. Translate to an algebraic expression: Half of the sum of two numbers.

CAUTION! The order in which we subtract and divide affects the answer! Answering $4 - h$ or $8 \div d$ in Examples 3(a) and 3(c) is incorrect.

EXAMPLE 4 Translate each phrase to an algebraic expression.
a) Half of some number
b) Seven pounds more than twice Owen's weight, in pounds
c) Six less than the product of two numbers
d) Nine times the difference of x and 10
e) Eighty-two percent of last year's enrollment

SOLUTION

Phrase	Variable(s)	Algebraic Expression
a) Half of some number	Let n represent the number.	$\frac{1}{2}n$, or $\frac{n}{2}$, or $n \div 2$
b) Seven pounds more than twice Owen's weight	Let w represent Owen's weight, in pounds.	$2w + 7$, or $7 + 2w$
c) Six less than the product of two numbers	Let m and n represent the numbers.	$mn - 6$
d) Nine times the difference of x and 10	The variable is already given.	$9(x - 10)$
e) Eighty-two percent of last year's enrollment	Let r represent last year's enrollment.	82% of r, or $0.82r$

YOUR TURN

C. Translating to Equations

The **equals** symbol, $=$, indicates that the expressions on either side of the equals sign represent the same number. An **equation** is a number sentence with the verb $=$.

It is important to be able to distinguish between expressions and equations. Compare the descriptions given in the following table.

Expression	Equation
No $=$ sign appears. Compare to an English *phrase*, like "The interesting book." May be of any length: x $3(x - 5) + 4y - 17(3 - y)$.	An $=$ sign appears. Compare to an English *sentence*, like "The book was interesting." May be of any length: $x = 7$ $3(x - 5) + 4y = 17(3 - y)$.

Although we do not study equations until Chapter 2, we can translate certain problem situations to equations now. The words "is the same as," "equal," "is," "are," "was," and "were" often translate to "$=$."

Words indicating equality ($=$): "is the same as," "equal," "is," "are," "was," "were," "represents"

When translating a problem to an equation, we first translate phrases to algebraic expressions, and then the entire statement to an equation containing those expressions.

EXAMPLE 5   Translate the following problem to an equation.

What number plus 478 is 1019?

SOLUTION   We let y represent the unknown number. The translation then comes almost directly from the English sentence.

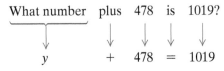

Note that "what number plus 478" translates to "$y + 478$" and "is" translates to "=."

5. Translate to an equation: What number times 12 is 672?

↩ YOUR TURN

Sometimes it helps to reword a problem before translating.

EXAMPLE 6   Translate the following problem to an equation.

The Beipanjiang Duge Bridge in China is the world's highest bridge. At 1854 ft, it is 574 ft higher than the Baluarte Bridge in Mexico. How high is the Baluarte Bridge?

Data: www.highestbridges.com

SOLUTION   Since one bridge is 574 ft *higher* than the other, its height is 574 ft *more than* the other's height. We let h represent the height, in feet, of the Baluarte Bridge and reword and translate as follows:

Digital representation of the Beipanjiang Duge Bridge superimposed on a photograph of its location

6. Translate to an equation: Valley College has 13 science instructors. There are 5 more science instructors than math instructors. How many math instructors are there?

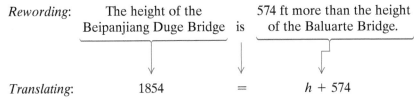

↩ YOUR TURN

When we translate a problem into mathematical language, we say that we **model** the problem. A **mathematical model** is a mathematical representation of a real-world situation. Note that the word *model* can be used as either a verb or a noun.

Information about a problem is often given as a set of numbers, called **data.** Sometimes data follow a pattern that can be modeled using an equation.

EXAMPLE 7   *Music.* The following table lists the amount charged for several purchases from an online music store. We let a represent the amount charged, in dollars, and n the number of songs. Find an equation giving a in terms of n.

A student's score on a test is the difference of 100 and x, where x is the number of points missed.

1. Translate to an algebraic expression that gives a student's score.

2. Evaluate the expression for $x = 11$.

Number of Songs Purchased, n	Amount Charged, a
2	$1.98
3	2.97
5	4.95
10	9.90

SOLUTION   To write an equation for a **in terms of** n means that a will be on one side of the equals sign and an expression involving n will be on the other side.

7. Suppose that an online music store charges $2.58 for 2 songs, $3.87 for 3 songs, and $12.90 for 10 songs. Using the same variables as those in Example 7, find an equation giving a in terms of n.

We look for a pattern in the data. Since the amount charged increases as the number of songs increases, we can try dividing the amount by the number of songs:

$$1.98/2 = 0.99; \qquad 4.95/5 = 0.99;$$
$$2.97/3 = 0.99; \qquad 9.90/10 = 0.99.$$

The quotient is the same, 0.99, for each pair of numbers. Thus each song costs $0.99. We reword and translate as follows:

Rewording: The amount charged   is   0.99   times   the number of songs.

Translating: a $=$ 0.99 $\cdot$ n

⟳ YOUR TURN

Technology Connection

Technology Connections are activities that make use of features that are common to most graphing calculators. In some cases, students may find the user's manual for their particular calculator helpful for exact keystrokes.

   Although all graphing calculators are not the same, most share the following characteristics.

Screen.   The large screen can show graphs and tables as well as the expressions entered. The screen has a different layout for different functions. Computations are performed in the **home screen**. On many calculators, the home screen is accessed by pressing **2ND** **(QUIT)**. The **cursor** shows location on the screen, and the **contrast** (set by **2ND** ⌃ or **2ND** ⌄) determines how dark the characters appear.

Keypad.   There are options written above the keys as well as on them. To access those above the keys, we press **2ND** or **ALPHA** and then the key. Expressions are generally entered as they would appear in print. For example, to evaluate $3xy + x$ for $x = 65$ and $y = 92$, we press 3 **(×)** 65 **(×)** 92 **(+)** 65 and then **ENTER**. The value of the expression, 18005, will appear at the right of the screen.

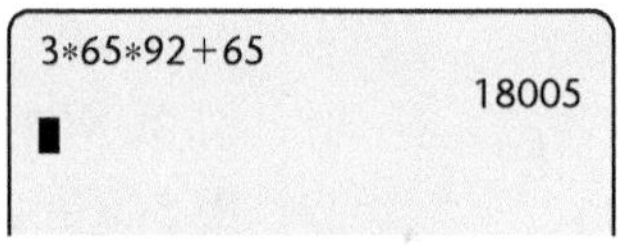

```
3*65*92+65
                      18005
■
```

Evaluate each of the following.
1. $27a - 18b$, for $a = 136$ and $b = 13$
2. $19xy - 9x + 13y$, for $x = 87$ and $y = 29$

[VIDEO]

Study Skills

Get the Facts

Throughout this textbook, you will find a feature called Study Skills. These tips are intended to help improve your math study skills. On the first day of class, we recommend that you complete this chart.

Instructor:
  Name _____
  Office hours and location _____
  Phone number _____
  E-mail address _____
Classmates:
  1. Name _____
  Phone number _____
  E-mail address _____
  2. Name _____
  Phone number _____
  E-mail address _____
Math lab on campus:
  Location _____
  Hours _____
  Phone number _____
  E-mail address _____
Tutoring:
  Campus location _____
  Hours _____
  E-mail address _____
Important supplements:
  (See the preface for a complete list of available supplements.)
  Supplements recommended by the instructor.
  _____
  _____

↘ **Chapter Resources:**
Translating for Success, p. 69; Collaborative Activity (Teamwork), p. 70

1.1   EXERCISE SET

Vocabulary and Reading Check

Choose from the following list of words to complete each statement. Not every word will be used.

constant	expression
equation	operation
evaluate	variable

1. In the expression $4 + x$, the number 4 is a(n) _____.

2. In the expression $4 + x$, the symbol + indicates the _____ of addition.

3. To _____ an algebraic expression, we substitute a number for each variable and carry out the operations.

4. A(n) _____ contains an equals sign.

Concept Reinforcement

Classify each of the following as either an expression or an equation.

5. $10n - 1$ **6.** $3x = 21$

7. $2x - 5 = 9$ **8.** $5(x + 2)$

9. $45 = a - 1$ **10.** $4a - 5b$

11. $2x - 3y = 8$ **12.** $r(t + 7) + 5$

A. Evaluating Algebraic Expressions

Evaluate.

13. $5a$, for $a = 9$ **14.** $11y$, for $y = 7$

15. $12 - r$, for $r = 4$ **16.** $t + 8$, for $t = 2$

17. $\dfrac{a}{b}$, for $a = 45$ and $b = 9$

18. $\dfrac{c + d}{3}$, for $c = 14$ and $d = 13$

19. $\dfrac{x + y}{4}$, for $x = 2$ and $y = 14$

20. $\dfrac{m}{n}$, for $m = 54$ and $n = 9$

21. $\dfrac{p - q}{7}$, for $p = 55$ and $q = 20$

22. $\dfrac{9m}{q}$, for $m = 6$ and $q = 18$

23. $\dfrac{5z}{y}$, for $z = 9$ and $y = 15$

24. $\dfrac{m - n}{2}$, for $m = 20$ and $n = 8$

Substitute to find the value of each expression.

25. *Hockey.*   The area of a rectangle with base b and height h is bh. A regulation hockey goal is 6 ft wide and 4 ft high. Find the area of the opening.

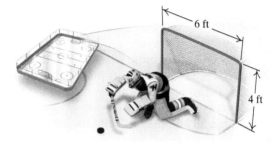

26. *Orbit Time.*   A communications satellite orbiting 300 mi above the earth travels about 27,000 mi in one orbit. The time, in hours, for an orbit is

$$\frac{27{,}000}{v},$$

where v is the velocity, in miles per hour. How long will an orbit take at a velocity of 1125 mph?

27. *Zoology.*   A great white shark has triangular teeth. Each tooth measures about 5 cm across the base and has a height of 6 cm. Find the surface area of the front side of one such tooth. (See Example 2.)

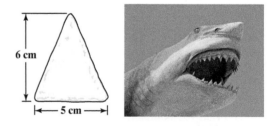

28. *Work Time.*   Justin takes three times as long to complete a job as Carl does. Suppose that t represents the time it takes Carl to complete the job. Then $3t$ represents the time it takes Justin. How long does it take Justin if Carl takes **(a)** 30 sec? **(b)** 90 sec? **(c)** 2 min?

29. *Area of a Parallelogram.*   The area of a parallelogram with base b and height h is bh. Edward

Tufte's sculpture *Spring Arcs* is in the shape of a parallelogram with base 67 ft and height 12 ft. What is the area of the parallelogram?

Data: edwardtufte.com

Spring Arcs *(2004), Edward Tufte. Solid stainless steel, footprint* 12' × 67'.

30. *Women's Softball.* A softball player's batting average is h/a, where h is the number of hits and a is the number of "at bats." In the 2015 Women's College World Series, Lauren Haeger of the Florida Gators had 8 hits in 14 at bats. What was her batting average? Round to the nearest thousandth.

B. Translating to Algebraic Expressions

Translate to an algebraic expression.

31. 5 more than Ron's age

32. The product of 4 and a

33. 6 times b

34. 7 more than Patti's weight

35. 9 less than c

36. 4 less than d

37. 6 increased by q

38. 11 increased by z

39. The difference of p and t

40. m subtracted from n

41. x less than y

42. 2 less than Kurt's age

43. x divided by w

44. The quotient of two numbers

45. The sum of the box's length and height

46. The sum of d and f

47. The product of 9 and twice m

48. Abby's speed minus twice the wind speed

49. Thirteen less than one quarter of some number

50. Four less than ten times a number

51. Five times the difference of two numbers

52. One third of the sum of two numbers

53. 64% of the women attending

54. 38% of a number

C. Translating to Equations

Translate each problem to an equation. Do not solve.

55. What number added to 73 is 201?

56. Seven times what number is 1596?

57. When 42 is multiplied by a number, the result is 2352. Find the number.

58. When 345 is added to a number, the result is 987. Find the number.

59. *Chess.* A chess board has 64 squares. If pieces occupy 19 squares, how many squares are unoccupied?

60. *Hours Worked.* A carpenter charges $35 per hour. How many hours did she work if she billed a total of $3640?

61. *Recycling.* Currently, Americans recycle 34.5% of all municipal solid waste. This is the same as recycling 87 million tons per year. What is the total amount of waste generated per year?

Data: Environmental Protection Agency

62. *Travel to Work.* For U.S. cities with populations greater than 5000, the longest average commute is 59.8 min in Indian Wells, Arizona. This is 51.2 min longer than the shortest average commute, which is in Fort Bliss, Texas. How long is the average commute in Fort Bliss?

Data: www.city-data.com

63. *Nutrition.* The number of grams f of dietary fiber recommended daily for children depends on the age a of the child, as shown in the following table. Find an equation for f in terms of a.

Age of Child, a (in years)	Grams of Dietary Fiber Recommended Daily, f
3	8
4	9
5	10
6	11
7	12
8	13

Data: The American Health Foundation

64. *Tuition.* The following table lists the tuition costs for students taking various numbers of hours of classes. Find an equation for the cost c of tuition for a student taking h hours of classes.

Number of Class Hours, h	Tuition, c
12	$1200
15	1500
18	1800
21	2100

65. *Postage Rates.* The U.S. Postal Service charges an extra fee for packages that must be processed by hand. The following table lists machinable and non-machinable costs for certain commercial packages. Find an equation for the nonmachinable cost n in terms of the machinable cost m.

Weight (in pounds)	Machinable Cost, m	Nonmachinable Cost, n
1	$4.17	$6.59
4	4.76	7.18
10	6.34	8.76

Data: pe.usps.gov

66. *Foreign Currency.* On Emily's trip to Italy, she used her debit card to withdraw money. The following table lists the amounts r that she received and the amounts s that were subtracted from her account. Find an equation for r in terms of s.

Amount Received, r (in U.S. dollars)	Amount Subtracted, s (in U.S. dollars)
$150	$153
75	78
120	123

67. *Number of Drivers.* The following table lists the number of vehicle miles v traveled annually per household by the number of drivers d in the household. Find an equation for v in terms of d.

Number of Drivers, d	Number of Vehicle Miles Traveled, v
1	10,000
2	20,000
3	30,000
4	40,000

Data: Energy Information Administration

68. *Meteorology.* The following table lists the number of centimeters of water w to which various amounts of snow s will melt under certain conditions. Find an equation for w in terms of s.

Depth of Snow, s (in centimeters)	Depth of Water, w (in centimeters)
120	12
135	13.5
160	16
90	9

B., C. Translating to Algebraic Expressions and Equations

In each of Exercises 69–76, match the phrase or sentence with the appropriate expression or equation from the column on the right.

69. ____ Twice the sum of two numbers

70. ____ Five less than a number is twelve.

71. ____ Twelve more than a number is five.

72. ____ Half of the product of two numbers

73. ____ Three times the sum of a number and five

74. ____ Twice the sum of two numbers is 48.

75. ____ One less than the product of two numbers is 48.

76. ____ Six more than the quotient of two numbers

a) $\dfrac{x}{y} + 6$

b) $2(x + y) = 48$

c) $\dfrac{1}{2} \cdot a \cdot b$

d) $t + 12 = 5$

e) $ab - 1 = 48$

f) $2(m + n)$

g) $3(t + 5)$

h) $x - 5 = 12$

To the student and the instructor: Writing exercises, denoted by 🗒, are meant to be answered using one or more sentences. Because answers to many writing exercises will vary, solutions are not listed in the answers at the back of the book.

77. What is the difference between a variable, a variable expression, and an equation?

78. What does it mean to evaluate an algebraic expression?

Synthesis

To the student and the instructor: Synthesis exercises are designed to challenge students to extend the concepts or skills studied in each section. Many synthesis exercises will require the assimilation of skills and concepts from several sections.

79. If the lengths of the sides of a square are doubled, is the area doubled? Why or why not?

80. Write a problem that translates to
$$2006 + t = 2014.$$

81. Signs of Distinction charges $120 per square foot for handpainted signs. The town of Belmar commissioned a triangular sign with a base of 3 ft and a height of 2.5 ft. How much will the sign cost?

82. Find the area that is shaded.

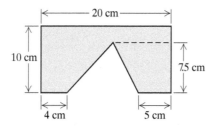

83. Evaluate $\dfrac{x - y}{3}$ when x is twice y and $x = 12$.

84. Evaluate $\dfrac{x + y}{2}$ when y is twice x and $x = 6$.

85. Evaluate $\dfrac{a + b}{4}$ when a is twice b and $a = 16$.

86. Evaluate $\dfrac{a - b}{3}$ when a is three times b and $a = 18$.

Answer each question with an algebraic expression.

87. If $w + 3$ is a whole number, what is the next whole number after it?

88. If $d + 2$ is an odd number, what is the preceding odd number?

Translate to an algebraic expression.

89. The perimeter of a rectangle with length l and width w (perimeter means distance around)

90. The perimeter of a square with side s (perimeter means distance around)

91. Ellie's race time, assuming that she took 5 sec longer than Dion and Dion took 3 sec longer than Molly. Assume that Molly's time was t seconds.

92. Kade's age 7 years from now if he is 2 years older than Monique and Monique is a years old

93. If the height of a triangle is doubled, is its area also doubled? Why or why not?

YOUR TURN ANSWERS: SECTION 1.1

1. 36 **2.** 150 in^2 **3.** Let n represent the number of students registered for the course; $n - 20$ **4.** Let x and y represent the numbers; $\frac{1}{2}(x + y)$ **5.** Let x represent the number; $x \cdot 12 = 672$ **6.** Let m represent the number of math instructors; $13 = m + 5$ **7.** $a = 1.29n$

1.2 The Commutative, Associative, and Distributive Laws

A. The Commutative Laws   **B.** The Associative Laws   **C.** The Distributive Law
D. The Distributive Law and Factoring

Study Skills

Learn by Example

The examples in each section are designed to prepare you for success with the exercise set. Study the step-by-step solutions of the examples, noting that color is used to indicate substitutions and to call attention to the new steps in multistep examples. The time you spend studying the examples will save you valuable time when you do your assignment.

The commutative, associative, and distributive laws discussed in this section enable us to write *equivalent expressions* that will simplify our work.

The expressions $4 + 4 + 4$, $3 \cdot 4$, and $4 \cdot 3$ all represent the same number, 12. Expressions that represent the same number are said to be **equivalent**. The equivalent expressions $t + 18$ and $18 + t$ are both translations of "eighteen more than a number." These expressions are equivalent because they represent the same number for any value of t. We can illustrate this by making some choices for t.

When $t = 3$, $t + 18 = 3 + 18 = 21$
          and $18 + t = 18 + 3 = 21.$
When $t = 40$, $t + 18 = 40 + 18 = 58$
          and $18 + t = 18 + 40 = 58.$

A. The Commutative Laws

Recall that changing the order in addition or multiplication does not change the result. Equations like $3 + 78 = 78 + 3$ and $5 \cdot 14 = 14 \cdot 5$ illustrate this idea and show that addition and multiplication are **commutative**.

> **THE COMMUTATIVE LAWS**
>
> *For Addition.*   For any numbers a and b,
>
> $a + b = b + a.$
>
> (Changing the order of addition does not affect the answer.)
>
> *For Multiplication.*   For any numbers a and b,
>
> $ab = ba.$
>
> (Changing the order of multiplication does not affect the answer.)

EXAMPLE 1   Use the commutative laws to write an expression equivalent to each of the following: **(a)** $y + 5$; **(b)** $9x$; **(c)** $7 + ab$.

SOLUTION

a) $y + 5$ is equivalent to $5 + y$ by the commutative law of addition.

b) $9x$ is equivalent to $x \cdot 9$ by the commutative law of multiplication.

c) $7 + ab$ is equivalent to $ab + 7$ by the commutative law of *addition*.

$7 + ab$ is also equivalent to $7 + ba$ by the commutative law of *multiplication*.

$7 + ab$ is also equivalent to $ba + 7$ by the two commutative laws, used together.

1. Use the commutative law of addition to write an expression equivalent to $7a + 3$.

 YOUR TURN

B. The Associative Laws

Parentheses can be used to indicate groupings. We generally simplify within the parentheses first. For example,

$$3 + (8 + 4) = 3 + 12 = 15$$

and

$$(3 + 8) + 4 = 11 + 4 = 15.$$

Similarly,

$$4 \cdot (2 \cdot 3) = 4 \cdot 6 = 24$$

and

$$(4 \cdot 2) \cdot 3 = 8 \cdot 3 = 24.$$

Note that, so long as only addition or only multiplication appears in an expression, changing the grouping does not change the result. Equations such as $3 + (7 + 5) = (3 + 7) + 5$ and $4(5 \cdot 3) = (4 \cdot 5)3$ illustrate that addition and multiplication are **associative**.

THE ASSOCIATIVE LAWS

For Addition. For any numbers a, b, and c,

$$a + (b + c) = (a + b) + c.$$

(Numbers can be grouped in any manner for addition.)

For Multiplication. For any numbers a, b, and c,

$$a \cdot (b \cdot c) = (a \cdot b) \cdot c.$$

(Numbers can be grouped in any manner for multiplication.)

EXAMPLE 2 Use an associative law to write an expression equivalent to each of the following: **(a)** $y + (z + 3)$; **(b)** $(8x)y$.

SOLUTION

a) $y + (z + 3)$ is equivalent to $(y + z) + 3$ by the associative law of addition.

b) $(8x)y$ is equivalent to $8(xy)$ by the associative law of multiplication.

YOUR TURN

2. Use the associative law of multiplication to write an expression equivalent to $37(mp)$.

When only addition or only multiplication is involved, parentheses do not change the result. For that reason, we sometimes omit them altogether. Thus,

$$x + (y + 7) = x + y + 7 \quad \text{and} \quad l(wh) = lwh.$$

A sum such as $(5 + 1) + (3 + 5) + 9$ can be simplified by pairing numbers that add to 10. The associative and commutative laws allow us to do this:

$$(5 + 1) + (3 + 5) + 9 = 5 + 5 + 9 + 1 + 3$$
$$= 10 + 10 + 3 = 23.$$

EXAMPLE 3 Use the commutative and/or the associative laws of addition to write two expressions equivalent to $(7 + x) + 3$. Then simplify.

SOLUTION

$$(7 + x) + 3 = (x + 7) + 3 \qquad \text{Using a commutative law;} \\ (x + 7) + 3 \text{ is one equivalent expression.}$$

$$= x + (7 + 3) \qquad \text{Using an associative law; } x + (7 + 3) \text{ is another equivalent expression.}$$

$$= x + 10 \qquad \text{Simplifying}$$

3. Use the commutative and/or the associative laws of multiplication to write two expressions equivalent to $2(x \cdot 3)$. Then simplify.

⟳ YOUR TURN

C. The Distributive Law

Student Notes

To remember the names *commutative, associative,* and *distributive,* first understand the concept. Next, use everyday life to link each word to the concept. For example, think of *commuting* to and from college as changing the order of appearance.

The *distributive law* is probably the single most important law for manipulating algebraic expressions. Unlike the commutative and the associative laws, the distributive law uses multiplication together with addition.

The distributive law relates expressions like $5(x + 2)$ and $5x + 10$, which involve both multiplication and addition. When two numbers are multiplied, the result is a **product**. The parts of the product are called **factors**. When two numbers are added, the result is a **sum**. The parts of the sum are called **terms**.

$5(x + 2)$ is a product. The factors are 5 and $(x + 2)$.

This factor is a sum. Its terms are x and 2.

$5x + 10$ is a sum. The terms are $5x$ and 10.

This term is a product. Its factors are 5 and x.

In general, a term is a number, a variable, or a product or a quotient of numbers and/or variables. Terms are separated by plus signs.

EXAMPLE 4 List the terms in the expression $3s + st + \dfrac{2s}{t}$.

SOLUTION Terms are separated by plus signs, so the terms in $3s + st + \dfrac{2s}{t}$ are $3s, st,$ and $\dfrac{2s}{t}$.

4. List the terms in the expression $5x + 3y$.

⟳ YOUR TURN

EXAMPLE 5 List the factors in the expression $x(3 + y)$.

SOLUTION Factors are parts of products, so the factors in $x(3 + y)$ are x and $(3 + y)$.

5. List the factors in the expression $4 \cdot x \cdot y$.

⟳ YOUR TURN

You have already used the distributive law although you may not have realized it at the time. To illustrate, try to multiply $3 \cdot 21$ mentally. Many people find the product, 63, by thinking of 21 as $20 + 1$ and then multiplying 20 by 3 and 1 by 3. The sum of the two products, $60 + 3$, is 63. Note that if the 3 does not multiply *both* 20 and 1, the result will not be correct.

We can compute $4(7 + 2)$ in two ways. As in the discussion of $3(20 + 1)$ above, to compute $4(7 + 2)$, we can multiply both 7 and 2 by 4 and add the results:

$$4(7 + 2) = 4 \cdot 7 + 4 \cdot 2 = 28 + 8 = 36.$$

By first adding inside the parentheses, we get the same result in a different way:

$$4(7 + 2) = 4(9) = 36.$$

Student Notes

The distributive law involves both addition *and* multiplication. Do not try to "distribute" when only multiplication is involved. For example, $5(3 \cdot x) = (5 \cdot 3) \cdot x = 15x$ (by the associative law of multiplication). *It is incorrect to write* $5(3 \cdot x) = (5 \cdot 3)(5 \cdot x)$.

> **THE DISTRIBUTIVE LAW**
>
> For any numbers a, b, and c,
>
> $$a(b + c) = ab + ac.$$
>
> (The product of a number and a sum can be written as the sum of two products.)

EXAMPLE 6   Multiply: $3(x + 2)$.

SOLUTION   We use the distributive law:

$$3(x + 2) = 3 \cdot x + 3 \cdot 2 \qquad \text{Using the distributive law}$$
$$= 3x + 6. \qquad \text{Note that } 3 \cdot x \text{ is the same as } 3x.$$

6. Multiply: $5(3 + y)$.

YOUR TURN

The distributive law can also be used when more than two terms are inside the parentheses.

EXAMPLE 7   Multiply: $6(s + 2 + 5w)$.

SOLUTION

$$6(s + 2 + 5w) = 6 \cdot s + 6 \cdot 2 + 6 \cdot 5w \qquad \text{Using the distributive law}$$
$$= 6s + 12 + (6 \cdot 5)w \qquad \text{Using the associative law for multiplication}$$
$$= 6s + 12 + 30w$$

7. Multiply: $4(5a + 6m + 1)$.

YOUR TURN

> **CAUTION!**   To use the distributive law for removing parentheses, be sure to multiply *each* term inside the parentheses by the multiplier outside. Thus,
>
> $$a(b + c) \neq ab + c \quad \text{but} \quad a(b + c) = ab + ac.$$

Because of the commutative law of multiplication, the distributive law can be used on the "right": $(b + c)a = ba + ca$.

EXAMPLE 8   Multiply: $(c + 4)5$.

SOLUTION

$$(c + 4)5 = c \cdot 5 + 4 \cdot 5 \qquad \text{Using the distributive law on the right}$$
$$= 5c + 20 \qquad \text{Using the commutative law; } c \cdot 5 = 5c$$

8. Multiply: $(11 + y)2$.

YOUR TURN

> **CAUTION!**   Note the differences between expressions such as $6(5w)$ and $6(5 + w)$:
>
> $6(5w)$
> - Only multiplication is involved.
> - Use an associative law.
> - $6(5w) = (6 \cdot 5)w = 30w$
>
> $6(5 + w)$
> - Both multiplication and addition are involved.
> - Use the distributive law.
> - $6(5 + w) = 6 \cdot 5 + 6 \cdot w = 30 + 6w$

↪ Check Your UNDERSTANDING

For each of the following expressions:

a) What operations are involved?

b) What law would be used to perform the multiplication?

c) Multiply.

1. $7(3 + x)$

2. $7(3x)$

---

9. Use the distributive law to factor $15x + 5$.

D. The Distributive Law and Factoring

If we use the distributive law in reverse, we have the basis of a process called **factoring**: $ab + ac = a(b + c)$. To **factor** an expression means to write an equivalent expression that is a product. Recall that the parts of the product are called **factors**. Note that "factor" can be used as either a verb or a noun. A **common factor** is a factor that appears in every term in an expression.

EXAMPLE 9 Use the distributive law to factor each of the following.

a) $3x + 3y$ 　　　　　　　　　　　　b) $7x + 21y + 7$

SOLUTION

a) By the distributive law,

$$3x + 3y = 3(x + y).$$ The common factor for $3x$ and $3y$ is 3.

b) $7x + 21y + 7 = 7 \cdot x + 7 \cdot 3y + 7 \cdot 1$ 　　The common factor is 7.

$$= 7(x + 3y + 1)$$ Using the distributive law. Be sure to include both the 1 and the common factor, 7.

↩ YOUR TURN

To check our factoring, we multiply to see if the original expression is obtained. For example, to check the **factorization** in Example 9(b), note that

$$7(x + 3y + 1) = 7 \cdot x + 7 \cdot 3y + 7 \cdot 1$$
$$= 7x + 21y + 7.$$

Since $7x + 21y + 7$ is what we started with in Example 9(b), we have a check.

1.2 EXERCISE SET

FOR EXTRA HELP   MyMathLab®

↪ Vocabulary and Reading Check

Choose from the following list of words to complete each statement. Not every word will be used.

associative 　　　　　factors
commutative 　　　　product
distributive 　　　　　sum
equivalent 　　　　　terms

1. _____ expressions represent the same number.

2. Changing the order of multiplication does not affect the answer. This is an example of a(n) _____ law.

3. The result of addition is called a(n) _____.

4. The numbers in a product are called _____.

↪ Concept Reinforcement

Determine whether each statement illustrates a commutative law, an associative law, or a distributive law.

5. $8 + x = x + 8$ 　　　　　**6.** $5b(c) = 5(bc)$

7. $x(y + z) = xy + xz$ 　　　**8.** $3(t + 4) = 3(4 + t)$

9. $5(x + 2) = (x + 2)5$ 　　　**10.** $2a + 2b = 2(a + b)$

A. The Commutative Laws

Use the commutative law of addition to write an equivalent expression.

11. $11 + t$ 　　　　　　　**12.** $a + 2$

13. $4 + 8x$ 　　　　　　　**14.** $ab + c$

15. $9x + 3y$ 　　　　　　 **16.** $3a + 7b$

17. $5(a + 1)$ 　　　　　　 **18.** $9(x + 5)$

Use the commutative law of multiplication to write an equivalent expression.

19. $7x$

20. xy

21. st

22. $13m$

23. $5 + ab$

24. $x + 3y$

25. $5(a + 1)$

26. $9(x + 5)$

B. The Associative Laws

Use the associative law of addition to write an equivalent expression.

27. $(x + 8) + y$

28. $(5 + m) + r$

29. $u + (v + 7)$

30. $x + (2 + y)$

31. $(ab + c) + d$

32. $(m + np) + r$

Use the associative law of multiplication to write an equivalent expression.

33. $(10x)y$

34. $(4u)v$

35. $2(ab)$

36. $9(7r)$

37. $3[2(a + b)]$

38. $5[x(2 + y)]$

A, B. The Commutative and Associative Laws

Use the commutative and/or the associative laws to write two equivalent expressions. Answers may vary.

39. $s + (t + 6)$

40. $7 + (v + w)$

41. $(17a)b$

42. $x(3y)$

Use the commutative and/or the associative laws to show why the expression on the left is equivalent to the expression on the right. Write a series of steps with labels, as in Example 3.

43. $(1 + x) + 2$ is equivalent to $x + 3$

44. $(2a)4$ is equivalent to $8a$

45. $(m \cdot 3)7$ is equivalent to $21m$

46. $4 + (9 + x)$ is equivalent to $x + 13$

C. The Distributive Law

List the terms in each expression.

47. $x + xyz + 1$

48. $9 + 17a + abc$

49. $2a + \dfrac{a}{3b} + 5b$

50. $3xy + 20 + \dfrac{4a}{b}$

51. $4x + 4y$

52. $14 + 2y$

List the factors in each expression.

53. $5n$

54. uv

55. $3(x + y)$

56. $(a + b)12$

57. $7 \cdot a \cdot b$

58. $m \cdot n \cdot 2$

59. $(a - b)(x - y)$

60. $(3 - a)(b + c)$

Multiply.

61. $2(x + 15)$

62. $3(x + 5)$

63. $4(1 + a)$

64. $7(1 + y)$

65. $10(9x + 6)$

66. $9(6m + 7)$

67. $5(r + 2 + 3t)$

68. $4(5x + 8 + 3p)$

69. $(a + b)2$

70. $(x + 2)7$

71. $(x + y + 2)5$

72. $(2 + a + b)6$

D. The Distributive Law and Factoring

Use the distributive law to factor each of the following. Check by multiplying.

73. $2a + 2b$

74. $5y + 5z$

75. $7 + 7y$

76. $13 + 13x$

77. $32x + 2$

78. $20a + 5$

79. $5x + 10 + 15y$

80. $3 + 27b + 6c$

81. $7a + 35b$

82. $3x + 24y$

83. $44x + 11y + 22z$

84. $14a + 56b + 7$

A, B, C. The Commutative, Associative, and Distributive Laws

Fill in each blank with the law that justifies that step.

85. $3(2 + x)$
$= 3(x + 2)$ _____
$= 3 \cdot x + 3 \cdot 2$ _____
$= 3x + 6$ Multiplying

86. $(y + 4)5$
$= 5(y + 4)$ _____
$= 5 \cdot y + 5 \cdot 4$ _____
$= 5y + 20$ Multiplying

87. $7(2x + 3y)$
$= 7(2x) + 7(3y)$ _____
$= (7 \cdot 2)x + (7 \cdot 3)y$ _____
$= 14x + 21y$ Multiplying

88. $(4a + 2)8$
$= 8(4a + 2)$ _____
$= 8(4a) + 8(2)$ _____
$= (8 \cdot 4)a + 8(2)$ _____
$= 32a + 16$ Multiplying

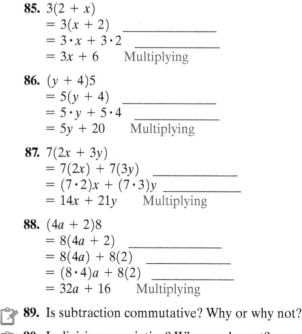 **89.** Is subtraction commutative? Why or why not?

90. Is division associative? Why or why not?

Synthesis

91. Give an example illustrating the distributive law, and identify the terms and the factors in your example. Explain how you can identify terms and factors in an expression.

92. Explain how the distributive, commutative, and associative laws can be used to show that $2(3x + 4y)$ is equivalent to $6x + 8y$.

Fill in each blank with the law that justifies that step.

93. $[2(x + 1)] + 3x$
$= [2 \cdot x + 2 \cdot 1] + 3x$ _____
$= [2x + 2] + 3x$ Multiplying
$= 2x + [2 + 3x]$ _____
$= 2x + [3x + 2]$ _____
$= [2x + 3x] + 2$ _____
$= [(2 + 3)x] + 2$ _____
$= [5x] + 2$ Adding
$= 5x + 2$

94. $12a + 4(b + 5)$
$= (4 \cdot 3)a + 4(b + 5)$ Writing 12 as a product
$= 4(3a) + 4(b + 5)$ _____
$= 4(3a) + 4(b) + 4(5)$ _____
$= 4(3a + b + 5)$ _____

Tell whether the expressions in each pairing are equivalent. Then explain why or why not.

95. $8 + 4(a + b)$ and $4(2 + a + b)$

96. $5(a \cdot b)$ and $5 \cdot a \cdot 5 \cdot b$

97. $7 \div 3m$ and $m \cdot 3 \div 7$

98. $(rt + st)5$ and $5t(r + s)$

99. $30y + x \cdot 15$ and $5[2(x + 3y)]$

100. $[c(2 + 3b)]5$ and $10c + 15bc$

101. Evaluate the expressions $3(2 + x)$ and $6 + x$ for $x = 0$. Do your results indicate that $3(2 + x)$ and $6 + x$ are equivalent? Why or why not?

102. Factor $15x + 40$. Then evaluate both $15x + 40$ and the factorization for $x = 4$. Do your results *guarantee* that the factorization is correct? Why or why not? (*Hint:* See Exercise 101.)

103. Aidan, Beth, and Cody consistently work more than 40 hours every week. The first 40 hours are paid at a regular pay rate of $10 per hour. The number of hours over 40 are paid at an overtime pay rate that is one and one-half times the regular pay rate. Each of the employees calculates the week's wages using a different formula.

- Aidan multiplies his overtime hours by 1.5, adds this to his regular hours, and then multiplies the sum by 10.
- Beth multiplies her regular hours by 10, then multiplies her overtime hours by 10 and then by 1.5. Finally, she adds the two amounts together.
- Cody multiplies his overtime hours by 15 and adds this to 400.

Let x represent the number of overtime hours worked in one week.

a) Write an algebraic expression for each method of calculating wages.

b) Use the commutative, associative, and distributive laws to show that all three methods of calculating wages yield the same total.

YOUR TURN ANSWERS: SECTION 1.2

1. $3 + 7a$ **2.** $(37m)p$ **3.** $2(3x)$; $(2 \cdot 3)x$; $6x$; answers may vary   **4.** $5x, 3y$ **5.** $4, x, y$ **6.** $15 + 5y$ **7.** $20a + 24m + 4$ **8.** $22 + 2y$ **9.** $5(3x + 1)$

Quick Quiz: Sections 1.1 and 1.2

To the student and the instructor: Beginning in the second section of each chapter, every exercise set contains a short quiz reviewing content already taught in the chapter. The numbers in brackets immediately following the directions or exercise indicate the section in which the skill was introduced. The answers to all quiz exercises appear at the back of the book. Continuous review of chapter content is excellent preparation for a chapter test.

1. Evaluate $x - y$ for $x = 17$ and $y = 8$.   [1.1]

2. Translate to an algebraic expression: Twice the sum of m and 3.   [1.1]

3. Translate to an equation. Do not solve. One-third of what number is 18?   [1.1]

4. Multiply: $3(x + 5y + 7)$.   [1.2]

5. Factor: $14a + 7t + 7$.   [1.2]

1.3 Fraction Notation

A. Factors and Prime Factorizations   **B.** Multiplication, Division, and Simplification of Fractions
C. Addition and Subtraction of Fractions

This section covers multiplication, addition, subtraction, and division with fractions, including fraction expressions that contain variables.

An example of **fraction notation** for a number is

$$\frac{2}{3},$$
← Numerator
← Denominator

The top number is called the **numerator**, and the bottom number is called the **denominator**.

A. Factors and Prime Factorizations

We first review how *natural numbers* are factored. **Natural numbers** can be thought of as the counting numbers:

$$1, 2, 3, 4, 5, \ldots.$$

(The dots indicate that the established pattern continues without ending.)

Since factors are parts of products, to factor a number, we express it as a product of two or more numbers. This product is called a **factorization**.

Several factorizations of 12 are

$$1 \cdot 12, \quad 2 \cdot 6, \quad 3 \cdot 4, \quad 2 \cdot 2 \cdot 3.$$

It is easy to overlook a factor of a number if the factorizations are not written methodically.

EXAMPLE 1   List all factors of 18.

SOLUTION   Beginning at 1, we check all natural numbers to see if they are factors of 18. If they are, we write the factorization. We stop when we have already included the next natural number in a factorization.

1 is a factor of every number. $1 \cdot 18$
2 is a factor of 18. $2 \cdot 9$
3 is a factor of 18. $3 \cdot 6$
4 is *not* a factor of 18.
5 is *not* a factor of 18.
6 is the next natural number, but we have already listed 6 as a factor in the product $3 \cdot 6$.

We stop at 6 because any natural number greater than 6 would be paired with a factor less than 6. (Remember that multiplication is commutative.)

We now write the factors of 18, going down the above list of factorizations writing the first factors and then up the list writing the second factors:

$$1, \quad 2, \quad 3, \quad 6, \quad 9, \quad 18.$$

Student Notes

If you are asked to "find the factors" of a number, your answer will be a list of numbers. If you are asked to "find a factorization" of a number, your answer will be a product. For example:

The factors of 12 are 1, 2, 3, 4, 6, 12.

A factorization of 12 is $2 \cdot 2 \cdot 3$.

1. List all factors of 54.

 YOUR TURN

Some numbers have only two different factors, the number itself and 1. Such numbers are called **prime**.

PRIME NUMBER

A *prime number* is a natural number that has exactly two different factors: the number itself and 1. The first several primes are 2, 3, 5, 7, 11, 13, 17, 19, and 23.

If a natural number other than 1 is not prime, we call it **composite**.

EXAMPLE 2 Label each number as prime, composite, or neither: 29, 4, 1.

SOLUTION

29 is prime. It has exactly two different factors, 29 and 1.

4 is not prime. It has three different factors, 1, 2, and 4. It is composite.

1 is not prime. It does not have two *different* factors. The number 1 is not considered composite. It is neither prime nor composite.

2. Label 21 as prime, composite, or neither.

YOUR TURN

Every composite number can be factored into a product of prime numbers. Such a factorization is called the **prime factorization** of that composite number.

EXAMPLE 3 Find the prime factorization of 36.

SOLUTION We first factor 36 in any way that we can, such as

$$36 = 4 \cdot 9.$$

Since 4 and 9 are not prime, we factor them:

$$36 = 4 \cdot 9$$
$$= 2 \cdot 2 \cdot 3 \cdot 3. \qquad \text{2 and 3 are both prime.}$$

3. Find the prime factorization of 100.

The prime factorization of 36 is $2 \cdot 2 \cdot 3 \cdot 3$.

YOUR TURN

Student Notes

When writing a factorization, you are writing an equivalent expression for the original number. Some students do this with a tree diagram:

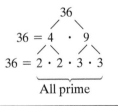

B. Multiplication, Division, and Simplification of Fractions

Recall from arithmetic that fractions are multiplied as follows.

MULTIPLICATION OF FRACTIONS

For any two fractions a/b and c/d,

$$\frac{a}{b} \cdot \frac{c}{d} = \frac{ac}{bd}.$$

(The numerator of the product is the product of the two numerators. The denominator of the product is the product of the two denominators.)

EXAMPLE 4 Multiply: **(a)** $\dfrac{2}{3} \cdot \dfrac{5}{7}$; **(b)** $\dfrac{4}{x} \cdot \dfrac{8}{y}$; **(c)** $9 \cdot \dfrac{7}{n}$.

SOLUTION We multiply numerators as well as denominators.

a) $\dfrac{2}{3} \cdot \dfrac{5}{7} = \dfrac{2 \cdot 5}{3 \cdot 7} = \dfrac{10}{21}$ 　　　　 **b)** $\dfrac{4}{x} \cdot \dfrac{8}{y} = \dfrac{4 \cdot 8}{x \cdot y} = \dfrac{32}{xy}$

4. Multiply: $\dfrac{2}{x} \cdot \dfrac{3}{7}$.

c) $9 \cdot \dfrac{7}{n} = \dfrac{9}{1} \cdot \dfrac{7}{n} = \dfrac{9 \cdot 7}{1 \cdot n} = \dfrac{63}{n}$ 　　 9 can be written $\dfrac{9}{1}$.

🔁 YOUR TURN

Any nonzero number divided by itself is 1.

> **FRACTION NOTATION FOR 1**
>
> For any number a, except 0,
>
> $$\dfrac{a}{a} = 1.$$

Two numbers whose product is 1 are **reciprocals**, or **multiplicative inverses**, of each other. All numbers, except zero, have reciprocals. For example,

the reciprocal of $\dfrac{2}{3}$ is $\dfrac{3}{2}$ because $\dfrac{2}{3} \cdot \dfrac{3}{2} = \dfrac{6}{6} = 1$;

the reciprocal of 9 is $\dfrac{1}{9}$ because $9 \cdot \dfrac{1}{9} = \dfrac{9}{1} \cdot \dfrac{1}{9} = \dfrac{9}{9} = 1$;

the reciprocal of $\dfrac{1}{4}$ is 4 because $\dfrac{1}{4} \cdot 4 = \dfrac{1}{4} \cdot \dfrac{4}{1} = \dfrac{4}{4} = 1$.

Reciprocals are used to rewrite division in an equivalent form that uses multiplication.

> **DIVISION OF FRACTIONS**
>
> To divide two fractions, multiply by the reciprocal of the divisor:
>
> $$\dfrac{a}{b} \div \dfrac{c}{d} = \dfrac{a}{b} \cdot \dfrac{d}{c}.$$

EXAMPLE 5 Divide: $\dfrac{1}{2} \div \dfrac{3}{5}$.

SOLUTION

$$\dfrac{1}{2} \div \dfrac{3}{5} = \dfrac{1}{2} \cdot \dfrac{5}{3}$$ 　　 $\dfrac{5}{3}$ is the reciprocal of $\dfrac{3}{5}$.

5. Divide: $\dfrac{3}{4} \div \dfrac{5}{7}$.

$$= \dfrac{5}{6}$$

🔁 YOUR TURN

Multiplying a number by 1 gives that same number because of the *identity property of* 1. A similar property can be stated for division.

Study Skills

Do the Exercises

- When you complete an odd-numbered exercise, you can check your answer at the back of the book. If an answer is incorrect, closely examine your work and, if necessary, consult your instructor for guidance.

- Do some even-numbered exercises, even if none are assigned. Because there are no answers given for them, you will gain practice doing exercises in a context similar to taking a quiz or a test. Check your answers later with a friend or your instructor.

THE IDENTITY PROPERTY OF 1

For any number a,

$$a \cdot 1 = 1 \cdot a = a.$$

(Multiplying a number by 1 gives that same number.) The number 1 is called the *multiplicative identity*.

For example, we can multiply $\frac{4}{5} \cdot \frac{6}{6}$ to find an expression equivalent to $\frac{4}{5}$. Since $\frac{6}{6} = 1$, the expression $\frac{4}{5} \cdot \frac{6}{6}$ is equivalent to $\frac{4}{5} \cdot 1$, or simply $\frac{4}{5}$. We have

$$\frac{4}{5} \cdot \frac{6}{6} = \frac{4 \cdot 6}{5 \cdot 6} = \frac{24}{30}.$$

Thus, $\frac{24}{30}$ is equivalent to $\frac{4}{5}$.

We reverse these steps by "removing a factor equal to 1"—in this case, $\frac{6}{6}$. By removing a factor that equals 1, we can *simplify* an expression like $\frac{24}{30}$ to an equivalent expression like $\frac{4}{5}$.

To simplify, we factor the numerator and the denominator, looking for the largest factor common to both. This is sometimes made easier by writing prime factorizations. After identifying common factors, we can express the fraction as a product of two fractions, one of which is in the form a/a.

EXAMPLE 6 Simplify: **(a)** $\dfrac{15}{40}$; **(b)** $\dfrac{36}{24}$.

SOLUTION

a) Note that 5 is a factor of both 15 and 40:

$$\frac{15}{40} = \frac{3 \cdot 5}{8 \cdot 5} \qquad \text{Factoring the numerator and the denominator, using the common factor, 5}$$

$$= \frac{3}{8} \cdot \frac{5}{5} \qquad \text{Rewriting as a product of two fractions; } \frac{5}{5} = 1$$

$$= \frac{3}{8} \cdot 1 = \frac{3}{8}. \qquad \text{Using the identity property of 1 (removing a factor equal to 1)}$$

b) $\dfrac{36}{24} = \dfrac{2 \cdot 2 \cdot 3 \cdot 3}{2 \cdot 2 \cdot 2 \cdot 3}$ Writing the prime factorizations and identifying common factors; 12/12 could also be used.

$$= \frac{3}{2} \cdot \frac{2 \cdot 2 \cdot 3}{2 \cdot 2 \cdot 3} \qquad \text{Rewriting as a product of two fractions; } \frac{2 \cdot 2 \cdot 3}{2 \cdot 2 \cdot 3} = 1$$

$$= \frac{3}{2} \cdot 1 = \frac{3}{2} \qquad \text{Using the identity property of 1}$$

6. Simplify: $\dfrac{35}{30}$.

YOUR TURN

It is always wise to check your result to see if any common factors of the numerator and the denominator remain. (This will not occur if prime factorizations are used correctly.) If common factors remain, repeat the process by removing another factor equal to 1 to simplify your result.

Student Notes

The following rules can help you quickly determine whether 2, 3, or 5 is a factor of a number.

2 is a factor of a number if the number is even (the ones digit is 0, 2, 4, 6, or 8).

3 is a factor of a number if the sum of its digits is divisible by 3.

5 is a factor of a number if its ones digit is 0 or 5.

"Canceling" is a shortcut that you may have used for removing a factor equal to 1 when working with fraction notation. With *great* concern, we mention it as a possible way to speed up your work. Canceling can be used only when removing common factors in numerators and denominators. Canceling *cannot* be used in sums or differences. Our concern is that "canceling" be used with understanding. Example 6(b) might have been done faster as follows:

$$\frac{36}{24} = \frac{2 \cdot 2 \cdot 3 \cdot \cancel{3}}{2 \cdot 2 \cdot 2 \cdot \cancel{3}} = \frac{3}{2}, \quad \text{or} \quad \frac{36}{24} = \frac{3 \cdot \cancel{12}}{2 \cdot \cancel{12}} = \frac{3}{2}, \quad \text{or} \quad \frac{\overset{3}{\overset{18}{\cancel{\cancel{36}}}}}{\underset{2}{\underset{12}{\cancel{\cancel{24}}}}} = \frac{3}{2}.$$

> **CAUTION!**   Unfortunately, canceling is often performed incorrectly:
>
> $$\frac{2+3}{2} \ne 3, \qquad \frac{4-1}{4-2} \ne \frac{1}{2}, \qquad \frac{15}{54} \ne \frac{1}{4}.$$
>
> The cancellations above are incorrect because the expressions canceled are *not* factors. For example, in $2 + 3$, the 2 and the 3 are terms, not factors. Only factors can be canceled. Correct simplifications are as follows:
>
> $$\frac{2+3}{2} = \frac{5}{2}, \qquad \frac{4-1}{4-2} = \frac{3}{2}, \qquad \frac{15}{54} = \frac{5 \cdot \cancel{3}}{18 \cdot \cancel{3}} = \frac{5}{18}.$$
>
> *Remember*: **If you can't factor, you can't cancel! If in doubt, don't cancel!**

Sometimes it is helpful to use 1 as a factor in the numerator or the denominator when simplifying.

EXAMPLE 7   Simplify: $\dfrac{9}{72}$.

SOLUTION

$$\frac{9}{72} = \frac{1 \cdot 9}{8 \cdot 9} \qquad \text{Factoring and using the identity property of 1 to write 9 as } 1 \cdot 9$$

$$= \frac{1 \cdot \cancel{9}}{8 \cdot \cancel{9}} = \frac{1}{8} \qquad \text{Simplifying by removing a factor equal to 1: } \tfrac{9}{9} = 1$$

7. Simplify: $\dfrac{6}{18}$.

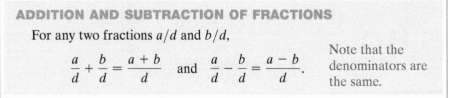 YOUR TURN

C. Addition and Subtraction of Fractions

When denominators are the same, fractions are added or subtracted by adding or subtracting numerators and keeping the same denominator.

> **CAUTION!**   When adding or subtracting fractions with the same denominator, add or subtract only the numerators. The denominator does not change.

ADDITION AND SUBTRACTION OF FRACTIONS

For any two fractions a/d and b/d,

$$\frac{a}{d} + \frac{b}{d} = \frac{a+b}{d} \quad \text{and} \quad \frac{a}{d} - \frac{b}{d} = \frac{a-b}{d}.$$

Note that the denominators are the same.

EXAMPLE 8   Add and simplify: $\dfrac{4}{8} + \dfrac{5}{8}$.

SOLUTION   The common denominator is 8. We add the numerators and keep the common denominator:

$$\frac{4}{8} + \frac{5}{8} = \frac{4+5}{8} = \frac{9}{8}.$$ You can think of this as $4 \cdot \frac{1}{8} + 5 \cdot \frac{1}{8} = 9 \cdot \frac{1}{8}$, or $\frac{9}{8}$.

8. Add and simplify: $\dfrac{2}{5} + \dfrac{4}{5}$.

↩ YOUR TURN

In arithmetic, we often write $1\frac{1}{8}$ rather than the "improper" fraction $\frac{9}{8}$. In algebra, $\frac{9}{8}$ is generally more useful and is quite "proper" for our purposes.

When denominators are different, we use the identity property of 1 and multiply to find a common denominator. Then we add, as in Example 8.

A common denominator is a number that is divisible by the denominator of each fraction. One method of finding a common denominator uses prime factorizations.

FINDING A COMMON DENOMINATOR USING PRIME FACTORIZATIONS

1. Find the prime factorization of each denominator.
2. Choose one factorization.
3. Multiply that factorization by any factors of the other denominator that it lacks.

Student Notes

Another way to find a common denominator is to list multiples of the larger denominator, stopping when the multiple is also a multiple of the smaller denominator. For example, in Example 9, multiples of 12 are 12, 24, 36, ... The first number in the list, 12, is not a multiple of 8, but the second multiple, 24, is also a multiple of 8. Thus the least common denominator is 24.

EXAMPLE 9   Add: $\dfrac{7}{8} + \dfrac{5}{12}$.

SOLUTION   We follow the steps listed above to find a common denominator.

1. $8 = 2 \cdot 2 \cdot 2$ Find the prime factorization of each denominator.
   $12 = 2 \cdot 2 \cdot 3$
2. $2 \cdot 2 \cdot 2$ Choose one factorization.
3. $2 \cdot 2 \cdot 2 \cdot 3$ The factorization of 8 is "missing" the factor 3 from the factorization of 12.

The common denominator is thus $2 \cdot 2 \cdot 2 \cdot 3$, or 24. It is divisible by both 8 and 12. We multiply both $\frac{7}{8}$ and $\frac{5}{12}$ by suitable forms of 1 to obtain two fractions with denominators of 24:

$$\frac{7}{8} + \frac{5}{12} = \frac{7}{8} \cdot \frac{3}{3} + \frac{5}{12} \cdot \frac{2}{2}$$ Multiplying by 1. Since $8 \cdot 3 = 24$, we multiply $\frac{7}{8}$ by $\frac{3}{3}$. Since $12 \cdot 2 = 24$, we multiply $\frac{5}{12}$ by $\frac{2}{2}$.

$$= \frac{21}{24} + \frac{10}{24}$$ Performing the multiplication

$$= \frac{31}{24}.$$ Adding fractions

9. Add: $\dfrac{9}{8} + \dfrac{4}{5}$.

↩ YOUR TURN

After adding, subtracting, multiplying, or dividing, we may still need to simplify the answer.

EXAMPLE 10 Perform the indicated operation and, if possible, simplify.

a) $\dfrac{7}{10} - \dfrac{1}{5}$ **b)** $8 \cdot \dfrac{5}{12}$ **c)** $\dfrac{\frac{5}{6}}{\frac{25}{9}}$

SOLUTION

a) Since one denominator, 10, is a multiple of the other denominator, 5, we use 10 as the common denominator.

$$\dfrac{7}{10} - \dfrac{1}{5} = \dfrac{7}{10} - \dfrac{1}{5}\cdot\dfrac{2}{2}$$ Since $5\cdot 2 = 10$, we multiply $\dfrac{1}{5}$ by $\dfrac{2}{2}$.

$$= \dfrac{7}{10} - \dfrac{2}{10}$$

$$= \dfrac{5}{10} = \dfrac{1\cdot 5}{2\cdot 5} = \dfrac{1}{2}$$ Removing a factor equal to 1: $\dfrac{5}{5} = 1$

b) $8 \cdot \dfrac{5}{12} = \dfrac{8\cdot 5}{12}$ Multiplying numerators and denominators. Think of 8 as $\frac{8}{1}$.

$$= \dfrac{2\cdot 2\cdot 2\cdot 5}{2\cdot 2\cdot 3}$$ Factoring; $\dfrac{4\cdot 2\cdot 5}{4\cdot 3}$ can also be used.

$$= \dfrac{2\cdot 2\cdot 2\cdot 5}{2\cdot 2\cdot 3}$$ Removing a factor equal to 1: $\dfrac{2\cdot 2}{2\cdot 2} = 1$

$$= \dfrac{10}{3}$$ Simplifying

c) $\dfrac{\frac{5}{6}}{\frac{25}{9}} = \dfrac{5}{6} \div \dfrac{25}{9}$ Rewriting horizontally. Remember that a fraction bar indicates division.

$$= \dfrac{5}{6}\cdot\dfrac{9}{25}$$ Multiplying by the reciprocal of $\frac{25}{9}$

$$= \dfrac{5\cdot 3\cdot 3}{2\cdot 3\cdot 5\cdot 5}$$ Writing as one fraction and factoring

$$= \dfrac{5\cdot 3\cdot 3}{2\cdot 3\cdot 5\cdot 5}$$ Removing a factor equal to 1: $\dfrac{5\cdot 3}{3\cdot 5} = 1$

$$= \dfrac{3}{10}$$ Simplifying

YOUR TURN

Check Your UNDERSTANDING

For each exercise, determine which of the following is the first step in performing the operation.

a) Find a common denominator.
b) Multiply numerators and then multiply denominators.
c) Multiply by the reciprocal of the divisor.

1. $\dfrac{2}{3}\cdot\dfrac{4}{5}$ **2.** $\dfrac{2}{3} + \dfrac{4}{5}$

3. $\dfrac{2}{3} \div \dfrac{4}{5}$ **4.** $\dfrac{9}{10} \div 2$

5. $\dfrac{5}{8} - \dfrac{1}{4}$ **6.** $\dfrac{3}{8}\cdot\dfrac{7}{2}$

10. Subtract and, if possible, simplify:

$$\dfrac{9}{20} - \dfrac{1}{4}.$$

Technology Connection

Many calculators can perform operations using fraction notation. You may find an option n/d on a key or in a **menu** of options that appears when a key is pressed. To select an item from a menu, we highlight its number and press **ENTER** or simply press the number of the item.

For some calculators, to add $\frac{2}{15} + \frac{7}{12}$, we first press **MATH** ⟩ ⌃ **ENTER** to choose the n/d option in the MATH NUM submenu. An empty fraction template will appear on the screen. We then press ② ⌄ ① ⑤ to enter the numerator and the denominator of the first fraction. Next, we press ⟩ to move outside of the fraction and then press ⊕ . To enter the second fraction, we press **MATH** ⟩ ⌃ **ENTER** again and then enter the numerator and the denominator of the second fraction. We press **ENTER** to calculate the result.

$$\frac{2}{15} + \frac{7}{12}$$
$$\frac{43}{60}$$

Since a fraction bar indicates division, we can also enter $\frac{2}{15} + \frac{7}{12}$ as 2/15 + 7/12. The answer is typically given in decimal notation. To convert this to fraction notation, we press **MATH** and select the FRAC option. The notation ANS ▶ FRAC shows that the calculator will convert .7166666667 to fraction notation.

```
2/15+7/12
            .7166666667
Ans▶Frac
            43/60
```

We see that $\frac{2}{15} + \frac{7}{12} = \frac{43}{60}$.

VIDEO

1.3 EXERCISE SET

⬥ Vocabulary and Reading Check

Choose from the following list of words to complete each statement. Not every word will be used.

add	multiply	prime
composite	numerator	reciprocal
denominator	opposite	

1. The top number in a fraction is called the

   _____.

2. A(n) _____ number has exactly two different factors.

3. To divide two fractions, multiply by the _____ of the divisor.

4. We need a common denominator in order to _____ fractions.

⬥ Concept Reinforcement

In each of Exercises 5–8, match the description with a number from the list on the right.

5. ____ A factor of 35   a) 2

6. ____ A number that has 3 as a factor   b) 7

7. ____ An odd composite number   c) 60

8. ____ The only even prime number   d) 65

To the student and the instructor: Beginning in this section, selected exercises are marked with the symbol Aha!. Students who pause to inspect an Aha! exercise should find the answer more readily than those who proceed mechanically. This is done to discourage rote memorization. Some later "Aha!" exercises in this exercise set are unmarked, to encourage students to always pause before working a problem.

A. Factors and Prime Factorizations

Label each of the following numbers as prime, composite, or neither.

9. 9	**10.** 15	**11.** 41	**12.** 49
13. 77	**14.** 37	**15.** 2	**16.** 1
17. 0	**18.** 16		

List all the factors of each number.

19. 50	**20.** 70	**21.** 42	**22.** 60

Find the prime factorization of each number. If the number is prime, state this.

23. 39	**24.** 34	**25.** 30
26. 98	**27.** 27	**28.** 54
29. 150	**30.** 56	**31.** 31

32. 180   **33.** 210   **34.** 79

35. 115   **36.** 143

B. Simplification of Fractions

Simplify.

37. $\dfrac{21}{35}$ **38.** $\dfrac{20}{26}$ **39.** $\dfrac{16}{56}$

40. $\dfrac{72}{27}$ **41.** $\dfrac{12}{48}$ **42.** $\dfrac{18}{84}$

43. $\dfrac{52}{13}$ **44.** $\dfrac{132}{11}$ **45.** $\dfrac{19}{76}$

46. $\dfrac{17}{51}$ **47.** $\dfrac{150}{25}$ **48.** $\dfrac{180}{36}$

49. $\dfrac{42}{50}$ **50.** $\dfrac{75}{80}$ **51.** $\dfrac{120}{82}$

52. $\dfrac{75}{45}$ **53.** $\dfrac{210}{98}$ **54.** $\dfrac{140}{350}$

B, C. Multiplication, Division, Addition, and Subtraction of Fractions

Perform the indicated operation and, if possible, simplify.

55. $\dfrac{1}{2} \cdot \dfrac{3}{5}$ **56.** $\dfrac{11}{10} \cdot \dfrac{8}{5}$ **57.** $\dfrac{9}{2} \cdot \dfrac{4}{3}$

Aha! **58.** $\dfrac{11}{12} \cdot \dfrac{12}{11}$ **59.** $\dfrac{1}{8} + \dfrac{3}{8}$ **60.** $\dfrac{1}{10} + \dfrac{7}{10}$

61. $\dfrac{4}{9} + \dfrac{13}{18}$ **62.** $\dfrac{4}{5} + \dfrac{8}{15}$ **63.** $\dfrac{3}{a} \cdot \dfrac{b}{7}$

64. $\dfrac{x}{5} \cdot \dfrac{y}{z}$ **65.** $\dfrac{4}{n} + \dfrac{6}{n}$ **66.** $\dfrac{9}{x} - \dfrac{5}{x}$

67. $\dfrac{3}{10} + \dfrac{8}{15}$ **68.** $\dfrac{7}{8} + \dfrac{5}{12}$ **69.** $\dfrac{11}{7} - \dfrac{4}{7}$ **1**

70. $\dfrac{12}{5} - \dfrac{2}{5}$ **2**   **71.** $\dfrac{13}{18} - \dfrac{4}{9}$ **72.** $\dfrac{13}{15} - \dfrac{11}{45}$

73. $\dfrac{11}{30} - \dfrac{2}{9}$ **74.** $\dfrac{5}{14} - \dfrac{5}{21}$ **75.** $\dfrac{7}{6} \div \dfrac{3}{5}$

76. $\dfrac{7}{5} \div \dfrac{10}{3}$ **77.** $12 \div \dfrac{4}{9}$ **27**   **78.** $\dfrac{9}{4} \div 9$

Aha! **79.** $\dfrac{7}{13} \div \dfrac{7}{13}$ **80.** $\dfrac{1}{10} \div \dfrac{1}{5}$ **81.** $\dfrac{\frac{2}{7}}{\frac{5}{3}}$

82. $\dfrac{\frac{3}{8}}{\frac{1}{5}}$ **83.** $\dfrac{9}{\frac{1}{2}}$ **84.** $\dfrac{\frac{3}{7}}{6}$

85. Under what circumstances would the sum of two fractions be easier to compute than the product of the same two fractions?

86. Under what circumstances would the product of two fractions be easier to compute than the sum of the same two fractions?

Synthesis

87. Bryce insists that $(2 + x)/8$ is equivalent to $(1 + x)/4$. What mistake do you think is being made and how could you demonstrate to Bryce that the two expressions are not equivalent?

88. *Research.* Mathematicians use computers to determine whether very large numbers are prime. Find the largest prime number currently known, and describe how it was found.

89. In the following table, the top number can be factored in such a way that the sum of the factors is the bottom number. For example, in the first column, 56 is factored as $7 \cdot 8$, since $7 + 8 = 15$, the bottom number. Find the missing numbers in each column.

Product	56	63	36	72	140	96	168
Factor	7						
Factor	8						
Sum	15	16	20	38	24	20	29

90. *Packaging.* Tritan Candies uses two sizes of boxes, 6 in. long and 8 in. long. These are packed end to end in larger cartons to be shipped. What is the shortest-length carton that will accommodate boxes of either size without any room left over? (Each carton must contain boxes of only one size; no mixing is allowed.)

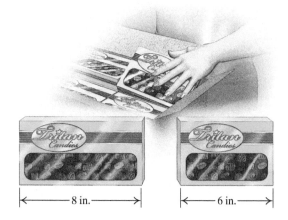

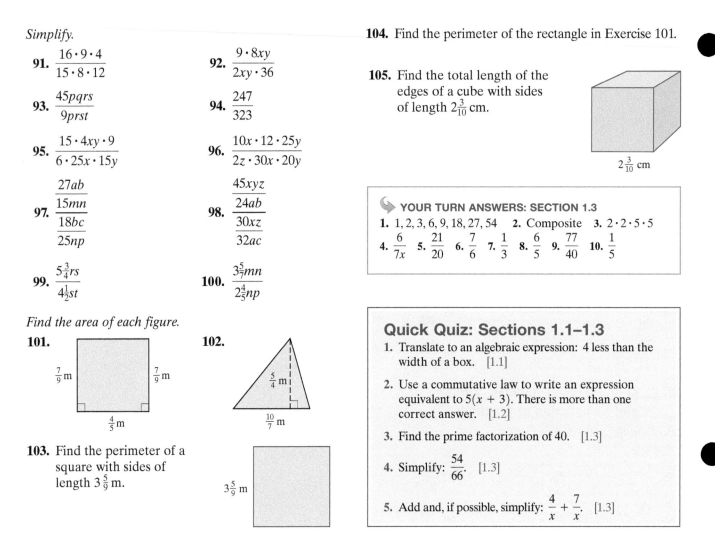

Simplify.

91. $\dfrac{16 \cdot 9 \cdot 4}{15 \cdot 8 \cdot 12}$

92. $\dfrac{9 \cdot 8xy}{2xy \cdot 36}$

93. $\dfrac{45pqrs}{9prst}$

94. $\dfrac{247}{323}$

95. $\dfrac{15 \cdot 4xy \cdot 9}{6 \cdot 25x \cdot 15y}$

96. $\dfrac{10x \cdot 12 \cdot 25y}{2z \cdot 30x \cdot 20y}$

97. $\dfrac{\frac{27ab}{15mn}}{\frac{18bc}{25np}}$

98. $\dfrac{\frac{45xyz}{24ab}}{\frac{30xz}{32ac}}$

99. $\dfrac{5\frac{3}{4}rs}{4\frac{1}{2}st}$

100. $\dfrac{3\frac{5}{7}mn}{2\frac{4}{5}np}$

Find the area of each figure.

101.

$\frac{7}{9}$ m (left side), $\frac{7}{9}$ m (right side), $\frac{4}{5}$ m (bottom)

102.

$\frac{5}{4}$ m (height), $\frac{10}{7}$ m (base)

103. Find the perimeter of a square with sides of length $3\frac{5}{9}$ m.

$3\frac{5}{9}$ m

104. Find the perimeter of the rectangle in Exercise 101.

105. Find the total length of the edges of a cube with sides of length $2\frac{3}{10}$ cm.

$2\frac{3}{10}$ cm

YOUR TURN ANSWERS: SECTION 1.3

1. $1, 2, 3, 6, 9, 18, 27, 54$ **2.** Composite   **3.** $2 \cdot 2 \cdot 5 \cdot 5$
4. $\dfrac{6}{7x}$ **5.** $\dfrac{21}{20}$ **6.** $\dfrac{7}{6}$ **7.** $\dfrac{1}{3}$ **8.** $\dfrac{6}{5}$ **9.** $\dfrac{77}{40}$ **10.** $\dfrac{1}{5}$

Quick Quiz: Sections 1.1–1.3

1. Translate to an algebraic expression: 4 less than the width of a box.   [1.1]

2. Use a commutative law to write an expression equivalent to $5(x + 3)$. There is more than one correct answer.   [1.2]

3. Find the prime factorization of 40.   [1.3]

4. Simplify: $\dfrac{54}{66}$.   [1.3]

5. Add and, if possible, simplify: $\dfrac{4}{x} + \dfrac{7}{x}$.   [1.3]

1.4   Positive and Negative Real Numbers

A. Whole Numbers and Integers   **B.** Rational Numbers   **C.** Real Numbers and Order   **D.** Absolute Value

A **set** is a collection of objects. The set containing 1, 3, and 7 is usually written $\{1, 3, 7\}$. In this section, we examine some important sets of numbers.

A. Whole Numbers and Integers

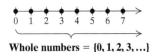

Whole numbers = {0, 1, 2, 3, ...}

The set of **whole numbers** can be written as $\{0, 1, 2, 3, \ldots\}$. We represent this set using dots on the number line, as shown at left.

To create the set of integers, we include all whole numbers, along with their *opposites*. To find the opposite of a number, we locate the number that is the same distance from 0 but on the other side of 0 on the number line. For example,

the opposite of 1 is negative 1, written -1;

and the opposite of 3 is negative 3, written -3.

Student Notes

It is not uncommon in mathematics for a symbol to have more than one meaning in different contexts. The symbol "−" in 5 − 3 indicates subtraction. The same symbol in −10 indicates the opposite of 10, or negative 10.

The **integers** consist of all whole numbers and their opposites. Note that, except for 0, opposites occur in pairs. Thus, 5 is the opposite of −5, just as −5 is the opposite of 5. The number 0 acts as its own opposite.

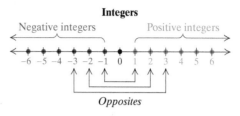

Opposites

SET OF INTEGERS

The set of integers = $\{\ldots, -4, -3, -2, -1, 0, 1, 2, 3, 4, \ldots\}$.

Integers are associated with many real-world problems and situations.

EXAMPLE 1 State which integer(s) corresponds to each situation.

a) In 2015, the per-capita household debt in Washington, D.C., was $76,080.

   Data: Federal Reserve Bank of New York

b) Badwater Basin in Death Valley, California, is 282 ft below sea level.

SOLUTION

a) The integer −76,080 corresponds to a debt of $76,080.

b) The integer −282 corresponds to 282 ft below sea level. The elevation is −282 ft.

1. State what integers correspond to this situation: The highest temperature ever recorded in Australia was 123 degrees Fahrenheit (°F) in Oodnadatta, South Australia. The coldest temperature ever recorded in Australia was 14°F below zero in Charlotte Pass, New South Wales.

YOUR TURN

B. Rational Numbers

A number like $\frac{5}{9}$, although built out of integers, is not itself an integer. Another set of numbers, the **rational numbers**, contains integers, fractions, and decimals. Some examples of rational numbers are

$$\frac{5}{9}, \quad -\frac{4}{7}, \quad 95, \quad -16, \quad 0, \quad \frac{-35}{8}, \quad 2.4, \quad -0.31.$$

The number $-\frac{4}{7}$ can be written as $\frac{-4}{7}$ or $\frac{4}{-7}$. Indeed, every number listed above can be written as an integer over an integer. For example, 95 can be written as $\frac{95}{1}$ and 2.4 can be written as $\frac{24}{10}$. In this manner, any *ratio*nal number can be expressed as the *ratio* of two integers. Rather than attempt to list all rational numbers, we use this idea of ratio to describe the set as follows.

> **SET OF RATIONAL NUMBERS**
>
> The set of rational numbers $= \left\{ \dfrac{a}{b} \,\middle|\, a \text{ and } b \text{ are integers and } b \neq 0 \right\}$.
>
> This is read "the set of all numbers a over b, such that a and b are integers and b does not equal zero."

You may have noted the following relationships among the sets or types of numbers.

- Natural numbers are counting numbers: $\{1, 2, 3, \ldots\}$.
- Whole numbers include all natural numbers and 0: $\{0, 1, 2, 3, \ldots\}$.
- Integers include all whole numbers and their opposites:

$$\{\ldots, -3, -2, -1, 0, 1, 2, 3, \ldots\}.$$

- Rational numbers are all ratios of integers, excluding those with denominators of 0. Rational numbers include all integers.

To *graph* a number is to mark its location on the number line.

EXAMPLE 2   Graph each of the following rational numbers: **(a)** $\frac{5}{2}$; **(b)** -3.2; **(c)** $\frac{11}{8}$.

SOLUTION

(a) Since $\frac{5}{2} = 2\frac{1}{2} = 2.5$, its graph is halfway between 2 and 3.

(b) -3.2 is $\frac{2}{10}$ of a unit to the left of -3.

(c) $\frac{11}{8} = 1\frac{3}{8} = 1.375$

2. Graph -1.1 on the number line.

⤵ YOUR TURN

Every rational number can be written using fraction notation or decimal notation.

EXAMPLE 3   Convert to decimal notation: $-\frac{5}{8}$.

SOLUTION   We first find decimal notation for $\frac{5}{8}$. Since $\frac{5}{8}$ means $5 \div 8$, we divide.

$$
\begin{array}{r}
0.6\,2\,5 \\
8\overline{)5.0\,0\,0} \\
\underline{4\,8} \\
2\,0 \\
\underline{1\,6} \\
4\,0 \\
\underline{4\,0} \\
0
\end{array}
$$

← The remainder is 0.

3. Convert $-\dfrac{7}{4}$ to decimal notation.

Thus, $\frac{5}{8} = 0.625$, so $-\frac{5}{8} = -0.625$.

⤵ YOUR TURN

Because the division in Example 3 ends with the remainder 0, we consider -0.625 a **terminating decimal**. If we are "bringing down" zeros and a remainder reappears, we have a **repeating decimal**, as shown in the next example.

EXAMPLE 4 Convert to decimal notation: $\frac{7}{11}$.

SOLUTION We divide:

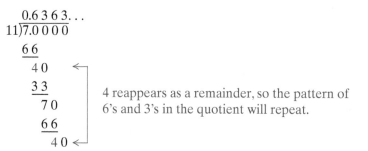

We abbreviate repeating decimals by writing a bar over the repeating part—in this case, $0.\overline{63}$. Thus, $\frac{7}{11} = 0.\overline{63}$.

4. Convert $-\frac{1}{6}$ to decimal notation.

YOUR TURN

Although we do not prove it here, the following is true.

Every rational number can be expressed as either a terminating decimal or a repeating decimal, and every terminating decimal or repeating decimal is a rational number.

C. Real Numbers and Order

Some numbers, when written in decimal form, neither terminate nor repeat. Such numbers are called **irrational numbers**. These numbers cannot be expressed as a ratio of two integers.

What sort of numbers are irrational? One example is π (the Greek letter *pi*, read "pie"), which is used to find the area and the circumference of a circle: $A = \pi r^2$ and $C = 2\pi r$.

Another irrational number, $\sqrt{2}$ (read "the square root of 2"), is the length of the diagonal of a square with sides of length 1. It is also the number that, when multiplied by itself, gives 2. No rational number can be multiplied by itself to get 2, although some approximations come close:

1.4 is an *approximation* of $\sqrt{2}$ because $(1.4)(1.4) = 1.96$;

1.41 is a better approximation because $(1.41)(1.41) = 1.9881$;

1.4142 is an even better approximation because $(1.4142)(1.4142) = 1.99996164$.

To approximate $\sqrt{2}$ on some calculators, we simply press ② and then √. With other calculators, we press √ ② **ENTER**. If needed, consult your manual.

EXAMPLE 5 Graph the real number $\sqrt{3}$ on the number line.

SOLUTION We use a calculator and approximate: $\sqrt{3} \approx 1.732$ ("$\approx$" means "approximately equals"). Then we locate this number on the number line.

5. Graph $\sqrt{5}$ on the number line.

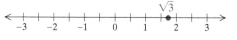

YOUR TURN

Not every square root is irrational. If the number under the radical sign √ is a perfect square, such as 4, 9, 16, or 25, the square root is rational. For example, $\sqrt{4} = 2$, $\sqrt{9} = 3$, $\sqrt{16} = 4$, and $\sqrt{25} = 5$.

The rational numbers and the irrational numbers together correspond to all the points on the number line and make up what is called the **real-number system**.

> **SET OF REAL NUMBERS**
>
> The set of real numbers = The set of all numbers corresponding to points on the number line.

The following figure shows the relationships among various kinds of numbers.

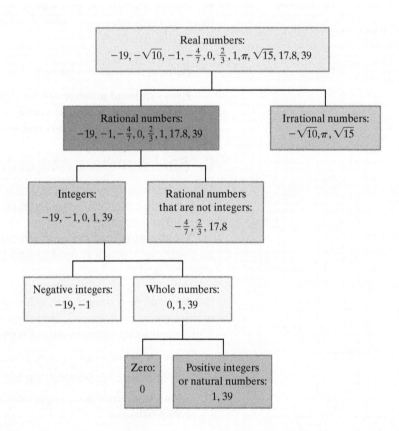

EXAMPLE 6   Which numbers in the following group are **(a)** whole numbers? **(b)** integers? **(c)** rational numbers? **(d)** irrational numbers? **(e)** real numbers?

$$-38, \quad -\frac{8}{5}, \quad 0, \quad 0.\overline{3}, \quad 4.5, \quad \sqrt{30}, \quad 52$$

SOLUTION

a) 0 and 52 are whole numbers.

b) $-38, 0,$ and 52 are integers.

c) $-38, -\frac{8}{5}, 0, 0.\overline{3}, 4.5,$ and 52 are rational numbers.

d) $\sqrt{30}$ is an irrational number.

e) $-38, -\frac{8}{5}, 0, 0.\overline{3}, 4.5, \sqrt{30},$ and 52 are real numbers.

6. Which numbers in the following group are integers?

$$5, \quad -15, \quad 0, \quad -3.6, \quad \frac{2}{3}$$

YOUR TURN

EXPLORING 🔍 **THE CONCEPT**

Use the number line shown at right to answer the following questions.

1. Graph the numbers 2 and 5.
   a) Which number is farther to the right on the number line?
   b) Which number is greater?
2. Graph the numbers 2 and −5.
   a) Which number is farther to the right on the number line?
   b) Which number is greater?
3. Graph the numbers −2 and −5.
   a) Which number is farther to the right on the number line?
   b) Which number is greater?
4. Why is 4 > 1, but −4 < −1?

ANSWERS
1. (a) 5; (b) 5   2. (a) 2; (b) 2   3. (a) −2; (b) −2
4. 4 is to the right of 1 on the number line, and −1 is to the right of −4 on the number line.

Real numbers are named in order on the number line, with larger numbers farther to the right. For any two numbers, the one to the left is less than the one to the right. We use the symbol **<** to mean "**is less than**." The sentence −8 < 6 means "−8 is less than 6." The symbol **>** means "**is greater than**." The sentence −3 > −7 means "−3 is greater than −7."

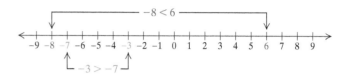

EXAMPLE 7 Use either < or > for ▓ to write a true sentence.

a) 2 ▓ 9

b) −3.45 ▓ 1.32

c) 6 ▓ −12

d) −18 ▓ −5

e) $\frac{7}{11}$ ▓ $\frac{5}{8}$

SOLUTION

a) Since 2 is to the left of 9 on the number line, we know that 2 is less than 9, so 2 < 9.

b) Since −3.45 is to the left of 1.32, we have −3.45 < 1.32.

c) Since 6 is to the right of −12, we have 6 > −12.

d) Since −18 is to the left of −5, we have −18 < −5.

e) We convert to decimal notation: $\frac{7}{11} = 0.\overline{63}$ and $\frac{5}{8} = 0.625$. Thus, $\frac{7}{11} > \frac{5}{8}$.

We also could have used a common denominator: $\frac{7}{11} = \frac{56}{88} > \frac{55}{88} = \frac{5}{8}$.

↩ YOUR TURN

7. Use either < or > for ▓ to write a true sentence: −8 ▓ −9.

Sentences like "$a < -5$" and "$-3 > -8$" are **inequalities**. It may be helpful to think of an inequality sign as "opening" toward the larger number, or as being widest next to the larger number.

Note that $a > 0$ means that a represents a positive real number and $a < 0$ means that a represents a negative real number.

Statements like $a \leq b$ and $b \geq a$ are also inequalities. We read $a \leq b$ as "a is **less than or equal to** b" and $a \geq b$ as "a is **greater than or equal to** b."

Student Notes

It is important to remember that just because an equation or inequality is written or printed, it is not necessarily *true*. For instance, 6 = 7 is an equation and 2 > 5 is an inequality. Of course, both statements are *false*.

EXAMPLE 8 Write a second inequality with the same meaning as $-11 \leq -3$.

SOLUTION Every inequality can be written in two ways:

$$-11 \leq -3 \quad \text{has the same meaning as} \quad -3 \geq -11.$$

If we read both inequalities, we say that

−11 is less than or equal to −3,   and

−3 is greater than or equal to −11.

These statements have the same meaning.

8. Write a second inequality with the same meaning as 5 > y.

↩ YOUR TURN

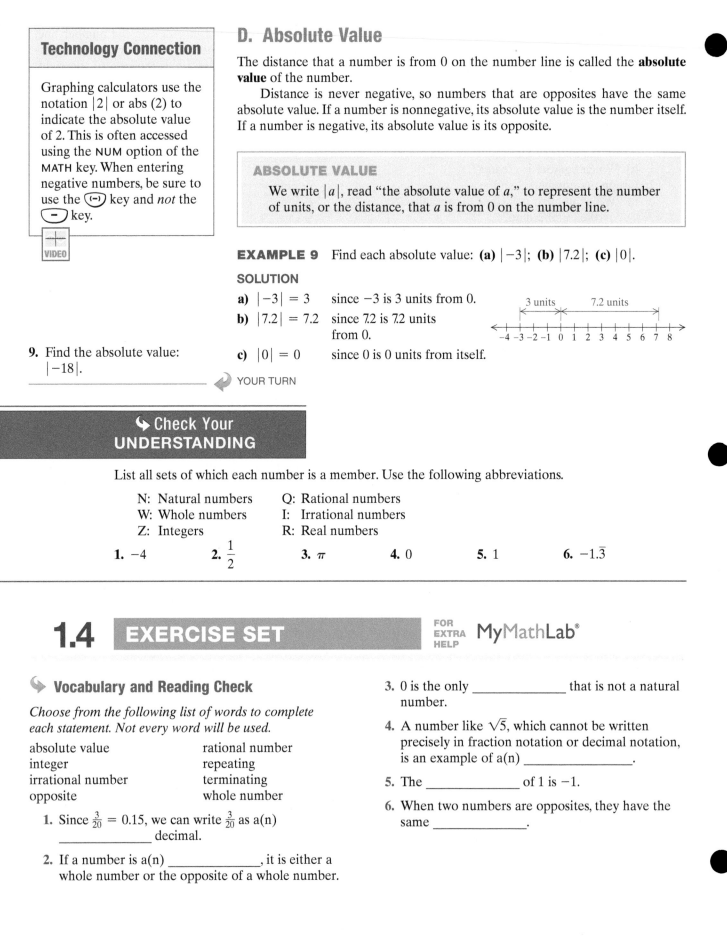

D. Absolute Value

The distance that a number is from 0 on the number line is called the **absolute value** of the number.

Distance is never negative, so numbers that are opposites have the same absolute value. If a number is nonnegative, its absolute value is the number itself. If a number is negative, its absolute value is its opposite.

> **ABSOLUTE VALUE**
>
> We write $|a|$, read "the absolute value of a," to represent the number of units, or the distance, that a is from 0 on the number line.

EXAMPLE 9 Find each absolute value: **(a)** $|-3|$; **(b)** $|7.2|$; **(c)** $|0|$.

SOLUTION

a) $|-3| = 3$ since -3 is 3 units from 0.

b) $|7.2| = 7.2$ since 7.2 is 7.2 units from 0.

c) $|0| = 0$ since 0 is 0 units from itself.

9. Find the absolute value: $|-18|$.

_____

⟳ YOUR TURN

↳ Check Your UNDERSTANDING

List all sets of which each number is a member. Use the following abbreviations.

N: Natural numbers Q: Rational numbers
W: Whole numbers I: Irrational numbers
Z: Integers R: Real numbers

1. -4 **2.** $\frac{1}{2}$ **3.** π **4.** 0   **5.** 1   **6.** $-1.\overline{3}$

1.4 EXERCISE SET

↳ Vocabulary and Reading Check

Choose from the following list of words to complete each statement. Not every word will be used.

absolute value   rational number
integer   repeating
irrational number   terminating
opposite   whole number

1. Since $\frac{3}{20} = 0.15$, we can write $\frac{3}{20}$ as a(n) _____ decimal.

2. If a number is a(n) _____, it is either a whole number or the opposite of a whole number.

3. 0 is the only _____ that is not a natural number.

4. A number like $\sqrt{5}$, which cannot be written precisely in fraction notation or decimal notation, is an example of a(n) _____.

5. The _____ of 1 is -1.

6. When two numbers are opposites, they have the same _____.

⤷ Concept Reinforcement

Translate each phrase to an algebraic expression.

7. The opposite of *n*

8. The absolute value of *x*

9. −10 is less than *x*

10. 6 is greater than or equal to *y*

A. Whole Numbers and Integers

State which real numbers correspond to each situation.

11. *Student Loans and Grants.* The maximum amount that an independent undergraduate first-year student can borrow with a Direct Stafford Loan is $9500. The maximum annual award for the Landscape Architecture Foundation Scholarship is $5000.

    Data: www.studentaid.ed.gov and www.collegescholarships.org

12. Using a NordicTrack exercise machine, Kylie burned 150 calories. She then drank an isotonic drink containing 65 calories.

13. *Record Temperature.* The highest temperature recorded in Alaska is 100 degrees Fahrenheit (°F) at Fort Yukon. The lowest temperature recorded in Alaska is 80°F below zero at Prospect Creek Camp.

    Data: www.netstate.com

14. The Dead Sea is 1349 ft below sea level, whereas Mt. Everest is 29,035 ft above sea level.

    Data: National Geographic

15. *Stock Market.* The Dow Jones Industrial Average is an indicator of the stock market. On September 29, 2008, the Dow Jones fell a record 777.68 points. On October 13, 2008, the Dow Jones gained a record 936.42 points.

    Data: CME Group Index Services, LLC; Dow Jones Indexes

16. Kittling County Volunteers received a technology grant of $10,000 and spent $4500 to update its website.

17. The New York Giants gained 8 yd on the first play. They lost 5 yd on the next play.

18. In the 2014 U.S. Open Championship, golfer Jordan Spieth finished 4 over par. In the 2015 U.S. Open Championship, he finished 5 under par.

    Data: PGA Tour Inc.

B. Rational Numbers

Graph each rational number on the number line.

19. −2

20. 5

21. −4.3

22. 3.87

23. $\frac{10}{3}$

24. $-\frac{17}{5}$

Basketball Statistics. Use the figure on p. 1 to answer Exercises 25 and 26.

25. Place the teams in order from lowest to highest by graphing their points-per-game differentials on the number line.

$$\xleftarrow{\hspace{0.3cm}}\underset{-7-6-5-4-3-2-101234567}{\rule{7cm}{0.4pt}}\xrightarrow{\hspace{0.3cm}}$$

26. Place the teams in order from lowest to highest by graphing their field-goal-percentage differentials on the number line.

$$\xleftarrow{\hspace{0.3cm}}\underset{-4-3-2-101234}{\rule{4.5cm}{0.4pt}}\xrightarrow{\hspace{0.3cm}}$$

Write decimal notation for each number.

27. $\frac{7}{8}$

28. $-\frac{1}{8}$

29. $-\frac{3}{4}$

30. $\frac{11}{6}$

31. $-\frac{7}{6}$

32. $-\frac{5}{12}$

33. $\frac{2}{3}$

34. $\frac{1}{4}$

35. $-\frac{1}{2}$

36. $-\frac{1}{9}$

Aha! 37. $\frac{13}{100}$

38. $-\frac{9}{20}$

C. Real Numbers and Order

To the student and the instructor: The calculator icon, 🖩, *indicates those exercises designed to be solved with a calculator.*

🖩 *Graph each irrational number on the number line.*

39. $\sqrt{5}$

40. $\sqrt{92}$

41. $-\sqrt{22}$

42. $-\sqrt{54}$

Write a true sentence using either < or >.

43. 5 ☐ 0

44. 8 ☐ −8

45. −9 ☐ 9

46. 0 ☐ −7

47. −8 ☐ −5

48. −4 ☐ −3

49. -5 ▢ -11

50. -3 ▢ -4

51. -12.5 ▢ -10.2

52. -10.3 ▢ -14.5

53. $\frac{5}{12}$ ▢ $\frac{11}{25}$

54. $-\frac{14}{17}$ ▢ $-\frac{27}{35}$

For each of the following, write a second inequality with the same meaning.

55. $-2 > x$

56. $a > 9$

57. $10 \le y$

58. $-12 \ge t$

For Exercises 59–64, consider the following list:

$$-83, \quad -4.7, \quad 0, \quad \tfrac{5}{9}, \quad 2.\overline{16}, \quad \pi, \quad \sqrt{17}, \quad 62.$$

59. List all rational numbers.

60. List all natural numbers.

61. List all integers.

62. List all irrational numbers.

63. List all real numbers.

64. List all nonnegative integers.

D. Absolute Value

Find each absolute value.

65. $|-58|$

66. $|-47|$

67. $|-12.2|$

68. $|4.3|$

69. $|\sqrt{2}|$

70. $|-456|$

71. $\left|-\frac{9}{7}\right|$

72. $|-\sqrt{3}|$

73. $|0|$

74. $\left|-\frac{3}{4}\right|$

75. $|x|$, for $x = -8$

76. $|a|$, for $a = -5$

77. Is every integer a rational number? Why or why not?

78. Is every integer a natural number? Why or why not?

Synthesis

79. Is the absolute value of a number always positive? Why or why not?

80. How many rational numbers are there between 0 and 1? Justify your answer.

81. Does "nonnegative" mean the same thing as "positive"? Why or why not?

List in order from least to greatest.

82. $13, -12, 5, -17$

83. $-23, 4, 0, -17$

84. $-\frac{2}{3}, \frac{1}{2}, -\frac{3}{4}, -\frac{5}{6}, \frac{3}{8}, \frac{1}{6}$

85. $\frac{4}{5}, \frac{4}{3}, \frac{4}{8}, \frac{4}{6}, \frac{4}{9}, \frac{4}{2}, -\frac{4}{3}$

Write a true sentence using either $<$, $>$, or $=$.

86. $|-5|$ ▢ $|-2|$

87. $|4|$ ▢ $|-7|$

88. $|-8|$ ▢ $|8|$

89. $|23|$ ▢ $|-23|$

Solve. Consider only integer replacements.

Aha! **90.** $|x| = 19$

91. $|x| < 3$

92. $2 < |x| < 5$

Given that $0.3\overline{3} = \frac{1}{3}$ and $0.6\overline{6} = \frac{2}{3}$, express each of the following as a ratio of two integers.

93. $0.1\overline{1}$

94. $0.9\overline{9}$

95. $5.5\overline{5}$

96. $7.7\overline{7}$

Translate to an inequality.

97. A number a is negative.

98. A number x is nonpositive.

99. The distance from x to 0 is no more than 10.

100. The distance from t to 0 is at least 20.

101. When Helga's calculator gives a decimal value for $\sqrt{2}$ and that value is promptly squared, the result is 2. Yet when that same decimal approximation is entered by hand and then squared, the result is not exactly 2. Why do you suppose this is?

102. Is the following statement true? Why or why not?

$$\sqrt{a^2} = |a| \quad \text{for any real number } a.$$

↪ YOUR TURN ANSWERS: SECTION 1.4

1. $123; -14$ **2.**   **3.** -1.75

4. $-0.1\overline{6}$ **5.**   **6.** $5, -15, 0$

7. $>$ **8.** $y < 5$ **9.** 18

Quick Quiz: Sections 1.1–1.4

1. Evaluate $12 - c$ for $c = 11$.   [1.1]

2. Factor: $5x + 5y + 15$.   [1.2]

3. Subtract and, if possible, simplify: $\frac{1}{2} - \frac{1}{3}$.   [1.3]

4. Divide and, if possible, simplify: $\frac{4}{5} \div \frac{2}{5}$.   [1.3]

5. Write a true sentence using either $<$ or $>$:

$0 \,\square\, -0.5$.   [1.4]

Mid-Chapter Review

An introduction to algebra involves learning some basic laws and terms.

Commutative Laws: $a + b = b + a$; $ab = ba$
Associative Laws: $a + (b + c) = (a + b) + c$; $a(bc) = (ab)c$
Distributive Law: $a(b + c) = ab + ac$

GUIDED SOLUTIONS

1. Evaluate $\dfrac{x - y}{3}$ for $x = 22$ and $y = 10$. [1.1]*

   Solution

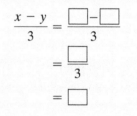

        Substituting

        Subtracting

        Dividing

2. Factor: $14x + 7$. [1.2]

   Solution

$$14x + 7 = \boxed{} \cdot 2x + \boxed{} \cdot 1$$

        Factoring each term using a common factor

$$= \boxed{} (2x + 1) \quad \text{Factoring out the common factor}$$

MIXED REVIEW

Evaluate. [1.1]

3. $x + y$, for $x = 3$ and $y = 12$

4. $\dfrac{2a}{5}$, for $a = 10$

Translate to an algebraic expression. [1.1]

5. 10 less than d

6. The product of 8 and the number of hours worked

7. Translate to an equation. Do not solve. [1.1]

Janine's class has 27 students. This is 5 fewer than the number originally enrolled. How many students originally enrolled in the class?

8. Determine whether 8 is a solution of $13t = 94$. [1.1]

9. Use the commutative law of addition to write an expression equivalent to $7 + 10x$. [1.2]

10. Use the associative law of multiplication to write an expression equivalent to $3(ab)$. [1.2]

Multiply. [1.2]

11. $4(2x + 8)$

12. $3(2m + 5n + 10)$

Factor. [1.2]

13. $18x + 15$

14. $9c + 12d + 3$

15. Find the prime factorization of 84. [1.3]

16. Simplify: $\frac{135}{315}$. [1.3]

Perform the indicated operation and, if possible, simplify. [1.3]

17. $\dfrac{11}{12} - \dfrac{3}{8}$

18. $\dfrac{8}{15} \div \dfrac{6}{11}$

19. Graph -2.5 on the number line. [1.4]

20. Write decimal notation for $-\frac{3}{20}$. [1.4]

Write a true sentence using either $<$ or $>$. [1.4]

21. $-16 \,\square\, -24$

22. $-\frac{3}{22} \,\square\, -\frac{2}{15}$

23. Write a second inequality with the same meaning as $x \geq 9$. [1.4]

Find the absolute value. [1.4]

24. $|-5.6|$

25. $|0|$

---

*The *section reference* [1.1] refers to Chapter 1, Section 1. The concept reviewed in Guided Solution 1 was developed in this section.

1.5 Addition of Real Numbers

A. Adding with the Number Line **B.** Adding Without the Number Line **C.** Problem Solving
D. Combining Like Terms

We now consider addition of real numbers. To gain understanding, we will use the number line first and then develop rules that allow us to add without the number line.

A. Adding with the Number Line

To add $a + b$ on the number line, we start at a and move according to b.

a) If b is positive, we move to the right (the positive direction).

b) If b is negative, we move to the left (the negative direction).

c) If b is 0, we stay at a.

EXAMPLE 1 Add: $-4 + 9$.

SOLUTION To add on the number line, we locate the first number, -4, and then move 9 units to the right. Note that it requires 4 units to reach 0. The difference between 9 and 4 is where we finish.

$$-4 + 9 = 5$$

Start at -4.

Move 9 units to the right.

1. Use the number line to add $-3 + 5$.

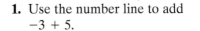

YOUR TURN

EXAMPLE 2 Add: $3 + (-5)$.

SOLUTION We locate the first number, 3, and then move 5 units to the left. Note that it requires 3 units to reach 0. The difference between 5 and 3 is 2, so we finish 2 units to the left of 0, at -2.

$$3 + (-5) = -2$$

Move 5 units to the left.

Start at 3.

2. Use the number line to add $1 + (-4)$.

YOUR TURN

EXAMPLE 3 Add: $-4 + (-3)$.

SOLUTION After locating -4, we move 3 units to the left. We finish a total of 7 units to the left of 0, at -7.

$$-4 + (-3) = -7$$

Move 3 units to the left.

Start at -4.

3. Use the number line to add $-2 + (-1)$.

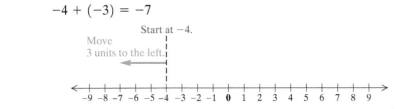

YOUR TURN

EXAMPLE 4   Add: $-5.2 + 0$.

SOLUTION   We locate -5.2 and move 0 units. Thus we finish where we started, at -5.2.

$$-5.2 + 0 = -5.2$$

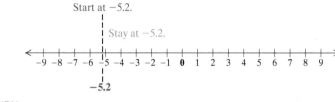

4. Use the number line to add $-4 + 0$.

YOUR TURN

From Examples 1–4, we can develop the following rules.

RULES FOR ADDITION OF REAL NUMBERS

1. *Positive numbers*: Add as usual. The answer is positive.
2. *Negative numbers*: Add absolute values and make the answer negative (see Example 3).
3. *A positive number and a negative number*: Subtract the smaller absolute value from the greater absolute value. Then:

   a) If the positive number has the greater absolute value, the answer is positive (see Example 1).
   b) If the negative number has the greater absolute value, the answer is negative (see Example 2).
   c) If the numbers have the same absolute value, the answer is 0.

4. *One number is zero*: The sum is the other number (see Example 4).

Rule 4 is known as the **identity property of 0**.

IDENTITY PROPERTY OF 0

For any real number a,

$$a + 0 = 0 + a = a.$$

(Adding 0 to a number gives that same number.) The number 0 is called the *additive identity*.

Student Notes

Parentheses are essential when a negative sign follows an operation.

B. Adding Without the Number Line

The rules listed above can be used without drawing the number line.

EXAMPLE 5   Add without using the number line.

a) $-12 + (-7)$ **b)** $-1.4 + 8.5$

c) $-36 + 21$ **d)** $1.5 + (-1.5)$

e) $-\frac{7}{8} + 0$ **f)** $\frac{2}{3} + \left(-\frac{5}{8}\right)$

5. Add without using the number line:

$$2.3 + (-9.1).$$

6. Add:

$$-3 + 10 + (-11) + (-1) + 4.$$

SOLUTION

a) $-12 + (-7) = -19$

Two negatives. *Think:* Add the absolute values, 12 and 7, to get 19. Make the answer *negative*, -19.

b) $-1.4 + 8.5 = 7.1$

A negative and a positive. *Think:* The difference of absolute values is $8.5 - 1.4$, or 7.1. The positive number has the greater absolute value, so the answer is *positive*, 7.1.

c) $-36 + 21 = -15$

A negative and a positive. *Think:* The difference of absolute values is $36 - 21$, or 15. The negative number has the greater absolute value, so the answer is *negative*, -15.

d) $1.5 + (-1.5) = 0$

A negative and a positive. *Think:* Since the numbers are opposites, they have the same absolute value and the answer is 0.

e) $-\dfrac{7}{8} + 0 = -\dfrac{7}{8}$

One number is zero. The sum is the other number, $-\frac{7}{8}$.

f) $\dfrac{2}{3} + \left(-\dfrac{5}{8}\right) = \dfrac{16}{24} + \left(-\dfrac{15}{24}\right)$

$$= \dfrac{1}{24}$$

This is similar to part (b) above. We find a common denominator and then add.

YOUR TURN

If we are adding several numbers, some positive and some negative, the commutative and associative laws allow us to add all the positive numbers, then add all the negative numbers, and then add the results. We can also add from left to right, if we prefer, but this is not always the easiest approach.

EXAMPLE 6   Add: $15 + (-2) + 7 + 14 + (-5) + (-12)$.

SOLUTION

$$15 + (-2) + 7 + 14 + (-5) + (-12)$$
$$= 15 + 7 + 14 + (-2) + (-5) + (-12) \qquad \text{Using the commutative law of addition}$$
$$= (15 + 7 + 14) + [(-2) + (-5) + (-12)] \qquad \text{Using the associative law of addition}$$
$$= 36 + (-19) \qquad \text{Adding the positives; adding the negatives}$$
$$= 17 \qquad \text{Adding a positive and a negative}$$

YOUR TURN

C. Problem Solving

EXAMPLE 7   *Credit-Card Interest Rates.*   During one four-week period, the national average credit-card interest rate dropped 0.09%, then rose 0.16%, stayed the same, and then dropped 0.02%. By how much did the national average credit-card interest rate change?

7. Refer to Example 7. In a different four-week period, the interest rate rose 0.01%, dropped 0.08%, dropped 0.01%, and then did not change. By how much did the interest rate change?

SOLUTION The problem translates to a sum:

Rewording: The 1st change plus the 2nd change plus the 3rd change plus the 4th change is the total change.

Translating: $-0.09 + 0.16 + 0 + (-0.02) = $ Total change

Adding from left to right, we have

$$-0.09 + 0.16 + 0 + (-0.02) = 0.07 + 0 + (-0.02) = 0.07 + (-0.02) = 0.05.$$

The national average credit-card interest rate rose 0.05% over the four-week period.

YOUR TURN

D. Combining Like Terms

When two terms have variable factors that are exactly the same, like $5a$ and $-7a$, the terms are called **like**, or **similar**, **terms**. Constants like 6 and 2 are also considered to be like terms. The distributive law enables us to **combine**, or **collect**, **like terms** in order to form equivalent expressions.

EXAMPLE 8 Combine like terms to form equivalent expressions.

a) $-7x + 9x$ **b)** $2a + (-3b) + (-5a) + 9b$

c) $6 + y + (-3.5y) + 2$

SOLUTION

a) $-7x + 9x = (-7 + 9)x$ Using the distributive law
$\qquad\qquad\quad = 2x$ Adding -7 and 9

b) $2a + (-3b) + (-5a) + 9b$
$\qquad = 2a + (-5a) + (-3b) + 9b$ Using the commutative law of addition
$\qquad = (2 + (-5))a + (-3 + 9)b$ Using the distributive law
$\qquad = -3a + 6b$ Adding

c) $6 + y + (-3.5y) + 2 = y + (-3.5y) + 6 + 2$ Using the commutative law of addition
$\qquad\qquad\qquad\qquad = (1 + (-3.5))y + 6 + 2$ Using the distributive law
$\qquad\qquad\qquad\qquad = -2.5y + 8$ Adding

YOUR TURN

↳ Check Your UNDERSTANDING

Add. Check using the number line.

1. $1 + 2$
2. $-1 + (-2)$
3. $-1 + 2$
4. $1 + (-2)$

8. Combine like terms to form an equivalent expression:

$$-9x + 5 + 13x + (-11).$$

1.5   EXERCISE SET

FOR EXTRA HELP   MyMathLab®

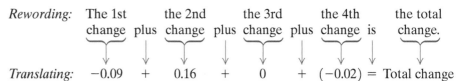

↳ Vocabulary and Reading Check

Choose from the following list of words to complete each statement. Words may be used more than once or not at all.

add	identity	like	negative
positive	subtract	zero	

1. To add $-3 + (-6)$, _____ 3 and 6 and make the answer _____.

2. To add $-1 + 8$, _____ 1 from 8 and make the answer _____.

3. To add $-11 + 5$, _____ 5 from 11 and make the answer _____.

4. The number 0 is called the additive _____.

5. The addition $-7 + 0 = -7$ illustrates the _____ property of 0.

6. The expressions $5x$ and $-9x$ are examples of _____ terms.

Concept Reinforcement

In each of Exercises 7–12, match the term with a like term from the column on the right.

7. ___ $8n$ a) $-3z$

8. ___ $7m$ b) $5x$

9. ___ 43 c) $2t$

10. ___ $28z$ d) $-4m$

11. ___ $-2x$ e) 9

12. ___ $-9t$ f) $-3n$

A. Adding with the Number Line

Add using the number line.

13. $5 + (-8)$ 14. $2 + (-5)$

15. $-6 + 10$ 16. $-3 + 8$

17. $-7 + 0$ 18. $-6 + 0$

19. $-3 + (-5)$ 20. $-4 + (-6)$

B. Adding Without the Number Line

Add. Do not use the number line except as a check.

21. $-6 + (-5)$ 22. $-8 + (-12)$

23. $10 + (-15)$ 24. $12 + (-22)$

25. $12 + (-12)$ 26. $17 + (-17)$

27. $-24 + (-17)$ 28. $-17 + (-25)$

29. $-13 + 13$ 30. $-31 + 31$

31. $20 + (-11)$ 32. $8 + (-5)$

33. $-36 + 0$ 34. $0 + (-74)$

35. $-3 + 14$ 36. $25 + (-6)$

37. $-24 + (-19)$ 38. $11 + (-9)$

39. $19 + (-19)$ 40. $-20 + (-6)$

41. $23 + (-5)$ 42. $-15 + (-7)$

43. $69 + (-85)$ 44. $-63 + 13$

45. $-3.6 + 2.8$ 46. $-6.5 + 4.7$

47. $-5.4 + (-3.7)$ 48. $-3.8 + (-9.4)$

49. $\frac{4}{5} + \left(\frac{-1}{5}\right)$ 50. $\frac{-2}{7} + \frac{3}{7}$

51. $\frac{-4}{7} + \frac{-2}{7}$ 52. $\frac{-5}{9} + \frac{-2}{9}$

53. $-\frac{2}{5} + \frac{1}{3}$ 54. $-\frac{4}{13} + \frac{1}{2}$

55. $\frac{-4}{9} + \frac{2}{3}$ 56. $\frac{1}{9} + \left(\frac{-1}{3}\right)$

57. $35 + (-14) + (-19) + (-5)$

58. $-28 + (-44) + 17 + 31 + (-94)$

Aha! 59. $-4.9 + 8.5 + 4.9 + (-8.5)$

60. $24 + 3.1 + (-44) + (-8.2) + 63$

C. Problem Solving

Solve. Write your answer as a complete sentence.

61. *Gasoline Prices.* During one month, the price of a gallon of 87-octane gasoline dropped 15¢, then dropped 3¢, and then rose 17¢. By how much did the price change during that period?

62. *Natural Gas Prices.* During one winter, the price of a gallon of natural gas dropped 2¢, then rose 25¢, and then dropped 43¢. By how much did the price change during that period?

63. *Telephone Bills.* Chloe's cell-phone bill for July was $82. She sent a check for $50 and then ran up $63 in charges for August. What was her new balance?

64. *Profits and Losses.* The following table lists the profits and losses of Premium Sales over a 3-year period. Find the profit or loss after this period of time.

Year	Profit or loss
2014	$-\$26,500$
2015	$-\$10,200$
2016	$+\$32,400$

65. *Yardage Gained.* In an intramural football game, the quarterback attempted passes with the following results.

First try	13-yd loss
Second try	12-yd gain
Third try	21-yd gain

Find the total gain (or loss).

66. *Account Balance.* Omari has $450 in a checking account. He writes a check for $530, makes a deposit of $75, and then writes a check for $90. What is the balance in the account?

67. *Lake Level.* During the course of a year, the water level of the Great Salt Lake dropped $\frac{2}{5}$ ft, rose $\frac{9}{10}$ ft, and dropped $\frac{6}{5}$ ft. By how much did the level change?

Data: U.S. Geological Survey

68. *Credit-Card Bills.* Logan's credit-card bill indicates that he owes $470. He sends a check to the credit-card company for $45, charges another $160 in merchandise, and then pays off another $500 of his bill. What is Logan's new balance?

69. *Peak Elevation.* The tallest mountain in the world, as measured from base to peak, is Mauna Kea in Hawaii. From a base 19,684 ft below sea

level, it rises 33,480 ft. What is the elevation of its peak?

Data: *Guinness World Records* 2007

70. *Class Size.* During the first two weeks of the semester, 5 students withdrew from Hailey's algebra class, 8 students were added to the class, and 4 students were dropped as "no-shows." By how many students did the original class size change?

D. Combining Like Terms

Combine like terms to form an equivalent expression.

71. $7a + 10a$

72. $3x + 8x$

73. $-3x + 12x$

74. $-2m + (-7m)$

75. $7m + (-9m)$

76. $-4x + 4x$

77. $-8y + (-2y)$

78. $10n + (-17n)$

79. $-3 + 8x + 4 + (-10x)$

80. $8a + 5 + (-a) + (-3)$

81. $6m + 9n + (-9n) + (-10m)$

82. $-11s + (-8t) + (-3s) + 8t$

83. $-4x + 6.3 + (-x) + (-10.2)$

84. $-7 + 10.5y + 13 + (-11.5y)$

Find the perimeter of, or the distance around, each figure.

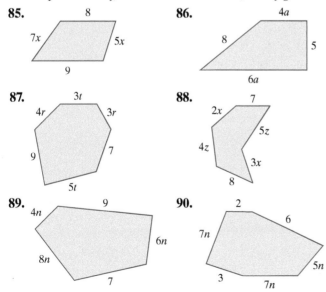

85.

86.

87.

88.

89.

90.

91. Explain in your own words why the sum of two negative numbers is negative.

92. Without performing the actual addition, explain why the sum of all integers from -10 to 10 is 0.

Synthesis

93. Under what circumstances will the sum of one positive number and several negative numbers be positive?

94. Is it possible to add real numbers without knowing how to calculate $a - b$ with a and b both nonnegative and $a \geq b$? Why or why not?

95. *Banking.* Travis had $257.33 in his checking account. After depositing $152 in the account and writing a check, his account was overdrawn by $42.37. What was the amount of the check?

96. *Comic-Book Values.* The value of a Spider-Man comic book dropped $12 and then rose $17.50 before settling at $61. What was the original value of the comic book?

Find the missing term or terms.

97. $4x +$ ___ $+ (-9x) + (-2y) = -5x - 7y$

98. $-3a + 9b +$ ___ $+ 5a = 2a - 6b$

99. $3m + 2n +$ ___ $+ (-2m) = 2n + (-6m)$

100. ___ $+ 9x + (-4y) + x = 10x - 7y$

Aha! **101.** $7t + 23 +$ ___ $+$ ___ $= 0$

102. *Geometry.* The perimeter of a rectangle is $7x + 10$. If the length of the rectangle is 5, express the width in terms of x.

103. *Golfing.* After five rounds of golf, a golf pro was 3 under par twice, 2 over par once, 2 under par once, and 1 over par once. On average, how far above or below par was the golfer?

> **YOUR TURN ANSWERS: SECTION 1.5**
> **1.** 2   **2.** -3 **3.** -3 **4.** -4 **5.** -6.8 **6.** -1
> **7.** The interest rate dropped 0.08%.   **8.** $4x - 6$

Quick Quiz: Sections 1.1–1.5

1. Multiply: $3(6a + 4c + 1)$.   [1.2]

2. Add and, if possible, simplify: $\frac{2}{10} + \frac{6}{5}$.   [1.3]

3. Multiply and, if possible, simplify: $\frac{2}{9} \cdot \frac{3}{8}$.   [1.3]

4. Find the absolute value: $|505|$.   [1.4]

5. Combine like terms: $-12 + 10 + 13x + (-20x)$.   [1.5]

1.6 Subtraction of Real Numbers

A. Opposites and Additive Inverses   **B.** Subtraction   **C.** Problem Solving

In arithmetic, when a number b is subtracted from another number a, the **difference**, $a - b$, is the number that when added to b gives a. For example, $8 - 5 = 3$ because $3 + 5 = 8$. We will use this approach to develop an efficient way of finding the value of $a - b$ for any real numbers a and b.

A. Opposites and Additive Inverses

Numbers such as 6 and -6 are *opposites*, or *additive inverses*, of each other. Whenever opposites are added, the result is 0; and whenever two numbers add to 0, those numbers are opposites.

EXAMPLE 1   Find the opposite of each number: **(a)** 34; **(b)** -8.3; **(c)** 0.

SOLUTION

a) The opposite of 34 is -34: $34 + (-34) = 0.$

b) The opposite of -8.3 is 8.3: $-8.3 + 8.3 = 0.$

c) The opposite of 0 is 0: $0 + 0 = 0.$

1. Find the opposite of -15.

YOUR TURN

To write the opposite of a number, we use the symbol $-$, as follows.

> **OPPOSITE**
>
> The *opposite*, or *additive inverse*, of a number a is written $-a$ (read "the opposite of a" or "the additive inverse of a").

Note that if we take a number, say 8, and find its opposite, -8, and then find the opposite of the result, we will have the original number, 8, again.

EXAMPLE 2   Find $-x$ and $-(-x)$ when $x = 16$.

SOLUTION

If $x = 16$, then $-x = -16$.   The opposite of 16 is -16.

If $x = 16$, then $-(-x) = -(-16) = 16$.   The opposite of the opposite of 16 is 16.

2. Find $-x$ and $-(-x)$ when $x = 3.5$.

YOUR TURN

> **THE OPPOSITE OF AN OPPOSITE**
>
> For any real number a,
>
> $$-(-a) = a.$$
>
> (The opposite of the opposite of a is a.)

3. Find $-x$ and $-(-x)$ when $x = -9$.

Student Notes

As you read mathematics, it is important to verbalize correctly the words and symbols to yourself. Consistently reading the expression $-x$ as "the opposite of x" is a good step in this direction.

EXAMPLE 3 Find $-x$ and $-(-x)$ when $x = -3$.

SOLUTION

If $x = -3$, then $-x = -(-3) = 3$. The opposite of -3 is 3.

Since $-(-x) = x$, it follows that $-(-(-3)) = -3$. Finding the opposite of an opposite

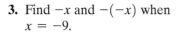 YOUR TURN

Note in Example 3 that an extra set of parentheses is used to show that we are substituting the negative number -3 for x. The notation $- -x$ is not used.

A symbol such as -8 is usually read "negative 8." It could be read "the additive inverse of 8," because the additive inverse of 8 is negative 8. It could also be read "the opposite of 8," because the opposite of 8 is -8.

A symbol like $-x$, which has a variable, should be read "the opposite of x" or "the additive inverse of x" and *not* "negative x," since to do so suggests that $-x$ represents a negative number. As we have seen, $-x$ is not always negative.

The symbol "$-$" is read differently depending on where it appears. For example, $-5 - (-x)$ should be read "negative five minus the opposite of x."

EXAMPLE 4 Write each of the following in words.

a) $2 - 8$ **b)** $t - (-4)$ **c)** $-6 - (-x)$

SOLUTION

a) $2 - 8$ is read "two minus eight."

b) $t - (-4)$ is read "t minus negative four."

4. Write $-y - 7$ in words.

c) $-6 - (-x)$ is read "negative six minus the opposite of x."

YOUR TURN

As we saw in Example 3, $-x$ can represent a positive number. This notation can be used to state the *law of opposites*.

THE LAW OF OPPOSITES

For any two numbers a and $-a$,

$$a + (-a) = 0.$$

(When opposites are added, their sum is 0.)

A negative number is said to have a "negative *sign*." A positive number is said to have a "positive *sign*." If we change a number to its opposite, or additive inverse, we say that we have "changed or reversed its sign."

EXAMPLE 5 Change the sign (find the opposite) of each number: **(a)** -3; **(b)** -10; **(c)** 14.

SOLUTION

a) When we change the sign of -3, we obtain 3.

b) When we change the sign of -10, we obtain 10.

5. Change the sign of 12.

c) When we change the sign of 14, we obtain -14.

YOUR TURN

B. Subtraction

Opposites are helpful when subtraction involves negative numbers. To see why, look for a pattern in the following:

Subtracting	*Reasoning*	*Adding the Opposite*
$9 - 5 = 4$	since $4 + 5 = 9$	$9 + (-5) = 4$
$5 - 8 = -3$	since $-3 + 8 = 5$	$5 + (-8) = -3$
$-6 - 4 = -10$	since $-10 + 4 = -6$	$-6 + (-4) = -10$
$-7 - (-10) = 3$	since $3 + (-10) = -7$	$-7 + 10 = 3$
$-7 - (-2) = -5$	since $-5 + (-2) = -7$	$-7 + 2 = -5$

The matching results suggest that we can subtract by adding the opposite of the number being subtracted. This can always be done and often provides the easiest way to subtract real numbers.

> **SUBTRACTION OF REAL NUMBERS**
>
> For any real numbers a and b,
>
> $$a - b = a + (-b).$$
>
> (To subtract, add the opposite, or additive inverse, of the number being subtracted.)

EXAMPLE 6   Subtract each of the following and then check with addition.

a) $2 - 6$ **b)** $4 - (-9)$ **c)** $-4.2 - (-3.6)$

d) $-1.8 - (-7.5)$ **e)** $\frac{1}{5} - \left(-\frac{3}{5}\right)$

SOLUTION

a) $2 - 6 = 2 + (-6) = -4$    The number being subtracted is 6. To subtract 6, we add the opposite of 6, which is -6. *Check:* $-4 + 6 = 2.$

b) $4 - (-9) = 4 + 9 = 13$ The number being subtracted is -9. To subtract -9, we add the opposite of -9, which is 9. *Check:* $13 + (-9) = 4.$

c) $-4.2 - (-3.6) = -4.2 + 3.6$ Adding the opposite of -3.6
$= -0.6$ *Check:* $-0.6 + (-3.6) = -4.2.$

d) $-1.8 - (-7.5) = -1.8 + 7.5$ Adding the opposite of -7.5
$= 5.7$ *Check:* $5.7 + (-7.5) = -1.8.$

e) $\dfrac{1}{5} - \left(-\dfrac{3}{5}\right) = \dfrac{1}{5} + \dfrac{3}{5}$ Adding the opposite of $-\dfrac{3}{5}$

$= \dfrac{1 + 3}{5}$ A common denominator exists so we add in the numerator.

$= \dfrac{4}{5}$

Check: $\dfrac{4}{5} + \left(-\dfrac{3}{5}\right) = \dfrac{4}{5} + \dfrac{-3}{5} = \dfrac{4 + (-3)}{5} = \dfrac{1}{5}.$

Technology Connection

On nearly all graphing calculators, it is essential to distinguish between the key for negation and the key for subtraction. To enter a negative number, we use (−); to subtract, we use (−).

6. Subtract: $-16 - (-20)$.

YOUR TURN

EXAMPLE 7 Simplify: $8 - (-4) - 2 - (-5) + 3$.

SOLUTION

$$8 - (-4) - 2 - (-5) + 3 = 8 + 4 + (-2) + 5 + 3 \quad \text{To subtract, we add the opposite.}$$
$$= 18$$

7. Simplify:

$-3 - (-1) + (-10) - 7$.

YOUR TURN

The terms of an algebraic expression are separated by plus signs. This means that the terms of $5x - 7y - 9$ are $5x, -7y$, and -9, since $5x - 7y - 9 = 5x + (-7y) + (-9)$.

EXAMPLE 8 Identify the terms of $4 - 2ab + 7a - 9$.

SOLUTION We have

$$4 - 2ab + 7a - 9 = 4 + (-2ab) + 7a + (-9), \quad \text{Rewriting as addition}$$

so the terms are $4, -2ab, 7a$, and -9.

8. Identify the terms of
$5a - 7x - 10 + y$.

YOUR TURN

EXAMPLE 9 Combine like terms.

a) $1 + 3x - 7x$ **b)** $-5a - 7b - 4a + 10b$

c) $4 - 3m - 9 + 2m$

SOLUTION

a) $1 + 3x - 7x = 1 + 3x + (-7x)$ Adding the opposite

$\left. \begin{array}{l} = 1 + (3 + (-7))x \\ = 1 + (-4)x \end{array} \right\}$ Using the distributive law. Try to do this mentally.

$= 1 - 4x$ Rewriting as subtraction to be more concise

b) $-5a - 7b - 4a + 10b = -5a + (-7b) + (-4a) + 10b$ Adding the opposite

$= -5a + (-4a) + (-7b) + 10b$ Using the commutative law of addition

$= -9a + 3b$ Combining like terms mentally

c) $4 - 3m - 9 + 2m = 4 + (-3m) + (-9) + 2m$ Rewriting as addition

$= 4 + (-9) + (-3m) + 2m$ Using the commutative law of addition

$= -5 + (-1m)$ We can write $-1m$ as $-m$.

$= -5 - m$

9. Combine like terms:

$x - 3y + 7y - 2x$.

YOUR TURN

↳ Check Your UNDERSTANDING

Rewrite each subtraction as an addition. Do not carry out the addition.

1. $6 - 8$

2. $-5 - (-1)$

3. $-10 - 7$

4. $13 - (-4)$

C. Problem Solving

The words "difference," "minus," and "less than" are examples of key words that translate to subtraction. Since subtraction is not commutative, the order in which we write a translation is important. Unless the context indicates otherwise, "the difference between 5 and 8" translates to $5 - 8$, and "subtract 10 from 3" means to calculate $3 - 10$.

We use subtraction to solve problems involving differences. These include problems that ask "How much more?" or "How much higher?"

EXAMPLE 10 *Record Elevations.* On March 26, 2012, James Cameron became the first person to make a solo visit to the deepest point of the ocean, descending 26,791 ft into the Marianas Trench. On October 24, 2014, Google senior vice-president Alan Eustace set a world record for parachuting from a height of 135,890 ft. What is the difference in elevation for these feats?

Data: www.space.com; www.gizmag.com

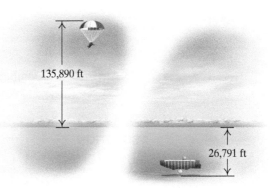

135,890 ft

26,791 ft

10. The lowest elevation in Asia, the Dead Sea, is 1349 ft below sea level. The highest elevation in Asia, Mount Everest, is 29,035 ft. Find the difference in elevation.

Data: *National Geographic*

SOLUTION To find the difference between two elevations, we always subtract the lower elevation from the higher elevation:

Higher elevation $-$ Lower elevation

$$135{,}890 \quad - \quad (-26{,}791)$$
$$= 135{,}890 + 26{,}791$$
$$= 162{,}681.$$

The parachute jump began 162,681 ft higher than the ocean dive ended.

↻ YOUR TURN

1.6   EXERCISE SET

FOR EXTRA HELP   MyMathLab®

↳ Vocabulary and Reading Check

In each of Exercises 1–4, two words appear under the blank. Choose the correct word to complete the statement.

1. The numbers 5 and -5 are ———— of each other.
   opposites/reciprocals

2. The number -100 has a negative ————.
   opposite/sign

3. We subtract by adding the ———— of the number being subtracted.
   opposite/reciprocal

4. The word ———— usually translates to subtraction.
   difference/quotient

⤷ Concept Reinforcement

In each of Exercises 5–12, match the expression with the appropriate expression from the column on the right.

5. ___ $-x$

6. ___ $12 - x$

7. ___ $12 - (-x)$

8. ___ $x - 12$

9. ___ $x - (-12)$

10. ___ $-x - 12$

11. ___ $-x - x$

12. ___ $-x - (-12)$

a) x minus negative twelve

b) The opposite of x minus x

c) The opposite of x minus twelve

d) The opposite of x

e) The opposite of x minus negative twelve

f) Twelve minus the opposite of x

g) Twelve minus x

h) x minus twelve

A. Opposites and Additive Inverses

Write each of the following in words.

13. $6 - 10$

14. $5 - 13$

15. $2 - (-12)$

16. $4 - (-1)$

17. $-x - y$

18. $-a - b$

19. $-3 - (-n)$

20. $-7 - (-m)$

Find the opposite, or additive inverse, of each number.

21. 51

22. -17

23. $-\frac{11}{3}$

24. $\frac{7}{2}$

25. -3.14

26. 48.2

Find $-x$ when x is each of the following.

27. -45

28. 26

29. $-\frac{14}{3}$

30. $\frac{1}{328}$

31. 0.101

32. 0

Find $-(-x)$ when x is each of the following.

33. 37

34. 29

35. $-\frac{2}{5}$

36. -9.1

Change the sign. (Find the opposite.)

37. -1

38. -7

39. 15

40. 10

B. Subtraction

Subtract.

41. $7 - 10$

42. $4 - 13$

43. $0 - 6$

44. $0 - 8$

45. $-4 - 3$

46. $-5 - 6$

47. $-9 - (-3)$

48. $-9 - (-5)$

Aha! **49.** $-8 - (-8)$

50. $-10 - (-10)$

51. $14 - 19$

52. $12 - 16$

53. $30 - 40$

54. $20 - 27$

55. $-9 - (-9)$

56. $-40 - (-40)$

57. $5 - 5$

58. $7 - 7$

59. $4 - (-4)$

60. $6 - (-6)$

61. $-7 - 4$

62. $-6 - 8$

63. $6 - (-10)$

64. $3 - (-12)$

65. $-4 - 15$

66. $-14 - 2$

67. $-6 - (-5)$

68. $-4 - (-1)$

69. $5 - (-12)$

70. $5 - (-6)$

71. $0 - (-3)$

72. $0 - (-5)$

73. $-5 - (-2)$

74. $-3 - (-1)$

75. $-7 - 14$

76. $-9 - 16$

77. $0 - 11$

78. $0 - 31$

79. $-8 - 0$

80. $-9 - 0$

81. $-52 - 8$

82. $-63 - 11$

83. $2 - 25$

84. $18 - 63$

85. $-4.2 - 3.1$

86. $-10.1 - 2.6$

87. $-1.3 - (-2.4)$

88. $-5.8 - (-7.3)$

89. $3.2 - 8.7$

90. $1.5 - 9.4$

91. $0.072 - 1$

92. $0.825 - 1$

93. $\frac{2}{11} - \frac{9}{11}$

94. $\frac{3}{7} - \frac{5}{7}$

95. $\frac{-1}{5} - \frac{3}{5}$

96. $\frac{-2}{9} - \frac{5}{9}$

97. $-\frac{2}{3} - \left(-\frac{1}{2}\right)$

98. $-\frac{1}{4} - \left(-\frac{2}{3}\right)$

Simplify.

99. $16 - (-12) - 1 - (-2) + 3$

100. $22 - (-18) + 7 + (-42) - 27$

101. $-31 + (-28) - (-14) - 17$

102. $-43 - (-19) - (-21) + 25$

103. $-34 - 28 + (-33) - 44$

104. $39 + (-88) - 29 - (-83)$

Aha! **105.** $-93 + (-84) - (-93) - (-84)$

106. $84 + (-99) + 44 - (-18) - 43$

Identify the terms in each expression.

107. $-3y - 8x$

108. $7a - 9b$

109. $9 - 5t - 3st$

110. $-4 - 3x + 2xy$

Combine like terms.

111. $10x - 13x$

112. $3a - 14a$

113. $7a - 12a + 4$

114. $-9x - 13x + 7$

115. $-8n - 9 + 7n$

116. $-7 + 9n - 8n$

117. $5 - 3x - 11$

118. $2 + 3a - 7$

119. $2 - 6t - 9 - 2t$

120. $-5 + 4b - 7 - 5b$

121. $5y + (-3x) - 9x + 1 - 2y + 8$

122. $14 - (-5x) + 2z - (-32) + 4z - 2x$

123. $13x - (-2x) + 45 - (-21) - 7x$

124. $8t - (-2t) - 14 - (-5t) + 53 - 9t$

C. Problem Solving

125. Subtract 32 from -8.

126. Subtract 19 from -7.

127. Subtract -25 from 18.

128. Subtract -31 from -5.

In each of Exercises 129–132, translate the phrase to mathematical language and simplify.

129. The difference between 3.8 and -5.2

130. The difference between -2.1 and -5.9

131. The difference between 114 and -79

132. The difference between 23 and -17

133. *Elevation.* The Jordan River begins in Lebanon at an elevation of 550 m above sea level and empties into the Dead Sea at an elevation of 400 m below sea level. During its 360-km length, by how many meters does it drop?

Data: Brittanica Online

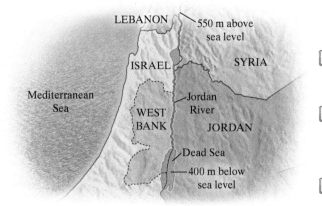

134. *Elevation.* The elevation of Mount Whitney, the highest peak in California, is 14,497 ft. Death Valley, California, is 282 ft below sea level. What is the difference in elevation?

Data: mount-whitney.com and infoplease.com

135. *Temperature Extremes.* The highest temperature ever recorded in the United States is 134°F in Greenland Ranch, California, on July 10, 1913. The lowest temperature ever recorded is -79.8°F in Prospect Creek, Alaska, on January 23, 1971. How much higher was the temperature in Greenland Ranch than that in Prospect Creek?

Data: infoplease.com

136. *Temperature Change.* In just 12 hr on February 21, 1918, the temperature in Granville, North Dakota, rose from -33°F to 50°F. By how much did the temperature change?

Data: weatherexplained.com

137. *Basketball.* A team's points differential is the difference between points scored and points allowed. The Chicago Bulls improved their points-per-game differential from -1.6 in a recent season to $+7.3$ in the following season. By how much did their differential change?

Data: espn.go.com

138. *Underwater Elevation.* The deepest point in the Pacific Ocean is the Challenger Deep in the Marianas Trench, with a depth of 10,994 m. The deepest point in the Atlantic Ocean is the Milwaukee Deep in the Puerto Rico Trench, with a depth of 8380 m. What is the difference in elevation of the two trenches?

Data: www.britannica.com

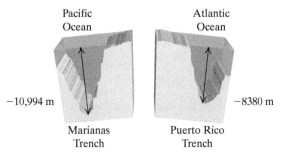

139. Lowell insists that if you can *add* real numbers, then you can also *subtract* real numbers. Do you agree? Why or why not?

140. Are the expressions $-a + b$ and $a + (-b)$ opposites of each other? Why or why not?

Synthesis

141. Explain the different uses of the symbol "$-$". Give examples of each and how they should be read.

142. If a and b are both negative, under what circumstances will $a - b$ be negative?

143. *Power Outages.* During the Northeast's electrical blackout of August 14, 2003, residents of Bloomfield, New Jersey, lost power at 4:00 P.M. One resident returned from vacation at 3:00 P.M. the following day to find the clocks in her apartment reading 8:00 A.M. At what time, and on what day, was power restored?

Tell whether each statement is true or false for all real numbers m and n. Use various replacements for m and n to support your answer.

144. If $m > n$, then $m - n > 0$.

145. If $m > n$, then $m + n > 0$.

146. If m and n are opposites, then $m - n = 0$.

147. If $m = -n$, then $m + n = 0$.

148. A gambler loses a wager and then loses "double or nothing" (meaning the gambler owes twice as much) twice more. After the three losses, the gambler's assets are $-\$20$. Explain how much the gambler originally bet and how the \$20 debt occurred.

149. List the keystrokes needed to compute $-9 - (-7)$.

150. If n is positive and m is negative, what is the sign of $n + (-m)$? Why?

YOUR TURN ANSWERS: SECTION 1.6

1. 15 **2.** $-3.5; 3.5$ **3.** $9; -9$ **4.** The opposite of y minus 7 **5.** -12 **6.** 4 **7.** -19 **8.** $5a, -7x, -10, y$
9. $-x + 4y$ **10.** 30,384 ft

Quick Quiz: Sections 1.1–1.6

1. Translate to an equation. Do not solve.

In the second quarter of 2011, a total of 9.25 million Apple iPad tablets were sold. This was 11% of the total PC sales. How many PCs were sold in all? [1.1]

Data: Apple; Business Insider

2. List all the factors of 52. [1.3]

3. Find the prime factorization of 52. [1.3]

4. Add: $-10 + (-19)$. [1.5]

5. Subtract: $-6 - (-11)$. [1.6]

1.7 Multiplication and Division of Real Numbers

A. Multiplication   B. Division

We now develop rules for multiplication and division of real numbers. Because multiplication and division are closely related, the rules are quite similar.

A. Multiplication

Student Notes

The multiplication $2 \cdot (-5)$ can be thought of as starting at 0 and *adding* -5 twice:

$$0 + (-5) + (-5) = -10.$$

We already know how to multiply two nonnegative numbers. To see how to multiply a positive number and a negative number, consider the following pattern in which multiplication is regarded as repeated addition:

This number → $4(-5) = (-5) + (-5) + (-5) + (-5) = -20$ ← This number
decreases by $3(-5) = (-5) + (-5) + (-5) = -15$ increases by
1 each time. $2(-5) = (-5) + (-5) = -10$ 5 each time.
$1(-5) = (-5) = -5$
$0(-5) = 0 = 0$

This pattern illustrates that the product of a negative number and a positive number is negative.

> **THE PRODUCT OF A NEGATIVE NUMBER AND A POSITIVE NUMBER**
>
> To multiply a positive number and a negative number, multiply their absolute values and make the result negative.

EXAMPLE 1   Multiply: **(a)** $8(-5)$; **(b)** $-\frac{1}{3} \cdot \frac{5}{7}$.

SOLUTION   The product of a negative number and a positive number is negative.

a) $8(-5) = -40$ *Think:* $8 \cdot 5 = 40$; make the answer negative.

b) $-\frac{1}{3} \cdot \frac{5}{7} = -\frac{5}{21}$ *Think:* $\frac{1}{3} \cdot \frac{5}{7} = \frac{5}{21}$; make the answer negative.

1. Multiply: $(-3)(10)$.

YOUR TURN

The pattern developed above also illustrates that the product of 0 and any real number is 0.

> **THE MULTIPLICATIVE PROPERTY OF ZERO**
>
> For any real number a,
>
> $$0 \cdot a = a \cdot 0 = 0.$$
>
> (The product of 0 and any real number is 0.)

EXAMPLE 2   Multiply: $173(-452)0$.

SOLUTION   We have

$$173(-452)0 = 173[(-452)0]$$ Because of the associative law of multiplication, we can multiply the last two factors first.

$$= 173[0]$$ Using the multiplicative property of zero

$$= 0.$$ Using the multiplicative property of zero again

Note that whenever 0 appears as a factor, the product is 0.

2. Multiply: $(-13)(0)(19)$.

YOUR TURN

Student Notes

The multiplication $(-2) \cdot (-5)$ can be thought of as starting at 0 and *subtracting* -5 twice:

$$0 - (-5) - (-5) = 0 + 5 + 5$$
$$= 10.$$

An alternative explanation is to write

$$(-2)(-5) = [(-1)(2)](-5)$$
$$= (-1)[(2)(-5)]$$
$$= -[-10]$$
$$= 10.$$

We can extend the above pattern still further to examine the product of two negative numbers.

This number → $2(-5) =$ $(-5) + (-5) = -10$ ← This number
decreases by $1(-5) =$ $(-5) = -5$ increases by 5
1 each time. $0(-5) =$ $0 = 0$ each time.
                $-1(-5) =$ $-(-5) = 5$
                $-2(-5) = -(-5) - (-5) = 10$

According to the pattern, the product of two negative numbers is positive.

> **THE PRODUCT OF TWO NEGATIVE NUMBERS**
>
> To multiply two negative numbers, multiply their absolute values. The answer is positive.

EXAMPLE 3 Multiply: **(a)** $(-6)(-8)$; **(b)** $(-1.2)(-3)$.

SOLUTION The product of two negative numbers is positive.

a) The absolute value of -6 is 6 and the absolute value of -8 is 8. Thus,

$$(-6)(-8) = 6 \cdot 8 \qquad \text{Multiplying absolute values. The answer is positive.}$$
$$= 48.$$

3. Multiply: $\left(-\dfrac{1}{2}\right)\left(-\dfrac{1}{3}\right)$.

b) $(-1.2)(-3) = (1.2)(3) \qquad$ Multiplying absolute values. The answer is positive.
$ = 3.6 \qquad$ Try to go directly to this step.

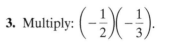 YOUR TURN

When three or more numbers are multiplied, we can order and group the numbers as we please, because of the commutative and associative laws.

EXAMPLE 4 Multiply: **(a)** $-3(-2)(-5)$; **(b)** $-4(-6)(-1)(-2)$.

SOLUTION

a) $-3(-2)(-5) = 6(-5) \qquad$ Multiplying the first two numbers. The product of two negatives is positive.

$ = -30 \qquad$ The product of a positive and a negative is negative.

b) $-4(-6)(-1)(-2) = 24 \cdot 2 \qquad$ Multiplying the first two numbers and the last two numbers

4. Multiply: $(-1)(-10)(-5)$.

$ = 48$

YOUR TURN

We can see the following pattern in the results of Example 4.

The product of an even number of negative numbers is positive.

The product of an odd number of negative numbers is negative.

When a number is multiplied by -1, the result is the opposite of that number. For example, $-1(7) = -7$ and $-1(-5) = 5$.

> **THE PROPERTY OF -1**
>
> For any real number a,
>
> $$-1 \cdot a = -a.$$
>
> (Negative one times a is the opposite of a.)

B. Division

Note that $a \div b$, or $\dfrac{a}{b}$, is the number, if one exists, that when multiplied by b gives a. For example, to show that $10 \div 2$ is 5, we need only note that $5 \cdot 2 = 10$. Thus division can always be checked with multiplication.

The rules for signs for division are the same as those for multiplication: The quotient of a positive number and a negative number is negative; the quotient of two negative numbers is positive.

> **RULES FOR MULTIPLICATION AND DIVISION**
> To multiply or divide two nonzero real numbers:
>
> 1. Using the absolute values, multiply or divide, as indicated.
> 2. If the signs are the same, the answer is positive.
> 3. If the signs are different, the answer is negative.

EXAMPLE 5 Divide, if possible, and check your answer.

a) $14 \div (-7)$

b) $\dfrac{-32}{-4}$

c) $\dfrac{-10}{2}$

d) $\dfrac{-17}{0}$

SOLUTION

a) $14 \div (-7) = -2$ *Think:* $14 \div 7 = 2$; the answer is negative.
   Check: $(-2)(-7) = 14$.

b) $\dfrac{-32}{-4} = 8$ *Think:* $32 \div 4 = 8$; the answer is positive.
   Check: $8(-4) = -32$.

c) $\dfrac{-10}{2} = -5$ *Think:* $10 \div 2 = 5$; the answer is negative.
   Check: $-5(2) = -10$.

d) $\dfrac{-17}{0}$ is **undefined**.   We look for a number that when multiplied by 0 gives -17. There is no such number because if 0 is a factor, the product is 0, not -17.

YOUR TURN

Student Notes

Try to regard "undefined" as a mathematical way of saying "we do not give any meaning to this expression."

5. Divide, if possible, and check your answer: $(-100) \div 25$.

Had Example 5(a) been written as $-14 \div 7$ or $-\frac{14}{7}$, rather than $14 \div (-7)$, the result would still have been -2. Thus from Examples 5(a)–5(c), we have the following:

$$\frac{-a}{b} = \frac{a}{-b} = -\frac{a}{b} \quad \text{and} \quad \frac{-a}{-b} = \frac{a}{b}.$$

EXAMPLE 6 Rewrite each of the following in two equivalent forms.

a) $\dfrac{5}{-2}$

b) $-\dfrac{3}{10}$

SOLUTION We use one of the properties just listed.

a) $\dfrac{5}{-2} = \dfrac{-5}{2}$ and $\dfrac{5}{-2} = -\dfrac{5}{2}$

b) $-\dfrac{3}{10} = \dfrac{-3}{10}$ and $-\dfrac{3}{10} = \dfrac{3}{-10}$

Since $\dfrac{-a}{b} = \dfrac{a}{-b} = -\dfrac{a}{b}$

6. Rewrite in two equivalent forms:
$$\dfrac{-5}{6}.$$

YOUR TURN

When a fraction contains a negative sign, it can be helpful to rewrite (or simply visualize) the fraction in an equivalent form.

EXAMPLE 7 Perform the indicated operation: **(a)** $\left(-\frac{4}{5}\right)\left(\frac{-7}{3}\right)$; **(b)** $-\frac{2}{7}+\frac{9}{-7}$.

SOLUTION

a) $\left(-\frac{4}{5}\right)\left(\frac{-7}{3}\right)=\left(-\frac{4}{5}\right)\left(-\frac{7}{3}\right)$ Rewriting $\frac{-7}{3}$ as $-\frac{7}{3}$

$\qquad\qquad\qquad\quad = \dfrac{28}{15}$ Try to go directly to this step.

b) Given a choice, we generally choose a positive denominator:

$\qquad -\dfrac{2}{7}+\dfrac{9}{-7}=\dfrac{-2}{7}+\dfrac{-9}{7}$ Rewriting both fractions with a common denominator of 7

$\qquad\qquad\qquad\quad = \dfrac{-11}{7}$, or $-\dfrac{11}{7}$.

7. Perform the indicated operation:

$\dfrac{-3}{4}-\left(\dfrac{7}{-4}\right).$

⟲ YOUR TURN

To divide with fraction notation, it is usually easiest to find a reciprocal and then multiply.

EXAMPLE 8 Find the reciprocal of each number, if it exists.

a) -27 **b)** $\dfrac{-3}{4}$ **c)** $-\dfrac{1}{5}$ **d)** 0

SOLUTION Two numbers are reciprocals of each other if their product is 1.

a) The reciprocal of -27 is $\frac{1}{-27}$. More often, this number is written as $-\frac{1}{27}$.

  Check: $(-27)\left(-\frac{1}{27}\right)=\frac{27}{27}=1.$

b) The reciprocal of $\frac{-3}{4}$ is $\frac{4}{-3}$, or, equivalently, $-\frac{4}{3}$.   *Check:* $\frac{-3}{4}\cdot\frac{4}{-3}=\frac{-12}{-12}=1.$

c) The reciprocal of $-\frac{1}{5}$ is -5.   *Check:* $-\frac{1}{5}(-5)=\frac{5}{5}=1.$

8. Find the reciprocal of $-\frac{10}{9}$, if it exists.

d) The reciprocal of 0 does not exist. To see this, recall that there is no number r for which $0\cdot r=1.$

⟲ YOUR TURN

EXAMPLE 9 Divide: **(a)** $-\frac{2}{3}\div\left(-\frac{5}{4}\right)$; **(b)** $-\frac{3}{4}\div\frac{3}{10}$.

SOLUTION We divide by multiplying by the reciprocal of the divisor.

a) $-\dfrac{2}{3}\div\left(-\dfrac{5}{4}\right)=-\dfrac{2}{3}\cdot\left(-\dfrac{4}{5}\right)=\dfrac{8}{15}$ Multiplying by the reciprocal

Be careful not to change the sign when taking a reciprocal!

9. Divide: $-\dfrac{5}{9}\div\dfrac{1}{15}.$

b) $-\dfrac{3}{4}\div\dfrac{3}{10}=-\dfrac{3}{4}\cdot\left(\dfrac{10}{3}\right)=-\dfrac{30}{12}=-\dfrac{5}{2}\cdot\dfrac{6}{6}=-\dfrac{5}{2}$ Removing a factor equal to 1: $\frac{6}{6}=1$

⟲ YOUR TURN

To divide negative numbers with decimal notation, it is usually easiest to carry out the division and then focus on the sign.

EXAMPLE 10 Divide: $27.9 \div (-3)$.

SOLUTION

$$27.9 \div (-3) = \frac{27.9}{-3} = -9.3 \qquad \text{Dividing: } 3\overline{)27.9.}^{\,9.3}$$
$$\text{The answer is negative.}$$

10. Divide: $-96 \div (-0.6)$.

YOUR TURN

In Example 5(d), we explained why we cannot divide -17 by 0. To see why *no* nonzero number b can be divided by 0, remember that $b \div 0$ would need to be the number that when multiplied by 0 gives b. That is, if $b \div 0 = r$, then $r \cdot 0 = b$. But since 0 times any number is 0, not b, we say that $b \div 0$ is **undefined** for $b \neq 0$.

In the special case of $0 \div 0$, we look for a number r such that $0 \div 0 = r$ and $r \cdot 0 = 0$. But, $r \cdot 0 = 0$ for *any* number r. For this reason, we say that $b \div 0$ is undefined for any choice of b. Sometimes we say that $0 \div 0$ is *indeterminate*.

Finally, note that $0 \div 7 = 0$ since $0 \cdot 7 = 0$. This can be written $0/7 = 0$. It is important not to confuse division *by* 0 with division *into* 0.

EXAMPLE 11 Divide, if possible: **(a)** $\frac{0}{-2}$; **(b)** $\frac{5}{0}$.

SOLUTION

a) $\dfrac{0}{-2} = 0$ We can divide 0 by a nonzero number.
       Check: $0(-2) = 0$.

11. Divide, if possible:

$0 \div (-12)$.

b) $\dfrac{5}{0}$ is undefined.   We cannot divide by 0.

YOUR TURN

> **DIVISION INVOLVING ZERO**
>
> For any real number a, $\dfrac{a}{0}$ is undefined, and for $a \neq 0$, $\dfrac{0}{a} = 0$.

↪ Check Your UNDERSTANDING

Determine whether each product or quotient is positive or negative. Do not perform the calculation.

1. $-12 \cdot 11$

2. $-3.2 \cdot (-48)$

3. $-\dfrac{2}{3} \div \left(-\dfrac{1}{2}\right)$

4. $-6 \cdot 3 \cdot (-4) \cdot (-5)$

5. $32.7 \div (-0.16)$

It is important *not* to confuse *opposite* with *reciprocal*. Keep in mind that the opposite, or additive inverse, of a number is what we add to the number in order to get 0. The reciprocal, or multiplicative inverse, is what we multiply the number by in order to get 1.

Compare the following.

Number	Opposite (Change the sign.)	Reciprocal (Invert but do not change the sign.)
$-\dfrac{3}{8}$	$\dfrac{3}{8}$ ←	$-\dfrac{8}{3}$ ← $\left(-\dfrac{3}{8}\right)\left(-\dfrac{8}{3}\right) = 1$
19	-19	$\dfrac{1}{19}$ $\quad -\dfrac{3}{8} + \dfrac{3}{8} = 0$
$\dfrac{18}{7}$	$-\dfrac{18}{7}$	$\dfrac{7}{18}$
-7.9	7.9	$-\dfrac{1}{7.9}$, or $-\dfrac{10}{79}$
0	0	Undefined

CONNECTING 🔗 THE CONCEPTS

The rules for multiplication and division of real numbers differ significantly from the rules for addition and subtraction. When simplifying an expression, look at the operation first to determine which set of rules to follow.

Addition

If the signs are the same, add absolute values. *The answer has the same sign as the numbers.*

If the signs are different, subtract absolute values. *The answer has the same sign as the number with the larger absolute value. If the absolute values are the same, the answer is 0.*

Subtraction

Add the opposite of the number being subtracted.

Multiplication

If the signs are the same, multiply absolute values. *The answer is positive.*

If the signs are different, multiply absolute values. *The answer is negative.*

Division

Multiply by the reciprocal of the divisor.

EXERCISES

Perform the indicated operation and, if possible, simplify.

1. $-8 + (-2)$ **2.** $-8 \cdot (-2)$

3. $-8 \div (-2)$ **4.** $-8 - (-2)$

5. $\dfrac{3}{5} - \dfrac{8}{5}$ **6.** $\dfrac{12}{5} + \left(\dfrac{-7}{5}\right)$

7. $(1.3)(-2.9)$ **8.** $-44.1 \div 6.3$

9. $-38 - (-38)$ **10.** $-46 - 46$

1.7   EXERCISE SET

FOR EXTRA HELP   MyMathLab®

🢒 Vocabulary and Reading Check

Choose from the following list of words to complete each statement. Words may be used more than once or not at all.

even	positive
negative	reciprocal
odd	undefined
opposite	zero

1. The product of two negative numbers is

_____.

2. The product of a(n) _____ number of negative numbers is negative.

3. Division by zero is _____.

4. To divide by a fraction, multiply by its

_____.

5. The _____ of a negative number is positive.

6. The _____ of a negative number is negative.

🢒 Concept Reinforcement

In each of Exercises 7–16, replace the blank with either 0 or 1 to match the description given.

7. The product of two reciprocals ____

8. The sum of a pair of opposites ____

9. The sum of a pair of additive inverses ____

10. The product of two multiplicative inverses ____

11. This number has no reciprocal. ____

12. This number is its own reciprocal. ____

13. This number is the multiplicative identity. ____

14. This number is the additive identity. ____

15. A nonzero number divided by itself ____

16. Division by this number is undefined. ____

A. Multiplication

Multiply.

17. $-4 \cdot 10$ **18.** $-5 \cdot 6$

19. $-8 \cdot 7$ **20.** $-9 \cdot 2$

21. $4 \cdot (-10)$ **22.** $9 \cdot (-5)$

23. $-9 \cdot (-8)$ **24.** $-10 \cdot (-11)$

25. $-19 \cdot (-10)$ **26.** $-12 \cdot (-10)$

27. $11 \cdot (-12)$ **28.** $15 \cdot (-43)$

29. $4.5 \cdot (-28)$ **30.** $-49 \cdot (-2.1)$

31. $-5 \cdot (-2.3)$ **32.** $-6 \cdot 4.8$

33. $(-25) \cdot 0$ **34.** $0 \cdot (-4.7)$

35. $\frac{2}{5} \cdot \left(-\frac{5}{7}\right)$ **36.** $\frac{5}{7} \cdot \left(-\frac{2}{3}\right)$

37. $-\frac{3}{8} \cdot \left(-\frac{2}{9}\right)$ **38.** $-\frac{5}{8} \cdot \left(-\frac{2}{5}\right)$

39. $(-5.3)(2.1)$ **40.** $(9.5)(-3.7)$

41. $-\frac{5}{9} \cdot \frac{3}{4}$ **42.** $-\frac{8}{3} \cdot \frac{9}{4}$

43. $3 \cdot (-7) \cdot (-2) \cdot 6$ **44.** $9 \cdot (-2) \cdot (-6) \cdot 7$

Aha! **45.** $27 \cdot (-34) \cdot 0$ **46.** $-43 \cdot (-74) \cdot 0$

47. $-\frac{1}{3} \cdot \frac{1}{4} \cdot \left(-\frac{3}{7}\right)$ **48.** $-\frac{1}{2} \cdot \frac{3}{5} \cdot \left(-\frac{2}{7}\right)$

49. $-2 \cdot (-5) \cdot (-3) \cdot (-5)$

50. $-3 \cdot (-5) \cdot (-2) \cdot (-1)$

51. $(-31) \cdot (-27) \cdot 0 \cdot (-13)$

52. $7 \cdot (-6) \cdot 5 \cdot (-4) \cdot 3 \cdot (-2) \cdot 1 \cdot 0$

53. $(-8)(-9)(-10)$

54. $(-7)(-8)(-9)(-10)$

55. $(-6)(-7)(-8)(-9)(-10)$

56. $(-5)(-6)(-7)(-8)(-9)(-10)$

B. Division

Divide, if possible, and check. If a quotient is undefined, state this.

57. $18 \div (-2)$ **58.** $\frac{24}{-3}$

59. $\frac{36}{-9}$ **60.** $26 \div (-13)$

61. $\frac{-56}{8}$ **62.** $\frac{-35}{-7}$

63. $\frac{-48}{-12}$ **64.** $-63 \div (-9)$

65. $-72 \div 8$ **66.** $\frac{-50}{25}$

67. $-10.2 \div (-2)$ **68.** $-2 \div 0.8$

69. $-100 \div (-11)$ **70.** $\frac{-64}{-7}$

71. $\frac{400}{-50}$ **72.** $-300 \div (-13)$

73. $\frac{48}{0}$ **74.** $\frac{0}{-5}$

75. $-4.8 \div 1.2$ **76.** $-3.9 \div 1.3$

77. $\frac{0}{-9}$ **78.** $0 \div 18$

Aha! **79.** $\frac{9.7(-2.8)0}{4.3}$ **80.** $\frac{(-4.9)(7.2)}{0}$

Write each expression in two equivalent forms, as in Example 6.

81. $\frac{-8}{3}$ **82.** $\frac{-10}{3}$

83. $\frac{29}{-35}$ **84.** $\frac{18}{-7}$

85. $-\frac{7}{3}$ **86.** $-\frac{4}{15}$

87. $\frac{-x}{2}$ **88.** $\frac{9}{-a}$

Find the reciprocal of each number, if it exists.

89. $-\frac{4}{5}$ **90.** $-\frac{13}{11}$

91. $\frac{51}{-10}$ **92.** $\frac{43}{-24}$

93. -10 **94.** 34

95. 4.3 **96.** -1.7

97. $\frac{-1}{4}$ **98.** $\frac{-1}{11}$

99. 0 **100.** -1

A, B. Addition, Subtraction, Multiplication, and Division

Perform the indicated operation and, if possible, simplify. If a quotient is undefined, state this.

101. $\left(\frac{-7}{4}\right)\left(-\frac{3}{5}\right)$ **102.** $\left(-\frac{5}{6}\right)\left(\frac{-1}{3}\right)$

103. $\frac{-3}{8} + \frac{-5}{8}$ **104.** $\frac{-4}{5} + \frac{7}{5}$

Aha! **105.** $\left(\frac{-9}{5}\right)\left(\frac{5}{-9}\right)$ **106.** $\left(-\frac{2}{7}\right)\left(\frac{5}{-8}\right)$

107. $\left(-\frac{3}{11}\right) - \left(-\frac{6}{11}\right)$ **108.** $\left(-\frac{4}{7}\right) - \left(-\frac{2}{7}\right)$

109. $\frac{7}{8} \div \left(-\frac{1}{2}\right)$ **110.** $\frac{3}{4} \div \left(-\frac{2}{3}\right)$

111. $\frac{9}{5} \cdot \frac{-20}{3}$ **112.** $\frac{-5}{12} \cdot \frac{7}{15}$

113. $\left(-\frac{18}{7}\right) + \left(\frac{3}{-7}\right)$ **114.** $\left(\frac{12}{-5}\right) + \left(-\frac{3}{5}\right)$

Aha! **115.** $-\frac{5}{9} \div \left(-\frac{5}{9}\right)$ **116.** $\frac{-5}{12} \div \frac{15}{7}$

117. $\frac{-3}{10} + \frac{2}{5}$ **118.** $\frac{-5}{9} + \frac{2}{3}$

119. $\frac{7}{10} \div \left(\frac{-3}{5}\right)$

120. $\left(\frac{-3}{5}\right) \div \frac{6}{15}$

121. $\frac{14}{-9} \div \frac{0}{3}$

122. $\frac{0}{-10} \div \frac{-3}{8}$

123. $\frac{2}{15} - \frac{7}{10}$

124. $-\frac{7}{20} - \frac{5}{16}$

125. Most calculators have a key, often appearing as **1/x**, for finding reciprocals. To use this key, we enter a number and then press **1/x** to find its reciprocal. What should happen if we key in a number and press the reciprocal key twice? Why?

126. Multiplication can be regarded as repeated addition. Using this idea and the number line, explain why $3 \cdot (-5) = -15$.

Synthesis

127. If two nonzero numbers are opposites of each other, are their reciprocals opposites of each other? Why or why not?

128. If two numbers are reciprocals of each other, are their opposites reciprocals of each other? Why or why not?

Translate to an algebraic expression or an equation.

129. The reciprocal of a sum

130. The sum of two reciprocals

131. The opposite of a sum

132. The sum of two opposites

133. A real number is its own opposite.

134. A real number is its own reciprocal.

135. Show that the reciprocal of a sum is *not* the sum of the two reciprocals.

136. Which real numbers are their own reciprocals?

137. Jenna is a meteorologist. On December 10, she notes that the temperature is $-3°\text{F}$ at 6:00 A.M. She predicts that the temperature will rise at a rate of 2° per hour for 3 hr, and then rise at a rate of 3° per hour for 6 hr. She also predicts that the temperature will then fall at a rate of 2° per hour for 3 hr, and then fall at a rate of 5° per hour for 2 hr. What is Jenna's temperature forecast for 8:00 P.M.?

Tell whether each expression represents a positive number or a negative number when m and n are negative.

138. $\dfrac{m}{-n}$

139. $\dfrac{-n}{-m}$

140. $-m \cdot \left(\dfrac{-n}{m}\right)$

141. $-\left(\dfrac{n}{-m}\right)$

142. $(m + n) \cdot \dfrac{m}{n}$

143. $(-n - m)\dfrac{n}{m}$

144. What must be true of m and n if $-mn$ is to be **(a)** positive? **(b)** zero? **(c)** negative?

145. The following is a proof that a positive number times a negative number is negative. Provide a reason for each step. Assume that $a > 0$ and $b > 0$.

$$a(-b) + ab = a[-b + b]$$
$$= a(0)$$
$$= 0$$

Therefore, $a(-b)$ is the opposite of ab.

146. Is it true that for any numbers a and b, if a is larger than b, then the reciprocal of a is smaller than the reciprocal of b? Why or why not?

147. *Research.* As her airplane was descending, Judy noticed that the onboard monitors indicated an altitude of 20,000 ft and an air temperature of $-15°\text{F}$. Find how many degrees air temperature drops for every 1000 ft of altitude. Then predict the temperature on the ground after Judy lands.

YOUR TURN ANSWERS: SECTION 1.7

1. -30 **2.** 0 **3.** $\frac{1}{6}$ **4.** -50 **5.** -4 **6.** $\frac{5}{-6}; -\frac{5}{6}$
7. 1 **8.** $-\frac{9}{10}$ **9.** $-\frac{25}{3}$ **10.** 160 **11.** 0

Quick Quiz: Sections 1.1–1.7

1. Factor: $22x + 11 + 33y$. [1.2]

2. Graph on the number line: -3.5. [1.4]

3. Simplify: $-3.1 + 1.5 + (-2.8) + (-1.7)$. [1.5]

4. Combine like terms: $3x - 7m - m - 4x$. [1.6]

5. Divide: $\dfrac{3}{8} \div \left(-\dfrac{3}{16}\right)$. [1.7]

1.8 Exponential Notation and Order of Operations

A. Exponential Notation   **B.** Order of Operations   **C.** Simplifying and the Distributive Law
D. The Opposite of a Sum

In this section, we learn how to use *exponential notation* and rules for the *order of operations* to perform certain algebraic manipulations.

A. Exponential Notation

A product like $3 \cdot 3 \cdot 3 \cdot 3$, in which the factors are the same, is called a **power**. Powers are often written in **exponential notation**. For

$$\underbrace{3 \cdot 3 \cdot 3 \cdot 3}_{4 \text{ factors}}, \quad \text{we write} \quad 3^4.$$

This is read "three to the fourth power," or simply, "three to the fourth." The number 4 is called an **exponent** and the number 3 a **base**. Because $3^4 = 81$, we say that 81 is a power of 3.

Expressions like s^2 and s^3 are usually read "*s* squared" and "*s* cubed," respectively. This comes from the fact that a square with sides of length *s* has an area *A* given by $A = s^2$ and a cube with sides of length *s* has a volume *V* given by $V = s^3$.

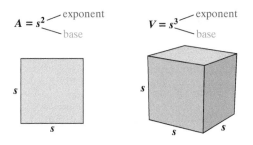

EXAMPLE 1   Write exponential notation for $10 \cdot 10 \cdot 10 \cdot 10 \cdot 10$.

SOLUTION

1. Write exponential notation for $(-9)(-9)(-9)$.

Exponential notation is 10^5.   5 is the exponent.
10 is the base.

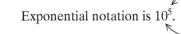   YOUR TURN

EXAMPLE 2   Simplify: **(a)** 5^2; **(b)** $(-5)^3$; **(c)** $(2n)^3$.

SOLUTION

a) $5^2 = 5 \cdot 5 = 25$ The base is 5. The exponent 2 indicates two factors of 5.

b) $(-5)^3 = (-5)(-5)(-5)$ The base is -5. The exponent 3 indicates three factors of -5.

$\qquad\qquad = 25(-5)$ Using the associative law of multiplication

$\qquad\qquad = -125$

c) $(2n)^3 = (2n)(2n)(2n)$ The base is $2n$.

$\qquad\qquad = 2 \cdot 2 \cdot 2 \cdot n \cdot n \cdot n$ Using the associative and commutative laws of multiplication

$\qquad\qquad = 8n^3$

2. Simplify: $(-2)^4$.

YOUR TURN

Student Notes

Although most scientific and graphing calculators follow the rules for order of operations when evaluating expressions, a few do not. Calculate $4 + 2 \times 5$ on your calculator. If the result is 30, your calculator does not follow the rules for order of operations. In this case, you will need to multiply 2×5 first and then add 4.

3. Simplify: $1 - 30 \div 3 + 7$.

YOUR TURN

To determine what the exponent 1 will mean, look for a pattern in the following:

$$7 \cdot 7 \cdot 7 \cdot 7 = 7^4$$
$$7 \cdot 7 \cdot 7 = 7^3$$
$$7 \cdot 7 = 7^2$$
$$? = 7^1$$

The exponent decreases by 1 each time.

The number of factors decreases by 1 each time. To extend the pattern, we say that

$$7 = 7^1.$$

EXPONENTIAL NOTATION

For any natural number n,

$$b^n \quad \text{means} \quad \overbrace{b \cdot b \cdot b \cdot b \cdots b}^{n \text{ factors}}.$$

B. Order of Operations

How should $4 + 2 \times 5$ be computed? If we multiply 2 by 5 and then add 4, the result is 14. If we add 2 and 4 first and then multiply by 5, the result is 30. Since these results differ, the order in which we perform operations matters. If grouping symbols such as parentheses ( ), brackets [ ], braces { }, absolute-value symbols | |, or fraction bars appear, they tell us what to do first. For example,

$$(4 + 2) \times 5 \quad \text{indicates} \quad 6 \times 5, \quad \text{resulting in 30,}$$

and

$$4 + (2 \times 5) \quad \text{indicates} \quad 4 + 10, \quad \text{resulting in 14.}$$

Besides grouping symbols, the following conventions exist for determining the order in which operations should be performed.

RULES FOR ORDER OF OPERATIONS

1. Simplify, if possible, within the innermost grouping symbols, ( ), [ ], { }, | |, and above or below any fraction bars.
2. Simplify all exponential expressions.
3. Perform all multiplications and divisions, working from left to right.
4. Perform all additions and subtractions, working from left to right.

Thus the correct way to compute $4 + 2 \times 5$ is to first multiply 2 by 5 and then add 4. The result is 14.

EXAMPLE 3 Simplify: $15 - 2 \cdot 5 + 3$.

SOLUTION When no groupings or exponents appear, we *always* multiply or divide before adding or subtracting:

$$15 - 2 \cdot 5 + 3 = 15 - 10 + 3 \qquad \text{Multiplying}$$
$$= 5 + 3 \Big\} \qquad \text{Subtracting and adding from}$$
$$= 8. \qquad \text{left to right}$$

Always calculate within parentheses first. When there are exponents and no parentheses, simplify powers before multiplying or dividing.

EXAMPLE 4   Simplify: **(a)** $(3 \cdot 4)^2$; **(b)** $3 \cdot 4^2$.

SOLUTION

a) $(3 \cdot 4)^2 = (12)^2$ Working within parentheses first

$\qquad\qquad = 144$

b) $3 \cdot 4^2 = 3 \cdot 16$ Simplifying the power

$\qquad\quad = 48$ Multiplying

Note that $(3 \cdot 4)^2 \neq 3 \cdot 4^2$.

4. Simplify: $(-5) \cdot 3^2$.

YOUR TURN

> **CAUTION!**   Example 4 illustrates that, in general, $(ab)^2 \neq ab^2$.

Student Notes

In $(-7)^2$, the base is -7. In -7^2, the base is 7.

Finding the opposite of a number is the same as multiplying the number by -1. Thus we evaluate expressions like $(-7)^2$ and -7^2 differently.

$(-7)^2 = 49$

1. *Parentheses*: The opposite of 7 is -7, or $-1 \cdot 7 = -7$.

2. *Exponents*: The square of -7 is 49.

$-7^2 = -49$

1. *Exponents*: The square of 7 is 49.

2. *Multiplication*: The opposite of 49 is -49, or $-1 \cdot 49 = -49$.

To simplify $-x^2$, it may help to write

$$-x^2 = (-1)x^2.$$

EXAMPLE 5   Evaluate for $x = 5$: **(a)** $(-x)^2$; **(b)** $-x^2$.

SOLUTION

a) $(-x)^2 = (-5)^2 = (-5)(-5) = 25$ The base is -5. We square the opposite of 5.

b) $-x^2 = (-1)x^2 = (-1)(5)^2 = (-1)(25) = -25$ The base is 5. We square 5 and then multiply by -1 (or find the opposite).

5. Evaluate $-x^2$ for $x = 10$.

YOUR TURN

> **CAUTION!**   Example 5 illustrates that, in general, $(-x)^2 \neq -x^2$.

Student Notes

When simplifying an expression, it is important to copy the entire expression on each line, not just the parts that have been simplified in a given step. As shown in Examples 6 and 7, each line should be equivalent to the line above it.

EXAMPLE 6   Evaluate $-15 \div 3(6 - a)^3$ for $a = 4$.

SOLUTION

$-15 \div 3(6 - a)^3 = -15 \div 3(6 - 4)^3$ Substituting 4 for a

$\qquad\qquad\qquad = -15 \div 3(2)^3$ Working within parentheses first

$\qquad\qquad\qquad = -15 \div 3 \cdot 8$ Simplifying the exponential expression

$\qquad\qquad\qquad = -5 \cdot 8$ Dividing and multiplying from left to right

$\qquad\qquad\qquad = -40$

6. Evaluate $-2 + 100 \div 2y$ for $y = 10$.

YOUR TURN

The symbols ( ), [ ], and { } are all used in the same way. Used inside or next to each other, they make it easier to locate the left and right sides of a grouping. When combinations of grouping symbols are used, we begin with the innermost grouping symbols and work to the outside.

EXAMPLE 7 Simplify: $8 \div 4 + 3[9 + 2(3 - 5)^3]$.

SOLUTION

$8 \div 4 + 3[9 + 2(3 - 5)^3] = 8 \div 4 + 3[9 + 2(-2)^3]$ Doing the calculations in the innermost grouping symbols first

$\qquad = 8 \div 4 + 3[9 + 2(-8)]$ $(-2)^3 = (-2)(-2)(-2)$
$\qquad\qquad\qquad\qquad\qquad\qquad\qquad\qquad = -8$

$\qquad = 8 \div 4 + 3[9 + (-16)]$

$\qquad = 8 \div 4 + 3[-7]$ Completing the calculations within the brackets

$\qquad = 2 + (-21)$ Multiplying and dividing from left to right

$\qquad = -19$

7. Simplify:

$\qquad 10 + 5[12 \div (2 - 8)]^2$.

YOUR TURN

EXAMPLE 8 Calculate: $\dfrac{12(9 - 7) + 4 \cdot 5}{2^4 + 3^2}$.

SOLUTION An equivalent expression with brackets is

$\qquad [12(9 - 7) + 4 \cdot 5] \div [2^4 + 3^2]$.   Here the grouping symbols are necessary.

In effect, we need to simplify the numerator, simplify the denominator, and then divide the results:

$$\frac{12(9 - 7) + 4 \cdot 5}{2^4 + 3^2} = \frac{12(2) + 4 \cdot 5}{16 + 9}$$

$$= \frac{24 + 20}{25} = \frac{44}{25}.$$

8. Calculate:

$\qquad \dfrac{28 \div 14 \cdot 2 - (6 - 1)}{2^2 + 6^2 \div (-3)}$.

YOUR TURN

C. Simplifying and the Distributive Law

Sometimes we cannot simplify within grouping symbols. When a sum or a difference is being grouped, we can use the distributive law to remove the grouping symbols.

EXAMPLE 9 Simplify: $5x - 9 + 2(4x + 5)$.

SOLUTION

$\qquad 5x - 9 + 2(4x + 5) = 5x - 9 + 8x + 10$ Using the distributive law
$\qquad\qquad\qquad\qquad\qquad\quad = 13x + 1$ Combining like terms

9. Simplify:

$\qquad 2x + 5(3x - 7) - 20$.

YOUR TURN

Now that exponents have been introduced, we can make our definition of *like*, or *similar*, terms more precise. **Like**, or **similar**, **terms** are either constant terms or terms containing the same variable(s) raised to the same power(s). Thus, 5 and -7, $19xy$ and $2yx$, as well as $4a^3b$ and a^3b are all pairs of like terms.

EXAMPLE 10   Simplify: $7x^2 + 3[x^2 + 2x] - 5x$.

SOLUTION

$$7x^2 + 3[x^2 + 2x] - 5x = 7x^2 + 3x^2 + 6x - 5x \qquad \text{Using the distributive law}$$

$$= 10x^2 + x \qquad \text{Combining like terms}$$

10. Simplify:

$2a^2 - ab + 10(a^2 + 6ab)$.

YOUR TURN

D. The Opposite of a Sum

An expression such as $-(x + y)$ indicates the *opposite*, or *additive inverse*, of the sum of x and y. When a sum within grouping symbols is preceded by a "$-$" symbol, we can multiply the sum by -1 and use the distributive law. In this manner, we can find an equivalent expression for the opposite of a sum.

EXAMPLE 11   Write an expression equivalent to $-(3x + 2y + 4)$ without using parentheses.

SOLUTION

$$-(3x + 2y + 4) = -1(3x + 2y + 4) \qquad \text{Using the property of } -1$$

$$= -1(3x) + (-1)(2y) + (-1)4 \qquad \text{Using the distributive law}$$

$$= -3x - 2y - 4 \qquad \text{Using the property of } -1$$

11. Write an expression equivalent to $-(2 + 5m + 10n)$ without using parentheses.

YOUR TURN

THE OPPOSITE OF A SUM

For any real numbers a and b,

$$-(a + b) = (-a) + (-b) = -a - b.$$

(The opposite of a sum is the sum of the opposites.)

To remove parentheses from an expression like $-(x - 7y + 5)$, we can first rewrite the subtraction as addition:

$$-(x - 7y + 5) = -(x + (-7y) + 5) \qquad \text{Rewriting as addition}$$

$$= -x + 7y - 5. \qquad \text{Taking the opposite of a sum}$$

This procedure is generally streamlined to one step in which we find the opposite by "removing parentheses and changing the sign of every term":

$$-(x - 7y + 5) = -x + 7y - 5.$$

EXAMPLE 12   Simplify: $3x - (4x + 2)$.

SOLUTION

$$3x - (4x + 2) = 3x + [-(4x + 2)] \qquad \text{Adding the opposite of } 4x + 2$$

$$= 3x + [-4x - 2] \qquad \text{Taking the opposite of } 4x + 2$$

$$= 3x + (-4x) + (-2)$$

$$= 3x - 4x - 2 \qquad \text{Try to go directly to this step.}$$

$$= -x - 2 \qquad \text{Combining like terms}$$

12. Simplify: $9 - (5x - 11)$.

YOUR TURN

We can also subtract expressions in parentheses using multiplication by -1.

EXAMPLE 13   Simplify: $5t^2 - 2t - (-4t^2 + 9t)$.

SOLUTION

$$
\begin{aligned}
5t^2 - 2t - (-4t^2 + 9t) \\
&= 5t^2 - 2t + [-(-4t^2 + 9t)] && \text{Adding the opposite of } -4t^2 + 9t \\
&= 5t^2 - 2t + [-1(-4t^2 + 9t)] && \text{Using the property of } -1: -a = -1 \cdot a \\
&= 5t^2 - 2t + [4t^2 - 9t] && \text{Using the distributive law} \\
&= 5t^2 - 2t + 4t^2 - 9t \\
&= 9t^2 - 11t && \text{Combining like terms}
\end{aligned}
$$

13. Simplify:

$10x - 3x^2 - (-x^2 + 12x)$.

YOUR TURN

Expressions such as $7 - 3(x + 2)$ can be simplified as follows:

$$
\begin{aligned}
7 - 3(x + 2) &= 7 + [-3(x + 2)] && \text{Adding the opposite of } 3(x + 2) \\
&= 7 + [-3x - 6] && \text{Multiplying } x + 2 \text{ by } -3 \\
&= 7 + (-3x) + (-6) \\
&= 1 + (-3x) && \text{Combining like terms} \\
&= 1 - 3x.
\end{aligned}
$$

EXAMPLE 14   Simplify: $3n - 2(4n - 5)$.

SOLUTION

$$
\begin{aligned}
3n - 2(4n - 5) &= 3n + [-2(4n - 5)] && \text{Adding the opposite of } 2(4n - 5) \\
&= 3n + [-8n + 10] && \text{Multiplying } 4n - 5 \text{ by } -2 \\
&= 3n + (-8n) + 10 \\
&= -5n + 10 && \text{Combining like terms}
\end{aligned}
$$

14. Simplify:

$6y^2 - 3(2y^2 - 6y - 7) + 10$.

YOUR TURN

Chapter Resources:
Collaborative Activity (Select the Symbols), p. 70; Decision Making: Connection, p. 71

In practice, we often multiply and remove parentheses in one step. Then Example 14 looks like

$$
\begin{aligned}
3n - 2(4n - 5) &= 3n - 8n + 10 \\
&= -5n + 10.
\end{aligned}
$$

EXAMPLE 15   Simplify: $7x^3 + 2 - [5(x^3 - 1) + 8]$.

SOLUTION

$$
\begin{aligned}
7x^3 + 2 - [5(x^3 - 1) + 8] &= 7x^3 + 2 - [5x^3 - 5 + 8] && \text{Removing parentheses} \\
&= 7x^3 + 2 - [5x^3 + 3] && \text{Combining like terms} \\
&= 7x^3 + 2 - 5x^3 - 3 && \text{Removing brackets} \\
&= 2x^3 - 1 && \text{Combining like terms}
\end{aligned}
$$

15. Simplify:

$2(x^2 + 3) - [6(x^2 - 1) + 5]$.

YOUR TURN

It is important to distinguish between the two tasks of **simplifying an expression** and **solving an equation**. In this chapter, we have not solved equations, but we have simplified expressions. This enabled us to write equivalent expressions that were simpler than the given expression.

↳ Check Your
UNDERSTANDING

Use the rules for order of operations to simplify each expression.

1. a) $7 - 3 \cdot 2$
b) $(7 - 3) \cdot 2$

2. a) $10 - 3^2$
b) $(10 - 3)^2$

3. a) $(-10)^2$
b) -10^2

4. a) $10 - 3 + 5$
b) $10 - (3 + 5)$

5. a) $100 \div (25 \cdot 4)$
b) $100 \div 25 \cdot 4$

6. a) $12 \div (4 + 2)$
b) $12 \div 4 + 2$

1.8 EXERCISE SET

FOR EXTRA HELP MyMathLab®

↳ Vocabulary and Reading Check

In each of Exercises 1–6, match the expression with the best illustration of that expression from the column on the right.

1. An exponent ____
2. A base ____
3. The square of a number ____
4. The cube of a number ____
5. Like terms ____
6. The opposite of a sum ____

a) 10^2
b) The 4 in 4^3
c) The 6 in x^6
d) $-(x + 3) = -x - 3$
e) $3y^4$ and $-y^4$
f) 8^3

↳ Concept Reinforcement

In each of Exercises 7–12, name the operation that should be performed first. Do not perform the calculations.

7. $4 + 8 \div 2 \cdot 2$
8. $7 - 9 + 15$
9. $5 - 2(3 + 4)$
10. $6 + 7 \cdot 3$
11. $18 - 2[4 + (3 - 2)]$
12. $\dfrac{5 - 6 \cdot 7}{2}$

A. Exponential Notation

Write exponential notation.

13. $x \cdot x \cdot x \cdot x \cdot x \cdot x$
14. $y \cdot y \cdot y \cdot y \cdot y \cdot y$
15. $(-5)(-5)(-5)$
16. $(-7)(-7)(-7)(-7)$
17. $3t \cdot 3t \cdot 3t \cdot 3t \cdot 3t$
18. $5m \cdot 5m \cdot 5m \cdot 5m \cdot 5m$
19. $2 \cdot n \cdot n \cdot n \cdot n$
20. $8 \cdot a \cdot a \cdot a$

Simplify.

21. 4^2
22. 5^3
23. $(-3)^2$
24. $(-7)^2$
25. -3^2
26. -7^2
27. 4^3
28. 9^1
29. $(-5)^4$
30. 5^4
31. 7^1
32. $(-1)^7$
33. $(-2)^5$
34. -2^5
35. $(3t)^4$
36. $(5t)^2$
37. $(-7x)^3$
38. $(-5x)^4$

B. Order of Operations

Simplify.

39. $5 + 3 \cdot 7$
40. $3 - 4 \cdot 2$
41. $10 \cdot 5 + 1 \cdot 1$
42. $19 - 5 \cdot 4 + 3$
43. $5 - 50 \div 5 \cdot 2$
44. $12 \div 3 + 18 \div 2$
Aha! 45. $14 \cdot 19 \div (19 \cdot 14)$
46. $18 - 6 \div 3 \cdot 2 + 7$
47. $3(-10)^2 - 8 \div 2^2$
48. $9 - 3^2 \div 9(-1)$
49. $8 - (2 \cdot 3 - 9)$
50. $(8 - 2 \cdot 3) - 9$
51. $(8 - 2)(3 - 9)$
52. $32 \div (-2)^2 \cdot 4$

53. $13(-10)^2 + 45 \div (-5)$

54. $2^4 + 2^3 - 10 \div (-1)^4$

55. $5 + 3(2 - 9)^2$

56. $9 - (3 - 5)^3 - 4$

57. $[2 \cdot (5 - 8)]^2 - 12$

58. $2^3 + 2^4 - 5[8 - 4(9 - 10)^2]$

59. $\dfrac{7 + 2}{5^2 - 4^2}$

60. $\dfrac{(5^2 - 3^2)^2}{2 \cdot 6 - 4}$

61. $8(-7) + |3(-4)|$

62. $|10(-5)| + 1(-1)$

63. $36 \div (-2)^2 + 4[5 - 3(8 - 9)^5]$

64. $-48 \div (7 - 9)^3 - 2[1 - 5(2 - 6) + 3^2]$

65. $\dfrac{7^2 - (-1)^7}{5 \cdot 7 - 4 \cdot 3^2 - 2^2}$

66. $\dfrac{(-2)^3 + 4^2}{2 \cdot 3 - 5^2 + 3 \cdot 7}$

67. $\dfrac{-3^3 - 2 \cdot 3^2}{8 \div 2^2 - (6 - |2 - 15|)}$

68. $\dfrac{(-5)^2 - 3 \cdot 5}{3^2 + 4 \cdot |6 - 7| \cdot (-1)^5}$

Evaluate.

69. $9 - 4x$, for $x = 7$

70. $1 + x^3$, for $x = -2$

71. $24 \div t^3$, for $t = -2$

72. $-100 \div a^2$, for $a = -5$

73. $45 \div a \cdot 5$, for $a = -3$

74. $50 \div 2 \cdot t$, for $t = 5$

75. $5x \div 15x^2$, for $x = 3$

76. $6a \div 12a^3$, for $a = 2$

77. $45 \div 3^2 x(x - 1)$, for $x = 3$

78. $-30 \div t(t + 4)^2$, for $t = -6$

79. $-x^2 - 5x$, for $x = -3$

80. $(-x)^2 - 5x$, for $x = -3$

81. $\dfrac{3a - 4a^2}{a^2 - 20}$, for $a = 5$

82. $\dfrac{a^3 - 4a}{a(a - 3)}$, for $a = -2$

C. Simplifying and the Distributive Law

Simplify.

83. $3x - 2 + 5(2x + 7)$

84. $8n + 3(5n + 2) + 1$

85. $2x^2 + 5(3x^2 - x) - 12x$

86. $6x^2 - x + 4(7x^2 + 3x)$

87. $9t - 7r + 2(3r + 6t)$

88. $4m - 9n + 3(2m - n)$

89. $5t^3 + t + 3(t - 2t^3)$

90. $8n^2 - 3n + 2(n - 4n^2)$

D. The Opposite of a Sum

Write an equivalent expression without using grouping symbols.

91. $-(9x + 1)$

92. $-(3x + 5)$

93. $-[-7n + 8]$

94. $-(6x - 7)$

95. $-(4a - 3b + 7c)$

96. $-[5n - m - 2p]$

97. $-(3x^2 + 5x - 1)$

98. $-(-9x^3 + 8x + 10)$

Simplify.

99. $8x - (6x + 7)$

100. $2a - (5a - 9)$

101. $2x - 7x - (4x - 6)$

102. $2a + 5a - (6a + 8)$

103. $15x - y - 5(3x - 2y + 5z)$

104. $4a - b - 4(5a - 7b + 8c)$

105. $3x^2 + 11 - (2x^2 + 5)$

106. $5x^4 + 3x - (5x^4 + 3x)$

107. $12a^2 - 3ab + 5b^2 - 5(-5a^2 + 4ab - 6b^2)$

108. $-8a^2 + 5ab - 12b^2 - 6(2a^2 - 4ab - 10b^2)$

109. $-7t^3 - t^2 - 3(5t^3 - 3t)$

110. $9t^4 + 7t - 5(9t^3 - 2t)$

111. $5(2x - 7) - [4(2x - 3) + 2]$

112. $3(6x - 5) - [3(1 - 8x) + 5]$

113. Some students use the mnemonic device PEMDAS to remember the rules for the order of operations. Explain what each letter represents and how the order of the letters in PEMDAS could lead a student to a wrong conclusion about the order of some operations.

114. Jake keys $18/2 \cdot 3$ into his calculator and expects the result to be 3. What mistake is he probably making?

Synthesis

115. Write the sentence $(-x)^2 \neq -x^2$ in words. Explain why $(-x)^2$ and $-x^2$ are not equivalent.

116. Write the sentence $-|x| \neq -x$ in words. Explain why $-|x|$ and $-x$ are not equivalent.

Simplify.

117. $5t - \{7t - [4r - 3(t - 7)] + 6r\} - 4r$

118. $z - \{2z - [3z - (4z - 5z) - 6z] - 7z\} - 8z$

119. $\{x - [f - (f - x)] + [x - f]\} - 3x$

120. Is it true that for all real numbers a and b,
$$ab = (-a)(-b)?$$
Why or why not?

121. Is it true that for all real numbers a, b, and c,
$$a|b - c| = ab - ac?$$
Why or why not?

If $n > 0$, $m > 0$, and $n \neq m$, classify each of the following as either true or false.

122. $-n + m = -(n + m)$

123. $m - n = -(n - m)$

124. $n(-n - m) = -n^2 + nm$

125. $-m(n - m) = -(mn + m^2)$

126. $-n(-n - m) = n(n + m)$

Evaluate.

Aha! **127.** $[x + 3(2 - 5x) \div 7 + x](x - 3)$, for $x = 3$

Aha! **128.** $[x + 2 \div 3x] \div [x + 2 \div 3x]$, for $x = -7$

129. $\dfrac{x^2 + 2^x}{x^2 - 2^x}$, for $x = 3$

130. $\dfrac{x^2 + 2^x}{x^2 - 2^x}$, for $x = 2$

131. In Mexico, between 500 B.C. and 600 A.D., the Mayans represented numbers using powers of 20 and certain symbols. For example, the symbols

represent $4 \cdot 20^3 + 17 \cdot 20^2 + 10 \cdot 20^1 + 0 \cdot 20^0$. Evaluate this number.

Data: National Council of Teachers of Mathematics, 1906 Association Drive, Reston, VA 22091

132. Examine the Mayan symbols and the numbers in Exercise 131. What number does

 , , and

each represent?

133. Calculate the volume of the tower shown below.

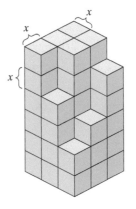

YOUR TURN ANSWERS: SECTION 1.8

1. $(-9)^3$ **2.** 16 **3.** -2 **4.** -45 **5.** -100 **6.** 498

7. 30 **8.** $\dfrac{1}{8}$ **9.** $17x - 55$ **10.** $12a^2 + 59ab$

11. $-2 - 5m - 10n$ **12.** $-5x + 20$ **13.** $-2x^2 - 2x$

14. $18y + 31$ **15.** $-4x^2 + 7$

Quick Quiz: Sections 1.1–1.8

1. Translate to an algebraic expression: Half of the sum of two numbers. [1.1]

2. Write an inequality with the same meaning as $x > -10$. [1.4]

3. Find decimal notation: $-\dfrac{11}{9}$. [1.4]

4. Multiply: $(-2.1)(-1.4)(-1)$. [1.7]

5. Remove parentheses and simplify:
$$5x - (2y - 3x) - 7y. \quad [1.8]$$

1. Twice the difference of a number and 11

2. The product of a number and 11 is 2.

3. Twice the difference of two numbers is 11.

4. The quotient of twice a number and 11

5. The quotient of 11 and the product of two numbers

Translating for Success

Use after Section 1.1.

Translate each phrase or sentence to an expression or an equation and match that translation with one of the choices A–O below. Do not solve.

A. $x = 0.2(11)$

B. $\dfrac{2x}{11}$

C. $2x + 2 = 11$

D. $2(11x + 2)$

E. $11x = 2$

F. $0.2x = 11$

G. $11(2x - y)$

H. $2(x - 11)$

I. $11 + 2x = 2$

J. $2x + y = 11$

K. $2(x - y) = 11$

L. $11(x + 2x)$

M. $2(x + y) = 11$

N. $2 + \dfrac{x}{11}$

O. $\dfrac{11}{xy}$

Answers on page A-3

An additional, animated version of this activity appears in MyMathLab. *To use MyMathLab, you need a course ID and a student access code. Contact your instructor for more information.*

6. Eleven times the sum of a number and twice the number

7. Twice the sum of two numbers is 11.

8. Two more than twice a number is 11.

9. Twice the sum of 11 times a number and 2

10. Twenty percent of some number is 11.

Collaborative Activity   *Teamwork*

Focus: Group problem solving; working collaboratively
Use after: Section 1.1
Time: 15 minutes
Group size: 2

Working and studying as a team often enables students to solve problems that are difficult to solve alone.

Activity

1. The left-hand column below lists the names of 12 colleges. A scrambled list of the names of their sports teams is on the right. As a group, without using outside references, match the names of the colleges to the teams.

1. University of Texas	a)	Antelopes
2. Western State College of Colorado	b)	Banana Slugs
3. University of North Carolina	c)	Sea Warriors
4. University of Massachusetts	d)	Gators
5. Hawaii Pacific University	e)	Mountaineers
6. University of Nebraska	f)	Sailfish
7. University of California, Santa Cruz	g)	Longhorns
8. University of Louisiana at Lafayette	h)	Tar Heels
9. Grand Canyon University	i)	Seawolves
10. Palm Beach Atlantic University	j)	Ragin' Cajuns
11. University of Alaska, Anchorage	k)	Cornhuskers
12. University of Florida	l)	Minutemen

2. After working for 5 min, confer with another group and reach mutual agreement.
3. Does the class agree on all 12 pairs?
4. Do you agree that group collaboration enhances our ability to solve problems?

Collaborative Activity   *Select the Symbols*

Focus: Order of operations
Use after: Section 1.8
Time: 15 minutes
Group size: 2

One way to master the rules for the order of operations is to insert symbols within a display of numbers in order to obtain a predetermined result. For example, the display

$$1 \quad 2 \quad 3 \quad 4 \quad 5$$

can be used to obtain the result 21 as follows:

$$(1 + 2) \div 3 + 4 \cdot 5.$$

Note that without an understanding of the rules for the order of operations, solving a problem of this sort is impossible.

Activity

1. Each group should prepare an exercise similar to the example shown here. (Exponents are not allowed.) To do so, first select five single-digit numbers for display. Then insert operations and grouping symbols and calculate the result.

2. Pair with another group. Each group should give the other its result along with its five-number display, and challenge the other group to insert symbols that will make the display equal the result given.

3. Share with the entire class the various mathematical statements developed by each group.

4. Are the answers always unique for any given set of numbers? Why or why not?

Decision Making & Connection   (*Use after Section 1.8.*)

Calorie Differential.   In order to lose weight, a person must burn more calories than he or she consumes. Burning 3500 excess calories will typically result in a weight loss of 1 lb. The difference between calorie consumption and calorie usage is called the *calorie differential*.

	Calorie Consumption	Calorie Usage	Calorie Differential
Monday	1900	2000	
Tuesday	2600	2100	
Wednesday	2400	2500	

1. The table at left shows Porter's calorie consumption and calorie usage for several days. Calculate the calorie differential for each day. If he is on a diet to lose weight, on which day(s) was he successful? If he is on a diet to gain weight, on which day(s) was he successful?

2. The Basal Metabolic Rate (BMR) is the number of calories a person uses at rest. One formula for BMR uses a person's weight in pounds w, height in inches h, and age in years a. Because of different body composition, there is one formula for women and another for men.

Women: $BMR = 655 + 4.35w + 4.7h - 4.7a$

Men: $BMR = 66 + 6.23w + 12.7h - 6.8a$

Calorie usage also depends on one's level of activity. To determine your total daily calorie needs to maintain your current weight, you can use a Harris Benedict Formula to multiply your BMR by the appropriate activity factor, as follows.

Activity Level	Daily Calorie Needs to Maintain Weight
Sedentary (little or no exercise)	BMR × 1.2
Lightly active (light exercise/sports 1–3 days/week)	BMR × 1.375
Moderately active (moderate exercise/sports 3–5 days/week)	BMR × 1.55
Very active (hard exercise/sports 6–7 days/week)	BMR × 1.725
Extra active (very hard exercise/sports and physical job)	BMR × 1.9

Data: bmi-calculator.net

If Porter is lightly active, weighs 180 lb, is 70 in. tall, and is 22 years old, what are his daily calorie needs in order to maintain his weight?

3. *Research.*   The number of calories burned by *extra* exercise can be added to a person's daily calorie needs in order to calculate total calorie usage. How many calories do you burn during exercise? Choose a favorite exercise, and find the number of calories burned per minute during that exercise.

4. Suppose you want to lose one pound. Follow the steps below to estimate how many calories you can consume each day if you exercise 30 min per day.

1) Use one of the formulas for BMR given above to estimate your Basal Metabolic Rate. Then use a Harris Benedict Formula to estimate your daily calorie needs to maintain your weight.

2) Calculate how many calories you would burn during 30 min of your favorite extra exercise.

3) Add the numbers found in steps (1) and (2) to calculate your daily calorie usage.

4) Divide −3500 by 7 to calculate the calorie differential needed each day in order to lose one pound in one week.

5) Add the number found in step (4) to the number found in step (3) to estimate the daily calorie consumption needed to lose one pound per week, assuming extra exercise.

Study Summary

KEY TERMS AND CONCEPTS	EXAMPLES	PRACTICE EXERCISES

SECTION 1.1: *Introduction to Algebra*

To **evaluate** an algebraic expression, substitute a number for each variable and carry out the operations. The result is a **value** of that expression.

Evaluate $\dfrac{x + y}{8}$ for $x = 15$ and $y = 9$.

$$\frac{x + y}{8} = \frac{15 + 9}{8} = \frac{24}{8} = 3$$

1. Evaluate $3 + 5c - d$ for $c = 3$ and $d = 10$.

To find the area of a rectangle, a triangle, or a parallelogram, evaluate the appropriate formula for the given values.

Find the area of a triangle with base 3.1 m and height 6 m.
$$A = \tfrac{1}{2}bh = \tfrac{1}{2}(3.1 \text{ m})(6 \text{ m})$$
$$= \tfrac{1}{2}(3.1)(6)(\text{m} \cdot \text{m}) = 9.3 \text{ m}^2$$

2. Find the area of a rectangle with length 8 ft and width $\frac{1}{2}$ ft.

Many problems can be solved by **translating** phrases to algebraic expressions and then forming an equation.

Translate to an equation. Do not solve.

   When 34 is subtracted from a number, the result is 13. What is the number?

Let n represent the number.

Rewording: $\underbrace{\text{34 subtracted from a number}}$ is 13

Translating: $n - 34$ $=$ 13

3. Translate to an equation. Do not solve.

78 is 92 less than some number. What is the number?

SECTION 1.2: *The Commutative, Associative, and Distributive Laws*

The Commutative Laws
$a + b = b + a$;
$ab = ba$

$3 + (-5) = -5 + 3$;
$8(10) = 10(8)$

4. Use the commutative law of addition to write an expression equivalent to $6 + 10n$.

The Associative Laws
$a + (b + c) = (a + b) + c$;
$a \cdot (b \cdot c) = (a \cdot b) \cdot c$

$-5 + (5 + 6) = (-5 + 5) + 6$;
$2 \cdot (5 \cdot 9) = (2 \cdot 5) \cdot 9$

5. Use the associative law of multiplication to write an expression equivalent to $3(ab)$.

The Distributive Law
$a(b + c) = ab + ac$

Multiply: $3(2x + 5y)$.
$$3(2x + 5y) = 3 \cdot 2x + 3 \cdot 5y = 6x + 15y$$
Factor: $16x + 24y + 8$.
$$16x + 24y + 8 = 8(2x + 3y + 1)$$

6. Multiply:
$$10(5m + 9n + 1).$$

7. Factor: $26x + 13$.

SECTION 1.3: *Fraction Notation*

A **prime** number has exactly two different factors, the number itself and 1. Natural numbers that have factors other than 1 and the number itself are **composite** numbers.

2, 3, 5, 7, 11, and 13 are the first six prime numbers. 4, 6, 8, 24, and 100 are examples of composite numbers.

8. Is 15 prime or composite?

The **prime factorization** of a composite number expresses that number as a product of prime numbers.

The prime factorization of 136 is $2 \cdot 2 \cdot 2 \cdot 17$.

9. Find the prime factorization of 84.

For any nonzero number a,
$$\frac{a}{a} = 1.$$

$\dfrac{15}{15} = 1$ and $\dfrac{2x}{2x} = 1.$ We assume $x \neq 0$.

10. Simplify: $\dfrac{t}{t}$.

Assume $t \neq 0$.

The Identity Property of 1
$a \cdot 1 = 1 \cdot a = a$

The number 1 is called the **multiplicative identity.**

$\dfrac{2}{3} = \dfrac{2}{3} \cdot \dfrac{5}{5}$ since $\dfrac{5}{5} = 1.$

11. Simplify: $\dfrac{9}{10} \cdot \dfrac{13}{13}$.

$\dfrac{a}{d} + \dfrac{b}{d} = \dfrac{a+b}{d}$

$\dfrac{a}{d} - \dfrac{b}{d} = \dfrac{a-b}{d}$

$\dfrac{a}{b} \cdot \dfrac{c}{d} = \dfrac{a \cdot c}{b \cdot d}$

$\dfrac{a}{b} \div \dfrac{c}{d} = \dfrac{a}{b} \cdot \dfrac{d}{c}$

$\dfrac{1}{6} + \dfrac{3}{8} = \dfrac{4}{24} + \dfrac{9}{24} = \dfrac{13}{24}$

$\dfrac{5}{12} - \dfrac{1}{6} = \dfrac{5}{12} - \dfrac{2}{12} = \dfrac{3}{12} = \dfrac{1 \cdot 3}{4 \cdot 3} = \dfrac{1}{4} \cdot \dfrac{3}{3} = \dfrac{1}{4} \cdot 1 = \dfrac{1}{4}$

$\dfrac{2}{5} \cdot \dfrac{7}{8} = \dfrac{2 \cdot 7}{5 \cdot 2 \cdot 4} = \dfrac{7}{20}$ Removing a factor equal to 1: $\frac{2}{2} = 1$

$\dfrac{10}{9} \div \dfrac{4}{15} = \dfrac{10}{9} \cdot \dfrac{15}{4} = \dfrac{2 \cdot 5 \cdot 3 \cdot 5}{3 \cdot 3 \cdot 2 \cdot 2} = \dfrac{25}{6}$ Removing a factor equal to 1: $\dfrac{2 \cdot 3}{2 \cdot 3} = 1$

Perform the indicated operation and, if possible, simplify.

12. $\dfrac{2}{3} + \dfrac{5}{6}$

13. $\dfrac{3}{4} - \dfrac{3}{10}$

14. $\dfrac{15}{14} \cdot \dfrac{35}{9}$

15. $15 \div \dfrac{3}{5}$

SECTION 1.4: *Positive and Negative Real Numbers*

Natural numbers:
$\{1, 2, 3, 4, \ldots\}$

1, 50, and 685 are examples of natural numbers.

Whole numbers:
$\{0, 1, 2, 3, 4, \ldots\}$

0, 37, and 14,615 are examples of whole numbers.

Integers:
$\{\ldots, -3, -2, -1, 0, 1, 2, 3, \ldots\}$

$-25, -2, 0, 1$, and 2000 are examples of integers.

Rational numbers:
$\left\{ \dfrac{a}{b} \middle| a \text{ and } b \text{ are integers and } b \neq 0 \right\}$

$\dfrac{1}{6}, \dfrac{-3}{7}, 0, 17, 0.758$, and $9.\overline{608}$ are examples of rational numbers.

The rational numbers and the **irrational numbers** make up the set of **real numbers.**

$\sqrt{7}$ and π are examples of irrational numbers.

16. Which of the following are integers?

$\dfrac{9}{10},\ 0,\ -15,\ \sqrt{2},\ \dfrac{30}{3}$

Every rational number can be written using fraction notation or decimal notation. When written in decimal notation, a rational number either **repeats** or **terminates.**

$-\dfrac{1}{16} = -0.0625$ This is a terminating decimal.

$\dfrac{5}{6} = 0.8333\ldots = 0.8\overline{3}$ This is a repeating decimal.

17. Find decimal notation:
$-\dfrac{10}{9}.$

Every real number corresponds to a point on the number line. For any two numbers, the one to the left on the number line is less than the one to the right. The symbol $<$ means "**is less than**" and the symbol $>$ means "**is greater than.**"

$$-3.1 \quad -\frac{1}{2} \quad \sqrt{2} \qquad 4$$

(number line from -4 to 4)

$$4 > -3.1 \qquad -\frac{1}{2} < \sqrt{2}$$

18. Write a true sentence using either $<$ or $>$:

$$-3 \quad\boxed{}\quad -4.$$

The **absolute value** of a number is the number of units that number is from 0 on the number line.

$|3| = 3$ since 3 is 3 units from 0.
$|-3| = 3$ since -3 is 3 units from 0.

19. Find the absolute value:

$$|-1.5|.$$

SECTION 1.5: *Addition of Real Numbers*

To **add** two real numbers, use the rules given in Section 1.5.

$-8 + (-3) = -11;$
$-8 + 3 = -5;$
$8 + (-3) = 5;$
$-8 + 8 = 0$

20. Add:

$$-15 + (-10) + 20.$$

The Identity Property of 0
$a + 0 = 0 + a = a$
The number 0 is called the **additive identity.**

$-35 + 0 = -35;$

$0 + \dfrac{2}{9} = \dfrac{2}{9}$

21. Add: $-2.9 + 0.$

SECTION 1.6: *Subtraction of Real Numbers*

The **opposite**, or **additive inverse**, of a number a is written $-a$. The opposite of the opposite of a is a.

$$-(-a) = a$$

Find $-x$ and $-(-x)$ when $x = -11$.
$\quad -x = -(-11) = 11;$
$\quad -(-x) = -(-(-11)) = -11 \qquad -(-x) = x$

22. Find $-(-x)$ when $x = -12$.

To **subtract** two real numbers, add the opposite of the number being subtracted.

$-10 - 12 = -10 + (-12) = -22;$
$-10 - (-12) = -10 + 12 = 2$

23. Subtract: $6 - (-9).$

The **terms** of an expression are separated by plus signs. **Like terms** either are constants or have the same variable factors. Like terms can be **combined** using the distributive law.

In the expression $-2x + 3y + 5x - 7y$:
$\quad$ The terms are $-2x, 3y, 5x,$ and $-7y.$
$\quad$ The like terms are $-2x$ and $5x$, and $3y$ and $-7y.$
Combining like terms gives

$$-2x + 3y + 5x - 7y = -2x + 5x + 3y - 7y$$
$$= (-2 + 5)x + (3 - 7)y$$
$$= 3x - 4y.$$

24. Combine like terms:
$3c + d - 10c - 2 + 8d.$

SECTION 1.7: *Multiplication and Division of Real Numbers*

To **multiply** or **divide** two real numbers, use the rules given in Section 1.7. Division by 0 is **undefined**.	$(-5)(-2) = 10;$ $30 \div (-6) = -5;$ $0 \div (-3) = 0;$ $-3 \div 0$ is undefined.	**25.** Multiply: $-3(-7)$. **26.** Divide: $10 \div (-2.5)$.

SECTION 1.8: *Exponential Notation and Order of Operations*

Exponential Notation Exponent $\overset{\text{Exponent}}{\searrow} \quad \overbrace{}^{n \text{ factors}}$ $b^n = b \cdot b \cdot b \cdots b$ $\nearrow$ Base	$6^2 = 6 \cdot 6 = 36;$ $(-6)^2 = (-6) \cdot (-6) = 36;$ $-6^2 = -(6 \cdot 6) = -36;$ $(6x)^2 = (6x) \cdot (6x) = 36x^2$	**27.** Evaluate: -10^2.
To perform multiple operations, use the rules for **order of operations** given in Section 1.8.	$\begin{aligned} -3 + (3 - 5)^3 \div 4(-1) &= -3 + (-2)^3 \div 4(-1) \\ &= -3 + (-8) \div 4(-1) \\ &= -3 + (-2)(-1) \\ &= -3 + 2 \\ &= -1 \end{aligned}$	**28.** Simplify: $120 \div (-10) \cdot 2 - 3(4 - 5)$.
The Opposite of a Sum For any real numbers a and b, $-(a + b) = -a - b.$	$-(2x - 3y) = -(2x) - (-3y) = -2x + 3y$	**29.** Write an equivalent expression without using grouping symbols: $-(-a + 2b - 3c)$.
Expressions containing parentheses can be simplified by removing parentheses using the distributive law.	Simplify: $3x^2 - 5(x^2 - 4xy + 2y^2) - 7y^2$. $\begin{aligned} 3x^2 &- 5(x^2 - 4xy + 2y^2) - 7y^2 \\ &= 3x^2 - 5x^2 + 20xy - 10y^2 - 7y^2 \\ &= -2x^2 + 20xy - 17y^2 \end{aligned}$	**30.** Simplify: $2m + n - 3(5 - m - 2n)$.

Review Exercises: Chapter 1

✎ Concept Reinforcement

In each of Exercises 1–10, classify the statement as either true or false.

1. $4x - 5y$ and $12 - 7a$ are both algebraic expressions containing two terms.   [1.2]*

2. $3t + 1 = 7$ and $8 - 2 = 9$ are both equations. [1.1]

3. The fact that $2 + x$ is equivalent to $x + 2$ is an illustration of the associative law for addition. [1.2]

4. The statement $4(a + 3) = 4 \cdot a + 4 \cdot 3$ illustrates the distributive law.   [1.2]

5. The number 2 is neither prime nor composite. [1.3]

6. Every irrational number can be written as a repeating decimal or a terminating decimal.   [1.4]

---

*The notation [1.2] refers to Chapter 1, Section 2.

7. Every natural number is a whole number and every whole number is an integer. [1.4]

8. The expressions $9r^2s$ and $5rs^2$ are like terms. [1.8]

9. The opposite of x, written $-x$, never represents a positive number. [1.6]

10. The number 0 has no reciprocal. [1.7]

Evaluate.

11. $8t$, for $t = 3$ [1.1]

12. $9 - y^2$, for $y = -5$ [1.8]

13. $-10 + a^2 \div (b + 1)$, for $a = 5$ and $b = -6$ [1.8]

Translate to an algebraic expression. [1.1]

14. 7 less than y

15. 10 more than the product of x and z

16. 15 times the difference of Brandt's speed and the wind speed

17. Translate to an equation. Do not solve. [1.1]

Backpacking burns twice as many calories per hour as housecleaning. If Katie burns 237 calories per hour housecleaning, how many calories per hour would she burn backpacking?

Data: www.myoptumhealth.com

18. The following table lists the number of calories that Kim burns when bowling for various lengths of time. Find an equation for the number of calories burned c when Kim bowls for t hours. [1.1]

Number of Hours Spent Bowling, t	Number of Calories Burned, c
$\frac{1}{2}$	100
2	400
$2\frac{1}{2}$	500

19. Use the commutative law of multiplication to write an expression equivalent to $3t + 5$. [1.2]

20. Use the associative law of addition to write an expression equivalent to $(2x + y) + z$. [1.2]

21. Use the commutative and associative laws to write three expressions equivalent to $4(xy)$. [1.2]

Multiply. [1.2]

22. $6(3x + 5y)$

23. $8(5x + 3y + 2)$

Factor. [1.2]

24. $21x + 15y$

25. $22a + 99b + 11$

26. Find the prime factorization of 56. [1.3]

Simplify. [1.3]

27. $\dfrac{20}{48}$

28. $\dfrac{18}{8}$

Perform the indicated operation and, if possible, simplify. [1.3]

29. $\dfrac{5}{12} + \dfrac{3}{8}$

30. $\dfrac{9}{16} \div 3$

31. $\dfrac{2}{3} - \dfrac{1}{15}$

32. $\dfrac{9}{10} \cdot \dfrac{6}{5}$

33. Tell which integers correspond to this situation. [1.4]

Becky borrowed $3600 to buy a used car. Clayton has $1350 in his savings account.

34. Graph on a number line: $\dfrac{-1}{3}$. [1.4]

35. Write an inequality with the same meaning as $-3 < x$. [1.4]

36. Write a true sentence using either $<$ or $>$: $-10 \;\square\; 0$. [1.4]

37. Find decimal notation: $-\dfrac{4}{9}$. [1.4]

38. Find the absolute value: $|-1|$. [1.4]

39. Find $-(-x)$ when x is -12. [1.6]

Simplify.

40. $-3 + (-7)$ [1.5]

41. $-\dfrac{2}{3} + \dfrac{1}{12}$ [1.5]

42. $-3.8 + 5.1 + (-12) + (-4.3) + 10$ [1.5]

43. $-2 - (-10)$ [1.6]

44. $-\dfrac{9}{10} - \dfrac{1}{2}$ [1.6]

45. $-2.7(3.4)$ [1.7]

46. $\dfrac{2}{3} \cdot \left(-\dfrac{3}{7}\right)$ [1.7]

47. $2 \cdot (-7) \cdot (-2) \cdot (-5)$ [1.7]

48. $35 \div (-5)$ [1.7]

49. $-5.1 \div 1.7$ [1.7]

50. $-\dfrac{3}{5} \div \left(-\dfrac{4}{15}\right)$ [1.7]

51. $120 - 6^2 \div 4 \cdot 8$ [1.8]

52. $(120 - 6^2) \div 4 \cdot 8$ [1.8]

53. $(120 - 6^2) \div (4 \cdot 8)$ [1.8]

54. $16 \div (-2)^3 - 5[3 - 1 + 2(4 - 7)]$ [1.8]

55. $|-3 \cdot 5 - 4 \cdot 8| - 3(-2)$ [1.8]

56. $\dfrac{4(18 - 8) + 7 \cdot 9}{9^2 - 8^2}$ [1.8]

Combine like terms.

57. $11a + 2b + (-4a) + (-3b)$ [1.5]

58. $7x - 3y - 11x + 8y$ [1.6]

59. Find the opposite of -7. [1.6]

60. Find the reciprocal of -7. [1.7]

61. Write exponential notation for $2x \cdot 2x \cdot 2x \cdot 2x$. [1.8]

62. Simplify: $(-5x)^3$. [1.8]

Remove parentheses and simplify. [1.8]

63. $2a - (5a - 9)$

64. $11x^4 + 2x + 8(x - x^4)$

65. $2n^2 - 5(-3n^2 + m^2 - 4mn) + 6m^2$

66. $8(x + 4) - 6 - [3(x - 2) + 4]$

Synthesis

67. Explain the difference between a constant and a variable. [1.1]

68. Explain the difference between a term and a factor. [1.2]

69. Describe at least three ways in which the distributive law was used in this chapter. [1.2]

70. Devise a rule for determining the sign of a negative number raised to a power. [1.8]

71. Evaluate $a^{50} - 20a^{25}b^4 + 100b^8$ for $a = 1$ and $b = 2$. [1.8]

72. If $0.090909 \ldots = \frac{1}{11}$ and $0.181818 \ldots = \frac{2}{11}$, what rational number is named by each of the following?
  a) $0.272727 \ldots$ [1.4]
  b) $0.909090 \ldots$ [1.4]

Simplify. [1.8]

73. $-|\frac{7}{8} - (-\frac{1}{2}) - \frac{3}{4}|$

74. $(|2.7 - 3| + 3^2 - |-3|) \div (-3)$

In each of Exercises 75–85, match the phrase with the most appropriate choice from the column on the right.

75. ___ A number is nonnegative. [1.4]

76. ___ The product of a number and its reciprocal is 1. [1.7]

77. ___ A number squared [1.8]

78. ___ A sum of squares [1.8]

79. ___ The opposite of an opposite is the original number. [1.6]

80. ___ The order in which numbers are added does not change the result. [1.2]

81. ___ A number is positive. [1.4]

82. ___ The absolute value of a product [1.4]

83. ___ A sum of a number and its reciprocal [1.7]

84. ___ The square of a sum [1.8]

85. ___ The absolute value of one number is less than the absolute value of another number. [1.4]

a) a^2

b) $a + b = b + a$

c) $a > 0$

d) $a + \dfrac{1}{a}$

e) $|ab|$

f) $(a + b)^2$

g) $|a| < |b|$

h) $a^2 + b^2$

i) $a \geq 0$

j) $a \cdot \dfrac{1}{a} = 1$

k) $-(-a) = a$

Test: Chapter 1   For step-by-step test solutions, access the Chapter Test Prep Videos in MyMathLab.

1. Evaluate $\dfrac{2x}{y}$ for $x = 10$ and $y = 5$.

2. Write an algebraic expression: Nine less than the product of two numbers.

3. Find the area of a triangle when the height h is 30 ft and the base b is 16 ft.

4. Use the commutative law of addition to write an expression equivalent to $3p + q$.

5. Use the associative law of multiplication to write an expression equivalent to $x \cdot (4 \cdot y)$.

6. Translate to an equation. Do not solve.

    About 1500 golden lion tamarins, an endangered species of monkey, live in the wild. This is 1050 more than live in zoos worldwide. How many golden lion tamarins live in zoos?

    Data: nationalzoo.si.edu

Multiply.

7. $7(5 + x)$

8. $-5(y - 2)$

Factor.

9. $11 + 44x$

10. $7x + 7 + 49y$

11. Find the prime factorization of 300.

12. Simplify: $\dfrac{24}{56}$.

Write a true sentence using either $<$ or $>$.

13. $-4 \,\square\, 0$ 14. $-3 \,\square\, -8$

Find the absolute value.

15. $\left|\dfrac{9}{4}\right|$ 16. $|-3.8|$

17. Find the opposite of $-\dfrac{2}{3}$.

18. Find the reciprocal of $-\dfrac{4}{7}$.

19. Find $-x$ when x is -10.

20. Write an inequality with the same meaning as $x \le -5$.

Perform the indicated operations and, if possible, simplify.

21. $3.1 - (-4.7)$

22. $-8 + 4 + (-7) + 3$

23. $-\dfrac{1}{8} - \dfrac{3}{4}$

24. $4 \cdot (-12)$

25. $-\dfrac{1}{2} \cdot \left(-\dfrac{4}{9}\right)$

26. $-\dfrac{3}{5} \div \left(-\dfrac{4}{5}\right)$

27. $4.864 \div (-0.5)$

28. $10 - 2(-16) \div 4^2 + |2 - 10|$

29. $9 + 7 - 4 - (-3)$

30. $256 \div (-16) \cdot 4$

31. $2^3 - 10[4 - 3(-2 + 18)]$

32. Combine like terms: $18y + 30a - 9a + 4y$.

33. Simplify: $(-2x)^4$.

Remove parentheses and simplify.

34. $4x - (3x - 7)$

35. $4(2a - 3b) + a - 7$

36. $3[5(y - 3) + 9] - 2(8y - 1)$

Synthesis

37. Find $\dfrac{5y - x}{2}$ when $x = 20$ and y is 4 less than half of x.

38. Insert one pair of parentheses to make the following a true statement:

    $$9 - 3 - 4 + 5 = 15.$$

39. Translate to an inequality: A number n is nonnegative.

40. Simplify: $a - \{3a - [4a - (2a - 4a)]\}$.

41. Classify the following as either true or false:

    $$a|b - c| = |ab| - |ac|.$$

Equations, Inequalities, and Problem Solving

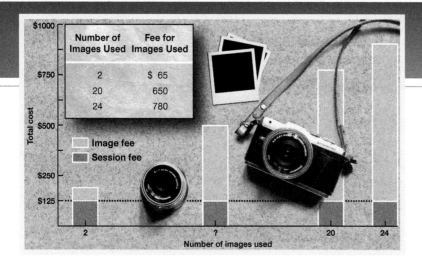

Number of Images Used	Fee for Images Used
2	$ 65
20	650
24	780

A Picture Is Worth a Thousand Words.

2.1 Solving Equations

2.2 Using the Principles Together

2.3 Formulas

MID-CHAPTER REVIEW

2.4 Applications with Percent

2.5 Problem Solving

2.6 Solving Inequalities

CONNECTING THE CONCEPTS

2.7 Solving Applications with Inequalities

CHAPTER RESOURCES

Translating for Success
Collaborative Activity
Decision Making: Connection

STUDY SUMMARY

REVIEW EXERCISES
CHAPTER TEST
CUMULATIVE REVIEW

When you hire a photographer for an event, the number of images that you can afford to buy may be limited. If a photographer charges a $125 session fee plus image fees as shown in the table above, how many images can you purchase for $500? (*See Example 3 in Section 2.5.*)

Whether it is the first photograph or the last print, math is used extensively throughout my photography.

Carlton Riffel, Owner, Riffel in Wilmington, North Carolina, uses math when editing images and calculating payments. He is also constantly adjusting camera settings using math to get the right exposure for his photographs.

ALF
Active Learning Figure
Explore the math using the Active Learning Figure in MyMathLab.

SA
Student Activity
Do the Student Activity in MyMathLab to see math in action.

Solving equations and inequalities is a recurring theme in much of mathematics. In this chapter, we study some of the principles used to solve equations and inequalities. We then use equations and inequalities to solve applied problems.

2.1 Solving Equations

A. Equations and Solutions   **B.** The Addition Principle   **C.** The Multiplication Principle
D. Selecting the Correct Approach

Solving equations is an essential part of problem solving in algebra. In this section, we study two of the most important principles used for this task.

A. Equations and Solutions

An equation is a number sentence stating that the expressions on either side of the equals sign represent the same number. Equations may be true, false, or neither true nor false.

The equation $8 \cdot 4 = 32$ is *true*.

The equation $7 - 2 = 4$ is *false*.

The equation $x + 6 = 13$ is *neither* true nor false, because we do not know what number x represents.

Equations like $x + 6 = 13$ are true for some replacements for x and are false for others.

> **SOLUTION OF AN EQUATION**
>
> Any replacement for a variable that makes an equation true is called a *solution* of the equation. To *solve* an equation means to find all of its solutions.

Student Notes

At the end of each example in this text, you will see YOUR TURN. This directs you to try an exercise similar to the corresponding example. The answers to those exercises appear at the end of each exercise set.

To determine whether a number is a solution of an equation, we substitute that number for the variable. If the values on both sides of the equals sign are the same, then the number that was substituted is a solution.

EXAMPLE 1   Determine whether 7 is a solution of $x + 6 = 13$.

SOLUTION   We have

$$\begin{array}{ll} x + 6 = 13 & \text{Writing the equation} \\ \overline{7 + 6 \ \big| \ 13} & \text{Substituting 7 for } x \\ 13 \stackrel{?}{=} 13 \quad \text{TRUE} & 13 = 13 \text{ is a true statement.} \end{array}$$

Since the left side and the right side are the same, 7 is a solution.

1. Determine whether 8 is a solution of $12 = 20 - t$.

YOUR TURN

> **CAUTION!**   Note that in Example 1 the solution is 7, not 13.

EXAMPLE 2 Determine whether -1 is a solution of $7x - 2 = 4x + 5$.

SOLUTION We have

$$
\begin{array}{c|c}
\multicolumn{2}{c}{7x - 2 = 4x + 5} \\
\hline
7(-1) - 2 & 4(-1) + 5 \\
-7 - 2 & -4 + 5 \\
-9 \overset{?}{=} 1 &
\end{array}
$$

Writing the equation
Substituting -1 for x
Carrying out calculations on both sides
FALSE   The statement $-9 = 1$ is false.

Since the left and the right sides differ, -1 is not a solution.

2. Determine whether -1 is a solution of $5 + 2n = 4 - n$.

YOUR TURN

B. The Addition Principle

Consider the equation

$$x = 7.$$

We can easily see that the solution of this equation is 7. Replacing x with 7, we get

$$7 = 7, \quad \text{which is true.}$$

In Example 1, we found that the solution of $x + 6 = 13$ is also 7. Although the solution of $x = 7$ may seem more obvious, because $x + 6 = 13$ and $x = 7$ have identical solutions, the equations are said to be **equivalent**.

Student Notes

Be sure to remember the difference between an expression and an equation. For example, $5a - 10$ and $5(a - 2)$ are *equivalent expressions* because they represent the same value for all replacements for a. The *equations* $5a = 10$ and $a = 2$ are *equivalent equations* because they have the same solution, 2.

> **EQUIVALENT EQUATIONS**
>
> Equations with the same solutions are called *equivalent equations*.

There are principles that enable us to begin with one equation and end up with an equivalent equation, like $x = 7$, for which the solution is obvious. One such principle concerns addition. The equation $a = b$ says that a and b stand for the same number. Suppose that this is true, and some number c is added to a. We get the same result if we add c to b, because a and b are the same number.

> **THE ADDITION PRINCIPLE**
>
> For any real numbers a, b, and c,
>
> $$a = b \quad \text{is equivalent to} \quad a + c = b + c.$$

To visualize the addition principle, consider a balance similar to one that a jeweler might use. When the two sides of the balance hold equal weight, the balance is level. If weight is then added or removed, equally, on both sides, the balance will remain level.

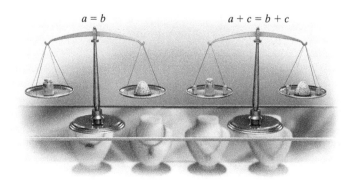

$a = b$ $a + c = b + c$

When using the addition principle, we often say that we "add the same number to both sides of an equation." We can also "subtract the same number from both sides," since subtraction can be regarded as the addition of an opposite.

EXAMPLE 3 Solve: $x + 5 = -7$.

SOLUTION We can add any number we like to both sides. Since -5 is the opposite, or additive inverse, of 5, we decide to add -5 to each side:

$$x + 5 = -7$$

$$x + 5 - 5 = -7 - 5 \qquad \text{Using the addition principle: adding } -5 \text{ to both sides or subtracting 5 from both sides}$$

$$x + 0 = -12 \qquad \text{Simplifying; } x + 5 - 5 = x + 5 + (-5)$$
$$= x + 0$$

$$x = -12. \qquad \text{Using the identity property of 0}$$

The equation $x = -12$ is equivalent to the equation $x + 5 = -7$ by the addition principle, so the solution of $x = -12$ is the solution of $x + 5 = -7$.

It is obvious that the solution of $x = -12$ is the number -12. To check the answer in the original equation, we substitute.

Check:

$$\begin{array}{c|c} x + 5 = -7 \\ \hline -12 + 5 & -7 \\ -7 \stackrel{?}{=} -7 & \text{TRUE} \qquad -7 = -7 \text{ is true.} \end{array}$$

The solution of the original equation is -12.

3. Solve: $x + 3 = -10$.
_____

⤶ YOUR TURN

In Example 3, note that because we added the *opposite*, or *additive inverse*, of 5, the left side of the equation simplified to x plus the *additive identity*, 0, or simply x. To solve $x + a = b$ for x, we add $-a$ to (or subtract a from) both sides.

EXAMPLE 4 Solve: $-6.5 = y - 8.4$.

SOLUTION The variable is on the right side this time. We can isolate y by adding 8.4 to each side:

$$-6.5 = y - 8.4 \qquad y - 8.4 \text{ can be regarded as } y + (-8.4).$$

$$-6.5 + 8.4 = y - 8.4 + 8.4 \qquad \text{Using the addition principle: Adding 8.4 to both sides "eliminates" } -8.4 \text{ on the right side.}$$

$$1.9 = y. \qquad y - 8.4 + 8.4 = y + (-8.4) + 8.4$$
$$= y + 0 = y$$

Student Notes

We can also think of "undoing" operations in order to isolate a variable. In Example 4, we began with $y - 8.4$ on the right side. To undo the subtraction, we *add* 8.4.

Check:

$$\begin{array}{c|c} -6.5 = y - 8.4 \\ \hline -6.5 & 1.9 - 8.4 \\ -6.5 \stackrel{?}{=} -6.5 & \text{TRUE} \qquad -6.5 = -6.5 \text{ is true.} \end{array}$$

The solution is 1.9.

4. Solve: $-8 = y - 19$.
_____

⤶ YOUR TURN

Note that the equations $a = b$ and $b = a$ have the same meaning. Thus, $-6.5 = y - 8.4$ could have been rewritten as $y - 8.4 = -6.5$.

C. The Multiplication Principle

A second principle for solving equations concerns multiplying. Suppose that a and b are equal. If a and b are multiplied by some number c, then ac and bc will also be equal.

> **THE MULTIPLICATION PRINCIPLE**
>
> For any real numbers a, b, and c, with $c \neq 0$,
>
> $$a = b \quad \text{is equivalent to} \quad a \cdot c = b \cdot c.$$

EXAMPLE 5 Solve: $\frac{5}{4}x = 10$.

SOLUTION We can multiply both sides by any nonzero number we like. Since $\frac{4}{5}$ is the reciprocal of $\frac{5}{4}$, we decide to multiply both sides by $\frac{4}{5}$:

$$\frac{5}{4}x = 10$$

$$\frac{4}{5} \cdot \frac{5}{4}x = \frac{4}{5} \cdot 10 \qquad \text{Using the multiplication principle: Multiplying both sides by } \frac{4}{5} \text{ "eliminates" the } \frac{5}{4} \text{ on the left.}$$

$$1 \cdot x = 8 \qquad \text{Simplifying}$$

$$x = 8. \qquad \text{Using the identity property of 1}$$

Check:

$$\begin{array}{c|c} \frac{5}{4}x = 10 \\ \hline \frac{5}{4} \cdot 8 & 10 \\ \frac{40}{4} & \\ 10 \stackrel{?}{=} 10 & \text{TRUE} \end{array} \qquad \begin{array}{l} \text{Think of 8 as } \frac{8}{1}. \\ \\ 10 = 10 \text{ is true.} \end{array}$$

The solution is 8.

5. Solve: $\dfrac{1}{5}x = -3$.

YOUR TURN

In Example 5, in order to get x alone on one side, we multiplied by the *reciprocal*, or *multiplicative inverse* of $\frac{5}{4}$. The simplified left side is x times the *multiplicative identity*, 1, or simply x.

Because division is the same as multiplying by a reciprocal, the multiplication principle also tells us that we can "divide both sides by the same nonzero number." That is,

$$\text{if } a = b, \text{ then } \quad \frac{1}{c} \cdot a = \frac{1}{c} \cdot b \quad \text{and} \quad \frac{a}{c} = \frac{b}{c} \qquad \text{(provided } c \neq 0).$$

In a product like $3x$, the multiplier 3 is called the **coefficient**. Examining the coefficient of the variable helps us to decide whether to multiply or to divide in order to solve an equation.

- When the coefficient is an integer or a decimal, it is usually easiest to divide both sides by the coefficient.
- When the coefficient is a fraction, it is usually easiest to multiply both sides by the reciprocal of the coefficient.

To illustrate these two approaches, let's compare the following solutions.

Coefficient a decimal

$$1.2x = 36$$

$$\frac{1.2x}{1.2} = \frac{36}{1.2} \qquad \text{Dividing by 1.2}$$

$$x = 30$$

Coefficient a fraction

$$\frac{2}{3}x = 14$$

$$\frac{3}{2} \cdot \frac{2}{3}x = \frac{3}{2} \cdot 14 \qquad \text{Multiplying by the reciprocal of } \frac{2}{3}$$

$$x = 21$$

Student Notes

In Example 6(a), we can think of undoing the multiplication $-4 \cdot x$ by *dividing* by -4.

EXAMPLE 6 Solve: **(a)** $-4x = 9$; **(b)** $-x = 5$.

SOLUTION

a) In $-4x = 9$, the coefficient of x is an integer, so we *divide* on both sides:

$$\frac{-4x}{-4} = \frac{9}{-4} \qquad \text{Using the multiplication principle: Dividing both sides by } -4 \text{ is the same as multiplying by } -\frac{1}{4}.$$

$$1 \cdot x = -\frac{9}{4} \qquad \text{Simplifying}$$

$$x = -\frac{9}{4}. \qquad \text{Using the identity property of 1}$$

Check:
$$\frac{-4x = 9}{-4\left(-\frac{9}{4}\right) \mid 9}$$
$$9 \stackrel{?}{=} 9 \quad \text{TRUE} \qquad 9 = 9 \text{ is true.}$$

The solution is $-\frac{9}{4}$.

b) To solve an equation like $-x = 5$, remember that when an expression is multiplied or divided by -1, its sign is changed. Here we divide both sides by -1 to change the sign of $-x$:

$$-x = 5 \qquad \text{Note that } -x = -1 \cdot x.$$

$$\frac{-x}{-1} = \frac{5}{-1} \qquad \text{Dividing both sides by } -1. \text{ (Multiplying by } -1 \text{ would also work. Note that the reciprocal of } -1 \text{ is } -1.)$$

$$x = -5. \qquad \text{Note that } \frac{-x}{-1} \text{ is the same as } \frac{x}{1}.$$

Check:
$$\frac{-x = 5}{-(-5) \mid 5}$$
$$5 \stackrel{?}{=} 5 \quad \text{TRUE} \qquad 5 = 5 \text{ is true.}$$

6. Solve: $-x = \dfrac{1}{3}$.

The solution is -5.

↩ YOUR TURN

MULTIPLYING AND DIVIDING BY -1

Multiplying a number by -1 changes its sign:

$$(-1)(x) = -x \quad \text{and} \quad (-1)(-x) = x.$$

Dividing a number by -1 changes its sign:

$$\frac{x}{-1} = -x \quad \text{and} \quad \frac{-x}{-1} = x.$$

Study Skills

Seeking Help?

A variety of resources are available to help make studying easier and more enjoyable.

- *Textbook supplements.* See the preface for a description of the supplements for this textbook.

- *Your college or university.* Your own school probably has many resources: a learning lab or tutoring center, study skills workshops or group tutoring sessions tailored for the course you are taking, or a bulletin board or network where you can locate a private tutor.

- *Your instructor.* Find out your instructor's office hours and visit when you need additional help. Many instructors also welcome student e-mail.

---

7. Solve: $\dfrac{4t}{7} = 6$.

↳ Check Your UNDERSTANDING

Match each equation with the step from the following list that would be used to most readily solve the equation.

a) Add 6 to both sides.

b) Subtract 6 from both sides.

c) Multiply both sides by 6.

d) Divide both sides by 6.

1. ____ $6x = 30$

2. ____ $x + 6 = 30$

3. ____ $\frac{1}{6}x = 30$

4. ____ $x - 6 = 30$

EXAMPLE 7   Solve: $\dfrac{2y}{9} = \dfrac{8}{3}$.

SOLUTION   To solve an equation like $\dfrac{2y}{9} = \dfrac{8}{3}$, we can rewrite the left side as $\dfrac{2}{9} \cdot y$ and then use the multiplication principle, multiplying by the reciprocal of $\dfrac{2}{9}$:

$$\frac{2y}{9} = \frac{8}{3}$$

$$\frac{2}{9} \cdot y = \frac{8}{3} \qquad \text{Rewriting } \frac{2y}{9} \text{ as } \frac{2}{9} \cdot y$$

$$\frac{9}{2} \cdot \frac{2}{9} \cdot y = \frac{9}{2} \cdot \frac{8}{3} \qquad \text{Multiplying both sides by } \frac{9}{2}$$

$$1y = \frac{3 \cdot 3 \cdot 2 \cdot 4}{2 \cdot 3} \qquad \text{Removing a factor equal to 1: } \frac{3 \cdot 2}{2 \cdot 3} = 1$$

$$y = 12.$$

Check:
$$\frac{2y}{9} = \frac{8}{3}$$

$$\begin{array}{c|c} \dfrac{2 \cdot 12}{9} & \dfrac{8}{3} \\[2mm] \dfrac{24}{9} & \\[2mm] \dfrac{8}{3} \stackrel{?}{=} \dfrac{8}{3} \text{ TRUE} & \end{array} \qquad \frac{8}{3} = \frac{8}{3} \text{ is true.}$$

The solution is 12.

↶ YOUR TURN

D. Selecting the Correct Approach

It is important that you be able to determine which principle should be used to solve a particular equation.

EXAMPLE 8   Solve: **(a)** $-8 + x = -3$; **(b)** $18 = 0.3t$.

SOLUTION

a) To undo addition of -8, we subtract -8 from (or add 8 to) both sides. Note that the opposite of *negative* 8 is *positive* 8.

$$-8 + x = -3$$
$$-8 + x + 8 = -3 + 8 \qquad \text{Using the addition principle}$$
$$x = 5$$

Check:
$$\begin{array}{c|c} -8 + x & = -3 \\ \hline -8 + 5 & -3 \\ -3 \stackrel{?}{=} -3 & \text{ TRUE} \end{array} \qquad -3 = -3 \text{ is true.}$$

The solution is 5.

b) To undo multiplication by 0.3, we either divide both sides by 0.3 or multiply both sides by $\frac{1}{0.3}$. Note that the reciprocal of *positive* 0.3 is *positive* $\frac{1}{0.3}$.

$$18 = 0.3t$$

$$\frac{18}{0.3} = \frac{0.3t}{0.3} \qquad \text{Using the multiplication principle}$$

$$60 = t \qquad \text{Simplifying}$$

Check: $\dfrac{18 = 0.3t}{18 \mid 0.3(60)}$

$18 \stackrel{?}{=} 18$ TRUE $18 = 18$ is true.

The solution is 60.

8. Solve: $2.6 + a = -0.46$.

 YOUR TURN

2.1 | EXERCISE SET

FOR EXTRA HELP   MyMathLab®

↪ Vocabulary and Reading Check

In each of Exercises 1–6, match the statement with the appropriate choice from the column on the right.

1. _____ The equations $x + 3 = 7$ and $6x = 24$

2. _____ The expressions $3(x - 2)$ and $3x - 6$

3. _____ A replacement that makes an equation true

4. _____ The role of 9 in $9ab$

5. _____ The principle used to solve $\frac{2}{3} \cdot x = -4$

6. _____ The principle used to solve $\frac{2}{3} + x = -4$

a) A coefficient

b) Equivalent expressions

c) Equivalent equations

d) The multiplication principle

e) The addition principle

f) A solution

A. Equations and Solutions

Determine whether the given number is a solution of the given equation.

7. $6 - x = -2$; 4

8. $6 - x = -2$; 8

9. $\frac{2}{3}t = 12$; 18

10. $\frac{2}{3}t = 12$; 8

11. $x + 7 = 3 - x$; -2

12. $-4 + x = 5x$; -1

13. $4 - \frac{1}{5}n = 8$; -20

14. $-3 = 5 - \dfrac{n}{2}$; 4

B. The Addition Principle

Solve. Don't forget to check!

15. $x + 10 = 21$

16. $t + 9 = 47$

17. $y + 7 = -18$

18. $x + 12 = -7$

19. $-6 = y + 25$

20. $-5 = x + 8$

21. $x - 18 = 23$

22. $x - 19 = 16$

23. $12 = -7 + y$

24. $15 = -8 + z$

25. $-5 + t = -11$

26. $-6 + y = -21$

27. $r + \frac{1}{3} = \frac{8}{3}$

28. $t + \frac{3}{8} = \frac{5}{8}$

29. $x - \frac{3}{5} = -\frac{7}{10}$

30. $x - \frac{2}{3} = -\frac{5}{6}$

31. $x - \frac{5}{6} = \frac{7}{8}$

32. $y - \frac{3}{4} = \frac{5}{6}$

33. $-\frac{1}{5} + z = -\frac{1}{4}$

34. $-\frac{2}{3} + y = -\frac{3}{4}$

35. $m - 2.8 = 6.3$

36. $y - 5.3 = 8.7$

37. $-9.7 = -4.7 + y$

38. $-7.8 = 2.8 + x$

C. The Multiplication Principle

Solve. Don't forget to check!

39. $8a = 56$

40. $6x = 72$

41. $84 = 7x$

42. $45 = 9t$

43. $-x = 38$

44. $100 = -x$

Aha! **45.** $-t = -8$

46. $-68 = -r$

47. $-7x = 49$

48. $-4x = 36$

49. $0.2m = 10$

50. $0.1n = 15$

51. $-1.2x = 0.24$

52. $-1.6x = 0.8$

53. $-1.3a = -10.4$

54. $-3.4t = -20.4$

55. $\dfrac{y}{8} = 11$

56. $\dfrac{a}{4} = 13$

57. $\dfrac{4}{5}x = 16$

58. $\dfrac{3}{4}x = 27$

59. $\dfrac{-x}{6} = 9$

60. $\dfrac{-t}{4} = 8$

61. $\dfrac{1}{9} = \dfrac{z}{-5}$

62. $\dfrac{2}{7} = \dfrac{x}{-3}$

Aha! **63.** $-\dfrac{3}{5}r = -\dfrac{3}{5}$

64. $-\dfrac{2}{5}y = -\dfrac{4}{15}$

65. $\dfrac{-3r}{2} = -\dfrac{27}{4}$

66. $\dfrac{5x}{7} = -\dfrac{10}{14}$

D. Selecting the Correct Approach

Solve. The icon 🖩 *indicates an exercise designed to provide practice using a calculator.*

67. $4.5 + t = -3.1$

68. $\frac{3}{4}x = 18$

69. $-8.2x = 20.5$

70. $t - 7.4 = -12.9$

71. $x - 4 = -19$

72. $y - 6 = -14$

73. $t - 3 = -8$

74. $t - 9 = -8$

75. $-12x = 14$

76. $-15x = 20$

77. $48 = -\frac{3}{8}y$

78. $14 = t + 27$

79. $a - \dfrac{1}{6} = -\dfrac{2}{3}$

80. $-\dfrac{x}{6} = \dfrac{2}{9}$

81. $-24 = \dfrac{8x}{5}$

82. $\dfrac{1}{5} + y = -\dfrac{3}{10}$

83. $-\frac{4}{3}t = -12$

84. $\frac{17}{35} = -x$

🖩 **85.** $-483.297 = -794.053 + t$

🖩 **86.** $-0.2344x = 2028.732$

📋 **87.** When solving an equation, how do you determine what number to add, subtract, multiply, or divide by on both sides of that equation?

📋 **88.** What is the difference between equivalent expressions and equivalent equations?

Skill Review

To the student and the instructor: Skill Review *exercises review skills previously studied in the text. The numbers in brackets immediately following the directions or exercise indicate the section in which the skill was introduced. The answers to all Skill Review exercises appear at the back of the book.*

89. Translate to an algebraic expression:

7 less than one-third of y. [1.1]

90. Multiply: $6(2x + 11)$. [1.2]

91. Factor: $35a + 55c + 5$. [1.2]

92. Graph on the number line: $-\frac{11}{5}$. [1.4]

Synthesis

📋 **93.** To solve $-3.5 = 14t$, Anika adds 3.5 to both sides. Will this form an equivalent equation? Will it help solve the equation? Explain.

📋 **94.** Explain why it is not necessary to state a subtraction principle: For any real numbers a, b, and c, $a = b$ is equivalent to $a - c = b - c$.

Solve for x. Assume a, c, m ≠ 0.

95. $mx = 11.6m$

96. $x - 4 + a = a$

97. $cx + 5c = 7c$

98. $c \cdot \dfrac{21}{a} = \dfrac{7cx}{2a}$

99. $7 + |x| = 30$

100. $ax - 3a = 5a$

101. If $t - 3590 = 1820$, find $t + 3590$.

102. If $n + 268 = 124$, find $n - 268$.

🖩 **103.** Chantal estimated her monthly business taxes to be \$225. As her last step, she multiplied by 0.3 when she should have divided by 0.3. What should the correct answer be?

📋 **104.** Are the equations $x = 5$ and $x^2 = 25$ equivalent? Why or why not?

➤ **YOUR TURN ANSWERS: SECTION 2.1**

1. Yes **2.** No **3.** -13 **4.** 11 **5.** -15 **6.** $-\frac{1}{3}$

7. $\frac{21}{2}$ **8.** -3.06

Prepare to Move On

Simplify. [1.8]

1. $3 \cdot 4 - 18$

2. $14 - 2(7 - 1)$

3. $4x + 10 - 5x$

4. $x - 2(3x - 7) + 14$

2.2   Using the Principles Together

A. Applying Both Principles   **B.** Combining Like Terms   **C.** Clearing Fractions and Decimals

The addition and multiplication principles, along with the properties and laws concerning real numbers, are our tools for solving equations.

A. Applying Both Principles

EXAMPLE 1   Solve: $5 + 3x = 17$.

SOLUTION   Were we to evaluate $5 + 3x$, the rules for the order of operations direct us to *first multiply* by 3 and *then add* 5. Because of this, we can isolate $3x$ and then x by reversing these operations: We *first subtract* 5 from both sides and *then divide* both sides by 3. Our goal is an equivalent equation of the form $x = a$.

$$5 + 3x = 17$$

$$5 + 3x - 5 = 17 - 5 \qquad \text{Using the addition principle: subtracting 5 from both sides (adding } -5)$$

$$5 + (-5) + 3x = 12 \qquad \text{Using a commutative law. Try to perform this step mentally.}$$

Isolate the x-term. $3x = 12$ Simplifying

$$\frac{3x}{3} = \frac{12}{3} \qquad \text{Using the multiplication principle: dividing both sides by 3 (multiplying by } \tfrac{1}{3})$$

Isolate x. $x = 4$ Simplifying. This is of the form $x = a$.

Check:
$$\begin{array}{c|c} 5 + 3x = 17 \\ \hline 5 + 3\cdot 4 & 17 \\ 5 + 12 & \\ 17 \stackrel{?}{=} 17 & \text{TRUE} \end{array}$$

We use the rules for order of operations: Find the product, $3 \cdot 4$, and then add.

The solution is 4.

YOUR TURN

1. Solve: $2x - 5 = 7$.

EXAMPLE 2   Solve: $\frac{4}{3}x - 7 = 1$.

SOLUTION   In $\frac{4}{3}x - 7$, we first multiply and then subtract. To reverse these steps, we first add 7 and then either divide by $\frac{4}{3}$ or multiply by $\frac{3}{4}$.

$$\frac{4}{3}x - 7 = 1$$

$$\frac{4}{3}x - 7 + 7 = 1 + 7 \qquad \text{Adding 7 to both sides}$$

$$\frac{4}{3}x = 8$$

$$\frac{3}{4}\cdot\frac{4}{3}x = \frac{3}{4}\cdot 8 \qquad \text{Multiplying both sides by } \tfrac{3}{4}$$

$$\left. \begin{array}{c} 1\cdot x = \dfrac{3\cdot 4\cdot 2}{4} \\[2mm] x = 6 \end{array} \right\} \quad \text{Simplifying}$$

Check:

$$
\begin{array}{c|c}
\frac{4}{3}x - 7 = 1 \\
\hline
\frac{4}{3} \cdot 6 - 7 & 1 \\
8 - 7 & \\
1 \stackrel{?}{=} 1 & \text{TRUE}
\end{array}
$$

The solution is 6.

2. Solve: $4 + \frac{2}{3}x = 2$.

↪ YOUR TURN

EXAMPLE 3 Solve: $45 - t = 13$.

SOLUTION We have

$$
\begin{aligned}
45 - t &= 13 \\
45 - t - 45 &= 13 - 45 && \text{Subtracting 45 from both sides} \\
\left.\begin{aligned} 45 + (-t) + (-45) &= 13 - 45 \\ 45 + (-45) + (-t) &= 13 - 45 \end{aligned}\right\} && \text{Try to do these steps mentally.} \\
-t &= -32 && \text{Try to go directly to this step.} \\
(-1)(-t) &= (-1)(-32) && \text{Multiplying both sides by } -1. \\
&&& \text{(Dividing by } -1 \text{ would also} \\
&&& \text{work.)} \\
t &= 32.
\end{aligned}
$$

Check:

$$
\begin{array}{c|c}
45 - t = 13 \\
\hline
45 - 32 & 13 \\
13 \stackrel{?}{=} 13 & \text{TRUE}
\end{array}
$$

The solution is 32.

3. Solve: $-3 - y = 8$.

↪ YOUR TURN

As our skills improve, certain steps can be streamlined.

EXAMPLE 4 Solve: $16.3 - 7.2y = -8.18$.

SOLUTION We have

$$
\begin{aligned}
16.3 - 7.2y &= -8.18 \\
16.3 - 7.2y - 16.3 &= -8.18 - 16.3 && \text{Subtracting 16.3 from both sides} \\
-7.2y &= -24.48 && \text{Simplifying} \\
\frac{-7.2y}{-7.2} &= \frac{-24.48}{-7.2} && \text{Dividing both sides by } -7.2 \\
y &= 3.4. && \text{Simplifying}
\end{aligned}
$$

Check:

$$
\begin{array}{c|c}
16.3 - 7.2y = -8.18 \\
\hline
16.3 - 7.2(3.4) & -8.18 \\
16.3 - 24.48 & \\
-8.18 \stackrel{?}{=} -8.18 & \text{TRUE}
\end{array}
$$

The solution is 3.4.

4. Solve: $0.8 = 2.8x + 5$.

↪ YOUR TURN

B. Combining Like Terms

When like terms appear on the same side of an equation, we combine them and then solve. When like terms appear on both sides of an equation, we can use the addition principle to rewrite all like terms on one side.

EXAMPLE 5 Solve.

a) $3x + 4x = -14$ **b)** $-x + 5 = -8x + 6$

c) $6x + 5 - 7x = 10 - 4x + 7$ **d)** $2 - 5(x + 5) = 3(x - 2) - 1$

SOLUTION

a) $3x + 4x = -14$

$$7x = -14 \qquad \text{Combining like terms}$$

$$\frac{7x}{7} = \frac{-14}{7} \qquad \text{Dividing both sides by 7}$$

$$x = -2 \qquad \text{Simplifying}$$

The check is left to the student. The solution is -2.

b) To solve $-x + 5 = -8x + 6$, we combine variable terms on one side and constant terms on the other. This can be done by subtracting 5 from both sides, to get all constant terms on the right, and adding $8x$ to both sides, to get all variable terms on the left. These steps can be performed in either order.

> Isolate variable terms on one side and constant terms on the other side.

$$-x + 5 = -8x + 6$$

$$-x + 8x + 5 = -8x + 8x + 6 \qquad \text{Adding } 8x \text{ to both sides}$$

$$7x + 5 = 6 \qquad \text{Simplifying}$$

$$7x + 5 - 5 = 6 - 5 \qquad \text{Subtracting 5 from both sides}$$

$$7x = 1 \qquad \text{Combining like terms}$$

$$\frac{7x}{7} = \frac{1}{7} \qquad \text{Dividing both sides by 7}$$

$$x = \frac{1}{7}$$

The check is left to the student. The solution is $\frac{1}{7}$.

Technology Connection

A **TABLE** feature lists the value of a variable expression for different choices of x. For example, to evaluate $6x + 5 - 7x$ for $x = 0, 1, 2, \ldots$, we first use (Y=) to enter $6x + 5 - 7x$ as y_1. We then use **2ND** (TBLSET) to specify TblStart $= 0$, ΔTbl $= 1$, and select **AUTO** twice. By pressing **2ND** (TBLSET), we can generate a table in which the value of $6x + 5 - 7x$ is listed for values of x starting at 0 and increasing by one's.

X	Y₁	
0	5	
1	4	
2	3	
3	2	
4	1	
5	0	
6	-1	
X = 0		

1. Create the above table on your graphing calculator. Scroll up and down to extend the table.

2. Enter $10 - 4x + 7$ as y_2. Your table should now have three columns.

3. For what x-value is y_1 the same as y_2? Compare this with the solution of Example 5(c). Is this a reliable way to solve equations? Why or why not?

VIDEO

c) $6x + 5 - 7x = 10 - 4x + 7$

$\quad\quad\quad -x + 5 = 17 - 4x$ Combining like terms on each side

$\quad -x + 5 + 4x = 17 - 4x + 4x$ Adding $4x$ to both sides

$\quad\quad\quad 5 + 3x = 17$ Simplifying. This is identical to Example 1.

$\quad 5 + 3x - 5 = 17 - 5$ Subtracting 5 from both sides

$\quad\quad\quad\quad 3x = 12$ Simplifying

$\quad\quad\quad\quad \dfrac{3x}{3} = \dfrac{12}{3}$ Dividing both sides by 3

$\quad\quad\quad\quad\quad x = 4$

Check:

$$\begin{array}{c|c} \multicolumn{2}{c}{6x + 5 - 7x = 10 - 4x + 7} \\ \hline 6 \cdot 4 + 5 - 7 \cdot 4 & 10 - 4 \cdot 4 + 7 \\ 24 + 5 - 28 & 10 - 16 + 7 \\ 1 \overset{?}{=} 1 & \text{TRUE} \end{array}$$

The solution is 4.

d) $2 - 5(x + 5) = 3(x - 2) - 1$

$\quad 2 - 5x - 25 = 3x - 6 - 1$ Using the distributive law. This is now similar to part (c) above.

$\quad\quad -5x - 23 = 3x - 7$ Combining like terms on each side

$\quad -5x - 23 + 7 = 3x - 7 + 7$ Adding 7 to both sides

$\quad\quad -5x - 16 = 3x$ Simplifying

$\quad -5x - 16 + 5x = 3x + 5x$ Adding $5x$ to both sides

$\quad\quad\quad -16 = 8x$

$\quad\quad\quad \dfrac{-16}{8} = \dfrac{8x}{8}$ Dividing both sides by 8

$\quad\quad\quad\quad -2 = x$ This is equivalent to $x = -2$.

The student can confirm that -2 checks and is the solution.

5. Solve:

$2x - (3 - x) = 7x - 1.$

↩ YOUR TURN

C. Clearing Fractions and Decimals

Equations are generally easier to solve when they do not contain fractions or decimals. The multiplication principle can be used to "clear" fractions or decimals, as shown here.

Clearing Fractions	Clearing Decimals
$\frac{1}{2}x + 5 = \frac{3}{4}$	$2.3x + 7 = 5.4$
$4\left(\frac{1}{2}x + 5\right) = 4 \cdot \frac{3}{4}$	$10(2.3x + 7) = 10 \cdot 5.4$
$2x + 20 = 3$	$23x + 70 = 54$

In each case, the resulting equation is equivalent to the original equation, but easier to solve.

> **AN EQUATION-SOLVING PROCEDURE**
>
> **1.** Use the multiplication principle to clear any fractions or decimals. (This is optional, but can ease computations. See Examples 6 and 7.)
> **2.** If necessary, use the distributive law to remove parentheses. Then combine like terms on each side, as needed. (See Example 5.)
> **3.** Use the addition principle, as needed, to isolate all variable terms on one side. Then combine like terms. (See Examples 1–7.)
> **4.** Multiply or divide to solve for the variable, using the multiplication principle. (See Examples 1–7.)
> **5.** Check all possible solutions in the original equation. (See Examples 1–4.)

The easiest way to clear an equation of fractions is to multiply *both sides* of the equation by the smallest, or *least*, common denominator of the fractions in the equation.

EXAMPLE 6   Solve: **(a)** $\frac{2}{3}x - \frac{1}{6} = 2x$; **(b)** $\frac{2}{5}(3x + 2) = 8$.

SOLUTION

a) We multiply both sides by 6, the least common denominator of $\frac{2}{3}$ and $\frac{1}{6}$.

$6\left(\dfrac{2}{3}x - \dfrac{1}{6}\right) = 6 \cdot 2x$	Multiplying both sides by 6
$6 \cdot \dfrac{2}{3}x - 6 \cdot \dfrac{1}{6} = 6 \cdot 2x$	
$4x - 1 = 12x$	Simplifying. Note that the fractions are cleared: $6 \cdot \frac{2}{3} = 4$, $6 \cdot \frac{1}{6} = 1$, and $6 \cdot 2 = 12$.
$4x - 1 - 4x = 12x - 4x$	Subtracting $4x$ from both sides
$-1 = 8x$	
$\dfrac{-1}{8} = \dfrac{8x}{8}$	Dividing both sides by 8
$-\dfrac{1}{8} = x$	$\dfrac{-1}{8} = -\dfrac{1}{8}$

> **CAUTION!** Be sure that the distributive law is used to multiply *all* the terms by 6.

The student can confirm that $-\frac{1}{8}$ checks and is the solution.

b) To solve $\frac{2}{5}(3x + 2) = 8$, we could first clear fractions, multiplying both sides by 5. For this equation, it is more efficient to multiply both sides by the reciprocal of $\frac{2}{5}$, which is $\frac{5}{2}$.

$\dfrac{5}{2} \cdot \dfrac{2}{5}(3x + 2) = \dfrac{5}{2} \cdot 8$	Multiplying both sides by $\frac{5}{2}$
$3x + 2 = 20$	Simplifying; $\frac{5}{2} \cdot \frac{2}{5} = 1$ and $\frac{5}{2} \cdot \frac{8}{1} = 20$
$3x + 2 - 2 = 20 - 2$	Subtracting 2 from both sides
$3x = 18$	
$\dfrac{3x}{3} = \dfrac{18}{3}$	Dividing both sides by 3
$x = 6$	

The student can confirm that 6 checks and is the solution.

6. Solve: $\frac{1}{4} - 5x = \frac{2}{3}x$.

YOUR TURN

Student Notes

Compare the steps of Examples 4 and 7. Note that the two different approaches yield the same solution. Using two approaches to solve a problem serves both as a check and as a valuable learning experience.

To clear an equation of decimals, we count the greatest number of decimal places in any one number. If the greatest number of decimal places is 1, we multiply both sides by 10; if it is 2, we multiply by 100; and so on.

EXAMPLE 7   Solve: $16.3 - 7.2y = -8.18$.

SOLUTION   The greatest number of decimal places in any one number is *two*. Multiplying by 100 will clear all decimals.

$$100(16.3 - 7.2y) = 100(-8.18) \qquad \text{Multiplying both sides by 100}$$
$$100(16.3) - 100(7.2y) = 100(-8.18) \qquad \text{Using the distributive law}$$
$$1630 - 720y = -818 \qquad \text{Simplifying}$$
$$1630 - 720y - 1630 = -818 - 1630 \qquad \text{Subtracting 1630 from both sides}$$
$$-720y = -2448 \qquad \text{Combining like terms}$$
$$\frac{-720y}{-720} = \frac{-2448}{-720} \qquad \text{Dividing both sides by } -720$$
$$y = 3.4$$

In Example 4, the same solution was found without clearing decimals. Finding the same answer in two ways is a good check. The solution is 3.4.

7. Solve: $0.8 = 2.8x + 5$.

⤶ YOUR TURN

↳ Check Your UNDERSTANDING

Match each equation with the step from the following list that could provide an appropriate first step in finding a solution.

a) Add 9 to both sides.

b) Subtract 9 from both sides.

c) Multiply both sides by 9.

d) Divide both sides by 9.

1. $\frac{1}{9} - x = \frac{2}{3}x$

2. $3x - 9 = 18$

3. $9 - 5x = 1$

4. $9(x - 6) = 90$

Determine the smallest positive number by which we can multiply in order to clear fractions.

5. $\frac{1}{3} + \frac{3}{4}t = 6$

6. $\frac{5}{6}x - \frac{2}{9} = \frac{1}{3}$

7. $\frac{1}{3}(4x - 7) = 10$

Determine the smallest power of 10 by which we can multiply in order to clear decimals.

8. $1.4 + 3.2n = 2.6$

9. $20 + 0.01t = 4$

10. $2.1x - 2.05 = 4$

2.2 EXERCISE SET

⟶ Vocabulary and Reading Check

Choose the principle or law from the following list that best completes each statement. Choices will be used more than once.

Addition principle   Distributive law
Multiplication principle

1. To isolate x in $x - 4 = 7$, we would use the
   _____.

2. To isolate x in $5x = 8$, we would use the
   _____.

3. To clear fractions or decimals, we use the
   _____.

4. To remove parentheses, we use the
   _____.

5. To solve $3x - 1 = 8$, we use the _____
   first.

6. To solve $5(x - 1) + 3(x + 7) = 2$, we use the
   _____first.

⟶ Concept Reinforcement

In each of Exercises 7–12, match the equation with an equivalent equation from the column on the right that could be the next step in finding a solution.

7. ____ $3x - 1 = 7$ a) $6x - 6 = 2$

8. ____ $4x + 5x = 12$ b) $4x + 2x = 3$

9. ____ $6(x - 1) = 2$ c) $3x = 7 + 1$

10. ____ $7x = 9$ d) $8x + 2x = 6 + 5$

11. ____ $4x = 3 - 2x$ e) $9x = 12$

12. ____ $8x - 5 = 6 - 2x$ f) $x = \frac{9}{7}$

A. Applying Both Principles

Solve.

13. $2x + 9 = 25$ 14. $5z + 2 = 57$

15. $7t - 8 = 27$ 16. $6x - 5 = 2$

17. $3x - 9 = 1$ 18. $5x - 9 = 41$

19. $8z + 2 = -54$ 20. $4x + 3 = -21$

21. $-37 = 9t + 8$ 22. $-39 = 1 + 5t$

23. $12 - t = 16$ 24. $9 - t = 21$

25. $-6z - 18 = -132$ 26. $-7x - 24 = -129$

27. $5.3 + 1.2n = 1.94$ 28. $6.4 - 2.5n = 2.2$

29. $32 - 7x = 11$ 30. $27 - 6x = 99$

31. $\frac{3}{5}t - 1 = 8$ 32. $\frac{2}{3}t - 1 = 5$

33. $6 + \frac{7}{2}x = -15$ 34. $6 + \frac{5}{4}x = -4$

35. $-\dfrac{4a}{5} - 8 = 2$ 36. $-\dfrac{8a}{7} - 2 = 4$

B. Combining Like Terms

Solve.

37. $6x + 10x = 18$ 38. $-3z + 8z = 45$

39. $4x - 6 = 6x$ 40. $7n = 2n + 4$

41. $2 - 5y = 26 - y$ 42. $6x - 5 = 7 + 2x$

43. $6x + 3 = 2x + 3$ 44. $5y + 3 = 2y + 15$

45. $5 - 2x = 3x - 7x + 25$

46. $10 - 3x = x - 2x + 40$

47. $7 + 3x - 6 = 3x + 5 - x$

48. $5 + 4x - 7 = 4x - 2 - x$

C. Clearing Fractions and Decimals

Clear fractions or decimals and solve.

49. $\frac{2}{3} + \frac{1}{4}t = 2$ 50. $-\frac{5}{6} + x = -\frac{1}{2} - \frac{2}{3}$

51. $\frac{2}{3} + 4t = 6t - \frac{2}{15}$ 52. $\frac{1}{2} + 4m = 3m - \frac{5}{2}$

53. $\frac{1}{3}x + \frac{2}{5} = \frac{4}{5} + \frac{3}{5}x - \frac{2}{3}$

54. $1 - \frac{2}{3}y = \frac{9}{5} - \frac{1}{5}y + \frac{3}{5}$

55. $2.1x + 45.2 = 3.2 - 8.4x$

56. $0.91 - 0.2z = 1.23 - 0.6z$

57. $0.76 + 0.21t = 0.96t - 0.49$

58. $1.7t + 8 - 1.62t = 0.4t - 0.32 + 8$

59. $\frac{2}{5}x - \frac{3}{2}x = \frac{3}{4}x + 3$ 60. $\frac{5}{16}y + \frac{3}{8}y = 2 + \frac{1}{4}y$

61. $\frac{1}{3}(2x - 1) = 7$ 62. $\frac{1}{5}(4x - 1) = 7$

A, B, C. Solving Linear Equations

Solve and check.

63. $7(2a - 1) = 21$ 64. $5(3 - 3t) = 30$

Aha! 65. $11 = 11(x + 1)$ 66. $9 = 3(5x - 2)$

67. $2(3 + 4m) - 6 = 48$ 68. $3(5 + 3m) - 8 = 7$

69. $2r + 8 = 6r + 10$

70. $3b - 2 = 7b + 4$

71. $4y - 4 + y + 24 = 6y + 20 - 4y$

72. $5y - 10 + y = 7y + 18 - 5y$

73. $19 - 3(2x - 1) = 7$

74. $5(d + 4) = 7(d - 2)$

75. $2(3t + 1) - 5 = t - (t + 2)$

76. $4x - (x + 6) = 5(3x - 1) + 8$

77. $19 - (2x + 3) = 2(x + 3) + x$

78. $13 - (2c + 2) = 2(c + 2) + 3c$

79. $\frac{3}{4}(3t - 4) = 15$

80. $\frac{3}{2}(2x + 5) = -\frac{15}{2}$

81. $\frac{1}{6}(\frac{3}{4}x - 2) = -\frac{1}{5}$

82. $\frac{2}{3}(\frac{7}{8} - 4x) - \frac{5}{8} = \frac{3}{8}$

83. $0.7(3x + 6) = 1.1 - (x - 3)$

84. $0.9(2x - 8) = 4 - (x + 5)$

85. $a + (a - 3) = (a + 2) - (a + 1)$

86. $0.8 - 4(b - 1) = 0.2 + 3(4 - b)$

87. Maggie solves $45 - t = 13$ (Example 3) by adding $t - 13$ to both sides. Is this approach preferable to the one used in Example 3? Why or why not?

88. Why must the rules for the order of operations be understood before solving the equations in this section?

Skill Review

89. Add: $\frac{2}{9} + \frac{1}{6}$. [1.3]

90. Multiply: $\frac{2}{7} \cdot \frac{7}{2}$. [1.3]

91. Find decimal notation: $-\frac{1}{9}$. [1.4]

92. Find the absolute value: $|-16|$. [1.4]

Synthesis

93. What procedure would you use to solve an equation like $0.23x + \frac{17}{3} = -0.8 + \frac{3}{4}x$? Could your procedure be streamlined? If so, how?

94. Ethan is determined to solve $3x + 4 = -11$ by first using the multiplication principle to "eliminate" the 3. How should he proceed and why?

95. Kyle estimates his long-distance driving time, in hours, by multiplying the number of miles that he will drive by $\frac{3}{200}$ and then adding $\frac{1}{4}$ for every major city he will drive through. If he estimates that a trip through two major cities will take 8 hr, how many miles will he be driving?

Some equations, like $3 = 7$ or $x + 2 = x + 5$, are never true and are called **contradictions**. *Other equations, like $7 = 7$ or $2x = 2x$, are always true and are called* **identities**. *Solve each of the following and if an identity or contradiction is found, state this.*

96. $2x = x + x$

97. $x + 5 + x = 2x$

98. $9x = 0$

99. $4x - x = 2x + x$

100. $x + 8 = 3 + x + 7$ Aha! **101.** $2|x| = -14$

102. $|3x| = 12$

Solve. Label any contradictions or identities.

103. $8.43x - 2.5(3.2 - 0.7x) = -3.455x + 9.04$

104. $0.008 + 9.62x - 42.8 = 0.944x + 0.0083 - x$

105. $-2[3(x - 2) + 4] = 4(5 - x) - 2x$

106. $0 = t - (-6) - (-7t)$

107. $3(x + 5) = 3(5 + x)$

108. $5(x - 7) = 3(x - 2) + 2x$

109. $2x(x + 5) - 3(x^2 + 2x - 1) = 9 - 5x - x^2$

110. $9 - 3x = 2(5 - 2x) - (1 - 5x)$

Aha! **111.** $[7 - 2(8 \div (-2))]x = 0$

112. $\dfrac{5x + 3}{4} + \dfrac{25}{12} = \dfrac{5 + 2x}{3}$

> **↪ YOUR TURN ANSWERS: SECTION 2.2**
> **1.** 6   **2.** -3 **3.** -11 **4.** -1.5 **5.** $-\frac{1}{2}$ **6.** $\frac{3}{68}$
> **7.** -1.5

Quick Quiz: Sections 2.1–2.2

1. Determine whether -5 is a solution of $7 - 3x = 22$. [2.1]

Solve and check.

2. $y + 7.5 = 2.1$ [2.1]

3. $\dfrac{2x}{5} = 10$ [2.1]

4. $2 - (x - 7) = 13$ [2.2]

5. $\dfrac{1}{10}r + \dfrac{1}{5} = \dfrac{1}{2}r$ [2.2]

Prepare to Move On

Evaluate. [1.8]

1. $3 - 5a$, for $a = 2$ **2.** $12 \div 4 \cdot t$, for $t = 5$

3. $7x - 2x$, for $x = -3$ **4.** $t(8 - 3t)$, for $t = -2$

2.3   Formulas

A. Evaluating Formulas   **B.** Solving for a Variable

An equation that shows a relationship between variable quantities uses two or more letters and is known as a **formula**. Most of the letters in this book are variables, but some are constants. For example, c in $E = mc^2$ represents the speed of light.

A. Evaluating Formulas

EXAMPLE 1 *Event Promotion.*   Event promoters use the formula

$$p = \frac{1.2x}{s}$$

to determine a ticket price p for an event with x dollars of expenses and anticipated ticket sales of s tickets. Grand Events expects expenses for a concert to be \$80,000 and anticipates selling 4000 tickets. What should the ticket price be?

Data: The Indianapolis Star, 2/27/03

1. Using the formula in Example 1, determine the ticket price when concert expenses are \$40,000 and anticipated sales are 2500 tickets.

SOLUTION   We substitute 80,000 for x and 4000 for s and calculate p:

$$p = \frac{1.2x}{s} = \frac{1.2(80,000)}{4000} = 24.$$

The ticket price should be \$24.

YOUR TURN

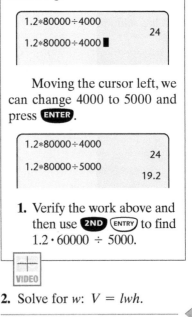

Technology Connection

Suppose that after calculating $1.2 \cdot 80{,}000 \div 4000$, we want to find $1.2 \cdot 80{,}000 \div 5000$. Pressing **2ND** **(ENTRY)** gives the following.

```
1.2*80000÷4000
              24
1.2*80000÷4000 ▇
```

Moving the cursor left, we can change 4000 to 5000 and press **ENTER**.

```
1.2*80000÷4000
              24
1.2*80000÷5000
            19.2
```

1. Verify the work above and then use **2ND** **(ENTRY)** to find $1.2 \cdot 60000 \div 5000$.

VIDEO

2. Solve for w: $V = lwh$.

B. Solving for a Variable

The **circumference** of a circle is the distance around the circle. The formula $C = 2\pi r$ gives the circumference C of a circle with radius r. For some circles (such as a circular pond), it is easier to measure the circumference than to measure the radius. In order to find the radius of a circle when we know its circumference, we can *solve* the formula for r.

To **solve** for a variable means to write an equivalent equation with that variable alone on one side of the equation. That letter should not appear at all on the other side of the equation.

EXAMPLE 2 *Circumference of a Circle.*   Solve for r: $C = 2\pi r$.

SOLUTION   We want r to appear alone on one side of the equation and not to appear at all on the other side.

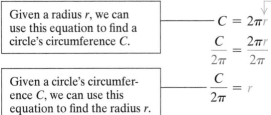

Given a radius r, we can use this equation to find a circle's circumference C.

$C = 2\pi r$ We want this variable alone.

$\dfrac{C}{2\pi} = \dfrac{2\pi r}{2\pi}$ Dividing both sides by 2π

$\dfrac{C}{2\pi} = r$

Given a circle's circumference C, we can use this equation to find the radius r.

YOUR TURN

To see how solving a formula is just like solving an equation, compare the following. In (A), we solve for x as usual; in (B), we show steps but do not simplify; and in (C), we *cannot* simplify because a, b, and c are unknown.

A. $5x + 2 = 12$

$5x = 12 - 2$

$5x = 10$

$x = \dfrac{10}{5} = 2$

B. $5x + 2 = 12$

$5x = 12 - 2$

$x = \dfrac{12 - 2}{5}$

C. $ax + b = c$

$ax = c - b$

$x = \dfrac{c - b}{a}$

EXAMPLE 3 *Motion.* The rate r at which an object moves is found by dividing distance d traveled by time t, or

$$r = \frac{d}{t}.$$

Solve for t.

SOLUTION We use the multiplication principle to clear fractions and then solve for t:

$r = \dfrac{d}{t}$ ⟵—— We want this variable alone.

$r \cdot t = \dfrac{d}{t} \cdot t$ Multiplying both sides by t

$rt = \dfrac{dt}{t}$ $\dfrac{d}{t} \cdot t = \dfrac{d}{t} \cdot \dfrac{t}{1} = \dfrac{dt}{t}$

$rt = d$ Removing a factor equal to 1: $t/t = 1$.
The equation is cleared of fractions.

$\dfrac{rt}{r} = \dfrac{d}{r}$ Dividing both sides by r

$t = \dfrac{d}{r}.$

This formula can be used to determine the time spent traveling when the distance and the rate are known.

YOUR TURN

EXAMPLE 4 Solve for y: $3x - 4y = 10$.

SOLUTION There is one term that contains y, so we begin by isolating that term on one side of the equation.

$3x - 4y = 10$ We want this variable alone.

$-4y = 10 - 3x$ Subtracting $3x$ from both sides

$-\frac{1}{4}(-4y) = -\frac{1}{4}(10 - 3x)$ Multiplying both sides by $-\frac{1}{4}$

$y = -\frac{10}{4} + \frac{3}{4}x$ Multiplying using the distributive law

$y = -\frac{5}{2} + \frac{3}{4}x$ Simplifying the fraction

YOUR TURN

Student Notes

When working with formulas, you should not interchange a lower-case letter with its associated uppercase letter; these letters may represent different quantities. For example, a formula for gravity uses both M and m.

3. Solve for x: $y = \dfrac{k}{x}$.

4. Solve for y: $x + 2y = 3$.

5. Solve for p:

$$T = 20(a + p) - 15.$$

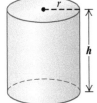

6. Solve for c: $D = 3c + 6cy$.

  YOUR TURN

EXAMPLE 5 *Nutrition.* The number of calories K needed each day by a moderately active woman who weighs w pounds, is h inches tall, and is a years old can be estimated using the formula

$$K = 917 + 6(w + h - a).*$$

Solve for w.

SOLUTION   We undo the operations by reversing the order in which they would be performed on the right side of the given equation:

$$K = 917 + 6(w + h - a) \qquad \text{We want } w \text{ alone.}$$

$$K - 917 = 6(w + h - a) \qquad \text{Subtracting 917 from both sides}$$

$$\frac{K - 917}{6} = w + h - a \qquad \text{Dividing both sides by 6}$$

$$\frac{K - 917}{6} + a - h = w. \qquad \begin{array}{l}\text{Adding } a \text{ and subtracting } h \text{ on} \\ \text{both sides}\end{array}$$

This formula can be used to estimate a woman's weight, if we know her age, height, and caloric needs.

  YOUR TURN

The above steps are similar to those used to solve equations with one variable. We use the addition and multiplication principles just as before. An important difference that we will see in the next example is that we may need to factor.

TO SOLVE A FORMULA FOR A GIVEN VARIABLE

   1. If the variable for which you are solving appears in a fraction, use the multiplication principle to clear fractions.
   2. Isolate the term(s) with the variable for which you are solving on one side of the equation.
   3. If two or more terms contain the variable for which you are solving, factor the variable out.
   4. Multiply or divide to solve for the variable in question.

We can also solve for a letter that represents a constant.

EXAMPLE 6 *Surface Area of a Right Circular Cylinder.* The formula $A = 2\pi rh + 2\pi r^2$ gives the surface area A of a right circular cylinder of height h and radius r. Solve for π.

SOLUTION   We have

$$A = 2\pi rh + 2\pi r^2 \qquad \text{We want this letter alone.}$$

$$A = \pi(2rh + 2r^2) \qquad \text{Factoring}$$

$$\frac{A}{2rh + 2r^2} = \pi. \qquad \begin{array}{l}\text{Dividing both sides by } 2rh + 2r^2\text{, or} \\ \text{multiplying both sides by } 1/(2rh + 2r^2)\end{array}$$

We can also write this as

$$\pi = \frac{A}{2rh + 2r^2}.$$

  YOUR TURN

---

*Based on information from M. Parker (ed.), *She Does Math*! (Washington, D.C.: Mathematical Association of America, 1995), p. 96.

↶ Check Your UNDERSTANDING

For each of the following, determine whether the formula has been solved for n.

1. $n = \dfrac{p}{t}$

2. $3m - 7(y - 2) = n$

3. $n = x - 4pn$

4. $\dfrac{1}{2}n = \dfrac{a + b + c}{2}$

CAUTION! Had we performed the following steps in Example 6, we would *not* have solved for π:

$$A = 2\pi rh + 2\pi r^2 \qquad \text{We want } \pi \text{ alone.}$$

$$A - 2\pi r^2 = 2\pi rh \qquad \text{Subtracting } 2\pi r^2 \text{ from both sides}$$
Two occurrences of π

$$\dfrac{A - 2\pi r^2}{2rh} = \pi. \qquad \text{Dividing both sides by } 2rh$$

The mathematics of each step is correct, but because π occurs on both sides of the formula, *we have not solved the formula for π.* Remember that the letter being solved for should be alone on one side of the equation, with no occurrence of that letter on the other side!

2.3   EXERCISE SET

FOR EXTRA HELP   MyMathLab®

↷ Vocabulary and Reading Check

In each of Exercises 1 and 2, classify the statement as either true or false.

1. All letters used in algebra represent variables.

2. If a letter appears on both sides of a formula, that formula is not solved for that letter.

In each of Exercises 3 and 4, three words appear under the blank. Choose the correct word to complete the statement.

3. The distance around a circle is its

_____
area/circumference/volume.

4. An equation that uses two or more letters to represent a relationship among quantities is a(n)

_____
constant/formula/variable.

A. Evaluating Formulas

Solve.

5. *Concerts.* The formula $d = 344t$ can be used to determine how far d, in meters, sound travels through room-temperature air in t seconds. Fans near the back of a large crowd at a concert will experience a time lag between the time they see a note played on stage and the time they hear that note. If the time lag is 0.9 sec, how far are the fans from the stage?

6. *Furnace Output.* Contractors in the Northeast use the formula $B = 30a$ to determine the minimum furnace output B, in British thermal units (Btu's), for a well-insulated house with a square feet of flooring.

Determine the minimum furnace output for an 1800-ft² house that is well insulated.
Data: U.S. Department of Energy

7. *College Enrollment.* At many colleges, the number of "full-time-equivalent" students F is given by

$$F = \dfrac{n}{15},$$

where n is the total number of credits for which students have enrolled in a given semester. Determine the number of full-time-equivalent students on a campus in which students registered for a total of 21,345 credits.

8. *Distance from a Storm.* The formula $M = \frac{1}{5}t$ can be used to determine how far M, in miles, you are from lightning when its thunder takes t seconds to reach your ears. If it takes 10 sec for the sound of thunder to reach you after you have seen the lightning, how far away is the storm?

9. *Federal Funds Rate.* The Federal Reserve Board sets a target f for the federal funds rate, that is, the interest rate that banks charge each other for overnight borrowing of Federal funds. This target rate can be estimated by

$$f = 8.5 + 1.4(I - U),$$

where I is the core inflation rate over the previous 12 months and U is the seasonally adjusted unemployment rate. If core inflation is 0.025 and unemployment is 0.044, what should the federal funds rate be?

Data: Greg Mankiw, Harvard University, www.gregmankiw.blogspot.com

10. *Calorie Density.* The calorie density D, in calories per ounce, of a food that contains c calories and weighs w ounces is given by

$$D = \frac{c}{w}.^*$$

Eight ounces of fat-free milk contains 84 calories. Find the calorie density of fat-free milk.

11. *Absorption of Ibuprofen.* When 400 mg of the painkiller ibuprofen is swallowed, the number of milligrams n in the bloodstream t hours later (for $0 \le t \le 6$) is estimated by

$$n = 0.5t^4 + 3.45t^3 - 96.65t^2 + 347.7t.$$

How many milligrams of ibuprofen remain in the blood 1 hr after 400 mg has been swallowed?

12. *Size of a League Schedule.* When all n teams in a league play every other team twice, a total of N games are played, where

$$N = n^2 - n.$$

If a soccer league has 7 teams and all teams play each other twice, how many games are played?

B. Solving for a Variable

In Exercises 13–52, solve each formula for the indicated letter.

13. $A = bh$, for b
(Area of parallelogram with base b and height h)

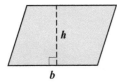

14. $A = bh$, for h

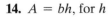
Data: Nutrition Action Healthletter, March 2000, p. 9. Center for Science in the Public Interest, Suite 300; 1875 Connecticut Ave NW, Washington, D.C. 20008.

15. $I = Prt$, for P
(Simple-interest formula, where I is interest, P is principal, r is interest rate, and t is time)

16. $I = Prt$, for t

17. $H = 65 - m$, for m
(To determine the number of heating degree days H for a day with m degrees Fahrenheit as the average temperature)

18. $d = h - 64$, for h
(To determine how many inches d above average an h-inch-tall woman is)

19. $P = 2l + 2w$, for l
(Perimeter of a rectangle of length l and width w)

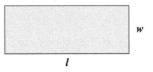

20. $P = 2l + 2w$, for w

21. $A = \pi r^2$, for π
(Area of a circle with radius r)

22. $A = \pi r^2$, for r^2

23. $A = \frac{1}{2}bh$, for h
(Area of a triangle with base b and height h)

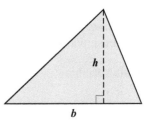

24. $A = \frac{1}{2}bh$, for b

25. $E = mc^2$, for c^2
(A relativity formula from physics)

26. $E = mc^2$, for m

27. $Q = \dfrac{c + d}{2}$, for d

28. $A = \dfrac{a + b + c}{3}$, for b

29. $p - q + r = 2$, for q
(Euler's formula from graph theory)

30. $p = \dfrac{r - q}{2}$, for q

31. $w = \dfrac{r}{f}$, for r

(To compute the wavelength w of a musical note with frequency f and speed of sound r)

32. $M = \dfrac{A}{s}$, for A

(To compute the Mach number M for speed A and speed of sound s)

33. $H = \dfrac{TV}{550}$, for T

(To determine the horsepower H of an airplane propeller with thrust T and airplane velocity V)

34. $P = \dfrac{ab}{c}$, for b

35. $F = \frac{9}{5}C + 32$, for C
(To convert the Celsius temperature C to the Fahrenheit temperature F)

36. $M = \frac{5}{9}n + 18$, for n

37. $2x - y = 1$, for y **38.** $3x - y = 7$, for y

39. $2x + 5y = 10$, for y **40.** $3x + 2y = 12$, for y

41. $4x - 3y = 6$, for y **42.** $5x - 4y = 8$, for y

43. $9x + 8y = 4$, for y **44.** $x + 10y = 2$, for y

45. $3x - 5y = 8$, for y **46.** $7x - 6y = 7$, for y

47. $z = 13 + 2(x + y)$, for x

48. $A = 115 + \frac{1}{2}(p + s)$, for s

49. $t = 27 - \frac{1}{4}(w - l)$, for l

50. $m = 19 - 5(x - n)$, for n

51. $A = at + bt$, for t

52. $S = rx + sx$, for x

53. *Area of a Trapezoid.* The formula
$$A = \tfrac{1}{2}ah + \tfrac{1}{2}bh$$
can be used to find the area A of a trapezoid with bases a and b and height h. Solve for h. (*Hint*: First clear fractions.)

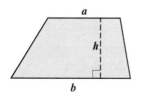

54. *Compounding Interest.* The formula
$$A = P + Prt$$
is used to find the amount A in an account when simple interest is added to an investment of P dollars (see Exercise 15). Solve for P.

55. *Chess Rating.* The formula
$$R = r + \dfrac{400(W - L)}{N}$$
is used to establish a chess player's rating R after that player has played N games, won W of them, and lost L of them. Here r is the average rating of the opponents. Solve for L.

Data: The U.S. Chess Federation

56. *Angle Measure.* The angle measure S of a sector of a circle is given by
$$S = \dfrac{360A}{\pi r^2},$$
where r is the radius, A is the area of the sector, and S is in degrees. Solve for r^2.

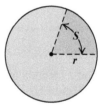

57. Julia has a formula that allows her to convert Celsius temperatures to Fahrenheit temperatures. She needs a formula for converting Fahrenheit temperatures to Celsius temperatures. What advice can you give her?

58. Under what circumstances would it be useful to solve $I = Prt$ for P? (See Exercise 15.)

Skill Review

Simplify.

59. $-2 + 5 - (-4) - 17$ [1.6]

60. $-98 \div \frac{1}{2}$ [1.7]

Aha! **61.** $4.2(-11.75)(0)$ [1.7]

62. $(-2)^5$ [1.8]

63. $20 \div (-4) \cdot 2 - 3$ [1.8]

64. $5|8 - (2 - 7)|$ [1.8]

Synthesis

65. The equations

$$P = 2l + 2w \quad \text{and} \quad w = \frac{P}{2} - l$$

are equivalent formulas involving the perimeter P, length l, and width w of a rectangle. Write a problem for which the second of the two formulas would be more useful.

66. While solving $2A = ah + bh$ for h, Eva writes

$$\frac{2A - ah}{b} = h.$$

What is her mistake?

67. A Harris–Benedict formula gives the number of calories K needed each day by a moderately active man who weighs w kilograms, is h centimeters tall, and is a years old as

$$K = 21.235w + 7.75h - 10.54a + 102.3.$$

If Janos is moderately active, weighs 80 kg, is 190 cm tall, and needs to consume 2852 calories per day, how old is he?

68. *Altitude and Temperature.* Air temperature drops about 1° Celsius (C) for each 100-m rise above ground level, up to 12 km. If the ground level temperature is t°C, find a formula for the temperature T at an elevation of h meters.

Data: *A Sourcebook of School Mathematics*, Mathematical Association of America, 1980

69. *Surface Area of a Cube.* The surface area A of a cube with side s is given by

$$A = 6s^2.$$

If a cube's surface area is 54 in², find the volume of the cube.

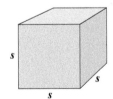

70. *Weight of a Fish.* An ancient fisherman's formula for estimating the weight of a fish is

$$w = \frac{lg^2}{800},$$

where w is the weight, in pounds, l is the length, in inches, and g is the girth (distance around the midsection), in inches. Estimate the girth of a 700-lb yellowfin tuna that is 8 ft long.

71. *Dosage Size.* Clark's rule for determining the size of a particular child's medicine dosage c is

$$c = \frac{w}{a} \cdot d,$$

where w is the child's weight, in pounds, and d is the usual adult dosage for an adult weighing a pounds. Solve for a.

Data: Olsen, June Looby, et al., *Medical Dosage Calculations*. Redwood City, CA: Addison-Wesley, 1995

Solve each formula for the given letter.

72. $\dfrac{y}{z} \div \dfrac{z}{t} = 1$, for y

73. $ac = bc + d$, for c

74. $qt = r(s + t)$, for t

75. $3a = c - a(b + d)$, for a

76. *Furnace Output.* The formula

$$B = 50a$$

is used in New England to estimate the minimum furnace output B, in Btu's, for an old, poorly insulated house with a square feet of flooring. Find an equation for determining the number of Btu's saved by insulating an old house. (See Exercise 6.)

77. Revise the formula in Exercise 67 so that a man's weight in pounds (2.2046 lb = 1 kg) and his height in inches (0.3937 in. = 1 cm) are used.

 YOUR TURN ANSWERS: SECTION 2.3

1. $19.20   **2.** $w = \dfrac{V}{lh}$ **3.** $x = \dfrac{k}{y}$ **4.** $y = -\frac{1}{2}x + \frac{3}{2}$

5. $p = \dfrac{T + 15}{20} - a$ **6.** $c = \dfrac{D}{3 + 6y}$

Quick Quiz: Sections 2.1–2.3

Solve and check.

1. $-x = -7$ [2.1]

2. $4(n - 7) - n = 36$ [2.2]

3. $1.2t - 0.05 = 3.2t$ [2.2]

4. $3(2x - 4) = 2 - 5(7 - x)$ [2.2]

5. Solve for y: $6x + 2y = 1$.   [2.3]

Prepare to Move On

Convert to decimal notation.   [1.4]

1. $\dfrac{1}{4}$ **2.** $\dfrac{9}{8}$ **3.** $\dfrac{2}{3}$ **4.** $\dfrac{5}{6}$

Mid-Chapter Review

We solve equations using the addition and multiplication principles.

For any real numbers a, b, and c:

a) $a = b$ is equivalent to $a + c = b + c$;

b) $a = b$ is equivalent to $ac = bc$, provided $c \neq 0$.

GUIDED SOLUTIONS

Solve. [2.2]*

1. $2x + 3 = 10$

Solution

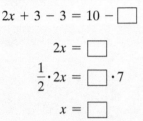

$2x + 3 - 3 = 10 - \boxed{}$ Using the addition principle

$2x = \boxed{}$ Simplifying

$\dfrac{1}{2} \cdot 2x = \boxed{} \cdot 7$ Using the multiplication principle

$x = \boxed{}$ Simplifying

2. $\frac{1}{2}(x - 3) = \frac{1}{3}(x - 4)$

Solution

$6 \cdot \frac{1}{2}(x - 3) = \boxed{} \cdot \frac{1}{3}(x - 4)$ Multiplying to clear fractions

$\boxed{}(x - 3) = \boxed{}(x - 4)$ The fractions are cleared.

$3x - \boxed{} = 2x - \boxed{}$ Multiplying

$3x - 9 + 9 = 2x - 8 + \boxed{}$ Using the addition principle

$3x = 2x + \boxed{}$ Simplifying

$3x - \boxed{} = 2x + 1 - 2x$ Using the addition principle

$x = \boxed{}$ Simplifying

MIXED REVIEW

Solve.

3. $x - 2 = -1$ [2.1]

4. $2 - x = -1$ [2.1]

5. $3t = 5$ [2.1]

6. $-\frac{3}{2}x = 12$ [2.1]

7. $\dfrac{y}{8} = 6$ [2.1]

8. $0.06x = 0.03$ [2.1]

9. $3x - 7x = 20$ [2.2]

10. $9x - 7 = 17$ [2.2]

11. $4(t - 3) - t = 6$ [2.2]

12. $8n - (3n - 5) = 5 - n$ [2.2]

13. $\frac{9}{10}y - \frac{7}{10} = \frac{21}{5}$ [2.2]

14. $2(t - 5) - 3(2t - 7) = 12 - 5(3t + 1)$ [2.2]

15. $\frac{2}{3}(x - 2) - 1 = -\frac{1}{2}(x - 3)$ [2.2]

Solve for the indicated variable. [2.3]

16. $E = wA$, for A

17. $Ax + By = C$, for y

18. $at + ap = m$, for a

19. $m = \dfrac{F}{a}$, for a

20. $v = \dfrac{b - f}{t}$, for b

---

*The notation [2.2] refers to Chapter 2, Section 2.

2.4 Applications with Percent

A. Converting Between Percent Notation and Decimal Notation   **B.** Solving Percent Problems

Percent problems arise so frequently in everyday life that often we are not even aware of them. In this section, we will solve some real-world percent problems. Before doing so, however, we need to review a few basics.

A. Converting Between Percent Notation and Decimal Notation

Oceans cover 70% of the earth's surface. This means that of every 100 square miles on the surface of the earth, 70 square miles is ocean. Thus, 70% is a ratio of 70 to 100.

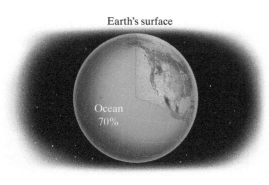

Earth's surface

Ocean 70%

Study Skills

Do It Again

Continual review of material is a key to success. Not only does review keep the material fresh for a final exam, in many courses (such as math), new concepts are developed using previous concepts. Take advantage of the review exercises built into this text. You may be surprised at the positive results you will see with just a few minutes of review each day.

1. Convert to decimal notation: 120%.

The percent symbol % means "per hundred." We can regard the percent symbol as part of a name for a number. For example,

$$70\% \quad \text{is defined to mean} \quad \frac{70}{100}, \quad \text{or} \quad 70 \times \frac{1}{100}, \quad \text{or} \quad 70 \times 0.01.$$

PERCENT NOTATION

$n\%$ means $\dfrac{n}{100}$,   or $n \times \dfrac{1}{100}$,   or $n \times 0.01$.

EXAMPLE 1   Convert to decimal notation: **(a)** 78%; **(b)** 1.3%.

SOLUTION

a) $78\% = 78 \times 0.01$ Replacing % with $\times 0.01$
$ = 0.78$

b) $1.3\% = 1.3 \times 0.01$ Replacing % with $\times 0.01$
$ = 0.013$

↩ YOUR TURN

As shown in Example 1, multiplication by 0.01 simply moves the decimal point two places to the left.

To convert from percent notation to decimal notation, move the decimal point two places to the left and drop the percent symbol.

EXAMPLE 2 Convert the percent notation in the following sentence to decimal notation: Plastic makes up 90% of all trash floating in the ocean.

Data: environment.nationalgeographic.com

SOLUTION

$$90\% = 90.0\% \qquad 0.90.0 \qquad 90\% = 0.90, \quad \text{or simply } 0.9$$

Move the decimal point two places to the left and drop the %.

2. Convert to decimal notation: 4%.

↪ YOUR TURN

The procedure used in Examples 1 and 2 can be reversed:

$$0.38 = 38 \times 0.01$$
$$= 38\%. \qquad \text{Replacing } \times 0.01 \text{ with } \%$$

To convert from decimal notation to percent notation, move the decimal point two places to the right and write a percent symbol.

EXAMPLE 3 Convert to percent notation: **(a)** 1.27; **(b)** $\frac{1}{4}$; **(c)** 0.3.

SOLUTION

a) We first move the decimal point two places to the right: 1.27.
and then write a % symbol: 127%   This is the same as multiplying 1.27 by 100 and writing %.

b) Note that $\frac{1}{4} = 0.25$. We move the decimal point two places to the right: 0.25.
and then write a % symbol: 25%   Multiplying by 100 and writing %

c) We first move the decimal point two places to the right (recall that 0.3 = 0.30): 0.30.
and then write a % symbol: 30%   Multiplying by 100 and writing %

3. Convert to percent notation: 0.37.

↪ YOUR TURN

B. Solving Percent Problems

In solving percent problems, we first *translate* the problem to an equation. Then we *solve* the equation. The key words in the translation are as follows.

KEY WORDS IN PERCENT TRANSLATIONS

"**Of**" translates to " · " or "×".   "**Is**" or "**Was**" translates to "=".

"**What**" translates to a variable.   "**%**" translates to "×$\frac{1}{100}$" or "× 0.01".

Student Notes

A way of checking answers is by estimating as follows:

$11\% \times 49 \approx 10\% \times 50 = 5.$

Since 5 is close to 5.39, our answer is reasonable.

4. What is 6% of 15?

EXAMPLE 4 What is 11% of 49?

SOLUTION

Translate: What is 11% of 49?

$$a = 0.11 \cdot 49$$

"of" means multiply;
$11\% = 0.11$

$$a = 5.39$$

Thus, 5.39 is 11% of 49. The answer is 5.39.

↩ YOUR TURN

EXAMPLE 5 3 is 16 percent of what?

SOLUTION

Translate: 3 is 16 percent of what?

$$3 = 0.16 \cdot y$$

$$\frac{3}{0.16} = y \qquad \text{Dividing both sides by 0.16}$$

$$18.75 = y$$

5. 5.2 is 8 percent of what?

Thus, 3 is 16 percent of 18.75. The answer is 18.75.

↩ YOUR TURN

EXAMPLE 6 What percent of $50 is $34?

SOLUTION

Translate: What percent of $50 is $34?

$$n \cdot 50 = 34$$

$$n = \frac{34}{50} \qquad \text{Dividing both sides by 50}$$

$$n = 0.68 = 68\% \qquad \begin{array}{l}\text{Converting}\\ \text{to percent}\\ \text{notation}\end{array}$$

6. What percent of 80 is 16?

Thus, $34 is 68% of $50. The answer is 68%.

↩ YOUR TURN

Examples 4–6 represent the three basic types of percent problems. Note that in all the problems, the following quantities are present:

- a percent, expressed in decimal notation in the translation,
- a base amount, referred to by the word "of" in the problem, and
- a percentage of the base, found by multiplying the base times the percent.

EXAMPLE 7 *Bicycling.* In 2012, the city of Chicago had designated 200 mi of bike lanes. The number of miles of bike lanes planned for Chicago by 2020 was 322.5% of the miles in 2012. If all the planned lanes are completed, how many miles of bike lanes will Chicago have in 2020?

Data: bicycling.com

SOLUTION We first reword and then translate. We let m = the number of miles of bike lanes in Chicago in 2020.

Rewording: What is 322.5% of 200?

Translating: m = 3.225 × 200

The letter is by itself on one side of the equation. To solve the equation, we need only multiply:

$$m = 3.225 \times 200 = 645.$$

Since 645 is 322.5% of 200, we have found that there will be 645 mi of bike lanes in Chicago in 2020.

YOUR TURN

EXAMPLE 8 *College Enrollment.* About 1.962 million U.S. students who graduated from high school in 2013 were attending college in the fall of 2013. This was 65.9% of all 2013 high school graduates, the lowest percentage in a decade. How many students graduated from high school in 2013?

Data: U.S. Bureau of Labor Statistics

SOLUTION Before translating the problem to mathematics, we reword and let S represent the total number of students, in millions, who graduated from high school in 2013.

Rewording: 1.962 is 65.9% of S.

Translating: 1.962 = 0.659 · S

$$\frac{1.962}{0.659} = S \qquad \text{Dividing both sides by 0.659}$$

$$2.977 \approx S \qquad \text{The symbol } \approx \text{ means } \textit{is approximately equal to.}$$

About 2.977 million U.S. students graduated from high school in 2013.

YOUR TURN

EXAMPLE 9 *Discounts.* During a recent promotion, Fitting Footwear reduced the price of a pair of running shoes from $139 to $114.

a) What percent of the regular price does the sale price represent?

b) What is the percent of discount?

SOLUTION

a) We reword and translate, using n for the unknown percent.

Rewording: What percent of 139 is 114?

Translating: n · 139 = 114

$$n = \frac{114}{139} \qquad \text{Dividing both sides by 139}$$

$$n \approx 0.82 = 82\% \qquad \text{Converting to percent notation}$$

The sale price is about 82% of the regular price.

---

7. Alicia presented her line of jewelry to 250 stores, and 25.6% of those stores placed an order. How many stores placed an order?

Student Notes

Always look for connections between examples. Here you should look for similarities between Examples 4 and 7 as well as between Examples 5 and 8 and between Examples 6 and 9.

---

8. At Valley Heights Community College, 252 first-year students enrolled in a Spanish class. This was 30% of the entire group. How many first-year students were there?

↳ Check Your UNDERSTANDING

Find each of the following.

1. 10% of 40

2. 50% of 40

3. 25% of 40

4. 100% of 40

5. 1% of 40

b) Since the original price of $139 represents 100% of the regular price, the sale price represents a discount of $(100 - 82)\%$, or 18%.

Alternatively, or as a check, we can find the amount of discount and then calculate the percent of discount, using x for the unknown percent.

Amount of discount: $\$139 - \$114 = \$25$

Rewording: What percent of 139 is 25?

Translating: x $\cdot$ 139 = 25

$$x = \frac{25}{139}$$ Dividing both sides by 139

$$x \approx 0.18 = 18\%$$ Converting to percent notation

We have a check. The percent of discount is 18%.

Chapter Resources:
Collaborative Activity, p. 145;
Decision Making: Connection, p. 145

9. Recently, Elm Cycles reduced the price of a BMW F 800 GS motorcycle from $15,000 to $13,500. What is the percent of discount?

YOUR TURN

2.4 EXERCISE SET

FOR EXTRA HELP   MyMathLab®

Vocabulary and Reading Check

Choose from the following list of words to complete each statement. Not every word will be used.

approximately	left	right
base	not	sale
decimal	percent	
hundred	retail	

1. To convert from percent notation to decimal notation, move the decimal point two places to the _____ and drop the percent symbol.

2. The percent symbol, %, means "per _____."

3. The expression 1.3% is written in _____ notation.

4. The word "of" in a percent problem generally refers to the _____ amount.

5. The _____ price is the original price minus the discount.

6. The symbol $\approx$ means "is _____ equal to."

Concept Reinforcement

In each of Exercises 7–16, match the question with the appropriate translation from the following list. Some choices are used more than once.

a) $a = (0.57)23$ **b)** $57 = 0.23y$
c) $n \cdot 23 = 57$ **d)** $n \cdot 57 = 23$
e) $23 = 0.57y$ **f)** $a = (0.23)57$

7. ____ What percent of 57 is 23?

8. ____ What percent of 23 is 57?

9. ____ 23 is 57% of what number?

10. ____ 57 is 23% of what number?

11. ____ 57 is what percent of 23?

12. ____ 23 is what percent of 57?

13. ____ What is 23% of 57?

14. ____ What is 57% of 23?

15. ____ 23% of what number is 57?

16. ____ 57% of what number is 23?

A. Converting Between Percent Notation and Decimal Notation

Convert the percent notation in each sentence to decimal notation.

17. *Education.* According to Scholastic, about 47% of children ages 6–17 believe that strong math skills are among the top three skills they should have.

Data: Kids & Family Reading Report

18. *Volunteering.* Of all Americans, 55% do volunteer work.

Data: The Nonprofit Almanac in Brief

19. *Sports and Calories.* Sports fans consume 5% fewer calories when their team wins.

Data: Southwest: The Magazine, March 2015, p. 20

20. If a NASCAR driver loses more than 3% body weight in sweat and does not replace those fluids, the driver's focus and reflexes start declining.

Data: CNN.com

21. Approximately 3.2% of U.S. adults are vegetarians.

Data: Vegetarian Times

22. *Gold.* Gold that is marked 10K is 41.6% gold.

23. *Energy Use.* Using landscape to shade a home's outdoor air-conditioning unit can reduce cooling costs by 10%.

Data: U.S. Department of Energy

24. Semisweet dark chocolate contains about 60% chocolate liquor.

Convert to decimal notation.

25. 6.25%   **26.** 8.375%

27. 0.2%   **28.** 0.8%

29. 175%   **30.** 250%

Convert the decimal notation in each sentence to percent notation.

31. *Video Games.* In a survey of 341 adult video game players, 0.79 never played educational games.

Data: Phan, Mikki H. (2011). "Video Gaming Trends: Action/ Adventure Games Are Most Popular." Software Usability Research Laboratory, usabilitynews.org

32. *Baseball Fans.* Baseball is the second most popular sport in the United States, with 0.17 of the adult population saying it is their favorite sport.

Data: Sports Business Daily, 1/25/11

33. *National Parks.* The four largest U.S. National Parks are in Alaska and make up 0.047 of the total area of U.S. protected areas.

Data: National Park Service

34. *Foreign Student Enrollment.* Of all foreign students studying in the United States, 0.019 are from Vietnam.

Data: iie.org

35. *Composition of the Sun.* The sun is 0.7 hydrogen.

36. *Jupiter's Atmosphere.* The atmosphere of Jupiter is 0.1 helium.

Convert to percent notation.

37. 0.0009   **38.** 0.0056

39. 1.06   **40.** 1.08

41. $\frac{3}{5}$ **42.** $\frac{3}{2}$

43. $\frac{8}{25}$ **44.** $\frac{5}{8}$

B. Solving Percent Problems

Solve.

45. What percent of 76 is 19?

46. What percent of 125 is 30?

47. 14 is 30% of what number?

48. 54 is 24% of what number?

49. 0.3 is 12% of what number?

50. 7 is 175% of what number?

51. What number is 1% of one million?

52. What number is 35% of 240?

53. What percent of 60 is 75?

Aha! **54.** What percent of 70 is 70?

55. What is 2% of 40?

56. What is 40% of 2?

Aha! **57.** 25 is what percent of 50?

58. 0.8 is 2% of what number?

59. What percent of 69 is 23?

60. What percent of 40 is 9?

Household Pets. In 2014, approximately 95.6 million cats lived as pets in the United States. The following circle graph shows the sources from which households obtain their pets. In each of Exercises 61–64, estimate how many cats came from the indicated source.

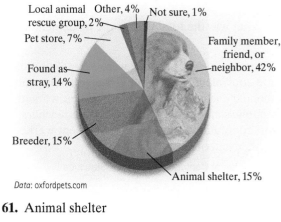

Pet Sources

Local animal rescue group, 2%
Other, 4%
Not sure, 1%
Pet store, 7%
Family member, friend, or neighbor, 42%
Found as stray, 14%
Breeder, 15%
Animal shelter, 15%

Data: oxfordpets.com

61. Animal shelter

62. Found as stray

63. Local animal rescue group

64. Family member, friend, or neighbor

65. *College Graduation.* To obtain his bachelor's degree in nursing, Cody must complete 125 credit hours of instruction. If he has completed 60% of his requirement, how many credits did Cody complete?

66. *College Graduation.* To obtain her bachelor's degree in journalism, Addie must complete 125 credit hours of instruction. If 20% of Addie's credit hours remain to be completed, how many credits does she still need to take?

67. *Batting Average.* In a recent season, Andrew McCutchen of the Pittsburgh Pirates had 172 hits. His batting average was 0.314, or 31.4%. That is, of the total number of at-bats, 31.4% were hits. How many at-bats did he have?

Data: Major League Baseball

68. *Pass Completions.* In a recent season, Peyton Manning of the Denver Broncos completed 395 passes. This was 66.2% of his attempts. How many passes did he attempt?

Data: National Football League

69. *Tipping.* Trent left a $4 tip for a meal that cost $25.

   a) What percent of the cost of the meal was the tip?
   b) What was the total cost of the meal including the tip?

70. *Tipping.* Selena left a $12.76 tip for a meal that cost $58.

   a) What percent of the cost of the meal was the tip?
   b) What was the total cost of the meal including the tip?

71. *Education.* In 2014, there were 3.5 million full-time equivalent teachers in U.S. elementary and secondary schools. Of these, 3.1 million worked in public schools. What percent of the teachers worked in public schools? What percent worked in other schools?

Data: National Center for Education Statistics

72. *Education.* In 2014, there were 54.8 million students enrolled in U.S. elementary and secondary schools. Of these, 49.8 million were enrolled in public schools. What percent of the students were in public schools? What percent were in other schools?

Data: National Center for Education Statistics

73. *Student Loans.* Glenn takes out a student loan for $2400. After a year, Glenn decides to pay off the interest, which is 6.80% of $2400. How much will he pay?

74. *Student Loans.* To finance her community college education, LaTonya takes out a student loan for $3500. After a year, LaTonya decides to pay off the interest, which is 4.50% of $3500. How much will she pay?

75. *Infant Health.* In a study of 300 pregnant women with "good-to-excellent" diets, 95% had babies in good or excellent health. How many women in this group had babies in good or excellent health?

76. *Infant Health.* In a study of 300 pregnant women with "poor" diets, 8% had babies in good or excellent health. How many women in this group had babies in good or excellent health?

77. *Cost of Self-Employment.* Because of additional taxes and fewer benefits, it has been estimated that a self-employed person must earn 20% more than a non–self-employed person performing the same task(s). If Tia earns $16 per hour working for Village Copy, how much would she need to earn on her own for a comparable income?

78. Refer to Exercise 77. Rik earns $18 per hour working for Round Edge stairbuilders. How much would Rik need to earn on his own for a comparable income?

79. *Budget Overruns.* The "Big Dig" in Boston, also known as the Central Artery/Tunnel Project, has been labeled the most expensive highway project in U.S. history. The original cost estimate for the project was $2.6 billion, and by the time the last major portion of the project was opened to vehicles, the cost was $12 billion over the original estimate. By what percent did the actual cost exceed the initial estimate?

Data: Associated Press, Dec. 20, 2003, cited by msnbc.msn.com: *Boston's "Big Dig" Opens to Public*; Construction Management Schools

80. *Swimming Records.* On December 18, 2009, Cesar Cielo Filho of Brazil set a world record by swimming 50 m on a long course (50-m pool) in 20.91 sec. This record is 0.65 sec longer than the record on the 50-m short course (25-m pool), which was set by Florent Manadou of France on December 5, 2014. By what percent is the record for a short course faster than the record for a long course?

Data: usaswimming.org

81. *Body Fat.* One author of this text exercises regularly at a local YMCA that recently offered a body-fat percentage test to its members. The device used measures the passage of a very low voltage of electricity through the body. The author's body-fat percentage was found to be 16.5%, and he weighs 191 lb. What part of his body weight, in pounds, is fat?

82. *Areas of Alaska and Arizona.* The area of Arizona is 19% of the area of Alaska. The area of Alaska is 586,400 mi². What is the area of Arizona?

83. *Spam e-Mail.* About 265 billion of the 294 billion e-mails sent each day are spam and viruses. What percent of e-mails are spam and viruses?

Data: Radicati Group

84. *Kissing and Colds.* In a medical study, it was determined that if 800 people kiss someone else who has a cold, only 56 will actually catch the cold. What percent is this?

85. *Calorie Content.* An 8-oz serving of Ocean Spray® Cranberry Juice Cocktail contains 120 calories. This is 2400% of the number of calories in an 8-oz serving of Diet Ocean Spray® Cranberry Juice Drink. How many calories are in an 8-oz serving of Diet Cranberry Juice Drink?

86. *Sodium Content.* Each serving of Planters® Lightly Salted Peanuts contains 95 mg of sodium. This is 50% of the sodium content in each serving of Planters® Dry Roasted Peanuts. How many milligrams of sodium are in a serving of Dry Roasted Peanuts?

87. How is the use of statistics in the following examples misleading?

a) A business explaining new restrictions on sick leave cited a recent survey indicating that 40% of all sick days were taken on Monday or Friday.

b) An advertisement urging summer installation of a security system quoted FBI statistics stating that over 26% of home burglaries occur between Memorial Day and Labor Day.

88. If Julian leaves a $12 tip for a $90 dinner, is he being generous, stingy, or neither? Explain.

Skill Review

89. Find the opposite of $-\frac{1}{3}$. [1.6]

90. Find the reciprocal of $-\frac{1}{3}$. [1.7]

91. Find $-(-x)$ for $x = -12$. [1.6]

92. Simplify: $(-3x)^2$. [1.8]

Synthesis

93. Campus Bookbuyers pays $30 for a book and sells it for $60. Is this a 100% markup or a 50% markup? Explain.

94. Recently in Brazil, women earned 30% less than men for the same work. As a publicity stunt to draw attention to this, a restaurant decided to charge male customers 30% more than women for food. If men were charged 30% more on all of their daily purchases, would this offset the inequality in pay? Why or why not?

95. The community of Bardville has 1332 left-handed females. If 48% of the community is female and 15% of all females are left-handed, how many people are in the community?

96. It has been determined that at the age of 10, a girl has reached 84.4% of her final adult height. Dana is 4 ft 8 in. at the age of 10. Predict her final adult height.

97. It has been determined that at the age of 15, a boy has reached 96.1% of his final adult height. Jaraan is 6 ft 4 in. at the age of 15. Predict his final adult height.

98. *Dropout Rate.* Between 2010 and 2012, the high school dropout rate in the United States decreased from 74 students per thousand to 66 students per thousand. Calculate the average percent by which the dropout rate decreased each year and use that percentage to estimate dropout rates for the United States in 2009 and in 2011.

Data: National Center for Education Statistics

99. *Photography.* A 6-in. by 8-in. photo is framed using a mat meant for a 5-in. by 7-in. photo. What percentage of the photo will be hidden by the mat?

100. Would it be better to receive a 5% raise and then, a year later, an 8% raise or the other way around? Why?

Aha!

101. Jorge is in the 30% tax bracket. This means that 30¢ of each dollar earned goes to taxes. Which would cost him the least: contributing $50 that is tax-deductible or contributing $40 that is not tax-deductible? Explain.

102. *Research.* If you have student loans, find the interest rates on those loans, whether the rates are fixed or variable, and how the interest is calculated. How much interest will you owe when the first payment is due?

YOUR TURN ANSWERS: SECTION 2.4

1. 1.2 **2.** 0.04 **3.** 37% **4.** 0.9 **5.** 65
6. 20% **7.** 64 stores **8.** 840 students **9.** 10%

Quick Quiz: Sections 2.1–2.4

Solve and check.

1. $6x - (5 - x) = 2x - 5$ [2.2]

2. $\dfrac{x}{2} - 10 = \dfrac{1}{3}$ [2.2]

3. Solve for p: $4p + mp = T$. [2.3]

4. Convert to decimal notation: 1.2%. [2.4]

5. 5 is 20% of what number? [2.4]

Prepare to Move On

Translate to an algebraic expression or equation. [1.1]

1. Twice the length plus twice the width

2. 5% of $180

3. The product of 10 and half of a

4. 10 more than three times a number

5. A board's width is 2 in. less than its length.

6. A number is four times as large as a second number.

2.5 Problem Solving

A. The Five Steps for Problem Solving   **B.** Percent Increase and Percent Decrease

Probably the most important use of algebra is as a tool for problem solving. In this section, we develop a five-step problem-solving approach that is used throughout the remainder of the text.

A. The Five Steps for Problem Solving

FIVE STEPS FOR PROBLEM SOLVING IN ALGEBRA

1. *Familiarize* yourself with the problem.
2. *Translate* to mathematical language. (This often means writing an equation.)
3. *Carry out* some mathematical manipulation. (This often means *solving* an equation.)
4. *Check* your possible answer in the original problem.
5. *State* the answer clearly, using a complete sentence.

Of the five steps, the most important is probably the first one: becoming familiar with the problem. Here are some hints for familiarization.

TO BECOME FAMILIAR WITH A PROBLEM

1. Read the problem carefully. Try to visualize the problem.
2. Reread the problem, perhaps aloud. Make sure you understand all words as well as any symbols or abbreviations.
3. List the information given and the question(s) to be answered. Choose a variable (or variables) to represent the unknown and specify exactly what the variable represents. For example, let L = length in centimeters, d = distance in miles, and so on.
4. Look for similarities between the problem and other problems you have already solved. Ask yourself what type of problem this is.
5. Find more information. Look up a formula in a book, at a library, or online. Consult a reference librarian or an expert in the field.
6. Make a table that organizes the information you have available. Look for patterns that may help in the translation.
7. Make a drawing and label it with known and unknown information, using specific units if given.
8. Think of a possible answer and check your guess. Note the manner in which your guess is checked.

EXAMPLE 1 *Hiking.* In 1957 at the age of 69, Emma "Grandma" Gatewood became the first woman to hike solo all 2100 mi of the Appalachian Trail—from Springer Mountain, Georgia, to Mount Katahdin, Maine. Gatewood repeated the feat in 1960 and again in 1963, becoming the first person to hike the trail three times. When Gatewood stood atop Big Walker Mountain, Virginia, she was three times as far from the northern end of the trail as from the southern end. At that point, how far was she from each end of the trail?

SOLUTION

1. **Familiarize.** We first note that the total distance that Gatewood hiked was 2100 mi. To gain some familiarity, let's suppose that Gatewood stood 600 mi from Springer Mountain. Three times 600 mi is 1800 mi. Since 600 mi + 1800 mi = 2400 mi and 2400 mi > 2100 mi, we see that our guess is too large. Rather than guess again, we let

   s = the distance, in miles, to the southern end

and

   $3s$ = the distance, in miles, to the northern end.

(We could also let n = the distance to the northern end, in miles, and $\frac{1}{3}n$ = the distance to the southern end, in miles.) It may be helpful to make a drawing.

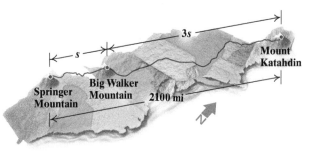

1. In 2005, Ken Looi of New Zealand covered 235.3 mi in 24 hr on his unicycle. After 8 hr, he was approximately twice as far from the finish line as he was from the start. How far had he traveled?

Data: *Guinness World Records*

2. Translate. From the drawing, we see that the lengths of the two parts of the trail must add up to 2100 mi. This leads to our translation.

Rewording: Distance to southern end plus distance to northern end is 2100 mi.

Translating: s + $3s$ = 2100

3. Carry out. We solve the equation:

$$s + 3s = 2100$$
$$4s = 2100 \qquad \text{Combining like terms}$$
$$s = 525. \qquad \text{Dividing both sides by 4}$$

4. Check. As predicted in the *Familiarize* step, s is less than 600 mi. If $s = 525$ mi, then $3s = 1575$ mi. Since 525 mi + 1575 mi = 2100 mi, we have a check.

5. State. Atop Big Walker Mountain, Gatewood stood 525 mi from Springer Mountain and 1575 mi from Mount Katahdin.

YOUR TURN

We can represent several *consecutive integers* using one variable.

	Examples	Algebraic Representation
Consecutive Integers (are 1 unit apart)	16, 17, 18, 19, 20; −31, −30, −29, −28	x, $x + 1$, $x + 2$, and so on
Consecutive Even Integers (are 2 units apart)	16, 18, 20, 22, 24; −52, −50, −48, −46	x, $x + 2$, $x + 4$, and so on
Consecutive Odd Integers (are 2 units apart)	21, 23, 25, 27, 29; −71, −69, −67, −65	x, $x + 2$, $x + 4$, and so on

EXAMPLE 2 *Interstate Mile Markers.* U.S. interstate highways post numbered markers at every mile to indicate location in case of an emergency. The sum of two consecutive mile markers on I-70 in Kansas is 559. Find the numbers on the markers.

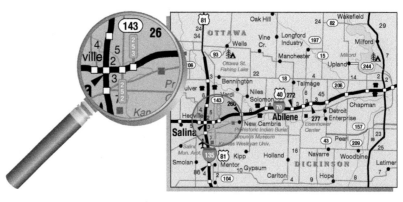

x	$x + 1$	Sum of x and $x + 1$
114	115	229
252	253	505
302	303	605

SOLUTION

1. Familiarize. The numbers on the mile markers are consecutive positive integers. Thus if we let $x =$ the smaller number, then $x + 1 =$ the larger number.

To become familiar with the problem, we can make a table, as shown at left. First, we guess a value for x; then we find $x + 1$. Finally, we add the two numbers and check the sum.

Study Skills

Set Reasonable Expectations

Do not be surprised if your success rate drops as you work through the exercises in this section. *This is normal.* Your success rate will increase as you gain experience with these types of problems and use some of the study skills already listed.

2. The sum of two consecutive mile markers on I-70 in Indiana is 337. (See Example 2.) Find the numbers on the markers.

From the table, we see that the first marker will be between 252 and 302. We could continue guessing and solve the problem this way, but let's develop our algebra skills.

2. **Translate.** We reword the problem and translate as follows.

   Rewording: First integer plus next integer is 559.

   Translating: x + $(x + 1)$ = 559

3. **Carry out.** We solve the equation:

$$x + (x + 1) = 559$$
$$2x + 1 = 559 \qquad \text{Using an associative law and combining like terms}$$
$$2x = 558 \qquad \text{Subtracting 1 from both sides}$$
$$x = 279. \qquad \text{Dividing both sides by 2}$$

If x is 279, then $x + 1$ is 280.

3. **Check.** Our possible answer is mile markers 279 and 280. These are consecutive positive integers and $279 + 280 = 559$, so the answers check.

4. **State.** The mile markers are 279 and 280.

YOUR TURN

EXAMPLE 3 *Photography.* Marissa has a budget of $500 for photographs for her company's new website. Several friends have recommended Fine Taste Photography for the job. Marissa has learned that Fine Taste charges $125 for a day session plus a fee for each image used. From her friends, she has gathered the following data. From the information given, how many images can Marissa buy without exceeding her budget?

Number of Images Used	Fee for Images Used
2	$ 65
20	650
24	780

SOLUTION

1. **Familiarize.** To determine how much each image costs, we divide each fee by the number of images. We find that the cost is $32.50 per image:

$$\$65 \div 2 = \$32.50,$$
$$\$650 \div 20 = \$32.50,$$
$$\$780 \div 24 = \$32.50.$$

Suppose that Marissa buys 14 images. Then the total cost is the fee for the session plus the fee for the images, or

Session fee plus cost per image times number of images,

$125 + $32.50 · 14

which is $580. Since this amount is greater than the $500 budgeted, we know that our guess of 14 images is too large. However, the process we used to check our guess can be used to translate the problem to mathematical language. We let n = the number of images that Marissa can buy for a total cost of $500.

2. Translate. We reword the problem and translate as follows.

Rewording: Session fee plus image fee is $500.

Translating: $125 \quad + \quad 32.50n \quad = \quad 500$

3. Carry out. We solve the equation:

$$125 + 32.50n = 500$$
$$32.50n = 375 \qquad \text{Subtracting 125 from both sides}$$
$$n \approx 11.54. \qquad \text{Dividing both sides by 32.50}$$

Since fraction parts of images do not make sense in the problem, we need to round the answer to a whole number. In this case, we must round *down* to avoid going over the budget. We have a possible solution of 11 images.

4. Check. The fee for 11 images is $11(\$32.50) = \$357.50$. The total cost is then $\$357.50 + \$125 = \$482.50$. If Marissa spends this amount, she will have $17.50 left, which is not enough to buy another image. Also note that our answer is fewer than 14 images, as we expected from the *Familiarize* step.

5. State. Marissa can buy 11 images from Fine Taste Photography without exceeding her budget.

3. Refer to Example 3. If Marissa's budget is increased to $1000, how many images can she buy without exceeding her budget?

YOUR TURN

EXAMPLE 4 *Perimeter of an NBA Court.* The perimeter of an NBA basketball court is 288 ft. The length is 44 ft longer than the width. Find the dimensions of the court.

Data: National Basketball Association

SOLUTION

1. Familiarize. Recall that the perimeter of a rectangle is twice the length plus twice the width. Suppose that the court were 30 ft wide. The length would then be $30 + 44$, or 74 ft, and the perimeter would be $2 \cdot 30$ ft $+ 2 \cdot 74$ ft, or 208 ft. This shows that in order for the perimeter to be 288 ft, the width must exceed 30 ft. Instead of guessing again, we let $w =$ the width of the court, in feet. Since the court is "44 ft longer than it is wide," we let $w + 44 =$ the length of the court, in feet.

Student Notes

Example 4 states that "the length is 44 ft longer than the width." Because "width" is used in the description of "length," we say that length is described *in terms of* width. When one quantity is described in terms of a second quantity, it is generally best to let the variable represent the second quantity.

Try to develop the habit of writing what each variable represents before writing an equation. In Example 4, you might write

width (in feet) $= w$,
length (in feet) $= w + 44$.

2. Translate. To translate, we use $w + 44$ as the length and 288 as the perimeter. To double the length, $w + 44$, we must use parentheses.

Rewording: Twice the length plus twice the width is 288 ft.

Translating: $2(w + 44) \quad + \quad 2w \quad = \quad 288$

3. Carry out. We solve the equation:

$$2(w + 44) + 2w = 288$$
$$2w + 88 + 2w = 288 \quad \text{Using the distributive law}$$
$$4w + 88 = 288 \quad \text{Combining like terms}$$
$$4w = 200$$
$$w = 50.$$

4. The perimeter of a standard high school basketball court is 268 ft. The length is 34 ft longer than the width. Find the dimensions of the court.

Data: Indiana High School Athletic Association

The dimensions appear to be $w = 50$ ft, and $l = w + 44 = 94$ ft.

4. Check. If the width is 50 ft and the length is 94 ft, then the court is 44 ft longer than it is wide. The perimeter is $2(50\,\text{ft}) + 2(94\,\text{ft}) = 100\,\text{ft} + 188\,\text{ft}$, or 288 ft, as specified. We have a check.

5. State. An NBA court is 50 ft wide and 94 ft long.

YOUR TURN

> *CAUTION!*   Always be sure to answer the original question completely. For instance, in Example 1 we are asked to find *two* numbers: the distances from the hiker to *each* end of the trail. Similarly, in Example 4 we are asked to find two dimensions, not just the width. Be sure to label each answer with the proper unit.

EXAMPLE 5 *Cross Section of a Roof.*   In a triangular gable end of a roof, the angle of the peak is twice as large as the angle on the back side of the house. The measure of the angle on the front side is 20° greater than the angle on the back side. How large are the angles?

SOLUTION

1. Familiarize. First, we make a drawing. Because the peak angle and the front angle are described in terms of the back angle, we label the measure of the back angle x. Then the measure of the front angle is $x + 20$, and the measure of the peak angle is $2x$.

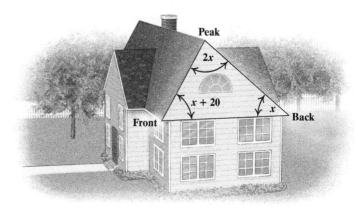

Student Notes

You may be expected to recall material that you have learned in an earlier course. If you have forgotten a formula, refresh your memory by consulting a textbook or a website.

2. Translate. To translate, we need to recall that the sum of the measures of the angles in any triangle is 180°.

Rewording:	Measure of back angle	+	measure of front angle	+	measure of peak angle	is	180°.
Translating:	x	+	$(x + 20)$	+	$2x$	=	180

3. Carry out. We solve:

$$x + (x + 20) + 2x = 180$$

$$4x + 20 = 180 \qquad \text{Combining like terms}$$

$$4x = 160 \qquad \text{Subtracting 20 from both sides}$$

$$x = 40. \qquad \text{Dividing both sides by 4}$$

The measures for the angles appear to be:

Back angle: $x = 40°,$

Front angle: $x + 20 = 40 + 20 = 60°,$

Peak angle: $2x = 2(40) = 80°.$

4. Check. Consider $40°, 60°,$ and $80°,$ as listed above. The measure of the front angle is $20°$ greater than the measure of the back angle, the measure of the peak angle is twice the measure of the back angle, and the sum, $40° + 60° + 80°,$ is $180°.$ These numbers check.

5. State. The measures of the angles are $40°, 60°,$ and $80°.$

5. The second angle of a triangle is 30° more than the first. The third angle is half as large as the first. Find the measures of the angles.

YOUR TURN

When working with motion problems, we often use tables, as well as the following motion formula.

MOTION FORMULA

$$d = r \cdot t$$

$$(\text{Distance} = \text{Rate} \times \text{Time})$$

EXAMPLE 6 *Motion.* Sharon drove for 3 hr on a highway and then for 1 hr on a side road. Her speed on the highway was 20 mph faster than her speed on the side road. If she traveled a total of 220 mi, how fast did she travel on the side road?

SOLUTION

1. **Familiarize.** After reading the problem carefully, we see that we are asked to find the rate of travel on the side road. Because Sharon's highway speed is described in terms of her speed on the side road, we let $x =$ Sharon's speed on the side road, in miles per hour.

   Since we are dealing with motion, we will use the motion formula $d = r \cdot t.$ We use the variables in this formula as headings for three columns in a table. The rows of the table correspond to the highway and the side road. We know the times traveled for each type of driving, and we have defined x to represent the speed on the side road. Thus we can fill in those entries in the table.

Road Type	d =	r	$\cdot$ t
Highway			3 hr
Side Road		x mph	1 hr

To fill in the remaining entries using the variable x, we know that Sharon's speed on the highway was 20 mph faster than her speed on the side road:

Highway speed $= (x + 20)$ mph.

Then, since $d = r \cdot t$, we multiply to find each distance:

Highway distance $= (x + 20)(3)$ mi;

Side-road distance $= x(1)$ mi.

Road Type	d	$=$	r	$\cdot$	t
Highway	$(x + 20)(3)$ mi		$(x + 20)$ mph		3 hr
Side Road	$x(1)$ mi		x mph		1 hr

2. **Translate.** We know that Sharon traveled a total of 220 mi. This gives us the translation to an equation.

Rewording: Highway distance plus side-road distance is total distance.

Translating: $(x + 20)(3)$ $+$ $x(1)$ $=$ 220

3. **Carry out.** We now have an equation to solve:

$$(x + 20)(3) + x(1) = 220$$
$$3x + 60 + x = 220 \qquad \text{Using the distributive law}$$
$$4x + 60 = 220 \qquad \text{Combining like terms}$$
$$4x = 160 \qquad \text{Subtracting 60 from both sides}$$
$$x = 40. \qquad \text{Dividing both sides by 4}$$

4. **Check.** Since x represents Sharon's speed on the side road, we have the following.

Road Type	Speed (Rate)	Time	Distance
Highway	$40 + 20 = 60$ mph	3 hr	$60(3) = 180$ mi
Side Road	40 mph	1 hr	$40(1) = 40$ mi

The total distance traveled is 40 mi + 180 mi, or 220 mi. The answer checks.

5. **State.** Sharon's speed on the side road was 40 mph.

↩ YOUR TURN

6. Erik drove to his first sales appointment at 60 mph. From there, he drove to his second appointment at 40 mph. It took him twice as long to drive to his first appointment as it did to his second. If he traveled a total of 80 mi, how long did it take him to drive from his first appointment to his second?

PROBLEM-SOLVING TIPS

1. The more problems you solve, the more your skills will improve.
2. Look for patterns when solving problems.
3. Clearly define variables before translating to an equation.
4. Consider the dimensions of the variables and constants in the equation. The variables that represent length should all be in the same unit, those that represent money should all be in dollars or all in cents, and so on.
5. Make sure that you have completely answered the original question using the appropriate units.

B. Percent Increase and Percent Decrease

Whenever a percent of a quantity is calculated and then added to or subtracted from that quantity, the result is a percent increase or a percent decrease. Some examples follow.

Percent Increase	Percent Decrease
Sales tax	Reduction in crimes
Salary increase	Lowered sale price
Population growth	Body weight loss

For each example of percent increase or percent decrease, there are three amounts:

The original amount,

The amount of increase or decrease,

The new amount.

Identifying each amount is important in solving percent increase or percent decrease problems. The following relationships exist among these amounts.

> The amount of increase is added to the original amount to get the new amount.
>
> The amount of decrease is subtracted from the original amount to get the new amount.
>
> The percent increase or percent decrease is calculated on the basis of the original amount.

EXAMPLE 7 *Fitness.* Through diet and exercise, Gabrielle's weight decreased from 150 lb to 145 lb. What was the percent decrease in her body weight?

SOLUTION We begin by identifying the three amounts. Then we recall that the percent of decrease is calculated on the basis of the original amount.

Student Notes

Since the original amount minus the amount of decrease is the new amount, another translation for Example 7 is

$$150 - p \cdot 150 = 145.$$

7. After he had recovered from a prolonged illness, Victor's weight increased from 140 lb to 150 lb. What was the percent increase in his body weight?

Original Weight	150 lb
New Weight	145 lb
Amount of Weight Lost	5 lb

Original amount − new amount
= 150 lb − 145 lb

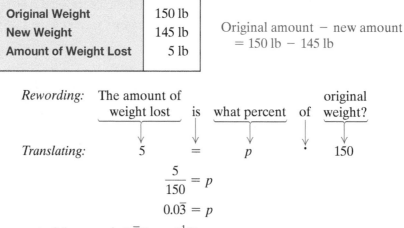

Rewording: The amount of weight lost is what percent of original weight?

Translating: 5 = p · 150

$$\frac{5}{150} = p$$

$$0.0\overline{3} = p$$

The percent of decrease is $3.\overline{3}\%$, or $3\frac{1}{3}\%$.

↪ YOUR TURN

EXAMPLE 8 *Charitable Organizations.* The tax-exempt Hope Food Pantry received a bill of $242.65 for food storage bags. The bill incorrectly included sales tax of 5.5%. How much does the organization actually owe?

↪ Check Your UNDERSTANDING

Complete the translation of each statement.

1. The sum of two consecutive odd numbers is 32.

   If x = the smaller number, then _____ = the larger number.

   The translation is
   $x +$ _____ $=$ __.

2. Including a 6% sales tax, Jaykob paid $36.57 for a sweatshirt.

   If x = the original price, then _____ = the sales tax. The translation is
   $x +$ _____ $=$ _____.

---

8. Casey is selling his collection of Native American pottery at an auction. He wants to be left with $1150 after paying a seller's premium of 8% on the final bid (hammer price) for the collection. What must the hammer price be in order for him to clear $1150?

SOLUTION

1. **Familiarize.** Sales tax is a percent increase and is calculated on the original price of an item. In this example, we are told the amount after the sales tax has been added, so we let x = the original price of the food storage bags. We can organize the information in a table.

Original Price (in dollars)	x
Amount of Sales Tax	5.5% of x, or $0.055x$
Price Including Sales Tax	$242.65

2. **Translate.** We reword the problem and translate as follows.

   Rewording: Original price plus sales tax is total price.

   Translating: x $+$ $0.055x$ $=$ 242.65

3. **Carry out.** We solve the equation:

$$x + 0.055x = 242.65$$
$$1x + 0.055x = 242.65$$
$$1.055x = 242.65 \quad \text{Combining like terms}$$
$$x = \frac{242.65}{1.055} \quad \text{Dividing both sides by 1.055}$$
$$x = 230.$$

4. **Check.** To check, we first find 5.5% of $230:

   5.5% of $230 = 0.055($230$) = $12.65.$ This is the amount of the sales tax.

   Next, we add the sales tax to find the total amount billed:

   $230 + $12.65 = $242.65.

   Since this is the amount that the organization was billed, our answer checks.

5. **State.** The organization owes $230.

↪ YOUR TURN

2.5 EXERCISE SET

↪ Vocabulary and Reading Check

1. List the steps in the five-step problem-solving approach in the correct order.

   1) _____ Carry out.

   2) _____ Check.

   3) _____ Familiarize.

   4) _____ State.

   5) _____ Translate.

Each of the following was part of a step in the solution to at least one example in this section. In the blank, write the name of the step during which each was done. Select from the five steps listed above.

2. _____ Solve an equation.

3. _____ Write the answer clearly.

4. _____ Make and check a guess.

5. _____ Reword the problem.

6. _____ Make a table.

7. _____ Recall a formula.

8. _____ Compare the answer with a prediction from an earlier step.

A. The Five Steps for Problem Solving

Solve. Even though you might find the answer quickly in another way, use the five-step approach in order to increase your skill of problem solving.

9. Three less than twice a number is 19. What is the number?

10. Two fewer than ten times a number is 78. What is the number?

11. Five times the sum of 3 and twice some number is 70. What is the number?

12. Twice the sum of 4 and three times some number is 34. What is the number?

13. *Kayaking.* On May 8, 2010, a team of kayakers set a record time of 5 hr 19 min 17 sec to kayak across the length of Loch Ness, Scotland, UK, a distance of 20.5 mi. When the kayakers were four times as far from the starting point as from the finish, how many miles had they traveled?

Data: *Guinness World Records*

14. *Sled-Dog Racing.* The Iditarod sled-dog race extends for 1049 mi from Anchorage to Nome. If a musher is twice as far from Anchorage as from Nome, how many miles has the musher traveled?

15. *Indy Car Racing.* In 2015, Juan Pablo Montoya won his second Indy 500, with a time of 3:05:57 for the 500-mi race. At one point, Montoya was 20 mi closer to the finish than to the start. How far had Montoya traveled at that point?

16. *NASCAR Racing.* In 2015, Jimmie Johnson won his fifth Autism Speaks 400 with a finish time of 3:23:16 for the 400-mi race. At one point, Johnson was 80 mi closer to the finish than to the start. How far had Johnson traveled at that point?

17. *Apartment Numbers.* The apartments in Erica's apartment house are consecutively numbered on each floor. The sum of her number and her next-door neighbor's number is 2409. What are the two numbers?

18. *Apartment Numbers.* The apartments in Brian's apartment house are numbered consecutively on each floor. The sum of his number and his next-door neighbor's number is 1419. What are the two numbers?

19. *Street Addresses.* The houses on the west side of Lincoln Avenue are consecutive odd numbers. Sam and Colleen are next-door neighbors and the sum of their house numbers is 572. Find their house numbers.

20. *Street Addresses.* The houses on the south side of Elm Street are consecutive even numbers. Wanda and Larry are next-door neighbors and the sum of their house numbers is 794. Find their house numbers.

21. The sum of three consecutive page numbers is 99. Find the numbers.

22. The sum of three consecutive page numbers is 60. Find the numbers.

23. *Longest Marriage.* As half of the world's longest-married couple, the woman was 2 years younger than her husband. Together, their ages totaled 206 years. How old were the man and the woman?

Data: *Guinness World Records*

24. *Oldest Bride.* The world's oldest bride was 19 years older than her groom. Together, their ages totaled 185 years. How old were the bride and the groom?

Data: *Guinness World Records*

25. *Movies.* Two of Hollywood's biggest box office flops were *The 13th Warrior* and *Mars Needs Moms.* Together, they lost $209.3 million, and *Mars Needs Moms* lost $12.7 million more than *The 13th Warrior* did. How much did each movie lose?

Data: xfinity.comcast.net

26. *e-Mail.* In a recent year, approximately 196.3 billion e-mails were sent each day. There were 21.1 billion more business e-mails sent each day than consumer e-mails. How many business e-mails and how many consumer e-mails were sent each day?

Data: radicati.com

27. *Page Numbers.* A book is lying open on a desk. The sum of the page numbers showing is 281. What are the page numbers?

28. *Perimeter of a Triangle.* The perimeter of a triangle is 195 mm. If the lengths of the sides are consecutive odd integers, find the length of each side.

29. A rectangular community garden is to be enclosed with 92 m of fencing. In order to allow for compost storage, the garden will be 4 m longer than it is wide. Determine the dimensions of the garden.

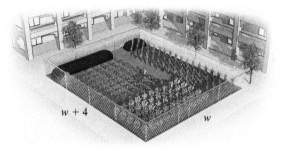

$w + 4$ w

30. *Hancock Building Dimensions.* The top of the John Hancock Building in Chicago is a rectangle whose length is 60 ft more than the width. The perimeter is 520 ft. Find the width and the length of the rectangle. Find the area of the rectangle.

31. *Two-by-four.* The perimeter of a cross section of a "two-by-four" piece of lumber is 10 in. The length is 2 in. longer than the width. Find the actual dimensions of the cross section of a two-by-four.

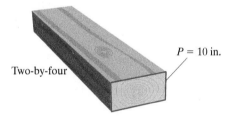

Two-by-four $P = 10$ in.

32. *Standard Billboard Sign.* A standard rectangular highway billboard sign has a perimeter of 124 ft. The length is 6 ft more than three times the width. Find the dimensions of the sign.

33. *Angles of a Triangle.* The second angle of an architect's triangle is three times as large as the first. The third angle is 30° more than the first. Find the measure of each angle.

34. *Angles of a Triangle.* The second angle of a triangular garden is four times as large as the first. The third angle is 45° less than the sum of the other two angles. Find the measure of each angle.

35. *Angles of a Triangle.* The second angle of a triangular kite is four times as large as the first. The third angle is 5° more than the sum of the other two angles. Find the measure of the second angle.

36. *Angles of a Triangle.* The second angle of a triangular building lot is three times as large as the first. The third angle is 10° more than the sum of the other two angles. Find the measure of the third angle.

37. *Rocket Sections.* A rocket is divided into three sections: the payload and navigation section in the top, the fuel section in the middle, and the rocket engine section in the bottom. The top section is one-sixth the length of the bottom section. The middle section is one-half the length of the bottom section. The total length is 240 ft. Find the length of each section.

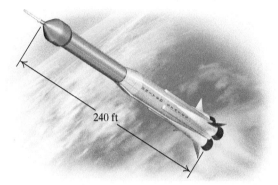

240 ft

38. *Gourmet Sandwiches.* Jenny, Demi, and Joel buy an 18-in. long gourmet sandwich and take it back to their apartment. Since they have different appetites, Jenny cuts the sandwich so that Demi gets half of what Jenny gets and Joel gets three-fourths of what Jenny gets. Find the length of each person's sandwich.

39. *Boating.* Gaston paddled for 3 hr upstream and then for 2 hr downstream. His speed upstream was 10 mph slower than his speed downstream. If he traveled a total of 30 mi, how fast did he travel downstream?

40. *Commuting.* Bjorn rode for $\frac{1}{2}$ hr on the train and then for $\frac{1}{3}$ hr on the bus. The speed of the train was 50 km/h faster than the speed of the bus. If he traveled a total of 37.5 km, what was the speed of the bus?

Aha! **41.** *Long-Distance Running.* Phoebe runs at 12 km/h and walks at 5 km/h. One afternoon, she ran and walked a total of 17 km. If she ran for the same length of time as she walked, for how long did she run?

42. *Driving.* Theodora drove on the Blue Ridge Parkway at 40 mph and then on an interstate highway at 70 mph. She drove three times as long on the Blue Ridge Parkway as she did on the interstate. If she drove a total of 285 mi, for how long did she drive on the interstate?

B. Percent Increase and Percent Decrease

43. *Endangered Species.* The population of gray wolves in Wisconsin increased from 570 in 2007 to 660 in 2014. Find the percent increase in the wolf population.

Data: Wisconsin Department of Natural Resources and U.S. Fish & Wildlife Service

44. *Monarch Butterflies.* In 2011, monarch butterflies overwintering in Mexico covered 4.02 hectares. This area decreased to 1.13 hectares in 2015. Find the percent decrease in the area occupied by overwintering monarch butterflies.

Data: monarchjointventure.org

45. *Town Budget.* The town of Maxwell decreased its annual budget from $1,600,000 to $1,400,000. Find the percent decrease in the budget.

46. *Employment.* The number of jobs in a state increased from 1,800,000 to 1,816,200. Find the percent increase in the number of jobs.

47. *Tax-Exempt Organizations.* A tax-exempt hospital received a bill of $1310.75 for linens. The bill incorrectly included sales tax of 7%. How much does the hospital owe?

48. *Automobile Tax.* Miles paid $5824, including 4% sales tax, for an out-of-state purchase of a car. In order to calculate the amount of sales tax he owes in his state, he must first determine the price of the car without sales tax. How much did the car cost before sales tax?

49. *Income Tax Deductions.* Jinney wants to deduct the sales tax that she paid this year on her state income tax return. She spent $4960.80 during the year, including 6% sales tax. How much sales tax did she pay?

50. *State Sales Tax.* The Tea Chest needs to send the sales tax that it has collected to the state. The total amount of sales, including 4% sales tax, is $7115.68. How much in sales tax must the store send to the state?

51. *Discount.* Raena paid $224 for a camera during a 30%-off sale. What was the regular price?

52. *Discount.* Raquel paid $68 for a digital picture frame during a 20%-off sale. What was the regular price?

53. *Job Change.* Bradley took a 15% pay cut when he changed jobs. If he makes $30,600 per year at his new job, what was the annual salary for his previous job?

54. *Retirement Account.* The value of Karen's retirement account decreased by 40% to $87,000. How much was her retirement account worth before the decrease?

Aha! **55.** *Couponing.* Through "extreme couponing," Marie saved 85% of her grocery bill. If she paid $15, what was the original amount of the bill?

56. *Couponing.* Elliot had a coupon for 12% off his first meal at a new restaurant. If he was charged $11 for the meal, what was the original price of the meal?

57. *Advertising.* A 30-sec television advertising slot during the 2016 Super Bowl cost $4.8 million. This was a 20% increase over the cost in 2013. How much did a 30-sec slot cost in 2013?

Data: bleacherreport.com

58. *Law Enforcement.* During the 2015 holiday season in California, officials arrested 350 individuals for driving under the influence of alcohol. This was a 23% increase from the number arrested during the same time in 2014. How many were arrested during the holidays in 2014?

Data: Times of San Diego

59. *Selling a Home.* The Brannons are planning to sell their home. If they want to be left with $117,500 after paying 6% of the selling price to a realtor as a commission, for how much must they sell their home?

60. *Law Enforcement.* In Charlotte, North Carolina, the number of annual crashes at targeted intersections fell 43.6% to 2591 crashes after red-light cameras were installed. How many crashes occurred annually before the cameras were installed?

Data: The National Campaign to Stop Red Light Running

A, B. Problem Solving

61. *Taxi Rates.* In Chicago, a taxi ride costs $3.25 plus $1.80 for each mile traveled. Debbie has budgeted $19 for a taxi ride (excluding tip). How far can she travel on her $19 budget?

Data: cityofchicago.org

62. *Taxi Fares.* In Las Vegas, taxi rides originating at the airport cost $5.45 plus $2.86 per additional mile. How far can Ashfaq travel for $24.95?

Data: Nevada Transportation Authority

63. *Truck Rentals.* Truck-Rite Rentals rents trucks at a daily rate of $39.95 plus 55¢ per mile. Concert Productions has budgeted $100 for renting a truck to haul equipment to an upcoming concert. How far can they travel in one day and stay within their budget? Round your answer to the nearest tenth of a mile.

64. *Truck Rentals.* Fine Line Trucks rents an 18-ft truck for $42 plus 35¢ per mile. Judy needs a truck for one day to deliver a shipment of plants. How far can she drive and stay within a budget of $70?

65. *Complementary Angles.* The sum of the measures of two *complementary* angles is 90°. If one angle measures 15° more than twice the measure of its complement, find the measure of each angle.

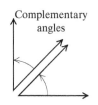
Complementary angles

66. *Complementary Angles.* Two angles are complementary. (See Exercise 65.) The measure of one angle is $1\frac{1}{2}$ times the measure of the other. Find the measure of each angle.

67. *Supplementary Angles.* The sum of the measures of two *supplementary* angles is 180°. If the measure of one angle is $3\frac{1}{2}$ times the measure of the other, find the measure of each angle.

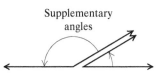
Supplementary angles

68. *Supplementary Angles.* Two angles are supplementary. (See Exercise 67.) If one angle measures 45° less than twice the measure of its supplement, find the measure of each angle.

69. *Copier Paper.* The perimeter of standard-size copier paper is 99 cm. The width is 6.3 cm less than the length. Find the length and the width.

70. *Stock Prices.* Sarah's investment in Jet Blue stock grew 28% to $448. How much did she originally invest?

71. *Savings Interest.* Janeka invested money in a savings account at a rate of 1% simple interest. After one year, she has $1555.40 in the account. How much did Janeka originally invest?

72. *Credit Cards.* The balance in Will's Mastercard® account grew 2%, to $870, in one month. What was his balance at the beginning of the month?

73. *Scrabble®.* In a single game on October 12, 2006, Michael Cresta and Wayne Yorra set three North American Scrabble records: the most points in one game by one player, the most total points in the game, and the most points on a single turn. Cresta scored 340 points more than Yorra, and together they scored 1320 points. What was Cresta's winning score?

Data: slate.com

74. *Bridge Construction.* The San Francisco–Oakland Bay Bridge consists of two spans connected by a tunnel. The East span is 556 ft longer than the West span, and the total length of the two spans is 19,796 ft. How long is the East span?

Data: historicbridges.org

75. *Cost of Food.* The equation $c = 1.2x + 32.94$ can be used to estimate the cost c of a Thanksgiving dinner for 10 people x years after 2000. Determine in what year the cost of a dinner for 10 people was $50.94.

Data: American Farm Bureau Federation

76. *Teacher Salaries.* We can use the equation $s = 1352x + 44,609$ to estimate the average salary s of teachers in public schools x years after 2000. In what year was the average salary $63,537?

Data: National Center for Education Statistics

77. *Cricket Chirps and Temperature.* The equation $T = \frac{1}{4}N + 40$ can be used to determine the temperature T, in degrees Fahrenheit, given the number of times N a cricket chirps per minute. Determine the number of chirps per minute for a temperature of 80°F.

78. *Race Time.* The equation $R = -0.028t + 20.8$ can be used to predict the world record in the 200-m dash, where R is the record in seconds and t is the number of years after 1920. In what year will the record be 18.0 sec?

79. Marcus claims that he can solve most of the problems in this section by guessing. Is there anything wrong with this approach? Why or why not?

80. When solving Exercise 24, Jamie used a to represent the bride's age and Ben used a to represent the groom's age. Is one of these approaches preferable to the other? Why or why not?

Skill Review

81. Multiply: $4(2n + 8t + 1)$. [1.8]

82. Factor: $12 + 18x + 21y$. [1.2]

83. Simplify: $x - 3[2x - 4(x - 1) + 2]$. [1.8]

84. Find the absolute value: $|0|$. [1.4]

Synthesis

85. Write a problem for a classmate to solve. Devise it so that the problem can be translated to the equation $x + (x + 2) + (x + 4) = 375$.

86. Write a problem for a classmate to solve. Devise it so that the solution is "Audrey can drive the rental truck for 50 mi without exceeding her budget."

87. *Discounted Dinners.* Kate's "Dining Card" entitles her to $10 off the price of a meal after a 15% tip has been added to the cost of the meal. If, after the discount, the bill is $32.55, how much did the meal originally cost?

88. *Test Scores.* Pam scored 78 on a test that had 4 fill-in questions worth 7 points each and 24 multiple-choice questions worth 3 points each. She had one fill-in question wrong. How many multiple-choice questions did Pam answer correctly?

89. *Gettysburg Address.* Abraham Lincoln's 1863 Gettysburg Address refers to the year 1776 as "four *score* and seven years ago." Determine what a score is.

90. One number is 25% of another. The larger number is 12 more than the smaller. What are the numbers?

91. A storekeeper goes to the bank to get $10 worth of change. She requests twice as many quarters as half dollars, twice as many dimes as quarters, three times as many nickels as dimes, and no pennies or dollars. How many of each coin did the storekeeper get?

92. *Perimeter of a Rectangle.* The width of a rectangle is three-fourths of the length. The perimeter of the rectangle becomes 50 cm when the length and the width are each increased by 2 cm. Find the length and the width.

93. *Discounts.* In exchange for opening a new credit account, Macy's Department Stores® subtracts 10% from all purchases made the day the account is established. Julio is opening an account and has a coupon for which he receives 10% off the first day's reduced price of a camera. If Julio's final price is $77.75, what was the price of the camera before the two discounts?

94. *Sharing Fruit.* Apples are collected in a basket for six people. One-third, one-fourth, one-eighth, and one-fifth of the apples are given to four people, respectively. The fifth person gets ten apples, and one apple remains for the sixth person. Find the original number of apples in the basket.

95. *eBay Purchases.* An eBay seller charges $9.99 for the first DVD purchased and $6.99 for all others. For shipping and handling, he charges the full shipping fee of $3 for the first DVD, one-half of the shipping charge for the second DVD, and one-third of the shipping charge per item for all remaining DVDs. The total cost of a shipment (excluding tax) was $45.45. How many DVDs were in the shipment?

96. *Winning Percentage.* In a basketball league, the Falcons won 15 of their first 20 games. In order to win 60% of the total number of games, how many more games must they play, assuming they win only half of the remaining games?

97. *Taxi Fares.* In New York City, a taxi ride costs $2.80 plus 50¢ per $\frac{1}{5}$ mile and 60¢ per minute stopped in traffic. Due to traffic, Mya's taxi took 20 min to complete what is, without traffic, a 10-min drive. If she is charged $23.80 for the ride, how far did Mya travel?

Data: NYC Taxi and Limousine Commission

98. *Test Scores.* Ella has an average score of 82 on three tests. Her average score on the first two tests is 85. What was the score on the third test?

99. A school purchases a piano and must choose between paying $2000 at the time of purchase or $2150 at the end of one year. Which option should the school select and why?

100. Annette claims the following problem has no solution: "The sum of the page numbers on facing pages is 191. Find the page numbers." Is she correct? Why or why not?

101. The perimeter of a rectangle is 101.74 cm. If the length is 4.25 cm longer than the width, find the dimensions of the rectangle.

102. The second side of a triangle is 3.25 cm longer than the first side. The third side is 4.35 cm longer than the second side. If the perimeter of the triangle is 26.87 cm, find the length of each side.

YOUR TURN ANSWERS: SECTION 2.5

1. About 78.4 mi   **2.** 168 and 169   **3.** 26 images
4. Width: 50 ft; length: 84 ft   **5.** 60°, 90°, 30°   **6.** $\frac{1}{2}$ hr
7. Approximately 7.1%, or $7\frac{1}{7}\%$ **8.** $1250

Quick Quiz: Sections 2.1–2.5

Solve and check.

1. $-\frac{2}{5}x = -\frac{5}{3}$ [2.1]

2. $3t - 8 - (9 - t) = 2(5t + 4) - 7$ [2.2]

3. Solve for v: $B = \dfrac{3}{v}$.   [2.3]

4. Caitlan has $500 in a savings account that is paying 2.1% interest per year. How much will she earn in a year?   [2.4]

5. One day Jordan spent a total of 4 hr studying American Literature, College Algebra, and Nonverbal Communication. He spent twice as much time studying algebra as he did studying literature, and he spent 30 min longer studying communication than he did studying literature. For how long did he study literature?   [2.5]

Prepare to Move On

Write a true sentence using either $<$ or $>$.   [1.4]

1. $-8 \,\square\, 1$ 2. $-2 \,\square\, -5$

Write a second inequality with the same meaning.   [1.4]

3. $x \geq -4$ 4. $5 > y$

2.6 Solving Inequalities

A. Solutions of Inequalities   **B.** Graphs of Inequalities   **C.** Set-Builder Notation and Interval Notation
D. Solving Inequalities Using the Addition Principle   **E.** Solving Inequalities Using the Multiplication
Principle   **F.** Using the Principles Together

Many real-world situations translate to *inequalities*. For example, a student might need to register for *at least* 12 credits; an elevator might be designed to hold *at most* 2000 pounds; a tax credit might be allowable for families with incomes of *less than* $40,000; and so on. Before solving applications like these, we must modify our equation-solving principles for solving inequalities.

A. Solutions of Inequalities

An inequality is a number sentence containing $>$ (is greater than), $<$ (is less than), $\geq$ (is greater than or equal to), $\leq$ (is less than or equal to), or $\neq$ (is not equal to). Inequalities may be true or false. For example,

$-3 \leq 5$ is *true* because $-3 < 5$ is true,

$-3 \leq -3$ is *true* because $-3 = -3$ is true, and

$-5 \geq 4$ is *false* because neither $-5 > 4$ nor $-5 = 4$ is true.

Inequalities like

$$-7 > x, \qquad t < 5, \qquad 5x - 2 \geq 9, \quad \text{and} \quad -3y + 8 \leq -7$$

are true for some replacements of the variable and false for others.

Any value for the variable that makes an inequality true is called a **solution**. The set of all solutions is called the **solution set**. When all solutions of an inequality have been found, we say that we have **solved** the inequality.

EXAMPLE 1   Determine whether the given number is a solution of $y \geq 6$:
(a) 6; **(b)** -4.

SOLUTION

a) Since $6 \geq 6$ is true, 6 is a solution.

1. Determine whether -1 is a solution of $x > -3$.

b) Since $-4 \geq 6$ is false, -4 is not a solution.

YOUR TURN

B. Graphs of Inequalities

Because the solutions of inequalities like $x < 2$ are too numerous to list, it is helpful to make a drawing, or a **graph**, that represents all the solutions. Graphs of inequalities in one variable can be drawn on the number line by shading all points that are solutions. Parentheses are used to indicate endpoints that *are not* solutions and brackets to indicate endpoints that *are* solutions.*

EXAMPLE 2   Graph each inequality: **(a)** $x < 2$; **(b)** $y \geq -3$; **(c)** $-2 < x \leq 3$.

SOLUTION

a) The solutions of $x < 2$ are those numbers less than 2. They are shown on the graph by shading all points to the left of 2. The parenthesis at 2 and the shading to its left indicate that 2 *is not* part of the graph, but numbers like 1.2 and 1.99 are.

b) The solutions of $y \geq -3$ are shown on the number line by shading all points to the right of -3. Since -3 is also a solution, the bracket at -3 indicates that -3 *is* part of the graph.

Student Notes

Note that $-2 < x < 3$ means $-2 < x$ *and* $x < 3$. Because of this, statements like $2 < x < 1$ make no sense—no number is both greater than 2 and less than 1.

c) The inequality $-2 < x \leq 3$ is read "-2 is less than x *and* x is less than or equal to 3," or "x is greater than -2 *and* less than or equal to 3." To be a solution of $-2 < x \leq 3$, a number must be a solution of both $-2 < x$ *and* $x \leq 3$. The number 1 is a solution, as are -0.5, 1.9, and 3. The parenthesis indicates that -2 *is not* a solution, whereas the bracket indicates that 3 *is* a solution. The other solutions are shaded.

2. Graph the inequality:

$$0 \leq t < 5.$$

YOUR TURN

---

*An alternative notation uses open dots to indicate endpoints that are not solutions and closed dots to indicate endpoints that are solutions. Using this notation, the solutions of $x < 2$ are graphed as 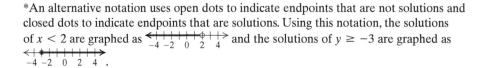 and the solutions of $y \geq -3$ are graphed as .

Consider the inequality and the graph in Example 2(a). Match each change in the inequality described in the left column with the corresponding change in the graph described in the right column.

| *Inequality* | *Graph* |

$$x < 2$$

1. Replace $<$ with $\leq$.
2. Replace $<$ with $>$.
3. Replace 2 with -1.

a) The) changes to a (and the shading moves to the right of the (.
b) The) changes to a].
c) The) moves to the left and the portion of the number line that is shaded changes.

ANSWERS
1. (b) 2. (a) 3. (c)

C. Set-Builder Notation and Interval Notation

To write the solution set of $x < 3$, we can use **set-builder notation:**

$$\{x \mid x < 3\}.$$

This is read "The set of all x such that x is less than 3."

Another way to write solutions of an inequality in one variable is to use **interval notation.** Interval notation uses parentheses, ( ), and brackets, [ ].

If a and b are real numbers with $a < b$, we define the **open interval (a, b)** as the set of all numbers x for which $a < x < b$. This means that x can be any number between a and b, but it cannot be either a or b.

The **closed interval $[a, b]$** is defined as the set of all numbers x for which $a \leq x \leq b$. **Half-open intervals $(a, b]$** and $[a, b)$ contain one endpoint and not the other.

We use the symbols ∞ and $-\infty$ to represent positive infinity and negative infinity, respectively. Thus the notation (a, ∞) represents the set of all real numbers greater than a, and $(-\infty, a)$ represents the set of all real numbers less than a.

Interval notation for a set of numbers corresponds to its graph.

CAUTION! Do not confuse the *interval* (a, b) with the *ordered pair* (a, b). The context in which the notation appears should make the meaning clear.

Student Notes

You may have noticed that for interval notation, the order of the endpoints mimics the number line, with the smaller number on the left and the larger on the right.

Also note which inequality signs correspond to brackets and which correspond to parentheses. The relationship could be written informally as

$$\geq \quad \leq \quad [\ ]$$
$$> \quad < \quad (\ ).$$

Interval Notation	Set-Builder Notation	Graph
(a, b)	$\{x \mid a < x < b\}$	← (———) → a b
$[a, b]$	$\{x \mid a \leq x \leq b\}$	← [———] → a b
$(a, b]$	$\{x \mid a < x \leq b\}$	← (———] → a b
$[a, b)$	$\{x \mid a \leq x < b\}$	← [———) → a b
(a, ∞)	$\{x \mid x > a\}$	← (———→ a
$[a, \infty)$	$\{x \mid x \geq a\}$	← [———→ a
$(-\infty, a)$	$\{x \mid x < a\}$	←———) → a
$(-\infty, a]$	$\{x \mid x \leq a\}$	←———] → a

EXAMPLE 3   Graph $t \geq -2$ on the number line and write the solution set using both set-builder notation and interval notation.

SOLUTION   Using set-builder notation, we write the solution set as $\{t \mid t \geq -2\}$.
Using interval notation, we write $[-2, \infty)$.
To graph the solution, we shade all numbers to the right of -2 and use a bracket to indicate that -2 is also a solution.

-7 -6 -5 -4 -3 -2 -1 0 1 2 3 4 5 6 7

3. Graph $y < 1$ on the number line and write the solution set using both set-builder notation and interval notation.

↪ YOUR TURN

D. Solving Inequalities Using the Addition Principle

We can visualize the addition principle for inequalities using a balance. When one side of the balance holds more weight than the other, the balance tips in that direction. If equal amounts of weight are then added to or subtracted from both sides of the balance, the balance remains tipped in the same direction.

The balance illustrates the idea that when a number, such as 2, is added to (or subtracted from) both sides of a true inequality, such as $3 < 7$, we get another true inequality:

$$3 + 2 < 7 + 2, \quad \text{or} \quad 5 < 9.$$

If we add -4 to both sides of $x + 4 < 10$, we get an *equivalent* inequality:

$$x + 4 + (-4) < 10 + (-4), \quad \text{or} \quad x < 6.$$

We say that $x + 4 < 10$ and $x < 6$ are **equivalent**, which means that both inequalities have the same solution set.

THE ADDITION PRINCIPLE FOR INEQUALITIES

For any real numbers a, b, and c:

$a < b$ is equivalent to $a + c < b + c$;
$a \leq b$ is equivalent to $a + c \leq b + c$;
$a > b$ is equivalent to $a + c > b + c$;
$a \geq b$ is equivalent to $a + c \geq b + c$.

As with equations, our goal is to isolate the variable on one side.

EXAMPLE 4   Solve $x + 2 > 8$ and then graph the solution. Write the solution set using both set-builder notation and interval notation.

SOLUTION   We use the addition principle, subtracting 2 from both sides:

$$x + 2 - 2 > 8 - 2 \qquad \text{Subtracting 2 from, or adding } -2 \text{ to, both sides}$$
$$x > 6.$$

Any number greater than 6 makes $x > 6$ true and is a solution of both that inequality and $x + 2 > 8$. Using set-builder notation, we write the solution set as $\{x \mid x > 6\}$. Using interval notation, we write the solution set as $(6, \infty)$. The graph is as follows:

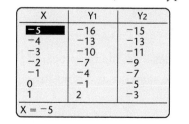

Because most inequalities have an infinite number of solutions, we cannot possibly check them all. A partial check can be made using one of the possible solutions. For this example, we can substitute any number greater than 6—say, 6.1—into the original inequality:

$$\frac{x + 2 > 8}{6.1 + 2 \mid 8}$$
$$8.1 \overset{?}{>} 8 \quad \text{TRUE} \qquad 8.1 > 8 \text{ is a true statement.}$$

Since $8.1 > 8$ is true, 6.1 is a solution. Any number greater than 6 is a solution.

4. Solve $t - 4 < 1$ and then graph the solution.

 YOUR TURN

Technology Connection

As a partial check of Example 5, we can let $y_1 = 3x - 1$ and $y_2 = 2x - 5$. We set TblStart $= -5$ and ΔTbl $= 1$ in the **TBLSET** menu to get the following table. By scrolling up or down, you can note that for $x \le -4$, we have $y_1 \le y_2$.

X	Y₁	Y₂
−5	−16	−15
−4	−13	−13
−3	−10	−11
−2	−7	−9
−1	−4	−7
0	−1	−5
1	2	−3

X = −5

---

5. Solve $4 + 5n \ge 4n - 1$ and then graph the solution.

 YOUR TURN

EXAMPLE 5 Solve $3x - 1 \le 2x - 5$ and then graph the solution. Write the solution set using both set-builder notation and interval notation.

SOLUTION We have

$$3x - 1 \le 2x - 5$$
$$3x - 1 + 1 \le 2x - 5 + 1 \qquad \text{Adding 1 to both sides}$$
$$3x \le 2x - 4 \qquad\qquad \text{Simplifying}$$
$$3x - 2x \le 2x - 4 - 2x \qquad \text{Subtracting } 2x \text{ from both sides}$$
$$x \le -4. \qquad\qquad \text{Simplifying}$$

The graph is as follows:

The student should check that numbers less than or equal to -4 are solutions. The solution set is $\{x \mid x \le -4\}$, or $(-\infty, -4]$.

E. Solving Inequalities Using the Multiplication Principle

There is a multiplication principle for inequalities similar to that for equations, but it must be modified when multiplying both sides by a negative number. Consider the true inequality

$$3 < 7.$$

If we multiply both sides by a *positive* number—say, 2—we get another true inequality:

$$3 \cdot 2 < 7 \cdot 2, \quad \text{or} \quad 6 < 14. \qquad \text{TRUE}$$

If we multiply both sides by a negative number—say, -2—we get a *false* inequality:

$$3 \cdot (-2) < 7 \cdot (-2), \quad \text{or} \quad -6 < -14. \qquad \text{FALSE}$$

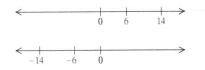

The fact that $6 < 14$ is true, but $-6 < -14$ is false, stems from the fact that the negative numbers, in a sense, *mirror* the positive numbers. Whereas 14 is to the *right* of 6 on the number line, the number -14 is to the *left* of -6. We must reverse the inequality symbol in $-6 < -14$ in order to get a true inequality:

$$-6 > -14. \quad \text{TRUE}$$

THE MULTIPLICATION PRINCIPLE FOR INEQUALITIES

For any real numbers a and b,
when c is a *positive* number,

$$a < b \quad \text{is equivalent to} \quad ac < bc, \quad \text{and}$$
$$a > b \quad \text{is equivalent to} \quad ac > bc;$$

when c is a *negative* number,

$$a < b \quad \text{is equivalent to} \quad ac > bc, \quad \text{and}$$
$$a > b \quad \text{is equivalent to} \quad ac < bc.$$

Similar statements hold for $\leq$ and $\geq$.

CAUTION!   When multiplying or dividing both sides of an inequality by a negative number, don't forget to reverse the inequality symbol!

EXAMPLE 6 Solve and graph each inequality: **(a)** $\frac{1}{4}x < 7$; **(b)** $-2y \leq 18$. Write the solution set using both set-builder notation and interval notation.

SOLUTION

a) $\frac{1}{4}x < 7$

$4 \cdot \frac{1}{4}x < 4 \cdot 7$ Multiplying both sides by 4, the reciprocal of $\frac{1}{4}$

$\qquad\qquad\qquad$ The symbol stays the same, since 4 is positive.

$\quad x < 28$ Simplifying

The solution set is $\{x \mid x < 28\}$, or $(-\infty, 28)$. The graph is shown at left.

b) $-2y \leq 18$

$\dfrac{-2y}{-2} \geq \dfrac{18}{-2}$ Multiplying both sides by $-\frac{1}{2}$, or dividing both sides by -2

$\qquad\qquad$ *At this step*, we reverse the inequality, because $-\frac{1}{2}$ is negative.

$\quad y \geq -9$ Simplifying

As a partial check, we substitute a number greater than -9, say -8, into the original inequality:

$$\frac{-2y \leq 18}{-2(-8) \overset{?}{\vert} 18}$$
$$16 \overset{?}{\leq} 18 \quad \text{TRUE} \quad 16 \leq 18 \text{ is a true statement.}$$

The solution set is $\{y \mid y \geq -9\}$, or $[-9, \infty)$. The graph is shown at left.

6. Solve and graph: $10 > -5x$.

YOUR TURN

F. Using the Principles Together

We use the addition and multiplication principles together to solve inequalities much as we did when solving equations.

↳ **Check Your**
UNDERSTANDING

Classify each pair of inequalities as "equivalent" or "not equivalent."

1. $x < -2$; $-2 > x$

2. $1 > x$; $x > 1$

3. $y + 1 < 7$; $y < 6$

4. $y - 1 < 7$; $y < 8$

5. $4t < 8$; $t < 2$

6. $-4t < 8$; $t < -2$

7. Solve: $12 - y \leq 3$.

EXAMPLE 7 Solve: $6 - 5y > 7$.

SOLUTION We have

$$6 - 5y > 7$$
$$-6 + 6 - 5y > -6 + 7 \qquad \text{Adding } -6 \text{ to both sides}$$
$$-5y > 1 \qquad \text{Simplifying}$$
$$-\tfrac{1}{5} \cdot (-5y) < -\tfrac{1}{5} \cdot 1 \qquad \text{Multiplying both sides by } -\tfrac{1}{5}, \text{ or dividing}$$
both sides by -5

Remember to reverse the inequality symbol!

$$y < -\tfrac{1}{5}. \qquad \text{Simplifying}$$

As a partial check, we substitute a number smaller than $-\tfrac{1}{5}$, say -1, into the original inequality:

$$\frac{6 - 5y > 7}{\begin{array}{c|c} 6 - 5(-1) & 7 \\ 6 - (-5) & \end{array}}$$
$$11 \overset{?}{>} 7 \quad \text{TRUE} \qquad 11 > 7 \text{ is a true statement.}$$

The solution set is $\{y \mid y < -\tfrac{1}{5}\}$, or $\left(-\infty, -\tfrac{1}{5}\right)$. The graph is shown at left.

↪ YOUR TURN

EXAMPLE 8 Solve: **(a)** $16.3 - 7.2p \leq -8.18$; **(b)** $1 - 3(x - 9) < 5(x + 6) - 2$.

SOLUTION

a) The greatest number of decimal places in any one number is *two*. Multiplying both sides by 100 will clear decimals. Then we proceed as before.

$$16.3 - 7.2p \leq -8.18$$
$$100(16.3 - 7.2p) \leq 100(-8.18) \qquad \text{Multiplying both sides by 100}$$
$$100(16.3) - 100(7.2p) \leq 100(-8.18) \qquad \text{Using the distributive law}$$
$$1630 - 720p \leq -818 \qquad \text{Simplifying}$$
$$1630 - 720p - 1630 \leq -818 - 1630 \qquad \text{Subtracting 1630 from both sides}$$
$$-720p \leq -2448 \qquad \text{Simplifying;}$$
$$-818 - 1630 = -2448$$

$$\frac{-720p}{-720} \geq \frac{-2448}{-720} \qquad \text{Dividing both sides by } -720$$

Remember to reverse the symbol!

$$p \geq 3.4$$

The solution set is $\{p \mid p \geq 3.4\}$, or $[3.4, \infty)$.

b)
$$1 - 3(x - 9) < 5(x + 6) - 2$$
$$1 - 3x + 27 < 5x + 30 - 2 \qquad \text{Using the distributive law to remove parentheses}$$
$$-3x + 28 < 5x + 28 \qquad \text{Simplifying}$$
$$-3x + 28 - 28 < 5x + 28 - 28 \qquad \text{Subtracting 28 from both sides}$$
$$-3x < 5x$$
$$-3x + 3x < 5x + 3x \qquad \text{Adding } 3x \text{ to both sides}$$
$$0 < 8x$$
$$0 < x \qquad \text{Dividing both sides by 8}$$

8. Solve: $\tfrac{1}{4} - \tfrac{2}{3}n \geq -\tfrac{1}{2}$.

The solution set is $\{x \mid 0 < x\}$, or $\{x \mid x > 0\}$, or $(0, \infty)$.

↪ YOUR TURN

CONNECTING 🔗 THE CONCEPTS

The procedure for solving inequalities is very similar to that used to solve equations. There are, however, two important differences.

- The multiplication principle for inequalities differs from the multiplication principle for equations: When we multiply or divide both sides of an inequality by a *negative* number, we must *reverse* the direction of the inequality.

- The solution set of an equation like those solved in this chapter typically consists of one number. The solution set of an inequality typically consists of a set of numbers and is written using either set-builder notation or interval notation.

Compare the following solutions.

Solve: $2 - 3x = x + 10$.

Solution

$2 - 3x = x + 10$	
$-3x = x + 8$	Subtracting 2 from both sides
$-4x = 8$	Subtracting x from both sides
$x = -2$	Dividing both sides by -4

The solution is -2.

Solve: $2 - 3x > x + 10$.

Solution

$2 - 3x > x + 10$	
$-3x > x + 8$	Subtracting 2 from both sides
$-4x > 8$	Subtracting x from both sides
$x < -2$	Dividing both sides by -4 and reversing the direction of the inequality symbol

The solution is $\{x \mid x < -2\}$, or $(-\infty, -2)$.

EXERCISES

Solve.

1. $x - 6 = 15$

2. $x - 6 \le 15$

3. $3x = -18$

4. $3x > -18$

5. $7 - 3x \ge 8$

6. $7 - 3x = 8$

7. $\dfrac{n}{6} - 6 = 5$

8. $\dfrac{n}{6} - 6 < 5$

9. $10 \ge -2(a - 5)$

10. $10 = -2(a - 5)$

2.6 EXERCISE SET

FOR EXTRA HELP   MyMathLab®

🢂 Vocabulary and Reading Check

Choose from the following list of words to complete each statement. Not every word will be used.

bracket   infinity   parenthesis
closed   interval   set-builder
half-open   open   solution

1. The number -2 is one _____ of the inequality $x < 0$.

2. The solution set $\{x \mid x \ge 10\}$ is an example of _____ notation.

3. The interval $[6, 10]$ is an example of a(n) _____ interval.

4. When graphing the solution of the inequality $-3 \le x \le 2$, place a(n) _____ at both ends of the interval.

🢂 Concept Reinforcement

Insert the symbol $<, >, \le,$ or $\ge$ to make each pair of inequalities equivalent.

5. $y - 10 \ge 4$; y ⬜ 14

6. $y + 6 \le 4$; y ⬜ -2

7. $3n > 90$; n ⬜ 30

8. $-7t \ge 56$; t ⬜ -8

9. $-5x \le 30$; x ⬜ -6

10. $4x < 12$; x ⬜ 3

11. $-2t > -14$; t ⬜ 7

12. $-3x < -15$; x ⬜ 5

A. Solutions of Inequalities

Determine whether each number is a solution of the given inequality.

13. $x > -4$

    a) 4   **b)** -6 **c)** -4

14. $t < 3$

    a) -3 **b)** 3   **c)** $2\frac{19}{20}$

15. $y \le 19$

    a) 18.99   **b)** 19.01   **c)** 19

16. $n \ge -4$

    a) 0   **b)** -4.1 **c)** -3.9

17. $c \ge -7$

    a) 0   **b)** $-5\frac{4}{5}$ **c)** $1\frac{1}{3}$

18. $m \le -2$

    a) $-1\frac{9}{10}$ **b)** 0   **c)** $-2\frac{1}{3}$

B. Graphs of Inequalities

Graph on the number line.

19. $y < 2$ **20.** $x \le 7$

21. $x \ge -1$ **22.** $t > -2$

23. $0 \le t$ **24.** $1 \le m$

25. $-5 \le x < 2$ **26.** $-3 < x \le 5$

27. $-4 < x < 0$ **28.** $0 \le x \le 5$

C. Set-Builder Notation and Interval Notation

Graph each inequality, and write the solution set using both set-builder notation and interval notation.

29. $y < 6$ **30.** $x > 4$

31. $x \ge -4$ **32.** $t \le 6$

33. $t > -3$ **34.** $y < -3$

35. $x \le -7$ **36.** $x \ge -6$

Describe each graph using both set-builder notation and interval notation.

37.
    -7 -6 -5 -4 -3 -2 -1 0 1 2 3 4 5 6 7

38.
    -7 -6 -5 -4 -3 -2 -1 0 1 2 3 4 5 6 7

39.
    -7 -6 -5 -4 -3 -2 -1 0 1 2 3 4 5 6 7

40.
    -7 -6 -5 -4 -3 -2 -1 0 1 2 3 4 5 6 7

41.
    -7 -6 -5 -4 -3 -2 -1 0 1 2 3 4 5 6 7

42.
    -7 -6 -5 -4 -3 -2 -1 0 1 2 3 4 5 6 7

43.
    -7 -6 -5 -4 -3 -2 -1 0 1 2 3 4 5 6 7

44.
    -7 -6 -5 -4 -3 -2 -1 0 1 2 3 4 5 6 7

D. Solving Inequalities Using the Addition Principle

Solve. Graph and write both set-builder notation and interval notation for each answer.

45. $y + 6 > 9$ **46.** $x + 8 \le -10$

47. $n - 6 < 11$ **48.** $n - 4 > -3$

49. $2x \le x - 9$ **50.** $3x \le 2x + 7$

51. $5 \ge t + 8$ **52.** $4 < t + 9$

53. $t - \frac{1}{8} > \frac{1}{2}$ **54.** $y - \frac{1}{3} > \frac{1}{4}$

55. $-9x + 17 > 17 - 8x$

56. $-8n + 12 > 12 - 7n$

Aha! **57.** $-23 < -t$ **58.** $19 < -x$

E. Solving Inequalities Using the Multiplication Principle

Solve. Graph and write both set-builder notation and interval notation for each answer.

59. $4x < 28$ **60.** $3x \ge 24$

61. $-24 > 8t$ **62.** $-16x < -64$

63. $1.8 \ge -1.2n$ **64.** $9 \le -2.5a$

65. $-2y \le \frac{1}{5}$ **66.** $-2x \ge \frac{1}{5}$

67. $-\frac{8}{5} > 2x$ **68.** $-\frac{5}{8} < -10y$

F. Using the Principles Together

Solve. Write both set-builder notation and interval notation for each answer.

69. $2 + 3x < 20$ **70.** $7 + 4y < 31$

71. $4t - 5 \le 23$ **72.** $15x - 7 \le -7$

73. $39 > 3 - 9x$ **74.** $5 > 5 - 7y$

75. $5 - 6y > 25$ **76.** $8 - 2y > 9$

77. $-3 < 8x + 7 - 7x$ **78.** $-5 < 9x + 8 - 8x$

79. $6 - 4y > 6 - 3y$ **80.** $7 - 8y > 5 - 7y$

81. $2.1x + 43.2 > 1.2 - 8.4x$

82. $0.96y - 0.79 \le 0.21y + 0.46$

83. $1.7t + 8 - 1.62t < 0.4t - 0.32 + 8$

84. $0.7n - 15 + n \ge 2n - 8 - 0.4n$

85. $\frac{x}{3} + 4 \le 1$ **86.** $\frac{2}{3} - \frac{x}{5} < \frac{4}{15}$

87. $3 < 5 - \dfrac{t}{7}$

88. $2 > 9 - \dfrac{x}{5}$

89. $4(2y - 3) \le -44$

90. $3(2y - 3) > 21$

91. $8(2t + 1) > 4(7t + 7)$

92. $3(t - 2) \ge 9(t + 2)$

93. $3(r - 6) + 2 < 4(r + 2) - 21$

94. $5(t + 3) + 9 \ge 3(t - 2) - 10$

95. $\frac{4}{5}(3x + 4) \le 20$

96. $\frac{2}{3}(2x - 1) \ge 10$

97. $\frac{2}{3}(\frac{7}{8} - 4x) - \frac{5}{8} < \frac{3}{8}$

98. $\frac{3}{4}(3x - \frac{1}{2}) - \frac{2}{3} < \frac{1}{3}$

99. Are the inequalities $x > -3$ and $x \ge -2$ equivalent? Why or why not?

100. Are the inequalities $t < -7$ and $t \le -8$ equivalent? Why or why not?

Skill Review

Simplify.   [1.8]

101. $5x - 2(3 - 6x)$

102. $8m - n - 3(2m + 5n)$

103. $x - 2[4y + 3(8 - x) - 1]$

104. $9x - 2\{4 - 5[6 - 2(x + 1) - x]\}$

Synthesis

105. Explain how it is possible for the graph of an inequality to consist of just one number. (*Hint*: See Example 2c.)

106. The statements of the addition and multiplication principles begin with *conditions* described for the variables. Explain the conditions given for each principle.

Solve.

Aha! **107.** $x < x + 1$

108. $6[4 - 2(6 + 3t)] > 5[3(7 - t) - 4(8 + 2t)] - 20$

109. $27 - 4[2(4x - 3) + 7] \ge 2[4 - 2(3 - x)] - 3$

Solve for x.

110. $\frac{1}{2}(2x + 2b) > \frac{1}{3}(21 + 3b)$

111. $-(x + 5) \ge 4a - 5$

112. $y < ax + b$ (Assume $a < 0$.)

113. $y < ax + b$ (Assume $a > 0$.)

114. Graph the solutions of $|x| < 3$ on the number line.

Aha! **115.** Determine the solution set of $|x| > -3$.

116. Determine the solution set of $|x| < 0$.

117. In order for a meal to be labeled "lowfat," it must have fewer than 3 g of fat per serving *and* the number of calories from fat must be 30% or fewer of the food's total number of calories.

  a) Each Chik Patties Breaded Veggie Pattie contains 150 calories per serving, and 54 of those calories come from fat. Can these patties be labeled "lowfat"?

  b) Cabot's 50% Reduced Fat Cheddar contains, as the name implies, 50% less fat than Cabot's regular cheddar cheese, but still cannot be labeled lowfat. What can you conclude about the fat content of a serving of Cabot's regular cheddar cheese?

↪ **YOUR TURN ANSWERS: SECTION 2.6**

1. Yes   **2.** ←|——|→
                      0   5
3. ←—)——→ $\{y \mid y < 1\}$, or $(-\infty, 1)$
      0 1
4. $\{t \mid t < 5\}$, or $(-\infty, 5)$ ←——)→
                                              0 5
5. $\{n \mid n \ge -5\}$, or $[-5, \infty)$ ←[——→
                                                 -5   0
6. $\{x \mid x > -2\}$, or $(-2, \infty)$ ←(——→
                                               -2 0
7. $\{y \mid y \ge 9\}$, or $[9, \infty)$ **8.** $\{n \mid n \le \frac{9}{8}\}$, or $(-\infty, \frac{9}{8}]$

Quick Quiz: Sections 2.1–2.6

Solve.

1. $1 - (d - 7) = 6d - 2(5d + 10)$ [2.2]

2. $3x - 7x > 10 - x$ [2.6]

3. Pernell left an 18% tip for his lunch service. If the tip was $2.70, what was the price of the lunch before the tip?   [2.4]

4. Melissa's dinner cost $20.89, including 8% sales tax. What was the price of the dinner before the tax was added?   [2.5]

5. Solve for c: $X = 12 - 5(d - c)$.   [2.3]

Prepare to Move On

Translate to an equation. Do not solve.   [1.1]

1. The area of a triangle is 5 m². The base is 3 m long. What is the height?

2. The perimeter of a triangle is 12 ft. One side is 1 ft longer than the shortest side. The third side is $\frac{1}{2}$ ft longer than the shortest side. How long is the shortest side of the triangle?

2.7 Solving Applications with Inequalities

A. Translating to Inequalities   **B.** Solving Problems

A. Translating to Inequalities

The five steps for problem solving can be used for problems involving inequalities. Before doing so, we list some important phrases to look for. Sample translations are listed as well.

Important Words	Sample Sentence	Definition of Variables	Translation
is at least	Kelby walks at least 1.5 mi a day.	Let k represent the length of Kelby's walk, in miles.	$k \geq 1.5$
is at most	At most 5 students dropped the course.	Let n represent the number of students who dropped the course.	$n \leq 5$
cannot exceed	The cost cannot exceed $12,000.	Let c represent the cost, in dollars.	$c \leq 12{,}000$
must exceed	The speed must exceed 40 mph.	Let s represent the speed, in miles per hour.	$s > 40$
is less than	Hamid's weight is less than 130 lb.	Let w represent Hamid's weight, in pounds.	$w < 130$
is more than	Boston is more than 200 mi away.	Let d represent the distance to Boston, in miles.	$d > 200$
is between	The film is between 90 min and 100 min long.	Let t represent the length of the film, in minutes.	$90 < t < 100$
minimum	Ned drank a minimum of 5 glasses of water a day.	Let w represent the number of glasses of water Ned drank.	$w \geq 5$
maximum	The maximum penalty is $100.	Let p represent the penalty, in dollars.	$p \leq 100$
no more than	Alan consumes no more than 1500 calories.	Let c represent the number of calories Alan consumes.	$c \leq 1500$
no less than	Patty scored no less than 80.	Let s represent Patty's score.	$s \geq 80$

The following phrases deserve special attention.

TRANSLATING "AT LEAST" AND "AT MOST"

The quantity x is at least some amount q: $x \geq q$.
(If x is *at least* q, it cannot be less than q.)

The quantity x is at most some amount q: $x \leq q$.
(If x is *at most* q, it cannot be more than q.)

B. Solving Problems

EXAMPLE 1 *Catering Costs.* To cater a party, Papa Roux charges a $50 setup fee plus $15 per person. The cost of Hotel Pharmacy's end-of-season softball party cannot exceed $450. How many people can attend the party?

SOLUTION

1. **Familiarize.** Suppose that 20 people were to attend the party. The cost would then be $50 + $15 · 20, or $350. This shows that more than 20 people could attend without exceeding $450. Instead of making another guess, we let n = the number of people in attendance.

2. **Translate.** The cost of the party will be $50 for the setup fee plus $15 times the number of people attending. We can reword as follows:

Rewording:	The setup fee	plus	the cost of the meals	cannot exceed	$450.
Translating:	50	+	$15 \cdot n$	$\leq$	450

3. **Carry out.** We solve for n:

$$50 + 15n \leq 450$$

$\qquad 15n \leq 400 \qquad$ Subtracting 50 from both sides

$\qquad n \leq \dfrac{400}{15} \qquad$ Dividing both sides by 15

$\qquad n \leq 26\dfrac{2}{3}. \qquad$ Simplifying

4. **Check.** The solution set is all numbers less than or equal to $26\frac{2}{3}$. Since n represents the number of people in attendance, we round to a whole number. Since the nearest whole number, 27, is not part of the solution set, we round *down* to 26. If 26 people attend, the cost will be $50 + $15 \cdot 26$, or $440, and if 27 attend, the cost will exceed $450.

5. **State.** At most 26 people can attend the party.

YOUR TURN

1. Refer to Example 1. Suppose that Hotel Pharmacy decides to use a different caterer. This caterer charges a $100 setup fee plus $12 per person. How many people can attend the party?

> **CAUTION!**   Solutions of problems should always be checked using the original wording of the problem. In some cases, answers might need to be whole numbers or integers or rounded off in a particular direction.

Some applications with inequalities involve *averages*, or *means*.

> **AVERAGE, OR MEAN**
>
> To find the **average**, or **mean**, of a set of numbers, add the numbers and then divide by the number of addends.

EXAMPLE 2   *Financial Aid.*   Full-time students in a health-care education program can receive financial aid and employee benefits from Covenant Health System by working at Covenant while attending school and also agreeing to work there after graduation. Students who work an average of at least 16 hr per week receive extra pay and part-time employee benefits. For the first three weeks of September, Dina worked 20 hr, 12 hr, and 14 hr. How many hours must she work during the fourth week in order to average at least 16 hr per week for the month?

Data: Covenant Health Systems

SOLUTION

1. **Familiarize.** Suppose that Dina works 10 hr during the fourth week. Her average for the month would be

$$\frac{20\text{ hr} + 12\text{ hr} + 14\text{ hr} + 10\text{ hr}}{4} = 14\text{ hr.}$$

There are 4 addends, so we divide by 4.

↳ Check Your UNDERSTANDING

Determine whether $5 satisfies the requirement in each statement.

1. The cost must exceed $5.
2. The cost cannot exceed $5.
3. No more than $5 per student can be spent.
4. The cost is no more than $6.
5. The price is no less than $6.

Chapter Resource:
Translating for Success, p. 144

2. Refer to Example 2. Suppose that for the first three weeks of October, Dina worked 12 hr, 14 hr, and 18 hr. How many hours must she work during the fourth week in order to average at least 16 hr per week for the month?

This shows that Dina must work more than 10 hr during the fourth week, if she is to average at least 16 hr of work per week. We let $x =$ the number of hours Dina works during the fourth week.

2. **Translate.** We reword the problem and translate as follows:

Rewording: The average number of hours Dina worked | should be at least | 16 hr.

Translating: $\dfrac{20 + 12 + 14 + x}{4} \geq 16$

3. **Carry out.** Because of the fraction, it is convenient to use the multiplication principle first:

$$\frac{20 + 12 + 14 + x}{4} \geq 16$$

$$4\left(\frac{20 + 12 + 14 + x}{4}\right) \geq 4 \cdot 16 \quad \text{Multiplying both sides by 4}$$

$$20 + 12 + 14 + x \geq 64$$

$$46 + x \geq 64 \quad \text{Simplifying}$$

$$x \geq 18. \quad \text{Subtracting 46 from both sides}$$

4. **Check.** As a partial check, we show that if Dina works 18 hr, she will average at least 16 hr per week:

$$\frac{20 + 12 + 14 + 18}{4} = \frac{64}{4} = 16. \quad \text{Note that 16 is at least 16.}$$

5. **State.** Dina will average at least 16 hr of work per week for September if she works at least 18 hr during the fourth week.

↪ YOUR TURN

2.7 EXERCISE SET

FOR EXTRA HELP MyMathLab®

↳ Vocabulary and Reading Check

In each of Exercises 1–4, two phrases appear under the blank. Choose the correct phrase to complete the statement.

1. If Matt's income is always $500 per week or more, then his income _____ $500.
   is at least/is at most

2. If Amy's entertainment budget for each month is $250, then her entertainment expenses _____ $250.
   must exceed/cannot exceed

3. If Lori works out 30 min or more every day, then her exercise time is _____ 30 min.
   no less than/no more than

4. If Marco must pay $20 or more toward his credit-card loan each month, then $20 is his _____ monthly payment.
   maximum/minimum

↳ Concept Reinforcement

In each of Exercises 5–12, match the sentence with one of the following:

$$a < b; \quad a \leq b; \quad b < a; \quad b \leq a.$$

5. a is at least b.
6. a exceeds b.
7. a is at most b.
8. a is exceeded by b.
9. b is no more than a.
10. b is no less than a.
11. b is less than a.
12. b is more than a.

A. Translating to Inequalities

Translate to an inequality.

13. A number is less than 10.

14. A number is greater than or equal to 4.

15. The temperature is at most −3°C.

16. A full-time student must take at least 12 credits of classes.

17. To rent a car, a driver must have a minimum of 5 years driving experience.

18. Focus-group sessions should last no more than 2 hr.

19. The age of the Mayan altar exceeds 1200 years.

20. The maximum safe exposure limit of formaldehyde is 2 parts per million.

21. Bianca earns no less than $12 per hour.

22. The cost of production of software cannot exceed $12,500.

23. Ireland gets between 1100 and 1600 hours of sunshine per year.

24. The cost of gasoline was between $2 and $4 per gallon.

B. Solving Problems

Use an inequality and the five-step approach to solve each problem.

25. *Furnace Repairs.*   RJ's Plumbing and Heating charges $55 plus $40 per hour for emergency service. Gary remembers being billed over $150 for an emergency call. How long was RJ's there?

26. *College Tuition.*   Vanessa's financial aid stipulates that her tuition not exceed $2500. If her local community college charges a $95 registration fee plus $675 per course, what is the greatest number of courses for which Vanessa can register?

27. *Graduate School.*   Unconditional acceptance into the Master of Business Administration (MBA) program at the University of Arkansas at Little Rock is awarded to students whose GMAT score plus 200 times the undergraduate grade point average is at least 1020. Robbin's GMAT score was 500. What must her grade point average be in order to be unconditionally accepted into the program?

Data: ualr.edu

28. *Car Payments.*   As a rule of thumb, debt payments (other than mortgages) should be less than 8% of a consumer's monthly gross income. Oliver makes $54,000 per year and has a $100 student-loan payment every month. What size car payment can he afford?

Data: money.cnn.com

29. *Quiz Average.*   Rod's quiz grades are 73, 75, 89, and 91. What scores on a fifth quiz will make his average quiz grade at least 85?

30. *Nutrition.*   Following the guidelines of the U.S. Department of Agriculture, Dale tries to eat at least 5 half-cup servings of vegetables each day. For the first six days of one week, she had 4, 6, 7, 4, 6, and 4 servings. How many servings of vegetables should Dale eat on Saturday, in order to average at least 5 servings per day for the week?

31. *College Course Load.*   To remain on financial aid, Millie must complete an average of at least 7 credits per quarter each year. In the first three quarters of a school year, Millie completed 5, 7, and 8 credits. How many credits of course work must Millie complete in the fourth quarter if she is to remain on financial aid?

32. *Music Lessons.*   Band members at Colchester Middle School are expected to average at least 20 min of practice time per day. One week Monroe practiced 15 min, 28 min, 30 min, 0 min, 15 min, and 25 min. How long must he practice on the seventh day if he is to meet expectations?

33. *Baseball.*   In order to qualify for a batting title, a major league baseball player must average at least 3.1 plate appearances per game. For the first nine games of the season, a player had 5, 1, 4, 2, 3, 4, 4, 3, and 2 plate appearances. How many plate appearances must the player have in the tenth game in order to average at least 3.1 per game?

Data: Major League Baseball

34. *Education.*   The Mecklenberg County Public Schools stipulate that a standard school day will average at least $5\frac{1}{2}$ hr, excluding meal breaks. For the first four days of one school week, bad weather resulted in school days of 4 hr, $6\frac{1}{2}$ hr, $3\frac{1}{2}$ hr, and $6\frac{1}{2}$ hr. How long must the Friday school day be in order to average at least $5\frac{1}{2}$ hr for the week?

Data: www.meck.k12.va.us

35. *Perimeter of a Triangle.*   One side of a triangle is 2 cm shorter than the base. The other side is 3 cm longer than the base. What lengths of the base will allow the perimeter to be greater than 19 cm?

36. *Perimeter of a Sign.*   The perimeter of a rectangular sign is not to exceed 50 ft. The length is to be twice the width. What widths will meet these conditions?

37. *Well Drilling.* All Seasons Well Drilling offers two plans. Under the "pay-as-you-go" plan, they charge $500 plus $8 per foot for a well of any depth. Under their "guaranteed-water" plan, they charge a flat fee of $4000 for a well that is guaranteed to provide adequate water for a household. For what depths would it save a customer money to use the pay-as-you-go plan?

38. *Cost of Road Service.* Rick's Automotive charges $50 plus $15 for each quarter hour when making a road call. Twin City Repair charges $70 plus $10 for each quarter hour. Under what circumstances would it be more economical for a motorist to call Rick's?

39. *Insurance-Covered Repairs.* Most insurance companies will replace a vehicle if an estimated repair exceeds 80% of the "blue-book" value of the vehicle. Michele's insurance company paid $8500 for repairs to her Subaru after an accident. What can be concluded about the blue-book value of the car?

40. *Insurance-Covered Repairs.* Following an accident, Jeff's Ford pickup was replaced by his insurance company because the damage was so extensive. Before the damage, the blue-book value of the truck was $21,000. How much would it have cost to repair the truck? (See Exercise 39.)

41. *Sizes of Packages.* The U.S. Postal Service defines a "package" as a parcel for which the sum of the length and the girth is less than 84 in. (Length is the longest side of a package and girth is the distance around the other two sides of the package.) A box has a fixed girth of 29 in. Determine (in terms of an inequality) those lengths for which the box is considered a "package."

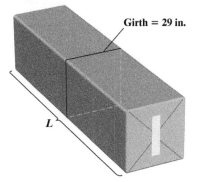

Girth = 29 in.

L

42. *Sizes of Envelopes.* Rhetoric Advertising is a direct-mail company. It determines that for a particular campaign, it can use any envelope with a fixed width of $3\frac{1}{2}$ in. and an area of at least $17\frac{1}{2}$ in². Determine (in terms of an inequality) those lengths that will satisfy the company constraints.

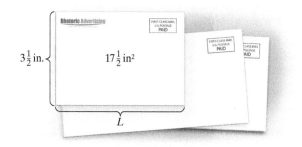

$3\frac{1}{2}$ in. $17\frac{1}{2}$ in²

L

43. *Body Temperature.* A person is considered to be feverish when his or her temperature is higher than 98.6°F. The formula $F = \frac{9}{5}C + 32$ can be used to convert Celsius temperatures C to Fahrenheit temperatures F. For which Celsius temperatures is a person considered feverish?

44. *Gold Temperatures.* Gold stays solid at Fahrenheit temperatures below 1945.4°. Determine (in terms of an inequality) those Celsius temperatures for which gold stays solid. Use the formula given in Exercise 43.

45. *Area of a Triangular Sign.* Zoning laws in Harrington prohibit displaying signs with areas exceeding 12 ft². If Flo's Marina is ordering a triangular sign with an 8-ft base, how tall can the sign be?

8 ft

FLO'S MARINA

?

46. *Area of a Triangular Flag.* As part of an outdoor education course, Trisha needs to make a bright-colored triangular flag with an area of at least 3 ft². What lengths can the triangle be if the base is $1\frac{1}{2}$ ft?

$1\frac{1}{2}$ ft

FORESTRY

?

47. *Fat Content in Foods.* Jif Creamy Reduced Fat peanut butter contains 12 g of fat per serving. In order for a food to be labeled "reduced fat," it must have at least 25% less fat than the regular item. What can you conclude about the number of grams of fat in a serving of the regular Jif peanut butter?
Data: Nutrition facts label

48. *Fat Content in Foods.* Reduced Fat Sargento® colby cheese contain 5 g of fat per serving. What can you conclude about the number of grams of fat in regular Sargento colby cheese? (See Exercise 47.)
Data: Nutrition facts label

49. *Pets.* In 2004, about 73.9 million dogs lived in U.S. households as pets. Since that time, the number of dogs as pets has risen an average of 1.1 million dogs per year. For what years can we expect more than 90 million dogs living as household pets?

50. *Pond Depth.* On July 1, Garrett's Pond was 25 ft deep. Since that date, the water level has dropped $\frac{2}{3}$ ft per week. For what dates will the water level not exceed 21 ft?

51. *Cell-Phone Budget.* Liam has budgeted $60 per month for his pay-as-you-go cell phone. He pays $1.99 for each day that he uses the phone plus 2¢ for each text message sent or received. If he uses the phone for 22 days, how many text messages can he send or receive?

52. *Banquet Costs.* The women's volleyball team can spend at most $700 for its awards banquet at a local restaurant. If the restaurant charges a $100 setup fee plus $24 per person, at most how many can attend?

53. *World Records in the Mile Run.* The formula
$$R = -0.0065t + 4.3259$$
can be used to predict the world record, in minutes, for the 1-mi run t years after 1900. Determine (in terms of an inequality) those years for which the world record will be less than 3.6 min.
Data: Information Please Database, Pearson Education, Inc.

54. *World Records in the Women's 1500-m Run.* The formula
$$R = -0.0026t + 4.0807$$
can be used to predict the world record, in minutes, for the 1500-m run t years after 1900. Determine (in terms of an inequality) those years for which the world record will be less than 3.7 min.
Data: Track and Field

55. *Toll Charges.* The equation
$$y = 0.122x + 0.912$$
can be used to estimate the cost y, in dollars, of driving x miles on the Pennsylvania Turnpike. For what mileages x will the cost be at most $14?
Data: turnpikeinfo.com

56. *Price of a Movie Ticket.* The average price of a movie ticket can be estimated by the equation
$$P = 0.237Y - 468.87,$$
where Y is the year and P is the average price, in dollars. For what years will the average price of a movie ticket be at least $9? (Include the year in which the $9 ticket first occurs.)
Data: National Association of Theatre Owners

57. If f represents Fran's age and t represents Todd's age, write a sentence that would translate to $t + 3 < f$.

58. Explain how the meanings of "Five more than a number" and "Five is more than a number" differ.

Skill Review

59. Use the commutative law of addition to write an expression equivalent to $xy + 7$. [1.2]

60. Find the prime factorization of 225. [1.3]

61. Change the sign of 18. [1.6]

62. Find $-(-x)$ for $x = -5$. [1.6]

Synthesis

63. Write a problem for a classmate to solve. Devise the problem so that the answer is "At most 18 passengers can go on the boat." Design the problem so that at least one number in the solution must be rounded down.

64. Write a problem for a classmate to solve. Devise the problem so that the answer is "The Rothmans can drive 90 mi without exceeding their truck rental budget."

65. *Ski Wax.* Green ski wax works best between 5° and 15° Fahrenheit. Determine those Celsius temperatures for which green ski wax works best. (See Exercise 43.)

66. *Parking Fees.* Mack's Parking Garage charges $4.00 for the first hour and $2.50 for each additional hour. For how long has a car been parked when the charge exceeds $16.50?

67. The area of a square can be no more than 64 cm². What lengths of a side will allow this?

Aha! **68.** The sum of two consecutive odd integers is less than 100. What is the largest pair of such integers?

69. *Frequent Buyer Bonus.* Alice's Books allows customers to select one free book for every 10 books purchased. The price of that book cannot exceed the average cost of the 10 books. Neoma has bought 9 books whose average cost is $12 per book. How much should her tenth book cost if she wants to select a $15 book for free?

70. *Parking Fees.* When asked how much the parking charge is for a certain car (see Exercise 66), Mack replies, "between 14 and 24 dollars." For how long has the car been parked?

71. *Grading.* After 9 quizzes, Blythe's average is 84. Is it possible for Blythe to improve her average by two points with the next quiz? Why or why not?

72. *Discount Card.* Barnes & Noble offers a member card for $25 per year. This card entitles a customer to a 40% discount off list price on hardcover bestsellers and a 10% discount on other eligible purchases. Describe two sets of circumstances for which an individual would save money by paying for a membership.

Data: Barnes & Noble

73. *Research.* Find a formula for the maximum mortgage payment, sometimes called front-end ratio, based on annual income. Using this formula, determine what annual salaries qualify for a $1000 monthly mortgage payment.

YOUR TURN ANSWERS: SECTION 2.7
1. At most 29 people **2.** At least 20 hr

Quick Quiz: Sections 2.1–2.7
Solve.

1. $3 - x \geq 1 - 2(5 - x)$ [2.6]

2. $0.03t - 2.1 = 10$ [2.2]

3. Kenneth saved $13.75 on a pair of running shoes that originally were priced at $68.75. What percent of the original price did he save? [2.4]

4. Two hours into a 72-km bicycle race, Jeannette realized that she had twice as far to ride as she had already ridden. At this point, how far was Jeannette from the finish line? [2.5]

5. On her first three history papers, Nara received scores of 88, 86, and 74. What scores on the fourth paper will make her average grade at least 80? [2.7]

Prepare to Move On
Graph on the number line. [1.4]
1. −12   2. 26

Simplify. [1.8]

3. $\dfrac{286 - 127}{6 - 4}$ 4. $\dfrac{1.3 - 9.8}{1 - 2}$

1. *Consecutive Integers.* The sum of two consecutive even integers is 102. Find the integers.

2. *Salary Increase.* After Susanna earned a 5% raise, her new salary was $25,750. What was her former salary?

3. *Dimensions of a Rectangle.* The length of a rectangle is 6 in. more than the width. The perimeter of the rectangle is 102 in. Find the length and the width.

4. *Population.* The population of Kelling Point is decreasing at a rate of 5% per year. The current population is 25,750. What was the population the previous year?

5. *Reading Assignment.* Quinn has 6 days to complete a 150-page reading assignment. How many pages must he read the first day so that he has no more than 102 pages left to read on the 5 remaining days?

Translating for Success

Use after Section 2.7.

Translate each word problem to an equation or an inequality and select a correct translation from A–O.

A. $0.05(25{,}750) = x$

B. $x + 2x = 102$

C. $2x + 2(x + 6) = 102$

D. $150 - x \le 102$

E. $x - 0.05x = 25{,}750$

F. $x + (x + 2) = 102$

G. $x + (x + 6) > 102$

H. $x + 5x = 150$

I. $x + 0.05x = 25{,}750$

J. $x + (2x + 6) = 102$

K. $x + (x + 1) = 102$

L. $102 + x > 150$

M. $0.05x = 25{,}750$

N. $102 + 5x > 150$

O. $x + (x + 6) = 102$

Answers on page A-7

An additional, animated version of this activity appears in MyMathLab. *To use MyMathLab, you need a course ID and a student access code. Contact your instructor for more information.*

6. *Numerical Relationship.* One number is 6 more than twice another. The sum of the numbers is 102. Find the numbers.

7. *Movie Collections.* Together, Mindy and Ken own 102 movies. If Ken owns 6 more than Mindy, how many does each have?

8. *Sales Commissions.* Kirk earns a commission of 5% on his sales. One year, he earned commissions totaling $25,750. What were his total sales for the year?

9. *Fencing.* Jess has 102 ft of fencing that he plans to use to enclose dog runs at two houses. The perimeter of one run is to be twice the perimeter of the other. Into what lengths should the fencing be cut?

10. *Quiz Scores.* Lupe has a total of 102 points on the first 6 quizzes in her sociology class. How many total points must she earn on the 5 remaining quizzes in order to have more than 150 points for the semester?

Collaborative Activity   *Sales and Discounts*

Focus: Applications and models using percent
Time: 15 minutes
Use after: Section 2.4
Group size: 3
Materials: Calculators are optional.

Often a store will reduce the price of an item by a fixed percentage. When the sale ends, the items are returned to their original prices. Suppose that a department store reduces all sporting goods 20%, all clothing 25%, and all electronics 10%.

Activity

1. Each group member should select one of the following items: a $50 basketball, an $80 jacket, or a $200 television. Fill in the first three columns of the first three rows of the chart below.

2. Apply the appropriate discount and determine the sale price of your item. Fill in the fourth column of the chart.

3. Next, find a multiplier for each item that can be used to convert the sale price back to the original price and fill in the remaining column of the chart. Does this multiplier depend on the price of the item?

4. Working as a group, compare the results of part (3) for all three items. Then develop a formula for a multiplier that will restore a sale price to its original price, p, after a discount r has been applied. Complete the fourth row of the table and check that your formula will duplicate the results of part (3).

5. Use the formula from part (4) to find the multiplier that a store would use to return an item to its original price after a "30% off" sale expires. Fill in the last line on the chart.

6. Inspect the last column of your chart. How can these multipliers be used to determine the percentage by which a sale price is increased when a sale ends?

Original Price, p	Discount, r	$1 - r$	Sale Price	Multiplier to convert back to p
p	r	$1 - r$		

Decision Making & Connection   *(Use after Section 2.4.)*

Cost of Self-Employment.   Because of additional taxes and fewer benefits, it has been estimated that a self-employed person must earn 20% more than a non–self-employed person performing the same work in order to have an equivalent income.

1. Barry is considering starting his own lawn-care business. He currently earns $16 per hour as a crew manager working for a large lawn-care company. He estimates that he can earn $18 per hour if he starts his own business, but he will lose his benefits. On the basis of these hourly rates alone, which option should Barry choose?

2. Develop a formula that gives the minimum amount A that a self-employed person must earn in order to make a salary equivalent to that of a non–self-employed person making x dollars.

3. *Research.*   Determine the extra costs that a self-employed person has. Consider self-employment taxes, insurance and vacation benefits, retirement benefits, and other costs. Based on your research, does the 20% rule seem accurate?

Study Summary

KEY TERMS AND CONCEPTS	EXAMPLES	PRACTICE EXERCISES

SECTION 2.1: *Solving Equations*

The Addition Principle for Equations

$a = b$ is equivalent to $a + c = b + c$.

Solve: $x + 5 = -2$.

$$x + 5 = -2$$
$$x + 5 + (-5) = -2 + (-5) \quad \text{Adding } -5 \text{ to both sides}$$
$$x = -7$$

1. Solve: $x - 8 = -3$.

The Multiplication Principle for Equations

$a = b$ is equivalent to $ac = bc$, for $c \neq 0$.

Solve: $-\frac{1}{3}x = 7$.

$$-\frac{1}{3}x = 7$$
$$(-3)\left(-\frac{1}{3}x\right) = (-3)(7) \quad \text{Multiplying both sides by } -3$$
$$x = -21$$

2. Solve: $\frac{1}{4}x = 1.2$.

SECTION 2.2: *Using the Principles Together*

When solving equations, we usually work in the reverse order of the order of operations.

Solve: $-3x - 7 = -8$.

$$-3x - 7 + 7 = -8 + 7 \quad \text{Adding 7 to both sides}$$
$$-3x = -1$$
$$\frac{-3x}{-3} = \frac{-1}{-3} \quad \text{Dividing both sides by } -3$$
$$x = \frac{1}{3}$$

3. Solve: $4 - 3x = 7$.

We can **clear fractions** by multiplying both sides of an equation by the least common multiple of the denominators in the equation.

We can **clear decimals** by multiplying both sides by a power of 10. If there is at most one decimal place in any one number, we multiply by 10. If there are at most two decimal places, we multiply by 100, and so on.

Solve: $\frac{1}{2}x - \frac{1}{3} = \frac{1}{6}x + \frac{2}{3}$.

$$6\left(\frac{1}{2}x - \frac{1}{3}\right) = 6\left(\frac{1}{6}x + \frac{2}{3}\right) \quad \text{Multiplying by 6, the least common multiple of the denominators}$$
$$6 \cdot \frac{1}{2}x - 6 \cdot \frac{1}{3} = 6 \cdot \frac{1}{6}x + 6 \cdot \frac{2}{3} \quad \text{Using the distributive law}$$
$$3x - 2 = x + 4 \quad \text{Simplifying}$$
$$2x = 6 \quad \text{Subtracting } x \text{ from and adding 2 to both sides}$$
$$x = 3$$

4. Solve: $\frac{1}{6}t - \frac{3}{4} = t - \frac{2}{3}$.

SECTION 2.3: *Formulas*

A **formula** uses letters to show a relationship among two or more quantities. Formulas can be solved for a given letter using the addition and multiplication principles.

Solve for y: $x = \frac{2}{5}y + 7$.

$$x = \frac{2}{5}y + 7 \quad \text{We are solving for } y.$$
$$x - 7 = \frac{2}{5}y \quad \text{Isolating the term containing } y$$
$$\frac{5}{2}(x - 7) = \frac{5}{2} \cdot \frac{2}{5}y \quad \text{Multiplying both sides by } \frac{5}{2}$$
$$\frac{5}{2}x - \frac{5}{2} \cdot 7 = 1 \cdot y \quad \text{Using the distributive law}$$
$$\frac{5}{2}x - \frac{35}{2} = y \quad \text{We have solved for } y.$$

5. Solve for c: $ac - bc = d$.

SECTION 2.4:   *Applications with Percent*

Key Words in Percent Translations

"Of" translates to "$\cdot$" or "$\times$"

"What" translates to a variable

"Is" or "Was" translates to "$=$"

"%" translates to "$\times \frac{1}{100}$" or "$\times 0.01$"

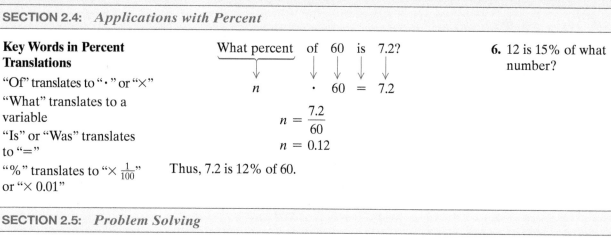

$$n = \frac{7.2}{60}$$

$$n = 0.12$$

Thus, 7.2 is 12% of 60.

6. 12 is 15% of what number?

SECTION 2.5:   *Problem Solving*

Five Steps for Problem Solving in Algebra

1. *Familiarize* yourself with the problem.
2. *Translate* to mathematical language. (This often means writing an equation or an inequality.)
3. *Carry out* some mathematical manipulation. (This often means *solving* an equation or an inequality.)
4. *Check* your possible answer in the original problem.
5. *State* the answer clearly.

The perimeter of a rectangle is 70 cm. The width is 5 cm longer than half the length. Find the length and the width.

1. **Familiarize.** The formula for the perimeter of a rectangle is $P = 2l + 2w$. We can describe the width in terms of the length: $w = \frac{1}{2}l + 5$.

2. **Translate.**

Rewording: Twice the length   plus   twice the width   is   the perimeter.

Translating: $2l$ $+$ $2(\frac{1}{2}l + 5) =$ 70

3. **Carry out.** Solve the equation:

$$2l + 2(\tfrac{1}{2}l + 5) = 70$$

$2l + l + 10 = 70$ — Using the distributive law

$3l + 10 = 70$ — Combining like terms

$3l = 60$ — Subtracting 10 from both sides

$l = 20.$ — Dividing both sides by 3

If $l = 20$, then $w = \frac{1}{2}l + 5 = \frac{1}{2} \cdot 20 + 5 = 10 + 5 = 15$.

4. **Check.** The width should be 5 cm longer than half the length. Since half the length is 10 cm, and 15 cm is 5 cm longer, this statement checks. The perimeter should be 70 cm. Since $2l + 2w = 2(20) + 2(15) = 40 + 30 = 70$, this statement also checks.

5. **State.** The length is 20 cm and the width is 15 cm.

7. Deborah rode a total of 120 mi in two bicycle fundraisers. One ride was 25 mi longer than the other. How long was each ride?

SECTION 2.6: *Solving Inequalities*

An **inequality** is any sentence containing $<, >, \leq, \geq,$ or $\neq$. Solution sets of inequalities can be **graphed** and written in **set-builder notation** or **interval notation.**

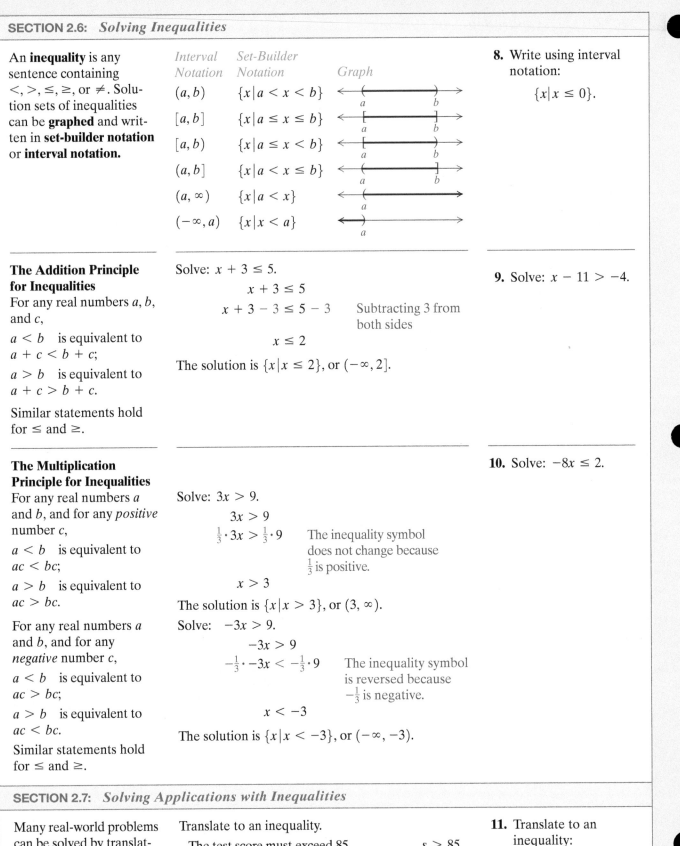

Interval Notation	Set-Builder Notation	Graph
(a, b)	$\{x \mid a < x < b\}$	
$[a, b]$	$\{x \mid a \leq x \leq b\}$	
$[a, b)$	$\{x \mid a \leq x < b\}$	
$(a, b]$	$\{x \mid a < x \leq b\}$	
(a, ∞)	$\{x \mid a < x\}$	
$(-\infty, a)$	$\{x \mid x < a\}$	

8. Write using interval notation:

$$\{x \mid x \leq 0\}.$$

The Addition Principle for Inequalities

For any real numbers $a, b,$ and $c,$

$a < b$ is equivalent to $a + c < b + c$;

$a > b$ is equivalent to $a + c > b + c$.

Similar statements hold for $\leq$ and $\geq$.

Solve: $x + 3 \leq 5$.

$$x + 3 \leq 5$$
$$x + 3 - 3 \leq 5 - 3 \qquad \text{Subtracting 3 from both sides}$$
$$x \leq 2$$

The solution is $\{x \mid x \leq 2\}$, or $(-\infty, 2]$.

9. Solve: $x - 11 > -4$.

The Multiplication Principle for Inequalities

For any real numbers a and $b,$ and for any *positive* number $c,$

$a < b$ is equivalent to $ac < bc$;

$a > b$ is equivalent to $ac > bc$.

For any real numbers a and $b,$ and for any *negative* number $c,$

$a < b$ is equivalent to $ac > bc$;

$a > b$ is equivalent to $ac < bc$.

Similar statements hold for $\leq$ and $\geq$.

Solve: $3x > 9$.

$$3x > 9$$
$$\tfrac{1}{3} \cdot 3x > \tfrac{1}{3} \cdot 9 \qquad \text{The inequality symbol does not change because } \tfrac{1}{3} \text{ is positive.}$$
$$x > 3$$

The solution is $\{x \mid x > 3\}$, or $(3, \infty)$.

Solve: $-3x > 9$.

$$-3x > 9$$
$$-\tfrac{1}{3} \cdot -3x < -\tfrac{1}{3} \cdot 9 \qquad \text{The inequality symbol is reversed because } -\tfrac{1}{3} \text{ is negative.}$$
$$x < -3$$

The solution is $\{x \mid x < -3\}$, or $(-\infty, -3)$.

10. Solve: $-8x \leq 2$.

SECTION 2.7: *Solving Applications with Inequalities*

Many real-world problems can be solved by translating the problem to an inequality and applying the five-step problem-solving strategy.

Translate to an inequality.

The test score must exceed 85.	$s > 85$
At most 15 volunteers greeted visitors.	$v \leq 15$
Ona makes no more than $100 per week.	$w \leq 100$
Herbs need at least 4 hr of sun per day.	$h \geq 4$

11. Translate to an inequality:

Luke runs no less than 3 mi per day.

Review Exercises: Chapter 2

Concept Reinforcement

In each of Exercises 1–8, classify the statement as either true or false.

1. $5x - 4 = 2x$ and $3x = 4$ are equivalent equations. [2.1]

2. $5 - 2t < 9$ and $t > 6$ are equivalent inequalities. [2.6]

3. Some equations have no solution. [2.1]

4. Consecutive odd integers are 2 units apart. [2.5]

5. For any number a, $a \le a$. [2.6]

6. The addition principle is always used before the multiplication principle. [2.2]

7. A 10% discount results in a sale price that is 90% of the original price. [2.4]

8. Often it is impossible to list all solutions of an inequality number by number. [2.6]

Solve.

9. $x + 9 = -16$ [2.1]

10. $-8x = -56$ [2.1]

11. $-\dfrac{x}{5} = 13$ [2.1]

12. $x - 0.1 = 1.01$ [2.1]

13. $-\frac{2}{3} + x = -\frac{1}{6}$ [2.1]

14. $4y + 11 = 5$ [2.2]

15. $5 - x = 13$ [2.2]

16. $3t + 7 = t - 1$ [2.2]

17. $7x - 6 = 25x$ [2.2]

18. $\frac{1}{4}x - \frac{5}{8} = \frac{3}{8}$ [2.2]

19. $14y = 23y - 17 - 10$ [2.2]

20. $0.22y - 0.6 = 0.12y + 3 - 0.8y$ [2.2]

21. $\frac{1}{4}x - \frac{1}{8}x = 3 - \frac{1}{16}x$ [2.2]

22. $6(4 - n) = 18$ [2.2]

23. $4(5x - 7) = -56$ [2.2]

24. $8(x - 2) = 4(x - 4)$ [2.2]

25. $-5x + 3(x + 8) = 16$ [2.2]

Solve each formula for the given letter. [2.3]

26. $C = \pi d$, for d

27. $V = \frac{1}{3}Bh$, for B

28. $5x - 2y = 10$, for y

29. $tx = ax + b$, for x

30. Find decimal notation for 1.2%. [2.4]

31. Find percent notation for $\frac{11}{25}$. [2.4]

32. What percent of 60 is 42? [2.4]

33. 49 is 35% of what number? [2.4]

Determine whether each number is a solution of $x \le -5$. [2.6]

34. -3 35. -7 36. 0

Graph on the number line. [2.6]

37. $5x - 6 < 2x + 3$ 38. $-2 < x \le 5$

39. $t > 0$

Solve. Write the answers in both set-builder notation and interval notation. [2.6]

40. $t + \frac{2}{3} \ge \frac{1}{6}$

41. $2 + 6y > 20$

42. $7 - 3y \ge 27 + 2y$

43. $-4y < 28$

44. $3 - 4x < 27$

45. $4 - 8x < 13 + 3x$

46. $13 \le -\frac{2}{3}t + 5$

47. $7 \le 1 - \frac{3}{4}x$

Solve.

48. The FACE clinic in Indianapolis, Indiana, places stray and feral cats in indoor residences and on farms as barn cats. In 2014, 280 cats were adopted into barn homes. This represented approximately 30% of the total number of cats placed. How many cats were adopted through the FACE clinic in 2014? [2.4]

   Data: "Farm Felines," Jon Shoulders, *Farm Indiana*, April 2015, pp. 24–29.

49. A 32-ft beam is cut into two pieces. One piece is 2 ft longer than the other. How long are the pieces? [2.5]

50. In 2012, a total of 294,000 students from China and India enrolled in U.S. colleges and universities. The number of Chinese students was 6000 less than twice the number of Indian students. How many Chinese students and how many Indian students enrolled in the United States in 2012?   [2.5]

Data: Open Doors Report

51. From 2014 to 2015, the number of international students studying in the United States increased by 14.18% to 1.13 million. How many new international students were there in 2014?   [2.4]

Data: U.S. Immigration and Customs Enforcement

52. The sum of two consecutive odd integers is 116. Find the integers.   [2.5]

53. The perimeter of a rectangle is 56 cm. The width is 6 cm less than the length. Find the width and the length.   [2.5]

54. After a 25% reduction, a picnic table is on sale for $120. What was the regular price?   [2.5]

55. The measure of the second angle of a triangle is 50° more than that of the first. The measure of the third angle is 10° less than twice the first. Find the measures of the angles.   [2.5]

56. Kathleen has budgeted an average of $95 per month for entertainment. For the first five months of the year, she has spent $98, $89, $110, $85, and $83. How much can Kathleen spend in the sixth month without exceeding her average entertainment budget?   [2.7]

57. To make copies of blueprints, Vantage Reprographics charges a $6 setup fee plus $4 per copy. Myra can spend no more than $65 for the copying. What number of copies will allow her to stay within budget?   [2.7]

Synthesis

58. How does the multiplication principle for equations differ from the multiplication principle for inequalities?   [2.1], [2.6]

59. Explain how checking the solutions of an equation differs from checking the solutions of an inequality.   [2.1], [2.6]

60. A study of sixth- and seventh-graders in Boston revealed that, on average, the students spent 3 hr 20 min per day watching TV or playing video and computer games. This represents 108% more than the average time spent reading or doing homework. How much time each day was spent, on average, reading or doing homework?   [2.4]

Data: Harvard School of Public Health

61. In June 2007, a team of Brazilian scientists exploring the Amazon measured its length as 65 mi longer than the Nile. If the combined length of both rivers is 8385 mi, how long is each river?   [2.5]

Data: news.nationalgeographic.com

Solve.

62. $2|n| + 4 = 50$ [1.4], [2.2]

63. $|3n| = 60$ [1.4], [2.1]

64. $y = 2a - ab + 3$, for a [2.3]

65. The Maryland Heart Center gives the following steps to calculate the number of fat grams needed daily by a moderately active woman. Write the steps as one formula relating the number of fat grams F to a woman's weight w, in pounds.   [2.3]

1) Calculate the total number of calories per day.
   _____ pounds × 12 calories = _____ total calories per day

2) Take the total number of calories and multiply by 30 percent.
   _____ calories per day × 0.30 = _____ calories from fat per day.

3) Take the number of calories from fat per day and divide by 9 (there are 9 calories per gram of fat).
   _____ calories from fat per day divided by 9 = _____ fat grams per day

Test: Chapter 2

For step-by-step test solutions, access the Chapter Test Prep Videos in MyMathLab.

Solve.

1. $t + 7 = 16$

2. $6x = -18$

3. $-\frac{4}{7}x = -28$

4. $3t + 7 = 2t - 5$

5. $\frac{1}{2}x - \frac{3}{5} = \frac{2}{5}$

6. $8 - y = 16$

7. $4.2x + 3.5 = 1.2 - 2.5x$

8. $4(x + 2) = 36$

9. $9 - 3x = 6(x + 4)$

10. $\frac{5}{6}(3x + 1) = 20$

Solve. Write the answers in both set-builder notation and interval notation.

11. $x + 6 > 1$

12. $14x + 9 > 13x - 4$

13. $-5y \geq 65$

14. $4n + 3 < -17$

15. $3 - 5x > 38$

16. $\frac{1}{2}t - \frac{1}{4} \leq \frac{3}{4}t$

17. $5 - 9x \geq 19 + 5x$

Solve each formula for the given letter.

18. $A = 2\pi rh$, for r

19. $w = \dfrac{P + l}{2}$, for l

20. Find decimal notation for 230%.

21. Find percent notation for 0.003.

22. What number is 18.5% of 80?

23. What percent of 75 is 33?

Graph on the number line.

24. $y < 4$

25. $-2 \leq x \leq 2$

Solve.

26. The perimeter of a rectangular calculator is 36 cm. The length is 4 cm greater than the width. Find the width and the length.

27. Kari is taking a 240-mi bicycle trip through Vermont. She has three times as many miles to go as she has already ridden. How many miles has she biked so far?

28. The perimeter of a triangle is 249 mm. If the sides are consecutive odd integers, find the length of each side.

29. By lowering the temperature of their electric hot-water heater from 140°F to 120°F, the Kellys' average electric bill dropped by 7% to $60.45. What was their electric bill before they lowered the temperature of their hot water?

30. Local light rail service in Denver, Colorado, costs $2.60 per trip (one way). A monthly pass costs $99. Gail is a student at Community College of Denver. Express as an inequality the number of trips per month that Gail should make if the pass is to save her money.

    Data: rtd-denver.com

Synthesis

Solve.

31. $c = \dfrac{2cd}{a - d}$, for d

32. $3|w| - 8 = 37$

33. Translate to an inequality.

    A plant marked "partial sun" needs at least 4 hr but no more than 6 hr of sun each day.

    Data: www.yardsmarts.com

34. A concert promoter had a certain number of tickets to give away. Five people got the tickets. The first got one-third of the tickets, the second got one-fourth of the tickets, and the third got one-fifth of the tickets. The fourth person got eight tickets, and there were five tickets left for the fifth person. Find the total number of tickets given away.

Cumulative Review: Chapters 1–2

Simplify.

1. $18 + (-30)$ [1.5]

2. $\frac{1}{2} - \left(-\frac{1}{4}\right)$ [1.6]

3. $-1.2(3.5)$ [1.7]

4. $-5 \div \left(-\frac{1}{2}\right)$ [1.7]

5. $150 - 10^2 \div 25 \cdot 4$ [1.8]

6. $(150 - 10^2) \div (25 \cdot 4)$ [1.8]

Remove parentheses and simplify.   [1.8]

7. $5x - (3x - 1)$

8. $3[4n - 5(2n - 1)] - 3(n - 7)$

9. Graph on the number line: $-\frac{5}{2}$.   [1.4]

10. Factor: $6x + 4y + 8z$.   [1.2]

Solve.

11. $4x - 7 = 3x + 9$ [2.1]

12. $\frac{2}{3}t + 7 = 13$ [2.2]

13. $9(2a - 1) = 4$ [2.2]

14. $12 - 3(5x - 1) = x - 1$ [2.2]

15. $3(x + 1) - 2 = 8 - 5(x + 7)$ [2.2]

Solve each formula for the given letter.   [2.3]

16. $\frac{1}{2}x = 2yz$, for z

17. $4x - 9y = 1$, for y

18. $an = p - rn$, for n

19. Find decimal notation for 183%.   [2.4]

20. Find percent notation for $\frac{3}{8}$.   [2.4]

21. Graph on the number line: $t > -\frac{5}{2}$.   [2.6]

Solve. Write the answer in both set-builder notation and interval notation.   [2.6]

22. $4t + 10 \leq 2$

23. $8 - t > 5$

24. $4 < 10 - \dfrac{x}{5}$

25. $4(2n - 3) \leq 2(5n - 8)$

Solve.

26. The total attendance at NCAA basketball games for a recent school year was 33 million. This was 31.25% less than the total attendance at NCAA football games during that year. What was the total attendance at NCAA football games during the school year?   [2.4]

Data: NCAA

27. The wavelength w, in meters per cycle, of a musical note is given by

$$w = \frac{r}{f},$$

where r is the speed of the sound, in meters per second, and f is the frequency, in cycles per second. The speed of sound in air is 344 m/sec. What is the wavelength of a note whose frequency in air is 24 cycles per second?   [2.3]

28. A 24-ft ribbon is cut into two pieces. One piece is 6 ft longer than the other. How long are the pieces?   [2.5]

29. Ariel has budgeted an average of $65 per month for bus fares. For the first five months of the year, she has spent $88, $15, $125, $50, and $60. How much can Ariel spend in the sixth month without exceeding her average bus-fare budget?   [2.7]

30. In 2010, about 285 million people had diabetes. The International Diabetes Foundation predicts that by 2030, 438 million people may have the disease. By what percent would the number of people with diabetes increase?   [2.5]

31. The second angle of a triangle is twice as large as the first. The third angle is 5° more than four times the first. Find the measure of the largest angle.   [2.5]

32. The length of a rectangular frame is 53 cm. For what widths would the perimeter be greater than 160 cm?   [2.7]

Synthesis

33. Simplify:

$$t - \{t - [3t - (2t - t) - t] - 4t\} - t.$$ [1.8]

34. Solve: $3|n| + 10 = 25$.   [2.2]

35. Lindy sold her Fender acoustic guitar on eBay using a third-party service. This service charges 35% of the first $500 of the selling price and 20% of the amount over $500. After these charges were deducted from the selling price, she received $745. For how much did her guitar sell?   [2.4], [2.5]

Introduction to Graphing

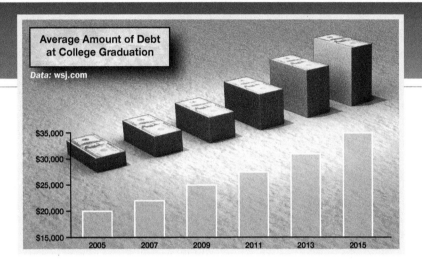

Average Amount of Debt at College Graduation

Data: wsj.com

$35,000
$30,000
$25,000
$20,000
$15,000

2005 2007 2009 2011 2013 2015

Congratulations! Now Here's the Bill.

3.1 Reading Graphs, Plotting Points, and Scaling Graphs

3.2 Graphing Linear Equations

3.3 Graphing and Intercepts

3.4 Rates

3.5 Slope

MID-CHAPTER REVIEW

3.6 Slope–Intercept Form

3.7 Point–Slope Form

CONNECTING THE CONCEPTS

CHAPTER RESOURCES

Visualizing for Success
Collaborative Activity
Decision Making: Connection

STUDY SUMMARY

REVIEW EXERCISES
CHAPTER TEST
CUMULATIVE REVIEW

For many college graduates, the excitement of earning a college degree is tempered by the arrival of the first bill for their student loan. Although student advisors and financial-aid officers work with students to make college education affordable, the average education debt has risen steadily, as indicated in the table shown. We can use these data to estimate the average education debt for years other than those shown. (*See Example 8 in Section 3.7.*)

In the financial aid office, we use math continually to help students work out the finances of their education.

Ben Burton, Chief Student Financial Officer at Ivy Tech Community College in Indianapolis, Indiana, uses math to calculate students' loan eligibility and the projected schedule and amount of their repayment, as well as their overall college debt.

ALF *Active Learning Figure* Explore the math using the Active Learning Figure in MyMathLab.

SA *Student Activity* Do the Student Activity in MyMathLab to see math in action.

We now begin our study of graphing. First, we examine graphs as they commonly appear in newspapers or magazines and develop terminology. Following that, we graph certain equations and study the connection between rate and slope. We will also learn how graphs are used as a problem-solving tool in many applications.

Our work in this chapter centers on equations that contain two variables.

3.1   Reading Graphs, Plotting Points, and Scaling Graphs

A. Problem Solving with Bar Graphs and Line Graphs   **B.** Points and Ordered Pairs
C. Numbering the Axes Appropriately

Today's print and electronic media make almost constant use of graphs. In this section, we consider problem solving with bar graphs and line graphs. Then we examine graphs that use a coordinate system.

A. Problem Solving with Bar Graphs and Line Graphs

A *bar graph* is a convenient way of showing comparisons. In every bar graph, certain categories, such as levels of education in the example below, are paired with certain numbers.

EXAMPLE 1 *Earnings.* The following bar graph shows median weekly earnings for full-time wage and salary workers ages 25 and older.

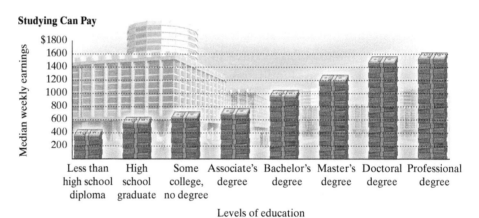

Studying Can Pay

Median weekly earnings: $1800, 1600, 1400, 1200, 1000, 800, 600, 400, 200

Levels of education: Less than high school diploma, High school graduate, Some college, no degree, Associate's degree, Bachelor's degree, Master's degree, Doctoral degree, Professional degree

Data: U.S. Bureau of Labor Statistics

a) Keagan plans to earn an associate's degree. What is the median income for workers with that level of education?

b) Isabella would like to make at least $1000 per week. What level of education should she pursue?

SOLUTION

a) Level of education is shown on the horizontal scale of the graph. We locate "associate's degree" and go to the top of that bar. Then we move horizontally from the top of the bar to the vertical scale, which shows earnings. We read there that Keagan can expect to make about $750 per week.

1. Use the bar graph in Example 1 to estimate by how much the median weekly earnings of a high school graduate exceeds that of a person who did not earn a high school diploma.

b) By moving up the vertical scale to $1000 and then moving horizontally, we see that the first bar to reach a height of $1000 or higher corresponds to a bachelor's degree. Thus Isabella should pursue a bachelor's, master's, doctoral, or professional degree in order to make at least $1000 per week.

↩ YOUR TURN

EXAMPLE 2 *Exercise and Pulse Rate.* The following *line graph* shows the relationship between a person's resting pulse rate and the number of months of regular exercise.* Note that the symbol ⚡ is used to indicate that counting on the vertical scale begins at 50.

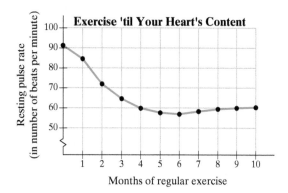

Since there are no points on the graph for pulse rates less than 50, that portion of the graph is omitted. This is indicated by the symbol ⚡.

a) How many months of regular exercise are required to lower the resting pulse rate as much as possible?

b) How many months of regular exercise are needed to achieve a resting pulse rate of 65 beats per minute?

SOLUTION

a) The lowest point on the graph is above 6. Thus, after 6 months of regular exercise, the resting pulse rate is lowered as much as possible.

b) To determine how many months of exercise are needed to lower a person's resting pulse rate to 65, we estimate 65 midway between 60 and 70 on the vertical scale. From that location, we move right until the line is reached. At that point, we move down to the horizontal scale and read the number of months required, as shown.

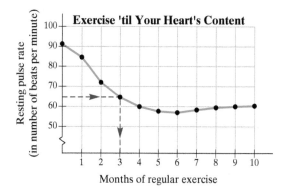

2. Use the graph in Example 2 to determine how many months of regular exercise are required before the resting pulse rate begins to rise.

The resting pulse rate is 65 beats per minute after 3 months of regular exercise.

↩ YOUR TURN

*Data from *Body Clock* by Dr. Martin Hughes (New York: Facts on File, Inc.), p. 60.

↪ **Check Your**
UNDERSTANDING

List the coordinates of each
point shown in the figure.

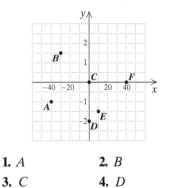

1. *A*   **2.** *B*
3. *C*   **4.** *D*
5. *E*   **6.** *F*

B. Points and Ordered Pairs

The line graph in Example 2 contains a collection of points. Each point pairs a number of months of exercise with a pulse rate. To create such a graph, we **graph**, or **plot**, pairs of numbers on a plane. This is done using two perpendicular number lines called **axes** (pronounced "ak-sēz"; singular, **axis**). The point at which the axes cross is called the **origin**. Arrows on the axes indicate the positive directions.

Consider the pair $(3, 4)$. The numbers in such a pair are called **coordinates**. The **first coordinate** in this case is 3, and the **second coordinate** is 4.* To plot, or graph, $(3, 4)$, we start at the origin, move horizontally to the 3, move up vertically 4 units, and then make a "dot." Thus, $(3, 4)$ is located above 3 on the first axis and to the right of 4 on the second axis.

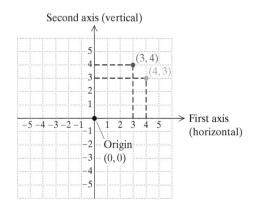

The point $(4, 3)$ is also plotted in the figure above. Note that $(3, 4)$ and $(4, 3)$ are different points. For this reason, coordinate pairs are called **ordered pairs**— the order in which the numbers appear is important.

EXAMPLE 3   Plot the point $(-3, 4)$.

SOLUTION   The first number, -3, is negative. Starting at the origin, we move 3 units in the negative horizontal direction (3 units to the left). The second number, 4, is positive, so we move 4 units in the positive vertical direction (up). The point $(-3, 4)$ is above -3 on the first axis and to the left of 4 on the second axis.

3. Plot the point $(-2, -4)$.

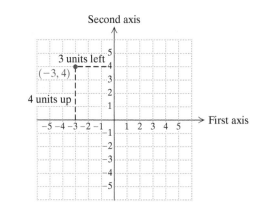

↪ YOUR TURN

To determine the coordinates of a given point, we see how far to the right or to the left of the origin the point is and how far above or below the origin it is. Note that the coordinates of the origin itself are $(0, 0)$.

*The first coordinate is called the *abscissa* and the second coordinate is called the *ordinate*. The plane is called the *Cartesian coordinate plane* after the French mathematician René Descartes (1595–1650).

EXAMPLE 4 Determine the coordinates of points $A, B, C, D, E, F,$ and G.

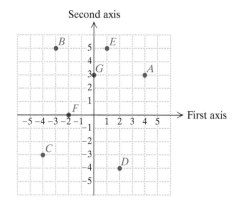

4. Determine the coordinates of point P.

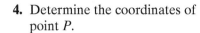
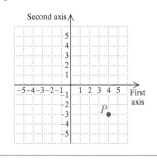

SOLUTION Point A is 4 units to the right of the origin and 3 units above the origin. Its coordinates are $(4, 3)$. The coordinates of the other points are as follows:

B: $(-3, 5)$; C: $(-4, -3)$; D: $(2, -4)$;

E: $(1, 5)$; F: $(-2, 0)$; G: $(0, 3)$.

YOUR TURN

The variables x and y are commonly used when graphing on a plane. Coordinates of ordered pairs are typically labeled in the form

$$(x\text{-coordinate}, y\text{-coordinate}).$$

The first, or horizontal, axis is labeled the x-axis, and the second, or vertical, axis is labeled the y-axis.

The horizontal axis and the vertical axis divide the plane into four regions, or **quadrants**, as indicated by Roman numerals in the following figure. Note that the point $(-4, 5)$ is in the second quadrant and the point $(5, -5)$ is in the fourth quadrant. The points $(3, 0)$ and $(0, 1)$ are on the axes and are not considered to be in any quadrant.

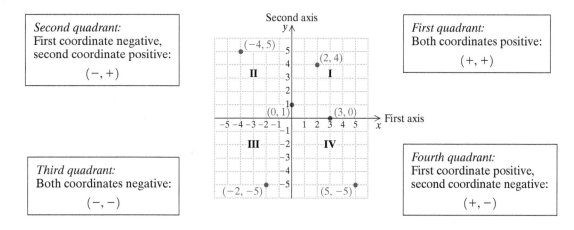

Second quadrant:
First coordinate negative, second coordinate positive:
$(-, +)$

First quadrant:
Both coordinates positive:
$(+, +)$

Third quadrant:
Both coordinates negative:
$(-, -)$

Fourth quadrant:
First coordinate positive, second coordinate negative:
$(+, -)$

Student Notes

The scale on the horizontal axis may be different from the scale on the vertical axis. Within each axis, however, the number of units that each mark represents must be constant.

C. Numbering the Axes Appropriately

We draw only part of the plane when we create a graph. Although it is standard to show the origin and parts of all quadrants, as in the graphs in Examples 3 and 4, sometimes it is more practical to show a different portion of the plane. Note, too, that on the graphs in Examples 3 and 4, the marks on the axes are one unit apart. In this case, we say that the **scale** of each axis is 1. Often it is necessary to use a different scale on one or both of the axes.

Technology Connection

The portion of the coordinate plane shown by a graphing calculator is called a **viewing window**. We indicate the **dimensions** of the window by setting a minimum x-value, a maximum x-value, a minimum y-value, and a maximum y-value. The **scale** by which we count must also be chosen. Window settings are often abbreviated in the form [L, R, B, T], with the letters representing **L**eft, **R**ight, **B**ottom, and **T**op endpoints. The window $[-10, 10, -10, 10]$ is called the **standard viewing window**. On most graphing calculators, a standard viewing window can be set up using an option in the ZOOM menu.

To set up a $[-100, 100, -5, 5]$ window, we press WINDOW and use the following settings. A scale of 10 for the x-axis is large enough for the marks on the x-axis to be distinct.

```
WINDOW
  Xmin=-100
  Xmax=100
  Xscl=10
  Ymin=-5
  Ymax=5
  Yscl=1
  Xres=1
```

VIDEO

When GRAPH is pressed, a graph extending from -100 to 100 along the x-axis (counted by 10's) and from -5 to 5 along the y-axis appears. When the arrow keys are pressed, a cursor can be moved, its coordinates appearing at the bottom of the window.

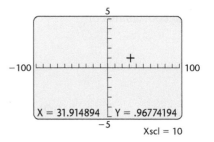

X = 31.914894   Y = .96774194   Xscl = 10

Set up the following viewing windows, choosing an appropriate scale for each axis. Then move the cursor and practice reading coordinates.

1. $[-10, 10, -10, 10]$
2. $[-5, 5, 0, 100]$
3. $[-1, 1, -0.1, 0.1]$

EXAMPLE 5   Use a grid 10 squares wide and 10 squares high to plot $(-34, 450)$, $(48, 95)$, and $(10, -200)$.

SOLUTION   Since x-coordinates vary from a low of -34 to a high of 48, the 10 horizontal squares must span $48 - (-34)$, or 82 units. Because 82 is not a multiple of 10, we round *up* to the next multiple of 10, which is 90. Dividing 90 by 10, we find that if each square is 9 units wide (has a scale of 9), we could represent all the x-values. However, since it is more convenient to count by 10's, we will instead use a scale of 10. Starting at 0, we count backward to -40 and forward to 60.

This is how we will arrange the x-axis.

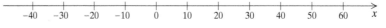

There is more than one correct way to cover the values from -34 to 48 using 10 increments. For instance, we could have counted from -60 to 90, using a scale of 15. In general, we try to use the smallest span and scale that will cover the given coordinates. Scales that are multiples of 2, 5, or 10 are especially convenient. The numbering always begins at the origin.

Since we must be able to show y-values from -200 to 450, the 10 vertical squares must span $450 - (-200)$, or 650 units. For convenience, we round 650 *up* to 700 and then divide by 10: $700 \div 10 = 70$. Using 70 as the scale, we count *down* from 0 until we pass -200 and *up* from 0 until we pass 450, as shown at left.

Next, we use the x-axis and the y-axis that we just developed to form a grid. In order for us to be able to place the axes correctly, the two 0's must

This is how we will arrange the y-axis.

y
490
420
350
280
210
140
70
0
-70
-140
-210

↪ **Chapter Resource:**
Collaborative Activity, p. 222

coincide where the axes cross. Finally, once the graph is numbered, we plot the points as shown.

5. Use a grid 10 squares wide and 10 squares high to plot $(15, -2)$, $(3, 60)$, and $(-20, 12)$. Scales may vary.

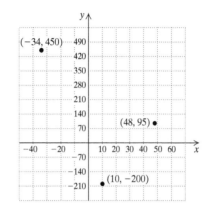

Each axis can have its own scale, as shown here.

↪ YOUR TURN

3.1 EXERCISE SET

FOR EXTRA HELP MyMathLab®

↪ Vocabulary and Reading Check

Match the letter from the graph with the most appropriate term. Not all letters will be used.

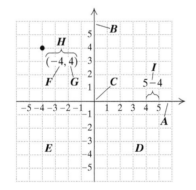

1. ___ Third quadrant

2. ___ Origin

3. ___ Second coordinate

4. ___ y-axis

5. ___ Ordered pair

6. ___ Scale

↪ Concept Reinforcement

In each of Exercises 7–10, match the set of coordinates with the graph below that would be the best for plotting the points.

7. ___ $(-9, 3)$, $(-2, -1)$, $(4, 5)$

8. ___ $(-2, -1)$, $(1, 5)$, $(7, 3)$

9. ___ $(-2, -9)$, $(2, 1)$, $(4, -6)$

10. ___ $(-2, -1)$, $(-9, 3)$, $(-4, -6)$

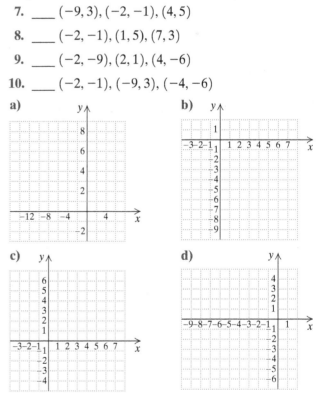

A. Problem Solving with Bar Graphs and Line Graphs

Driving under the Influence. A blood-alcohol level of 0.08% or higher makes driving illegal in the United States. The following bar graph shows how many drinks a person of a certain weight would need to consume in 1 hr in order to reach a blood-alcohol level of 0.08%. Note that a 12-oz beer, a 5-oz glass of wine, or a cocktail containing $1\frac{1}{2}$ oz of distilled liquor all count as one drink.

Data: soberup.com and vsa.vassar.edu

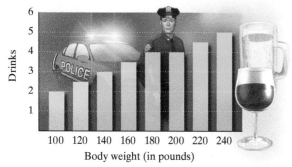

Friends Don't Let Friends Drive Drunk

11. Approximately how many drinks would a 100-lb person have consumed in 1 hr to reach a blood-alcohol level of 0.08%?

12. Approximately how many drinks would a 160-lb person have consumed in 1 hr to reach a blood-alcohol level of 0.08%?

13. What can you conclude about the weight of someone who has consumed 3 drinks in 1 hr without reaching a blood-alcohol level of 0.08%?

14. What can you conclude about the weight of someone who has consumed 4 drinks in 1 hr without reaching a blood-alcohol level of 0.08%?

Education Degrees. The following line graph shows the percentage of bachelor's degrees conferred in education-related fields for various years.

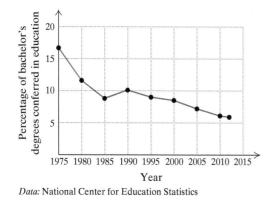

Data: National Center for Education Statistics

15. Approximately what percentage of bachelor's degrees were conferred in education in 1980?

16. Approximately what percentage of bachelor's degrees were conferred in education in 2012?

17. In which two years were approximately 9% of bachelor's degrees conferred in education?

18. In which two years were approximately 10% of bachelor's degrees conferred in education?

19. In which year after 1985 was the percentage of bachelor's degrees conferred in education the highest?

20. Between which years did the percentage of bachelor's degrees conferred in education increase?

B. Points and Ordered Pairs

Plot each group of points.

21. $(1, 2), (-2, 3), (4, -1), (-5, -3), (4, 0), (0, -2)$

22. $(-2, -4), (4, -3), (5, 4), (-1, 0), (-4, 4), (0, 5)$

23. $(4, 4), (-2, 4), (5, -3), (-5, -5), (0, 4), (0, -4),$ $(-4, 0), (0, 0)$

24. $(2, 5), (-1, 3), (3, -2), (-2, -4), (0, 0), (0, -5),$ $(5, 0), (-5, 0)$

25. *Text Messaging.* Listed below are estimates of the number of text messages sent in the United States for various years. Make a line graph of the data.

Year	Number of Monthly Text Messages (in billions)
2009	161
2010	247
2011	367
2012	423
2013	498
2014	561

Data: CTIA—The Wireless Association®

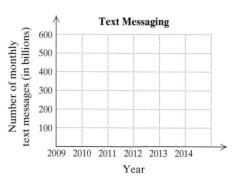

26. *Ozone Layer.* Make a line graph of the data in the following table.

Year	Minimum Daily Ozone Level in the Southern Polar Region (in Dobson Units)
2009	96
2010	118
2011	95
2012	124
2013	116
2014	114

Data: National Aeronautics and Space Administration

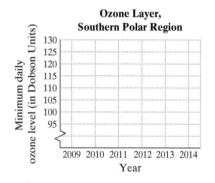

In Exercises 27–30, determine the coordinates of points *A, B, C, D,* and *E.*

27.

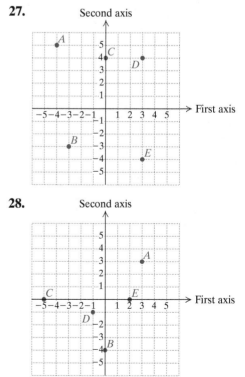

28.

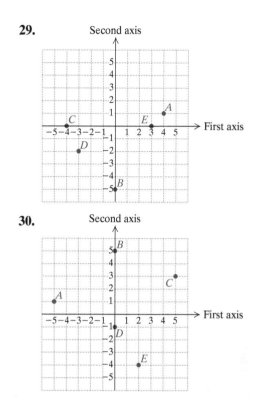

29.

30.

In which quadrant or on which axis is each point located?

31. $(7, -2)$	**32.** $(-1, -4)$	**33.** $(-4, -3)$
34. $(1, -5)$	**35.** $(0, -3)$	**36.** $(6, 0)$
37. $(-4.9, 8.3)$	**38.** $(7.5, 2.9)$	**39.** $\left(-\frac{5}{2}, 0\right)$
40. $(0, 2.8)$	**41.** $(160, 2)$	**42.** $\left(-\frac{1}{2}, 2000\right)$

43. In which quadrants are the first coordinates positive?

44. In which quadrants are the second coordinates negative?

45. In which quadrants do both coordinates have the same sign?

46. In which quadrants do the first and second coordinates have opposite signs?

C. Numbering the Axes Appropriately

In Exercises 47–56, use a grid 10 squares wide and 10 squares tall to plot the given coordinates. Choose your scale carefully. Scales may vary.

47. $(-75, 5), (-18, -2), (9, -4)$

48. $(-13, 3), (48, -1), (62, -4)$

49. $(-1, 83), (-5, -14), (5, 37)$

50. $(2, -79), (4, -25), (-4, 12)$

51. $(-10, -4), (-16, 7), (3, 15)$

52. $(5, -16), (-7, -4), (12, 3)$

53. $(-100, -5), (350, 20), (800, 37)$

54. $(750, -8), (-150, 17), (400, 32)$

55. $(-83, 491), (-124, -95), (54, -238)$

56. $(738, -89), (-49, -6), (-165, 53)$

57. The following graph was included in a mailing sent by Agway® to their oil customers in 2000. What information is missing from the graph and why might the graph be considered misleading?

Residential Fuel Oil and Natural Gas Prices

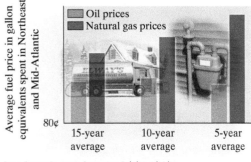

Data: Energy Research Center Inc. *3/1/99–2/29/00

58. What do all points plotted on the vertical axis of a graph have in common?

Skill Review

Calculate each of the following.

59. $-\frac{1}{2} + \frac{1}{3}$ [1.5]

60. $-2.6 - 9.1$ [1.6]

61. $3(-40)$ [1.7]

62. $-100 \div (-20)$ [1.7]

63. $(-1)(-2)(-3)(-4)$ [1.7]

64. $3 + (-4) + (-6) + 10$ [1.5]

Synthesis

65. Describe what the result would be if the first and second coordinates of every point in the following graph of an arrow were interchanged.

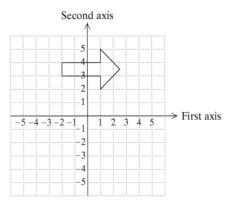

66. The graph accompanying Example 2 flattens out. Why do you think this occurs?

67. In which quadrant(s) could a point be located if its coordinates are opposites of each other?

68. In which quadrant(s) could a point be located if its coordinates are reciprocals of each other?

69. The points $(-1, 1), (4, 1),$ and $(4, -5)$ are three corners of a rectangle. Find the coordinates of the fourth corner.

70. The pairs $(-2, -3), (-1, 2),$ and $(4, -3)$ can serve as three (of four) vertices for three different parallelograms. Find the fourth vertex of each parallelogram.

71. Graph eight points such that the sum of the coordinates in each pair is 7. Answers may vary.

72. Find the perimeter of a rectangle if three of its vertices are $(5, -2), (-3, -2),$ and $(-3, 3)$.

73. Find the area of a triangle whose vertices have coordinates $(0, 9), (0, -4),$ and $(5, -4)$.

Sorting Solid Waste. *Circle graphs, or pie charts, are often used to show what percent of the whole each item in a group represents. Use the following pie chart to answer Exercises 74–77.*

Sorting Solid Waste

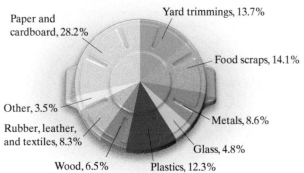

Data: U.S. Environmental Protection Agency

74. In 2012, Americans generated 250.9 million tons of waste. How much of the waste was glass?

75. In 2012, the average American generated 4.38 lb of waste per day. How much of that was yard trimmings?

76. Americans are recycling about 34.1% of all glass that is in the waste stream. How much glass did Americans recycle in 2012? (See Exercise 74.)

77. Americans are recycling about 13.5% of all yard trimmings. What amount of yard trimmings did the average American recycle per day in 2012? (Use the information in Exercise 75.)

Coordinates on the Globe. Coordinates can also be used to describe a location on a sphere: 0° latitude is the equator and 0° longitude is a line from the North Pole to the South Pole through France and Algeria. In the following figure, hurricane Clara is at a point about 260 mi northwest of Bermuda near latitude 36.0° North, longitude 69.0° West.

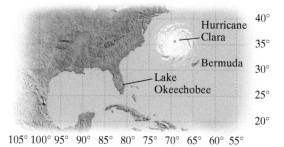

78. Approximate the latitude and the longitude of Bermuda.

79. Approximate the latitude and the longitude of Lake Okeechobee.

80. In the *Star Trek* science-fiction series, a three-dimensional coordinate system is used to locate objects in space. If the center of a planet is used as the origin, how many "quadrants" will exist? Why? If possible, sketch a three-dimensional coordinate system and label each "quadrant."

81. *Research.* Find the minimum daily ozone layer for the Southern Polar Region for the most recent year. (See Exercise 26.) Extend the line graph in Exercise 26 to include this information. How would you describe the trends shown by this graph?

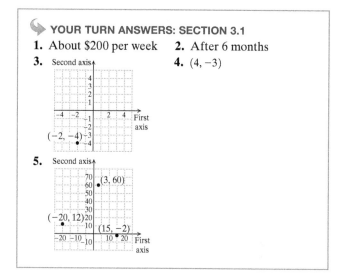

YOUR TURN ANSWERS: SECTION 3.1

1. About $200 per week   **2.** After 6 months
3.
4. (4, −3)
5.

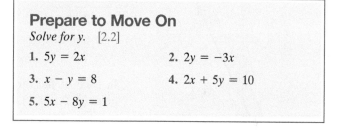

Prepare to Move On

Solve for y. [2.2]

1. $5y = 2x$ **2.** $2y = -3x$

3. $x - y = 8$ **4.** $2x + 5y = 10$

5. $5x - 8y = 1$

3.2 | Graphing Linear Equations

A. Solutions of Equations   **B.** Graphing Linear Equations   **C.** Applications

Not only do graphs help us to visualize data, but they can also be used to represent solutions of equations.

A. Solutions of Equations

When an equation contains two variables, solutions are ordered pairs. Each number in the pair replaces a letter in the equation. Unless stated otherwise, the first number in each pair replaces the variable that occurs first alphabetically.

EXAMPLE 1 Determine whether each of the following pairs is a solution of $4b - 3a = 22$: **(a)** $(2, 7)$; **(b)** $(1, 6)$.

Student Notes

It is sometimes—but not always—necessary to use parentheses when substituting for a variable. It is always safe to do so, and developing the habit of using parentheses when substituting may reduce errors.

1. Determine whether the ordered pair $(2, -3)$ is a solution of $6c - d = 15$.

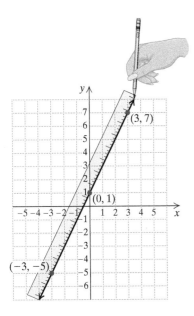

2. Show that the pairs $(2, -3)$ and $(0, -5)$ are solutions of $x - y = 5$. Then graph the two points to find another pair that is a solution.

SOLUTION

a) We substitute 2 for a and 7 for b (alphabetical order of variables):

$$4b - 3a = 22$$

$$\frac{4(7) - 3(2) \quad\big|\quad 22}{\qquad 28 - 6 \quad\big|}$$

$$22 \overset{?}{=} 22 \quad \text{TRUE}$$

Since $22 = 22$ is *true*, the pair $(2, 7)$ *is* a solution.

b) In this case, we replace a with 1 and b with 6:

$$4b - 3a = 22$$

$$\frac{4(6) - 3(1) \quad\big|\quad 22}{\qquad 24 - 3 \quad\big|}$$

$$21 \overset{?}{=} 22 \quad \text{FALSE} \qquad 21 \neq 22$$

Since $21 = 22$ is *false*, the pair $(1, 6)$ is *not* a solution.

YOUR TURN

EXAMPLE 2 Show that the pairs $(3, 7)$, $(0, 1)$, and $(-3, -5)$ are solutions of $y = 2x + 1$. Then graph the three points to find another pair that is a solution.

SOLUTION To show that a pair is a solution, we substitute, replacing x with the first coordinate and y with the second coordinate of each pair:

$y = 2x + 1$	$y = 2x + 1$	$y = 2x + 1$
$7 \mid 2 \cdot 3 + 1$	$1 \mid 2 \cdot 0 + 1$	$-5 \mid 2(-3) + 1$
$\mid 6 + 1$	$\mid 0 + 1$	$\mid -6 + 1$
$7 \overset{?}{=} 7$ TRUE	$1 \overset{?}{=} 1$ TRUE	$-5 \overset{?}{=} -5$ TRUE

In each of the three cases, the substitution results in a true equation. Thus the pairs $(3, 7)$, $(0, 1)$, and $(-3, -5)$ are all solutions. We graph them as shown at left.

The three points appear to "line up," or are *collinear*. Will other points that line up with these points also represent solutions of $y = 2x + 1$? To find out, we draw a line passing through $(-3, -5)$, $(0, 1)$, and $(3, 7)$. The line appears to pass through $(2, 5)$. Let's check to see if this pair is a solution of $y = 2x + 1$:

$$y = 2x + 1$$

$$\frac{5 \quad\big|\quad 2 \cdot 2 + 1}{\quad\big|\quad 4 + 1}$$

$$5 \overset{?}{=} 5 \quad \text{TRUE}$$

We see that $(2, 5)$ *is* a solution.

YOUR TURN

In Example 2, any point on the line represents a solution of $y = 2x + 1$. The line is called the *graph* of the equation and it represents *all* solutions of the equation.

> ### SOLUTIONS OF EQUATIONS IN TWO VARIABLES
> - Solutions of equations in two variables are ordered pairs.
> - Numbers in ordered pairs replace variables in equations in alphabetical order, unless stated otherwise.
> - To graph an equation is to make a drawing that represents its solutions.
> - Every solution of an equation is represented by a point on its graph, and every point on the graph of an equation represents a solution.

B. Graphing Linear Equations

Equations like $y = 2x + 1$ or $4b - 3a = 22$ are said to be **linear** because the **graph** of each equation is a line. In general, any equation that can be written in the form $y = mx + b$ or $Ax + By = C$ (where $m, b, A, B,$ and C are constants and A and B are not both 0) is linear.

Linear equations can be graphed as follows.

> **TO GRAPH A LINEAR EQUATION**
> 1. Select a value for one coordinate and calculate the corresponding value of the other coordinate. Form an ordered pair. This pair is one solution of the equation.
> 2. Repeat step (1) to find a second ordered pair. A third ordered pair serves as a check when all three points line up.
> 3. Plot the ordered pairs and draw a straight line passing through the points. The line represents all solutions of the equation.

EXAMPLE 3   Graph: $y = -3x + 1$.

SOLUTION   Since $y = -3x + 1$ is in the form $y = mx + b$, the equation is linear and the graph is a straight line. We select a convenient value for x, compute y, and form an ordered pair. Then we repeat the process for other choices of x.

If $x = 2$,   then $y = -3 \cdot 2 + 1 = -5$,   and $(2, -5)$ is a solution.
If $x = 0$,   then $y = -3 \cdot 0 + 1 = 1$,   and $(0, 1)$ is a solution.
If $x = -1$,   then $y = -3(-1) + 1 = 4$,   and $(-1, 4)$ is a solution.

Results are often listed in a table, as shown below. The points corresponding to each pair are then plotted.

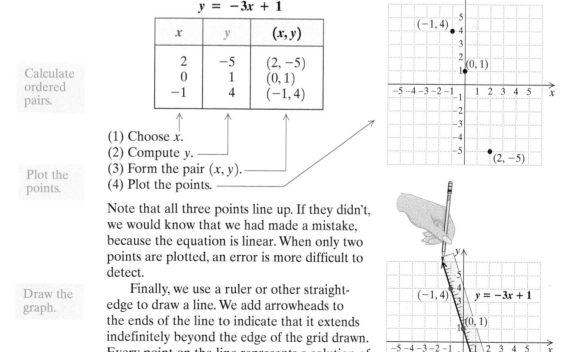

$$y = -3x + 1$$

x	y	(x, y)
2	−5	$(2, -5)$
0	1	$(0, 1)$
−1	4	$(-1, 4)$

Calculate ordered pairs.

(1) Choose x.
(2) Compute y.
(3) Form the pair (x, y).
(4) Plot the points.

Plot the points.

Note that all three points line up. If they didn't, we would know that we had made a mistake, because the equation is linear. When only two points are plotted, an error is more difficult to detect.

Draw the graph.

Finally, we use a ruler or other straight-edge to draw a line. We add arrowheads to the ends of the line to indicate that it extends indefinitely beyond the edge of the grid drawn. Every point on the line represents a solution of $y = -3x + 1$.

3. Graph: $y = -2x + 3$.

YOUR TURN

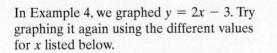

EXPLORING THE CONCEPT

In Example 4, we graphed $y = 2x - 3$. Try graphing it again using the different values for x listed below.

1. Complete the following table.
2. Plot the points on graph paper and draw the graph.
3. Does the choice of x-values affect the graph?

x	y	(x, y)
-1		
2		
3		

ANSWERS
1. y: $-5, 1, 3$; (x, y): $(-1, -5), (2, 1), (3, 3)$
2. Same graph as in Example 4
3. No

4. Graph: $y = 3x - 1$.
_____ YOUR TURN

EXAMPLE 4   Graph: $y = 2x - 3$.

SOLUTION   We select some convenient x-values and compute y-values.

If $x = 0$,   then $y = 2 \cdot 0 - 3 = -3$,   and $(0, -3)$ is a solution.
If $x = 1$,   then $y = 2 \cdot 1 - 3 = -1$,   and $(1, -1)$ is a solution.
If $x = 4$,   then $y = 2 \cdot 4 - 3 = 5$,   and $(4, 5)$ is a solution.

$$y = 2x - 3$$

x	y	(x, y)
0	-3	$(0, -3)$
1	-1	$(1, -1)$
4	5	$(4, 5)$

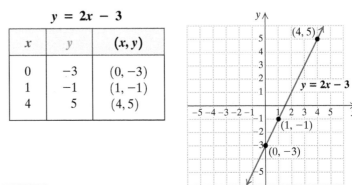

ALF
Active Learning Figure

EXAMPLE 5   Graph: $4x + 2y = 12$.

SOLUTION   To form ordered pairs, we can replace either variable with a number and then calculate the other coordinate:

If $y = 0$, we have $4x + 2 \cdot 0 = 12$
$$4x = 12$$
$$x = 3,$$

so $(3, 0)$ is a solution.

If $x = 0$, we have $4 \cdot 0 + 2y = 12$
$$2y = 12$$
$$y = 6,$$

so $(0, 6)$ is a solution.

If $y = 2$, we have $4x + 2 \cdot 2 = 12$
$$4x + 4 = 12$$
$$4x = 8$$
$$x = 2,$$

so $(2, 2)$ is a solution.

$$4x + 2y = 12$$

x	y	(x, y)
3	0	$(3, 0)$
0	6	$(0, 6)$
2	2	$(2, 2)$

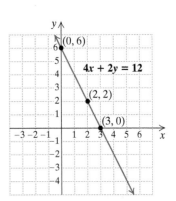

5. Graph: $2x + 3y = 6$.
_____ YOUR TURN

Note that in Examples 3 and 4 the variable y is isolated on one side of the equation. This generally simplifies calculations, so it is important to be able to solve for y before graphing.

EXAMPLE 6   Graph $3y = 2x$ by first solving for y.

SOLUTION   To isolate y, we divide both sides by 3, or multiply both sides by $\frac{1}{3}$:

$$3y = 2x$$

$$\tfrac{1}{3} \cdot 3y = \tfrac{1}{3} \cdot 2x \qquad \text{Using the multiplication principle to multiply both sides by } \tfrac{1}{3}$$

$$\left.\begin{array}{l} 1y = \tfrac{2}{3} \cdot x \\[4pt] y = \tfrac{2}{3} x. \end{array}\right\} \quad \text{Simplifying}$$

Because all the equations above are equivalent, we can use $y = \frac{2}{3}x$ to draw the graph of $3y = 2x$.

To graph $y = \frac{2}{3}x$, we can select x-values that are multiples of 3. This allows us to avoid fractions when corresponding y-values are computed.

$$\left.\begin{array}{lll} \text{If } x = 3, & \text{then } y = \tfrac{2}{3} \cdot 3 = 2. \\ \text{If } x = -3, & \text{then } y = \tfrac{2}{3}(-3) = -2. \\ \text{If } x = 6, & \text{then } y = \tfrac{2}{3} \cdot 6 = 4. \end{array}\right\} \quad \begin{array}{l} \text{Note that when multiples of 3 are} \\ \text{substituted for } x, \text{ the } y\text{-coordinates} \\ \text{are not fractions.} \end{array}$$

The following table lists these solutions. Next, we plot the points and see that they form a line. Finally, we draw and label the line.

$3y = 2x$, or $y = \frac{2}{3}x$

x	y	(x, y)
3	2	$(3, 2)$
-3	-2	$(-3, -2)$
6	4	$(6, 4)$

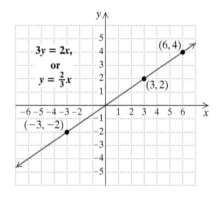

6. Graph $3y = 5x$ by first solving for y.

YOUR TURN

EXAMPLE 7   Graph $x + 5y = -10$ by first solving for y.

SOLUTION   We have

$$x + 5y = -10$$

$$5y = -x - 10 \qquad \text{Adding } -x \text{ to both sides}$$

$$y = \tfrac{1}{5}(-x - 10) \qquad \text{Multiplying both sides by } \tfrac{1}{5}$$

$$y = -\tfrac{1}{5}x - 2. \qquad \text{Using the distributive law}$$

CAUTION!   It is very important to multiply *both* $-x$ and -10 by $\frac{1}{5}$.

Thus, $x + 5y = -10$ is equivalent to $y = -\frac{1}{5}x - 2$. If we now choose x-values that are multiples of 5, we can avoid fractions when calculating the corresponding y-values.

If $x = 5$, then $y = -\frac{1}{5} \cdot 5 - 2 = -1 - 2 = -3$.

If $x = 0$, then $y = -\frac{1}{5} \cdot 0 - 2 = 0 - 2 = -2$.

If $x = -5$, then $y = -\frac{1}{5}(-5) - 2 = 1 - 2 = -1$.

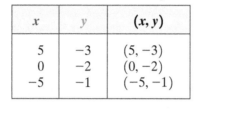

$x + 5y = -10$, or $y = -\frac{1}{5}x - 2$

x	y	(x, y)
5	−3	(5, −3)
0	−2	(0, −2)
−5	−1	(−5, −1)

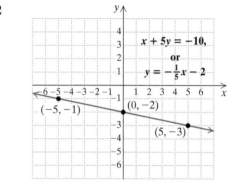

7. Graph $x + 2y = -6$ by first solving for y.

YOUR TURN

C. Applications

Linear equations appear in many real-life situations.

EXAMPLE 8 *Fuel Cost.* The amount of fuel that an engine burns while idling depends on the size of the engine. The table at left lists the annual cost of idling for 5 min per day for various engine sizes. These costs are based on a gasoline price of $3.75 per gallon. The annual idling cost can be approximated by

$$c = 19a - 4,$$

where c is the annual cost of idling 5 min per day and a is the engine size, in liters. Graph the equation and then use the graph to estimate the annual idling cost for a 5.3-L engine.

SOLUTION We graph the equation by selecting values for a and then calculating the associated values of c.

If $a = 2$, then $c = 19(2) - 4 = 34$.

If $a = 4$, then $c = 19(4) - 4 = 72$.

If $a = 6$, then $c = 19(6) - 4 = 110$.

a	c
2	34
4	72
6	110

Engine Size (in liters)	Annual Idling Cost
1.8	$ 30
2.0	34
3.5	62
4.3	75
5.7	109
6.2	112

Data: U.S. Department of Energy, quoted in www.greenenergytimes.org, October 15, 2011

Because we are *selecting* values for a and *calculating* values for c, we represent a on the horizontal axis and c on the vertical axis. Counting by 0.5 horizontally and by 10 vertically will allow us to plot all three pairs, as shown on the left below.

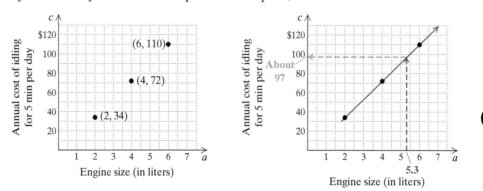

8. Use the graph in Example 8 to estimate the annual cost of idling for 5 min per day for a 2.5-L engine.

Since the three points are collinear, our calculations are probably correct. We draw a line, as shown on the right at the bottom of the preceding page. We begin the line at $a = 1.8$, since that is the smallest engine size given in the table above. To estimate the idling cost for a 5.3-L engine, we locate the point on the line that is above 5.3 and then find the value on the c-axis that corresponds to that point. The idling cost for a 5.3-L engine is about $97.

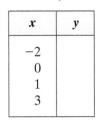 YOUR TURN

↪Check Your UNDERSTANDING

> **CAUTION!** When the coordinates of a point are read from a graph, as in Example 8, values should not be considered exact.

1. Complete the following table to find four solutions of the equation $2x + y = 5$.

x	y
-2	
0	
1	
3	

Many equations in two variables have graphs that are not straight lines. Three such *nonlinear* graphs are shown below. As before, each graph represents the solutions of the given equation.

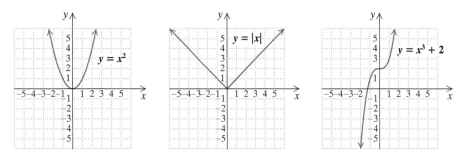

2. Use the table in Exercise 1 to list four ordered pairs that are on the graph of $2x + y = 5$.

Technology Connection

Most graphing calculators require that y be alone on one side before the equation is entered. For example, to graph $5y + 4x = 13$, we would first solve for y. The student can check that solving for y yields $y = -\frac{4}{5}x + \frac{13}{5}$.

We press (Y=), enter $-\frac{4}{5}x + \frac{13}{5}$ as Y1, and press (GRAPH). The graph is shown here in the standard viewing window $[-10, 10, -10, 10]$.

There are many graphing calculator apps available for mobile devices. These can also be used to graph equations and to study characteristics of graphs. The graph of $y = -\frac{4}{5}x + \frac{13}{5}$ is shown below in one such app.

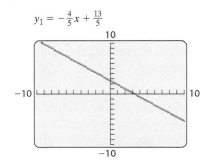

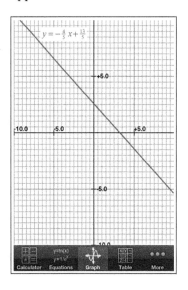

Using a graphing calculator, graph each of the following. Select the "standard" $[-10, 10, -10, 10]$ window.

1. $y = -5x + 6.5$
2. $y = 3x + 4.5$
3. $7y - 4x = 22$
4. $5y + 11x = -20$
5. $2y - x^2 = 0$
6. $y + x^2 = 8$

VIDEO

3.2 EXERCISE SET

FOR EXTRA HELP   MyMathLab®

Vocabulary and Reading Check

In each of Exercises 1–6, classify the statement as either true or false.

1. A linear equation in two variables has at most one solution.

2. Every solution of $y = 3x - 7$ is an ordered pair.

3. The graph of $y = 3x - 7$ represents all solutions of the equation.

4. If a point is on the graph of $y = 3x - 7$, the corresponding ordered pair is a solution of the equation.

5. To find a solution of $y = 3x - 7$, we can choose any value for x and calculate the corresponding y-value.

6. The graph of every equation is a straight line.

A. Solutions of Equations

Determine whether each equation has the given ordered pair as a solution.

7. $y = 4x - 7$; $(2, 1)$

8. $y = 5x + 8$; $(0, 8)$

9. $3y + 4x = 19$; $(5, 1)$

10. $5x - 3y = 15$; $(0, 5)$

11. $4m - 5n = 7$; $(3, -1)$

12. $3q - 2p = -8$; $(1, -2)$

In each of Exercises 13–20, an equation and two ordered pairs are given. Show that each pair is a solution and graph the two pairs to find another solution. Answers may vary.

13. $y = x + 3$; $(-1, 2), (4, 7)$

14. $y = x - 2$; $(3, 1), (-2, -4)$

15. $y = \frac{1}{2}x + 3$; $(4, 5), (-2, 2)$

16. $y = \frac{1}{2}x - 1$; $(6, 2), (0, -1)$

17. $y + 3x = 7$; $(2, 1), (4, -5)$

18. $2y + x = 5$; $(-1, 3), (7, -1)$

19. $4x - 2y = 10$; $(0, -5), (4, 3)$

20. $6x - 3y = 3$; $(1, 1), (-1, -3)$

B. Graphing Linear Equations

Graph each equation.

21. $y = x + 1$

22. $y = x - 1$

23. $y = -x$

24. $y = x$

25. $y = 2x$

26. $y = -3x$

27. $y = 2x + 2$

28. $y = 3x - 2$

29. $y = -\frac{1}{2}x$

30. $y = \frac{1}{4}x$

31. $y = \frac{1}{3}x - 4$

32. $y = \frac{1}{2}x + 1$

33. $x + y = 4$

34. $x + y = -5$

35. $x - y = -2$

36. $y - x = 3$

37. $x + 2y = -6$

38. $x + 2y = 8$

39. $y = -\frac{2}{3}x + 4$

40. $y = \frac{3}{2}x + 1$

41. $4x = 3y$

42. $2x = 5y$

43. $5x - y = 0$

44. $3x - 5y = 0$

45. $6x - 3y = 9$

46. $8x - 4y = 12$

47. $6y + 2x = 8$

48. $8y + 2x = -4$

C. Applications

49. *Student Aid.* The average award a of federal student financial assistance per student is approximated by
$$a = 0.08t + 2.5,$$
where a is in thousands of dollars and t is the number of years after 1994. Graph the equation and use the graph to estimate the average amount of federal student aid per student in 2018.

Data: U.S. Department of Education, Office of Postsecondary Education

50. *Value of a Color Copier.* The value of Duplio-graphic's color copier is given by
$$v = -0.68t + 3.4,$$
where v is the value, in thousands of dollars, t years from the date of purchase. Graph the equation and use the graph to estimate the value of the copier after $2\frac{1}{2}$ years.

51. *Fuel Efficiency.* A typical tractor-trailer will move 18 tons of air per mile at 55 mph. Air resistance increases with speed, causing fuel efficiency to decrease at higher speeds. At highway speeds, a certain truck's fuel efficiency t, in miles per gallon (mpg), can be approximated by
$$t = -0.1s + 13.1,$$
where s is the speed of the truck, in miles per hour (mph). Graph the equation and then use the graph to estimate the fuel efficiency at 66 mph.

Data: Kenworth Truck Co.

52. *Increasing Life Expectancy.* A smoker is 15 times more likely to die of lung cancer than a nonsmoker. An ex-smoker who stopped smoking t years ago is w times more likely to die of lung cancer than a nonsmoker, where

$$w = 15 - t.$$

Graph the equation and use the graph to estimate how much more likely it is for Sandy to die of lung cancer than Polly, if Polly never smoked and Sandy quit $2\frac{1}{2}$ years ago.

Data: *Body Clock* by Dr. Martin Hughes, p. 60. New York: Facts on File, Inc.

53. *Photobook Pricing.* The price p, in dollars, of an 8-in. by 8-in. hardcover photobook is given by

$$p = 0.7n + 16,$$

where n is the number of pages in the photobook. Graph the equation and use the graph to estimate the price of a photobook containing 25 pages.

Data: www.shutterfly.com

54. *Value of Computer Software.* The value v of a shopkeeper's inventory software program, in hundreds of dollars, is given by

$$v = -\tfrac{3}{4}t + 6,$$

where t is the number of years since the shopkeeper first bought the program. Graph the equation and use the graph to estimate what the program is worth 4 years after it was first purchased.

55. *Apps.* In the first two years after the iPad was launched, the number of apps available for that device could be estimated by

$$a = 0.23d - 7,$$

where a is the number of apps, in thousands, and d is the number of days after the iPad was launched. Graph the equation and use the graph to estimate the number of apps available 400 days after the iPad was launched.

Data: PadGadget Apps Tracker and Apple App Store

56. *Record Temperature Drop.* On January 22, 1943, the temperature T, in degrees Fahrenheit, in Spearfish, South Dakota, could be approximated by

$$T = -2m + 54,$$

where m is the number of minutes since 9:00 A.M. that morning. Graph the equation and use the graph to estimate the temperature at 9:15 A.M.

Data: National Oceanic Atmospheric Administration

57. *Cost of College.* The cost T, in hundreds of dollars, of tuition and fees at many community colleges can be approximated by

$$T = \tfrac{5}{4}c + 2,$$

where c is the number of credits for which a student registers. Graph the equation and use the graph to estimate the cost of tuition and fees when a student registers for 4 three-credit courses.

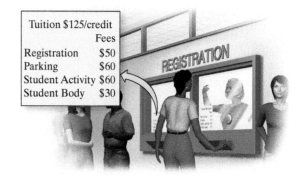

58. *Cost of College.* The tuition cost C, in thousands of constant 2014 dollars, of a year at a public two-year college can be approximated by

$$C = \tfrac{1}{12}t + 2.5,$$

where t is the number of years after 2004. Graph the equation and use the graph to predict the cost of a year at a public two-year college in 2020.

Data: *Trends in College Pricing* 2014, College Board Advocacy and Policy Center

59. The equations $3x + 4y = 8$ and $y = -\tfrac{3}{4}x + 2$ are equivalent. Which equation would be easier to graph and why?

60. Suppose that a linear equation is graphed by plotting three points and that the three points line up. Does this *guarantee* that the equation is being correctly graphed? Why or why not?

Skill Review

Solve and check. [2.2]

61. $5x + 3(2 - x) = 12$

62. $3(y - 5) - 8y = 6$

Solve. [2.3]

63. $A = \dfrac{T + Q}{2}$, for Q

64. $pq + p = w$, for p

65. $Ax + By = C$, for y

66. $\dfrac{y - k}{m} = x - h$, for y

Synthesis

67. Janice consistently makes the mistake of plotting the x-coordinate of an ordered pair using the y-axis, and the y-coordinate using the x-axis. How will Janice's incorrect graph compare with the correct graph?

68. Explain how the graph in Example 8 can be used to determine the engine size for which the annual idling cost is $100.

69. *Bicycling.* Long Beach Island in New Jersey is a long, narrow, flat island. For exercise, Laura routinely bikes to the northern tip of the island and back. Because of steady wind, she uses one gear going north and another for her return. Laura's bike has 21 gears and the sum of the two gears used on her ride is always 24. Write and graph an equation that represents the different pairings of gears that Laura might use. Note that there are no fraction gears on a bicycle.

In each of Exercises 70–73, find an equation for the graph shown.

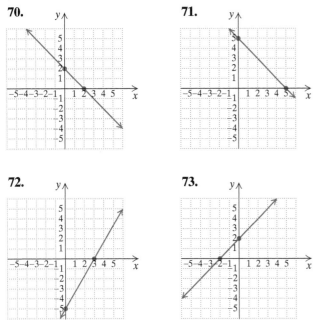

70.

71.

72.

73.

74. Translate to an equation:

  d dimes and *n* nickels total $1.75.

  Then graph the equation and use the graph to determine three different combinations of dimes and nickels that total $1.75. (See also Exercise 87.)

75. Translate to an equation:

  d $25 dinners and *l* $5 lunches total $225.

  Then graph the equation and use the graph to determine three different combinations of lunches and dinners that total $225 (see also Exercise 87).

Use the suggested x-values $-3, -2, -1, 0, 1, 2,$ *and* 3 *to graph each equation.*

76. $y = |x|$ Aha! **77.** $y = -|x|$

Aha! **78.** $y = |x| - 2$ **79.** $y = x^2$

80. $y = x^2 + 1$

For each of Exercises 81–86, use a graphing calculator to graph the equation. Use a $[-10, 10, -10, 10]$ *window.*

81. $y = -2.8x + 3.5$

82. $y = 4.5x + 2.1$

83. $y = 2.8x - 3.5$

84. $y = -4.5x - 2.1$

85. $y = x^2 + 4x + 1$

86. $y = -x^2 + 4x - 7$

87. Study each graph in Exercises 74 and 75. Does *every* point on the graph represent a solution of the associated problem? Why or why not?

88. *Research.* Find the size of the engine in your car or a friend's car. (See Example 8.)

  a) Use the results of Example 8 to estimate the annual cost of idling this engine for 5 min per day.

  b) The data in Example 8 were based on a gasoline price of $3.75 per gallon. Estimate the annual idling cost of this engine at current gasoline prices using the formula

  $$x = c \cdot p \div 3.75,$$

  where x is the current idling cost, c is the cost calculated in part (a), and p is the current gasoline price per gallon.

89. Exercise 51 discusses fuel efficiency. If fuel costs $3.50 per gallon, how much money will a truck driver save on a 500-mi trip by driving at 55 mph instead of 70 mph? How many gallons of fuel will be saved?

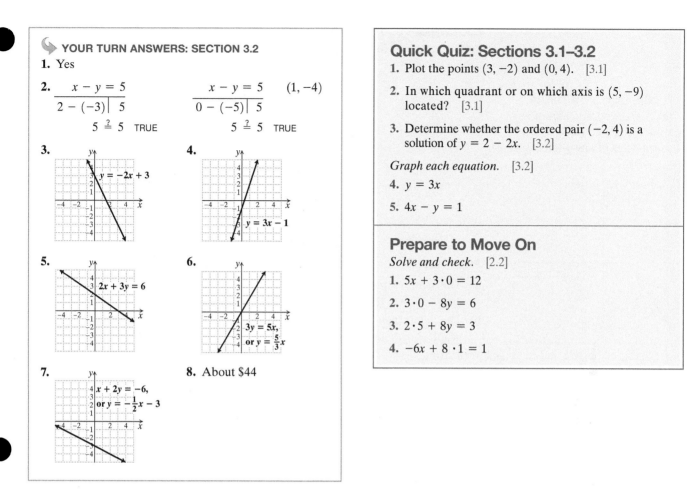

↳ **YOUR TURN ANSWERS: SECTION 3.2**

1. Yes

2.
$$\frac{x - y = 5}{2 - (-3) \mid 5}$$
$$5 \overset{?}{=} 5 \quad \text{TRUE}$$

$$\frac{x - y = 5}{0 - (-5) \mid 5} \quad (1, -4)$$
$$5 \overset{?}{=} 5 \quad \text{TRUE}$$

3. $y = -2x + 3$

4. $y = 3x - 1$

5. $2x + 3y = 6$

6. $3y = 5x$, or $y = \frac{5}{3}x$

7. $x + 2y = -6$, or $y = -\frac{1}{2}x - 3$

8. About $44

Quick Quiz: Sections 3.1–3.2

1. Plot the points $(3, -2)$ and $(0, 4)$. [3.1]

2. In which quadrant or on which axis is $(5, -9)$ located? [3.1]

3. Determine whether the ordered pair $(-2, 4)$ is a solution of $y = 2 - 2x$. [3.2]

Graph each equation. [3.2]

4. $y = 3x$

5. $4x - y = 1$

Prepare to Move On

Solve and check. [2.2]

1. $5x + 3 \cdot 0 = 12$

2. $3 \cdot 0 - 8y = 6$

3. $2 \cdot 5 + 8y = 3$

4. $-6x + 8 \cdot 1 = 1$

3.3 Graphing and Intercepts

A. Intercepts   **B.** Using Intercepts to Graph   **C.** Graphing Horizontal Lines or Vertical Lines

Unless a line is horizontal or vertical, it will cross both axes. Often, finding the points at which a line crosses the axes provides a quick way to graph its equation.

A. Intercepts

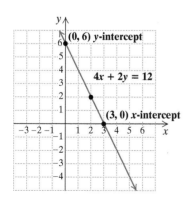

To graph $4x + 2y = 12$, we can plot $(3, 0)$, $(0, 6)$, and $(2, 2)$ and then draw the line.

- The point at which a graph crosses the y-axis is called the **y-intercept**. In the figure shown here, the y-intercept is $(0, 6)$. **The x-coordinate of a y-intercept is always 0.**

- The point at which a graph crosses the x-axis is called the **x-intercept**. In the figure shown here, the x-intercept is $(3, 0)$. **The y-coordinate of an x-intercept is always 0.**

Some curves have more than one y-intercept or more than one x-intercept.

EXAMPLE 1 For the graph shown below, **(a)** give the coordinates of any x-intercepts and **(b)** give the coordinates of any y-intercepts.

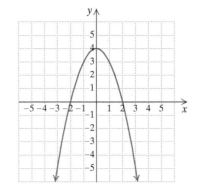

1. For the graph shown below, **(a)** give the coordinates of any x-intercepts and **(b)** give the coordinates of any y-intercepts.

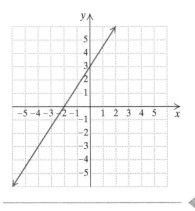

SOLUTION

a) The x-intercepts are the points at which the graph crosses the x-axis. For the graph shown, the x-intercepts are $(-2, 0)$ and $(2, 0)$.

b) The y-intercept is the point at which the graph crosses the y-axis. For the graph shown, the y-intercept is $(0, 4)$.

YOUR TURN

B. Using Intercepts to Graph

It is important to know how to locate the intercepts of a graph from the equation being graphed.

> **TO FIND INTERCEPTS**
>
> To find the y-intercept(s) of an equation's graph, replace x with 0 and solve for y. Write the y-intercept as an ordered pair $(0, b)$.
>
> To find the x-intercept(s) of an equation's graph, replace y with 0 and solve for x. Write the x-intercept as an ordered pair $(a, 0)$.

EXAMPLE 2 Find the y-intercept and the x-intercept of the graph of $2x + 4y = 20$.

SOLUTION To find the y-intercept, we let $x = 0$ and solve for y:

$$2 \cdot 0 + 4y = 20 \qquad \text{Replacing } x \text{ with } 0$$
$$4y = 20$$
$$y = 5.$$

Thus the y-intercept is $(0, 5)$.

To find the x-intercept, we let $y = 0$ and solve for x:

$$2x + 4 \cdot 0 = 20 \qquad \text{Replacing } y \text{ with } 0$$
$$2x = 20$$
$$x = 10.$$

2. Find the y-intercept and the x-intercept of the graph of $x + 3y = 6$.

Thus the x-intercept is $(10, 0)$.

YOUR TURN

Since two points are sufficient to graph a line, intercepts can be used to graph linear equations.

EXAMPLE 3   Graph $2x + 4y = 20$ using intercepts. Find a third point as a check.

SOLUTION   In Example 2, we showed that the y-intercept is $(0, 5)$ and the x-intercept is $(10, 0)$. Before drawing a line, we plot a third point as a check. We substitute any convenient value for x and solve for y.
   If we let $x = 5$, then

$$2 \cdot 5 + 4y = 20 \qquad \text{Substituting 5 for } x$$
$$10 + 4y = 20$$
$$4y = 10 \qquad \text{Subtracting 10 from both sides}$$
$$y = \tfrac{10}{4}, \text{ or } 2\tfrac{1}{2}. \qquad \text{Solving for } y; \left(5, 2\tfrac{1}{2}\right) \text{ must be on the graph.}$$

The point $\left(5, 2\tfrac{1}{2}\right)$ appears to line up with the intercepts, so our work is probably correct. To finish, we draw and label the line.

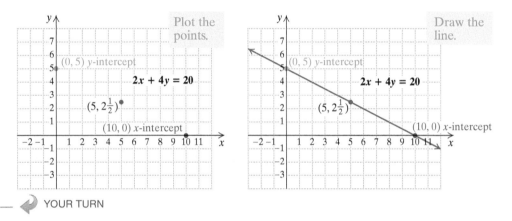

3. Graph $3x + 5y = 15$ using intercepts.

⟳ YOUR TURN

Note that when we solved for the y-intercept, we replaced x with 0 and simplified $2x + 4y = 20$ to $4y = 20$. Thus, to find the y-intercept, we can momentarily ignore the x-term and solve the remaining equation.
   In a similar manner, when we solved for the x-intercept, we simplified $2x + 4y = 20$ to $2x = 20$. Thus, to find the x-intercept, we can momentarily ignore the y-term and then solve this remaining equation.

EXAMPLE 4   Graph $3x - 2y = 60$ using intercepts. Find a third point as a check.

SOLUTION   To find the y-intercept, we let $x = 0$. This amounts to temporarily ignoring the x-term and then solving:

$$-2y = 60 \qquad \text{For } x = 0, \text{ we have } 3 \cdot 0 - 2y, \text{ or simply } -2y.$$
$$y = -30.$$

The y-intercept is $(0, -30)$.
   To find the x-intercept, we let $y = 0$. This amounts to temporarily disregarding the y-term and then solving:

$$3x = 60 \qquad \text{For } y = 0, \text{ we have } 3x - 2 \cdot 0, \text{ or simply } 3x.$$
$$x = 20.$$

The x-intercept is $(20, 0)$.
   To find a third point, we can replace x with 4 and solve for y:

$$3 \cdot 4 - 2y = 60 \qquad \text{Numbers other than 4 can be used for } x.$$
$$12 - 2y = 60$$
$$-2y = 48$$
$$y = -24. \qquad \text{This means that } (4, -24) \text{ is on the graph.}$$

Study Skills

Create Your Own Glossary

Understanding terminology is essential for success in any math course. Try writing your own glossary toward the back of your notebook. Often, simply writing out a word's definition can help you remember what the word means.

In order for us to graph all three points, the y-axis of our graph must go down to at least -30 and the x-axis must go up to at least 20. Using a scale of 5 units per square allows us to display both intercepts and $(4, -24)$, as well as the origin.

The point $(4, -24)$ appears to line up with the intercepts, so we draw and label the line, as shown below.

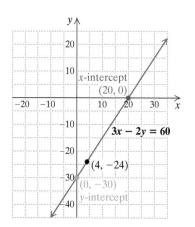

4. Graph $2x - 5y = 30$ using intercepts.

YOUR TURN

Technology Connection

When an equation is graphed using a graphing calculator, we may not always see both intercepts. For example, if $y = -0.8x + 17$ is graphed in the window $[-10, 10, -10, 10]$, neither intercept is visible, as shown in the graph on the left below.

To better view intercepts, we can change the window dimensions or we can zoom out. On many graphing calculator apps, we can zoom in or out using the touch screen. On a graphing calculator without a touch screen, the ZOOM feature allows us to reduce or magnify a graph or a portion of a graph. Before zooming, we must set the ZOOM *factors* in the memory of the ZOOM key. If we zoom out with factors set at 5, both intercepts are visible but the axes are heavily drawn, as shown in the graph in the middle below.

This suggests that the *scales* of the axes should be changed. To do this, we use the WINDOW menu and set Xscl to 5 and Yscl to 5. The resulting graph has tick marks 5 units apart and clearly shows both intercepts, as shown in the graph on the right below. Other choices for Xscl and Yscl can also be made.

Graph each equation so that both intercepts can be easily viewed. Zoom or adjust the window settings so that tick marks can be clearly seen on both axes.

1. $y = -0.72x - 15$ **2.** $y - 2.13x = 27$
3. $5x + 6y = 84$ **4.** $2x - 7y = 150$
5. $19x - 17y = 200$ **6.** $6x + 5y = 159$

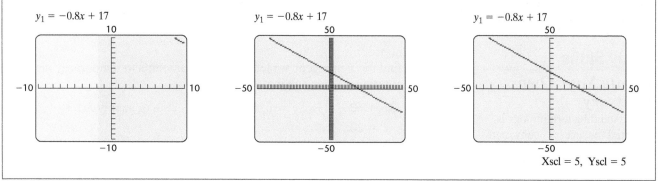

C. Graphing Horizontal Lines or Vertical Lines

The equations graphed in Examples 3 and 4 are in the form $Ax + By = C$. We have already stated that any equation in the form $Ax + By = C$ is linear, provided A and B are not both zero. What if A or B (but not both) is zero? We will

Student Notes

A horizontal line (other than $y = 0$) has no x-intercept. A vertical line (other than $x = 0$) has no y-intercept. All other lines intersect both the x-axis and the y-axis. If a line goes through the origin, the x-intercept and the y-intercept are the same point, $(0, 0)$.

find that when A is zero, there is no x-term and the graph is a horizontal line. We will also find that when B is zero, there is no y-term and the graph is a vertical line.

EXAMPLE 5   Graph: $y = 3$.

SOLUTION   We can regard the equation $y = 3$ as $0 \cdot x + y = 3$. No matter what number we choose for x, we find that y must be 3 if the equation is to be solved. Consider the following table.

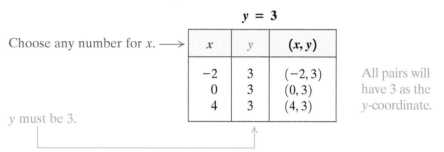

Choose any number for x. $\longrightarrow$

$y = 3$

x	y	(x, y)
-2	3	$(-2, 3)$
0	3	$(0, 3)$
4	3	$(4, 3)$

All pairs will have 3 as the y-coordinate.

y must be 3.

When we plot $(-2, 3)$, $(0, 3)$, and $(4, 3)$ and connect the points, we obtain a horizontal line. Any pair of the form $(x, 3)$ is a solution, so the line is parallel to the x-axis with y-intercept $(0, 3)$. Note that the graph of $y = 3$ has no x-intercept.

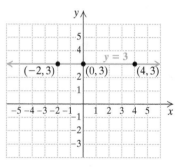

5. Graph: $y = 2$.

YOUR TURN

EXAMPLE 6   Graph: $x = -4$.

SOLUTION   We can regard the equation $x = -4$ as $x + 0 \cdot y = -4$. We create a table with all -4's in the x-column.

Student Notes

Sometimes students draw horizontal lines when they should be drawing vertical lines and vice versa. To avoid this mistake, first locate the correct number on the axis whose label is given. Thus, to graph $x = 2$, we locate 2 on the x-axis and then draw a line perpendicular to that axis at that point. Note that the graph of $x = 2$ on a plane is a line, whereas the graph of $x = 2$ on the number line is a point.

$x = -4$

x must be -4. $\longrightarrow$

x	y	(x, y)
-4	-5	$(-4, -5)$
-4	1	$(-4, 1)$
-4	3	$(-4, 3)$

All pairs will have -4 as the x-coordinate.

Any number can be used for y.

When we plot $(-4, -5)$, $(-4, 1)$, and $(-4, 3)$ and connect the points, we obtain a vertical line. Any ordered pair of the form $(-4, y)$ is a solution. The line is parallel to the y-axis with x-intercept $(-4, 0)$. Note that the graph of $x = -4$ has no y-intercept.

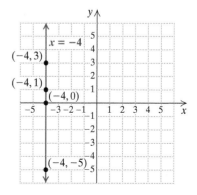

6. Graph: $x = -2$.

YOUR TURN

➥ Check Your UNDERSTANDING

Draw a line with the given intercepts.

1. x-intercept: $(-2, 0)$; y-intercept: $(0, 4)$

2. x-intercept: $(3, 0)$; no y-intercept

3. no x-intercept; y-intercept: $(0, -3)$

LINEAR EQUATIONS IN ONE VARIABLE

The graph of $y = b$ is a horizontal line, with y-intercept $(0, b)$.

The graph of $x = a$ is a vertical line, with x-intercept $(a, 0)$.

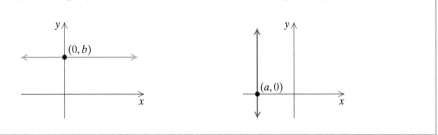

EXAMPLE 7 Write an equation for each graph.

a)

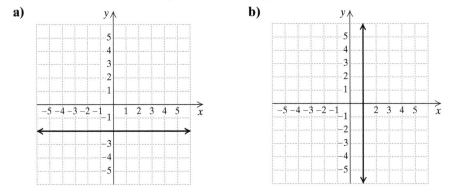

b)

7. Write an equation for the following graph.

SOLUTION

a) Note that every point on the horizontal line passing through $(0, -2)$ has -2 as the y-coordinate. Thus the equation of the line is $y = -2$.

b) Note that every point on the vertical line passing through $(1, 0)$ has 1 as the x-coordinate. Thus the equation of the line is $x = 1$.

🔄 YOUR TURN

3.3 EXERCISE SET

FOR EXTRA HELP   MyMathLab®

➥ Vocabulary and Reading Check

Choose from the following list of words to complete each statement. Words will be used more than once.

  x-intercept
  y-intercept

1. A horizontal line has a(n) _____.

2. A vertical line has a(n) _____.

3. To find a(n) _____, replace x with 0 and solve for y.

4. To find a(n) _____, replace y with 0 and solve for x.

5. The point $(-3, 0)$ could be a(n) _____ of a graph.

6. The point $(0, 7)$ could be a(n) _____ of a graph.

● ⤷ **Concept Reinforcement**

In each of Exercises 7–12, match the phrase with the appropriate choice from the column on the right.

7. ___ A vertical line

8. ___ A horizontal line

9. ___ A *y*-intercept

10. ___ An *x*-intercept

11. ___ A third point as a check

12. ___ Use a scale of 10 units per square.

a) $2x + 5y = 100$

b) $(3, -2)$

c) $(1, 0)$

d) $(0, 2)$

e) $y = 3$

f) $x = -4$

A. Intercepts

*For Exercises 13–20, list **(a)** the coordinates of the y-intercept and **(b)** the coordinates of all x-intercepts.*

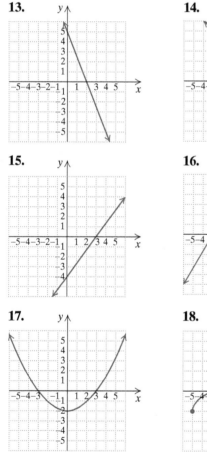

13.

14.

15.

16.

17.

18.

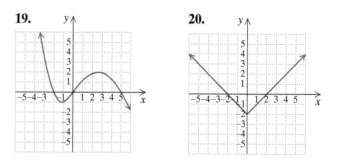

19.

20.

*For Exercises 21–30, list **(a)** the coordinates of any y-intercept and **(b)** the coordinates of any x-intercept. Do not graph.*

21. $3x + 5y = 15$

22. $2x + 7y = 14$

23. $9x - 2y = 36$

24. $10x - 3y = 60$

25. $-4x + 5y = 80$

26. $-5x + 6y = 100$

27. $x = 12$

28. $y = 10$

29. $y = -9$

30. $x = -5$

B. Using Intercepts to Graph

Find the intercepts. Then graph.

31. $3x + 5y = 15$

32. $2x + y = 6$

33. $x + 2y = 4$

34. $2x + 5y = 10$

35. $-x + 2y = 8$

36. $-x + 3y = 9$

37. $3x + y = 9$

38. $2x - y = 8$

39. $y = 2x - 6$

40. $y = -3x + 6$

41. $5x - 10 = 5y$

42. $3x - 9 = 3y$

43. $2x - 5y = 10$

44. $2x - 3y = 6$

45. $6x + 2y = 12$

46. $4x + 5y = 20$

47. $4x + 3y = 16$

48. $3x + 2y = 8$

49. $2x + 4y = 1$

50. $3x - 6y = 1$

51. $5x - 3y = 180$

52. $10x + 7y = 210$

53. $y = -30 + 3x$

54. $y = -40 + 5x$

55. $-4x = 20y + 80$

56. $60 = 20x - 3y$

57. $y - 3x = 0$

58. $x + 2y = 0$

C. Graphing Horizontal Lines or Vertical Lines

Graph.

59. $y = 1$

60. $y = 4$

61. $x = 3$

62. $x = 6$

63. $y = -2$

64. $y = -4$

65. $x = -1$

66. $x = -6$

67. $y = -15$

68. $x = 20$

69. $y = 0$

70. $y = \frac{3}{2}$

71. $x = -\frac{5}{2}$

72. $x = 0$

73. $-4x = -100$

74. $12y = -360$

75. $35 + 7y = 0$

76. $-3x - 24 = 0$

Write an equation for each graph.

77.

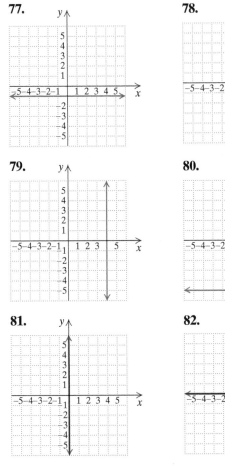

78.

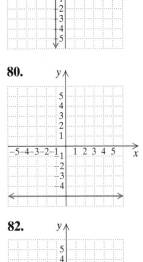

79.

80.

81.

82.

83. Explain in your own words why the graph of $y = 8$ is a horizontal line.

84. Explain in your own words why the graph of $x = -4$ is a vertical line.

Skill Review

Translate to an algebraic expression. [1.1]

85. 7 less than d

86. 5 more than w

87. The sum of 7 and four times a number

88. The product of 3 and a number

89. Twice the sum of two numbers

90. Half of the sum of two numbers

Synthesis

91. Describe what the graph of $x + y = C$ will look like for any choice of C.

92. If the graph of a linear equation passes through (r, s) and $(-r, -s)$, what can you conclude about its x- and y-intercepts? Why?

93. Write an equation for the x-axis.

94. Write an equation of the line parallel to the x-axis and passing through $(3, 5)$.

95. Write an equation of the line parallel to the y-axis and passing through $(-2, 7)$.

96. Find the coordinates of the point of intersection of the graphs of $y = x$ and $y = 6$.

97. Find the coordinates of the point of intersection of the graphs of the equations $x = -3$ and $y = 4$.

98. Write an equation of the line shown in Exercise 13.

99. Write an equation of the line shown in Exercise 16.

100. Find the value of C such that the graph of $3x + C = 5y$ has an x-intercept of $(-4, 0)$.

101. Find the value of C such that the graph of $4x = C - 3y$ has a y-intercept of $(0, -8)$.

102. For A and B nonzero, the graphs of $Ax + D = C$ and $By + D = C$ will be parallel to an axis. Explain why.

103. Find the x-intercept of the graph of $Ax + D = C$.

In each of Exercises 104–109, find the intercepts of the equation algebraically. Then adjust the window and scale so that the intercepts can be checked graphically with no further window adjustments.

104. $3x + 2y = 50$

105. $2x - 7y = 80$

106. $y = 1.3x - 15$

107. $y = 0.2x - 9$

108. $25x - 20y = 1$

109. $50x + 25y = 1$

110. Draw a graph illustrating the following data. What portion of the graph is horizontal? How is this reflected in the data?

Month	President's Approval Rating
February	60%
March	50
April	40
May	40
June	40
July	40
August	60

YOUR TURN ANSWERS: SECTION 3.3

1. (a) $(-2, 0)$; **(b)** $(0, 3)$ **2.** *y*-intercept: $(0, 2)$; *x*-intercept: $(6, 0)$

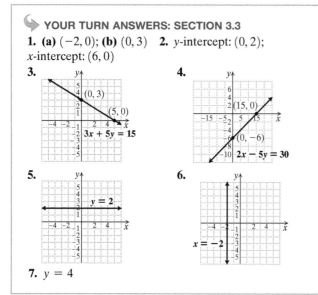

3.

4.

5.

6.

7. $y = 4$

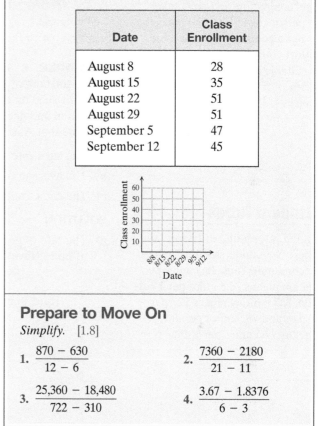

Quick Quiz: Sections 3.1–3.3

Graph.

1. $y = 2x - 5$ [3.2]   **2.** $y = 4$ [3.3]

3. $x = -1$ [3.3]   **4.** $3x - 4y = 12$ [3.3]

5. Listed below are class enrollments for the three weeks prior to and the three weeks following the beginning of the semester. Make a line graph of the data. [3.1]

Date	Class Enrollment
August 8	28
August 15	35
August 22	51
August 29	51
September 5	47
September 12	45

Prepare to Move On

Simplify. [1.8]

1. $\dfrac{870 - 630}{12 - 6}$

2. $\dfrac{7360 - 2180}{21 - 11}$

3. $\dfrac{25{,}360 - 18{,}480}{722 - 310}$

4. $\dfrac{3.67 - 1.8376}{6 - 3}$

3.4 Rates

A. Rates of Change   **B.** Visualizing Rates

A. Rates of Change

Because graphs make use of two axes, they allow us to see how two quantities change with respect to each other. The ratio of the vertical change to the horizontal change is used to represent this type of change and is referred to as a *rate*.

> **RATE**
>
> A *rate* is a ratio that indicates how two quantities change with respect to each other.

Student Notes

In the rates found in Example 1, there is an understood "1" in the second unit. Thus, 26.5 miles per gallon means 26.5 miles per 1 gallon, $27 per day means $27 per 1 day, and 79.5 miles per day means 79.5 miles per 1 day.

1. Refer to Exercise 1. Find the average daily rate of gas consumption, in gallons per day.

Rates occur often in everyday life:

A website that grows by 10,000 visitors over a period of 2 months has an average *growth rate* of $\frac{10,000}{2}$, or 5000, visitors per month.

A vehicle traveling 150 mi in 3 hr is moving at a *rate* of $\frac{150}{3}$, or 50, mph (miles per hour).

A class of 25 students pays a total of $93.75 to visit a museum. The *rate* is $\frac{\$93.75}{25}$, or $3.75, per student.

> **CAUTION!** To calculate a rate, it is important to keep track of the units being used.

EXAMPLE 1 *Car Rental.* On January 3, Alisha rented a Chevrolet Cruze with a full tank of gas and 9312 mi on the odometer. On January 7, she returned the car with 9630 mi on the odometer.* If the rental agency charged Alisha $108 for the rental and needed 12 gal of gas to fill up the gas tank, find the following rates. Assume that Alisha did not purchase any gas during the rental time.

a) The car's rate of gas mileage, in miles per gallon

b) The average cost of the rental, in dollars per day

c) The car's average rate of travel, in miles per day

SOLUTION

a) The rate of gas mileage, in miles per gallon, is found by dividing the number of miles traveled by the number of gallons used for that amount of driving:

$$\text{Rate, in miles per gallon} = \frac{9630 \text{ mi} - 9312 \text{ mi}}{12 \text{ gal}} \qquad \boxed{\text{The word "per" indicates division.}}$$

$$= \frac{318 \text{ mi}}{12 \text{ gal}}$$

$$= 26.5 \text{ mi/gal} \qquad \text{Dividing}$$

$$= 26.5 \text{ miles per gallon.}$$

b) The average cost of the rental, in dollars per day, is found by dividing the cost of the rental by the number of days:

$$\text{Rate, in dollars per day} = \frac{108 \text{ dollars}}{4 \text{ days}} \qquad \begin{array}{l}\text{From January 3}\\\text{to January 7 is}\\ 7 - 3 = 4 \text{ days.}\end{array}$$

$$= 27 \text{ dollars/day}$$

$$= \$27 \text{ per day.}$$

c) The car's average rate of travel, in miles per day, is found by dividing the number of miles traveled by the number of days:

$$\text{Rate, in miles per day} = \frac{318 \text{ mi}}{4 \text{ days}} \qquad \begin{array}{l} 9630 \text{ mi} - 9312 \text{ mi} = 318 \text{ mi;}\\ \text{From January 3 to January 7 is}\\ 7 - 3 = 4 \text{ days.}\end{array}$$

$$= 79.5 \text{ mi/day}$$

$$= 79.5 \text{ miles per day.}$$

YOUR TURN

---

*For all problems concerning rentals, assume that the pickup time was later in the day than the return time so that no late fees were applied.

> **CAUTION!** Units are a vital part of real-world problems. They must be considered in the translation of a problem and included in every answer.

Many problems involve a rate of travel, or *speed*. The **speed** of an object is found by dividing the distance traveled by the time required to travel that distance.

EXAMPLE 2 *Transportation.* An Atlantic City Express bus makes regular trips between Paramus and Atlantic City, New Jersey. At 6:00 P.M., the bus is at mileage marker 40 on the Garden State Parkway, and at 8:00 P.M., it is at marker 170. Find the average speed of the bus.

SOLUTION Speed is the distance traveled divided by the time spent traveling.

$$\text{Bus speed} = \frac{\text{Distance traveled}}{\text{Time spent traveling}}$$

$$= \frac{\text{Change in mileage}}{\text{Change in time}}$$

$$= \frac{130 \text{ mi}}{2 \text{ hr}} \qquad \begin{array}{l} 170 \text{ mi} - 40 \text{ mi} = 130 \text{ mi};\\ 8{:}00 \text{ P.M.} - 6{:}00 \text{ P.M.} = 2 \text{ hr} \end{array}$$

$$= 65 \frac{\text{mi}}{\text{hr}}$$

$$= 65 \text{ miles per hour,} \qquad \begin{array}{l} \text{This } \textit{average} \text{ speed does not}\\ \text{or 65 mph} \qquad\qquad\quad \text{indicate by how much the bus}\\ \text{speed may vary along the route.} \end{array}$$

2. A hummingbird flew across Shannon's yard, a distance of 250 ft, in 5 sec. Find the average speed of the hummingbird.

↩ YOUR TURN

B. Visualizing Rates

Graphs allow us to visualize a rate of change. As a rule, the quantity listed in the numerator appears on the vertical axis and the quantity listed in the denominator appears on the horizontal axis.

EXAMPLE 3 *Library Holdings.* In 2011, U.S. public libraries spent an average of $0.40 per capita on electronic materials. Here, "per capita" means "per person living in the library's legal service area." This amount was increasing at a rate of $0.50 per year. Draw a graph to represent this information.

Data: publiclibrariesonline.org

SOLUTION To label the axes, note that the rate is given as $0.50 per year, or

$$0.50 \frac{\text{dollars}}{\text{year}}. \quad \begin{array}{l} \leftarrow \text{Numerator: vertical axis}\\ \leftarrow \text{Denominator: horizontal axis} \end{array}$$

We list *Per-capita spending on electronic materials* on the vertical axis and *Year* on the horizontal axis. (See the figure on the left at the top of the following page.)

Next, we select a scale for each axis that allows us to plot the information. If we count by increments of $0.50 on the vertical axis, we can show $0.40 for 2011 and increasing amounts for later years. On the horizontal axis, we can count by increments of 1 year. (See the figure in the middle at the top of the following page.)

↳ Check Your UNDERSTANDING

At 3.00 P.M., Lisa passed the 2-mi mark on a walking trail. At 3:45 P.M., she passed the 5-mi mark.

1. Between 3:00 P.M. and 3:45 P.M., how many minutes did Lisa walk?

2. Between the 2-mi mark and the 5-mi mark, how many miles did Lisa walk?

3. What is Lisa's walking rate?

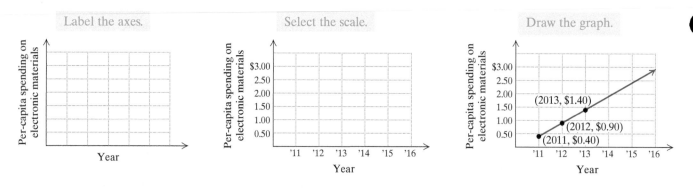

Label the axes.   Select the scale.   Draw the graph.

We are given that libraries spent an average of $0.40 per capita on electronic materials in 2011. This gives us one point on the graph: (2011, $0.40).

We use the rate of change to find more points on the graph.

Spending in 2011: $0.40

Spending in 2012: $0.40 + $0.50 = $0.90

Spending in 2013: $0.90 + $0.50 = $1.40

These calculations give us two more points on the graph: (2012, $0.90) and (2013, $1.40). After plotting the three points, we draw a line through them. This gives us the graph on the right above.

3. In 2013, U.S. consumers spent, on average, $1078 on gifts. This amount was decreasing at a rate of $15 per year. Draw a graph to represent this information.

YOUR TURN

EXAMPLE 4 *Banking.* Nadia prepared the following graph from data collected on a recent day at a local bank.

a) What rate can be determined from the graph?

b) What is that rate?

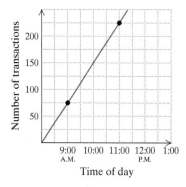

4. Refer to Example 4. Nadia prepared the following graph for data at a different bank location. What is the rate?

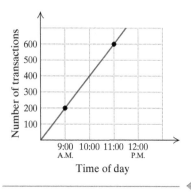

SOLUTION

a) Because the vertical axis shows the number of transactions and the horizontal axis lists the time in hour-long increments, we can find the rate *Number of transactions per hour.*

b) The points (9:00, 75) and (11:00, 225) are both on the graph. This tells us that in the 2 hours between 9:00 and 11:00, there were 225 − 75 = 150 transactions. Thus the average rate is

$$\frac{225 \text{ transactions} - 75 \text{ transactions}}{11:00 - 9:00} = \frac{150 \text{ transactions}}{2 \text{ hours}}$$

$$= 75 \text{ transactions per hour.}$$

YOUR TURN

3.4 EXERCISE SET

FOR EXTRA HELP **MyMathLab®**

Vocabulary and Reading Check

Classify each of the following statements as either true or false.

1. A rate is a ratio.

2. In the phrase "meters per second," the word "per" indicates division.

3. To find the speed of an object, divide the time by the distance traveled.

4. A vertical change in a graph should be used in the numerator of the associated rate of change.

Concept Reinforcement

For each of Exercises 5–8, fill in the missing units for the rate.

5. If Eva biked 100 miles in 5 hours, her average rate was 20 _____.

6. If it took Lauren 18 hours to read 6 chapters, her average rate was 3 _____.

7. If Denny's ticket cost $300 for a 150-mile flight, his average rate was 2 _____.

8. If Marc made 8 cakes using 20 cups of flour, his average rate was $2\frac{1}{2}$ _____.

A. Rates of Change

Solve. For Exercises 9–16, round answers to the nearest cent, where appropriate.

9. *Car Rentals.* Late on June 5, Gaya rented a Honda Odyssey with a full tank of gas and 13,741 mi on the odometer. On June 8, she returned the vehicle with 14,131 mi on the odometer. The rental agency charged Gaya $118 for the rental and needed 13 gal of gas to fill up the tank.

    a) Find the vehicle's rate of gas consumption, in miles per gallon.
    b) Find the average cost of the rental, in dollars per day.
    c) Find the average rate of travel, in miles per day.
    d) Find the rental rate, in cents per mile.

10. *Car Rentals.* On February 10, Oscar rented a Chevy Tahoe with a full tank of gas and 13,091 mi on the odometer. On February 12, he returned the vehicle with 13,322 mi on the odometer. The rental agency charged $92 for the rental and needed 14 gal of gas to fill the tank.

    a) Find the vehicle's rate of gas consumption, in miles per gallon.
    b) Find the average cost of the rental, in dollars per day.
    c) Find the average rate of travel, in miles per day.
    d) Find the rental rate, in cents per mile.

11. *Bicycle Rentals.* At 9:00, Jodi rented a mountain bike from The Bike Rack. She returned the bicycle at 11:00, after cycling 14 mi. Jodi paid $15 for the rental.

    a) Find Jodi's average speed, in miles per hour.
    b) Find the rental rate, in dollars per hour.
    c) Find the rental rate, in dollars per mile.

12. *Bicycle Rentals.* At 2:00, Braden rented a mountain bike from Slickrock Cycles. He returned the bike at 5:00, after cycling 18 mi. Braden paid $24 for the rental.

    a) Find Braden's average speed, in miles per hour.
    b) Find the rental rate, in dollars per hour.
    c) Find the rental rate, in dollars per mile.

13. *Proofreading.* Sergei began proofreading at 9:00 A.M., starting at the top of page 93. He worked until 2:00 P.M. that day and finished page 195. He billed the publishers $110 for the day's work.

    a) Find the rate of pay, in dollars per hour.
    b) Find the average proofreading rate, in number of pages per hour.
    c) Find the rate of pay, in dollars per page.

14. *Temporary Help.* A typist for Kelly Services reports to 3E's Properties for work at 10:00 A.M. and leaves at 6:00 P.M. after having typed from the end of page 8 to the end of page 50 of a proposal. 3E's pays $120 for the typist's services.

    a) Find the rate of pay, in dollars per hour.
    b) Find the average typing rate, in number of pages per hour.
    c) Find the rate of pay, in dollars per page.

15. *National Debt.* The U.S. federal budget debt was $15,041 billion in 2011 and $18,884 billion in 2016. Find the average rate at which the debt was increasing.

    Data: U.S. Office of Management and Budget

16. *Four-Year-College Tuition.* The average amount of room, board, and tuition at a four-year college was $22,092 in 2011 and $23,600 in 2015. Find the average rate at which room, board, and tuition are increasing.

Data: U.S. National Center for Education Statistics

17. *Elevators.* At 2:38, Lara entered an elevator on the 34th floor of the Regency Hotel. At 2:40, she stepped off at the 5th floor.

a) Find the elevator's average rate of travel, in number of floors per minute.

b) Find the elevator's average rate of travel, in seconds per floor.

18. *Snow Removal.* By 1:00 P.M., Shani had already shoveled 2 driveways, and by 6:00 P.M. that day, the number was up to 7.

a) Find Shani's average shoveling rate, in number of driveways per hour.

b) Find Shani's average shoveling rate, in hours per driveway.

19. *Mountaineering.* The fastest ascent of Mt. Everest was accomplished by the Sherpa guide Pemba Dorje of Nepal in 2004. Pemba Dorje climbed from base camp, elevation 17,700 ft, to the summit, elevation 29,029 ft, in 8 hr 10 min.

Data: *Guinness World Records*

a) Find Pemba Dorje's average rate of ascent, in feet per minute.

b) Find Pemba Dorje's average rate of ascent, in minutes per foot.

20. *Pottery.* Master potter Mark Byles holds the record for the most clay pots thrown in 1 hr. He made 150 clay flower pots during the allotted hour.

Data: *Guinness World Records*

a) Find Mark's average pot-throwing rate, in pots per minute.

b) Find Mark's average pot-throwing rate, in minutes per pot.

B. Visualizing Rates

In each of Exercises 21–30, draw a linear graph to represent the given information. Be sure to label and number the axes appropriately. (See Example 3.)

21. *Salaries in Public Schools.* In 2011, the average hourly wage in the United States for an instructional teacher aide was $13.55, and the wage was increasing at a rate of $0.23 per year.

Data: U.S. Census Bureau, *Statistical Abstract of the United States,* 2012

22. *Salaries in Public Schools.* In 2011, the average annual salary in the United States for a public-school technology administrator was $91,000, and the salary was increasing at a rate of $1700 per year.

Data: U.S. Census Bureau, *Statistical Abstract of the United States,* 2012

23. *Amount of Paper Sent to Landfills.* In 2008, approximately 28 million tons of paper went to landfills in the United States, and the amount was decreasing at a rate of 2 million tons per year.

Data: www.paperrecycles.org

24. *Law Enforcement.* In 2012, approximately 9 million property crimes were reported in the United States, and the number was decreasing at a rate of 0.15 million crimes per year.

Data: FBI UCS Annual Crime Reports, compiled by DisasterCenter.com

25. *Train Travel.* At 3:00 P.M., the Boston–Washington Metroliner had traveled 230 mi and was cruising at a rate of 90 miles per hour.

26. *Plane Travel.* At 4:00 P.M., the Seattle–Los Angeles shuttle had traveled 400 mi and was cruising at a rate of 300 miles per hour.

27. *Wages.* By 2:00 P.M., Diane had earned $50. She continued earning money at a rate of $15 per hour.

28. *Wages.* By 3:00 P.M., Arnie had earned $70. He continued earning money at a rate of $12 per hour.

29. *Telephone Bills.* Roberta's phone bill was already $7.50 when she made a call for which she was charged at a rate of $0.10 per minute.

30. *Telephone Bills.* At 3:00 P.M., Theo's phone bill was $6.50 and increasing at a rate of 7¢ per minute.

In Exercises 31–40, use the graph provided to calculate a rate of change in which the units of the horizontal axis are used in the denominator.

31. *Call Center.* The following graph shows data from a technical assistance call center. At what rate are calls being handled?

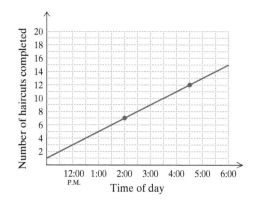

32. *Hairdresser.* Eve's Custom Cuts has a graph displaying data from a recent day of work. At what rate does Eve work?

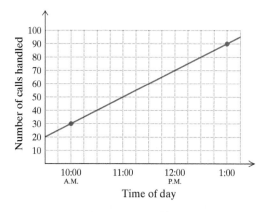

33. *Train Travel.* The following graph shows data from a recent train ride from Chicago to St. Louis. At what rate did the train travel?

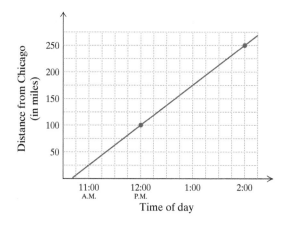

34. *Train Travel.* The following graph shows data from a recent train ride from Denver to Kansas City. At what rate did the train travel?

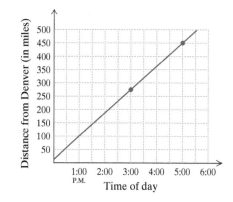

35. *Cost of a Telephone Call.* The following graph shows data from a recent phone call between the United States and Netherlands. At what rate was the customer being billed?

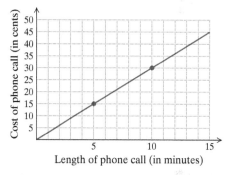

36. *Cost of a Telephone Call.* The following graph shows data from a recent phone call between the United States and South Korea. At what rate was the customer being billed?

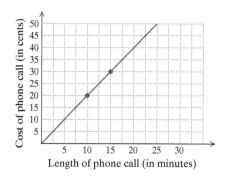

37. *Population.* The following graph shows data regarding the population of Lithuania. At what rate was the population changing?

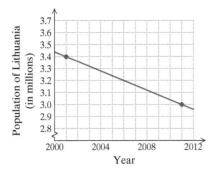

38. *Depreciation of an Office Machine.* Data regarding the value of a particular color copier is represented in the following graph. At what rate is the value changing?

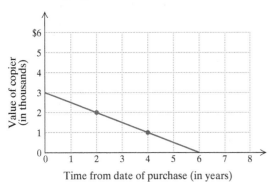

39. *Gas Mileage.* The following graph shows data for a Honda Insight (hybrid) driven on city streets. At what rate was the vehicle consuming gas?

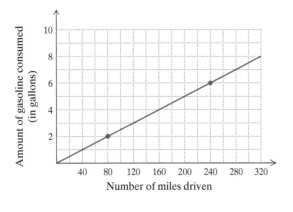

40. *Gas Mileage.* The following graph shows data for a Ford Mustang driven on highways. At what rate was the vehicle consuming gas?

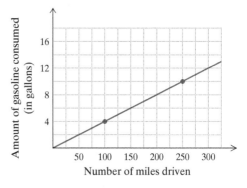

In Exercises 41–46, match each description with the most appropriate graph from the choices below. Scales are intentionally omitted. Assume that of the three sports listed, swimming is the slowest and biking is the fastest.

41. ____ Robin trains for triathlons by running, biking, and then swimming every Saturday.

42. ____ Gene trains for triathlons by biking, running, and then swimming every Sunday.

43. ____ Shirley trains for triathlons by swimming, biking, and then running every Sunday.

44. ____ Evan trains for triathlons by swimming, running, and then biking every Saturday.

45. ____ Angie trains for triathlons by biking, swimming, and then running every Sunday.

46. ____ Mick trains for triathlons by running, swimming, and then biking every Saturday.

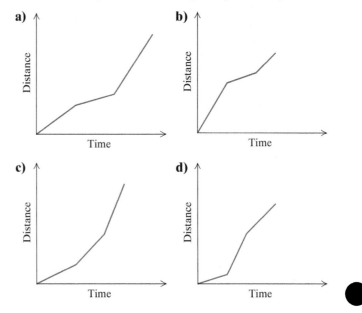

e)

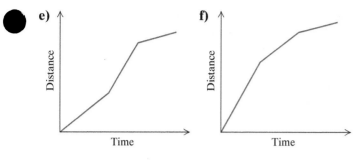

f)

47. What does a negative rate of travel indicate? Explain.

48. Explain how to convert from kilometers per hour to meters per second.

Skill Review

49. Find the prime factorization of 150. [1.3]

50. Graph on the number line: -3.5. [1.4]

51. Find decimal notation: $-\dfrac{11}{8}$. [1.4]

52. Find the absolute value: $\left| -\dfrac{1}{3} \right|$. [1.4]

53. Find the opposite of $\dfrac{3}{2}$. [1.6]

54. Find the reciprocal of $\dfrac{3}{2}$. [1.7]

Synthesis

55. How would the graphs of Jon's and Jenny's total earnings compare in each of the following situations?

    a) Jon earns twice as much per hour as Jenny.
    b) Jon and Jenny earn the same hourly rate, but Jenny received a bonus for a cost-saving suggestion.
    c) Jon is paid by the hour, and Jenny is paid a weekly salary.

56. Write an exercise similar to those in Exercises 9–20 for a classmate to solve. Design the problem so that the solution is "The motorcycle's rate of gas consumption was 65 miles per gallon."

57. *Aviation.* A Boeing 737 airplane climbs from sea level to a cruising altitude of 31,500 ft at a rate of 6300 ft/min. After cruising for 3 min, the jet is forced to land, descending at a rate of 3500 ft/min. Represent the flight with a graph in which altitude is measured on the vertical axis and time on the horizontal axis.

58. *Wages with Commissions.* Each salesperson at Mike's Bikes is paid $140 per week plus 13% of all sales up to $2000, and then 20% on any sales in excess of $2000. Draw a graph in which sales are measured on the horizontal axis and wages on the vertical axis. Then use the graph to estimate the wages paid when a salesperson sells $2700 in merchandise in one week.

59. *Taxi Fares.* The driver of a New York City Yellow Cab recently charged $2.50 plus 50¢ for each fifth of a mile traveled. Draw a graph that could be used to determine the cost of a fare.

60. *Gas Mileage.* Suppose that a Honda motorcycle travels twice as far as a Honda Insight on the same amount of gas. (See Exercise 39.) Draw a graph that reflects this information.

61. *Aviation.* Tim's F-16 jet is moving forward at a deck speed of 95 mph aboard an aircraft carrier that is traveling 39 mph in the same direction. How fast is the jet traveling, in minutes per mile, with respect to the sea?

62. *Navigation.* In 3 sec, Penny walks 24 ft to the bow (front) of a tugboat. The boat is cruising at a rate of 5 ft/sec. What is Penny's rate of travel with respect to land?

63. *Running.* Anne ran from the 4-km mark to the 7-km mark of a 10-km race in 15.5 min. At this rate, how long would it take Anne to run a 5-mi race?

64. *Running.* Jerod ran from the 2-mi marker to the finish line of a 5-mi race in 25 min. At this rate, how long would it take Jerod to run a 10-km race?

65. Trevor picks apples twice as fast as Doug. By 4:30, Doug had already picked 4 bushels of apples. Fifty minutes later, his total reached $5\frac{1}{2}$ bushels. Find Trevor's picking rate. Give your answer in bushels per hour.

66. At 3:00 P.M., Carrie and Chad had already made 46 candles. By 5:00 P.M., the total had reached 100 candles. Assuming a constant production rate, at what time did they make their 82nd candle?

YOUR TURN ANSWERS: SECTION 3.4

1. 3 gal per day   **2.** 50 ft per sec

3.
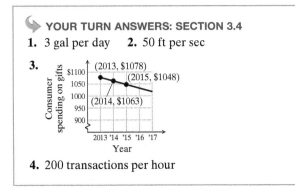

4. 200 transactions per hour

Quick Quiz: Sections 3.1–3.4

1. In which quadrant or on which axis is $(0, -15)$ located?   [3.1]

2. Determine whether the ordered pair $(-3, -1)$ is a solution of the equation $2a - 3b = -9$.   [3.2]

3. Graph: $y = -3$.   [3.3]

4. Graph: $y = -3x$.   [3.2]

5. In 2008, there were about 39,200 commercial pilots in the United States. The number of pilots is expected to increase to 46,500 by 2018. Find the rate of increase, in number of pilots per year.   [3.4]

Data: U.S. Department of Labor, Bureau of Labor Statistics

Prepare to Move On
Simplify.

1. $-2 - (-7)$ [1.6]   **2.** $-9 - (-3)$ [1.6]

3. $\dfrac{5 - (-4)}{-2 - 7}$ [1.8]   **4.** $\dfrac{8 - (-4)}{2 - 11}$ [1.8]

5. $\dfrac{-4 - 8}{11 - 2}$ [1.8]   **6.** $\dfrac{-5 - (-3)}{4 - 6}$ [1.8]

7. $\dfrac{-6 - (-6)}{-2 - 7}$ [1.8]   **8.** $\dfrac{-3 - 5}{-1 - (-1)}$ [1.8]

3.5   Slope

A. Rate and Slope   **B.** Horizontal Lines and Vertical Lines   **C.** Applications

---

A *rate* is a measure of how two quantities change with respect to each other. In this section, we discuss how rate is related to the slope of a line.

A. Rate and Slope

Gary can replace his employees' laptop computers every two years, or he can replace them every three years if he chooses a more durable model. The following tables list the costs of replacing the computers using both programs.

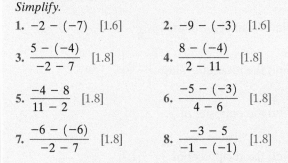

Two-Year Replacement	
Years Since Start of Program	Cost of Computers
0	$ 0
2	3,000
4	6,000
6	9,000
8	12,000

Three-Year Replacement	
Years Since Start of Program	Cost of Computers
0	$ 0
3	4,000
6	8,000
9	12,000
12	16,000

We now graph the pairs of numbers listed in the tables, using the horizontal axis for the number of years since the start of the program and the vertical axis for the cost of the computers.

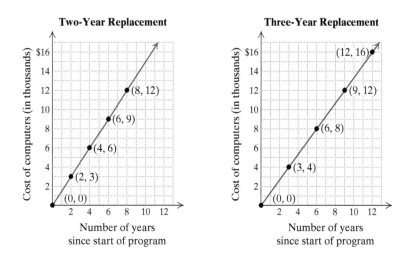

By comparing the cost of the computers over a specified period of time, we can compare the two rates. Note that the rate of the two-year replacement program is greater so its graph is steeper.

Replacement Program	Cost of Computers	Rate
Two-year	$3000 every 2 years	$\dfrac{3 \text{ thousand dollars}}{2 \text{ years}} = \dfrac{3}{2}$ thousand dollars per year, or $1500/year
Three-year	$4000 every 3 years	$\dfrac{4 \text{ thousand dollars}}{3 \text{ years}} = \dfrac{4}{3}$ thousand dollars per year, or $1333.33/year

The rates $\frac{3}{2}$ and $\frac{4}{3}$ can also be found using the coordinates of any two points that are on each line.

EXAMPLE 1   Use the graph of the Two-Year Replacement program above to find the cost of computers per year.

SOLUTION   We can use the points $(6, 9)$ and $(8, 12)$ to find the rate for the two-year program. To do so, remember that these coordinates tell us that after 6 years, the computer cost is $9 thousand, and after 8 years, the computer cost is $12 thousand. In the 2 years between the 6-year and 8-year points, $12 − $9, or $3 thousand, was spent. Thus we have

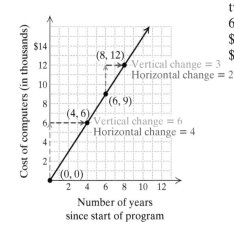

$$\text{Two-year rate} = \frac{\text{change in cost of computers}}{\text{corresponding change in time}}$$

$$= \frac{12 - 9 \text{ thousand dollars}}{8 - 6 \text{ years}}$$

$$= \frac{3 \text{ thousand dollars}}{2 \text{ years}}$$

$$= \frac{3}{2} \text{ thousand dollars per year, or } \$1500/\text{year}.$$

Because the line is straight, the same rate is found using *any* pair of points on the line. For example, using $(0, 0)$ and $(4, 6)$, we have

$$\text{Two-year rate} = \frac{6 - 0 \text{ thousand dollars}}{4 - 0 \text{ years}} = \frac{6 \text{ thousand dollars}}{4 \text{ years}}$$

$$= \frac{3}{2} \text{ thousand dollars per year, or } \$1500/\text{year.}$$

1. Use the graph of the Three-Year Replacement program to find the cost of computers per year.

YOUR TURN

When the axes of a graph are simply labeled x and y, the ratio of vertical change to horizontal change is the rate at which y is changing with respect to x. This ratio is a measure of a line's slant, or **slope**.

> The rate is always the vertical change divided by the corresponding horizontal change.

Consider a line passing through $(2, 3)$ and $(6, 5)$, as shown below. We find the ratio of vertical change, or *rise*, to horizontal change, or *run*, as follows:

$$\frac{\text{Ratio of vertical change}}{\text{to horizontal change}} = \frac{\text{change in } y}{\text{change in } x} = \frac{\text{rise}}{\text{run}}$$

$$= \frac{5 - 3}{6 - 2}$$

$$= \frac{2}{4}, \text{ or } \frac{1}{2}.$$

Note that these calculations can be performed without viewing a graph.

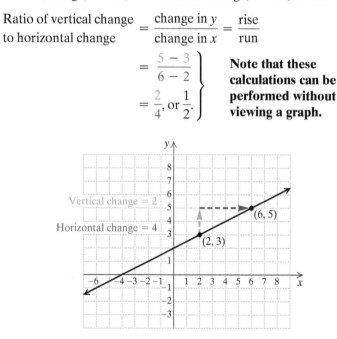

Thus the y-coordinates of points on this line increase at a rate of 2 units for every 4-unit increase in x, which is 1 unit for every 2-unit increase in x, or $\frac{1}{2}$ unit for every 1-unit increase in x. The slope of the line is $\frac{1}{2}$.

In the definition of *slope* below, the *subscripts* 1 and 2 are used to distinguish point 1 from point 2. The slightly lowered 1's and 2's are not exponents but are used to denote x-values (and y-values) for different points.

Student Notes

The notation x_1 is read "x sub one."

The *slope* of the line containing points (x_1, y_1) and (x_2, y_2) is given by

$$m = \frac{\text{change in } y}{\text{change in } x} = \frac{\text{rise}}{\text{run}} = \frac{y_2 - y_1}{x_2 - x_1}.$$

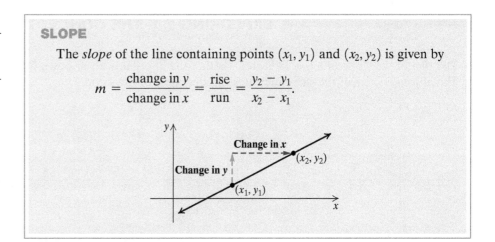

EXAMPLE 2 Find the slope of the line containing the points $(-4, 3)$ and $(2, -6)$.

SOLUTION From $(-4, 3)$ to $(2, -6)$, the change in y, or rise, is $-6 - 3$, or -9. The change in x, or run, is $2 - (-4)$, or 6. Thus,

$$\begin{aligned}
\text{Slope} &= \frac{\text{change in } y}{\text{change in } x} \\
&= \frac{\text{rise}}{\text{run}} \\
&= \frac{-6 - 3}{2 - (-4)} \\
&= \frac{-9}{6} \\
&= -\frac{9}{6}, \text{ or } -\frac{3}{2}.
\end{aligned}$$

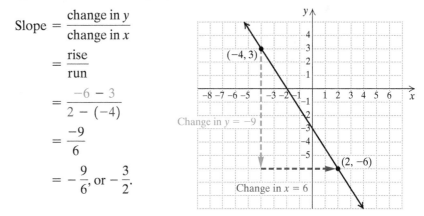

The graph of the line is shown above for reference.

YOUR TURN

2. Find the slope of the line containing the points $(2, 7)$ and $(-6, 0)$.

Student Notes

You may wonder which point should be regarded as (x_1, y_1) and which should be (x_2, y_2). To see that the math works out the same either way, perform both calculations on your own.

CAUTION! When we use the formula

$$m = \frac{y_2 - y_1}{x_2 - x_1},$$

it makes no difference which point is considered (x_1, y_1). What matters is that we subtract the y-coordinates in the same order that we subtract the x-coordinates.

To illustrate, we reverse *both* of the subtractions in Example 2. The slope is still $-\frac{3}{2}$:

$$\text{Slope} = \frac{\text{change in } y}{\text{change in } x} = \frac{3 - (-6)}{-4 - 2} = \frac{9}{-6} = -\frac{3}{2}.$$

EXPLORING 🔍 THE CONCEPT

The *sign* of the slope of a line indicates whether the line slants up or down from left to right.
The *absolute value* of the slope indicates the steepness of the slope.

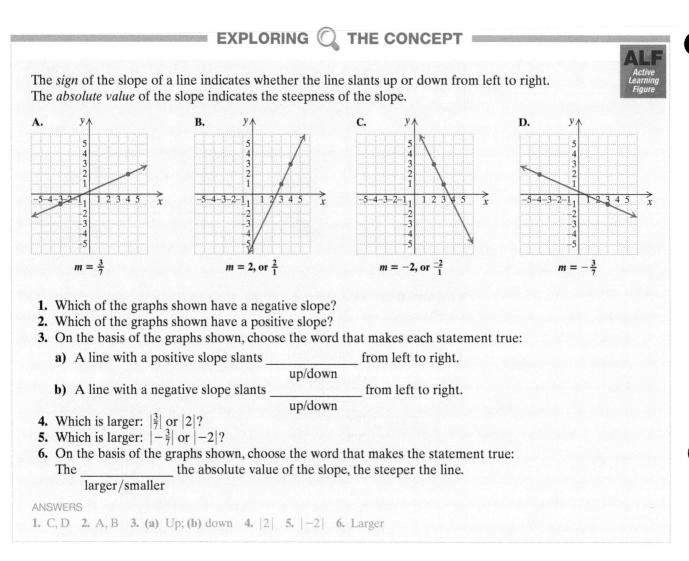

A. $m = \frac{3}{7}$ B. $m = 2$, or $\frac{2}{1}$ C. $m = -2$, or $\frac{-2}{1}$ D. $m = -\frac{3}{7}$

1. Which of the graphs shown have a negative slope?
2. Which of the graphs shown have a positive slope?
3. On the basis of the graphs shown, choose the word that makes each statement true:
    a) A line with a positive slope slants _____ from left to right.
            up/down
    b) A line with a negative slope slants _____ from left to right.
            up/down
4. Which is larger: $\left|\frac{3}{7}\right|$ or $|2|$?
5. Which is larger: $\left|-\frac{3}{7}\right|$ or $|-2|$?
6. On the basis of the graphs shown, choose the word that makes the statement true:
    The _____ the absolute value of the slope, the steeper the line.
        larger/smaller

ANSWERS
1. C, D   2. A, B   3. (a) Up; (b) down   4. $|2|$ 5. $|-2|$ 6. Larger

---

✎ Check Your
UNDERSTANDING

The points $(2, 7)$ and $(5, 3)$ are on a line.

1. Moving from $(2, 7)$ to $(5, 3)$, what is the horizontal distance and direction?

2. Moving from $(2, 7)$ to $(5, 3)$, what is the vertical distance and direction?

3. What is the rise?

4. What is the run?

5. Is the slope positive, negative, zero, or undefined?

6. What is the slope of the line?

> A line with positive slope slants up from left to right, and a line with negative slope slants down from left to right. The larger the absolute value of the slope, the steeper the line.

B. Horizontal Lines and Vertical Lines

What about the slope of a horizontal line or a vertical line?

EXAMPLE 3   Find the slope of the line given by $y = 4$.

SOLUTION   Consider the points $(2, 4)$ and $(-3, 4)$, which are on the line. The change in y, or the rise, is $4 - 4$, or 0. The change in x, or the run, is $-3 - 2$, or -5. Thus,

$$m = \frac{4 - 4}{-3 - 2}$$
$$= \frac{0}{-5}$$
$$= 0.$$

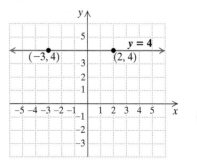

3. Find the slope of the line
$y = 2$.

_____

Any two points on a horizontal line have the same y-coordinate. Thus the change in y is 0, so the slope is 0.

YOUR TURN

> **A horizontal line has slope 0.**

EXAMPLE 4 Find the slope of the line given by $x = -3$.

SOLUTION Consider the points $(-3, 4)$ and $(-3, -2)$, which are on the line. The change in y, or the rise, is $-2 - 4$, or -6. The change in x, or the run, is $-3 - (-3)$, or 0. Thus,

$$m = \frac{-2 - 4}{-3 - (-3)}$$

$$= \frac{-6}{0}. \quad \text{(undefined)}$$

Since division by 0 is not defined, the slope of this line is not defined. The answer to a problem of this type is "The slope of this line is undefined."

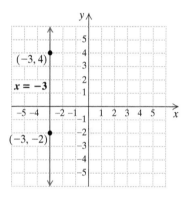

4. Find the slope of the line
$x = -1$.

_____

YOUR TURN

> **The slope of a vertical line is undefined.**

C. Applications

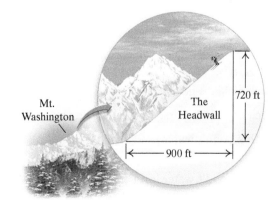

Road grade $\frac{a}{b}$ (expressed as a percent)

Slope has many real-world applications, ranging from car speed to production rate. Slope can also measure steepness. For example, numbers like 2%, 3%, and 6% are often used to represent the **grade** of a road, a measure of a road's steepness. That is, since $3\% = \frac{3}{100}$, a 3% grade means that for every horizontal distance of 100 ft, the road rises or drops 3 ft. The concept of grade also occurs in skiing or snowboarding, where a 7% grade is considered very tame, but a 70% grade is considered steep.

EXAMPLE 5 *Skiing.* Among the steepest skiable terrain in North America, the Headwall on Mount Washington, in New Hampshire, drops 720 ft over a horizontal distance of 900 ft. Find the grade of the Headwall.

Mt. Washington

The Headwall

720 ft

900 ft

5. A mountain road rises 250 ft over a horizontal distance of 4000 ft. Find the grade of the road.

SOLUTION The grade of the Headwall is its slope, expressed as a percent:

$$m = \frac{720}{900} = \frac{8}{10} = 80\%.$$ Grade is slope expressed as a percent.

YOUR TURN

Carpenters use slope when designing stairs, ramps, or roof pitches. Another application occurs in the engineering of a dam—the force or strength of a river depends on how much the river drops over a specified distance.

3.5   EXERCISE SET

FOR EXTRA HELP   MyMathLab®

Vocabulary and Reading Check

Choose from the following list the expression or word that best completes each statement.

x	negative
y	positive
$x_2 - x_1$	rise
$y_2 - y_1$	run
change in x	undefined
change in y	zero

1. Slope is the rate at which _____ is changing with respect to _____.

2. The slope of a line can be expressed in terms of *change* as follows: $\underline{\quad\quad}$.

3. The slope of a line can be expressed using *rise* and *run* as follows: $\underline{\quad\quad}$.

4. The slope of the line containing (x_1, y_1) and (x_2, y_2) is given by $\underline{\quad\quad}$.

5. If a line slants up from left to right, the sign of its slope is _____ , and if a line slants down from left to right, the sign of its slope is _____.

6. The slope of a horizontal line is _____, and the slope of a vertical line is _____.

Concept Reinforcement

In each of Exercises 7–14, state whether the rate is positive, negative, or zero.

7. The rate at which a child's height changes

8. The rate at which an elderly person's height changes

9. The rate at which a pond's water level changes during a drought

10. The rate at which a pond's water level changes during the rainy season

11. The rate at which a person's I.Q. changes during his or her sleep

12. The rate at which the number of people in attendance at a basketball game changes in the moments before the opening tipoff

13. The rate at which the number of people in attendance at a basketball game changes in the moments after the final buzzer sounds

14. The rate at which the number of U.S. Senators changes

A. Rate and Slope

15. *Blogging.* Find the rate at which a professional blogger is paid.

Data: readwriteweb.com

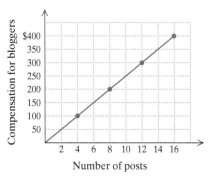

16. *Fitness.* Find the rate at which a runner burns calories.

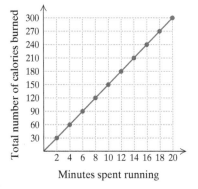

17. *Retail Sales.* Find the rate of change of the total retail sales of bookstores.

Data: U.S. Census Bureau

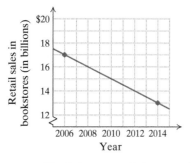

18. *Employment.* Find the rate of change of the number of news reporters and correspondents employed in the United States.

Data: U.S. Department of Labor, Bureau of Labor Statistics

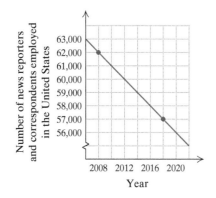

19. *College Admission Tests.* Find the rate of change of SAT critical reading scores with respect to family income.

Data: The College Board, College-Bound Seniors 2011 Total Group Profile Report

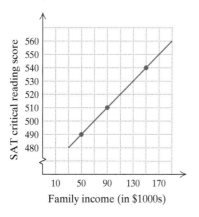

20. *Long-Term Care.* Find the rate of change of Medicaid spending on long-term care.

Data: The Henry J. Kaiser Family Foundation

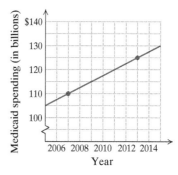

21. *Meteorology.* Find the rate of change of the temperature in Spearfish, Montana, on January 22, 1943, as shown below.

Data: National Oceanic Atmospheric Administration

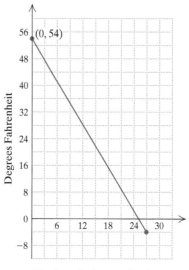

Number of minutes after 9 A.M.

22. Find the rate of change of the birth rate among teenagers reported in the United States.

Data: U.S. National Center for Health Statistics

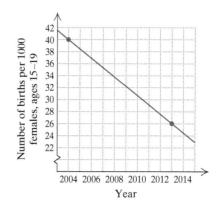

Find the slope, if it is defined, of each line. If the slope is undefined, state this.

23.

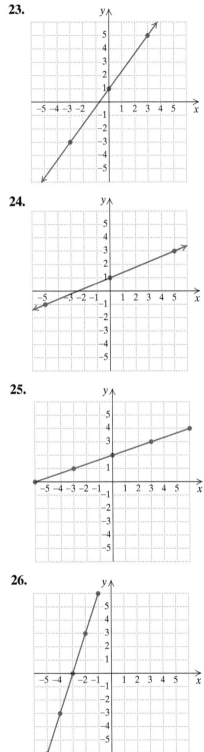

24.

25.

26.

27.

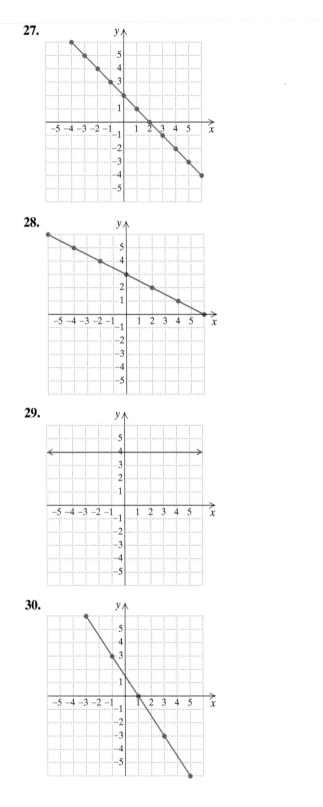

28.

29.

30.

31.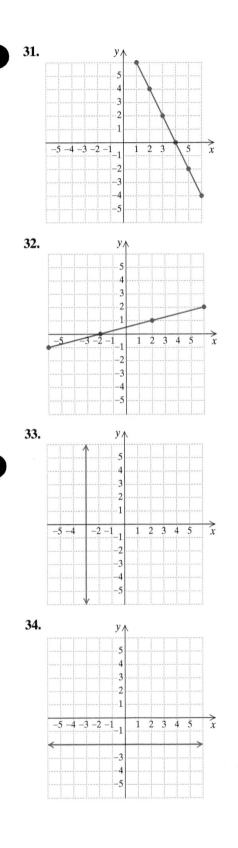

32.

33.

34.

35.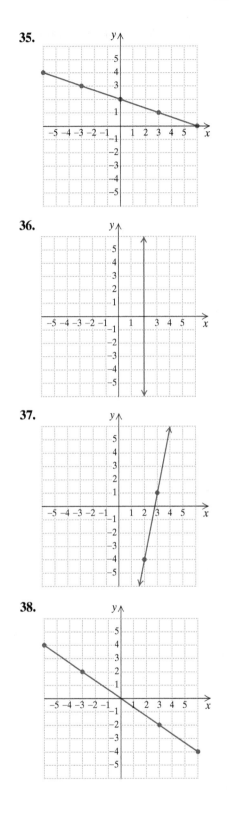

36.

37.

38.

Find the slope of the line containing each given pair of points. If the slope is undefined, state this.

39. $(1, 3)$ and $(5, 8)$ **40.** $(1, 8)$ and $(6, 9)$

41. $(-2, 4)$ and $(3, 0)$ **42.** $(-4, 2)$ and $(2, -3)$

43. $(-4, 0)$ and $(5, 6)$ **44.** $(3, 0)$ and $(6, 9)$

45. $(0, 7)$ and $(-3, 10)$ **46.** $(0, 9)$ and $(-5, 0)$

47. $(-2, 3)$ and $(-6, 5)$ **48.** $(-1, 4)$ and $(5, -8)$

Aha! **49.** $\left(-2, \frac{1}{2}\right)$ and $\left(-5, \frac{1}{2}\right)$ **50.** $(-5, -1)$ and $(2, -3)$

51. $(5, -4)$ and $(2, -7)$ **52.** $(-10, 3)$ and $(-10, 4)$

53. $(6, -4)$ and $(6, 5)$ **54.** $(5, -2)$ and $(-4, -2)$

B. Horizontal Lines and Vertical Lines

Find the slope of each line whose equation is given. If the slope is undefined, state this.

55. $y = 5$ **56.** $y = 13$

57. $x = -8$ **58.** $x = 18$

59. $x = 9$ **60.** $x = -7$

61. $y = -10$ **62.** $y = -4$

C. Applications

63. *Surveying.* Lick Skillet Road, near Boulder, Colorado, climbs 230 m over a horizontal distance of 1600 m. What is the grade of the road?

64. *Navigation.* Capital Rapids drops 54 ft vertically over a horizontal distance of 1080 ft. What is the slope of the rapids?

65. *Construction.* Part of New Valley rises 28 ft over a horizontal distance of 80 ft, and is too steep to build on. What is the slope of the land?

66. *Engineering.* At one point, Yellowstone's Beartooth Highway rises 315 ft over a horizontal distance of 4500 ft. Find the grade of the road.

67. *Carpentry.* Find the slope (or pitch) of the roof.

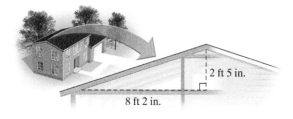

2 ft 5 in.

8 ft 2 in.

68. *Exercise.* Find the slope (or grade) of the treadmill.

RUNRITE

0.4 ft

5 ft

69. *Bicycling.* To qualify as a rated climb on the Tour de France, a grade must average at least 4%. The ascent of Dooley Mountain, Oregon, part of the Elkhorn Classic, begins at 3500 ft and climbs to 5400 ft over a horizontal distance of 37,000 ft. What is the grade of the road? Would it qualify as a rated climb if it were part of the Tour de France?

Data: barkercityherald.com

70. *Accessible Design.* According to the U.S. Department of Justice's 2010 ADA Statistics for Accessible Design, handrails are required for any ramps for which there is a 1-ft vertical rise or more for every 20-ft horizontal run. The ramps in the McCormick Tribune Campus Center of the Illinois Institute of Technology in Chicago, Illinois, do not have handrails. What must be true of the grade of the ramps?

Data: twistersifter.com

71. Explain why the order in which coordinates are subtracted to find slope does not matter so long as y-coordinates and x-coordinates are subtracted in the same order.

72. If one line has a slope of -3 and another has a slope of 2, which line is steeper? Why?

Skill Review

73. Multiply: $3(4 + a)$. [1.2]

74. Factor: $14 + 35x$. [1.2]

75. Find $-x$ when x is -15. [1.6]

76. Write another inequality with the same meaning as $x \geq 3$. [1.4]

77. Write exponential notation for $5t \cdot 5t \cdot 5t$. [1.8]

78. Simplify: $(-2y)^4$. [1.8]

Synthesis

79. The points $(-4, -3)$, $(1, 4)$, $(4, 2)$, and $(-1, -5)$ are vertices of a quadrilateral. Using the slopes, explain why the quadrilateral is a parallelogram.

80. Which is steeper and why: a ski slope that is $50°$ or one with a grade of 100%?

81. The plans shown below are for a skateboard "Fun Box." For the ramps labeled A, find the slope or grade.

Data: www.heckler.com

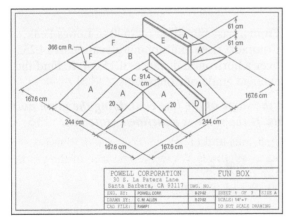

82. A line passes through $(4, -7)$ and never enters the first quadrant. What numbers could the line have for its slope?

83. A line passes through $(2, 5)$ and never enters the second quadrant. What numbers could the line have for its slope?

84. *Architecture.* Architects often use the equation $x + y = 18$ to determine the height y, in inches, of the riser of a step when the tread is x inches wide. (See Exercise 70.) Express the slope of stairs designed with this equation without using the variable y.

In Exercises 85 and 86, the slope of the line is $-\frac{2}{3}$, but the numbering on one axis is missing. How many units should each tick mark on that unnumbered axis represent?

85.

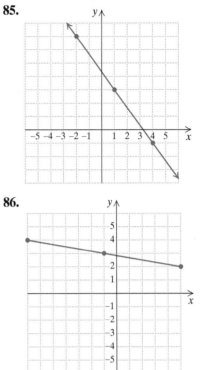

86.

YOUR TURN ANSWERS: SECTION 3.5
1. $\frac{4}{3}$ thousand dollars per year, or $1333.33 per year
2. $\frac{7}{8}$ **3.** 0 **4.** Undefined **5.** 6.25%

Quick Quiz: Sections 3.1–3.5
Graph.
1. $x = -2$ [3.3]   **2.** $3x - y = 3$ [3.3]

3. $y = 2x - 4$ [3.2]

4. Find the slope of the line containing $(9, -4)$ and $(10, -5)$. [3.5]

5. Find the slope of the line given by $y = 21$. If the slope is undefined, state this. [3.5]

Prepare to Move On
Solve for y. [2.3]
1. $2x + 3y = 7$ **2.** $3x - 4y = 8$

3. $ax + by = c$ **4.** $ax - by = c$

Mid-Chapter Review

We can plot points and graph equations on a *Cartesian coordinate plane.*

- A point is represented by an *ordered pair.*
- The graph of an equation represents all of its solutions.
- Equations can be *linear* or *nonlinear.*
- The *slope* of a line represents a rate of change: $\text{Slope} = m = \dfrac{\text{change in } y}{\text{change in } x} = \dfrac{y_2 - y_1}{x_2 - x_1}.$

GUIDED SOLUTIONS

1. Find the *y*-intercept and the *x*-intercept of the graph of $y - 3x = 6$. [3.3]

Solution

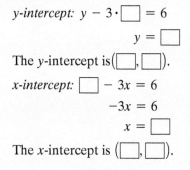

y-intercept: $y - 3 \cdot \boxed{} = 6$

$y = \boxed{}$

The *y*-intercept is $(\boxed{}, \boxed{})$.

x-intercept: $\boxed{} - 3x = 6$

$-3x = 6$

$x = \boxed{}$

The *x*-intercept is $(\boxed{}, \boxed{})$.

2. Find the slope of the line containing the points $(1, 5)$ and $(3, -1)$. [3.5]

Solution

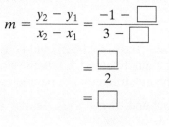

$m = \dfrac{y_2 - y_1}{x_2 - x_1} = \dfrac{-1 - \boxed{}}{3 - \boxed{}}$

$= \dfrac{\boxed{}}{2}$

$= \boxed{}$

MIXED REVIEW

3. Plot the point $(0, -3)$. [3.1]

4. In which quadrant is the point $(4, -15)$ located? [3.1]

5. Determine whether $(-2, -3)$ is a solution of $y = 5 - x$. [3.2]

Graph by hand.

6. $y = x - 3$ [3.2]

7. $y = -3x$ [3.2]

8. $3x - y = 2$ [3.2]

9. $4x - 5y = 20$ [3.3]

10. $y = -2$ [3.3]

11. $x = 1$ [3.3]

12. By the end of June, Conservalot Builders had winterized 10 homes. By the end of August, they had winterized a total of 38 homes. Find the rate at which the company was winterizing homes. [3.4]

13. From a base elevation of 9600 ft, Longs Peak, Colorado, rises to a summit elevation of 14,255 ft over a horizontal distance of 15,840 ft. Find the average grade of Longs Peak. [3.5]

Find the slope of the line containing the given pair of points. If the slope is undefined, state this. [3.5]

14. $(-5, -2)$ and $(1, 8)$

15. $(1, 2)$ and $(4, -7)$

16. $(0, 0)$ and $(0, -2)$

17. $(6, -3)$ and $(2, -3)$

18. What is the slope of the line given by $y = 4$? [3.5]

19. What is the slope of the line given by $x = -7$? [3.5]

20. Find the *x*-intercept and the *y*-intercept of the line given by $2y - 3x = 12$. [3.3]

3.6 Slope–Intercept Form

A. Using the *y*-intercept and the Slope to Graph a Line   **B.** Equations in Slope–Intercept Form
C. Graphing and Slope–Intercept Form   **D.** Slope and Parallel Lines

If we know the slope and the *y*-intercept of a line, it is possible to graph the line. In this section, we will discover that a line's slope and its *y*-intercept can be determined directly from the line's equation, provided the equation is written in a certain form.

A. Using the *y*-intercept and the Slope to Graph a Line

Last year, Gary spent $4000 upgrading the local area network (LAN) for his company's computer system. Now he plans to spend $3000 every two years to update his employees' laptop computers. We can make a table and draw a graph showing his costs.

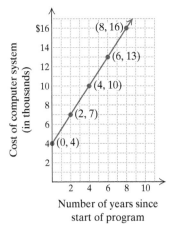

Years Since Start of Program	Cost of Computers (in thousands)
0	$ 4
2	7
4	10
6	13
8	16

The slope of the line is the rate of change in computer cost per year;

$$\text{Slope} = \frac{\text{change in } y}{\text{change in } x} = \frac{\text{rise}}{\text{run}} = \frac{y_2 - y_1}{x_2 - x_1},$$

where (x_1, y_1) and (x_2, y_2) are any two points on the graphed line. Here we select $(0, 4)$ and $(2, 7)$:

$$\text{Slope} = \frac{\text{change in } y}{\text{change in } x} = \frac{7 - 4}{2 - 0} = \frac{3}{2}. \qquad \text{The rate of change is } \tfrac{3}{2} \text{ thousand dollars per year, or \$1500/year.}$$

Knowing that the slope is $\frac{3}{2}$, we could have drawn the graph by plotting $(0, 4)$ and from there moving 3 units *up* and 2 units *to the right*. This would have located the point $(2, 7)$. Using $(0, 4)$ and $(2, 7)$, we can then draw the line. This is the method used in the next example.

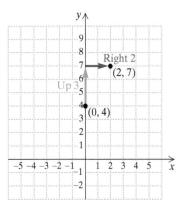

EXAMPLE 1 Draw a line that has slope $\frac{1}{4}$ and y-intercept $(0, 2)$.

SOLUTION We plot $(0, 2)$ and from there move 1 unit *up* and 4 units *to the right*. This locates the point $(4, 3)$. We plot $(4, 3)$ and draw a line passing through $(0, 2)$ and $(4, 3)$, as shown on the right below.

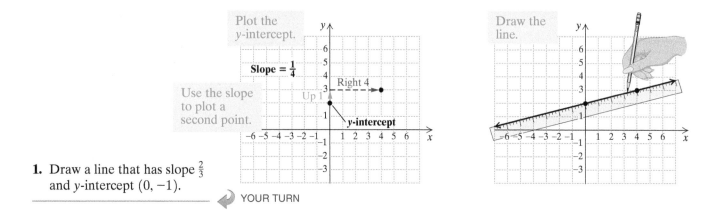

1. Draw a line that has slope $\frac{2}{3}$ and y-intercept $(0, -1)$.

YOUR TURN

B. Equations in Slope–Intercept Form

It is possible to read the slope and the y-intercept of a line directly from its equation. To find the y-intercept of an equation's graph, we replace x with 0 and solve the resulting equation for y. Compare the following.

$$y = 2x + 3 \qquad\qquad y = mx + b$$
$$\quad = 2 \cdot 0 + 3 = 0 + 3 = 3 \qquad\qquad = m \cdot 0 + b = 0 + b = b$$

The y-intercept of the graph of $y = 2x + 3$ is $(0, 3)$.

The y-intercept of the graph of $y = mx + b$ is $(0, b)$.

To calculate the slope of the graph of $y = 2x + 3$, we need two ordered pairs that are solutions of the equation. The y-intercept $(0, 3)$ is one pair; a second pair, $(1, 5)$, can be found by substituting 1 for x. We then have

$$\text{Slope} = \frac{\text{change in } y}{\text{change in } x} = \frac{5 - 3}{1 - 0} = \frac{2}{1} = 2.$$

Note that the slope, 2, is also the x-coefficient in $y = 2x + 3$. It can be similarly shown that the graph of any equation of the form $y = mx + b$ has slope m (see Exercise 87).

THE SLOPE–INTERCEPT EQUATION

The equation $y = mx + b$ is called the *slope–intercept equation*. The graph of $y = mx + b$ has slope m and y-intercept $(0, b)$.

The equation of any nonvertical line can be written in this form. The use of the letter m for slope may have its origin in the French verb *monter*, to climb.

EXAMPLE 2 Find the slope and the y-intercept of each line whose equation is given.

a) $y = \frac{4}{5}x - 8$ **b)** $2x + y = 5$ **c)** $4x - 4y = 7$

Student Notes

To write an equation "in the form $y = mx + b$" means to solve the equation for y, so that "$y =$" is followed by the term containing x and then the constant term. It may help to write your equation below the general form and align the y, the $=$, and the x. For Example 2(b), you could write

$$y = mx + b$$
$$y = -2x + 5.$$

From this form, you can see that $m = -2$ and $b = 5$.

SOLUTION

a) We rewrite $y = \frac{4}{5}x - 8$ as $y = \frac{4}{5}x + (-8)$. Now we simply read the slope and the y-intercept from the equation:

$$y = \tfrac{4}{5}x + (-8).$$

The slope is $\frac{4}{5}$. The y-intercept is $(0, -8)$.

b) We first solve for y to find an equivalent equation in the form $y = mx + b$:

$$2x + y = 5$$
$$y = -2x + 5. \qquad \text{Adding } -2x \text{ to both sides}$$

The slope is -2. The y-intercept is $(0, 5)$.

c) We rewrite the equation in the form $y = mx + b$:

$$4x - 4y = 7$$
$$-4y = -4x + 7 \qquad \text{Adding } -4x \text{ to both sides}$$
$$y = -\tfrac{1}{4}(-4x + 7) \qquad \text{Multiplying both sides by } -\tfrac{1}{4}$$
$$y = x - \tfrac{7}{4} \qquad \text{Using the distributive law}$$
$$y = 1 \cdot x + \left(-\tfrac{7}{4}\right). \qquad \text{The (unwritten) coefficient of } x \text{ is 1.}$$

The slope is 1. The y-intercept is $\left(0, -\tfrac{7}{4}\right)$.

2. Find the slope and the y-intercept of the line given by $y = -\frac{1}{3}x - 7$.

YOUR TURN

EXAMPLE 3 A line has slope $-\frac{12}{5}$ and y-intercept $(0, 11)$. Find an equation of the line.

SOLUTION We use the slope–intercept equation, substituting $-\frac{12}{5}$ for m and 11 for b:

$$y = mx + b = -\tfrac{12}{5}x + 11.$$

The desired equation is $y = -\frac{12}{5}x + 11$.

3. A line has slope 4 and y-intercept $(0, -1)$. Find an equation of the line.

YOUR TURN

EXAMPLE 4 *Online College Enrollment.* The following graph shows the number of students, in millions, taking at least one online college course for various years. Determine an equation for the graph.

Online College Enrollment

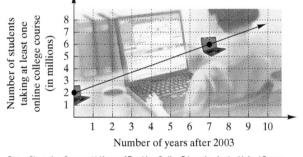

Number of students taking at least one online college course (in millions)

Number of years after 2003

Data: Changing Course: 10 Years of Tracking Online Education in the United States, Babson Survey Research Group

4. Determine an equation for the following graph.

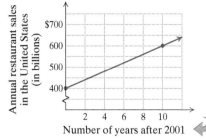

Annual restaurant sales in the United States (in billions)

$700
600
500
400

2   4   6   8   10
Number of years after 2001

Data: National Restaurant Association

SOLUTION If we know the *y*-intercept and the slope, we can use slope–intercept form to write an equation for the line. From the graph, we see that $(0, 2)$ is the *y*-intercept and that the line passes through $(7, 6)$. We calculate the slope:

$$m = \frac{\text{change in } y}{\text{change in } x} = \frac{6 - 2}{7 - 0} = \frac{4}{7}.$$

The desired equation is $y = \frac{4}{7}x + 2$,   Using $\frac{4}{7}$ for *m* and 2 for *b*

where *y* is the number of students, in millions, taking at least one online college course *x* years after 2003.

YOUR TURN

C. Graphing and Slope–Intercept Form

In Example 1, we drew a graph, knowing only the slope and the *y*-intercept. In Example 2, we determined the slope and the *y*-intercept of a line by examining its equation. We now combine the two procedures to develop a quick way to graph a linear equation.

EXAMPLE 5   Graph: **(a)** $y = \frac{3}{4}x + 5$; **(b)** $2x + 3y = 3$.

SOLUTION   To graph each equation, we plot the *y*-intercept and find additional points using the slope.

a) We can read the slope and the *y*-intercept from the equation $y = \frac{3}{4}x + 5$:

Determine the slope and the *y*-intercept.

    Slope: $\frac{3}{4}$;   *y*-intercept: $(0, 5)$.

Plot the *y*-intercept.

We plot the *y*-intercept $(0, 5)$. This gives us one point on the line. Starting at $(0, 5)$, we use the slope $\frac{3}{4}$ to find another point.

Use the slope to find a second point.

- We move 3 units *up* since the numerator (change in *y*) is *positive*.
- We move 4 units *to the right* since the denominator (change in *x*) is *positive*.

This gives us a second point on the line, $(4, 8)$.
    We can find a third point on the line by rewriting the slope $\frac{3}{4}$ as $\frac{-3}{-4}$, since these fractions are equivalent. Now, starting again at $(0, 5)$, we use the slope $\frac{-3}{-4}$ to find another point.

Use the slope to find a third point.

- We move 3 units *down* since the numerator (change in *y*) is *negative*.
- We move 4 units *to the left* since the denominator (change in *x*) is *negative*.

This gives us a third point on the line, $(-4, 2)$.

Draw the line.

    Finally, we draw the line. As a partial check, note that the slope is positive, as we would expect.

Student Notes

Recall the following:

$$\frac{3}{4} = \frac{-3}{-4};$$

$$-\frac{3}{4} = \frac{-3}{4} = \frac{3}{-4};$$

$$2 = \frac{2}{1} \quad \text{and} \quad -2 = \frac{-2}{1}.$$

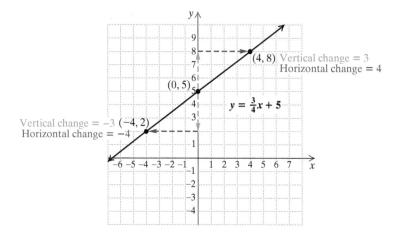

Vertical change = 3
Horizontal change = 4

$(4, 8)$

$(0, 5)$

$y = \frac{3}{4}x + 5$

Vertical change = −3 $(-4, 2)$
Horizontal change = −4

Consider $y = \frac{2}{3}x + 1$.

1. Is the equation written in slope–intercept form?

2. With reference to the form $y = mx + b$, what is m for this equation?

3. What is the slope of the graph of the equation?

4. For a horizontal run of 3 units, how much does the graph of the equation rise vertically?

5. With reference to the form $y = mx + b$, what is b for this equation?

6. What is the y-intercept of the graph of the equation?

7. Is the graph of the equation a straight line?

5. Graph: $y = -\frac{1}{2}x - 2$.

b) To graph $2x + 3y = 3$, we first rewrite it to find the slope and the y-intercept:

$$2x + 3y = 3$$
$$3y = -2x + 3 \qquad \text{Adding } -2x \text{ to both sides}$$
$$y = \frac{1}{3}(-2x + 3) \qquad \text{Multiplying both sides by } \frac{1}{3}$$
$$y = -\frac{2}{3}x + 1. \qquad \text{Using the distributive law}$$

We plot the y-intercept, $(0, 1)$.

The slope is $-\frac{2}{3}$. For graphing, we think of this slope as $\frac{-2}{3}$ or $\frac{2}{-3}$. Starting at $(0, 1)$, we use the slope $\frac{-2}{3}$ to find a second point.

• We move 2 units *down* because the numerator is *negative*.
• We move 3 units *to the right* because the denominator is *positive*.

We plot the new point, $(3, -1)$.

Now, starting at $(3, -1)$ and again using the slope $\frac{-2}{3}$, we move to a third point, $(6, -3)$.

Alternatively, we can start at $(0, 1)$ and use the slope $\frac{2}{-3}$.

• We move 2 units *up* because the numerator is *positive*.
• We move 3 units *to the left* because the denominator is *negative*.

This leads to another point on the graph, $(-3, 3)$.

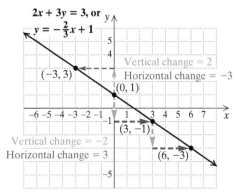

It is important to be able to use both $\frac{2}{-3}$ and $\frac{-2}{3}$ to draw the graph.

⟍ YOUR TURN

Student Notes

The signs of the numerator and the denominator of the slope indicate whether to move up, down, left, or right. Compare the following slopes.

$\frac{1}{2} \begin{array}{l} \leftarrow 1 \text{ unit up} \\ \leftarrow 2 \text{ units right} \end{array}$

$\frac{-1}{-2} \begin{array}{l} \leftarrow 1 \text{ unit down} \\ \leftarrow 2 \text{ units left} \end{array}$

$\frac{-1}{2} \begin{array}{l} \leftarrow 1 \text{ unit down} \\ \leftarrow 2 \text{ units right} \end{array}$

$\frac{1}{-2} \begin{array}{l} \leftarrow 1 \text{ unit up} \\ \leftarrow 2 \text{ units left} \end{array}$

D. Slope and Parallel Lines

Two lines are parallel if they lie in the same plane and do not intersect no matter how far they are extended. If two lines are vertical, they are parallel. How can we tell if nonvertical lines are parallel? The answer is simple: We look at their slopes.

> **SLOPE AND PARALLEL LINES**
>
> Two different lines are parallel if they have the same slope or if both lines are vertical.

EXAMPLE 6 Determine whether the graphs of

$$y = -3x + 4 \quad \text{and} \quad 6x + 2y = -10$$

are parallel.

SOLUTION We compare the slopes of the two lines to determine whether the graphs are parallel.

One of the two equations given is in slope–intercept form:

$$y = -3x + 4. \qquad \text{The slope is } -3 \text{ and the } y\text{-intercept is } (0, 4).$$

To find the slope of the other line, we need to rewrite the other equation in slope–intercept form:

$$6x + 2y = -10$$
$$2y = -6x - 10 \qquad \text{Adding } -6x \text{ to both sides}$$
$$y = -3x - 5. \qquad \text{The slope is } -3 \text{ and the } y\text{-intercept is } (0, -5).$$

6. Determine whether the graphs of $2x - y = 3$ and $y = -2x - 7$ are parallel.

_____

Since both lines have slope -3 but different y-intercepts, the graphs are parallel. There is no need for us to actually graph either equation.

◁ YOUR TURN

3.6 EXERCISE SET

FOR EXTRA HELP   MyMathLab®

◈ Vocabulary and Reading Check

Classify each of the following as either "slope" or "y-intercept."

1. m for $y = mx + b$

2. $(0, b)$ for $y = mx + b$

3. The point at which a graph crosses the y-axis

4. A value that is the same for parallel lines

5. $(0, 5)$ for $y = x + 5$

6. 1 for $y = x + 5$

◈ Concept Reinforcement

In each of Exercises 7–12, match the phrase with the most appropriate choice from the following list.

a) $\left(0, \frac{3}{4}\right)$ **b)** 2
c) $(0, -3)$ **d)** $\frac{2}{3}$
e) $(0, -2)$ **f)** 3

7. ____ The slope of the graph of $y = 3x - 2$

8. ____ The slope of the graph of $y = 2x - 3$

9. ____ The slope of the graph of $y = \frac{2}{3}x + 3$

10. ____ The y-intercept of the graph of $y = 2x - 3$

11. ____ The y-intercept of the graph of $y = 3x - 2$

12. ____ The y-intercept of the graph of $y = \frac{2}{3}x + \frac{3}{4}$

A. Using the y-intercept and the Slope to Graph a Line

Draw a line that has the given slope and y-intercept.

13. Slope $\frac{2}{3}$; y-intercept $(0, 1)$

14. Slope $\frac{3}{5}$; y-intercept $(0, -1)$

15. Slope $\frac{5}{3}$; y-intercept $(0, -2)$

16. Slope $\frac{1}{2}$; y-intercept $(0, 0)$

17. Slope $-\frac{1}{3}$; y-intercept $(0, 5)$

18. Slope $-\frac{4}{5}$; y-intercept $(0, 6)$

19. Slope 2; y-intercept $(0, 0)$

20. Slope -2; y-intercept $(0, -3)$

21. Slope -3; y-intercept $(0, 2)$

22. Slope 3; y-intercept $(0, 4)$

Aha! **23.** Slope 0; y-intercept $(0, -5)$

24. Slope 0; y-intercept $(0, 1)$

B. Equations in Slope–Intercept Form

Find the slope and the y-intercept of the line from the given equation.

25. $y = -\frac{2}{7}x + 5$ **26.** $y = -\frac{3}{8}x + 4$

27. $y = \frac{1}{3}x + 7$ **28.** $y = \frac{4}{5}x + 1$

29. $y = \frac{9}{5}x - 4$ **30.** $y = -\frac{9}{10}x - 5$

31. $-3x + y = 7$ **32.** $-4x + y = 7$

33. $4x + 2y = 8$ **34.** $3x + 4y = 12$

Aha! **35.** $y = 3$ **36.** $y - 3 = 5$

37. $2x - 5y = -8$ **38.** $12x - 6y = 9$

39. $9x - 8y = 0$ **40.** $7x = 5y$

Find the slope–intercept equation of the line with the indicated slope and y-intercept.

41. Slope 5; y-intercept $(0, 7)$

42. Slope -4; y-intercept $\left(0, -\frac{3}{5}\right)$

43. Slope $\frac{7}{8}$; y-intercept $(0, -1)$

44. Slope $\frac{5}{7}$; y-intercept $(0, 4)$

45. Slope $-\frac{5}{3}$; y-intercept $(0, -8)$

46. Slope $\frac{3}{4}$; y-intercept $(0, -35)$

Aha! **47.** Slope 0; y-intercept $\left(0, \frac{1}{3}\right)$

48. Slope 7; y-intercept $(0, 0)$

Determine an equation for each graph shown.

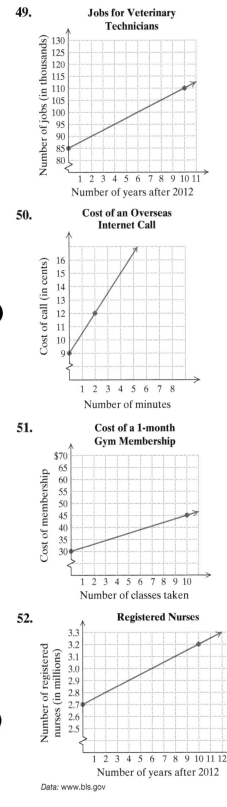

49. Jobs for Veterinary Technicians

50. Cost of an Overseas Internet Call

51. Cost of a 1-month Gym Membership

52. Registered Nurses

Data: www.bls.gov

C. Graphing and Slope–Intercept Form

Graph.

53. $y = \frac{2}{3}x + 2$

54. $y = -\frac{2}{3}x - 3$

55. $y = -\frac{2}{3}x + 3$

56. $y = \frac{2}{3}x - 2$

57. $y = \frac{3}{2}x + 3$

58. $y = \frac{3}{2}x - 2$

59. $y = -\frac{4}{3}x + 3$

60. $y = -\frac{3}{2}x - 2$

61. $2x + y = 1$

62. $3x + y = 2$

63. $3x + y = 0$

64. $2x + y = 0$

65. $4x + 5y = 15$

66. $2x + 3y = 9$

67. $x - 4y = 12$

68. $x + 5y = 20$

D. Slope and Parallel Lines

Determine whether each pair of equations represents parallel lines.

69. $y = \frac{3}{4}x + 6$,
$y = \frac{3}{4}x - 2$

70. $y = \frac{1}{3}x - 2$,
$y = -\frac{1}{3}x + 1$

71. $y = 2x - 5$,
$4x + 2y = 9$

72. $y = -3x + 1$,
$6x + 2y = 8$

73. $3x + 4y = 8$,
$7 - 12y = 9x$

74. $3x = 5y - 2$,
$10y = 4 - 6x$

75. Can a horizontal line be graphed using the method of Example 5? Why or why not?

76. Can a vertical line be graphed using the method of Example 5? Why or why not?

Skill Review

Solve.

77. Mr. and Mrs. Sturgis left a $3 tip for a meal that cost $25. What percent of the cost of the meal was the tip? [2.4]

78. Irniq is writing a 1000-word essay. He has three times as many words to write as he has already written. How many words has he written? [2.5]

79. In December, the balance in Dalila's college savings account grew 15%, to $2760. What was her balance at the beginning of the month? [2.5]

80. The perimeter of a hospital's rectangular flower garden is 140 ft. The width is 30 ft less than the length. Find the width and the length. [2.5]

Synthesis

81. Explain how it is possible for an incorrect graph to be drawn, even after plotting three points that line up.

82. Which would you prefer, and why: graphing an equation of the form $y = mx + b$ or graphing an equation of the form $Ax + By = C$?

83. Show that the slope of the line given by $y = mx + b$ is m. (*Hint*: Substitute both 0 and 1 for x to form two ordered pairs. Then use the formula, Slope = change in y/change in x.)

84. Write an equation of the line with the same slope as the line given by $5x + 2y = 8$ and the same y-intercept as the line given by $3x - 7y = 10$.

85. Write an equation of the line parallel to the line given by $-4x + 8y = 5$ and having the same y-intercept as the line given by $4x - 3y = 0$.

86. Write an equation of the line parallel to the line given by $3x - 2y = 8$ and having the same y-intercept as the line given by $2y + 3x = -4$.

87. Write an equation of the line parallel to the line given by $4x + 5y = 9$ and having the same y-intercept as the line given by $2x + 3y = 12$.

Two lines are perpendicular if either the product of their slopes is −1, or one line is vertical and the other horizontal. For Exercises 88–93, determine whether each pair of equations represents perpendicular lines.

88. $3y = 5x - 3$,
$3x + 5y = 10$

89. $y + 3x = 10$,
$2x - 6y = 18$

90. $3x + 5y = 10$,
$15x + 9y = 18$

91. $10 - 4y = 7x$,
$7y + 21 = 4x$

92. $x = 5$,
$y = \frac{1}{2}$

93. $y = -2x$,
$x = \frac{1}{2}$

94. Write an equation of the line perpendicular to the line given by $2x + 3y = 7$ (see Exercises 88–93) and having the same y-intercept as the line given by $5x + 2y = 10$.

95. Write an equation of the line perpendicular to the line given by $3x - 5y = 8$ (see Exercises 88–93) and having the same y-intercept as the line given by $2x + 4y = 12$.

96. Write an equation of the line perpendicular to the line given by $3x - 2y = 9$ (see Exercises 88–93) and having the same y-intercept as the line given by $2x + 5y = 0$.

97. Write an equation of the line perpendicular to the line given by $2x + 5y = 6$ (see Exercises 88–93) that passes through $(2, 6)$. (*Hint*: Draw a graph.)

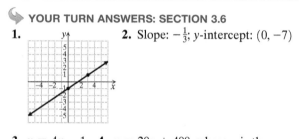 **98.** Graph $y_1 = -\frac{3}{4}x - 2$, $y_2 = -\frac{1}{5}x - 2$, $y_3 = -\frac{3}{4}x - 5$, and $y_4 = -\frac{1}{5}x - 5$ using the SIMULTANEOUS mode. Then match each line with the corresponding equation. Check using TRACE.

YOUR TURN ANSWERS: SECTION 3.6

1. 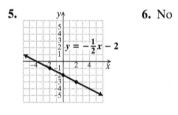 **2.** Slope: $-\frac{1}{3}$; y-intercept: $(0, -7)$

3. $y = 4x - 1$ **4.** $y = 20x + 400$, where y is the amount of U.S. restaurant sales, in billions of dollars, and x is the number of years after 2001

5.

$y = -\frac{1}{2}x - 2$

6. No

Quick Quiz: Sections 3.1–3.6

Graph.

1. $y = 3$ [3.3] **2.** $y = \frac{1}{2}x - 4$ [3.6]

3. In which quadrant or on which axis is $(-12, -0.02)$ located? [3.1]

4. List the coordinates of the x- and y-intercepts of the graph of $y - x = 7$. [3.3]

5. Find the slope of the graph of $y - x = 7$. [3.6]

Prepare to Move On

Solve. [2.3]

1. $y - k = m(x - h)$, for y

2. $y - 9 = -2(x + 4)$, for y

Simplify. [1.6]

3. $-10 - (-3)$ **4.** $8 - (-5)$

5. $-4 - 5$

3.7 Point–Slope Form

A. Writing Equations Using Point–Slope Form **B.** Graphing and Point–Slope Form

C. Estimations and Predictions Using Two Points

Specifying a line's slope and one point through which the line passes enables us to draw the line. In this section, we study how this same information can be used to produce an equation of the line.

A. Writing Equations Using Point–Slope Form

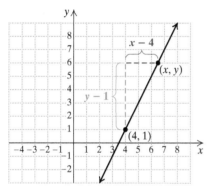

Consider a line with slope 2 passing through $(4, 1)$, as shown in the figure at left. In order for another point (x, y) to be on this line, the coordinates (x, y) must satisfy the slope equation

$$\frac{y - 1}{x - 4} = 2.$$

Pairs like $(5, 3)$ and $(3, -1)$ are on the line and are solutions of this equation, since

$$\frac{3 - 1}{5 - 4} = 2 \quad \text{and} \quad \frac{-1 - 1}{3 - 4} = 2.$$

Note, however, that $(4, 1)$ is not itself a solution of the equation:

$$\frac{1 - 1}{4 - 4} \neq 2.$$

To avoid this difficulty, we use the multiplication principle.

$$(x - 4) \cdot \frac{y - 1}{x - 4} = 2(x - 4) \qquad \text{Multiplying both sides by } x - 4$$

$$y - 1 = 2(x - 4) \qquad \text{Removing a factor equal to 1: } \frac{x - 4}{x - 4} = 1$$

Every point on the line is a solution of this equation. This is considered a **point–slope equation** of the line shown. **A point–slope equation can be written anytime a line's slope and a point on the line are known.**

Student Notes

You can remember the point–slope equation by calculating the slope m of a line containing a *general* point (x, y) and a *specific* point (x_1, y_1):

$$m = \frac{y - y_1}{x - x_1}.$$

Multiplying by $x - x_1$, we have

$$m(x - x_1) = y - y_1.$$

---

1. Write a point–slope equation for the line with slope $\frac{1}{2}$ that contains the point $(-2, 1)$.

THE POINT–SLOPE EQUATION

The equation $y - y_1 = m(x - x_1)$ is called the *point–slope equation* for the line with slope m that contains the point (x_1, y_1).

When we are using the point–slope equation, x and y remain as variables, and m, x_1, and y_1 are replaced by the slope and the coordinates of a point on the line.

EXAMPLE 1 Write a point–slope equation for the line with slope $-\frac{4}{3}$ that contains the point $(1, -6)$.

SOLUTION We substitute $-\frac{4}{3}$ for m, 1 for x_1, and -6 for y_1:

$$y - y_1 = m(x - x_1) \qquad \text{Using the point–slope equation}$$

$$y - (-6) = -\frac{4}{3}(x - 1). \qquad \text{Substituting}$$

YOUR TURN

Student Notes

There are several forms in which a line's equation can be written. For instance, as shown in Example 2, $y - 1 = 2(x - 3), y - 1 = 2x - 6$, and $y = 2x - 5$ all are equations of the same line.

2. Write the slope–intercept equation for the line with slope -3 that contains the point $(5, 7)$.

Student Notes

Using $(4, -7)$ instead of $(2, -1)$ for the point in Example 3 gives a point–slope equation of

$$y - (-7) = -3(x - 4).$$

This still leads to the slope–intercept equation

$$y = -3x + 5.$$

Since there are many points on a line, there are many point–slope equations for a line. There is only one y-intercept, so there is only one slope–intercept equation for a line.

3. Write the slope–intercept equation for the line containing the points $(10, 3)$ and $(2, -5)$.

EXAMPLE 2   Write the slope–intercept equation for the line with slope 2 that contains the point $(3, 1)$.

SOLUTION   There are two parts to this solution. First, we write an equation in point–slope form:

$$y - y_1 = m(x - x_1)$$
$$y - 1 = 2(x - 3). \qquad \text{Substituting}$$

 Write in point–slope form.

Next, we find an equivalent equation of the form $y = mx + b$:

$$y - 1 = 2(x - 3)$$
$$y - 1 = 2x - 6 \qquad \text{Using the distributive law}$$
$$y = 2x - 5. \qquad \text{Adding 1 to both sides to get slope–intercept form}$$

Write in slope–intercept form.

YOUR TURN

EXAMPLE 3   Write the slope–intercept equation for the line containing the points $(2, -1)$ and $(4, -7)$.

SOLUTION   As in Example 2, we first write an equation in point–slope form. This time, however, the slope is not given, so we use the two points given to calculate the slope:

$$m = \frac{y_2 - y_1}{x_2 - x_1} = \frac{-7 - (-1)}{4 - 2} = \frac{-6}{2} = -3.$$

Next, we use the slope and one of the given points to write an equation in point–slope form:

$$y - y_1 = m(x - x_1)$$
$$y - (-1) = -3(x - 2). \qquad \text{Substituting}$$

Finally, we find an equivalent equation of the form $y = mx + b$:

$$y - (-1) = -3(x - 2)$$
$$y + 1 = -3x + 6 \qquad \text{Simplifying}$$
$$y = -3x + 5. \qquad \text{Writing in slope–intercept form}$$

Had we used $(4, -7)$ when finding an equation in point–slope form, the final answer in slope–intercept form would have been the same. (See the Student Notes at left.)

YOUR TURN

TO WRITE AN EQUATION FOR A LINE

1. Determine the slope and a point on the line. If two points are given, use the points to find the slope and choose one of the points to use to write the equation.
2. Use the slope and the point to write a point–slope equation for the line. *Each point on a line will give a different point–slope equation.*
3. If desired, write the equation in slope–intercept form. If the y-intercept is given, you can go directly to this step. *There is only one slope–intercept equation for a line.*

B. Graphing and Point–Slope Form

When we know a line's slope and a point that is on the line, we can draw the graph. For example, the information given in the statement of Example 2 is sufficient for drawing a graph.

EXAMPLE 4   Graph the line with slope 2 that passes through $(3, 1)$.

SOLUTION   We plot $(3, 1)$, move *up* 2 and *to the right* 1 (since $2 = \frac{2}{1}$), and draw the line.

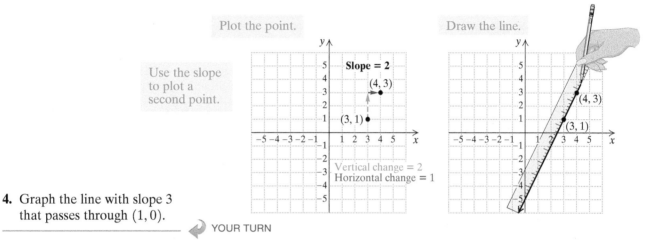

Use the slope to plot a second point.

4. Graph the line with slope 3 that passes through $(1, 0)$.

YOUR TURN

EXAMPLE 5   Graph: $y - 2 = 3(x - 4)$.

SOLUTION   Since $y - 2 = 3(x - 4)$ is in point–slope form, we know that the line has slope 3, or $\frac{3}{1}$, and passes through the point $(4, 2)$. We plot $(4, 2)$ and then find a second point by moving *up* 3 units and *to the right* 1 unit. The line can then be drawn, as shown below.

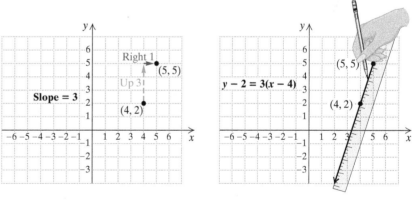

5. Graph: $y - 1 = 2(x - 3)$.

YOUR TURN

EXAMPLE 6   Graph: $y + 4 = -\frac{5}{2}(x + 3)$.

SOLUTION   Once an equation is in point–slope form, $y - y_1 = m(x - x_1)$, we can proceed much as we did in Example 5. To find an equivalent equation in point–slope form, we subtract opposites instead of adding:

$$y + 4 = -\tfrac{5}{2}(x + 3)$$
$$y - (-4) = -\tfrac{5}{2}(x - (-3)).$$

$y - (-4) = y + 4$ and
$x - (-3) = x + 3$. This is now in
point–slope form.

Study Skills

Understand Your Mistakes

When your instructor returns a graded quiz, test, or assignment, it is important that you review and understand what your mistakes were. Take advantage of the opportunity to learn from your mistakes.

---

6. Graph: $y + 2 = -\frac{1}{3}(x + 1)$.

From this last equation, $y - (-4) = -\frac{5}{2}(x - (-3))$, we see that the line passes through $(-3, -4)$ and has slope $-\frac{5}{2}$, or $\frac{5}{-2}$.

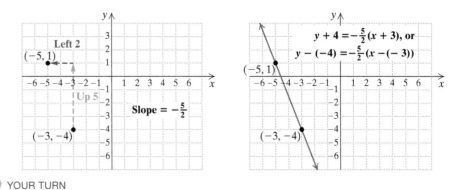

YOUR TURN

C. Estimations and Predictions Using Two Points

We can estimate real-life quantities that are not already known by using two points with known coordinates. When the unknown point is located *between* the two points, this process is called **interpolation**. If a graph passing through the known points is *extended* to predict future values, the process is called **extrapolation**.

EXAMPLE 7 *Education Debt.* The average education debt per borrower at college graduation nearly doubled from 2005 to 2015.

a) Examine the data from 2007 and 2015. Let x represent the number of years after 2005 and y the average debt. Then use the two points to determine the slope–intercept equation of the line.

b) Use the equation found in part (a) to estimate the average education debt at college graduation in 2010 and in 2020.

SOLUTION

a) We first draw and label a horizontal axis to display the number of years after 2005 and a vertical axis to display the average education debt, in thousands of dollars. Next, we number the axes, choosing scales that include both the given values and the values to be estimated.

Since $x =$ the number of years after 2005, we plot $(2, 22)$ and $(10, 35)$ and draw a line passing through both points.

Year	Average Education Debt at College Graduation
2005	$20,000
2007	22,000
2009	25,000
2011	27,500
2013	31,000
2015	35,000

Data: wsj.com

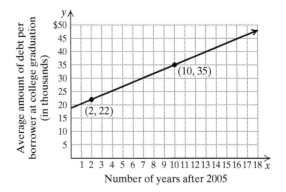

To find an equation for the line, we first calculate its slope:

$$m = \frac{\text{change in } y}{\text{change in } x} = \frac{35 - 22}{10 - 2} = \frac{13}{8} = 1.625.$$

Chapter Resources:
Visualizing for Success, p. 221;
Decision Making: Connection, p. 222

The average education debt increased at a rate of $1.625 thousand (or $1625) per year.

We can use either of the given points to write an equation of the line. Choosing $(10, 35)$, we write an equation in point–slope form and then an equivalent equation in slope–intercept form:

$y - y_1 = m(x - x_1)$	Writing the general point–slope equation
$y - 35 = 1.625(x - 10)$	Substituting. This is a point–slope equation.
$y - 35 = 1.625 - 16.25$	Using the distributive law
$y = 1.625x + 18.75.$	Adding 35 to both sides. This is slope–intercept form.

b) To estimate the average education debt at college graduation in 2010, we substitute 5 for x in the slope–intercept equation:

$$y = 1.625 \cdot 5 + 18.75 = 26.875. \qquad \text{2010 is 5 years after 2005.}$$

In 2010, the average education debt was about $26.875 thousand (or $26,875). Because 2010 is *between* 2005 and 2015, this is *interpolation*.

To estimate the average education debt at college graduation in 2020, we substitute 15 for x in the slope–intercept equation:

$$y = 1.625 \cdot 15 + 18.75 = 43.125. \qquad \text{2020 is 15 years after 2005.}$$

In 2020, the average education debt will be about $43.125 thousand (or $43,125). Because 2020 is *beyond* 2015, this is *extrapolation*. This estimate assumes that the trend indicated by the data continues.

The following graph confirms the estimates found above.

7. The data used for Example 7 do not all lie on a straight line. Thus a different pair of points may yield a different equation for the line used to model the application.

a) The data from 2009 and 2015 correspond to the points $(4, 25)$ and $(10, 35)$. Find an equation for the line containing these points and then rewrite it in slope–intercept form.

b) Use the equation found in part (a) to estimate the average education debt at college graduation in 2010 and in 2020.

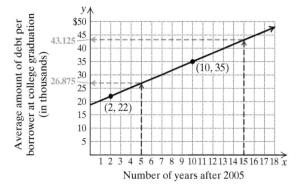

YOUR TURN

↳ Check Your
UNDERSTANDING

For each description of a line, determine what numbers you would substitute for $m, x_1,$ and y_1 in $y - y_1 = m(x - x_1)$.

1. The line has slope 2 and contains $(4, 8)$.

2. The line contains $(-1, 6)$ and has slope $\frac{1}{5}$.

3. The line has slope -3 and contains $(0, 7)$.

4. The line contains $(-4, 3)$ and $(2, 0)$.

CONNECTING 🔗 THE CONCEPTS

Any line can be described by an infinite number of equivalent equations. We write the equation in the form that is most useful for us. For example, all four of the equations shown at right describe the same line.

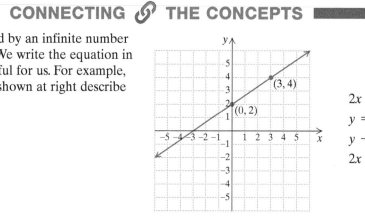

$$2x - 3y = -6;$$
$$y = \tfrac{2}{3}x + 2;$$
$$y - 4 = \tfrac{2}{3}(x - 3);$$
$$2x + 6 = 3y$$

Form of a Linear Equation	Example	Uses
Standard form: $Ax + By = C$	$2x - 3y = -6$	Finding x- and y-intercepts Graphing using intercepts
Slope–intercept form: $y = mx + b$	$y = \dfrac{2}{3}x + 2$	Finding slope and y-intercept Graphing using slope and y-intercept Writing an equation given slope and y-intercept Entering equations into a graphing calculator
Point–slope form: $y - y_1 = m(x - x_1)$	$y - 4 = \dfrac{2}{3}(x - 3)$	Finding slope and a point on the line Graphing using slope and a point on the line Writing an equation given slope and a point on the line

EXERCISES

Tell whether each equation is in standard form, slope–intercept form, point–slope form, or none of these.

1. $y = -\tfrac{1}{2}x - 7$

2. $5x - 8y = 10$

3. $x = y + 2$

4. $\tfrac{1}{2}x + \tfrac{1}{3}y = 5$

5. $y - 2 = 5(x - (-1))$

6. $3y + 7 = x$

Write each equation in standard form.

7. $2x = 5y + 10$

8. $y = 2x + 7$

Write each equation in slope–intercept form.

9. $2x - 7y = 8$

10. $y + 5 = -(x + 3)$

3.7 EXERCISE SET

Vocabulary and Reading Check

Classify each of the following statements as either true or false.

1. The equation $y - 7 = 3(x - 5)$ is written in point–slope form.

2. The graph of $y - 7 = 3(x - 5)$ has slope 3.

3. The graph of $y - 7 = 3(x - 5)$ contains the point $(7, 5)$.

4. Estimating an unknown point that lies between two known points is called interpolating.

Concept Reinforcement

In each of Exercises 5–12, match the given information about a line with the appropriate equation from the column on the right.

5. ____ Slope 5; includes $(2, 3)$

6. ____ Slope 5; includes $(3, 2)$

7. ____ Slope -5; includes $(2, 3)$

8. ____ Slope -5; includes $(3, 2)$

9. ____ Slope -5; includes $(-2, -3)$

10. ____ Slope 5; includes $(-2, -3)$

11. ____ Slope -5; includes $(-3, -2)$

12. ____ Slope 5; includes $(-3, -2)$

a) $y + 3 = 5(x + 2)$

b) $y - 2 = 5(x - 3)$

c) $y + 2 = 5(x + 3)$

d) $y - 3 = -5(x - 2)$

e) $y + 3 = -5(x + 2)$

f) $y + 2 = -5(x + 3)$

g) $y - 3 = 5(x - 2)$

h) $y - 2 = -5(x - 3)$ Aha!

A. Writing Equations Using Point–Slope Form

Write a point–slope equation for the line with the given slope that contains the given point.

13. $m = 3$; $(1, 6)$

14. $m = 2$; $(3, 7)$

15. $m = \frac{3}{5}$; $(2, 8)$

16. $m = \frac{2}{3}$; $(4, 1)$

17. $m = -4$; $(3, 1)$

18. $m = -5$; $(6, 2)$

19. $m = \frac{3}{2}$; $(5, -4)$

20. $m = -\frac{4}{3}$; $(7, -1)$

21. $m = -\frac{5}{4}$; $(-2, 6)$

22. $m = \frac{7}{2}$; $(-3, 4)$

23. $m = -2$; $(-4, -1)$

24. $m = -3$; $(-2, -5)$

25. $m = 1$; $(-2, 8)$

26. $m = -1$; $(-3, 6)$

Write the slope–intercept equation for the line with the given slope that contains the given point.

27. $m = 4$; $(3, 5)$

28. $m = 3$; $(6, 2)$

29. $m = \frac{7}{4}$; $(4, -2)$

30. $m = \frac{8}{3}$; $(3, -4)$

31. $m = -2$; $(-3, 7)$

32. $m = -3$; $(-2, 1)$

33. $m = -4$; $(-2, -1)$

34. $m = -5$; $(-1, -4)$

35. $m = \frac{2}{3}$; $(5, 6)$

36. $m = \frac{3}{2}$; $(7, 4)$

Aha! 37. $m = -\frac{5}{6}$; $(0, 4)$

38. $m = -\frac{3}{4}$; $(0, 5)$

Write the slope–intercept equation for the line containing the given pair of points.

39. $(2, 3)$ and $(4, 1)$

40. $(6, 8)$ and $(3, 5)$

41. $(-3, 1)$ and $(3, 5)$

42. $(-3, 4)$ and $(3, 1)$

43. $(5, 0)$ and $(0, -2)$

44. $(-2, 0)$ and $(0, 3)$

45. $(-4, -1)$ and $(1, 9)$

46. $(-3, 5)$ and $(-1, -3)$

B. Graphing and Point–Slope Form

For each point–slope equation given, state the slope and a point on the graph.

47. $y - 9 = \frac{2}{7}(x - 8)$

48. $y - 3 = 9(x - 2)$

49. $y + 2 = -5(x - 7)$

50. $y - 4 = -\frac{2}{9}(x + 5)$

51. $y - 4 = -\frac{5}{3}(x + 2)$

52. $y + 7 = -4(x - 9)$

53. $y = \frac{4}{7}x$

54. $y = 3x$

55. Graph the line with slope $\frac{4}{3}$ that passes through the point $(1, 2)$.

56. Graph the line with slope $\frac{2}{5}$ that passes through the point $(3, 4)$.

57. Graph the line with slope $-\frac{3}{4}$ that passes through the point $(2, 5)$.

58. Graph the line with slope $-\frac{3}{2}$ that passes through the point $(1, 4)$.

Graph.

59. $y - 5 = \frac{1}{3}(x - 2)$

60. $y - 2 = \frac{1}{2}(x - 1)$

61. $y - 1 = -\frac{1}{4}(x - 3)$

62. $y - 1 = -\frac{1}{2}(x - 3)$

63. $y + 2 = \frac{2}{3}(x - 1)$

64. $y - 1 = \frac{3}{4}(x + 5)$

65. $y + 4 = 3(x + 1)$

66. $y + 3 = 2(x + 1)$

67. $y - 4 = -2(x + 1)$

68. $y + 3 = -1(x - 4)$

69. $y + 4 = 3(x + 2)$

70. $y + 3 = -(x + 2)$

71. $y + 1 = -\frac{3}{5}(x - 2)$

72. $y - 2 = -\frac{2}{3}(x + 1)$

C. Estimations and Predictions Using Two Points

In Exercises 73–80, assume that the data are linear.

73. *High Cholesterol.*   The percentage of U.S. women ages 60 and older with high cholesterol fell from about 27% in 2002 to about 20% in 2010.

Data: CDC/NCHS, National Health and Nutrition Examination Survey

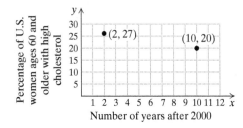

a) Find a linear equation that fits the data.
b) Estimate the percentage of U.S. women ages 60 and older with high cholesterol in 2008.
c) Estimate the percentage of U.S. women ages 60 and older with high cholesterol in 2015.

74. *Customer Satisfaction.*   The American customer satisfaction index scores of McDonald's restaurants grew from 61 in 2002 to 71 in 2014.

Data: American Customer Satisfaction Index at www.statista.com

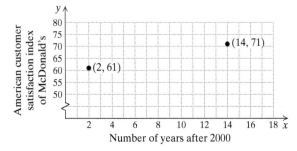

a) Find a linear equation that fits the data.
b) Estimate the customer satisfaction index of McDonald's in 2012.
c) Estimate the customer satisfaction index of McDonald's in 2017.

75. *Social Cost of Carbon.*   The Social Cost of Carbon (SCC) is an estimate of the environmental damages—such as increased health costs, flooding risk, agricultural losses, energy costs, and more—resulting from adding a metric ton of carbon dioxide to the atmosphere. In 2015, and using 2007 dollars and a moderate rate of inflation, the U.S. Environmental Protection

Agency estimated that the SCC in 2020 will be $42 per ton and in 2030 will be $50 per ton.

Data: epa.gov

a) Find an equation for the line containing the given data points. Let x represent the number of years after 2007 and y the SCC in dollars per ton, assuming a moderate rate of inflation.
b) Estimate the SCC in 2040 and in 2050, assuming a moderate rate of inflation.

76. *Social Cost of Carbon.*   The Social Cost of Carbon (SCC) estimates the environmental damages resulting from adding a metric ton of carbon dioxide to the atmosphere. (See Exercise 75.) In 2015, using 2007 dollars and assuming an accelerated rate of climate change, the U.S. Environmental Protection Agency estimated that the SCC in 2020 will be $123 per ton and in 2030 will be $152 per ton.

Data: epa.gov

a) Find an equation for the line containing the given data points. Let x represent the number of years after 2007 and y the SCC in dollars per ton, assuming an accelerated rate of climate change.
b) Estimate the SCC in 2040 and in 2050, assuming an accelerated rate of climate change.

77. *Video and Computer Games.*   The revenue from the sale of physical video and computer games decreased from $10.05 billion in 2010 to $5.47 billion in 2014. Let R represent the revenue from the sale of physical video and computer games, in billions of dollars, and t the number of years after 2008, the year in which revenue began to decrease.

Data: www.statista.com

a) Find a linear equation that fits the data.
b) Estimate the revenue from the sale of physical video and computer games in 2016.
c) If the trend continues, in what year will there be no sales of physical video and computer games?

78. *Newspapers.*   The average number of issues of newspapers distributed daily in the United States decreased from 54.6 million per day in 2004 to 29.1 million per day in 2014. Let N represent the number of newspaper issues distributed daily and t the number of years after 2000.

Data: Alliance for Audited Media

a) Find a linear equation that fits the data.
b) Estimate the number of issues distributed daily in 2018.
c) If the trend continues, in what year will no newspapers be distributed?

79. *Urban Population.* The percent of the U.S. population that lives in metropolitan areas was approximately 79% in 2000 and 80.7% in 2010.

Data: U.S. Census Bureau

a) Define variables x and y and find an equation for the line containing the given data points. Answers may vary.

b) Estimate the percent of the U.S. population that lived in metropolitan areas in 2005, and estimate the percent living in metropolitan areas in 2015.

80. *Aging Population.* The number of U.S. residents over the age of 65 was approximately 36.3 million in 2004 and 46.2 million in 2014.

Data: U.S. Census Bureau

a) Define variables x and y and find an equation for the line containing the given data points. Answers may vary.

b) Estimate the number of U.S. residents over the age of 65 in 2006 and in 2016.

81. Can equations for horizontal or vertical lines be written in point–slope form? Why or why not?

82. Describe a situation in which it is easier to graph the equation of a line in point–slope form than in slope–intercept form.

Skill Review

Solve.

83. $\frac{3}{8}x = -24$ [2.1]

84. $6 - x = -3$ [2.2]

85. $\frac{t}{3} = 6$ [2.1]

86. $\frac{1}{2}n - \frac{1}{3} = \frac{1}{6}n + \frac{3}{2}$ [2.2]

87. $2(x - 7) > 5x + 3$ [2.6] **88.** $10 - x \le 12$ [2.6]

Synthesis

89. Describe a procedure that can be used to write the slope–intercept equation for any nonvertical line passing through two given points.

90. Any nonvertical line has many equations in point–slope form, but only one in slope–intercept form. Why is this?

Graph.

Aha! **91.** $y - 3 = 0(x - 52)$ **92.** $y + 4 = 0(x + 93)$

Write the slope–intercept equation for each line shown.

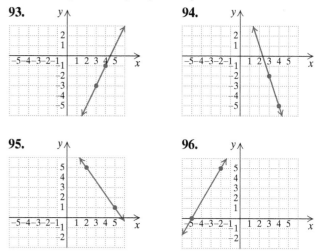

93.   **94.**

95.   **96.**

97. Write a point–slope equation of the line passing through $(-4, 7)$ that is parallel to the line given by $2x + 3y = 11$.

98. Write a point–slope equation of the line passing through $(3, -1)$ that is parallel to the line given by $4x - 5y = 9$.

Aha! **99.** Write an equation of the line parallel to the line given by $y = 3 - 4x$ that passes through $(0, 7)$.

100. Write the slope–intercept equation of the line that has the same y-intercept as the line $x - 3y = 6$ and contains the point $(5, -1)$.

101. Write the slope–intercept equation of the line that contains the point $(-1, 5)$ and is parallel to the line passing through $(2, 7)$ and $(-1, -3)$.

102. Write the slope–intercept equation of the line that has x-intercept $(-2, 0)$ and is parallel to $4x - 8y = 12$.

Another form of a linear equation is the double-intercept *form:* $\frac{x}{a} + \frac{y}{b} = 1$. *From this form, we can read the* x-intercept $(a, 0)$ *and the* y-intercept $(0, b)$ *directly.*

103. Find the x-intercept and the y-intercept of the graph of $\frac{x}{2} + \frac{y}{5} = 1$.

104. Find the x-intercept and the y-intercept of the graph of $\frac{x}{10} - \frac{y}{3} = 1$.

105. Write the equation $4y - 3x = 12$ in double-intercept form and find the intercepts.

106. Write the equation $6x + 5y = 30$ in double-intercept form and find the intercepts.

 107. Why is slope–intercept form more useful than point–slope form when using a graphing calculator? How can point–slope form be modified so that it is more easily used with graphing calculators?

108. If data do not lie exactly on a straight line, more than one equation can be used to model the data, as illustrated in Example 7 and Your Turn Exercise 7. A line of "best fit" can be found using methods such as *linear regression*. On a graphing calculator, enter x-coordinates in L_1 and y-coordinates in L_2 using the STAT EDIT menu. Then in the STAT CALC menu, choose the Lin Reg option. Use linear regression to find an equation that fits the data in the table in Example 7.

Quick Quiz: Sections 3.1–3.7
Graph.

1. $y - 3 = 2x$ [3.2] **2.** $x + 7 = 8$ [3.3]

3. $y = -3x + 2$ [3.6] **4.** $3x - 6y = 6$ [3.3]

5. $y - 1 = 2(x + 1)$ [3.7]

Prepare to Move On
Simplify. [1.8]

1. $(-5)^3$ **2.** $(-2)^6$

3. -2^6 **4.** $3 \cdot 2^4 - 5 \cdot 2^3$

5. $(5 - 7)^2(3 - 2 \cdot 2)$

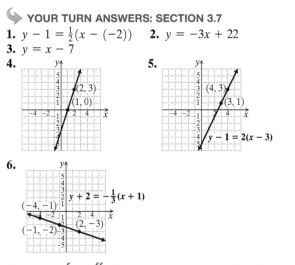

➤ **YOUR TURN ANSWERS: SECTION 3.7**

1. $y - 1 = \frac{1}{2}(x - (-2))$ **2.** $y = -3x + 22$

3. $y = x - 7$

4.

5.

6.

7. (a) $y = \frac{5}{3}x + \frac{55}{3}$, where y is the average education debt per borrower, in thousands of dollars, y years after 2005; **(b)** about \$26.667 thousand (or \$26,667); about \$43.333 thousand (or \$43,333)

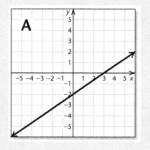

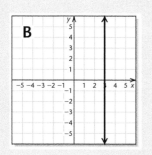

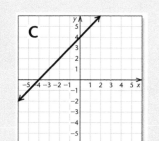

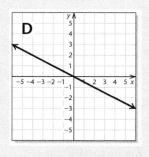

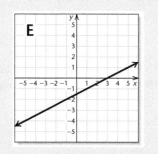

Visualizing for Success

Use after Section 3.7.

Match each equation with its graph.

1. $y = x + 4$

2. $y = 2x$

3. $y = 3$

4. $x = 3$

5. $y = -\frac{1}{2}x$

6. $2x - 3y = 6$

7. $y = -3x - 2$

8. $3x + 2y = 6$

9. $y - 3 = 2(x - 1)$

10. $y + 2 = \frac{1}{2}(x + 1)$

Answers on page A-17

An additional, animated version of this activity appears in MyMathLab. *To use MyMathLab, you need a course ID and a student access code. Contact your instructor for more information.*

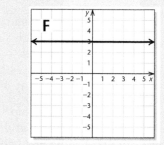

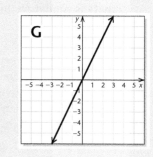

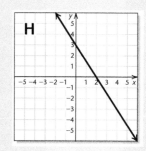

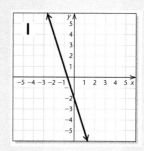

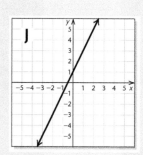

Collaborative Activity   *You Sank My Battleship!*

Focus: Graphing points; logical questioning
Use after: Section 3.1
Time: 15–25 minutes
Group size: 3–5
Materials: Graph paper

In the game Battleship®, a player places a miniature ship on a grid that only that player can see. An opponent guesses at coordinates that might "hit" the "hidden" ship. The following activity is similar to this game.

Activity

1. Using only integers from −10 to 10 (inclusive), one group member should secretly record the coordinates of a point on a slip of paper. (This point is the hidden "battleship.")

2. The other group members can then ask up to 10 "yes/no" questions in an effort to determine the coordinates of the secret point. Be sure to phrase each question mathematically (for example, "Is the *x*-coordinate negative?").

3. The group member who selected the point should answer each question. On the basis of the answer given, another group member should cross out the points no longer under consideration. All group members should check that this is done correctly.

4. If the hidden point has not been determined after 10 questions have been answered, the secret coordinates should be revealed.

5. Repeat parts (1)–(4) until everyone has had the opportunity to select the hidden point and answer questions.

Decision Making & Connection

Depreciation.   From the minute a new car is driven out of a dealership, it *depreciates*, or drops in value with the passing of time. The Kelley Blue Book Market Report is a periodic listing of the values of used cars. The following data are taken from two such reports from the 2015 and the 2016 Central Editions.

Car	Lending Value, July 2015	Lending Value, January 2016
2010 Mazda 3	$8100	$6325
2010 Hyundai Elantra	7925	7200
2010 Nissan Sentra	8225	6625

1. Assuming that the values are dropping linearly, draw a line representing the value of each car. Draw all three lines on the same graph, using different colored pens or pencils if possible. Let the horizontal axis represent the time, in years, after January 2015, and the vertical axis the value of each car.

(Use after Section 3.7.)

2. At what rate is each car depreciating and how are the different rates illustrated in the graph of part (1)?

3. Find a linear equation that models the value of each car. Let *x* represent the number of years after January 2015, and *y* the value.

4. Use the equations found in part (3) to estimate the value of each car in January 2017.

5. If you had planned to buy one of the three cars in July 2015 and sell it in January 2017, which one would you have purchased, and why?

6. *Research.*   Find the trade-in value of your car or of that of a friend. Use the trade-in value and the price originally paid for the car to estimate the rate of depreciation. Then find a linear equation that models the value of the car and use it to estimate its value one year from now.

Study Summary

KEY TERMS AND CONCEPTS	EXAMPLES	PRACTICE EXERCISES

SECTION 3.1: *Reading Graphs, Plotting Points, and Scaling Graphs*

Ordered pairs can be **plotted** or **graphed** using a **coordinate system** that uses two **axes**, which are most often labeled x and y. The axes intersect at the **origin**, $(0, 0)$, and divide a plane into four **quadrants**.

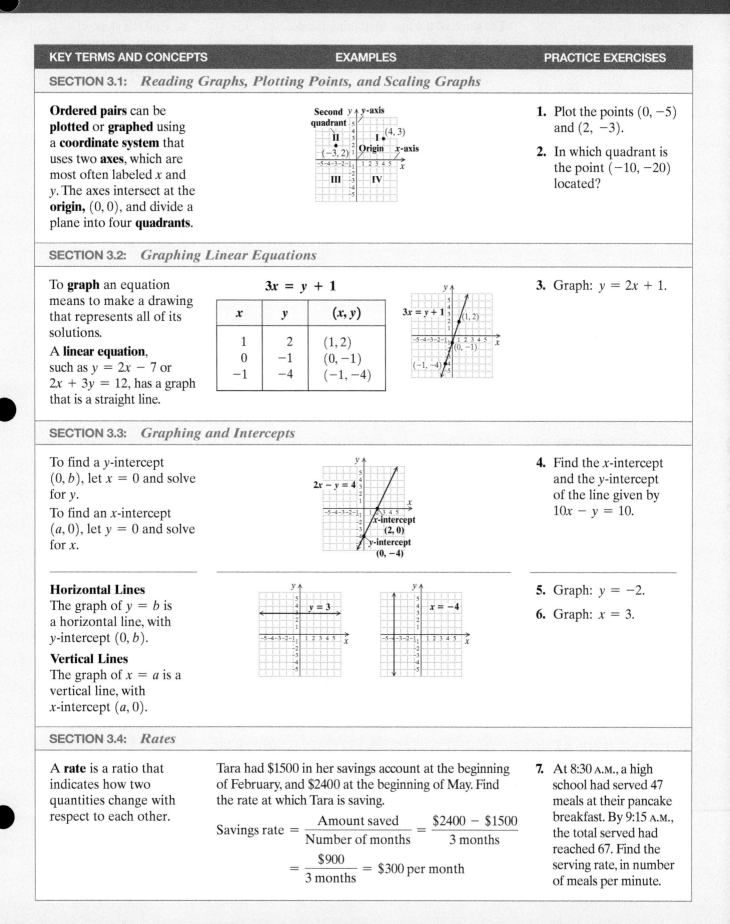

1. Plot the points $(0, -5)$ and $(2, -3)$.

2. In which quadrant is the point $(-10, -20)$ located?

SECTION 3.2: *Graphing Linear Equations*

To **graph** an equation means to make a drawing that represents all of its solutions.

A **linear equation**, such as $y = 2x - 7$ or $2x + 3y = 12$, has a graph that is a straight line.

$$3x = y + 1$$

x	y	(x, y)
1	2	$(1, 2)$
0	-1	$(0, -1)$
-1	-4	$(-1, -4)$

3. Graph: $y = 2x + 1$.

SECTION 3.3: *Graphing and Intercepts*

To find a y-intercept $(0, b)$, let $x = 0$ and solve for y.

To find an x-intercept $(a, 0)$, let $y = 0$ and solve for x.

4. Find the x-intercept and the y-intercept of the line given by $10x - y = 10$.

Horizontal Lines
The graph of $y = b$ is a horizontal line, with y-intercept $(0, b)$.

Vertical Lines
The graph of $x = a$ is a vertical line, with x-intercept $(a, 0)$.

5. Graph: $y = -2$.

6. Graph: $x = 3$.

SECTION 3.4: *Rates*

A **rate** is a ratio that indicates how two quantities change with respect to each other.

Tara had $1500 in her savings account at the beginning of February, and $2400 at the beginning of May. Find the rate at which Tara is saving.

$$\text{Savings rate} = \frac{\text{Amount saved}}{\text{Number of months}} = \frac{\$2400 - \$1500}{3 \text{ months}}$$

$$= \frac{\$900}{3 \text{ months}} = \$300 \text{ per month}$$

7. At 8:30 A.M., a high school had served 47 meals at their pancake breakfast. By 9:15 A.M., the total served had reached 67. Find the serving rate, in number of meals per minute.

SECTION 3.5: *Slope*

Slope

$$\text{Slope} = m = \frac{\text{change in } y}{\text{change in } x}$$

$$= \frac{\text{rise}}{\text{run}} = \frac{y_2 - y_1}{x_2 - x_1}$$

The slope of the line containing the points $(-1, -4)$ and $(2, -6)$ is

$$m = \frac{-6 - (-4)}{2 - (-1)} = \frac{-2}{3} = -\frac{2}{3}.$$

8. Find the slope of the line containing the points $(1, 4)$ and $(-9, 3)$.

The slope of a horizontal line is 0.

The slope of a vertical line is undefined.

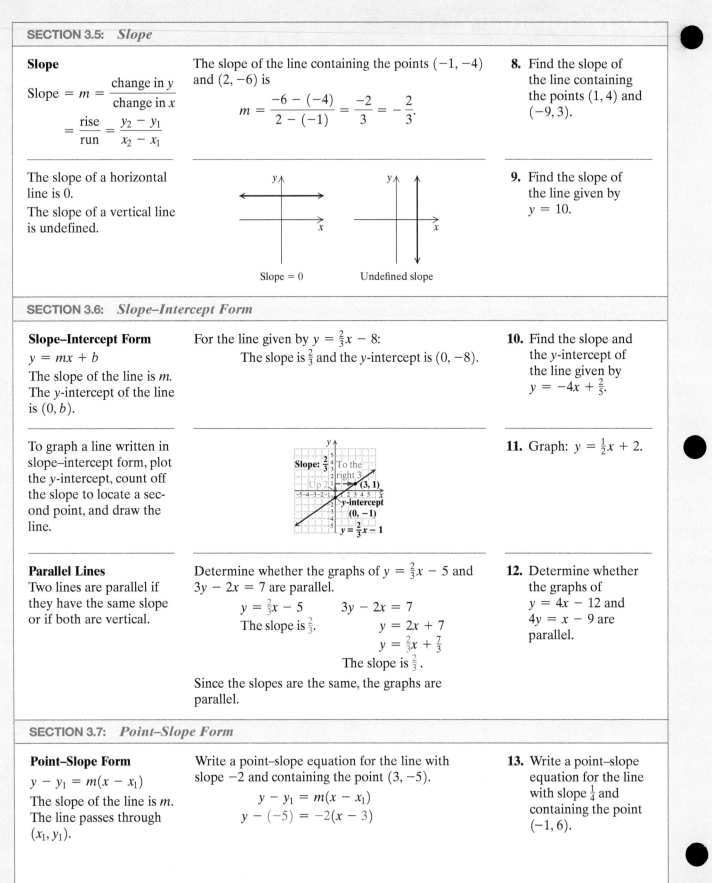

Slope = 0   Undefined slope

9. Find the slope of the line given by $y = 10$.

SECTION 3.6: *Slope–Intercept Form*

Slope–Intercept Form

$y = mx + b$

The slope of the line is m.
The y-intercept of the line is $(0, b)$.

For the line given by $y = \frac{2}{3}x - 8$:

The slope is $\frac{2}{3}$ and the y-intercept is $(0, -8)$.

10. Find the slope and the y-intercept of the line given by $y = -4x + \frac{2}{5}$.

To graph a line written in slope–intercept form, plot the y-intercept, count off the slope to locate a second point, and draw the line.

Slope: $\frac{2}{3}$ To the right 3

Up 2   (3, 1)

y-intercept (0, −1)

$y = \frac{2}{3}x - 1$

11. Graph: $y = \frac{1}{2}x + 2$.

Parallel Lines
Two lines are parallel if they have the same slope or if both are vertical.

Determine whether the graphs of $y = \frac{2}{3}x - 5$ and $3y - 2x = 7$ are parallel.

$$y = \frac{2}{3}x - 5 \qquad 3y - 2x = 7$$

The slope is $\frac{2}{3}$.

$$3y = 2x + 7$$
$$y = \frac{2}{3}x + \frac{7}{3}$$

The slope is $\frac{2}{3}$.

Since the slopes are the same, the graphs are parallel.

12. Determine whether the graphs of
$y = 4x - 12$ and
$4y = x - 9$ are parallel.

SECTION 3.7: *Point–Slope Form*

Point–Slope Form

$y - y_1 = m(x - x_1)$

The slope of the line is m.
The line passes through (x_1, y_1).

Write a point–slope equation for the line with slope -2 and containing the point $(3, -5)$.

$$y - y_1 = m(x - x_1)$$
$$y - (-5) = -2(x - 3)$$

13. Write a point–slope equation for the line with slope $\frac{1}{4}$ and containing the point $(-1, 6)$.

Review Exercises: Chapter 3

⤷ Concept Reinforcement

Classify each of the following statements as either true or false.

1. Not every ordered pair lies in one of the four quadrants. [3.1]

2. The equation of a vertical line cannot be written in slope–intercept form. [3.6]

3. Equations for lines written in slope–intercept form appear in the form $Ax + By = C$. [3.6]

4. Every horizontal line has an x-intercept. [3.3]

5. A line's slope is a measure of rate. [3.5]

6. A positive rate of ascent means that an airplane is flying increasingly higher above the earth. [3.4]

7. Any two points on a line can be used to determine the slope of a nonvertical line. [3.5]

8. Knowing a line's slope is enough to write the equation of the line. [3.6]

9. Knowing two points on a line is enough to write the equation of the line. [3.7]

10. Parallel lines that are not vertical have the same slope. [3.6]

The following bar graph shows the number of volunteers in the United States by age group. [3.1]

Data: U.S. Department of Labor, Bureau of Labor Statistics

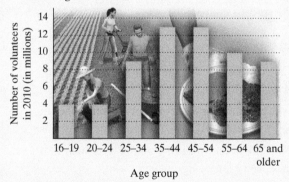

Volunteering in America

11. Approximately how many more people ages 35–44 volunteered than those ages 25–34?

12. About 2% of volunteers ages 16–19 volunteered in environmental organizations. How many people ages 16–19 volunteered in environmental organizations?

Plot each point. [3.1]

13. $(5, -1)$ 14. $(2, 3)$ 15. $(-4, 0)$

In which quadrant is each point located? [3.1]

16. $(-8, -7)$ 17. $(15.3, -13.8)$ 18. $\left(-\frac{1}{2}, \frac{1}{10}\right)$

Find the coordinates of each point in the figure. [3.1]

19. A 20. B 21. C

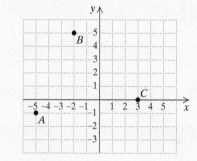

22. Use a grid 10 squares wide and 10 squares high to plot $(-65, -2)$, $(-10, 6)$, and $(25, 7)$. Choose the scale carefully. [3.1]

23. Determine whether the equation $y = 2x + 7$ has the given ordered pair as a solution: **(a)** $(3, 1)$; **(b)** $(-3, 1)$. [3.2]

24. Show that the ordered pairs $(0, -3)$ and $(2, 1)$ are solutions of the equation $2x - y = 3$. Then use the graph of the two points to determine another solution. Answers may vary. [3.2]

Graph.

25. $y = x - 5$ [3.2]   26. $y = -\frac{1}{4}x$ [3.2]

27. $y = -x + 4$ [3.2]   28. $4x + y = 3$ [3.2]

29. $4x + 5 = 3$ [3.3]   30. $5x - 2y = 10$ [3.3]

31. $y = 6$ [3.3]   32. $y = \frac{2}{3}x - 5$ [3.6]

33. $2x + y = 4$ [3.3]

34. $y + 2 = -\frac{1}{2}(x - 3)$ [3.7]

35. *Organic Gardening.* The number of U.S. households g, in millions, that use only all-natural fertilizer and pest control can be approximated by

   $g = 1.75t + 5$,

   where t is the number of years after 2004. Graph the equation and use the graph to estimate the number of households using natural garden products in 2018. [3.2]

   Data: The National Gardening Association.

36. At 4:00 P.M., Jesse's Honda Civic was at mile marker 17 of Interstate 290 in Chicago. At 4:45 P.M., the car was at mile marker 23. [3.4]

   a) Find Jesse's driving rate, in number of miles per minute.

   b) Find Jesse's driving rate, in number of minutes per mile.

37. *Gas Mileage.* The following graph shows data for the gas consumption of a 4-cylinder Ford Explorer driven on city streets. At what rate was the vehicle consuming gas? [3.4]

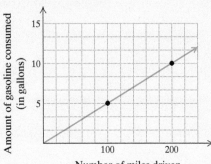

Find the slope of each line. [3.5]

38.

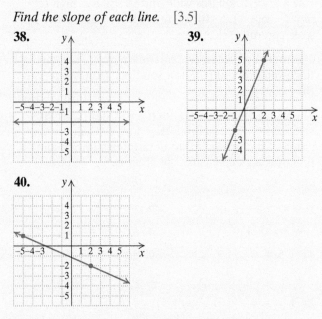

39.

40.

Find the slope of the line containing the given pair of points. If it is undefined, state this. [3.5]

41. $(-2, 5)$ and $(3, -1)$

42. $(6, 5)$ and $(-2, 5)$

43. $(-3, 0)$ and $(-3, 5)$

44. $(-8.3, 4.6)$ and $(-9.9, 1.4)$

45. Find the x-intercept and the y-intercept of the line given by $5x - 8y = 80$. [3.3]

46. Find the slope and the y-intercept of the line given by $3x + 5y = 45$. [3.6]

47. *Architecture.* To meet federal standards, a wheelchair ramp cannot rise more than 1 ft over a horizontal distance of 12 ft. Express this slope as a grade. [3.5]

48. Write the slope–intercept equation of the line with slope $\frac{3}{8}$ and y-intercept $(0, 7)$. [3.6]

49. Write a point–slope equation for the line with slope $-\frac{1}{3}$ that contains the point $(-2, 9)$. [3.7]

50. Write the slope–intercept equation for the line with slope 5 that contains the point $(3, -10)$. [3.7]

51. Write the slope–intercept equation for the line containing the points $(-2, 5)$ and $(3, 10)$. [3.7]

52. *Level of Education.* The percentage of the U.S. population with a bachelor's degree or higher increased from 27.7% in 2004 to 32.1% in 2014. [3.7]

   Data: U.S. Census Bureau

   a) Assuming linear growth, find an equation that fits the data. Let p represent the percentage of the U.S. population with a bachelor's degree or higher t years after 2000.

   b) Estimate the percentage in 2012.

   c) Predict the percentage in 2020.

Synthesis

53. Can two perpendicular lines share the same y-intercept? Why or why not? [3.3], [3.6]

54. Is it possible for a graph to have only one intercept? Why or why not? [3.3]

55. Find the value of m in $y = mx + 3$ such that $(-2, 5)$ is on the graph. [3.2]

56. Find the area and the perimeter of a rectangle for which $(-2, 2)$, $(7, 2)$, and $(7, -3)$ are three of the vertices. [3.1]

57. Find three solutions of $y = 4 - |x|$. [3.2]

Test: Chapter 3

For step-by-step test solutions, access the Chapter Test Prep Videos in MyMathLab®.

Income. *The following line graph shows data for median household income in the United States, adjusted for inflation.*

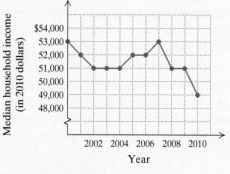

Data: U.S. Census Bureau

1. By how much did median household income decrease from 2007 to 2010?

2. In which year or years shown was median household income the highest?

In which quadrant is each point located?

3. $(-2, -10)$

4. $(-1.6, 2.3)$

Find the coordinates of each point in the figure.

5. A

6. B

7. C

Graph.

8. $y = 2x - 1$

9. $2x - 4y = -8$

10. $y + 1 = 6$

11. $y = \frac{3}{4}x$

12. $2x - y = 3$

13. $x = -1$

14. $y + 4 = -\frac{1}{2}(x - 3)$

Find the slope of the line containing each pair of points. If it is undefined, state this.

15. $(3, -2)$ and $(4, 3)$

16. $(-5, 6)$ and $(-1, -3)$

17. $(4, 7)$ and $(4, -8)$

18. *Running.* Jon reached the 3-km mark of a race at 2:15 P.M. and the 6-km mark at 2:24 P.M. What was his running rate?

19. At one point Filbert Street, the steepest street in San Francisco, drops 63 ft over a horizontal distance of 200 ft. Find the road grade.

20. Find the *x*-intercept and the *y*-intercept of the line given by $5x - y = 30$.

21. Find the slope and the *y*-intercept of the line given by $y - 8x = 10$.

22. Write the slope–intercept equation of the line with slope $-\frac{1}{3}$ and *y*-intercept $(0, -11)$.

23. *Aerobic Exercise.* A person's target heart rate is the number of beats per minute that brings the most aerobic benefit to his or her heart. The target heart rate for a 20-year-old is 150 beats per minute; for a 60-year-old, it is 120 beats per minute.

    a) Graph the data and determine an equation for the related line. Let a = age and r = target heart rate, in number of beats per minute.

    b) Calculate the target heart rate for a 36-year-old.

Synthesis

24. Write an equation of the line that is parallel to the graph of $2x - 5y = 6$ and has the same *y*-intercept as the graph of $3x + y = 9$.

25. A diagonal of a square connects the points $(-3, -1)$ and $(2, 4)$. Find the area and the perimeter of the square.

26. List the coordinates of three other points that are on the same line as $(-2, 14)$ and $(17, -5)$. Answers may vary.

Cumulative Review: Chapters 1–3

1. Evaluate $\dfrac{x}{5y}$ for $x = 70$ and $y = 2$. [1.1]

2. Multiply: $6(2a - b + 3)$. [1.2]

3. Factor: $8x - 4y + 2$. [1.2]

4. Find the prime factorization of 54. [1.3]

5. Find decimal notation: $-\dfrac{3}{20}$. [1.4]

6. Find decimal notation: 36.7%. [2.4]

Simplify.

7. $\dfrac{3}{5} - \dfrac{5}{12}$ [1.3]

8. $3.4 + (-0.8)$ [1.5]

9. $(-2)(-1.4)(2.6)$ [1.7]

10. $\dfrac{3}{8} \div \left(-\dfrac{9}{10}\right)$ [1.7]

11. $1 - [32 \div (4 + 2^2)]$ [1.8]

12. $3(x - 1) - 2[x - (2x + 7)]$ [1.8]

Solve.

13. $\dfrac{5}{3}x = -45$ [2.1]

14. $3x - 7 = 41$ [2.2]

15. $\dfrac{3}{4} = \dfrac{-n}{8}$ [2.1]

16. $14 - 5x = 2x$ [2.2]

17. $3(5 - x) = 2(3x + 4)$ [2.2]

18. $\dfrac{1}{4}x - \dfrac{2}{3} = \dfrac{3}{4} + \dfrac{1}{3}x$ [2.2]

19. $x - 28 < 20 - 2x$ [2.6]

20. Solve $A = 2\pi rh + \pi r^2$ for h. [2.3]

21. In which quadrant is the point $(3, -1)$ located? [3.1]

22. Graph on the number line: $-1 < x \le 2$. [2.6]

23. Use a grid 10 squares wide and 10 squares high to plot $(-150, -40)$, $(40, -7)$, and $(0, 6)$. Choose the scale carefully. [3.1]

Graph.

24. $x = 3$ [3.3]

25. $2x - 5y = 10$ [3.3]

26. $y = -2x + 1$ [3.2]

27. $y = \dfrac{2}{3}x$ [3.2]

28. $2y - 5 = 3$ [3.3]

29. Find the slope and the y-intercept of the line given by $3x - y = 2$. [3.6]

30. Find the slope of the line containing the points $(-4, 1)$ and $(2, -1)$. [3.5]

31. Write an equation of the line with slope $\dfrac{2}{7}$ and y-intercept $(0, -4)$. [3.6]

32. Write a point–slope equation of the line with slope $-\dfrac{3}{8}$ that contains the point $(-6, 4)$. [3.7]

33. A 150-lb person will burn 240 calories per hour when riding a bicycle at 6 mph. The same person will burn 410 calories per hour when cycling at 12 mph. [3.7]
    Data: American Heart Association

    a) Graph the data and determine an equation for the related line. Let $r =$ the rate at which the person is cycling and $c =$ the number of calories burned per hour. Use the horizontal axis for r.

    b) Use the equation of part (a) to estimate the number of calories burned per hour by a 150-lb person cycling at 10 mph.

34. In 2013, the mean annual earnings of individuals with a high school diploma was $30,000. This was about 62% of the mean annual earnings of those with a bachelor's degree. What were the mean annual earnings of individuals with a bachelor's degree in 2013? [2.4]
    Data: U.S. Census Bureau

35. In order to qualify for availability pay, a criminal investigator must average at least 2 hr of unscheduled duty per workday. For the first four days of one week, Alayna worked 1, 0, 3, and 2 unscheduled hours. How many unscheduled hours must she work on Friday in order to qualify for availability pay? [2.7]
    Data: U.S. Department of Justice

Synthesis

36. Anya's salary at the end of a year is $26,780. This reflects a 4% salary increase in February and then a 3% cost-of-living increase in June. What was her salary at the beginning of the year? [2.5]

Solve. If no solution exists, state this.

37. $4|x| - 13 = 3$ [1.4], [2.2]

38. $\dfrac{2 + 5x}{4} = \dfrac{11}{28} + \dfrac{8x + 3}{7}$ [2.2]

39. $5(7 + x) = (x + 6)5$ [2.2]

Polynomials

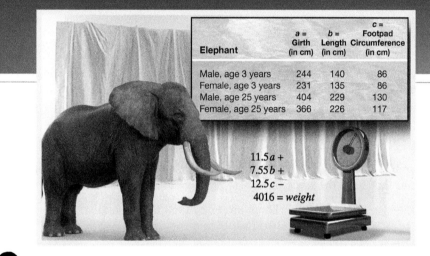

Elephant	a = Girth (in cm)	b = Length (in cm)	c = Footpad Circumference (in cm)
Male, age 3 years	244	140	86
Female, age 3 years	231	135	86
Male, age 25 years	404	229	130
Female, age 25 years	366	226	117

$$11.5a + 7.55b + 12.5c - 4016 = weight$$

It's Not Easy to Weigh an Elephant.

4.1 Exponents and Their Properties

4.2 Negative Exponents and Scientific Notation

CONNECTING THE CONCEPTS

4.3 Polynomials

4.4 Addition and Subtraction of Polynomials

MID-CHAPTER REVIEW

4.5 Multiplication of Polynomials

4.6 Special Products

4.7 Polynomials in Several Variables

4.8 Division of Polynomials

CHAPTER RESOURCES

Visualizing for Success
Collaborative Activity
Decision Making: Connection

STUDY SUMMARY

REVIEW EXERCISES
CHAPTER TEST
CUMULATIVE REVIEW

Wildlife experts have developed formulas for estimating the weight of an elephant using measurements that are much easier to obtain than the weight itself. One such formula uses the girth of the elephant, measured at the heart, the length of the elephant, and the circumference of the footpad of the elephant to estimate the elephant's weight. We can use that formula to estimate the weight of the elephants described in the table above. (*See Exercises 15 and 16 in Exercise Set 4.7.*)

Data: "How Much Does That Elephant Weigh?" by Mark MacAllister on fieldtripearth.org

As a field biologist, having mathematical skill is incredibly important in order for me to translate the ordinary behavior of animals into a scientific story.

Shermin de Silva, Director of Uda Walawe Elephant Research Project, uses a variety of aspects of math, including algebra, geometry, and calculus, to study how elephants behave and move in their environment, as well as to keep track of changes in their population size for conservation.

 ALF *Active Learning Figure* Explore the math using the Active Learning Figure in MyMathLab.

 SA *Student Activity* Do the Student Activity in MyMathLab to see math in action.

Algebraic expressions such as $16t^2$, $5a^2 - 3ab$, and $3x^2 - 7x + 5$ are called *polynomials*. Polynomials occur frequently in applications and appear in most branches of mathematics. Thus learning to add, subtract, multiply, and divide polynomials is an important part of nearly every course in elementary algebra. The focus of this chapter is finding equivalent expressions, not solving equations.

4.1 Exponents and Their Properties

A. Multiplying Powers with Like Bases **B.** Dividing Powers with Like Bases **C.** Zero as an Exponent
D. Raising a Power to a Power **E.** Raising a Product or a Quotient to a Power

Before beginning our study of polynomials, we must develop some rules for working with exponents.

A. Multiplying Powers with Like Bases

An expression like a^3 means $a \cdot a \cdot a$. We can use this fact to find the product of two expressions that have the same base:

$$a^3 \cdot a^2 = (a \cdot a \cdot a)(a \cdot a) \qquad \text{There are three factors in } a^3 \text{ and}$$
$$\text{two factors in } a^2$$
$$a^3 \cdot a^2 = a \cdot a \cdot a \cdot a \cdot a \qquad \text{Using an associative law}$$
$$a^3 \cdot a^2 = a^5.$$

Note that the exponent in a^5 is the sum of the exponents in $a^3 \cdot a^2$. That is, $3 + 2 = 5$. Similarly,

$$b^4 \cdot b^3 = (b \cdot b \cdot b \cdot b)(b \cdot b \cdot b)$$
$$b^4 \cdot b^3 = b^7, \quad \text{where } 4 + 3 = 7.$$

Adding the exponents gives the correct result.

THE PRODUCT RULE

For any number a and any positive integers m and n,

$$a^m \cdot a^n = a^{m+n}.$$

(To multiply powers with the same base, keep the base and add the exponents.)

EXAMPLE 1 Multiply and simplify each of the following. (Here "simplify" means express the product as one base to a power whenever possible.)

a) $2^3 \cdot 2^8$

b) $5^3 \cdot 5^8 \cdot 5$

c) $(r + s)^7 (r + s)^6$

d) $(a^3 b^2)(a^3 b^5)$

SOLUTION

a) $2^3 \cdot 2^8 = 2^{3+8}$ Adding exponents: $a^m \cdot a^n = a^{m+n}$

        $= 2^{11}$

> **CAUTION!** The base is unchanged: $2^3 \cdot 2^8 \neq 4^{11}$.

b) $5^3 \cdot 5^8 \cdot 5 = 5^3 \cdot 5^8 \cdot 5^1$ Recall that $x^1 = x$ for any number x.

          $= 5^{3+8+1}$ Adding exponents

          $= 5^{12}$

> **CAUTION!** $5^{12} \neq 5 \cdot 12$.

c) $(r + s)^7 (r + s)^6 = (r + s)^{7+6}$ The base here is $r + s$.

             $= (r + s)^{13}$

> **CAUTION!** $(r + s)^{13} \neq r^{13} + s^{13}$.

d) $(a^3 b^2)(a^3 b^5) = a^3 b^2 a^3 b^5$ Using an associative law

              $= a^3 a^3 b^2 b^5$ Using a commutative law

              $= a^6 b^7$ Adding exponents: $a^3 a^3 = a^6$ and $b^2 b^5 = b^7$

1. Multiply and simplify:

     $x^5 \cdot x \cdot x^4$.

YOUR TURN

B. Dividing Powers with Like Bases

Recall that any expression that is divided or multiplied by 1 is unchanged. This, together with the fact that any nonzero number divided by itself is 1, leads to a rule for division:

$$\frac{a^5}{a^2} = \frac{a \cdot a \cdot a \cdot a \cdot a}{a \cdot a} = \frac{a \cdot a \cdot a}{1} \cdot \frac{a \cdot a}{a \cdot a}$$

$$\frac{a^5}{a^2} = \frac{a \cdot a \cdot a}{1} \cdot 1$$

$$\frac{a^5}{a^2} = a \cdot a \cdot a = a^3.$$

Note that the exponent in a^3 is the difference of the exponents in $\dfrac{a^5}{a^2}$. Similarly,

$$\frac{x^4}{x^3} = \frac{x \cdot x \cdot x \cdot x}{x \cdot x \cdot x} = \frac{x}{1} \cdot \frac{x \cdot x \cdot x}{x \cdot x \cdot x} = \frac{x}{1} \cdot 1 = x = x^1.$$

Subtracting the exponents gives the correct result.

> **THE QUOTIENT RULE**
>
> For any nonzero number a and any positive integers m and n for which $m > n$,
>
> $$\frac{a^m}{a^n} = a^{m-n}.$$
>
> (To divide powers with the same base, subtract the exponent of the denominator from the exponent of the numerator.)

EXAMPLE 2   Divide and simplify. (Here "simplify" means express the quotient as one base to an exponent whenever possible.)

a) $\dfrac{7^9}{7^4}$ **b)** $\dfrac{(5a)^{12}}{(5a)^4}$ **c)** $\dfrac{4p^5 q^7}{6p^2 q}$

SOLUTION

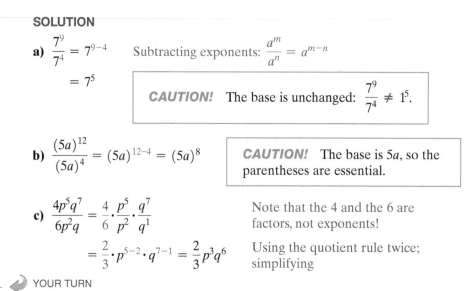

a) $\dfrac{7^9}{7^4} = 7^{9-4}$ Subtracting exponents: $\dfrac{a^m}{a^n} = a^{m-n}$

$= 7^5$

> **CAUTION!** The base is unchanged: $\dfrac{7^9}{7^4} \neq 1^5$.

b) $\dfrac{(5a)^{12}}{(5a)^4} = (5a)^{12-4} = (5a)^8$

> **CAUTION!** The base is $5a$, so the parentheses are essential.

2. Divide and simplify:

$$\dfrac{(x+p)^6}{(x+p)}.$$

c) $\dfrac{4p^5q^7}{6p^2q} = \dfrac{4}{6} \cdot \dfrac{p^5}{p^2} \cdot \dfrac{q^7}{q^1}$ Note that the 4 and the 6 are factors, not exponents!

$= \dfrac{2}{3} \cdot p^{5-2} \cdot q^{7-1} = \dfrac{2}{3}p^3q^6$ Using the quotient rule twice; simplifying

YOUR TURN

C. Zero as an Exponent

The quotient rule can be used to help determine what 0 means as an exponent. Consider a^4/a^4, with $a \neq 0$. Since the numerator and the denominator are the same,

$$\dfrac{a^4}{a^4} = 1.$$ We assume $a \neq 0$.

On the other hand, using the quotient rule would give us

$$\dfrac{a^4}{a^4} = a^{4-4} = a^0.$$ Subtracting exponents

Since $a^0 = a^4/a^4 = 1$, this suggests that $a^0 = 1$ for any nonzero value of a.

> **THE EXPONENT ZERO**
>
> For any real number a, with $a \neq 0$,
>
> $$a^0 = 1.$$
>
> (Any nonzero number raised to the 0 power is 1.)

Note that in the above box, 0^0 is not defined. For this text, we will assume that expressions like a^m do not represent 0^0.

Recall that in the rules for order of operations, simplifying exponential expressions is done before multiplying.

EXAMPLE 3 Simplify: **(a)** 1948^0; **(b)** $(-9)^0$; **(c)** $(3x)^0$; **(d)** $3x^0$; **(e)** $(-1)9^0$; **(f)** -9^0.

SOLUTION

a) $1948^0 = 1$ Any nonzero number raised to the 0 power is 1.

b) $(-9)^0 = 1$ Any nonzero number raised to the 0 power is 1. The base here is -9.

c) $(3x)^0 = 1$, for any $x \neq 0$.   The parentheses indicate that the base is $3x$.

d) $3x^0 = 3 \cdot x^0$ The base is x.

$= 3 \cdot 1$ $x^0 = 1$, for any $x \neq 0$

$= 3$

e) $(-1)9^0 = (-1)1 = -1$ The base here is 9.

f) -9^0 is read "the opposite of 9^0" and is equivalent to $(-1)9^0$:

$$-9^0 = (-1)9^0 = (-1)1 = -1.$$

Note from parts (b), (e), and (f) that $-9^0 = (-1)9^0$ and $-9^0 \neq (-9)^0$.

3. Simplify: 99^0.

 YOUR TURN

D. Raising a Power to a Power

Consider an expression like $(7^2)^4$:

$$(7^2)^4 = (7^2)(7^2)(7^2)(7^2)$$ There are four factors of 7^2.
$$(7^2)^4 = (7 \cdot 7)(7 \cdot 7)(7 \cdot 7)(7 \cdot 7)$$ We could also use the product rule.
$$(7^2)^4 = 7 \cdot 7 \cdot 7 \cdot 7 \cdot 7 \cdot 7 \cdot 7 \cdot 7$$ Using an associative law
$$(7^2)^4 = 7^8.$$

Note that the exponent in 7^8 is the product of the exponents in $(7^2)^4$. Similarly,

$$(y^5)^3 = y^5 \cdot y^5 \cdot y^5$$ There are three factors of y^5.
$$(y^5)^3 = (y \cdot y \cdot y \cdot y \cdot y)(y \cdot y \cdot y \cdot y \cdot y)(y \cdot y \cdot y \cdot y \cdot y)$$
$$(y^5)^3 = y^{15}.$$

Once again, we get the same result if we multiply exponents:

$$(y^5)^3 = y^{5 \cdot 3} = y^{15}.$$

Student Notes

There are several rules for manipulating exponents in this section. One way to remember them all is to replace variables with small numbers (other than 1) and see what the results suggest. For example, multiplying $2^2 \cdot 2^3$ and examining the result is a fine way of reminding yourself that $a^m \cdot a^n = a^{m+n}$.

> **THE POWER RULE**
>
> For any number a and any whole numbers m and n,
>
> $$(a^m)^n = a^{mn}.$$
>
> (To raise a power to a power, multiply the exponents and leave the base unchanged.)

Remember that for this text we assume that 0^0 is not considered.

EXAMPLE 4   Simplify: $(m^2)^5$.

SOLUTION

$$(m^2)^5 = m^{2 \cdot 5}$$ Multiplying exponents: $(a^m)^n = a^{mn}$
$$= m^{10}$$

4. Simplify: $(3^8)^{10}$.

 YOUR TURN

E. Raising a Product or a Quotient to a Power

When an expression inside parentheses is raised to a power, the inside expression is the base. Let's compare $2a^3$ and $(2a)^3$:

$2a^3 = 2 \cdot a \cdot a \cdot a$;   The base is a. $(2a)^3 = (2a)(2a)(2a)$ The base is $2a$.
$$(2a)^3 = (2 \cdot 2 \cdot 2)(a \cdot a \cdot a)$$
$$(2a)^3 = 2^3 a^3$$
$$(2a)^3 = 8a^3.$$

We see that $2a^3$ and $(2a)^3$ are *not* equivalent. Note too that $(2a)^3$ can be simplified by cubing each factor in $2a$. This leads to the following rule for raising a product to a power.

CAUTION! The rule $(ab)^n = a^n b^n$ applies only to *products* raised to a power, not to sums or differences. For example, $(3 + 4)^2 \neq 3^2 + 4^2$ since $49 \neq 9 + 16$. Similarly, $(5 + x)^2 \neq 5^2 + x^2$.

RAISING A PRODUCT TO A POWER

For any numbers a and b and any whole number n,

$$(ab)^n = a^n b^n.$$

(To raise a product to a power, raise each factor to that power.)

EXAMPLE 5   Simplify: **(a)** $(4a)^3$; **(b)** $(-5x^4)^2$; **(c)** $(a^7 b)^2(a^3 b^4)$.

SOLUTION

a) $(4a)^3 = 4^3 a^3 = 64a^3$ Raising each factor to the third power and simplifying

b) $(-5x^4)^2 = (-5)^2(x^4)^2$ Raising each factor to the second power. Parentheses are important here.

$\quad\quad = 25x^8$ Simplifying $(-5)^2$ and using the power rule

c) $(a^7 b)^2(a^3 b^4) = (a^7)^2 b^2 a^3 b^4$ Raising a product to a power

$\quad\quad = a^{14} b^2 a^3 b^4$ Multiplying exponents

$\quad\quad = a^{17} b^6$ Adding exponents

5. Simplify: $(-2y^4)^3$.

↪ YOUR TURN

There is a similar rule for raising a quotient to a power.

RAISING A QUOTIENT TO A POWER

For any numbers a and b, $b \neq 0$, and any whole number n,

$$\left(\frac{a}{b}\right)^n = \frac{a^n}{b^n}.$$

(To raise a quotient to a power, raise the numerator to the power and divide by the denominator to the power.)

EXAMPLE 6   Simplify: **(a)** $\left(\dfrac{x}{5}\right)^2$; **(b)** $\left(\dfrac{5}{a^4}\right)^3$; **(c)** $\left(\dfrac{3a^4}{b^3}\right)^2$.

SOLUTION

a) $\left(\dfrac{x}{5}\right)^2 = \dfrac{x^2}{5^2} = \dfrac{x^2}{25}$ Squaring the numerator and the denominator

b) $\left(\dfrac{5}{a^4}\right)^3 = \dfrac{5^3}{(a^4)^3}$ Raising a quotient to a power

$\quad\quad = \dfrac{125}{a^{4\cdot 3}} = \dfrac{125}{a^{12}}$ Using the power rule and simplifying

c) $\left(\dfrac{3a^4}{b^3}\right)^2 = \dfrac{(3a^4)^2}{(b^3)^2}$ Raising a quotient to a power

$\quad\quad = \dfrac{3^2(a^4)^2}{b^{3\cdot 2}} = \dfrac{9a^8}{b^6}$ Raising a product to a power and using the power rule

6. Simplify: $\left(\dfrac{-3}{a^3}\right)^2$.

↪ YOUR TURN

⤷ Check Your
UNDERSTANDING

In the following summary of definitions and rules, we assume that no denominators are 0 and that 0^0 is not considered.

Each of the following statements is an example of one of the properties listed in the box at right. For each statement, name the corresponding property.

1. $(a^6)^4 = a^{24}$

2. $(5x)^7 = 5^7 x^7$

3. $m^6 \cdot m^4 = m^{10}$

4. $\dfrac{p^9}{p^3} = p^6$

5. $\left(\dfrac{c}{3}\right)^8 = \dfrac{c^8}{3^8}$

DEFINITIONS AND PROPERTIES OF EXPONENTS

For any whole numbers m and n,

1 as an exponent:	$a^1 = a$
0 as an exponent:	$a^0 = 1$
The Product Rule:	$a^m \cdot a^n = a^{m+n}$
The Quotient Rule:	$\dfrac{a^m}{a^n} = a^{m-n}$
The Power Rule:	$(a^m)^n = a^{mn}$
Raising a product to a power:	$(ab)^n = a^n b^n$
Raising a quotient to a power:	$\left(\dfrac{a}{b}\right)^n = \dfrac{a^n}{b^n}$

4.1 EXERCISE SET

FOR EXTRA HELP **MyMathLab®**

⤷ **Vocabulary and Reading Check**

Complete the sentence using the most appropriate phrase from the column on the right. Each phrase is used once.

1. To raise a product to a power, ____

2. To raise a quotient to a power, ____

3. To raise a power to a power, ____

4. To divide powers with the same base, ____

5. Any nonzero number raised to the 0 power ____

6. To multiply powers with the same base, ____

7. To square a fraction, ____

8. To square a product, ____

a) keep the base and add the exponents.

b) multiply the exponents and leave the base unchanged.

c) square the numerator and square the denominator.

d) square each factor.

e) raise each factor to that power.

f) raise the numerator to the power and divide by the denominator to the power.

g) is one.

h) subtract the exponent of the denominator from the exponent of the numerator.

⤷ **Concept Reinforcement**

Identify the base and the exponent in each expression.

9. $(2x)^5$

10. $(x + 1)^0$

11. $2x^3$

12. $-y^6$

13. $\left(\dfrac{4}{y}\right)^7$

14. $(-5x)^4$

A. Multiplying Powers with Like Bases

Simplify.

15. $d^3 \cdot d^{10}$

16. $8^4 \cdot 8^3$

17. $a^6 \cdot a$

18. $y^7 \cdot y^9$

19. $6^5 \cdot 6^{10}$

20. $t^0 \cdot t^{16}$

21. $(3y)^4 (3y)^8$

22. $(2t)^8 (2t)^{17}$

23. $(5p)^0 (5p)^1$

24. $(8n)(8n)^9$

25. $(x + 3)^5 (x + 3)^8$

26. $(m - 3)^4 (m - 3)^5$

27. $(a^2 b^7)(a^3 b^2)$

28. $(a^8 b^3)(a^4 b)$

29. $r^3 \cdot r^7 \cdot r^0$

30. $s^4 \cdot s^5 \cdot s^2$

31. $(mn^5)(m^3 n^4)$

32. $(a^3 b)(ab)^4$

B. Dividing Powers with Like Bases

Simplify. Assume that no denominator is 0 and that 0^0 is not considered.

33. $\dfrac{7^5}{7^2}$

34. $\dfrac{4^7}{4^3}$

35. $\dfrac{t^8}{t}$

36. $\dfrac{x^7}{x}$

37. $\dfrac{(5a)^7}{(5a)^6}$

38. $\dfrac{(3m)^9}{(3m)^8}$

Aha! **39.** $\dfrac{(x+y)^8}{(x+y)^8}$

40. $\dfrac{(9x)^{10}}{(9x)^2}$

41. $\dfrac{(r+s)^{12}}{(r+s)^4}$

42. $\dfrac{(a-b)^4}{(a-b)^3}$

43. $\dfrac{12d^9}{15d^2}$

44. $\dfrac{10n^7}{15n^3}$

45. $\dfrac{8a^9b^7}{2a^2b}$

46. $\dfrac{12r^{10}s^7}{4r^2s}$

47. $\dfrac{x^{12}y^9}{x^0y^2}$

48. $\dfrac{a^{10}b^{12}}{a^2b^0}$

C. Zero as an Exponent

Simplify.

49. t^0 when $t=15$

50. y^0 when $y=38$

51. $5x^0$ when $x=-22$

52. $7m^0$ when $m=1.7$

53. 7^0+4^0

54. $(8+5)^0$

55. $(-3)^1-(-3)^0$

56. $(-4)^0-(-4)^1$

D. Raising a Power to a Power

Simplify.

57. $(x^3)^{11}$

58. $(a^5)^8$

59. $(5^8)^4$

60. $(2^5)^2$

61. $(t^{20})^4$

62. $(x^{25})^6$

E. Raising a Product or a Quotient to a Power

Simplify. Assume that no denominator is 0 and that 0^0 is not considered.

63. $(10x)^2$

64. $(5a)^2$

65. $(-2a)^3$

66. $(-3x)^3$

67. $(-5n^7)^2$

68. $(-4m^4)^2$

69. $(a^2b)^7$

70. $(xy^4)^9$

71. $(r^5t)^3(r^2t^8)$

72. $(a^4b^6)(a^2b)^5$

73. $(2x^5)^3(3x^4)$

74. $(5x^3)^2(2x^7)$

75. $\left(\dfrac{x}{5}\right)^3$

76. $\left(\dfrac{2}{a}\right)^4$

77. $\left(\dfrac{7}{6n}\right)^2$

78. $\left(\dfrac{4x}{3}\right)^3$

79. $\left(\dfrac{a^3}{b^8}\right)^6$

80. $\left(\dfrac{x^5}{y^2}\right)^7$

81. $\left(\dfrac{x^2y}{z^3}\right)^4$

82. $\left(\dfrac{a^4}{b^2c}\right)^5$

83. $\left(\dfrac{a^3}{-2b^5}\right)^4$

84. $\left(\dfrac{x^5}{-3y^3}\right)^4$

85. $\left(\dfrac{5x^7y}{-2z^4}\right)^3$

86. $\left(\dfrac{-4p^5}{3m^2n^3}\right)^3$

Aha! **87.** $\left(\dfrac{4x^3y^5}{3z^7}\right)^0$

88. $\left(\dfrac{5a^7}{2b^5c}\right)^0$

89. Explain in your own words why $-5^2 \neq (-5)^2$.

90. Under what circumstances should exponents be added?

Skill Review

Solve.

91. $-\dfrac{x}{7}=3$ [2.1]

92. $3x-2 \leq 5-x$ [2.6]

93. $\frac{1}{2}x+\frac{1}{3}=\frac{1}{6}x$ [2.2]

94. $6(x-10)=3[4(x-5)]$ [2.2]

95. $8-2(n-7)>9-(3-n)$ [2.6]

96. $8(5-y)=32$ [2.2]

Synthesis

97. Under what conditions does a^n represent a negative number? Why?

98. Using the quotient rule, explain why 9^0 is 1.

99. Suppose that the width of a square is three times the width of a second square. How do the areas of the squares compare? Why?

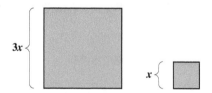

100. Suppose that the width of a cube is twice the width of a second cube. How do the volumes of the cubes compare? Why?

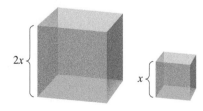

Find a value of the variable that shows that the two expressions are not *equivalent. Answers may vary.*

101. $3x^2$; $(3x)^2$

102. $(a + 5)^2$; $a^2 + 5^2$

103. $\dfrac{t^6}{t^2}$; t^3

104. $\dfrac{a + 7}{7}$; a

Simplify.

105. $y^{4x} \cdot y^{2x}$

106. $a^{10k} \div a^{2k}$

107. $\dfrac{x^{5t}(x^t)^2}{(x^{3t})^2}$

108. $\dfrac{\left(\frac{1}{2}\right)^3 \left(\frac{2}{3}\right)^4}{\left(\frac{5}{6}\right)^3}$

109. Solve for x:

$$\frac{t^{26}}{t^x} = t^x.$$

Replace ▨ *with* $>, <,$ *or* $=$ *to write a true sentence.*

110. 3^5 ▨ 3^4

111. 4^2 ▨ 4^3

112. 4^3 ▨ 5^3

113. 4^3 ▨ 3^4

114. 9^7 ▨ 3^{13}

115. 25^8 ▨ 125^5

 Use the fact that $10^3 \approx 2^{10}$ to estimate each of the following powers of 2. Then compute the power of 2 with a calculator and find the difference between the exact value and the approximation.

116. 2^{14}

117. 2^{22}

118. 2^{26}

119. 2^{31}

🖩 *In computer science,* 1 *KB of memory refers to* 1 *kilobyte, or* 1×10^3 *bytes, of memory. This is really an approximation of* 1×2^{10} *bytes (since computer memory actually uses powers of 2).*

120. The TI-84 Plus graphing calculator has 480 KB of "FLASH ROM." How many bytes is this?

121. The TI-84 Plus Silver Edition graphing calculator has 1.5 MB (megabytes) of FLASH ROM, where 1 MB is 1000 KB (see Exercise 120). How many bytes of FLASH ROM does this calculator have?

> 👉 **YOUR TURN ANSWERS: SECTION 4.1**
> **1.** x^{10} **2.** $(x + p)^5$ **3.** 1   **4.** 3^{80} **5.** $-8y^{12}$
> **6.** $\dfrac{9}{a^6}$

> **Prepare to Move On**
> *Perform the indicated operations.*
> **1.** $-10 - 14$ [1.6]   **2.** $-3(5)$ [1.7]
> **3.** $-16 + 5$ [1.5]   **4.** $12 - (-4)$ [1.6]
> **5.** $-3 + (-11)$ [1.5]   **6.** $-8 - (-12)$ [1.6]

4.2 Negative Exponents and Scientific Notation

A. Negative Integers as Exponents   **B.** Scientific Notation
C. Multiplying and Dividing Using Scientific Notation

We now attach a meaning to negative exponents. Once we understand both positive exponents and negative exponents, we can study a method for writing numbers known as *scientific notation*.

A. Negative Integers as Exponents

Let's define negative exponents so that the existing rules that apply to whole-number exponents will hold for all integer exponents. To do so, consider a^{-5} and the rule for adding exponents:

$$a^{-5} = a^{-5} \cdot 1 \qquad \text{Using the identity property of 1}$$

$$= \frac{a^{-5}}{1} \cdot \frac{a^5}{a^5} \qquad \text{Writing 1 as } \frac{a^5}{a^5} \text{ and } a^{-5} \text{ as } \frac{a^{-5}}{1}$$

$$= \frac{a^{-5+5}}{a^5} \qquad \text{Adding exponents}$$

$$= \frac{1}{a^5}. \qquad -5 + 5 = 0 \text{ and } a^0 = 1$$

This leads to our definition of negative exponents.

NEGATIVE EXPONENTS

For any nonzero real number a and any integer n,

$$a^{-n} = \frac{1}{a^n}.$$

(The numbers a^{-n} and a^n are reciprocals of each other.)

Study Skills

Connect the Dots

Whenever possible, look for connections between concepts covered in different sections or chapters. For example, two sections may both discuss exponents, or two chapters may both cover polynomials.

---

1. Express using positive exponents and, if possible, simplify: $7x^{-2}$.

EXAMPLE 1 Express using positive exponents and, if possible, simplify.

a) m^{-3} **b)** 4^{-2} **c)** $(-3)^{-2}$ **d)** ab^{-1}

SOLUTION

a) $m^{-3} = \dfrac{1}{m^3}$ m^{-3} is the reciprocal of m^3.

b) $4^{-2} = \dfrac{1}{4^2} = \dfrac{1}{16}$ 4^{-2} is the reciprocal of 4^2. Note that $4^{-2} \neq 4(-2)$.

c) $(-3)^{-2} = \dfrac{1}{(-3)^2} = \dfrac{1}{(-3)(-3)} = \dfrac{1}{9}$ $(-3)^{-2}$ is the reciprocal of $(-3)^2$.
Note that $(-3)^{-2} \neq -\dfrac{1}{3^2}$.

d) $ab^{-1} = a\left(\dfrac{1}{b^1}\right) = a\left(\dfrac{1}{b}\right) = \dfrac{a}{b}$ b^{-1} is the reciprocal of b^1. Note that the base is b, not ab.

YOUR TURN

CAUTION!   A negative exponent does not, in itself, indicate that an expression is negative. As shown in Example 1,

$$4^{-2} \neq 4(-2) \quad \text{and} \quad (-3)^{-2} \neq -\frac{1}{3^2}.$$

The following is another way to illustrate why negative exponents are defined as they are.

On this side, we divide by 5 at each step.		On this side, the exponents decrease by 1.
	$125 = 5^3$	
	$25 = 5^2$	
	$5 = 5^1$	
	$1 = 5^0$	
	$\dfrac{1}{5} = 5^?$	
	$\dfrac{1}{25} = 5^?$	

To continue the pattern, it follows that

$$\frac{1}{5} = \frac{1}{5^1} = 5^{-1}, \qquad \frac{1}{25} = \frac{1}{5^2} = 5^{-2}, \quad \text{and, in general,} \quad \frac{1}{a^n} = a^{-n}.$$

EXAMPLE 2   Express $\dfrac{1}{x^7}$ using negative exponents.

2. Express $\dfrac{1}{10^7}$ using negative exponents.

SOLUTION   We know that $\dfrac{1}{a^n} = a^{-n}$. Thus, $\dfrac{1}{x^7} = x^{-7}$.

YOUR TURN

The rules for exponents still hold when exponents are negative.

EXAMPLE 3 Simplify. Do not use negative exponents in the answer.

a) $t^5 \cdot t^{-2}$

b) $(5x^2y^{-3})^4$

c) $\dfrac{x^{-4}}{x^{-5}}$

d) $\dfrac{1}{t^{-5}}$

e) $\dfrac{s^{-3}}{t^{-5}}$

f) $\dfrac{-10x^{-3}y}{5x^2y^5}$

Student Notes

Unless the exponent is large, we usually evaluate exponential expressions where possible. Thus in Example 3(b), we write 625 instead of 5^4.

SOLUTION

a) $t^5 \cdot t^{-2} = t^{5+(-2)} = t^3$ Adding exponents

b) $(5x^2y^{-3})^4 = 5^4(x^2)^4(y^{-3})^4$ Raising each factor to the fourth power

$$= 625x^8y^{-12} = \frac{625x^8}{y^{12}}$$

c) $\dfrac{x^{-4}}{x^{-5}} = x^{-4-(-5)} = x^1 = x$ We subtract exponents even if the exponent in the denominator is negative.

d) Since $\dfrac{1}{a^n} = a^{-n}$, we have $\dfrac{1}{t^{-5}} = t^{-(-5)} = t^5$.

e) $\dfrac{s^{-3}}{t^{-5}} = s^{-3} \cdot \dfrac{1}{t^{-5}}$

$$= \frac{1}{s^3} \cdot t^5 = \frac{t^5}{s^3}$$ Using the result from part (d) above

3. Simplify:

$$\frac{3x^{-2}}{x^{10}}.$$

Do not use negative exponents in the answer.

f) $\dfrac{-10x^{-3}y}{5x^2y^5} = \dfrac{-10}{5} \cdot \dfrac{x^{-3}}{x^2} \cdot \dfrac{y^1}{y^5}$ Note that the -10 and 5 are factors.

$$= -2 \cdot x^{-3-2} \cdot y^{1-5}$$ Using the quotient rule twice; simplifying

$$= -2x^{-5}y^{-4} = \frac{-2}{x^5y^4}$$

⮌ YOUR TURN

The result from Example 3(e) can be generalized.

> **FACTORS AND NEGATIVE EXPONENTS**
>
> For any nonzero real numbers a and b and any integers m and n,
>
> $$\frac{a^{-n}}{b^{-m}} = \frac{b^m}{a^n}.$$
>
> (A factor can be moved to the other side of the fraction bar if the sign of the exponent is changed.)

EXAMPLE 4 Simplify: $\dfrac{-15x^{-7}}{5y^2z^{-4}}$.

SOLUTION We can move the factors x^{-7} and z^{-4} to the other side of the fraction bar if we change the sign of each exponent:

$$\frac{-15x^{-7}}{5y^2z^{-4}} = \frac{-15}{5} \cdot \frac{x^{-7}}{y^2z^{-4}}$$ We can simply divide the constant factors.

$$= -3 \cdot \frac{z^4}{y^2x^7}$$

$$= \frac{-3z^4}{x^7y^2}.$$

4. Simplify: $\dfrac{12a^{-1}}{4bc^{-3}}$.

⮌ YOUR TURN

Another way to change the sign of the exponent is to take the reciprocal of the base. To understand why this is true, note that

$$\left(\frac{s}{t}\right)^{-5} = \frac{s^{-5}}{t^{-5}} = \frac{t^5}{s^5} = \left(\frac{t}{s}\right)^5.$$

This often provides the easiest way to simplify an expression containing a negative exponent.

RECIPROCALS AND NEGATIVE EXPONENTS

For any nonzero real numbers a and b and any integer n,

$$\left(\frac{a}{b}\right)^{-n} = \left(\frac{b}{a}\right)^n.$$

(A base to a power is equal to the reciprocal of the base raised to the opposite power.)

EXAMPLE 5   Simplify: $\left(\dfrac{x^4}{2y}\right)^{-3}$.

SOLUTION

$$\left(\frac{x^4}{2y}\right)^{-3} = \left(\frac{2y}{x^4}\right)^3 \qquad \text{Taking the reciprocal of the base and changing the sign of the exponent}$$

$$= \frac{(2y)^3}{(x^4)^3} \qquad \text{Raising a quotient to a power by raising both the numerator and the denominator to the power}$$

$$= \frac{2^3 y^3}{x^{12}} \qquad \text{Raising a product to a power; using the power rule in the denominator}$$

$$= \frac{8y^3}{x^{12}} \qquad \text{Cubing 2}$$

5. Simplify: $\left(\dfrac{3x}{y^5}\right)^{-2}$.

↩ YOUR TURN

B. Scientific Notation

Scientific notation provides a useful way of writing the very large or very small numbers that occur in science. The following are examples of scientific notation.

The mass of the earth:

6.0×10^{24} kilograms (kg) $= 6{,}000{,}000{,}000{,}000{,}000{,}000{,}000{,}000$ kg

The mass of a hydrogen atom:

1.7×10^{-24} g $= 0.0000000000000000000000017$ g

SCIENTIFIC NOTATION

Scientific notation for a number is an expression of the type

$$N \times 10^m,$$

where N is at least 1 but less than 10 (that is, $1 \le N < 10$), N is expressed in decimal notation, and m is an integer.

Student Notes

Definitions are usually written as concisely as possible, so that every phrase included is important. The definition for scientific notation states that $1 \le N < 10$. Thus, 2.68×10^5 is written in scientific notation, but 26.8×10^5 and 0.268×10^5 are *not* written in scientific notation.

↳ Check Your UNDERSTANDING

Choose from the column on the right the equivalent expression.

1. x^{-6} **a)** $6x$

2. $\dfrac{1}{x^{-6}}$ **b)** $-6x$

          c) x^6

3. $-x^{-6}$ **d)** $\dfrac{1}{x^6}$

          e) $-\dfrac{1}{x^6}$

State whether each of the following is written in scientific notation.

4. 1.007×10^{-18}

5. 0.25×10^6

6. 21.37×10^{12}

⬤ **6.** Convert to decimal notation:

    8.04×10^{-3}.

---

7. Write in scientific notation:

    600,000.

Converting from scientific notation to decimal notation involves multiplying by a power of 10. Consider the following.

Scientific Notation	Multiplication	Decimal Notation
4.52×10^2	4.52×100	452.
4.52×10^1	4.52×10	45.2
4.52×10^0	4.52×1	4.52
4.52×10^{-1}	4.52×0.1	0.452
4.52×10^{-2}	4.52×0.01	0.0452

We generally perform this multiplication mentally. Thus to convert $N \times 10^m$ to decimal notation, we move the decimal point as follows.

• When m is positive, we move the decimal point m places to the right.
• When m is negative, we move the decimal point $|m|$ places to the left.

EXAMPLE 6 Convert to decimal notation: **(a)** 7.893×10^5; **(b)** 4.7×10^{-8}.

SOLUTION

a) Since the exponent is positive, the decimal point moves to the right:

    7.89300. $7.893 \times 10^5 = 789,300$ The decimal point moves 5 places to the right.

    5 places

b) Since the exponent is negative, the decimal point moves to the left:

    0.00000004.7 $4.7 \times 10^{-8} = 0.000000047$ The decimal point moves 8 places to the left.

    8 places

YOUR TURN

To convert from decimal notation to scientific notation, this procedure is reversed.

EXAMPLE 7 Write in scientific notation: **(a)** 83,000; **(b)** 0.0327.

SOLUTION

a) We need to find m such that $83,000 = 8.3 \times 10^m$. To change 8.3 to 83,000 requires moving the decimal point 4 places to the right. This can be accomplished by multiplying by 10^4. Thus,

    $83,000 = 8.3 \times 10^4$.   This is scientific notation.

b) We need to find m such that $0.0327 = 3.27 \times 10^m$. To change 3.27 to 0.0327 requires moving the decimal point 2 places to the left. This can be accomplished by multiplying by 10^{-2}. Thus,

    $0.0327 = 3.27 \times 10^{-2}$.   This is scientific notation.

YOUR TURN

In scientific notation, positive exponents are used to represent large numbers and negative exponents are used to represent small numbers between 0 and 1.

C. Multiplying and Dividing Using Scientific Notation

Products and quotients of numbers written in scientific notation are found using the rules for exponents.

EXAMPLE 8   Simplify. Write the answer in scientific notation.

a) $(1.8 \times 10^9) \cdot (2.3 \times 10^{-4})$ **b)** $(3.41 \times 10^5) \div (1.1 \times 10^{-3})$

SOLUTION

a) $(1.8 \times 10^9) \cdot (2.3 \times 10^{-4})$

$\qquad = 1.8 \times 2.3 \times 10^9 \times 10^{-4}$ Using the associative and commutative laws

$\qquad = 4.14 \times 10^{9+(-4)}$ Adding exponents

$\qquad = 4.14 \times 10^5$

b) $(3.41 \times 10^5) \div (1.1 \times 10^{-3})$

$\qquad = \dfrac{3.41 \times 10^5}{1.1 \times 10^{-3}}$

$\qquad = \dfrac{3.41}{1.1} \times \dfrac{10^5}{10^{-3}}$

$\qquad = 3.1 \times 10^{5-(-3)}$ Subtracting exponents

$\qquad = 3.1 \times 10^8$

 Chapter Resource:
Decision Making: Connection, p. 294

8. Simplify:

$(2.2 \times 10^{-8}) \times (2.5 \times 10^{-10})$.

 YOUR TURN

Technology Connection

A key labeled (10ˣ), (^), or (EE) is used to enter scientific notation into a calculator. Sometimes this is a secondary function, meaning that another key— often labeled SHIFT or **2ND**—must be pressed first.

To check Example 9(a), we press

3.1 (EE) 5 (×) 4.5 (EE) ((-)) 3 **ENTER**.

The result that appears represents 1.395×10^3. On some calculators, the mode SCI must be selected in order to display scientific notation.

```
3.1E5∗4.5E−3
                    1.395E3
```

Calculate each of the following.

1. $(3.8 \times 10^9) \cdot (4.5 \times 10^7)$
2. $(2.9 \times 10^{-8}) \div (5.4 \times 10^6)$
3. $(9.2 \times 10^7) \div (2.5 \times 10^{-9})$

When a problem is stated using scientific notation, we generally use scientific notation for the answer. This may require an additional conversion.

EXAMPLE 9   Simplify. Write the answer in scientific notation.

a) $(3.1 \times 10^5) \cdot (4.5 \times 10^{-3})$
b) $(7.2 \times 10^{-7}) \div (8.0 \times 10^6)$

SOLUTION

a) We have

$(3.1 \times 10^5) \cdot (4.5 \times 10^{-3}) = 3.1 \times 4.5 \times 10^5 \times 10^{-3}$

$\qquad\qquad\qquad\qquad\qquad = 13.95 \times 10^2.$

Our answer is not yet in scientific notation because 13.95 is not between 1 and 10. We convert to scientific notation as follows:

$13.95 \times 10^2 = 1.395 \times 10^1 \times 10^2$ Substituting 1.395×10^1 for 13.95

$\qquad\qquad = 1.395 \times 10^3.$ Adding exponents

b) $(7.2 \times 10^{-7}) \div (8.0 \times 10^6) = \dfrac{7.2 \times 10^{-7}}{8.0 \times 10^6} = \dfrac{7.2}{8.0} \times \dfrac{10^{-7}}{10^6}$

$= 0.9 \times 10^{-13}$

$= 9.0 \times 10^{-1} \times 10^{-13}$ Substituting 9.0×10^{-1} for 0.9

9. Simplify: $\dfrac{1.2 \times 10^6}{3.0 \times 10^{-11}}$.

$= 9.0 \times 10^{-14}$ Adding exponents

↩ YOUR TURN

■ **CONNECTING 🔗 THE CONCEPTS** ■

Definitions and Properties of Exponents

The following summary assumes that no denominators are 0 and that 0^0 is not considered. For any integers m and n,

1 as an exponent: $a^1 = a$

0 as an exponent: $a^0 = 1$

Negative exponents: $a^{-n} = \dfrac{1}{a^n},$

$\dfrac{a^{-n}}{b^{-m}} = \dfrac{b^m}{a^n},$

$\left(\dfrac{a}{b}\right)^{-n} = \left(\dfrac{b}{a}\right)^n$

The Product Rule: $a^m \cdot a^n = a^{m+n}$

The Quotient Rule: $\dfrac{a^m}{a^n} = a^{m-n}$

The Power Rule: $(a^m)^n = a^{mn}$

Raising a product to a power: $(ab)^n = a^n b^n$

Raising a quotient to a power: $\left(\dfrac{a}{b}\right)^n = \dfrac{a^n}{b^n}$

EXERCISES

Simplify. Do not use negative exponents in the answer.

1. $x^4 x^{10}$

2. $x^{-4} x^{-10}$

3. $\dfrac{x^{-4}}{x^{10}}$

4. $\dfrac{x^4}{x^{-10}}$

5. $\left(x^{-4}\right)^{-10}$

6. $\left(x^4\right)^{10}$

7. $\dfrac{1}{c^{-8}}$

8. c^{-8}

9. $\left(\dfrac{a^3}{b^4}\right)^5$

10. $\left(\dfrac{a^3}{b^4}\right)^{-5}$

4.2 EXERCISE SET

FOR EXTRA HELP MyMathLab®

↳ Vocabulary and Reading Check

Classify each of the following statements as either true or false.

1. A negative exponent indicates a reciprocal.

2. The expressions 3^{-2} and -3^2 are equivalent.

3. A positive exponent of the base 10 in scientific notation indicates a number greater than or equal to 10.

4. The number 18.68×10^{12} is written in scientific notation.

⮕ Concept Reinforcement

Match each expression with an equivalent expression from the column on the right.

5. ____ $\left(\dfrac{x^3}{y^2}\right)^{-2}$ a) $\dfrac{y^6}{x^9}$

6. ____ $\left(\dfrac{y^2}{x^3}\right)^{-2}$ b) $\dfrac{x^9}{y^6}$

7. ____ $\left(\dfrac{y^{-2}}{x^{-3}}\right)^{-3}$ c) $\dfrac{y^4}{x^6}$

8. ____ $\left(\dfrac{x^{-3}}{y^{-2}}\right)^{-3}$ d) $\dfrac{x^6}{y^4}$

State whether scientific notation for each of the following numbers would include a positive power of 10 or a negative power of 10.

9. The length of an Olympic marathon, in centimeters

10. The thickness of a cat's whisker, in meters

11. The mass of a hydrogen atom, in grams

12. The mass of a pickup truck, in grams

13. The time between leap years, in seconds

14. The time between a bird's heartbeats, in hours

A. Negative Integers as Exponents

Express using positive exponents. Then, if possible, simplify.

15. 2^{-3} 16. 10^{-5} 17. $(-2)^{-6}$

18. $(-3)^{-4}$ 19. t^{-9} 20. x^{-7}

21. $8x^{-3}$ 22. xy^{-9} 23. $\dfrac{1}{a^{-8}}$

24. $\dfrac{1}{z^{-6}}$ 25. 7^{-1} 26. 3^{-1}

27. $3a^8b^{-6}$ 28. $5a^{-7}b^4$ 29. $\left(\dfrac{x}{2}\right)^{-5}$

30. $\left(\dfrac{a}{2}\right)^{-4}$ 31. $\dfrac{z^{-4}}{3x^5}$ 32. $\dfrac{y^{-5}}{x^{-3}}$

Express using negative exponents.

33. $\dfrac{1}{9^2}$ 34. $\dfrac{1}{5^2}$ 35. $\dfrac{1}{y^3}$

36. $\dfrac{1}{t^4}$ 37. $\dfrac{1}{t}$ 38. $\dfrac{1}{8}$

Simplify. Do not use negative exponents in the answer.

39. $2^{-5} \cdot 2^8$ 40. $5^{-8} \cdot 5^{10}$

41. $x^{-3} \cdot x^{-9}$ 42. $x^{-4} \cdot x^{-7}$

43. $t^{-3} \cdot t$ 44. $y^{-5} \cdot y$

45. $(5a^{-2}b^{-3})(2a^{-4}b)$ 46. $(3a^{-5}b^{-7})(2ab^{-2})$

47. $(n^{-5})^3$ 48. $(m^{-5})^{10}$

49. $(t^{-3})^{-6}$ 50. $(a^{-4})^{-7}$

51. $(mn)^{-7}$ 52. $(ab)^{-9}$

53. $(3x^{-4})^2$ 54. $(2a^{-5})^3$

55. $(5r^{-4}t^3)^2$ 56. $(4x^5y^{-6})^3$

57. $\dfrac{t^{12}}{t^{-2}}$ 58. $\dfrac{x^7}{x^{-2}}$

59. $\dfrac{y^{-7}}{y^{-3}}$ 60. $\dfrac{z^{-6}}{z^{-2}}$

61. $\dfrac{15y^{-7}}{3y^{-10}}$ 62. $\dfrac{-12a^{-5}}{2a^{-8}}$

63. $\dfrac{2x^6}{x}$ 64. $\dfrac{3x}{x^{-1}}$

65. $\dfrac{-15a^{-7}}{10b^{-9}}$ 66. $\dfrac{12x^{-6}}{8y^{-10}}$

Aha!   67. $\dfrac{t^{-7}}{t^{-7}}$ 68. $\dfrac{a^{-5}}{b^{-7}}$

69. $\dfrac{3t^4}{s^{-2}u^{-4}}$ 70. $\dfrac{5x^{-8}}{y^{-3}z^2}$

71. $(x^4y^5)^{-3}$ 72. $(t^5x^3)^{-4}$

73. $(3m^{-5}n^{-3})^{-2}$ 74. $(2y^{-4}z^{-2})^{-3}$

75. $(a^{-5}b^7c^{-2})(a^{-3}b^{-2}c^6)$

76. $(x^3y^{-4}z^{-5})(x^{-4}y^{-2}z^9)$

77. $\left(\dfrac{a^4}{3}\right)^{-2}$ 78. $\left(\dfrac{y^2}{2}\right)^{-2}$

79. $\left(\dfrac{m^{-1}}{n^{-4}}\right)^3$ 80. $\left(\dfrac{x^2y}{z^{-5}}\right)^3$

81. $\left(\dfrac{2a^2}{3b^4}\right)^{-3}$ 82. $\left(\dfrac{a^2b}{2d^3}\right)^{-5}$

Aha!   83. $\left(\dfrac{5x^{-2}}{3y^{-2}z}\right)^0$ 84. $\left(\dfrac{4a^3b^{-2}}{5c^{-3}}\right)^1$

85. $\dfrac{-6a^3b^{-5}}{-3a^7b^{-8}}$ 86. $\dfrac{12x^{-2}y^4}{-3xy^{-7}}$

87. $\dfrac{10x^{-4}yz^7}{8x^7y^{-3}z^{-3}}$ 88. $\dfrac{9a^6b^{-4}c^7}{27a^{-4}b^5c^9}$

B. Scientific Notation

Convert to decimal notation.

89. 4.92×10^3 **90.** 8.13×10^4

91. 8.92×10^{-3} **92.** 7.26×10^{-4}

93. 9.04×10^8 **94.** 1.35×10^7

95. 3.497×10^{-6} **96.** 9.043×10^{-3}

Convert to scientific notation.

97. 36,000,000   **98.** 27,400

99. 0.00583   **100.** 0.0814

101. 78,000,000,000   **102.** 3,700,000,000,000

103. 0.000001032   **104.** 0.00000008

C. Multiplying and Dividing Using Scientific Notation

Multiply or divide, as indicated, and write scientific notation for the result.

105. $(3 \times 10^5)(2 \times 10^8)$

106. $(3.1 \times 10^7)(2.1 \times 10^{-4})$

107. $(3.8 \times 10^9)(6.5 \times 10^{-2})$

108. $(7.1 \times 10^{-7})(8.6 \times 10^{-5})$

109. $(8.7 \times 10^{-12})(4.5 \times 10^{-5})$

110. $(4.7 \times 10^5)(6.2 \times 10^{-12})$

111. $\dfrac{8.5 \times 10^8}{3.4 \times 10^{-5}}$

112. $\dfrac{5.6 \times 10^{-2}}{2.5 \times 10^5}$

113. $(4.0 \times 10^3) \div (8.0 \times 10^8)$

114. $(1.5 \times 10^{-3}) \div (1.6 \times 10^{-6})$

115. $\dfrac{7.5 \times 10^{-9}}{2.5 \times 10^{12}}$ **116.** $\dfrac{3.0 \times 10^{-2}}{6.0 \times 10^{10}}$

117. Without performing actual computations, explain why 3^{-29} is smaller than 2^{-29}.

118. Explain why each of the following is not scientific notation: 12.6×10^8; $4.8 \times 10^{1.7}$; 0.207×10^{-5}.

Skill Review

Simplify.

119. $\frac{1}{6} - \frac{1}{3}$ [1.6]   **120.** $x - (x - 7)$ [1.8]

121. $(-2a)^5$ [1.8]   **122.** $|7 - 8|$ [1.8]

123. $24 \div 6 \cdot 2 - 3[4(3 - 1) - 7]$ [1.8]

124. $(-3)(-1)(-5)(-1)(-1)$ [1.7]

Synthesis

125. Explain why $(-17)^{-8}$ is positive.

126. What requirements must be met in order for x^{-n} to represent a negative integer? Why?

127. Explain why scientific notation cannot be used without an understanding of the rules for exponents.

128. Write the reciprocal of 1.25×10^{-6} in scientific notation.

129. Write $8^{-3} \cdot 32 \div 16^2$ as a power of 2.

130. Write $81^3 \div 27 \cdot 9^2$ as a power of 3.

Simplify each of the following. Do not use a calculator.

Aha! **131.** $\dfrac{125^{-4}(25^2)^4}{125}$ **132.** $(13^{-12})^2 \cdot 13^{25}$

133. $[(5^{-3})^2]^{-1}$ **134.** $5^0 - 5^{-1}$

135. $3^{-1} + 4^{-1}$

136. Determine whether each of the following is true for all pairs of integers m and n and all positive numbers x and y.

a) $x^m \cdot y^n = (xy)^{mn}$
b) $x^m \cdot y^m = (xy)^{2m}$
c) $(x - y)^m = x^m - y^m$

Solve. Write scientific notation for each answer.

137. *Ecology.* In one year, a large tree can remove from the air the same amount of carbon dioxide produced by a car traveling 500 mi. If New York City contains approximately 600,000 trees, how many miles of car traffic can those trees clean in one year?

Data: Colorado Tree Coalition; New York City Department of Parks and Recreations

138. *Computer Technology.* One gigabyte is 1 billion bytes of information, and a terabyte is 1000 gigabytes. Intel Corp. has developed silicon-based connections that use lasers to move data at a rate of 50 gigabytes per second. The printed collection of the U.S. Library of Congress contains 10 terabytes of information. How long would it take to copy the Library of Congress using these connections?

Data: spie.org; newworldencyclopedia.org

139. *Astronomy.* The diameter of the Milky Way galaxy is approximately 5.88×10^{17} mi. The distance that light travels in one year, or one light-year, is 5.88×10^{12} mi. How many light-years is it from one end of the galaxy to the other?

140. *Coral Reefs.* There are 10 million bacteria per square centimeter of coral in a coral reef. The coral reefs near the Hawaiian Islands cover 14,000 km². How many bacteria are there in Hawaii's coral reefs?

Data: livescience.com; U.S. Geological Survey

141. *Biology.* A human hair is about 4×10^{-5} m in diameter. A strand of DNA is 2 nanometers in diameter. How many strands of DNA laid side by side would it take to equal the width of a human hair?

↘ **YOUR TURN ANSWERS: SECTION 4.2**

1. $\dfrac{7}{x^2}$ **2.** 10^{-7} **3.** $\dfrac{3}{x^{12}}$ **4.** $\dfrac{3c^3}{ab}$ **5.** $\dfrac{y^{10}}{9x^2}$

6. 0.00804 **7.** 6×10^5 **8.** 5.5×10^{-18} **9.** 4.0×10^{16}

Quick Quiz: Sections 4.1–4.2

Simplify.

1. $(2n^2m)^3(5nm^4)^2$ [4.1] **2.** $6^{-12} \cdot 6$ [4.2]

3. $(2x^2y^{-1})^{-3}$ [4.2] **4.** 289^0 [4.1]

5. Convert 30,070,000 to scientific notation. [4.2]

Prepare to Move On

Combine like terms. [1.6]

1. $9x + 2y - x - 2y$ **2.** $5a - 7b - 8a + b$

3. $-3x + (-2) - 5 - (-x)$ **4.** $2 - t - 3t - r - 7$

Evaluate. [1.8]

5. $4 + x^3$, for $x = 10$

6. $-x^2 - 5x + 3$, for $x = -2$

4.3 | Polynomials

A. Terms **B.** Types of Polynomials **C.** Degree and Coefficients **D.** Combining Like Terms
E. Evaluating Polynomials and Applications

We now examine an important algebraic expression known as a *polynomial.* Certain polynomials have appeared earlier in this text so you already have some experience working with them.

A. Terms

At this point, we have seen a variety of algebraic expressions like

$$3a^2b^4, \quad 2l + 2w, \quad \text{and} \quad 5x^2 + x - 2.$$

Within these expressions, $3a^2b^4$, $2l$, $2w$, $5x^2$, x, and -2 are examples of *terms.* A **term** can be a number (like -2), a variable (like x), a product of numbers and/or variables (like $3a^2b^4$), or a quotient of numbers and/or variables (like $7/t$).

If a term is a product of constants and/or variables, it is called a **monomial**. A term, but not a monomial, can include division by a variable. A **polynomial** is a monomial or a sum of monomials.

Examples of monomials: $7, \quad t, \quad 2l, \quad 2w, \quad 5x^3y, \quad \frac{3}{7}a^5$

Examples of polynomials: $4x - 7, \quad \frac{2}{3}t^2, \quad -5n^2 + m - 1, \quad x, \quad 0$

Note that $4x - 7$ is a polynomial because it can be written as $4x + (-7)$. When a polynomial is written as a sum of monomials, each monomial is called a *term of the polynomial*.

EXAMPLE 1   Identify the terms of the polynomial $3t^4 - 5t^6 - 4t + 2$.

SOLUTION   The terms are $3t^4$, $-5t^6$, $-4t$, and 2. We can see this by rewriting all subtractions as additions of opposites:

$$3t^4 - 5t^6 - 4t + 2 = 3t^4 + (-5t^6) + (-4t) + 2.$$

These are the terms of the polynomial.

1. Identify the terms of the polynomial $3x - 6 + 5x^4$.

YOUR TURN

B. Types of Polynomials

A polynomial with two terms is called a **binomial**, and one with three terms is called a **trinomial**. Polynomials with four or more terms have no special name.

Monomials	Binomials	Trinomials	No Special Name
$4x^2$	$2x + 4$	$3t^3 + 4t + 7$	$4x^3 - 5x^2 + xy - 8$
9	$3a^5 + 6bc$	$6x^7 - 8z^2 + 4$	$z^5 + 2z^4 - z^3 + 7z + 3$
$-7a^{19}b^5$	$-9x^7 - 6$	$4x^2 - 6x - \frac{1}{2}$	$4x^6 - 3x^5 + x^4 - x^3 + 2x - 1$

The following algebraic expressions are *not* polynomials:

(1) $\dfrac{x + 3}{x - 4}$,   **(2)** $5x^3 - 2x^2 + \dfrac{1}{x}$,   **(3)** $\dfrac{1}{x^3 - 2}$.

Expressions (1) and (3) are not polynomials because they represent quotients, not sums. Expression (2) is not a polynomial because $1/x$ is not a monomial.

Student Notes

Example 2(c) illustrates that the degree of a constant, such as 7, is 0. Another way to understand this is to write 7 as $7 \cdot 1 = 7x^0$. Then the degree is the exponent 0.

C. Degree and Coefficients

The **degree of a monomial** is the number of variable factors in the monomial. Thus the degree of $7t^2$ is 2 because $7t^2$ has two variable factors: $7t^2 = 7 \cdot t \cdot t$. When there is only one variable, the degree is the exponent of that variable.

EXAMPLE 2   Determine the degree of each monomial: **(a)** $8x^4$; **(b)** $3x$; **(c)** 7.

SOLUTION

a) The degree of $8x^4$ is 4. x^4 represents 4 variable factors: $x \cdot x \cdot x \cdot x$.

b) The degree of $3x$ is 1.   There is 1 variable factor.

2. Determine the degree of the term $12t$.

c) The degree of 7 is 0.   There is no variable factor.

YOUR TURN

The degree of a constant, such as 7, is 0 since there are no variable factors. The monomial 0 is an exception, since $0 = 0x = 0x^2 = 0x^3$ and so on. We say that the monomial 0 has *no* degree. Note that a term of a polynomial is a monomial, so that we can also refer to the degree of a term.

The part of a term that is a constant factor is the **coefficient** of that term. Thus the coefficient of $3x$ is 3, and the coefficient for the term 7 is simply 7.

EXAMPLE 3   Identify the coefficient of each term in the polynomial

$$4x^3 - 7x^2 + x - 8.$$

SOLUTION

The coefficient of $4x^3$ is 4.

The coefficient of $-7x^2$ is -7.

The coefficient of the third term is 1, since $x = 1x$.

The coefficient of -8 is simply -8.

3. Identify the coefficient of each term in the polynomial $y^5 - y^2 + 12y - 9.$

YOUR TURN

The **leading term** of a polynomial is the term of highest degree. Its coefficient is called the **leading coefficient**, and its degree is referred to as the **degree of the polynomial**. To see how this terminology is used, consider the polynomial

$$3x^2 - 8x^3 + 5x^4 + 7x - 6.$$

The *terms* are	$3x^2$,	$-8x^3$,	$5x^4$,	$7x$,	and	-6.
The *coefficients* are	3,	-8,	5,	7,	and	-6.
The *degrees of the term* are	2,	3,	4,	1,	and	0.

The *leading term* is $5x^4$ and the *leading coefficient* is 5.

The *degree of the polynomial* is 4.

D. Combining Like Terms

Like, or *similar*, *terms* are either constant terms or terms containing the same variable(s) raised to the same power(s). For example, the like terms in

$$4x^3 + 5x - 7x^2 + 2x^3 + x^2$$

are $4x^3$ and $2x^3$ as well as $-7x^2$ and x^2.

Often we can simplify polynomials by *combining*, or *collecting*, like terms.

EXAMPLE 4   Write an equivalent expression by combining like terms.

a) $2x^3 + 6x^3$

b) $5x^2 + 7 + 2x^4 - 6x^2 - 11 - 2x^4$

c) $7a^3 - 5a^2 + 9a^3 + a^2$

d) $2.3y^5 - 4.6y^3 - 8.1y^5 + 3.7y^2$

e) $\frac{2}{3}x^4 - x^3 - \frac{1}{6}x^4 + \frac{2}{5}x^3 - \frac{3}{10}x^3$

SOLUTION

Student Notes

Remember that when we combine like terms, we are not solving equations, but are forming equivalent expressions.

a) $2x^3 + 6x^3 = (2 + 6)x^3$ Using the distributive law

$= 8x^3$

b) $5x^2 + 7 + 2x^4 - 6x^2 - 11 - 2x^4$

$= 5x^2 - 6x^2 + 2x^4 - 2x^4 + 7 - 11$

$= (5 - 6)x^2 + (2 - 2)x^4 + (7 - 11)$ ⎫

$= -1x^2 + 0x^4 + (-4)$ ⎬ These steps are often done mentally.

$= -x^2 - 4$ ⎭

c) $7a^3 - 5a^2 + 9a^3 + a^2 = 7a^3 - 5a^2 + 9a^3 + 1a^2$ $a^2 = 1 \cdot a^2 = 1a^2$

$= 16a^3 - 4a^2$ $7a^3 + 9a^3 = 16a^3;$ $-5a^2 + 1a^2 = -4a^2$

d) $2.3y^5 - 4.6y^3 - 8.1y^5 + 3.7y^2$

$= 2.3y^5 - 8.1y^5 - 4.6y^3 + 3.7y^2$ There is only one pair of like terms.

$= -5.8y^5 - 4.6y^3 + 3.7y^2$ $2.3 - 8.1 = 2.3 + (-8.1) = -5.8$

4. Write an equivalent expression by combining like terms:

$$2n^2 - 6n + n^2 + 6n.$$

e) $\frac{2}{3}x^4 - x^3 - \frac{1}{6}x^4 + \frac{2}{5}x^3 - \frac{3}{10}x^3 = \left(\frac{2}{3} - \frac{1}{6}\right)x^4 + \left(-1 + \frac{2}{5} - \frac{3}{10}\right)x^3$

$$= \left(\frac{4}{6} - \frac{1}{6}\right)x^4 + \left(-\frac{10}{10} + \frac{4}{10} - \frac{3}{10}\right)x^3$$

$$= \frac{3}{6}x^4 - \frac{9}{10}x^3$$

$$= \frac{1}{2}x^4 - \frac{9}{10}x^3$$

YOUR TURN

Note in Example 4 that the solutions are written so that the term of highest degree appears first, followed by the term of next highest degree, and so on. This is known as **descending order**. If the degrees increase from left to right, then we say that the polynomial is written in **ascending order**.

E. Evaluating Polynomials and Applications

When each variable in a polynomial is replaced with a number, the polynomial then represents a number, or *value*, that can be calculated using the rules for order of operations.

EXAMPLE 5 Evaluate $-x^2 + 3x + 9$ for $x = -2$.

SOLUTION For $x = -2$, we have

Substitute. $\quad -x^2 + 3x + 9 = -(-2)^2 + 3(-2) + 9$ The negative sign in front of x^2 remains.

$$= -(4) + 3(-2) + 9 \quad \text{Evaluating the exponential expression}$$

Simplify. $\quad = -4 + (-6) + 9 \quad$ Multiplying

$$= -10 + 9 = -1. \quad \text{Adding}$$

5. Evaluate $t^3 - 5t^2 - 10t + 1$ for $t = -1$.

YOUR TURN

6. Use the polynomial given in Example 6 to estimate the power generated by a 20-mph wind.

EXAMPLE 6 *Renewable Energy Sources.* The number of watts of power P generated by a particular wind turbine at a wind speed of x miles per hour can be approximated by the polynomial

$$P = 0.0157x^3 + 0.1163x^2 - 1.3396x + 3.7063.$$

Estimate the power generated by a 10-mph wind.

Data: QST, November 2006

SOLUTION To find the power generated by a 10-mph wind, we evaluate the polynomial for $x = 10$:

$$P = 0.0157x^3 + 0.1163x^2 - 1.3396x + 3.7063$$

$$= 0.0157(10)^3 + 0.1163(10)^2 - 1.3396(10) + 3.7063$$

$$= 15.7 + 11.63 - 13.396 + 3.7063$$

$$= 17.6403.$$

Since this value is approximate, we round the answer. A 10-mph wind will generate about 17.6 watts.

YOUR TURN

Sometimes, a graph can be used to estimate the value of a polynomial visually.

EXAMPLE 7 *Renewable Energy Sources.*   In the following figure, the polynomial from Example 6 has been graphed by evaluating it for several choices of x. Use the graph to estimate the power generated by a 30-mph wind.

Check Your UNDERSTANDING

1. How many terms are in the polynomial $x^3 + 4x^2 - 7x + 5$?

2. What is the coefficient of the term $4x^2$?

3. What is the degree of the term $4x^2$?

4. Which terms in the expression

$8x^3 + 4x^2 + 7x^2 + 8x$

are like terms?

5. Evaluate $-x^2$ for $x = -10$.

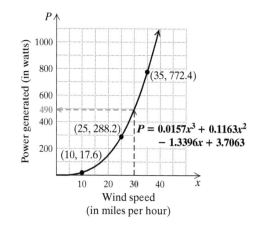

The first graph shows a curve with points $(10, 17.6)$, $(25, 288.2)$, and $(35, 772.4)$, labeled with $P = 0.0157x^3 + 0.1163x^2 - 1.3396x + 3.7063$. Axes: Power generated (in watts) vs. Wind speed (in miles per hour).

SOLUTION   To estimate the power generated by a 30-mph wind, we locate 30 on the horizontal axis of the graph. From there, we move vertically until we meet the curve. From that point, we move horizontally to the P-axis.

The second graph shows the same curve with a dashed line at $P = 490$ corresponding to $x = 30$. Points $(10, 17.6)$, $(25, 288.2)$, $(35, 772.4)$ are labeled, with $P = 0.0157x^3 + 0.1163x^2 - 1.3396x + 3.7063$.

7. Use the graph given in Example 7 to estimate the power generated by a 15-mph wind.

We see that the power generated by a 30-mph wind is approximately 490 watts. (For $x = 30$, the value of $0.0157x^3 + 0.1163x^2 - 1.3396x + 3.7063$ is approximately 490.)

YOUR TURN

Technology Connection

One way to evaluate a polynomial is to use the TRACE key. For example, to evaluate $0.0157x^3 + 0.1163x^2 - 1.3396x + 3.7063$ in Example 7 for $x = 30$, we can graph the polynomial $y = 0.0157x^3 + 0.1163x^2 - 1.3396x + 3.7063$. We then use TRACE and enter an x-value of 30.

The value of the polynomial appears as y, and the cursor appears at $(30, 492.0883)$. The VALUE option of the CALC menu works in a similar way.

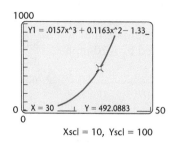

Y1 = .0157x^3 + 0.1163x^2 - 1.33_
X = 30   Y = 492.0883
Xscl = 10, Yscl = 100

1. Use TRACE or CALC VALUE to find the value of $0.0157x^3 + 0.1163x^2 - 1.3396x + 3.7063$ for $x = 40$.

4.3  EXERCISE SET

FOR EXTRA HELP MyMathLab®

⤳ Vocabulary and Reading Check

Match the description in the left-hand column with the most appropriate algebraic expression from the column on the right.

1. ____ A polynomial with four terms
2. ____ A polynomial with 7 as its leading coefficient
3. ____ A trinomial written in descending order
4. ____ A polynomial with degree 5
5. ____ A binomial with degree 7
6. ____ A monomial of degree 0
7. ____ An expression with two terms that is not a binomial
8. ____ An expression with three terms that is not a trinomial

a) $8x^3 + \dfrac{2}{x^2}$

b) $5x^4 + 3x^3 - 4x + 7$

c) $\dfrac{3}{x} - 6x^2 + 9$

d) $8t - 4t^5$

e) 5

f) $6x^2 + 7x^4 - 2x^3$

g) $4t - 2t^7$

h) $3t^2 + 4t + 7$

⤳ Concept Reinforcement

Determine whether each expression is a polynomial.

9. $3x - 7$

10. $-2x^5 + 9 - 7x^2$

11. $\dfrac{x^2 + x + 1}{x^3 - 7}$

12. -10

13. $\frac{1}{4}x^{10} - 8.6$

14. $\dfrac{3}{x^4} - \dfrac{1}{x} + 13$

A. Terms

Identify the terms of each polynomial.

15. $8x^3 - 11x^2 + 6x + 1$

16. $5a^3 + 4a^2 - a - 7$

17. $-t^6 - 3t^3 + 9t - 4$

18. $n^5 - 4n^3 + 2n - 8$

B. Types of Polynomials

Classify each polynomial as a monomial, a binomial, a trinomial, or a polynomial with no special name.

19. $x^2 - 23x + 17$

20. $-9x^2$

21. $x^3 - 7x + 2x^2 - 4$

22. $t^3 + 4$

23. $y + 8$

24. $3x^8 + 12x^3 - 9$

25. 17

26. $2x^4 - 7x^3 + x^2 + x - 6$

C. Degree and Coefficients

27. Complete the following table for the polynomial $7x^2 + 8x^5 - 4x^3 + 6 - \frac{1}{2}x^4$.

Term	Coefficient	Degree of the Term	Degree of the Polynomial
		5	
$-\frac{1}{2}x^4$			
	-4		
		2	
	6		

28. Complete the following table for the polynomial $-3x^4 + 6x^3 - 2x^2 + 8x + 7$.

Term	Coefficient	Degree of the Term	Degree of the Polynomial
	-3		
$6x^3$			
		2	
		1	
	7		

Determine the coefficient and the degree of each term in each polynomial.

29. $8x^4 + 2x$

30. $9a^3 - 4a^2$

31. $9t^2 - 3t + 4$

32. $7x^4 + 5x - 3$

33. $x^4 - x^3 + 4x - 3$

34. $2a^5 + a^2 + 8a + 10$

For each of the following polynomials, (a) list the degree of each term; (b) determine the leading term and the leading coefficient; and (c) determine the degree of the polynomial.

35. $5t + t^3 + 8t^4$

36. $1 + 6n + 4n^2$

37. $3a^2 - 7 + 2a^4$

38. $9x^4 + x^2 + x^7 - 12$

39. $8 + 6x^2 - 3x - x^5$

40. $9a - a^4 + 3 + 2a^3$

D. Combining Like Terms

Combine like terms. Write all answers in descending order.

41. $5n^2 + n + 6n^2$

42. $5a + 7a^2 + 3a$

43. $3a^4 - 2a + 2a + a^4$

44. $9b^5 + 3b^2 - 2b^5 - 3b^2$

45. $4b^3 + 5b + 7b^3 + b^2 - 6b$

46. $6x^2 + 2x^4 - 2x^2 - x^4 - 4x^2 + x$

47. $10x^2 + 2x^3 - 3x^3 - 4x^2 - 6x^2 - x^4$

48. $12t^6 - t^3 + 8t^6 + 4t^3 - t^7 - 3t^3$

49. $\frac{1}{5}x^4 + 7 - 2x^2 + 3 - \frac{2}{15}x^4 + 2x^2$

50. $\frac{1}{6}x^3 + 3x^2 - \frac{1}{3}x^3 + 7 + x^2 - 10$

51. $8.3a^2 + 3.7a - 8 - 9.4a^2 + 1.6a + 0.5$

52. $1.4y^3 + 2.9 - 7.7y - 1.3y - 4.1 + 9.6y^3$

E. Evaluating Polynomials and Applications

Evaluate each polynomial for $x = 3$ and for $x = -3$.

53. $-4x + 9$

54. $-6x + 5$

55. $2x^2 - 3x + 7$

56. $4x^2 - 6x + 9$

57. $-3x^3 + 7x^2 - 4x - 8$

58. $-2x^3 - 3x^2 + 4x + 2$

59. $2x^4 - \frac{1}{9}x^3$

60. $\frac{1}{3}x^4 - 2x^3$

61. *Skydiving.* During the first 13 sec of a jump, the number of feet that a skydiver falls in t seconds is approximated by the polynomial

$$11.12t^2.$$

In 2009, 108 U.S. skydivers fell headfirst in formation from a height of 18,000 ft. How far had they fallen 10 sec after having jumped from the plane?

Data: www.telegraph.co.uk

62. *Skydiving.* For jumps that exceed 13 sec, the polynomial $173t - 369$ can be used to approximate the distance, in feet, that a skydiver has fallen in t seconds. Approximately how far has a skydiver fallen 20 sec after having jumped from a plane?

Circumference. The circumference of a circle of radius r is given by the polynomial $2\pi r$, where π is an irrational number. For an approximation of π, use 3.14.

63. Find the circumference of a circle with radius 10 cm.

64. Find the circumference of a circle with radius 5 ft.

Area of a Circle. The area of a circle of radius r is given by the polynomial πr^2. Use 3.14 for π.

65. Find the area of a circle with radius 7 m.

66. Find the area of a circle with radius 6 ft.

67. *Kayaking.* The distance $s(t)$, in feet, traveled by a body falling freely from rest in t seconds is approximated by $s(t) = 16t^2$. On March 4, 2009, Brazilian kayaker Pedro Olivia set a world record waterfall descent on the Rio Sacre in Brazil. He was airborne for 2.9 sec. How far did he drop?

Data: www.telegraph.co.uk

68. *SCAD Diving.* The SCAD thrill ride is a 2.5-sec free fall into a net. How far does the diver fall? (See Exercise 67.)

Data: "What is SCAD?", www.scadfreefall.co.uk

69. *Stacking Spheres.* In 2004, the journal *Annals of Mathematics* accepted a proof of the so-called Kepler Conjecture: that the most efficient way to pack spheres is in the shape of a square pyramid. The number N of balls in the stack is given by the polynomial

$$N = \frac{1}{3}x^3 + \frac{1}{2}x^2 + \frac{1}{6}x,$$

where x is the number of layers. Use both the function and the figure to find $N(3)$. Then calculate the number of oranges in a pyramid with 5 layers.

Data: The New York Times 4/6/04

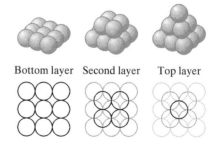

Bottom layer   Second layer   Top layer

70. *Stacking Cannonballs.* The function in Exercise 69 was discovered by Thomas Harriot, assistant to Sir Walter Raleigh, when preparing for an expedition at sea. How many cannonballs did they pack if there were 10 layers to their pyramid?

Data: *The New York Times* 4/7/04

Veterinary Science. *Gentamicin is an antibiotic frequently used by veterinarians. The concentration, in micrograms per milliliter (mcg/mL), of Gentamicin in a horse's bloodstream t hours after injection can be approximated by the polynomial*

$$C = -0.005t^4 + 0.003t^3 + 0.35t^2 + 0.5t.$$

Use the following graph for Exercises 71 and 72.

Data: Michele Tulis, DVM, telephone interview

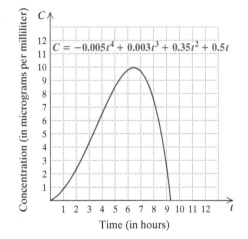

71. Estimate the concentration, in mcg/mL, of Gentamicin in the bloodstream 2 hr after injection.

72. Estimate the concentration, in mcg/mL, of Gentamicin in the bloodstream 4 hr after injection.

73. Explain how it is possible for a term to not be a monomial.

74. Why is it important to understand the rules for order of operations when evaluating polynomials?

Skill Review

Graph.

75. $y = 3x$ [3.2]

76. $y = \frac{1}{2}x - 5$ [3.6]

77. $3x - y = 3$ [3.3]

78. $y = -1$ [3.3]

79. $x = 2$ [3.3]

80. $y + 2 = -2(x - 1)$ [3.7]

Synthesis

81. Suppose that the coefficients of a polynomial are all integers and the polynomial is evaluated for some integer. Must the value of the polynomial then also be an integer? Why or why not?

82. Is it easier to evaluate a polynomial before or after like terms have been combined? Why?

83. Construct a polynomial in x (meaning that x is the variable) of degree 5 with four terms, with coefficients that are consecutive even integers. Write in descending order.

Simplify.

84. $\frac{9}{2}x^8 + \frac{1}{9}x^2 + \frac{1}{2}x^9 + \frac{9}{2}x + \frac{9}{2}x^9 + \frac{8}{9}x^2 + \frac{1}{2}x - \frac{1}{2}x^8$

85. $(3x^2)^3 + 4x^2 \cdot 4x^4 - x^4(2x)^2 + ((2x)^2)^3 - 100x^2(x^2)^2$

86. A polynomial in x has degree 3. The coefficient of x^2 is 3 less than the coefficient of x^3. The coefficient of x is three times the coefficient of x^2. The remaining constant is 2 more than the coefficient of x^3. The sum of the coefficients is -4. Find the polynomial.

87. Use the graph for Exercises 71 and 72 to determine the times for which the concentration of Gentamicin is 5 mcg/mL.

88. *Path of the Olympic Arrow.* The Olympic flame at the 1992 Summer Olympics in Spain was lit by a flaming arrow. As the arrow moved d meters horizontally from the archer, its height h, in meters, was approximated by the polynomial

$$-0.0064d^2 + 0.8d + 2.$$

Complete the table for the choices of d given. Then plot the points and draw a graph representing the path of the arrow.

d	$-0.0064d^2 + 0.8d + 2$
0	
30	
60	
90	
120	

▦ *Semester Averages.* *Professor Sakima calculates a student's average for her course using*

$$A = 0.3q + 0.4t + 0.2f + 0.1h,$$

with q, t, f, and h representing a student's quiz average, test average, final exam score, and homework average, respectively. In Exercises 89 and 90, find the given student's course average rounded to the nearest tenth.

89. Beth: quizzes: 60, 85, 72, 91; final exam: 84; tests: 89, 93, 90; homework: 88

90. Cameron: quizzes: 95, 99, 72, 79; final exam: 91; tests: 68, 76, 92; homework: 86

🖥 **91.** *Research.* Find out how grades are calculated in one or more of your classes. Write a polynomial that will determine a student's final course average. See Exercises 89 and 90 for examples.

92. *Volume of a Display.* The number of spheres in a triangular pyramid with x layers is given by

$$N(x) = \tfrac{1}{6}x^3 + \tfrac{1}{2}x^2 + \tfrac{1}{3}x.$$

The volume of a sphere of radius r is given by

$$V(r) = \tfrac{4}{3}\pi r^3.$$

where π can be approximated as 3.14.

Greta's Chocolate has a window display of truffles piled in a triangular pyramid formation 5 layers deep. If the diameter of each truffle is 3 cm, find the volume of chocolate in the display.

93. Complete the table for the given choices of t. Then plot the points and connect them with a smooth curve representing the graph of the polynomial.

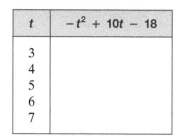

t	$-t^2 + 10t - 18$
3	
4	
5	
6	
7	

👆 **YOUR TURN ANSWERS: SECTION 4.3**
1. $3x, -6, 5x^4$ **2.** 1 **3.** $1, -1, 12, -9$ **4.** $3n^2$ **5.** 5
6. Approximately 149 watts **7.** Approximately 60 watts

Quick Quiz: Sections 4.1–4.3
Simplify.

1. $(-3x^5 y^4)^2$ [4.1] **2.** $\dfrac{48a^5 b}{-3ab}$ [4.1]

3. List the coefficients of the polynomial $x^6 - 6x + \tfrac{1}{2}$. [4.3]

4. Evaluate $-x - 3x^2 + x^3$ for $x = -10$. [4.3]

5. Evaluate t^0 for $t = -12.5$. [4.1]

Prepare to Move On
Simplify. [1.8]

1. $2x + 5 - (x + 8)$ **2.** $3x - 7 - (5x - 1)$

3. $4a + 3 - (-2a + 6)$ **4.** $\tfrac{1}{2}t - \tfrac{1}{4} - \left(\tfrac{3}{2}t + \tfrac{3}{4}\right)$

5. $4t^4 + 8t - (5t^4 - 9t)$

6. $0.1a^2 + 5 - (-0.3a^2 + a - 6)$

4.4 Addition and Subtraction of Polynomials

A. Addition of Polynomials **B.** Opposites of Polynomials **C.** Subtraction of Polynomials
D. Problem Solving

A. Addition of Polynomials

To add two polynomials, we write a plus sign between them and combine like terms.

Student Notes

In Example 1, the parentheses indicate the beginning and the end of each polynomial.

EXAMPLE 1 Write an equivalent expression by adding.

a) $(-5x^3 + 6x - 1) + (4x^3 + 3x^2 + 2)$

b) $\left(\frac{2}{3}x^4 + 3x^2 - 7x + \frac{1}{2}\right) + \left(-\frac{1}{3}x^4 + 5x^3 - 3x^2 + 3x - \frac{1}{2}\right)$

SOLUTION

a) $(-5x^3 + 6x - 1) + (4x^3 + 3x^2 + 2)$

$\quad = -5x^3 + 6x - 1 + 4x^3 + 3x^2 + 2$ Writing without parentheses

$\quad = -5x^3 + 4x^3 + 3x^2 + 6x - 1 + 2$ Using the commutative and associative laws

$\quad = (-5 + 4)x^3 + 3x^2 + 6x + (-1 + 2)$ Combining like terms; using the distributive law

$\quad = -x^3 + 3x^2 + 6x + 1$ Note that $-1x^3 = -x^3$.

b) $\left(\frac{2}{3}x^4 + 3x^2 - 7x + \frac{1}{2}\right) + \left(-\frac{1}{3}x^4 + 5x^3 - 3x^2 + 3x - \frac{1}{2}\right)$

$\quad = \left(\frac{2}{3} - \frac{1}{3}\right)x^4 + 5x^3 + (3 - 3)x^2 + (-7 + 3)x + \left(\frac{1}{2} - \frac{1}{2}\right)$ Combining like terms

$\quad = \frac{1}{3}x^4 + 5x^3 - 4x$

1. Write an equivalent expression by adding:

$(-y^3 + 6y^2 - 4y) + (y^2 + 2y).$

↩ YOUR TURN

EXAMPLE 2 Add: $(2 - 3x + x^2) + (-5 + 7x - 3x^2 + x^3)$.

SOLUTION We have

$(2 - 3x + x^2) + (-5 + 7x - 3x^2 + x^3)$

$\quad = (2 - 5) + (-3 + 7)x + (1 - 3)x^2 + x^3$ You might do this step mentally.

$\quad = -3 + 4x - 2x^2 + x^3.$ Then you would write only this.

2. Add:

$(-1 - 4n + n^2) +$
$(7 + 10n + 3n^2).$

↩ YOUR TURN

In the polynomials of the last example, the terms are arranged according to degree, from least to greatest, in *ascending order*. As a rule, answers are written in ascending order when the polynomials in the original problem are given in ascending order. If the polynomials in the original problem are given in descending order, the answer is usually written in descending order.

To add using columns, we write the polynomials one under the other, listing like terms under one another and leaving space for any missing terms.

> **CAUTION!** Note that equations like those in Examples 1 and 2 are written to show how one expression can be rewritten in an equivalent form. This is very different from solving an equation.

EXAMPLE 3 Write in columns and add:

$\quad 9x^5 - 2x^3 + 6x^2 + 3 \quad$ and $\quad 5x^4 - 7x^2 + 6 \quad$ and $\quad 3x^6 - 5x^5 + x^2 + 5.$

SOLUTION We arrange the polynomials with like terms in columns.

$$
\begin{array}{llll}
9x^5 & -2x^3 + 6x^2 + & 3 & \\
& 5x^4 \qquad -7x^2 + & 6 & \text{We leave spaces for missing terms.} \\
\underline{3x^6 - 5x^5 \qquad\qquad\quad + 1x^2 +} & 5 & \text{Writing } x^2 \text{ as } 1x^2 \\
3x^6 + 4x^5 + 5x^4 - 2x^3 \qquad\qquad + & 14 & \text{Adding}
\end{array}
$$

The answer is $3x^6 + 4x^5 + 5x^4 - 2x^3 + 14.$

3. Add using columns:

$12x^3 + 6x - 7 \quad$ and

$x^3 - 5x^2 + 6x - 1.$

↩ YOUR TURN

B. Opposites of Polynomials

The opposite of the polynomial $x^2 - 3x + 5$ is written $-(x^2 - 3x + 5)$. We can remove the parentheses and the negative sign by changing the sign of every term within the parentheses.

> ### THE OPPOSITE OF A POLYNOMIAL
>
> To find an equivalent polynomial for the *opposite*, or *additive inverse*, of a polynomial, change the sign of every term. This is the same as multiplying the polynomial by −1.

EXAMPLE 4   Write two equivalent expressions for the opposite of $4x^5 - 7x^3 - 8x + \frac{5}{6}$.

SOLUTION

i) $-\left(4x^5 - 7x^3 - 8x + \frac{5}{6}\right)$ This is one way to write the opposite of $4x^5 - 7x^3 - 8x + \frac{5}{6}$.

ii) $-4x^5 + 7x^3 + 8x - \frac{5}{6}$ Changing the sign of every term

Thus, $-\left(4x^5 - 7x^3 - 8x + \frac{5}{6}\right)$ and $-4x^5 + 7x^3 + 8x - \frac{5}{6}$ are equivalent. Both expressions represent the opposite of $4x^5 - 7x^3 - 8x + \frac{5}{6}$.

4. Write two equivalent expressions for the opposite of $-a^4 - 5a^2 + 10$.

YOUR TURN

EXAMPLE 5   Simplify: $-\left(-7x^4 - \frac{5}{9}x^3 + 8x^2 - x + 67\right)$.

SOLUTION   We have

$$-\left(-7x^4 - \frac{5}{9}x^3 + 8x^2 - x + 67\right) = 7x^4 + \frac{5}{9}x^3 - 8x^2 + x - 67.$$

The same result can be found by multiplying by −1:

$$-(-7x^4 - \frac{5}{9}x^3 + 8x^2 - x + 67)$$
$$= -1(-7x^4) + (-1)\left(-\frac{5}{9}x^3\right) + (-1)(8x^2) + (-1)(-x) + (-1)67$$
$$= 7x^4 + \frac{5}{9}x^3 - 8x^2 + x - 67.$$

5. Simplify:

$$-(3.5a^4 - \frac{2}{3}a^3 - 7a + 9).$$

YOUR TURN

C. Subtraction of Polynomials

We can now subtract one polynomial from another by adding the opposite of the polynomial being subtracted. Recall that $5 - 3 = 5 + (-3)$ and, as a rule, $a - b = a + (-b)$.

EXAMPLE 6   Write an equivalent expression by subtracting.

a) $(9x^5 + x^3 - 2x^2 + 4) - (-2x^5 + x^4 - 4x^3 - 3x^2)$
b) $(7x^5 + x^3 - 9x) - (3x^5 - 4x^3 + 5)$

SOLUTION

a) $(9x^5 + x^3 - 2x^2 + 4) - (-2x^5 + x^4 - 4x^3 - 3x^2)$
$$= 9x^5 + x^3 - 2x^2 + 4 + 2x^5 - x^4 + 4x^3 + 3x^2 \quad \text{Adding the opposite}$$
$$= 11x^5 - x^4 + 5x^3 + x^2 + 4 \quad \text{Combining like terms}$$

b) $(7x^5 + x^3 - 9x) - (3x^5 - 4x^3 + 5)$
$$= 7x^5 + x^3 - 9x + (-3x^5) + 4x^3 - 5 \quad \text{Adding the opposite}$$
$$= 7x^5 + x^3 - 9x - 3x^5 + 4x^3 - 5 \quad \text{Try to go directly to this step.}$$
$$= 4x^5 + 5x^3 - 9x - 5 \quad \text{Combining like terms}$$

6. Write an equivalent expression by subtracting:

$$(y^3 - 7y^2 - 9) - (-3y^3 - y^2 + 11).$$

YOUR TURN

To subtract using columns, we first replace the coefficients in the polynomial being subtracted with their opposites. We then add as before.

EXAMPLE 7 Write in columns and subtract:

$$(5x^2 - 3x + 6) - (9x^2 - 5x - 3).$$

SOLUTION

i) $\quad 5x^2 - 3x + 6$ Writing like terms in columns
$\quad \underline{-(9x^2 - 5x - 3)}$

ii) $\quad 5x^2 - 3x + 6$
$\quad \underline{-9x^2 + 5x + 3}$ Changing signs and removing parentheses
                      You might start with this step.

iii) $\quad 5x^2 - 3x + 6$
$\quad \underline{-9x^2 + 5x + 3}$
$\quad -4x^2 + 2x + 9$ Adding

YOUR TURN

7. Write in columns and subtract:

$$(12x^2 + x - 1) - (4x^2 + 10x - 8).$$

If you can do so without error, you can mentally find the opposite of each term being subtracted, and write the answer. Lining up like terms is important and may require leaving some blank space.

EXAMPLE 8 Write in columns and subtract:

$$(x^3 + x^2 - 12) - (-2x^3 + x^2 - 3x + 6).$$

SOLUTION We have

$$\begin{array}{r} x^3 + x^2 \qquad\; - 12 \\ \underline{-(-2x^3 + x^2 - 3x + 6)} \\ 3x^3 \qquad\quad + 3x - 18 \end{array}$$ Leaving a blank space for the missing term

YOUR TURN

8. Write in columns and subtract:

$$(5x^4 - 3x^3 - x^2 + 6) - (-4x^3 + 9x - 2).$$

D. Problem Solving

EXAMPLE 9 Find a polynomial for the sum of the areas of rectangles A, B, C, and D.

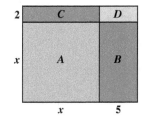

SOLUTION

1. **Familiarize.** Recall that the area of a rectangle is the product of its length and width.

2. **Translate.** We translate the problem to mathematical language. The sum of the areas is a sum of products. We find each product and then add:

$$\underbrace{\text{Area of } A}_{} \text{ plus } \underbrace{\text{area of } B}_{} \text{ plus } \underbrace{\text{area of } C}_{} \text{ plus } \underbrace{\text{area of } D}_{}$$

$$x \cdot x \qquad + \qquad 5 \cdot x \qquad + \qquad 2 \cdot x \qquad + \qquad 2 \cdot 5.$$

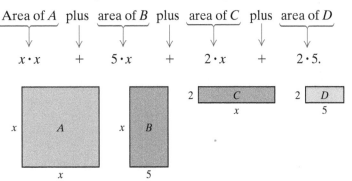

9. Find a polynomial for the sum of the areas of rectangles $A, B, C,$ and $D.$

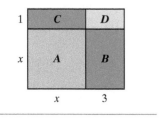

3. Carry out. We simplify and combine like terms:

$$x^2 + 5x + 2x + 10 = x^2 + 7x + 10.$$

4. Check. A partial check is to replace x with a number—say, 3. Then we evaluate $x^2 + 7x + 10$ and compare that result with an alternative calculation:

$$3^2 + 7 \cdot 3 + 10 = 9 + 21 + 10 = 40.$$

When we substitute 3 for x and calculate the total area by regarding the figure as one large rectangle, we should also get 40:

$$\text{Total area} = (x + 5)(x + 2) = (3 + 5)(3 + 2) = 8 \cdot 5 = 40.$$

Our check is only partial, since it is possible for an incorrect answer to equal 40 when evaluated for $x = 3$. This would be unlikely, especially if a second choice of x—say, $x = 5$—also checks. We leave that check to the student.

5. State. A polynomial for the sum of the areas is $x^2 + 7x + 10.$

⟳ YOUR TURN

EXAMPLE 10   A 16-ft wide round fountain is built in a square city park that measures x ft by x ft. Find a polynomial for the remaining area of the park.

SOLUTION

1. Familiarize. We make a drawing of the square park and the circular fountain, and let x represent the length of a side of the park.

The area of a square of side s is given by $A = s^2$, and the area of a circle of radius r is given by $A = \pi r^2$. Note that a circle with a diameter of 16 ft has a radius of 8 ft.

2. Translate. We reword the problem and translate as follows.

Rewording: Area of park   minus   area of fountain   is   area left over.

Translating: $x\text{ ft} \cdot x\text{ ft}$ $-$ $\pi \cdot 8\text{ ft} \cdot 8\text{ ft}$ $=$ Area left over

3. Carry out. We carry out the multiplication:

$$x^2\text{ ft}^2 - 64\pi\text{ ft}^2 = (x^2 - 64\pi)\text{ ft}^2 = \text{Area left over.}$$

4. Check. As a partial check, note that the units in the answer are square feet (ft^2), a measure of area, as expected.

5. State. The remaining area of the park is $(x^2 - 64\pi)\text{ ft}^2.$

⟳ YOUR TURN

Write an equivalent expression without parentheses.

1. $x^2 - (x)$

2. $x^2 - (-x)$

3. $x^2 - (x + 1)$

4. $x^2 - (x - 1)$

5. $x^2 - (-x + 1)$

6. $x^2 - (-x - 1)$

10. Consider the fountain and park described in Example 10. If the fountain is 20 ft wide, find a polynomial for the remaining area of the park.

Technology Connection

To check polynomial addition or subtraction, we can let $y_1 =$ the expression before the addition or subtraction has been performed and $y_2 =$ the simplified sum or difference. If the addition or subtraction is correct, y_1 will equal y_2 and $y_2 - y_1$ will be 0. We enter $y_2 - y_1$ as y_3, using **VARS**. Below is a check of Example 6(b) in which

$$y_1 = (7x^5 + x^3 - 9x) - (3x^5 - 4x^3 + 5),$$
$$y_2 = 4x^5 + 5x^3 - 9x - 5,$$

and $y_3 = y_2 - y_1$.

We graph only y_3. If indeed y_1 and y_2 are equivalent, then y_3 should equal 0. This means its graph should coincide with the x-axis. The TRACE or TABLE feature

can confirm that y_3 is always 0, or we can select y_3 to be drawn bold in the ⟨ Y= ⟩ window.

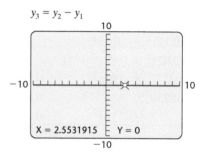

1. Use a graphing calculator to check Examples 1, 2, and 6.

4.4 EXERCISE SET

◆ Vocabulary and Reading Check

Choose from the following list the word or words to complete each statement. Words may be used more than once or not at all.

ascending	like	opposite
descending	missing	sign

1. To subtract a polynomial, add the _____ of the polynomial.

2. When the terms of a polynomial are arranged according to degree, from least to greatest, the polynomial is written in _____ order.

3. To write an equivalent expression for the _____ of a polynomial, we change the _____ of every term.

4. To add or subtract polynomials using columns, we write _____ terms in columns and leave spaces for _____ terms.

◆ Concept Reinforcement

For each of Exercises 5–8, replace ▨ with the correct expression or operation sign.

5. $(3x^2 + 2) + (6x^2 + 7) = (3 + 6)\,▨ + (2 + 7)$

6. $(5t - 6) + (4t + 3) = (5 + 4)t + (\,▨ + 3)$

7. $(9x^3 - x^2) - (3x^3 + x^2) = 9x^3 - x^2 - 3x^3 \,▨\, x^2$

8. $(-2n^3 + 5) - (n^2 - 2) = -2n^3 + 5 - n^2 \,▨\, 2$

A. Addition of Polynomials

Add.

9. $(3x + 2) + (x + 7)$

10. $(x + 1) + (12x + 10)$

11. $(2t + 7) + (-8t + 1)$

12. $(4t - 3) + (-11t + 2)$

13. $(x^2 + 6x + 3) + (-4x^2 - 5)$

14. $(x^2 - 5x + 4) + (8x - 9)$

15. $(7t^2 - 3t - 6) + (2t^2 + 4t + 9)$

16. $(8a^2 + 4a - 7) + (6a^2 - 3a - 1)$

17. $(4m^3 - 7m^2 + m - 5) + (4m^3 + 7m^2 - 4m - 2)$

18. $(5n^3 - n^2 + 4n + 11) + (2n^3 - 4n^2 + n - 11)$

19. $(3 + 6a + 7a^2 + a^3) + (4 + 7a - 8a^2 + 6a^3)$

20. $(7 + 4t - 5t^2 + 6t^3) + (2 + t + 6t^2 - 4t^3)$

21. $(3x^6 + 2x^4 - x^3 + 5x) + (-x^6 + 3x^3 - 4x^2 + 7x^4)$

22. $(4x^5 - 6x^3 - 9x + 1) + (3x^4 + 6x^3 + 9x^2 + x)$

23. $\left(\frac{3}{5}x^4 + \frac{1}{2}x^3 - \frac{2}{3}x + 3\right) + \left(\frac{2}{5}x^4 - \frac{1}{4}x^3 - \frac{3}{4}x^2 - \frac{1}{6}x\right)$

24. $\left(\frac{1}{3}x^9 + \frac{1}{5}x^5 - \frac{1}{2}x^2 + 7\right) + \left(-\frac{1}{5}x^9 + \frac{1}{4}x^4 - \frac{3}{5}x^5\right)$

25. $(5.3t^2 - 6.4t - 9.1) + (4.2t^3 - 1.8t^2 + 7.3)$

26. $(4.9a^3 + 3.2a^2 - 5.1a) + (2.1a^2 - 3.7a + 4.6)$

27. $-4x^3 + 8x^2 + 3x - 2$
$\underline{ - 4x^2 + 3x + 2}$

28. $-3x^4 + 6x^2 + 2x - 4$
$\underline{ - 3x^2 + 2x + 4}$

29. $0.05x^4 + 0.12x^3 - 0.5x^2$
$ - 0.02x^3 + 0.02x^2 + 2x$
$1.5x^4 + 0.01x^2 + 0.15$
$ 0.25x^3 + 0.85$
$\underline{-0.25x^4 + 10x^2 - 0.04}$

30. $0.15x^4 + 0.10x^3 - 0.9x^2$
$ - 0.01x^3 + 0.01x^2 + x$
$1.25x^4 + 0.11x^2 + 0.01$
$ 0.27x^3 + 0.99$
$\underline{-0.35x^4 + 15x^2 - 0.03}$

B. Opposites of Polynomials

Write two equivalent expressions for the opposite of each polynomial, as in Example 4.

31. $-3t^3 + 4t^2 - 7$

32. $-x^3 - 5x^2 + 2x$

33. $x^4 - 8x^3 + 6x$

34. $5a^3 + 2a - 17$

Simplify.

35. $-(3a^4 - 5a^2 + 1.2)$

36. $-(-6a^3 + 0.2a^2 - 7)$

37. $-\left(-4x^4 + 6x^2 + \frac{3}{4}x - 8\right)$

38. $-\left(3x^5 - 2x^3 - \frac{3}{5}x^2 + 16\right)$

C. Subtraction of Polynomials

Subtract.

39. $(3x + 1) - (5x + 8)$

40. $(7x + 3) - (3x + 2)$

41. $(-9t + 12) - (t^2 + 3t - 1)$

42. $(a^2 - 3a - 2) - (2a^2 - 6a - 2)$

43. $(4a^2 + a - 7) - (3 - 8a^3 - 4a^2)$

44. $(-4x^2 + 2x) - (-5x^2 + 2x^3 + 3)$

Aha! **45.** $(7x^3 - 2x^2 + 6) - (6 - 2x^2 + 7x^3)$

46. $(8x^5 + 3x^4 + x - 1) - (8x^5 + 3x^4 - 1)$

47. $(3 + 5a + 3a^2 - a^3) - (2 + 4a - 9a^2 + 2a^3)$

48. $(7 + t - 5t^2 + 2t^3) - (1 + 2t - 4t^2 + 5t^3)$

49. $\left(\frac{5}{8}x^3 - \frac{1}{4}x - \frac{1}{3}\right) - \left(-\frac{1}{2}x^3 + \frac{1}{4}x - \frac{1}{3}\right)$

50. $\left(\frac{1}{5}x^3 + 2x^2 - \frac{3}{10}\right) - \left(-\frac{2}{5}x^3 + 2x^2 + \frac{7}{1000}\right)$

51. $(0.07t^3 - 0.03t^2 + 0.01t) - (0.02t^3 + 0.04t^2 - 1)$

52. $(0.9a^3 + 0.2a - 5) - (0.7a^4 - 0.3a - 0.1)$

53. $x^3 + 3x^2 + 1$
$\underline{-(x^3 + x^2 - 5)}$

54. $x^2 + 5x + 6$
$\underline{-(x^2 + 2x + 1)}$

55. $4x^4 - 2x^3$
$\underline{-(7x^4 + 6x^3 + 7x^2)}$

56. $5x^4 + 6x^3 - 9x^2$
$\underline{-(-6x^4 + x^2)}$

D. Problem Solving

57. Solve.

    a) Find a polynomial for the sum of the areas of the rectangles shown in the figure.

    b) Find the sum of the areas when $x = 5$ and $x = 7$.

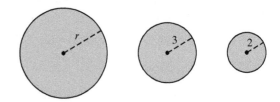

58. Solve. Leave the answers in terms of π.

    a) Find a polynomial for the sum of the areas of the circles shown in the figure.

    b) Find the sum of the areas when $r = 5$ and $r = 11.3$.

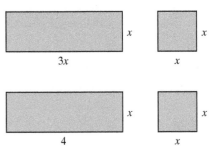

In each of Exercises 59 and 60, find a polynomial for the perimeter of the figure.

59.

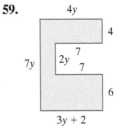

60.

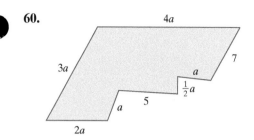

Find two algebraic expressions for the area of each figure. First, regard the figure as one large rectangle, and then regard the figure as a sum of four smaller rectangles.

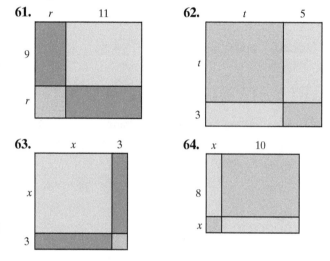

61. **62.**

63. **64.**

In each of Exercises 65–68, find a polynomial for the shaded area of the figure.

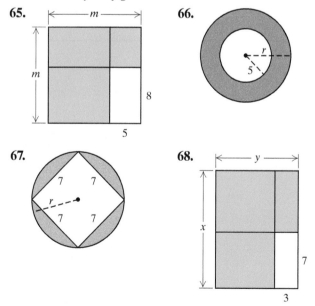

65. **66.**

67. **68.**

69. A 2-ft by 6-ft bath enclosure is installed in a new bathroom measuring *x* ft by *x* ft. Find a polynomial for the remaining floor area.

70. A 5-ft by 7-ft Jacuzzi™ is installed on an outdoor deck measuring *y* ft by *y* ft. Find a polynomial for the remaining area of the deck.

71. A 12-ft wide round patio is laid in a garden measuring *z* ft by *z* ft. Find a polynomial for the remaining area of the garden.

72. A 10-ft wide round water trampoline is floating in a pool measuring *x* ft by *x* ft. Find a polynomial for the remaining surface area of the pool.

73. A 12-m by 12-m mat includes a circle of diameter *d* meters for wrestling. Find a polynomial for the area of the mat outside the wrestling circle.

74. A 2-m by 3-m rug is spread inside a tepee that has a diameter of *x* meters. Find a polynomial for the area of the tepee's floor that is not covered.

75. Explain why parentheses are used in the statement of the solution of Example 10: $(x^2 - 64\pi)$ ft^2.

76. Is the sum of two trinomials always a trinomial? Why or why not?

Skill Review

Find the slope–intercept equation for each line described.

77. Slope $\frac{1}{3}$; *y*-intercept $(0, 2)$ [3.6]

78. Slope 4; contains the point $(-6, 0)$ [3.7]

79. Slope 0; contains the point $(4, 10)$ [3.7]

80. Slope -3; contains the point $(1, 5)$ [3.7]

81. Slope $-\frac{4}{7}$; contains the point $(0, -4)$ [3.6]

82. Contains the points $(-2, 0)$ and $(4, -8)$ [3.7]

Synthesis

83. What can be concluded about two polynomials whose sum is zero?

84. Which, if any, of the commutative, associative, and distributive laws are needed for adding polynomials? Why?

Simplify.

85. $(6t^2 - 7t) + (3t^2 - 4t + 5) - (9t - 6)$

86. $(3x^2 - 4x + 6) - (-2x^2 + 4) + (-5x - 3)$

87. $4(x^2 - x + 3) - 2(2x^2 + x - 1)$

88. $3(2y^2 - y - 1) - (6y^2 - 3y - 3)$

89. $(345.099x^3 - 6.178x) - (94.508x^3 - 8.99x)$

In each of Exercises 90–93, find a polynomial for the surface area of the right rectangular solid.

90.

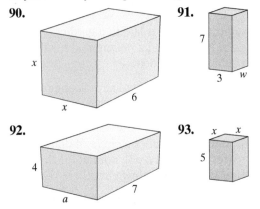

91.

92.

93.

94. Find a polynomial for the total length of all edges in the figure appearing in Exercise 93.

95. *Total Profit.* Hadley Electronics is marketing a new digital camera. Total revenue is the total amount of money taken in. The firm determines that when it sells x cameras, its total revenue is given by

$$R = 175x - 0.4x^2.$$

Total cost is the total cost of producing x cameras. Hadley Electronics determines that the total cost of producing x cameras is given by

$$C = 5000 + 0.6x^2.$$

The total profit P is

$$(\text{Total Revenue}) - (\text{Total Cost}) = R - C.$$

a) Find a polynomial for total profit.
b) What is the total profit on the production and sale of 75 cameras?
c) What is the total profit on the production and sale of 120 cameras?

96. Does replacing each occurrence of the variable x in $4x^7 - 6x^3 + 2x$ with its opposite result in the opposite of the polynomial? Why or why not?

YOUR TURN ANSWERS: SECTION 4.4

1. $-y^3 + 7y^2 - 2y$ **2.** $6 + 6n + 4n^2$
3. $13x^3 - 5x^2 + 12x - 8$
4. $-(-a^4 - 5a^2 + 10); a^4 + 5a^2 - 10$
5. $-3.5a^4 + \frac{2}{3}a^3 + 7a - 9$ **6.** $4y^3 - 6y^2 - 20$
7. $8x^2 - 9x + 7$ **8.** $5x^4 + x^3 - x^2 - 9x + 8$
9. $x^2 + 4x + 3$ **10.** $(x^2 - 100\pi)\,\text{ft}^2$

Quick Quiz: Sections 4.1–4.4

Simplify.

1. $\left(\dfrac{-2x^3}{3y^5}\right)^{-2}$ [4.2]   **2.** $(y + 3)^5(y + 3)^7$ [4.1]

3. Determine the degree of $5n - n^4 + 7 + 6n^2$.   [4.3]

4. Combine like terms and write in descending order: $2y - 7y^2 - y^3 + y^2$.   [4.3]

5. Subtract: $(5y^2 - y + 7) - (-y - 3)$.   [4.4]

Prepare to Move On

Simplify.

1. $2(x^2 - x + 3)$ [1.8]

2. $-5(3x^2 - 2x - 7)$ [1.8]

3. $x^2 \cdot x^6$ [4.1]   **4.** $y^6 \cdot y$ [4.1]

5. $2n \cdot n^2$ [4.1]   **6.** $-6n^4 \cdot n^8$ [4.1]

Mid-Chapter Review

Properties of exponents allow us to simplify exponential expressions and to work with polynomials.

$$a^1 = a, \qquad a^m \cdot a^n = a^{m+n}, \qquad (ab)^n = a^n b^n,$$

$$a^0 = 1, \qquad (a^m)^n = a^{mn}, \qquad \left(\frac{a}{b}\right)^n = \frac{a^n}{b^n}$$

$$a^{-n} = \frac{1}{a^n}, \qquad \frac{a^m}{a^n} = a^{m-n},$$

GUIDED SOLUTIONS

1. Simplify: $\dfrac{x^{-3}y}{x^{-4}y^7}$. [4.2]

Solution

$$\frac{x^{-3}y}{x^{-4}y^7} = x^{-3-(\boxed{})}y^{\boxed{}-7} = x^{\boxed{}}y^{\boxed{}} = \frac{\boxed{}}{\boxed{}}$$

2. Subtract: $(x^2 + 7x - 12) - (3x^2 - 6x - 1)$. [4.4]

Solution

$$(x^2 + 7x - 12) - (3x^2 - 6x - 1) = x^2 + 7x - 12 \,\boxed{}\, 3x^2 \,\boxed{}\, 6x \,\boxed{}\, 1 \qquad \text{Adding the opposite}$$

$$= \boxed{}x^2 + \boxed{}x - \boxed{} \qquad \text{Combining like terms}$$

MIXED REVIEW

Simplify. Do not use negative exponents in the answer.

3. $(x^2 y^5)^8$ [4.1]

4. $(4x)^0$ [4.1]

5. $\dfrac{3a^{11}}{12a}$ [4.1]

6. $\dfrac{-48ab^7}{18ab^6}$ [4.1]

7. $5x^{-2}$ [4.2]

8. $(a^{-2}bc^3)^{-1}$ [4.2]

9. $\left(\dfrac{2a^2}{3}\right)^{-3}$ [4.2]

10. $\dfrac{8m^{-2}n^3}{12m^4n^7}$ [4.2]

11. Convert to decimal notation: 1.89×10^{-6}. [4.2]

12. Convert to scientific notation: 27,000,000,000. [4.2]

13. Determine the degree of $8t^2 - t^3 + 5$. [4.3]

14. Combine like terms:

$$3a^2 - 6a - a^2 + 7 + a - 10. \quad [4.3]$$

Perform the indicated operation and simplify.

15. $(3x^2 - 2x + 6) + (5x - 3)$ [4.4]

16. $(9x + 6) - (2x - 1)$ [4.4]

17. $(4x^2 - x - 7) - (10x^2 - 3x + 5)$ [4.4]

18. $(t^9 + 3t^6 - 8t^2) + (5t^7 - 3t^6 + 8t^2)$ [4.4]

19. $(3a^4 - 9a^3 - 7) - (4a^3 + 13a^2 - 3)$ [4.4]

20. $(x^4 - 2x^2 - \frac{1}{2}x) - (x^5 - x^4 + \frac{1}{2}x)$ [4.4]

4.5   Multiplication of Polynomials

A. Multiplying Monomials   **B.** Multiplying a Monomial and a Polynomial   **C.** Multiplying Any Two Polynomials

We now multiply polynomials using techniques based on the distributive, associative, and commutative laws and the rules for exponents.

A. Multiplying Monomials

Consider $(3x)(4x)$. We multiply as follows:

$$
\begin{aligned}
(3x)(4x) &= 3 \cdot x \cdot 4 \cdot x && \text{Using an associative law} \\
&= 3 \cdot 4 \cdot x \cdot x && \text{Using a commutative law} \\
&= (3 \cdot 4) \cdot (x \cdot x) && \text{Using an associative law} \\
&= 12x^2.
\end{aligned}
$$

> **TO MULTIPLY MONOMIALS**
>
> To find an equivalent expression for the product of two monomials, multiply the coefficients and then multiply any variables using the product rule for exponents.

Student Notes

Remember that when we compute $(3 \cdot 5)(2 \cdot 4)$, each factor is used only once, even if we change the order:

$$
\begin{aligned}
(3 \cdot 5)(2 \cdot 4) &= (5 \cdot 2)(3 \cdot 4) \\
&= 10 \cdot 12 = 120.
\end{aligned}
$$

In the same way,

$$
\begin{aligned}
(3 \cdot x)(2 \cdot x) &= (3 \cdot 2)(x \cdot x) \\
&= 6x^2.
\end{aligned}
$$

Some students mistakenly "reuse" a factor.

---

1. Multiply to form an equivalent expression: $(-6t^3)(9t)$.

EXAMPLE 1   Multiply to form an equivalent expression.

a) $(5x)(6x)$ **b)** $(-a)(3a)$ **c)** $(7x^5)(-4x^3)$

SOLUTION

a) $(5x)(6x) = (5 \cdot 6)(x \cdot x)$ Multiplying the coefficients; multiplying the variables

$\qquad\qquad\quad = 30x^2$ Simplifying

b) $(-a)(3a) = (-1a)(3a)$ Writing $-a$ as $-1a$ can ease calculations.

$\qquad\qquad\quad = (-1 \cdot 3)(a \cdot a)$ Using an associative law and a commutative law

$\qquad\qquad\quad = -3a^2$

c) $(7x^5)(-4x^3) = [7(-4)](x^5 \cdot x^3)$

$\qquad\qquad\qquad\quad = -28x^{5+3}$

$\qquad\qquad\qquad\quad = -28x^8$ } Using the product rule for exponents

YOUR TURN

B. Multiplying a Monomial and a Polynomial

To find an equivalent expression for the product of a monomial, such as $5x$, and a polynomial, such as $2x^2 - 3x + 4$, we use the distributive law.

EXAMPLE 2   Multiply: **(a)** $x(x + 3)$; **(b)** $5x(2x^2 - 3x + 4)$.

SOLUTION

a) $x(x + 3) = x \cdot x + x \cdot 3$ Using the distributive law

$\qquad\qquad = x^2 + 3x$

b) $5x(2x^2 - 3x + 4) = (5x)(2x^2) - (5x)(3x) + (5x)(4)$ Using the distributive law

$$= 10x^3 - 15x^2 + 20x$$ Performing the three multiplications

2. Multiply: $3a(5a^3 - a + 11)$.

YOUR TURN

The product in Example 2(a) can be visualized as the area of a rectangle with width x and length $x + 3$. Note that the total area can be expressed as $x(x + 3)$ or, by adding the two smaller areas, $x^2 + 3x$.

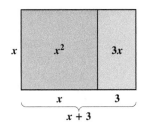

THE PRODUCT OF A MONOMIAL AND A POLYNOMIAL

To multiply a monomial and a polynomial, multiply each term of the polynomial by the monomial.

Try to do this mentally, when possible. Remember that we multiply coefficients and, when the bases match, add exponents.

EXAMPLE 3 Multiply: $2x^2(x^3 - 7x^2 + 10x - 4)$.

SOLUTION

$$Think: \quad \underbrace{2x^2 \cdot x^3} - \underbrace{2x^2 \cdot 7x^2} + \underbrace{2x^2 \cdot 10x} - \underbrace{2x^2 \cdot 4}$$

3. Multiply:

$5y^3(2y^3 - 8y^2 + y - 5)$.

$$2x^2(x^3 - 7x^2 + 10x - 4) = 2x^5 - 14x^4 + 20x^3 - 8x^2$$

YOUR TURN

C. Multiplying Any Two Polynomials

Before considering the product of *any* two polynomials, let's look at products of two binomials. To multiply, we again begin by using the distributive law. This time, however, it is a *binomial* rather than a monomial that is being distributed.

EXAMPLE 4 Multiply each pair of binomials.

a) $x + 5$ and $x + 4$ **b)** $4x - 3$ and $x - 2$

SOLUTION

a) $(x + 5)(x + 4) = (x + 5)x + (x + 5)4$ Using the distributive law

$$= x(x + 5) + 4(x + 5)$$ Using the commutative law for multiplication

$$= x \cdot x + x \cdot 5 + 4 \cdot x + 4 \cdot 5$$ Using the distributive law (twice)

$$= x^2 + 5x + 4x + 20$$ Multiplying the monomials

$$= x^2 + 9x + 20$$ Combining like terms

b) $(4x - 3)(x - 2) = (4x - 3)x - (4x - 3)2$ Using the distributive law

$\qquad = x(4x - 3) - 2(4x - 3)$ Using the commutative law for multiplication. This step is often omitted.

$\qquad = x \cdot 4x - x \cdot 3 - 2 \cdot 4x - 2(-3)$ Using the distributive law (twice)

$\qquad = 4x^2 - 3x - 8x + 6$ Multiplying the monomials

$\qquad = 4x^2 - 11x + 6$ Combining like terms

4. Multiply $x + 3$ and $5x - 1$.

 YOUR TURN

To visualize the product in Example 4(a), consider a rectangle of length $x + 5$ and width $x + 4$. The total area can be expressed as $(x + 5)(x + 4)$ or, by adding the four smaller areas, $x^2 + 5x + 4x + 20$.

Study Skills

Take a Peek Ahead

Try to at least glance at the next section of material that will be covered in class. This will make it easier to concentrate on your instructor's lecture instead of trying to write everything down.

Let's consider the product of a binomial and a trinomial. Again we make repeated use of the distributive law.

EXAMPLE 5   Multiply: $(x^2 + 2x - 3)(x + 4)$.

SOLUTION

$(x^2 + 2x - 3)(x + 4)$

$\qquad = (x^2 + 2x - 3)x + (x^2 + 2x - 3)4$ Using the distributive law

$\qquad = x(x^2 + 2x - 3) + 4(x^2 + 2x - 3)$ Using the commutative law

$\qquad = x \cdot x^2 + x \cdot 2x - x \cdot 3 + 4 \cdot x^2 + 4 \cdot 2x - 4 \cdot 3$ Using the distributive law (twice)

5. Multiply:

$(2t^2 - t + 5)(t + 3)$.

$\qquad = x^3 + 2x^2 - 3x + 4x^2 + 8x - 12$ Multiplying the monomials

$\qquad = x^3 + 6x^2 + 5x - 12$ Combining like terms

 YOUR TURN

To use columns for long multiplication, we multiply each term in the top row by every term in the bottom row. We write like terms in columns, and then add the results. Such multiplication is similar to multiplying whole numbers.

$$
\begin{array}{r}
321 \\
\times\, 12 \\
\hline
642 \\
321 \\
\hline
3852
\end{array}
\qquad
\begin{array}{r}
300 + 20 + 1 \\
\times 10 + 2 \\
\hline
600 + 40 + 2 \\
3000 + 200 + 10 \\
\hline
3000 + 800 + 50 + 2
\end{array}
$$

Multiplying the top row by 2

Multiplying the top row by 10

Adding

EXAMPLE 6 Multiply: $(5x^4 - 2x^2 + 3x)(x^2 + 2x)$.

SOLUTION

$$
\begin{array}{r}
5x^4 - 2x^2 + 3x \\
x^2 + 2x \\
\hline
10x^5 \quad\quad - 4x^3 + 6x^2 \\
5x^6 \quad\quad - 2x^4 + 3x^3 \\
\hline
5x^6 + 10x^5 - 2x^4 - x^3 + 6x^2
\end{array}
$$

Note that each polynomial is written in descending order.
Multiplying the top row by $2x$
Multiplying the top row by x^2
Combining like terms
Line up like terms in columns.

6. Multiply using columns:

$(3x^3 - x^2 + 5)(x - 4)$.

↩ YOUR TURN

THE PRODUCT OF TWO POLYNOMIALS

To multiply two polynomials P and Q, select one of the polynomials—say, P. Then multiply each term of P by every term of Q and combine like terms.

Sometimes we multiply horizontally, while still aligning like terms as we write the product.

EXAMPLE 7 Multiply: $(2x^3 + 3x^2 - 4x + 6)(3x + 5)$.

SOLUTION

$$
\begin{aligned}
&\overbrace{\hspace{3cm}}^{\text{Multiplying by } 3x} \\
(2x^3 + 3x^2 - 4x + 6)(3x + 5) = 6x^4 + {}& 9x^3 - 12x^2 + 18x \\
+ {}& \underbrace{10x^3 + 15x^2 - 20x + 30}_{\text{Multiplying by } 5} \\
= 6x^4 + {}& 19x^3 + 3x^2 - 2x + 30
\end{aligned}
$$

7. Multiply:

$(5y^3 - 4y^2 + y + 2)(3y + 1)$.

↩ YOUR TURN

Checking by Evaluating

How can we be certain that our multiplication (or addition or subtraction) of polynomials is correct? One check is to simply review our calculations. A different type of check makes use of the fact that equivalent expressions have the same value when evaluated for the same replacement. Thus a quick, partial, check of Example 7 can be made by selecting a convenient replacement for x (say, 1) and comparing the values of the expressions $(2x^3 + 3x^2 - 4x + 6)(3x + 5)$ and $6x^4 + 19x^3 + 3x^2 - 2x + 30$:

$$
\begin{aligned}
(2x^3 + 3x^2 - 4x + 6)(3x + 5) &= (2 \cdot 1^3 + 3 \cdot 1^2 - 4 \cdot 1 + 6)(3 \cdot 1 + 5) \\
&= (2 + 3 - 4 + 6)(3 + 5) \\
&= 7 \cdot 8 = 56;
\end{aligned}
$$

$$
\begin{aligned}
6x^4 + 19x^3 + 3x^2 - 2x + 30 &= 6 \cdot 1^4 + 19 \cdot 1^3 + 3 \cdot 1^2 - 2 \cdot 1 + 30 \\
&= 6 + 19 + 3 - 2 + 30 \\
&= 28 - 2 + 30 = 56.
\end{aligned}
$$

Since the value of both expressions is 56, the multiplication in Example 7 is very likely correct.

It is possible, by chance, for two expressions that are not equivalent to share the same value when evaluated. For this reason, checking by evaluating is only a partial check. Consult your instructor for the checking approach that he or she prefers.

↩ Check Your UNDERSTANDING

Match the expression with the correct result from the column on the right. Choices may be used more than once.

1. ____ $3x^2 \cdot 2x^4$ **a)** $6x^8$

2. ____ $3x^8 + 5x^8$ **b)** $8x^6$

3. ____ $4x^3 \cdot 2x^5$ **c)** $6x^6$

4. ____ $3x^5 \cdot 2x^3$ **d)** $8x^8$

5. ____ $4x^6 + 2x^6$

6. ____ $4x^4 \cdot 2x^2$

Technology Connection

Tables can also be used to check polynomial multiplication. To illustrate, we can check Example 7 by entering $y_1 = (2x^3 + 3x^2 - 4x + 6)(3x + 5)$ and $y_2 = 6x^4 + 19x^3 + 3x^2 - 2x + 30$.

When $\boxed{\text{TABLE}}$ is then pressed, we are shown two columns of values—one for y_1 and one for y_2. If our multiplication is correct, the columns of values will match.

X	Y₁	Y₂
−3	36	36
−2	−10	−10
−1	22	22
0	30	30
1	56	56
2	286	286
3	1050	1050

X = −3

1. Form a table and scroll up and down to check Example 6.
2. Check Example 7 by letting

$$y_1 = (2x^3 + 3x^2 - 4x + 6)(3x + 5),$$
$$y_2 = 6x^4 + 19x^3 + 3x^2 - 2x + 30,$$

and

$$y_3 = y_2 - y_1.$$

Then check that y_3 is always 0.

4.5 EXERCISE SET

FOR EXTRA HELP   MyMathLab®

⤷ Vocabulary and Reading Check

For each step, write the letter that gives an explanation of what was done on that step.

a) Combining like terms
b) Multiplying monomials
c) Using the commutative law for multiplication
d) Using the distributive law

$(2x + 3)(5x - 4)$
$= (2x + 3)(5x) - (2x + 3)(4)$ (d)
$= (5x)(2x + 3) - 4(2x + 3)$ 1. ___
$= (5x)(2x) + (5x)(3) - 4(2x) - 4(3)$ 2. ___
$= 10x^2 + 15x - 8x - 12$ 3. ___
$= 10x^2 + 7x - 12$ 4. ___

A. Multiplying Monomials

Multiply.

5. $(3x^5)7$ **6.** $2x^3 \cdot 11$

7. $(-x^3)(x^4)$ **8.** $(-x^2)(-x)$

9. $(-x^6)(-x^2)$ **10.** $(-x^5)(x^3)$

11. $4t^2(9t^2)$ **12.** $(6a^8)(3a^2)$

13. $(0.3x^3)(-0.4x^6)$ **14.** $(-0.1x^6)(0.2x^4)$

15. $\left(-\frac{1}{4}x^4\right)\left(\frac{1}{5}x^8\right)$ **16.** $\left(-\frac{1}{5}x^3\right)\left(-\frac{1}{3}x\right)$

17. $(-5n^3)(-1)$ **18.** $19t^2 \cdot 0$

19. $(-4y^5)(6y^2)(-3y^3)$ **20.** $7x^2(-2x^3)(2x^6)$

B. Multiplying a Monomial and a Polynomial

Multiply.

21. $5x(4x + 1)$ **22.** $3x(2x - 7)$

23. $(a - 9)3a$ **24.** $(a - 7)4a$

25. $x^2(x^3 + 1)$ **26.** $-2x^3(x^2 - 1)$

27. $-3n(2n^2 - 8n + 1)$

28. $4n(3n^3 - 4n^2 - 5n + 10)$

29. $-5t^2(3t + 6)$

30. $7t^2(2t + 1)$

31. $\frac{2}{3}a^4\left(6a^5 - 12a^3 - \frac{5}{8}\right)$

32. $\frac{3}{4}t^5\left(8t^6 - 12t^4 + \frac{12}{7}\right)$

33. $0.2x^3(1.2x^2 - 0.5x)$

34. $0.01x(230x^5 + 0.12x^3)$

C. Multiplying Any Two Polynomials

Multiply. Write the answer in the same order as the given polynomials.

35. $(x + 3)(x + 4)$

36. $(x + 7)(x + 3)$

37. $(t + 7)(t - 3)$

38. $(t - 4)(t + 3)$

39. $(a - 0.6)(a - 0.7)$

40. $(a - 0.4)(a - 0.8)$

41. $(x + 3)(x - 3)$

42. $(x + 6)(x - 6)$

43. $(4 - x)(7 - 2x)$

44. $(5 + x)(5 + 2x)$

45. $\left(t + \frac{3}{2}\right)\left(t + \frac{4}{3}\right)$

46. $\left(a - \frac{2}{5}\right)\left(a + \frac{5}{2}\right)$

47. $\left(\frac{1}{4}a + 2\right)\left(\frac{3}{4}a - 1\right)$

48. $\left(\frac{2}{5}t - 1\right)\left(\frac{3}{5}t + 1\right)$

Draw and label rectangles similar to those following Examples 2 and 4 to illustrate each product.

49. $x(x + 5)$ **50.** $x(x + 2)$

51. $(x + 1)(x + 2)$ **52.** $(x + 3)(x + 1)$

53. $(x + 5)(x + 3)$ **54.** $(x + 4)(x + 6)$

Multiply and check. Write the answer in descending order.

55. $(x^2 - x + 3)(x + 1)$

56. $(x^2 + x - 7)(x + 2)$

57. $(2a + 5)(a^2 - 3a + 2)$

58. $(3t - 4)(t^2 - 5t + 1)$

59. $(y^2 - 7)(3y^4 + y + 2)$

60. $(a^2 + 4)(5a^3 - 3a - 1)$

Aha! **61.** $(3x + 2)(7x + 4x + 1)$

62. $(4x - 5x - 3)(1 + 2x^2)$

63. $\left(3x + \frac{1}{2}\right)(x^2 - 4x - 2)$

64. $\left(6x^2 - 2x - \frac{1}{2}\right)(2x^2 - 1)$

65. $(1.2a^3 - 0.2a^2 + 5a)(0.4a^2 - 10)$

66. $(1.5a - 0.3)(0.2a^2 - 4a + 8)$

67. $(x^2 + 5x - 1)(x^2 - x + 3)$

68. $(x^2 - 3x + 2)(x^2 + x + 1)$

69. $\left(5t^2 - t + \frac{1}{2}\right)(2t^2 + t - 4)$

70. $(2t^2 - 5t - 4)\left(3t^2 - t + \frac{1}{2}\right)$

71. $(x + 1)(x^3 + 7x^2 + 5x + 4)$

72. $(x + 2)(x^3 + 5x^2 + 9x + 3)$

73. Is it possible to understand polynomial multiplication without understanding the distributive law? Why or why not?

74. The polynomials
$$(a + b + c + d) \quad \text{and} \quad (r + s + m + p)$$
are multiplied. Without performing the multiplication, determine how many terms the product will contain. Provide a justification for your answer.

Skill Review

75. Solve $A = \dfrac{b + c}{2}$ for c.   [2.3]

76. What percent of 30 is 24?   [2.4]

77. Graph $t \geq -3$ on the number line.   [2.6]

78. In which quadrant or on what axis is the point $\left(-\frac{1}{2}, 0\right)$ located?   [3.1]

79. Find the slope of the line containing the points $(2, 3)$ and $(5, -1)$.   [3.5]

80. Find the slope and the y-intercept of the line given by $x - 3y = 5$.   [3.6]

Synthesis

81. Under what conditions will the product of two binomials be a trinomial?

82. Explain how the following figure can be used to show that $(x + 3)^2 \neq x^2 + 9$.

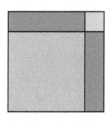

Find a polynomial for the shaded area of each figure.

83.

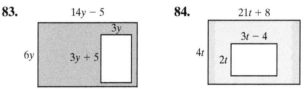

$14y - 5$, $3y$, $6y$, $3y + 5$

84.

$21t + 8$, $3t - 4$, $4t$, $2t$

For each figure, determine what the missing number must be in order for the figure to have the given area.

85. Area is $x^2 + 8x + 15$ **86.** Area is $x^2 + 7x + 10$

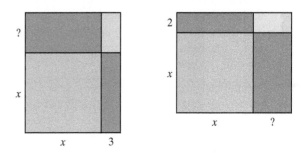

87. A box with a square bottom and no top is to be made from a 12-in.-square piece of cardboard. Squares with side x are cut out of the corners and the sides are folded up. Find the polynomials for the volume and the outside surface area of the box.

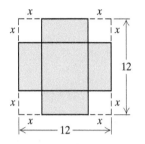

88. Find a polynomial for the volume of the solid shown below.

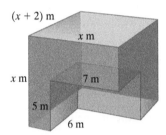

89. An open wooden box is a cube with side x cm. The box, including its bottom, is made of wood that is 1 cm thick. Find a polynomial for the interior volume of the cube.

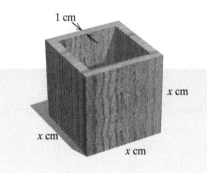

90. A side of a cube is $(x + 2)$ cm long. Find a polynomial for the volume of the cube.

91. A sphere of radius x is enclosed in a plastic cube that is just large enough for the sphere.
  a) Find a polynomial for the amount of air that is in the cube outside the sphere.
  b) The diameter of a baseball is 2.9 in. If a baseball is enclosed in the plastic cube, how much air is in the box?

92. A rectangular garden is twice as long as it is wide and is surrounded by a sidewalk that is 4 ft wide.

The area of the sidewalk is 256 ft². Find the dimensions of the garden.

Compute and simplify.

93. $(x + 3)(x + 6) + (x + 3)(x + 6)$

Aha! **94.** $(x - 2)(x - 7) - (x - 7)(x - 2)$

95. $(x + 5)^2 - (x - 3)^2$

96. $(x + 2)(x + 4)(x - 5)$

97. $(x - 3)^3$

Aha! **98.** Extend the pattern and simplify
$$(x - a)(x - b)(x - c)(x - d) \cdots (x - z).$$

99. Use a graphing calculator to check your answers to Exercises 21, 43, and 55. Use graphs, tables, or both, as directed by your instructor.

YOUR TURN ANSWERS: SECTION 4.5

1. $-54t^4$ **2.** $15a^4 - 3a^2 + 33a$
3. $10y^6 - 40y^5 + 5y^4 - 25y^3$ **4.** $5x^2 + 14x - 3$
5. $2t^3 + 5t^2 + 2t + 15$ **6.** $3x^4 - 13x^3 + 4x^2 + 5x - 20$
7. $15y^4 - 7y^3 - y^2 + 7y + 2$

Quick Quiz: Sections 4.1–4.5

1. Multiply: $-2x^4(3x^5 - 8x + 12)$. [4.5]

2. Add: $(2x^4 + 7x^2 - 10x) + (8x^4 - 7x^2 - 5x + 1)$. [4.4]

3. Subtract: $(6a^2 - 8a - 3) - (-a^2 + 2a - 1)$. [4.4]

4. Simplify: $(x^2y)(5xy)^3$. [4.1]

5. Classify $x^7 + x - 3$ as a monomial, a binomial, a trinomial, or a polynomial with no special name. [4.3]

Prepare to Move On

Simplify.

1. 0.7^2 [1.8] **2.** $(7x^3)^2$ [4.1]

3. $\left(-\dfrac{3}{2}\right)^2$ [1.8] **4.** $2(3x)\left(\dfrac{1}{4}\right)$ [1.8]

5. $2 \cdot 5a \cdot 0.2$ [1.8] **6.** $\left(\dfrac{1}{3}t^4\right)^2$ [4.1]

4.6 Special Products

A. Products of Two Binomials   **B.** Multiplying Sums and Differences of Two Terms
C. Squaring Binomials   **D.** Multiplications of Various Types

Patterns that we note in certain polynomial products allow us to compute such products quickly.

A. Products of Two Binomials

We can find the product $(x + 5)(x + 4)$ by using the distributive law a total of three times. Note that each term in $x + 5$ is multiplied by each term in $x + 4$:

$$(x + 5)(x + 4) = x \cdot x + x \cdot 4 + 5 \cdot x + 5 \cdot 4$$
$$= x^2 + 4x + 5x + 20$$
$$= x^2 + 9x + 20.$$

Note that $x \cdot x$ is found by multiplying the *First* terms of the binomials, $x \cdot 4$ is found by multiplying the *Outer* terms of the two binomials, $5 \cdot x$ is the product of the *Inner* terms of the two binomials, and $5 \cdot 4$ is the product of the *Last* terms of the binomials:

$$\begin{array}{cccc} \text{First} & \text{Outer} & \text{Inner} & \text{Last} \\ \text{terms} & \text{terms} & \text{terms} & \text{terms} \end{array}$$
$$(x + 5)(x + 4) = x \cdot x + 4 \cdot x + 5 \cdot x + 5 \cdot 4.$$

To remember this method for multiplying, we use the initials **FOIL**.

THE FOIL METHOD

To multiply two binomials, $A + B$ and $C + D$, multiply the First terms AC, the Outer terms AD, the Inner terms BC, and then the Last terms BD. Then combine like terms, if possible.

$$(A + B)(C + D) = AC + AD + BC + BD$$

1. Multiply First terms: AC.
2. Multiply Outer terms: AD.
3. Multiply Inner terms: BC.
4. Multiply Last terms: BD.
      ↓
   FOIL

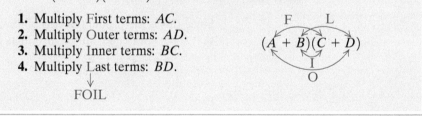

Because addition is commutative, the individual multiplications can be performed in any order. Both FLOI and FIOL yield the same result as FOIL, but FOIL is most easily remembered and most widely used.

EXAMPLE 1   Form an equivalent expression by multiplying: $(x + 8)(x^2 + 5)$.

SOLUTION

$$
\begin{array}{cccc}
\text{F} & \text{L} & \text{F} & \text{O} \quad \text{I} \quad \text{L}
\end{array}
$$

$$
(x + 8)(x^2 + 5) = x^3 + 5x + 8x^2 + 40 \qquad \text{There are no like terms.}
$$
$$
= x^3 + 8x^2 + 5x + 40 \qquad \text{Writing in descending order}
$$

1. Form an equivalent expression by multiplying:

$(t + 3)(t^2 - 5)$.

YOUR TURN

After multiplying, remember to combine any like terms.

EXAMPLE 2   Multiply to form an equivalent expression.

a) $(x + 7)(x + 4)$ **b)** $(y + 3)(y - 2)$

c) $(4t^3 + 5t)(3t^2 - 2)$ **d)** $(3 - 4x)(7 - 5x^3)$

SOLUTION

a) $(x + 7)(x + 4) = x^2 + 4x + 7x + 28 \qquad$ Using FOIL
$$
= x^2 + 11x + 28 \qquad \text{Combining like terms}
$$

b) $(y + 3)(y - 2) = y^2 - 2y + 3y - 6 \qquad$ Using FOIL
$$
= y^2 + y - 6
$$

c) $(4t^3 + 5t)(3t^2 - 2) = 12t^5 - 8t^3 + 15t^3 - 10t \qquad$ Using FOIL and the rules for exponents
$$
= 12t^5 + 7t^3 - 10t
$$

d) $(3 - 4x)(7 - 5x^3) = 21 - 15x^3 - 28x + 20x^4$
$$
= 21 - 28x - 15x^3 + 20x^4
$$

2. Multiply to form an equivalent expression: $(x - 1)(x - 9)$.

In general, if the original binomials are written in *ascending* order, the answer is also written that way.

YOUR TURN

B. Multiplying Sums and Differences of Two Terms

Consider the product of the sum and the difference of the same two terms, such as

$$(x + 5)(x - 5).$$

Since this is the product of two binomials, we can use FOIL. In doing so, we find that the "outer" and "inner" products are opposites:

$$
(x + 5)(x - 5) = x^2 - 5x + 5x - 25 \qquad \text{The "outer" and "inner" terms "drop out." Their sum is zero.}
$$
$$
= x^2 - 25.
$$

Because opposites always add to zero, for products like $(x + 5)(x - 5)$, we write the result directly without using FOIL.

THE PRODUCT OF A SUM AND A DIFFERENCE

The product of the sum and the difference of the same two terms is the square of the first term minus the square of the second term:

$$(A + B)(A - B) = \underline{A^2 - B^2}.$$

This is called a *difference of squares*.

Student Notes

Multiplying the binomials in Example 3 using FOIL may reinforce your understanding. In each case, the "outer" and "inner" products are opposites and add to 0. Using $(A + B)(A - B) = A^2 - B^2$ simply skips that step.

3. Multiply: $(y - 10)(y + 10)$.

EXAMPLE 3 Multiply.

a) $(x + 4)(x - 4)$ b) $(5 + 2w)(5 - 2w)$ c) $(3a^4 - 5)(3a^4 + 5)$

SOLUTION

$$(A + B)(A - B) = A^2 - B^2$$

a) $(x + 4)(x - 4) = x^2 - 4^2$ Saying the words can help: "The square of the first term, x^2, minus the square of the second, 4^2"

$$= x^2 - 16 \quad \text{Simplifying}$$

b) $(5 + 2w)(5 - 2w) = 5^2 - (2w)^2$

$$= 25 - 4w^2 \quad \text{Squaring both 5 and } 2w$$

c) $(3a^4 - 5)(3a^4 + 5) = (3a^4)^2 - 5^2$

$$= 9a^8 - 25 \quad (3a^4)^2 = 3^2 \cdot (a^4)^2 = 9 \cdot a^{4 \cdot 2} = 9a^8$$

YOUR TURN

C. Squaring Binomials

Consider the square of a binomial, such as $(x + 3)^2$. This can be expressed as $(x + 3)(x + 3)$. Since this is the product of two binomials, we can use FOIL. But again, this type of product occurs so often that a faster method has been developed. Look for a pattern in the following:

a) $(x + 3)^2 = (x + 3)(x + 3)$

$$= x^2 + 3x + 3x + 9$$
$$= x^2 + 6x + 9;$$

b) $(a - 7)^2 = (a - 7)(a - 7)$

$$= a^2 - 7a - 7a + 49$$
$$= a^2 - 14a + 49.$$

Perhaps you noticed that in each product the "outer" product and the "inner" product are identical. The other two terms, the "first" product and the "last" product, are squares.

> **THE SQUARE OF A BINOMIAL**
>
> The square of a binomial is the square of the first term, plus twice the product of the two terms, plus the square of the last term:
>
> $$(A + B)^2 = A^2 + 2AB + B^2;$$
> $$(A - B)^2 = A^2 - 2AB + B^2.$$
>
> These are called *perfect-square trinomials.**

EXAMPLE 4 Write an equivalent expression for each square of a binomial.

a) $(x + 7)^2$ b) $(t - 5)^2$
c) $(3a + 0.4)^2$ d) $(5x - 3x^4)^2$

---

*Another name for these is *trinomial squares*.

Student Notes

Note that each of the expressions in Example 4 is a product of two binomials; for example, $(x + 7)^2 = (x + 7)(x + 7)$. You may wish to also multiply using FOIL to reinforce your understanding that the "outer" product and the "inner" product are identical.

$$(A + B)^2 = A^2 + 2 \cdot A \cdot B + B^2$$

a) $(x + 7)^2 = x^2 + 2 \cdot x \cdot 7 + 7^2$ Saying the words can help: "The square of the first term, x^2, plus twice the product of the terms, $2 \cdot x \cdot 7$, plus the square of the second term, 7^2"

$$= x^2 + 14x + 49$$

b) $(t - 5)^2 = t^2 - 2 \cdot t \cdot 5 + 5^2$
$$= t^2 - 10t + 25$$

c) $(3a + 0.4)^2 = (3a)^2 + 2 \cdot 3a \cdot 0.4 + 0.4^2$
$$= 9a^2 + 2.4a + 0.16$$

d) $(5x - 3x^4)^2 = (5x)^2 - 2 \cdot 5x \cdot 3x^4 + (3x^4)^2$
$$= 25x^2 - 30x^5 + 9x^8$$ Using the rules for exponents

4. Multiply: $(x^2 - 3)^2$.

YOUR TURN

CAUTION! Although the square of a product is the product of the squares, the square of a sum is *not* the sum of the squares. That is, $(AB)^2 = A^2B^2$, but

The term $2AB$ is missing.

$$(A + B)^2 \ne A^2 + B^2.$$

To confirm this inequality, note that

$$(7 + 5)^2 = 12^2 = 144,$$

whereas

$$7^2 + 5^2 = 49 + 25 = 74, \quad \text{and} \quad 74 \ne 144.$$

Geometrically, $(A + B)^2$ can be viewed as the area of a square with sides of length $A + B$:

$$(A + B)(A + B) = (A + B)^2.$$

This is equal to the sum of the areas of the four smaller regions:

$$A^2 + AB + AB + B^2 = A^2 + 2AB + B^2.$$

Thus,

$$(A + B)^2 = A^2 + 2AB + B^2.$$

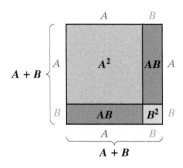

Note that the areas A^2 and B^2 do not fill the area $(A + B)^2$. Two additional areas, AB and AB, are needed.

Check Your
UNDERSTANDING

For each product given, answer the following questions.

a) Which of the following patterns could be used to find the product?

$$(A + B)(A - B) = A^2 - B^2,$$
$$(A + B)^2 = A^2 + 2AB + B^2,$$
$$(A - B)^2 = A^2 - 2AB + B^2$$

b) What expression corresponds to A? What expression corresponds to B?

1. $(x - 2)^2$

2. $(4y + 7)(4y - 7)$

3. $(3a + \frac{1}{4})^2$

4. $(2x - 3)(2x - 3)$

5. $(2p - 5)(2p + 5)$

EXPLORING 🔍 THE CONCEPT

Match each of the following squares with the appropriate square of a binomial and perfect-square trinomial.

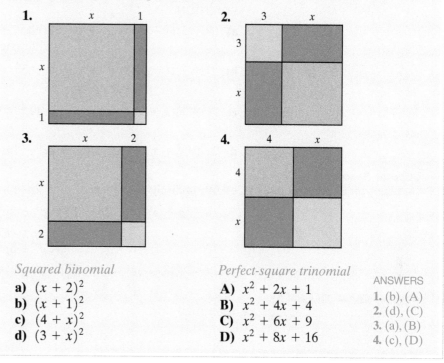

Squared binomial

a) $(x + 2)^2$
b) $(x + 1)^2$
c) $(4 + x)^2$
d) $(3 + x)^2$

Perfect-square trinomial

A) $x^2 + 2x + 1$
B) $x^2 + 4x + 4$
C) $x^2 + 6x + 9$
D) $x^2 + 8x + 16$

ANSWERS
1. (b), (A)
2. (d), (C)
3. (a), (B)
4. (c), (D)

D. Multiplications of Various Types

Recognizing patterns often helps when new problems are encountered. To simplify a new multiplication problem, always examine what type of product it is so that the best method for finding that product can be used. To do this, ask yourself questions similar to the following.

> **MULTIPLYING TWO POLYNOMIALS**
>
> **1.** Is the multiplication the product of a monomial and a polynomial? If so, multiply each term of the polynomial by the monomial.
>
> **2.** Is the multiplication the product of two binomials? If so:
>
> **a)** Is it the product of the sum and the difference of the *same* two terms? If so, use the pattern
> $$(A + B)(A - B) = A^2 - B^2.$$
>
> **b)** Is the product the square of a binomial? If so, use the pattern
> $$(A + B)(A + B) = (A + B)^2 = A^2 + 2AB + B^2$$
> or
> $$(A - B)(A - B) = (A - B)^2 = A^2 - 2AB + B^2.$$
>
> **c)** If neither (a) nor (b) applies, use FOIL.
>
> **3.** Is the multiplication the product of two polynomials other than those above? If so, multiply each term of one by every term of the other. Use columns if you wish.

EXAMPLE 5 Multiply.

a) $(x + 3)(x - 3)$ **b)** $(t + 7)(t - 5)$
c) $(x + 7)(x + 7)$ **d)** $2x^3(9x^2 + x - 7)$
e) $(p + 3)(p^2 + 2p - 1)$ **f)** $\left(3x - \frac{1}{4}\right)^2$

SOLUTION

a) $(x + 3)(x - 3) = x^2 - 9$ This is the product of the sum and the difference of the same two terms.

b) $(t + 7)(t - 5) = t^2 - 5t + 7t - 35$ Using FOIL
$$= t^2 + 2t - 35$$

c) $(x + 7)(x + 7) = x^2 + 14x + 49$ This is the square of a binomial, $(x + 7)^2$.

d) $2x^3(9x^2 + x - 7) = 18x^5 + 2x^4 - 14x^3$ Multiplying each term of the trinomial by the monomial

Chapter Resource:
Visualizing for Success, p. 293

e) We multiply each term of $p^2 + 2p - 1$ by every term of $p + 3$:

$(p + 3)(p^2 + 2p - 1) = p^3 + 2p^2 - p$ Multiplying by p
$\qquad\qquad\qquad\qquad\quad + 3p^2 + 6p - 3$ Multiplying by 3
$\qquad\qquad\qquad\quad = p^3 + 5p^2 + 5p - 3.$ Combining like terms

f) $\left(3x - \frac{1}{4}\right)^2 = 9x^2 - 2(3x)\left(\frac{1}{4}\right) + \frac{1}{16}$ Squaring a binomial
$\qquad\qquad\quad = 9x^2 - \frac{3}{2}x + \frac{1}{16}$

5. Multiply: $(4x + 7)(4x - 7)$.

YOUR TURN

4.6   EXERCISE SET

FOR EXTRA HELP   MyMathLab®

Vocabulary and Reading Check

Classify each of the following statements as either true or false.

1. FOIL is simply a memory device for finding the product of two binomials.

2. The polynomial $x^2 + 49$ is an example of a perfect-square trinomial.

3. Once FOIL is used, it is always possible to combine like terms.

4. The square of $A + B$ is not the sum of the squares of A and B.

A. Products of Two Binomials

Multiply.

5. $(x^2 + 2)(x + 3)$ **6.** $(x - 5)(x^2 - 6)$

7. $(t^4 - 2)(t + 7)$ **8.** $(n^3 + 8)(n - 4)$

9. $(y + 2)(y - 3)$ **10.** $(a + 2)(a + 2)$

11. $(3x + 2)(3x + 5)$ **12.** $(4x + 1)(2x + 7)$

13. $(5x - 3)(x + 4)$ **14.** $(4x - 5)(4x + 5)$

15. $(3 - 2t)(5 - t)$ **16.** $(7 - a)(4 - 3a)$

17. $(x^2 + 3)(x^2 - 7)$ **18.** $(x^2 + 2)(x^2 - 8)$

B. Multiplying Sums and Differences of Two Terms

Multiply.

19. $\left(p - \frac{1}{4}\right)\left(p + \frac{1}{4}\right)$ **20.** $\left(q + \frac{1}{3}\right)\left(q - \frac{1}{3}\right)$

21. $(x + 0.3)(x - 0.3)$ **22.** $(x - 0.1)(x + 0.1)$

23. $(10x^2 + 3)(10x^2 - 3)$ **24.** $(5 - 4x^5)(5 + 4x^5)$

25. $(1 - 5t^3)(1 + 5t^3)$ **26.** $(x^{10} + 3)(x^{10} - 3)$

C. Squaring Binomials

Find an equivalent expression for the following squares of binomials.

27. $(t - 2)^2$ **28.** $(a - 10)^2$

29. $(x + 10)(x + 10)$ **30.** $(x + 12)(x + 12)$

31. $(3x + 2)^2$ **32.** $(5x + 1)^2$

33. $(1 - 10a)^2$ **34.** $(7 - 6p)^2$

35. $(x^3 + 12)^2$ **36.** $(x^5 - 4)^2$

D. Multiplications of Various Types

Multiply. Try to recognize the type of product before multiplying.

37. $(x^2 + 3)(x^3 - 1)$ **38.** $(x^4 - 3)(2x + 1)$

39. $(x + 8)(x - 8)$ **40.** $(x + 1)(x - 1)$

41. $(-3n + 2)(n + 7)$

42. $(-m + 5)(2m - 9)$

43. $(x + 3)^2$

44. $(2x - 1)^2$

45. $(7x^3 - 1)^2$

46. $(5x^3 + 2)^2$

47. $(9a^3 + 1)(9a^3 - 1)$

48. $(t^2 - 0.2)(t^2 + 0.2)$

49. $(x^4 + 0.1)(x^4 - 0.1)$

50. $(a^3 + 5)(a^3 - 5)$

51. $\left(t - \frac{3}{4}\right)\left(t + \frac{3}{4}\right)$

52. $\left(m - \frac{2}{3}\right)\left(m + \frac{2}{3}\right)$

53. $(1 - 3t)(1 + 5t^2)$

54. $(1 + 2t)(1 - 3t^2)$

55. $\left(a - \frac{2}{5}\right)^2$

56. $\left(t - \frac{1}{5}\right)^2$

57. $(t^4 + 3)^2$

58. $(a^3 + 6)^2$

59. $(5x - 9)(9x + 5)$

60. $(7x - 2)(2x - 7)$

61. $7n^3(2n^2 - 1)$

62. $5m^3(4 - 3m^2)$

63. $(a - 3)(a^2 + 2a - 4)$

64. $(x^2 - 5)(x^2 + x - 1)$

65. $(7 - 3x^4)(7 - 3x^4)$

66. $(x - 4x^3)^2$

67. $(2 - 3x^4)^2$

68. $(5 - 2t^3)^2$

69. $(5t + 6t^2)^2$

70. $(3p^2 - p)^2$

71. $5x(x^2 + 6x - 2)$

72. $6x(-x^5 + 6x^2 + 9)$

73. $(q^5 + 1)(q^5 - 1)$

74. $(p^4 + 2)(p^4 - 2)$

75. $3t^2(5t^3 - t^2 + t)$

76. $-5x^3(x^2 + 8x - 9)$

77. $(6x^4 - 3x)^2$

78. $(8a^3 + 5)(8a^3 - 5)$

79. $(9a + 0.4)(2a^3 + 0.5)$

80. $(2a - 0.7)(8a^3 - 0.5)$

81. $\left(\frac{1}{5} - 6x^4\right)\left(\frac{1}{5} + 6x^4\right)$

82. $\left(3 + \frac{1}{2}t^5\right)\left(3 + \frac{1}{2}t^5\right)$

83. $(a + 1)(a^2 - a + 1)$

84. $(x - 5)(x^2 + 5x + 25)$

In each of Exercises 85–92, find the total area of all shaded rectangles.

85.
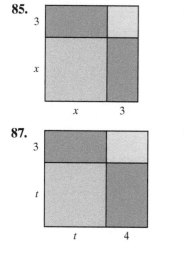

86.
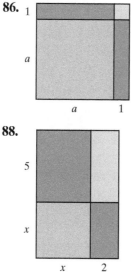

87.

88.

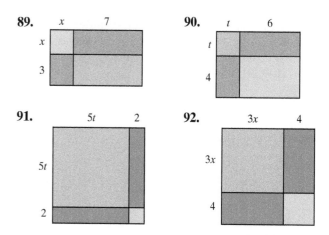

89.

90.

91.

92.

Draw and label rectangles similar to those in Exercises 85–92 to illustrate each of the following.

93. $(x + 5)^2$

94. $(x + 8)^2$

95. $(3 + x)^2$

96. $(7 + t)^2$

97. $(2t + 1)^2$

98. $(3a + 2)^2$

99. Kristi feels that since she can find the product of any two binomials using FOIL, she needn't study the other special products. What advice would you give her?

100. Under what conditions is the product of two binomials a binomial?

Skill Review

Solve.

101. *Energy Use.* Bailey's refrigerator, upright freezer, and washing machine together use 1200 kilowatt-hours per year (kWh/year) of electricity. Her refrigerator uses three times as much energy as her washing machine, and her freezer uses six times as much energy as her washing machine. How much energy is used by each appliance? [2.5]

Data: ftc.gov/appliancedata

102. *Energy Credit.* During some years, U.S. tax law allowed a tax credit of 30% of the price of a geothermal heat pump, with a maximum credit of $2000. For what prices of a heat pump could a 30% credit be taken? [2.7]

Solve. [2.3]

103. $3ab = c$, for a

104. $ax - by = c$, for x

Synthesis

105. By writing $19 \cdot 21$ as $(20 - 1)(20 + 1)$, Blair can find the product mentally. How do you think he does this?

106. The product $(A + B)^2$ can be regarded as the sum of the areas of four regions (as shown following Example 4). How might one visually represent $(A + B)^3$? Why?

Multiply.

Aha! **107.** $(4x^2 + 9)(2x + 3)(2x - 3)$

108. $(9a^2 + 1)(3a - 1)(3a + 1)$

Aha! **109.** $(3t - 2)^2(3t + 2)^2$

110. $(5a + 1)^2(5a - 1)^2$

111. $(t^3 - 1)^4(t^3 + 1)^4$

112. $(32.41x + 5.37)^2$

Calculate as the difference of squares.

113. 18×22 [*Hint:* $(20 - 2)(20 + 2)$.]

114. 93×107

Solve.

115. $(x + 2)(x - 5) = (x + 1)(x - 3)$

116. $(2x + 5)(x - 4) = (x + 5)(2x - 4)$

Find a polynomial for the total shaded area in each figure.

117.

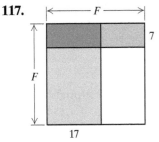

118.

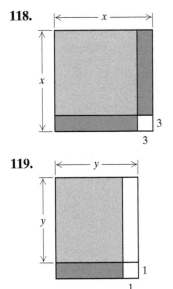

119.

120. Find $(10 - 2x)^2$ by subtracting the white areas from 10^2.

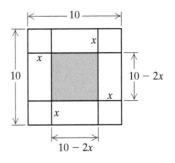

121. Find $(y - 2)^2$ by subtracting the white areas from y^2.

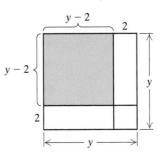

122. Find three consecutive integers for which the sum of the squares is 65 more than three times the square of the smallest integer.

123. Use a graphing calculator and the method developed in Section 4.5 to check your answers to Exercises 22, 47, and 83.

YOUR TURN ANSWERS: SECTION 4.6

1. $t^3 + 3t^2 - 5t - 15$ **2.** $x^2 - 10x + 9$ **3.** $y^2 - 100$
4. $x^4 - 6x^2 + 9$ **5.** $16x^2 - 49$

Quick Quiz: Sections 4.1–4.6

Multiply.

1. $(10a^3 b^{10})(-2ab^2)$ [4.1]   **2.** $n \cdot n^{13} \cdot n^0$ [4.1]

3. $(x - 3)(4x^2 - x + 2)$ [4.5]

4. $(10x^5 + 1)(10x^5 - 1)$ [4.6]

5. $(3a + 4)^2$ [4.6]

Prepare to Move On

Evaluate. [1.1]

1. $\dfrac{2m}{n}$, for $m = 10$ and $n = 5$

2. $x - y$, for $x = 37$ and $y = 18$

Combine like terms. [1.6]

3. $2a - 6c - c + 10a$ **4.** $4y + 7y + 2y - w$

4.7 Polynomials in Several Variables

A. Evaluating Polynomials   **B.** Like Terms and Degree   **C.** Addition and Subtraction
D. Multiplication

Thus far, the polynomials that we have studied have had only one variable. Polynomials such as

$$5x + x^2y - 3y + 7, \quad 9ab^2c - 2a^3b^2 + 8a^2b^3, \quad \text{and} \quad 4m^2 - 9n^2$$

contain two or more variables. In this section, we will add, subtract, multiply, and evaluate such **polynomials in several variables**.

A. Evaluating Polynomials

To evaluate a polynomial in two or more variables, we substitute numbers for the variables. Then we compute, using the rules for order of operations.

EXAMPLE 1   Evaluate the polynomial $4 + 3x + xy^2 + 8x^3y^3$ for $x = -2$ and $y = 5$.

SOLUTION   We substitute -2 for x and 5 for y:

$$4 + 3x + xy^2 + 8x^3y^3 = 4 + 3(-2) + (-2) \cdot 5^2 + 8(-2)^3 \cdot 5^3$$
$$= 4 - 6 - 50 - 8000 = -8052.$$

1. Evaluate the polynomial $x^2y - 3y^2 - 2xy$ for $x = 4$ and $y = -3$.

YOUR TURN

EXAMPLE 2   *Surface Area of a Right Circular Cylinder.*   The surface area of a right circular cylinder is given by the polynomial

$$2\pi rh + 2\pi r^2,$$

where h is the height and r is the radius of the base. A 12-oz can has a height of 4.7 in. and a radius of 1.2 in. Approximate its surface area to the nearest tenth of a square inch.

SOLUTION   We evaluate the polynomial for $h = 4.7$ in. and $r = 1.2$ in. If 3.14 is used to approximate π, we have

$$2\pi rh + 2\pi r^2 \approx 2(3.14)(1.2 \text{ in.})(4.7 \text{ in.}) + 2(3.14)(1.2 \text{ in.})^2$$
$$= 2(3.14)(1.2 \text{ in.})(4.7 \text{ in.}) + 2(3.14)(1.44 \text{ in}^2)$$
$$= 35.4192 \text{ in}^2 + 9.0432 \text{ in}^2 = 44.4624 \text{ in}^2.$$

If the π key of a calculator is used, we have

$$2\pi rh + 2\pi r^2 \approx 2(3.141592654)(1.2 \text{ in.})(4.7 \text{ in.})$$
$$+ 2(3.141592654)(1.2 \text{ in.})^2$$
$$\approx 44.48495197 \text{ in}^2.$$

2. Use the formula given in Example 2 to find the surface area of a right circular cylinder with a height of 10 cm and a radius of 3 cm. Round to the nearest tenth of a square centimeter.

Note that the unit in the answer (square inches) is a unit of area. The surface area is about 44.5 in^2 (square inches).

YOUR TURN

B. Like Terms and Degree

Recall that the degree of a monomial is the number of variable factors in the term. For example, the degree of $5x^2$ is 2 because there are two variable factors in $5 \cdot x \cdot x$. Similarly, the degree of $5a^2b^4$ is 6 because there are 6 variable factors in $5 \cdot a \cdot a \cdot b \cdot b \cdot b \cdot b$. Note that 6 can be found by adding the exponents 2 and 4. The degree of a polynomial is the degree of the term of highest degree.

EXAMPLE 3 Identify the degree of each term and the degree of the polynomial

$$9x^2y^3 - 14xy^2z^3 + xy + 4y + 5x^2 + 7.$$

SOLUTION

Term	Degree	Degree of the Polynomial
$9x^2y^3$	5	
$-14xy^2z^3$	6	
xy	2	
$4y$	1	6
$5x^2$	2	
7	0	

3. Identify the degree of each term and the degree of the polynomial $2x^2y + 9y - 6 + 4x^3y^5$.

YOUR TURN

Note in Example 3 that although both xy and $5x^2$ have degree 2, they are *not* like terms. *Like*, or *similar, terms* either have exactly the same variables with exactly the same exponents or are constants. For example,

$8a^4b^7$ and $5b^7a^4$ are like terms

and

-17 and 3 are like terms,

but

$-2x^2y$ and $9xy^2$ are *not* like terms.

As always, combining like terms is based on the distributive law.

EXAMPLE 4 Combine like terms to form equivalent expressions.

a) $9x^2y + 3xy^2 - 5x^2y - xy^2$
b) $7ab - 5ab^2 + 3ab^2 + 6a^3 + 9ab - 11a^3 + b - 1$

SOLUTION

a) $9x^2y + 3xy^2 - 5x^2y - xy^2 = (9 - 5)x^2y + (3 - 1)xy^2$
$= 4x^2y + 2xy^2$ Try to go directly to this step.

b) $7ab - 5ab^2 + 3ab^2 + 6a^3 + 9ab - 11a^3 + b - 1$
$= -5a^3 - 2ab^2 + 16ab + b - 1$ We choose to write descending powers of a. Other, equivalent, forms can also be used.

4. Combine like terms:

$2p^2x - x^2 + 9px^2 - 5p^2x.$

YOUR TURN

Student Notes

Always read the problem carefully. The difference between

$$(-5x^3 - 3y) + (8x^3 + 4x^2)$$

and

$$(-5x^3 - 3y)(8x^3 + 4x^2)$$

is enormous. To avoid wasting time working on an incorrectly copied exercise, be sure to double-check that you have written the correct problem in your notebook.

5. Add: $(3a^2b - b^2 + 7ab) + (4ab - 7a^2b)$.

6. Subtract:

$(12x^3 - 3xy^2 - xy) - (-4xy^2 + xy - 5x^3)$.

7. Multiply:

$(ab - 2a)(5a^2 + 3ab^2 + b^3)$.

C. Addition and Subtraction

The same procedures used for adding or subtracting polynomials in one variable are used to add or subtract polynomials in several variables.

EXAMPLE 5 Add.

a) $(-5x^3 + 3y - 5y^2) + (8x^3 + 4x^2 + 7y^2)$

b) $(5ab^2 - 4a^2b + 5a^3 + 2) + (3ab^2 - 2a^2b + 3a^3b - 5)$

SOLUTION

a) $(-5x^3 + 3y - 5y^2) + (8x^3 + 4x^2 + 7y^2)$

$$= (-5 + 8)x^3 + 4x^2 + 3y + (-5 + 7)y^2 \qquad \text{Try to do this step mentally.}$$

$$= 3x^3 + 4x^2 + 3y + 2y^2$$

b) $(5ab^2 - 4a^2b + 5a^3 + 2) + (3ab^2 - 2a^2b + 3a^3b - 5)$

$$= 8ab^2 - 6a^2b + 5a^3 + 3a^3b - 3$$

YOUR TURN

EXAMPLE 6 Subtract:

$$(4x^2y + x^3y^2 + 3x^2y^3 + 6y) - (4x^2y - 6x^3y^2 + x^2y^2 - 5y).$$

SOLUTION   We find the opposite of $4x^2y - 6x^3y^2 + x^2y^2 - 5y$ and then add.

$$(4x^2y + x^3y^2 + 3x^2y^3 + 6y) - (4x^2y - 6x^3y^2 + x^2y^2 - 5y)$$

$$= 4x^2y + x^3y^2 + 3x^2y^3 + 6y - 4x^2y + 6x^3y^2 - x^2y^2 + 5y$$

$$= 7x^3y^2 + 3x^2y^3 - x^2y^2 + 11y \qquad \text{Combining like terms}$$

YOUR TURN

D. Multiplication

To multiply polynomials in several variables, multiply each term of one polynomial by every term of the other.

EXAMPLE 7 Multiply: $(3x^2y - 2xy + 3y)(xy + 2y)$.

SOLUTION

$$
\begin{array}{r}
3x^2y - 2xy + 3y \\
xy + 2y \\
\hline
6x^2y^2 - 4xy^2 + 6y^2 \qquad \text{Multiplying by } 2y \\
3x^3y^2 - 2x^2y^2 + 3xy^2 \qquad \text{Multiplying by } xy \\
\hline
3x^3y^2 + 4x^2y^2 - xy^2 + 6y^2 \qquad \text{Adding}
\end{array}
$$

YOUR TURN

↳ Check Your UNDERSTANDING

Answer each of the following questions with reference to the polynomial

$$3ax^2 - 2xy^3 - axy + x^3y + 7ax^2.$$

1. What are the variables in the polynomial?

2. How many terms are in the polynomial?

3. Which term has a coefficient of -1?

4. Which terms are of degree 3?

5. List any pairs of like terms.

Using patterns for special products, such as the square of a binomial, can speed up our work.

EXAMPLE 8 Multiply.

a) $(p + 5w)(2p - 3w)$ **b)** $(3x + 2y)^2$
c) $(a^3 - 7a^2b)^2$ **d)** $(3x^2y + 2y)(3x^2y - 2y)$
e) $(-2x^3y^2 + 5t)(2x^3y^2 + 5t)$ **f)** $(2x + 3 - 2y)(2x + 3 + 2y)$

SOLUTION

$$\begin{array}{cccc} \text{F} & \text{O} & \text{I} & \text{L} \end{array}$$

a) $(p + 5w)(2p - 3w) = 2p^2 - 3pw + 10pw - 15w^2$
$$= 2p^2 + 7pw - 15w^2 \quad \text{Combining like terms}$$

$$(A + B)^2 = A^2 + 2 \cdot A \cdot B + B^2$$

b) $(3x + 2y)^2 = (3x)^2 + 2(3x)(2y) + (2y)^2 \quad \text{Squaring a binomial}$
$$= 9x^2 + 12xy + 4y^2$$

$$(A - B)^2 = A^2 - 2 \cdot A \cdot B + B^2$$

c) $(a^3 - 7a^2b)^2 = (a^3)^2 - 2(a^3)(7a^2b) + (7a^2b)^2 \quad \text{Squaring a binomial}$
$$= a^6 - 14a^5b + 49a^4b^2 \quad \text{Using the rules for exponents}$$

$$(A + B)(A - B) = A^2 - B^2$$

d) $(3x^2y + 2y)(3x^2y - 2y) = (3x^2y)^2 - (2y)^2 \quad \begin{array}{l}\text{Multiplying the sum and}\\ \text{the difference of two terms}\end{array}$
$$= 9x^4y^2 - 4y^2 \quad \begin{array}{l}\text{Using the rules for}\\ \text{exponents}\end{array}$$

e) $(-2x^3y^2 + 5t)(2x^3y^2 + 5t) = (5t - 2x^3y^2)(5t + 2x^3y^2) \quad \begin{array}{l}\text{Using the com-}\\ \text{mutative law for}\\ \text{addition twice}\end{array}$
$$= (5t)^2 - (2x^3y^2)^2 \quad \begin{array}{l}\text{Multiplying the sum and}\\ \text{the difference of two}\\ \text{terms}\end{array}$$
$$= 25t^2 - 4x^6y^4$$

$$(A - B)(A + B) = A^2 - B^2$$

f) $(2x + 3 - 2y)(2x + 3 + 2y) = (2x + 3)^2 - (2y)^2 \quad \begin{array}{l}\text{Multiplying}\\ \text{a sum and a}\\ \text{difference}\end{array}$
$$= 4x^2 + 12x + 9 - 4y^2 \quad \begin{array}{l}\text{Squaring a}\\ \text{binomial}\end{array}$$

YOUR TURN

In Example 8, we recognized patterns that might not be obvious, particularly in parts (e) and (f). In part (e), we *can* use FOIL, and in part (f), we *can* use long multiplication, but doing so is slower. By carefully inspecting a problem before "jumping in," we can save ourselves considerable work. At least one instructor refers to this as "working smart" instead of "working hard."*

---

*Thanks to Pauline Kirkpatrick of Wharton County Junior College for this language.

Technology Connection

One way to evaluate the polynomial in Example 1 for $x = -2$ and $y = 5$ is to store -2 to X and 5 to Y and enter the polynomial.

```
-2 → X
              -2
5 → Y
               5
4+3X+XY²+8X³Y³
           -8052
■
```

Evaluate.

1. $3x^2 - 2y^2 + 4xy + x$, for $x = -6$ and $y = 2.3$

2. $a^2b^2 - 8c^2 + 4abc + 9a$, for $a = 11, b = 15$, and $c = -7$

8. Multiply:

$(5x - 2xy)(x^2 - 2y)$.

Chapter Resource:
Collaborative Activity, p. 294

4.7 EXERCISE SET

FOR EXTRA HELP MyMathLab®

⤷ Vocabulary and Reading Check

Consider the polynomial $2x^2y + 7xy^3 + 5x^2y$. *Choose from the list on the right the phrase that describes the expression on the left as it relates to the polynomial.*

1. ____ $2x^2y$ and $5x^2y$
2. ____ x, y
3. ____ 4
4. ____ 3

a) The degree of the polynomial
b) Like terms
c) The number of terms in the polynomial
d) The variables in the polynomial

Elephant	Girth, *g* (in centimeters)	Length, *l* (in centimeters)	Footpad Circumference, *c* (in centimeters)
Male, age 3 yr	244	140	86
Female, age 3 yr	231	135	86
Male, age 25 yr	404	229	130
Female, age 25 yr	366	226	117

Data: "How Much Does That Elephant Weigh?" by Mark MacAllister on fieldtripearth.org

⤷ Concept Reinforcement

Each of the expressions in Exercises 5–10 can be regarded as either (a) the square of a binomial, (b) the product of the sum and the difference of the same two terms, or (c) neither of the above. Select the appropriate choice for each expression.

5. $(3x + 5y)^2$
6. $(4x - 9y)(4x + 9y)$
7. $(5a + 6b)(-6b + 5a)$
8. $(4a - 3b)(4a - 3b)$
9. $(r - 3s)(5r + 3s)$
10. $(2x - 7y)(7y - 2x)$

A. Evaluating Polynomials

Evaluate each polynomial for $x = 5$ *and* $y = -2$.

11. $x^2 - 2y^2 + 3xy$

12. $x^2 + 5y^2 - 4xy$

Evaluate each polynomial for $x = 2$, $y = -3$, *and* $z = -4$.

13. $xy^2z - z$
14. $xy - x^2z + yz^2$

Zoology. *The polynomial* $11.5g + 7.55l + 12.5c - 4016$ *can be used to estimate the weight of an elephant, in kilograms, given the girth g of the elephant at the heart, the length l of the elephant, and the circumference c of the elephant's footpad. Use the formula and the data from the following table for Exercises 15 and 16.*

15. Estimate the weight of the 3-year-old female elephant described in the table. Round your estimate to the nearest kilogram.

16. Estimate the weight of the 25-year-old male elephant described in the table. Round your estimate to the nearest kilogram.

Surface Area of a Silo. *A silo is a structure that is shaped like a right circular cylinder with a half sphere on top. The surface area of a silo of height h and radius r (including the area of the base) is given by the polynomial* $2\pi rh + \pi r^2$.

17. A coffee grinder is shaped like a silo, with a height of 7 in. and a radius of $1\frac{1}{2}$ in. Find the surface area of the coffee grinder. Use 3.14 for π.

18. A $1\frac{1}{2}$-oz bottle of roll-on deodorant has a height of 4 in. and a radius of $\frac{3}{4}$ in. Find the surface area of the bottle if the bottle is shaped like a silo. Use 3.14 for π.

Altitude of a Launched Object. *The altitude of an object, in meters, is given by the polynomial*

$$h + vt - 4.9t^2,$$

where h is the height, in meters, at which the launch occurs, v is the initial upward speed (or velocity), in meters per second, and t is the number of seconds for which the object is airborne.

19. A bocce ball is thrown upward with an initial speed of 18 m/sec by a person atop the Leaning Tower of Pisa, which is 50 m above the ground. How high will the ball be 2 sec after it has been thrown?

50 m

20. A golf ball is launched upward with an initial speed of 30 m/sec by a golfer on the third level of The Golf Club at Chelsea Piers, Manhattan, which is 10 m above the ground. How high above the ground will the ball be after 3 sec?

B. Like Terms and Degree

Identify the degree of each term of each polynomial. Then find the degree of the polynomial.

21. $3x^2y - 5xy + 2y^2 - 11$

22. $xy^3 + 7x^3y^2 - 6xy^4 + 2$

23. $7 - abc + a^2b + 9ab^2$

24. $3p - pq - 7p^2q^3 - 8pq^6$

Combine like terms.

25. $3r + s - r - 7s$

26. $9a + b - 8a - 5b$

27. $5xy^2 - 2x^2y + x + 3x^2$

28. $m^3 + 2m^2n - 3m^2 + 3mn^2$

29. $6u^2v - 9uv^2 + 3vu^2 - 2v^2u + 11u^2$

30. $3x^2 + 6xy + 3y^2 - 5x^2 - 10xy$

31. $5a^2c - 2ab^2 + a^2b - 3ab^2 + a^2c - 2ab^2$

32. $3s^2t + r^2t - 9ts^2 - st^2 + 5t^2s - 7tr^2$

C. Addition and Subtraction

Add or subtract, as indicated.

33. $(6x^2 - 2xy + y^2) + (5x^2 - 8xy - 2y^2)$

34. $(7r^3 + rs - 5r^2) - (2r^3 - 3rs + r^2)$

35. $(3a^4 - 5ab + 6ab^2) - (9a^4 + 3ab - ab^2)$

36. $(2r^2t - 5rt + rt^2) - (7r^2t + rt - 5rt^2)$

Aha! **37.** $(5r^2 - 4rt + t^2) + (-6r^2 - 5rt - t^2) + (-5r^2 + 4rt - t^2)$

38. $(2x^2 - 3xy + y^2) + (-4x^2 - 6xy - y^2) + (4x^2 + 6xy + y^2)$

39. $(x^3 - y^3) - (-2x^3 + x^2y - xy^2 + 2y^3)$

40. $(a^3 + b^3) - (-5a^3 + 2a^2b - ab^2 + 3b^3)$

41. $(2y^4x^3 - 3y^3x) + (5y^4x^3 - y^3x) - (9y^4x^3 - y^3x)$

42. $(5a^2b - 7ab^2) - (3a^2b + ab^2) + (a^2b - 2ab^2)$

43. Subtract $7x + 3y$ from the sum of $4x + 5y$ and $-5x + 6y$.

44. Subtract $5a + 2b$ from the sum of $2a + b$ and $3a - 4b$.

D. Multiplication

Multiply.

45. $(4c - d)(3c + 2d)$ **46.** $(5x + y)(2x - 3y)$

47. $(xy - 1)(xy + 5)$

48. $(ab + 3)(ab - 5)$

49. $(2a - b)(2a + b)$

50. $(a - 3b)(a + 3b)$

51. $(5rt - 2)(4rt - 3)$

52. $(3xy - 1)(4xy + 2)$

53. $(m^3n + 8)(m^3n - 6)$

54. $(9 - u^2v^2)(2 - u^2v^2)$

55. $(6x - 2y)(5x - 3y)$

56. $(7a - 6b)(5a + 4b)$

57. $(pq + 0.1)(-pq + 0.1)$

58. $(rt + 0.2)(-rt + 0.2)$

59. $(x + h)^2$

60. $(a - r)^2$

61. $(4a - 5b)^2$

62. $(2x + 5y)^2$

63. $(ab + cd^2)(ab - cd^2)$

64. $(p^3 - 5q)(p^3 + 5q)$

65. $(2xy + x^2y + 3)(xy + y^2)$

66. $(5cd - c^2 - d^2)(2c - c^2d)$

Aha! **67.** $(a + b - c)(a + b + c)$

68. $(x + y + 2z)(x + y - 2z)$

69. $[a + b + c][a - (b + c)]$

70. $(a + b + c)(a - b - c)$

Find the total area of each figure.

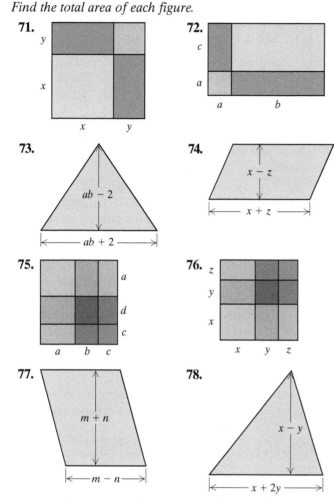

71.

72.

73.

74.

75.

76.

77.

78.

Draw and label rectangles similar to those in Exercises 71, 72, 75, and 76 to illustrate each product.

79. $(r + s)(u + v)$ **80.** $(m + r)(n + v)$

81. $(a + b + c)(a + d + f)$

82. $(r + s + t)^2$

83. Is it possible for a polynomial in 4 variables to have a degree less than 4? Why or why not?

84. A fourth-degree monomial is multiplied by a third-degree monomial. What is the degree of the product? Explain your reasoning.

Skill Review

Simplify. [1.8]

85. $-16 + 20 \div 2^2 \cdot 5 - 3$

86. $2 - (3 - 8)^2 \div (-5)$

87. $2[3 - 4(5 - 6)^2 - 1] + 10$

88. $-6|-2 - (-1)| + 3(-5)$

89. $2a - 3(-x - 5a + 7)$

90. $4(y + 7) - (y - 2) + (2y - 9)$

Synthesis

91. The concept of "leading term" was intentionally not discussed in this section. Why not?

92. Explain how it is possible for the sum of two trinomials in several variables to be a binomial in one variable.

Find a polynomial for the shaded area in each figure. (Leave results in terms of π where appropriate.)

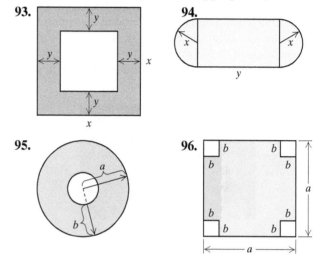

93.

94.

95.

96.

97. Find a polynomial for the total volume of the following figure.

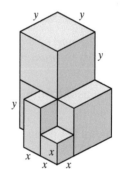

98. Find the shaded area in the following figure using each of the approaches given below. Then check that both answers match.

a) Find the shaded area by subtracting the area of the unshaded square from the total area of the figure.

b) Find the shaded area by adding the areas of the three shaded rectangles.

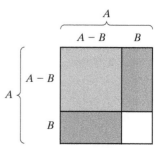

Find a polynomial for the surface area of each solid object shown. (Leave results in terms of π.)

99.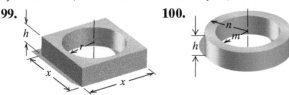

100.

101. The observatory at Danville University is shaped like a silo that is 40 ft high and 30 ft wide (see Exercise 17). The Heavenly Bodies Astronomy Club is to paint the exterior of the observatory using paint that covers 250 ft² per gallon. How many gallons should they purchase? Explain your reasoning.

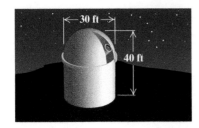

102. Multiply: $(x + a)(x - b)(x - a)(x + b)$.

Spreadsheet applications allow values for cells to be calculated from values in other cells. For example, if the cell C1 contains the formula

$$= A1 + 2 * B1,$$

the value in C1 will be the sum of the value in A1 and twice the value in B1. This formula is a polynomial in the two variables A1 and B1.

103. The cell D4 contains the formula

$$= 2 * A4 + 3 * B4.$$

What is the value in D4 if the value in A4 is 5 and the value in B4 is 10?

104. The cell D6 contains the formula

$$= A1 - 0.2 * B1 + 0.3 * C1.$$

What is the value in D6 if the value in A1 is 10, the value in B1 is −3, and the value in C1 is 30?

105. *Interest Compounded Annually.* An amount of money P that is invested at the yearly interest rate r grows to the amount $P(1 + r)^t$ after t years. Find a polynomial that can be used to determine the amount to which P will grow after 2 years.

106. *Yearly Depreciation.* An investment P that drops in value at the yearly rate r drops in value to

$$P(1 - r)^t$$

after t years. Find a polynomial that can be used to determine the value to which P has dropped after 2 years.

107. Suppose that $10,400 is invested at 4.5%, compounded annually. How much is in the account at the end of 5 years? (See Exercise 105.)

108. A $90,000 investment in computer hardware is depreciating at a yearly rate of 12.5%. How much is the investment worth after 4 years? (See Exercise 106.)

109. *Research.* Find the interest rates on several accounts offered by your bank or another local bank.

    a) Suppose that $10,000 is invested in each account. Compare the amounts in each account at the end of 10 years if the interest were compounded annually. (See Exercise 105.)

    b) Find the formula used to calculate the amount in an account if the interest is compounded more often than once a year. Use the compounding period from the bank you chose to compare amounts in each account after 10 years.

YOUR TURN ANSWERS: SECTION 4.7

1. −51 **2.** 245.0 cm² (using the π key)
3. Degrees of terms: 3, 1, 0, 8; degree of polynomial: 8
4. $-3p^2x - x^2 + 9px^2$
5. $-4a^2b - b^2 + 11ab$ **6.** $17x^3 + xy^2 - 2xy$
7. $5a^3b + 3a^2b^3 + ab^4 - 10a^3 - 6a^2b^2 - 2ab^3$
8. $5x^3 - 10xy - 2x^3y + 4xy^2$

Quick Quiz: Sections 4.1–4.7

1. Evaluate $-2x^3 - 3x^2 - 5x + 17$ for $x = -2$. [4.3]

2. Subtract:
$(3.1x^2 + 2.7x - 1.1) - (-x^2 + 5.6x - 7.2)$. [4.4]

3. Multiply: $(x + 3)(5x^3 - x + 4)$. [4.5]

4. Multiply: $(a^2 + 10)(a^2 - 10)$. [4.6]

5. Evaluate $xz - z^2 - 2y$ for $x = -2, y = 3$, and $z = -5$. [4.7]

Prepare to Move On

Subtract. [4.4]

1. $x^2 - 3x - 7$
   $-(\quad\quad 5x - 3)$

2. $2x^3 \quad\quad - x + 3$
   $-(\quad\quad x^2 \quad - 1)$

3. $3x^2 + x + 5$
   $-(3x^2 + 3x)$

4. $4x^3 - 3x^2 + x$
   $-(4x^3 - 8x^2)$

4.8 Division of Polynomials

A. Dividing by a Monomial   **B.** Dividing by a Binomial

In this section, we study division of polynomials. Polynomial division is similar to division in arithmetic, and the same terms are used.

$$\text{Divisor} \longrightarrow 3\overline{)7} \begin{array}{l} 2 \longleftarrow \text{Quotient} \\ 7 \longleftarrow \text{Dividend} \\ \underline{-6} \\ 1 \longleftarrow \text{Remainder} \end{array}$$

A. Dividing by a Monomial

We first consider division by a monomial. When dividing a monomial by a monomial, we subtract exponents when bases are the same. For example,

$$\frac{42a^2b^5}{-3ab^2} = \frac{42}{-3}a^{2-1}b^{5-2} \qquad \text{Recall that } a^m/a^n = a^{m-n}.$$

$$= -14ab^3.$$

> **CAUTION!** The coefficients are divided but the exponents are subtracted.

To divide a polynomial by a monomial, we note that since

$$\frac{A}{C} + \frac{B}{C} = \frac{A+B}{C},$$

it follows that

$$\frac{A+B}{C} = \frac{A}{C} + \frac{B}{C}. \qquad \text{Switching the left side and the right side of the equation}$$

This is actually how we perform divisions like $86 \div 2$:

$$\frac{86}{2} = \frac{80+6}{2} = \frac{80}{2} + \frac{6}{2} = 40 + 3.$$

Similarly, to divide a polynomial by a monomial, we divide each term by the monomial:

$$\frac{80x^5 + 6x^7}{2x^3} = \frac{80x^5}{2x^3} + \frac{6x^7}{2x^3}$$

$$= \frac{80}{2}x^{5-3} + \frac{6}{2}x^{7-3} \qquad \begin{array}{l}\text{Dividing coefficients and} \\ \text{subtracting exponents}\end{array}$$

$$= 40x^2 + 3x^4.$$

EXAMPLE 1   Divide $x^4 + 15x^3 - 6x^2$ by $3x$.

SOLUTION   We divide each term of $x^4 + 15x^3 - 6x^2$ by $3x$:

$$\frac{x^4 + 15x^3 - 6x^2}{3x} = \frac{x^4}{3x} + \frac{15x^3}{3x} - \frac{6x^2}{3x}$$

$$= \frac{1}{3}x^{4-1} + \frac{15}{3}x^{3-1} - \frac{6}{3}x^{2-1} \qquad \begin{array}{l}\text{Dividing coefficients} \\ \text{and subtracting} \\ \text{exponents}\end{array}$$

$$= \frac{1}{3}x^3 + 5x^2 - 2x. \qquad \text{This is the quotient.}$$

To check, we multiply our answer by $3x$, using the distributive law:

$$3x\left(\frac{1}{3}x^3 + 5x^2 - 2x\right) = 3x \cdot \frac{1}{3}x^3 + 3x \cdot 5x^2 - 3x \cdot 2x$$

$$= x^4 + 15x^3 - 6x^2.$$

This is the polynomial that was being divided, so our answer, $\frac{1}{3}x^3 + 5x^2 - 2x$, checks.

1. Divide $t^3 - 4t^2 + 12t$ by $4t$.

YOUR TURN

EXAMPLE 2   Divide and check: $(10a^5b^4 - 2a^3b^2 + 6a^2b) \div (-2a^2b)$.

SOLUTION   We have

$$\frac{10a^5b^4 - 2a^3b^2 + 6a^2b}{-2a^2b} = \frac{10a^5b^4}{-2a^2b} - \frac{2a^3b^2}{-2a^2b} + \frac{6a^2b}{-2a^2b}$$

> We divide coefficients and subtract exponents.

$$= -\frac{10}{2}a^{5-2}b^{4-1} - \left(-\frac{2}{2}\right)a^{3-2}b^{2-1} + \left(-\frac{6}{2}\right)a^{2-2}b^{1-1}$$

$$= -5a^3b^3 + ab - 3.$$

Check: $-2a^2b(-5a^3b^3 + ab - 3)$
$$= -2a^2b\,(-5a^3b^3) + (-2a^2b)(ab) + (-2a^2b)(-3)$$
$$= 10a^5b^4 - 2a^3b^2 + 6a^2b$$

2. Divide and check:

$(-3x^2y + 9x^3y^2 - 6x^2y^2) \div (3x^2y)$.

Our answer, $-5a^3b^3 + ab - 3$, checks.

YOUR TURN

B. Dividing by a Binomial

The divisors in Examples 1 and 2 have just one term. For divisors with more than one term, we use long division, much as we do in arithmetic.

EXAMPLE 3   Divide $x^2 + 5x + 6$ by $x + 3$.

SOLUTION   We begin by dividing x^2 by x:

> Divide   the first term, x^2, by the first term in the divisor: $x^2/x = x$. Ignore the term 3 for the moment.

$$
\begin{array}{r}
x \\
x + 3 \overline{) x^2 + 5x + 6} \\
-(x^2 + 3x) \\
\hline
2x.
\end{array}
$$

> Multiply: $x(x + 3) = x^2 + 3x$.

> Subtract   both x^2 and $3x$: $x^2 + 5x - (x^2 + 3x) = 2x.$

Now we "bring down" the next term—in this case, 6. The current remainder, $2x + 6$, now becomes the focus of our division. We divide $2x$ by x.

Student Notes

Since long division of polynomials requires many steps, we recommend that you double-check each step of your work as you move forward.

$$
\begin{array}{r}
x + 2 \\
x + 3 \overline{) x^2 + 5x + 6} \\
-(x^2 + 3x) \\
\hline
2x + 6 \\
-(2x + 6) \\
\hline
0
\end{array}
$$

> Divide $2x$ by x: $2x/x = 2$.

> Multiply: $2(x + 3) = 2x + 6$.

> Subtract: $(2x + 6) - (2x + 6) = 0$.

The quotient is $x + 2$. The notation R 0 indicates a remainder of 0, although a remainder of 0 is generally not listed in an answer.

Check: To check, we multiply the quotient by the divisor and add any remainder to see if we get the dividend:

$$\overbrace{(x + 3)}^{\text{Divisor}} \quad \overbrace{(x + 2)}^{\text{Quotient}} \quad + \quad \overbrace{0}^{\text{Remainder}} \quad = \quad \overbrace{x^2 + 5x + 6.}^{\text{Dividend}}$$

3. Divide $x^2 - 3x - 10$ by $x - 5$.

Our answer, $x + 2$, checks.

◀ YOUR TURN

EXAMPLE 4   Divide: $(2x^2 + 5x - 1) \div (2x - 1)$.

SOLUTION   We begin by dividing $2x^2$ by $2x$:

> Divide the first term by the first term: $2x^2/(2x) = x$.

$$\begin{array}{r} x \\ 2x - 1 \overline{) 2x^2 + 5x - 1} \\ \underline{-(2x^2 - x)} \\ 6x \end{array}$$

CAUTION!   Write the parentheses around the polynomial being subtracted to remind you to subtract all its terms.

> Multiply: $x(2x - 1) = 2x^2 - x$.

> Subtract by changing signs and adding: $(2x^2 + 5x) - (2x^2 - x) = 6x$.

Now, we bring down the -1 and divide $6x - 1$ by $2x - 1$, focusing on the $6x$ and the $2x$.

$$\begin{array}{r} x + 3 \\ 2x - 1 \overline{) 2x^2 + 5x - 1} \\ \underline{-(2x^2 - x)} \\ 6x - 1 \\ \underline{-(6x - 3)} \\ 2 \end{array}$$

The division procedure ends when the degree of the remainder is less than that of the divisor.

> Divide $6x$ by $2x$: $6x/(2x) = 3$.

> Multiply: $3(2x - 1) = 6x - 3$.

> Subtract. Note that $-1 - (-3) = -1 + 3 = 2$.

The answer is $x + 3$ with R 2.
   Another way to write $x + 3$ R 2 is as

$$\text{Quotient} \quad \underbrace{x + 3}_{} + \underbrace{\frac{2 \leftarrow \text{Remainder}}{2x - 1}}_{\text{Divisor}}.$$

(This is the way answers will be given at the back of the book.)

Check: To check, we multiply the divisor by the quotient and add the remainder:

4. Divide:

$(2x^2 - 3x + 5) \div (2x + 3)$.

$$(2x - 1)(x + 3) + 2 = 2x^2 + 5x - 3 + 2$$
$$= 2x^2 + 5x - 1. \quad \text{Our answer checks.}$$

◀ YOUR TURN

When we are dividing polynomials, it is important that both the divisor and the dividend are written in descending order. Any missing terms in the dividend should be written in, using 0 for the coefficients.

EXAMPLE 5 Divide each of the following.

a) $(x^3 + 1) \div (x + 1)$ **b)** $(9a^2 + a^3 - 5) \div (a^2 - 1)$

SOLUTION

a)
$$
\begin{array}{r}
x^2 - x + 1 \\
x + 1 \overline{)x^3 + 0x^2 + 0x + 1} \\
\end{array}
$$
$\longleftarrow$ Writing in the missing terms

$-(x^3 + x^2)$

$-x^2 + 0x$ $\longleftarrow$ Subtracting $x^3 + x^2$ from $x^3 + 0x^2$ and bringing down the $0x$

$-(-x^2 - x)$

$x + 1$ $\longleftarrow$ Subtracting $-x^2 - x$ from $-x^2 + 0x$ and bringing down the 1

$-(x + 1)$

0

The answer is $x^2 - x + 1$.

Check: $(x + 1)(x^2 - x + 1) = x^3 - x^2 + x + x^2 - x + 1$
$$= x^3 + 1.$$

b) We rewrite the problem in descending order.

$$(a^3 + 9a^2 - 5) \div (a^2 - 1).$$

Thus,

$$
\begin{array}{r}
a + 9 \\
a^2 - 1 \overline{)a^3 + 9a^2 + 0a - 5} \\
\end{array}
$$
$\longleftarrow$ Writing in the missing term

$-(a^3 \quad - a)$

$9a^2 + a - 5$ $\longleftarrow$ Subtracting $a^3 - a$ from $a^3 + 9a^2 + 0a$ and bringing down the -5

$-(9a^2 \quad - 9)$

$a + 4$ The degree of the remainder is less than the degree of the divisor, so we are finished.

c) The answer is $a + 9 + \dfrac{a + 4}{a^2 - 1}$.

Check: $(a^2 - 1)(a + 9) + a + 4 = a^3 + 9a^2 - a - 9 + a + 4$
$$= a^3 + 9a^2 - 5.$$

5. Divide: $(2x^2 - 5) \div (x - 1)$.

YOUR TURN

↳ Check Your UNDERSTANDING

Place the dividend and divisor appropriately, making sure that they are written in the correct form. Do not carry out the division.

1. $(x^2 - 8x + 12) \div (x - 2)$ **2.** $(x^3 - 1) \div (x - 1)$

3. $(x - 3x^3 + 2) \div (3 + x)$

4.8 EXERCISE SET

FOR EXTRA HELP MyMathLab®

Vocabulary and Reading Check

Use the words from the following list to label the numbered expressions from the division shown.

dividend
divisor
quotient
remainder

$$\begin{array}{r} ②x + 2 \\ ①\,x - 3\overline{)x^2 - x + 9}\;③ \\ \underline{-(x^2 - 3x)} \\ 2x + 9 \\ \underline{-(2x - 6)} \\ 15\;④ \end{array}$$

1. _____

2. _____

3. _____

4. _____

A. Dividing by a Monomial

Divide and check.

5. $\dfrac{40x^6 - 25x^3}{5}$

6. $\dfrac{16a^5 - 24a^2}{8}$

7. $\dfrac{u - 2u^2 + u^7}{u}$

8. $\dfrac{50x^5 - 7x^4 + 2x}{x}$

9. $(18t^3 - 24t^2 + 6t) \div (3t)$

10. $(20t^3 - 15t^2 + 30t) \div (5t)$

11. $(42x^5 - 36x^3 + 9x^2) \div (6x^2)$

12. $(24x^6 + 18x^4 + 8x^3) \div (4x^3)$

13. $(32t^5 + 16t^4 - 8t^3) \div (-8t^3)$

14. $(36t^6 - 27t^5 - 9t^2) \div (-9t^2)$

15. $\dfrac{8x^2 - 10x + 1}{2x}$

16. $\dfrac{9x^2 + 3x - 2}{3x}$

17. $\dfrac{5x^3y + 10x^5y^2 + 15x^2y}{5x^2y}$

18. $\dfrac{12a^3b^2 + 4a^4b^5 + 16ab^2}{4ab^2}$

19. $\dfrac{9r^2s^2 + 3r^2s - 6rs^2}{-3rs}$

20. $\dfrac{4x^4y - 8x^6y^2 + 12x^8y^6}{4x^4y}$

B. Dividing by a Binomial

Divide.

21. $(x^2 - 8x + 12) \div (x - 2)$

22. $(x^2 + 2x - 15) \div (x + 5)$

23. $(t^2 - 10t - 20) \div (t - 5)$

24. $(t^2 + 8t - 15) \div (t + 4)$

25. $(2x^2 + 11x - 5) \div (x + 6)$

26. $(3x^2 - 2x - 13) \div (x - 2)$

27. $\dfrac{t^3 + 27}{t + 3}$

28. $\dfrac{a^3 + 8}{a + 2}$

29. $\dfrac{a^2 - 21}{a - 5}$

30. $\dfrac{t^2 - 13}{t - 4}$

31. $(6x^2 - x - 15) \div (2x + 3)$

32. $(10x^2 + x - 3) \div (2x - 1)$

33. $(5x^2 - 16x) \div (5x - 1)$

34. $(3x^2 - 7x + 1) \div (3x - 1)$

35. $(6a^2 + 17a + 8) \div (2a + 5)$

36. $(10a^2 + 19a + 9) \div (2a + 3)$

37. $\dfrac{2t^3 - 9t^2 + 11t - 3}{2t - 3}$

38. $\dfrac{8t^3 - 22t^2 - 5t + 12}{4t + 3}$

39. $(x^3 - x^2 + x - 1) \div (x - 1)$

40. $(t^3 - t^2 + t - 1) \div (t + 1)$

41. $(t^4 + 4t^2 + 3t - 6) \div (t^2 + 5)$

42. $(t^4 - 2t^2 + 4t - 5) \div (t^2 - 3)$

43. $(4x^4 - 3 - x - 4x^2) \div (2x^2 - 3)$

44. $(x + 6x^4 - 4 - 3x^2) \div (1 + 2x^2)$

45. How is the distributive law used when dividing a polynomial by a binomial?

46. On an assignment, Selina *incorrectly* writes

$$\frac{12x^3 - 6x}{3x} = 4x^2 - 6x.$$

What mistake do you think she is making and how might you convince her that a mistake has been made?

Skill Review

Graph.

47. $y = -\frac{2}{3}x + 4$ [3.6]   **48.** $8x = 4y$ [3.2]

49. Find the slope and the y-intercept of the line given by $2y = 8x + 7$. [3.6]

50. Find the slope–intercept form of the line containing the points $(6, 3)$ and $(-2, -7)$. [3.7]

Synthesis

51. Explain how to form trinomials for which division by $x - 5$ results in a remainder of 3.

52. Under what circumstances will the quotient of two binomials have more than two terms?

Divide.

53. $(10x^{9k} - 32x^{6k} + 28x^{3k}) \div (2x^{3k})$

54. $(45a^{8k} + 30a^{6k} - 60a^{4k}) \div (3a^{2k})$

55. $(6t^{3h} + 13t^{2h} - 4t^h - 15) \div (2t^h + 3)$

56. $(x^4 + a^2) \div (x + a)$

57. $(5a^3 + 8a^2 - 23a - 1) \div (5a^2 - 7a - 2)$

58. $(15y^3 - 30y + 7 - 19y^2) \div (3y^2 - 2 - 5y)$

59. Divide the sum of $4x^5 - 14x^3 - x^2 + 3$ and $2x^5 + 3x^4 + x^3 - 3x^2 + 5x$ by $3x^3 - 2x - 1$.

60. Divide $5x^7 - 3x^4 + 2x^2 - 10x + 2$ by the sum of $(x - 3)^2$ and $5x - 8$.

If the remainder is 0 when one polynomial is divided by another, the divisor is a factor *of the dividend. Find the value(s) of c for which $x - 1$ is a factor of each polynomial.*

61. $x^2 - 4x + c$

62. $2x^2 - 3cx - 8$

63. $c^2x^2 + 2cx + 1$

64. *Business.* A company's **revenue** R from an item is defined as the price paid per item times the quantity of items sold.

a) Easy on the Eyes sells high-quality reproductions of original watercolors. Find an expression for the price paid per reproduction if the revenue from the sale of q reproductions is $(80q - q^2)$ dollars.

b) Find an expression for the price paid per reproduction if one more reproduction is sold but the revenue remains $(80q - q^2)$ dollars.

65. The volume of a cube is $(a^3 + 3a^2 + 3a + 1)$ cm³.

a) Find the length of one side of the cube.

b) Find the area of one side of the cube.

YOUR TURN ANSWERS: SECTION 4.8

1. $\frac{1}{4}t^2 - t + 3$ **2.** $-1 + 3xy - 2y$ **3.** $x + 2$

4. $x - 3 + \dfrac{14}{2x + 3}$ **5.** $2x + 2 + \dfrac{-3}{x - 1}$

Quick Quiz: Sections 4.1–4.8

1. Determine the leading term of the polynomial $3t^2 - t^5 + 10 + 2t^4$. [4.3]

2. Multiply: $3x^4(-5x^2 - 6x + 1)$. [4.5]

3. Multiply: $(\frac{1}{3}a^2 + 6)^2$. [4.6]

4. Add: $(a^2b - 2ab + 5ab^2) + (3ab^2 - a^2b - 8ab)$. [4.7]

5. Divide: $(16n^4 - 8n^3 + 4n) \div (4n)$. [4.8]

Prepare to Move On

List the factors in each expression. [1.2]

1. $3(x + 7)$

2. $(x - 3)(x + 10)$

3. $x(x + 1)(2x - 5)$

Factor. [1.2]

4. $9x + 15y$

5. $14a + 7c + 7$

1

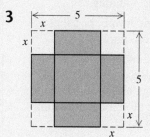

2

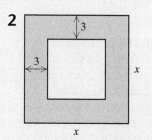

3

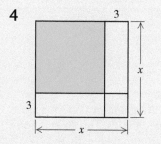

4

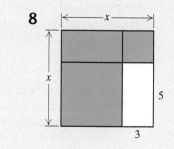

5

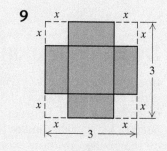

Visualizing for Success

Use after Section 4.6.

In each of Exercises 1–10, find two algebraic expressions for the shaded area of the figure from the list below.

A. $9 - 4x^2$

B. $x^2 - (x - 6)^2$

C. $(x + 3)(x - 3)$

D. $10^2 + 2^2$

E. $8x + 15$

F. $(x + 5)(x + 3) - x^2$

G. $x^2 - 6x + 9$

H. $(3 - 2x)^2 + 4x(3 - 2x)$

I. $(x + 3)^2$

J. $(5x + 3)^2 - 25x^2$

K. $(5 - 2x)^2 + 4x(5 - 2x)$

L. $x^2 - 9$

M. 104

N. $x^2 - 15$

O. $12x - 36$

P. $30x + 9$

Q. $(x - 5)(x - 3) + 3(x - 5) + 5(x - 3)$

R. $(x - 3)^2$

S. $25 - 4x^2$

T. $x^2 + 6x + 9$

Answers on page A-24

An additional, animated version of this activity appears in MyMathLab. To use MyMathLab, you need a course ID and a student access code. Contact your instructor for more information.

6

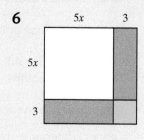

7

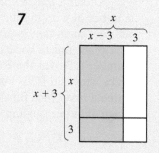

8

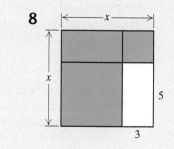

9

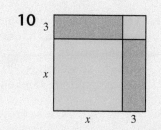

10

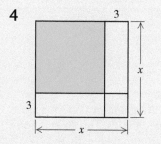

Collaborative Activity   *Finding the Magic Number*

Focus: Evaluating polynomials in several
    variables
Use after: Section 4.7
Time: 15–25 minutes
Group size: 3
Materials: A coin for each person

Can you determine the requirements for your
baseball team to clinch first place?

 A team's *magic number* is the combined
number of wins by that team and losses by
another team that guarantee the leading team a
first-place finish. For example, if the Cubs' magic
number is 3 over the Reds, any combination of
Cubs wins and Reds losses that totals 3 will guarantee a first-place finish for the Cubs. A team's
magic number is computed using the polynomial

$$G - P - L + 1,$$

where G is the length of the season, in games,
P is the number of games that the leading team
has played, and L is the total number of games
that the second-place team has lost minus the total
number of games that the leading team has lost.

Activity

1. The standings below are from a fictitious
   league. Each group should calculate the
   Jaguars' magic number with respect to the
   Catamounts as well as the Jaguars' magic

number with respect to the Wildcats. Assume
that the schedule is 162 games long.

	W	L
Jaguars	92	64
Catamounts	90	66
Wildcats	89	66

2. Each group member should play the role of
   one of the teams, using coin tosses to simulate
   the remaining games. If a group member correctly predicts the side (heads or tails) that
   comes up, the coin toss represents a win for
   that team. Should the other side appear, the
   toss represents a loss. Assume that these games
   are against other (unlisted) teams in the league.
   Each group member should perform three coin
   tosses and then update the standings.

3. Recalculate the two magic numbers, using
   the updated standings from part (2).

4. Slowly—one coin toss at a time—play out the
   remainder of the season. Record all wins and
   losses, update the standings, and recalculate the
   magic numbers each time all three group members have completed a round of coin tosses.

5. Examine the work in part (4) and explain
   why a magic number of 0 indicates that a
   team has been eliminated from contention.

Decision Making ⌗ Connection   *(Use after Section 4.2.)*

Buying a Smartphone. One decision to make
when purchasing a smartphone is the amount of
memory you need. In order to determine that, you
need to know what kind of files you plan to store
on the phone. For these exercises, use the following equivalents: 1 megabyte = 1 million bytes and
1 gigabyte (GB) = 1 billion bytes.

1. Anna is considering a smartphone with 64 GB of
   memory. She plans to use the phone primarily for
   viewing movies. The average size of the movies she
   wants to download is 800 megabytes. How many
   movies could she store on the device?

2. Nate has a collection of 2000 songs. The average
   size of each song is 10 megabytes. The smartphone
   he is considering can be purchased with 16 GB,
   32 GB, or 64 GB. What size phone should he buy?

3. *Research.* Determine the average size of a game
   file. If Nate instead had a collection of 2000 games,
   what size phone should he buy?

4. On the basis of your own use of media, estimate the
   amount of memory you would need to purchase.

Study Summary

KEY TERMS AND CONCEPTS	EXAMPLES	PRACTICE EXERCISES

SECTION 4.1: *Exponents and Their Properties*

1 as an exponent:	$a^1 = a$	$3^1 = 3$	*Simplify.*
0 as an exponent:	$a^0 = 1$	$3^0 = 1$	**1.** 6^1 **2.** $(-5)^0$
The Product Rule:	$a^m \cdot a^n = a^{m+n}$	$3^5 \cdot 3^9 = 3^{5+9} = 3^{14}$	
The Quotient Rule:	$\dfrac{a^m}{a^n} = a^{m-n}$	$\dfrac{3^7}{3} = 3^{7-1} = 3^6$	**3.** $x^5 x^{11}$ **4.** $\dfrac{8^9}{8^2}$
The Power Rule:	$(a^m)^n = a^{mn}$	$(3^4)^2 = 3^{4\cdot2} = 3^8$	**5.** $(y^5)^3$
Raising a product to a power:	$(ab)^n = a^n b^n$	$(3x^5)^4 = 3^4(x^5)^4 = 81x^{20}$	**6.** $(x^3 y)^{10}$
Raising a quotient to a power:	$\left(\dfrac{a}{b}\right)^n = \dfrac{a^n}{b^n}$	$\left(\dfrac{3}{x}\right)^6 = \dfrac{3^6}{x^6}$	**7.** $\left(\dfrac{x^2}{7}\right)^5$

SECTION 4.2: *Negative Exponents and Scientific Notation*

$a^{-n} = \dfrac{1}{a^n};$ $\dfrac{a^{-n}}{b^{-m}} = \dfrac{b^m}{a^n}$	$3^{-2} = \dfrac{1}{3^2} = \dfrac{1}{9};$ $\dfrac{3^{-7}}{x^{-5}} = \dfrac{x^5}{3^7}$	*Write without negative exponents.* **8.** 10^{-1} **9.** $\dfrac{x^{-1}}{y^{-3}}$
Scientific notation: $N \times 10^m, 1 \le N < 10.$	$4100 = 4.1 \times 10^3;$ $5 \times 10^{-3} = 0.005$	**10.** Convert to scientific notation: 0.000904. **11.** Convert to decimal notation: 6.9×10^5.

SECTION 4.3: *Polynomials*

A **polynomial** is a monomial or a sum of monomials.

When a polynomial is written as a sum of monomials, each monomial is a **term** of the polynomial.

The **degree of a term** of a polynomial is the number of variable factors in that term.

The **coefficient** of a term is the part of the term that is a constant factor.

The **leading term** of a polynomial is the term of highest degree.

The **leading coefficient** is the coefficient of the leading term.

The **degree of the polynomial** is the degree of the leading term.

Polynomial: $10x - x^3 + 4x^5 + 7$

Term	$10x$	$-x^3$	$4x^5$	7
Degree of Term	1	3	5	0
Coefficient of Term	10	-1	4	7
Leading Term	$4x^5$			
Leading Coefficient	4			
Degree of Polynomial	5			

For Exercises 12–17, consider the polynomial $x^2 - 10 + 5x - 8x^6$.

12. List the terms of the polynomial.

13. What is the degree of the term $5x$?

14. What is the coefficient of the term x^2?

15. What is the leading term of the polynomial?

16. What is the leading coefficient of the polynomial?

17. What is the degree of the polynomial?

A **monomial** has one term.
A **binomial** has two terms.
A **trinomial** has three terms.

Monomial: $4x^3$
Binomial: $x^2 - 5$
Trinomial: $3t^3 + 2t - 10$

18. Classify the polynomial
$$8x - 3 - x^4$$
as a monomial, a binomial, or a trinomial.

Like terms, or **similar terms,** are either constant terms or terms containing the same variable(s) raised to the same power(s). These can be **combined** within a polynomial.

Combine like terms:
$$3y^4 + 6y^2 - 7 - y^4 - 6y^2 + 8.$$
$3y^4 + 6y^2 - 7 - y^4 - 6y^2 + 8$
$= \underline{3y^4 - y^4} + \underline{6y^2 - 6y^2} - \underline{7 + 8}$
$= \quad 2y^4 \quad + \quad 0 \quad + \quad 1$
$= 2y^4 + 1$

19. Combine like terms:
$3x^2 + 5x - 10x + x.$

To **evaluate** a polynomial, replace the variable with a number. The **value** is calculated using the rules for order of operations.

Evaluate $t^3 - 2t^2 - 5t + 1$ for $t = -2$.
$t^3 - 2t^2 - 5t + 1$
$= (-2)^3 - 2(-2)^2 - 5(-2) + 1$
$= -8 - 2(4) - (-10) + 1$
$= -8 - 8 + 10 + 1$
$= -5$

20. Evaluate $2 - 3x - x^2$ for $x = -1$.

SECTION 4.4:   *Addition and Subtraction of Polynomials*

Add polynomials by combining like terms.

$(2x^2 - 3x + 7) + (5x^3 + 3x - 9)$
$= 2x^2 + (-3x) + 7 + 5x^3 + 3x + (-9)$
$= 5x^3 + 2x^2 - 2$

21. Add:
$(9x^2 - 3x) + (4x - x^2).$

Subtract polynomials by adding the opposite of the polynomial being subtracted.

$(2x^2 - 3x + 7) - (5x^3 + 3x - 9)$
$= 2x^2 - 3x + 7 + (-5x^3 - 3x + 9)$
$= 2x^2 - 3x + 7 - 5x^3 - 3x + 9$
$= -5x^3 + 2x^2 - 6x + 16$

22. Subtract:
$(9x^2 - 3x) - (4x - x^2).$

SECTION 4.5:   *Multiplication of Polynomials*

Multiply polynomials by multiplying each term of one polynomial by each term of the other.

$(x + 2)(x^2 - x - 1)$
$= x \cdot x^2 - x \cdot x - x \cdot 1 + 2 \cdot x^2$
$\quad - 2 \cdot x - 2 \cdot 1$
$= x^3 - x^2 - x + 2x^2 - 2x - 2$
$= x^3 + x^2 - 3x - 2$

23. Multiply:
$(x - 1)(x^2 - x - 2).$

SECTION 4.6:   *Special Products*

FOIL (First, Outer, Inner, Last):
$(A + B)(C + D)$
$\quad = AC + AD + BC + BD$

$$\begin{array}{c} \overset{F \qquad L}{\overbrace{}} \\ (A + B)(C + D) \\ \underset{O}{\underbrace{}} \\ I \end{array}$$

$(x + 3)(x - 2) = x^2 - 2x + 3x - 6$
$\qquad\qquad\qquad = x^2 + x - 6$

24. Multiply:
$(x + 4)(2x + 3).$

The product of a sum and a difference: $(A + B)(A - B) = A^2 - B^2$ $A^2 - B^2$ is a **difference of squares.**	$(t^3 + 5)(t^3 - 5) = (t^3)^2 - 5^2$ $\qquad = t^6 - 25$	**25.** Multiply: $(5 + 3x)(5 - 3x)$.
The square of a binomial: $(A + B)^2 = A^2 + 2AB + B^2;$ $(A - B)^2 = A^2 - 2AB + B^2$ $A^2 + 2AB + B^2$ and $A^2 - 2AB + B^2$ are called **perfect-square trinomials.**	$(5x + 3)^2 = (5x)^2 + 2(5x)(3) + 3^2$ $\qquad = 25x^2 + 30x + 9;$ $(5x - 3)^2 = (5x)^2 - 2(5x)(3) + 3^2$ $\qquad = 25x^2 - 30x + 9$	*Multiply.* **26.** $(x + 9)^2$ **27.** $(8x - 1)^2$

SECTION 4.7: *Polynomials in Several Variables*

To **evaluate** a polynomial, replace each variable with a number and simplify.	Evaluate $4 - 3xy + x^2y$ for $x = 5$ and $y = -1$. $4 - 3xy + x^2y = 4 - 3(5)(-1)$ $\qquad\qquad + (5)^2(-1)$ $\qquad = 4 - (-15) + (-25)$ $\qquad = -6$	**28.** Evaluate $xy - y^2 - 4x$ for $x = -2$ and $y = 3$.
The **degree** of a term is the number of variable factors in the term or the sum of the exponents of the variables.	The degree of $-19x^3yz^2$ is 6.	**29.** What is the degree of $4mn^5$?
Add, subtract, and multiply polynomials in several variables in the same way as polynomials in one variable.	$(3xy^2 - 4x^2y + 5xy) + (xy - 6x^2y)$ $\quad = 3xy^2 - 10x^2y + 6xy;$ $(3xy^2 - 4x^2y + 5xy) - (xy - 6x^2y)$ $\quad = 3xy^2 + 2x^2y + 4xy;$ $(2a^2b + 3a)(5a^2b - a)$ $\quad = 10a^4b^2 + 13a^3b - 3a^2$	**30.** Add: $(3cd^2 + 2c) +$ $(4cd - 9c)$. **31.** Subtract: $(8pw - p^2w) -$ $(p^2w + 8pw)$. **32.** Multiply: $(7xy - x^2)^2$.

SECTION 4.8: *Division of Polynomials*

To divide a polynomial by a monomial, divide each term by the monomial. Divide coefficients and subtract exponents.	$\dfrac{3t^5 - 6t^4 + 4t^2 + 9t}{3t}$ $= \dfrac{3t^5}{3t} - \dfrac{6t^4}{3t} + \dfrac{4t^2}{3t} + \dfrac{9t}{3t}$ $= t^4 - 2t^3 + \frac{4}{3}t + 3$	**33.** Divide: $\dfrac{4y^5 - 8y^3 + 16y^2}{4y^2}$.
To divide a polynomial by a binomial, use long division.	Divide: $(x^2 + 5x - 2) \div (x - 3)$. $\quad\quad\quad\ x + 8$ $x - 3 \overline{)\, x^2 + 5x - 2}$ $\quad\ \underline{-(x^2 - 3x)}$ $\quad\quad\quad\ 8x - 2$ $\quad\quad\quad \underline{-(8x - 24)}$ $\quad\quad\quad\quad\quad 22$ The result is $x + 8 + \dfrac{22}{x - 3}$.	**34.** Divide: $(x^2 - x + 4) \div$ $(x + 1)$.

Review Exercises: Chapter 4

⬥ Concept Reinforcement

Classify each of the following statements as either true or false.

1. When two polynomials that are written in descending order are added, the result is generally written in descending order. [4.4]

2. The product of the sum and the difference of the same two terms is a difference of squares. [4.6]

3. When a binomial is squared, the result is a perfect-square trinomial. [4.6]

4. FOIL can be used whenever two polynomials are being multiplied. [4.6]

5. The degree of a polynomial cannot exceed the value of the polynomial's leading coefficient. [4.3]

6. Scientific notation is used only for extremely large numbers. [4.2]

7. FOIL can be used with polynomials in several variables. [4.7]

8. A positive number raised to a negative exponent can never represent a negative number. [4.2]

Simplify. [4.1]

9. $n^3 \cdot n^8 \cdot n$

10. $(7x)^8 \cdot (7x)^2$

11. $t^6 \cdot t^0$

12. $\dfrac{4^5}{4^2}$

13. $\dfrac{(a + b)^4}{(a + b)^4}$

14. $\dfrac{-18c^9d^3}{2c^5d}$

15. $(-2xy^2)^3$

16. $(2x^3)(-3x)^2$

17. $(a^2b)(ab)^5$

18. $\left(\dfrac{2t^5}{3s^4}\right)^2$

19. Express using a positive exponent: 8^{-6}. [4.2]

20. Express using a negative exponent: $\dfrac{1}{a^9}$. [4.2]

Simplify. Do not use negative exponents in the answer. [4.2]

21. $4^5 \cdot 4^{-7}$

22. $\dfrac{6a^{-5}b}{3a^8b^{-8}}$

23. $(w^3)^{-5}$

24. $(2x^{-3}y)^{-2}$

25. $\left(\dfrac{2x}{y}\right)^{-3}$

26. Convert to decimal notation: 4.7×10^8. [4.2]

27. Convert to scientific notation: 0.0000109. [4.2]

Multiply or divide and write scientific notation for the result. [4.2]

28. $(3.8 \times 10^4)(5.5 \times 10^{-1})$

29. $\dfrac{1.28 \times 10^{-8}}{2.5 \times 10^{-4}}$

30. Identify the terms of the polynomial:
$$-4y^5 + 7y^2 - 3y - 2. \quad [4.3]$$

31. List the coefficients of the terms in the polynomial:
$$7n^4 - \tfrac{5}{6}n^2 - 4n + 10. \quad [4.3]$$

*For each polynomial, **(a)** list the degree of each term; **(b)** determine the leading term and the leading coefficient; and **(c)** determine the degree of the polynomial.* [4.3]

32. $4t^2 + 6 + 15t^5$

33. $-2x^5 + 7 - 3x^2 + x$

Classify each polynomial as a monomial, a binomial, a trinomial, or a polynomial with no special name. [4.3]

34. $4x^3 - 5x + 3$

35. $4 - 9t^3 - 7t^4 + 10t^2$

36. $7y^2$

Combine like terms and write in descending order. [4.3]

37. $-4t^3 + 2t + 4t^3 + 8 - t - 9$

38. $-a + \tfrac{1}{3} + 20a^5 - 1 - 6a^5 - 2a^2$

Evaluate each polynomial for $x = -2$. [4.3]

39. $9x - 6$

40. $x^2 - 3x + 6$

Add or subtract, as indicated. [4.4]

41. $(8x^4 - x^3 + x - 4) + (x^5 + 7x^3 - 3x - 5)$

42. $(5a^5 - 2a^3 - 9a^2) + (2a^5 + a^3) + (-a^5 - 3a^2)$

43. $(3x^5 - 4x^4 + 2x^2 + 3) - (2x^5 - 4x^4 + 3x^3 + 4x^2 - 5)$

44. $\begin{array}{r} -\tfrac{3}{4}x^4 + \tfrac{1}{2}x^3 \qquad\qquad + \tfrac{7}{8} \\ -\tfrac{1}{4}x^3 - x^2 - \tfrac{7}{4}x \\ +\tfrac{3}{2}x^4 \qquad + \tfrac{2}{3}x^2 \qquad - \tfrac{1}{2} \\ \hline \end{array}$

45. $\begin{array}{r} 2x^5 \qquad - x^3 \qquad + x + 3 \\ -(3x^5 - x^4 + 4x^3 + 2x^2 - x + 3) \\ \hline \end{array}$

46. The length of a rectangle is 3 m greater than its width.

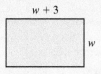

w + 3

w

a) Find a polynomial for the perimeter. [4.4]
b) Find a polynomial for the area. [4.5]

Multiply.

47. $5x^2(-6x^3)$ [4.5]   **48.** $(7x + 1)^2$ [4.6]

49. $(a - 7)(a + 4)$ [4.6]

50. $(d - 8)(d + 8)$ [4.6]

51. $(4x^2 - 5x + 1)(3x - 2)$ [4.5]

52. $3t^2(5t^3 - 2t^2 + 4t)$ [4.5]

53. $(2a + 9)(2a - 9)$ [4.6]

54. $(x - 0.8)(x - 0.5)$ [4.6]

55. $(x^4 - 2x + 3)(x^3 + x - 1)$ [4.5]

56. $(4y^3 - 5)^2$ [4.6]

57. $(2t^2 + 3)(t^2 - 7)$ [4.6]

58. $\left(a - \frac{1}{2}\right)\left(a + \frac{2}{3}\right)$ [4.6]

59. $(-7 + 2n)(7 + 2n)$ [4.6]

60. Evaluate $2 - 5xy + y^2 - 4xy^3 + x^6$ for $x = -1$ and $y = 2$. [4.7]

Identify the coefficient and the degree of each term of each polynomial. Then find the degree of the polynomial. [4.7]

61. $x^5y - 7xy + 9x^2 - 8$

62. $a^3b^8c^2 - c^{22} + a^5c^{10}$

Combine like terms. [4.7]

63. $u + 3v - 5u + v - 7$

64. $6m^3 + 3m^2n + 4mn^2 + m^2n - 5mn^2$

Add or subtract, as indicated. [4.7]

65. $(4a^2 - 10ab - b^2) + (-2a^2 - 6ab + b^2)$

66. $(6x^3y^2 - 4x^2y - 6x) - (-5x^3y^2 + 4x^2y + 6x^2 - 6)$

Multiply. [4.7]

67. $(2x + 5y)(x - 3y)$

68. $(5ab - cd^2)^2$

69. Find a polynomial for the shaded area. [4.7]

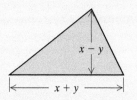

x − y

x + y

Divide. [4.8]

70. $(3y^5 - y^2 + 12y) \div (3y)$

71. $(6x^3 - 5x^2 - 13x + 13) \div (2x + 3)$

72. $\dfrac{t^4 + t^3 + 2t^2 - t - 3}{t + 1}$

Synthesis

73. Explain why $5x^3$ and $(5x)^3$ are not equivalent expressions. [4.1]

74. A binomial is squared and the result, written in descending order, is $x^2 - 6x + 9$. Is it possible to determine what binomial was squared? Why or why not? [4.6]

75. Determine, without performing the multiplications, the degree of each product. [4.5]
a) $(x^5 - 6x^2 + 3)(x^4 + 3x^3 + 7)$
b) $(x^7 - 4)^4$

76. Simplify:
$$(-3x^5 \cdot 3x^3 - x^6(2x)^2 + (3x^4)^2 + (2x^2)^4 - 20x^2(x^3)^2)^2. \text{[4.1], [4.3]}$$

77. A polynomial has degree 4. There is no x^2-term. The coefficient of x^4 is two times the coefficient of x^3. The coefficient of x is 3 less than the coefficient of x^4. The remaining coefficient is 7 less than the coefficient of x. The sum of the coefficients is 15. Find the polynomial. [4.3]

78. Multiply: $[(x - 5) - 4x^3][(x - 5) + 4x^3]$. [4.6]

79. Solve: $(x - 7)(x + 10) = (x - 4)(x - 6)$. [2.2], [4.6]

80. *Blood Donors.* Every 4–6 weeks, Jordan donates 1.14×10^6 cubic millimeters (two pints) of whole blood, from which platelets are removed and the blood returned to his body. In one cubic millimeter of blood, there are about 2×10^5 platelets. Approximate the number of platelets in Jordan's typical donation. [4.2]

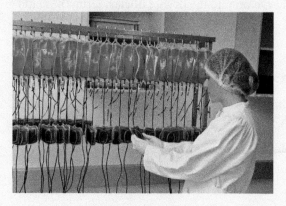

Test: Chapter 4   For step-by-step test solutions, access the Chapter Test Prep Videos in MyMathLab.

Simplify.

1. $x^7 \cdot x \cdot x^5$

2. $\dfrac{3^8}{3^7}$

3. $\dfrac{(3m)^4}{(3m)^4}$

4. $(t^5)^9$

5. $(-3y^2)^3$

6. $(5x^4y)(-2x^5y)^3$

7. $\dfrac{24a^7b^4}{20a^2b}$

8. $\left(\dfrac{4p}{5q^3}\right)^2$

9. Express using a positive exponent: y^{-7}.

10. Express using a negative exponent: $\dfrac{1}{5^6}$.

Simplify.

11. $t^{-4} \cdot t^{-5}$

12. $\dfrac{9x^3y^2}{3x^8y^{-3}}$

13. $(2a^3b^{-1})^{-4}$

14. $\left(\dfrac{ab}{c}\right)^{-3}$

15. Convert to scientific notation: 3,060,000,000.

16. Convert to decimal notation: 5×10^{-8}.

Multiply or divide, as indicated, and write scientific notation for the result.

17. $\dfrac{5.6 \times 10^6}{3.2 \times 10^{-11}}$

18. $(2.4 \times 10^5)(5.4 \times 10^{16})$

19. Classify $4x^2y - 7y^3$ as a monomial, a binomial, a trinomial, or a polynomial with no special name.

20. Identify the coefficient of each term of the polynomial:
$$3x^5 - x + \tfrac{1}{9}.$$

21. Determine the degree of each term, the leading term and the leading coefficient, and the degree of the polynomial:
$$2t^3 - t + 7t^5 + 4.$$

22. Evaluate $x^2 + 5x - 1$ for $x = -3$.

Combine like terms and write in descending order.

23. $y^2 - 3y - y + \tfrac{3}{4}y^2$

24. $3 - x^2 + 8x + 5x^2 - 6x - 2x + 4x^3$

Add or subtract, as indicated.

25. $(3x^5 + 5x^3 - 5x^2 - 3) + (x^5 + x^4 - 3x^2 + 2x - 4)$

26. $\left(x^4 + \tfrac{2}{3}x + 5\right) + \left(4x^4 + 5x^2 + \tfrac{1}{3}x\right)$

27. $(5a^4 + 3a^3 - a^2 - 2a - 1) - (7a^4 - a^2 - a + 6)$

28. $(t^3 - 0.3t^2 - 20) - (t^4 - 1.5t^3 + 0.3t^2 - 11)$

Multiply.

29. $-2x^2(3x^2 - 3x - 5)$

30. $\left(x - \tfrac{1}{3}\right)^2$

31. $(5t - 7)(5t + 7)$

32. $(3b + 5)(2b - 1)$

33. $(x^6 - 4)(x^8 + 4)$

34. $(8 - y)(6 + 5y)$

35. $(2x + 1)(3x^2 - 5x - 3)$

36. $(8a^3 + 3)^2$

37. Evaluate $2x^2y - 3y^2$ for $x = -3$ and $y = 2$.

38. Combine like terms:
$$2x^3y - y^3 + xy^3 + 8 - 6x^3y - x^2y^2 + 11.$$

39. Subtract:
$$(8a^2b^2 - ab + b^3) - (-6ab^2 - 7ab - ab^3 + 5b^3).$$

40. Multiply: $(3x^5 - y)(3x^5 + y)$.

Divide.

41. $(12x^4 + 9x^3 - 15x^2) \div (3x^2)$

42. $(6x^3 - 8x^2 - 14x + 13) \div (x + 2)$

Synthesis

43. The height of a box is 1 less than its length, and the length is 2 more than its width. Express the volume in terms of the length.

44. Simplify: $2^{-1} - 4^{-1}$.

45. Every day about 265 billion spam e-mails are sent. If each spam e-mail wastes 4 sec of the recipient's time, how many hours are wasted each day due to spam?

Data: Radicati Group

Cumulative Review: Chapters 1–4

1. Evaluate $\dfrac{2x + y}{5}$ for $x = 12$ and $y = 6$. [1.1]

2. Evaluate $5x^2y - xy + y^2$ for $x = -1$ and $y = -2$. [4.7]

Simplify.

3. $\frac{1}{15} - \frac{2}{9}$ [1.6]

4. $2 - [10 - (5 + 12 \div 2^2 \cdot 3)]$ [1.8]

5. $2y - (y - 7) + 3$ [1.8]

6. $t^4 \cdot t^7 \cdot t$ [4.1]

7. $\dfrac{-100x^6y^8}{25xy^5}$ [4.1]

8. $(2a^2b)(5ab^3)^2$ [4.1]

9. Factor: $10a - 6b + 12$. [1.2]

10. Determine the degree of the polynomial
$-x^4 + 5x^3 + 3x^6 - 1$. [4.3]

11. In which quadrant is $(-2, 5)$ located? [3.1]

Graph.

12. $3y + 2x = 0$ [3.2]   13. $3y - 2x = 12$ [3.3]

14. $3y = 2$ [3.3]   15. $3y = 2x + 9$ [3.6]

16. Find the slope and the y-intercept of the line given by $y = \frac{1}{10}x + \frac{3}{8}$. [3.6]

17. Find the slope of the line containing the points $(2, 3)$ and $(-6, 8)$. [3.5]

18. Write an equation of the line with slope $-\frac{2}{3}$ and y-intercept $(0, -10)$. [3.6]

Solve.

19. $\frac{1}{6}n = -\frac{2}{3}$ [2.1]

20. $5y + 7 = 8y - 1$ [2.2]

21. $2 - (x - 7) = 8 - 4(x + 5)$ [2.2]

22. $-\frac{1}{2}t \le 4$ [2.6]

23. $3x - 5 > 9x - 8$ [2.6]

24. Solve $c = \dfrac{5pq}{2t}$ for t. [2.3]

Add or subtract, as indicated.

25. $(2u^2v - uv^2 + uv) + (3u^2 - v^2u + 5vu^2)$ [4.7]

26. $(2x^5 - x^4 - x) - (x^5 - x^4 + x)$ [4.4]

Multiply.

27. $8x^3(-2x^2 - 6x + 7)$ [4.5]

28. $(x - 2)(x^2 + x - 5)$ [4.5]

29. $(4t^2 + 3)^2$ [4.6]

30. $(\frac{1}{2}x + 1)(\frac{1}{2}x - 1)$ [4.6]

31. $(2r^2 + s)(3r^2 - 4s)$ [4.7]

32. Divide: $(x^2 - x + 3) \div (x - 1)$. [4.8]

Simplify. Do not use negative exponents in the answer. [4.2]

33. 7^{-10}

34. $(3x^{-7}y^{-2})^{-1}$

35. In 2015, China and Germany together had installed solar panels capable of producing about 83 thousand megawatts of electricity. China's solar capacity was 3 thousand megawatts greater than that of Germany. What was China's solar capacity? [2.5]

Data: businessgreen.com

36. In 2013, U.S. electric utilities used coal to generate 1.5 trillion kilowatt hours (kWh) of electricity. This was 39% of the total amount of electricity generated in 2013. How much electricity was generated in 2013? [2.4]

Data: U.S. Energy Information Administration

37. While studying in the United States under a student visa, Ana can work on campus no more than 20 hr per week. For the first 4 days of one week, she worked 3, 2, 5, and 6 hr. How many hours can she work on the fifth day without violating this restriction? [2.7]

38. U.S. retail inventory shrink, or loss due to theft and error, increased from $34.5 billion in 2011 to $44 billion in 2014. Find the rate of change of inventory shrink. [3.4]

Data: National Retail Federation

Synthesis

Solve. If no solution exists, state this.

39. $3x - 2(x + 6) = 4(x - 3)$ [2.2]

40. $x - (2x - 1) = 3x - 4(x + 1) + 10$ [2.2]

Simplify.

41. $7^{-1} + 8^0$ [4.2]

42. $-2x^5(x^7) + (x^3)^4 - (4x^5)^2(-x^2)$ [4.1], [4.3]

Polynomials and Factoring

Waves in a Wind Sea

Waves in a Wind Sea

5.1 Introduction to Factoring

5.2 Factoring Trinomials of the Type $x^2 + bx + c$

5.3 Factoring Trinomials of the Type $ax^2 + bx + c$

5.4 Factoring Perfect-Square Trinomials and Differences of Squares

MID-CHAPTER REVIEW

5.5 Factoring: A General Strategy

5.6 Solving Quadratic Equations by Factoring

CONNECTING THE CONCEPTS

5.7 Solving Applications

CHAPTER RESOURCES

Translating for Success
Collaborative Activities
Decision Making: Connection

STUDY SUMMARY

REVIEW EXERCISES
CHAPTER TEST
CUMULATIVE REVIEW

Ocean surface waves resulting from wind can reach heights of 100 ft as measured from the base of the trough. A stronger wind results in higher waves, but, as the graph indicates, the relationship between wind speed and wave height is not linear. We can use a *quadratic equation* to model wave height given wind speed and to predict wave heights for other wind speeds. (**See Exercise 29 in Exercise Set 5.7.**)

Mastery of mathematical skills is essential for all of my investigations into the impact of humans on the aquatic environment.

Ashanti Johnson, Chemical Oceanographer at University of Texas in Arlington, Texas, uses math to study events impacting marine environments ranging from the Arctic Circle to the coastlines of Georgia, Florida, and Puerto Rico.

ALF
Active Learning Figure
Explore the math using the Active Learning Figure in MyMathLab.

Student Activity
Do the Student Activity in MyMathLab to see math in action.

Factoring is multiplying reversed. Thus factoring polynomials requires a solid understanding of how to multiply polynomials. Factoring is an important skill that will be used to solve equations and simplify other types of expressions found later in the study of algebra.

5.1 Introduction to Factoring

A. Factoring Monomials   **B.** Factoring When Terms Have a Common Factor   **C.** Factoring by Grouping

Just as a number like 15 can be factored as $3 \cdot 5$, a polynomial like $x^2 + 7x$ can be factored as $x(x + 7)$. In both cases, we ask ourselves, "What was multiplied to obtain the given result?" The situation is much like a popular television game show in which an "answer" is given and participants must find the "question" to which the answer corresponds.

Study Skills

You've Got Mail

Many students overlook an excellent way to get questions answered—email. If your instructor makes his or her email address available, consider using it to get help. Often, just the act of writing out your question brings clarity. You might also take a photo of your work and email it to your instructor. Be sure to allow some time for your instructor to reply.

FACTORING

To *factor* a polynomial is to find an equivalent expression that is a product. An equivalent expression of this type is called a *factorization* of the polynomial.

A. Factoring Monomials

To factor a monomial, we find monomials whose product is equivalent to the original monomial. For example, $20x^2$ can be factored as $2 \cdot 10x^2$, $4x \cdot 5x$, or $10x \cdot 2x$, as well as several other ways. To check, we multiply.

EXAMPLE 1   Find three factorizations of $15x^3$.

SOLUTION

a) $15x^3 = (3 \cdot 5)(x \cdot x^2)$ Thinking of how 15 and x^3 can each be factored
$ = (3x)(5x^2)$ The factors are $3x$ and $5x^2$. *Check*: $3x \cdot 5x^2 = 15x^3$.

b) $15x^3 = (3 \cdot 5)(x^2 \cdot x)$
$ = (3x^2)(5x)$ The factors are $3x^2$ and $5x$. *Check*: $3x^2 \cdot 5x = 15x^3$.

c) $15x^3 = ((-5)(-3))x^3$
$ = (-5)(-3x^3)$ The factors are -5 and $-3x^3$.
 *Check*: $(-5)(-3x^3) = 15x^3$.

We see that $(3x)(5x^2)$, $(3x^2)(5x)$, and $(-5)(-3x^3)$ are all factorizations of $15x^3$. Other factorizations exist as well.

1. Find three factorizations of $30a^3$.

YOUR TURN

B. Factoring When Terms Have a Common Factor

To factor a polynomial with two or more terms, we use the distributive law with the sides of the equation switched.

Multiply: $3(x + 2y - z) = 3 \cdot x + 3 \cdot 2y - 3 \cdot z$
$ = 3x + 6y - 3z$

Factor: $3x + 6y - 3z = 3 \cdot x + 3 \cdot 2y - 3 \cdot z$
$ = 3(x + 2y - z)$

In the factorization above, note that since 3 appears as a factor of $3x$, $6y$, and $-3z$, it is a *common factor* for all the terms of $3x + 6y - 3z$.

We generally factor out the *largest* common factor.

EXAMPLE 2 Factor to form an equivalent expression: $8a - 12$.

SOLUTION We write the prime factorization of both terms, lining up common factors in columns:

The prime factorization of $8a$ is $2 \cdot 2 \cdot 2 \cdot a$;

The prime factorization of 12 is $2 \cdot 2 \cdot \qquad 3$.

Since both factorizations include two factors of 2, the largest common factor is $2 \cdot 2$, or 4:

$$8a - 12 = 4 \cdot 2a - 4 \cdot 3 \qquad \text{4 is a factor of } 8a \text{ and of 12.}$$
$$8a - 12 = 4(2a - 3). \qquad \text{Try to go directly to this step.}$$

Check: $4(2a - 3) = 4 \cdot 2a - 4 \cdot 3 = 8a - 12$, as expected.

2. Factor to form an equivalent expression: $12x - 30$.

The factorization of $8a - 12$ is $4(2a - 3)$.

YOUR TURN

CAUTION! $2 \cdot 2 \cdot 2 \cdot a - 2 \cdot 2 \cdot 3$ is a factorization of the *terms* of $8a - 12$ but not of the polynomial itself. The factorization of $8a - 12$ is $4(2a - 3)$.

A common factor may contain a variable.

EXAMPLE 3 Factor: $24x^5 + 30x^2$.

SOLUTION

The prime factorization of $24x^5$ is $2 \cdot 2 \cdot 2 \cdot 3 \cdot \quad x \cdot x \cdot x \cdot x \cdot x.$

The prime factorization of $30x^2$ is $2 \cdot \qquad 3 \cdot 5 \cdot x \cdot x.$

The largest common factor is $2 \cdot 3 \cdot x \cdot x$, or $6x^2$.

$$24x^5 + 30x^2 = 6x^2 \cdot 4x^3 + 6x^2 \cdot 5 \qquad \text{Factoring each term}$$
$$= 6x^2(4x^3 + 5) \qquad \text{Factoring out } 6x^2$$

Check: $6x^2(4x^3 + 5) = 6x^2 \cdot 4x^3 + 6x^2 \cdot 5 = 24x^5 + 30x^2$, as expected.

3. Factor: $30x^4 + 40x^3$.

The factorization of $24x^5 + 30x^2$ is $6x^2(4x^3 + 5)$.

YOUR TURN

The largest common factor of a polynomial is the largest common factor of the coefficients times the largest common factor of the variable(s) in all the terms. Suppose in Example 3 that you did not recognize the *largest* common factor, and removed only part of it, as follows:

$$24x^5 + 30x^2 = 2x^2 \cdot 12x^3 + 2x^2 \cdot 15 \qquad 2x^2 \text{ is a common factor.}$$
$$= 2x^2(12x^3 + 15). \qquad 12x^3 + 15 \text{ itself contains a common factor, 3.}$$

Note that $12x^3 + 15$ still has a common factor, 3. To find the largest common factor, we extend the above factoring, as follows, until no more common factors exist:

$$24x^5 + 30x^2 = 2x^2[3(4x^3 + 5)] \qquad \text{Factoring } 12x^3 + 15. \text{ Remember to rewrite the first common factor, } 2x^2.$$

$$= 6x^2(4x^3 + 5). \qquad \text{Using an associative law; } 2x^2 \cdot 3 = 6x^2$$

Since $4x^3 + 5$ cannot be factored any further, we say that we have factored *completely*. **When we are directed simply to factor, it is understood that we should always factor completely.**

EXAMPLE 4 Factor: $12x^5 - 15x^4 + 27x^3$.

SOLUTION

The prime factorization of $12x^5$ is $2 \cdot 2 \cdot 3 \cdot \quad x \cdot x \cdot x \cdot x \cdot x.$
The prime factorization of $15x^4$ is $\quad 3 \cdot \quad 5 \cdot x \cdot x \cdot x \cdot x.$
The prime factorization of $27x^3$ is $\quad 3 \cdot 3 \cdot 3 \cdot \quad x \cdot x \cdot x.$

The largest common factor is $3 \cdot x \cdot x \cdot x$, or $3x^3$.

$$12x^5 - 15x^4 + 27x^3 = 3x^3 \cdot 4x^2 - 3x^3 \cdot 5x + 3x^3 \cdot 9$$
$$= 3x^3(4x^2 - 5x + 9)$$

Since $4x^2 - 5x + 9$ has no common factor, we are done, except for a check:

$$3x^3(4x^2 - 5x + 9) = 3x^3 \cdot 4x^2 - 3x^3 \cdot 5x + 3x^3 \cdot 9$$
$$= 12x^5 - 15x^4 + 27x^3,$$

4. Factor: $8a^5 + 12a^3 - 6a^2$.

as expected. The factorization of $12x^5 - 15x^4 + 27x^3$ is $3x^3(4x^2 - 5x + 9)$.

YOUR TURN

Note in Examples 3 and 4 that the *largest* common variable factor is the *smallest* power of x in the original polynomial.

With practice, we can determine the largest common factor without writing the prime factorization of each term. Then, to factor, we write the largest common factor and parentheses and then fill in the parentheses. It is customary for the leading coefficient of the polynomial inside the parentheses to be positive.

EXAMPLE 5 Factor: **(a)** $8r^5s^2 + 16rs^5$; **(b)** $-3xy + 6xz - 3x$.

SOLUTION

a) $8r^5s^2 + 16rs^5 = 8rs^2(r^4 + 2s^3)$ Try to go directly to this step.

The largest common factor is $8rs^2$. $\begin{cases} 8r^5s^2 = 2 \cdot 2 \cdot 2 \cdot \quad r \cdot r^4 \cdot s^2 \\ 16rs^5 = 2 \cdot 2 \cdot 2 \cdot 2 \cdot r \cdot \quad s^2 \cdot s^3 \end{cases}$

Check: $8rs^2(r^4 + 2s^3) = 8r^5s^2 + 16rs^5$.

Student Notes

The 1 in $(y - 2z + 1)$ plays an important role in Example 5(b). Without the 1, the term $-3x$ would not appear in the check.

b) $-3xy + 6xz - 3x = -3x(y - 2z + 1)$ Note that either $-3x$ or $3x$ can be the largest common factor.

We generally factor out a negative when the first coefficient is negative. The way we factor can depend on the situation in which we are working. We might also factor as follows:

$$-3xy + 6xz - 3x = 3x(-y + 2z - 1).$$

The checks are left to the student.

5. Factor: $12a^2b^4 - 8a^3b^3$.

YOUR TURN

In some texts, the largest common factor is referred to as the *greatest* common factor. We have avoided this language because, as shown in Example 5(b), the largest common factor may represent a negative value that is actually *less* than other common factors.

> **TIPS FOR FACTORING**
> 1. Factor out the largest common factor, if one exists.
> 2. The common factor multiplies a polynomial with the same number of terms as the original polynomial.
> 3. Factoring can always be checked by multiplying. Multiplication should yield the original polynomial.

C. Factoring by Grouping

Sometimes algebraic expressions contain a common factor with two or more terms.

EXAMPLE 6   Factor: $x^2(x + 1) + 2(x + 1)$.

SOLUTION   The binomial $x + 1$ is a factor of both $x^2(x + 1)$ and $2(x + 1)$. Thus, $x + 1$ is a common factor:

$$x^2(x + 1) + 2(x + 1) = (x + 1)x^2 + (x + 1)2$$

Using a commutative law twice. Try to do this step mentally.

$$= (x + 1)(x^2 + 2).$$

Factoring out the common factor, $x + 1$

To check, we could simply reverse the above steps.
The factorization is $(x + 1)(x^2 + 2)$.

6. Factor: $x(x - 5) + 3(x - 5)$.

YOUR TURN

Some polynomials with four terms have a common binomial factor. In order to identify that factor, we regroup into two groups of two terms each. Factoring common factors from each group separately may reveal a common binomial factor. This method, known as **factoring by grouping**, can be tried on any polynomial with four terms. (Factoring by grouping also works for some polynomials with more than four terms.)

EXAMPLE 7   Factor: $5x^3 - x^2 + 15x - 3$.

SOLUTION

$$5x^3 - x^2 + 15x - 3 = (5x^3 - x^2) + (15x - 3)$$

Grouping as two binomials

$$= x^2(5x - 1) + 3(5x - 1)$$

Factoring each binomial

$$= (5x - 1)(x^2 + 3)$$

Factoring out the common factor, $5x - 1$

Check: $(5x - 1)(x^2 + 3) = 5x \cdot x^2 + 5x \cdot 3 - 1 \cdot x^2 - 1 \cdot 3$ Using FOIL
$$= 5x^3 - x^2 + 15x - 3.$$

7. Factor: $x^3 + 3x^2 + 4x + 12$.

YOUR TURN

EXAMPLE 8   Factor by grouping.

a) $2x^3 + 8x^2 + x + 4$ **b)** $4t^3 - 15 + 20t^2 - 3t$

SOLUTION

a) $2x^3 + 8x^2 + x + 4 = (2x^3 + 8x^2) + (x + 4)$

$\qquad\qquad\qquad = 2x^2(x + 4) + 1(x + 4)$ Factoring $2x^3 + 8x^2$ to find a common binomial factor. Writing the 1 helps with the next step.

> **CAUTION!**   Be sure to include the term 1.

$\qquad\qquad\qquad = (x + 4)(2x^2 + 1)$ Factoring out the common factor, $x + 4$. The 1 is essential in the factor $2x^2 + 1$.

Check: $(x + 4)(2x^2 + 1) = x \cdot 2x^2 + x \cdot 1 + 4 \cdot 2x^2 + 4 \cdot 1$ Using FOIL

$\qquad\qquad\qquad\qquad\quad = 2x^3 + x + 8x^2 + 4$

$\qquad\qquad\qquad\qquad\quad = 2x^3 + 8x^2 + x + 4.$ Using a commutative law

The factorization is $(x + 4)(2x^2 + 1)$.

b) When we try grouping $4t^3 - 15 + 20t^2 - 3t$ as

$\qquad (4t^3 - 15) + (20t^2 - 3t),$

we are unable to factor $4t^3 - 15$. When this happens, we can rearrange the polynomial and try a different grouping:

$4t^3 - 15 + 20t^2 - 3t = 4t^3 + 20t^2 - 3t - 15$ Using the commutative law to rearrange the terms

$\qquad\qquad\qquad\qquad = 4t^2(t + 5) - 3(t + 5)$ By factoring out -3, we see that $t + 5$ is a common factor.

$\qquad\qquad\qquad\qquad = (t + 5)(4t^2 - 3).$ To check, we use FOIL.

8. Factor by grouping:

$\qquad 3x^3 - 3x^2 + 2x - 2.$

↩ YOUR TURN

To reverse the order of subtraction, we can factor out -1.

> **FACTORING OUT -1 TO REVERSE SUBTRACTION**
>
> $b - a = -1(a - b) = -(a - b)$

EXAMPLE 9   Factor: $t^5 - 5t^4 + 10 - 2t$.

SOLUTION

$t^5 - 5t^4 + 10 - 2t = (t^5 - 5t^4) + (10 - 2t)$ Grouping

$\qquad\qquad\qquad\quad = t^4(t - 5) + 2(5 - t)$ Factoring each binomial

$\qquad\qquad\qquad\quad = t^4(t - 5) + 2(-1)(t - 5)$ Factoring out -1 to reverse $5 - t$

$\qquad\qquad\qquad\quad = t^4(t - 5) - 2(t - 5)$ Simplifying

$\qquad\qquad\qquad\quad = (t - 5)(t^4 - 2)$ Factoring out $t - 5$

The check is left to the student.

9. Factor: $2x^3 - 16x^2 + 8 - x$.

↩ YOUR TURN

Checking by Evaluating

Chapter Resource:
Decision Making: Connection,
p. 362

One way to check a factorization is to multiply. A second type of check uses the fact that equivalent expressions have the same value when evaluated for the same replacement. Thus a quick, partial check of Example 8(a) can be made by using a convenient replacement for x (say, 1) and evaluating both $2x^3 + 8x^2 + x + 4$ and $(x + 4)(2x^2 + 1)$:

$$2 \cdot 1^3 + 8 \cdot 1^2 + 1 + 4 = 2 + 8 + 1 + 4 = 15;$$
$$(1 + 4)(2 \cdot 1^2 + 1) = 5 \cdot 3 = 15.$$

Since the value of both expressions is the same, the factorization is probably correct. Evaluating for several values will make the check more certain.

Technology Connection

A partial check of a factorization can be performed using a table or a graph. To check Example 8(a), we let

$$y_1 = 2x^3 + 8x^2 + x + 4 \quad \text{and} \quad y_2 = (x + 4)(2x^2 + 1).$$

Then we set up a table in AUTO mode. If the factorization is correct, the values of y_1 and y_2 will be the same regardless of the table settings used.

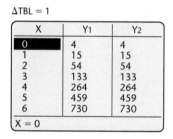

ΔTBL = 1

X	Y₁	Y₂
0	4	4
1	15	15
2	54	54
3	133	133
4	264	264
5	459	459
6	730	730

X = 0

We can also graph $y_1 = 2x^3 + 8x^2 + x + 4$ and $y_2 = (x + 4)(2x^2 + 1)$. If the graphs appear to coincide, the factorization is probably correct. The TRACE feature can be used to confirm this. This approach is illustrated using a graphing calculator app.

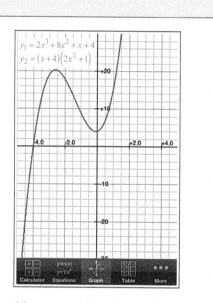

Use a table or a graph to determine whether each of the following factorizations is correct.

1. $x^2 - 7x - 8 = (x - 8)(x + 1)$
2. $10x^2 + 37x + 7 = (5x - 1)(2x + 7)$
3. $12x^2 - 17x - 5 = (6x + 1)(2x - 5)$
4. $12x^2 - 17x - 5 = (4x + 1)(3x - 5)$
5. $x^2 - 4 = (x - 2)(x - 2)$
6. $x^2 - 4 = (x + 2)(x - 2)$

↳ Check Your UNDERSTANDING

Determine the largest common factor for each pair of terms.

1. $16x$ and 24
2. $6x^3$ and $8x^4$
3. $a^2b^3c^5$ and a^4bc^6
4. $30u^2v^5$ and $40u^3v^3$

Determine whether each of the following could be a factorization of a polynomial.

5. $x(x - 3)$
6. $(x + 2)(x - 1)$
7. $x \cdot x + 3 \cdot x$
8. $x(x + 1) - 7$

5.1   EXERCISE SET

FOR EXTRA HELP   MyMathLab®

⮕ Vocabulary and Reading Check

Classify each of the following statements as either true or false.

1. The largest common factor of a polynomial always has the same degree as the polynomial itself.

2. When the leading coefficient of a polynomial is negative, we generally factor out a common factor with a negative coefficient.

3. A polynomial is not prime if it contains a common factor other than 1 or −1.

4. All polynomials with four terms can be factored by grouping.

⮕ Concept Reinforcement

In each of Exercises 5–12, match the phrase with the most appropriate choice from the column on the right.

5. ____ A factorization of $35a^2b$

6. ____ A factor of $35a^2b$

7. ____ A common factor of $5x + 10$ and $4x + 8$

8. ____ A factorization of $3x^4 - 9x^2$

9. ____ A factorization of $9x^4 - 3x^2$

10. ____ A common factor of $2x + 10$ and $4x + 8$

11. ____ A factor of $3a + 6a^2$

12. ____ A factorization of $3a + 6a^2$

a) $3a(1 + 2a)$

b) $x + 2$

c) $3x^2(3x^2 - 1)$

d) $1 + 2a$

e) $3x^2(x^2 - 3)$

f) $5a^2$

g) 2

h) $7a \cdot 5ab$

A.   Factoring Monomials

Find three factorizations for each monomial. Answers may vary.

13. $14x^3$

14. $22x^3$

15. $-15a^4$

16. $-8t^5$

17. $25t^5$

18. $9a^4$

B.   Factoring When Terms Have a Common Factor

Factor. Remember to use the largest common factor and to check by multiplying. Factor out a negative factor if the first coefficient is negative.

19. $8x + 24$

20. $10x + 50$

21. $2x^2 + 2x - 8$

22. $6x^2 + 3x - 15$

23. $3t^2 + t$

24. $2t^2 + t$

25. $-5y^2 - 10y$

26. $-4y^2 - 12y$

27. $x^3 + 6x^2$

28. $5x^4 - x^2$

29. $16a^4 - 24a^2$

30. $25a^5 + 10a^3$

31. $-6t^6 + 9t^4 - 4t^2$

32. $-10t^5 + 15t^4 + 9t^3$

33. $6x^8 + 12x^6 - 24x^4 + 30x^2$

34. $10x^4 - 30x^3 - 50x - 20$

35. $x^5y^5 + x^4y^3 + x^3y^3 - x^2y^2$

36. $x^9y^6 - x^7y^5 + x^4y^4 + x^3y^3$

37. $-35a^3b^4 + 10a^2b^3 - 15a^3b^2$

38. $-21r^5t^4 - 14r^4t^6 + 21r^3t^6$

C.   Factoring by Grouping

Factor.

39. $n(n - 6) + 3(n - 6)$

40. $b(b + 5) + 3(b + 5)$

41. $x^2(x + 3) - 7(x + 3)$

42. $3z^2(2z + 9) + (2z + 9)$

43. $y^2(2y - 9) + (2y - 9)$

44. $x^2(x - 7) - 3(x - 7)$

Factor by grouping, if possible, and check.

45. $x^3 + 2x^2 + 5x + 10$

46. $z^3 + 3z^2 + 7z + 21$

47. $9n^3 - 6n^2 + 3n - 2$

48. $10x^3 - 25x^2 + 2x - 5$

49. $4t^3 - 20t^2 + 3t - 15$

50. $8a^3 - 2a^2 + 12a - 3$

51. $7x^3 + 5x^2 - 21x - 15$

52. $5x^3 + 4x^2 - 10x - 8$

53. $6a^3 + 7a^2 + 6a + 7$

54. $7t^3 - 5t^2 + 7t - 5$

55. $2x^3 - 12x^2 - x + 6$

56. $x^3 - x^2 - 2x + 5$

57. $p^3 + p^2 - 3p + 10$

58. $a^3 - 3a^2 + 6 - 2a$

59. $y^3 + 8y^2 - 2y - 16$

60. $3x^3 + 18x^2 - 5x - 25$

61. $2x^3 + 36 - 8x^2 - 9x$

62. $20g^3 + 5 - 4g^2 - 25g$

63. In answering a factoring problem, Taylor says the largest common factor is $-5x^2$ and Kimber says the largest common factor is $5x^2$. Can they both be correct? Why or why not?

64. Write a two-sentence paragraph in which the word "factor" is used at least once as a noun and once as a verb.

Skill Review

Simplify.

65. $\frac{2}{5} \div \frac{10}{3}$ [1.3]

66. $-1 + 20 \div 2^2 \cdot 5$ [1.8]

67. $(2xy^{-4})^{-1}$ [4.2]

68. $(a^2b^3c)(a^5b^4c^2)$ [4.1]

69. $(3x^2 - x - 3) - (-x^2 - 6x + 10)$ [4.4]

70. $(w^2y + yz - w^2) + (w^2 - 2w^2y + wz)$ [4.7]

Synthesis

71. Suresh factors $12x^2y - 18xy^2$ as $6xy \cdot 2x - 6xy \cdot 3y$. Is this the factorization of the polynomial? Why or why not?

72. Azrah recognizes that evaluating usually provides only a partial check of her factoring. Because of this, she often performs a second check with a different replacement value. Is this a good idea? Why or why not?

Factor, if possible.

73. $4x^5 + 6x^2 + 6x^3 + 9$

74. $x^6 + x^2 + x^4 + 1$

75. $2x^4 + 2x^3 - 4x^2 - 4x$

76. $x^3 + x^2 - 2x + 2$

Aha! **77.** $5x^5 - 5x^4 + x^3 - x^2 + 3x - 3$

Aha! **78.** $ax^2 + 2ax + 3a + x^2 + 2x + 3$

79. Write a trinomial of degree 7 for which $8x^2y^3$ is the largest common factor. Answers may vary.

80. Kris and Tina are each calculating the amount of sheet metal needed to form a circular canister with a radius of 5 in. and a height of 20 in. Kris uses the formula $A = 2\pi rh + 2\pi r^2$. Tina adds $5 + 20$, and then multiplies that sum by 10 and then by π. Do the methods give the same result? Which seems easier?

YOUR TURN ANSWERS: SECTION 5.1
1. $(3a)(10a^2), (6a^2)(5a), (-3)(-10a^3)$; answers may vary **2.** $6(2x - 5)$ **3.** $10x^3(3x + 4)$
4. $2a^2(4a^3 + 6a - 3)$ **5.** $4a^2b^3(3b - 2a)$
6. $(x - 5)(x + 3)$ **7.** $(x + 3)(x^2 + 4)$
8. $(x - 1)(3x^2 + 2)$ **9.** $(x - 8)(2x^2 - 1)$

Prepare to Move On
Multiply. [4.6]

1. $(x + 2)(x + 7)$ **2.** $(x - 2)(x - 7)$

3. $(x + 2)(x - 7)$ **4.** $(x - 2)(x + 7)$

List all factors of each number. [1.3]

5. 60 **6.** 18

5.2 Factoring Trinomials of the Type $x^2 + bx + c$

A. When the Constant Term Is Positive **B.** When the Constant Term Is Negative **C.** Prime Polynomials
D. Factoring Completely

We now learn how to factor trinomials like

$$x^2 + 5x + 4 \quad \text{or} \quad x^2 + 3x - 10,$$

for which no common factor exists and the leading coefficient is 1. Recall that when factoring, we are writing an equivalent expression that is a product. For these trinomials, the factors will be binomials.

Study Skills

Leave a Trail

Your supporting work is not "scrap" work. Most instructors regard your reasoning as extremely important. Try to organize your supporting work so that your instructor (and you as well) can follow your steps. If you abandon a line of work, a simple × through it will allow you and others to review it if desired.

Compare the following multiplications:

$$
\begin{array}{cccc}
\text{F} & \text{O} & \text{I} & \text{L} \\
\downarrow & \downarrow & \downarrow & \downarrow
\end{array}
$$

$$(x + 2)(x + 5) = x^2 + 5x + 2x + 2 \cdot 5$$
$$= x^2 + 7x + 10;$$

$$(x - 2)(x - 5) = x^2 - 5x - 2x + (-2)(-5)$$
$$= x^2 - 7x + 10;$$

$$(x + 3)(x - 7) = x^2 - 7x + 3x + 3(-7)$$
$$= x^2 - 4x - 21;$$

$$(x - 3)(x + 7) = x^2 + 7x - 3x + (-3)7$$
$$= x^2 + 4x - 21.$$

Note that for all four products:

- The product of the two binomials is a trinomial.
- The coefficient of x in the trinomial is the sum of the constant terms in the binomials.
- The constant term in the trinomial is the product of the constant terms in the binomials.

These observations lead to a method for factoring certain trinomials. We first consider those with a positive constant term, as in the first two products above.

A. When the Constant Term Is Positive

To factor a polynomial like $x^2 + 7x + 10$, we think of FOIL in reverse. The x^2 suggests that the first term of each binomial factor is x. Next, we look for numbers p and q such that

$$x^2 + 7x + 10 = (x + p)(x + q).$$

To get the middle term and the last term of the trinomial, we need two numbers, p and q, whose product is 10 and whose sum is 7. Those numbers are 2 and 5. Thus the factorization is

$$(x + 2)(x + 5).$$

Check: $(x + 2)(x + 5) = x^2 + 5x + 2x + 10$
$$= x^2 + 7x + 10.$$

EXAMPLE 1   Factor to form an equivalent expression:

$$x^2 + 5x + 6.$$

SOLUTION   Think of FOIL in reverse. The first term of each factor is x:

$$(x + \quad)(x + \quad).$$

To complete the factorization, we need a constant term for each binomial. The constants must have a product of 6 and a sum of 5. We list some pairs of numbers that multiply to 6 and then check the sum of each pair of factors.

A GEOMETRIC APPROACH TO EXAMPLE 1

The product of two binomials can be regarded as the sum of the areas of four rectangles. Thus we can regard the factoring of $x^2 + 5x + 6$ as a search for p and q so that the sum of areas A, B, C, and D is $x^2 + 5x + 6$.

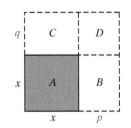

Note that area D is the product of p and q. In order for area D to be 6, p and q must be either 1 and 6 or 2 and 3. We illustrate both below.

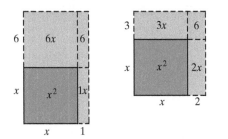

When p and q are 1 and 6, the total area is $x^2 + 7x + 6$, but when p and q are 2 and 3, as shown on the right, the total area is $x^2 + 5x + 6$, as desired. Thus the factorization of $x^2 + 5x + 6$ is $(x + 2)(x + 3)$.

Student Notes

It is important to be able to systematically list all pairs of factors of a number. For the factorizations in this section, the order of the numbers within each pair is not important.

Pairs of Factors of 6	Sums of Factors
1, 6	7
2, 3	5 ←
−1, −6	−7
−2, −3	−5

The numbers that we are looking for are 2 and 3.

Every pair has a product of 6.   *One* pair has a sum of 5.

Since

$$2 \cdot 3 = 6 \quad \text{and} \quad 2 + 3 = 5,$$

the factorization of $x^2 + 5x + 6$ is $(x + 2)(x + 3)$.

Check: $(x + 2)(x + 3) = x^2 + 3x + 2x + 6 = x^2 + 5x + 6.$

Thus, $(x + 2)(x + 3)$ is a product that is equivalent to $x^2 + 5x + 6$.

Note that since 5 and 6 are both positive, when factoring $x^2 + 5x + 6$ we need not have listed negative factors of 6. Note too that changing the signs of the factors changes only the sign of the sum (see the table above).

1. Factor to form an equivalent expression: $x^2 + 6x + 8$.

⟲ YOUR TURN

At the beginning of this section, we considered the multiplication $(x - 2)(x - 5)$. For this product, the resulting trinomial, $x^2 - 7x + 10$, has a positive constant term but a negative coefficient of x. This is because the *product* of two negative numbers is always positive, whereas the *sum* of two negative numbers is always negative.

TO FACTOR $x^2 + bx + c$ WHEN c IS POSITIVE

When the constant term c of a trinomial is positive, look for two numbers with the same sign. Select pairs of numbers with the sign of b, the coefficient of the middle term.

$$x^2 - 7x + 10 = (x - 2)(x - 5); \qquad b \text{ is negative; } c \text{ is positive.}$$

$$x^2 + 7x + 10 = (x + 2)(x + 5) \qquad b \text{ is positive; } c \text{ is positive.}$$

EXAMPLE 2 Factor: $y^2 - 8y + 12$.

SOLUTION Since the constant term is positive and the coefficient of the middle term is negative, we look for a factorization of 12 in which both factors are negative. Their sum must be -8.

Pairs of Factors of 12	Sums of Factors
−1, −12	−13
−2, −6	−8 ←
−3, −4	−7

We need a sum of -8. The numbers we need are -2 and -6.

2. Factor: $a^2 - 8a + 15$.

The factorization of $y^2 - 8y + 12$ is $(y - 2)(y - 6)$. The check is left to the student.

⟲ YOUR TURN

B. When the Constant Term Is Negative

As we saw in two of the multiplications at the beginning of this section, the product of two binomials can have a negative constant term:

$$(x + 3)(x - 7) = x^2 - 4x - 21$$

and

$$(x - 3)(x + 7) = x^2 + 4x - 21.$$

It is important to note that when the signs of the constants in the binomials are reversed, only the sign of the middle term of the trinomial changes.

EXAMPLE 3 Factor: $x^2 - 8x - 20$.

SOLUTION The constant term, -20, must be expressed as the product of a negative number and a positive number. Since the sum of these two numbers must be negative (specifically, -8), the negative number must have the greater absolute value.

Pairs of Factors of −20	Sums of Factors
1, −20	−19
2, −10	−8 ←——— The numbers we need are 2 and −10.
4, −5	−1

The numbers that we are looking for are 2 and -10.

Check: $(x + 2)(x - 10) = x^2 - 10x + 2x - 20$
$$= x^2 - 8x - 20.$$

The factorization of $x^2 - 8x - 20$ is $(x + 2)(x - 10)$.

3. Factor: $t^2 - 2t - 15$.

YOUR TURN

TO FACTOR $x^2 + bx + c$ WHEN c IS NEGATIVE

When the constant term c is negative, look for a positive number and a negative number that multiply to c. Select pairs of numbers for which the number with the larger absolute value has the sign of b, the coefficient of the middle term.

$$x^2 - 4x - 21 = (x + 3)(x - 7); \qquad b \text{ is negative; } c \text{ is negative.}$$

$$x^2 + 4x - 21 = (x - 3)(x + 7) \qquad b \text{ is positive; } c \text{ is negative.}$$

Student Notes

Writing a trinomial in descending order will help you to identify b and c.

EXAMPLE 4 Factor: $t^2 - 24 + 5t$.

SOLUTION We first write the trinomial in descending order: $t^2 + 5t - 24$. The factorization of the constant term, -24, must have one positive factor and one negative factor. The sum of the factors must be 5, so the positive factor must have the larger absolute value. Thus we consider only pairs of factors in which the positive factor has the larger absolute value.

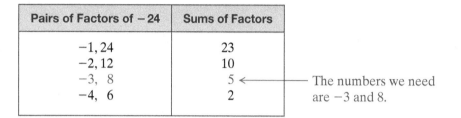

Pairs of Factors of -24	Sums of Factors	
$-1, 24$	23	
$-2, 12$	10	
$-3,\ 8$	5 ←	The numbers we need are -3 and 8.
$-4,\ 6$	2	

4. Factor: $t^2 - 28 + 3t$.

The factorization is $(t - 3)(t + 8)$. The check is left to the student.

YOUR TURN

Polynomials in two or more variables, such as $a^2 + 4ab - 21b^2$, are factored in a similar manner.

EXAMPLE 5 Factor: $a^2 + 4ab - 21b^2$.

SOLUTION We look for numbers p and q such that

$$a^2 + 4ab - 21b^2 = (a + pb)(a + qb).$$

Our thinking is much the same as if we were factoring $x^2 + 4x - 21$. We look for factors of -21 whose sum is 4. Those factors are -3 and 7. Thus,

$$a^2 + 4ab - 21b^2 = (a - 3b)(a + 7b).$$

Check: $(a - 3b)(a + 7b) = a^2 + 7ab - 3ba - 21b^2$
$$= a^2 + 4ab - 21b^2.$$

5. Factor: $x^2 - 3xy - 18y^2$.

The factorization of $a^2 + 4ab - 21b^2$ is $(a - 3b)(a + 7b)$.

YOUR TURN

C. Prime Polynomials

EXAMPLE 6 Factor: $x^2 - x + 5$.

SOLUTION Since 5 has very few factors, we can easily check all possibilities.

Pairs of Factors of 5	Sums of Factors
$5,\ 1$	6
$-5, -1$	-6

6. Factor: $x^2 + x + 1$.

Since there are no factors whose sum is -1, the polynomial is *not* factorable into binomials.

YOUR TURN

In this text, a polynomial like $x^2 - x + 5$ that cannot be factored using rational numbers is said to be **prime**. In more advanced courses, other types of numbers are considered. There, polynomials like $x^2 - x + 5$ can be factored and are not considered prime.

D. Factoring Completely

Often factoring requires two or more steps. Remember, when told to factor, we should *factor completely*. This means that the final factorization should contain only prime polynomials.

EXAMPLE 7   Factor: $-2x^3 + 20x^2 - 50x$.

SOLUTION   *Always* look first for a common factor. This time there is one. Since the leading coefficient is negative, we begin by factoring out $-2x$:

$$-2x^3 + 20x^2 - 50x = -2x(x^2 - 10x + 25).$$

Now consider $x^2 - 10x + 25$. Since the constant term is positive and the coefficient of the middle term is negative, we look for a factorization of 25 in which both factors are negative. Their sum must be -10.

Pairs of Factors of 25	Sums of Factors
$-25, -1$	-26
$-5, -5$	-10 ⟵——— The numbers we need are -5 and -5.

The factorization of $x^2 - 10x + 25$ is $(x - 5)(x - 5)$, or $(x - 5)^2$. The factorization of the original polynomial also includes the factor $-2x$.

> **CAUTION!**   When factoring involves more than one step, be careful to write out the *entire* factorization.

Check: $-2x(x - 5)(x - 5) = -2x[x^2 - 10x + 25]$ Multiplying binomials
$$= -2x^3 + 20x^2 - 50x.$$ Using the distributive law

The factorization of $-2x^3 + 20x^2 - 50x$ is $-2x(x - 5)(x - 5)$, or $-2x(x - 5)^2$.

YOUR TURN

Once any common factors have been factored out, the following summary can be used to factor $x^2 + bx + c$.

> **TO FACTOR $x^2 + bx + c$**
>
> **1.** Find a pair of factors that have c as their product and b as their sum.
>
>   **a)** When c is positive, both factors will have the same sign as b.
>   **b)** When c is negative, one factor will be positive and the other will be negative. Select the factors such that the factor with the larger absolute value has the same sign as b.
>
> **2.** Check by multiplying.

Note that each polynomial has a unique factorization (except for the order in which the factors are written).

Student Notes

Whenever a new set of parentheses is created while factoring, check the expression inside the parentheses to see if it can be factored further.

Chapter Resource:
Collaborative Activity
(Visualizing Factoring), p. 361

---

7. Factor: $3x^3 + 30x^2 + 27x$.

↳ Check Your UNDERSTANDING

For each polynomial, list all pairs of positive factors of the constant term.

1. $x^2 + 13x + 30$

2. $x^2 + 10x + 9$

3. $a^2 + 12a + 11$

4. $t^2 + 14t + 48$

5.2 EXERCISE SET

FOR EXTRA HELP MyMathLab®

Vocabulary and Reading Check

Classify each of the following statements as either true or false.

1. Whenever the product of a pair of factors is negative, the factors have the same sign.

2. Anytime the sum of a negative number and a positive number is negative, the negative number has the greater absolute value.

3. When factoring any polynomial, it is always best to look first for a common factor.

4. When we factor a trinomial such as $x^2 + 6x + 5$, we try to write it as the product of two binomials.

Concept Reinforcement

For Exercises 5–10, assume that $x^2 + bx + c$ can be factored as $(x + p)(x + q)$. Complete each sentence by replacing each blank with either "positive" or "negative."

5. If b is positive and c is positive, then p will be _____ and q will be _____.

6. If b is negative and c is positive, then p will be _____ and q will be _____.

7. If p is negative and q is negative, then b must be _____ and c must be _____.

8. If p is positive and q is positive, then b must be _____ and c must be _____.

9. If b, c, and p are all negative, then q must be _____.

10. If b and c are negative and p is positive, then q must be _____.

A, B, C, D. Factoring $x^2 + bx + c$

Factor completely. Remember to look first for a common factor. Check by multiplying. If a polynomial is prime, state this.

11. $x^2 + 8x + 16$
12. $x^2 + 9x + 20$
13. $x^2 + 11x + 10$
14. $y^2 + 8y + 7$
15. $t^2 - 9t + 14$
16. $a^2 - 9a + 20$
17. $b^2 - 5b + 4$
18. $z^2 - 8z + 7$
19. $d^2 - 7d + 10$
20. $x^2 - 8x + 15$
21. $x^2 - 2x - 15$
22. $x^2 - x - 42$
23. $x^2 + 2x - 15$
24. $x^2 + x - 42$
25. $2x^2 - 14x - 36$
26. $3y^2 - 9y - 84$
27. $-x^3 + 6x^2 + 16x$
28. $-x^3 + x^2 + 42x$

29. $4y - 45 + y^2$
30. $7x - 60 + x^2$
31. $x^2 - 72 + 6x$
32. $-2x - 99 + x^2$
33. $-5b^2 - 35b + 150$
34. $-c^4 - c^3 + 56c^2$
35. $x^5 - x^4 - 2x^3$
36. $2a^2 - 4a - 70$
37. $x^2 + 5x + 10$
38. $x^2 + 11x + 18$
39. $32 + 12t + t^2$
40. $y^2 - y + 1$
41. $x^2 + 20x + 99$
42. $x^2 + 20x + 100$
43. $3x^3 - 63x^2 - 300x$
44. $2x^3 - 40x^2 + 192x$
45. $-4x^2 - 40x - 100$
46. $-2x^2 + 42x + 144$
47. $y^2 - 20y + 96$
48. $144 - 25t + t^2$
49. $-a^6 - 9a^5 + 90a^4$
50. $-a^4 - a^3 + 132a^2$
51. $t^2 + \frac{2}{3}t + \frac{1}{9}$
52. $x^2 - \frac{2}{5}x + \frac{1}{25}$
53. $11 + w^2 - 4w$
54. $6 + p^2 + 2p$
55. $p^2 - 7pq + 10q^2$
56. $a^2 - 2ab - 3b^2$
57. $m^2 + 5mn + 5n^2$
58. $x^2 - 11xy + 24y^2$
59. $s^2 - 4st - 12t^2$
60. $b^2 + 8bc - 20c^2$
61. $6a^{10} + 30a^9 - 84a^8$
62. $5a^8 - 20a^7 - 25a^6$

63. Without multiplying $(x - 17)(x - 18)$, explain why it cannot possibly be a factorization of $x^2 + 35x + 306$.

64. Shari factors $x^3 - 8x^2 + 15x$ as $(x^2 - 5x)(x - 3)$. Is she wrong? Why or why not? What advice would you offer?

Skill Review

Solve.

65. $\frac{y}{2} = -5$ [2.1]
66. $x - 0.05 = 1.08$ [2.1]
67. $3x + 7 = 12$ [2.2]
68. $9 - 5x = 4 - 3x$ [2.2]
69. $x - (2x - 5) = 16$ [2.2]
70. $\frac{1}{3}x - \frac{1}{6} = \frac{1}{2}$ [2.2]

Synthesis

71. When searching for a factorization, why do we list pairs of numbers with the correct *product* instead of pairs of numbers with the correct *sum*?

72. When factoring $x^2 + bx + c$ with a large value of c, Riley begins by writing out the prime factorization of c. What is the advantage of doing this?

73. Find all integers b for which $a^2 + ba - 50$ can be factored.

74. Find all integers m for which $y^2 + my + 50$ can be factored.

Factor completely.

75. $y^2 - 0.2y - 0.08$

76. $x^2 + \frac{1}{2}x - \frac{3}{16}$

77. $-\frac{1}{3}a^3 + \frac{1}{3}a^2 + 2a$

78. $-a^7 + \frac{25}{7}a^5 + \frac{30}{7}a^6$

79. $x^{2m} + 11x^m + 28$

80. $-t^{2n} + 7t^n - 10$

Aha! **81.** $(a + 1)x^2 + (a + 1)3x + (a + 1)2$

82. $ax^2 - 5x^2 + 8ax - 40x - (a - 5)9$
(*Hint:* See Exercise 81.)

83. Find the volume of a cube if its surface area is $(6x^2 + 36x + 54)$ square meters.

Find a polynomial in factored form for the shaded area in each figure. (Use π in your answers where appropriate.)

84.

85.

86.

87.

88.

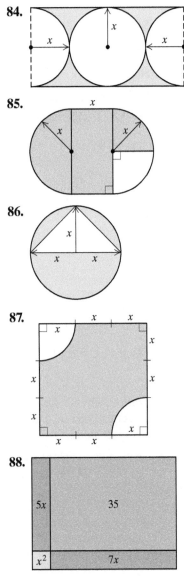

89.

90. A census taker asks a woman, "How many children do you have?"

"Three," she answers.

"What are their ages?"

She responds, "The product of their ages is 36. The sum of their ages is the house number next door."

The math-savvy census taker walks next door, reads the house number, appears puzzled, and returns to the woman, asking, "Is there something you forgot to tell me?"

"Oh yes," says the woman. "I'm sorry. The oldest child is playing a video game."

The census taker records the three ages, thanks the woman for her time, and leaves.

How old is each child? Explain how you reached this conclusion. (*Hint:* Consider factorizations.)

YOUR TURN ANSWERS: SECTION 5.2

1. $(x + 2)(x + 4)$ **2.** $(a - 3)(a - 5)$
3. $(t + 3)(t - 5)$ **4.** $(t + 7)(t - 4)$
5. $(x - 6y)(x + 3y)$ **6.** Not factorable, or prime
7. $3x(x + 9)(x + 1)$

Quick Quiz: Sections 5.1–5.2

Factor.

1. $12x^4 - 30x^3 + 6x^2$ [5.1]

2. $3x^3 - 12x^2 + 5x - 20$ [5.1]

3. $x^2 - 3x + 2$ [5.2]

4. $a^2 + 2ab - 8b^2$ [5.2]

5. $3x^2 + 15x + 18$ [5.2]

Prepare to Move On

Multiply. [4.6]

1. $(2x + 3)(3x + 4)$ **2.** $(2x + 3)(3x - 4)$

3. $(2x - 3)(3x + 4)$ **4.** $(2x - 3)(3x - 4)$

5. $(5x - 1)(x - 7)$ **6.** $(x + 6)(3x - 5)$

5.3 Factoring Trinomials of the Type $ax^2 + bx + c$

A. Factoring with FOIL **B.** The Grouping Method

In this section, we learn to factor trinomials in which the leading, or x^2, coefficient is not 1. First, we will use another FOIL-based method and then we will use an alternative method that involves factoring by grouping. Use the method that you prefer or the one recommended by your instructor.

A. Factoring with FOIL

Before factoring trinomials of the type $ax^2 + bx + c$, consider the following:

$$\overset{\text{F}\qquad\text{O}\qquad\text{I}\qquad\text{L}}{(2x + 5)(3x + 4) = 6x^2 + 8x + 15x + 20}$$
$$= 6x^2 + 23x + 20.$$

To factor $6x^2 + 23x + 20$, we could reverse the multiplication and look for two binomials whose product is this trinomial. We see from above that:

- the product of the First terms must be $6x^2$;
- the product of the Outer terms plus the product of the Inner terms must be $23x$; and
- the product of the Last terms must be 20.

How can such a factorization be found without first seeing the corresponding multiplication? Our first approach relies on trial and error and FOIL.

TO FACTOR $ax^2 + bx + c$ USING FOIL

1. Make certain that all common factors have been removed. If any remain, factor out the largest common factor.
2. Find two First terms whose product is ax^2:

$$(\,\blacksquare\, x + \;)(\,\blacksquare\, x + \;) = ax^2 + bx + c.$$
$$\underline{\hspace{4cm}} \text{FOIL}$$

3. Find two Last terms whose product is c:

$$(\,\blacksquare\, x + \blacksquare\,)(\,\blacksquare\, x + \blacksquare\,) = ax^2 + bx + c.$$
$$\underline{\hspace{4cm}} \text{FOIL}$$

4. Check by multiplying to see if the sum of the Outer and Inner products is bx. If it is not, repeat steps (2) and (3) until the correct combination is found.

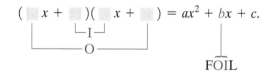

If no correct combination exists, state that the polynomial is prime.

EXAMPLE 1 Factor: $3x^2 - 10x - 8$.

SOLUTION

1. First, check for a common factor. In this case, there is none (other than 1 or -1).

2. Find two **First** terms whose product is $3x^2$. The only possibilities for the **First** terms are $3x$ and x:

$$(3x + \quad)(x + \quad).$$

3. Find two **Last** terms whose product is -8. There are four pairs of factors of -8 and each can be listed in two ways:

$$\begin{array}{cc} -1, & 8 \\ 1, & -8 \\ -2, & 4 \\ 2, & -4 \end{array} \quad \text{and} \quad \begin{array}{cc} 8, & -1 \\ -8, & 1 \\ 4, & -2 \\ -4, & 2. \end{array}$$

> **CAUTION!** For this factoring method, list all pairs of factors in two ways. Since the **First** terms differ, changing the order of the factors of -8 results in a different product.

4. Knowing that all **First** and **Last** products will check, systematically inspect the **Outer** and **Inner** products resulting from steps (2) and (3). Look for the combination in which the sum of the products is the middle term, $-10x$.

Pair of Factors	Corresponding Trial	Product	
$-1,\ 8$	$(3x - 1)(x + 8)$	$3x^2 + 24x - x - 8$ $= 3x^2 + 23x - 8$	Wrong middle term
$1, -8$	$(3x + 1)(x - 8)$	$3x^2 - 24x + x - 8$ $= 3x^2 - 23x - 8$	Wrong middle term
$-2,\ 4$	$(3x - 2)(x + 4)$	$3x^2 + 12x - 2x - 8$ $= 3x^2 + 10x - 8$	Wrong middle term
$2, -4$	$(3x + 2)(x - 4)$	$3x^2 - 12x + 2x - 8$ $= 3x^2 - 10x - 8$	Correct middle term!
$8, -1$	$(3x + 8)(x - 1)$	$3x^2 - 3x + 8x - 8$ $= 3x^2 + 5x - 8$	Wrong middle term
$-8,\ 1$	$(3x - 8)(x + 1)$	$3x^2 + 3x - 8x - 8$ $= 3x^2 - 5x - 8$	Wrong middle term
$4, -2$	$(3x + 4)(x - 2)$	$3x^2 - 6x + 4x - 8$ $= 3x^2 - 2x - 8$	Wrong middle term
$-4,\ 2$	$(3x - 4)(x + 2)$	$3x^2 + 6x - 4x - 8$ $= 3x^2 + 2x - 8$	Wrong middle term

The correct factorization is $(3x + 2)(x - 4)$.

1. Factor: $3x^2 - 13x - 10$.

 YOUR TURN

Two observations can be made from Example 1. First, we listed all possible trials even though we generally stop after finding the correct factorization. We did this to show that **each trial differs only in the middle term of the product.** Second, note that **only the sign of the middle term changes when the signs in the binomials are reversed.**

A GEOMETRIC APPROACH TO EXAMPLE 2

The factoring of $10x^2 + 37x + 7$ can be regarded as a search for r and s so that the sum of areas A, B, C, and D is $10x^2 + 37x + 7$.

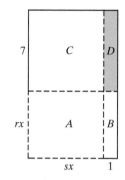

Because A must be $10x^2$, the product rs must be 10. Only when r is 2 and s is 5 will the sum of areas B and C be $37x$ (see below).

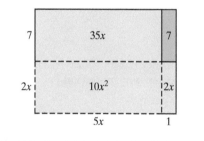

2. Factor: $14x^2 + 13x + 3$.

YOUR TURN

EXAMPLE 2 Factor: $10x^2 + 37x + 7$.

SOLUTION

1. There is no factor (other than 1 or -1) common to all three terms.

2. Because $10x^2$ factors as $10x \cdot x$ or $5x \cdot 2x$, we have two possibilities:

$$(10x + \ )(x + \ ) \quad \text{or} \quad (5x + \ )(2x + \ ).$$

3. There are two pairs of factors of 7 and each can be listed in two ways:

$$\begin{matrix} 1, \ 7 \\ -1, -7 \end{matrix} \quad \text{and} \quad \begin{matrix} 7, \ 1 \\ -7, -1. \end{matrix}$$

4. Look for **O**uter and **I**nner products for which the sum is the middle term. Because all coefficients in $10x^2 + 37x + 7$ are positive, we need consider only those combinations involving positive factors of 7.

Trial   *Product*

$(10x + 1)(x + 7)$ $10x^2 + 70x + 1x + 7$
                    $= 10x^2 + 71x + 7$ Wrong middle term

$(10x + 7)(x + 1)$ $10x^2 + 10x + 7x + 7$
                    $= 10x^2 + 17x + 7$ Wrong middle term

$(5x + 7)(2x + 1)$ $10x^2 + 5x + 14x + 7$
                    $= 10x^2 + 19x + 7$ Wrong middle term

$(5x + 1)(2x + 7)$ $10x^2 + 35x + 2x + 7$
                    $= 10x^2 + 37x + 7$ Correct middle term!

The correct factorization is $(5x + 1)(2x + 7)$.

EXAMPLE 3 Factor: $24x^3 - 76x^2 + 40x$.

SOLUTION

1. First, we factor out the largest common factor, $4x$:

$$4x(6x^2 - 19x + 10).$$

2. Next, we factor $6x^2 - 19x + 10$. Since $6x^2$ can be factored as $3x \cdot 2x$ or $6x \cdot x$, we have two possibilities:

$$(3x + \ )(2x + \ ) \quad \text{or} \quad (6x + \ )(x + \ ).$$

3. The constant term, 10, can be factored as $1 \cdot 10, 2 \cdot 5, (-1)(-10)$, and $(-2)(-5)$. Since the middle term is negative, we need consider only the factorizations with negative factors:

$$\begin{matrix} -1, -10 \\ -2, \ -5 \end{matrix} \quad \text{and} \quad \begin{matrix} -10, -1 \\ -5, -2. \end{matrix}$$

4. The two possibilities from step (2) and the four possibilities from step (3) give $2 \cdot 4$, or 8, possible factorizations to consider.

We first try these factors with $(3x + \quad)(2x + \quad)$, looking for **O**uter and **I**nner products for which the sum is $-19x$. If none gives the correct factorization, then we will consider $(6x + \quad)(x + \quad)$.

Trial	Product
$(3x - 1)(2x - 10)$	$6x^2 - 30x - 2x + 10$
	$= 6x^2 - 32x + 10$ Wrong middle term
$(3x - 10)(2x - 1)$	$6x^2 - 3x - 20x + 10$
	$= 6x^2 - 23x + 10$ Wrong middle term
$(3x - 2)(2x - 5)$	$6x^2 - 15x - 4x + 10$
	$= 6x^2 - 19x + 10$ Correct middle term!
$(3x - 5)(2x - 2)$	$6x^2 - 6x - 10x + 10$
	$= 6x^2 - 16x + 10$ Wrong middle term

The factorization of $6x^2 - 19x + 10$ is $(3x - 2)(2x - 5)$, but do not forget the common factor! The factorization of $24x^3 - 76x^2 + 40x$ is

$$4x(3x - 2)(2x - 5).$$

3. Factor: $12n^4 + 10n^3 - 12n^2$.

⟳ YOUR TURN

In Example 3, look again at the possibility $(3x - 5)(2x - 2)$. Without multiplying, we can reject such a possibility. To see why, note that

$$(3x - 5)(2x - 2) = (3x - 5)2(x - 1).$$

The expression $2x - 2$ has a common factor, 2. But we removed the *largest* common factor in step (1). If $2x - 2$ were one of the factors, then 2 would be *another* common factor in addition to the original, $4x$. Thus, $(2x - 2)$ cannot be part of the factorization of $6x^2 - 19x + 10$. Similar reasoning can be used to reject $(3x - 1)(2x - 10)$ as a possible factorization.

Once the largest common factor is factored out, none of the remaining factors can have a common factor.

Student Notes

Keep your work organized so that you can see what you have already considered. For example, when factoring $6x^2 - 19x + 10$, we can list all possibilities and cross out those in which a common factor appears:

$\cancel{(3x - 1)(2x - 10)}$
$(3x - 10)(2x - 1)$
$(3x - 2)(2x - 5)$
$\cancel{(3x - 5)(2x - 2)}$
$(6x - 1)(x - 10)$
$\cancel{(6x - 10)(x - 1)}$
$\cancel{(6x - 2)(x - 5)}$
$(6x - 5)(x - 2)$

By being organized and not erasing, we can see that there are only four possible factorizations.

TIPS FOR FACTORING $ax^2 + bx + c$

To factor $ax^2 + bx + c$ $(a > 0)$:

- Make sure that any common factor has been factored out.
- Once the largest common factor has been factored out of the original trinomial, no binomial factor can contain a common factor (other than 1 or -1).
- If c is positive, then the signs in both binomial factors must match the sign of b.
- Reversing the signs in the binomials reverses the sign of the middle term of their product.
- Organize your work so that you can keep track of which possibilities you have checked.
- Remember to include the largest common factor—if there is one—in the final factorization.
- Check by multiplying.

EXAMPLE 4 Factor: $10x + 8 - 3x^2$.

SOLUTION Before attempting to factor $10x + 8 - 3x^2$, we write it in descending order:

$$10x + 8 - 3x^2 = -3x^2 + 10x + 8. \qquad \text{Using the commutative law to write descending order}$$

Although we could now factor $-3x^2 + 10x + 8$, the procedure is easier if we have a positive leading coefficient. This can be found by factoring out -1:

$$-3x^2 + 10x + 8 = -1(3x^2 - 10x - 8) \qquad \text{Factoring out } -1 \text{ changes the signs of the coefficients.}$$

$$= -1(3x + 2)(x - 4). \qquad \text{Using the result from Example 1}$$

The factorization of $10x + 8 - 3x^2$ is $-1(3x + 2)(x - 4)$.

4. Factor: $9y - 10 - 2y^2$.

YOUR TURN

B. The Grouping Method

Another method of factoring trinomials of the type $ax^2 + bx + c$ is known as the *grouping method*. The grouping method relies on rewriting $ax^2 + bx + c$ in the form $ax^2 + px + qx + c$ and then factoring by grouping. To develop this method, consider the following*:

$$(2x + 5)(3x + 4) = 2x \cdot 3x + 2x \cdot 4 + 5 \cdot 3x + 5 \cdot 4 \qquad \text{Using FOIL}$$

$$= 2 \cdot 3 \cdot x^2 + 2 \cdot 4x + 5 \cdot 3x + 5 \cdot 4$$

$$= 2 \cdot 3 \cdot x^2 + (2 \cdot 4 + 5 \cdot 3)x + 5 \cdot 4$$

$$\qquad\qquad\quad \underset{a}{\uparrow} \qquad\qquad\quad \underset{b}{\uparrow} \qquad\qquad \underset{c}{\uparrow}$$

$$= 6x^2 \quad + \quad 23x \quad + \quad 20.$$

Note that reversing these steps shows that $6x^2 + 23x + 20$ can be rewritten as $6x^2 + 8x + 15x + 20$ and then factored by grouping. Note that the numbers that add to b (in this case, $2 \cdot 4$ and $5 \cdot 3$) also multiply to ac (in this case, $2 \cdot 3 \cdot 5 \cdot 4$).

TO FACTOR $ax^2 + bx + c$ USING THE GROUPING METHOD

1. Factor out the largest common factor, if one exists.
2. Multiply the leading coefficient a and the constant c.
3. Find a pair of factors of ac whose sum is b.
4. Rewrite the middle term, bx, as a sum or a difference using the factors found in step (3).
5. Factor by grouping.
6. Include any common factor from step (1) and check by multiplying.

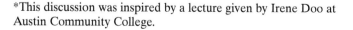

*This discussion was inspired by a lecture given by Irene Doo at Austin Community College.

EXAMPLE 5   Factor: $3x^2 - 10x - 8$.

SOLUTION

1. First, we note that there is no common factor (other than 1 or -1).

2. We multiply the leading coefficient, 3, and the constant, -8:

   $$3(-8) = -24.$$

3. We next look for a factorization of -24 in which the sum of the factors is the coefficient of the middle term, -10.

Pairs of Factors of -24	Sums of Factors
1, -24	-23
-1, 24	23
2, -12	-10 ←
-2, 12	10
3, -8	-5
-3, 8	5
4, -6	-2
-4, 6	2

$2 + (-12) = -10$

We normally stop listing pairs of factors once we have found the one we need.

4. Next, we express the middle term as a sum or a difference using the factors found in step (3):

   $$-10x = 2x - 12x.$$

5. We now factor by grouping as follows:

   $$3x^2 - 10x - 8 = 3x^2 + 2x - 12x - 8$$

   Substituting $2x - 12x$ for $-10x$. We could also use $-12x + 2x$.

   $$= x(3x + 2) - 4(3x + 2)$$
   $$= (3x + 2)(x - 4).$$

   Factoring by grouping

6. *Check:* $(3x + 2)(x - 4) = 3x^2 - 12x + 2x - 8 = 3x^2 - 10x - 8.$

The factorization of $3x^2 - 10x - 8$ is $(3x + 2)(x - 4)$.

5. Factor: $2x^2 - 3x - 20$.

YOUR TURN

EXAMPLE 6   Factor: $8x^3 + 22x^2 - 6x$.

SOLUTION

1. We factor out the largest common factor, $2x$:

   $$8x^3 + 22x^2 - 6x = 2x(4x^2 + 11x - 3).$$

2. We multiply the leading coefficient, 4, and the constant term, -3:

   $$4(-3) = -12.$$

3. We next look for factors of -12 that add to 11.

Pairs of Factors of -12	Sums of Factors
1, -12	-11
-1, 12	11 ←
.	.
.	.
.	.

Since $-1 + 12 = 11$, there is no need to list other pairs of factors.

4. We then rewrite the $11x$ in $4x^2 + 11x - 3$ using the results of step (3):

$$11x = -1x + 12x, \quad \text{or} \quad 11x = 12x - 1x.$$

5. Next, we factor by grouping:

$$4x^2 + 11x - 3 = 4x^2 - 1x + 12x - 3 \qquad \text{Rewriting the middle term; } 12x - 1x \text{ could also be used.}$$

$$\begin{aligned} &= x(4x - 1) + 3(4x - 1) \\ &= (4x - 1)(x + 3). \end{aligned} \Bigg\} \quad \text{Factoring by grouping}$$

6. The factorization of $4x^2 + 11x - 3$ is $(4x - 1)(x + 3)$. But don't forget the common factor, $2x$. The factorization of the original trinomial is

$$2x(4x - 1)(x + 3).$$

6. Factor: $30n^3 + 5n^2 - 60n$.

YOUR TURN

EXAMPLE 7 Factor: $6r^2 + 19rw + 15w^2$.

SOLUTION In order for the product of the first terms to be $6r^2$ and the product of the last terms to be $15w^2$, the binomial factors will be of the form

$$(\square r + \square w)(\square r + \square w).$$

We find the missing numbers using the same procedure that we used in the previous examples.

1. There is no common factor (other than 1 or -1).

2. We multiply the coefficients 6 and 15: $6 \cdot 15 = 90$.

3. We next look for factors of 90 that add to 19.

Pairs of Factors of 90	Sums of Factors
1, 90	91
2, 45	47
3, 15	18
5, 18	23
6, 15	21
9, 10	19 ← 9 + 10 = 19

4. We rewrite $19rw$ in $6r^2 + 19rw + 15w^2$ using the results of step (3):

$$19rw = 9rw + 10rw.$$

5. Next, we factor by grouping:

$$6r^2 + 19rw + 15w^2 = 6r^2 + 9rw + 10rw + 15w^2 \qquad \text{Rewriting the middle term}$$

$$\begin{aligned} &= 3r(2r + 3w) + 5w(2r + 3w) \\ &= (2r + 3w)(3r + 5w). \end{aligned} \Bigg\} \quad \text{Factoring by grouping}$$

6. *Check:* $(2r + 3w)(3r + 5w) = 6r^2 + 10rw + 9rw + 15w^2$

$$= 6r^2 + 19rw + 15w^2$$

The factorization of $6r^2 + 19rw + 15w^2$ is $(2r + 3w)(3r + 5w)$.

7. Factor: $12x^2 - 8xy - 15y^2$.

YOUR TURN

↳ Check Your UNDERSTANDING

Complete each factorization.

1. $3x^2 + 13x - 10$
$= (3x - 2)(\square + \square)$

2. $10x^2 - 23x + 12$
$= (2x - 3)(\square - \square)$

3. $10x^2 + 14x + 4$
$= 2(x + 1)(\square + \square)$

4. $-2x^2 - 5x + 18$
$= -(x - 2)(\square + \square)$

5.3 EXERCISE SET

FOR EXTRA HELP   MyMathLab®

➧ Vocabulary and Reading Check

Match each polynomial in Exercises 1–4 with the statement that is true when factoring that polynomial.

a) Factor out a common factor first.

b) The only possibilities for the first terms of the binomial factors are $5x$ and x.

c) Both of the constants in the binomial factors must be negative.

d) The polynomial is prime.

1. ____ $11x^2 + 3x + 1$

2. ____ $12x^2 + 18x + 21$

3. ____ $5x^2 - 14x - 3$

4. ____ $2x^2 - 7x + 5$

A, B. Factoring Trinomials of the Type $ax^2 + bx + c$

Factor completely. If a polynomial is prime, state this.

5. $2x^2 + 7x - 4$

6. $3x^2 + x - 4$

7. $3x^2 - 17x - 6$

8. $5x^2 - 19x - 4$

9. $4t^2 + 12t + 5$

10. $6t^2 + 17t + 7$

11. $15a^2 - 14a + 3$

12. $10a^2 - 11a + 3$

13. $6x^2 + 17x + 12$

14. $6x^2 + 19x + 10$

15. $6x^2 - 10x - 4$

16. $5t^3 - 21t^2 + 18t$

17. $7t^3 + 15t^2 + 2t$

18. $15t^2 + 20t - 75$

19. $10 - 23x + 12x^2$

20. $-20 + 31x - 12x^2$

21. $-35x^2 - 34x - 8$

22. $28x^2 + 38x - 6$

23. $4 + 6t^2 - 13t$

24. $9 + 8t^2 - 18t$

25. $25x^2 + 40x + 16$

26. $49t^2 + 42t + 9$

27. $20y^2 + 59y - 3$

28. $25a^2 - 23a - 2$

29. $14x^2 + 73x + 45$

30. $35x^2 - 57x - 44$

31. $-2x^2 + 15 + x$

32. $2t^2 - 19 - 6t$

33. $-6x^2 - 33x - 15$

34. $-12x^2 - 28x + 24$

35. $10a^2 - 8a - 18$

36. $20y^2 - 25y + 5$

37. $12x^2 + 68x - 24$

38. $6x^2 + 21x + 15$

39. $4x + 1 + 3x^2$

40. $-9 + 18x^2 + 21x$

Factor. Use factoring by grouping even though it may seem reasonable to first combine like terms.

41. $x^2 + 3x - 2x - 6$

42. $x^2 + 4x - 2x - 8$

43. $8t^2 - 6t - 28t + 21$

44. $35t^2 - 40t + 21t - 24$

45. $6x^2 + 4x + 15x + 10$

46. $3x^2 - 2x + 3x - 2$

47. $2y^2 + 8y - y - 4$

48. $7n^2 + 35n - n - 5$

49. $6a^2 - 8a - 3a + 4$

50. $10a^2 - 4a - 5a + 2$

Factor completely. If a polynomial is prime, state this.

51. $16t^2 + 23t + 7$

52. $9t^2 + 14t + 5$

53. $-9x^2 - 18x - 5$

54. $-16x^2 - 32x - 7$

55. $10x^2 + 30x - 70$

56. $10a^2 + 25a - 15$

57. $18x^3 + 21x^2 - 9x$

58. $6x^3 - 4x^2 - 10x$

59. $89x + 64 + 25x^2$

60. $47 - 42y + 9y^2$

61. $168x^3 + 45x^2 + 3x$

62. $144x^5 - 168x^4 + 48x^3$

63. $-14t^4 + 19t^3 + 3t^2$

64. $-70a^4 + 68a^3 - 16a^2$

65. $132y + 32y^2 - 54$

66. $220y + 60y^2 - 225$

67. $2a^2 - 5ab + 2b^2$

68. $3p^2 - 16pq - 12q^2$

69. $8s^2 + 22st + 14t^2$

70. $10s^2 + 4st - 6t^2$

71. $27x^2 - 72xy + 48y^2$

72. $-30a^2 - 87ab - 30b^2$

73. $-24a^2 + 34ab - 12b^2$

74. $15a^2 - 5ab - 20b^2$

75. $19x^3 - 3x^2 + 14x^4$

76. $10x^5 - 2x^4 + 22x^3$

77. $18a^7 + 8a^6 + 9a^8$

78. $40a^8 + 16a^7 + 25a^9$

79. Asked to factor $2x^2 - 18x + 36$, Juan *incorrectly* answers
$$2x^2 - 18x + 36 = 2(x^2 + 9x + 18)$$
$$= 2(x + 3)(x + 6).$$
If this were a 10-point quiz question, how many points would you deduct? Why?

80. Asked to factor $4x^2 + 28x + 48$, Therese *incorrectly* answers
$$4x^2 + 28x + 48 = (2x + 6)(2x + 8)$$
$$= 2(x + 3)(x + 4).$$
If this were a 10-point quiz question, how many points would you deduct? Why?

Skill Review

Graph.

81. $2x - 5y = 10$ [3.3]

82. $-5x = 10$ [3.3]

83. $y = \frac{2}{3}x - 1$ [3.6]

84. $y - 2 = -2(x + 4)$ [3.7]

85. $\frac{1}{2}y = 1$ [3.3]

86. $x = -y$ [3.2]

Synthesis

87. Explain how you would prove to a fellow student that a given trinomial is prime.

88. For the trinomial $ax^2 + bx + c$, suppose that a is the product of three different prime factors and c is the product of another two prime factors. How many possible factorizations (like those in Example 1) exist? Explain your reasoning.

Factor. If a polynomial is prime, state this.

89. $18x^2y^2 - 3xy - 10$

90. $8x^2y^3 + 10xy^2 + 2y$

91. $9a^2b^3 + 25ab^2 + 16$

92. $-9t^{10} - 12t^5 - 4$

93. $16t^{10} - 8t^5 + 1$

94. $9a^2b^2 - 15ab - 2$

95. $-15x^{2m} + 26x^m - 8$

96. $-20x^{2n} - 16x^n - 3$

97. $3a^{6n} - 2a^{3n} - 1$

98. $a^{2n+1} - 2a^{n+1} + a$

99. $7(t - 3)^{2n} + 5(t - 3)^n - 2$

100. $3(a + 1)^{n+1}(a + 3)^2 - 5(a + 1)^n(a + 3)^3$

101. Kara bought a number of sections of decorative fencing, all of equal length, to fence in a rectangular 1500-ft^2 yard. For each of the shorter sides of the rectangle, she used 4 sections plus 2 ft cut from another section. For each of the longer sides, she used 7 sections plus 1 ft cut from another section. How long was each section of fencing?

Quick Quiz: Sections 5.1–5.3
Factor completely. If a polynomial is prime, state this.

1. $6a^2b^3 - 9ab^4 + 15a^3b^5$ [5.1]

2. $x^3 - 7x^2 + 12x$ [5.2]

3. $p^2 + p - 7$ [5.2]

4. $4a^2 + 16a + 15$ [5.3]

5. $6x^2 - x - 5$ [5.3]

Prepare to Move On
Multiply. [4.6]

1. $(x - 2)^2$ **2.** $(x + 2)^2$

3. $(x + 2)(x - 2)$ **4.** $(5t - 3)^2$

5. $(4a + 1)^2$ **6.** $(2n + 7)(2n - 7)$

5.4   Factoring Perfect-Square Trinomials and Differences of Squares

A. Recognizing Perfect-Square Trinomials   **B.** Factoring Perfect-Square Trinomials
C. Recognizing Differences of Squares   **D.** Factoring Differences of Squares   **E.** Factoring Completely

Reversing the rules for special products provides us with shortcuts for factoring certain polynomials.

A. Recognizing Perfect-Square Trinomials

Some trinomials are squares of binomials. For example, $x^2 + 10x + 25$ is the square of the binomial $x + 5$, because

$$(x + 5)^2 = x^2 + 2 \cdot x \cdot 5 + 5^2 = x^2 + 10x + 25.$$

A trinomial that is the square of a binomial is called a **perfect-square trinomial**.
We can square binomials using the following special-product rule:

$$(A + B)^2 = A^2 + 2AB + B^2;$$
$$(A - B)^2 = A^2 - 2AB + B^2.$$

Reading the right sides first, we can use these equations to factor perfect-square trinomials. In order for a trinomial to be the square of a binomial, it must have the following:

1. Two expressions, A^2 and B^2, must be squares, such as

   $$4, \quad x^2, \quad 81m^2, \quad 16t^2.$$

2. Neither A^2 nor B^2 is being subtracted.

3. The remaining term is either $2 \cdot A \cdot B$ or $-2 \cdot A \cdot B$, where A and B are the **square roots** of A^2 and B^2.

Because $3^2 = 9$, we say that 3 is the *square root* of 9. Also, 2 is the square root of 4, 4 is the square root of 16, and so on.
   Note that in order for an expression to be a square, its coefficient must be a perfect square and the power(s) of the variable(s) must be even.

EXAMPLE 1   Determine whether each of the following is a perfect-square trinomial.

a) $x^2 + 6x + 9$ **b)** $t^2 - 8t - 9$ **c)** $16x^2 + 49 - 56x$

SOLUTION

a) **1.** Two expressions, x^2 and 9, are squares.

    2. Neither x^2 nor 9 is being subtracted.

    3. The remaining term, $6x$, is $2 \cdot x \cdot 3$, where x and 3 are the square roots of x^2 and 9.

    Thus, $x^2 + 6x + 9$ *is* a perfect-square trinomial.

b) Both t^2 and 9 are squares. However, since 9 is being subtracted, $t^2 - 8t - 9$ *is not* a perfect-square trinomial.

c) To see if $16x^2 + 49 - 56x$ is a perfect-square trinomial, it helps to first write it in descending order:

$$16x^2 - 56x + 49.$$

Student Notes

If you're not already quick to recognize the squares that represent $1^2, 2^2, 3^2, \ldots, 12^2$, this would be a good time to memorize these numbers.

Study Skills

Fill in Your Blanks

Don't hesitate to write out any missing steps that you'd like to see included. For instance, in Example 1(c), we state (in red) that $16x^2$ is a square. To solidify your understanding, you may want to write $4x \cdot 4x = 16x^2$ in the margin of your text.

1. Determine whether $100 - 20x + x^2$ is a perfect-square trinomial.

Next, note that:

1. Two expressions, $16x^2$ and 49, are squares.
2. Neither $16x^2$ nor 49 is being subtracted.
3. Twice the product of the square roots, $2 \cdot 4x \cdot 7$, is $56x$. The remaining term, $-56x$, is the opposite of this product.

Thus, $16x^2 + 49 - 56x$ *is* a perfect-square trinomial.

⤵ YOUR TURN

B. Factoring Perfect-Square Trinomials

To factor perfect-square trinomials, we recognize the following patterns.

> **FACTORING A PERFECT-SQUARE TRINOMIAL**
>
> $A^2 + 2AB + B^2 = (A + B)^2;$
> $A^2 - 2AB + B^2 = (A - B)^2$

Student Notes

The following diagram may help clarify the factoring in Examples 2(a) and 2(b):

$$x^2 + 2 \cdot x \cdot 3 + 3^2$$
$$\downarrow \quad \downarrow \quad \sqrt{}$$
$$= (x + 3)^2;$$

$$t^2 - 2 \cdot t \cdot 7 + 7^2$$
$$\downarrow \quad \downarrow \quad \sqrt{}$$
$$= (t - 7)^2.$$

Each factorization uses the square roots of the squared terms and the sign of the remaining term. To verify these equations, you should compute $(A + B)(A + B)$ and $(A - B)(A - B)$.

EXAMPLE 2   Factor.

a) $x^2 + 6x + 9$ **b)** $t^2 + 49 - 14t$ **c)** $100x^2 - 180x + 81$

SOLUTION

a) $x^2 + 6x + 9 = x^2 + 2 \cdot x \cdot 3 + 3^2 = (x + 3)^2$ The sign of the middle term is positive.

$$A^2 + 2 \; A \; B + B^2 = (A + B)^2$$

b) $t^2 + 49 - 14t = t^2 - 14t + 49$ Using a commutative law to write in descending order

$$= t^2 - 2 \cdot t \cdot 7 + 7^2 = (t - 7)^2$$

$$A^2 - 2 \; A \; B + B^2 = (A - B)^2$$

c) $100x^2 - 180x + 81 = (10x)^2 - 2 \cdot 10x \cdot 9 + 9^2 = (10x - 9)^2$ $(10x)^2 = 100x^2$

$$A^2 \quad - 2 \; A \quad B + B^2 = (A \; - \; B)^2$$

2. Factor: $x^2 + 2x + 1$.

The checks are left to the student.

⤵ YOUR TURN

Polynomials in more than one variable can also be perfect-square trinomials.

EXAMPLE 3   Factor: $4p^2 - 12pt + 9t^2$.

SOLUTION   We have

$$4p^2 - 12pt + 9t^2 = (2p)^2 - 2(2p)(3t) + (3t)^2$$ $4p^2$ and $9t^2$ are squares.

$$= (2p - 3t)^2.$$ The sign of the middle term is negative.

Check: $(2p - 3t)(2p - 3t) = 4p^2 - 12pt + 9t^2.$

3. Factor: $100a^2 + 20ab + b^2$.

The factorization is $(2p - 3t)^2$.

⤵ YOUR TURN

EXAMPLE 4 Factor: $-75m^3 - 60m^2 - 12m$.

SOLUTION *Always* look first for a common factor. This time there is one. We factor out $-3m$ so that the leading coefficient of the polynomial inside the parentheses is positive:

Factor out the common factor.

$$-75m^3 - 60m^2 - 12m = -3m[25m^2 + 20m + 4] \qquad 25m^2 = (5m)^2$$
$$= -3m[(5m)^2 + 2(5m)(2) + 2^2]$$
$$= -3m(5m + 2)^2.$$

Factor the perfect-square trinomial.

Check: $-3m(5m + 2)^2 = -3m(5m + 2)(5m + 2)$
$$= -3m(25m^2 + 20m + 4)$$
$$= -75m^3 - 60m^2 - 12m.$$

The factorization is $-3m(5m + 2)^2$.

4. Factor: $-4n^3 + 12n^2 - 9n$.

YOUR TURN

C. Recognizing Differences of Squares

Any expression that can be written in the form $A^2 - B^2$, like $x^2 - 9$, is called a **difference of squares**. In order for a binomial to be a difference of squares, it must have the following.

1. There must be two expressions, both squares, such as

$$25, \quad t^2, \quad 4x^2, \quad 1, \quad x^6, \quad 49y^8, \quad 100x^2y^2.$$

2. The terms in the binomial must have different signs.

EXAMPLE 5 Determine whether each of the following is a difference of squares.

a) $9x^2 - 64$ **b)** $25 - t^3$ **c)** $-4x^{10} + 36$

SOLUTION

a) 1. The first expression is a square: $9x^2 = (3x)^2$.
The second expression is a square: $64 = 8^2$.

   2. The terms have different signs.

Thus, $9x^2 - 64$ is a difference of squares, $(3x)^2 - 8^2$.

b) 1. The expression t^3 is not a square.

Thus, $25 - t^3$ is not a difference of squares.

c) 1. The expressions $4x^{10}$ and 36 are squares: $4x^{10} = (2x^5)^2$ and $36 = 6^2$.

   2. The terms have different signs.

5. Determine whether $-x^2 + 49$ is a difference of squares.

Thus, $-4x^{10} + 36$ is a difference of squares, $6^2 - (2x^5)^2$. It is helpful to rewrite $-4x^{10} + 36$ in the equivalent form $36 - 4x^{10}$ before factoring.

YOUR TURN

D. Factoring Differences of Squares

To factor a difference of squares, we reverse the special product

$$(A + B)(A - B) = A^2 - AB + AB - B^2 = A^2 - B^2.$$

> **FACTORING A DIFFERENCE OF SQUARES**
> $$A^2 - B^2 = (A + B)(A - B)$$

Once we have identified the expressions that are playing the roles of A and B, the factorization can be written directly.

EXAMPLE 6 Factor: **(a)** $x^2 - 4$; **(b)** $1 - 9p^2$; **(c)** $s^6 - 16t^{10}$; **(d)** $50x^2 - 8x^8$.

SOLUTION

a) $x^2 - 4 = x^2 - 2^2 = (x + 2)(x - 2)$

$A^2 - B^2 = (A + B)(A - B)$

b) $1 - 9p^2 = 1^2 - (3p)^2 = (1 + 3p)(1 - 3p)$

$A^2 - B^2 = (A + B)(A - B)$

c) $s^6 - 16t^{10} = (s^3)^2 - (4t^5)^2$ Using the rules for powers

$A^2 - B^2$

$= (s^3 + 4t^5)(s^3 - 4t^5)$ Try to go directly to this step.

$(A + B) (A - B)$

d) *Always* look first for a common factor. This time there is one, $2x^2$:

> Factor out the common factor.

$50x^2 - 8x^8 = 2x^2(25 - 4x^6)$

> Factor the difference of squares.

$= 2x^2[5^2 - (2x^3)^2]$ Recognizing $A^2 - B^2$. Try to do this mentally.

$= 2x^2(5 + 2x^3)(5 - 2x^3).$

Check: $2x^2(5 + 2x^3)(5 - 2x^3) = 2x^2(25 - 4x^6)$
$= 50x^2 - 8x^8.$

The factorization of $50x^2 - 8x^8$ is $2x^2(5 + 2x^3)(5 - 2x^3)$.

6. Factor: $a^4 - 9b^2$.

YOUR TURN

CAUTION! A difference of squares is *not* the square of the difference; that is,

$A^2 - B^2 \neq (A - B)^2.$ To see this, note that
$(A - B)^2 = A^2 - 2AB + B^2.$

EXAMPLE 7 Factor, if possible: $x^2 + 16$.

SOLUTION There is no common factor. Although both x^2 and 16 are squares, $x^2 + 16$ is a *sum* of squares, not a *difference* of squares. This polynomial cannot be factored and is thus prime.

7. Factor, if possible: $a^2 + 25$.

YOUR TURN

CAUTION! There is no general formula for factoring a sum of squares. In particular,

$A^2 + B^2 \neq (A + B)^2 \quad \text{and} \quad A^2 + B^2 \neq (A + B)(A - B).$

E. Factoring Completely

Sometimes, as in Examples 4 and 6(d), a *complete* factorization requires two or more steps. Factoring is complete only when no factor can be factored further.

EXAMPLE 8   Factor: $y^4 - 16$.

SOLUTION   We have

Factor a difference of squares.

Factor another difference of squares.

$$
\begin{aligned}
y^4 - 16 &= (y^2)^2 - 4^2 && \text{Recognizing } A^2 - B^2 \\
&= (y^2 + 4)(y^2 - 4) && \text{Note that } y^2 - 4 \text{ is not prime.} \\
&= (y^2 + 4)(y + 2)(y - 2). && \text{Note that } y^2 - 4 \text{ is itself a} \\
&&& \text{difference of squares.}
\end{aligned}
$$

Check: $(y^2 + 4)(y + 2)(y - 2) = (y^2 + 4)(y^2 - 4)$
$$= y^4 - 16.$$

The factorization is $(y^2 + 4)(y + 2)(y - 2)$.

8. Factor: $t^4 - 81$.

YOUR TURN

Note in Example 8 that the factor $y^2 + 4$ is a sum of squares that cannot be factored further.

As you proceed through the exercises, these suggestions may prove helpful.

TIPS FOR FACTORING

1. *Always* look first for a common factor! If there is one, factor it out.
2. Be alert for perfect-square trinomials and for binomials that are differences of squares. Once recognized, they can be factored without trial and error.
3. Always factor completely.
4. Check by multiplying.

↳ Check Your UNDERSTANDING

Each of the following is a perfect-square trinomial of the form $A^2 + 2AB + B^2$ or $A^2 - 2AB + B^2$. Determine the values of A and B in each one.

1. $x^2 + 10x + 25$

2. $y^2 - 2y + 1$

3. $4x^2 - 12x + 9$

4. $25c^2 + 60c + 36$

Each of the following is a difference of squares of the form $A^2 - B^2$. Determine the values of A and B in each one.

5. $x^2 - 25$

6. $1 - 4y^2$

7. $m^4 - 16$

8. $100n^4 - 81$

5.4 EXERCISE SET

FOR EXTRA HELP MyMathLab®

↳ Vocabulary and Reading Check

Identify each of the following as either a perfect-square trinomial, a difference of squares, a prime polynomial, or none of these.

1. $4x^2 + 49$ **2.** $x^2 - 64$

3. $t^2 - 100$ **4.** $x^2 - 5x + 4$

5. $9x^2 + 6x + 1$ **6.** $a^2 - 8a + 16$

7. $2t^2 + 10t + 6$ **8.** $25x^2 - 3$

9. $16t^2 - 25$ **10.** $4r^2 + 20r + 25$

A. Recognizing Perfect-Square Trinomials

Determine whether each of the following is a perfect-square trinomial.

11. $x^2 + 18x + 81$ **12.** $x^2 - 16x + 64$

13. $x^2 - 10x - 25$ **14.** $x^2 - 14x - 49$

15. $x^2 - 3x + 9$ **16.** $x^2 + 4x + 4$

17. $9x^2 + 25 - 30x$ **18.** $36x^2 + 16 - 24x$

B. Factoring Perfect-Square Trinomials

Factor completely. Remember to look first for a common factor and to check by multiplying. If a polynomial is prime, state this.

19. $x^2 + 16x + 64$ **20.** $x^2 + 10x + 25$

21. $x^2 - 10x + 25$ **22.** $x^2 - 16x + 64$

23. $5p^2 + 20p + 20$ **24.** $3p^2 - 12p + 12$

25. $1 - 2t + t^2$ **26.** $1 + t^2 + 2t$

27. $18x^2 + 12x + 2$ **28.** $25x^2 + 10x + 1$

29. $49 - 56y + 16y^2$ **30.** $75 - 60m + 12m^2$

31. $-x^5 + 18x^4 - 81x^3$ **32.** $-2x^2 + 40x - 200$

33. $2n^3 + 40n^2 + 200n$ **34.** $x^3 + 24x^2 + 144x$

35. $20x^2 + 100x + 125$ **36.** $27m^2 - 36m + 12$

37. $49 - 42x + 9x^2$ **38.** $64 - 112x + 49x^2$

39. $16x^2 + 24x + 9$ **40.** $2a^2 + 28a + 98$

41. $2 + 20x + 50x^2$ **42.** $9x^2 + 30x + 25$

43. $9p^2 + 12px + 4x^2$ **44.** $x^2 - 3xy + 9y^2$

45. $a^2 - 12ab + 49b^2$

46. $25m^2 - 20mn + 4n^2$

47. $-64m^2 - 16mn - n^2$

48. $-81p^2 + 18pw - w^2$

49. $-32s^2 + 80st - 50t^2$

50. $-36a^2 - 96ab - 64b^2$

C. Recognizing Differences of Squares

Determine whether each of the following is a difference of squares.

51. $x^2 - 100$ **52.** $x^2 + 49$

53. $n^4 + 1$ **54.** $n^4 - 81$

55. $-1 + 64t^2$ **56.** $-12 + 25t^2$

D. Factoring Differences of Squares

Factor completely. Remember to look first for a common factor. If a polynomial is prime, state this.

57. $x^2 - 25$ **58.** $x^2 - 36$

59. $p^2 - 9$ **60.** $q^2 + 1$

61. $-49 + t^2$ **62.** $-64 + m^2$

63. $6a^2 - 24$ **64.** $x^2 - 8x + 16$

65. $49x^2 - 14x + 1$ **66.** $3t^2 - 3$

67. $200 - 2t^2$ **68.** $98 - 8w^2$

69. $-80a^2 + 45$ **70.** $25x^2 - 4$

71. $5t^2 - 80$ **72.** $-4t^2 + 64$

73. $8x^2 - 162$ **74.** $24x^2 - 54$

75. $36x - 49x^3$ **76.** $16x - 81x^3$

77. $49a^4 - 20$ **78.** $25a^4 - 9$

E. Factoring Completely

Factor completely.

79. $t^4 - 1$

80. $x^4 - 16$

81. $-3x^3 + 24x^2 - 48x$

82. $-2a^4 + 36a^3 - 162a^2$

83. $75t^3 - 27t$ **84.** $80s^4 - 45s^2$

85. $a^8 - 2a^7 + a^6$ **86.** $x^8 - 8x^7 + 16x^6$

87. $10a^2 - 10b^2$ **88.** $6p^2 - 6q^2$

89. $16x^4 - y^4$ **90.** $98x^2 - 32y^2$

91. $18t^2 - 8s^2$ **92.** $a^4 - 81b^4$

93. Explain in your own words how to determine whether a polynomial is a perfect-square trinomial.

94. Explain in your own words how to determine whether a polynomial is a difference of squares.

Skill Review

Perform the indicated operation.

95. $(2x^3 - x + 3) + (x^2 + x - 5)$ [4.4]

96. $(3t^2 - 2t - 5) - (t^2 - 8t + 6)$ [4.4]

97. $(2x^2 + y)(3x^2 - y)$ [4.7]

98. $-5x(-x^2 + 3x - 7)$ [4.5]

99. $(21x^3 - 3x^2 + 9x) \div (3x)$ [4.8]

100. $(x^2 - x - 10) \div (x - 5)$ [4.8]

Synthesis

101. Leon concludes that since
$$x^2 - 9 = (x - 3)(x + 3),$$
it must follow that
$$x^2 + 9 = (x + 3)(x - 3).$$
What mistake(s) is he making?

102. Write directions that would enable someone to construct a polynomial that contains a perfect-square trinomial, a difference of squares, and a common factor.

Factor completely. If a polynomial is prime, state this.

103. $x^8 - 2^8$

104. $3x^2 - \frac{1}{3}$

105. $18x^3 - \frac{8}{25}x$

106. $0.81t - t^3$

107. $(y - 5)^4 - z^8$

108. $x^2 - \left(\frac{1}{x}\right)^2$

109. $-x^4 + 8x^2 + 9$

110. $-16x^4 + 96x^2 - 144$

Aha! 111. $(y + 3)^2 + 2(y + 3) + 1$

112. $49(x + 1)^2 - 42(x + 1) + 9$

113. $27p^3 - 45p^2 - 75p + 125$

114. $a^{2n} - 49b^{2n}$

115. $81 - b^{4k}$

116. $9b^{2n} + 12b^n + 4$

117. Subtract $(x^2 + 1)^2$ from $x^2(x + 1)^2$.

Factor by grouping. Look for a grouping of three terms that is a perfect-square trinomial.

118. $t^2 + 4t + 4 - 25$

119. $y^2 + 6y + 9 - x^2 - 8x - 16$

Find c such that each polynomial is the square of a binomial.

120. $cy^2 + 6y + 1$

121. $cy^2 - 24y + 9$

122. Find the only positive value of a for which $x^2 + a^2x + a^2$ factors into $(x + a)^2$.

123. Show that the difference of the squares of two consecutive integers is the sum of the integers. (*Hint:* Use x for the smaller number.)

124. Use a geometric approach, similar to that shown on p. 312, to show that $x^2 + 6x + 9$ is a perfect square.

⇱ YOUR TURN ANSWERS: SECTION 5.4

1. Yes   **2.** $(x + 1)^2$ **3.** $(10a + b)^2$
4. $-n(2n - 3)^2$ **5.** Yes   **6.** $(a^2 + 3b)(a^2 - 3b)$
7. Prime   **8.** $(t^2 + 9)(t + 3)(t - 3)$

Quick Quiz: Sections 5.1–5.4

Factor completely. If a polynomial is prime, state this.

1. $2x^3 + 6x^2 - x - 3$ [5.1]

2. $x^2 - 24x + 80$ [5.2]

3. $9a^3 - 6a^2 - 8a$ [5.3]

4. $y^2 + 16y + 64$ [5.4]

5. $3z^4 - 3$ [5.4]

Prepare to Move On

Simplify.   [4.1]

1. $(2x^2y^4)^3$ 2. $(-5x^2y)^3$

Multiply.   [4.5]

3. $(x - 1)^3$ 4. $(p + t)^3$

Mid-Chapter Review

The following is a good strategy to follow when you encounter a mixed set of factoring problems.

1. Factor out any common factor.
2. If there are *two* terms, determine whether it is a difference of squares. If so, factor using the pattern
   $$A^2 - B^2 = (A + B)(A - B).$$
3. If there are *three* terms, determine whether it is a perfect-square trinomial. If so, factor using the pattern
   $$A^2 + 2AB + B^2 = (A + B)^2 \quad \text{or} \quad A^2 - 2AB + B^2 = (A - B)^2.$$
4. If there are *three* terms and it is not a perfect-square trinomial, try factoring using FOIL or by grouping.
5. If there are *four* terms, try factoring by grouping.

GUIDED SOLUTIONS

Factor completely.

1. $12x^3y - 8xy^2 + 24x^2y = \square\,(3x^2 - 2y + 6x)$ [5.1]   Factoring out the largest common factor. No further factorization is possible.

2. $3a^3 - 3a^2 - 90a = 3a\left(\square - \square - \square\right)$ Factoring out the largest common factor

$\qquad\qquad\qquad = 3a\left(a - \square\right)\left(a + \square\right)$ [5.2]   Factoring the trinomial

MIXED REVIEW

Factor completely. If a polynomial is prime, state this.

3. $6x^5 - 18x^2$ [5.1]

4. $x^2 + 10x + 16$ [5.2]

5. $2x^2 + 13x - 7$ [5.3]

6. $x^3 + 3x^2 + 2x + 6$ [5.1]

7. $64n^2 - 9$ [5.4]

8. $x^2 - 2x - 5$ [5.2]

9. $6p^2 - 6t^2$ [5.4]

10. $b^2 - 14b + 49$ [5.4]

11. $12x^2 - x - 1$ [5.3]

12. $a - 10a^2 + 25a^3$ [5.4]

13. $10x^4 - 10$ [5.4]

14. $t^2 + t - 10$ [5.2]

15. $15d^2 - 30d + 75$ [5.1]

16. $15p^2 + 16px + 4x^2$ [5.3]

17. $-2t^3 - 10t^2 - 12t$ [5.2]

18. $10c^2 + 20c + 10$ [5.4]

19. $5 + 3x - 2x^2$ [5.3]

20. $2m^3n - 10m^2n - 6mn + 30n$ [5.1]

5.5 Factoring: A General Strategy

A. Choosing the Right Method

Thus far, each section in this chapter has examined one or two different methods for factoring polynomials. In practice, when the need for factoring a polynomial arises, we must decide on our own which method to use. The following guidelines provide an approach for this.

TO FACTOR A POLYNOMIAL

A. Always look for a common factor. If there is one, factor out the largest common factor. If the leading term is negative, then factor out a negative common factor. Be sure to include the common factor in your final answer.

B. Then look at the number of terms.

Two terms: If you have a difference of squares, factor accordingly:

$$A^2 - B^2 = (A + B)(A - B).$$

Three terms: If the trinomial is a perfect-square trinomial, factor accordingly:

$$A^2 + 2AB + B^2 = (A + B)^2;$$
$$A^2 - 2AB + B^2 = (A - B)^2.$$

If it is not a perfect-square trinomial, try using FOIL or grouping.

Four terms: Try factoring by grouping.

C. Always *factor completely*. When a factor can itself be factored, be sure to factor it. Remember that some polynomials, like $x^2 + 9$, are prime.

D. Check.

We can always check a factorization by multiplying. Another way to check a factorization is to evaluate both the original polynomial and the factorization for one or more choices of the variable(s). Since the polynomial and its factorization are equivalent expressions, they have the same value for any replacements of the variable(s). This check becomes more certain if it is done for more than one replacement. In fact, for a polynomial in one variable of degree n, a check of $n + 1$ replacements is sufficient to show that the factorization is correct.

Student Notes

Quickly checking both the leading term and the constant term of a trinomial to see if they are squares can save you time. If they aren't both squares and there is no common factor, the trinomial can't possibly be a perfect-square trinomial.

A. Choosing the Right Method

EXAMPLE 1 Factor: $x^2 - 20x + 100$.

SOLUTION

A. We look first for a common factor. There is none.

B. This polynomial is a perfect-square trinomial. We factor it accordingly:

$$x^2 - 20x + 100 = x^2 - 2 \cdot x \cdot 10 + 10^2 \qquad \text{Try to do this step mentally.}$$
$$= (x - 10)^2, \text{ or } (x - 10)(x - 10).$$

1. Factor: $a^2 + 2a + 1$.

Study Skills

Keep Your Focus

When you are studying with someone else, it can be very tempting to talk about topics not related to what you need to study. If you see that this may happen, explain to your partner(s) that you enjoy the conversation, but would enjoy it more later—after the work has been completed.

2. Factor: $5x^3 + 15x - x^2 - 3$.

3. Factor: $-3t^3 + 39t^2 - 66t$.

C. Nothing can be factored further, so we have factored completely.

D. We check by multiplying: $(x - 10)(x - 10) = x^2 - 20x + 100$.

The factorization is $(x - 10)(x - 10)$, or $(x - 10)^2$.

YOUR TURN

EXAMPLE 2 Factor: $2x^3 + 10x^2 + x + 5$.

SOLUTION

A. We look for a common factor. There is none.

B. Because there are four terms, we try factoring by grouping:

$$2x^3 + 10x^2 + x + 5$$
$$= (2x^3 + 10x^2) + (x + 5) \quad \text{Separating into two binomials}$$
$$= 2x^2(x + 5) + 1(x + 5) \quad \text{Factoring out the largest common factor from each binomial. The 1 serves as an aid.}$$
$$= (x + 5)(2x^2 + 1). \quad \text{Factoring out the common factor, } x + 5$$

C. Nothing can be factored further, so we have factored completely.

D. As a partial check, we evaluate the original polynomial and the factorization for $x = 2$:

$$2x^3 + 10x^2 + x + 5 = 2(2)^3 + 10(2)^2 + 2 + 5 = 2 \cdot 8 + 10 \cdot 4 + 2 + 5 = 63;$$
$$(x + 5)(2x^2 + 1) = (2 + 5)(2(2)^2 + 1) = 7 \cdot 9 = 63.$$

The values are the same for $x = 2$. If we were to evaluate both expressions for three additional values of x, the check would be complete.

The factorization is $(x + 5)(2x^2 + 1)$.

YOUR TURN

EXAMPLE 3 Factor: $-n^5 + 2n^4 + 35n^3$.

SOLUTION

A. We note that there is a common factor, $-n^3$:

$$-n^5 + 2n^4 + 35n^3 = -n^3(n^2 - 2n - 35).$$

B. The factor $n^2 - 2n - 35$ is not a perfect-square trinomial. We factor it using trial and error:

$$-n^5 + 2n^4 + 35n^3 = -n^3(n^2 - 2n - 35)$$
$$= -n^3(n - 7)(n + 5).$$

C. Nothing can be factored further, so we have factored completely.

D. We check by multiplying:

$$-n^3(n - 7)(n + 5) = -n^3(n^2 - 2n - 35)$$
$$= -n^5 + 2n^4 + 35n^3.$$

The factorization is $-n^3(n - 7)(n + 5)$.

YOUR TURN

EXAMPLE 4   Factor: $6x^2y^4 - 21x^3y^5 + 3x^2y^6$.

SOLUTION

A. We first factor out the largest common factor, $3x^2y^4$:

$$6x^2y^4 - 21x^3y^5 + 3x^2y^6 = 3x^2y^4(2 - 7xy + y^2).$$

B. The constant term in $2 - 7xy + y^2$ is not a square, so we do not have a perfect-square trinomial. Note that x appears only in $-7xy$. The product of a form like $(1 - y)(2 - y)$ has no x in the middle term. Thus, $2 - 7xy + y^2$ cannot be factored.

C. Nothing can be factored further, so we have factored completely.

D. We check by multiplying:

$$3x^2y^4(2 - 7xy + y^2) = 6x^2y^4 - 21x^3y^5 + 3x^2y^6.$$

The factorization is $3x^2y^4(2 - 7xy + y^2)$.

4. Factor:

$$12p^3w - 18p^2w^2 + 6p^2w.$$

YOUR TURN

EXAMPLE 5   Factor: $98x^3 + 280x^2 + 200x$.

SOLUTION

A. We first factor out the largest common factor, $2x$:

$$98x^3 + 280x^2 + 200x = 2x(49x^2 + 140x + 100).$$

B. The trinomial in the parentheses is a perfect-square trinomial: $49x^2 = (7x)^2$, $100 = (10)^2$, and $140x = 2 \cdot 7x \cdot 10$. We factor it accordingly:

$$98x^3 + 280x^2 + 200x = 2x(49x^2 + 140x + 100)$$
$$= 2x(7x + 10)^2.$$

C. Nothing can be factored further, so we have factored completely.

D. As a partial check, we evaluate the original polynomial and the factorization for $x = 1$:

$$98x^3 + 280x^2 + 200x = 98(1)^3 + 280(1)^2 + 200(1) = 98 + 280 + 200 = 578;$$
$$2x(7x + 10)^2 = 2(1)[(7(1) + 10)^2] = 2(17)^2 = 2(289) = 578.$$

The values are the same for $x = 1$. If we were to evaluate both expressions for three additional values of x, the check would be complete.

5. Factor:

$$75a^4 + 30a^3 + 3a^2.$$

The factorization is $2x(7x + 10)^2$.

YOUR TURN

EXAMPLE 6   Factor: $-10m^2 - mn + 3n^2$.

SOLUTION

A. We first look for a common factor. Since the first term is negative, we factor out -1:

$$-1(10m^2 + mn - 3n^2).$$

B. There are three terms in the parentheses. Since none is a square, the trinomial is not a perfect square. We factor using FOIL. (We could also use grouping.) There are two possibilities for the first terms in the binomial factors:

$$(2m + \quad)(5m + \quad) \quad \text{or} \quad (m + \quad)(10m + \quad).$$

The possibilities for the second terms in the binomials are

$$\begin{matrix} n, -3n \\ -n, \quad 3n \end{matrix} \quad \text{and} \quad \begin{matrix} -3n, \quad n \\ 3n, -n. \end{matrix}$$

We check the possible combinations and find that the factorization of $10m^2 + mn - 3n^2$ is $(2m - n)(5m + 3n)$. Thus,

$$-10m^2 - mn + 3n^2 = -1(2m - n)(5m + 3n).$$

C. Nothing can be factored further, so we have factored completely.

D. We check by multiplying:

$$-1(2m - n)(5m + 3n) = -1(10m^2 + mn - 3n^2)$$
$$= -10m^2 - mn + 3n^2.$$

The factorization is $-1(2m - n)(5m + 3n)$, or $-(2m - n)(5m + 3n)$.

6. Factor: $-9x^2 - 37xy - 4y^2$.

⟲ YOUR TURN

EXAMPLE 7 Factor: $x^2y^2 + 7xy + 12$.

SOLUTION

A. We first look for a common factor. There is none.

B. Since only one term is a square, we do not have a perfect-square trinomial. We use trial and error, treating the product xy as a single variable:

$$(xy + \quad)(xy + \quad).$$

We factor the last term, 12. All the signs are positive, so we consider only positive factors. Possibilities are 1, 12 and 2, 6 and 3, 4. The pair 3, 4 gives a sum of 7 for the coefficient of the middle term. Thus,

$$x^2y^2 + 7xy + 12 = (xy + 3)(xy + 4).$$

C. Nothing can be factored further, so we have factored completely.

D. *Check:* $(xy + 3)(xy + 4) = x^2y^2 + 7xy + 12$.

The factorization is $(xy + 3)(xy + 4)$.

⟲ YOUR TURN

⤷ **Chapter Resource:**
Collaborative Activity (Matching Factorizations), p. 362

7. Factor: $c^2t^2 - 9ct + 20$.

Compare the variables appearing in Example 6 with those in Example 7. Note that if the leading term contains one variable and a different variable is in the last term, as in Example 6, each factor contains two variable terms. When two variables appear in the leading term and no variables appear in the last term, as in Example 7, each factor contains just one variable term.

EXAMPLE 8 Factor: $a^4 - 16b^4$.

SOLUTION

A. We look first for a common factor. There is none.

B. There are two terms. Since $a^4 = (a^2)^2$ and $16b^4 = (4b^2)^2$, we see that we have a difference of squares. Thus,

$$a^4 - 16b^4 = (a^2 + 4b^2)(a^2 - 4b^2).$$

C. The factor $(a^2 - 4b^2)$ is itself a difference of squares. Thus,

$$a^4 - 16b^4 = (a^2 + 4b^2)(a + 2b)(a - 2b). \qquad \text{Factoring } a^2 - 4b^2$$

D. *Check:* $(a^2 + 4b^2)(a + 2b)(a - 2b) = (a^2 + 4b^2)(a^2 - 4b^2)$
$$= a^4 - 16b^4.$$

The factorization is $(a^2 + 4b^2)(a + 2b)(a - 2b)$.

⟲ YOUR TURN

⤷ **Check Your UNDERSTANDING**

For each polynomial, determine **(a)** whether there is a common factor (other than 1 or -1) and **(b)** the number of terms.

1. $3x^2 - 5x^2 + 6x$

2. $ax^2 - ay^2$

3. $ax - by + ay - bx$

4. $24m^2 - 18mn - 32n^2$

5. $x^2y^2 - xy - 12$

8. Factor: $16 - t^4$.

5.5 EXERCISE SET

FOR EXTRA HELP MyMathLab®

↪ Vocabulary and Reading Check

Choose from the following list of terms to complete each statement. Not every term will be used.

common factor   multiplying
dividing   perfect-square trinomial
grouping   prime polynomial

1. As a first step when factoring polynomials, always check for a(n) _____.

2. When factoring a trinomial, if two terms are not squares, it cannot be a(n) _____.

3. If a polynomial has four terms and no common factor, it may be possible to factor by _____.

4. It is always possible to check a factorization by _____.

A. Factoring Polynomials

Factor completely. If a polynomial is prime, state this.

5. $5a^2 - 125$
6. $10c^2 - 810$

7. $y^2 + 49 - 14y$
8. $a^2 + 25 + 10a$

9. $3t^2 + 16t + 21$
10. $8t^2 + 31t - 4$

11. $x^3 + 18x^2 + 81x$
12. $x^3 - 24x^2 + 144x$

13. $x^3 - 5x^2 - 25x + 125$

14. $x^3 + 3x^2 - 4x - 12$

15. $27t^3 - 3t$
16. $98t^2 - 18$

17. $9x^3 + 12x^2 - 45x$
18. $20x^3 - 4x^2 - 72x$

19. $t^2 + 25$
20. $4x^2 + 20x - 144$

21. $6y^2 + 18y - 240$
22. $4n^2 + 81$

23. $-2a^6 + 8a^5 - 8a^4$
24. $-x^5 - 14x^4 - 49x^3$

25. $5x^5 - 80x$
26. $4x^4 - 64$

27. $t^4 - 9$
28. $9 + t^8$

29. $-x^6 + 2x^5 - 7x^4$
30. $-x^5 + 4x^4 - 3x^3$

31. $p^2 - w^2$
32. $a^2b^2 - c^2$

33. $ax^2 + ay^2$
34. $12n^2 + 24n^3$

35. $2\pi rh + 2\pi r^2$
36. $4\pi r^2 + 2\pi r$

Aha! 37. $(a + b)5a + (a + b)3b$

38. $5c(a^3 + b) - (a^3 + b)$

39. $x^2 + x + xy + y$

40. $n^2 + 2n + np + 2p$

41. $160a^2m^4 - 10a^2$

42. $32t^4 - 162y^4$

43. $a^2 - 2a - ay + 2y$

44. $2x^2 - 4x + xz - 2z$

45. $3x^2 + 13xy - 10y^2$

46. $-x^2 - y^2 - 2xy$

47. $8m^3n - 32m^2n^2 + 24mn$

48. $a^2 - 7a - 6$

49. $\frac{9}{16} - y^2$
50. $\frac{1}{36}a^2 - m^2$

51. $4b^2 + a^2 - 4ab$
52. $7p^4 - 7q^4$

53. $16x^2 + 24xy + 9y^2$

54. $6a^2b^3 + 12a^3b^2 - 3a^4b^2$

55. $m^2 - 5m + 8$
56. $25z^2 + 10zy + y^2$

57. $10x^2 - 11x - 6$
58. $24x^2 - 47x - 21$

59. $a^4b^4 - 16$
60. $a^5 - 4a^4b - 5a^3b^2$

61. $80cd^2 - 36c^2d + 4c^3$
62. $2p^2 + pq + q^2$

63. $3b^2 + 17ab - 6a^2$
64. $2mn - 360n^2 + m^2$

65. $-12 - x^2y^2 - 8xy$

66. $m^2n^2 - 4mn - 32$

67. $14t + 8t^2 - 15$

68. $8y - 15 + 12y^2$

69. $5p^2t^2 + 25pt - 30$

70. $a^4b^3 + 2a^3b^2 - 15a^2b$

71. $4ab^5 - 32b^4 + a^2b^6$
72. $-60 + 52x - 8x^2$

73. $x^6 + x^5y - 2x^4y^2$
74. $2s^6t^2 + 10s^3t^3 + 12t^4$

75. $36a^2 - 15a + \frac{25}{16}$
76. $a^2 + 2a^2bc + a^2b^2c^2$

77. $\frac{1}{81}x^2 - \frac{8}{27}x + \frac{16}{9}$
78. $\frac{1}{4}a^2 + \frac{1}{3}ab + \frac{1}{9}b^2$

79. $1 - 16x^{12}y^{12}$
80. $b^4a - 81a^5$

81. $4a^2b^2 + 12ab + 9$
82. $9c^2 + 6cd + d^2$

83. $z^4 + 6z^3 - 6z^2 - 36z$
84. $t^5 - 2t^4 + 5t^3 - 10t^2$

85. $x^3 + 5x^2 - x - 5$
86. $x^3 + 3x^2 - 16x - 48$

87. Kelly factored $16 - 8x + x^2$ as $(x - 4)^2$, while Tony factored it as $(4 - x)^2$. Are they both correct? Why or why not?

88. Describe in your own words or draw a diagram representing a strategy for factoring polynomials.

Skill Review

Solve.

89. There were approximately 1.91 billion smartphone users worldwide in 2015. This was 42.9% of all mobile phone users. How many mobile phone users were there in 2015? [2.4]

Data: emarketer.com

90. The number of billionaires in China increased from 120 in 2009 to 335 in 2015. What was the average rate of increase? [3.4]

Data: sify.com and forbes.com

91. In 2015, U.S. shoppers spent $33.9 billion on gifts for Mother's Day and for Father's Day combined. They spent $8.5 billion more for Mother's Day than for Father's Day. How much did shoppers spend for each holiday? [2.5]

Data: National Retail Federation

92. Shelby plans to spend no more than $50 on roses for a Mother's Day gift. If the shipping for the flowers is $15 and each rose costs $6.50, how many roses can she purchase? [2.7]

Synthesis

93. There are third-degree polynomials in x that we are not yet able to factor, despite the fact that they are not prime. Explain how such a polynomial could be created.

94. Describe a method that could be used to find a binomial of degree 16 that can be expressed as the product of prime binomial factors.

Factor.

95. $36 - 12x + x^2 - a^2$

96. $100 + 20t + t^2 - 4y^2$

97. $-(x^5 + 7x^3 - 18x)$ **98.** $18 + a^3 - 9a - 2a^2$

99. $-x^4 + 7x^2 + 18$ **100.** $-3a^4 + 15a^2 - 12$

Aha! **101.** $y^2(y + 1) - 4y(y + 1) - 21(y + 1)$

102. $y^2(y - 1) - 2y(y - 1) + (y - 1)$

103. $(y + 4)^2 + 2x(y + 4) + x^2$

104. $6(x - 1)^2 + 7y(x - 1) - 3y^2$

105. $2(a + 3)^4 - (a + 3)^3(b - 2) - (a + 3)^2(b - 2)^2$

106. $5(t - 1)^5 - 6(t - 1)^4(s - 1) + (t - 1)^3(s - 1)^2$

107. $49x^4 + 14x^2 + 1 - 25x^6$

YOUR TURN ANSWERS: SECTION 5.5

1. $(a + 1)^2$ **2.** $(5x - 1)(x^2 + 3)$
3. $-3t(t - 11)(t - 2)$ **4.** $6p^2w(2p - 3w + 1)$
5. $3a^2(5a + 1)^2$ **6.** $-1(9x + y)(x + 4y)$, or
$-(9x + y)(x + 4y)$ **7.** $(ct - 5)(ct - 4)$
8. $(4 + t^2)(2 + t)(2 - t)$

Quick Quiz: Sections 5.1–5.5

Factor.

1. $10m^2 - 90$ [5.4]   **2.** $a + a^2 - 6$ [5.2]

3. $12x^3 - 6x^2 - 10x + 5$ [5.1]

4. $8m^2n^2 - 16mn + 8$ [5.4]

5. $6c^3 - 5c^2 - 6c$ [5.3]

Prepare to Move On

Solve. [2.2]

1. $8x - 9 = 0$ **2.** $2x + 7 = 0$

3. $3 - x = 0$ **4.** $3x + 5 = 0$

5. $4x - 1 = 0$ **6.** $22 - 2x = 0$

5.6 Solving Quadratic Equations by Factoring

A. The Principle of Zero Products   **B.** Factoring to Solve Equations

When we factor a polynomial, we are forming an *equivalent expression*. We now use our factoring skills to *solve equations*. We already know how to solve linear equations like $x + 2 = 7$ and $2x = 9$. The equations we will learn to solve in this section contain a variable raised to a power greater than 1 and will usually have more than one solution.

Second-degree equations like $4t^2 - 7 = 2$ and $x^2 + 6x + 5 = 0$ are called **quadratic equations**.

> **QUADRATIC EQUATION**
>
> A *quadratic equation* is an equation equivalent to one of the form
>
> $$ax^2 + bx + c = 0,$$
>
> where a, b, and c are constants, with $a \neq 0$.

In order to solve quadratic equations, we need to develop a new principle.

A. The Principle of Zero Products

Suppose we are told that the product of two numbers is 6. From this alone, it is impossible to know the value of either number—the product could be $2 \cdot 3$, $6 \cdot 1$, $12 \cdot \frac{1}{2}$, and so on. However, if we are told that the product of two numbers is 0, we know that at least one of the two factors must itself be 0. For example, if $(x + 3)(x - 2) = 0$, we can conclude that either $x + 3$ is 0 or $x - 2$ is 0.

> **THE PRINCIPLE OF ZERO PRODUCTS**
>
> An equation $AB = 0$ is true if and only if $A = 0$ or $B = 0$, or both. (A product is 0 if and only if at least one factor is 0.)

EXAMPLE 1   Solve: $(x + 3)(x - 2) = 0$.

SOLUTION   We look for all values of x that make the equation true. The equation tells us that the product of $x + 3$ and $x - 2$ is 0. In order for any product to be 0, at least one factor must be 0. Thus we look for any value of x for which $x + 3 = 0$, as well as any value of x for which $x - 2 = 0$, that is, either

$$x + 3 = 0 \quad or \quad x - 2 = 0. \qquad \text{Using the principle of zero products. There are two equations to solve.}$$

We solve each equation:

$$x + 3 = 0 \quad or \quad x - 2 = 0$$
$$x = -3 \quad or \quad x = 2.$$

Both -3 and 2 should be checked in the original equation.

Check: For -3:

$$\frac{(x + 3)(x - 2) = 0}{(-3 + 3)(-3 - 2) \;\Big|\; 0}$$

The factor $x + 3$ is 0 when $x = -3$. $\longrightarrow 0(-5)$

$$0 \stackrel{?}{=} 0 \quad \text{TRUE}$$

For 2:

$$\frac{(x + 3)(x - 2) = 0}{(2 + 3)(2 - 2) \;\Big|\; 0}$$

The factor $x - 2$ is 0 when $x = 2$. $\longleftarrow 5(0)$

$$0 \stackrel{?}{=} 0 \quad \text{TRUE}$$

The solutions are -3 and 2.

YOUR TURN

When we are using the principle of zero products, the word "or" is meant to emphasize that any one of the factors could be the one that represents 0.

Study Skills

Identify the Highlights

If you haven't already tried one, consider using a highlighter as you read. By highlighting sentences or phrases that you find especially important, you will make it easier to review important material in the future.

1. Solve: $(x - 1)(x + 5) = 0$.

EXAMPLE 2 Solve: $3(5x + 1)(x - 7) = 0$.

SOLUTION The factors in this equation are $3, 5x + 1$, and $x - 7$. Since the factor 3 is constant, the only way in which $3(5x + 1)(x - 7)$ can be 0 is for one of the other factors to be 0, that is,

$$5x + 1 = 0 \quad or \quad x - 7 = 0 \qquad \text{Using the principle of zero products}$$
$$5x = -1 \quad or \qquad x = 7 \qquad \text{Solving the two equations separately}$$
$$x = -\tfrac{1}{5} \quad or \qquad x = 7. \qquad \begin{array}{l} 5x + 1 = 0 \text{ when } x = -\tfrac{1}{5}; \\ x - 7 = 0 \text{ when } x = 7 \end{array}$$

Check: For $-\tfrac{1}{5}$:

$$3(5x + 1)(x - 7) = 0$$

$$\begin{array}{c|c} 3\left(5\left(-\tfrac{1}{5}\right) + 1\right)\left(-\tfrac{1}{5} - 7\right) & 0 \\ 3(-1 + 1)\left(-7\tfrac{1}{5}\right) & \\ 3(0)\left(-7\tfrac{1}{5}\right) & \\ 0 \stackrel{?}{=} 0 & \text{TRUE} \end{array}$$

For 7:

$$3(5x + 1)(x - 7) = 0$$

$$\begin{array}{c|c} 3(5(7) + 1)(7 - 7) & 0 \\ 3(35 + 1)0 & \\ 0 \stackrel{?}{=} 0 & \text{TRUE} \end{array}$$

The solutions are $-\tfrac{1}{5}$ and 7.

2. Solve:
$$6(x + 2)(3x + 5) = 0.$$

YOUR TURN

The constant factor 3 in Example 2 is never 0 and is not a solution of the equation. However, a variable factor such as x or t *can* equal 0, and must be considered when using the principle of zero products.

EXAMPLE 3 Solve: $7t(t - 5) = 0$.

SOLUTION We have

$$7 \cdot t(t - 5) = 0 \qquad \text{The factors are } 7, t, \text{ and } t - 5.$$
$$t = 0 \quad or \quad t - 5 = 0 \qquad \text{Using the principle of zero products}$$
$$t = 0 \quad or \qquad t = 5. \qquad \begin{array}{l}\text{Solving. Note that the constant factor, } 7, \\ \text{is never 0.}\end{array}$$

The solutions are 0 and 5. The check is left to the student.

3. Solve: $10a(a + 9) = 0$.

YOUR TURN

B. Factoring to Solve Equations

By factoring and using the principle of zero products, we can now solve a variety of quadratic equations.

EXAMPLE 4 Solve: $x^2 + 5x + 6 = 0$.

SOLUTION We first factor the polynomial, and then use the principle of zero products:

$$x^2 + 5x + 6 = 0$$
$$(x + 2)(x + 3) = 0 \qquad \text{Factoring}$$
$$x + 2 = 0 \quad or \quad x + 3 = 0 \qquad \text{Using the principle of zero products}$$
$$x = -2 \quad or \qquad x = -3.$$

Check: For -2:

$$x^2 + 5x + 6 = 0$$

$$\frac{(-2)^2 + 5(-2) + 6 \;\big|\; 0}{}$$
$$4 - 10 + 6$$
$$-6 + 6$$
$$0 \overset{?}{=} 0 \quad \text{TRUE}$$

For -3:

$$x^2 + 5x + 6 = 0$$

$$\frac{(-3)^2 + 5(-3) + 6 \;\big|\; 0}{}$$
$$9 - 15 + 6$$
$$-6 + 6$$
$$0 \overset{?}{=} 0 \quad \text{TRUE}$$

The solutions are -2 and -3.

4. Solve: $t^2 - 8t + 15 = 0$.
_____

YOUR TURN

Student Notes

Checking for a common factor is an important step that is often overlooked. In Example 5, the quadratic expression must be factored. If we "divide both sides by x," we will not find the solution 0.

5. Solve: $n^2 - 8n = 0$.
_____

EXAMPLE 5 Solve: $x^2 + 7x = 0$.

SOLUTION Although there is no constant term, because of the x^2-term, the equation is still quadratic. We factor and use the principle of zero products:

$$x^2 + 7x = 0$$
$$x(x + 7) = 0 \qquad \text{Factoring out the largest common factor, } x$$
$$x = 0 \quad or \quad x + 7 = 0 \qquad \text{Using the principle of zero products}$$
$$x = 0 \quad or \qquad x = -7.$$

The solutions are 0 and -7. The check is left to the student.

YOUR TURN

> **CAUTION!** We *must* have 0 on one side of the equation before the principle of zero products can be used. Get all nonzero terms on one side and 0 on the other.

EXAMPLE 6 Solve: **(a)** $x^2 - 8x = -16$; **(b)** $4t^2 = 25$.

SOLUTION

a) We first add 16 to get 0 on one side:

$$x^2 - 8x = -16$$
$$x^2 - 8x + 16 = 0 \qquad \text{Adding 16 to both sides to get 0 on one side}$$
$$(x - 4)(x - 4) = 0 \qquad \text{Factoring}$$
$$x - 4 = 0 \quad or \quad x - 4 = 0 \qquad \text{Using the principle of zero products}$$
$$x = 4 \quad or \qquad x = 4.$$

There is only one solution, 4. The check is left to the student.

b) We have

Get 0 on one side.

Factor.

Use the principle of zero products.

Solve each equation separately.

$$4t^2 = 25$$
$$4t^2 - 25 = 0 \qquad \text{Subtracting 25 from both sides to get 0 on one side}$$
$$(2t - 5)(2t + 5) = 0 \qquad \text{Factoring a difference of squares}$$
$$2t - 5 = 0 \quad or \quad 2t + 5 = 0$$
$$2t = 5 \quad or \qquad 2t = -5 \;\Big\}$$
$$t = \tfrac{5}{2} \quad or \qquad t = -\tfrac{5}{2}. \;\Big]$$
Solving the two equations separately

6. Solve: $a^2 = -10a - 25$.
_____

The solutions are $\tfrac{5}{2}$ and $-\tfrac{5}{2}$. The check is left to the student.

YOUR TURN

↳ Check Your UNDERSTANDING

For each equation, use the principle of zero products to write two or three linear equations—one for each factor that contains a variable. Do not solve.

1. $(x + 4)(x - 5) = 0$

2. $(2x - 7)(3x + 4) = 0$

3. $x(x - 3) = 0$

4. $x(x + 7)(x - 9) = 0$

5. $3(x + 6)(2x + 1) = 0$

7. Solve:

$$(3x + 4)(x - 1) = -2.$$

EXAMPLE 7 Solve: $(x + 3)(2x - 1) = 9$.

SOLUTION Be careful with an equation like this! Since we need 0 on one side, we first multiply the product on the left and then subtract 9 from both sides.

$(x + 3)(2x - 1) = 9$	This is not a product equal to 0.
$2x^2 + 5x - 3 = 9$	Multiplying on the left
$2x^2 + 5x - 3 - 9 = 9 - 9$	Subtracting 9 from both sides to get 0 on one side
$2x^2 + 5x - 12 = 0$	Combining like terms
$(2x - 3)(x + 4) = 0$	Factoring. Now we have a product equal to 0.
$2x - 3 = 0 \quad or \quad x + 4 = 0$	Using the principle of zero products
$2x = 3 \quad or \quad x = -4$	
$x = \frac{3}{2} \quad or \quad x = -4$	

Check: For $\frac{3}{2}$:

$$\frac{(x + 3)(2x - 1) = 9}{\left(\frac{3}{2} + 3\right)\left(2 \cdot \frac{3}{2} - 1\right) \,\bigg|\, 9}$$

$$\left(\frac{9}{2}\right)(2) \,\bigg|\,$$

$$9 \overset{?}{=} 9 \quad \text{TRUE}$$

For -4:

$$\frac{(x + 3)(2x - 1) = 9}{(-4 + 3)(2(-4) - 1) \,\bigg|\, 9}$$

$$(-1)(-9) \,\bigg|\,$$

$$9 \overset{?}{=} 9 \quad \text{TRUE}$$

The solutions are $\frac{3}{2}$ and -4.

↩ YOUR TURN

EXPLORING 🔍 THE CONCEPT

ALF
Active Learning Figure

Since an x-intercept is on the x-axis, the y-coordinate of an x-intercept will be 0. In order to find any x-intercepts of the graph of an equation, we can replace y with 0 and solve for x.

Graphs of quadratic equations of the form $y = ax^2 + bx + c$ are shaped as shown below. Each x-intercept represents a solution of $ax^2 + bx + c = 0$. For example, since 3 is a solution of $x^2 - 2x - 3 = 0$, then $(3, 0)$ is an x-intercept of the graph of $y = x^2 - 2x - 3$.

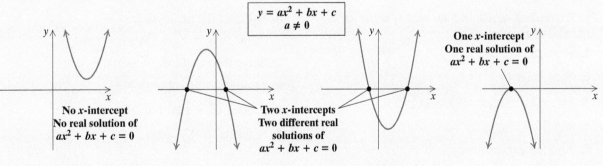

$$y = ax^2 + bx + c$$
$$a \neq 0$$

No x-intercept
No real solution of
$ax^2 + bx + c = 0$

Two x-intercepts
Two different real
solutions of
$ax^2 + bx + c = 0$

One x-intercept
One real solution of
$ax^2 + bx + c = 0$

1. The x-intercepts of the graph of $y = x^2 + x - 6$ are shown on the graph at right. What are the solutions of $x^2 + x - 6 = 0$?

2. The solutions of $x^2 + 4x - 5 = 0$ are -5 and 1. What are the x-intercepts of the graph of $y = x^2 + 4x - 5$?

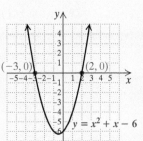

ANSWERS

1. $-3, 2$ **2.** $(-5, 0), (1, 0)$

Because every x-intercept has a y-coordinate of 0, we can find any x-intercepts of the graph of an equation in x and y by letting $y = 0$ and solving for x.

EXAMPLE 8 Find the x-intercepts of the graph of the equation shown. (The grid is intentionally not included.)

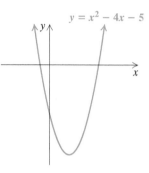

$$y = x^2 - 4x - 5$$

8. Find the x-intercepts of the graph of the equation shown.

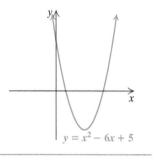

$$y = x^2 - 6x + 5$$

SOLUTION To find the x-intercepts, we let $y = 0$ and solve for x:

$$0 = x^2 - 4x - 5 \qquad \text{Substituting 0 for } y$$
$$0 = (x - 5)(x + 1) \qquad \text{Factoring}$$
$$x - 5 = 0 \quad or \quad x + 1 = 0 \qquad \text{Using the principle of zero products}$$
$$x = 5 \quad or \qquad x = -1. \qquad \text{Solving for } x$$

The x-intercepts are $(5, 0)$ and $(-1, 0)$. The check is left to the student.

YOUR TURN

CONNECTING 🔗 THE CONCEPTS

Recall that an *equation* is a statement that two *expressions* are equal. When we simplify expressions, combine expressions, and form equivalent expressions, each result is an expression. When we are asked to solve an equation, the result is one or more numbers. Remember to read the directions to an exercise carefully so you do not attempt to "solve" an expression.

EXERCISES

For Exercises 1–4, tell whether each is either an expression or an equation.

1. $x^2 - 25$

2. $x^2 - 25 = 0$

3. $x(x + 3) - 2(2x - 7) - (x - 5)$

4. $x = 10$

5. Add the expressions:
$$(2x^3 - 5x + 1) + (x^2 - 3x - 1).$$

6. Subtract the expressions:
$$(x^2 - x - 5) - (3x^2 - x + 6).$$

7. Solve the equation: $t^2 - 100 = 0$.

8. Multiply: $(3a - 2)(2a - 5)$.

9. Factor: $n^2 - 10n + 9$.

10. Solve: $x^2 + 16 = 10x$.

Technology Connection

A graphing calculator allows us to solve polynomial equations even when an equation cannot be solved by factoring. For example, to solve $x^2 - 3x - 5 = 0$, we can let $y_1 = x^2 - 3x - 5$ and $y_2 = 0$. Selecting a bold line type to the left of y_2 in the ⟨ Y= ⟩ window makes the line easier to see. Using the INTERSECT option of the CALC menu, we select the two graphs in which we are interested, along with a guess. The graphing calculator displays the nearest point of intersection.

An alternative method uses only y_1 and the ZERO option of the CALC menu. This option requires you to enter an x-value to the left of each x-intercept as a LEFT BOUND. An x-value to the right of the x-intercept is then entered as a RIGHT BOUND. Finally, a GUESS value between the two bounds is entered and the x-intercept, or ZERO, is displayed.

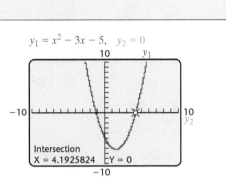

$y_1 = x^2 - 3x - 5, \quad y_2 = 0$

Intersection
X = 4.1925824   Y = 0

Use a graphing calculator to find the solutions, if they exist, accurate to two decimal places.

1. $x^2 + 4x - 3 = 0$
2. $x^2 - 5x - 2 = 0$
3. $x^2 + 13.54x + 40.95 = 0$
4. $x^2 - 4.43x + 6.32 = 0$
5. $1.235x^2 - 3.409x = 0$

VIDEO

5.6 EXERCISE SET

◆ Vocabulary and Reading Check

For each of Exercises 1–4, match the phrase with the most appropriate choice from the column on the right.

1. ____ The name of equations of the type $ax^2 + bx + c = 0$, with $a \neq 0$

   a) 2

   b) 0

2. ____ The maximum number of solutions of a quadratic equation

   c) Quadratic

   d) The principle of zero products

3. ____ The idea that $A \cdot B = 0$ if and only if $A = 0$ or $B = 0$

4. ____ The number that a product must equal before the principle of zero products is used

A. The Principle of Zero Products

Solve using the principle of zero products.

5. $(x + 2)(x + 9) = 0$

6. $(x + 3)(x + 10) = 0$

7. $(x + 1)(x - 8) = 0$

8. $(x + 5)(x - 4) = 0$

9. $(2t - 3)(t + 6) = 0$

10. $(5t - 8)(t - 1) = 0$

11. $4(7x - 1)(10x - 3) = 0$

12. $6(4x - 3)(2x + 9) = 0$

13. $x(x - 7) = 0$

14. $x(x + 2) = 0$

15. $\left(\frac{2}{3}x - \frac{12}{11}\right)\left(\frac{7}{4}x - \frac{1}{12}\right) = 0$

16. $\left(\frac{1}{9} - 3x\right)\left(\frac{1}{5} + 2x\right) = 0$

17. $6n(3n + 8) = 0$

18. $10n(4n - 5) = 0$

19. $(20 - 0.4x)(7 - 0.1x) = 0$

20. $(1 - 0.05x)(1 - 0.3x) = 0$

B. Factoring to Solve Equations

Solve by factoring and using the principle of zero products.

21. $x^2 - 7x + 6 = 0$

22. $x^2 - 6x + 5 = 0$

23. $x^2 + 4x - 21 = 0$

24. $x^2 - 7x - 18 = 0$

25. $n^2 + 11n + 18 = 0$

26. $n^2 + 8n + 15 = 0$

27. $x^2 - 10x = 0$

28. $x^2 + 8x = 0$

29. $6t + t^2 = 0$

30. $3t - t^2 = 0$

31. $x^2 - 36 = 0$

32. $x^2 - 100 = 0$

33. $4t^2 = 49$

34. $9t^2 = 25$

35. $0 = 25 + x^2 + 10x$

36. $0 = 6x + x^2 + 9$

37. $64 + x^2 = 16x$

38. $x^2 + 1 = 2x$

39. $4t^2 = 8t$

40. $12t = 3t^2$

41. $4y^2 = 7y + 15$

42. $12y^2 - 5y = 2$

43. $(x - 7)(x + 1) = -16$

44. $(x + 2)(x - 7) = -18$

45. $15z^2 + 7 = 20z + 7$

46. $14z^2 - 3 = 21z - 3$

47. $36m^2 - 9 = 40$

48. $81x^2 - 5 = 20$

49. $(x + 3)(3x + 5) = 7$

50. $(x - 1)(5x + 4) = 2$

51. $3x^2 - 2x = 9 - 8x$

52. $x^2 - 2x = 18 + 5x$

53. $(6a + 1)(a + 1) = 21$

54. $(2t + 1)(4t - 1) = 14$

55. Use this graph to solve $x^2 - 3x - 4 = 0$.

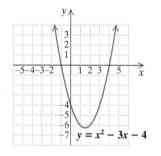

56. Use this graph to solve $x^2 + x - 6 = 0$.

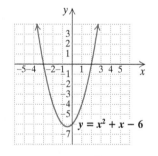

57. Use this graph to solve $-x^2 - x + 6 = 0$.

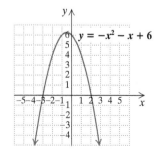

58. Use this graph to solve $-x^2 + 2x + 3 = 0$.

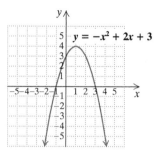

Find the x-intercepts for the graph of each equation. Grids are intentionally not included.

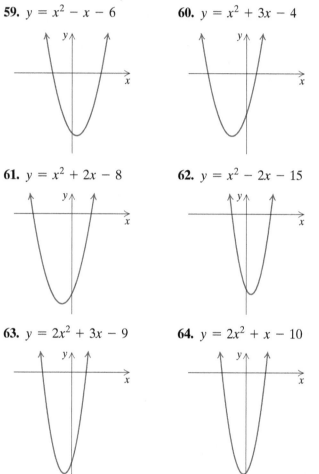

59. $y = x^2 - x - 6$

60. $y = x^2 + 3x - 4$

61. $y = x^2 + 2x - 8$

62. $y = x^2 - 2x - 15$

63. $y = 2x^2 + 3x - 9$

64. $y = 2x^2 + x - 10$

65. The equation $x^2 + 1 = 0$ has no real-number solutions. What implications does this have for the graph of $y = x^2 + 1$?

66. What is the difference between a quadratic polynomial and a quadratic equation?

Skill Review

67. Find the opposite of -65. [1.6]

68. Find the reciprocal of $\frac{3}{4}$. [1.7]

69. Find the absolute value of -1.65. [1.4]

70. Write 0.00068 in scientific notation. [4.2]

71. Write $-\frac{2}{3}$ in decimal notation. [1.4]

72. Write $\frac{5}{4}$ in percent notation. [2.4]

Synthesis

73. What is wrong with solving $x^2 = 3x$ by dividing both sides of the equation by x?

74. When the principle of zero products is used to solve a quadratic equation, will there always be two different solutions? Why or why not?

Solve.

75. $(2x - 11)(3x^2 + 29x + 56) = 0$

76. $(4x + 1)(15x^2 - 7x - 2) = 0$

77. Find an equation with integer coefficients that has the given numbers as solutions. For example, 3 and -2 are solutions of $x^2 - x - 6 = 0$.

  a) $-4, 5$ **b)** $-1, 7$ **c)** $\frac{1}{4}, 3$
  d) $\frac{1}{2}, \frac{1}{3}$ **e)** $\frac{2}{3}, \frac{3}{4}$ **f)** $-1, 2, 3$

Solve.

78. $16(x - 1) = x(x + 8)$

79. $a(9 + a) = 4(2a + 5)$

80. $(t - 5)^2 = 2(5 - t)$

81. $-x^2 + \frac{9}{25} = 0$

82. $a^2 = \frac{49}{100}$

Aha! 83. $(t + 1)^2 = 9$

84. $\frac{27}{25}x^2 = \frac{1}{3}$

85. $x^3 - 2x^2 - x + 2 = 0$

86. $x^3 + 4x^2 - 9x - 36 = 0$

87. For each equation on the left, find an equivalent equation on the right.

  a) $x^2 + 10x - 2 = 0$ $4x^2 + 8x + 36 = 0$
  b) $(x - 6)(x + 3) = 0$ $(2x + 8)(2x - 5) = 0$
  c) $5x^2 - 5 = 0$ $9x^2 - 12x + 24 = 0$
  d) $(2x - 5)(x + 4) = 0$ $(x + 1)(5x - 5) = 0$
  e) $x^2 + 2x + 9 = 0$ $x^2 - 3x - 18 = 0$
  f) $3x^2 - 4x + 8 = 0$ $2x^2 + 20x - 4 = 0$

88. Explain how to construct an equation that has seven solutions.

89. Explain how the graph in Exercise 57 can be used to visualize the solutions of
$$-x^2 - x + 6 = 4.$$

Use a graphing calculator to find the solutions of each equation. Round solutions to the nearest hundredth.

90. $-x^2 + 0.63x + 0.22 = 0$

91. $x^2 - 9.10x + 15.77 = 0$

92. $6.4x^2 - 8.45x - 94.06 = 0$

93. $x^2 + 13.74x + 42.00 = 0$

94. $0.84x^2 - 2.30x = 0$

95. $1.23x^2 + 4.63x = 0$

96. $x^2 + 1.80x - 5.69 = 0$

97. The square of the sum of two consecutive integers is 225. What are the integers?

98. The sum of the squares of two consecutive integers is 313. What are the integers?

↳ YOUR TURN ANSWERS: SECTION 5.6

1. $-5, 1$ **2.** $-2, -\frac{5}{3}$ **3.** $-9, 0$ **4.** $3, 5$
5. $0, 8$ **6.** -5 **7.** $-1, \frac{2}{3}$ **8.** $(1, 0)$ and $(5, 0)$

Quick Quiz: Sections 5.1–5.6

1. Factor: $x^2 - 2x - 8$. [5.2]

2. Solve: $x^2 - 2x - 8 = 0$. [5.6]

3. Solve: $a^2 = 100$. [5.6]

4. Factor: $a^2 - 100$. [5.4]

5. Factor: $2a^2 + 40a + 200$. [5.4]

Prepare to Move On

Translate to an algebraic expression. [1.1]

1. The square of the sum of two consecutive integers

2. The product of two consecutive integers

Solve. [2.5]

3. The first angle of a triangle is four times as large as the second. The measure of the third angle is 30° less than that of the second. How large are the angles?

4. A rectangular table top is twice as long as it is wide. The perimeter of the table is 192 in. What are the dimensions of the table?

5.7 Solving Applications

A. Applications   **B.** The Pythagorean Theorem

A. Applications

We can use the five-step problem-solving process and our methods of solving quadratic equations to solve new types of problems.

EXAMPLE 1 *Race Numbers.* Terry and Jody registered their boats in the Lakeport Race at the same time. The racing numbers assigned to their boats were consecutive numbers, the product of which was 156. Find the numbers.

SOLUTION

1. **Familiarize.** Consecutive numbers are one apart, like 49 and 50. Let $x =$ the first boat number; then $x + 1 =$ the next boat number.

2. **Translate.** We reword the problem before translating:

   Rewording: The first boat number times the next boat number is 156.

   Translating: $x \cdot (x + 1) = 156$

3. **Carry out.** We solve the equation as follows:

$$x(x + 1) = 156$$
$$x^2 + x = 156 \qquad \text{Multiplying}$$
$$x^2 + x - 156 = 0 \qquad \text{Subtracting 156 to get 0 on one side}$$
$$(x - 12)(x + 13) = 0 \qquad \text{Factoring}$$
$$x - 12 = 0 \quad or \quad x + 13 = 0 \qquad \text{Using the principle of zero products}$$
$$x = 12 \quad or \quad x = -13. \qquad \text{Solving each equation}$$

4. **Check.** The solutions of the equation are 12 and -13. Since race numbers are not negative, -13 must be rejected. On the other hand, if x is 12, then $x + 1$ is 13 and $12 \cdot 13 = 156$. Thus the solution 12 checks.

5. **State.** The boat numbers for Terry and Jody were 12 and 13.

1. Refer to Example 1. Suppose that the product of the racing numbers of the boats was 132. Find the numbers.

YOUR TURN

EXAMPLE 2 *Manufacturing.* Wooden Work, Ltd., builds cutting boards that are twice as long as they are wide. The most popular board that Wooden Work makes has an area of 800 cm². What are the dimensions of the board?

SOLUTION

1. **Familiarize.** We first make a drawing. Recall that the area of any rectangle is Length · Width. We let x = the width of the board, in centimeters. The length is then $2x$, since the board is twice as long as it is wide.

$2x$ x

2. **Translate.** We reword and translate as follows:

Rewording: The area of the rectangle   is   800 cm².

Translating: $2x \cdot x$ =   800

3. **Carry out.** We solve the equation as follows:

$$2x \cdot x = 800$$
$$2x^2 = 800$$
$$2x^2 - 800 = 0 \quad \text{Subtracting 800 to get 0 on one side of the equation}$$
$$2(x^2 - 400) = 0 \quad \text{Factoring out a common factor of 2}$$
$$2(x - 20)(x + 20) = 0 \quad \text{Factoring a difference of squares}$$
$$x - 20 = 0 \quad or \quad x + 20 = 0 \quad \text{Using the principle of zero products}$$
$$x = 20 \quad or \quad x = -20. \quad \text{Solving each equation}$$

4. **Check.** The solutions of the equation are 20 and −20. Since the width must be positive, −20 cannot be a solution. To check 20 cm, we note that if the width is 20 cm, then the length is $2 \cdot 20$ cm = 40 cm and the area is 20 cm · 40 cm = 800 cm². Thus the solution 20 checks.

5. **State.** The cutting board is 20 cm wide and 40 cm long.

2. Refer to Example 2. Wooden Work also builds a cutting board that has an area of 450 cm². This board is also twice as long as it is wide. What are the dimensions of this cutting board?

YOUR TURN

EXAMPLE 3 *Dimensions of a Leaf.* Each leaf of one particular *Philodendron* species is approximately a triangle. A typical leaf has an area of 320 in². If the leaf is 12 in. longer than it is wide, find the length and the width of the leaf.

SOLUTION

1. **Familiarize.** The formula for the area of a triangle is Area = $\frac{1}{2} \cdot$ (base) · (height). We let b = the width, in inches, of the triangle's base and $b + 12$ = the height, in inches.

2. **Translate.** We reword and translate as follows:

Rewording: The area of the leaf   is   320 in².

Translating: $\frac{1}{2} \cdot b(b + 12)$ =   320

3. Carry out. We solve the equation as follows:

$$\frac{1}{2} \cdot b \cdot (b + 12) = 320$$

$$\frac{1}{2}(b^2 + 12b) = 320 \qquad \text{Multiplying}$$

$$b^2 + 12b = 640 \qquad \text{Multiplying by 2 to clear fractions}$$

$$b^2 + 12b - 640 = 0 \qquad \text{Subtracting 640 to get 0 on one side}$$

$$(b + 32)(b - 20) = 0 \qquad \text{Factoring}$$

$$b + 32 = 0 \quad or \quad b - 20 = 0 \qquad \text{Using the principle of zero products}$$

$$b = -32 \quad or \qquad b = 20.$$

3. Like the *Philodendron* in Example 3, each leaf of the *Triangular Leaf Senecio* approximates a triangle. One leaf of this plant is 6 in. longer than it is wide and has an area of 8 in². Find the length and the width of the leaf.

4. Check. The width must be positive, so −32 cannot be a solution. Suppose that the base is 20 in. The height would be 20 + 12, or 32 in., and the area $\frac{1}{2}(20)(32)$, or 320 in². These numbers check in the original problem.

5. State. The leaf is 32 in. long and 20 in. wide.

YOUR TURN

Number of Minutes t After Injection	Number of Micrograms N of Epinephrine in the Bloodstream
2	160
5	250
8	160

EXAMPLE 4 *Medicine.* For certain people suffering an extreme allergic reaction, the drug epinephrine (adrenaline) is sometimes prescribed. The number of micrograms N of epinephrine in an adult's bloodstream t minutes after 250 micrograms have been injected is shown in the table at left for several values of t and can be approximated by

$$-10t^2 + 100t = N.$$

How long after an injection will there be about 210 micrograms of epinephrine in the bloodstream?

Data: Chohan, Naina, Rita M. Doyle, and Patricia Nayle (eds.), *Nursing Handbook*, 21st ed. Springhouse, PA: Springhouse Corporation, 2001

SOLUTION

1. Familiarize. To familiarize ourselves with this problem, we could calculate N for different choices of t. We leave this to the student. Note that there may be two solutions, one on each side of the time at which the drug's effect peaks.

2. Translate. To find the length of time after injection when 210 micrograms are in the bloodstream, we replace N with 210 in the formula above:

$$-10t^2 + 100t = 210. \qquad \text{Substituting 210 for } N. \text{ This is now an equation in one variable.}$$

3. Carry out. We solve the equation as follows:

$$-10t^2 + 100t = 210$$

$$-10t^2 + 100t - 210 = 0 \qquad \text{Subtracting 210 from both sides to get 0 on one side}$$

$$-10(t^2 - 10t + 21) = 0 \qquad \text{Factoring out the largest common factor, } -10$$

$$-10(t - 3)(t - 7) = 0 \qquad \text{Factoring}$$

$$t - 3 = 0 \quad or \quad t - 7 = 0 \qquad \text{Using the principle of zero products}$$

$$t = 3 \quad or \qquad t = 7.$$

4. Refer to Example 4. How long after an injection will there be about 160 micrograms of epinephrine in the bloodstream?

4. Check. Since $-10 \cdot 3^2 + 100 \cdot 3 = -90 + 300 = 210$, the number 3 checks. Since $-10 \cdot 7^2 + 100 \cdot 7 = -490 + 700 = 210$, the number 7 also checks.

5. State. There will be 210 micrograms of epinephrine in the bloodstream approximately 3 minutes and 7 minutes after injection.

YOUR TURN

As a visual check for Example 4, we can either let $y_1 = -10x^2 + 100x$ and $y_2 = 210$, or let $y_1 = -10x^2 + 100x - 210$ and $y_2 = 0$. In either case, the points of intersection occur at $x = 3$ and $x = 7$, as shown.

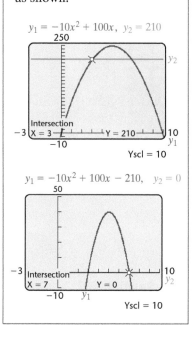

B. The Pythagorean Theorem

The following problems involve the Pythagorean theorem, which relates the lengths of the sides of a *right* triangle. A triangle is a **right triangle** if it has a 90°, or *right*, angle. The side opposite the 90° angle is called the **hypotenuse**. The other sides are called **legs**.

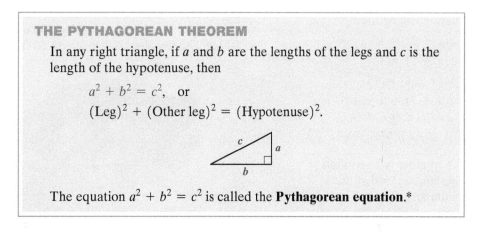

THE PYTHAGOREAN THEOREM

In any right triangle, if a and b are the lengths of the legs and c is the length of the hypotenuse, then

$$a^2 + b^2 = c^2, \quad \text{or}$$
$$(\text{Leg})^2 + (\text{Other leg})^2 = (\text{Hypotenuse})^2.$$

The equation $a^2 + b^2 = c^2$ is called the **Pythagorean equation.***

The Pythagorean theorem is named for the Greek mathematician Pythagoras (569?–500? B.C.). We can think of this relationship as adding areas.

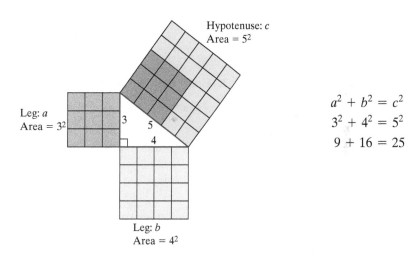

$$a^2 + b^2 = c^2$$
$$3^2 + 4^2 = 5^2$$
$$9 + 16 = 25$$

If we know the lengths of any two sides of a right triangle, we can use the Pythagorean equation to determine the length of the third side.

---

*The *converse* of the Pythagorean theorem is also true. That is, if $a^2 + b^2 = c^2$, then the triangle is a right triangle.

↳ Check Your UNDERSTANDING

Choose from the following list the equation that best models each problem. Not every equation will be used.

a) $x(x + 2) = 15$

b) $x(x + 1) = 15$

c) $\frac{1}{2}x(x + 1) = 15$

d) $x(x + 2) = x + 4$

e) $x = x^2 - 2$

f) $x^2 + (x + 2)^2 = (x + 4)^2$

1. A number is 2 less than its square. Find all such numbers.

2. The product of two consecutive odd integers is 15. Find the integers.

3. The base of a triangle is 1 ft longer than its height. Find the base and the height if the triangle has an area of $15\ \text{ft}^2$.

4. The sides of a right triangle are consecutive even integers. Find the length of each side of the triangle.

↳ **Chapter Resource:**
Translating for Success, p. 360

5. Refer to Example 5. Another brace on the bridge is also 50 ft long, but the horizontal distance that it spans is 34 ft longer than the height that it reaches at the center of the bridge. Find both distances.

EXAMPLE 5 *Bridge Design.*   A 50-ft diagonal brace on a bridge connects a support under the road surface at the center of the bridge to a side support on the bridge. The horizontal distance that it spans is 10 ft longer than the height that it reaches at the center of the bridge. Find both distances.

SOLUTION

1. **Familiarize.** We first make a drawing. The diagonal brace and the missing distances form the hypotenuse and the legs of a right triangle. We let x = the length of the vertical leg. Then $x + 10$ = the length of the horizontal leg. The hypotenuse has length 50 ft.

2. **Translate.** Since the triangle is a right triangle, we can use the Pythagorean theorem:

$$a^2 + b^2 = c^2$$
$$x^2 + (x + 10)^2 = 50^2. \quad \text{Substituting}$$

3. **Carry out.** We solve the equation as follows:

$$x^2 + (x^2 + 20x + 100) = 2500 \qquad \text{Squaring}$$
$$2x^2 + 20x + 100 = 2500 \qquad \text{Combining like terms}$$
$$2x^2 + 20x - 2400 = 0 \qquad \text{Subtracting 2500 to get 0 on one side}$$
$$2(x^2 + 10x - 1200) = 0 \qquad \text{Factoring out a common factor}$$
$$2(x + 40)(x - 30) = 0 \qquad \text{Factoring. A calculator could be helpful here.}$$
$$x + 40 = 0 \quad or \quad x - 30 = 0 \qquad \text{Using the principle of zero products}$$
$$x = -40 \quad or \qquad x = 30.$$

4. **Check.** The integer -40 cannot be a length of a side because it is negative. If the length is 30 ft, then $x + 10 = 40$, and $30^2 + 40^2 = 900 + 1600 = 2500$, which is 50^2. So the solution 30 checks.

5. **State.** The height that the brace reaches at the center of the bridge is 30 ft, and the distance that it reaches to the middle of the bridge is 40 ft.

↻ YOUR TURN

5.7 EXERCISE SET

FOR EXTRA HELP MyMathLab®

↳ Vocabulary and Reading Check

Complete each statement with the correct variable expression or number.

1. If x is an integer, then the next consecutive integer is _____.

2. If the length of a rectangle is twice its width and its width is w, then its length is _____.

3. A right triangle contains a(n) _____ angle.

4. If the legs of a right triangle have lengths a and b and the hypotenuse is length c, then _____.

A. Applications

Solve. Use the five-step problem-solving approach.

5. A number is 6 less than its square. Find all such numbers.

6. A number is 30 less than its square. Find all such numbers.

7. *Parking-Space Numbers.* The product of two consecutive parking space numbers is 132. Find the numbers.

8. *Page Numbers.* The product of the page numbers on two facing pages of a book is 420. Find the page numbers.

9. The product of two consecutive even integers is 168. Find the integers.

10. The product of two consecutive odd integers is 195. Find the integers.

11. *Construction.* The front porch on Trent's new home is five times as long as it is wide. If the area of the porch is 320 ft^2, find the dimensions.

12. *Furnishings.* The work surface of Anita's desk is a rectangle that is twice as long as it is wide. If the area of the desktop is 18 ft^2, find the length and the width of the desk.

13. A photo is 5 cm longer than it is wide. Find the length and the width if the area is 84 cm^2.

14. An envelope is 4 cm longer than it is wide. The area is 96 cm^2. Find the length and the width.

15. *Dimensions of a Sail.* The height of the jib sail on a Lightning sailboat is 5 ft greater than the length of its "foot." If the area of the sail is 42 ft^2, find the length of the foot and the height of the sail.

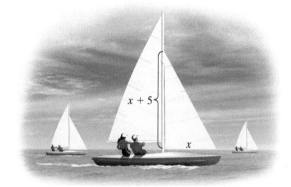

16. *Dimensions of a Sail.* The height of a triangular mainsail on a sailboat is 4 ft longer than the base of the sail. If the area of the sail is 38.5 ft^2, find the height and the base of the sail.

17. *Road Design.* A triangular traffic island has a base that is half as long as its height. Find the base and the height if the island has an area of 64 ft^2.

18. *Tent Design.* The triangular entrance to a tent is two-thirds as wide as it is tall. The area of the entrance is 12 ft^2. Find the height and the base.

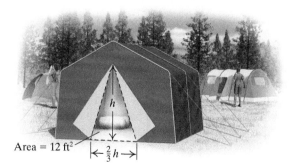

Area = 12 ft^2 $\frac{2}{3}h$

19. *Cabin Design.* The front of an A-frame cabin in a national park is in the shape of a triangle, with an area of 60 ft². If the height is 1 ft less than twice the base, find the base and the height of the front of the cabin.

20. *Flower Garden Design.* The demonstration flower garden at a state fair is in the shape of a triangle. The base of the triangle is 6 m longer than twice the height. If the area is 28 m², find the base and the height of the flower garden.

Games in a League's Schedule. In a sports league of x teams in which all teams play each other twice, the total number N of games played is given by $x^2 - x = N$. Use this formula for Exercises 21 and 22.

21. The Colchester Youth Soccer League plays a total of 56 games, with all teams playing each other twice. How many teams are in the league?

22. The teams in a women's softball league play each other twice, for a total of 132 games. How many teams are in the league?

Number of Handshakes. The number of possible handshakes H within a group of n people is given by $H = \frac{1}{2}(n^2 - n)$. Use this formula for Exercises 23–26.

23. At a meeting, there are 12 people. How many handshakes are possible?

24. At a party, there are 25 people. How many handshakes are possible?

25. *High-Fives.* After winning the championship, all Dallas Maverick teammates exchanged hugs. Altogether there were 66 hugs. How many players were there?

26. *Toasting.* During a toast at a party, there were 105 "clicks" of glasses. How many people took part in the toast?

27. *Medicine.* For many people suffering from constricted bronchial muscles, the drug Albuterol is prescribed. The number of micrograms A of Albuterol in a person's bloodstream t minutes after 200 micrograms have been inhaled can be approximated by

$$A = -50t^2 + 200t.$$

How long after an inhalation will there be about 150 micrograms of Albuterol in the bloodstream?

Data: Chohan, Naina, Rita M. Doyle, and Patricia Nayle (eds.), *Nursing Handbook*, 21st ed. Springhouse, PA: Springhouse Corporation, 2001

28. *Medicine.* For adults with certain heart conditions, the drug Primacor (milrinone lactate) is prescribed. The number of milligrams M of Primacor in the bloodstream of a 132-lb patient t hours after a 3-mg dose has been injected can be approximated by

$$M = -\frac{1}{2}t^2 + \frac{5}{2}t.$$

How long after an injection will there be about 2 mg in the bloodstream?

Data: Chohan, Naina, Rita M. Doyle, and Patricia Nayle (eds.), *Nursing Handbook*, 21st ed. Springhouse, PA: Springhouse Corporation, 2001

29. *Wave Height.* The height of waves in a storm, measured from the base of the trough, depends on the speed of the wind. Assuming that the wind has no obstructions for a long distance, the maximum wave height H for a wind speed x can be approximated by

$$H = 0.03x^2 + 0.6x - 6,$$

where H is in feet and x is in knots (nautical miles per hour). For what wind speed would the maximum wave height be 3 ft?

Data: Smith, Craig B., *Extreme Waves*, Joseph Henry Press, 2006

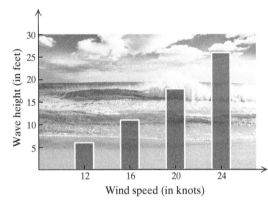

30. *Cabinet Making.* Dovetail Woodworking determines that the revenue R, in thousands of dollars, from the sale of x sets of cabinets is given by $R(x) = 2x^2 + x$. If the cost C, in thousands of dollars, of producing x sets of cabinets is given by $C(x) = x^2 - 2x + 10$, how many sets must be produced and sold in order for the company to break even?

B. The Pythagorean Theorem

31. *Construction.* The diagonal braces in a fire tower are 15 ft long and span a horizontal distance of 12 ft. How high does each brace reach vertically?

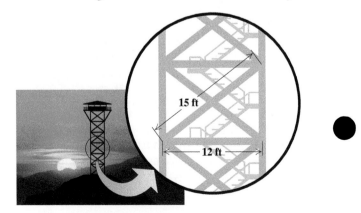

15 ft

12 ft

32. *Roadway Design.* Elliott Street is 24 ft wide when it ends at Main Street in Brattleboro, Vermont. A 40-ft long diagonal crosswalk allows pedestrians to cross Main Street to or from either corner of Elliott Street (see the figure). Determine the width of Main Street.

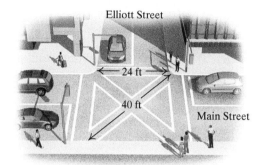

Elliott Street

24 ft

40 ft

Main Street

33. *Archaeology.* Archaeologists have discovered that the 18th-century garden of the Charles Carroll House in Annapolis, Maryland, was a right triangle. One leg of the triangle was formed by a 400-ft long sea wall. The hypotenuse of the triangle was 200 ft longer than the other leg. What were the dimensions of the garden?

Data: www.bsos.umd.edu

34. *Guy Wire.* The height of a wind power assessment tower is 5 m shorter than the guy wire that supports it. If the guy wire is anchored 15 m from the foot of the antenna, how tall is the antenna?

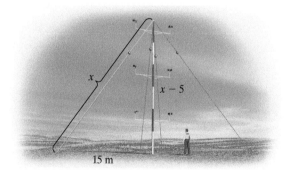

x

$x - 5$

15 m

35. *Parking Lot Design.* A rectangular parking lot is 50 ft longer than it is wide. Determine the dimensions of the parking lot if it measures 250 ft diagonally.

36. *Right Triangle.* One leg of a right triangle is 7 cm shorter than the other leg. The length of the hypotenuse is 13 cm. Find the length of each side.

37. *Carpentry.* In order to build a deck at a right angle to their house, Lucinda and Felipe place a stake in the ground a precise distance from the back wall of their house. This stake will combine with two marks on the

house to form a right triangle. From a course in geometry, Lucinda remembers that there are three consecutive integers that can work as sides of a right triangle. Find the sides of that triangle.

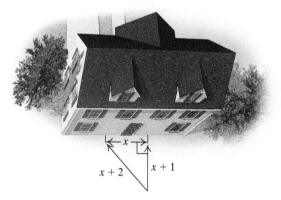

x

$x + 2$

$x + 1$

38. *Right Triangle.* The lengths of the two longest sides of a right triangle are consecutive odd integers. The shorter leg is 7 m shorter than the longer leg. Find the lengths of the sides of the triangle.

39. *Right Triangle.* The longest side of a right triangle is 1 ft longer than three times the length of the shortest side. The other side of the triangle is 1 ft shorter than three times the length of the shortest side. Find the lengths of the sides of the triangle.

40. *Right Triangle.* The longest side of a right triangle is 5 yd shorter than six times the length of the shortest side. The other side of the triangle is 5 yd longer than five times the length of the shortest side. Find the lengths of the sides of the triangle.

A. Applications

41. *Architecture.* An architect has allocated a rectangular space of 264 ft² for a square dining room and a 10-ft wide kitchen, as shown in the figure. Find the dimensions of each room.

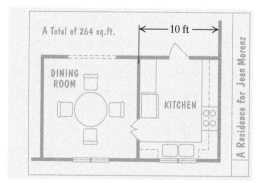

A Total of 264 sq.ft.

10 ft

DINING ROOM

KITCHEN

A Residence for Jean Morenz

42. *Design.* A window panel for a sun porch consists of a 7-ft tall rectangular window stacked above a square window. The windows have the same width. If the total area of the window panel is 18 ft², find the dimensions of each window.

7 ft

Height of a Rocket. For Exercises 43–46, assume that a water rocket is launched upward with an initial velocity of 48 ft/sec. Its height h, in feet, after t seconds, is given by $h = 48t - 16t^2$.

43. Determine the height of the rocket $\frac{1}{2}$ sec after it has been launched.

44. Determine the height of the rocket 2.5 sec after it has been launched.

45. When will the rocket be exactly 32 ft above the ground?

46. When will the rocket crash into the ground?

47. Do we now have the ability to solve *any* problem that translates to a quadratic equation? Why or why not?

48. Write a problem for a classmate to solve such that only one of two solutions of a quadratic equation can be used as an answer.

Skill Review

Solve. Label any contradictions or identities.

49. $3(x - 2) = 5x + 2(3 - x)$ [2.2]

50. $9x - 2[4(x + 1)] = 4(x - 2 + x)$ [2.2]

51. $10x + 3 = 5(2x + 1) - 2$ [2.2]

Solve. Write the answers in both set-builder notation and interval notation.

52. $3 - y \geq 7$ [2.6]

53. $\frac{1}{3}x + \frac{1}{2} < \frac{1}{6}$ [2.6]

54. Solve $x + 2y = 7$ for y.   [2.3]

Synthesis

The converse of the Pythagorean theorem is also true. That is, if $a^2 + b^2 = c^2$, then the triangle is a right triangle (where a and b are the lengths of the legs and c is the length of the hypotenuse). Use this result to answer Exercises 55 and 56.

55. An archaeologist has straight rods of 3 ft, 4 ft, and 5 ft. Explain how she could draw a 7-ft by 9-ft rectangle on a piece of land being excavated.

56. Explain how straight rods of 5 cm, 12 cm, and 13 cm can be used to draw a right triangle that has two 45° angles.

57. *Sailing.* The mainsail of a Lightning sailboat is a right triangle in which the hypotenuse is called the leech. If a 24-ft tall mainsail has a leech length of 26 ft and if Dacron® sailcloth costs $1.50 per square foot, find the cost of the fabric for a new mainsail.

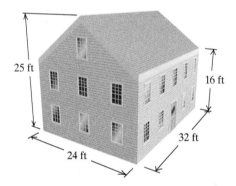

26 ft

24 ft

58. *Roofing.* A *square* of shingles covers 100 ft² of surface area. How many squares will be needed to reshingle the house shown?

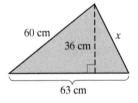

25 ft

16 ft

32 ft

24 ft

59. Solve for x.

60 cm

36 cm

x

63 cm

60. *Pool Sidewalk.* A cement walk of uniform width is built around a 20-ft by 40-ft rectangular pool. The total area of the pool and the walk is 1500 ft². Find the width of the walk.

61. *Folding Sheet Metal.* An open rectangular gutter is made by turning up the sides of a piece of metal 20 in. wide, as shown. The area of the cross section of the gutter is 48 in². Find the possible depths of the gutter.

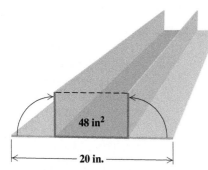

48 in²

20 in.

62. Find a polynomial for the shaded area surrounding the square in the following figure.

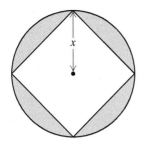

x

63. *Telephone Service.* Use the information in the following figure to determine the height of the telephone pole.

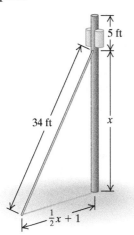

5 ft

34 ft

x

$\frac{1}{2}x + 1$

64. *Dimensions of a Closed Box.* The total surface area of a closed box is 350 m². The box is 9 m high and has a square base and lid. Find the length of a side of the base.

65. The maximum length of a postcard that can be mailed with one postcard stamp is $\frac{7}{4}$ in. longer than the maximum width. The maximum area is $\frac{51}{2}$ in². Find the maximum length and the maximum width of a postcard.

Data: USPS

Medicine. For certain people with acid reflux, the drug Pepcid (famotidine) is used. The number of milligrams N of Pepcid in an adult's bloodstream t hours after a 20-mg tablet has been swallowed can be approximated by

$$N = -0.009t(t - 12)^3.$$

Use a graphing calculator with the window $[-1, 13, -1, 25]$ *and the* TRACE *feature to answer Exercises 66–68.*

Data: Chohan, Naina, Rita M. Doyle, and Patricia Nayle (eds.), *Nursing Handbook*, 21st ed. Springhouse, PA: Springhouse Corporation, 2001

66. Approximately how long after a tablet has been swallowed will there be 18 mg in the bloodstream?

67. Approximately how long after a tablet has been swallowed will there be 10 mg in the bloodstream?

68. Approximately how long after a tablet has been swallowed will the peak dosage in the bloodstream occur?

69. *Research.* Find an online calculator that asks for the width, length, and height of a roof and calculates the roof area.

a) The roof shown in Exercise 58 is a gable roof. Enter the dimensions of this roof in the online calculator and compare your answer with the one calculated for you.

b) Develop a formula that could be used by the online calculator to find the roof area.

YOUR TURN ANSWERS: SECTION 5.7

1. 11 and 12   **2.** 15 cm wide and 30 cm long
3. 2 in. wide and 8 in. long   **4.** Approximately 2 min and 8 min after injection   **5.** Horizontal distance spanned: 48 ft; height at center of bridge: 14 ft

---

Quick Quiz: Sections 5.1–5.7

1. Factor: $3x^3 - 15x^2 - 7x + 35$. [5.1]

2. Solve: $20x^2 = 11x + 3$. [5.6]

3. Factor: $10x^2 - 17x + 3$. [5.3]

4. Factor: $6x^8 - 6$. [5.4]

5. The hypotenuse of a right triangle is 1 ft longer than the length of the longer leg. The length of the shorter leg is 9 ft. Find the length of each side. [5.7]

Prepare to Move On

Simplify. [1.3]

1. $\frac{24}{28}$ **2.** $\frac{90}{88}$ **3.** $\frac{124}{155}$

Divide, if possible. [1.7]

4. $\frac{0}{7}$ **5.** $\frac{13}{0}$

1. Angle Measures. The degree measures of the angles of a triangle are three consecutive integers. Find the measures of the angles.

2. Rectangle Dimensions. The area of a rectangle is 3604 ft². The length is 15 ft longer than the width. Find the dimensions of the rectangle.

3. Sales Tax. Claire paid $3604 for a used pickup truck. This included 6% for sales tax. How much did the truck cost before tax?

4. Wire Cutting. A 180-m wire is cut into three pieces. The third piece is 2 m longer than the first. The second is two-thirds as long as the first. How long is each piece?

5. Perimeter. The perimeter of a rectangle is 240 ft. The length is 2 ft greater than the width. Find the length and the width.

Translating for Success

Use after Section 5.7.

Translate each word problem to an equation and select a correct translation from equations A–O.

A. $2x \cdot x = 288$

B. $x(x + 60) = 7021$

C. $59 = x \cdot 60$

D. $x^2 + (x + 15)^2 = 3604$

E. $x^2 + (x + 70)^2 = 130^2$

F. $0.06x = 3604$

G. $2(x + 2) + 2x = 240$

H. $\frac{1}{2}x(x - 1) = 1770$

I. $x + \frac{2}{3}x + (x + 2) = 180$

J. $0.59x = 60$

K. $x + 0.06x = 3604$

L. $2x^2 + x = 288$

M. $x(x + 15) = 3604$

N. $x^2 + 60 = 7021$

O. $x + (x + 1) + (x + 2) = 180$

Answers on page A-29

An additional, animated version of this activity appears in MyMathLab. *To use MyMathLab, you need a course ID and a student access code. Contact your instructor for more information.*

6. Cell-Phone Tower. A guy wire on a cell-phone tower is 130 ft long and is attached to the top of the tower. The height of the tower is 70 ft longer than the distance from the point on the ground where the wire is attached to the bottom of the tower. Find the height of the tower.

7. Sales Meeting Attendance. PTQ Corporation held a sales meeting in Tucson. Of the 60 employees, 59 of them attended the meeting. What percent attended the meeting?

8. Dimensions of a Pool. A rectangular swimming pool is twice as long as it is wide. The area of the surface is 288 ft². Find the dimensions of the pool.

9. Dimensions of a Triangle. The height of a triangle is 1 cm less than the length of the base. The area of the triangle is 1770 cm². Find the height and the length of the base.

10. Width of a Rectangle. The length of a rectangle is 60 ft longer than the width. Find the width if the area of the rectangle is 7021 ft².

Collaborative Activity   *Visualizing Factoring*

Focus: Visualizing factoring
Use after: Section 5.2
Time: 20–30 minutes
Group size: 3
Materials: Graph paper and scissors

Factoring a polynomial like $x^2 + 11x + 10$ can be thought of as determining the length and the width of a rectangle that has area $x^2 + 11x + 10$.

Activity

1. a) To factor $x^2 + 11x + 10$ geometrically, the group needs to cut out shapes like those below to represent x^2, $11x$, and 10. This can be done by either tracing the figures below or by selecting a value for x, say 4, and using the squares on the graph paper to cut out the following:

   x^2:   Using the value selected for x, cut out a square that is x units on each side.

   $11x$:   Using the value selected for x, cut out a rectangle that is 1 unit wide and x units long. Repeat this to form 11 such strips.

   10:   Cut out two rectangles with whole-number dimensions and an area of 10. One should be 2 units by 5 units and the other 1 unit by 10 units.

   b) The group, working together, should then attempt to use one of the two rectangles with area 10, along with all of the other shapes, to piece together one large rectangle. Only one of the rectangles with area 10 will work.

   c) From the large rectangle formed in part (b), use the length and the width to determine the factorization of $x^2 + 11x + 10$. Where do the dimensions of the rectangle representing 10 appear in the factorization?

2. Repeat step (1) above, but this time use the other rectangle with area 10, and use only 7 of the 11 strips, along with the x^2-shape. Piece together the shapes to form one large rectangle. What factorization do the dimensions of this rectangle suggest?

3. Cut out rectangles with area 12 and use the above approach to factor $x^2 + 8x + 12$. What dimensions should be used for the rectangle with area 12?

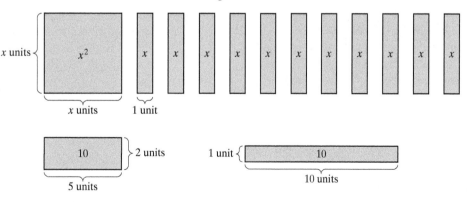

Collaborative Activity   *Matching Factorizations**

Focus: Factoring
Use after: Section 5.5
Time: 20 minutes
Group size: Begin with the entire class. If there is an odd number of students, the instructor should participate.
Materials: Prepared sheets of paper, pins or tape. On half of the sheets, the instructor writes a polynomial. On the remaining sheets, the instructor writes the factorization of those polynomials. The polynomials and factorizations are similar; for example,

$$x^2 - 2x - 8, \quad (x - 2)(x - 4),$$
$$x^2 - 6x + 8, \quad (x - 1)(x - 8),$$
$$x^2 - 9x + 8, \quad (x + 2)(x - 4)$$

Activity

1. As class members enter the room, the instructor pins or tapes either a polynomial or a factorization to the back of each student. Class members are told only whether their sheet of paper contains a polynomial or a factorization.

2. After all students are wearing a sheet of paper, they should mingle with one another, attempting to match up their factorization with the appropriate polynomial or vice versa. They may ask questions of one another that relate to factoring and polynomials. Answers to the questions should be yes or no. For example, a legitimate question might be "Is my last term negative?" or "Do my factors have opposite signs?"

3. The game is over when all factorization/polynomial pairs have "found" one another.

---

*Thanks to Jann MacInnes of Florida Community College at Jacksonville–Kent Campus for suggesting this activity.

Decision Making & Connection   *(Use after Section 5.1.)*

Designing a Tournament. The number of games in a tournament varies widely depending on the type of tournament being played.

In a *single elimination* tournament, a player or team is eliminated after one loss.

In a *double elimination* tournament, a player or team is eliminated after losing twice.

In a *round-robin* tournament, each team plays every other team once, and a winner is chosen on the basis of the win–loss records.

In a *double round-robin* tournament, each team plays every other team twice.

Other types of tournaments include ladder and pyramid tournaments.

1. The number of games G required for a round-robin tournament with n teams is given by $G = \frac{1}{2}n^2 - \frac{1}{2}n$. If 15 basketball teams enter a round-robin tournament, how many games must be played?

2. The number of games G required for a double round-robin tournament with n teams is given by $G = n^2 - n$. If 15 basketball teams enter a double round-robin tournament, how many games must be played?

3. Ideally, in a round-robin tournament, all the teams play at the same time. If there is an even number of teams, then there will be $n/2$ games occurring simultaneously during each round. If there is an odd number of teams, then $(n - 1)/2$ games will occur simultaneously, with one team resting for each round. Use this information and the formula given in Exercise 1 to find a formula for the number of rounds needed for an even number of teams and the number for an odd number of teams.

4. *Research.* Find the number of games required in a single elimination tournament with n teams.

5. *Research.* Find the number of games required in a double elimination tournament with n teams. Is there a minimum number or a maximum number?

6. If you were planning a chess tournament, what type of tournament would you choose? How many players would you invite to participate?

Study Summary

KEY TERMS AND CONCEPTS	EXAMPLES	PRACTICE EXERCISES

SECTION 5.1: *Introduction to Factoring*

To **factor** a polynomial means to write it as a product of polynomials. Always begin by factoring out the **largest common factor**.	$12x^4 - 30x^3 = 6x^3(2x - 5)$	**1.** Factor: $12x^4 - 18x^3 + 30x$.
Some polynomials with four terms can be **factored by grouping**.	$3x^3 - x^2 - 6x + 2 = x^2(3x - 1) - 2(3x - 1)$ $\qquad\qquad\qquad = (3x - 1)(x^2 - 2)$	**2.** Factor: $2x^3 - 6x^2 - x + 3$.

SECTION 5.2: *Factoring Trinomials of the Type $x^2 + bx + c$*

Some trinomials of the type $x^2 + bx + c$ can be factored by reversing the steps of FOIL.	Factor: $x^2 - 11x + 18$.	**3.** Factor: $x^2 - 7x - 18$.

Pairs of Factors of 18	Sums of Factors
$-1, -18$	-19
$-2, \ -9$	-11

The factorization is $(x - 2)(x - 9)$.

SECTION 5.3: *Factoring Trinomials of the Type $ax^2 + bx + c$*

One method for factoring trinomials of the type $ax^2 + bx + c$ is a FOIL-based method.	Factor: $6x^2 - 5x - 6$. The factors will be in the form $(3x + \ )(2x + \ )$ or $(6x + \ )(x + \ )$. We list all pairs of factors of -6, and check possible products by multiplying those possibilities that do not contain a common factor. $(3x - 2)(2x + 3) = 6x^2 + 5x - 6,$ $(3x + 2)(2x - 3) = 6x^2 - 5x - 6 \longleftarrow$ This is the correct product. The factorization is $(3x + 2)(2x - 3)$.	**4.** Factor: $6x^2 + x - 2$.

Another method for factoring trinomials of the type $ax^2 + bx + c$ involves factoring by grouping.

Factor: $6x^2 - 5x - 6$.

We multiply the leading coefficient and the constant term: $6(-6) = -36$. Look for factors of -36 that add to -5.

Pairs of Factors of −36	Sums of Factors
1, −36	−35
2, −18	−16
3, −12	−9
4, −9	−5

Rewrite $-5x$ as $4x - 9x$ and factor by grouping:
$$6x^2 - 5x - 6 = 6x^2 + 4x - 9x - 6$$
$$= 2x(3x + 2) - 3(3x + 2) = (3x + 2)(2x - 3).$$

5. Factor:
$8x^2 - 22x + 15$.

SECTION 5.4: *Factoring Perfect-Square Trinomials and Differences of Squares*

Factoring a Perfect-Square Trinomial
$A^2 + 2AB + B^2 = (A + B)^2$;
$A^2 - 2AB + B^2 = (A - B)^2$

Factor: $y^2 + 100 - 20y$.
$$A^2 - 2AB + B^2 = (A - B)^2$$
$$y^2 + 100 - 20y = y^2 - 20y + 100 = (y - 10)^2$$

6. Factor:
$100n^2 + 81 + 180n$.

Factoring a Difference of Squares
$A^2 - B^2 = (A + B)(A - B)$

Factor: $9t^2 - 1$.
$$A^2 - B^2 = (A + B)(A - B)$$
$$9t^2 - 1 = (3t + 1)(3t - 1)$$

7. Factor:
$144t^2 - 25$.

SECTION 5.5: *Factoring: A General Strategy*

A general strategy for factoring polynomials can be found in Section 5.6.

Factor: $5x^5 - 80x$.
$5x^5 - 80x$
$$= 5x(x^4 - 16) \quad \text{5x is the largest common factor.}$$
$$= 5x(x^2 + 4)(x^2 - 4) \quad x^4 - 16 \text{ is a difference of squares.}$$
$$= 5x(x^2 + 4)(x + 2)(x - 2) \quad x^2 - 4 \text{ is also a difference of squares.}$$

Check:
$$5x(x^2 + 4)(x + 2)(x - 2) = 5x(x^2 + 4)(x^2 - 4)$$
$$= 5x(x^4 - 16)$$
$$= 5x^5 - 80x.$$

8. Factor:
$3x^3 - 36x^2 + 108x$.

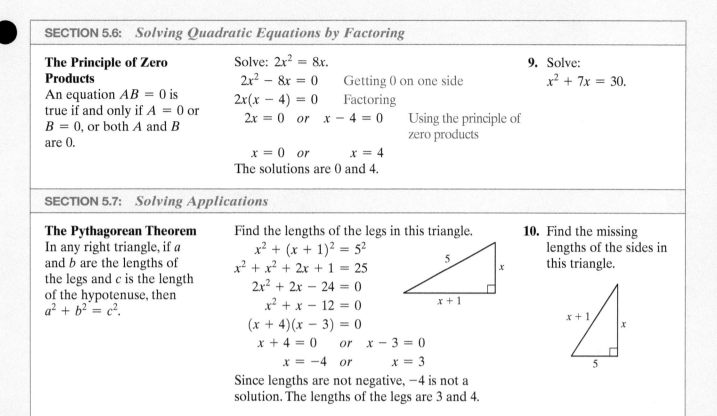

SECTION 5.6: *Solving Quadratic Equations by Factoring*

The Principle of Zero Products
An equation $AB = 0$ is true if and only if $A = 0$ or $B = 0$, or both A and B are 0.

Solve: $2x^2 = 8x$.
$$2x^2 - 8x = 0 \quad \text{Getting 0 on one side}$$
$$2x(x - 4) = 0 \quad \text{Factoring}$$
$$2x = 0 \quad or \quad x - 4 = 0 \quad \text{Using the principle of zero products}$$
$$x = 0 \quad or \quad x = 4$$
The solutions are 0 and 4.

9. Solve:
$x^2 + 7x = 30.$

SECTION 5.7: *Solving Applications*

The Pythagorean Theorem
In any right triangle, if a and b are the lengths of the legs and c is the length of the hypotenuse, then $a^2 + b^2 = c^2$.

Find the lengths of the legs in this triangle.
$$x^2 + (x + 1)^2 = 5^2$$
$$x^2 + x^2 + 2x + 1 = 25$$
$$2x^2 + 2x - 24 = 0$$
$$x^2 + x - 12 = 0$$
$$(x + 4)(x - 3) = 0$$
$$x + 4 = 0 \quad or \quad x - 3 = 0$$
$$x = -4 \quad or \quad x = 3$$
Since lengths are not negative, -4 is not a solution. The lengths of the legs are 3 and 4.

10. Find the missing lengths of the sides in this triangle.

Review Exercises: Chapter 5

⤷ Concept Reinforcement

Classify each of the following statements as either true or false.

1. The largest common variable factor is the largest power of the variable in the polynomial.　[5.1]

2. A prime polynomial has no common factor other than 1 or -1.　[5.2]

3. Every perfect-square trinomial can be expressed as a binomial squared.　[5.4]

4. Every binomial can be regarded as a difference of squares.　[5.4]

5. Every quadratic equation has two different solutions.　[5.6]

6. The principle of zero products can be applied whenever a product equals 0.　[5.6]

7. In a right triangle, the hypotenuse is always longer than either leg.　[5.7]

8. The Pythagorean theorem can be applied to any triangle that has an angle measuring at least 90°.　[5.7]

Find three factorizations of each monomial.　[5.1]

9. $20x^3$

10. $-18x^5$

Factor completely. If a polynomial is prime, state this.

11. $12x^4 - 18x^3$　[5.1]

12. $100t^2 - 1$　[5.4]

13. $x^2 + x - 12$　[5.2]

14. $x^2 + 14x + 49$　[5.4]

15. $12x^3 + 12x^2 + 3x$　[5.4]

16. $6x^3 + 9x^2 + 2x + 3$　[5.1]

17. $6a^2 + a - 5$　[5.3]

18. $25t^2 + 9 - 30t$　[5.4]

19. $81a^4 - 1$　[5.4]

20. $9x^3 + 12x^2 - 45x$　[5.3]

21. $2x^3 - 250$　[5.5]

22. $x^4 + 4x^3 - 2x - 8$　[5.1]

23. $a^2b^4 - 64$　[5.4]

24. $-8x^6 + 32x^5 - 4x^4$ [5.1]

25. $75 + 12x^2 - 60x$ [5.4]

26. $y^2 + 9$ [5.4]

27. $-t^3 + t^2 + 42t$ [5.2]

28. $4x^2 - 25$ [5.4]

29. $n^2 - 60 - 4n$ [5.2]

30. $5z^2 - 30z + 10$ [5.1]

31. $4t^2 + 13t + 10$ [5.3]

32. $2t^2 - 7t - 4$ [5.3]

33. $7x^3 + 35x^2 + 28x$ [5.2]

34. $-6x^3 + 150x$ [5.4]

35. $15 - 8x + x^2$ [5.2]

36. $3x + x^2 + 5$ [5.2]

37. $x^2y^2 + 6xy - 16$ [5.2]

38. $12a^2 + 84ab + 147b^2$ [5.4]

39. $m^2 + 5m + mt + 5t$ [5.1]

40. $6r^2 + rs - 15s^2$ [5.3]

Solve. [5.6]

41. $(x - 9)(x + 11) = 0$

42. $x^2 + 2x - 35 = 0$

43. $16x^2 = 9$

44. $3x^2 + 2 = 5x$

45. $(x + 1)(x - 2) = 4$

46. $9t - 15t^2 = 0$

47. $3x^2 + 3 = 6x$

48. The square of a number is 12 more than the number. Find all such numbers. [5.7]

49. A stone is tossed down from a cliff at an initial velocity of 4 ft/sec. The distance s that the stone falls after t seconds is given by $s = 4t + 16t^2$. If the cliff is 420 ft tall, in how many seconds will the stone reach the bottom of the cliff? [5.7]

50. Find the x-intercepts of the graph of $y = 2x^2 - 3x - 5$. [5.6]

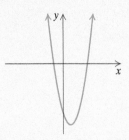

51. The front of a house is a triangle that is as wide as it is tall. Its area is 98 ft^2. Find the height and the base. [5.7]

52. Josh needs to add a diagonal brace to his LEGO® robot. The brace must span a height of 8 holes and a width of 6 holes. How long should the brace be? [5.7]

Synthesis

53. On a quiz, Celia writes the factorization of $4x^2 - 100$ as $(2x - 10)(2x + 10)$. If this were a 10-point question, how many points would you give Celia? Why? [5.4]

54. How does solving quadratic equations differ from solving linear equations? [5.6]

Solve.

55. The pages of a book measure 15 cm by 20 cm. Margins of equal width surround the printing on each page and constitute one-half of the area of the page. Find the width of the margins. [5.7]

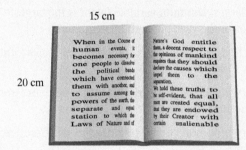

15 cm

20 cm

When in the Course of human events, it becomes necessary for one people to dissolve the political bands which have connected them with another, and to assume among the powers of the earth, the separate and equal station to which the Laws of Nature and of

Nature's God entitle them, a decent respect to the opinions of mankind requires that they should declare the causes which impel them to the separation. We hold these truths to be self-evident, that all men are created equal, that they are endowed by their Creator with certain unalienable

56. The cube of a number is the same as twice the square of the number. Find the number. [5.7]

57. The length of a rectangle is two times its width. When the length is increased by 20 cm and the width is decreased by 1 cm, the area is 160 cm^2. Find the original length and width. [5.7]

Solve. [5.6]

58. $(x - 2)2x^2 + x(x - 2) - (x - 2)15 = 0$

Aha! **59.** $x^2 + 25 = 0$

Test: Chapter 5   For step-by-step test solutions, access the Chapter Test Prep Videos in MyMathLab®.

Factor completely. If a polynomial is prime, state this.

1. $x^2 - 13x + 36$

2. $x^2 + 25 - 10x$

3. $6y^2 - 8y^3 + 4y^4$

4. $x^3 + x^2 + 2x + 2$

5. $t^7 - 3t^5$

6. $a^3 + 3a^2 - 4a$

7. $28x - 48 + 10x^2$

8. $4t^2 - 25$

9. $-6m^3 - 9m^2 - 3m$

10. $3w^2 - 75$

11. $45r^2 + 60r + 20$

12. $3x^4 - 48$

13. $49t^2 + 36 + 84t$

14. $x^4 + 2x^3 - 3x - 6$

15. $x^2 + 3x + 6$

16. $6t^3 + 9t^2 - 15t$

17. $3m^2 - 9mn - 30n^2$

Solve.

18. $x^2 - 6x + 5 = 0$

19. $2x^2 - 7x = 15$

20. $4t - 10t^2 = 0$

21. $25t^2 = 1$

22. $x(x - 1) = 20$

23. Find the x-intercepts of the graph of
    $y = 3x^2 - 5x - 8$.

24. The length of a rectangle is 6 m more than the width. The area of the rectangle is 40 m². Find the length and the width.

25. The number of possible handshakes H within a group of n people is given by $H = \frac{1}{2}(n^2 - n)$. At a meeting, everyone shook hands once with everyone else. If there were 45 handshakes, how many people were at the meeting?

26. A mason wants to be sure she has a right corner in a building's foundation. She marks a point 3 ft from the corner along one wall and another point 4 ft from the corner along the other wall. If the corner is a right angle, what should the distance be between the two marked points?

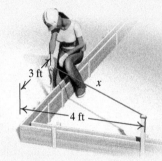

Synthesis

27. *Dimensions of an Open Box.* A rectangular piece of cardboard is twice as long as it is wide. A 4-cm square is cut out of each corner, and the sides are turned up to make a box with an open top. The volume of the box is 616 cm³. Find the original dimensions of the cardboard.

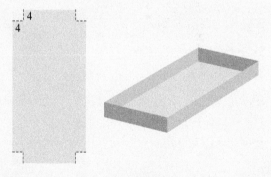

28. Factor: $(a + 3)^2 - 2(a + 3) - 35$.

29. Solve: $20x(x + 2)(x - 1) = 5x^3 - 24x - 14x^2$.

Cumulative Review: Chapters 1–5

Simplify. Do not use negative exponents in the answer.

1. $\frac{3}{8} \div \frac{3}{4}$ [1.3]

2. $\frac{3}{8} + \frac{3}{4}$ [1.3]

3. $-2 + (20 \div 4)^2 - 6 \cdot (-1)^3$ [1.8]

4. $(3x^2y^3)^{-2}$ [4.2]

5. $(t^2)^3 \cdot t^4$ [4.1]

6. $(3x^4 - 2x^2 + x - 7) + (5x^3 + 2x^2 - 3)$ [4.4]

7. $\dfrac{3t^3s^{-1}}{12t^{-5}s}$ [4.2]

8. $\left(\dfrac{-2x^2y}{3z^4}\right)^3$ [4.1]

9. Evaluate $-x$ for $x = -8$.   [1.6]

10. Determine the leading term of the polynomial
$4x^3 - 6x^2 - x^4 + 7$.   [4.3]

11. Divide: $(8x^4 - 20x^3 + 2x^2 - 4x) \div (4x)$.   [4.8]

Multiply.

12. $-4t^8(t^3 - 2t - 5)$ [4.5]

13. $(3x - 5)^2$ [4.6]

14. $(10x^5 + y)(10x^5 - y)$ [4.7]

Factor completely.

15. $c^2 - 1$ [5.4]

16. $5x + 5y + 10x^2 + 10xy$ [5.1]

17. $4r^2 - 4rt + t^2$ [5.4]

18. $10y^2 + 40$ [5.1]

19. $x^2y - 3xy + 2y$ [5.2]

20. $12x^2 - 5xy - 2y^2$ [5.3]

Solve.

21. $\frac{1}{3} + 2x = \frac{1}{2}$ [2.2]

22. $8y - 6(y - 2) = 3(2y + 7)$ [2.2]

23. $3x - 7 \geq 4 - 8x$ [2.6]

24. $x^2 + x = 12$ [5.6]

25. $3x^2 = 12$ [5.6]

26. $3x^2 = 12x$ [5.6]

27. Solve $a = bc + dc$ for c.   [2.3]

28. Find the slope of the line containing the points $(6, 7)$ and $(-2, 7)$.   [3.5]

29. Write the slope–intercept equation for the line with slope 5 that contains the point $\left(-\frac{1}{3}, 0\right)$.   [3.7]

Graph.

30. $4(x + 1) = 8$ [3.3]

31. $x + y = 5$ [3.3]

32. $y = \frac{3}{2}x - 2$ [3.6]

33. $3x + 5y = 10$ [3.6]

Solve.

34. On average, men talk 97 min more per month on cell phones than do women. The sum of men's average minutes and women's average minutes is 647 min. What is the average number of minutes per month that men talk on cell phones?   [2.5]

Data: International Communications Research for Cingular Wireless

35. A 13-ft ladder is placed against a building in such a way that the distance from the top of the ladder to the ground is 7 ft more than the distance from the bottom of the ladder to the building. Find both distances.   [5.7]

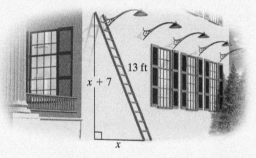

36. Donna's quiz grades are 8, 3, 5, and 10. What scores on the fifth quiz will make her average quiz grade at least 7?   [2.7]

37. Approximately 28 million people worldwide had paid subscriptions for digital music services in 2013. This was an increase from 8 million people in 2010.   [3.7]

Data: ifpi.org

a) Graph the data and determine an equation for the related line. Let d represent the number of people with paid digital music service subscriptions and t the number of years after 2010.

b) Use the equation of part (a) to estimate the number of people with paid digital music service subscriptions in 2016.

Synthesis

38. Solve $x = \dfrac{abx}{2 - b}$ for b.   [2.3]

39. Solve: $6x^3 + 4x^2 = 2x$.   [5.6]

Rational Expressions and Equations

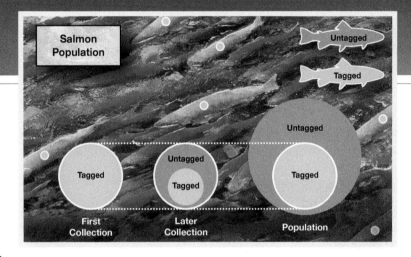

Salmon Population

Untagged

Tagged

Untagged

Tagged

Tagged

Untagged

Tagged

Tagged

First Collection

Later Collection

Population

Counting Can Be an UpHill Battle.

6.1 Rational Expressions

6.2 Multiplication and Division

6.3 Addition, Subtraction, and Least Common Denominators

6.4 Addition and Subtraction with Unlike Denominators

MID-CHAPTER REVIEW

6.5 Complex Rational Expressions

6.6 Rational Equations

CONNECTING THE CONCEPTS

6.7 Applications Using Rational Equations and Proportions

CHAPTER RESOURCES

Translating for Success
Collaborative Activity
Decision Making: Connection

STUDY SUMMARY

REVIEW EXERCISES
CHAPTER TEST
CUMULATIVE REVIEW

Determining the size of a wildlife population can be challenging, since it is usually unreasonable or impossible to collect all of the population. To estimate such a population, naturalists often use a method that relies on *proportions*. They begin by tagging a number of the animals in a population. These are then released and allowed to mix throughout the entire population. When a second collection is made, some animals will have been tagged and some will not. Reasoning that the ratio of tagged to untagged animals is approximately the same in this second collection as it is in the entire population, as illustrated in the graph shown, naturalists write and solve a proportion to estimate the size of the population. (*See Example 8 in Section 6.7.*)

I use math in studying the impact of climate change, water pollution, and invasive species on endangered species, water quality, and marine habitats.

Melanie Okoro, Water Quality Specialist/Regional AIS Coordinator, NOAA Fisheries West Coast Region in Sacramento, California, uses math to visualize data: a critical component of science communication.

 ALF *Active Learning Figure* Explore the math using the Active Learning Figure in MyMathLab.

 SA *Student Activity* Do the Student Activity in MyMathLab to see math in action.

R ational expressions are similar to fractions in arithmetic. In this chapter, we learn how to simplify, add, subtract, multiply, and divide rational expressions. These skills are then used to solve equations that arise from real-life problems.

6.1 Rational Expressions

A. Restricting Replacement Values   **B.** Simplifying Rational Expressions   **C.** Factors That Are Opposites

Just as a rational number is any number that can be written as a quotient of two integers, a **rational expression** is any expression that can be written as a quotient of two polynomials. The following are examples of rational expressions:

$$\frac{7}{3}, \quad \frac{5}{x+6}, \quad \frac{t^2 - 5t + 6}{4t^2 - 7}.$$

Rational expressions are examples of *algebraic fractions*. They are also examples of *fraction expressions*.

A. Restricting Replacement Values

Because rational expressions may contain variables in a denominator, and because division by 0 is undefined, we must avoid certain replacement values. For example, in the expression

$$\frac{x+5}{x-7},$$

when x is replaced with 7, the denominator is 0, and the expression is undefined:

$$\frac{x+5}{x-7} = \frac{7+5}{7-7} = \frac{12}{0}. \qquad \text{Division by 0 is undefined.}$$

When x is replaced with a number other than 7, such as 6, the expression *is* defined because the denominator is not 0:

$$\frac{x+5}{x-7} = \frac{6+5}{6-7} = \frac{11}{-1} = -11.$$

The expression is also defined when $x = -5$:

$$\frac{x+5}{x-7} = \frac{-5+5}{-5-7} = \frac{0}{-12} = 0. \qquad \text{0 divided by a nonzero number is 0.}$$

Any replacement for the variable that makes the *denominator* 0 will cause an expression to be undefined.

EXAMPLE 1   Find all numbers for which the rational expression

$$\frac{x+4}{x^2 - 3x - 10}$$

is undefined.

Technology Connection

To check Example 1 with a graphing calculator, let $y_1 = x^2 - 3x - 10$ and $y_2 = (x + 4)/y_1$ and use the TABLE feature. Since $x^2 - 3x - 10$ is 0 for $x = 5$, it is impossible to evaluate y_2 for $x = 5$. Scrolling up in the table would show that the same is true for $x = -2$.

TBLSTART = 0   ΔTBL = 1

X	Y₁	Y₂
0	−10	−.4
1	−12	−.4167
2	−12	−.5
3	−10	−.7
4	−6	−1.333
5	0	ERROR
6	8	1.25

X = 5

1. Find all numbers for which the rational expression $\dfrac{x^2 - x - 2}{x^2 - 25}$ is undefined.

Student Notes

When using a graphing calculator or tutorial software, you may need to use parentheses around the numerator and around the denominator of rational expressions. For example,

$$5/3x \quad \text{means} \quad \frac{5}{3}x$$

and

$$5/(3x) \quad \text{means} \quad \frac{5}{3x}.$$

SOLUTION To determine which numbers make the rational expression undefined, we set the *denominator* equal to 0 and solve:

$$x^2 - 3x - 10 = 0 \qquad \text{We set the denominator equal to 0.}$$
$$(x - 5)(x + 2) = 0 \qquad \text{Factoring}$$
$$x - 5 = 0 \quad or \quad x + 2 = 0 \qquad \text{Using the principle of zero products}$$
$$x = 5 \quad or \qquad x = -2. \qquad \text{Solving each equation}$$

Check:

For $x = 5$:

$$\frac{x + 4}{x^2 - 3x - 10} = \frac{5 + 4}{5^2 - 3 \cdot 5 - 10}$$
$$= \frac{9}{25 - 15 - 10} = \frac{9}{0}. \qquad \text{This expression is undefined, as expected.}$$

For $x = -2$:

$$\frac{x + 4}{x^2 - 3x - 10} = \frac{-2 + 4}{(-2)^2 - 3(-2) - 10}$$
$$= \frac{2}{4 + 6 - 10} = \frac{2}{0}. \qquad \text{This expression is undefined, as expected.}$$

Thus, $\dfrac{x + 4}{x^2 - 3x - 10}$ is undefined for $x = 5$ and $x = -2$.

YOUR TURN

B. Simplifying Rational Expressions

A rational expression is said to be *simplified* when the numerator and the denominator have no factors (other than 1) in common. To simplify a rational expression that contains variables, we use the same process we would use to simplify $\frac{15}{40}$:

$$\frac{15}{40} = \frac{3 \cdot 5}{8 \cdot 5} \qquad \text{Factoring the numerator and the denominator. Note the common factor, 5.}$$
$$= \frac{3}{8} \cdot \frac{5}{5} \qquad \text{Rewriting as a product of two fractions}$$
$$= \frac{3}{8} \cdot 1 \qquad \frac{5}{5} = 1$$
$$= \frac{3}{8}. \qquad \text{Using the identity property of 1 to remove the factor 1}$$

Similar steps are followed when simplifying rational expressions: We factor and remove a factor equal to 1, using the fact that

$$\frac{ab}{cb} = \frac{a}{c} \cdot \frac{b}{b} \quad \text{and} \quad \frac{a}{c} \cdot \frac{b}{b} = \frac{a}{c}.$$

Student Notes

In Examples 2 and 3, our first step is to *factor* the numerator and *factor* the denominator. This is necessary in order to remove a *factor* equal to 1.

2. Simplify: $\dfrac{15ab^2}{20b}$.

EXAMPLE 2 Simplify: $\dfrac{8x^2}{24x}$.

SOLUTION

$$\frac{8x^2}{24x} = \frac{8 \cdot x \cdot x}{3 \cdot 8 \cdot x}$$ Factoring the numerator and the denominator. Note that the greatest common factor is $8 \cdot x$.

$$= \frac{x}{3} \cdot \frac{8x}{8x}$$ Rewriting as a product of two rational expressions

$$= \frac{x}{3} \cdot 1 \qquad \frac{8x}{8x} = 1$$

$$= \frac{x}{3}$$ Removing the factor 1

YOUR TURN

We say that $\dfrac{8x^2}{24x}$ *simplifies to* $\dfrac{x}{3}$. In more advanced courses, we would say that $8x^2/(24x)$ simplifies to $x/3$, *with the restriction that $x \neq 0$.* In the work that follows, we assume that all denominators are nonzero.

EXAMPLE 3 Simplify: $\dfrac{5a + 15}{10}$.

SOLUTION

$$\frac{5a + 15}{10} = \frac{5(a + 3)}{5 \cdot 2}$$ Factoring the numerator and the denominator. The greatest common factor is 5.

$$= \frac{5}{5} \cdot \frac{a + 3}{2}$$ Rewriting as a product of two rational expressions

$$= 1 \cdot \frac{a + 3}{2} \qquad \frac{5}{5} = 1$$

$$= \frac{a + 3}{2}$$ Removing the factor 1

3. Simplify: $\dfrac{16}{2x + 10}$.

YOUR TURN

The result in Example 3 can be partially checked using a replacement for a—say, $a = 2$.

Student Notes

Checking by evaluating is only a partial check. To see why this check is not foolproof, see Exercise 69.

Original expression:
$$\frac{5a + 15}{10} = \frac{5 \cdot 2 + 15}{10}$$
$$= \frac{25}{10} = \frac{5}{2}$$ The results are the same.

Simplified expression:
$$\frac{a + 3}{2} = \frac{2 + 3}{2}$$
$$= \frac{5}{2}$$

If we do not get the same result when evaluating both expressions, we know that a mistake has been made. For example, if $(5a + 15)/10$ is *incorrectly* simplified as $(a + 15)/2$ and we evaluate using $a = 2$, we have the following.

Original expression:
$$\frac{5a + 15}{10} = \frac{5 \cdot 2 + 15}{10}$$
$$= \frac{5}{2}$$ The results are different.

Incorrectly simplified expression:
$$\frac{a + 15}{2} = \frac{2 + 15}{2}$$
$$= \frac{17}{2}$$

This demonstrates that a mistake has been made.

4. Simplify: $\dfrac{x^2 - 5x}{x^2 - 3x - 10}$.

Sometimes the common factor has two or more terms.

EXAMPLE 4 Simplify.

a) $\dfrac{6x - 12}{7x - 14}$

b) $\dfrac{18t^2 + 6t}{6t^2 + 15t}$

c) $\dfrac{x^2 + 3x + 2}{x^2 - 1}$

SOLUTION

a) $\dfrac{6x - 12}{7x - 14} = \dfrac{6(x - 2)}{7(x - 2)}$ Factoring the numerator and the denominator. The greatest common factor is $x - 2$.

$\phantom{\dfrac{6x - 12}{7x - 14}} = \dfrac{6}{7} \cdot \dfrac{x - 2}{x - 2}$ Rewriting as a product of two rational expressions

$\phantom{\dfrac{6x - 12}{7x - 14}} = \dfrac{6}{7} \cdot 1$ $\dfrac{x - 2}{x - 2} = 1$

$\phantom{\dfrac{6x - 12}{7x - 14}} = \dfrac{6}{7}$ Removing the factor 1

b) $\dfrac{18t^2 + 6t}{6t^2 + 15t} = \dfrac{3t \cdot 2(3t + 1)}{3t(2t + 5)}$ Factoring the numerator and the denominator. The greatest common factor is $3t$.

$\phantom{\dfrac{18t^2 + 6t}{6t^2 + 15t}} = \dfrac{3t}{3t} \cdot \dfrac{2(3t + 1)}{2t + 5}$ Rewriting as a product of two rational expressions

$\phantom{\dfrac{18t^2 + 6t}{6t^2 + 15t}} = 1 \cdot \dfrac{2(3t + 1)}{2t + 5}$ $\dfrac{3t}{3t} = 1$

$\phantom{\dfrac{18t^2 + 6t}{6t^2 + 15t}} = \dfrac{2(3t + 1)}{2t + 5}$ Removing the factor 1. The numerator and the denominator have no common factor so the simplification is complete.

c) $\dfrac{x^2 + 3x + 2}{x^2 - 1} = \dfrac{(x + 1)(x + 2)}{(x + 1)(x - 1)}$ Factoring; $x + 1$ is the greatest common factor.

$\phantom{\dfrac{x^2 + 3x + 2}{x^2 - 1}} = \dfrac{x + 1}{x + 1} \cdot \dfrac{x + 2}{x - 1}$ Rewriting as a product of two rational expressions

$\phantom{\dfrac{x^2 + 3x + 2}{x^2 - 1}} = 1 \cdot \dfrac{x + 2}{x - 1}$ $\dfrac{x + 1}{x + 1} = 1$

$\phantom{\dfrac{x^2 + 3x + 2}{x^2 - 1}} = \dfrac{x + 2}{x - 1}$ Removing the factor 1

YOUR TURN

Canceling is a shortcut that can be used—and easily *misused*—to simplify rational expressions. If done with care and understanding, canceling streamlines the process of removing a factor equal to 1. Example 4(c) could have been streamlined as follows:

$$\dfrac{x^2 + 3x + 2}{x^2 - 1} = \dfrac{\cancel{(x + 1)}(x + 2)}{\cancel{(x + 1)}(x - 1)}$$ When a factor equal to 1 is noted, it is "canceled": $\dfrac{x + 1}{x + 1} = 1.$

$$\phantom{\dfrac{x^2 + 3x + 2}{x^2 - 1}} = \dfrac{x + 2}{x - 1}.$$ Simplifying

> **CAUTION!** Canceling is often used incorrectly:
>
> $$\frac{\cancel{x} + 7}{\cancel{x} + 3}; \qquad \frac{a^2 - \cancel{5}}{\cancel{5}}; \qquad \frac{6x^2 + 5\cancel{x} + 1}{4x^2 - 3\cancel{x}}.$$
>
> Incorrect! Incorrect! Incorrect!
>
> None of the above cancellations removes a factor equal to 1. Factors are parts of products. For example, in $x \cdot 7$, x and 7 are factors, but in $x + 7$, x and 7 are terms, *not* factors. *Only factors can be canceled.*

EXAMPLE 5 Simplify: $\dfrac{3x^2 - 2x - 1}{x^2 - 3x + 2}$.

SOLUTION We factor the numerator and the denominator and look for common factors:

$$\frac{3x^2 - 2x - 1}{x^2 - 3x + 2} = \frac{(3x + 1)(x - 1)}{(x - 2)(x - 1)} \qquad \begin{array}{l}\text{Try to visualize this as} \\[4pt] \dfrac{3x + 1}{x - 2} \cdot \dfrac{x - 1}{x - 1}.\end{array}$$

$$= \frac{3x + 1}{x - 2}. \qquad \begin{array}{l}\text{Removing a factor equal to 1:} \\[4pt] \dfrac{x - 1}{x - 1} = 1\end{array}$$

5. Simplify: $\dfrac{n^2 - 16}{2n^2 + 7n - 4}$.

↪ YOUR TURN

C. Factors That Are Opposites

Consider

$$\frac{x - 4}{8 - 2x}, \quad \text{or, equivalently,} \quad \frac{x - 4}{2(4 - x)}.$$

At first glance, the numerator and the denominator do not appear to have any common factors. But $x - 4$ and $4 - x$ are opposites, or additive inverses, of each other. Thus we can find a common factor by factoring out -1 in either expression.

EXAMPLE 6 Simplify $\dfrac{x - 4}{8 - 2x}$ and check by evaluating.

SOLUTION We have

$$\frac{x - 4}{8 - 2x} = \frac{x - 4}{2(4 - x)} \qquad \text{Factoring}$$

$$= \frac{x - 4}{2(-1)(x - 4)} \qquad \text{Note that } 4 - x = -x + 4 = -1(x - 4).$$

$$= \frac{x - 4}{-2(x - 4)} \qquad \begin{array}{l}\text{Had we originally factored out } -2\text{, we} \\ \text{could have gone directly to this step.}\end{array}$$

$$= \frac{1}{-2} \cdot \frac{x - 4}{x - 4} \qquad \begin{array}{l}\text{Rewriting as a product. It is important to} \\ \text{write the 1 in the numerator.}\end{array}$$

$$= -\frac{1}{2}. \qquad \begin{array}{l}\text{Removing a factor equal to 1:} \\ (x - 4)/(x - 4) = 1\end{array}$$

↪ Check Your UNDERSTANDING

Simplify.

1. $\dfrac{3 \cdot x^2}{3 \cdot y}$

2. $\dfrac{2t(t - 6)}{2t \cdot t}$

3. $\dfrac{(x + 3)(x - 2)}{(x + 3)(x - 1)}$

4. $\dfrac{(y + 4)(y - 7)}{(y - 4)(y + 4)}$

5. $\dfrac{a - 2}{-1(a - 2)}$

As a partial check, note that for any choice of x other than 4, the value of the rational expression is $-\frac{1}{2}$. For example, if $x = 5$, then

$$\frac{x - 4}{8 - 2x} = \frac{5 - 4}{8 - 2 \cdot 5}$$

$$= \frac{1}{-2} = -\frac{1}{2}.$$

6. Simplify: $\dfrac{x - 2}{2 - x}$.

⟳ YOUR TURN

6.1 EXERCISE SET

FOR EXTRA HELP MyMathLab®

⟳ Vocabulary and Reading Check

Choose the word written under the blank that best completes the statement.

1. A rational expression can be written as a
_____ of two polynomials.
difference/quotient

2. A rational expression is undefined when the
_____ is zero.
denominator/numerator

3. A rational expression is simplified when the numerator and the denominator have no _____
factors/terms

(other than 1) in common.

4. When we cancel, we remove _____ .
a factor equal to 1/a restricted value

⟳ Concept Reinforcement

In each of Exercises 5–8, match the rational expression with the list of numbers in the column on the right for which the rational expression is undefined.

5. ___ $\dfrac{3t}{(t + 1)(t - 4)}$

6. ___ $\dfrac{2t + 7}{(2t - 1)(3t + 4)}$

7. ___ $\dfrac{a + 7}{a^2 - a - 12}$

8. ___ $\dfrac{m - 3}{m^2 - 2m - 15}$

a) $-1, 4$

b) $-3, 5$

c) $-\dfrac{4}{3}, \dfrac{1}{2}$

d) $-3, 4$

A. Restricting Replacement Values

List all numbers for which each rational expression is undefined.

9. $\dfrac{18}{-11x}$

10. $\dfrac{13}{-5t}$

11. $\dfrac{y - 3}{y + 5}$

12. $\dfrac{a + 6}{a - 10}$

13. $\dfrac{t - 5}{3t - 15}$

14. $\dfrac{x^2 - 4}{5x + 10}$

15. $\dfrac{x^2 - 25}{x^2 - 3x - 28}$

16. $\dfrac{p^2 - 9}{p^2 - 7p + 10}$

17. $\dfrac{t^2 + t - 20}{2t^2 + 11t - 6}$

18. $\dfrac{x^2 + 2x + 1}{3x^2 - x - 14}$

B. Simplifying Rational Expressions

Simplify. Show all steps.

19. $\dfrac{50a^2b}{40ab^3}$

20. $\dfrac{-24x^4y^3}{6x^7y}$

21. $\dfrac{6t + 12}{6t - 18}$

22. $\dfrac{5n - 30}{5n + 5}$

23. $\dfrac{21t - 7}{24t - 8}$

24. $\dfrac{10n + 25}{8n + 20}$

25. $\dfrac{a^2 - 9}{a^2 + 4a + 3}$

26. $\dfrac{a^2 + 5a + 6}{a^2 - 9}$

Simplify, if possible. Then check by evaluating, as in Example 6.

27. $\dfrac{-36x^8}{54x^5}$

28. $\dfrac{45a^4}{30a^6}$

29. $\dfrac{-2y + 6}{-8y}$

30. $\dfrac{4x - 12}{6x}$

31. $\dfrac{6a^2 + 3a}{7a^2 + 7a}$

32. $\dfrac{-4m^2 + 4m}{-8m^2 + 12m}$

33. $\dfrac{t^2 - 16}{t^2 - t - 20}$

34. $\dfrac{a^2 - 4}{a^2 + 5a + 6}$

35. $\dfrac{3a^2 + 9a - 12}{6a^2 - 30a + 24}$

36. $\dfrac{2t^2 - 6t + 4}{4t^2 + 12t - 16}$

37. $\dfrac{x^2 - 8x + 16}{x^2 - 16}$

38. $\dfrac{x^2 - 25}{x^2 + 10x + 25}$

39. $\dfrac{t^2 - 1}{t + 1}$

40. $\dfrac{a^2 - 1}{a - 1}$

41. $\dfrac{y^2 + 4}{y + 2}$

42. $\dfrac{m^2 + 9}{m + 3}$

43. $\dfrac{5x^2 + 20}{10x^2 + 40}$

44. $\dfrac{6x^2 + 54}{4x^2 + 36}$

45. $\dfrac{y^2 + 6y}{2y^2 + 13y + 6}$

46. $\dfrac{t^2 + 2t}{2t^2 + t - 6}$

47. $\dfrac{4x^2 - 12x + 9}{10x^2 - 11x - 6}$

48. $\dfrac{4x^2 - 4x + 1}{6x^2 + 5x - 4}$

C. Factors That Are Opposites

Simplify.

49. $\dfrac{10 - x}{x - 10}$

50. $\dfrac{x - 8}{8 - x}$

51. $\dfrac{7t - 14}{2 - t}$

52. $\dfrac{3 - n}{5n - 15}$

53. $\dfrac{a - b}{4b - 4a}$

54. $\dfrac{2p - 2q}{q - p}$

55. $\dfrac{3x^2 - 3y^2}{2y^2 - 2x^2}$

56. $\dfrac{7a^2 - 7b^2}{3b^2 - 3a^2}$

Aha! **57.** $\dfrac{7s^2 - 28t^2}{28t^2 - 7s^2}$

58. $\dfrac{9m^2 - 4n^2}{4n^2 - 9m^2}$

59. Explain how simplifying is related to the identity property of 1.

60. If a rational expression is undefined for $x = 5$ and $x = -3$, what is the degree of the denominator? Why?

Skill Review

Factor.

61. $3x^3 + 15x^2 + x + 5$ [5.1]

62. $3x^2 + 16x + 5$ [5.3]

63. $18y^4 - 27y^3 + 3y^2$ [5.1]

64. $25a^2 - 16b^2$ [5.4]

65. $m^3 - 8m^2 + 16m$ [5.4]

66. $5x^2 - 35x + 60$ [5.2]

Synthesis

67. Luke *incorrectly* simplifies

$$\dfrac{x^2 + x - 2}{x^2 + 3x + 2} \quad \text{as} \quad \dfrac{x - 1}{x + 2}.$$

He then checks his simplification by evaluating both expressions for $x = 1$. Use this situation to explain why evaluating is not a foolproof check.

68. How could you convince someone that $a - b$ and $b - a$ are opposites of each other?

Simplify.

69. $\dfrac{16y^4 - x^4}{(x^2 + 4y^2)(x - 2y)}$

70. $\dfrac{(x - 1)(x^4 - 1)(x^2 - 1)}{(x^2 + 1)(x - 1)^2(x^4 - 2x^2 + 1)}$

71. $\dfrac{x^5 - 2x^3 + 4x^2 - 8}{x^7 + 2x^4 - 4x^3 - 8}$

72. $\dfrac{10t^4 - 8t^3 + 15t - 12}{8 - 10t + 12t^2 - 15t^3}$

73. $\dfrac{(t^4 - 1)(t^2 - 9)(t - 9)^2}{(t^4 - 81)(t^2 + 1)(t + 1)^2}$

74. $\dfrac{(t + 2)^3(t^2 + 2t + 1)(t + 1)}{(t + 1)^3(t^2 + 4t + 4)(t + 2)}$

75. $\dfrac{x^3 + 6x^2 - 4x - 24}{x^2 + 4x - 12}$

76. $\dfrac{10x^2 - 10}{5x^3 - 30x^2 - 5x + 30}$

77. $\dfrac{(x^2 - y^2)(x^2 - 2xy + y^2)}{(x + y)^2(x^2 - 4xy - 5y^2)}$

78. $\dfrac{x^4 - y^4}{(y - x)^4}$

79. Select any number x, multiply by 2, add 5, multiply by 5, subtract 25, and divide by 10. What do you get? Explain how this procedure can be used for a number trick.

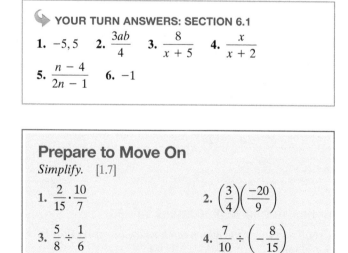

YOUR TURN ANSWERS: SECTION 6.1

1. $-5, 5$ **2.** $\dfrac{3ab}{4}$ **3.** $\dfrac{8}{x + 5}$ **4.** $\dfrac{x}{x + 2}$

5. $\dfrac{n - 4}{2n - 1}$ **6.** -1

Prepare to Move On

Simplify.   [1.7]

1. $\dfrac{2}{15} \cdot \dfrac{10}{7}$

2. $\left(\dfrac{3}{4}\right)\left(\dfrac{-20}{9}\right)$

3. $\dfrac{5}{8} \div \dfrac{1}{6}$

4. $\dfrac{7}{10} \div \left(-\dfrac{8}{15}\right)$

6.2 Multiplication and Division

A. Multiplication   **B.** Division

Multiplication and division of rational expressions are similar to multiplication and division with fractions. In this section, we again assume that all denominators are nonzero.

A. Multiplication

Recall that to multiply fractions, we multiply numerator times numerator and denominator times denominator. Rational expressions are multiplied in a similar way.

> **THE PRODUCT OF TWO RATIONAL EXPRESSIONS**
>
> To multiply rational expressions, form the product of the numerators and the product of the denominators:
>
> $$\frac{A}{B} \cdot \frac{C}{D} = \frac{AC}{BD}.$$
>
> Then factor and, if possible, simplify the result.

For example,

$$\frac{3}{5} \cdot \frac{8}{11} = \frac{3 \cdot 8}{5 \cdot 11} \quad \text{and} \quad \frac{x}{3} \cdot \frac{x+2}{y} = \frac{x(x+2)}{3y}.$$

Fraction bars are grouping symbols, so parentheses are needed when writing some products.

EXAMPLE 1 Multiply. (Write the product as a single rational expression.) Then simplify, if possible, by removing a factor equal to 1.

a) $\dfrac{5a^3}{4} \cdot \dfrac{2}{5a}$

b) $(x^2 - 3x - 10) \cdot \dfrac{x+4}{x^2 - 10x + 25}$

c) $\dfrac{10x + 20}{2x^2 - 3x + 1} \cdot \dfrac{x^2 - 1}{5x + 10}$

SOLUTION

a)
$$\frac{5a^3}{4} \cdot \frac{2}{5a} = \frac{5a^3(2)}{4(5a)} \qquad \text{Forming the product of the numerators and the product of the denominators}$$

$$= \frac{5 \cdot a \cdot a \cdot a \cdot 2}{2 \cdot 2 \cdot 5 \cdot a} \qquad \text{Factoring the numerator and the denominator}$$

$$= \frac{\cancel{5} \cdot \cancel{a} \cdot a \cdot a \cdot \cancel{2}}{\cancel{2} \cdot 2 \cdot \cancel{5} \cdot \cancel{a}} \;\Bigg\}$$

$$= \frac{a^2}{2} \qquad\qquad\qquad \text{Removing a factor equal to 1: } \frac{2 \cdot 5 \cdot a}{2 \cdot 5 \cdot a} = 1$$

Student Notes

Note in Example 1 that when we multiply the rational expressions, we form the product of the numerators and the product of the denominators, but we do not carry out the multiplication. Our next step is to factor the numerator and factor the denominator, and this is easier to do if the multiplication has not been carried out.

1. Multiply and, if possible, simplify:

$$\frac{t + 2}{2t} \cdot \frac{10t^2}{t^2 + t - 2}.$$

b) $(x^2 - 3x - 10) \cdot \dfrac{x + 4}{x^2 - 10x + 25}$

$$= \frac{x^2 - 3x - 10}{1} \cdot \frac{x + 4}{x^2 - 10x + 25}$$
  Writing $x^2 - 3x - 10$ as a rational expression

$$= \frac{(x^2 - 3x - 10)(x + 4)}{1(x^2 - 10x + 25)}$$
  Multiplying the numerators and the denominators

$$= \frac{(x - 5)(x + 2)(x + 4)}{(x - 5)(x - 5)}$$
  Factoring the numerator and the denominator

$$= \frac{\cancel{(x - 5)}(x + 2)(x + 4)}{\cancel{(x - 5)}(x - 5)}$$
  Removing a factor equal to 1: $\dfrac{x - 5}{x - 5} = 1$

$$= \frac{(x + 2)(x + 4)}{x - 5}$$

c) $\dfrac{10x + 20}{2x^2 - 3x + 1} \cdot \dfrac{x^2 - 1}{5x + 10}$

$$= \frac{(10x + 20)(x^2 - 1)}{(2x^2 - 3x + 1)(5x + 10)}$$
  Multiply.

$$= \frac{5(2)(x + 2)(x + 1)(x - 1)}{(x - 1)(2x - 1)5(x + 2)}$$
  Factor.   Try to go directly to this step.

$$= \frac{\cancel{5}(2)(x + 2)(x + 1)\cancel{(x - 1)}}{\cancel{(x - 1)}(2x - 1)\cancel{5}\cancel{(x + 2)}}$$
  Simplify. $\dfrac{5(x + 2)(x - 1)}{5(x + 2)(x - 1)} = 1$

$$= \frac{2(x + 1)}{2x - 1}$$

↪ YOUR TURN

Because our results are often used in problems that require factored form, there is no need to routinely multiply out the numerator or the denominator.

B. Division

As with fractions, reciprocals of rational expressions are found by interchanging the numerator and the denominator. For example,

the reciprocal of $\dfrac{3x}{x + 5}$ is $\dfrac{x + 5}{3x}$ and   the reciprocal of $\dfrac{1}{y - 8}$ is $y - 8$.

Perform the indicated operation.

1. $\dfrac{a}{b} \cdot \dfrac{c}{d}$

2. $\dfrac{a}{b} \div \dfrac{c}{d}$

3. $\dfrac{a}{b} \cdot c$

4. $\dfrac{a}{b} \div c$

> **THE QUOTIENT OF TWO RATIONAL EXPRESSIONS**
>
> To divide by a rational expression, multiply by its reciprocal:
>
> $$\frac{A}{B} \div \frac{C}{D} = \frac{A}{B} \cdot \frac{D}{C} = \frac{AD}{BC}.$$
>
> Then factor and, if possible, simplify.

EXAMPLE 2 Divide: **(a)** $\dfrac{x}{5} \div \dfrac{7}{y}$; **(b)** $(x + 2) \div \dfrac{x - 1}{x + 3}$.

SOLUTION

a) $\dfrac{x}{5} \div \dfrac{7}{y} = \dfrac{x}{5} \cdot \dfrac{y}{7}$ Multiplying by the reciprocal of the divisor

$= \dfrac{xy}{35}$ Multiplying rational expressions

b) $(x + 2) \div \dfrac{x - 1}{x + 3} = \dfrac{x + 2}{1} \cdot \dfrac{x + 3}{x - 1}$ Multiplying by the reciprocal of the divisor. Writing $x + 2$ as $\dfrac{x + 2}{1}$ can be helpful.

$= \dfrac{(x + 2)(x + 3)}{x - 1}$

2. Divide:

$\dfrac{x + 2}{x - 2} \div \dfrac{x - 3}{x + 1}$.

YOUR TURN

As usual, we should simplify when possible.

EXAMPLE 3 Divide and, if possible, simplify.

a) $\dfrac{x^2 - 2x - 3}{x^2 - 4} \div \dfrac{x + 1}{x + 5}$

b) $\dfrac{a^2 + 3a + 2}{a^2 + 4} \div (5a^2 + 10a)$

SOLUTION

a) $\dfrac{x^2 - 2x - 3}{x^2 - 4} \div \dfrac{x + 1}{x + 5}$

$= \dfrac{x^2 - 2x - 3}{x^2 - 4} \cdot \dfrac{x + 5}{x + 1}$ Multiply by the reciprocal.

$= \dfrac{(x - 3)(x + 1)(x + 5)}{(x - 2)(x + 2)(x + 1)}$ Multiply. Factor.

$= \dfrac{(x - 3)(x + 1)(x + 5)}{(x - 2)(x + 2)(x + 1)}$ Simplify.

$= \dfrac{(x - 3)(x + 5)}{(x - 2)(x + 2)}$ $\dfrac{x + 1}{x + 1} = 1$

b) $\dfrac{a^2 + 3a + 2}{a^2 + 4} \div (5a^2 + 10a)$

$= \dfrac{a^2 + 3a + 2}{a^2 + 4} \cdot \dfrac{1}{5a^2 + 10a}$ Multiplying by the reciprocal of the divisor

$= \dfrac{(a + 2)(a + 1)}{(a^2 + 4)5a(a + 2)}$ Multiplying rational expressions and factoring

$= \dfrac{(a + 2)(a + 1)}{(a^2 + 4)5a(a + 2)}$

$= \dfrac{a + 1}{(a^2 + 4)5a}$ Removing a factor equal to 1: $\dfrac{a + 2}{a + 2} = 1$

YOUR TURN

Technology Connection

To perform a partial check of Example 3(a), we enter the original expression as y_1 and the simplified expression as y_2. If the n/d mode is not available, we enclose each numerator, each denominator, and each rational expression in parentheses.

Comparing values of y_1 and y_2, we see that the simplification is probably correct.

X	Y₁	Y₂
-5	ERROR	0
-4	-.5833	-.5833
-3	-2.4	-2.4
-2	ERROR	ERROR
-1	ERROR	5.3333
0	3.75	3.75
1	4	4

X = -5

1. Check Example 3(b).
2. Why are there 3 ERROR messages shown for y_1 on the screen above, and only 1 for y_2?

3. Divide and, if possible, simplify:

$\dfrac{x + 1}{x^2 - 1} \div \dfrac{x + 1}{x^2 - 2x + 1}$.

6.2 EXERCISE SET

FOR EXTRA HELP   MyMathLab®

Vocabulary and Reading Check

Choose from the following list the correct ending for each statement.

a) interchange the numerator and the denominator.
b) multiply by its reciprocal.
c) multiply numerators and multiply denominators.
d) remove a factor equal to 1.

1. To simplify rational expressions, _____

2. To multiply rational expressions, _____

3. To find a reciprocal, _____

4. To divide by a rational expression, _____

Concept Reinforcement

In each of Exercises 5–10, match the product or quotient with an equivalent expression from the column on the right.

5. ____ $\dfrac{x}{2} \cdot \dfrac{5}{y}$

6. ____ $\dfrac{x}{2} \div \dfrac{5}{y}$

7. ____ $x \cdot \dfrac{5}{y}$

8. ____ $\dfrac{x}{2} \div y$

9. ____ $x \div \dfrac{5}{y}$

10. ____ $\dfrac{5}{y} \div \dfrac{x}{2}$

a) $\dfrac{5x}{y}$

b) $\dfrac{xy}{5}$

c) $\dfrac{xy}{10}$

d) $\dfrac{5x}{2y}$

e) $\dfrac{x}{2y}$

f) $\dfrac{10}{xy}$

A. Multiplication

Multiply. Leave each answer in factored form.

11. $\dfrac{3x}{8} \cdot \dfrac{x+2}{5x-1}$

12. $\dfrac{2x}{7} \cdot \dfrac{3x+5}{x-1}$

13. $\dfrac{a-4}{a+6} \cdot \dfrac{a+2}{a+6}$

14. $\dfrac{a+3}{a+6} \cdot \dfrac{a+3}{a-1}$

15. $\dfrac{n-4}{n^2+4} \cdot \dfrac{n+4}{n^2-4}$

16. $\dfrac{t+3}{t^2-2} \cdot \dfrac{t+3}{t^2-4}$

Multiply and, if possible, simplify. Leave any numerators and denominators with more than one term in factored form.

17. $\dfrac{8t^3}{5t} \cdot \dfrac{3}{4t}$

18. $\dfrac{18}{a^5} \cdot \dfrac{2a^2}{3a}$

19. $\dfrac{3c}{d^2} \cdot \dfrac{8d}{6c^3}$

20. $\dfrac{3x^2y}{2} \cdot \dfrac{4}{xy^3}$

21. $\dfrac{y^2-16}{4y+12} \cdot \dfrac{y+3}{y-4}$

22. $\dfrac{m^2-n^2}{4m+4n} \cdot \dfrac{m+n}{m-n}$

23. $\dfrac{x^2-3x-10}{(x-2)^2} \cdot \dfrac{x-2}{x-5}$

24. $\dfrac{t+2}{t-2} \cdot \dfrac{t^2-5t+6}{(t+2)^2}$

25. $\dfrac{n^2-6n+5}{n+6} \cdot \dfrac{n-6}{n^2+36}$

26. $\dfrac{a+2}{a-2} \cdot \dfrac{a^2+4}{a^2+5a+4}$

27. $\dfrac{a^2-9}{a^2} \cdot \dfrac{7a}{a^2+a-12}$

28. $\dfrac{x^2+10x-11}{9x} \cdot \dfrac{x^3}{x+11}$

29. $\dfrac{y^2-y}{y^2+5y+4} \cdot (y+4)$

30. $(n-3) \cdot \dfrac{n^2+4n}{n^2-5n+6}$

31. $\dfrac{4v-8}{5v} \cdot \dfrac{15v^2}{4v^2-16v+16}$

32. $\dfrac{4a^2}{3a^2-12a+12} \cdot \dfrac{3a-6}{2a}$

33. $\dfrac{t^2+2t-3}{t^2+4t-5} \cdot \dfrac{t^2-3t-10}{t^2+5t+6}$

34. $\dfrac{x^2+5x+4}{x^2-6x+8} \cdot \dfrac{x^2+5x-14}{x^2+8x+7}$

35. $\dfrac{12y+12}{5y+25} \cdot \dfrac{3y^2-75}{8y^2-8}$

36. $\dfrac{9t^2-900}{5t^2-20} \cdot \dfrac{5t+10}{3t-30}$

Aha! 37. $\dfrac{x^2+4x+4}{(x-1)^2} \cdot \dfrac{x^2-2x+1}{(x+2)^2}$

38. $\dfrac{x^2+7x+12}{x^2+6x+8} \cdot \dfrac{4-x^2}{x^2+x-6}$

39. $\dfrac{t^2-4t+4}{2t^2-7t+6} \cdot \dfrac{2t^2+7t-15}{t^2-10t+25}$

40. $\dfrac{5y^2 - 4y - 1}{3y^2 + 5y - 12} \cdot \dfrac{y^2 + 6y + 9}{y^2 - 2y + 1}$

41. $(10x^2 - x - 2) \cdot \dfrac{4x^2 - 8x + 3}{10x^2 - 11x - 6}$

42. $\dfrac{2x^2 - 5x + 3}{6x^2 - 5x - 1} \cdot (6x^2 + 13x + 2)$

43. $\dfrac{49x^2 - 25}{4x - 14} \cdot \dfrac{6x^2 - 13x - 28}{28x - 20}$

44. $\dfrac{9t^2 - 4}{8t^2 - 10t + 3} \cdot \dfrac{10t - 5}{3t - 2}$

45. $\dfrac{8x^2 + 14xy - 15y^2}{3x^3 - x^2y} \cdot \dfrac{3x - y}{4xy - 3y^2}$

46. $\dfrac{2x^2 - xy}{6x^2 + 7xy + 2y^2} \cdot \dfrac{9x^2 - 6xy - 8y^2}{3xy - 4y^2}$

B. Division

Find the reciprocal of each expression.

47. $\dfrac{2x}{9}$

48. $\dfrac{3 - x}{x^2 + 4}$

49. $a^4 + 3a$

50. $\dfrac{1}{a^2 - b^2}$

Divide and, if possible, simplify. Leave any numerators and denominators with more than one term in factored form.

51. $\dfrac{x}{4} \div \dfrac{5}{x}$

52. $\dfrac{5}{x} \div \dfrac{x}{12}$

53. $\dfrac{a^5}{b^4} \div \dfrac{a^2}{b}$

54. $\dfrac{x^5}{y^2} \div \dfrac{x^2}{y}$

55. $\dfrac{t - 3}{6} \div \dfrac{t + 1}{8}$

56. $\dfrac{10}{a + 3} \div \dfrac{15}{a}$

57. $\dfrac{4y - 8}{y + 2} \div \dfrac{y - 2}{y^2 - 4}$

58. $\dfrac{x^2 - 1}{x} \div \dfrac{x + 1}{2x - 2}$

59. $\dfrac{a}{a - b} \div \dfrac{b}{b - a}$

60. $\dfrac{x - y}{6} \div \dfrac{y - x}{3}$

61. $(n^2 + 5n + 6) \div \dfrac{n^2 - 4}{n + 3}$

62. $(v^2 - 1) \div \dfrac{(v + 1)(v - 3)}{v^2 + 9}$

63. $\dfrac{a + 2}{a - 1} \div \dfrac{3a + 6}{a - 5}$

64. $\dfrac{t - 3}{t + 2} \div \dfrac{4t - 12}{t + 1}$

65. $(2x - 1) \div \dfrac{2x^2 - 11x + 5}{4x^2 - 1}$

66. $(a + 7) \div \dfrac{3a^2 + 14a - 49}{a^2 + 8a + 7}$

67. $\dfrac{w^2 - 14w + 49}{2w^2 - 3w - 14} \div \dfrac{3w^2 - 20w - 7}{w^2 - 6w - 16}$

68. $\dfrac{2m^2 + 59m - 30}{m^2 - 10m + 25} \div \dfrac{2m^2 - 21m + 10}{m^2 + m - 30}$

69. $\dfrac{c^2 + 10c + 21}{c^2 - 2c - 15} \div (5c^2 + 32c - 21)$

70. $\dfrac{z^2 - 2z + 1}{z^2 - 1} \div (4z^2 - z - 3)$

71. $\dfrac{-3 + 3x}{16} \div \dfrac{x - 1}{5}$

72. $\dfrac{-4 + 2x}{15} \div \dfrac{x - 2}{3}$

73. $\dfrac{x - 1}{x + 2} \div \dfrac{1 - x}{4 + x^2}$

74. $\dfrac{-12 + 4x}{12} \div \dfrac{6 - 2x}{6}$

75. $\dfrac{x - y}{x^2 + 2xy + y^2} \div \dfrac{x^2 - y^2}{x^2 - 5xy + 4y^2}$

76. $\dfrac{a^2 - b^2}{a^2 - 4ab + 4b^2} \div \dfrac{a^2 - 3ab + 2b^2}{a - 2b}$

77. Why is it important to insert parentheses when multiplying rational expressions such as

$$\dfrac{x + 2}{5x - 7} \cdot \dfrac{3x - 1}{x + 4}?$$

78. As a first step in dividing $\dfrac{x}{3}$ by $\dfrac{7}{x}$, Jan canceled the x's.

Explain why this is incorrect, and show the correct division.

Skill Review

Graph.

79. $y = \frac{1}{2}x - 5$ [3.5]

80. $3x - 2y = 12$ [3.3]

81. $3(x - 1) = 4$ [3.6]

82. $y - 2 = -(x + 4)$ [3.7]

83. $3y = 5x$ [3.2]

84. $\frac{1}{2}y = 2$ [3.6]

Synthesis

85. Is the reciprocal of a product the product of the two reciprocals? Why or why not?

86. Explain why the quotient

$$\frac{x + 3}{x - 5} \div \frac{x - 7}{x + 1}$$

is undefined for $x = 5, x = -1$, and $x = 7$, but *is* defined for $x = -3$.

87. Find the reciprocal of $2\frac{1}{3}x$.

88. Find the reciprocal of $7.25x$.

Simplify.

89. $(x - 2a) \div \dfrac{a^2x^2 - 4a^4}{a^2x + 2a^3}$

90. $\dfrac{3x^2 - 2xy - y^2}{x^2 - y^2} \div (3x^2 + 4xy + y^2)^2$

91. $\dfrac{3a^2 - 5ab - 12b^2}{3ab + 4b^2} \div (3b^2 - ab)^2$

Aha! **92.** $\dfrac{a^2 - 3b}{a^2 + 2b} \cdot \dfrac{a^2 - 2b}{a^2 + 3b} \cdot \dfrac{a^2 + 2b}{a^2 - 3b}$

93. $\dfrac{z^2 - 8z + 16}{z^2 + 8z + 16} \div \dfrac{(z - 4)^5}{(z + 4)^5} \div \dfrac{3z + 12}{z^2 - 16}$

94. $\dfrac{(t + 2)^3}{(t + 1)^3} \div \dfrac{t^2 + 4t + 4}{t^2 + 2t + 1} \cdot \dfrac{t + 1}{t + 2}$

95. $\dfrac{a^4 - 81b^4}{a^2c - 6abc + 9b^2c} \cdot \dfrac{a + 3b}{a^2 + 9b^2} \div \dfrac{a^2 + 6ab + 9b^2}{(a - 3b)^2}$

96. $\dfrac{3y^3 + 6y^2}{y^2 - y - 12} \div \dfrac{y^2 - y}{y^2 - 2y - 8} \cdot \dfrac{y^2 + 5y + 6}{y^2}$

97. $\dfrac{xy - 2x + y - 2}{xy + 4x - y - 4} \cdot \dfrac{xy + y + 4x + 4}{xy - y - 2x + 2}$

98. $\dfrac{ab^2 + 2b^2 + a + 2}{ab - a - 3b + 3} \cdot \dfrac{ab^2 - 3b^2 - a + 3}{ab + a + 2b + 2}$

99. $\dfrac{3x^2 - 12x + bx - 4b}{4x^2 - 16x - bx + 4b} \div \dfrac{3bx + b^2 + 6x + 2b}{4bx - b^2 + 8x - 2b}$

100. $\dfrac{2x^2y - xy^2 + 6x^2 - 3xy}{3y^2 - xy + 9y - 3x} \div \dfrac{2x^3 - x^2y + 8x^2 - 4xy}{3xy - x^2 + 6y - 2x}$

101. $\dfrac{8n^2 - 10n + 3}{4n^2 - 4n - 3} \cdot \dfrac{6n^2 - 5n - 6}{6n^2 + 7n - 5} \div \dfrac{12n^2 - 17n + 6}{6n^2 + 7n - 5}$

102. $\dfrac{2p^2 - p - 6}{16p^2 - 25} \cdot \dfrac{12p^2 + 13p - 35}{4p^2 + 12p + 9} \div \dfrac{12p^2 + 43p + 35}{2p^2 - p - 6}$

103. Use a graphing calculator to check that

$$\frac{x - 1}{x^2 + 2x + 1} \div \frac{x^2 - 1}{x^2 - 5x + 4}$$

is equivalent to

$$\frac{x^2 - 5x + 4}{(x + 1)^3}.$$

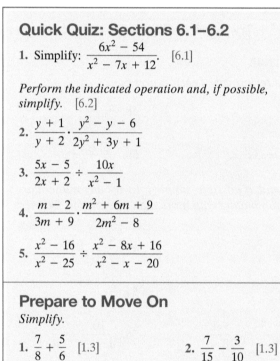

YOUR TURN ANSWERS: SECTION 6.2

1. $\dfrac{5t}{t - 1}$ **2.** $\dfrac{(x + 2)(x + 1)}{(x - 2)(x - 3)}$ **3.** $\dfrac{x - 1}{x + 1}$

Quick Quiz: Sections 6.1–6.2

1. Simplify: $\dfrac{6x^2 - 54}{x^2 - 7x + 12}$.   [6.1]

Perform the indicated operation and, if possible, simplify. [6.2]

2. $\dfrac{y + 1}{y + 2} \cdot \dfrac{y^2 - y - 6}{2y^2 + 3y + 1}$

3. $\dfrac{5x - 5}{2x + 2} \div \dfrac{10x}{x^2 - 1}$

4. $\dfrac{m - 2}{3m + 9} \cdot \dfrac{m^2 + 6m + 9}{2m^2 - 8}$

5. $\dfrac{x^2 - 16}{x^2 - 25} \div \dfrac{x^2 - 8x + 16}{x^2 - x - 20}$

Prepare to Move On

Simplify.

1. $\dfrac{7}{8} + \dfrac{5}{6}$ [1.3]   **2.** $\dfrac{7}{15} - \dfrac{3}{10}$ [1.3]

3. $2x^2 - x + 1 - (x^2 - x - 2)$ [4.4]

4. $3x^2 + x - 7 - (5x^2 + 5x - 8)$ [4.4]

6.3   **Addition, Subtraction, and Least Common Denominators**

A. Addition When Denominators Are the Same   **B.** Subtraction When Denominators Are the Same
C. Least Common Multiples and Denominators

A. Addition When Denominators Are the Same

Recall that to add fractions having the same denominator, like $\frac{2}{7}$ and $\frac{3}{7}$, we add the numerators and keep the common denominator: $\frac{2}{7} + \frac{3}{7} = \frac{5}{7}$. The same procedure is used for all rational expressions that share a common denominator. In this section, we assume that all denominators are nonzero.

> **THE SUM OF TWO RATIONAL EXPRESSIONS**
>
> To add when the denominators are the same, add the numerators and keep the common denominator:
> $$\frac{A}{B} + \frac{C}{B} = \frac{A + C}{B}.$$

Study Skills

Visualize the Steps

If you have completed all assignments and are studying for a quiz or a test, don't feel that you need to redo every assigned problem. A more productive use of your time would be to work through one problem of each type. Then read through the other problems, visualizing the steps that lead to a solution. When you are unsure of how to solve a problem, work that problem in its entirety, seeking outside help as needed.

EXAMPLE 1   Add. Simplify the result, if possible.

a) $\dfrac{4}{a} + \dfrac{3 + a}{a}$

b) $\dfrac{2x^2 + 3x - 7}{2x + 1} + \dfrac{x^2 + x - 8}{2x + 1}$

c) $\dfrac{x - 5}{x^2 - 9} + \dfrac{2}{x^2 - 9}$

SOLUTION

a) $\dfrac{4}{a} + \dfrac{3 + a}{a} = \dfrac{7 + a}{a}$ When the denominators are alike, add the numerators and keep the common denominator.

b) $\dfrac{2x^2 + 3x - 7}{2x + 1} + \dfrac{x^2 + x - 8}{2x + 1} = \dfrac{(2x^2 + 3x - 7) + (x^2 + x - 8)}{2x + 1}$

$= \dfrac{3x^2 + 4x - 15}{2x + 1}$ Combining like terms

$= \dfrac{(3x - 5)(x + 3)}{2x + 1}$ Factoring. There are no common factors, so we cannot simplify further.

c) $\dfrac{x - 5}{x^2 - 9} + \dfrac{2}{x^2 - 9} = \dfrac{x - 3}{x^2 - 9}$ Combining like terms in the numerator: $x - 5 + 2 = x - 3$

$= \dfrac{x - 3}{(x - 3)(x + 3)}$ Factoring

$= \dfrac{1 \cdot (x - 3)}{(x - 3)(x + 3)}$ Removing a factor equal to 1: $\dfrac{x - 3}{x - 3} = 1$

$= \dfrac{1}{x + 3}$

1. Add. Simplify the result, if possible.

$\dfrac{3x + 1}{x + 2} + \dfrac{x + 7}{x + 2}$

YOUR TURN

B. Subtraction When Denominators Are the Same

When two fractions have the same denominator, we subtract one numerator from the other and keep the common denominator, as in $\frac{5}{7} - \frac{2}{7} = \frac{3}{7}$. The same procedure is used with rational expressions.

> **THE DIFFERENCE OF TWO RATIONAL EXPRESSIONS**
>
> To subtract when the denominators are the same, subtract the second numerator from the first numerator and keep the common denominator:
>
> $$\frac{A}{B} - \frac{C}{B} = \frac{A - C}{B}.$$

> *CAUTION!* The fraction bar under a numerator is a grouping symbol, just like parentheses. Thus, when a numerator is subtracted, it is important to subtract *every* term in that numerator.

EXAMPLE 2 Subtract and, if possible, simplify.

a) $\dfrac{3x}{x + 2} - \dfrac{x - 5}{x + 2}$

b) $\dfrac{x^2}{x - 4} - \dfrac{x + 12}{x - 4}$

SOLUTION

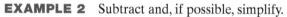

a) $\dfrac{3x}{x + 2} - \dfrac{x - 5}{x + 2} = \dfrac{3x - (x - 5)}{x + 2}$ The parentheses are needed to make sure that we subtract both terms.

$= \dfrac{3x - x + 5}{x + 2}$ Removing the parentheses and changing signs (using the distributive law)

$= \dfrac{2x + 5}{x + 2}$ Combining like terms

b) $\dfrac{x^2}{x - 4} - \dfrac{x + 12}{x - 4} = \dfrac{x^2 - (x + 12)}{x - 4}$ Remember the parentheses!

$= \dfrac{x^2 - x - 12}{x - 4}$ Removing parentheses (using the distributive law)

$= \dfrac{(x - 4)(x + 3)}{x - 4}$ Factoring, in hopes of simplifying

$= \dfrac{(x - 4)(x + 3)}{x - 4}$ Removing a factor equal to 1: $\dfrac{x - 4}{x - 4} = 1$

$= x + 3$

2. Subtract and, if possible, simplify:

$$\frac{3a}{2a - 1} - \frac{a - 1}{2a - 1}.$$

YOUR TURN

C. Least Common Multiples and Denominators

Thus far, every pair of rational expressions that we have added or subtracted shared a common denominator. To add or subtract rational expressions that have different denominators, we must first find equivalent rational expressions that *do* have a common denominator.

In algebra, we find a common denominator much as we do in arithmetic. Recall that to add $\frac{1}{12}$ and $\frac{7}{30}$, we first identify the smallest number that contains both 12 and 30 as factors. Such a number, the **least common multiple (LCM)** of the denominators, is then used as the **least common denominator (LCD)**.

Let's find the LCM of 12 and 30 using a method that can also be used with polynomials. We begin by writing the prime factorizations of 12 and 30:

$$12 = 2 \cdot 2 \cdot 3;$$
$$30 = 2 \cdot 3 \cdot 5.$$

The LCM must include the factors of each number, so it must include each prime factor the greatest number of times that it appears in either of the factorizations. To find the LCM for 12 and 30, we select one factorization, say

$$2 \cdot 2 \cdot 3,$$

and note that because it lacks a factor of 5, it does not contain the entire factorization of 30. If we multiply $2 \cdot 2 \cdot 3$ by 5, every prime factor occurs just often enough to contain both 12 and 30 as factors.

12 is a factor of the LCM.

$$\text{LCM} = 2 \cdot 2 \cdot 3 \cdot 5$$

30 is a factor of the LCM.

Note that each prime factor—2, 3, and 5—is used the greatest number of times that it appears in either of the individual factorizations. The factor 2 occurs twice and the factors 3 and 5 once each.

Student Notes

The terms *least common multiple* and *least common denominator* are similar enough that they may be confusing. We find the LCM of polynomials, and we find the LCD of rational expressions; the LCD of two rational expressions is the LCM of their denominators.

TO FIND THE LEAST COMMON DENOMINATOR (LCD)

1. Write the prime factorization of each denominator.
2. Select one of the factorizations and inspect it to see if it completely contains the other factorization.

   a) If it does, it represents the LCM of the denominators.
   b) If it does not, multiply that factorization by any factors of the other denominator that it lacks. The final product is the LCM of the denominators.

The LCD is the LCM of the denominators. It should contain each factor the greatest number of times that it occurs in any of the individual factorizations.

EXAMPLE 3 Find the LCD of $\dfrac{5}{36x^2}$ and $\dfrac{7}{24x}$.

SOLUTION

1. We begin by writing the prime factorizations of $36x^2$ and $24x$:

$$36x^2 = 2 \cdot 2 \cdot 3 \cdot 3 \cdot x \cdot x;$$
$$24x = 2 \cdot 2 \cdot 2 \cdot 3 \cdot x.$$

2. We select the factorization of $36x^2$. (We could have selected $24x$ instead.) Except for a third factor of 2, this factorization contains the entire factorization of $24x$. Thus we multiply $36x^2$ by a third factor of 2.

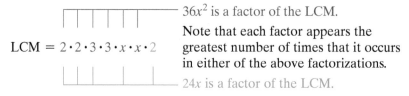

$$LCM = 2 \cdot 2 \cdot 3 \cdot 3 \cdot x \cdot x \cdot 2$$

$36x^2$ is a factor of the LCM.

Note that each factor appears the greatest number of times that it occurs in either of the above factorizations.

$24x$ is a factor of the LCM.

3. Find the LCD of

$$\frac{25}{12a^3} \quad \text{and} \quad \frac{8}{45a^2}.$$

The LCM of the denominators is thus $2^3 \cdot 3^2 \cdot x^2$, or $72x^2$. The LCD of the expressions is $72x^2$.

YOUR TURN

Let's now add $\dfrac{1}{12}$ and $\dfrac{7}{30}$:

$$\frac{1}{12} + \frac{7}{30} = \frac{1}{2 \cdot 2 \cdot 3} + \frac{7}{2 \cdot 3 \cdot 5}. \qquad \begin{array}{l}\text{The least common denominator} \\ \text{(LCD) is } 2 \cdot 2 \cdot 3 \cdot 5.\end{array}$$

We found above that the LCD is $2 \cdot 2 \cdot 3 \cdot 5$, or 60. To get the LCD, we see that the first denominator, 12, needs a factor of 5, and the second denominator, 30, needs another factor of 2. We multiply $\frac{1}{12}$ by 1, using $\frac{5}{5}$, and we multiply $\frac{7}{30}$ by 1, using $\frac{2}{2}$. Since $a \cdot 1 = a$, for any number a, the values of the fractions are not changed.

$$\frac{1}{12} + \frac{7}{30} = \frac{1}{2 \cdot 2 \cdot 3} \cdot \frac{5}{5} + \frac{7}{2 \cdot 3 \cdot 5} \cdot \frac{2}{2} \qquad \frac{5}{5} = 1 \text{ and } \frac{2}{2} = 1$$

$$= \frac{5}{60} + \frac{14}{60} \qquad \begin{array}{l}\text{Both denominators are now} \\ \text{the LCD.}\end{array}$$

$$= \frac{19}{60} \qquad \begin{array}{l}\text{Adding the numerators and} \\ \text{keeping the LCD}\end{array}$$

Expressions like $\dfrac{5}{36x^2}$ and $\dfrac{7}{24x}$ are added in much the same manner. In Example 3, we found that the LCD is $2 \cdot 2 \cdot 2 \cdot 3 \cdot 3 \cdot x \cdot x$, or $72x^2$. To obtain equivalent expressions with this LCD, we multiply each expression by 1, using the missing factors of the LCD to write 1:

$$\frac{5}{36x^2} + \frac{7}{24x} = \frac{5}{2 \cdot 2 \cdot 3 \cdot 3 \cdot x \cdot x} + \frac{7}{2 \cdot 2 \cdot 2 \cdot 3 \cdot x}$$

$$= \frac{5}{2 \cdot 2 \cdot 3 \cdot 3 \cdot x \cdot x} \cdot \frac{2}{2} + \frac{7}{2 \cdot 2 \cdot 2 \cdot 3 \cdot x} \cdot \frac{3 \cdot x}{3 \cdot x}$$

The LCD requires another factor of 2.

The LCD requires additional factors of 3 and x.

$$= \frac{10}{72x^2} + \frac{21x}{72x^2} \qquad \text{Both denominators are now the LCD.}$$

$$= \frac{21x + 10}{72x^2}.$$

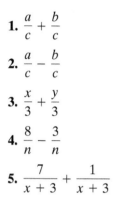

↳ Check Your UNDERSTANDING

Perform the indicated operation.

1. $\dfrac{a}{c} + \dfrac{b}{c}$

2. $\dfrac{a}{c} - \dfrac{b}{c}$

3. $\dfrac{x}{3} + \dfrac{y}{3}$

4. $\dfrac{8}{n} - \dfrac{3}{n}$

5. $\dfrac{7}{x+3} + \dfrac{1}{x+3}$

4. Find the least common multiple of $2x^2 + 5x - 3$ and $x^2 + 6x + 9$.

You now have the "big picture" of why LCMs are needed when adding rational expressions. For the remainder of this section, we will practice finding LCMs and rewriting rational expressions so that they have the LCD as the denominator. In the next section, we will return to the addition and subtraction of rational expressions.

EXAMPLE 4 For each pair of polynomials, find the least common multiple.

a) $15a$ and $35b$

b) $21x^3y^6$ and $7x^5y^2$

c) $x^2 + 5x - 6$ and $x^2 - 1$

SOLUTION

a) We write the prime factorizations and then construct the LCM, starting with the factorization of $15a$.

$$15a = 3 \cdot 5 \cdot a$$
$$35b = 5 \cdot 7 \cdot b$$

LCM $= 3 \cdot 5 \cdot a \cdot 7 \cdot b$

$15a$ is a factor of the LCM.

Each factor appears the greatest number of times that it occurs in either of the above factorizations.

$35b$ is a factor of the LCM.

The LCM is $3 \cdot 5 \cdot a \cdot 7 \cdot b$, or $105ab$.

b) $21x^3y^6 = 3 \cdot 7 \cdot x \cdot x \cdot x \cdot y \cdot y \cdot y \cdot y \cdot y \cdot y$ Try to visualize the factors of x and y mentally.

$$7x^5y^2 = 7 \cdot x \cdot x \cdot x \cdot x \cdot x \cdot y \cdot y$$

LCM $= 3 \cdot 7 \cdot x \cdot x \cdot x \cdot y \cdot y \cdot y \cdot y \cdot y \cdot y \cdot x \cdot x$

We start with the prime factorization of $21x^3y^6$.

We multiply by the factors of $7x^5y^2$ that are lacking.

Note that we used the highest power of each factor in $21x^3y^6$ and $7x^5y^2$. The LCM is $21x^5y^6$.

c) $x^2 + 5x - 6 = (x - 1)(x + 6)$
   $x^2 - 1 = (x - 1)(x + 1)$

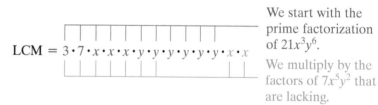

LCM $= (x - 1)(x + 6)(x + 1)$

We start with the factorization of $x^2 + 5x - 6$.

We multiply by the factor of $x^2 - 1$ that is missing.

The LCM is $(x - 1)(x + 6)(x + 1)$. There is no need to multiply this out.

↩ YOUR TURN

The procedure above can be used to find the LCM of three or more polynomials as well. We factor each polynomial and then construct the LCM using each factor the greatest number of times that it appears in any one factorization.

Student Notes

If you prefer, the LCM for a group of three polynomials can be found by finding the LCM of two of them and then finding the LCM of that result and the remaining polynomial.

EXAMPLE 5 For each group of polynomials, find the LCM.

a) $12x$, $16y$, and $8xyz$

b) $x^2 + 4$, $x + 1$, and 5

SOLUTION

a)
$$12x = 2 \cdot 2 \cdot 3 \cdot x$$
$$16y = 2 \cdot 2 \cdot 2 \cdot 2 \cdot y$$
$$8xyz = 2 \cdot 2 \cdot 2 \cdot x \cdot y \cdot z$$

$$\text{LCM} = 2 \cdot 2 \cdot 3 \cdot x \cdot 2 \cdot 2 \cdot y \cdot z$$

We start with the factorization of $12x$. Any one of the three could be chosen.

We multiply by the factors of $16y$ that are missing.

We then multiply by the factor of $8xyz$ that is missing.

The LCM is $2^4 \cdot 3 \cdot xyz$, or $48xyz$.

5. Find the LCM of $x^2 + 5x$, $x^2 - 25$, and $5x$.

b) Since $x^2 + 4$, $x + 1$, and 5 are not factorable, the LCM is their product: $5(x^2 + 4)(x + 1)$.

YOUR TURN

To add or subtract rational expressions with different denominators, we first write equivalent expressions that have the LCD. To do this, we multiply each rational expression by a carefully constructed form of 1.

EXAMPLE 6 Find equivalent expressions that have the LCD:

$$\frac{x + 3}{x^2 + 5x - 6}, \quad \frac{x + 7}{x^2 - 1}.$$

SOLUTION From Example 4(c), we know that the LCD is

$$(x + 6)(x - 1)(x + 1).$$

The factor of the LCD that is missing from $x^2 + 5x - 6$ is $x + 1$. We multiply by 1 using $(x + 1)/(x + 1)$:

$$\frac{x + 3}{x^2 + 5x - 6} = \frac{x + 3}{(x + 6)(x - 1)} \cdot \frac{x + 1}{x + 1} = \frac{(x + 3)(x + 1)}{(x + 6)(x - 1)(x + 1)}.$$

The factor of the LCD that is missing from $x^2 - 1$ is $x + 6$. We multiply by 1 using $(x + 6)/(x + 6)$:

6. Find equivalent expressions that have the LCD:

$$\frac{5}{x^2 + x}, \quad \frac{2}{3x}.$$

$$\frac{x + 7}{x^2 - 1} = \frac{x + 7}{(x + 1)(x - 1)} \cdot \frac{x + 6}{x + 6} = \frac{(x + 7)(x + 6)}{(x + 1)(x - 1)(x + 6)}.$$

Both rational expressions now have the same denominator. We leave the results in factored form.

YOUR TURN

6.3 EXERCISE SET

FOR EXTRA HELP MyMathLab®

Vocabulary and Reading Check

Use one or more words to complete each of the following statements.

1. To add two rational expressions when the denominators are the same, add _____ and keep the common _____.

2. When a numerator is being subtracted, use parentheses to make sure that every _____ in that numerator is subtracted.

3. The least common multiple of two denominators is usually referred to as the _____ and is abbreviated _____.

4. The least common denominator of two rational expressions must contain every _____ that is in either denominator.

A, B. Addition and Subtraction When Denominators Are the Same

Perform the indicated operation. Simplify, if possible.

5. $\dfrac{3}{t} + \dfrac{5}{t}$

6. $\dfrac{8}{y^2} + \dfrac{2}{y^2}$

7. $\dfrac{x}{12} + \dfrac{2x+5}{12}$

8. $\dfrac{a}{7} + \dfrac{3a-4}{7}$

9. $\dfrac{4}{a+3} + \dfrac{5}{a+3}$

10. $\dfrac{5}{x+2} + \dfrac{8}{x+2}$

11. $\dfrac{11}{4x-7} - \dfrac{3}{4x-7}$

12. $\dfrac{9}{2x+3} - \dfrac{5}{2x+3}$

13. $\dfrac{3y+8}{2y} - \dfrac{y+1}{2y}$

14. $\dfrac{5+3t}{4t} - \dfrac{2t+1}{4t}$

15. $\dfrac{5x+7}{x+3} + \dfrac{x+11}{x+3}$

16. $\dfrac{3x+4}{x-1} + \dfrac{2x-9}{x-1}$

17. $\dfrac{5x+7}{x+3} - \dfrac{x+11}{x+3}$

18. $\dfrac{3x+4}{x-1} - \dfrac{2x-9}{x-1}$

19. $\dfrac{a^2}{a-4} + \dfrac{a-20}{a-4}$

20. $\dfrac{x^2}{x+5} + \dfrac{7x+10}{x+5}$

21. $\dfrac{y^2}{y+2} - \dfrac{5y+14}{y+2}$

22. $\dfrac{t^2}{t-3} - \dfrac{8t-15}{t-3}$

Aha! 23. $\dfrac{t^2-5t}{t-1} + \dfrac{5t-t^2}{t-1}$

24. $\dfrac{y^2+6y}{y+2} + \dfrac{2y+12}{y+2}$

25. $\dfrac{x-6}{x^2+5x+6} + \dfrac{9}{x^2+5x+6}$

26. $\dfrac{x-5}{x^2-4x+3} + \dfrac{2}{x^2-4x+3}$

27. $\dfrac{3a^2+14}{a^2+5a-6} - \dfrac{13a}{a^2+5a-6}$

28. $\dfrac{2a^2+15}{a^2-7a+12} - \dfrac{11a}{a^2-7a+12}$

29. $\dfrac{t^2-5t}{t^2+6t+9} + \dfrac{4t-12}{t^2+6t+9}$

30. $\dfrac{y^2-7y}{y^2+8y+16} + \dfrac{6y-20}{y^2+8y+16}$

31. $\dfrac{2y^2+3y}{y^2-7y+12} - \dfrac{y^2+4y+6}{y^2-7y+12}$

32. $\dfrac{3a^2+7}{a^2-2a-8} - \dfrac{7+3a^2}{a^2-2a-8}$

33. $\dfrac{3-2x}{x^2-6x+8} + \dfrac{7-3x}{x^2-6x+8}$

34. $\dfrac{1-2t}{t^2-5t+4} + \dfrac{4-3t}{t^2-5t+4}$

35. $\dfrac{x-9}{x^2+3x-4} - \dfrac{2x-5}{x^2+3x-4}$

36. $\dfrac{5-3x}{x^2-2x+1} - \dfrac{x+1}{x^2-2x+1}$

C. Least Common Multiples and Denominators

Find the LCM.

37. 15, 36   38. 18, 30   39. 8, 9

40. 12, 15   41. 6, 12, 15   42. 8, 32, 50

Find the LCM.

43. $18t^2, 6t^5$

44. $8x^5, 24x^2$

45. $15a^4b^7, 10a^2b^8$

46. $6a^2b^7, 9a^5b^2$

47. $2(y-3), 6(y-3)$

48. $4(x-1), 8(x-1)$

49. $x^2-2x-15, x^2-9$

50. $t^2-4, t^2+7t+10$

51. t^3+4t^2+4t, t^2-4t

52. y^3-y^2, y^4-y^2

53. $6xz^2, 8x^2y, 15y^3z$

54. $12s^3t, 15sv^2, 6t^4v$

55. $a+1, (a-1)^2, a^2-1$

56. $x-2, (x+2)^2, x^2-4$

57. $2n^2+n-1, 2n^2+3n-2$

58. $m^2-2m-3, 2m^2+3m+1$

Aha! 59. $t-3, t+3, t^2-9$

60. $a-5, a^2-10a+25$

61. $6x^3 - 24x^2 + 18x,\ 4x^5 - 24x^4 + 20x^3$

62. $9x^3 - 9x^2 - 18x,\ 6x^5 - 24x^4 + 24x^3$

Find equivalent expressions that have the LCD.

63. $\dfrac{5}{6t^4},\ \dfrac{s}{18t^2}$

64. $\dfrac{7}{10y^2},\ \dfrac{x}{5y^6}$

65. $\dfrac{7}{3x^4y^2},\ \dfrac{4}{9xy^3}$

66. $\dfrac{3}{2a^2b},\ \dfrac{7}{8ab^2}$

67. $\dfrac{2x}{x^2 - 4},\ \dfrac{4x}{x^2 + 5x + 6}$

68. $\dfrac{5x}{x^2 - 9},\ \dfrac{2x}{x^2 + 11x + 24}$

69. Explain why the product of two numbers is not always their least common multiple.

70. If the LCM of two numbers is their product, what can you conclude about the two numbers?

Skill Review

Solve.

71. $2x - 7 = 5x + 3$ [2.2]

72. $\frac{1}{3}x < \frac{2}{5}x - 1$ [2.6]

73. $x^2 - 8x = 20$ [5.6]

74. $-x \geq 16 + x$ [2.6]

75. $2x^2 + 4x + 2 = 0$ [5.6]

76. $2 - 3(x - 7) = 15$ [2.2]

Synthesis

77. If the LCM of two third-degree polynomials is a sixth-degree polynomial, what can be concluded about the two polynomials?

78. If the LCM of a binomial and a trinomial is the trinomial, what relationship exists between the two expressions?

Perform the indicated operations. Simplify, if possible.

79. $\dfrac{6x - 1}{x - 1} + \dfrac{3(2x + 5)}{x - 1} + \dfrac{3(2x - 3)}{x - 1}$

80. $\dfrac{2x + 11}{x - 3} \cdot \dfrac{3}{x + 4} + \dfrac{-1}{4 + x} \cdot \dfrac{6x + 3}{x - 3}$

81. $\dfrac{x^2}{3x^2 - 5x - 2} - \dfrac{2x}{3x + 1} \cdot \dfrac{1}{x - 2}$

82. $\dfrac{x + y}{x^2 - y^2} + \dfrac{x - y}{x^2 - y^2} - \dfrac{2x}{x^2 - y^2}$

African Artistry. In Southeast Mozambique, the design of every woven handbag, or gipatsi *(plural,* sipatsi*) is created by repeating two or more geometric patterns.*

Each pattern encircles the bag, sharing the strands of fabric with any pattern above or below. The length, or period, of each pattern is the number of strands required to construct the pattern. For a gipatsi to be considered beautiful, each individual pattern must fit a whole number of times around the bag.

Data: Gerdes, Paulus, *Women, Art and Geometry in Southern Africa.* Asmara, Eritrea: Africa World Press, Inc., p. 5

83. A weaver is using two patterns to create a gipatsi. Pattern A is 10 strands long, and pattern B is 3 strands long. What is the smallest number of strands that can be used to complete the gipatsi?

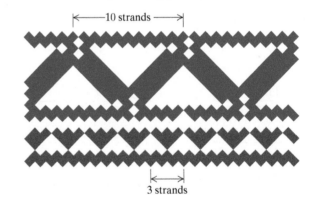

84. A weaver is using a four-strand pattern, a six-strand pattern, and an eight-strand pattern. What is the smallest number of strands that can be used to complete the gipatsi?

85. For technical reasons, the number of strands is generally a multiple of 4. Answer Exercise 85 with this additional requirement in mind.

Find the LCM.

86. 80, 96, 108

87. $4x^2 - 25,\ 6x^2 - 7x - 20,\ (9x^2 + 24x + 16)^2$

88. $9n^2 - 9,\ (5n^2 - 10n + 5)^2,\ 15n - 15$

89. *Printers.* The Brother MFC7360N laser printer can print 24 pages per minute. The Canon Pixma MX922 laser printer can print 15 pages per minute. If both machines begin printing at the same instant, how long will it be until they again begin printing a page at exactly the same time?

Data: toptenreviews.com

90. *Vitamins.* Dianara takes two vitamin D_3 capsules and four fish-oil capsules each day. The vitamin D_3 comes in a jar of 18 and the fish oil in a jar of 120. If she started both on the same day, and if she replaces each bottle as soon as it is empty, how long will it be before she replaces both bottles on the same day?

91. *Bus Schedules.* Beginning at 5:00 A.M., a hotel shuttle bus leaves Salton Airport every 15 min, and the downtown shuttle bus leaves the airport every 25 min. What time will it be when both shuttles again leave at the same time?

92. *Appliances.* Dishwashers last an average of 9 years, garbage disposals an average of 12 years, and gas ranges an average of 15 years. If an apartment house is equipped with new dishwashers, garbage disposals, and gas ranges in 2020, in what year will all three appliances need to be replaced at once?

Data: National Association of Home Builders/Bank of America Home Equity Study of Life Expectancy of Home Components

93. Explain how evaluating can be used to perform a partial check on the result of Example 1(c):
$$\frac{x-5}{x^2-9}+\frac{2}{x^2-9}=\frac{1}{x+3}.$$

94. On p. 385, the second step in finding an LCD is to select one of the factorizations of the denominators. Does it matter which one is selected? Why or why not?

95. *Research.* Determine the average life of three or more appliances in your home. If all these appliances were new when your home was built, when can you expect them all to fail in the same year?

Quick Quiz: Sections 6.1–6.3

1. Simplify: $\frac{x+3}{x^2+10x+21}$. [6.1]

2. Find the LCM: x^2-6x, x^2-7x+6. [6.3]

Perform the indicated operation and, if possible, simplify.

3. $\frac{x+1}{x^2+10x+21}\cdot\frac{x+7}{x^2-1}$ [6.2]

4. $\frac{2t-10}{9t^2}\div\frac{t^3-7t^2+10t}{6t}$ [6.2]

5. $\frac{x+3}{x-2}-\frac{x-5}{x-2}$ [6.3]

Prepare to Move On

Using the identity $-\frac{a}{b}=\frac{-a}{b}=\frac{a}{-b}$, *write each number in two equivalent forms.* [1.7]

1. $-\frac{5}{8}$ **2.** $\frac{4}{-11}$

Write an equivalent expression without parentheses. [1.8]

3. $-(x-y)$ **4.** $-(3-a)$

Multiply and simplify. [1.8]

5. $-1(2x-7)$ **6.** $-1(a-b)$

6.4 Addition and Subtraction with Unlike Denominators

A. Adding and Subtracting with LCDs   **B.** When Factors Are Opposites

A. Adding and Subtracting with LCDs

We now know how to rewrite two rational expressions in equivalent forms that use the LCD. Once rational expressions share a common denominator, they can be added or subtracted. We assume that all denominators are nonzero.

> **TO ADD OR SUBTRACT RATIONAL EXPRESSIONS HAVING DIFFERENT DENOMINATORS**
> 1. Find the LCD.
> 2. Multiply each rational expression by a form of 1 made up of the factors of the LCD missing from that expression's denominator.
> 3. Add or subtract the numerators, as indicated. Write the sum or the difference over the LCD.
> 4. If possible, factor and simplify.

EXAMPLE 1 Add: $\dfrac{5x^2}{8} + \dfrac{7x}{12}$.

SOLUTION

1. First, we find the LCD:

$$\left.\begin{array}{l} 8 = 2 \cdot 2 \cdot 2 \\ 12 = 2 \cdot 2 \cdot 3 \end{array}\right\} \quad \text{LCD} = 2 \cdot 2 \cdot 2 \cdot 3, \text{ or } 24.$$

2. The denominator 8 must be multiplied by 3 in order to obtain the LCD. The denominator 12 must be multiplied by 2 in order to obtain the LCD. Thus we multiply the first expression by $\frac{3}{3}$ and the second expression by $\frac{2}{2}$ to get the LCD:

$$\frac{5x^2}{8} + \frac{7x}{12} = \frac{5x^2}{2 \cdot 2 \cdot 2} + \frac{7x}{2 \cdot 2 \cdot 3}$$

$$= \frac{5x^2}{2 \cdot 2 \cdot 2} \cdot \frac{3}{3} + \frac{7x}{2 \cdot 2 \cdot 3} \cdot \frac{2}{2} \qquad \text{Multiplying by a form of 1 to get the LCD}$$

$$= \frac{15x^2}{24} + \frac{14x}{24}.$$

3. Next, we add the numerators: $\dfrac{15x^2}{24} + \dfrac{14x}{24} = \dfrac{15x^2 + 14x}{24}.$

4. We factor in hope of simplifying: $\dfrac{x(15x + 14)}{24}.$

   The expression cannot be simplified.

↩ YOUR TURN

EXAMPLE 2 Subtract: $\dfrac{7}{8x} - \dfrac{5}{12x^2}$.

SOLUTION We follow the four steps shown above. First, we find the LCD:

$$\left.\begin{array}{l} 8x = 2 \cdot 2 \cdot 2 \cdot x \\ 12x^2 = 2 \cdot 2 \cdot 3 \cdot x \cdot x \end{array}\right\} \quad \text{LCD} = 2 \cdot 2 \cdot 3 \cdot x \cdot x \cdot 2, \text{ or } 24x^2.$$

The denominator $8x$ must be multiplied by $3x$ in order to obtain the LCD. The denominator $12x^2$ must be multiplied by 2 in order to obtain the LCD. Thus we multiply by $\dfrac{3x}{3x}$ and $\dfrac{2}{2}$ to get the LCD. Then we subtract and, if possible, simplify.

$$\frac{7}{8x} - \frac{5}{12x^2} = \frac{7}{8x} \cdot \frac{3x}{3x} - \frac{5}{12x^2} \cdot \frac{2}{2}$$

$$= \frac{21x}{24x^2} - \frac{10}{24x^2} \qquad \boxed{\begin{array}{l} \textit{CAUTION!} \quad \text{Do not simplify} \\ \textit{these } \text{rational expressions or} \\ \text{you will lose the LCD.} \end{array}}$$

$$= \frac{21x - 10}{24x^2} \qquad \begin{array}{l} \text{This cannot be simplified,} \\ \text{so we are done.} \end{array}$$

↩ YOUR TURN

1. Add: $\dfrac{2x}{9} + \dfrac{5}{12}$.

2. Subtract: $\dfrac{3}{10x^2} - \dfrac{5}{4x}$.

After adding or subtracting numerators, we factor, if possible, in hope of simplifying. We also leave the simplified answer in factored form.

EXAMPLE 3 Add: $\dfrac{2a}{a^2 - 1} + \dfrac{1}{a^2 + a}$.

SOLUTION First, we find the LCD:

Find the LCD.

$$\left.\begin{array}{l} a^2 - 1 = (a - 1)(a + 1) \\ a^2 + a = a(a + 1). \end{array}\right\} \quad \text{LCD} = (a - 1)(a + 1)a$$

We multiply by a form of 1 to get the LCD in each expression:

Write each expression with the LCD.

$$\dfrac{2a}{a^2 - 1} + \dfrac{1}{a^2 + a} = \dfrac{2a}{(a - 1)(a + 1)} \cdot \dfrac{a}{a} + \dfrac{1}{a(a + 1)} \cdot \dfrac{a - 1}{a - 1}$$

Multiplying by $\dfrac{a}{a}$ and $\dfrac{a - 1}{a - 1}$ to get the LCD

$$= \dfrac{2a^2}{(a - 1)(a + 1)a} + \dfrac{a - 1}{a(a + 1)(a - 1)}$$

Add numerators.

$$= \dfrac{2a^2 + a - 1}{a(a - 1)(a + 1)} \qquad \text{Adding numerators}$$

Simplify.

$$= \dfrac{(2a - 1)(a + 1)}{a(a - 1)(a + 1)} \qquad \text{Simplifying by factoring and removing a factor equal to 1:}$$

$$= \dfrac{2a - 1}{a(a - 1)}. \qquad \dfrac{a + 1}{a + 1} = 1$$

3. Add: $\dfrac{4x}{x^2 - 25} + \dfrac{x}{x + 5}$.

YOUR TURN

EXAMPLE 4 Subtract.

a) $\dfrac{x + 4}{x - 2} - \dfrac{x - 7}{x + 5}$

b) $\dfrac{x}{x^2 + 5x + 6} - \dfrac{2}{x^2 + 3x + 2}$

Student Notes

As you can see, adding or subtracting rational expressions can involve many steps. Therefore, it is wise to double-check each step of your work as you work through each problem. Waiting to inspect your work at the end of each problem is usually a less efficient use of your time.

SOLUTION

a) First, we find the LCD. It is just the product of the denominators:

$$\text{LCD} = (x - 2)(x + 5).$$

We multiply by a form of 1 to get the LCD in each expression. Then we subtract and try to simplify.

$$\dfrac{x + 4}{x - 2} - \dfrac{x - 7}{x + 5} = \dfrac{x + 4}{x - 2} \cdot \dfrac{x + 5}{x + 5} - \dfrac{x - 7}{x + 5} \cdot \dfrac{x - 2}{x - 2}$$

$$= \dfrac{x^2 + 9x + 20}{(x - 2)(x + 5)} - \dfrac{x^2 - 9x + 14}{(x - 2)(x + 5)} \qquad \text{Multiplying out numerators (but not denominators)}$$

$$= \dfrac{x^2 + 9x + 20 - (x^2 - 9x + 14)}{(x - 2)(x + 5)} \qquad \text{Parentheses are important.}$$

$$= \dfrac{x^2 + 9x + 20 - x^2 + 9x - 14}{(x - 2)(x + 5)} \qquad \text{Removing parentheses and subtracting every term}$$

$$= \dfrac{18x + 6}{(x - 2)(x + 5)}$$

$$= \dfrac{6(3x + 1)}{(x - 2)(x + 5)} \qquad \text{We cannot simplify.}$$

b) $\dfrac{x}{x^2 + 5x + 6} - \dfrac{2}{x^2 + 3x + 2}$

Find the LCD. $= \dfrac{x}{(x + 2)(x + 3)} - \dfrac{2}{(x + 2)(x + 1)}$ Factoring denominators. The LCD is $(x + 2)(x + 3)(x + 1)$.

$= \dfrac{x}{(x + 2)(x + 3)} \cdot \dfrac{x + 1}{x + 1} - \dfrac{2}{(x + 2)(x + 1)} \cdot \dfrac{x + 3}{x + 3}$

Write each expression with the LCD. $= \dfrac{x^2 + x}{(x + 2)(x + 3)(x + 1)} - \dfrac{2x + 6}{(x + 2)(x + 3)(x + 1)}$

Subtract numerators. $= \dfrac{x^2 + x - (2x + 6)}{(x + 2)(x + 3)(x + 1)}$ Don't forget the parentheses!

$= \dfrac{x^2 + x - 2x - 6}{(x + 2)(x + 3)(x + 1)}$ Remember to subtract each term in $2x + 6$.

$= \dfrac{x^2 - x - 6}{(x + 2)(x + 3)(x + 1)}$ Combining like terms in the numerator

Simplify. $= \dfrac{\cancel{(x + 2)}(x - 3)}{\cancel{(x + 2)}(x + 3)(x + 1)}$

4. Subtract:

$\dfrac{x}{x^2 + 3x + 2} - \dfrac{3}{x^2 - x - 2}.$ $= \dfrac{x - 3}{(x + 3)(x + 1)}$ Factoring and simplifying; $\dfrac{x + 2}{x + 2} = 1$

↪ YOUR TURN

B. When Factors Are Opposites

A special case arises when one denominator is the opposite of the other. When this occurs, we can multiply either expression by 1 using $\dfrac{-1}{-1}$.

EXAMPLE 5   Add: $\dfrac{3}{8a} + \dfrac{1}{-8a}$.

SOLUTION

$\dfrac{3}{8a} + \dfrac{1}{-8a} = \dfrac{3}{8a} + \dfrac{-1}{-1} \cdot \dfrac{1}{-8a}$

> When denominators are opposites, we multiply one rational expression by $-1/-1$ to get the LCD.

$= \dfrac{3}{8a} + \dfrac{-1}{8a} = \dfrac{2}{8a}$

5. Add: $\dfrac{t}{2} + \dfrac{3}{-2}$.

$= \dfrac{2 \cdot 1}{2 \cdot 4a} = \dfrac{1}{4a}$ Simplifying by removing a factor equal to 1: $\dfrac{2}{2} = 1$

↪ YOUR TURN

Expressions of the form $a - b$ and $b - a$ are opposites of each other. When either of these binomials is multiplied by -1, the result is the other binomial:

$\left.\begin{array}{l} -1(a - b) = -a + b = b + (-a) = b - a; \\ -1(b - a) = -b + a = a + (-b) = a - b. \end{array}\right\}$ Multiplication by -1 reverses the order in which subtraction occurs.

EXAMPLE 6 Subtract: $\dfrac{5x}{x-7} - \dfrac{3x}{7-x}$.

SOLUTION

$$\dfrac{5x}{x-7} - \dfrac{3x}{7-x} = \dfrac{5x}{x-7} - \dfrac{-1}{-1} \cdot \dfrac{3x}{7-x} \qquad \text{Note that } x-7 \text{ and } 7-x \text{ are opposites.}$$

$$= \dfrac{5x}{x-7} - \dfrac{-3x}{x-7} \qquad \text{Performing the multiplication.}$$
$$\textit{Note: } -1(7-x) = -7+x = x-7.$$

$$= \dfrac{5x-(-3x)}{x-7} \left. \vphantom{\dfrac{5x+3x}{x-7}} \right\}$$
$$= \dfrac{5x+3x}{x-7} \qquad \begin{array}{l}\text{Subtracting. The parentheses are}\\ \text{important.}\end{array}$$

$$= \dfrac{8x}{x-7} \qquad \text{Simplifying}$$

6. Subtract:

$$\dfrac{x}{x-5} - \dfrac{7}{5-x}.$$

↻ YOUR TURN

Sometimes, after factoring to find the LCD, we find a factor in one denominator that is the opposite of a factor in the other denominator. When this happens, multiplication by $-1/-1$ can again be helpful.

EXAMPLE 7 Perform the indicated operations and simplify.

a) $\dfrac{x}{x^2-25} + \dfrac{3}{5-x}$

b) $\dfrac{x+9}{x^2-4} + \dfrac{6-x}{4-x^2} - \dfrac{1+x}{x^2-4}$

SOLUTION

a)
$$\dfrac{x}{x^2-25} + \dfrac{3}{5-x} = \dfrac{x}{(x-5)(x+5)} + \dfrac{3}{5-x} \qquad \text{Factoring}$$

$$= \dfrac{x}{(x-5)(x+5)} + \dfrac{3}{5-x} \cdot \dfrac{-1}{-1} \qquad \begin{array}{l}\text{Multiplication by}\\ -1/-1 \text{ changes}\\ 5-x \text{ to } x-5.\end{array}$$

$$= \dfrac{x}{(x-5)(x+5)} + \dfrac{-3}{x-5} \qquad (5-x)(-1) = x-5$$

$$= \dfrac{x}{(x-5)(x+5)} + \dfrac{-3}{(x-5)} \cdot \dfrac{x+5}{x+5} \qquad \begin{array}{l}\text{The LCD is}\\ (x-5)(x+5).\end{array}$$

$$= \dfrac{x}{(x-5)(x+5)} + \dfrac{-3x-15}{(x-5)(x+5)}$$

$$= \dfrac{-2x-15}{(x-5)(x+5)}$$

Student Notes

Your answer may differ slightly from the answer found at the back of the book and still be correct. For example, an equivalent answer to Example 7(a) is $-\dfrac{2x+15}{(x-5)(x+5)}$:

$$\dfrac{-2x-15}{(x-5)(x+5)} = \dfrac{-(2x+15)}{(x-5)(x+5)}$$
$$= -\dfrac{2x+15}{(x-5)(x+5)}.$$

Before reworking an exercise, be sure that your answer is indeed incorrect.

Technology Connection

The TABLE feature can be used to check addition or subtraction of rational expressions. Below we check Example 7(a), using

$$y_1 = \dfrac{x}{x^2-25} + \dfrac{3}{5-x} \quad \text{and} \quad y_2 = \dfrac{-2x-15}{(x-5)(x+5)}.$$

Because the values for y_1 and y_2 match, we have a check.

ΔTBL = 1

X	Y₁	Y₂
1	.70833	.70833
2	.90476	.90476
3	1.3125	1.3125
4	2.5556	2.5556
5	ERROR	ERROR
6	−2.455	−2.455
7	−1.208	−1.208

X = 1

Write an equivalent expression with the given denominator.

1. $\dfrac{3}{x + 4} = \dfrac{\square}{(x + 4)(x - 5)}$

2. $\dfrac{1}{a^2 b} = \dfrac{\square}{2a^3 b^5}$

3. $\dfrac{1}{2 - y} = \dfrac{\square}{y - 2}$

7. Perform the indicated operations and simplify:

$$\dfrac{x + 4}{x + 1} - \dfrac{x}{3 - x} - \dfrac{4x}{x^2 - 2x - 3}.$$

↘ YOUR TURN

b) Since $4 - x^2$ is the opposite of $x^2 - 4$, multiplying the second rational expression by $-1/-1$ will lead to a common denominator:

$$\dfrac{x + 9}{x^2 - 4} + \dfrac{6 - x}{4 - x^2} - \dfrac{1 + x}{x^2 - 4} = \dfrac{x + 9}{x^2 - 4} + \dfrac{6 - x}{4 - x^2} \cdot \dfrac{-1}{-1} - \dfrac{1 + x}{x^2 - 4}$$

$$= \dfrac{x + 9}{x^2 - 4} + \dfrac{x - 6}{x^2 - 4} - \dfrac{1 + x}{x^2 - 4}$$

$$= \dfrac{(x + 9) + (x - 6) - (1 + x)}{x^2 - 4} \quad \begin{array}{l}\text{Adding and}\\ \text{subtracting}\\ \text{numerators}\end{array}$$

$$= \dfrac{x + 9 + x - 6 - 1 - x}{x^2 - 4}$$

$$= \dfrac{x + 2}{x^2 - 4}$$

$$= \dfrac{(x + 2) \cdot 1}{(x + 2)(x - 2)} \quad \left.\begin{array}{l}\\ \\ \end{array}\right\} \text{ Simplifying}$$

$$= \dfrac{1}{x - 2}.$$

6.4 EXERCISE SET

FOR EXTRA HELP   MyMathLab®

↘ **Vocabulary and Reading Check**

In Exercises 1– 4, the four steps for adding rational expressions with different denominators are listed. Fill in the missing word or words for each step.

1. To add or subtract when the denominators are different, first find the _____.

2. Multiply each rational expression by a form of 1 made up of the factors of the LCD that are _____ from that expression's _____.

3. Add or subtract the _____, as indicated. Write the sum or the difference over the _____.

4. _____, if possible.

A. Adding and Subtracting with LCDs

Perform the indicated operation. Simplify, if possible.

5. $\dfrac{3}{x^2} + \dfrac{5}{x}$

6. $\dfrac{6}{x} + \dfrac{7}{x^2}$

7. $\dfrac{1}{6r} - \dfrac{3}{8r}$

8. $\dfrac{4}{9t} - \dfrac{7}{6t}$

9. $\dfrac{3}{uv^2} + \dfrac{4}{u^3 v}$

10. $\dfrac{8}{cd^2} + \dfrac{1}{c^2 d}$

11. $\dfrac{-2}{3xy^2} - \dfrac{6}{x^2 y^3}$

12. $\dfrac{8}{9t^3} - \dfrac{5}{6t^2}$

13. $\dfrac{x + 3}{8} + \dfrac{x - 2}{6}$

14. $\dfrac{x - 4}{9} + \dfrac{x + 5}{12}$

15. $\dfrac{x - 2}{6} - \dfrac{x + 1}{3}$

16. $\dfrac{a + 2}{2} - \dfrac{a - 4}{4}$

17. $\dfrac{a + 3}{15a} + \dfrac{2a - 1}{3a^2}$

18. $\dfrac{5a + 1}{2a^2} + \dfrac{a + 2}{6a}$

19. $\dfrac{4z - 9}{3z} - \dfrac{3z - 8}{4z}$

20. $\dfrac{x - 1}{4x} - \dfrac{2x + 3}{x}$

21. $\dfrac{3c + d}{cd^2} + \dfrac{c - d}{c^2 d}$

22. $\dfrac{u + v}{u^2 v} + \dfrac{2u + v}{uv^2}$

23. $\dfrac{4x + 2t}{3xt^2} - \dfrac{5x - 3t}{x^2 t}$

24. $\dfrac{5x + 3y}{2x^2 y} - \dfrac{3x + 4y}{xy^2}$

25. $\dfrac{3}{x - 2} + \dfrac{3}{x + 2}$

26. $\dfrac{5}{x - 1} + \dfrac{5}{x + 1}$

27. $\dfrac{t}{t + 3} - \dfrac{1}{t - 1}$

28. $\dfrac{y}{y - 3} + \dfrac{12}{y + 4}$

29. $\dfrac{3}{x + 1} + \dfrac{2}{3x}$

30. $\dfrac{2}{x + 5} + \dfrac{3}{4x}$

31. $\dfrac{3}{2t^2 - 2t} - \dfrac{5}{2t - 2}$

32. $\dfrac{8}{3t^2 - 15t} - \dfrac{3}{2t - 10}$

33. $\dfrac{3a}{a^2 - 9} + \dfrac{a}{a + 3}$

34. $\dfrac{5p}{p^2 - 16} + \dfrac{p}{p - 4}$

35. $\dfrac{6}{z + 4} - \dfrac{2}{3z + 12}$

36. $\dfrac{t}{t - 3} - \dfrac{5}{4t - 12}$

37. $\dfrac{5}{q - 1} + \dfrac{2}{(q - 1)^2}$

38. $\dfrac{3}{w + 2} + \dfrac{7}{(w + 2)^2}$

39. $\dfrac{3a}{4a - 20} + \dfrac{9a}{6a - 30}$

40. $\dfrac{4a}{5a - 10} + \dfrac{3a}{10a - 20}$

41. $\dfrac{y}{y - 1} - \dfrac{y - 1}{y}$

42. $\dfrac{x + 4}{x} + \dfrac{x}{x + 4}$

43. $\dfrac{6}{a^2 + a - 2} + \dfrac{4}{a^2 - 4a + 3}$

44. $\dfrac{x}{x^2 + 2x + 1} + \dfrac{1}{x^2 + 5x + 4}$

45. $\dfrac{x}{x^2 + 9x + 20} - \dfrac{4}{x^2 + 7x + 12}$

46. $\dfrac{x}{x^2 + 5x + 6} - \dfrac{2}{x^2 + 3x + 2}$

47. $\dfrac{3z}{z^2 - 4z + 4} + \dfrac{10}{z^2 + z - 6}$

48. $\dfrac{3}{x^2 - 9} + \dfrac{2}{x^2 - x - 6}$

Aha! 49. $\dfrac{-7}{x^2 + 25x + 24} - \dfrac{0}{x^2 + 11x + 10}$

50. $\dfrac{x}{x^2 + 17x + 72} - \dfrac{1}{x^2 + 15x + 56}$

51. $3 + \dfrac{4}{2x + 1}$

52. $2 + \dfrac{1}{5 - x}$

53. $3 - \dfrac{2}{4 - x}$

54. $4 - \dfrac{3}{3x + 2}$

B. When Factors Are Opposites

Perform the indicated operation. Simplify, if possible.

55. $\dfrac{5x}{4} - \dfrac{x - 2}{-4}$

56. $\dfrac{x}{6} - \dfrac{2x - 3}{-6}$

Aha! 57. $\dfrac{x}{x - 5} + \dfrac{x}{5 - x}$

58. $\dfrac{y}{y - 2} - \dfrac{y}{2 - y}$

59. $\dfrac{y^2}{y - 3} + \dfrac{9}{3 - y}$

60. $\dfrac{t^2}{t - 2} + \dfrac{4}{2 - t}$

61. $\dfrac{c - 5}{c^2 - 64} - \dfrac{5 - c}{64 - c^2}$

62. $\dfrac{b - 4}{b^2 - 49} + \dfrac{b - 4}{49 - b^2}$

63. $\dfrac{4 - p}{25 - p^2} + \dfrac{p + 1}{p - 5}$

64. $\dfrac{y + 2}{y - 7} + \dfrac{3 - y}{49 - y^2}$

65. $\dfrac{x}{x - 4} - \dfrac{3}{16 - x^2}$

66. $\dfrac{x}{3 - x} - \dfrac{2}{x^2 - 9}$

67. $\dfrac{a}{a^2 - 1} + \dfrac{2a}{a - a^2}$

68. $\dfrac{3x + 2}{3x + 6} + \dfrac{x}{4 - x^2}$

69. $\dfrac{4x}{x^2 - y^2} - \dfrac{6}{y - x}$

70. $\dfrac{4 - a^2}{a^2 - 9} - \dfrac{a - 2}{3 - a}$

A, B. Adding and Subtracting

Perform the indicated operations. Simplify, if possible.

71. $\dfrac{x - 3}{2 - x} - \dfrac{x + 3}{x + 2} + \dfrac{x + 6}{4 - x^2}$

72. $\dfrac{t - 5}{1 - t} - \dfrac{t + 4}{t + 1} + \dfrac{t + 2}{t^2 - 1}$

73. $\dfrac{2x + 5}{x + 1} + \dfrac{x + 7}{x + 5} - \dfrac{5x + 17}{(x + 1)(x + 5)}$

74. $\dfrac{x + 5}{x + 3} + \dfrac{x + 7}{x + 2} - \dfrac{7x + 19}{(x + 3)(x + 2)}$

75. $\dfrac{1}{x + y} + \dfrac{1}{x - y} - \dfrac{2x}{x^2 - y^2}$

76. $\dfrac{2r}{r^2 - s^2} + \dfrac{1}{r + s} - \dfrac{1}{r - s}$

77. $\dfrac{1}{x^2 + 7x + 12} - \dfrac{2}{x^2 + 4x + 3} + \dfrac{3}{x^2 + 5x + 4}$

78. $\dfrac{4}{x^2 + x - 2} + \dfrac{2}{x^2 - 4x + 3} - \dfrac{5}{x^2 - x - 6}$

79. What is the advantage of using the *least* common denominator—rather than just *any* common denominator—when adding or subtracting rational expressions?

80. Describe a procedure that can be used to add any two rational expressions.

Skill Review

Simplify.

81. $3 - 12 \div (-4)$ [1.8]

82. $|-6 - 1|$ [1.8]

83. $(1.2 \times 10^8)(2.5 \times 10^6)$ [4.2]

84. $(2a^3 b^{-5})(-3ab^4)$ [4.2]

85. $(3a^{-1}b)^{-2}$ [4.2]

86. $-(-12)$ [1.6]

Synthesis

87. How could you convince someone that

$$\frac{1}{3-x} \quad \text{and} \quad \frac{1}{x-3}$$

are opposites of each other?

88. Are parentheses as important for adding rational expressions as they are for subtracting rational expressions? Why or why not?

Write expressions for the perimeter and the area of each rectangle.

89.

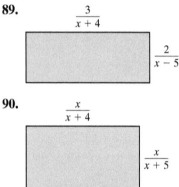

$$\frac{3}{x+4}$$

$$\frac{2}{x-5}$$

90.

$$\frac{x}{x+4}$$

$$\frac{x}{x+5}$$

Perform the indicated operations. Simplify, if possible.

91. $\dfrac{x^2}{3x^2-5x-2} - \dfrac{2x}{3x+1} \cdot \dfrac{1}{x-2}$

92. $\dfrac{2x+11}{x-3} \cdot \dfrac{3}{x+4} + \dfrac{2x+1}{4+x} \cdot \dfrac{3}{3-x}$

93. $\dfrac{2x-16}{x^2-x-2} - \dfrac{x+2}{x^2-5x+6} - \dfrac{x+6}{x^2-2x-3}$

94. $\dfrac{2x+8}{x^2+3x+2} - \dfrac{x-2}{x^2+5x+6} - \dfrac{x+2}{x^2+4x+3}$

Aha! **95.** $\left(\dfrac{x}{x+7} - \dfrac{3}{x+2}\right)\left(\dfrac{x}{x+7} + \dfrac{3}{x+2}\right)$

96. $\dfrac{1}{ay-3a+2xy-6x} - \dfrac{xy+ay}{a^2-4x^2}\left(\dfrac{1}{y-3}\right)^2$

97. $\left(\dfrac{a}{a-b} + \dfrac{b}{a+b}\right)\left(\dfrac{1}{3a+b} + \dfrac{2a+6b}{9a^2-b^2}\right)$

98. $\dfrac{2x^2+5x-3}{2x^2-9x+9} + \dfrac{x+1}{3-2x} + \dfrac{4x^2+8x+3}{x-3} \cdot \dfrac{x+3}{9-4x^2}$

99. Express

$$\frac{a-3b}{a-b}$$

as a sum of two rational expressions with denominators that are opposites of each other. Answers may vary.

100. Use a graphing calculator to check the answer to Exercise 29.

101. Why does the word ERROR appear in the table displayed in the Technology Connection on p. 395?

↳ YOUR TURN ANSWERS: SECTION 6.4

1. $\dfrac{8x+15}{36}$ **2.** $\dfrac{-25x+6}{20x^2}$ **3.** $\dfrac{x(x-1)}{(x-5)(x+5)}$

4. $\dfrac{x-6}{(x-2)(x+2)}$ **5.** $\dfrac{t-3}{2}$ **6.** $\dfrac{x+7}{x-5}$

7. $\dfrac{2(x+2)}{x+1}$

Quick Quiz: Sections 6.1–6.4

Perform the indicated operation and, if possible, simplify.

1. $\dfrac{x^2-64}{10x^2} \cdot \dfrac{5x}{2x+16}$ [6.2]

2. $\dfrac{2x-3}{x-4} - \dfrac{x}{x+2}$ [6.4]

3. $\dfrac{x^2-6}{x-3} - \dfrac{x}{x-3}$ [6.3]

4. $\dfrac{2t}{2t+1} + \dfrac{t}{t-1}$ [6.4]

5. $\dfrac{2x^2-x-1}{x^2-4x+4} \div \dfrac{x-2}{x-1}$ [6.2]

Prepare to Move On

Divide and, if possible, simplify.

1. $\dfrac{\frac{3}{4}}{\frac{5}{6}}$ [1.3]

2. $\dfrac{\frac{8}{15}}{\frac{9}{10}}$ [1.3]

3. $\dfrac{2x+6}{x-1} \div \dfrac{3x+9}{x-1}$ [6.2]

4. $\dfrac{x^2-9}{x^2-4} \div \dfrac{x^2+6x+9}{x^2+4x+4}$ [6.2]

Mid-Chapter Review

The process of adding and subtracting rational expressions is significantly different from multiplying and dividing. The first thing to note when combining rational expressions is the operation sign.

Operation	Need Common Denominator?	Procedure	Tips and Cautions
Addition	Yes	Write equivalent expressions with a common denominator. Add numerators. Keep denominator.	Do not simplify after writing with the LCD. Instead, simplify after adding the numerators.
Subtraction	Yes	Write equivalent expressions with a common denominator. Subtract numerators. Keep denominator.	Use parentheses around the numerator being subtracted. Simplify after subtracting the numerators.
Multiplication	No	Multiply numerators. Multiply denominators.	Do not perform the polynomial multiplication. Instead, factor and try to simplify.
Division	No	Multiply by the reciprocal of the divisor.	Begin by rewriting as a multiplication using the reciprocal of the divisor.

GUIDED SOLUTIONS

1. Divide: $\dfrac{a^2}{a-10} \div \dfrac{a^2+5a}{a^2-100}$. Simplify, if possible. [6.2]

Solution

$$\dfrac{a^2}{a-10} \div \dfrac{a^2+5a}{a^2-100}$$

$$= \dfrac{a^2}{a-10} \cdot \dfrac{\square}{\square} \qquad \text{Multiplying by the reciprocal of the divisor}$$

$$= \dfrac{a \cdot a \cdot (a+10) \cdot \square}{(a-10) \cdot a \cdot \square} \qquad \text{Multiplying and factoring}$$

$$= \dfrac{\square}{\square} \cdot \dfrac{a(a+10)}{a+5} \qquad \text{Factoring out a factor equal to 1}$$

$$= \dfrac{a(a+10)}{a+5} \qquad \text{Simplifying}$$

2. Add: $\dfrac{2}{x} + \dfrac{1}{x^2+x}$. Simplify, if possible. [6.4]

Solution

$$\dfrac{2}{x} + \dfrac{1}{x^2+x}$$

$$= \dfrac{2}{x} + \dfrac{1}{x\,\square} \qquad \text{Factoring denominators. The LCD is } x(x+1).$$

$$= \dfrac{2}{x} \cdot \dfrac{\square}{\square} + \dfrac{1}{x(x+1)} \qquad \text{Multiplying by 1 to get the LCD in the first denominator}$$

$$= \dfrac{\square}{x(x+1)} + \dfrac{1}{x(x+1)} \qquad \text{Multiplying}$$

$$= \dfrac{\square}{x(x+1)} \qquad \text{Adding numerators. We cannot simplify.}$$

MIXED REVIEW

Perform the indicated operation and, if possible, simplify.

3. $\dfrac{3}{5x} + \dfrac{2}{x^2}$ [6.4]

4. $\dfrac{3}{5x} \cdot \dfrac{2}{x^2}$ [6.2]

5. $\dfrac{3}{5x} \div \dfrac{2}{x^2}$ [6.2]

6. $\dfrac{3}{5x} - \dfrac{2}{x^2}$ [6.4]

7. $\dfrac{2x - 6}{5x + 10} \cdot \dfrac{x + 2}{6x - 12}$ [6.2]

8. $\dfrac{2}{x - 5} \div \dfrac{6}{x - 5}$ [6.2]

9. $\dfrac{x}{x + 2} - \dfrac{1}{x - 1}$ [6.4]

10. $\dfrac{2}{x + 3} + \dfrac{3}{x + 4}$ [6.4]

11. $\dfrac{5}{2x - 1} + \dfrac{10x}{1 - 2x}$ [6.4]

12. $\dfrac{3}{x - 4} - \dfrac{2}{4 - x}$ [6.4]

13. $\dfrac{(x - 2)(2x + 3)}{(x + 1)(x - 5)} \div \dfrac{(x - 2)(x + 1)}{(x - 5)(x + 3)}$ [6.2]

14. $\dfrac{a}{6a - 9b} - \dfrac{b}{4a - 6b}$ [6.4]

15. $\dfrac{x^2 - 16}{x^2 - x} \cdot \dfrac{x^2}{x^2 - 5x + 4}$ [6.2]

16. $\dfrac{x + 1}{x^2 - 7x + 10} + \dfrac{3}{x^2 - x - 2}$ [6.4]

17. $\dfrac{3u^2 - 3}{4} \div \dfrac{4u + 4}{3}$ [6.2]

18. $(t^2 + t - 20) \cdot \dfrac{t + 5}{t - 4}$ [6.2]

19. $\dfrac{a^2 - 2a + 1}{a^2 - 4} \div (a^2 - 3a + 2)$ [6.2]

20. $\dfrac{2x - 7}{x} - \dfrac{3x - 5}{2}$ [6.4]

6.5   Complex Rational Expressions

A. Using Division to Simplify   **B.** Multiplying by the LCD

A **complex rational expression** is a rational expression that has one or more rational expressions within its numerator or denominator. Here are some examples:

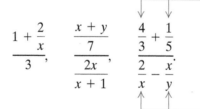

$$\dfrac{1 + \dfrac{2}{x}}{3}, \qquad \dfrac{\dfrac{x + y}{7}}{\dfrac{2x}{x + 1}}, \qquad \dfrac{\dfrac{4}{3} + \dfrac{1}{5}}{\dfrac{2}{x} - \dfrac{x}{y}}.$$

These are rational expressions within the complex rational expression.

To simplify a complex rational expression is to write an equivalent expression that is no longer complex. We will consider two methods for simplifying complex rational expressions. We assume that all denominators are nonzero.

A. Using Division to Simplify (Method 1)

Our first method for simplifying complex rational expressions involves rewriting the expression as a quotient of two rational expressions.

> **TO SIMPLIFY A COMPLEX RATIONAL EXPRESSION BY DIVIDING**
>
> 1. Add or subtract, as needed, to get a single rational expression in the numerator.
> 2. Add or subtract, as needed, to get a single rational expression in the denominator.
> 3. Divide the numerator by the denominator (invert the divisor and multiply).
> 4. If possible, simplify by removing a factor equal to 1.

The key here is to express a complex rational expression as one rational expression divided by another.

EXAMPLE 1 Simplify: $\dfrac{\dfrac{x}{x-3}}{\dfrac{4}{5x-15}}$.

SOLUTION Here the numerator and the denominator are already single rational expressions. This allows us to start by dividing (step 3):

$$\dfrac{\dfrac{x}{x-3}}{\dfrac{4}{5x-15}} = \dfrac{x}{x-3} \div \dfrac{4}{5x-15} \qquad \text{Rewriting with a division symbol}$$

$$= \dfrac{x}{x-3} \cdot \dfrac{5x-15}{4} \qquad \begin{array}{l}\text{Multiplying by the reciprocal of the}\\\text{divisor (inverting and multiplying)}\end{array}$$

$$= \dfrac{x}{x-3} \cdot \dfrac{5(x-3)}{4} \qquad \begin{array}{l}\text{Factoring and removing a factor}\\\text{equal to 1: } \dfrac{x-3}{x-3} = 1\end{array}$$

$$= \dfrac{5x}{4}.$$

1. Simplify: $\dfrac{\dfrac{1}{x^2+x}}{\dfrac{2}{5x+5}}$.

YOUR TURN

EXAMPLE 2 Simplify.

a) $\dfrac{\dfrac{5}{2a}+\dfrac{1}{a}}{\dfrac{1}{4a}-\dfrac{5}{6}}$

b) $\dfrac{\dfrac{x^2}{y}-\dfrac{5}{x}}{xz}$

SOLUTION

a) $\dfrac{\dfrac{5}{2a}+\dfrac{1}{a}}{\dfrac{1}{4a}-\dfrac{5}{6}} = \dfrac{\dfrac{5}{2a}+\dfrac{1}{a}\cdot\dfrac{2}{2}}{\dfrac{1}{4a}\cdot\dfrac{3}{3}-\dfrac{5}{6}\cdot\dfrac{2a}{2a}}$

⟵ Multiplying by 1 to get the LCD, $2a$, for the numerator of the complex rational expression

⟵ Multiplying by 1 to get the LCD, $12a$, for the denominator of the complex rational expression

1. Form a single rational expression in the numerator.

2. Form a single rational expression in the denominator.

$$= \dfrac{\dfrac{5}{2a}+\dfrac{2}{2a}}{\dfrac{3}{12a}-\dfrac{10a}{12a}} = \dfrac{\dfrac{7}{2a}}{\dfrac{3-10a}{12a}} \qquad \begin{array}{l}\text{⟵ Adding}\\[4pt]\text{⟵ Subtracting}\end{array}$$

3. Divide the numerator by the denominator.

$$= \dfrac{7}{2a} \div \dfrac{3-10a}{12a} \qquad \begin{array}{l}\text{Rewriting with a division symbol. This is often}\\\text{done mentally.}\end{array}$$

$$= \dfrac{7}{2a} \cdot \dfrac{12a}{3-10a} \qquad \begin{array}{l}\text{Multiplying by the reciprocal of the divisor}\\\text{(inverting and multiplying)}\end{array}$$

4. Simplify.

$$= \dfrac{7}{2a} \cdot \dfrac{2a\cdot 6}{3-10a} \qquad \text{Removing a factor equal to 1: } \dfrac{2a}{2a} = 1$$

$$= \dfrac{42}{3-10a}$$

b) $\dfrac{\dfrac{x^2}{y} - \dfrac{5}{x}}{xz} = \dfrac{\dfrac{x^2}{y} \cdot \dfrac{x}{x} - \dfrac{5}{x} \cdot \dfrac{y}{y}}{xz}$ ← Multiplying by 1 to get the LCD, xy, for the numerator of the complex rational expression

$= \dfrac{\dfrac{x^3}{xy} - \dfrac{5y}{xy}}{xz}$

$= \dfrac{\dfrac{x^3 - 5y}{xy}}{xz}$ ← Subtracting

← If you prefer, write xz as $\dfrac{xz}{1}$.

$= \dfrac{x^3 - 5y}{xy} \div (xz)$ Rewriting with a division symbol

$= \dfrac{x^3 - 5y}{xy} \cdot \dfrac{1}{xz}$ Multiplying by the reciprocal of the divisor (inverting and multiplying)

$= \dfrac{x^3 - 5y}{x^2 yz}$

YOUR TURN

2. Simplify: $\dfrac{\dfrac{1}{t} - \dfrac{1}{5}}{\dfrac{t-5}{t}}$.

↳ **Check Your**
UNDERSTANDING

For each question, refer to the following complex rational expression:

$$\dfrac{\dfrac{1}{a} + \dfrac{a}{3}}{\dfrac{a^2}{a+3}}.$$

1. What is the numerator of the complex rational expression?

2. What is the denominator of the complex rational expression?

3. List all the rational expressions within the complex rational expression.

4. List the denominators of all the rational expressions within the complex rational expression.

5. What is the LCD of the three rational expressions within the complex rational expression?

B. Multiplying by the LCD (Method 2)

A second method for simplifying complex rational expressions relies on multiplying by a carefully chosen expression that is equal to 1. This multiplication by 1 will result in an expression that is no longer complex.

> **TO SIMPLIFY A COMPLEX RATIONAL EXPRESSION BY MULTIPLYING BY THE LCD**
>
> **1.** Find the LCD of *all* rational expressions within the complex rational expression.
> **2.** Multiply the complex rational expression by an expression equal to 1. Write 1 as the LCD over itself: LCD/LCD.
> **3.** Simplify. No fraction expressions should remain within the complex rational expression.
> **4.** Factor and, if possible, simplify the remaining expression.

EXAMPLE 3 Simplify: $\dfrac{\dfrac{1}{2} + \dfrac{3}{4}}{\dfrac{5}{6} - \dfrac{3}{8}}$.

SOLUTION

1. The LCD of $\frac{1}{2}, \frac{3}{4}, \frac{5}{6}$, and $\frac{3}{8}$ is 24.

2. We multiply by an expression equal to 1:

$\dfrac{\dfrac{1}{2} + \dfrac{3}{4}}{\dfrac{5}{6} - \dfrac{3}{8}} = \dfrac{\dfrac{1}{2} + \dfrac{3}{4}}{\dfrac{5}{6} - \dfrac{3}{8}} \cdot \dfrac{24}{24}$ Multiplying by an expression equal to 1, using the LCD: $\dfrac{24}{24} = 1$

Student Notes

Pay careful attention to the different ways in which Method 1 and Method 2 use an LCD. In Method 1, one LCD is found in the numerator and one in the denominator. In Method 2, *all* rational expressions are considered in order to find one LCD.

3. Simplify: $\dfrac{\dfrac{5}{2} - \dfrac{4}{3}}{\dfrac{2}{3} + \dfrac{3}{4}}$.

3. Using the distributive law, we perform the multiplication:

$$\frac{\dfrac{1}{2} + \dfrac{3}{4}}{\dfrac{5}{6} - \dfrac{3}{8}} \cdot \frac{24}{24} = \frac{\left(\dfrac{1}{2} + \dfrac{3}{4}\right)24}{\left(\dfrac{5}{6} - \dfrac{3}{8}\right)24}$$

← Multiplying the numerator by 24
Don't forget the parentheses!
← Multiplying the denominator by 24

$$= \frac{\dfrac{1}{2}(24) + \dfrac{3}{4}(24)}{\dfrac{5}{6}(24) - \dfrac{3}{8}(24)} \qquad \text{Using the distributive law}$$

$$= \frac{\dfrac{24}{2} + \dfrac{3(24)}{4}}{\dfrac{5(24)}{6} - \dfrac{3(24)}{8}}$$

$$= \frac{12 + 18}{20 - 9}, \text{ or } \frac{30}{11}. \qquad \text{Simplifying}$$

4. The result, $\frac{30}{11}$, cannot be factored or simplified, so we are done.

YOUR TURN

Multiplying like this effectively clears fractions in both the numerator and the denominator of the complex rational expression.

EXAMPLE 4 Simplify.

a) $\dfrac{\dfrac{5}{2a} + \dfrac{1}{a}}{\dfrac{1}{4a} - \dfrac{5}{6}}$

b) $\dfrac{1 - \dfrac{1}{x}}{1 - \dfrac{1}{x^2}}$

SOLUTION

1. Find the LCD.

2. Multiply by LCD/LCD.

3., 4. Simplify.

a) The denominators within the complex expression are $2a$, a, $4a$, and 6, so the LCD is $12a$. We multiply by 1 using $(12a)/(12a)$:

$$\frac{\dfrac{5}{2a} + \dfrac{1}{a}}{\dfrac{1}{4a} - \dfrac{5}{6}} = \frac{\dfrac{5}{2a} + \dfrac{1}{a}}{\dfrac{1}{4a} - \dfrac{5}{6}} \cdot \frac{12a}{12a} = \frac{\dfrac{5}{2a}(12a) + \dfrac{1}{a}(12a)}{\dfrac{1}{4a}(12a) - \dfrac{5}{6}(12a)}$$

Using the distributive law

$$= \frac{\dfrac{5(12a)}{2a} + \dfrac{12a}{a}}{\dfrac{12a}{4a} - \dfrac{5(12a)}{6}}$$

$$= \frac{\dfrac{5 \cdot 2 \cdot 6 \cdot a}{2 \cdot a} + \dfrac{12 \cdot a}{a}}{\dfrac{4 \cdot 3 \cdot a}{4 \cdot a} - \dfrac{5 \cdot 2 \cdot 6 \cdot a}{6}}$$

All fractions have been cleared.

$$= \frac{30 + 12}{3 - 10a} = \frac{42}{3 - 10a}.$$

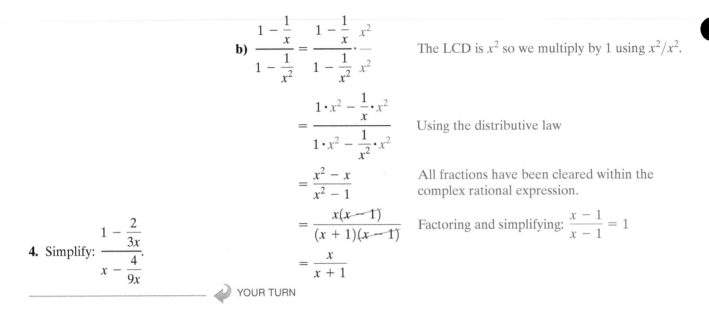

b) $\dfrac{1 - \dfrac{1}{x}}{1 - \dfrac{1}{x^2}} = \dfrac{1 - \dfrac{1}{x}}{1 - \dfrac{1}{x^2}} \cdot \dfrac{x^2}{x^2}$ The LCD is x^2 so we multiply by 1 using x^2/x^2.

$= \dfrac{1 \cdot x^2 - \dfrac{1}{x} \cdot x^2}{1 \cdot x^2 - \dfrac{1}{x^2} \cdot x^2}$ Using the distributive law

$= \dfrac{x^2 - x}{x^2 - 1}$ All fractions have been cleared within the complex rational expression.

$= \dfrac{x(x - 1)}{(x + 1)(x - 1)}$ Factoring and simplifying: $\dfrac{x - 1}{x - 1} = 1$

$= \dfrac{x}{x + 1}$

4. Simplify: $\dfrac{1 - \dfrac{2}{3x}}{x - \dfrac{4}{9x}}$.

 YOUR TURN

It is important to understand both of the methods studied in this section. Sometimes, as in Example 1, the complex rational expression is either given as—or easily written as—a quotient of two rational expressions. In these cases, Method 1 (using division) is probably the easier method to use. Other times, as in Example 4, it is not difficult to find the LCD of all denominators in the complex rational expression. When this occurs, it is usually easier to use Method 2 (multiplying by the LCD). The more practice you get using both methods, the better you will be at selecting the easier method for any given problem.

6.5 EXERCISE SET

FOR EXTRA HELP MyMathLab®

◆ Vocabulary and Reading Check

Consider the expression

$$\dfrac{\dfrac{2}{3} + \dfrac{1}{x}}{\dfrac{5}{x}}.$$

Choose from the following list the word (or words) that best completes each statement, with reference to the above expression. Not all words in the list will be used.

complex
denominator
least common denominator
numerator
opposite
reciprocal

1. The expression given above is a(n) _____ rational expression.

2. The expression $\dfrac{5}{x}$ is the _____ of the above expression.

3. The _____ of the rational expressions within the expression above is $3x$.

4. To simplify, we can multiply the numerator by the _____ of $\dfrac{5}{x}$.

◆ Concept Reinforcement

*Each of Exercises 5–8 shows a complex rational expression and the first step taken to simplify that expression. Indicate for each which method is being used: **(a)** using division to simplify (Method 1) or **(b)** multiplying by the LCD (Method 2).*

5. _____ $\dfrac{\dfrac{1}{x} + \dfrac{1}{2}}{\dfrac{1}{3} - \dfrac{1}{x}} = \dfrac{\dfrac{1}{x} + \dfrac{1}{2}}{\dfrac{1}{3} - \dfrac{1}{x}} \cdot \dfrac{6x}{6x}$

6. _____ $\dfrac{\dfrac{1}{x} + \dfrac{1}{2}}{\dfrac{1}{3} - \dfrac{1}{x}} = \dfrac{\dfrac{1}{x} \cdot \dfrac{2}{2} + \dfrac{1}{2} \cdot \dfrac{x}{x}}{\dfrac{1}{3} \cdot \dfrac{x}{x} - \dfrac{1}{x} \cdot \dfrac{3}{3}}$

7. ___ $\dfrac{\dfrac{x-1}{x}}{\dfrac{x^2}{x^2-1}} = \dfrac{x-1}{x} \div \dfrac{x^2}{x^2-1}$

8. ___ $\dfrac{\dfrac{x-1}{x}}{\dfrac{x^2}{x^2-1}} = \dfrac{\dfrac{x-1}{x}}{\dfrac{x^2}{x^2-1}} \cdot \dfrac{x(x+1)(x-1)}{x(x+1)(x-1)}$

A, B. Complex Rational Expressions

Simplify. Use either method or the method specified by your instructor.

9. $\dfrac{\dfrac{1}{2}+\dfrac{1}{3}}{\dfrac{1}{4}-\dfrac{1}{6}}$

10. $\dfrac{\dfrac{2}{5}-\dfrac{1}{10}}{\dfrac{7}{20}-\dfrac{4}{15}}$

11. $\dfrac{1+\dfrac{1}{4}}{2+\dfrac{3}{4}}$

12. $\dfrac{3+\dfrac{1}{4}}{1+\dfrac{1}{2}}$

13. $\dfrac{\dfrac{x}{4}+x}{\dfrac{4}{x}+x}$

14. $\dfrac{\dfrac{1}{c}+2}{\dfrac{1}{c}-5}$

15. $\dfrac{\dfrac{x+2}{x-1}}{\dfrac{x+4}{x-3}}$

16. $\dfrac{\dfrac{x-1}{x+3}}{\dfrac{x-6}{x+2}}$

17. $\dfrac{\dfrac{10}{t}}{\dfrac{2}{t^2}-\dfrac{5}{t}}$

18. $\dfrac{\dfrac{5}{x}-\dfrac{2}{x^2}}{\dfrac{2}{x^2}}$

19. $\dfrac{\dfrac{2a-5}{3a}}{\dfrac{a-7}{6a}}$

20. $\dfrac{\dfrac{a+5}{a^2}}{\dfrac{a-2}{3a}}$

21. $\dfrac{\dfrac{x}{6}-\dfrac{3}{x}}{\dfrac{1}{3}+\dfrac{1}{x}}$

22. $\dfrac{\dfrac{2}{x}+\dfrac{x}{4}}{\dfrac{3}{4}-\dfrac{2}{x}}$

23. $\dfrac{\dfrac{1}{s}-\dfrac{1}{5}}{\dfrac{s-5}{s}}$

24. $\dfrac{\dfrac{1}{9}-\dfrac{1}{n}}{\dfrac{n+9}{9}}$

25. $\dfrac{\dfrac{1}{t^2}+1}{\dfrac{1}{t}-1}$

26. $\dfrac{2+\dfrac{1}{x}}{2-\dfrac{1}{x^2}}$

27. $\dfrac{\dfrac{x^2}{x^2-y^2}}{\dfrac{x}{x+y}}$

28. $\dfrac{\dfrac{a^2-b^2}{ab}}{\dfrac{a-b}{b}}$

29. $\dfrac{\dfrac{7}{c^2}+\dfrac{4}{c}}{\dfrac{6}{c}-\dfrac{3}{c^3}}$

30. $\dfrac{\dfrac{4}{t^3}-\dfrac{1}{t^2}}{\dfrac{3}{t}+\dfrac{5}{t^2}}$

31. $\dfrac{\dfrac{2}{7a^4}-\dfrac{1}{14a}}{\dfrac{3}{5a^2}+\dfrac{2}{15a}}$

32. $\dfrac{\dfrac{5}{4x^3}-\dfrac{3}{8x}}{\dfrac{3}{2x}+\dfrac{3}{4x^3}}$

Aha! 33. $\dfrac{\dfrac{x}{5y^3}+\dfrac{3}{10y}}{\dfrac{3}{10y}+\dfrac{x}{5y^3}}$

34. $\dfrac{\dfrac{a}{6b^3}+\dfrac{4}{9b^2}}{\dfrac{5}{6b}-\dfrac{1}{9b^3}}$

35. $\dfrac{\dfrac{3}{ab^4}+\dfrac{4}{a^3b}}{\dfrac{5}{a^3b}-\dfrac{3}{ab}}$

36. $\dfrac{\dfrac{2}{x^2y}+\dfrac{3}{xy^2}}{\dfrac{3}{xy^2}+\dfrac{2}{x^2y}}$

37. $\dfrac{t-\dfrac{9}{t}}{t+\dfrac{4}{t}}$

38. $\dfrac{s+\dfrac{2}{s}}{s-\dfrac{3}{s}}$

39. $\dfrac{y+y^{-1}}{y-y^{-1}}$

40. $\dfrac{x-x^{-1}}{x+x^{-1}}$

$\left(\text{Hint: } y^{-1}=\dfrac{1}{y}.\right)$

41. $\dfrac{\dfrac{1}{a-h}-\dfrac{1}{a}}{h}$

42. $\dfrac{\dfrac{1}{x+h}-\dfrac{1}{x}}{h}$

43. $\dfrac{\dfrac{x^{-1}+y^{-1}}{x^2-y^2}}{xy}$

44. $\dfrac{\dfrac{a^{-1}+b^{-1}}{a^2-b^2}}{ab}$

45. $\dfrac{t+5+\dfrac{3}{t}}{t+2+\dfrac{1}{t}}$

46. $\dfrac{a+3+\dfrac{2}{a}}{a+2+\dfrac{5}{a}}$

47. $\dfrac{x-2-\dfrac{1}{x}}{x-5-\dfrac{4}{x}}$

48. $\dfrac{x-3-\dfrac{2}{x}}{x-4-\dfrac{3}{x}}$

49. $\dfrac{\dfrac{a^2-4}{a^2+3a+2}}{\dfrac{a^2-5a-6}{a^2-6a-7}}$

50. $\dfrac{\dfrac{x^2-x-12}{x^2-2x-15}}{\dfrac{x^2+8x+12}{x^2-5x-14}}$

51. $\dfrac{\dfrac{x}{x^2+3x-4}-\dfrac{1}{x^2+3x-4}}{\dfrac{x}{x^2+6x+8}+\dfrac{3}{x^2+6x+8}}$

52. $\dfrac{\dfrac{x}{x^2+5x-6}+\dfrac{6}{x^2+5x-6}}{\dfrac{x}{x^2-5x+4}-\dfrac{2}{x^2-5x+4}}$

53. Is it possible to simplify complex rational expressions without knowing how to divide rational expressions? Why or why not?

54. Why is the distributive law important when simplifying complex rational expressions?

Skill Review

Factor.

55. $6x^3 - 9x^2 - 4x + 6$ [5.1]

56. $12a^2b + 4ab^2 - 8ab$ [5.1]

57. $30n^3 - 3n^2 - 9n$ [5.3]

58. $25a^2 - 40ab + 16b^2$ [5.4]

59. $n^4 - 1$ [5.4]

60. $p^2w^2 - 2pw - 120$ [5.2]

Synthesis

61. Which of the two methods presented would you use to simplify Exercise 36? Why?

62. Which of the two methods presented would you use to simplify Exercise 19? Why?

In each of Exercises 63–66, find all x-values for which the given expression is undefined.

63. $\dfrac{\dfrac{x-5}{x-6}}{\dfrac{x-7}{x-8}}$

64. $\dfrac{\dfrac{x+1}{x+2}}{\dfrac{x+3}{x+4}}$

65. $\dfrac{\dfrac{2x+3}{5x+4}}{\dfrac{3}{7} - \dfrac{x^2}{21}}$

66. $\dfrac{\dfrac{3x-5}{2x-7}}{\dfrac{4x}{5} - \dfrac{8}{15}}$

67. Use multiplication by the LCD (Method 2) to show that

$$\frac{A}{B} \div \frac{C}{D} = \frac{A}{B} \cdot \frac{D}{C}.$$

(*Hint:* Begin by forming a complex rational expression.)

68. The formula

$$\frac{P\left(1 + \dfrac{i}{12}\right)^2}{\dfrac{\left(1 + \dfrac{i}{12}\right)^2 - 1}{\dfrac{i}{12}}},$$

where P is a loan amount and i is an interest rate, arises in certain business situations. Simplify this expression. (*Hint:* Expand the binomials.)

Simplify.

69. $\dfrac{\dfrac{x}{x+5} + \dfrac{3}{x+2}}{\dfrac{2}{x+2} - \dfrac{x}{x+5}}$

70. $\dfrac{\dfrac{z}{1 - \dfrac{z}{2+2z}} - 2z}{\dfrac{2z}{5z-2} - 3}$

Aha! **71.** $\left[\dfrac{\dfrac{x-1}{x-1} - 1}{\dfrac{x+1}{x-1} + 1}\right]^5$

72. $1 + \dfrac{1}{1 + \dfrac{1}{1 + \dfrac{1}{x}}}$

73. $\dfrac{1 - \dfrac{25}{x^2}}{1 + \dfrac{2}{x} - \dfrac{15}{x^2}}$

74. $\dfrac{1 - \dfrac{1}{y} + \dfrac{1}{y^2} - \dfrac{1}{y^3}}{1 - \dfrac{1}{y^4}}$

75. Find the simplified form for the reciprocal of

$$\frac{2}{x-1} - \frac{1}{3x-2}.$$

76. Under what circumstance(s) will there be no restrictions on the variable appearing in a complex rational expression?

YOUR TURN ANSWERS: SECTION 6.5

1. $\dfrac{5}{2x}$ **2.** $-\dfrac{1}{5}$ **3.** $\dfrac{14}{17}$ **4.** $\dfrac{3}{3x+2}$

Quick Quiz: Sections 6.1–6.5

Simplify.

1. $\dfrac{8x^2 - 20x - 12}{4x^2 - 14x + 6}$ [6.1]

2. $\dfrac{\dfrac{2}{x} + \dfrac{2}{3}}{\dfrac{1}{x^2}}$ [6.5]

Perform the indicated operation and, if possible, simplify.

3. $\dfrac{a+2}{a+1} \cdot \dfrac{a+3}{a+2}$ [6.2]

4. $\dfrac{4}{x-3} - \dfrac{5x-2}{x+1}$ [6.4] **5.** $\dfrac{x+2}{2x} + \dfrac{x+1}{3x}$ [6.4]

Prepare to Move On

Solve.

1. $(x-1)7 - (x+1)9 = 4(x+2)$ [2.2]

2. $\dfrac{5}{9} - \dfrac{2x}{3} = \dfrac{5x}{6} + \dfrac{4}{3}$ [2.2] **3.** $x^2 - 7x + 12 = 0$ [5.6]

4. $x^2 + 13x - 30 = 0$ [5.6]

6.6 Rational Equations

A. Solving Rational Equations

Our study of rational expressions allows us to solve a type of equation that we could not have solved prior to this chapter.

A. Solving Rational Equations

A **rational equation** is an equation containing one or more rational expressions, often with the variable in a denominator. Here are some examples:

$$\frac{2}{3} + \frac{5}{6} = \frac{x}{9}, \qquad t + \frac{7}{t} = -5, \qquad \frac{x^2}{x-1} = \frac{1}{x-1}.$$

TO SOLVE A RATIONAL EQUATION

1. List any restrictions that exist. Numbers that make a denominator equal 0 can never be solutions.
2. Clear the equation of fractions by multiplying both sides by the LCM of the denominators.
3. Solve the resulting equation using the addition principle, the multiplication principle, and the principle of zero products, as needed.
4. Check the possible solution(s) in the original equation.

When clearing an equation of fractions, we use the terminology LCM instead of LCD because the LCM is not used as a denominator in this setting.

EXAMPLE 1 Solve: $\dfrac{x}{6} - \dfrac{x}{8} = \dfrac{1}{12}$.

SOLUTION Because no variable appears in a denominator, no restrictions exist. The LCM of 6, 8, and 12 is 24, so we multiply both sides by 24:

$$24\left(\frac{x}{6} - \frac{x}{8}\right) = 24 \cdot \frac{1}{12} \qquad \begin{array}{l}\text{Using the multiplication principle} \\ \text{to multiply both sides by the LCM.} \\ \text{Parentheses are important!}\end{array}$$

$$24 \cdot \frac{x}{6} - 24 \cdot \frac{x}{8} = 24 \cdot \frac{1}{12} \qquad \text{Using the distributive law}$$

Be sure to multiply *each* term by the LCM.

$$\left.\begin{array}{c}\dfrac{24x}{6} - \dfrac{24x}{8} = \dfrac{24}{12} \\ 4x - 3x = 2\end{array}\right\} \qquad \begin{array}{l}\text{Simplifying. Note that all fractions have} \\ \text{been cleared. If fractions remain, we have} \\ \text{either made a mistake or have not used the} \\ \text{LCM of the denominators.}\end{array}$$

$$x = 2.$$

Study Skills

Does More Than One Solution Exist?

Keep in mind that many problems—in math and elsewhere—have more than one solution. When asked to solve an equation, we are expected to find any and all solutions of the equation.

Check:
$$\frac{x}{6} - \frac{x}{8} = \frac{1}{12}$$

$$\frac{\frac{2}{6} - \frac{2}{8}}{\frac{1}{3} - \frac{1}{4}} \mid \frac{1}{12}$$

$$\frac{4}{12} - \frac{3}{12}$$

$$\frac{1}{12} \stackrel{?}{=} \frac{1}{12} \quad \text{TRUE}$$

This checks, so the solution is 2.

1. Solve: $\dfrac{t}{3} + \dfrac{t}{5} = 1$.

YOUR TURN

Recall that the multiplication principle states that $a = b$ is equivalent to $a \cdot c = b \cdot c$, *provided c is not 0*. To clear fractions in rational equations, we often must multiply by a variable expression. For example, to clear the fraction in $1/x = 5$, we multiply both sides of the equation by x. Since x *could* represent 0, the new equation may not be equivalent to the original equation. For this reason, **checking in the original equation is essential**.

EXAMPLE 2   Solve.

a) $\dfrac{2}{3x} + \dfrac{1}{x} = 10$ **b)** $x + \dfrac{6}{x} = -5$

SOLUTION

List restrictions.

a) Note that in $\dfrac{2}{3x} + \dfrac{1}{x} = 10$, if $x = 0$, both denominators are 0. We list this restriction:

$$x \neq 0.$$

We now clear the equation of fractions and solve, using the LCM of the denominators, $3x$.

Clear fractions.

$$3x\left(\frac{2}{3x} + \frac{1}{x}\right) = 3x \cdot 10 \qquad \begin{array}{l}\text{Using the multiplication principle} \\ \text{to multiply both sides by the LCM.} \\ \textit{Don't forget the parentheses!}\end{array}$$

$$3x \cdot \frac{2}{3x} + 3x \cdot \frac{1}{x} = 3x \cdot 10 \qquad \text{Using the distributive law}$$

$$2 + 3 = 30x \qquad \begin{array}{l}\text{Removing factors equal to 1:} \\ (3x)/(3x) = 1 \text{ and } x/x = 1. \text{ This clears} \\ \text{all fractions.}\end{array}$$

Solve.

$$5 = 30x$$

$$\frac{5}{30} = x, \quad \text{so } x = \frac{1}{6}. \qquad \begin{array}{l}\text{Dividing both sides by 30, or} \\ \text{multiplying both sides by } 1/30\end{array}$$

Since $\frac{1}{6} \neq 0$, and 0 is the only restricted value, $\frac{1}{6}$ *should* check.

Check.   *Check:*
$$\frac{2}{3x} + \frac{1}{x} = 10$$

$$\frac{\dfrac{2}{3 \cdot \frac{1}{6}} + \dfrac{1}{\frac{1}{6}}}{} \mid 10$$

$$\frac{2}{\frac{1}{2}} + \frac{1}{\frac{1}{6}}$$

$$2 \cdot \frac{2}{1} + 1 \cdot \frac{6}{1}$$

$$4 + 6$$

$$10 \stackrel{?}{=} 10 \quad \text{TRUE}$$

The solution is $\frac{1}{6}$.

b) To solve $x + \dfrac{6}{x} = -5$, note that if $x = 0$, the expression $\dfrac{6}{x}$ is undefined. We list the restriction:

$$x \neq 0.$$

We now clear the equation of fractions and solve:

$$x\left(x + \frac{6}{x}\right) = x(-5) \qquad \text{Multiplying both sides by the LCM, } x. \textit{Don't forget the parentheses!}$$

$$x \cdot x + \cancel{x} \cdot \frac{6}{\cancel{x}} = -5x \qquad \text{Using the distributive law}$$

$$x^2 + 6 = -5x \qquad \text{Removing a factor equal to 1: } x/x = 1. \text{ We are left with a quadratic equation.}$$

$$x^2 + 5x + 6 = 0 \qquad \text{Using the addition principle to add } 5x \text{ to both sides}$$

$$(x + 3)(x + 2) = 0 \qquad \text{Factoring}$$

$$x + 3 = 0 \quad or \quad x + 2 = 0 \qquad \text{Using the principle of zero products}$$

$$x = -3 \quad or \quad x = -2 \qquad \text{The only restricted value is 0, so both answers should check.}$$

Check: For -3:   For -2:

$$
\begin{array}{c|c}
x + \dfrac{6}{x} = -5 & \\
\hline
-3 + \dfrac{6}{-3} & -5 \\
-3 - 2 & \\
& -5 \overset{?}{=} -5 \quad \text{TRUE}
\end{array}
$$

$$
\begin{array}{c|c}
x + \dfrac{6}{x} = -5 & \\
\hline
-2 + \dfrac{6}{-2} & -5 \\
-2 - 3 & \\
& -5 \overset{?}{=} -5 \quad \text{TRUE}
\end{array}
$$

Both of these check, so there are two solutions, -3 and -2.

YOUR TURN

EXAMPLE 3   Solve.

a) $1 + \dfrac{3x}{x + 2} = \dfrac{-6}{x + 2}$ **b)** $\dfrac{3}{x - 5} + \dfrac{1}{x + 5} = \dfrac{2}{x^2 - 25}$

c) $\dfrac{x^2}{x - 1} = \dfrac{1}{x - 1}$

SOLUTION

a) The only denominator in $1 + \dfrac{3x}{x + 2} = \dfrac{-6}{x + 2}$ is $x + 2$. To find restrictions, we set this equal to 0 and solve:

$$x + 2 = 0$$
$$x = -2.$$

If $x = -2$, the rational expressions are undefined. We list the restriction:

$$x \neq -2.$$

Student Notes

Not all checking is performed to find errors in computation. For these equations, the solution process itself can introduce numbers that do not check.

2. Solve: $\dfrac{2}{3} - \dfrac{1}{n} = \dfrac{7}{3n}$.

↳ Check Your UNDERSTANDING

For each equation, determine the simplest expression that can be used to multiply both sides of the equation in order to clear fractions. (Do not solve.)

1. $6 + \dfrac{1}{x} = x$

2. $\dfrac{x}{3} + \dfrac{1}{4x} = \dfrac{1}{6}$

3. $\dfrac{3}{y} - \dfrac{1}{y + 5} = \dfrac{y}{y^2 + 5y}$

Technology Connection

We can use a table to check possible solutions of rational equations. Consider the equation in Example 3(c) and the possible solutions that were found, 1 and −1. To check these solutions, we enter

$$y_1 = \frac{x^2}{x - 1} \text{ and } y_2 = \frac{1}{x - 1}.$$

After setting Indpnt to Ask and Depend to Auto in the TBLSET menu, we display the table and enter $x = 1$. The ERROR messages indicate that 1 is not a solution because it is not an allowable replacement for x in the equation. Next, we enter $x = -1$. Since y_1 and y_2 have the same value, we know that the equation is true when $x = -1$, and thus −1 is a solution.

X	Y₁	Y₂
1	ERROR	ERROR
−1	−.5	−.5

X =

Use a graphing calculator to check the possible solutions of Examples 3(a) and 3(b).

We clear fractions using the LCM $x + 2$ and solve.

$$(x + 2)\left(1 + \frac{3x}{x + 2}\right) = (x + 2)\frac{-6}{x + 2}$$

Multiplying both sides by the LCM. *Don't forget the parentheses!*

$$(x + 2) \cdot 1 + (x + 2)\frac{3x}{x + 2} = (x + 2)\frac{-6}{x + 2}$$

Using the distributive law; removing a factor equal to 1: $(x + 2)/(x + 2) = 1$

$$x + 2 + 3x = -6$$
$$4x + 2 = -6$$
$$4x = -8$$
$$x = -2.$$

Above, we stated that $x \neq -2$.

Because of the above restriction, −2 must be rejected as a solution. The check below simply confirms this.

Check:

$$1 + \frac{3x}{x + 2} = \frac{-6}{x + 2}$$

$$\begin{array}{c|c} 1 + \dfrac{3(-2)}{-2 + 2} & \dfrac{-6}{-2 + 2} \\ 1 + \dfrac{-6}{0} \overset{?}{=} \dfrac{-6}{0} \end{array}$$ FALSE

The equation has no solution.

b) The denominators in $\dfrac{3}{x - 5} + \dfrac{1}{x + 5} = \dfrac{2}{x^2 - 25}$ are $x - 5, x + 5$, and $x^2 - 25$. Setting them equal to 0 and solving, we find that the rational expressions are undefined when $x = 5$ or $x = -5$. We list the restrictions:

$$x \neq 5, \qquad x \neq -5.$$

We clear fractions using the LCM, $(x - 5)(x + 5)$, and solve.

$$(x - 5)(x + 5)\left(\frac{3}{x - 5} + \frac{1}{x + 5}\right) = (x - 5)(x + 5)\frac{2}{(x - 5)(x + 5)}$$

$$\frac{(x - 5)(x + 5)3}{x - 5} + \frac{(x - 5)(x + 5)}{x + 5} = \frac{2(x - 5)(x + 5)}{(x - 5)(x + 5)}$$

Using the distributive law

$$(x + 5)3 + (x - 5) = 2$$

Removing factors equal to 1: $(x - 5)/(x - 5) = 1$, $(x + 5)/(x + 5) = 1$, and $\dfrac{(x - 5)(x + 5)}{(x - 5)(x + 5)} = 1$

$$3x + 15 + x - 5 = 2$$ Using the distributive law
$$4x + 10 = 2$$
$$4x = -8$$
$$x = -2.$$ −2 is not restricted, so it *should* check.

The student can check to confirm that −2 is the solution.

Student Notes

When solving an equation like
$\dfrac{x^2}{x-1} = \dfrac{1}{x-1}$ in Example 3(c),
we can simply equate numerators
since the denominators are equal.
We must still note restrictions first.

3. Solve:
$$\frac{6}{x^2 - 4} = \frac{x+1}{x^2 - 2x}.$$

c) To solve $\dfrac{x^2}{x-1} = \dfrac{1}{x-1}$, note that if $x = 1$, the denominators are 0. We list the restriction:

$$x \neq 1.$$

We clear fractions using the LCM, $x - 1$, and solve.

$(x-1) \cdot \dfrac{x^2}{x-1} = (x-1) \cdot \dfrac{1}{x-1}$	Multiplying both sides by $x - 1$, the LCM
$x^2 = 1$	Removing a factor equal to 1: $(x-1)/(x-1) = 1$
$x^2 - 1 = 0$	Subtracting 1 from both sides
$(x-1)(x+1) = 0$	Factoring
$x - 1 = 0 \quad or \quad x + 1 = 0$	Using the principle of zero products
$x = 1 \quad or \quad x = -1$	Above, we stated that $x \neq 1$.

Because of the above restriction, 1 must be rejected as a solution. The student should check in the original equation that -1 *does* check. The solution is -1.

YOUR TURN

CONNECTING 🔗 THE CONCEPTS

An equation contains an equals sign; an expression does not. Be careful not to confuse simplifying an expression with solving an equation. When expressions are simplified, the result is an equivalent expression. When equations are solved, the result is a solution. Compare the following.

Simplify: $\dfrac{x-1}{6x} + \dfrac{4}{9}$.

> The equals signs indicate that all the expressions are equivalent.

SOLUTION

$$\frac{x-1}{6x} + \frac{4}{9} = \frac{x-1}{6x} \cdot \frac{3}{3} + \frac{4}{9} \cdot \frac{2x}{2x}$$

$$= \frac{3x - 3}{18x} + \frac{8x}{18x} \qquad \text{Writing with the LCD, } 18x$$

$$= \frac{11x - 3}{18x} \qquad \text{The result is an expression.}$$

The expressions

$$\frac{11x - 3}{18x} \quad \text{and} \quad \frac{x-1}{6x} + \frac{4}{9}$$

are equivalent.

Solve: $\dfrac{x-1}{6x} = \dfrac{4}{9}$.

SOLUTION

$$\frac{x-1}{6x} = \frac{4}{9} \qquad \boxed{\text{Each line is an equivalent equation.}}$$

$$18x \cdot \frac{x-1}{6x} = 18x \cdot \frac{4}{9} \qquad \text{Multiplying by the LCM, } 18x$$

$$3 \cdot 6x \cdot \frac{x-1}{6x} = 2 \cdot 9 \cdot x \cdot \frac{4}{9}$$

$$3(x-1) = 2x \cdot 4$$

$$3x - 3 = 8x$$

$$-3 = 5x$$

$$-\frac{3}{5} = x \qquad \text{The result is a solution.}$$

The solution is $-\dfrac{3}{5}$.

(continued)

EXERCISES

Tell whether each exercise contains either an expression or an equation. Then simplify the expression or solve the equation, as indicated.

1. Add and, if possible, simplify: $\dfrac{2}{5n} + \dfrac{3}{2n-1}$.

2. Solve: $\dfrac{3}{y} - \dfrac{1}{4} = \dfrac{1}{y}$.

3. Solve: $\dfrac{5}{x+3} = \dfrac{3}{x+2}$.

4. Multiply and, if possible, simplify:
$$\dfrac{8t+8}{2t^2+t-1} \cdot \dfrac{t^2-1}{t^2-2t+1}.$$

5. Subtract and, if possible, simplify:
$$\dfrac{2a}{a+1} - \dfrac{4a}{1-a^2}.$$

6. Solve: $\dfrac{20}{x} = \dfrac{x}{5}$.

6.6 EXERCISE SET

FOR EXTRA HELP MyMathLab®

⤵ Vocabulary and Reading Check

Classify each of the following statements as either true or false.

1. Every rational equation has at least one solution.

2. It is possible for a rational equation to have more than one solution.

3. When both sides of an equation are multiplied by a variable expression, the result is not always an equivalent equation.

4. All the equation-solving principles studied thus far may be needed when solving a rational equation.

A. Solving Rational Equations

Solve. If no solution exists, state this.

5. $\dfrac{3}{5} - \dfrac{2}{3} = \dfrac{x}{6}$

6. $\dfrac{5}{8} - \dfrac{3}{5} = \dfrac{x}{10}$

7. $\dfrac{1}{8} + \dfrac{1}{12} = \dfrac{1}{t}$

8. $\dfrac{1}{6} + \dfrac{1}{10} = \dfrac{1}{t}$

9. $\dfrac{x}{6} - \dfrac{6}{x} = 0$

10. $\dfrac{x}{7} - \dfrac{7}{x} = 0$

11. $\dfrac{2}{x} = \dfrac{5}{x} - \dfrac{1}{4}$

12. $\dfrac{3}{t} = \dfrac{4}{t} - \dfrac{1}{5}$

13. $\dfrac{5}{3t} + \dfrac{3}{t} = 1$

14. $\dfrac{3}{4x} + \dfrac{5}{x} = 1$

15. $\dfrac{12}{x} = \dfrac{x}{3}$

16. $\dfrac{x}{2} = \dfrac{18}{x}$

17. $y + \dfrac{4}{y} = -5$

18. $n + \dfrac{3}{n} = -4$

19. $\dfrac{n+2}{n-6} = \dfrac{1}{2}$

20. $\dfrac{a-4}{a+6} = \dfrac{1}{3}$

21. $x + \dfrac{12}{x} = -7$

22. $x + \dfrac{8}{x} = -9$

23. $\dfrac{3}{x-4} = \dfrac{5}{x+1}$

24. $\dfrac{1}{x+3} = \dfrac{4}{x-1}$

25. $\dfrac{a}{6} - \dfrac{a}{10} = \dfrac{1}{6}$

26. $\dfrac{t}{8} - \dfrac{t}{12} = \dfrac{1}{8}$

27. $\dfrac{x+1}{3} - 1 = \dfrac{x-1}{2}$

28. $\dfrac{x+2}{5} - 1 = \dfrac{x-2}{4}$

29. $\dfrac{y+3}{y-3} = \dfrac{6}{y-3}$

30. $\dfrac{3}{a+7} = \dfrac{a+10}{a+7}$

31. $\dfrac{3}{x+4} = \dfrac{5}{x}$

32. $\dfrac{2}{x+3} = \dfrac{7}{x}$

33. $\dfrac{n+1}{n+2} = \dfrac{n-3}{n+1}$

34. $\dfrac{n+2}{n-3} = \dfrac{n+1}{n-2}$

35. $\dfrac{5}{t-2} + \dfrac{3t}{t-2} = \dfrac{4}{t^2-4t+4}$

36. $\dfrac{4}{t-3} + \dfrac{2t}{t-3} = \dfrac{12}{t^2-6t+9}$

37. $\dfrac{x}{x+5} - \dfrac{5}{x-5} = \dfrac{14}{x^2-25}$

38. $\dfrac{5}{x+1} + \dfrac{2x}{x^2-1} = \dfrac{1}{x+1}$

39. $\dfrac{3}{x-3} + \dfrac{5}{x+2} = \dfrac{5x}{x^2-x-6}$

40. $\dfrac{2}{x-2} + \dfrac{1}{x+4} = \dfrac{x}{x^2+2x-8}$

41. $\dfrac{5}{t-3} - \dfrac{30}{t^2-9} = 1$

42. $\dfrac{1}{y + 3} + \dfrac{1}{y - 3} = \dfrac{1}{y^2 - 9}$

43. $\dfrac{7}{6 - a} = \dfrac{a + 1}{a - 6}$

44. $\dfrac{t - 12}{t - 10} = \dfrac{1}{10 - t}$

Aha! **45.** $\dfrac{-2}{x + 2} = \dfrac{x}{x + 2}$

46. $\dfrac{3}{2x - 6} = \dfrac{x}{2x - 6}$

47. $\dfrac{5}{3x + 3} + \dfrac{1}{2x - 2} = \dfrac{1}{x^2 - 1}$

48. $\dfrac{4}{x^2 - 4} + \dfrac{1}{x + 2} + \dfrac{2}{3x - 6} = 0$

49. When solving rational equations, why do we multiply each side by the LCM of the denominators?

50. Explain the difference between adding rational expressions and solving rational equations.

Skill Review

51. Find the x-intercept and the y-intercept of the line given by $6x - y = 18$. [3.3]

52. Find the slope of the line containing the points $(6, 0)$ and $(-3, -1)$. [3.5]

53. Find the slope and the y-intercept of the line given by $2x + y = 5$. [3.6]

54. Determine whether the graphs of the following equations are parallel:

$y = 2 - x,$
$3x + 3y = 7.$ [3.6]

55. Write the slope–intercept equation of the line with slope $\frac{1}{3}$ and y-intercept $(0, -2)$. [3.6]

56. Write a point–slope equation for the line with slope -4 that contains the point $(1, -5)$. [3.7]

Synthesis

57. Describe a method that can be used to create rational equations that have no solution.

58. How can a graph be used to determine how many solutions an equation has?

Solve.

59. $1 + \dfrac{x - 1}{x - 3} = \dfrac{2}{x - 3} - x$

60. $\dfrac{4}{y - 2} + \dfrac{3}{y^2 - 4} = \dfrac{5}{y + 2} + \dfrac{2y}{y^2 - 4}$

61. $\dfrac{12 - 6x}{x^2 - 4} = \dfrac{3x}{x + 2} - \dfrac{3 - 2x}{2 - x}$

62. $\dfrac{x}{x^2 + 3x - 4} + \dfrac{x + 1}{x^2 + 6x + 8} = \dfrac{2x}{x^2 + x - 2}$

63. $7 - \dfrac{a - 2}{a + 3} = \dfrac{a^2 - 4}{a + 3} + 5$

64. $\dfrac{x^2}{x^2 - 4} = \dfrac{x}{x + 2} - \dfrac{2x}{2 - x}$

65. $\dfrac{1}{x - 1} + x - 5 = \dfrac{5x - 4}{x - 1} - 6$

66. $\dfrac{5 - 3a}{a^2 + 4a + 3} - \dfrac{2a + 2}{a + 3} = \dfrac{3 - a}{a + 1}$

67. $\dfrac{\frac{1}{x} + 1}{x} = \dfrac{\frac{1}{x}}{2}$

68. $\dfrac{\frac{1}{3}}{x} = \dfrac{1 - \frac{1}{x}}{x}$

69. Use a graphing calculator to check your answers to Exercises 13, 21, 31, and 59.

70. The reciprocal of a number is the number itself. What is the number?

Quick Quiz: Sections 6.1–6.6

1. Multiply and, if possible, simplify:

$$\dfrac{x^2 - 12x + 11}{x + 7} \cdot \dfrac{7}{x - 1}. \quad [6.2]$$

2. Find the LCM: $x^3 - 4x, x^2 - x - 2$. [6.3]

3. Subtract and, if possible, simplify: $\dfrac{3}{2n} - \dfrac{n - 1}{n + 2}$. [6.4]

4. Simplify: $\dfrac{\frac{x - 2}{x + 1}}{\frac{x + 2}{x + 1}}$. [6.5]

5. Solve: $\dfrac{1}{x} - \dfrac{2}{x - 4} = \dfrac{3}{x^2 - 4x}$. [6.6]

Prepare to Move On

Solve.

1. The sum of two consecutive odd numbers is 276. Find the numbers. [2.5]

2. The product of two consecutive even integers is 48. Find the numbers. [5.7]

3. The height of a triangle is 3 cm longer than its base. If the area of the triangle is 54 cm², find the measurements of the base and the height. [5.7]

4. Between June 9 and June 24, Seth's beard grew 0.9 cm. Find the rate at which Seth's beard grows. [3.4]

6.7   Applications Using Rational Equations and Proportions

A. Problems Involving Work   **B.** Problems Involving Motion   **C.** Problems Involving Proportions

In many areas of study, applications involving rates, proportions, or reciprocals translate to rational equations.

A. Problems Involving Work

EXAMPLE 1   Brian and Reba volunteer in a community garden. Brian can mulch the garden alone in 8 hr and Reba can mulch the garden alone in 10 hr. How long will it take the two of them, working together, to mulch the garden?

SOLUTION

1. **Familiarize.** This *work problem* is a type of problem we have not yet encountered. Work problems are often *incorrectly* translated to mathematical language in several ways.

   a) Add the times together: 8 hr + 10 hr = 18 hr.   ⟵ Incorrect

   This cannot be the correct approach since it should not take Brian and Reba longer to do the job together than it takes either of them working alone.

   b) Average the times: (8 hr + 10 hr)/2 = 9 hr.   ⟵ Incorrect

   Again, this is longer than it would take Brian to do the job alone.

   c) Assume that each person does half the job.   ⟵ Incorrect

   If each person does half the job, Brian would be finished with his half in 4 hr, and Reba with her half in 5 hr. Since they are working together, Brian would continue to help Reba after completing his half. This does tell us that the job will take between 4 hr and 5 hr when they work together.

   Each incorrect approach began with the time it takes each worker to do the job. The correct approach instead focuses on the *rate* of work, or the amount of the job that each person completes in 1 hr.

   Since it takes Brian 8 hr to mulch the entire garden, in 1 hr he mulches $\frac{1}{8}$ of the garden. Since it takes Reba 10 hr to mulch the entire garden, in 1 hr she mulches $\frac{1}{10}$ of the garden. Together, they mulch $\frac{1}{8} + \frac{1}{10} = \frac{5}{40} + \frac{4}{40} = \frac{9}{40}$ of the garden per hour. The rates are thus

   Brian: $\frac{1}{8}$ of the garden per hour,

   Reba: $\frac{1}{10}$ of the garden per hour,

   Together: $\frac{9}{40}$ of the garden per hour.

   We are looking for the time required to mulch 1 entire garden.

	Fraction of the Garden Mulched		
Time	By Brian	By Reba	Together
1 hr	$\frac{1}{8}$	$\frac{1}{10}$	$\frac{1}{8} + \frac{1}{10}$, or $\frac{9}{40}$
2 hr	$\frac{1}{8} \cdot 2$	$\frac{1}{10} \cdot 2$	$\left(\frac{1}{8} + \frac{1}{10}\right)2$, or $\frac{9}{40} \cdot 2$, or $\frac{9}{20}$
3 hr	$\frac{1}{8} \cdot 3$	$\frac{1}{10} \cdot 3$	$\left(\frac{1}{8} + \frac{1}{10}\right)3$, or $\frac{9}{40} \cdot 3$, or $\frac{27}{40}$
t hr	$\frac{1}{8} \cdot t$	$\frac{1}{10} \cdot t$	$\left(\frac{1}{8} + \frac{1}{10}\right)t$, or $\frac{9}{40} \cdot t$

2. **Translate.** From the table, we see that t must be some number for which

$$\underbrace{\text{Fraction of garden}}_{\text{done by Brian in } t \text{ hr}} \quad \overbrace{\frac{1}{8} \cdot t}^{} + \overbrace{\frac{1}{10} \cdot t}^{} = 1, \quad \underbrace{\text{Fraction of garden}}_{\text{done by Reba in } t \text{ hr}}$$

or

$$\frac{t}{8} + \frac{t}{10} = 1.$$

3. **Carry out.** We solve the equation:

$$\frac{t}{8} + \frac{t}{10} = 1 \qquad \text{The LCD is 40.}$$

$$40\left(\frac{t}{8} + \frac{t}{10}\right) = 40 \cdot 1 \qquad \text{Multiplying to clear fractions}$$

$$\frac{40t}{8} + \frac{40t}{10} = 40$$

$$5t + 4t = 40 \qquad \text{Simplifying}$$

$$9t = 40$$

$$t = \frac{40}{9}, \text{ or } 4\frac{4}{9}.$$

4. **Check.** In $\frac{40}{9}$ hr, Brian mulches $\frac{1}{8} \cdot \frac{40}{9}$, or $\frac{5}{9}$, of the garden and Reba mulches $\frac{1}{10} \cdot \frac{40}{9}$, or $\frac{4}{9}$, of the garden. Together, they mulch $\frac{5}{9} + \frac{4}{9}$, or 1 garden. The fact that our solution is between 4 hr and 5 hr (see part (c) of the *Familiarize* step above) is also a check.

5. **State.** It will take $4\frac{4}{9}$ hr for Brian and Reba, working together, to mulch the garden.

1. Refer to Example 1. Suppose that Brian can pick the produce in the garden alone in 60 min and Reba can pick the produce alone in 45 min. How long will it take the two of them, working together, to pick the produce?

YOUR TURN

THE WORK PRINCIPLE

If

$a = $ the time needed for A to complete the work alone,

$b = $ the time needed for B to complete the work alone, and

$t = $ the time needed for A and B to complete the work together,

then

$$\frac{t}{a} + \frac{t}{b} = 1.$$

The following are equivalent equations that can also be used:

$$\left(\frac{1}{a} + \frac{1}{b}\right)t = 1, \quad \frac{1}{a} \cdot t + \frac{1}{b} \cdot t = 1, \quad \text{and} \quad \frac{1}{a} + \frac{1}{b} = \frac{1}{t}.$$

B. Problems Involving Motion

Problems dealing with distance, rate (or speed), and time are called **motion problems**. To translate them, we use either the basic motion formula, $d = rt$, or the formulas $r = d/t$ or $t = d/r$, which can be derived from $d = rt$.

EXAMPLE 2   On her road bike, Olivia bikes 5 km/h faster than Jason does on his mountain bike. In the time it takes Olivia to travel 50 km, Jason travels 40 km. Find the speed of each bicyclist.

SOLUTION

1. **Familiarize.** Let's make a guess and check it.

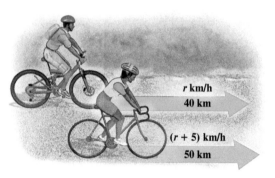

r km/h
40 km

(*r* + 5) km/h
50 km

$$\begin{array}{lll} \textit{Guess:} & \text{Jason's speed:} & 10 \text{ km/h} \\ & \text{Olivia's speed:} & 10 + 5, \text{ or } 15 \text{ km/h} \\ & \text{Jason's time:} & 40/10 = 4 \text{ hr} \\ & \text{Olivia's time:} & 50/15 = 3\tfrac{1}{3} \text{ hr} \end{array} \left. \begin{array}{l} \\ \\ \end{array} \right\} \begin{array}{l} \text{The times are} \\ \text{not the same.} \end{array}$$

Our guess is wrong, but we can make some observations. If Jason's speed $= r$, in kilometers/hour, then Olivia's speed $= r + 5$. Jason's travel time is the same as Olivia's travel time.

We can also make a sketch and label it to help us visualize the situation.

2. **Translate.** We organize the information in a table. By looking at how we checked our guess, we see that we can fill in the **Time** column of the table using the formula *Time = Distance/Rate*.

	Distance	Speed	Time
Jason's Mountain Bike	40	r	$40/r$
Olivia's Road Bike	50	$r + 5$	$50/(r + 5)$

Since we know that the times are the same, we can write an equation:

$$\frac{40}{r} = \frac{50}{r + 5}.$$

3. **Carry out.** We solve the equation:

$$\frac{40}{r} = \frac{50}{r + 5} \qquad \text{The LCM is } r(r + 5).$$

$$r(r + 5)\frac{40}{r} = r(r + 5)\frac{50}{r + 5} \qquad \text{Multiplying to clear fractions}$$

$$40r + 200 = 50r \qquad \text{Simplifying}$$

$$200 = 10r$$

$$20 = r.$$

2. Peter can drive 25 mph faster on the highway than he can on county roads. In the time it would take Peter to drive 70 mi on county roads, he could drive 120 mi on the highway. How fast can he drive on each type of road?

4. Check. If our answer checks, Jason's mountain bike is going 20 km/h and Olivia's road bike is going $20 + 5 = 25$ km/h.

Traveling 50 km at 25 km/h, Olivia is riding for $\frac{50}{25} = 2$ hr. Traveling 40 km at 20 km/h, Jason is riding for $\frac{40}{20} = 2$ hr. Our answer checks since the two times are the same.

5. State. Olivia's speed is 25 km/h, and Jason's speed is 20 km/h.

YOUR TURN

ALF
Active Learning Figure

EXPLORING THE CONCEPT

Motion problems are often much simpler to solve if the information is organized in a table. For each motion problem, fill in the missing entries in the table using the list of options given below.

1. Tara runs 1 km/h faster than Cassie. Tara can run 20 km in the same time that it takes Cassie to run 18 km. Find the speed of each runner.

	Distance	Speed	Time
Tara	20	**(a)** ____	**(b)** ____
Cassie	**(c)** ____	r	**(d)** ____

Options: 18 $r + 1$ $\dfrac{20}{r+1}$ $\dfrac{18}{r}$

2. Damon rode 50 mi to a state park at a certain speed. Had he been able to ride 3 mph faster, the trip would have been $\frac{1}{4}$ hr shorter. How fast did he ride?

	Distance	Speed	Time
Actual Trip	**(a)** ____	r	**(b)** ____
Faster Trip	50	**(c)** ____	**(d)** ____

Options: 50 $r + 3$ $\dfrac{50}{r+3}$ $\dfrac{50}{r}$

ANSWERS

1. (a) $r + 1$; (b) $\dfrac{20}{r+1}$; (c) 18; (d) $\dfrac{18}{r}$

2. (a) 50; (b) $\dfrac{50}{r}$; (c) $r + 3$; (d) $\dfrac{50}{r+3}$

C. Problems Involving Proportions

A **ratio** of two quantities is their quotient. For example, 37% is the ratio of 37 to 100, or $\frac{37}{100}$. A **proportion** is an equation stating that two ratios are equal.

> **PROPORTION**
>
> An equality of ratios,
>
> $$\frac{A}{B} = \frac{C}{D},$$
>
> is called a *proportion*. The fractions within a proportion are said to be *proportional* to each other.

In geometry, if two triangles are **similar**, then their corresponding angles have the same measure and their corresponding sides are *proportional*. To illustrate, if triangle *ABC* is similar to triangle *RST*, then angles *A* and *R* have the same measure, angles *B* and *S* have the same measure, angles *C* and *T* have the same measure, and

$$\frac{a}{r} = \frac{b}{s} = \frac{c}{t}.$$

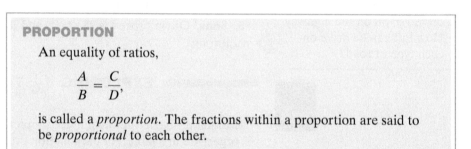

EXAMPLE 3 *Similar Triangles.* Triangles *ABC* and *XYZ* are similar. Solve for *z* if $x = 10$, $a = 8$, and $c = 5$.

SOLUTION We make a drawing, write a proportion, and then solve. Note that side *a* is always opposite angle *A*, side *x* is always opposite angle *X*, and so on.

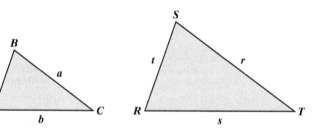

We have

$$\frac{z}{5} = \frac{10}{8}$$ The proportions $\frac{5}{z} = \frac{8}{10}$, $\frac{5}{8} = \frac{z}{10}$, or

$$\frac{8}{5} = \frac{10}{z}$$ could also be used.

$$40 \cdot \frac{z}{5} = 40 \cdot \frac{10}{8}$$ Multiplying both sides by the LCM, 40

$$8z = 50$$ Simplifying

$$z = \frac{50}{8}, \text{ or } 6.25.$$

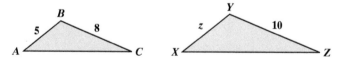
YOUR TURN

↳ Check Your UNDERSTANDING

Refer to the following pair of similar triangles to complete each proportion.

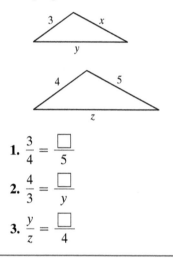

1. $\dfrac{3}{4} = \dfrac{\square}{5}$

2. $\dfrac{4}{3} = \dfrac{\square}{y}$

3. $\dfrac{y}{z} = \dfrac{\square}{4}$

---

3. Triangles *QRS* and *MNP* are similar. Solve for *r* if $m = 5$, $n = 8$, and $q = 4$.

EXAMPLE 4 *Architecture.* A *blueprint* is a scale drawing of a building representing an architect's plans. Ellia is adding 12 ft to the length of an apartment and needs to indicate the addition on an existing blueprint. If a 10-ft long bedroom is represented by $2\frac{1}{2}$ in. on the blueprint, how much longer should Ellia make the drawing in order to represent the addition?

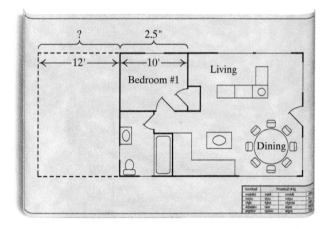

Student Notes

In Example 4, the ratios in the proportion both have "inches on drawing" in the numerator and "feet in real life" in the denominator. We would obtain the same solution if we wrote ratios with feet in the numerator and inches in the denominator. What is important is that each quantity occupy the same position in both ratios.

4. Refer to Example 4. Suppose that the width of the apartment is represented on the blueprint by 4 in. How wide is the apartment?

 Chapter Resources:
Translating for Success, p. 426;
Collaborative Activity, p. 427;
Decision Making: Connection,
p. 427

5. Refer to Example 5. Suppose that Brent sent or received 534 messages in 6 days. At this rate, how many text messages would he send or receive in 30 days?

SOLUTION We let w represent the width, in inches, of the addition that Ellia is drawing. Because the drawing must have the correct proportions, we have

Inches on drawing ⟶ $\dfrac{w}{12} = \dfrac{2.5}{10}$. ⟵ Inches on drawing
Feet in real life ⟶ $\phantom{\dfrac{w}{12}}$ ⟵ Feet in real life

To solve for w, we multiply both sides by the LCM of the denominators, 60:

$$60 \cdot \frac{w}{12} = 60 \cdot \frac{2.5}{10}$$
$$5w = 6 \cdot 2.5 \qquad \text{Simplifying}$$
$$w = \tfrac{15}{5}, \text{ or } 3.$$

Ellia should make the blueprint 3 in. longer.

YOUR TURN

EXAMPLE 5 *Text Messaging.* Brent sent or received 384 messages in 8 days. At this rate, how many text messages would he send or receive in 30 days?

SOLUTION We let $x =$ the number of text messages Brent would send or receive in 30 days. We form a proportion in which the ratio of the number of text messages to the number of days is expressed in two ways:

Number of text messages ⟶ $\dfrac{384}{8} = \dfrac{x}{30}$. ⟵ Number of text messages
Number of days ⟶ $\phantom{\dfrac{384}{8}}$ ⟵ Number of days

To solve for x, we multiply both sides by the LCM of the denominators, 120:

$$120 \cdot \frac{384}{8} = 120 \cdot \frac{x}{30}$$
$$15 \cdot 8 \cdot \frac{384}{8} = 4 \cdot 30 \cdot \frac{x}{30}$$
$$15 \cdot 384 = 4x$$
$$5760 = 4x$$
$$1440 = x.$$

At this rate, Brent will send or receive 1440 text messages in 30 days.

YOUR TURN

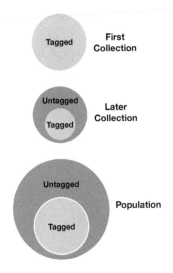

EXAMPLE 6   *Wildlife Population.*   To determine the number of brook trout in River Denys, Cape Breton, Nova Scotia, a team of volunteers and professionals caught and marked 1190 brook trout. Later, they captured 915 brook trout, of which 24 were marked. Estimate the number of brook trout in River Denys.

Data: www.gov.ns.ca

SOLUTION   We let $T =$ the brook trout population in River Denys. If we assume that the percentage of marked trout in the second group of trout captured is the same as the percentage of marked trout in the entire river, we can form a proportion in which this percentage is expressed in two ways:

Trout originally marked $\longrightarrow$ $\dfrac{1190}{T} = \dfrac{24}{915}.$ $\longleftarrow$ Marked trout in second group
Entire population $\longrightarrow$ $\phantom{\dfrac{1190}{T}}$ $\longleftarrow$ Total trout in second group

To solve for T, we multiply by the LCM, $915T$:

$$915T \cdot \frac{1190}{T} = 915T \cdot \frac{24}{915}$$ Multiplying both sides by $915T$

$$915 \cdot 1190 = 24T$$ Removing factors equal to 1: $T/T = 1$ and $915/915 = 1$

$$\frac{915 \cdot 1190}{24} = T \text{ or } T \approx 45{,}369.$$ Dividing both sides by 24

There are about 45,369 brook trout in the river.

YOUR TURN

6. Refer to Example 6. Suppose that the team captured 875 brook trout, of which 35 were marked. Estimate the number of brook trout in the River Denys.

6.7   EXERCISE SET

FOR EXTRA HELP   MyMathLab®

⮞ Vocabulary and Reading Check

Classify each of the following statements as either true or false.

1. In order to find the time that it would take two people to complete a job working together, we average the time that it takes each of them to complete the job working separately.

2. To find the rate at which two people work together, we add the rates at which they work separately.

3. Distance equals rate times time.

4. Rate equals distance divided by time.

5. Time equals distance divided by rate.

6. If two triangles are similar, their corresponding sides are of equal length.

⮞ Concept Reinforcement

Find each rate.

7. If Sandy can decorate a cake in 2 hr, what is her hourly rate?

8. If Eric can decorate a cake in 3 hr, what is his hourly rate?

9. If Sandy can decorate a cake in 2 hr and Eric can decorate the same cake in 3 hr, what is their hourly rate, working together?

10. If Lisa and Mark can mow a lawn together in 1 hr, what is their hourly rate?

11. If Lisa can mow a lawn by herself in 3 hr, what is her hourly rate?

12. If Lisa and Mark can mow a lawn together in 1 hr, and Lisa can mow the same lawn by herself in 3 hr, what is Mark's hourly rate, working alone?

A. Problems Involving Work

13. *Volunteerism.*   It takes Kelby 10 hr per week to prepare food for delivery to senior citizens. Natalie can do the same job in 15 hr. How long would it take Kelby and Natalie to prepare the food working together?

14. *Painting.*   Oliver can paint the porch on the Heppner's Victorian house in 20 hr. It would take Pat 30 hr to paint the porch. How long would it take them to paint the porch if they worked together?

15. *Home Restoration.* Bryan can refinish the floor of an apartment in 8 hr. Armando can refinish the floor in 6 hr. How long will it take them, working together, to refinish the floor?

16. *Custom Embroidery.* Chandra can embroider logos on a team's sweatshirts in 6 hr. Traci, a new employee, needs 9 hr to complete the same job. Working together, how long will it take them to do the job?

17. *Filling a Pool.* The San Paulo community swimming pool can be filled in 12 hr if water enters through a pipe alone or in 30 hr if water enters through a hose alone. If water is entering through both the pipe and the hose, how long will it take to fill the pool?

18. *Filling a Tank.* A community water tank can be filled in 18 hr by the town office well alone and in 22 hr by the high school well alone. How long will it take to fill the tank if both wells are working?

19. *Pumping Water.* An ABS Robusta 300 TS sump pump can remove water from Martha's flooded basement in 70 min. The Little Giant 1-A sump pump can complete the same job in 30 min. How long would it take the two pumps together to pump out the basement?

Data: www.shoppumps.com

20. *Air Purifying.* The Blueair 203 air purifier can clean the air in a family room in 12 min. The Blueair 205 can clean the air in a room of the same size in 10 min. How long would it take the two machines together to clean the air in such a room?

Data: usa.blueair.com

21. *Printing.* The Canon imageRUNNER 2545 can copy Sunil's dissertation in 5 min. The Xerox 4112 can copy the same document in 2 min. If the two machines work together, how long would they take to copy the dissertation?

Data: canon.com; xerox.com

22. *Shredding.* The Destroyit 2404SC Strip Cut Shredder can shred a week's worth of paper from New Law Office in 7 min. The Martin Yale 45CC4 Cross Cut Paper Shredder can shred the same amount of paper in 10 min. How long would it take the shredders, working together, to shred a week's worth of paper?

Data: shredderwarehouse.com

B. Problems Involving Motion

23. *Train Speeds.* A CSX freight train is traveling 14 km/h slower than an AMTRAK passenger train. The CSX train travels 330 km in the same time that it takes the AMTRAK train to travel 400 km. Find their speeds. Complete the following table as part of the familiarization.

	Distance =	Rate ·	Time
	Distance (in km)	**Speed (in km/h)**	**Time (in hours)**
CSX	330		
AMTRAK	400	r	$\dfrac{400}{r}$

24. *Speed of Travel.* A loaded Roadway truck is moving 40 mph faster than a New York Railways freight train. In the time that it takes the train to travel 150 mi, the truck travels 350 mi. Find their speeds. Complete the following table as part of the familiarization.

	Distance =	Rate ·	Time
	Distance (in miles)	**Speed (in miles per hour)**	**Time (in hours)**
Truck	350	r	$\dfrac{350}{r}$
Train	150		

25. *Driving Speed.* Sean's Camaro travels 15 mph faster than Rita's Harley. In the same time that Sean travels 156 mi, Rita travels 120 mi. Find their speeds.

26. *Bicycle Speed.* Ada bicycles 5 km/h slower than Elin. In the same time that it takes Ada to ride 48 km, Elin can ride 63 km. How fast does each bicyclist travel?

27. *Walking Speed.* Baruti walks 4 km/h faster than Tau. In the time that it takes Tau to walk 7.5 km, Baruti walks 13.5 km. Find their speeds.

28. *Cross-Country Skiing.* Luca cross-country skis 4 km/h faster than Lea. In the time that it takes Lea to ski 18 km, Luca skis 24 km. Find their speeds.

Aha! **29.** *Tractor Speed.* Manley's tractor is just as fast as Caledonia's. It takes Manley 1 hr more than it takes Caledonia to drive to town. If Manley is 20 mi from town and Caledonia is 15 mi from town, how long does it take Caledonia to drive to town?

30. *Boat Speed.* Tory and Emilio's motorboats travel at the same speed. Tory pilots her boat 40 km before docking. Emilio continues for another 2 hr, traveling a total of 100 km before docking. How long did it take Tory to navigate the 40 km?

31. *Train Travel.* A freight train covered 120 mi at a certain speed. Had the train been able to travel 10 mph faster, the trip would have been 2 hr shorter. How fast did the train go?

32. *Moped Speed.* Dexter rode 60 mi on his moped at a certain speed. Had he been able to travel 5 mph faster, the trip would have taken 1 hr less time. How fast did Dexter go?

C. Problems Involving Proportions

Geometry. For each pair of similar triangles, find the value of the indicated letter.

33. *b*

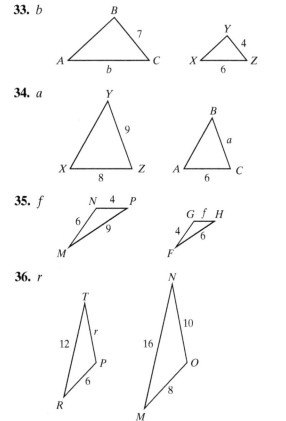

34. *a*

35. *f*

36. *r*

Graphing. Find the indicated length.

37. *r*

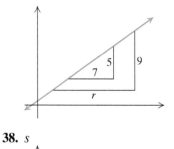

38. *s*

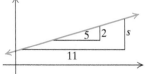

Architecture. Use the following blueprint to find the indicated length.

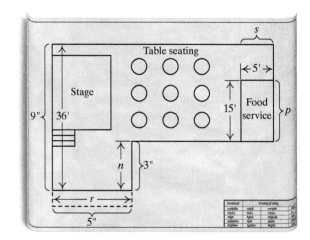

39. *p*, in inches on blueprint

40. *s*, in inches on blueprint

41. *r*, in feet on actual building

42. *n*, in feet on actual building

Find the indicated length.

43. *l*

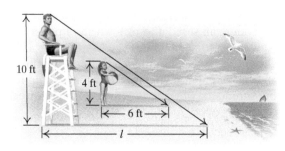

44. *h*

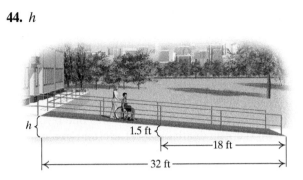

Flight Mechanics. *The wing aspect ratio for a bird or an airplane is the ratio of the wing span to the wing width. Generally, higher-aspect ratios are more efficient during low-speed flying. Use the following table for Exercises 45 and 46.*

Bird or Airplane	Wing Span	Wing Width
Grey heron	180 cm	24 cm
House sparrow	10 in.	1.8 in.
Albatross	9 ft	0.6 ft
Concorde (retired in 2003)	84 ft	47 ft
Boeing 747-400	64 m	8 m

45. Herons and storks, both waders, have the same wing aspect ratios. A white stork has a wing span of 200 cm. What is the wing width of a white stork?

46. The Piper PA-28 Cherokee airplane, with a wing span of 35 ft, has the same wing aspect ratio as a house sparrow. What is the width of a wing of the Piper airplane?

47. *Spending Habits.* In the first 8 days of September, Felicia spent $17.40 on coffee. At this rate, how much will she spend in September? (September has 30 days.)

48. *Burning Calories.* The average 140-lb adult burns about 380 calories bicycling 10 mi at a moderate rate. How far should the average 140-lb adult ride in order to burn 100 calories?

Data: www.livestrong.com

Aha! **49.** *Photography.* Aziza snapped 234 photos over a period of 14 days. At this rate, how many would she take in 42 days?

50. *Mileage.* The Toyota Prius is a hybrid car that travels approximately 204 mi on 4 gal of gas. Find the amount of gas required for a 714-mi trip.

Data: toyota.com

51. *Light Bulbs.* A sample of 220 compact fluorescent light bulbs contained 8 defective bulbs. How many defective bulbs would you expect in a batch of 1430 bulbs?

52. *Flash Drives.* A sample of 150 flash drives contained 7 defective drives. How many defective flash drives would you expect in a batch of 2700 flash drives?

53. *Veterinary Science.* The amount of water needed by a small dog depends on its weight. A moderately active 8-lb Shih Tzu needs approximately 12 oz of water per day. How much water does a moderately active 5-lb Bolognese require each day?

Data: www.smalldogsparadise.com

54. *Miles Driven.* Emmanuel is allowed to drive his leased car for 45,000 mi in 4 years without penalty. In the first $1\frac{1}{2}$ years, Emmanuel has driven 16,000 mi. At this rate, will he exceed the mileage allowed for 4 years?

55. *Environmental Science.* To determine the number of humpback whales in a pod, a marine biologist, using tail markings, identifies 27 members of the pod. Several weeks later, 40 whales from the pod are randomly sighted. Of the 40 sighted, 12 are from the 27 originally identified. Estimate the number of whales in the pod.

56. *Fox Population.* To determine the number of foxes in King County, a naturalist catches, tags, and then releases 25 foxes. Later, 36 foxes are caught; 4 of them have tags. Estimate the fox population of the county.

57. Is it correct to assume that two workers will complete a task twice as quickly as one person working alone? Why or why not?

58. If two triangles are exactly the same shape and size, are they similar? Why or why not?

Skill Review

Perform the indicated operation and simplify.

59. $(x^3 - 3x - 7) - (x^2 - 4x + 8)$ [4.4]

60. $(2x^3 - 7)(x + 3)$ [4.5]

61. $(3y^2z - 2yz^2 + y^2) + (4yz^2 + 5y^2 - 6yz)$ [4.7]

62. $(6a^2b^3 + 12ab^2 - 3a^2b) \div (3ab)$ [4.8]

63. $(8n^3 + 3)(8n^3 - 3)$ [4.6]

64. $(x^3 - x + 7) \div (x - 1)$ [4.8]

Synthesis

65. Two steamrollers are paving a parking lot. Working together, will the two steamrollers take less than half as long as the slower steamroller would working alone? Why or why not?

66. Two fuel lines are filling a freighter with oil. Will the faster fuel line take more or less than twice as long to fill the freighter by itself? Why?

67. *Filling a Bog.* The Norwich cranberry bog can be filled in 9 hr and drained in 11 hr. How long will it take to fill the bog if the drainage gate is left open?

68. *Filling a Tub.* Gretchen's hot tub can be filled in 10 min and drained in 8 min. How long will it take to empty a full tub if the water is left on?

69. Show that the four equations in the box labeled "The Work Principle" in this section are equivalent.

70. *Quilting.* Ricki and Maura work together and sew a quilt in 4 hr. Working alone, Maura would need 6 hr more than Ricki to sew a quilt. How long would it take each of them working alone?

71. *Wiring.* Janet can wire a house in 28 hr. Linus can wire a house in 34 hr. How long will it take Janet and Linus, working together, to wire *two* houses?

72. *Grading.* Alma can grade a batch of placement exams in 3 hr. Kevin can grade a batch in 4 hr. If they work together to grade a batch of exams, what percentage of the exams will have been graded by Alma?

73. According to the U.S. Census Bureau, Population Division, in February 2016, there was one birth every 8 sec, one death every 11 sec, and one new international migrant every 27 sec. How many seconds does it take for a net gain of one person?

74. *Home Maintenance.* Fuel used in many chain saws is made by pouring a 3.2-oz bottle of 2-cycle oil into 160 oz of gasoline. Gus accidentally poured 5.6 oz of 2-cycle oil into 200 oz of gasoline. How much more oil or gasoline should he add in order for the fuel to have the proper ratio of oil to gasoline?

75. At what time after 4:00 will the minute hand and the hour hand of a clock first be in the same position?

76. At what time after 10:30 will the hands of a clock first be perpendicular?

Average speed is defined as total distance divided by total time.

77. Ferdaws drove 200 km. For the first 100 km of the trip, she drove at a speed of 40 km/h. For the second half of the trip, she traveled at a speed of 60 km/h. What was the average speed of the entire trip? (It was *not* 50 km/h.)

78. For the first 50 mi of a 100-mi trip, Garry drove 40 mph. What speed would he have to travel for the last half of the trip so that the average speed for the entire trip would be 45 mph?

79. *Commuting.* To reach an appointment 50 mi away, Dr. Wright allowed 1 hr. After driving 30 mi, she realized that her speed would have to be increased 15 mph for the remainder of the trip. What was her speed for the first 30 mi?

80. Given that
$$\frac{A}{B} = \frac{C}{D},$$
write three other proportions using $A, B, C,$ and D.

81. *Distances.* The shadow from a 40-ft cliff just reaches across a water-filled quarry at the same time that a 6-ft tall diver casts a 10-ft shadow. How wide is the quarry?

82. Simplest fraction notation for a rational number is $\frac{9}{17}$. Find an equivalent ratio where the sum of the numerator and the denominator is 104.

83. If two triangles are similar, are their areas and perimeters proportional? Why or why not?

84. Are the equations

$$\frac{A + B}{B} = \frac{C + D}{D} \quad \text{and} \quad \frac{A}{B} = \frac{C}{D}$$

equivalent? Why or why not?

YOUR TURN ANSWERS: SECTION 6.7

1. $25\frac{5}{7}$ min
2. County roads: 35 mph; highway: 60 mph
3. $\frac{32}{5}$, or 6.4   **4.** 16 ft
5. 2670 text messages   **6.** 29,750 brook trout

Quick Quiz: Sections 6.1–6.7

1. Simplify: $\dfrac{x^2y^2 - 5x^2y + 6x^2}{7x^3y - 7x^3}$. [6.1]

2. Divide and, if possible, simplify:

$$\frac{7x + 21}{x - 4} \div \frac{x + 3}{5x - 20}. \quad [6.2]$$

3. Add and, if possible, simplify:

$$\frac{1}{x^2 + 3x + 2} + \frac{1}{x^2 - x - 2}. \quad [6.4]$$

4. Solve: $\dfrac{7}{x} = \dfrac{3}{x + 4}$. [6.6]

5. Scot can prepare a tray of appetizers in 75 min. Steve can prepare the same tray in 60 min. How long would it take them, working together, to prepare the tray of appetizers? [6.7]

Prepare to Move On

Graph each pair of equations on the same set of axes.

1. $y = 2x - 1,$
  $y = x + 3$ [3.6]

2. $x + y = 3,$
  $x - y = 2$ [3.3]

3. $y = \frac{1}{2}x + 2,$
  $y = \frac{1}{2}x - 1$ [3.6]

4. $x + y = 2,$
  $2x + 2y = 4$ [3.3]

1. **Advertising.** In 2014, advertisers worldwide spent $16.1 billion on social-network advertising. This was a 45.3% increase over the amount spent in 2013. How much was spent in 2013?

   Data: cmocouncil.org

2. **Bicycling.** The speed of one bicyclist is 2 km/h faster than the speed of another bicyclist. The first bicyclist travels 60 km in the same amount of time that it takes the second to travel 50 km. Find the speed of each bicyclist.

3. **Filling Time.** A swimming pool can be filled in 5 hr by hose A alone and in 6 hr by hose B alone. How long would it take to fill the tank if both hoses were working?

4. **Payroll.** In 2016, the total payroll for Kraftside Productions was $16.1 million. Of this amount, 45.3% was paid to employees working on an assembly line. How much money was paid to assembly-line workers?

Translating for Success

Use after Section 6.7.

Translate each word problem to an equation and select a correct translation from equations A–O.

A. $2x + 2(x + 1) = 613$

B. $x^2 + (x + 1)^2 = 613$

C. $\dfrac{60}{x + 2} = \dfrac{50}{x}$

D. $x = 45.3\% \cdot 16.1$

E. $\dfrac{197}{7} = \dfrac{x}{30}$

F. $x + (x + 1) = 613$

G. $\dfrac{7}{197} = \dfrac{x}{30}$

H. $x^2 + (x + 2)^2 = 612$

I. $x^2 + (x + 1)^2 = 612$

J. $\dfrac{50}{x + 2} = \dfrac{60}{x}$

K. $x + 45.3\% \cdot x = 16.1$

L. $\dfrac{5 + 6}{2} = t$

M. $x^2 + (x + 1)^2 = 452$

N. $\dfrac{1}{5} + \dfrac{1}{6} = \dfrac{1}{t}$

O. $x^2 + (x + 2)^2 = 452$

Answers on page A-35

An additional, animated version of this activity appears in MyMathLab. *To use MyMathLab, you need a course ID and a student access code. Contact your instructor for more information.*

5. **Cycling Distance.** A bicyclist traveled 197 mi in 7 days. At this rate, how many miles could the cyclist travel in 30 days?

6. **Sides of a Square.** If the sides of a square are increased by 2 ft, the area of the original square plus the area of the enlarged square is 452 ft^2. Find the length of a side of the original square.

7. **Consecutive Integers.** The sum of two consecutive integers is 613. Find the integers.

8. **Sums of Squares.** The sum of the squares of two consecutive odd integers is 612. Find the integers.

9. **Sums of Squares.** The sum of the squares of two consecutive integers is 613. Find the integers.

10. **Rectangle Dimensions.** The length of a rectangle is 1 ft longer than its width. Find the dimensions of the rectangle if the perimeter is 613 ft.

Collaborative Activity   *Sharing the Workload*

Focus: Modeling, estimation, and work problems
Use after: Section 6.7
Time: 10–15 minutes
Group size: 3
Materials: Paper, pencils, textbooks, and a watch

Many tasks can be done by two people working together. If both people work at the same rate, each does half the task, and the project is completed in half the time that it takes each person individually. However, when the work rates differ, the faster worker performs more than half of the task.

Activity

1. The project is to write the numbers from 1 to 100 on a sheet of paper. Two of the members in each group should write the numbers, one working slowly and one working quickly. The third group member should record the time required for each to write the numbers.

2. Using the times from step (1), calculate how long it will take the two workers, working together, to complete the task.

3. Next, have the same workers as in step (1) — working at the same speeds as in step (1) — perform the task together. To do this, one person should begin writing with 1, while the other worker, using a separate sheet of paper, begins with 100 and writes the numbers counting backward. The third member is again the timekeeper and should note when the two workers have written all the numbers.

4. Compare the actual experimental time from part (3) with the time predicted by the model in part (2). List reasons that might account for any discrepancy.

5. Let t_1, t_2, and t_3 represent the times required for the first worker, the second worker, and the two workers together, respectively, to complete a task. Then develop a model that can be used to find t_3 when t_1 and t_2 are known.

Decision Making & Connection   *(Use after Section 6.7.)*

Phone Costs. If you talk or text rarely, a prepaid cell phone may be more cost-effective than one with a monthly bill. Customers using a prepaid phone load the phone before using it with a dollar amount of their choice. Each use of the phone is charged against that dollar amount. When that amount is used up, the customer must load the phone before again using it.

Xavier loads his prepaid phone with $20 at the beginning of one month, hoping that amount will last until the beginning of the next month. Under Xavier's prepaid plan, each text message costs 5¢, voice calls cost 10¢ per minute, and a 7-day pass for up to 1 GB of data costs $10.

1. Suppose that Xavier communicates only with text messages.

    a) How many messages can he send or receive before reloading the phone?

    b) At the end of the 8th day of the month, Xavier had sent or received 110 text messages. At that rate, will he need to reload his phone before the end of the month? Why or why not?

2. Suppose that Xavier communicates only with voice calls.

    a) How many minutes can he talk before reloading the phone?

    b) At the end of the 12th day of the month, Xavier had talked for 75 min. At that rate, will he need to reload his phone before the end of the month? Why or why not?

3. After several months, Xavier has talked on the phone for an average of one hour per month and sent or received an average of 600 text messages per month. How much is he spending on his phone, on average, each month?

4. Xavier can move to a plan with unlimited talk and text for $30 per month. If he still plans to talk for an average of one hour per month, for how many text messages would the monthly plan save him money?

5. Because of a schedule change, Xavier needs to add data usage to his phone for two weeks out of every month. He can move to a monthly plan with unlimited talk and text plus 3 GB of data for $50 per month. Describe an average monthly combination of data, talk, and text for which the prepaid plan is still cost-effective for him.

6. *Research.* Estimate your monthly voice, text, and data usage. Find both a prepaid phone plan and a monthly service plan that are available in your area. Which is the better plan for you?

Study Summary

KEY TERMS AND CONCEPTS	EXAMPLES	PRACTICE EXERCISES

SECTION 6.1: *Rational Expressions*

A **rational expression** can be written as a quotient of two polynomials and is undefined when the denominator is 0. We simplify rational expressions by removing a factor equal to 1.

$$\frac{x^2 - 3x - 4}{x^2 - 1} = \frac{(x + 1)(x - 4)}{(x + 1)(x - 1)}$$

$$= \frac{x - 4}{x - 1} \quad \frac{x + 1}{x + 1} = 1$$

1. Simplify:
$$\frac{3x^2 - 6x + 3}{x^2 - 4x + 3}.$$

SECTION 6.2: *Multiplication and Division*

The Product of Two Rational Expressions

$$\frac{A}{B} \cdot \frac{C}{D} = \frac{AC}{BD}$$

$$\frac{5v + 5}{v - 2} \cdot \frac{2v^2 - 8v + 8}{v^2 - 1}$$

$$= \frac{5(v + 1) \cdot 2(v - 2)(v - 2)}{(v - 2)(v + 1)(v - 1)}$$

$$= \frac{10(v - 2)}{v - 1} \quad \frac{(v + 1)(v - 2)}{(v + 1)(v - 2)} = 1$$

2. Multiply and, if possible, simplify:
$$\frac{10a + 20}{a^2 - 4} \cdot \frac{a^2 - a - 2}{4a}.$$

The Quotient of Two Rational Expressions

$$\frac{A}{B} \div \frac{C}{D} = \frac{A}{B} \cdot \frac{D}{C} = \frac{AD}{BC}$$

$$(x^2 - 5x - 6) \div \frac{x^2 - 1}{x + 6}$$

$$= \frac{x^2 - 5x - 6}{1} \cdot \frac{x + 6}{x^2 - 1}$$

$$= \frac{(x - 6)(x + 1)(x + 6)}{(x + 1)(x - 1)}$$

$$= \frac{(x - 6)(x + 6)}{x - 1} \quad \frac{x + 1}{x + 1} = 1$$

3. Divide and, if possible, simplify:
$$\frac{t^2 - 100}{t^2 - 2t} \div \frac{t^2 + 8t - 20}{t^3}.$$

SECTION 6.3: *Addition, Subtraction, and Least Common Denominators*

The Sum of Two Rational Expressions

$$\frac{A}{B} + \frac{C}{B} = \frac{A + C}{B}$$

$$\frac{7x + 8}{x + 1} + \frac{4x + 3}{x + 1} = \frac{7x + 8 + 4x + 3}{x + 1}$$

$$= \frac{11x + 11}{x + 1}$$

$$= \frac{11(x + 1)}{x + 1}$$

$$= 11 \quad \frac{x + 1}{x + 1} = 1$$

4. Add and, if possible, simplify:
$$\frac{2x + 4}{x + 5} + \frac{x - 5}{x + 5}.$$

The Difference of Two Rational Expressions

$$\frac{A}{B} - \frac{C}{B} = \frac{A - C}{B}$$

$$\frac{7x + 8}{x + 1} - \frac{4x + 3}{x + 1} = \frac{7x + 8 - (4x + 3)}{x + 1}$$

$$= \frac{7x + 8 - 4x - 3}{x + 1}$$

$$= \frac{3x + 5}{x + 1}$$

5. Subtract and, if possible, simplify:
$$\frac{4x + 1}{3x - 7} - \frac{x + 8}{3x - 7}.$$

To find the **least common multiple, LCM**, of a set of polynomials, write the prime factorizations of the polynomials. The LCM contains each factor the greatest number of times that it occurs in any of the individual factorizations.	Find the LCM of $m^2 - 5m + 6$ and $m^2 - 4m + 4$. $\left.\begin{array}{l} m^2 - 5m + 6 = (m - 2)(m - 3) \\ m^2 - 4m + 4 = (m - 2)(m - 2) \end{array}\right\}$ Factoring each expression $\textbf{LCM} = (m - 2)(m - 2)(m - 3)$	**6.** Find the LCM of $a^2 - 25$ and $a^2 - 6a + 5$.

SECTION 6.4: *Addition and Subtraction with Unlike Denominators*

To add or subtract rational expressions with different denominators, first rewrite the expressions as equivalent expressions with a common denominator. The **least common denominator, LCD**, is the LCM of the denominators.	$\dfrac{2x}{x^2 - 16} + \dfrac{x}{x - 4}$ $= \dfrac{2x}{(x + 4)(x - 4)} + \dfrac{x}{x - 4}$ The LCD is $(x + 4)(x - 4)$. $= \dfrac{2x}{(x + 4)(x - 4)} + \dfrac{x}{x - 4} \cdot \dfrac{x + 4}{x + 4}$ $= \dfrac{2x}{(x + 4)(x - 4)} + \dfrac{x^2 + 4x}{(x + 4)(x - 4)}$ $= \dfrac{x^2 + 6x}{(x + 4)(x - 4)} = \dfrac{x(x + 6)}{(x + 4)(x - 4)}$	**7.** Subtract and, if possible, simplify: $\dfrac{3x - 1}{x - 1} - \dfrac{x + 1}{x - 2}.$

SECTION 6.5: *Complex Rational Expressions*

Complex rational expressions contain one or more rational expressions within the numerator and/or the denominator. They can be simplified either by using division or by multiplying by a form of 1 to clear the fractions.	**Using division to simplify:** $\dfrac{\dfrac{1}{6} - \dfrac{1}{x}}{\dfrac{6 - x}{6}} = \dfrac{\dfrac{1}{6} \cdot \dfrac{x}{x} - \dfrac{1}{x} \cdot \dfrac{6}{6}}{\dfrac{6 - x}{6}} = \dfrac{\dfrac{x - 6}{6x}}{\dfrac{6 - x}{6}}$ Forming a single rational expression in the numerator $= \dfrac{x - 6}{6x} \div \dfrac{6 - x}{6} = \dfrac{x - 6}{6x} \cdot \dfrac{6}{6 - x}$ $= \dfrac{6(x - 6)}{6x(-1)(x - 6)} = \dfrac{1}{-x} = -\dfrac{1}{x}$ $\dfrac{6(x - 6)}{6(x - 6)} = 1$ **Multiplying by 1 to simplify:** $\dfrac{\dfrac{4}{x}}{\dfrac{3}{x} + \dfrac{2}{x^2}} = \dfrac{\dfrac{4}{x}}{\dfrac{3}{x} + \dfrac{2}{x^2}} \cdot \dfrac{x^2}{x^2}$ The LCD of all the denominators is x^2; multiplying by $\dfrac{x^2}{x^2}$ $= \dfrac{\dfrac{4}{x} \cdot \dfrac{x^2}{1}}{\left(\dfrac{3}{x} + \dfrac{2}{x^2}\right) \cdot \dfrac{x^2}{1}}$ $= \dfrac{\dfrac{4 \cdot x \cdot x}{x}}{\dfrac{3 \cdot x \cdot x}{x} + \dfrac{2 \cdot x^2}{x^2}} = \dfrac{4x}{3x + 2}$	**8.** Simplify: $\dfrac{1 - \dfrac{2}{3x}}{x - \dfrac{4}{9x}}.$

SECTION 6.6: *Rational Equations*

To Solve a Rational Equation

1. List any restrictions.
2. Clear the equation of fractions.
3. Solve the resulting equation.
4. Check the possible solution(s) in the original equation.

Solve: $\dfrac{2}{x+1} = \dfrac{1}{x-2}$.   The restrictions are $x \neq -1$ and $x \neq 2$.

$$\frac{2}{x+1} = \frac{1}{x-2}$$

$$(x+1)(x-2) \cdot \frac{2}{x+1} = (x+1)(x-2) \cdot \frac{1}{x-2}$$

$$2(x-2) = x+1$$
$$2x - 4 = x + 1$$
$$x = 5$$

Check: Since $\dfrac{2}{5+1} = \dfrac{1}{5-2}$, the solution is 5.

9. Solve: $\dfrac{1}{3} + \dfrac{1}{6} = \dfrac{1}{t}$.

SECTION 6.7: *Applications Using Rational Equations and Proportions*

The Work Principle

If a = the time needed for A to complete the work alone,

b = the time needed for B to complete the work alone,

and

t = the time needed for A and B to complete the work together, then:

$$\frac{t}{a} + \frac{t}{b} = 1;$$

$$\left(\frac{1}{a} + \frac{1}{b}\right)t = 1;$$

$$\frac{1}{a} \cdot t + \frac{1}{b} \cdot t = 1;$$

$$\frac{1}{a} + \frac{1}{b} = \frac{1}{t}.$$

Brian and Reba volunteer in a community garden. Brian can mulch the garden alone in 8 hr and Reba can mulch the garden alone in 10 hr. How long would it take them, working together, to mulch the garden?

If t = the time, in hours, that it takes Brian and Reba to do the job working together, then

$$\frac{1}{8} \cdot t + \frac{1}{10} \cdot t = 1 \qquad \text{Using the work principle}$$

$$t = 4\tfrac{4}{9}\,\text{hr.} \qquad \text{Solving the equation}$$

See Example 1 in Section 6.7 for a complete solution of this problem.

10. It takes Kenesha 3 hr to stain a large bookshelf. It takes Fletcher 2 hr to stain the same size bookshelf. How long would it take them, working together, to stain the bookshelf?

The Motion Formula

$d = r \cdot t$,

$r = \dfrac{d}{t}$,

or

$t = \dfrac{d}{r}$.

On her road bike, Olivia bikes 5 km/h faster than Jason does on his mountain bike. In the time it takes Olivia to travel 50 km, Jason travels 40 km. Find the speed of each bicyclist.

If r = Jason's speed, in kilometers/hour, then $r + 5$ = Olivia's speed. Using $t = d/r$ and the fact that the times are equal, we have

$$\frac{40}{r} = \frac{50}{r+5}$$

$$r = 20. \qquad \text{Solving the equation}$$

Olivia's speed is 25 km/h, and Jason's speed is 20 km/h. See Example 3 in Section 6.7 for a complete solution of this problem.

11. Jerry jogs 5 mph faster than he walks. In the time it would take Jerry to walk 8 mi, he could jog 18 mi. Find how fast Jerry jogs and how fast he walks.

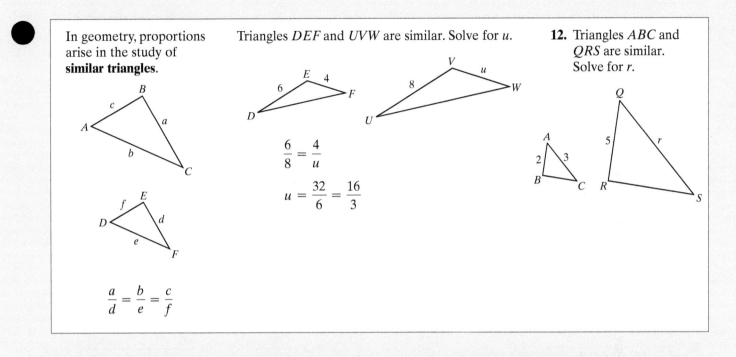

In geometry, proportions arise in the study of **similar triangles**.

Triangles *DEF* and *UVW* are similar. Solve for *u*.

$$\frac{6}{8} = \frac{4}{u}$$

$$u = \frac{32}{6} = \frac{16}{3}$$

$$\frac{a}{d} = \frac{b}{e} = \frac{c}{f}$$

12. Triangles *ABC* and *QRS* are similar. Solve for *r*.

Review Exercises: Chapter 6

➦ Concept Reinforcement

Classify each of the following statements as either true or false.

1. Every rational expression can be simplified. [6.1]

2. The expression $(t - 3)/(t^2 - 4)$ is undefined for $t = 2$. [6.1]

3. The expression $(t - 3)/(t^2 - 4)$ is undefined for $t = 3$. [6.1]

4. To multiply rational expressions, a common denominator is never required. [6.2]

5. To divide rational expressions, a common denominator is never required. [6.2]

6. To add rational expressions, a common denominator is never required. [6.3]

7. To subtract rational expressions, a common denominator is never required. [6.3]

8. The number 0 can never be a solution of a rational equation. [6.6]

List all numbers for which each expression is undefined. [6.1]

9. $\dfrac{17}{-x^2}$

10. $\dfrac{x - 5}{x^2 - 36}$

11. $\dfrac{x^2 + 3x + 2}{x^2 + x - 30}$

12. $\dfrac{-6}{(t + 2)^2}$

Simplify. [6.1]

13. $\dfrac{3x^2 - 9x}{3x^2 + 15x}$

14. $\dfrac{14x^2 - x - 3}{2x^2 - 7x + 3}$

15. $\dfrac{6y^2 - 36y + 54}{4y^2 - 36}$

16. $\dfrac{5x^2 - 20y^2}{2y - x}$

Multiply or divide and, if possible, simplify. [6.2]

17. $\dfrac{a^2 - 36}{10a} \cdot \dfrac{2a}{a + 6}$

18. $\dfrac{6y - 12}{2y^2 + 3y - 2} \cdot \dfrac{y^2 - 4}{8y - 8}$

19. $\dfrac{16 - 8t}{3} \div \dfrac{t - 2}{12t}$

20. $\dfrac{4x^4}{x^2 - 1} \div \dfrac{2x^3}{x^2 - 2x + 1}$

21. $\dfrac{x^2 + 1}{x - 2} \cdot \dfrac{2x + 1}{x + 1}$

22. $(t^2 + 3t - 4) \div \dfrac{t^2 - 1}{t + 4}$

Find the LCM. [6.3]

23. $10a^3b^8, \quad 12a^5b$

24. $x^2 - x, \quad x^5 - x^3, \quad x^4$

25. $y^2 - y - 2, \quad y^2 - 4$

Add or subtract and, if possible, simplify.

26. $\dfrac{x+6}{x+3} + \dfrac{9-4x}{x+3}$ [6.3]

27. $\dfrac{6x-3}{x^2-x-12} - \dfrac{2x-15}{x^2-x-12}$ [6.3]

28. $\dfrac{3x-1}{2x} - \dfrac{x-3}{x}$ [6.4]

29. $\dfrac{2a+4b}{5ab^2} - \dfrac{5a-3b}{a^2b}$ [6.4]

30. $\dfrac{y^2}{y-2} + \dfrac{6y-8}{2-y}$ [6.4]

31. $\dfrac{t}{t+1} + \dfrac{t}{1-t^2}$ [6.4]

32. $\dfrac{d^2}{d-2} + \dfrac{4}{2-d}$ [6.4]

33. $\dfrac{1}{x^2-25} - \dfrac{x-5}{x^2-4x-5}$ [6.4]

34. $\dfrac{3x}{x+2} - \dfrac{x}{x-2} + \dfrac{8}{x^2-4}$ [6.4]

35. $\dfrac{3}{4t} + \dfrac{3}{3t+2}$ [6.4]

Simplify. [6.5]

36. $\dfrac{\dfrac{1}{z}+1}{\dfrac{1}{z^2}-1}$

37. $\dfrac{\dfrac{5}{2x^2}}{\dfrac{3}{4x}+\dfrac{4}{x^3}}$

38. $\dfrac{\dfrac{c}{d}-\dfrac{d}{c}}{\dfrac{1}{c}+\dfrac{1}{d}}$

Solve. [6.6]

39. $\dfrac{3}{x} - \dfrac{1}{4} = \dfrac{1}{2}$

40. $\dfrac{3}{x+4} = \dfrac{1}{x-1}$

41. $x + \dfrac{6}{x} = -7$

42. $1 = \dfrac{2}{x-1} + \dfrac{2}{x+2}$

Solve. [6.7]

43. Jackson can sand the oak floors and stairs in a two-story home in 12 hr. Charis can do the same job in 9 hr. How long would it take if they worked together? (Assume that two sanders are available.)

44. Jennifer's home is 105 mi from her college dorm, and Elizabeth's home is 93 mi away. One Friday afternoon, they left school at the same time and arrived at their homes at the same time. If Jennifer drove 8 mph faster than Elizabeth, how fast did each drive?

45. To estimate the harbor seal population in Bristol Bay, scientists radio-tagged 33 seals. Several days later, they collected a sample of 40 seals, and 24 of them were tagged. Estimate the seal population of the bay.

46. Triangles *ABC* and *XYZ* are similar. Find the value of *x*.

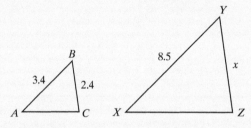

47. A sample of 30 weather-alert radios contained 4 defective ones. How many defective radios would you expect to find in a batch of 540?

Synthesis

48. For what procedures in this chapter is the LCM of denominators used to clear fractions? [6.5], [6.6]

49. A student always uses the common denominator found by multiplying the denominators of the expressions being added. How could this approach be improved? [6.3]

Simplify.

50. $\dfrac{2a^2+5a-3}{a^2} \cdot \dfrac{5a^3+30a^2}{2a^2+7a-4} \div \dfrac{a^2+6a}{a^2+7a+12}$ [6.2]

51. $\dfrac{12a}{(a-b)(b-c)} - \dfrac{2a}{(b-a)(c-b)}$ [6.4]

Aha! 52. $\dfrac{5(x-y)}{(x-y)(x+2y)} - \dfrac{5(x-3y)}{(x+2y)(x-3y)}$ [6.3]

53. The last major-league baseball player to average at least 4 hits in every 10 at bats was Ted Williams in 1941. Suppose that Miguel Cabrera currently has 153 hits after 395 at bats. If he is assured 125 more at bats, what percentage of those must be hits if he is to average 4 hits for every 10 at bats? [6.7]

Test: Chapter 6   For step-by-step test solutions, access the Chapter Test Prep Videos in MyMathLab.

List all numbers for which each expression is undefined.

1. $\dfrac{2-x}{5x}$

2. $\dfrac{x^2+x-30}{x^2-3x+2}$

3. Simplify: $\dfrac{6x^2+17x+7}{2x^2+7x+3}$.

Multiply or divide and, if possible, simplify.

4. $\dfrac{t^2-9}{12t}\cdot\dfrac{8t^2}{t^2-4t+3}$

5. $\dfrac{25y^2-1}{9y^2-6y}\div\dfrac{5y^2+9y-2}{3y^2+y-2}$

6. $\dfrac{4a^2+1}{4a^2-1}\div\dfrac{4a^2}{4a^2+4a+1}$

7. $(x^2+6x+9)\cdot\dfrac{(x-3)^2}{x^2-9}$

8. Find the LCM:

$$y^2-9,\qquad y^2+10y+21,\qquad y^2+4y-21.$$

Add or subtract, and, if possible, simplify.

9. $\dfrac{2+x}{x^3}+\dfrac{7-4x}{x^3}$

10. $\dfrac{5-t}{t^2+1}-\dfrac{t-3}{t^2+1}$

11. $\dfrac{2x-4}{x-3}+\dfrac{x-1}{3-x}$

12. $\dfrac{2x-4}{x-3}-\dfrac{x-1}{3-x}$

13. $\dfrac{7}{t-2}+\dfrac{4}{t}$

14. $\dfrac{y}{y^2+6y+9}+\dfrac{1}{y^2+2y-3}$

15. $\dfrac{1}{x-1}+\dfrac{4}{x^2-1}-\dfrac{2}{x^2-2x+1}$

Simplify.

16. $\dfrac{9-\dfrac{1}{y^2}}{3-\dfrac{1}{y}}$

17. $\dfrac{\dfrac{x}{8}-\dfrac{8}{x}}{\dfrac{1}{8}+\dfrac{1}{x}}$

Solve.

18. $\dfrac{1}{t}+\dfrac{1}{3t}=\dfrac{1}{2}$

19. $\dfrac{15}{x}-\dfrac{15}{x-2}=-2$

20. Kopy Kwik has 2 copiers. One can produce a year-end report in 20 min. The other can produce the same document in 30 min. How long would it take both machines, working together, to produce the report?

21. The average 140-lb adult burns about 320 calories walking 4 mi at a moderate speed. How far should the average 140-lb adult walk in order to burn 100 calories?

Data: www.walking.about.com

22. Ryan drives 20 km/h faster than Alicia. In the same time that Alicia drives 225 km, Ryan drives 325 km. Find the speed of each car.

Synthesis

23. Simplify: $1-\dfrac{1}{1-\dfrac{1}{1-\dfrac{1}{a}}}$.

24. The square of a number is the opposite of the number's reciprocal. Find the number.

Cumulative Review: Chapters 1–6

1. Use the commutative law of multiplication to write an expression equivalent to $a + bc$. [1.2]

2. Evaluate $-x^2$ for $x = 5$. [1.8]

3. Evaluate $(-x)^2$ for $x = 5$. [1.8]

4. Simplify: $-3[2(x - 3) - (x + 5)]$. [1.8]

Solve.

5. $4(y - 5) = -2(y - 2)$ [2.2]

6. $x^2 + 11x + 10 = 0$ [5.6]

7. $49 = x^2$ [5.6]

8. $\frac{4}{9}t + \frac{2}{3} = \frac{1}{3}t - \frac{2}{9}$ [2.2]

9. $\frac{4}{x} + x = 5$ [6.6]

10. $\frac{2}{x - 3} = \frac{5}{3x + 1}$ [6.6]

11. $2x^2 + 7x = 4$ [5.6]

12. $4(x + 7) < 5(x - 3)$ [2.6]

13. $\frac{2}{x^2 - 9} + \frac{5}{x - 3} = \frac{3}{x + 3}$ [6.6]

14. Solve $3a - b + 9 = c$ for b. [2.3]

Graph. [3.2], [3.3], [3.6]

15. $y = \frac{3}{4}x + 5$

16. $x = -3$

17. $4x + 5y = 20$

18. $y = 6$

19. Find the slope of the line containing the points $(1, 5)$ and $(2, 3)$. [3.5]

20. Find the slope and the y-intercept of the line given by $2x - 4y = 1$. [3.6]

21. Write the slope–intercept equation of the line with slope $-\frac{5}{8}$ and y-intercept $(0, -4)$. [3.6]

Simplify.

22. $\frac{x^{-5}}{x^{-3}}$ [4.2]

23. $-(2a^2b^7)^2$ [4.1]

24. Subtract: $(-8y^2 - y + 2) - (y^3 - 6y^2 + y - 5)$. [4.4]

Multiply.

25. $(2x^2 - 1)(x^3 + x - 3)$ [4.5]

26. $(6x - 5y)^2$ [4.6]

27. $(3n + 2)(n - 5)$ [4.6]

28. $(2x^3 + 1)(2x^3 - 1)$ [4.6]

Factor.

29. $6x - 2x^2 - 24x^4$ [5.1]

30. $16x^2 - 81$ [5.4]

31. $t^2 - 10t + 24$ [5.2]

32. $8x^2 + 10x + 3$ [5.3]

33. $6x^2 - 28x + 16$ [5.3]

34. $25t^2 + 40t + 16$ [5.4]

35. $x^2y^2 - xy - 20$ [5.2]

36. $x^4 + 2x^3 - 3x - 6$ [5.1]

Simplify.

37. $\frac{4t - 20}{t^2 - 16} \cdot \frac{t - 4}{t - 5}$ [6.2]

38. $\frac{x^2 - 1}{x^2 - x - 2} \div \frac{x - 1}{x - 2}$ [6.2]

39. $\frac{5ab}{a^2 - b^2} + \frac{a + b}{a - b}$ [6.4]

40. $\frac{x + 2}{4 - x} - \frac{x + 3}{x - 4}$ [6.4]

41. $\dfrac{1 + \dfrac{2}{x}}{1 - \dfrac{4}{x^2}}$ [6.5]

42. Divide:
    $(15x^4 - 12x^3 + 6x^2 + 2x + 18) \div (x + 3)$. [4.8]

Solve.

43. For each order, alibris.com charges a standard shipping fee of $1.00 plus $2.99 per book. The shipping cost for Dae's book order was $45.85. How many books did she order? [2.5]

    Data: alibris.com

44. Nikki is laying out two square flower gardens in a client's lawn. Each side of one garden is 2 ft longer than each side of the smaller garden. Together, the area of the gardens is 340 ft². Find the length of a side of the smaller garden. [5.7]

45. It takes Wes 25 min to file a week's worth of receipts. Corey, a new employee, takes 75 min to do the same job. How long would it take if they worked together? [6.7]

46. Rachel burned 450 calories in a workout. She burned twice as many in her aerobics session as she did doing calisthenics. How many calories did she burn doing calisthenics? [2.5]

Synthesis

47. Solve: $\frac{1}{3}|n| + 8 = 56$. [1.4], [2.2]

48. Solve: $x(x^2 + 3x - 28) - 12(x^2 + 3x - 28) = 0$. [5.6]

49. Solve: $\frac{2}{x - 3} \cdot \frac{3}{x + 3} - \frac{4}{x^2 - 7x + 12} = 0$. [6.6]

Systems and More Graphing

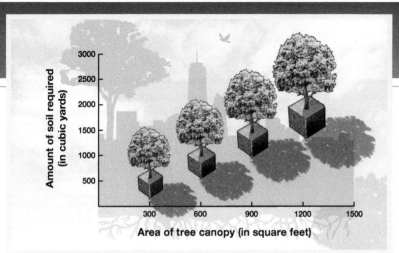

A Tree Can Do a Lot For You!

According to the Colorado Tree Coalition, besides being attractive, trees in urban areas remove air pollutants, produce oxygen, reduce topsoil erosion, reduce energy costs by lowering air temperature in the summer, extend the life of paved surfaces by shading them, and even increase traffic safety. Because of these benefits, continual effort is made in many cities to increase the urban tree canopy. This effort involves more than simply planting trees; urban designers must allow sufficient space for trees when planning sidewalks, streets, and other structures. Data like those illustrated above can be used to estimate the soil volume requirements for a given tree. (*See Example 3 in Section 7.7.*)

7.1 Systems of Equations and Graphing

7.2 Systems of Equations and Substitution

7.3 Systems of Equations and Elimination

CONNECTING THE CONCEPTS

7.4 More Applications Using Systems

MID-CHAPTER REVIEW

7.5 Linear Inequalities in Two Variables

7.6 Systems of Linear Inequalities

7.7 Direct Variation and Inverse Variation

CHAPTER RESOURCES

Visualizing for Success
Collaborative Activity
Decision Making: Connection

STUDY SUMMARY

REVIEW EXERCISES
CHAPTER TEST
CUMULATIVE REVIEW

There is hardly anything that I do that doesn't involve the use of a form of math somewhere in the process.

Syd Knight, a campus planner and site designer in Charlottesville, Virginia, uses models of college growth and cost estimates for campus planning, as well as geometry and trigonometry in site design. Although computers can "crunch the numbers," the designer still must understand what the computer is doing and the math behind its processes.

 ALF *Active Learning Figure* Explore the math using the Active Learning Figure in MyMathLab.

 SA *Student Activity* Do the Student Activity in MyMathLab to see math in action.

any problems translate into two or more equations that must all be true simultaneously. In this chapter, we use a graphing method and two algebraic methods to solve such *systems of equations*. These three methods are then used in a variety of applications and as an aid in solving systems of inequalities.

7.1 | Systems of Equations and Graphing

A. Solutions of Systems   **B.** Solving Systems of Equations by Graphing

A. Solutions of Systems

It is often easier to translate a real-world situation to a *system of two equations* that use two variables, or unknowns, than it is to represent the situation with one equation using one variable. Let's see how the following problem can be represented by two equations using two unknowns:

l, or *w* + 44

w

The perimeter of an NBA basketball court is 288 ft. The length is 44 ft longer than the width. Find the dimensions of the court.

Data: National Basketball Association

If we let w = the width of the court, in feet, and l = the length of the court, in feet, the problem translates to the following system of equations:

$$2l + 2w = 288, \qquad \text{The perimeter is 288 ft.}$$
$$l = w + 44. \qquad \text{The length is 44 ft more than the width.}$$

Both equations must be true. A solution of this system is an ordered pair of the form (l, w) for which $2l + 2w = 288$ *and* $l = w + 44$.

> **Variables are listed in alphabetical order within an ordered pair unless stated otherwise.**

A solution of a system of equations makes *both* equations true. We say that the equations are solved *simultaneously*.

EXAMPLE 1   Consider the system from above:

$$2l + 2w = 288,$$
$$l = w + 44.$$

Determine whether each pair is a solution of the system: **(a)** (94, 50); **(b)** (90, 46).

SOLUTION

a) We check by substituting (alphabetically) 94 for *l* and 50 for *w*:

$$\begin{array}{c|c} 2l + 2w = 288 \\ \hline 2 \cdot 94 + 2 \cdot 50 & 288 \\ 188 + 100 & \\ 288 \stackrel{?}{=} 288 & \text{TRUE} \end{array} \qquad \begin{array}{c|c} l = w + 44 \\ \hline 94 & 50 + 44 \\ 94 \stackrel{?}{=} 94 & \text{TRUE} \end{array}$$

Since (94, 50) checks in *both* equations, it is a solution of the system.

1. Determine whether $(1, -3)$ is a solution of the system

$$x + y = -2,$$
$$x - y = 4.$$

b) We substitute 90 for l and 46 for w:

$2l + 2w = 288$	
$2 \cdot 90 + 2 \cdot 46$	288
$180 + 92$	
$272 \overset{?}{=} 288$	FALSE

$l = w + 44$	
90	$46 + 44$
$90 \overset{?}{=} 90$	TRUE

Since $(90, 46)$ is not a solution of *both* equations, it is not a solution of the system.

YOUR TURN

> A **system of equations** is a set of two or more equations that are to be solved simultaneously.
>
> A **solution** of a system makes all of the equations true.
>
> To **solve** a system of two equations is to find all ordered pairs (if any exist) for which both equations are true.

B. Solving Systems of Equations by Graphing

A graph of an equation is a set of points representing its solution set. Each point on the graph corresponds to an ordered pair that is a solution of the equation. By graphing two equations using one set of axes, we can identify a solution of both equations by looking for a point of intersection.

Student Notes

Because it is so critical to accurately identify any point(s) of intersection, use graph paper, a ruler or other straightedge, and a sharp pencil to graph each line as neatly as possible.

EXAMPLE 2 Solve this system of equations by graphing:

$$x + y = 7,$$
$$y = 3x - 1.$$

SOLUTION We graph the equations using any method studied earlier. The equation $x + y = 7$ can be graphed quickly using the intercepts, $(0, 7)$ and $(7, 0)$. The equation $y = 3x - 1$ is in slope–intercept form, so it can be graphed by plotting its y-intercept, $(0, -1)$, and "counting off" a slope of 3.

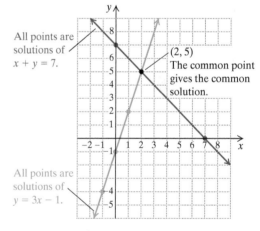

All points are solutions of $x + y = 7$.

$(2, 5)$
The common point gives the common solution.

All points are solutions of $y = 3x - 1$.

The "apparent" solution of the system, $(2, 5)$, should be checked in both equations.

Check:

$x + y = 7$	
$2 + 5$	7
$7 \overset{?}{=} 7$	TRUE

$y = 3x - 1$	
5	$3 \cdot 2 - 1$
$5 \overset{?}{=} 5$	TRUE

Since it checks in both equations, $(2, 5)$ is a solution of the system.

YOUR TURN

2. Solve by graphing:

$$x + y = 5,$$
$$y = 2x + 2.$$

A system of equations that has at least one solution, like the systems in Examples 1 and 2, is said to be **consistent**. A system for which there is no solution is said to be **inconsistent**.

EXAMPLE 3 Solve this system of equations by graphing:

$$y = \tfrac{5}{2}x + 4,$$
$$y = \tfrac{5}{2}x - 3.$$

SOLUTION Both equations are in slope–intercept form so it is easy to see that both lines have the same slope, $\tfrac{5}{2}$. The y-intercepts differ so the lines are parallel, as shown in the figure at right.

Because the lines are parallel, there is no point of intersection. Thus the system is inconsistent and has no solution.

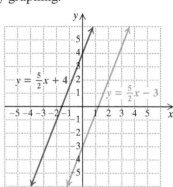

YOUR TURN

3. Solve by graphing:

$$x + y = 1,$$
$$x + y = 3.$$

↳ Check Your UNDERSTANDING

Use the slope and the y-intercept of each line to determine which of the following systems corresponds to each description.

a) $y = -3x + 7,$
$y = -3x + 9$

b) $y = x + 2,$
$y = -x + 2$

c) $y = 3x + 1,$
$y = 3x + 1$

1. The graphs of the equations are parallel.

2. The graphs of the equations are the same line.

3. The graphs of the equations intersect at one point.

4. The system has one solution.

5. The system has no solution.

6. The system has an infinite number of solutions.

---

4. Solve by graphing:

$$y = \tfrac{1}{2}x - 1,$$
$$x - 2y = 2.$$

Sometimes both equations in a system have the same graph.

EXAMPLE 4 Solve this system of equations by graphing:

$$2x + 3y = 6,$$
$$-8x - 12y = -24.$$

SOLUTION Graphing the equations, we see that both represent the same line. This can also be seen by solving each equation for y, obtaining the equivalent slope–intercept form for both equations, $y = -\tfrac{2}{3}x + 2$. Because the equations are equivalent, any solution of one equation is a solution of the other. We show four such solutions.

We check one solution, $(0, 2)$, in each of the original equations.

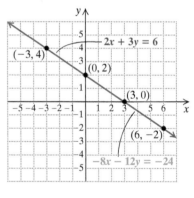

$$\begin{array}{c|c} 2x + 3y = 6 \\ \hline 2(0) + 3(2) & 6 \\ 0 + 6 & \\ & 6 \overset{?}{=} 6 \quad \text{TRUE} \end{array}$$

$$\begin{array}{c|c} -8x - 12y = -24 \\ \hline -8(0) - 12(2) & -24 \\ 0 - 24 & \\ & -24 \overset{?}{=} -24 \quad \text{TRUE} \end{array}$$

On your own, check that $(3, 0)$ is also a solution of the system. If two points are solutions, the lines coincide and all points on the line are solutions.

Since a solution exists, the system is consistent. We state that there is an infinite number of solutions.

YOUR TURN

When one equation can be obtained by multiplying both sides of another equation by a nonzero constant, the two equations are **dependent**.

The equations in Example 4 are dependent, but those in Examples 2 and 3 are **independent**. For systems of two equations, when two equations are dependent, they are equivalent. For systems containing more than two equations, the definition of dependent is slightly different and it is possible for dependent equations to not be equivalent. Such systems are beyond the scope of this text.

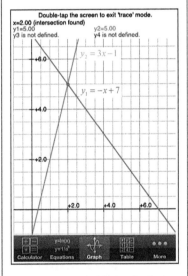
When a system of two linear equations in two variables is graphed, one of the following must occur.

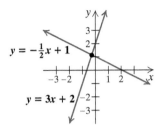 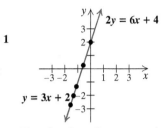

Graphs intersect at one point.
The system is *consistent* and has one solution. Since neither equation is a multiple of the other, the equations are *independent*.

Graphs are parallel.
The system is *inconsistent* because there is no solution. Since neither equation is a multiple of the other, the equations are *independent*.

Equations have the same graph.
The system is *consistent* and has an infinite number of solutions. Since one equation is a multiple of the other, the equations are *dependent*.

Algebraic 🔗 Graphical Connection

Let's take an algebraic–graphical look at equation solving.

Consider the equation $2x - 5 = 3$. Solving algebraically, we have

$$2x - 5 = 3$$
$$2x = 8 \qquad \text{Adding 5 to both sides}$$
$$x = 4. \qquad \text{Dividing both sides by 2}$$

To solve $2x - 5 = 3$ graphically, we graph the equations $y = 2x - 5$ and $y = 3$, as shown at right. The point of intersection, $(4, 3)$, indicates that when x is 4, the value of $2x - 5$ is 3. Thus the solution of $2x - 5 = 3$ is 4.

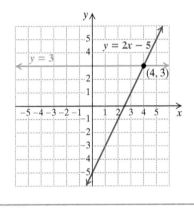

EXAMPLE 5 Solve graphically: $5 - x = x - 1$.

SOLUTION We graph $y = 5 - x$ and $y = x - 1$, as shown. The graphs intersect at $(3, 2)$, indicating that for the x-value 3 both $5 - x$ and $x - 1$ share the same value (in this case, 2). As a check, note that $5 - 3 = 3 - 1$ is true. The solution is 3.

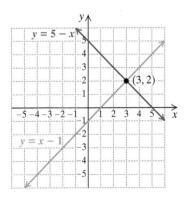

5. Solve graphically:

$$2x - 4 = x + 1.$$

YOUR TURN

Student Notes

Referring to Example 5, note that the solution of the *equation*

$$5 - x = x - 1$$

is the *x-coordinate*, 3, of the point of intersection. Compare this with the solution of the *system*

$$y = 5 - x,$$
$$y = x - 1,$$

which is the *ordered pair*, $(3, 2)$, corresponding to the point of intersection.

Although graphing lets us "see" the solution of a system, it does not always allow us to find a precise solution. For example, the solution of the system

$$3x + 7y = 5,$$
$$6x - 7y = 1$$

is $\left(\frac{2}{3}, \frac{3}{7}\right)$, but finding that precise solution from a graph—*even with a computer or a graphing calculator*—can be difficult. Fortunately, systems like this can be solved accurately using algebraic methods discussed in later sections.

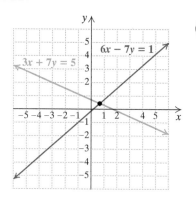

7.1   EXERCISE SET

FOR EXTRA HELP   MyMathLab®

⬥ Vocabulary and Reading Check

Choose the correct word(s) to complete each statement.

1. A solution of a system of two equations is an ordered pair that is a solution of _____
   both/at least one
   equation(s).

2. A solution of a system of two equations can be found by identifying where the two graphs _____.
   intersect/cross the *y*-axis

3. A system of equations that has at least one solution is said to be _____.
   inconsistent/consistent

4. When one equation in a system can be obtained by multiplying both sides of another equation in the system by a nonzero constant, the equations are said to be _____.
   dependent/independent

A. Solutions of Systems

Determine whether each ordered pair is a solution of the system of equations. Use alphabetical order of the variables.

5. $(2, 5)$; $2x + 3y = 19,$
            $3x - y = 1$

6. $(1, 4)$; $5x - 2y = -3,$
            $7x - 3y = -5$

7. $(3, 2)$; $3b - 2a = 0,$
            $b + 2a = 15$

8. $(2, -2)$; $b + 2a = 2,$
             $b - a = -4$

9. $(-15, 20)$; $3x + 2y = -5,$
               $4y + 5x = 5$

10. $(-2, -1)$; $r - 3t = 1,$
               $r + 2t = 0$

B. Solving Systems of Equations by Graphing

Solve each system of equations by graphing. If there is no solution or an infinite number of solutions, state this.

11. $x + y = 4,$
    $x - y = 2$

12. $x - y = 3,$
    $x + y = 7$

13. $y = -2x + 5,$
    $x + y = 4$

14. $y = 2x - 5,$
    $x + y = 4$

15. $y = -2x - 1,$
    $y = 2 - x$

16. $y = -3x + 1,$
    $y = 3 - x$

17. $4x - 20 = 5y,$
    $8x - 10y = 12$

18. $6x + 12 = 2y,$
    $6 - y = -3x$

19. $x = 4,$
    $y = -1$

20. $x = -6,$
    $y = 1$

21. $2x + y = 8,$
    $x - y = 7$

22. $3x + y = 4,$
    $x - y = 4$

23. $y - x = 5,$
    $x + 2y = 4$

24. $y - x = 8,$
    $x + 2y = 1$

25. $x + 2y = 7,$
    $3x + 6y = 21$

26. $x + 3y = 6,$
    $4x + 12y = 24$

27. $2x = 3y - 6,$
    $x = 3y$

28. $3y - 9 = 6x,$
    $y = x$

Aha! 29. $y = \frac{1}{5}x + 4,$
        $2y = \frac{2}{5}x + 8$

30. $y = \frac{1}{3}x + 2,$
    $y = \frac{1}{3}x - 7$

31. $4x + y = 2,$
    $x = \frac{1}{2}y + 5$

32. $3x - y = 1,$
    $x = \frac{1}{5}y + 1$

33. $2x - 3y = 5,$
    $x - 2y = 6$

34. $3x + 4y = 8,$
    $x + 2y = 10$

35. $3x + 2y = 1,$
    $2x + 5y = -14$

36. $4x + 2y = -2,$
    $5x + 4y = 5$

37. $x = \frac{1}{3}y,$
$\quad y = 6$

38. $x = \frac{1}{2}y,$
$\quad x = 3$

Solve graphically.

39. $2x - 1 = 3$

40. $3x - 1 = 2$

41. $x - 4 = 6 - x$

42. $x - 3 = 1 - x$

43. $2x - 1 = -x + 5$

44. $-x + 4 = 2x - 5$

45. $\frac{1}{2}x + 3 = -\frac{1}{2}x - 1$

46. $\frac{3}{2}x + 5 = -\frac{1}{2}x + 1$

47. Is it possible for a system of two linear equations to have exactly two solutions? Why or why not?

48. Suppose that the graphs of both lines in a system of two equations have the same slope. What must be true of the solution of the system?

Skill Review

Add or subtract and, if possible, simplify.

49. $-12.1 + 0.68$ [1.5]

50. $\frac{2}{3} - \left(-\frac{1}{4}\right)$ [1.6]

51. $(2x^2 - 3x - 7) + (x^3 + 5x + 8)$ [4.4]

52. $(xy^2 - xy + y^2) - (xy^2 + 3x^2y - y^2)$ [4.7]

53. $\dfrac{m + 3}{m} + \dfrac{2m}{m - 1}$ [6.4]

54. $\dfrac{1}{a - b} - \dfrac{3}{b - a}$ [6.4]

Synthesis

55. Suppose that the equations in a system of two linear equations are dependent. Does it follow that the system is consistent? Why or why not?

56. Explain why slope–intercept form can be especially useful when solving systems of equations by graphing.

57. Which of the systems in Exercises 11–38 contain dependent equations?

58. Which of the systems in Exercises 11–38 are inconsistent?

59. Write an equation that can be paired with $5x + 2y = 3$ to form a system that has $(-1, 4)$ as the solution. Answers may vary.

60. Solve by graphing:
$$4x - 8y = -7,$$
$$2x + 3y = 7.$$
(*Hint*: Use four squares per unit on your graph.)

61. The solution of the following system is $(2, -3)$. Find A and B.
$$Ax - 3y = 13,$$
$$x - By = 14$$

62. *Printing Costs.* In order to print flyers for his new business, Henri plans to purchase an inkjet printer for $50 or a laserjet printer for $230. An inkjet

cartridge costs $20 and will print 200 flyers, and a laserjet cartridge costs $80 and will print 8000 flyers.

a) Create cost equations for each method of printing flyers.

b) Graph both cost equations on the same set of axes, with cost, in dollars, on the vertical axis.

c) Use the graph to determine how many flyers Henri must print in order for the costs to be the same for both printers.

63. *Video Viewing.* In 2015, American adults spent, on average, about 1 hr 15 min per day viewing video content using a mobile device. This number was increasing at a rate of 14 daily minutes per year. That same year, American adults spent, on average, about 4 hr 15 min per day viewing video programming on television. This number was decreasing at a rate of 5 daily minutes per year.

Data: emarketer.com

a) Write two equations for v, the daily video time, in minutes, for adults on mobile devices and on television t years after 2015.

b) Use a graphing calculator to determine the year in which the daily video time on mobile devices and on television will be the same.

c) Is the total amount of video viewing time increasing or decreasing, and at what rate?

d) *Research.* Find the number of hours that adults currently spend each day viewing video content on mobile devices and on television. Determine whether the equations found in part (a) accurately predict current viewing habits.

64. Use a graphing calculator to solve the system
$$y = 1.2x - 32.7,$$
$$y = -0.7x + 46.15.$$

65. Use a graphing calculator to solve
$$1.3x - 4.9 = 6.3 - 3.7x.$$

YOUR TURN ANSWERS: SECTION 7.1

1. Yes **2.** $(1, 4)$ **3.** No solution **4.** Infinite number of solutions **5.** 5

Prepare to Move On

Solve. [2.2]

1. $3x - (4 - 2x) = 9$

2. $7x - 2(4 + 3x) = 5$

3. $2(8 - 5y) - y = 4$

4. $3(5 - 2y) - 4y = 8$

5. Solve $3x - 4y = 2$ for y. [2.3]

6. Solve $2x + 3y = -1$ for x. [2.3]

7.2 | Systems of Equations and Substitution

A. The Substitution Method   **B.** Problem Solving

A. The Substitution Method

One method for solving systems is known as the **substitution method**. It uses algebra instead of graphing and is thus considered an *algebraic* method.

EXAMPLE 1 Solve the system

$$x + y = 7, \quad \textbf{(1)}$$
$$y = 3x - 1. \quad \textbf{(2)}$$

We have numbered the equations (1) and (2) for easy reference.

SOLUTION The second equation says that y and $3x - 1$ represent the same value. Thus, in the first equation, we can substitute $3x - 1$ for y:

$$x + y = 7, \quad \text{Equation (1)}$$
$$x + 3x - 1 = 7. \quad \text{Substituting } 3x - 1 \text{ (from equation 2) for } y \text{ (in equation 1)}$$

The equation $x + 3x - 1 = 7$ has only one variable, for which we now solve:

$$4x - 1 = 7 \quad \text{Combining like terms}$$
$$4x = 8 \quad \text{Adding 1 to both sides}$$
$$x = 2. \quad \text{Dividing both sides by 4}$$

We have found the x-value of the solution. To find the y-value, we return to the original pair of equations. Substituting into either equation will give us the y-value. We choose equation (1):

$$x + y = 7 \quad \text{Equation (1)}$$
$$2 + y = 7 \quad \text{Substituting 2 for } x$$
$$y = 5. \quad \text{Subtracting 2 from both sides}$$

The ordered pair $(2, 5)$ appears to be a solution. We check:

$$\frac{x + y = 7}{2 + 5 \mid 7}$$
$$7 \overset{?}{=} 7 \quad \text{TRUE}$$

$$\frac{y = 3x - 1}{5 \mid 3 \cdot 2 - 1}$$
$$5 \overset{?}{=} 5 \quad \text{TRUE}$$

Since $(2, 5)$ checks, it is the solution. We can also check by graphing $x + y = 7$ and $y = 3x - 1$ on the same set of axes, as shown at left.

YOUR TURN

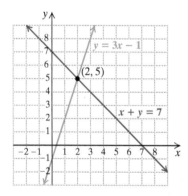

1. Solve the system

$$x = 2y - 1,$$
$$x - y = 3.$$

> *CAUTION!* A solution of a system of equations in two variables is an ordered *pair* of numbers. Once you have solved for one variable, don't forget the other!

EXAMPLE 2 Solve:

$$x = 3 - 2y, \quad \textbf{(1)}$$
$$y - 3x = 5. \quad \textbf{(2)}$$

SOLUTION   We substitute $3 - 2y$ for x in the second equation:

$$y - 3x = 5 \qquad \text{Equation (2)}$$
$$y - 3(3 - 2y) = 5. \qquad \text{Substituting } 3 - 2y \text{ for } x. \text{ The parentheses are} \\ \text{very important.}$$

Now we solve for y:

$$y - 9 + 6y = 5 \qquad \text{Using the distributive law}$$
$$\left. \begin{array}{r} 7y - 9 = 5 \\ 7y = 14 \\ y = 2. \end{array} \right\} \quad \text{Solving for } y$$

Next, we substitute 2 for y in equation (1) of the original system:

$$x = 3 - 2y \qquad \text{Equation (1)}$$
$$= 3 - 2 \cdot 2 \qquad \text{Substituting 2 for } y$$
$$= -1. \qquad \text{Simplifying}$$

We check the ordered pair $(-1, 2)$.

Check:

$x = 3 - 2y$	
-1	$3 - 2 \cdot 2$
	$3 - 4$
$-1 \overset{?}{=} -1$	TRUE

$y - 3x = 5$	
$2 - 3(-1)$	5
$2 + 3$	
$5 \overset{?}{=} 5$	TRUE

The pair $(-1, 2)$ is the solution. A graph is shown at left as another check.

↪ YOUR TURN

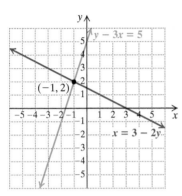

2. Solve:

$$2x - y = 5,$$
$$y = 1 - x.$$

Sometimes neither equation has a variable alone on one side. In that case, we solve one equation for one of the variables and then proceed as before.

EXAMPLE 3   Solve:

$$x - 2y = 6, \qquad \textbf{(1)}$$
$$3x + 2y = 4. \qquad \textbf{(2)}$$

SOLUTION   We can solve either equation for either variable. Since the coefficient of x is 1 in equation (1), it is easiest to solve that equation for x:

$$x - 2y = 6 \qquad \text{Equation (1)}$$
$$x = 6 + 2y. \qquad \textbf{(3)} \quad \text{Adding } 2y \text{ to both sides}$$

We substitute $6 + 2y$ for x in equation (2) of the original pair and solve for y:

$$3x + 2y = 4 \qquad \text{Equation (2)}$$
$$3(6 + 2y) + 2y = 4 \qquad \text{Substituting } 6 + 2y \text{ for } x$$

> Remember to use parentheses when you substitute.

$$18 + 6y + 2y = 4 \qquad \text{Using the distributive law}$$
$$18 + 8y = 4 \qquad \text{Combining like terms}$$
$$8y = -14 \qquad \text{Subtracting 18 from both sides}$$
$$y = \frac{-14}{8} = -\frac{7}{4}. \qquad \text{Dividing both sides by 8 and} \\ \text{simplifying}$$

Technology Connection

To check Example 3 with a graphing calculator, we must first solve each equation for y. When we do so, equation (1) becomes $y = (6 - x)/(-2)$ and equation (2) becomes $y = (4 - 3x)/2$.

1. Use the **INTERSECT** option of the **CALC** menu to determine the solution of the system.

2. What happens when parentheses are deleted from the two equations above?

To find x, we can substitute $-\frac{7}{4}$ for y in equation (1), (2), or (3). Because it is generally easier to use an equation that has already been solved for a specific variable, we decide to use equation (3):

$$x = 6 + 2y = 6 + 2\left(-\tfrac{7}{4}\right) = 6 - \tfrac{7}{2} = \tfrac{12}{2} - \tfrac{7}{2} = \tfrac{5}{2}.$$

We check the ordered pair $\left(\frac{5}{2}, -\frac{7}{4}\right)$ in the original equations.

Check:

$$\begin{array}{c|c}
\multicolumn{2}{c}{x - 2y = 6} \\
\hline
\frac{5}{2} - 2\left(-\frac{7}{4}\right) & 6 \\
\frac{5}{2} + \frac{7}{2} & \\
\frac{12}{2} & \\
6 \overset{?}{=} 6 & \text{TRUE}
\end{array}$$

$$\begin{array}{c|c}
\multicolumn{2}{c}{3x + 2y = 4} \\
\hline
3 \cdot \frac{5}{2} + 2\left(-\frac{7}{4}\right) & 4 \\
\frac{15}{2} - \frac{7}{2} & \\
\frac{8}{2} & \\
4 \overset{?}{=} 4 & \text{TRUE}
\end{array}$$

3. Solve:

$$3x + \ y = 1,$$
$$2x + 3y = 5.$$

Since $\left(\frac{5}{2}, -\frac{7}{4}\right)$ checks, it is the solution.

YOUR TURN

Some systems have no solution and some have an infinite number of solutions.

EXAMPLE 4   Solve each system.

a) $y = \frac{5}{2}x + 4,$ **(1)**   **b)** $2y = 6x + 4,$ **(1)**
     $y = \frac{5}{2}x - 3$ **(2)** $y = 3x + 2$ **(2)**

SOLUTION

a) As we see from the graph at left, the lines are parallel and the system has no solution. Let's see what happens if we try to solve this system by substituting $\frac{5}{2}x - 3$ for y in the first equation:

$$y = \tfrac{5}{2}x + 4 \qquad \text{Equation (1)}$$
$$\tfrac{5}{2}x - 3 = \tfrac{5}{2}x + 4 \qquad \text{Substituting } \tfrac{5}{2}x - 3 \text{ for } y$$
$$-3 = 4. \qquad \text{Subtracting } \tfrac{5}{2}x \text{ from both sides}$$

When we subtract $\frac{5}{2}x$ from both sides, we obtain a *false* equation.

When solving algebraically leads to a false equation, we state that the system has no solution and thus is inconsistent.

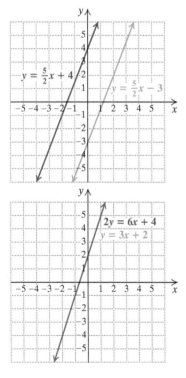

b) As we see from the graph at left, the lines coincide, so the system has an infinite number of solutions. If we use substitution to solve the system, we can substitute $3x + 2$ for y in equation (1):

$$2y = 6x + 4 \qquad \text{Equation (1)}$$
$$2(3x + 2) = 6x + 4 \qquad \text{Substituting } 3x + 2 \text{ for } y$$
$$6x + 4 = 6x + 4$$
$$4 = 4. \qquad \text{Subtracting } 6x \text{ from both sides}$$

This last equation is always true.

4. Solve:

$$y = 1 - 2x,$$
$$2x + y = 3.$$

When the algebraic solution of a system of two equations leads to an equation that is true for all real numbers, we state that the system has an infinite number of solutions.

YOUR TURN

Study Skills

This Looks Familiar

When a topic arises that seems familiar, be alert to any new or extended material that is included. View a familiar topic as a chance to sharpen existing skills and to fill in any "holes" in your understanding of the material.

↳ Check Your UNDERSTANDING

1. The solution of a system of equations is $x = 10$ and $a = 8$. Write the solution as an ordered pair.

2. In the solution of the following system, $x = 2$. What is the value of y?

$$3x - 2y = 7,$$
$$5x + 4y = 8$$

3. After substitution, an equation simplifies to $4 = 6$. How many solutions does the system have?

4. After substitution, an equation simplifies to $9 = 9$. How many solutions does the system have?

Chapter Resource:
Collaborative Activity, p. 484

TO SOLVE A SYSTEM OF TWO EQUATIONS USING SUBSTITUTION

1. Solve for a variable in either one of the equations if neither equation already has a variable isolated.
2. Substitute the result of step (1) in the *other* equation for the variable isolated in step (1).
3. Solve the equation from step (2).
4. Substitute the solution from step (3) into one of the other equations to solve for the other variable.
5. Check that the ordered pair resulting from steps (3) and (4) checks in both of the original equations.

Note: Obtaining a false equation indicates that there is no solution. Obtaining an equation that is always true indicates that there is an infinite number of solutions.

B. Problem Solving

EXAMPLE 5 *Supplementary Angles.* Two angles are supplementary. One angle measures 30° more than twice the other. Find the measures of the two angles.

SOLUTION

1. **Familiarize.** Recall that two angles are supplementary if the sum of their measures is 180°. We could try to guess a solution, but instead we make a drawing and translate. Let x and y represent the measures of the two angles, in degrees.

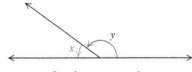
Supplementary angles

2. **Translate.** Since we are told that the angles are supplementary, one equation is

$$x + y = 180. \quad \textbf{(1)}$$

The second sentence can be rephrased and translated as follows:

Rewording: One angle is 30° more than two times the other.

Translating: $y = 2x + 30$ **(2)**

We now have a system of two equations in two unknowns:

$$x + y = 180, \quad \textbf{(1)}$$
$$y = 2x + 30. \quad \textbf{(2)}$$

3. **Carry out.** We substitute $2x + 30$ for y in equation (1) and solve for x:

$$
\begin{array}{ll}
x + y = 180 & \text{Equation (1)} \\
x + (2x + 30) = 180 & \text{Substituting} \\
3x + 30 = 180 & \\
3x = 150 & \text{Subtracting 30 from both sides} \\
x = 50. & \text{Dividing both sides by 3}
\end{array}
$$

Substituting 50 for x in equation (2) then gives us

$$
\begin{array}{ll}
y = 2x + 30 & \text{Equation (2)} \\
y = 2 \cdot 50 + 30 & \text{Substituting 50 for } x \\
y = 130.
\end{array}
$$

5. Two angles are complementary. (The sum of their angles is 90°.) One angle measures twice the other. Find the measures of the two angles.

4. Check. If one angle is 50° and the other is 130°, then the sum of the measures is 180°. Thus the angles are supplementary. If 30° is added to twice the measure of the smaller angle, we have $2 \cdot 50° + 30°$, or 130°, which is the measure of the other angle. The numbers check.

5. State. One angle measures 50° and the other 130°.

↪ YOUR TURN

7.2 EXERCISE SET

FOR EXTRA HELP MyMathLab®

↪ Vocabulary and Reading Check

Classify each of the following statements as either true or false.

1. When using the substitution method, we must solve for the variables in the order in which they occur alphabetically.

2. The substitution method often requires us to first solve for a variable, much as we did when solving for a letter in a formula.

3. When solving a system of equations algebraically leads to a false equation, the system has no solution.

4. When solving a system of two equations algebraically leads to an equation that is always true, the system has an infinite number of solutions.

A. The Substitution Method

Solve each system using the substitution method. If a system has no solution or an infinite number of solutions, state this.

5. $x + y = 9,$
    $y = x + 1$

6. $x + y = 5,$
    $x = y + 1$

7. $x = y + 1,$
    $x + 2y = 4$

8. $y = x - 3,$
    $3x + y = 5$

9. $y = 5x - 1,$
    $y - 3x = 1$

10. $y = 2x - 5,$
    $2y - x = 2$

11. $a = -4b,$
    $a + 5b = 5$

12. $r = -3s,$
    $r + 4s = 10$

13. $x = y - 5,$
    $2x + 5y = 4$

14. $x = y - 6,$
    $3x + 2y = 2$

15. $x = 2y + 1,$
    $3x - 6y = 2$

16. $y = 3x - 1,$
    $6x - 2y = 2$

17. $s + t = -5,$
    $s - t = 3$

18. $s - t = 2,$
    $s + t = -4$

19. $x - y = 5,$
    $x + 2y = 7$

20. $y - 2x = -6,$
    $2y - x = 5$

21. $x - 2y = 7,$
    $3x - 21 = 6y$

22. $x - 4y = 3,$
    $2x - 6 = 8y$

23. $y = 2x + 5,$
    $-2y = -4x - 10$

24. $y = -2x + 3,$
    $3y = -6x + 9$

25. $4x - y = -3,$
    $2x + 5y = 2$

26. $2x + 3y = -2,$
    $2x - y = 9$

27. $a - b = 6,$
    $3a - 2b = 12$

28. $x - y = -3,$
    $2x + 3y = -6$

29. $s = \frac{1}{2}r,$
    $3r - 4s = 10$

30. $x = \frac{1}{2}y,$
    $2x + y = 12$

31. $x - 3y = 7,$
    $-4x + 12y = 28$

32. $8x + 2y = 6,$
    $y = 3 - 4x$

33. $x - 2y = 5,$
    $2y - 3x = 1$

34. $x - 3y = -1,$
    $5y - 2x = 4$

Aha! **35.** $2x - y = 0,$
    $2x - y = -2$

36. $5x = y - 3,$
    $5x = y + 5$

B. Problem Solving

Solve using a system of equations.

37. The sum of two numbers is 63. One number is 7 more than the other. Find the numbers.

38. The sum of two numbers is 74. One number is 6 more than the other. Find the numbers.

39. Find two numbers for which the sum is 51 and the difference is 13.

40. Find two numbers for which the sum is 51 and the difference is 5.

41. The difference between two numbers is 2. Three times the larger number plus one-half the smaller is 34. What are the numbers?

42. The difference between two numbers is 11. Twice the smaller number plus three times the larger is 93. What are the numbers?

43. *Supplementary Angles.* Two angles are supplementary. One angle is 15° more than twice the other. Find the measure of each angle.

44. *Supplementary Angles.* Two angles are supplementary. One angle is 8° less than three times the other. Find the measure of each angle.

45. *Complementary Angles.* Two angles are complementary. One angle is 3° less than one-half of the other. Find the measure of each angle. (*Complementary angles* are pairs of angles for which the sum is 90°.)

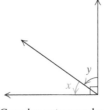

Complementary angles

46. *Complementary Angles.* Two angles are complementary. One angle is 42° more than one-half of the other. Find the measure of each angle.

47. *Billboards.* As an advertisement, the Meiji Seika Kaisha confectionary factory in Takatsuki, Osaka Prefecture, Japan, built a giant billboard in the shape of a chocolate bar. The perimeter of the billboard was 388 m, and the length was 138 m more than the width. Find the length and the width.

Data: www.worldrecordsacademy.org

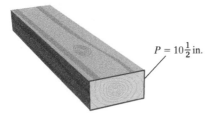

48. *Lumber.* The perimeter of a cross section of a piece of lumber is $10\frac{1}{2}$ in. The length is twice the width. Find the actual dimensions of the cross section of the lumber.

$P = 10\frac{1}{2}$ in.

49. *Dimensions of Wyoming.* The state of Wyoming is a rectangle with a perimeter of 1280 mi. The width is 90 mi less than the length. Find the length and the width.

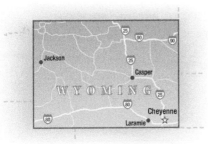

50. *Dimensions of Colorado.* The state of Colorado is roughly in the shape of a rectangle whose perimeter is 1300 mi. The width is 110 mi less than the length. Find the length and the width.

51. *Soccer.* The perimeter of a soccer field is 280 yd. The width is 5 yd more than half the length. Find the length and the width.

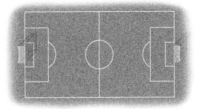

52. *Lacrosse.* The perimeter of a lacrosse field is 340 yd. The length is 10 yd less than twice the width. Find the length and the width.

53. *Racquetball.* The height of the front wall of a standard racquetball court is four times the width of the service zone. Together, these measurements total 25 ft. Find the height and the width.

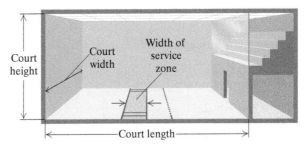

54. *Racquetball.* A regulation racquetball court has a perimeter of 120 ft, with a length that is twice the width. Find the length and the width of a court.

55. *Library Circulation.* In 2013, the patrons of U.S. public libraries, checked out a total of 951 million print and electronic books. They checked out 787 million more print books than electronic books. How many print books and how many electronic books did patrons check out in 2013?

Data: publiclibrariesonline.org

56. *Music Streaming.* The total number of on-demand music streams in the United States in the first six months of 2016 was 208.9 billion. There were 18.3 billion more audio streams than there were video streams. How many audio streams and how many video streams were there in the first six months of 2016?

Data: Nielsen.com

57. Craig solves every system of two equations (in *x* and *y*) by first solving for *y* in the first equation and then substituting into the second equation. Is he using the best approach? Why or why not?

58. Describe two advantages of the substitution method over the graphing method for solving systems of equations.

Skill Review

Multiply or divide and, if possible, simplify.

59. $\frac{2}{15} \div \left(-\frac{1}{3}\right)$ [1.7]

60. $(-1.3)(-2.5)$ [1.7]

61. $(2x + 1)(x^2 - 3)$ [4.6]

62. $(a^3 + b)^2$ [4.7]

63. $(x^2 - x + 10) \div (x - 5)$ [4.8]

64. $\dfrac{x^2 - 1}{5x^2} \div \dfrac{x^2 - x - 2}{x^2 + 2x}$ [6.2]

Synthesis

65. How can Kiara tell by inspection that the following system has no solution?

$$x = 2y - 1,$$
$$x = 2y + 3$$

66. Under what circumstances can a system of equations be solved more easily by graphing than by substitution?

Solve by the substitution method.

67. $\dfrac{1}{6}(a + b) = 1,$

$\dfrac{1}{4}(a - b) = 2$

68. $\dfrac{x}{5} - \dfrac{y}{2} = 3,$

$\dfrac{x}{4} + \dfrac{3y}{4} = 1$

69. $y + 5.97 = 2.35x,$
$2.14y - x = 4.88$

70. $a + 4.2b = 25.1,$
$9a - 1.8b = 39.78$

71. *Age at Marriage.* Trudy is 20 years younger than Dennis. She feels that she needs to be 7 more than half of Dennis's age before they can marry. What is the youngest age at which Trudy can marry Dennis and honor this requirement?

Exercises 72 and 73 contain systems of three equations in three variables. A solution is an ordered triple of the form (x, y, z). Use the substitution method to solve.

72. $x + y + z = 4,$
$x - 2y - z = 1,$
$y = -1$

73. $x + y + z = 180,$
$x = z - 70,$
$2y - z = 0$

74. *Softball.* The perimeter of a softball diamond is two-thirds of the perimeter of a baseball diamond. Together, the two perimeters measure 200 yd. Find the distance between the bases in each sport.

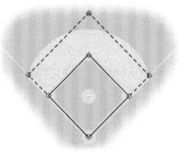

75. Solve Example 3 by first solving for 2*y* in equation (1) and then substituting for 2*y* in equation (2). Is this method easier than the procedure used in Example 3? Why or why not?

76. Write a system of two linear equations that can be solved more quickly—but still precisely—by a graphing calculator than by substitution. Time yourself using both methods to solve the system.

YOUR TURN ANSWERS: SECTION 7.2

1. $(7, 4)$ **2.** $(2, -1)$ **3.** $\left(-\frac{2}{7}, \frac{13}{7}\right)$ **4.** No solution
5. $30°, 60°$

Quick Quiz: Sections 7.1–7.2

1. Determine whether $(-1, -2)$ is a solution of this system of equations:

$$x - y = 1,$$
$$x - 2y = 3. \quad [7.1]$$

Solve by graphing. If there is no solution or an infinite number of solutions, state this. [7.1]

2. $y = x - 3,$
$x + y = 5$

3. $x - 2y = 4,$
$y = \frac{1}{2}x + 3$

Solve using the substitution method. If there is no solution or an infinite number of solutions, state this. [7.2]

4. $y = 2x - 1,$
$x - 3y = 8$

5. $2x + y = 1,$
$2y = 2 - 4x$

Prepare to Move On

Simplify. [1.8]

1. $3(4x + 2y) - 5(2x + y)$

2. $5(2x + 3y) - 3(7x + 5y)$

3. $3(8x + 1 - 2y) - 8(3x + 1)$

4. $2(3x - 4y) + 4(5x + 2y)$

7.3 | Systems of Equations and Elimination

A. The Elimination Method   **B.** Problem Solving

One way to solve the system of equations

$$2x + 3y = 13, \quad \textbf{(1)}$$
$$4x - 3y = 17 \quad \textbf{(2)}$$

is by substitution. Solving equation (1) for y, we find that $y = \frac{13}{3} - \frac{2}{3}x$. We then use the expression $\frac{13}{3} - \frac{2}{3}x$ in equation (2) as the replacement for y:

$$4x - 3\left(\frac{13}{3} - \frac{2}{3}x\right) = 17.$$

Although this approach works, another method, *elimination*, allows us to solve this system without using fraction notation.

A. The Elimination Method

The **elimination method** for solving systems of equations makes use of the addition principle. To see how it works, we use it to solve the system above.

EXAMPLE 1 Solve the system

$$2x + 3y = 13, \quad \textbf{(1)}$$
$$4x - 3y = 17. \quad \textbf{(2)}$$

SOLUTION According to equation (2), $4x - 3y$ and 17 are the same number. Thus we can add $4x - 3y$ to the left side and 17 to the right side of equation (1):

$$2x + 3y = 13 \quad \textbf{(1)}$$
$$\underline{4x - 3y = 17} \quad \textbf{(2)}$$
$$6x + 0y = 30. \quad \text{Adding. Note that } y \text{ has been "eliminated."}$$

The resulting equation has just one variable:

$$6x = 30.$$

Dividing both sides of this equation by 6, we find that $x = 5$.

Next, we substitute 5 for x in either of the original equations:

$$2x + 3y = 13 \qquad \text{Equation (1)}$$
$$2 \cdot 5 + 3y = 13 \qquad \text{Substituting 5 for } x$$
$$10 + 3y = 13$$
$$3y = 3$$
$$y = 1. \qquad \text{Solving for } y$$

We check the ordered pair $(5, 1)$. The graph shown at left also serves as a check.

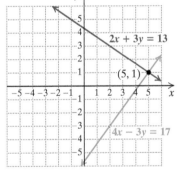

Check:

$2x + 3y = 13$	
$2(5) + 3(1)$	13
$10 + 3$	
$13 \stackrel{?}{=} 13$ TRUE	

$4x - 3y = 17$	
$4(5) - 3(1)$	17
$20 - 3$	
$17 \stackrel{?}{=} 17$ TRUE	

Since $(5, 1)$ checks in both equations, it is the solution.

1. Solve the system
$$3x - y = 1,$$
$$2x + y = 4.$$

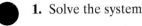 YOUR TURN

In Example 1, we used addition to eliminate the variable y because two terms, $-3y$ in equation (2) and $3y$ in equation (1), are opposites. When a system has no pair of terms that are opposites, we can multiply one or both of the equations by appropriate numbers to create a pair of terms that are opposites.

EXAMPLE 2 Solve:

$$2x + 3y = 8, \quad \textbf{(1)}$$
$$x + 3y = 7. \quad \textbf{(2)}$$

SOLUTION Adding these equations as they now appear will not eliminate a variable. However, if the $3y$ were $-3y$ in one equation, we could eliminate y. We multiply both sides of equation (2) by -1 to find an equivalent equation that contains $-3y$, and then add:

$$\begin{array}{ll} 2x + 3y = 8 & \text{Equation (1)} \\ \underline{-x - 3y = -7} & \text{Multiplying both sides of equation (2) by } -1 \\ x = 1. & \text{Adding; } 3y + (-3y) = 0 \end{array}$$

Next, we substitute 1 for x in either of the original equations:

$$\begin{array}{ll} x + 3y = 7 & \text{Equation (2)} \\ 1 + 3y = 7 & \text{Substituting 1 for } x \\ \left.\begin{array}{l} 3y = 6 \\ y = 2. \end{array}\right\} & \text{Solving for } y \end{array}$$

We can check the ordered pair $(1, 2)$. The graph shown at left is also a check.

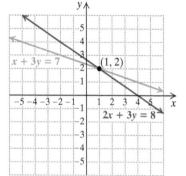

Check:

$$\begin{array}{c|c} 2x + 3y = 8 & \\ \hline 2 \cdot 1 + 3 \cdot 2 & 8 \\ 2 + 6 & \\ & 8 \stackrel{?}{=} 8 \quad \text{TRUE} \end{array} \qquad \begin{array}{c|c} x + 3y = 7 & \\ \hline 1 + 3 \cdot 2 & 7 \\ 1 + 6 & \\ & 7 \stackrel{?}{=} 7 \quad \text{TRUE} \end{array}$$

Since $(1, 2)$ checks in both equations, it is the solution.

2. Solve:

$$4x + 3y = 5,$$
$$4x + y = 7.$$

⤶ YOUR TURN

When deciding which variable to eliminate, we inspect the coefficients in both equations. If one coefficient is a multiple of the coefficient of the same variable in the other equation, that is the easier variable to eliminate.

EXAMPLE 3 Solve:

$$3x + 6y = -6, \quad \textbf{(1)}$$
$$5x - 2y = 14. \quad \textbf{(2)}$$

SOLUTION No terms are opposites, but if both sides of equation (2) are multiplied by 3 $\left(\text{or if both sides of equation (1) are multiplied by } \tfrac{1}{3}\right)$, the coefficients of y will be opposites. Note that 6 is the LCM of 2 and 6:

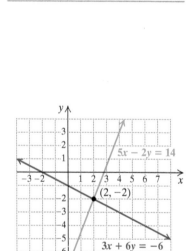

$$\begin{array}{ll} 3x + 6y = -6 & \text{Equation (1)} \\ \underline{15x - 6y = 42} & \text{Multiplying both sides of equation (2) by 3} \\ 18x = 36 & \text{Adding; } 6y + (-6y) = 0 \\ x = 2. & \text{Solving for } x \end{array}$$

We then substitute 2 for x in either equation (1) or equation (2):

$$3 \cdot 2 + 6y = -6 \qquad \text{Substituting 2 for } x \text{ in equation (1)}$$

$$\left. \begin{array}{r} 6 + 6y = -6 \\ 6y = -12 \\ y = -2. \end{array} \right\} \quad \text{Solving for } y$$

3. Solve:

$$2x + 5y = 4,$$
$$6x - 3y = 4.$$

We leave it to the student to confirm that $(2, -2)$ checks and is the solution. The graph in the margin at the bottom of the preceding page also serves as a check.

YOUR TURN

Sometimes both equations must be multiplied.

EXAMPLE 4   Solve:

$$3y + 1 + 2x = 0, \qquad \textbf{(1)}$$
$$5x = 7 - 4y. \qquad \textbf{(2)}$$

SOLUTION   It is often helpful to write both equations in the form $Ax + By = C$ before attempting to eliminate a variable:

$$2x + 3y = -1, \qquad \textbf{(3)} \qquad \text{Subtracting 1 from both sides and rearranging the terms of the first equation}$$

$$5x + 4y = 7. \qquad \textbf{(4)} \qquad \text{Adding } 4y \text{ to both sides of equation (2)}$$

Since neither coefficient of x is a multiple of the other and neither coefficient of y is a multiple of the other, we use the multiplication principle with *both* equations. Note that we can eliminate the x-term by multiplying both sides of equation (3) by 5 and both sides of equation (4) by -2:

> Multiply to get terms that are opposites.

$$\begin{array}{r} 10x + 15y = -5 \qquad \text{Multiplying both sides of equation (3) by 5} \\ \underline{-10x - 8y = -14} \qquad \text{Multiplying both sides of equation (4) by } -2 \\ 7y = -19 \qquad \text{Adding} \\ y = \tfrac{-19}{7} = -\tfrac{19}{7}. \qquad \text{Dividing by 7} \end{array}$$

> Solve for one variable.

We substitute $-\frac{19}{7}$ for y in equation (3):

> Substitute.

$$\begin{array}{ll} 2x + 3y = -1 & \text{Equation (3)} \\ 2x + 3\left(-\tfrac{19}{7}\right) = -1 & \text{Substituting } -\tfrac{19}{7} \text{ for } y \\ 2x - \tfrac{57}{7} = -1 & \\ 2x = -1 + \tfrac{57}{7} & \text{Adding } \tfrac{57}{7} \text{ to both sides} \\ 2x = -\tfrac{7}{7} + \tfrac{57}{7} = \tfrac{50}{7} & \\ x = \tfrac{50}{7} \cdot \tfrac{1}{2} = \tfrac{25}{7}. & \text{Solving for } x \end{array}$$

> Solve for the other variable.

We check the ordered pair $\left(\frac{25}{7}, -\frac{19}{7}\right)$.

> Check in both equations.

Check:

$$\begin{array}{c|c} \hline 3y + 1 + 2x = 0 \\ \hline 3\left(-\tfrac{19}{7}\right) + 1 + 2 \cdot \tfrac{25}{7} \; \Big| \; 0 \\ -\tfrac{57}{7} + \tfrac{7}{7} + \tfrac{50}{7} \; \Big| \\ 0 \stackrel{?}{=} 0 \quad \text{TRUE} \end{array}$$

$$\begin{array}{c|c} \hline 5x = 7 - 4y \\ \hline 5 \cdot \tfrac{25}{7} \; \Big| \; 7 - 4\left(-\tfrac{19}{7}\right) \\ \tfrac{125}{7} \; \Big| \; \tfrac{49}{7} + \tfrac{76}{7} \\ \tfrac{125}{7} \stackrel{?}{=} \tfrac{125}{7} \quad \text{TRUE} \end{array}$$

> State the solution as an ordered pair.

4. Solve:

$$2x = 5y + 4,$$
$$3x + 2y = -13.$$

The solution is $\left(\frac{25}{7}, -\frac{19}{7}\right)$.

YOUR TURN

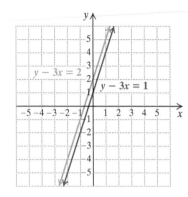

5. Solve:

$$y = 2x + 1,$$
$$2x - y = 3.$$

_____

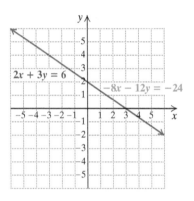

6. Solve:

$$x - 4y = 2,$$
$$2x = 8y + 4.$$

_____

7. Solve:

$$0.1x + 0.3y = 0.2,$$
$$x - 0.2y = 0.8.$$

_____

Next, we consider a system with no solution and see what happens when the elimination method is used.

EXAMPLE 5 Solve:

$$y - 3x = 2, \quad \textbf{(1)}$$
$$y - 3x = 1. \quad \textbf{(2)}$$

SOLUTION To eliminate y, we multiply both sides of equation (2) by -1. Then we add:

$$
\begin{array}{ll}
y - 3x = 2 & \\
\underline{-y + 3x = -1} & \text{Multiplying both sides of equation (2) by } -1 \\
 0 = 1. & \text{Adding. Note that this is a } \textit{false} \text{ equation.}
\end{array}
$$

Note that in eliminating y, we eliminated x as well. The resulting equation, $0 = 1$, is false for any pair (x, y), so there is *no solution*. The graph at left confirms this.

↩ YOUR TURN

Sometimes there is an infinite number of solutions.

EXAMPLE 6 Solve:

$$2x + 3y = 6, \quad \textbf{(1)}$$
$$-8x - 12y = -24. \quad \textbf{(2)}$$

SOLUTION To eliminate x, we multiply both sides of equation (1) by 4 and then add:

$$
\begin{array}{ll}
8x + 12y = 24 & \text{Multiplying both sides of equation (1) by 4} \\
\underline{-8x - 12y = -24} & \\
 0 = 0. & \text{Adding. Note that this equation is } \textit{always} \text{ true.}
\end{array}
$$

Again, we have eliminated *both* variables. This time, however, the resulting equation, $0 = 0$, is always true, indicating that the equations are dependent. Such a system has an infinite number of solutions. The graph at left confirms this.

↩ YOUR TURN

When decimals or fractions appear, we can first multiply to clear them. Then we proceed as before.

EXAMPLE 7 Solve:

$$\tfrac{1}{2}x + \tfrac{3}{4}y = 2, \quad \textbf{(1)}$$
$$x + 3y = 7. \quad \textbf{(2)}$$

SOLUTION The number 4 is the LCM of the denominators in equation (1). Thus we multiply both sides of equation (1) by 4 to clear fractions:

$$
\begin{array}{ll}
4\left(\tfrac{1}{2}x + \tfrac{3}{4}y\right) = 4 \cdot 2 & \text{Multiplying both sides of equation (1) by 4} \\
4 \cdot \tfrac{1}{2}x + 4 \cdot \tfrac{3}{4}y = 8 & \text{Using the distributive law} \\
2x + 3y = 8. &
\end{array}
$$

The resulting system is

$$2x + 3y = 8, \quad \text{This equation is equivalent to equation (1).}$$
$$x + 3y = 7.$$

As we saw in Example 2, the solution of this system is $(1, 2)$.

↩ YOUR TURN

1. Adding in the following system will result in the elimination of what variable?

$$3x - 2y = 7,$$
$$3x + 2y = 5$$

2. In the following system, what number should equation (1) be multiplied by in order to eliminate x when adding?

$$5x - 2y = 7, \quad \textbf{(1)}$$
$$10x + 4y = 3 \quad \textbf{(2)}$$

3. In order to eliminate y by adding in the following system, we can multiply equation (1) by 3 and equation (2) by what number?

$$3x + 2y = 5, \quad \textbf{(1)}$$
$$5x + 3y = 8 \quad \textbf{(2)}$$

Student Notes

Before diving into a problem, spend some time deciding which variable to eliminate. Always try to select the approach that will minimize your calculations

8. The purchase price of an Epson SureColor P400 printer is $530, and the ink for each 11-in. by 14-in. photo costs $2.21. The purchase price of a Canon Pixma Pro-100 printer is $394, and the ink for each 11-in. by 14-in. photo costs $2.38. For what number of 11-in. by 14-in. photos will the total cost (without paper) be the same?

Data: www.computershopper.com

B. Problem Solving

EXAMPLE 8 *Printing Costs.* The purchase price of a Brother HL-L6200DW printer is $199, and the ink for each black-and-white page costs 1.3¢. The purchase price of an OKI B412dn is $166, and the ink for each black-and-white page costs 1.8¢. For what number of pages will the total cost (without paper) be the same?

Data: www.computershopper.com

1. **Familiarize.** Some costs are given in dollars and some in cents. Before calculating any costs, we convert cents to dollars: 1.3¢ = $0.013 and 1.8¢ = $0.018.

 To become familiar with this problem, we make and check a guess of 100 pages. If the Brother is purchased, the cost would be $199 + $0.013 · 100, or $200.30. If the OKI is purchased, the cost would be $166 + $0.018 · 100, or $167.80. Since $200.30 ≠ $167.80, our guess is incorrect. However, from the check, we can see how equations can be written to model the situation. We let p = the number of pages printed and c = the total cost (without paper) of printing those pages.

2. **Translate.** We reword the problem and translate as follows:

	Total cost using the Brother	is	the purchase price	plus	$0.013	times	the number of pages.
Rewording:							
Translating:	c	$=$	199	$+$	0.013	$\cdot$	p

	Total cost using the OKI	is	the purchase price	plus	$0.018	times	the number of pages.
Rewording:							
Translating:	c	$=$	166	$+$	0.018	$\cdot$	p

We now have the system of equations

$$c = 199 + 0.013p,$$
$$c = 166 + 0.018p.$$

3. **Carry out.** To solve the system, we multiply the second equation by -1 and add to eliminate c:

$$\begin{aligned} c &= 199 + 0.013p \\ -c &= -166 - 0.018p \\ \hline 0 &= 33 - 0.005p. \end{aligned}$$

We can now solve for p:

$$0.005p = 33 \qquad \text{Adding } 0.005p \text{ to both sides}$$
$$p = 6600. \qquad \text{Dividing both sides by } 0.005$$

4. **Check.** To print 6600 pages using the Brother would cost

$$\$199 + \$0.013 \cdot 6600 = \$199 + \$85.80, \text{ or } \$284.80.$$

To print 6600 pages using the OKI would cost

$$\$166 + \$0.018 \cdot 6600 = \$166 + \$118.80, \text{ or } \$284.80.$$

The costs are the same.

5. **State.** For 6600 pages, the total costs using the two printers are the same.

↻ YOUR TURN

CONNECTING 🔗 THE CONCEPTS

We now have three distinctly different ways to solve a system of equations. Each method has certain strengths and weaknesses.

Method	Strengths	Weaknesses
Graphical	Solutions are displayed visually. Works with any system that can be graphed.	For some systems, only approximate solutions can be found graphically. The graph drawn may not be large enough to show the solution.
Substitution	Always yields exact solutions. Easy to use when a variable is alone on one side of an equation.	Substitution introduces extensive computations with fractions when solving more complicated systems. Solutions are not graphically displayed.
Elimination	Always yields exact solutions. Can often be used to avoid working with fractions.	Solutions are not graphically displayed.

EXERCISES

Solve using the best method for each system.

1. $x = y$,
   $x + y = 4$

2. $x + y = 5$,
   $x - y = 3$

3. $y = x + 1$,
   $2x + y = 6$

4. $2x + 3y = 6$,
   $2x - 3y = 2$

5. $y = 2x + 1$,
   $y = \frac{1}{2}x - 2$

6. $y = \frac{2}{3}x - 5$,
   $y = \frac{2}{3}x + 1$

7.3   EXERCISE SET

FOR EXTRA HELP MyMathLab®

👆 Vocabulary and Reading Check

Following are the steps used to solve a system of equations using elimination. Choose the word from the following list that best completes each step.

both   ordered
eliminate   solve
opposites   variable

1. Multiply to get terms that are _____.

2. Add to _____ one variable.

3. _____ for the remaining variable.

4. Substitute and solve for the _____ that was eliminated.

5. Check the values for the variables in _____ equations.

6. State the answer as a(n) _____ pair.

A. The Elimination Method

Solve using the elimination method. If a system has no solution or an infinite number of solutions, state this.

7. $x - y = 3$,
   $x + y = 13$

8. $x + y = 5$,
   $x - y = 1$

9. $x + y = 6$,
   $-x + 3y = -2$

10. $x + y = 6$,
   $-x + 2y = 15$

11. $4x + y = 5,$
$2x - y = 7$

12. $2x - y = 3,$
$3x + y = -8$

13. $5a + 4b = 7,$
$-5a + b = 8$

14. $7c + 4d = 16,$
$c - 4d = -4$

15. $8x - 5y = -9,$
$3x + 5y = -2$

16. $3a - 3b = -15,$
$-3a - 3b = -3$

17. $3a - 6b = 8,$
$-3a + 6b = -8$

18. $8x + 3y = 4,$
$-8x - 3y = -4$

19. $-x - y = 3,$
$2x - y = -3$

20. $x - y = 1,$
$3x - y = -5$

21. $x + 3y = 19,$
$x - y = -1$

22. $3x - y = 8,$
$x + 2y = 5$

23. $8x - 3y = -6,$
$5x + 6y = 75$

24. $x - y = 3,$
$2x - 3y = -1$

25. $2w - 3z = -1,$
$-4w + 6z = 5$

26. $7p + 5q = 2,$
$8p - 9q = 17$

27. $4a + 6b = -1,$
$a - 3b = 2$

28. $x + 9y = 1,$
$2x - 6y = 10$

29. $3y = x,$
$5x + 14 = y$

30. $5a = 2b,$
$2a + 11 = 3b$

31. $4x - 10y = 13,$
$-2x + 5y = 8$

32. $2p + 5q = 9,$
$3p - 2q = 4$

33. $2n - 15 - 10m = 40,$
$28 = n - 4m$

34. $30y + 14 + x = 0,$
$41 = 5y - 2x$

35. $3x + 5y = 4,$
$-2x + 3y = 10$

36. $2x + y = 13,$
$4x + 2y = 23$

37. $0.06x + 0.05y = 0.07,$
$0.4x - 0.3y = 1.1$

38. $x - \frac{3}{2}y = 13,$
$\frac{3}{2}x - y = 17$

39. $x + \frac{9}{2}y = \frac{15}{4},$
$\frac{9}{10}x - y = \frac{9}{20}$

40. $1.8x - 2y = 0.9,$
$0.04x + 0.18y = 0.15$

B. Problem Solving

Solve.

41. *Car Rentals.* The University of Minnesota maintains a fleet of vehicles for lease to university departments. To transport equipment, the School of Drama needs to lease either a cargo van or a utility vehicle for one month. The cargo van costs $303 for one month plus 29¢ per mile. The utility vehicle costs $388 for one month plus 18¢ per mile. For what mileage is the cost the same?

Data: pts.umn.edu

42. *RV Rentals.* Greenwood RV (recreational vehicle) Rentals rents a 29-ft motorhome for $225 per night plus 32¢ per mile. Cruise America RV Rentals rents a similar RV for $170 per night plus 35¢ per mile. For what mileage is the cost the same for a one-night rental?

Data: Greenwood RV Rentals, LLC; Cruise America RV Rentals and Sales

43. *Complementary Angles.* Two angles are complementary. The sum of the measure of one angle and one-half of the measure of the other is 74°. Find the measure of each angle.

44. *Complementary Angles.* Two angles are complementary. The sum of the measure of one angle and one-fourth of the measure of the other is 78°. Find the measure of each angle.

45. *Copier Costs.* EthanElliott Enterprises plans to lease a new copier. The copier sales representative suggested two copiers that would meet their needs. The Sharp MX-C301W leases for $90.25 per month plus 1.5¢ per page, and the Sharp MX-3050N leases for $160.65 per month plus 1.3¢ per page. For what number of pages copied per month will the cost of the two copiers be the same?

Data: G5

46. *International Calling Plans.* Sydney plans to spend one month in Ireland as part of an archaeology internship. Her cell-phone carrier offers two plans that fit her needs while overseas. The Silver plan costs $60 for the month plus 50¢ per minute for calls. The Gold plan costs $120 for the month plus 35¢ per minute for calls. For what number of minutes will the cost of the two plans be the same?

Data: att.com

47. *Supplementary Angles.* Two angles are supplementary. The difference of the measures of the angles is 106°. Find the measure of each angle.

48. *Supplementary Angles.* Two angles are supplementary. The difference of the measures of the angles is 90°. Find the measure of each angle.

49. *Planting Grapes.* South Wind Vineyards uses 820 acres to plant Chardonnay and Riesling grapes. The vintner knows the profits will be greatest by planting 140 more acres of Chardonnay than Riesling. How many acres of each type of grape should be planted?

50. *Baking.* Maple Branch Bakers sells 175 loaves of bread each day—some white and the rest whole-wheat. Because of a regular order from a local sandwich shop, Maple Branch consistently bakes 9 more loaves of white bread than whole-wheat. How many loaves of each type of bread do they bake?

51. *Framing.* Angel has 18 ft of molding from which he needs to make a rectangular mirror frame. The frame is to be twice as long as it is wide. What are the dimensions of the frame?

52. *Gardening.* Patrice has 108 ft of fencing for a rectangular garden. If the garden's length is to be $1\frac{1}{2}$ times its width, what should the garden's dimensions be?

53. Describe a method that could be used for writing a system that contains dependent equations.

54. Describe a method that could be used for writing an inconsistent system of equations.

Skill Review

Solve.

55. $x - (5 - x) = 3(x + 1)$ **56.** $9t^2 = 1$ [5.6]
   [2.2]

57. $3a + 5 \geq 4 - 6a$ [2.6]   **58.** $3x^2 + 5 = 16x$ [5.6]

59. $x + \dfrac{1}{x} = 2$ [6.6]   **60.** $\dfrac{2}{x+1} - \dfrac{1}{x-2} = 2$
   [6.6]

Synthesis

61. If a system has an infinite number of solutions, does it follow that *any* ordered pair is a solution? Why or why not?

62. Explain how the multiplication and addition principles are used in this section. Then count the number of times that these principles are used in Example 4.

Solve using substitution, elimination, or graphing.

63. $y = 3x + 4$,
   $3 + y = 2(y - x)$

64. $x + y = 7$,
   $3(y - x) = 9$

65. $0.05x + y = 4$,
   $\dfrac{x}{2} + \dfrac{y}{3} = 1\dfrac{1}{3}$

66. $2(5a - 5b) = 10$,
   $-5(2a + 6b) = 10$

Aha! **67.** $y = -\frac{2}{7}x + 3$,
   $y = \frac{4}{5}x + 3$

68. $y = \frac{2}{5}x - 7$,
   $y = \frac{2}{5}x + 4$

Solve for x and y.

69. $ax + by + c = 0$,
   $ax + cy + b = 0$

70. $y = ax + b$,
   $y = x + c$

71. *Caged Rabbits and Pheasants.* Several ancient Chinese books included problems that can be solved by translating to systems of equations. *Arithmetical Rules in Nine Sections* is a book of 246 problems compiled by a Chinese mathematician, Chang Tsang, who died in 152 B.C. One of the problems is: Suppose that there are a number of rabbits and pheasants confined in a cage. In all, there are 35 heads and 94 feet. How many rabbits and how many pheasants are there? Solve the problem.

72. *Age.* Patrick's age is 20% of his mother's age. Twenty years from now, Patrick's age will be 52% of his mother's age. How old are Patrick and his mother now?

73. *Age.* If 5 is added to a man's age and the total is divided by 5, the result will be his daughter's age. Five years ago, the man's age was eight times his daughter's age. Find their present ages.

74. *Dimensions of a Triangle.* When the base of a triangle is increased by 1 ft and the height is increased by 2 ft, the height changes from being two-thirds of the base to being four-fifths of the base. Find the original dimensions of the triangle.

YOUR TURN ANSWERS: SECTION 7.3
1. $(1, 2)$ **2.** $(2, -1)$ **3.** $\left(\frac{8}{9}, \frac{4}{9}\right)$ **4.** $(-3, -2)$
5. No solution   **6.** Infinite number of solutions
7. $\left(\frac{7}{8}, \frac{3}{8}\right)$ **8.** 800 photos

Quick Quiz: Sections 7.1–7.3

1. Solve by graphing:

   $x - y = 3$,
   $y = \frac{1}{2}x - 2$.   [7.1]

2. Solve using substitution:

   $3x + y = 5$,
   $x - y = 2$.   [7.2]

3. Solve using elimination:

   $3x - 4y = 8$,
   $5x + 4y = 2$.   [7.3]

Solve using any method.   [7.1], [7.2], [7.3]

4. $2x = y - 1$,   5. $x + y = 5$,
   $x + 3y = 8$ $x - y = 3$

Prepare to Move On

Convert to decimal notation.   [2.4]

1. 12.2%   **2.** 0.5%

Solve.   [2.4]

3. What percent of 65 is 26?

4. What number is 17% of 18?

Translate to an algebraic expression.   [1.1], [2.4]

5. 12% of the number of liters

6. 10.5% of the number of pounds

7.4 More Applications Using Systems

A. Total-Value Problems   **B.** Mixture Problems

The five steps for problem solving and the methods for solving systems of equations can be used in a variety of applications.

A. Total-Value Problems

EXAMPLE 1 *Basketball Scores.* In the final game of the 2016 NBA playoffs, the Cleveland Cavaliers scored 93 points. Of those, 21 points were from free throws. The remaining 72 points were the result of a combination of 33 two-point and three-point baskets. How many baskets of each type were made?

Data: National Basketball Association

SOLUTION

1. **Familiarize.** Suppose that of the 33 baskets, 20 were two-pointers and 13 were three-pointers. These 33 baskets would then amount to a total of

$$20 \cdot 2 + 13 \cdot 3 = 40 + 39 = 79 \text{ points.}$$ The total should be 72.

Although our guess is incorrect, checking the guess has familiarized us with the problem. We let $w =$ the number of two-pointers made and $r =$ the number of three-pointers made.

2. **Translate.** Since a total of 33 baskets was made, we must have

$$w + r = 33.$$

To find a second equation, we reword some information and focus on the number of points scored, just as when we checked our guess above.

	The points scored		the points scored		
Rewording:	from two-pointers	plus	from three-pointers	totaled	72.
Translating:	$w \cdot 2$	$+$	$r \cdot 3$	$=$	72

The problem has been translated to the following system of equations:

$$w + r = 33, \quad \textbf{(1)}$$
$$2w + 3r = 72. \quad \textbf{(2)}$$

3. **Carry out.** For purposes of review, we solve by substitution. First, we solve equation (1) for w:

$$w + r = 33 \qquad \text{Equation (1)}$$
$$w = 33 - r. \quad \textbf{(3)} \qquad \text{Solving for } w$$

Next, we replace w in equation (2) with $33 - r$:

$$2w + 3r = 72 \qquad \text{Equation (2)}$$
$$2(33 - r) + 3r = 72 \qquad \text{Substituting } 33 - r \text{ for } w$$
$$66 - 2r + 3r = 72$$
$$\left. \begin{array}{l} 66 + r = 72 \\ r = 6. \end{array} \right\} \quad \text{Solving for } r$$

We find w by substituting 6 for r in equation (3):

$$w = 33 - r = 33 - 6 = 27.$$

Cavaliers 93, Wariors 89

CLEVELAND (93)

James 9–24 8–10 27, Irving 10–23 4–4 26,

Smith 5–13 0–0 12, Thompson 3–3 3–4 9,

Love 3–9 3–4 9, Jefferson 1–4 0–0 2,

Shumpert 1–3 3–3 6, Williams 1–3 0–0 2,

Totals 33–82 21–25 93.

GOLDEN STATE (89)

Green 11–15 4–4 32, Thompson 6–17 0–0 14,

Curry 6–19 1–1 17, Barnes 3–10 2–2 10,

Ezeli 0–4 0–0 0, Iguodala 2–6 0–2 4,

Livingston 3–7 2–2 8, Varejao 0–1 1–2 1,

Speights 0–2 0–0 0, Barbosa 1–2 0–0 3,

Totals 32–83 10–13 89.

Study Skills

Putting Math to Use

One excellent way to study math is to use it in your everyday life. The concepts of this section can be easily reinforced if you look for real-life situations in which systems of equations can be used. Keep this in mind when you are shopping at a grocery or hardware store and buying multiple items of two or more types.

1. In a recent game, Kristaps Porzingis of the New York Knicks scored 20 points on a combination of 9 two-point and three-point baskets. How many shots of each type were made?

Data: National Basketball Association

4. Check. If the Cavaliers made 27 two-pointers and 6 three-pointers, they would have made 33 shots, for a total of

$$27 \cdot 2 + 6 \cdot 3 = 54 + 18 = 72 \text{ points.}$$

The numbers check.

5. State. The Cavaliers made 27 two-pointers and 6 three-pointers.

YOUR TURN

EXAMPLE 2   At UCLA, a student commuter parking permit costs $231 per quarter. A clean-fuel parking permit costs $189 per quarter. One afternoon, 52 students bought permits of these two types at the parking services office. If a total of $11,214 was collected from these students, how many of each type of permit was purchased?

Data: ucla.edu

SOLUTION

1. Familiarize. When faced with a new problem, it is often useful to compare it to a similar problem that you have already solved. Here instead of counting two-point and three-point baskets, as in Example 1, we are counting commuter permits and clean-fuel permits. We let c = the number of commuter permits and f = the number of clean-fuel permits purchased.

2. Translate. Organizing the information in a table can be helpful.

	Commuter Permits	Clean-Fuel Permits	Total
Number of Permits	c	f	52
Cost per Permit	$231	$189	
Amount Paid	$231c$	$189f$	11,214

Since 52 permits were purchased, we have

$$c + f = 52. \quad \textbf{(1)}$$

To find a second equation, we focus on the amount of money paid:

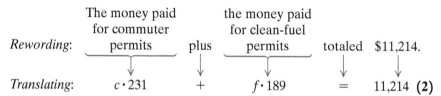

Rewording: The money paid for commuter permits plus the money paid for clean-fuel permits totaled $11,214.

Translating: $c \cdot 231$ $+$ $f \cdot 189$ $=$ 11,214 **(2)**

We can now write a system of equations:

$$c + f = 52, \quad \textbf{(1)}$$
$$231c + 189f = 11,214. \quad \textbf{(2)}$$

3. Carry out. The system can be solved using elimination:

$$
\begin{array}{ll}
-231c - 231f = -12{,}012 & \text{Multiplying both sides} \\
 & \text{of equation (1) by } -231 \\
\underline{231c + 189f = 11{,}214} & \text{Equation (2)} \\
-42f = -798 & \text{Adding} \\
f = 19. & \text{Dividing both sides by } -42
\end{array}
$$

2. Refer to Example 2. Suppose that on another afternoon, 28 students purchased either commuter or clean-fuel permits. If a total of $6048 was collected from these students, how many of each type of permit was purchased?

We substitute 19 for f and solve for c:

$$c + f = 52 \qquad \text{Using equation (1)}$$
$$c + 19 = 52$$
$$c = 33. \qquad \text{Subtracting 19 from both sides}$$

4. Check. If $c = 33$ and $f = 19$, the total number of permits is 52. The students paid $33(\$231)$, or $7623, for the commuter permits and $19(\$189)$, or $3591, for the clean-fuel permits. The total paid was therefore $7623 + $3591, or $11,214. The numbers check.

5. State. The students purchased 33 commuter permits and 19 clean-fuel permits.

YOUR TURN

B. Mixture Problems

EXAMPLE 3 *Blending Coffees.* The Roasted Bean's best-selling coffee, Perfect Morning Blend, sells for $13.35 per pound. This blend is a mix of bold organic Brazilian beans that sell for $13.95 per pound and mellow fair-trade Mexican beans that sell for $11.95 per pound. How much Brazilian coffee and how much Mexican coffee should be mixed in order to make a 50-lb batch of Perfect Morning Blend?

SOLUTION

1. **Familiarize.** This problem seems similar to Example 2. We have two different prices per pound for two types of coffee. We let $b = $ the number of pounds of organic Brazilian coffee and $m = $ the number of pounds of fair-trade Mexican coffee used.

   We do *not* have a total value for the batch of Perfect Morning Blend, but we can calculate that value by multiplying 50 lb by $13.35 per pound.

2. **Translate.** Since a 50-lb batch is being made, we must have

$$b + m = 50. \qquad \textbf{(1)}$$

The value of the beans that make up the blend must be the same as the value of the blend.

	The value of the Brazilian beans	plus	the value of the Mexican beans	is	the value of the Perfect Morning Blend.
Rewording:					
Translating:	$b \cdot 13.95$	$+$	$m \cdot 11.95$	$=$	$50 \cdot 13.35$	$\textbf{(2)}$

This information can be organized in a table.

	Brazilian Beans	Mexican Beans	Perfect Morning Blend	
Number of Pounds	b	m	50	$\longrightarrow b + m = 50$
Price per Pound	13.95	11.95	13.35	
Value of Beans	$13.95b$	$11.95m$	$50 \cdot 13.35$, or 667.50	$\longrightarrow 13.95b + 11.95m = 667.50$

We have translated to a system of equations:

$$b + m = 50, \quad \textbf{(1)}$$
$$13.95b + 11.95m = 667.50. \quad \textbf{(2)}$$

3. **Carry out.** When equation (1) is solved for b, we have $b = 50 - m$. We then substitute $50 - m$ for b in equation (2):

$13.95(50 - m) + 11.95m = 667.50$	Solving by substitution
$697.50 - 13.95m + 11.95m = 667.50$	Using the distributive law
$-2m = -30$	Combining like terms; subtracting 697.50 from both sides
$m = 15.$	Dividing both sides by -2

If $m = 15$, we see from equation (1) that $b = 35$.

4. **Check.** If 35 lb of Brazilian beans and 15 lb of Mexican beans are mixed, a 50-lb blend will result. The value of 35 lb of Brazilian beans is $35(\$13.95)$, or $\$488.25$. The value of 15 lb of Mexican beans is $15(\$11.95)$, or $\$179.25$, so the value of the blend is $\$488.25 + \$179.25 = \$667.50$. A 50-lb blend priced at $\$13.35$ per pound is also worth $\$667.50$, so our answer checks.

5. **State.** The Perfect Morning Blend should be made by combining 35 lb of organic Brazilian beans with 15 lb of fair-trade Mexican beans.

3. Refer to Example 3. The Roasted Bean uses the same Brazilian and Mexican beans to form a Quiet Afternoon Blend that sells for $12.35 per pound. How much of each type of bean should be mixed in order to make a 50-lb batch of Quiet Afternoon Blend?

⤺ YOUR TURN

EXAMPLE 4 *Paint Colors.* At a local "paint swap," Madison found large supplies of Skylite Pink (12.5% red pigment) and MacIntosh Red (20% red pigment). How many gallons of each color should be mixed in order to create a 15-gal batch of Summer Rose (17% red pigment)?

SOLUTION

1. **Familiarize.** This problem is similar to Example 3. Instead of mixing two types of coffee and keeping an eye on the price of the mixture, we are mixing two types of paint and keeping an eye on the amount of pigment in the mixture.

To visualize this problem, think of the pigment as a solid that, given time, would settle to the bottom of each can. Let's guess that 3 gal of Skylite Pink and 12 gal of MacIntosh Red are mixed. How much pigment would be in the mixture? The Skylite Pink would contribute 12.5% of 3 gal, or 0.375 gal of pigment, and the MacIntosh Red would contribute 20% of 12 gal, or 2.4 gal of pigment. Thus the 15-gal mixture would contain $0.375 + 2.4 = 2.775$ gal of pigment. Since Madison needs the 15 gal of Summer Rose to be 17% pigment, and since 17% of 15 gal is 2.55 gal, our guess is incorrect. Rather than guess again, we let $p = $ the number of gallons of Skylite Pink used and $m = $ the number of gallons of MacIntosh Red used.

Student Notes

When solving a new problem, see if it is like a problem that you have already solved. You can then modify the earlier approach to fit the problem.

p gal

m gal

12.5% pigment

20% pigment

$m + p$ gal

17% pigment

Sara mixes *x* lb of raisins with *y* lb of peanuts to make 10 lb of trail mix. The value of the raisins is $7/lb, the value of the peanuts is $4/lb, and the value of the trail mix is $5/lb.

1. What is the value of 10 lb of trail mix?

2. What is the value of *x* lb of raisins?

3. What is the value of *y* lb of peanuts?

4. If Sara uses 5 lb of raisins and 5 lb of peanuts, would her trail mix have a value of $5/lb?

5. In order to achieve the correct value of the trail mix, should Sara use more raisins than peanuts, or more peanuts than raisins? Why?

4. Refer to Example 4. Madison also plans to use Skylite Pink and MacIntosh Red in order to create a 10-gal batch of Spring Blush (14% red pigment). How many gallons of each should be mixed?

2. **Translate.** As in Example 3, the information given can be arranged in a table.

	Skylite Pink	MacIntosh Red	Summer Rose
Amount of Paint (in gallons)	*p*	*m*	15
Percent Pigment	12.5%	20%	17%
Amount of Pigment (in gallons)	0.125*p*	0.2*m*	0.17 × 15, or 2.55

A system of two equations can be formed by reading across the first and third rows of the table. Since Madison needs 15 gal of mixture, we must have

$$p + m = 15. \quad \leftarrow \text{Total amount of paint}$$

Madison also needs 2.55 gal of pigment. Since the amount of pigment in the Summer Rose paint comes from the pigment in the Skylite Pink and the MacIntosh Red paint, we have

$$0.125p + 0.2m = 2.55. \quad \leftarrow \text{Total amount of pigment}$$

We have translated to a system of equations:

$$p + \quad m = 15, \qquad \textbf{(1)}$$
$$0.125p + 0.2m = 2.55. \qquad \textbf{(2)}$$

3. **Carry out.** We note that $-5(0.2m) = -m$, so we multiply both sides of equation (2) by -5 to eliminate *m* (other approaches will also work):

$$\begin{array}{rcl} p + m = & 15 \\ -0.625p - m = & -12.75 \quad \text{Multiplying both sides of equation (2) by } -5 \\ \hline 0.375p \quad\quad = & 2.25 \\ p = & 6. \quad \text{Dividing both sides by } 0.375 \end{array}$$

If $p = 6$, we see from equation (1) that $m = 9$.

4. **Check.** Clearly, 6 gal of Skylite Pink and 9 gal of MacIntosh Red do add up to a total of 15 gal. To determine whether the mixture is the right color, Summer Rose, we calculate the amount of pigment in the mixture:

$$0.125 \cdot 6 + 0.2 \cdot 9 = 0.75 + 1.8 = 2.55 \text{ gal.}$$

Since 2.55 is indeed 17% of 15, the mixture is the correct color.

5. **State.** Madison needs 6 gal of Skylite Pink and 9 gal of MacIntosh Red in order to make 15 gal of Summer Rose.

↩ YOUR TURN

Re-examine Examples 1–4, looking for similarities. Examples 3 and 4 are often called *mixture problems*, but they have much in common with Examples 1 and 2.

7.4 EXERCISE SET

FOR EXTRA HELP   MyMathLab®

➷ Vocabulary and Reading Check

Match each statement with the most appropriate translation from the following list.

a) $x + y = 5$
b) $x + y = 12$
c) $x + y = 30$
d) $2x + 3y = 30$
e) $0.1x + 0.05y = 4$
f) $5x + 4y = 140$

1. _____ Benita purchased a total of 12 pens and notebooks.

2. _____ Benita spent $30 on the purchase of pens costing $2 each and notebooks costing $3 each.

3. _____ Chaz is mixing a 30-lb batch of peanuts and raisins.

4. _____ A mixture of peanuts costing $5 per pound and raisins costing $4 per pound is worth a total of $140.

5. _____ Roy is mixing a 5-lb batch of two kinds of fertilizer.

6. _____ A mixture of fertilizer that is 10% nitrogen with a fertilizer that is 5% nitrogen contains a total of 4 lb of nitrogen.

A. Total-Value Problems

Solve. Use the five steps for problem solving.

7. *Basketball Scoring.*   In a recent game, Kevin Durant of the Oklahoma City Thunder scored 24 points on a combination of 11 two-point and three-point baskets. How many shots of each type were made?

   Data: National Basketball Association

8. *Basketball Scoring.*   In a recent game, the Orlando Magic scored 68 points on a combination of 30 two-point and three-point baskets. How many shots of each type were made?

   Data: National Basketball Association

9. *Basketball Scoring.*   During the seven games of the 2016 NBA finals, the Cleveland Cavaliers scored 582 of their points on a combination of 263 two-point and three-point baskets. How many shots of each type were made?

   Data: National Basketball Association

10. *Basketball Scoring.*   During the seven games of the 2016 NBA finals, MVP LeBron James scored 177 points on a combination of 82 two-point and three-point baskets. How many shots of each type did he make?

   Data: National Basketball Association

11. *College Credits.*   Each course at Pease County Community College is worth either 3 or 4 credits. The members of the women's swim team are taking a total of 27 courses that are worth a total of 89 credits. How many 3-credit courses and how many 4-credit courses are being taken?

12. *College Credits.*   Each course at Mt. Regis College is worth either 3 or 4 credits. The members of the men's soccer team are taking a total of 48 courses that are worth a total of 155 credits. How many 3-credit courses and how many 4-credit courses are being taken?

13. *Returnable Bottles.*   As part of a fundraiser, the Cobble Hill Daycare collected 430 returnable bottles and cans, some worth 5 cents each and the rest worth 10 cents each. If the total value of the cans and bottles was $26.20, how many 5-cent bottles or cans and how many 10-cent bottles or cans were collected?

14. *Ice Cream Cones.*   A busload of campers stopped at a dairy stand for ice cream. They ordered 40 cones, some soft-serve at $2.25 and the rest hard-pack at $2.75. If the total bill was $96, how many of each type of cone were ordered?

15. *Yellowstone Park Admissions.*   Entering Yellowstone National Park costs $25 for a car and $20 for a motorcycle. One summer day, 5950 cars or motorcycles entered and paid a total of $137,650. How many motorcycles entered?

   Data: National Park Service

16. *Disneyland Admissions.* A two-day Park Hopper ticket to Disneyland costs $223 for children ages 3–9 and $235 for adults and children ages 10 and older. A bus full of 23 kindergartners and parents paid a total of $5201 for their tickets. How many children and how many adults were on the bus?

Data: disneyland.disney.go.com

17. *Museum Admissions.* The National Underground Railroad Freedom Center in Cincinnati, Ohio, charges an admission fee of $15 for adults and $10.50 for students. One day, a total of $7312.50 was collected from 642 adults and students. How many adult admissions were there?

Data: www.freedomcenter.org

18. *Zoo Admissions.* The Bronx Zoo charges an admission fee of $33.95 for adults and $23.95 for children. One July day, a total of $26,092 was collected from 960 admissions. How many adult admissions were there?

Data: Bronx Zoo

19. *International Messaging.* AT&T recently charged 25¢ to send a text internationally and 50¢ to send a picture or a video. During one month, Talya sent a total of 46 text, picture, and video messages internationally. If her bill for the messages was $15.00, how many texts did she send?

Data: m.att.com

20. *Music Lessons.* Jillian charges $25 for a private guitar lesson and $18 for a group guitar lesson. One day in August, Jillian earned $265 from 12 students. How many studen ts of each type did Jillian teach?

21. *Autoharp Strings.* Anna purchased 32 strings for her autoharp. Wrapped strings cost $4.49 each and unwrapped strings cost $2.99 each. If she paid a total of $107.68 for the strings, how many of each type did she buy?

22. *Gasoline Purchases.* On his trip home from college, Gil bought gas first for $3.59 per gallon and then for $3.27 per gallon. If he purchased a total of 20 gal for $69.24, how many gallons did he buy at each price?

B. Mixture Problems

23. *Coffee Blends.* Michelle and Gerry mix decaffeinated Sumatra coffee costing $14.95 per pound with regular Sumatra coffee costing $13.95 per pound. Last month they made 8 lb of the blend for $118.10. How much of each type of coffee did they use?

24. *Grass Seed.* The Seed Superstore recommends a blend of fescue and Kentucky bluegrass for a Midwestern shady lawn. Brock and Miriam paid $70.25 for a 5-lb mixture of Spartan II fescue seed and America bluegrass seed. If the fescue cost $11.95 per pound and the bluegrass cost $18.95 per pound, how many pounds of each did they buy?

Data: www.seedsuperstore.com

25. *Catering.* Gloria plans to mix cranberry juice cocktail costing $1.89 per liter with sparkling water costing $2.09 per liter to make 16 L of punch for an office party. If the punch costs $1.94 per liter, how many liters of each will she use?

26. *Mixed Nuts.* A vendor wishes to mix peanuts worth $2.40 per kilogram with Brazil nuts worth $4.80 per kilogram in order to make 480 kg of a mixture worth $4.12 per kilogram. How much of each should be used?

27. *Acid Mixtures.* Jerome's experiment requires him to mix a 50%-acid solution with an 80%-acid solution to create 200 mL of a 68%-acid solution. How much 50%-acid solution and how much 80%-acid solution should he use? Complete the following table as part of the *Translate* step.

Type of Solution	50%-Acid	80%-Acid	68%-Acid Mix
Amount of Solution	x	y	
Percent Acid	50%		68%
Amount of Acid in Solution		$0.8y$	

28. *Production.* Streakfree window cleaner is 12% alcohol and Sunstream window cleaner is 30% alcohol. How much of each should be used in order to make 90 oz of a cleaner that is 20% alcohol?

29. *Chemistry.* E-Chem Testing has a solution that is 80% base and another that is 30% base. A technician needs 200 L of a solution that is 62% base. The 200 L will be prepared by mixing the two solutions on hand. How much of each should be used?

30. *Horticulture.* A solution containing 28% fungicide is to be mixed with a solution containing 40% fungicide to make 300 L of a solution containing 36% fungicide. How much of each solution should be used?

31. *Octane Ratings.* When a tanker delivers gas to a gas station, it brings only two grades of gasoline, the highest and the lowest, filling two large underground tanks. If you purchase a middle grade, the pump's computer mixes the high grade and the low grade appropriately. How much 87-octane gas and how much 93-octane gas should be blended in order to make 12 gal of 91-octane gas?

Data: Champlain Electric and Petroleum Equipment

32. *Octane Ratings.* Refer to Exercise 31. If a gas station is supplied with 87-octane gas and 95-octane gas, how much of each should be blended in order to make 10 gal of 93-octane gas?

A, B. Mixed Applications

33. *Printing.* Using some pages that hold 1300 words per page and others that hold 1850 words per page, a typesetter is able to completely fill 12 pages with an 18,350-word document. How many pages of each kind were used?

34. *Coin Value.* A collection of quarters and nickels is worth $2.50. There are 26 coins in all. How many of each are there?

35. *Basketball Scoring.* Wilt Chamberlain once scored a record 100 points on a combination of 64 foul shots (each worth one point) and two-pointers. How many shots of each type did he make?

36. *Basketball Scoring.* Blake Griffin recently scored 22 points on a combination of 13 foul shots and two-pointers. How many shots of each type did he make?

37. *Aromatherapy.* Mallory has a bottle of Calm Morning fragrance that is 3% lavender essential oil and a bottle of Relaxed Evening fragrance that is 1% lavender essential oil. How much of each fragrance should be used in order to create 100 mL (milliliters) of Balanced Day fragrance that is 1.5% lavender essential oil?

38. *Cough Syrup.* Dr. Zeke's cough syrup is 2% alcohol. Vitabrite cough syrup is 5% alcohol. How much of each type should be used in order to prepare an 80-oz batch of cough syrup that is 3% alcohol?

Aha! **39.** *Nutrition.* New England Natural Bakers Muesli gets 20% of its calories from fat. Breadshop Supernatural granola gets 35% of its calories from fat. How much of each type should be used in order to create a 40-lb mixture that gets 27.5% of its calories from fat?

Data: Onion River Cooperative, Burlington VT

40. *Textile Production.* DRG Outdoor Products uses one insulation that is 20% goose down and another that is 34% goose down. How many pounds of each should be used in order to create 50 lb of insulation that is 25% goose down?

41. Why might fraction answers be acceptable on problems like Examples 3 and 4, but not on problems like Examples 1 and 2?

42. Write a problem for a classmate to solve by translating to a system of two equations in two unknowns.

Skill Review

Graph.

43. $x + y = 3$ [3.3]

44. $y = \frac{1}{2}x + 2$ [3.6]

45. $3x - y = 2$ [3.2]

46. $y = \frac{1}{2}$ [3.3]

47. $x = -3$ [3.3]

48. $2x = 4 - y$ [3.2]

Synthesis

49. In Exercise 36, suppose that some of Griffin's 13 baskets were three-pointers. Could the problem still be solved? Why or why not?

50. In Exercise 34, suppose that some of the 26 coins may have been half-dollars. Could the problem still be solved? Why or why not?

51. *Coffee.* Kona coffee, grown only in Hawaii, is highly desired and quite expensive. To create a blend of beans that is 30% Kona, the Brewtown Beanery is adding pure Kona beans to a 45-lb sack of Colombian beans. How many pounds of Kona should be added to the 45 lb of Colombian?

52. *Chemistry.* A tank contains 8000 L of a solution that is 40% acid. How much water should be added in order to make a solution that is 30% acid?

53. *Automobile Maintenance.* The radiator in Candy's Honda Accord contains 6.3 L of antifreeze and water. This mixture is 30% antifreeze. How much should be drained and replaced with pure antifreeze so that the mixture will be 50% antifreeze?

6.3 liters

54. *Investing.* One year Shannon made $114 from two investments: $1100 was invested at one yearly rate and $1800 at a rate that was 1.5% higher. Find the two rates of interest.

55. *Octane Rating.* Many cars need gasoline with an octane rating of at least 87. After mistakenly putting 5 gal of 85-octane gas in her empty gas tank, Deanna plans to add 91-octane gas until the mixture's octane rating is 87. How much 91-octane gas should she add?

56. *Sporting-Goods Prices.* Together, a bat, ball, and glove cost $99.00. The bat costs $9.95 more than the ball, and the glove costs $65.45 more than the bat. How much does each cost?

57. *Overtime Pay.* Juanita is paid "time and a half" for any hours that she works in excess of 40 hr per week. The week before Christmas she worked 55 hr and was paid $812.50. What was her normal hourly wage?

58. *Investing.* Khalid invested $54,000, part of it at 4% and the rest at 4.5%. The total yield after one year is $2305. How much was invested at each rate?

59. *Payroll.* Ace Engineering pays a total of $325 per hour when employing some workers at $20 per hour and others at $25 per hour. When the number of $20 workers is increased by 50% and the number of $25 workers is decreased by 20%, the cost per hour is $400. How many workers were originally employed at each rate?

60. *Dairy Farming.* The Benson Cooperative Creamery has 1000 gal of milk that is 4.6% butterfat. How much skim milk (no butterfat) should be added to make milk that is 2% butterfat?

61. A two-digit number is six times the sum of its digits. The tens digit is 1 more than the ones digit. Find the number.

62. The sum of the digits of a two-digit number is 12. When the digits are reversed, the number is decreased by 18. Find the original number.

63. *Literature.* In Lewis Carroll's *Through the Looking Glass*, Tweedledum says to Tweedledee, "The sum of your weight and twice mine is 361 pounds." Then Tweedledee says to Tweedledum, "Contrariwise, the sum of your weight and twice mine is 362 pounds." Find the weights of Tweedledum and Tweedledee.

⤷ **YOUR TURN ANSWERS: SECTION 7.4**
1. Two-pointers: 7; three-pointers: 2 **2.** Commuter permits: 18; clean-fuel permits: 10 **3.** Brazilian: 10 lb; Mexican: 40 lb **4.** Skylite Pink: 8 gal; MacIntosh Red: 2 gal

---

Quick Quiz: Sections 7.1–7.4

1. Solve by graphing:
$$y = -\tfrac{2}{3}x + 1,$$
$$x - y = 4. \quad [7.1]$$

2. Solve using substitution:
$$2x + y = 7,$$
$$x + 3y = 5. \quad [7.2]$$

3. Solve using elimination:
$$x + 2y = 3,$$
$$3x - 2y = 5. \quad [7.3]$$

4. The perimeter of a rectangle is 40 ft. The length is three times the width. Find the length and the width of the rectangle. [7.2]

5. Kerianne ordered 30 sandwiches for lunch during a staff meeting. In her order were vegetarian sandwiches costing $6.50 each and turkey sandwiches costing $7.50 each. The total cost of the sandwiches was $217. How many sandwiches of each type did she buy? [7.4]

Prepare to Move On

Graph each solution set on the number line. [2.6]

1. $x + 2 < 5$ **2.** $5 - 3x > 20$

3. $13 - 6x \le 1$ **4.** $6 > -\tfrac{1}{2}x + 3$

Mid-Chapter Review

A system of two equations is a pair of equations that are to be solved simultaneously.

- We can solve systems of two equations by graphing or by using substitution or elimination.
- A system of two equations may have one solution, no solution, or an infinite number of solutions.

GUIDED SOLUTIONS

1. Solve using substitution: $x - y = 3$,
$\qquad\qquad\qquad y = 2x + 1$.   [7.2]

2. Solve using elimination: $8x - 11y = 12$,
$\qquad\qquad\qquad 3x + 11y = 10$.   [7.3]

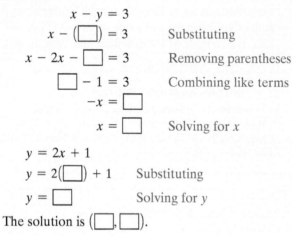

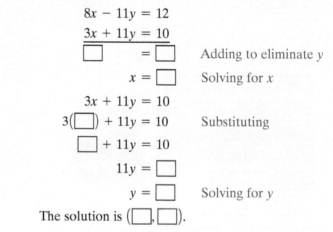

MIXED EXERCISES

For Exercises 3–14, if there is no solution or an infinite number of solutions, state this.

Solve by graphing.   [7.1]

3. $y = x - 3$,
$y = \frac{1}{2}x + 1$

4. $y - 3x = 1$,
$y = 3x - 5$

Solve using substitution.   [7.2]

5. $x = y - 5$,
$2x + y = 7$

6. $2y = x + 7$,
$3x + 2y = 5$

7. $x - 3y = 6$,
$2y = x - 1$

8. $x = 5 - 2y$,
$4y + 2x = 7$

Solve using elimination.   [7.3]

9. $2x + 3y = 1$,
$x - 3y = 8$

10. $3x - 4y = 1$,
$2x + 2y = 5$

11. $x + y = 3$,
$2x + 2y = 6$

12. $3y = 5x + 4$,
$2y = 3x - 1$

Solve using any method.   [7.1], [7.2], [7.3]

13. $6x + 5y = 15$,
$3y - 6x = 1$

14. $2x + 7y = 0$,
$9x - 10y = 0$

15. $\frac{1}{2}x + \frac{1}{5}y = 9$,
$\frac{1}{3}x - \frac{4}{15}y = -2$

16. $0.1x - 0.5y = -2.8$,
$0.2x = 0.3y$

Solve.

17. Two angles are complementary. One angle is 12° more than twice the other. Find the measure of each angle.   [7.2]

18. *Prepaid Cards.*   The Rush prepaid Visa® card charges a one-time fee of $3.95 plus a monthly maintenance fee of $7.95 per month. The Vision prepaid Visa® card charges a one-time fee of $9.95 plus a monthly maintenance fee of $4.95 per month. For what number of months will the cost of the two cards be the same?   [7.3]

Data: rushcard.com and visionprepaid.com

19. In a recent game, the Toronto Raptors scored 74 points on a combination of 34 two-point and three-point baskets. How many shots of each type were made?   [7.3]

Data: www.nba.com

20. Allen needs to mix paint containing 10% blue pigment with paint containing 20% blue pigment in order to make 8 gal of paint containing 17.5% blue pigment. How much of each should he use?   [7.4]

7.5 | Linear Inequalities in Two Variables

A. Graphing Linear Inequalities   **B.** Linear Inequalities in One Variable

As is the case for linear equations, the solutions of *linear inequalities* like $5x + 4y < 13$ or $y \geq \frac{2}{3}x - 5$ can be represented graphically.

A. Graphing Linear Inequalities

The solutions of an inequality in one variable, such as $x < 2$, can be graphed on the number line. To graph such an inequality, we shade part of the number line, indicating the endpoint with either a parenthesis or a bracket. Since the inequality symbol in $x < 2$ is $<$, we indicate the endpoint with a parenthesis.

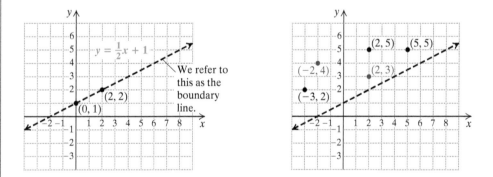

We graph inequalities like $y > \frac{1}{2}x + 1$ using a plane instead of the number line. We shade a region of the plane and indicate the *boundary line* of that region with either a dashed line or a solid line, depending on the inequality symbol used.

EXAMPLE 1   Graph: $y > \frac{1}{2}x + 1$.

SOLUTION   To graph the inequality, we first graph the line $y = \frac{1}{2}x + 1$. This line is a *boundary line* for the graph of the inequality. We find the equation of the boundary line by replacing $>$ with $=$ in the inequality.

Since the symbol $>$ is used, the solutions of $y = \frac{1}{2}x + 1$ are *not* part of the solutions of the inequality. To indicate this, we draw the line dashed.

The plane is now split in two. If we consider the coordinates of a few points above the line, we will find that all represent solutions of $y > \frac{1}{2}x + 1$.

Here is a check for the points $(2, 3)$ and $(-2, 4)$:

$$
\begin{array}{c|c}
y > \frac{1}{2}x + 1 & y > \frac{1}{2}x + 1 \\
\hline
3 \;\Big|\; \frac{1}{2} \cdot 2 + 1 & 4 \;\Big|\; \frac{1}{2}(-2) + 1 \\
\;\Big|\; 1 + 1 & \;\Big|\; -1 + 1 \\
3 \overset{?}{>} 2 \quad \text{TRUE} & 4 \overset{?}{>} 0 \quad \text{TRUE}
\end{array}
$$

The student can check that *any* point on the same side of the dashed line as $(2, 3)$ or $(-2, 4)$ is a solution. If one point in a region represents a solution of an inequality, then *all* points in that region represent solutions. Thus the graph of $y > \frac{1}{2}x + 1$, shown on the following page, consists of all points in the shaded

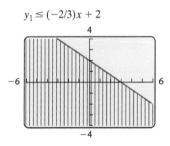
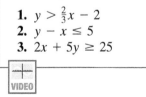
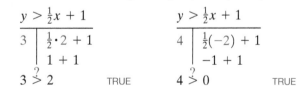

We refer to this as the boundary line.

Study Skills

Map Out Your Day

As the semester winds down and projects are due, it becomes more critical than ever that you manage your time wisely. If you aren't already doing so, consider writing out an hour-by-hour schedule for each day.

region. Furthermore, note that for any inequality of the form $y > mx + b$ or $y \geq mx + b$, we shade the region *above* the boundary line.

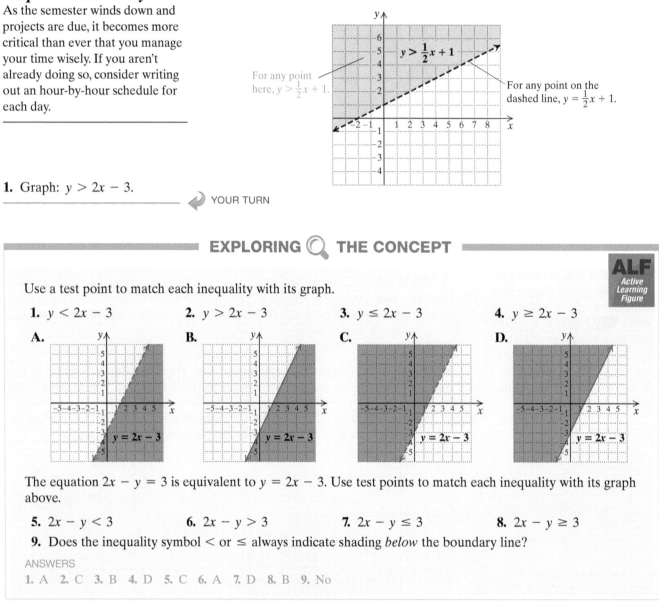

1. Graph: $y > 2x - 3$.

YOUR TURN

EXPLORING 🔍 THE CONCEPT

ALF
Active Learning Figure

Use a test point to match each inequality with its graph.

1. $y < 2x - 3$ **2.** $y > 2x - 3$ **3.** $y \leq 2x - 3$ **4.** $y \geq 2x - 3$

A.   **B.**   **C.**   **D.**

The equation $2x - y = 3$ is equivalent to $y = 2x - 3$. Use test points to match each inequality with its graph above.

5. $2x - y < 3$ **6.** $2x - y > 3$ **7.** $2x - y \leq 3$ **8.** $2x - y \geq 3$

9. Does the inequality symbol $<$ or $\leq$ always indicate shading *below* the boundary line?

ANSWERS
1. A **2.** C **3.** B **4.** D **5.** C **6.** A **7.** D **8.** B **9.** No

---

TO GRAPH A LINEAR INEQUALITY

1. Draw the boundary line by replacing the inequality symbol with an equals sign and graphing the resulting equation. If the inequality symbol is $<$ or $>$, the line is dashed. If the symbol is $\leq$ or $\geq$, the line is solid.
2. Shade the region on one side of the boundary line. To determine which side, select a point not on the line as a test point. If that point's coordinates are a solution of the inequality, shade the region containing the point. If not, shade the other region.
3. Inequalities of the form $y < mx + b$ or $y \leq mx + b$ are shaded below the boundary line. Inequalities of the form $y > mx + b$ or $y \geq mx + b$ are shaded above the boundary line.

Student Notes

A mistake in checking your test point could lead you to shade the wrong region of the plane. For this reason, you may want to use a second test point to assure yourself that the correct region is shaded.

EXAMPLE 2   Graph: $2x + 3y \leq 6$.

SOLUTION   First, we graph the boundary line $2x + 3y = 6$. This can be done either by using the intercepts, $(0, 2)$ and $(3, 0)$, or by finding slope–intercept form, $y = -\frac{2}{3}x + 2$. Since the inequality contains the symbol $\leq$, we draw a solid boundary line to include all pairs on the line as part of the solution. The graph of $2x + 3y \leq 6$ also includes the region either above or below the line. By using a "test point" that is clearly above or below the line, we can determine which region to shade. The origin, $(0, 0)$, is often a convenient test point, so long as it does not lie on the boundary line:

$$\frac{2x + 3y \leq 6}{2 \cdot 0 + 3 \cdot 0 \;|\; 6}$$
$$0 \overset{?}{\leq} 6 \quad \text{TRUE}$$

The point $(0, 0)$ is a solution and it is in the region below the boundary line. Thus this region, along with the line itself, represents the solution.

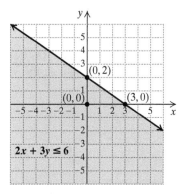

The original inequality is equivalent to $y \leq -\frac{2}{3}x + 2$. Note that for any inequality of the form $y \leq mx + b$ or $y < mx + b$, we shade the region *below* the boundary line.

2. Graph: $4x + y \geq 4$.

⤷ YOUR TURN

B. Linear Inequalities in One Variable

EXAMPLE 3   Graph $y \leq -2$ on a plane.

SOLUTION   We graph $y = -2$ as a solid line to indicate that all points on the line are solutions. Again, we select $(0, 0)$ as a test point. It may help to write $y \leq -2$ as $y \leq 0 \cdot x - 2$:

$$\frac{y \leq 0 \cdot x - 2}{0 \;|\; 0 \cdot 0 - 2}$$
$$0 \overset{?}{\leq} -2 \quad \text{FALSE}$$

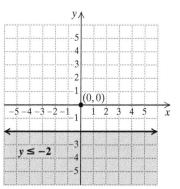

Since $(0, 0)$ is *not* a solution, we do not shade the region in which it appears. Instead, we shade below the boundary line as shown. The solution consists of all ordered pairs with y-coordinates less than or equal to -2.

3. Graph $y > 1$ on a plane.

⤷ YOUR TURN

EXAMPLE 4   Graph $x < 3$ on a plane.

SOLUTION   We graph $x = 3$ using a dashed line. To determine which region to shade, we again use the test point $(0, 0)$. It may help to write $x < 3$ as $x + 0 \cdot y < 3$:

$$\frac{x + 0 \cdot y < 3}{0 + 0 \cdot 0 \;\big|\; 3}$$
$$0 \overset{?}{<} 3 \quad \text{TRUE}$$

Since $(0, 0)$ is a solution, we shade to the left of the boundary line. The solution consists of all ordered pairs with first coordinates less than 3.

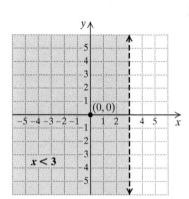

4. Graph $x \geq -1$ on a plane.

_____   ↩ YOUR TURN

↩ Check Your UNDERSTANDING

For the graph of each inequality, **(a)** determine whether the boundary line is dashed or solid, and **(b)** determine whether $(0, 0)$ is in the solution set.

1. $x + y < 1$ **2.** $2x \leq y + 3$
3. $y < -\frac{1}{2}x - 7$ **4.** $y > -4$
5. $x \geq 1$ **6.** $2x - 7 \geq 6y$

7.5   EXERCISE SET

FOR EXTRA HELP   MyMathLab®

↩ Vocabulary and Reading Check

Classify each of the following statements as either true or false.

1. To graph a linear inequality, we first graph a line.

2. To determine which region to shade, we must test at least two points.

3. To graph an inequality of the form $y < mx + b$, we first draw a solid boundary line.

4. To graph an inequality of the form $y \geq mx + b$, we shade above the boundary line.

A, B. Graphing Linear Inequalities

5. Determine whether $(-2, -6)$ is a solution of $2x + y < -10$.

6. Determine whether $(8, -1)$ is a solution of $2y - x \leq -10$.

Aha! **7.** Determine whether $\left(\frac{1}{3}, \frac{9}{10}\right)$ is a solution of $2y + 3x \geq -1$.

8. Determine whether $(-2, 3)$ is a solution of $x + 0 \cdot y > -1$.

Graph on a plane.

9. $y \leq x - 1$ **10.** $y \leq x + 5$
11. $y < x + 4$ **12.** $y < x - 2$
13. $y \geq x - 1$ **14.** $y \geq x - 3$
15. $y \leq 3x + 2$ **16.** $y \leq 2x - 1$
17. $x + y \leq 4$ **18.** $x + y \leq 6$
19. $x - y > 7$ **20.** $x - y > 5$
21. $y \geq 1 - 2x$ **22.** $y > 2 - 3x$
23. $y + 2x > 0$ **24.** $y + 3x \geq 0$
25. $x \geq 4$ **26.** $x > -4$
27. $y > -1$ **28.** $y \leq 4$
29. $y < 0$ **30.** $y \geq -5$
31. $x \leq -2$ **32.** $x < 5$
33. $\frac{1}{3}x - y < -5$ **34.** $y - \frac{1}{2}x \leq -1$
35. $2x + 3y \leq 12$ **36.** $5x + 4y \geq 20$
37. $3x - 2y \geq -6$ **38.** $2x - 5y \leq -10$

39. Examine the solution of Example 2. Why is the point $(4.5, -1)$ *not* a good choice for a test point?

40. Why is $(0, 0)$ such a "convenient" test point to use?

Skill Review

Factor completely.

41. $x^2 - 3x - 40$ [5.2]

42. $6y^4 - 6$ [5.4]

43. $25a^2 + 10a + 1$ [5.4]

44. $6x^2 - x - 2$ [5.3]

45. $x^3 - 3x^2 + x - 3$ [5.1]

46. $36x^4 - 42x^3 + 72x$ [5.1]

Synthesis

47. When graphing a linear inequality, if $(0, 0)$ is on the boundary line, Cyrus chooses a test point that is on an axis. Why is this a convenient choice?

48. Without drawing a graph, explain why the graph of $3x + y < 0$ is shaded *below* the boundary line but the graph of $3x - y < 0$ is shaded *above* the boundary line.

49. *Elevators.* Many Otis elevators have a capacity of 1000 lb. Suppose that c children, each weighing 75 lb, and a adults, each weighing 150 lb, are on an elevator. Find and graph an inequality that asserts that the elevator is overloaded.

50. *Hockey Wins and Losses.* A hockey team needs at least 60 points for the season in order to make the playoffs. Suppose that the Coyotes finish with w wins, each worth 2 points, and t ties, each worth 1 point. Find and graph an inequality that indicates that the Coyotes made the playoffs.

Find an inequality for each graph.

51.

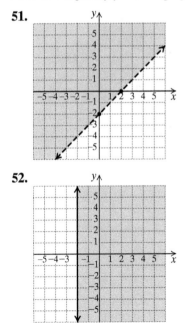

52.

Graph on a plane. (Hint: Use several test points.)

53. $xy \le 0$

54. $xy \ge 0$

55. Graph: $y + 3x \le 4.9$.

56. Graph: $0.7x - y \le 2.3$.

↘ **YOUR TURN ANSWERS: SECTION 7.5**

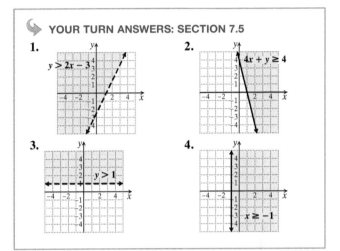

1. $y > 2x - 3$

2. $4x + y \ge 4$

3. $y > 1$

4. $x \ge -1$

Quick Quiz: Sections 7.1–7.5

1. Solve by graphing:

$$y = 2x + 1,$$
$$x + y = 1. \quad [7.1]$$

2. Solve using substitution:

$$4x + y = 2,$$
$$x + 3y = 2. \quad [7.2]$$

3. Solve using elimination:

$$3x + 2y = 5,$$
$$5x - 3y = 2. \quad [7.3]$$

4. Kristi walked and ran 20 mi in 4 hr. It takes her 15 min to walk 1 mi and 10 min to run 1 mi. How many miles did she walk and how many did she run? [7.4]

5. Graph: $y > x - 1$. [7.5]

Prepare to Move On

1. Solve for y: $2x - 4y = 7$. [2.3]

2. Determine whether $(-2, 3)$ is a solution of both $3x + y \le 0$ and $y - x > 4$. [7.5]

3. Determine whether $(-5, -6)$ is a solution of both $2x - 3y < 0$ and $x - y > 0$. [7.5]

7.6 Systems of Linear Inequalities

A. Graphing Systems of Inequalities

We now consider **systems of linear inequalities** in two variables, such as

$$x + y \leq 3,$$
$$x - y < 3.$$

A solution of a system of inequalities makes *both* inequalities true.

When systems of equations are solved graphically, we search for points common to both lines. To solve a system of inequalities graphically, we again look for points common to both graphs. This is accomplished by graphing each inequality and determining where the graphs overlap, or intersect.

A. Graphing Systems of Inequalities

EXAMPLE 1 Graph the solutions of the system

$$x + y \leq 3,$$
$$x - y < 3.$$

SOLUTION To graph $x + y \leq 3$, we draw the graph of $x + y = 3$ using a solid line. We graph the line using the intercepts $(3, 0)$ and $(0, 3)$ (see the graph on the left below). Since $(0, 0)$ is a solution of $x + y \leq 3$, we shade (in red) all points on that side of the line. The arrows near the ends of the lines also indicate the region that contains solutions.

Next, we superimpose the graph of $x - y < 3$, using a dashed line for $x - y = 3$ and again using $(0, 0)$ as a test point. Since $(0, 0)$ is again a solution, we shade (in blue) the region on the same side of the dashed line as $(0, 0)$.

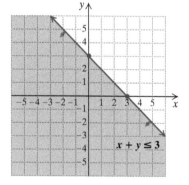

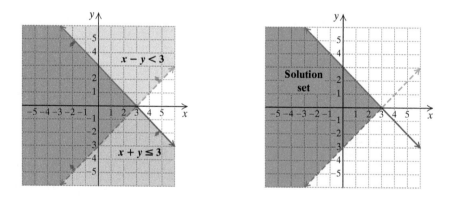

1. Graph the solutions of the system

$$x + y \geq 3,$$
$$x - y > 3.$$

The solution set of the system is the region shaded purple along with the purple portion of the line $x + y = 3$.

YOUR TURN

EXAMPLE 2 Graph the solutions of the system

$$x \geq 2,$$
$$x - 3y < 6.$$

SOLUTION We graph $x \geq 2$ using blue and $x - 3y < 6$ using red. The solution set is the purple region along with the purple portion of the solid line.

Student Notes

Unless you have ready access to different colored pencils, using arrows (as in the examples) or patterned shadings to indicate shaded regions is a good idea.

2. Graph the solutions of the system

$$y < 2,$$
$$x + y \leq 1.$$

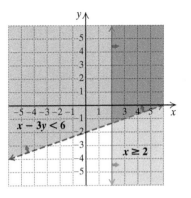
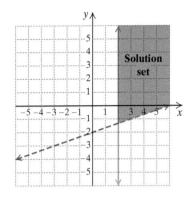

YOUR TURN

EXAMPLE 3 Graph the solutions of the system

$$x - 2y < 0,$$
$$-2x + y > 2.$$

SOLUTION We graph $x - 2y < 0$ using red and $-2x + y > 2$ using blue. The region that is purple is the solution set of the system since the coordinates of those points make both inequalities true.

Chapter Resource:
Visualizing for Success, p. 483

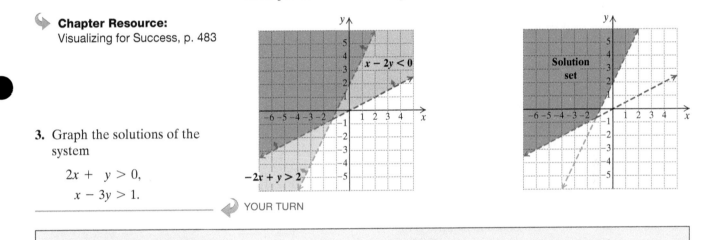

3. Graph the solutions of the system

$$2x + y > 0,$$
$$x - 3y > 1.$$

YOUR TURN

Technology Connection

Many graphing calculators can display systems of linear inequalities. To do this, we enter each inequality, indicating the side that should be shaded. The graphing calculator automatically uses a different pattern for each region.

$$y \leq -\tfrac{2}{3}x + 5, \quad \blacktriangle y_1 = (-2/3)x + 5,$$
$$y \geq \tfrac{1}{2}x - 3, \quad \blacktriangledown y_2 = .5x - 3,$$
$$y \leq 3x + 4; \quad \blacktriangle y_3 = 3x + 4$$

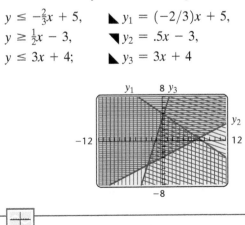

If your calculator has an Inequalz application, these inequalities can be entered as

$$y_1 \leq (-2/3)x + 5,$$
$$y_2 \geq .5x - 3,$$
$$y_3 \leq 3x + 4.$$

You may also be able to choose to shade only the intersection of the inequality.

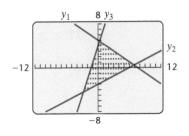

↳ Check Your
UNDERSTANDING

The graph at right represents the solution set of a system of inequalities. Determine whether each of the following is a solution of the system.

1. $(0, 0)$ **2.** $(2, 3)$

3. $(2, -3)$ **4.** $(-3, 0)$

5. $(-4, 0)$ **6.** $(3, 0)$

7. $(0, 4)$ **8.** $(-3, 1)$

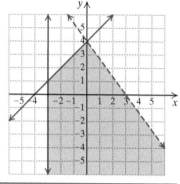

7.6   EXERCISE SET

FOR EXTRA HELP   MyMathLab®

↳ **Vocabulary and Reading Check**

Classify each of the following statements as either true or false.

1. If an ordered pair is a solution of one inequality in a system, then it must be a solution of the system.

2. The graph of a system of inequalities may contain part of a line.

3. We can sometimes use the same test point to graph both inequalities in a system.

4. The solution set of a system of inequalities is the intersection of the solution sets of the individual inequalities.

A. Graphing Systems of Inequalities

Graph the solutions of each system.

5. $x + y \leq 3$,
$x - y \leq 5$

6. $x + y \leq 7$,
$x - y \leq 4$

7. $x + y < 6$,
$x + y > 0$

8. $y - 2x > 1$,
$y - 2x < 3$

9. $x > 3$,
$x + y \leq 4$

10. $y \geq -1$,
$x > 3 + y$

11. $y \geq x$,
$y \leq 1 - x$

12. $y > 3x - 2$,
$y < -x + 4$

13. $x \geq -3$,
$y \leq 2$

14. $x \leq 4$,
$y \geq -1$

15. $x \leq 0$,
$y \leq 0$

16. $x \geq 0$,
$y \geq 0$

17. $2x - 3y \geq 9$,
$2y + x > 6$

18. $3x - 2y \geq 8$,
$2x + y > 6$

19. $y > \frac{1}{2}x - 2$,
$x + y \leq 1$

20. $2y - 3x \leq 4$,
$\frac{2}{3}x + y > 4$

21. $x + y \leq 5$,
$x \geq 0$,
$y \geq 0$,
$y \leq 3$

22. $x + 2y \leq 8$,
$x \leq 6$,
$x \geq 0$,
$y \geq 0$

23. $y - x \geq 1$,
$y - x \leq 3$,
$x \leq 5$,
$x \geq 2$

24. $x - 2y \leq 0$,
$y - 2x \leq 2$,
$x \leq 2$,
$y \leq 2$

25. $y \leq x$,
$x \geq -2$,
$x \leq -y$

26. $y > 0$,
$2y + x \geq -6$,
$x + 2 \leq 2y$

27. In Example 1, why was one line a solid line and the other dashed?

28. Explain why the solution set of a system of inequalities is indicated by the region where the individual graphs overlap.

Skill Review

Simplify.

29. $\dfrac{x^2 - x - 6}{x^2 - 6x + 9}$ [6.1]

30. $3y - (8 - y)$ [1.8]

31. $y^6 \cdot y^{-10}$ [4.2]

32. $(-3a^2)^2$ [4.1]

33. $2^3 - 24 \div 2 \cdot (-6)$ [1.8]

34. $|3 - 4(1 - 2^2)|$ [1.8]

Synthesis

35. In Example 2, is the point $\left(2, -\frac{4}{3}\right)$ part of the solution set? Why or why not?

36. Will a system of linear inequalities always have a solution? Why or why not?

Graph the solutions of each system. If no solution exists, state this.

37. $3r + 6t \geq 36,$
$\quad 2r + 3t \geq 21,$
$\quad 5r + 3t \geq 30,$
$\quad\quad\quad t \geq 0,$
$\quad\quad\quad r \geq 0$

38. $2x + 5y \geq 18,$
$\quad 4x + 3y \geq 22,$
$\quad 2x + y \geq 8,$
$\quad\quad\quad x \geq 0,$
$\quad\quad\quad y \geq 0$

39. $x + 3y \leq 6,$
$\quad 2x + y \geq 4,$
$\quad 3x + 9y \geq 18$

40. $2x + 5y \geq 10,$
$\quadx - 3y \leq 6,$
$\quad 4x + 10y \leq 20$

Aha! 41. $2x + 3y \leq 1,$
$\quad 4x + 6y > 9,$
$\quad 5x - 2y \leq 8,$
$\quad\quad\quad x \leq 12,$
$\quad\quad\quad y \geq -15$

42. $5x - 4y \geq 8,$
$\quad 2x + 3y < 9,$
$\quad\quad\quad 2y \geq -8,$
$\quad\quad\quad x \leq -5,$
$\quad\quad\quad y < -6$

43. *Quilting.* The International Plowing Match and Country Festival, held in Crosby, Ontario, sponsors several quilting competitions. The maximum perimeter allowed for the Theme Quilt is 200 in. Find and graph a system of inequalities that indicates that a quilt's length and width meet this requirement.

Data: www.ipm2012.ca

44. *Architecture.* Most architects agree that the sum of a step's riser r and tread t, in inches, should not be less than 17 in. In Kennebunk, Maine, the maximum riser height for nonresidential buildings is 7 in., and the minimum tread width is 11 in. Find and graph a system of inequalities that indicates that a stair design has met the architect's and the town's requirements.

Data: www.kennebunkmaine.org

45. Use a graphing calculator to solve Exercise 18.

46. Use a graphing calculator to solve Exercise 17.

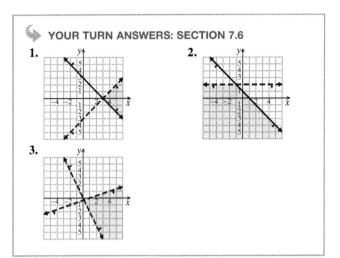

YOUR TURN ANSWERS: SECTION 7.6

1.
2.
3.

Quick Quiz: Sections 7.1–7.6

1. Solve by graphing:
$$y = 2x - 1,$$
$$2x + y = 3. \quad [7.1]$$

2. Graph the solutions of the system
$$x - y < 2,$$
$$2x + y \geq 4. \quad [7.6]$$

3. Graph: $3x - 2y \geq 12.$ [7.5]

Solve using either substitution or elimination. If the system has no solution or an infinite number of solutions, state this.   [7.2], [7.3]

4. $x - y = 2,$
$\quad 2x + y = 4$

5. $3x + y = 5,$
$\quad 2x + y = 1$

Prepare to Move On

Solve.   [2.1]

1. $210 = k \cdot 10$

2. $0.4 = k \cdot 0.5$

3. $3 = k \cdot 7$

4. $5 = \dfrac{k}{8}$

5. $100 = \dfrac{k}{0.16}$

7.7 Direct Variation and Inverse Variation

A. Equations of Direct Variation   **B.** Problem Solving with Direct Variation
C. Equations of Inverse Variation   **D.** Problem Solving with Inverse Variation

Many problems lead to equations of the form $y = kx$ or $y = k/x$, where k is a constant. Such equations are called *equations of variation*.

A. Equations of Direct Variation

A bicycle tour following the Underground Railroad route is traveling at a rate of 12 mph. In 1 hr, it goes 12 mi. In 2 hr, it goes 24 mi. In 3 hr, it goes 36 mi, and so on. In the following graph, we use the number of hours as the first coordinate and the number of miles traveled as the second coordinate:

$$(1, 12), \quad (2, 24), \quad (3, 36), \quad (4, 48), \quad \text{and so on.}$$

Note that the second coordinate is always 12 times the first.

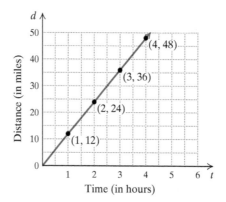

In this example, distance is a constant multiple of time, so we say that there is **direct variation** and that distance **varies directly** as time. The **equation of variation** is $d = 12t$.

DIRECT VARIATION

When a situation translates to an equation described by $y = kx$, with k a constant, we say that y *varies directly* as x. The equation $y = kx$ is called an *equation of direct variation*.

Note that for $k > 0$, any equation of the form $y = kx$ indicates that as x increases, y increases as well.

The terminologies

"y varies as x,"

"y is directly proportional to x," and

"y is proportional to x"

also imply direct variation. The constant k—that is, the **constant of proportionality**, or the **variation constant**—can be found if one pair of values of x and y is known. When k is known, other pairs can be determined.

EXAMPLE 1 If y varies directly as x, and $y = 2$ when $x = 5$, find the equation of variation.

SOLUTION The words "y varies directly as x" indicate direct variation:

$$y = kx \qquad \text{"}y \text{ varies directly as } x\text{" means } y = kx.$$
$$2 = k \cdot 5 \qquad \text{Substituting to solve for } k$$
$$\tfrac{2}{5} = k, \quad \text{or} \quad k = 0.4. \qquad \text{Dividing both sides by 5}$$

Thus the equation of variation is $y = 0.4x$.

1. If y varies directly as x, and $y = 3$ when $x = 15$, find the equation of variation.

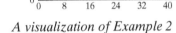

YOUR TURN

When y varies directly as x, the constant of proportionality is also the slope of the associated graph—the rate at which y changes with respect to x.

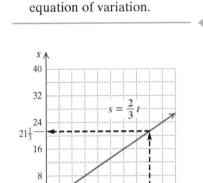

A visualization of Example 2

EXAMPLE 2 Find an equation in which s varies directly as t, and $s = 10$ when $t = 15$. Then find the value of s when $t = 32$.

SOLUTION We have

$$s = kt \qquad \text{We are told "}s \text{ varies directly as } t.\text{"}$$
$$10 = k \cdot 15 \qquad \text{Substituting 10 for } s \text{ and 15 for } t$$
$$\tfrac{10}{15} = k, \quad \text{or} \quad k = \tfrac{2}{3}. \qquad \text{Solving for } k$$

Thus the equation of variation is $s = \tfrac{2}{3}t$. When $t = 32$, we have

$$s = \tfrac{2}{3}t \qquad k = \tfrac{2}{3}$$
$$= \tfrac{2}{3} \cdot 32 \qquad \text{Substituting 32 for } t \text{ in the equation of variation}$$
$$= \tfrac{64}{3}, \text{ or } 21\tfrac{1}{3}.$$

The value of s is $21\tfrac{1}{3}$ when $t = 32$.

2. Find an equation in which s varies directly as t, and $s = 12$ when $t = 2$. Then find the value of s when $t = 5$.

YOUR TURN

B. Problem Solving with Direct Variation

In applications, it is often necessary to find an equation of variation and then use it to find other values, much as we did in Example 2.

EXAMPLE 3 The leaves and branches of a tree are called its *crown*. The *canopy* of a tree is the (roughly) circular shadow from above the crown onto the ground. The volume V of soil required for a tree to thrive varies directly as the area A of its canopy. From the table at left, we see that a tree with a canopy area of 640 ft² requires a soil volume of 1000 ft³. What soil volume is required for a tree with a canopy area of 800 ft²?

Area of Tree Canopy (in square feet)	Soil Volume Required (in cubic feet)
320	500
480	750
640	1000
1200	1875

Data: Center for Urban Forest Research

Canopy area A

Soil volume V

SOLUTION

1., 2. Familiarize and **Translate.** The words "vary directly" indicate direct variation. Using the variables provided, we see that an equation $V = kA$ applies.

3. Carry out. We find an equation of variation:

$$V = kA$$
$$1000 = k(640) \qquad \text{Substituting}$$
$$\frac{1000}{640} = k$$
$$\frac{25}{16} = k.$$

The equation of variation is $V = \frac{25}{16}A$. When $A = 800$, we have

$$V = \frac{25}{16}A \qquad k = \frac{25}{16}$$
$$= \frac{25}{16}(800) \qquad \text{Substituting 800 for } A$$
$$= 1250.$$

3. The time t that it takes to download a file varies directly as the size s of the file. Using a broadband Internet connection, Abriana can download a 4-MB (megabyte) song in 10 sec. How long would it take her to download a 210-MB television show?

4. Check. To check, you might note from the table above that 800 is between 640 and 1200 in the first column, and 1250 is between the corresponding values of 1000 and 1875 in the second column. Also, the ratios 1000/640 and 1250/800 are both 25/16.

5. State. A tree with a canopy area of 800 ft^2 requires a soil volume of 1250 ft^3.

↪ YOUR TURN

C. Equations of Inverse Variation

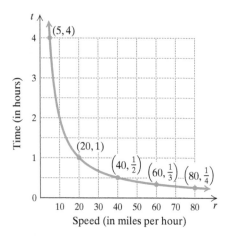

A car is traveling a distance of 20 mi. At a speed of 5 mph, the trip will take 4 hr. At 20 mph, it will take 1 hr. At 40 mph, it will take $\frac{1}{2}$ hr, and so on. This determines a set of pairs of numbers:

$$(5, 4), \qquad (20, 1), \qquad \left(40, \tfrac{1}{2}\right), \quad \text{and so on.}$$

Note that the product of speed and time for each of these pairs is 20. Note too that as the speed *increases*, the time *decreases*.

    In this case, the product of speed and time is constant so we say that there is **inverse variation** and that time **varies inversely** as speed. The equation of variation is

$$rt = 20 \text{ (a constant),} \quad \text{which is more commonly expressed as} \quad t = \frac{20}{r}.$$

> **INVERSE VARIATION**
>
> When a situation translates to an equation described by $y = k/x$, with k a constant, we say that y *varies inversely* as x. The equation $y = k/x$ is called an *equation of inverse variation*.

    Note that for $k > 0$, any equation of the form $y = k/x$ indicates that as x increases, y decreases.

    The terminology

        "y is inversely proportional to x"

also implies inverse variation. The constant k is again called the *constant of proportionality*, or the *variation constant*.

Student Notes

When inverse variation is indicated, the first step is to write $y = k/x$.

4. If y varies inversely as x, and $y = 5$ when $x = 20$, find the equation of variation.

Determine the type of variation.

Write the general form of the equation.

Substitute.

Solve for k.

Write the equation of variation.

Solve.

EXAMPLE 4 If y varies inversely as x, and $y = 145$ when $x = 0.8$, find the equation of variation.

SOLUTION The words "y varies inversely as x" indicate inverse variation:

$$y = \frac{k}{x} \qquad \text{"}y \text{ varies inversely as } x\text{" means } y = \frac{k}{x}.$$

$$145 = \frac{k}{0.8} \qquad \text{Substituting to solve for } k$$

$$(0.8)145 = k \qquad \text{Multiplying both sides by } 0.8$$

$$116 = k.$$

The equation of variation is $y = \dfrac{116}{x}$.

YOUR TURN

D. Problem Solving with Inverse Variation

Often in applications, we must decide what kind of variation, if any, applies.

EXAMPLE 5 *Community Service.* Maxwell Elementary School needs 40 volunteers to each donate 8 hr per month to read one-on-one with a child. If 50 people actually volunteer, how many hours per month would each person need to donate?

SOLUTION

1. **Familiarize.** What kind of variation applies to this situation? It seems reasonable that the greater the number of people volunteering, the less time each will need to donate. The total number of volunteer hours will be constant and inverse variation applies. If $T =$ the number of hours each person must volunteer and $N =$ the number of volunteers, then as N increases, T decreases.

2. **Translate.** Since inverse variation applies, we have

$$T = \frac{k}{N}.$$

3. **Carry out.** We find an equation of variation:

$$T = \frac{k}{N}$$

$$8 = \frac{k}{40} \qquad \text{Substituting 8 for } T \text{ and 40 for } N$$

$$40 \cdot 8 = k \qquad \text{Multiplying both sides by 40}$$

$$320 = k.$$

The equation of variation is $T = \dfrac{320}{N}$. When $N = 50$, we have

$$T = \frac{320}{50} \qquad \text{Substituting 50 for } N$$

$$T = 6.4.$$

5. A hospital needs 50 volunteers to each donate 10 hr per month. If only 40 people actually volunteer, how many hours per month does each person need to donate?

4. Check. To check, note that 40 volunteers would donate a total of $(40)(8)$, or 320 hr per month. If each of 50 volunteers donated 6.4 hr per month, the total volunteer time would be $(50)(6.4)$, which is also 320 hr. Also, as the number of volunteers increases from 40 to 50, the time each needs to donate per month decreases from 8 hr to 6.4 hr.

5. State. Each of the 50 volunteers would donate 6.4 hr each month.

↩ YOUR TURN

TO SOLVE VARIATION PROBLEMS

1. Determine from the problem whether direct variation or inverse variation applies.

2. Find the equation of variation for the problem.

   a) Write an equation of the form $y = kx$ for direct variation or $y = k/x$ for inverse variation.
   b) Substitute known values.
   c) Solve for k.
   d) Rewrite the equation from part (a), replacing k with its value.

3. Use the equation of variation, as needed, to find unknown values.

↩ **Chapter Resource:**
Decision Making: Connection, p. 484

↩ Check Your
UNDERSTANDING

Determine whether each equation represents direct variation or inverse variation.

1. $y = \dfrac{10}{x}$

2. $y = \frac{1}{10}x$

3. $y = 7.3x$

4. $y = 4/x$

5. $xy = 2$

6. $\dfrac{x}{y} = 2$

7.7   EXERCISE SET

FOR EXTRA HELP MyMathLab®

↩ **Vocabulary and Reading Check**

Match each description of variation with the appropriate equation of variation from the list on the right.

1. _____ y varies directly as x

2. _____ y varies inversely as x

3. _____ t is inversely proportional to s

4. _____ t is directly proportional to s

5. _____ a is proportional to c

6. _____ a varies inversely as c

   a) $y = kx$

   b) $a = kc$

   c) $t = ks$

   d) $a = \dfrac{k}{c}$

   e) $t = \dfrac{k}{s}$

   f) $y = \dfrac{k}{x}$

↩ **Concept Reinforcement**

Determine whether each situation reflects either direct variation or inverse variation.

7. Two copiers can complete the job in 3 hr, whereas three copiers require only 2 hr.

8. Fionna wrote 5 invitations in 30 min and 8 invitations in 48 min.

9. Tomas found 8 flaws in 100 shirts and 24 flaws in 300 shirts.

10. Three neighbors collected debris along a mile of roadway in 40 min while another group of five neighbors collected a mile's worth of debris in 24 min.

A. Equations of Direct Variation

For each of the following, find an equation of variation in which y varies directly as x and the following are true.

11. $y = 40$, when $x = 8$

12. $y = 30$, when $x = 3$

13. $y = 1.75$, when $x = 0.25$

14. $y = 3.2$, when $x = 0.08$

15. $y = 0.3$, when $x = 0.5$

16. $y = 1.2$, when $x = 1.1$

17. $y = 200$, when $x = 300$

18. $y = 500$, when $x = 60$

C. Equations of Inverse Variation

For each of the following, find an equation of variation in which y varies inversely as x and the following are true.

19. $y = 10$, when $x = 12$

20. $y = 8$, when $x = 3$

21. $y = 0.25$, when $x = 4$

22. $y = 0.25$, when $x = 8$

23. $y = 50$, when $x = 0.4$

24. $y = 42$, when $x = 50$

25. $y = 42$, when $x = 5$

26. $y = 0.2$, when $x = 5$

B, D. Problem Solving with Variation

Solve.

27. *Wages.* Chardeé's pay P varies directly as the number of hours worked H. For 15 hr of work, the pay is $135. Find the pay for 23 hr of work.

28. *Oatmeal Servings.* The number of servings S of oatmeal varies directly as the net weight W of the container purchased. A 42-oz box of oatmeal contains 30 servings. How many servings does a 63-oz box of oatmeal contain?

29. *Energy Use.* The number of kilowatt-hours (kWh) of electricity n used by an air conditioner varies directly as its capacity c, measured in tons. A 2-ton air conditioner uses 18.5 kWh in 10 hr. How many kilowatt-hours would a 3-ton air conditioner use in 10 hr?

Data: www.lccc.net

30. *Gas Volume.* The volume V of a gas varies inversely as the pressure P on it. The volume of a gas is 200 cm^3 (cubic centimeters) under a pressure of 32 kg/cm^2. What will be its volume under a pressure of 20 kg/cm^2?

31. *Gold.* The karat rating of a gold object varies directly as the percentage of gold in the object. A 14-karat gold chain is 58.25% gold. What is the percentage of gold in a 24-karat gold chain?

32. *Calories.* The number of calories in food varies directly as its weight. A cereal contains 120 calories in a 25-gram serving. How many calories are in a 35-gram serving?

33. *Job Completion.* The number of workers n required to complete a job varies inversely as the time t spent working. It takes 5 hr for 6 people to clean a gym. How long would it take 15 people to complete the job?

34. *Electricity.* A thermistor senses temperature changes by measuring changes in resistance. The resistance R in a PTC thermistor varies directly as the temperature T. A PTC thermistor has a resistance of 5000 ohms at a temperature of 25°C. What is the temperature if the resistance is 6000 ohms?

35. *Musical Tones.* The frequency, or pitch, of a musical tone varies inversely as its wavelength. The U.S. standard concert A has a pitch of 440 vibrations per second and a wavelength of 2.4 ft. The E above concert A has a frequency of 660 vibrations per second. What is its wavelength?

36. *Filling Time.* The time required to fill a pool varies inversely as the rate of flow into the pool. A tanker can fill a pool in 90 min at a rate of 1200 L/min. How long would it take to fill the pool at a rate of 2000 L/min?

Aha! **37.** *Chartering a Boat.* The cost per person of chartering a boat is inversely proportional to the number of people who are chartering the boat. If it costs $70 per person when 40 people charter the Sea Otter V, how many people chartered the boat if the cost were $140 per person?

38. *Answering Questions.* The number of minutes m that a student should allow for each question on a quiz is inversely proportional to the number of questions n on the quiz. If a 16-question quiz allows students 2.5 min per question, how many questions would appear on a quiz in which students are allowed 4 min per question?

State whether each situation represents direct variation, inverse variation, or neither. Give reasons for your answers.

39. The cost of mailing an overnight package in the United States and the distance that it travels

40. A runner's speed in a race and the time it takes to run the race

41. The weight of a turkey and its cooking time

42. The number of plays that it takes to go 80 yd for a touchdown and the average gain per play

Skill Review

43. 16 is 8% of what number? [2.4]

44. Solve $a = \frac{1}{3}bc$ for c. [2.3]

45. Find the slope of the line containing the points $(-3, 6)$ and $(-1, -7)$. [3.5]

46. Write the slope–intercept equation of the line with slope $-\frac{2}{5}$ and y-intercept $(0, 6)$. [3.6]

47. Convert to scientific notation: 0.00307. [4.2]

48. Find the LCM of $a^3 - a$, $a^2 - 2a + 1$, and a^2. [6.3]

Synthesis

49. If x varies inversely as y and y varies inversely as z, how does x vary with regard to z? Why?

50. If a varies directly as b and b varies inversely as c, how does a vary with regard to c? Why?

Write an equation of variation for each situation. Use k as the variation constant in each equation.

51. *Wind Energy.* The power P in a windmill varies directly as the cube of the wind speed v.

52. *Acoustics.* The square of the pitch P of a vibrating string varies directly as the tension t on the string.

53. *Ecology.* In a stream, the amount of salt S carried varies directly as the sixth power of the speed of the stream v.

54. *Lighting.* The intensity of illumination I from a light source varies inversely as the square of the distance d from the source.

55. *Crowd Size.* The number of people attending a First Night celebration N is directly proportional to the temperature T and inversely proportional to the percentage chance of rain or snow, P.

Write an equation of variation for each situation. Include a value for the variation constant in each equation.

56. *Geometry.* The volume V of a sphere varies directly as the cube of the radius r.

57. *Geometry.* The perimeter P of an equilateral octagon varies directly as the length S of a side.

58. *Geometry.* The circumference C of a circle varies directly as the diameter d.

59. *Geometry.* The area A of a circle varies directly as the square of the length of the radius r.

60. *Research.* Find examples that use the phrases "directly proportional to" and "inversely proportional to." Explain how your examples use these phrases consistently or inconsistently with their definitions in this section.

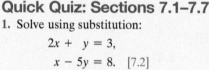

YOUR TURN ANSWERS: SECTION 7.7

1. $y = \frac{1}{5}x$, or $y = 0.2x$ **2.** $s = 6t$; 30

3. 525 sec, or 8 min 45 sec   **4.** $y = \dfrac{100}{x}$

5. 12.5 hr per month

Quick Quiz: Sections 7.1–7.7

1. Solve using substitution:
$$2x + y = 3,$$
$$x - 5y = 8. \quad [7.2]$$

2. Solve using elimination:
$$3x - 4y = 3,$$
$$2x + 2y = 5. \quad [7.3]$$

3. For an office reception, Mariah wants to provide 10 lb of a nut mixture that is 20% peanuts. She can buy mixtures that are 10% peanuts and mixtures that are 50% peanuts. How much of each mixture should she use? [7.4]

4. Graph: $y > x - 3$. [7.5]

5. If y varies inversely as x, and $y = 9$ when $x = 6$, find the equation of variation. [7.7]

Prepare to Move On

Simplify each square.

1. 10^2 [1.8]

2. $(-10)^2$ [1.8]

3. $(-5x)^2$ [4.1]

4. $(s^3t^4)^2$ [4.1]

5. $(x + y)^2$ [4.6]

6. $(2a - b)^2$ [4.6]

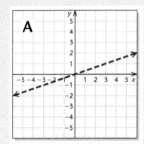

A

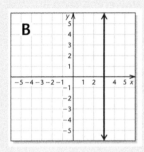

B

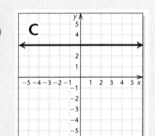

C

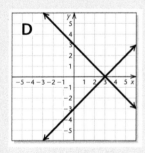

D

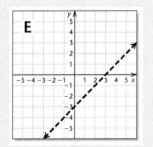

E

Visualizing for Success

Use after Section 7.6.

Match each equation or inequality or system of equations or inequalities with its graph.

1. $x + y = 3$

2. $x = 3$

3. $y < x - 3$

4. $y = \frac{1}{3}x$

5. $y = 3$

6. $x - y = 3$,
   $x + y = 3$

7. $y < \frac{1}{3}x$

8. $y \geq x - 3$,
   $3x - y > 1$

9. $x - y < 3$

10. $y = \frac{1}{3}x$,
   $y = 3x$

Answers on page A-40

An additional, animated version of this activity appears in MyMathLab. *To use MyMathLab, you need a course ID and a student access code. Contact your instructor for more information.*

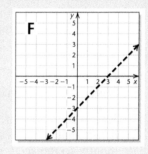

F

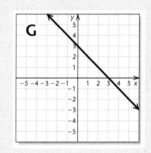

G

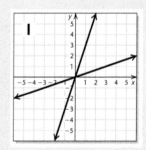

H

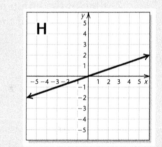

I

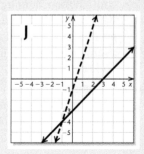

J

Collaborative Activity   *Conserving Energy and Money*

Focus: Systems of linear equations (two variables)
Use after: Section 7.2
Time: 20 minutes
Group size: 2
Materials: Graph paper

Jean, a condo owner, has an old electric water heater that consumes $110 of electricity per month. To replace it with a new gas water heater that consumes only $35 of gas per month will cost $550 plus $250 for installation. Jean wants to know how long it will take for the new gas water heater to "pay for itself," in other words, to save as much as it costs.

Activity

1. The "break-even" point occurs when the costs of the two water heaters are equal. Determine, by "guessing and checking," the number of months before Jean breaks even.

2. One of the group members should create a cost equation of the form $y = mx + b$ for the electric heater, where y is the cost, in dollars, and x is the time, in months. Note that when $x = 0$, the cost is $0. He or she should also graph the equation on the graph paper provided.

3. The second group member should create a cost equation of the form $y = mx + b$ for the gas heater, where y is the cost, in dollars, and x is the time, in months. This equation should be graphed on the same graph used in part (2).

4. Working together, the group should determine the coordinates of the point of intersection of the two lines. This is the break-even point. Which coordinate indicates the number of months before Jean breaks even? What does the other coordinate indicate? Compare the answer found graphically with the estimate made in part (1).

Decision Making & Connection   *(Use after Section 7.7.)*

Renting an Apartment. When deciding whether or not to rent an apartment, or which apartment to rent, there are many considerations besides the amount of monthly rent.

1. Most landlords require a renter to have a minimum monthly income that varies directly with the monthly apartment rent. Suppose that, for a monthly rent of $650, a landlord requires the renter to have a monthly take-home pay of $1625. What would the income requirement be for an apartment that rents for $850 per month?

2. Many landlords require a renter to pay for at least some of the utilities. For apartments with similar construction, heating and cooling costs vary directly with the size of the apartment. Suppose that, for a 1200-ft² apartment, the heating and cooling bill averages $60 per month. What would be the heating and cooling bill for an 800-ft² apartment?

3. The costs listed in the following table are for apartments of the same size but in different complexes. Write an equation that models the cost of each

apartment, and determine the time required for the costs to be the same.

	Cobblefield Downs	Grassymeadow Park
Rent	$900 per month	$625 per month
Credit Check	$30 before move-in	$30 before move-in
Gas	Included	$50 new account fee plus $30 per month
Internet	$70 installation plus $40 per month	$45 per month
Furniture	Included	$1100

4. *Research.* Choose an apartment in your area and find all costs associated with moving in. Write an equation that models the cost of the apartment for a rental period of t months.

Study Summary

KEY TERMS AND CONCEPTS	EXAMPLES	PRACTICE EXERCISES

SECTION 7.1: *Systems of Equations and Graphing*

A solution of a **system of two equations** is an ordered pair that makes both equations true. The intersection of the graphs of the equations gives the solution of the system.

A **consistent** system has at least one solution; an **inconsistent** system has no solution.

When one equation in a system of two equations is a nonzero multiple of the other, the equations are **dependent** and there is an infinite number of solutions. Otherwise, the equations are **independent**.

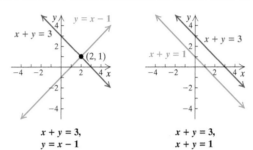

$x + y = 3,$
$y = x - 1$

$x + y = 3,$
$x + y = 1$

The graphs intersect at (2, 1). The graphs do not intersect.
The solution is (2, 1). There is no solution.
The system is *consistent*. The system is *inconsistent*.
The equations are *independent*. The equations are *independent*.

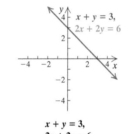

$x + y = 3,$
$2x + 2y = 6$

The graphs are the same.
There is an infinite number of solutions.
The system is *consistent*.
The equations are *dependent*.

1. Solve by graphing. If there is no solution or an infinite number of solutions, state this.

$y = -\frac{1}{2}x + 3,$
$x - 2y = 6$

SECTION 7.2: *Systems of Equations and Substitution*

To use the **substitution method,** we solve one equation for a variable and substitute the resulting expression for that variable in the other equation.

Solve:

$2x + 3y = 8,$
$x = y + 1.$

Substitute and solve for y:
$2(y + 1) + 3y = 8$
$2y + 2 + 3y = 8$
$5y + 2 = 8$
$5y = 6$
$y = \frac{6}{5}.$

Substitute and solve for x:
$x = y + 1$
$x = \frac{6}{5} + 1$
$x = \frac{6}{5} + \frac{5}{5}$
$x = \frac{11}{5}.$

The solution is $\left(\frac{11}{5}, \frac{6}{5}\right).$

2. Solve using substitution. If there is no solution or an infinite number of solutions, state this.

$y = 3 - 2x,$
$3x + 2y = 5$

SECTION 7.3: *Systems of Equations and Elimination*

| To use the **elimination method,** we add to eliminate a variable. | Solve: $$4x - 2y = 6,$$ $$3x + y = 7.$$ Eliminate y and solve for x: $$4x - 2y = 6$$ $$\underline{6x + 2y = 14}$$ $$10x = 20$$ $$x = 2.$$ The solution is $(2, 1)$. Substitute and solve for y: $$3x + y = 7$$ $$3 \cdot 2 + y = 7$$ $$6 + y = 7$$ $$y = 1.$$ | **3.** Solve using elimination. If there is no solution or an infinite number of solutions, state this. $$2x - 3y = 4,$$ $$4x + 2y = 3$$ |

SECTION 7.4: *More Applications Using Systems*

| **Total-value problems** and **mixture problems** are two types of problems that are often translated readily into a system of equations. | At a local "paint swap," Madison found large supplies of Skylite Pink (12.5% red pigment) and MacIntosh Red (20% red pigment). How many gallons of each color should be mixed in order to create a 15-gal batch of Summer Rose (17% red pigment)? We let p = the number of gallons of Skylite Pink and m = the number of gallons of MacIntosh Red and translate the problem to a system of equations: $$p + m = 15,$$ $$0.125p + 0.2m = 0.17(15).$$ Solving, we find that $p = 6$ and $m = 9$. For a complete solution, see Example 4 in Section 7.4. | **4.** An industrial cleaning solution that is 40% nitric acid is added to a solution that is 15% nitric acid in order to create 2 L of a solution that is 25% nitric acid. How much 40%-acid and how much 15%-acid should be used? |

SECTION 7.5: *Linear Inequalities in Two Variables*

| The solutions of a **linear inequality** like $y \le \frac{1}{2}x - 1$ are ordered pairs that make the inequality true. The graph of a linear inequality is a region in the plane bounded by a solid line if the inequality symbol is $\le$ or $\ge$ and by a dashed line if the inequality symbol is $<$ or $>$. A test point not on the boundary line is used to determine which half-plane to shade as the graph of the solution. | Graph: $y \le \frac{1}{2}x - 1$. **1.** Graph $y = \frac{1}{2}x - 1$, using a solid line. **2.** Test $(0,0)$. Since $0 \le \frac{1}{2}(0) - 1$, or $0 \le -1$, is *false*, shade the region that does *not* contain $(0,0)$. | **5.** Graph: $y < 2x + 3$. |

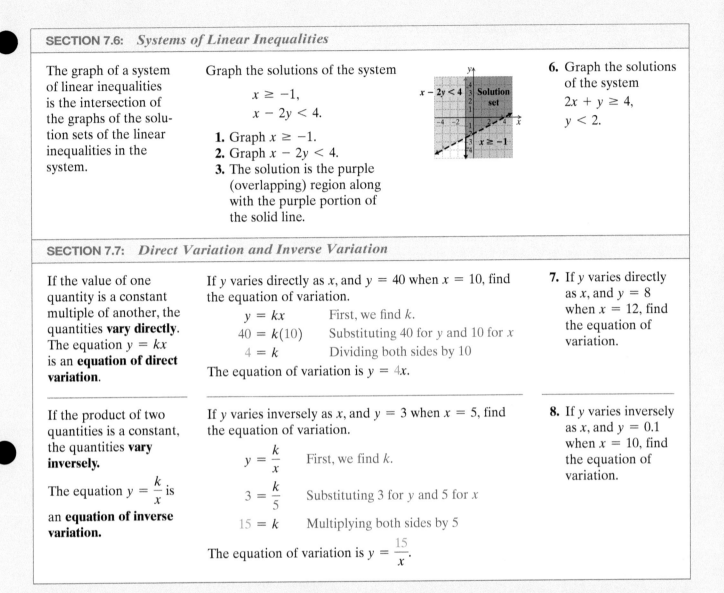

SECTION 7.6: *Systems of Linear Inequalities*

The graph of a system of linear inequalities is the intersection of the graphs of the solution sets of the linear inequalities in the system.	Graph the solutions of the system $$x \geq -1,$$ $$x - 2y < 4.$$ **1.** Graph $x \geq -1$. **2.** Graph $x - 2y < 4$. **3.** The solution is the purple (overlapping) region along with the purple portion of the solid line.	**6.** Graph the solutions of the system $$2x + y \geq 4,$$ $$y < 2.$$

SECTION 7.7: *Direct Variation and Inverse Variation*

If the value of one quantity is a constant multiple of another, the quantities **vary directly**. The equation $y = kx$ is an **equation of direct variation**.	If y varies directly as x, and $y = 40$ when $x = 10$, find the equation of variation. $y = kx$ — First, we find k. $40 = k(10)$ — Substituting 40 for y and 10 for x $4 = k$ — Dividing both sides by 10 The equation of variation is $y = 4x$.	**7.** If y varies directly as x, and $y = 8$ when $x = 12$, find the equation of variation.
If the product of two quantities is a constant, the quantities **vary inversely**. The equation $y = \dfrac{k}{x}$ is an **equation of inverse variation**.	If y varies inversely as x, and $y = 3$ when $x = 5$, find the equation of variation. $y = \dfrac{k}{x}$ — First, we find k. $3 = \dfrac{k}{5}$ — Substituting 3 for y and 5 for x $15 = k$ — Multiplying both sides by 5 The equation of variation is $y = \dfrac{15}{x}$.	**8.** If y varies inversely as x, and $y = 0.1$ when $x = 10$, find the equation of variation.

Review Exercises: Chapter 7

➤ Concept Reinforcement

Complete each statement.

1. The solution of a system of two equations in two variables is a(n) _____ pair of numbers. [7.1]

2. When both equations in a system are equivalent to each other, there is a(n) _____ number of solutions. [7.1]

3. When the equations in a system have graphs that are _____, the system has no solution. [7.1]

4. When substituting an expression for a variable, use _____ around the expression being substituted. [7.2]

5. To use the elimination method, the _____ of one of the variables must be made to be (if they are not already) opposites of each other. [7.3]

6. The graph of an inequality of the form $y < mx + b$ is shaded _____ the line $y = mx + b$. [7.5]

7. The graph of the solutions of a system of linear inequalities is the _____ of the graphs of the inequalities. [7.6]

8. If the product of x and y is constant, then y varies _____ as x. [7.7]

Determine whether each ordered pair is a solution of the system of equations. [7.1]

9. $(2, -5)$;
$x - 2y = 12,$
$2x - y = 1$

10. $(-3, 1)$;
$3b + a = 0,$
$a + 5b = 2$

Solve by graphing. If there is no solution or an infinite number of solutions, state this. [7.1]

11. $y = 4x - 1,$
$y = x + 2$

12. $x - y = 8,$
$x + y = 4$

13. $3x - 4y = 8,$
$4y - 3x = 6$

14. $2x + y = 3,$
$4x + 2y = 6$

Solve using the substitution method. If there is no solution or an infinite number of solutions, state this. [7.2]

15. $y = 4 - x,$
$3x + 4y = 21$

16. $x + 2y = 6,$
$2x + y = 8$

17. $x + y = 5,$
$y = 2 - x$

18. $x + y = 6,$
$y = 3 - 2x$

Solve using the elimination method. If there is no solution or an infinite number of solutions, state this. [7.3]

19. $3x - 2y = 0,$
$2x + 2y = 50$

20. $x - y = 8,$
$2x + y = 7$

21. $x - \frac{1}{3}y = -\frac{13}{3},$
$3x - y = -13$

22. $4x + 3y = -1,$
$2x + 9y = 2$

Solve using any appropriate method. If there is no solution or an infinite number of solutions, state this. [7.1], [7.2], [7.3]

23. $5x - 2y = 7,$
$4x - 3y = 14$

24. $-x - y = -5,$
$2x - y = 4$

25. $x - 2y = 5,$
$3x + 4y = 10$

26. $3x - y = 5,$
$6x = 2y + 10$

27. $2x + 5y = 8,$
$3x + 4y = 10$

28. $-4x + 6y = -10,$
$6x - 9y = 12$

Solve. [7.4]

29. *Perimeter of a Garden.* Hassam is designing a rectangular garden with a perimeter of 66 ft. The length of the garden is 1 ft longer than three times the width. Find the dimensions of the garden.

30. *Basketball.* In a recent NBA game, Monta Ellis scored 21 points on a combination of 10 two-point and three-point baskets. How many shots of each type were made?

Data: National Basketball Association

31. *Business.* F3 Sales spent $11,150 to purchase tablet computers for its 35 employees. Some employees requested tablets costing $250 and some requested tablets costing $350. How many of each type did F3 purchase?

32. *Fat Content.* Café Rich instant flavored coffee gets 40% of its calories from fat. Café Light coffee gets 25% of its calories from fat. How much of each brand of coffee should be mixed in order to make 200 g of instant coffee with 30% of its calories from fat?

Graph on a plane. [7.5]

33. $x \le y$

34. $x - 2y \ge 4$

35. $y > -2$

36. $y \ge \frac{2}{3}x - 5$

37. $2x + y < 1$

38. $x < 4$

Graph the solutions of each system. [7.6]

39. $x \ge 1,$
$y \le -1$

40. $x - y > 2,$
$x + y < 1$

41. If y varies directly as x, and $y = 81$ when $x = 3$, find the equation of variation. [7.7]

42. *Cubicle Space.* The size of a cubicle c in an office space varies inversely as the number of workers w in that space. When there are 10 workers in Office A-213, each worker has 180 ft^2 of space. How much space will each worker have if there are 12 workers in that space? [7.7]

Synthesis

43. Explain why any solution of a system of equations is a point of intersection of the graphs of each equation in the system. [7.1]

44. Nigel sketches the boundary lines of a system of two linear inequalities and notes that the lines are parallel. Since there is no point of intersection, he concludes that the solution set is empty. What is wrong with this conclusion? [7.6]

45. The solution of the following system is $(6, 2)$. Find C and D.

$2x - Dy = 6,$
$Cx + 4y = 14$ [7.1]

46. Solve using the substitution method:

$x - y + 2z = -3,$
$2x + y - 3z = 11,$
$z = -2.$ [7.2]

47. Solve:

$$3(x - y) = 4 + x,$$
$$x = 5y + 2. \quad [7.2]$$

48. For a two-digit number, the sum of the ones digit and the tens digit is 6. When the digits are reversed, the new number is 18 more than the original number. Find the original number.   [7.4]

49. An administrative assistant agrees to a compensation package of $42,000 plus a computer. After 7 months, she leaves the company. At that point, her compensation package consisted of the computer and $23,750. What was the value of the computer?   [7.4]

Test: Chapter 7   For step-by-step test solutions, access the Chapter Test Prep Videos in MyMathLab.

1. Determine whether $(3, -2)$ is a solution of the following system of equations:

$$2x + y = 4,$$
$$5x - 6y = 27.$$

Solve by graphing. If there is no solution or an infinite number of solutions, state this.

2. $y = -2x + 5,$
$y = 4x - 1$

3. $2y - x = 7,$
$2x - 4y = 4$

Solve using the substitution method. If there is no solution or an infinite number of solutions, state this.

4. $2x + 11 = y,$
$3x + 2y = 1$

5. $4x + y = 5,$
$2x + y = 4$

Solve using the elimination method. If there is no solution or an infinite number of solutions, state this.

6. $x - y = 1,$
$2x + y = 8$

7. $\frac{3}{2}x - y = 24,$
$2x + \frac{3}{2}y = 15$

8. $4x + 5y = 5,$
$6x + 7y = 7$

Solve using any appropriate method. If there is no solution or an infinite number of solutions, state this.

9. $x = 5y - 10,$
$15y = 3x + 30$

10. $2x + 3y = 8,$
$3x + 2y = 5$

Solve.

11. *Complementary Angles.*   Two angles are complementary. One angle is 2° less than three times the other. Find the angles.

12. *Cooking.*   A chef has one vinaigrette that is 40% vinegar and another that is 25% vinegar. How much of each is needed in order to make 60 L of a vinaigrette that is 30% vinegar?

13. *Taxi Fares.*   A New York City taxi recently cost $2.50 plus $2.00 per mile. A Boston taxi cost $2.20 plus $2.80 per mile. For what distance will the cost of the two taxis be the same?

Graph on a plane.

14. $y > x - 1$

15. $2x - y \le 4$

16. $y < -2$

Graph the solutions of each system.

17. $y \ge x - 5,$
$y < \frac{1}{2}x$

18. $x + y \le 4,$
$x \ge 0,$
$y \ge 0$

19. If y varies inversely as x, and $y = 9$ when $x = 2$, find the equation of variation.

20. *Hospital Costs.*   A hospital's linen cost varies directly with the number of patients. Wellview Hospital has a daily linen cost of $1178 when there are 124 patients in the hospital. What is the daily linen cost when there are 140 patients in the hospital?

Synthesis

21. You are in line at a ticket window. There are two more people ahead of you in line than there are behind you. In the entire line, there are three times as many people as there are behind you. How many are in the line?

22. Graph on a plane: $|x| \le 5$.

23. Find the numbers C and D such that $(-2, 3)$ is a solution of the system

$$Cx - 4y = 7,$$
$$3x + Dy = 8.$$

Cumulative Review: Chapters 1–7

1. Evaluate $9t \div 6t^3$ for $t = -2$. [1.8]

Simplify. [1.8]

2. $3x^2 - 2(-5x^2 + y) + y$

3. $40 - 8^2 \div 4 \cdot 4$

4. $\dfrac{|-2 \cdot 5 - 3 \cdot 4|}{5^2 - 2 \cdot 7}$

Solve.

5. $2x + 1 = 5(2 - x)$ [2.2]

6. $t + \dfrac{6}{t} = 5$ [6.6]

7. $2y + 9 \le 5y + 11$ [2.6]

8. $n^2 = 100$ [5.6]

9. $3x + y = 5,$
$y = x + 1$ [7.2]

10. $\dfrac{4}{x - 1} = \dfrac{3}{x + 2}$ [6.6]

11. $6x^2 = x + 2$ [5.6]

12. $3x + 2y = 4,$
$5x - 4y = 1$ [7.3]

Solve each formula. [2.3]

13. $t = \frac{1}{3}pq$, for p

14. $A = \dfrac{r + s}{2}$, for s

Add or subtract, as indicated, to form an equivalent expression. Simplify, if possible.

15. $(8a^2 - 6a - 7) + (3a^3 + 6a - 7)$ [4.4]

16. $\dfrac{2x + 1}{x + 2} + \dfrac{x + 5}{x + 2}$ [6.3]

17. $\dfrac{m + n}{2m + n} + \dfrac{n}{m - n}$ [6.4]

18. $(8a^2 - 6a - 7) - (3a^3 + 6a - 7)$ [4.4]

19. $\dfrac{4p - q}{p + q} - \dfrac{10p - q}{p + q}$ [6.3]

20. $\dfrac{x + 5}{x - 2} - \dfrac{x - 1}{2 - x}$ [6.4]

Multiply.

21. $(5a^2 + b)(5a^2 - b)$ [4.7]

22. $(2n + 5)^2$ [4.6]

23. $(8t^2 + 5)(t^3 + 4)$ [4.6]

24. $\dfrac{x^2 - 2x + 1}{x^2 - 4} \cdot \dfrac{x^2 + 4x + 4}{x^2 - 3x + 2}$ [6.2]

Divide.

25. $(2x^2 - 5x - 3) \div (x - 3)$ [4.8]

26. $\dfrac{x^2 - x}{4x^2 + 8x} \div \dfrac{x^2 - 1}{2x}$ [6.2]

Factor completely.

27. $x^3 + x^2 + 2x + 2$ [5.1]

28. $m^4 - 1$ [5.4]

29. $2x^3 + 18x^2 + 40x$ [5.2]

30. $x^4 - 6x^2 + 9$ [5.4]

31. $10x^2 - 29x + 10$ [5.3]

Graph.

32. $2x + y = 6$ [3.3]

33. $y = -2x + 1$ [3.6]

34. $2x = 10$ [3.3]

35. $x = 2y$ [3.2]

36. $x < 2y$ [7.5]

37. $x + y \le 1,$
$x - y > 2$ [7.6]

38. Find the x-intercept and the y-intercept of the line given by $10x - 15y = 60$. [3.3]

39. Write the slope–intercept equation for the line with slope -2 and y-intercept $\left(0, \frac{4}{7}\right)$. [3.6]

Simplify.

40. $\dfrac{3x^{-2}}{6x^{-12}}$ [4.2]

41. $\left(\dfrac{a^2 b}{a^3 b^4}\right)^{-1}$ [4.2]

Solve.

42. Each nurse practitioner at the Midway Clinic is required to see an average of at least 40 patients per day. During the first 4 workdays of one week, Michael saw 50, 35, 42, and 38 patients. How many patients must he see on the fifth day in order to meet his requirement? [2.7]

43. A snowmobile is traveling 40 mph faster than a dog sled. In the same time that the dog sled travels 24 mi, the snowmobile travels 104 mi. Find the speeds of the sled and the snowmobile. [6.7]

44. Tia and Avery live next door to each other on Meachin Street. Their house numbers are consecutive odd numbers, and the product of their house numbers is 143. Find the house numbers. [5.7]

45. Calls from the United States to Egypt made using the Rising Sun prepaid calling card cost 7.9¢ per minute plus a maintenance fee of 99¢ per week. Calls made using the Laya card cost 10.9¢ per minute plus a maintenance fee of 79¢ per week. For what number of minutes of calls to Egypt per week will the cost of the two cards be the same? [7.4]

Data: callingcards.com

46. A newspaper uses self-employed copywriters to write its advertising copy. If they use 12 copywriters, each person works 35 hr per week. How many hours per week would each person work if the newspaper uses 10 copywriters? [7.7]

Synthesis

47. Solve $t = px - qx$ for x. [2.3]

48. Write the slope–intercept equation of the line that contains the point $(-2, 3)$ and is parallel to the line $2x - y = 7$. [3.7]

49. Graph on a plane: $y \ge x^2$. [7.5]

Chapter 8

Radical Expressions and Equations

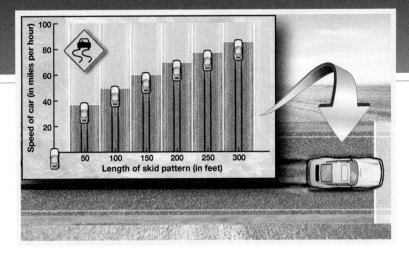

I Wasn't Going That Fast!

n order to help reconstruct vehicle crashes, crime scene analysts and forensic investigators use skid patterns on the road surface to estimate the speed that a car was traveling when the brakes were applied. The car's speed is related both to the length of the skid and to a drag factor describing the road surface and can be modeled by a *radical equation.* **(See Exercise 92 in Exercise Set 8.2 and Exercises 41 and 42 in Exercise Set 8.5.)**

8.1 Introduction to Square Roots and Radical Expressions

8.2 Multiplying and Simplifying Radical Expressions

8.3 Quotients Involving Square Roots

8.4 Radical Expressions with Several Terms

MID-CHAPTER REVIEW

8.5 Radical Equations

CONNECTING THE CONCEPTS

8.6 Applications Using Right Triangles

8.7 Higher Roots and Rational Exponents

CHAPTER RESOURCES

Translating for Success
Collaborative Activity
Decision Making: Connection

STUDY SUMMARY

REVIEW EXERCISES
CHAPTER TEST
CUMULATIVE REVIEW

Using mathematics is a critically important part of my job.

Jody Kasper, Chief of Police, Northhampton, Massachusetts, uses math to calculate vehicle speeds after crashes, to digitally map crime scenes, and to calculate fines for tickets. In addition, she manages a $5.9 million budget each year.

ALF
Active Learning Figure
Explore the math using the Active Learning Figure in MyMathLab.

SA
Student Activity
Do the Student Activity in MyMathLab to see math in action.

Y ou may already be familiar with the notion of *square roots.* For example, 3 is a square root of 9 because $3^2 = 9$. In this chapter, we look at square roots of polynomials and rational expressions. We will learn to manipulate such *radical expressions* and to solve *radical equations.*

8.1 Introduction to Square Roots and Radical Expressions

A. Square Roots **B.** Radicands and Radical Expressions **C.** Irrational Numbers
D. Square Roots and Absolute Value **E.** Problem Solving

We begin our study of radical expressions by examining square roots of numbers, square roots of variable expressions, and an application involving a formula.

A. Square Roots

To find the square of a number, we multiply that number by itself. When the process is reversed, we say that we are looking for a number's *square root.* For example, since $5^2 = 25$, we say that

25 is the square of 5   and   5 is a square root of 25.

> **SQUARE ROOT**
> The number c is a *square root* of a if $c^2 = a$.

We say that 5 is *a* square root of 25 because every positive number has two square roots. The square roots of 25 are 5 and -5 because $5^2 = 25$ and $(-5)^2 = 25$.

EXAMPLE 1 Find the square roots of 49.

SOLUTION The square roots of 49 are 7 and -7. To check, note that $7^2 = 49$ and $(-7)^2 = (-7)(-7) = 49$.

1. Find the square roots of 81.

YOUR TURN

We use a **radical sign**, $\sqrt{}$, to indicate the nonnegative square root, or **principal square root**, of a number. Thus the square roots of 49 are 7 and -7, but $\sqrt{49} = 7$. Note that $\sqrt{-49} \neq -7$. To represent the negative square root of 49, we write $-\sqrt{49}$. When we say *the* square root (singular) of 49, we mean the principal, or positive, square root.

EXAMPLE 2 Find each of the following: **(a)** $\sqrt{225}$; **(b)** $-\sqrt{64}$.

SOLUTION

a) The principal square root of 225 is its positive square root, so $\sqrt{225} = 15$.

b) The symbol $-\sqrt{64}$ represents the opposite of $\sqrt{64}$. Since $\sqrt{64} = 8$, we have $-\sqrt{64} = -8$.

2. Find $\sqrt{81}$.

YOUR TURN

B. Radicands and Radical Expressions

A **radical expression** is an algebraic expression that contains at least one radical sign. Here are some examples:

$$\sqrt{14}, \qquad 8 + \sqrt{2x}, \qquad \sqrt{t^2 + 4}, \qquad \sqrt{\dfrac{x^2 - 5}{2}}.$$

The **radicand** in a radical expression is the expression under the radical sign.

EXAMPLE 3 Identify the radicand in each expression: **(a)** $\sqrt{x}$; **(b)** $3\sqrt{y^2 - 5}$.

SOLUTION

a) In $\sqrt{x}$, the radicand is x.

b) In $3\sqrt{y^2 - 5}$, the radicand is $y^2 - 5$.

YOUR TURN

3. Identify the radicand in the expression $y\sqrt{5x}$.

The square of any nonzero real number is always positive. For example, $8^2 = 64$ and $(-11)^2 = 121$. Therefore, the following expressions are not real numbers:

$$\sqrt{-100}, \qquad \sqrt{-49}, \qquad -\sqrt{-3}.$$

C. Irrational Numbers

Numbers like $\sqrt{2}$ are real but not rational and are called *irrational*. An irrational number cannot be written as the *ratio* of two integers. A number that is the square of some rational number, like 64 or $\frac{9}{25}$, is called a *perfect square*. The square root of a perfect square is always rational; the square root of a nonnegative number that is not a perfect square is irrational.

EXAMPLE 4 Classify each of the following numbers as either rational or irrational.

a) $\sqrt{10}$ **b)** $\sqrt{25}$ **c)** $-\sqrt{9}$

SOLUTION

a) $\sqrt{10}$ is irrational, since 10 is not a perfect square.

b) $\sqrt{25}$ is rational, since 25 is a perfect square: $\sqrt{25} = 5$.

c) $-\sqrt{9}$ is rational, since 9 is a perfect square: $-\sqrt{9} = -3$.

YOUR TURN

4. Classify $\sqrt{8}$ as either rational or irrational.

In the following list, the irrational numbers appear in red: $\sqrt{1}$, $\sqrt{2}$, $\sqrt{3}$, $\sqrt{4}$, $\sqrt{5}$, $\sqrt{6}$, $\sqrt{7}$, $\sqrt{8}$, $\sqrt{9}$, $\sqrt{10}$, $\sqrt{11}$, $\sqrt{12}$, $\sqrt{13}$, $\sqrt{14}$, $\sqrt{15}$, $\sqrt{16}$, $\sqrt{17}$, $\sqrt{18}$, $\sqrt{19}$, $\sqrt{20}$, $\sqrt{21}$, $\sqrt{22}$, $\sqrt{23}$, $\sqrt{24}$, $\sqrt{25}$.

We use a calculator to find decimal approximations of irrational square roots.

EXAMPLE 5 Use a calculator to approximate $\sqrt{10}$ to three decimal places.

SOLUTION Calculators vary in their methods of operation. On some, we press $\boxed{\sqrt{}}$ and then the number; on others, we enter the number and then press $\boxed{\sqrt{}}$.

$$\sqrt{10} \approx 3.162277660 \qquad \text{Using a calculator with a 10-digit display}$$

With most graphing calculators, we press **2ND** $\boxed{\sqrt{}}$, the radicand (in this case 10), **)**, and then **ENTER**.

Decimal representation of an irrational number is nonrepeating and nonending. Rounding to three decimal places, we have $\sqrt{10} \approx 3.162$.

YOUR TURN

5. Use a calculator to approximate $\sqrt{2}$ to three decimal places.

D. Square Roots and Absolute Value

Let's compare $\sqrt{(-5)^2}$ and $\sqrt{5^2}$:

$$\sqrt{(-5)^2} = \sqrt{25} = 5, \quad |-5| = 5;$$
$$\sqrt{5^2} = \sqrt{25} = 5, \quad |5| = 5.$$

Note that squaring a number and then taking the square root is the same as taking the absolute value of the number. The principal square root of the square of A is the absolute value of A:

For any real number A, $\sqrt{A^2} = |A|$.

EXAMPLE 6   Simplify $\sqrt{(3x)^2}$ given that x can represent any real number.

SOLUTION   To simplify $\sqrt{(3x)^2}$, note that if x represents a negative number, say, -2, the result is *not* $3x$. To see this, note that $\sqrt{(3(-2))^2} \neq 3(-2)$. Since the principal square root is always positive, to write $\sqrt{(3x)^2} = 3x$ would be incorrect. Instead, we write

$$\sqrt{(3x)^2} = |3x|, \text{ or } 3|x|. \quad \text{Note that } 3x \text{ will be negative for } x < 0.$$

6. Simplify $\sqrt{(x-3)^2}$ given that x can represent any real number.

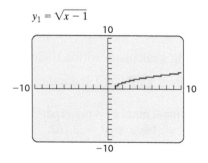 YOUR TURN

In many cases, it can be assumed that radicands that are variable expressions do not represent the square of a negative number. When this assumption is made, absolute-value symbols are not needed:

For $A \geq 0$, $\sqrt{A^2} = A$.

EXAMPLE 7   Simplify each expression. Assume that all variables represent nonnegative numbers.

a) $\sqrt{(8x)^2}$ **b)** $\sqrt{(t+2)^2}$

SOLUTION

a) $\sqrt{(8x)^2} = 8x$ Since $8x$ is assumed to be nonnegative, $|8x| = 8x$.

b) $\sqrt{(t+2)^2} = t + 2$ Since $t + 2$ is assumed to be nonnegative, $|t + 2| = t + 2$.

7. Simplify $\sqrt{n^2}$. Assume that n is a nonnegative number.

YOUR TURN

Technology Connection

Graphing equations that contain radical expressions often involves approximating irrational numbers. Also, since the square root of a negative number is not real, such graphs may not exist for all choices of x. For example, the graph of $y = \sqrt{x-1}$ does not exist for $x < 1$.

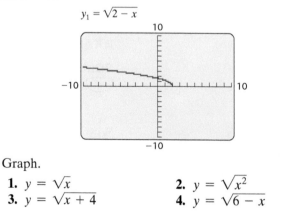

Similarly, the graph of $y = \sqrt{2-x}$ does not exist for $x > 2$.

$y_1 = \sqrt{x-1}$

$y_1 = \sqrt{2-x}$

Graph.

1. $y = \sqrt{x}$ **2.** $y = \sqrt{x^2}$
3. $y = \sqrt{x+4}$ **4.** $y = \sqrt{6-x}$

Chapter Resource:
Collaborative Activity, p. 536

8. Refer to Example 8. Find the number of spaces needed when an average of 90 cars arrive during peak hours.

Student Notes

In Example 8, the number 16.393 is rounded *up* to 17 because of the real-world situation being modeled. In another situation, a number may be rounded *down*. Always make sure that your answers make sense in the real world.

E. Problem Solving

Radical expressions often appear in applications.

EXAMPLE 8 *Parking-Lot Arrival Spaces.* The attendants at a downtown parking lot use spaces to leave cars before they are taken to long-term parking stalls. The required number N of such spaces is approximated by the formula

$$N = 2.5\sqrt{A},$$

where A is the average number of arrivals in peak hours. Find the number of spaces needed when an average of 43 cars arrive during peak hours.

SOLUTION We substitute 43 into the formula. We use a calculator to find an approximation:

$$N = 2.5\sqrt{43}$$
$$\approx 2.5(6.557) \qquad \text{Rounding } \sqrt{43} \text{ to } 6.557$$
$$\approx 16.393 \approx 17.$$

Note that we round *up* to 17 spaces because rounding down would create some overcrowding. Thus, for an average of 43 arrivals, 17 spaces are needed.

YOUR TURN

Calculator Note. In Example 8, we rounded $\sqrt{43}$. Generally, when using a calculator, we round *at the end* of our work. This can change the answer:

$$N = 2.5\sqrt{43} \approx 2.5(6.557438524) = 16.39359631 \approx 16.394.$$

Note the discrepancy in the third decimal place. When using a calculator for approximation, be aware of possible variations in answers. You may get answers that differ slightly from those at the back of the book. Answers to the exercises have been found by rounding at the end of the calculations.

Check Your UNDERSTANDING

Classify each of the following statements as either true or false.

1. 16 has two square roots.

2. $\sqrt{16}$ represents 4 or -4.

3. $\sqrt{16}$ is rational.

4. 100 is the square of 10.

5. 4 is the square root of 2.

6. $\sqrt{25} = 5$

7. $\sqrt{-25} = -5$

8. $-\sqrt{25} = -5$

8.1   EXERCISE SET

FOR EXTRA HELP   MyMathLab®

Vocabulary and Reading Check

In each of Exercises 1–6, match the phrase with the most appropriate choice from the column on the right.

1. ____ The name for an expression written under a radical
2. ____ The name for an algebraic expression that contains at least one radical sign
3. ____ The name for a number that is real but not rational
4. ____ The name for a positive number's positive square root
5. ____ The sign of $3x$ when x represents a negative number
6. ____ The sign of $\sqrt{A^2}$ when A represents a negative number

a) Irrational
b) Negative
c) Radicand
d) Positive
e) Radical expression
f) Principal square root

Concept Reinforcement

Classify each of the following statements as either true or false. Do not use a calculator.

7. $\sqrt{37}$ is between 6 and 7.
8. $\sqrt{75}$ is between 7 and 8.
9. $\sqrt{150}$ is between 11 and 12.
10. $\sqrt{103}$ is between 10 and 11.

A. Square Roots

Find the square roots of each number.

11. 100   12. 4
13. 36   14. 9
15. 1   16. 121
17. 144   18. 169

Simplify.

19. $\sqrt{100}$ 20. $\sqrt{9}$
21. $-\sqrt{1}$ 22. $-\sqrt{49}$
23. $\sqrt{0}$ 24. $-\sqrt{81}$
25. $-\sqrt{121}$ 26. $\sqrt{400}$
27. $\sqrt{900}$ 28. $\sqrt{441}$
29. $-\sqrt{144}$ 30. $-\sqrt{625}$

B. Radicands and Radical Expressions

Identify the radicand in each expression.

31. $\sqrt{10a}$ 32. $\sqrt{x^2 y}$
33. $5\sqrt{t^3 - 2}$ 34. $8\sqrt{x^2 + 2}$
35. $x^2 y\sqrt{\dfrac{7}{x + y}}$ 36. $ab^2\sqrt{\dfrac{a}{a - b}}$

C. Irrational Numbers

Classify each number as either rational or irrational.

37. $\sqrt{4}$ 38. $\sqrt{15}$
39. $\sqrt{11}$ 40. $\sqrt{12}$
41. $\sqrt{32}$ 42. $\sqrt{64}$
43. $-\sqrt{16}$ 44. $-\sqrt{144}$
Aha! 45. $-\sqrt{19^2}$ 46. $-\sqrt{22}$

Use a calculator to approximate each of the following numbers. Round to three decimal places.

47. $\sqrt{8}$ 48. $\sqrt{6}$
49. $\sqrt{15}$ 50. $\sqrt{19}$
51. $\sqrt{83}$ 52. $\sqrt{43}$

D. Square Roots and Absolute Value

Simplify. Assume that x can represent any real number.

53. $\sqrt{x^2}$ 54. $\sqrt{(7x)^2}$
55. $\sqrt{(10x)^2}$ 56. $\sqrt{(x - 1)^2}$
57. $\sqrt{(x + 7)^2}$ 58. $\sqrt{(4 - x)^2}$
59. $\sqrt{(5 - 2x)^2}$ 60. $\sqrt{(3x + 1)^2}$

Simplify. Assume that all variables represent nonnegative numbers.

61. $\sqrt{x^2}$ 62. $\sqrt{t^2}$
63. $\sqrt{(5y)^2}$ 64. $\sqrt{(3a)^2}$
65. $\sqrt{16t^2}$ 66. $\sqrt{25x^2}$
67. $\sqrt{(n + 7)^2}$ 68. $\sqrt{(a + 2)^2}$

E. Problem Solving

Parking Spaces. *Solve. Use the formula $N = 2.5\sqrt{A}$ of Example 8.*

69. Find the number of spaces needed when the average number of arrivals is **(a)** 36; **(b)** 29.

70. Find the number of spaces needed when the average number of arrivals is **(a)** 49; **(b)** 53.

🔲 *Hang Time.* An athlete's hang time *(time airborne for a jump)*, *T, in seconds, is given by* $T = 0.144\sqrt{V}$,* *where V is the athlete's vertical leap, in inches.*

71. Russell Westbrook of the Oklahoma City Thunder can jump 36.5 in. vertically. Find his hang time.

   Data: verticalshockprogram.com

72. LeBron James of the Cleveland Cavaliers can jump 44 in. vertically. Find his hang time.

   Data: verticalshockprogram.com

73. Which is the more exact way to write the square root of 12: 3.464101615 or $\sqrt{12}$? Why?

74. What is the difference between "*the* square root of 10" and "*a* square root of 10"?

Skill Review

Solve.

75. $3(x - 1) - 5 = 2 - (x + 7)$ [2.2]

76. $\dfrac{3}{4} = \dfrac{5}{x}$ [6.6]

77. $5 - 6n < 2(n - 7)$ [2.6]

78. $2x^2 - x = 15$ [5.6]

79. $x + y = 4,$
$2x - y = 7$ [7.3]

80. $\dfrac{1}{x} + \dfrac{2}{x + 1} = 2$ [6.6]

Synthesis

81. Explain in your own words why $\sqrt{A^2} \neq A$ when A is negative.

82. One number has only one square root. What is this number and why is it unique in this regard?

Simplify, if possible.

83. $\sqrt{\frac{1}{100}}$

84. $\sqrt{0.01}$

85. $\sqrt{(-7)^2}$

86. $\sqrt{-7^2}$

87. $\sqrt{3^2 + 4^2}$

88. $\sqrt{\sqrt{81}}$

Aha! **89.** Between what two consecutive integers is $-\sqrt{33}$?

90. Find at least one number that is the square of an integer and the cube of a different integer.

---

*Based on an article by Peter Brancazio, "The Mechanics of a Slam Dunk," *Popular Mechanics*, November 1991. Courtesy of Peter Brancazio, Brooklyn College.

Solve. If no real-number solution exists, state this.

91. $\sqrt{t^2} = 4$

92. $\sqrt{y^2} = -5$

93. $-\sqrt{x^2} = -3$

94. $a^2 = 36$

95. For which values of x is $\sqrt{x - 10}$ not a real number?

96. For which values of x is $\sqrt{10 - x}$ not a real number?

Simplify. Assume that all variables represent positive numbers.

97. $\sqrt{\dfrac{144x^8}{36y^6}}$

98. $\sqrt{\dfrac{y^{12}}{8100}}$

99. $\sqrt{\dfrac{400}{m^{16}}}$

100. $\sqrt{\dfrac{p^2}{3600}}$

101. Use the graph of $y = \sqrt{x}$, shown below, to estimate each of the following to the nearest tenth: **(a)** $\sqrt{3}$; **(b)** $\sqrt{5}$; **(c)** $\sqrt{7}$. Be sure to check by multiplying.

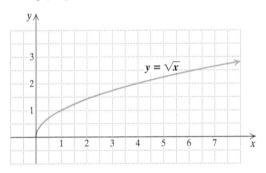

Speed of Sound. The speed V of sound traveling through air, in feet per second, is given by

$$V = \dfrac{1087\sqrt{273 + t}}{16.52},$$

where t is the temperature, in degrees Celsius. Using a calculator, find the speed of sound through air at each of the following temperatures. Round to the nearest tenth.

102. 28°C

103. 5°C

104. −10°C

105. Use a graphing calculator to draw the graphs of $y_1 = \sqrt{x - 2}$, $y_2 = \sqrt{x + 7}$, $y_3 = 5 + \sqrt{x}$, and $y_4 = -4 + \sqrt{x}$. If possible, graph all four equations using the SIMULTANEOUS mode and a $[-10, 10, -10, 10]$ window. Then determine which equation corresponds to each curve.

106. For which values of t is $\sqrt{(t + 4)^2} = t + 4$ false?

107. For which values of a is $\sqrt{(a - 1)^2} = a - 1$ false?

108. What restrictions on a, if any, are needed in order for
$$\sqrt{64a^{16}} = 8a^8$$
to be true? Explain how you found your answer.

Prepare to Move On

Simplify. [4.1]

1. $(t^{10})^2$ 2. $2a \cdot 25a^{12}$

3. $9x^6 \cdot 5x$ 4. $3m \cdot (10m^4)^2$

5. $2y \cdot (5y^7)^2$

YOUR TURN ANSWERS: SECTION 8.1

1. $-9, 9$ 2. 9   3. $5x$ 4. Irrational   5. 1.414
6. $|x - 3|$ 7. n 8. 24 spaces

8.2   Multiplying and Simplifying Radical Expressions

A. Multiplying   **B.** Simplifying and Factoring   **C.** Simplifying Square Roots of Powers
D. Multiplying and Simplifying

We now learn to multiply and simplify radical expressions. For this section, we assume that no radicands were formed by squaring negative quantities. Because of this assumption, we will not need to use absolute-value symbols when simplifying.

A. Multiplying

To see how to multiply with radical notation, consider the following:

$$\sqrt{9} \cdot \sqrt{4} = 3 \cdot 2 = 6; \qquad \text{This is a product of square roots.}$$
$$\sqrt{9 \cdot 4} = \sqrt{36} = 6. \qquad \text{This is the square root of a product.}$$

Note that $\sqrt{9} \cdot \sqrt{4} = \sqrt{9 \cdot 4}$. This is generalized in the following rule.

Student Notes

Every statement in a definition or a rule is important and should be carefully understood. In the product rule for square roots, the statement "for any real numbers $\sqrt{A}$ and $\sqrt{B}$" means that A and B must be nonnegative. The product rule does not hold, for example, for $\sqrt{-16} \cdot \sqrt{-25}$.

THE PRODUCT RULE FOR SQUARE ROOTS

For any real numbers $\sqrt{A}$ and $\sqrt{B}$,
$$\sqrt{A} \cdot \sqrt{B} = \sqrt{A \cdot B}.$$

(To multiply square roots, multiply the radicands and take the square root.)

EXAMPLE 1 Multiply: **(a)** $\sqrt{5}\,\sqrt{7}$; **(b)** $\sqrt{6}\,\sqrt{6}$; **(c)** $\sqrt{\frac{2}{3}}\sqrt{\frac{7}{5}}$; **(d)** $\sqrt{2x}\,\sqrt{3y}$.

SOLUTION

a) $\sqrt{5}\,\sqrt{7} = \sqrt{5 \cdot 7} = \sqrt{35}$

b) $\sqrt{6}\,\sqrt{6} = \sqrt{6 \cdot 6} = \sqrt{36} = 6$ Try to do this one directly: $\sqrt{6}\,\sqrt{6} = 6$.

c) $\sqrt{\dfrac{2}{3}}\,\sqrt{\dfrac{7}{5}} = \sqrt{\dfrac{2}{3} \cdot \dfrac{7}{5}} = \sqrt{\dfrac{14}{15}}$

d) $\sqrt{2x}\,\sqrt{3y} = \sqrt{6xy}$ Note that in order for $\sqrt{2x}$ and $\sqrt{3y}$ to be real numbers, we must have $x, y \geq 0$.

1. Multiply: $\sqrt{10}\sqrt{x}$.

YOUR TURN

B. Simplifying and Factoring

To factor a square root, we can use the product rule in reverse. That is,

$$\sqrt{AB} = \sqrt{A}\,\sqrt{B}.$$

This property is especially useful when a radicand contains a perfect square as a factor. For example, the radicand in $\sqrt{50}$ is not a perfect square, but one of its factors, 25, *is* a perfect square. Thus,

$$\sqrt{50} = \sqrt{25 \cdot 2} \qquad \text{25 is a perfect-square factor of 50.}$$
$$= \sqrt{25} \cdot \sqrt{2} \qquad \sqrt{25} \text{ is a rational factor of } \sqrt{50}.$$
$$= 5\sqrt{2}. \qquad \text{Simplifying } \sqrt{25}; \sqrt{2} \text{ cannot be simplified further.}$$

If we did not easily see that 50 contains a perfect square as a factor, we could proceed by writing the prime factorization of 50:

$$\sqrt{50} = \sqrt{2 \cdot 5 \cdot 5} \qquad \text{Factoring into prime factors}$$
$$= \sqrt{2} \cdot \sqrt{5 \cdot 5} \qquad \text{Grouping pairs of like factors;}$$
$$\qquad\qquad\qquad\quad 5 \cdot 5 \text{ is a perfect square.}$$
$$= \sqrt{2} \cdot 5.$$

To avoid any uncertainty as to what is under the radical sign, it is customary to write the radical factor last. Thus, $\sqrt{50} = 5\sqrt{2}$.

A radical expression, like $\sqrt{6}$, in which the radicand has no perfect-square factors, is considered to be in simplified form.

> **SIMPLIFIED FORM OF A SQUARE ROOT**
>
> A radical expression for a square root is simplified when its radicand has no factor other than 1 that is a perfect square.

A variable raised to an even power, such as t^2 or x^6, is also a perfect square.

Simplifying is always easiest if the *largest* perfect-square factor is identified in the first step.

EXAMPLE 2 Simplify by factoring. Assume that all variables represent non-negative numbers.

a) $\sqrt{18}$ **b)** $\sqrt{a^2 b}$ **c)** $\sqrt{196 t^2 u}$

SOLUTION

a) $\sqrt{18} = \sqrt{9 \cdot 2} \qquad \text{9 is a perfect-square factor of 18.}$
$\qquad\quad = \sqrt{9}\,\sqrt{2} \qquad \sqrt{9} \text{ is a rational factor of } \sqrt{18}.$
$\qquad\quad = 3\sqrt{2} \qquad\quad \sqrt{2} \text{ cannot be simplified further.}$

b) $\sqrt{a^2 b} = \sqrt{a^2}\,\sqrt{b} \qquad$ Identifying a perfect-square factor and factoring into a product of radicals
$\qquad\quad\; = a\sqrt{b} \qquad\qquad$ No absolute-value signs are necessary since a is assumed to be nonnegative.

c) We may not recognize that $196 = 14^2$. Let's suppose that we noticed only that 4 and t^2 are perfect-square factors of $196 t^2 u$:

$$\sqrt{196 t^2 u} = \sqrt{4} \cdot \sqrt{t^2} \cdot \sqrt{49u} \qquad \text{Identifying perfect-square factors}$$
$$= 2t\sqrt{49u}. \qquad\qquad \text{We assume } t \geq 0. \text{ This is not simplified completely.}$$

Because the radicand still contains a perfect-square factor, we have not simplified completely. To finish simplifying, we rewrite the product and simplify $\sqrt{49u}$:

$$\sqrt{196t^2u} = 2t\sqrt{49u} \qquad \text{Rewriting the last step above}$$
$$= 2t\sqrt{49}\sqrt{u} \qquad \text{Identifying a perfect-square factor of } 49u$$
$$= 2t \cdot 7\sqrt{u}$$
$$= 14t\sqrt{u}.$$

2. Simplify by factoring: $\sqrt{20}$.

↩ YOUR TURN

EXAMPLE 3　Evaluate $\sqrt{b^2 - 4ac}$ for $a = 3, b = 6,$ and $c = -5$.

SOLUTION　We have

$$\sqrt{b^2 - 4ac} = \sqrt{6^2 - 4 \cdot 3(-5)} \qquad \text{Substituting}$$
$$= \sqrt{36 - 12(-5)}$$
$$= \sqrt{36 + 60} \qquad\qquad \sqrt{36 + 60} \neq \sqrt{36} + \sqrt{60}$$
$$= \sqrt{96}$$
$$= \sqrt{2 \cdot 2 \cdot 2 \cdot 2 \cdot 2 \cdot 3} \qquad \text{The prime factorization always}$$
$$\qquad\qquad\qquad\qquad\qquad \text{allows us to identify the largest}$$
$$\qquad\qquad\qquad\qquad\qquad \text{perfect-square factor.}$$
$$= \sqrt{16} \cdot \sqrt{6}$$

3. Evaluate $\sqrt{b^2 - 4ac}$ for $a = 4, b = 2,$ and $c = -1$.

$$= 4\sqrt{6}.$$

↩ YOUR TURN

C. Simplifying Square Roots of Powers

To take the square root of an even power such as x^{10}, note that $x^{10} = (x^5)^2$. Then

$$\sqrt{x^{10}} = \sqrt{(x^5)^2} = x^5. \qquad \text{Remember that we assume } x \geq 0.$$

The exponent of the square root is half the exponent of the radicand. That is,

$$\sqrt{x^{10}} = x^5. \longleftarrow \tfrac{1}{2}(10) = 5$$

EXAMPLE 4　Simplify: **(a)** $\sqrt{x^6}$; **(b)** $\sqrt{p^{12}}$; **(c)** $\sqrt{t^{22}}$.

SOLUTION

a) $\sqrt{x^6} = \sqrt{(x^3)^2} = x^3$　　Half of 6 is 3.

b) $\sqrt{p^{12}} = \sqrt{(p^6)^2} = p^6$　　Half of 12 is 6.

4. Simplify: $\sqrt{t^{16}}$.

c) $\sqrt{t^{22}} = \sqrt{(t^{11})^2} = t^{11}$　　Half of 22 is 11.

↩ YOUR TURN

If a radicand is an odd power, we can simplify by factoring. For square roots of powers, after we have simplified, the radicand never contains an exponent greater than 1.

EXAMPLE 5　Simplify: **(a)** $\sqrt{x^9}$; **(b)** $\sqrt{32p^{19}}$.

SOLUTION

a) $\sqrt{x^9} = \sqrt{x^8 \cdot x}$　　x^8 is the largest perfect-square factor of x^9.

$$= \sqrt{x^8}\sqrt{x}$$
$$= x^4\sqrt{x}$$

CAUTION!　The square root of x^9 is *not* x^3.

b) $\sqrt{32p^{19}} = \sqrt{16p^{18} \cdot 2p}$ 16 is the largest perfect-square factor of 32; p^{18} is the largest perfect-square factor of p^{19}.

$$= \sqrt{16}\sqrt{p^{18}}\sqrt{2p}$$

$$= 4p^9\sqrt{2p}$$ Simplifying. We assume that p is positive. Since $2p$ has no perfect-square factor, we are done.

5. Simplify: $\sqrt{50x^5}$.

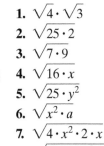 YOUR TURN

D. Multiplying and Simplifying

Sometimes we can simplify after multiplying. To do so, we again try to identify any perfect-square factors of the radicand.

EXAMPLE 6   Multiply and, if possible, simplify. Remember that all variables are assumed to represent nonnegative numbers.

a) $\sqrt{2}\,\sqrt{14}$ **b)** $\sqrt{5t}\,\sqrt{6t}$

c) $\sqrt{n^2}\,\sqrt{n^3}$ **d)** $\sqrt{2x^8}\,\sqrt{9x^3}$

SOLUTION

a) $\sqrt{2}\sqrt{14} = \sqrt{2 \cdot 14}$ Multiplying

$$= \sqrt{2 \cdot 2 \cdot 7}$$ Writing the prime factorization

$$= \sqrt{2^2}\sqrt{7}$$ Note that $2 \cdot 2$, or 4, is a perfect-square factor.

$$= 2\sqrt{7}$$ Simplifying

b) $\sqrt{5t}\,\sqrt{6t} = \sqrt{5 \cdot 2 \cdot 3 \cdot t^2}$ Multiplying

$$= \sqrt{t^2}\sqrt{30} \;\Big\}$$ Simplifying

$$= t\sqrt{30}$$

c) $\sqrt{n^2}\sqrt{n^3} = \sqrt{n^5}$ Multiplying

$$= \sqrt{n^4 \cdot n}$$ n^4 is a perfect square.

$$= \sqrt{n^4}\sqrt{n}$$ Factoring

$$= n^2\sqrt{n}$$ Simplifying

Note that both $\sqrt{n^2}$ and $\sqrt{n^3}$ can be simplified. If we had simplified before multiplying, the result would be the same.

d) Before multiplying, note that we can simplify both $\sqrt{2x^8}$ and $\sqrt{9x^3}$.

$$\sqrt{2x^8}\sqrt{9x^3} = \sqrt{2 \cdot x^8}\sqrt{9 \cdot x^2 \cdot x}$$ $x^8, 9$, and x^2 are perfect squares.

$$= \sqrt{2}\sqrt{x^8}\sqrt{9}\sqrt{x^2}\sqrt{x}$$

$$= \sqrt{2} \cdot x^4 \cdot 3 \cdot x \cdot \sqrt{x}$$ Simplifying

$$= 3 \cdot x^4 \cdot x \cdot \sqrt{2} \cdot \sqrt{x}$$ Using a commutative law

$$= 3x^5\sqrt{2x}$$ Multiplying

We could also begin by multiplying radicands. The result would be the same.

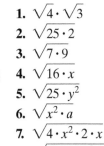 YOUR TURN

↳ Check Your UNDERSTANDING

Simplify, Assume that all variables represent nonnegative numbers.

1. $\sqrt{4} \cdot \sqrt{3}$

2. $\sqrt{25 \cdot 2}$

3. $\sqrt{7 \cdot 9}$

4. $\sqrt{16 \cdot x}$

5. $\sqrt{25 \cdot y^2}$

6. $\sqrt{x^2 \cdot a}$

7. $\sqrt{4 \cdot x^2 \cdot 2 \cdot x}$

8. $\sqrt{25 \cdot 3 \cdot a^2 \cdot b^2 \cdot b}$

6. Multiply and, if possible, simplify:

$$\sqrt{6x}\sqrt{3x}.$$

8.2 EXERCISE SET

FOR EXTRA HELP MyMathLab®

Vocabulary and Reading Check

Classify each of the following statements as either true or false.

1. The largest perfect-square factor of 200 is 100.

2. If a radicand of a square-root expression contains a perfect-square factor, then the expression is not simplified.

3. The expression $5\sqrt{x^3}$ cannot be simplified further.

4. If x is nonnegative, then the square root of x^{102} is x^{51}.

Concept Reinforcement

In each of Exercises 5–12, match the expression with the equivalent expression from the column on the right. Assume that all variables represent nonnegative numbers.

5. ___ $\sqrt{3} \cdot \sqrt{7}$ a) $7 \cdot 3$

6. ___ $\sqrt{9} \cdot \sqrt{7}$ b) $\sqrt{(a^3)^2}$

7. ___ $\sqrt{49 \cdot 3}$ c) $a^4 \cdot \sqrt{a}$

8. ___ $\sqrt{49 \cdot 9}$ d) $7 \cdot \sqrt{3}$

9. ___ $\sqrt{a^2}$ e) a^2

10. ___ $\sqrt{a^9}$ f) $\sqrt{21}$

11. ___ $\sqrt{a^6}$ g) a

12. ___ $\sqrt{a^4}$ h) $3\sqrt{7}$

A. Multiplying

Multiply.

13. $\sqrt{2}\sqrt{5}$ 14. $\sqrt{3}\sqrt{11}$ 15. $\sqrt{4}\sqrt{3}$

16. $\sqrt{2}\sqrt{9}$ 17. $\sqrt{\frac{3}{5}}\sqrt{\frac{7}{8}}$ 18. $\sqrt{\frac{3}{8}}\sqrt{\frac{1}{5}}$

19. $\sqrt{10}\sqrt{10}$ 20. $\sqrt{15}\sqrt{15}$ 21. $\sqrt{25}\sqrt{3}$

22. $\sqrt{36}\sqrt{2}$ 23. $\sqrt{2}\sqrt{x}$ 24. $\sqrt{3}\sqrt{a}$

25. $\sqrt{7}\sqrt{2a}$ 26. $\sqrt{5}\sqrt{7t}$ 27. $\sqrt{3x}\sqrt{7y}$

28. $\sqrt{3m}\sqrt{5n}$ 29. $\sqrt{3a}\sqrt{2bc}$ 30. $\sqrt{3x}\sqrt{yz}$

B. Simplifying and Factoring

Simplify by factoring. Assume that all variables represent nonnegative numbers.

31. $\sqrt{12}$ 32. $\sqrt{28}$ 33. $\sqrt{75}$

34. $\sqrt{45}$ 35. $\sqrt{500}$ 36. $\sqrt{300}$

37. $\sqrt{16t}$ 38. $\sqrt{64a}$ 39. $\sqrt{20z}$

40. $\sqrt{40m}$ 41. $\sqrt{100y^2}$ 42. $\sqrt{9x^2}$

43. $\sqrt{13x^2}$ 44. $\sqrt{29t^2}$ 45. $\sqrt{27b^2}$

46. $\sqrt{125a^2}$ 47. $\sqrt{144x^2y}$ 48. $\sqrt{256u^2v}$

In Exercises 49–54, evaluate $\sqrt{b^2 - 4ac}$ for the values of a, b, and c given.

49. $a = 2$, $b = 4$, $c = -1$

50. $a = 3$, $b = 2$, $c = -4$

51. $a = 1$, $b = 4$, $c = 4$

52. $a = 1$, $b = -3$, $c = -10$

53. $a = 3$, $b = -6$, $c = -4$

54. $a = 1$, $b = -8$, $c = -3$

C. Simplifying Square Roots of Powers

Simplify. Assume that all variables represent nonnegative numbers.

55. $\sqrt{a^{18}}$ 56. $\sqrt{t^{20}}$ 57. $\sqrt{x^{16}}$

58. $\sqrt{p^{14}}$ 59. $\sqrt{r^5}$ 60. $\sqrt{a^7}$

61. $\sqrt{t^{15}}$ 62. $\sqrt{x^{25}}$ 63. $\sqrt{40a^3}$

64. $\sqrt{250y^3}$ 65. $\sqrt{45x^5}$ 66. $\sqrt{20b^7}$

67. $\sqrt{200p^{25}}$ 68. $\sqrt{99m^{33}}$

D. Multiplying and Simplifying

Multiply and, if possible, simplify.

69. $\sqrt{2}\sqrt{10}$ 70. $\sqrt{3}\sqrt{6}$

71. $\sqrt{3} \cdot \sqrt{27}$ 72. $\sqrt{2} \cdot \sqrt{8}$

73. $\sqrt{3x}\sqrt{12y}$ 74. $\sqrt{2c}\sqrt{50d}$

75. $\sqrt{17}\sqrt{17x}$ 76. $\sqrt{11}\sqrt{11x}$

77. $\sqrt{10b}\sqrt{50b}$ 78. $\sqrt{6a}\sqrt{18a}$

Aha! 79. $\sqrt{12t}\sqrt{12t}$ 80. $\sqrt{10a}\sqrt{10a}$

81. $\sqrt{ab}\sqrt{ac}$ 82. $\sqrt{xy}\sqrt{xz}$

83. $\sqrt{x^7}\sqrt{x^{10}}$ 84. $\sqrt{y^3}\sqrt{y^9}$

85. $\sqrt{7m^5}\sqrt{14m}$ 86. $\sqrt{15m^7}\sqrt{5m}$

87. $\sqrt{x^2y^3}\sqrt{xy^4}$ 88. $\sqrt{x^3y^2}\sqrt{xy}$

89. $\sqrt{6ab}\sqrt{12a^2b^5}$ 90. $\sqrt{5xy^2}\sqrt{10x^2y^3}$

B. Simplifying: Applications

91. *Water Flow.* The required water flow f from a fire hose, in number of gallons per minute, is given by

$$f = 400\sqrt{p},$$

where p is the population, in thousands, of a community. What is the required flow for a community with a population of 25,000? of 100,000?

Data: inetdocs.loudoun.gov

92. *Speed of a Skidding Car.* The formula

$$r = \sqrt{30dL}$$

can be used to approximate the speed r, in miles per hour, of a car that has left a skid pattern that is L feet long on a road with drag factor d.

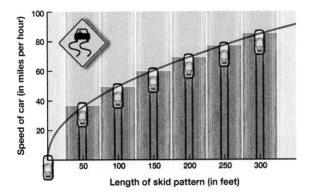

a) A common drag factor for a dry asphalt road is 0.8. Simplify the formula for a typical dry asphalt road.

b) What was the speed of a car that left skid marks of 150 ft on a dry road?

93. *Sound Design.* In order to match the sound heard from speakers with the sound heard from a stage, acoustic engineers delay the sound signal sent to the speakers. The speed of sound changes with temperature and can be estimated using the formula

$$s = 21.9\sqrt{5t + 2457},$$

where s is the speed of sound, in number of feet per second, and t is the temperature, in degrees Fahrenheit.

During one large outdoor concert, the air temperature was 81°F. What was the speed of sound? Give your answer both in simplified radical form and as an approximation to the nearest foot per second.

94. *Falling Object.* The number of seconds t that it takes for an object to fall d meters when thrown down at a velocity of 9.5 meters per second (m/sec) can be estimated by

$$t = 0.45\sqrt{d + 4.6} - 1.$$

A rock is thrown at 9.5 m/sec from the overlook in Great Bluffs State Park 175.4 m above the Mississippi River in Minnesota. After how many seconds will the rock hit the water? Give your answer both in simplified radical form and as an approximation to the nearest tenth of a second.

95. How would you convince someone that the following equation is not true?

$$\sqrt{x^2 - 49} = x^2 - \sqrt{49} = x - 7$$

This is not true.

96. Explain why the rules for manipulating exponents are important when simplifying radical expressions.

Skill Review

Simplify.

97. $\frac{2}{3} - \left(-\frac{1}{12}\right)$ [1.6]

98. $-\frac{9}{10} \div \left(-\frac{3}{5}\right)$ [1.7]

99. $\dfrac{38x^4y^7}{34x^2y}$ [4.1]

100. $\dfrac{3x^2 - 8x + 5}{x^2 - 1}$ [6.1]

101. $-6 - 4^3 \div (-2) \cdot 4 + (1 - 3)^2$ [1.8]

102. $\dfrac{a^{-10}}{a^{-11}}$ [4.2]

Synthesis

103. Explain why $\sqrt{16x^4} = 4x^2$, but $\sqrt{4x^{16}} \ne 2x^4$.

104. Simplify $\sqrt{49}$, $\sqrt{490}$, $\sqrt{4900}$, $\sqrt{49,000}$, and $\sqrt{490,000}$; then describe the pattern you see.

Simplify.

105. $\sqrt{0.01}$

106. $\sqrt{0.49}$

107. $\sqrt{0.0625}$

108. $\sqrt{0.000001}$

Use the proper symbol ($>, <,$ or $=$) between each pair of values to form a true sentence. Do not use a calculator.

109. $4\sqrt{14}$ ▢ 15

110. $3\sqrt{11}$ ▢ $7\sqrt{2}$

111. $\sqrt{450}$ ▢ $15\sqrt{2}$

112. 16 ▢ $\sqrt{15}\sqrt{17}$

113. 8 ▢ $\sqrt{15} + \sqrt{17}$

114. $5\sqrt{7}$ ▢ $4\sqrt{11}$

Multiply and, if possible, simplify.

115. $\sqrt{54(x + 1)}\sqrt{6y(x + 1)^2}$

116. $\sqrt{18(x - 2)}\sqrt{20(x - 2)^3}$

117. $\sqrt{x^9}\sqrt{2x}\sqrt{10x^5}$

118. $\sqrt{2^{109}}\sqrt{x^{306}}\sqrt{x^{11}}$

Fill in the blank.

119. $\sqrt{21x^9} \cdot \underline{\hspace{1cm}} = 7x^{14}\sqrt{6x^7}$

120. $\sqrt{35x^7} \cdot \underline{\hspace{1cm}} = 5x^5\sqrt{14x^3}$

Simplify.

121. $\sqrt{x^{16n}}$

122. $\sqrt{0.04x^{4n}}$

123. Simplify $\sqrt{y^n}$, when n is an even natural number.

124. Simplify $\sqrt{y^n}$, when n is an odd whole number greater than or equal to 3.

> 🢂 **YOUR TURN ANSWERS: SECTION 8.2**
> **1.** $\sqrt{10x}$ **2.** $2\sqrt{5}$ **3.** $2\sqrt{5}$ **4.** t^8 **5.** $5x^2\sqrt{2x}$
> **6.** $3x\sqrt{2}$

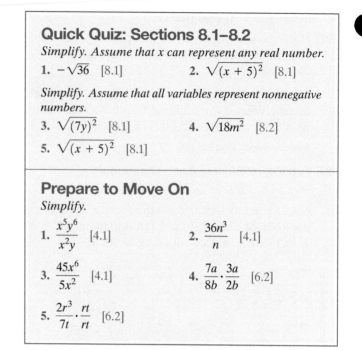

Quick Quiz: Sections 8.1–8.2

Simplify. Assume that x can represent any real number.

1. $-\sqrt{36}$ [8.1]   **2.** $\sqrt{(x+5)^2}$ [8.1]

Simplify. Assume that all variables represent nonnegative numbers.

3. $\sqrt{(7y)^2}$ [8.1]   **4.** $\sqrt{18m^2}$ [8.2]

5. $\sqrt{(x+5)^2}$ [8.1]

Prepare to Move On

Simplify.

1. $\dfrac{x^5y^6}{x^2y}$ [4.1]   **2.** $\dfrac{36n^3}{n}$ [4.1]

3. $\dfrac{45x^6}{5x^2}$ [4.1]   **4.** $\dfrac{7a}{8b} \cdot \dfrac{3a}{2b}$ [6.2]

5. $\dfrac{2r^3}{7t} \cdot \dfrac{rt}{rt}$ [6.2]

8.3 Quotients Involving Square Roots

A. Dividing Radical Expressions   **B.** Rationalizing Denominators with One Term

In this section, we divide radical expressions and simplify quotients containing radicals. Again, we assume that absolute-value signs are not needed when simplifying.

Study Skills

Review Material on Your Own

Never hesitate to review earlier material. For example, if you feel unsure about how to multiply with fraction notation, review that material before studying any new material that involves multiplication of fractions. Doing (and checking) some practice problems from an earlier section is also a good way to sharpen skills that you may not have used for a while.

A. Dividing Radical Expressions

To see how to divide with radical notation, consider the following:

$$\frac{\sqrt{100}}{\sqrt{4}} = \frac{10}{2} = 5 \quad \text{since } \sqrt{100} = 10 \text{ and } \sqrt{4} = 2;$$

$$\sqrt{\frac{100}{4}} = \sqrt{25} = 5 \quad \text{since } \frac{100}{4} = 25 \text{ and } \sqrt{25} = 5.$$

Both results are the same, so we see that $\dfrac{\sqrt{100}}{\sqrt{4}} = \sqrt{\dfrac{100}{4}}$.

> **THE QUOTIENT RULE FOR SQUARE ROOTS**
>
> For any real numbers $\sqrt{A}$ and $\sqrt{B}$, with $B \neq 0$,
>
> $$\frac{\sqrt{A}}{\sqrt{B}} = \sqrt{\frac{A}{B}}.$$
>
> (To divide two square roots, divide the radicands and take the square root.)

EXAMPLE 1 Divide and simplify: **(a)** $\dfrac{\sqrt{27}}{\sqrt{3}}$; **(b)** $\dfrac{\sqrt{8a^7}}{\sqrt{2a}}$.

SOLUTION

a) $\dfrac{\sqrt{27}}{\sqrt{3}} = \sqrt{\dfrac{27}{3}}$ We can now simplify the radicand: $\dfrac{27}{3} = 9$.

$\quad\quad = \sqrt{9} = 3$

b) $\dfrac{\sqrt{8a^7}}{\sqrt{2a}} = \sqrt{\dfrac{8a^7}{2a}}$ Now $\dfrac{8a^7}{2a}$ can be simplified.

$\quad\quad = \sqrt{4a^6} = 2a^3$ We assume $a > 0$.

1. Divide and simplify: $\dfrac{\sqrt{48}}{\sqrt{3}}$.

YOUR TURN

The quotient rule for square roots can also be read from right to left:

$$\sqrt{\dfrac{A}{B}} = \dfrac{\sqrt{A}}{\sqrt{B}}.$$

EXAMPLE 2 Simplify by taking square roots in the numerator and the denominator separately.

a) $\sqrt{\dfrac{36}{25}}$ 　　　　　 **b)** $\sqrt{\dfrac{1}{16}}$ 　　　　　 **c)** $\sqrt{\dfrac{49}{t^2}}$

SOLUTION

a) $\sqrt{\dfrac{36}{25}} = \dfrac{\sqrt{36}}{\sqrt{25}} = \dfrac{6}{5}$ Taking the square root of the numerator and the square root of the denominator. This is sometimes done mentally, in one step.

2. Simplify by taking square roots in the numerator and the denominator separately:

$\sqrt{\dfrac{a^2}{100}}$.

b) $\sqrt{\dfrac{1}{16}} = \dfrac{\sqrt{1}}{\sqrt{16}} = \dfrac{1}{4}$ Taking the square root of the numerator and the square root of the denominator

c) $\sqrt{\dfrac{49}{t^2}} = \dfrac{\sqrt{49}}{\sqrt{t^2}} = \dfrac{7}{t}$ We assume $t > 0$.

YOUR TURN

Sometimes a rational expression can be simplified to one that has a perfect-square numerator and/or a perfect-square denominator.

EXAMPLE 3 Simplify: **(a)** $\sqrt{\dfrac{14}{50}}$; **(b)** $\sqrt{\dfrac{48x^3}{3x^7}}$.

SOLUTION

a) $\sqrt{\dfrac{14}{50}} = \sqrt{\dfrac{7 \cdot 2}{25 \cdot 2}}$

$\quad\quad = \sqrt{\dfrac{7 \cdot 2}{25 \cdot 2}}$ Removing a factor equal to 1: $\dfrac{2}{2} = 1$

$\quad\quad = \dfrac{\sqrt{7}}{\sqrt{25}} = \dfrac{\sqrt{7}}{5}$ $\sqrt{\dfrac{A}{B}} = \dfrac{\sqrt{A}}{\sqrt{B}}$

b) $\sqrt{\dfrac{48x^3}{3x^7}} = \sqrt{\dfrac{16 \cdot 3x^3}{x^4 \cdot 3x^3}}$ Removing a factor equal to 1: $\dfrac{3x^3}{3x^3} = 1$

3. Simplify: $\sqrt{\dfrac{12x}{3x^{11}}}$.

$\quad\quad = \dfrac{\sqrt{16}}{\sqrt{x^4}} = \dfrac{4}{x^2}$ We assume $x \neq 0$.

YOUR TURN

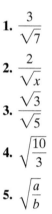

↳ Check Your
UNDERSTANDING

Write a form of 1 by which each expression can be multiplied in order to rationalize the denominator.

1. $\dfrac{3}{\sqrt{7}}$

2. $\dfrac{2}{\sqrt{x}}$

3. $\dfrac{\sqrt{3}}{\sqrt{5}}$

4. $\sqrt{\dfrac{10}{3}}$

5. $\sqrt{\dfrac{a}{b}}$

---

4. Rationalize the denominator:

$$\sqrt{\dfrac{x}{3}}.$$

B. Rationalizing Denominators with One Term

The expressions

$$\dfrac{1}{\sqrt{2}} \quad \text{and} \quad \dfrac{\sqrt{2}}{2}$$

are equivalent, but the second expression does not have an irrational number in the denominator. We can **rationalize the denominator** of a radical expression by multiplying by a carefully chosen form of 1.

EXAMPLE 4   Rationalize each denominator: **(a)** $\dfrac{3}{\sqrt{2}}$; **(b)** $\sqrt{\dfrac{5}{a}}$.

SOLUTION

a) We multiply by 1, using the fact that $\sqrt{2} \cdot \sqrt{2} = 2$ to choose the form of 1:

$$\dfrac{3}{\sqrt{2}} = \dfrac{3}{\sqrt{2}} \cdot \dfrac{\sqrt{2}}{\sqrt{2}} \qquad \text{Multiplying by 1, using the denominator, } \sqrt{2}, \text{ to write 1}$$

$$= \dfrac{3\sqrt{2}}{2}. \qquad \sqrt{2} \cdot \sqrt{2} = 2. \text{ The denominator is now rational.}$$

b) $\sqrt{\dfrac{5}{a}} = \dfrac{\sqrt{5}}{\sqrt{a}}$ The square root of a quotient is the quotient of the square roots.

$$= \dfrac{\sqrt{5}}{\sqrt{a}} \cdot \dfrac{\sqrt{a}}{\sqrt{a}} \qquad \text{Multiplying by 1, using the denominator, } \sqrt{a}, \text{ to write 1}$$

$$= \dfrac{\sqrt{5a}}{a} \qquad \sqrt{a} \cdot \sqrt{a} = a. \text{ We assume } a > 0.$$

↩ YOUR TURN

It is usually easiest to rationalize a denominator after the expression has been simplified.

EXAMPLE 5   Rationalize each denominator: **(a)** $\dfrac{\sqrt{2}}{\sqrt{45}}$; **(b)** $\sqrt{\dfrac{7a^2}{12a}}$.

Student Notes

Be careful when writing radical expressions. There is a *big* difference between the expressions

$$\dfrac{\sqrt{10}}{15} \quad \text{and} \quad \sqrt{\dfrac{10}{15}}.$$

SOLUTION

a) $\dfrac{\sqrt{2}}{\sqrt{45}} = \dfrac{\sqrt{2}}{\sqrt{9}\sqrt{5}}$ Simplifying the denominator. Note that 9 is a perfect square.

$$= \dfrac{\sqrt{2}}{3\sqrt{5}}$$

$$= \dfrac{\sqrt{2}}{3\sqrt{5}} \cdot \dfrac{\sqrt{5}}{\sqrt{5}} \qquad \text{Multiplying by 1, using } \sqrt{5} \text{ to write 1}$$

$$= \dfrac{\sqrt{10}}{3 \cdot 5} = \dfrac{\sqrt{10}}{15}$$

b) $\sqrt{\dfrac{7a^2}{12a}} = \sqrt{\dfrac{7a}{12}}$ Simplifying the radicand; $a^2/a = a$

$$= \dfrac{\sqrt{7a}}{\sqrt{12}} \qquad \begin{array}{l}\text{The square root of a quotient is the quotient} \\ \text{of the square roots.}\end{array}$$

$$= \dfrac{\sqrt{7a}}{\sqrt{4}\sqrt{3}} \qquad \begin{array}{l}\text{Simplifying the denominator. Note that} \\ \text{4 is a perfect square.}\end{array}$$

$$= \dfrac{\sqrt{7a}}{2\sqrt{3}}$$

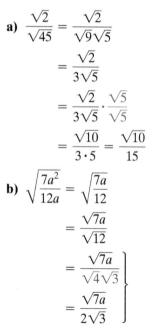

$$= \frac{\sqrt{7a}}{2\sqrt{3}} \cdot \frac{\sqrt{3}}{\sqrt{3}} \quad \text{Multiplying by 1}$$

$$= \frac{\sqrt{21a}}{2 \cdot 3}$$

5. Rationalize the denominator:

$$\frac{\sqrt{5}}{\sqrt{12}}.$$

$$= \frac{\sqrt{21a}}{6}$$

⟲ YOUR TURN

CAUTION! Our solutions in Example 5 cannot be simplified any further. A common mistake is to remove a factor equal to 1 that does not exist. For example, $\frac{\sqrt{10}}{15}$ *cannot* be simplified to $\frac{\sqrt{2}}{3}$ because $\sqrt{10}$ and 15 do not share a common factor.

8.3 EXERCISE SET

FOR EXTRA HELP MyMathLab®

◆ Vocabulary and Reading Check

Classify each of the following statements as either true or false.

1. To divide one square root by another, we can divide one radicand by the other and then take the square root.

2. The square root of a quotient can be found by dividing the square root of the numerator by the square root of the denominator.

3. To rationalize the denominator of a fraction, we square both the numerator and the denominator.

4. It is usually easiest to rationalize a denominator before the expression has been simplified.

A. Dividing Radical Expressions

Simplify. Assume that all variables represent positive numbers.

5. $\dfrac{\sqrt{500}}{\sqrt{5}}$

6. $\dfrac{\sqrt{50}}{\sqrt{2}}$

7. $\dfrac{\sqrt{40}}{\sqrt{10}}$

8. $\dfrac{\sqrt{72}}{\sqrt{2}}$

9. $\dfrac{\sqrt{55}}{\sqrt{5}}$

10. $\dfrac{\sqrt{18}}{\sqrt{3}}$

11. $\dfrac{\sqrt{5}}{\sqrt{20}}$

12. $\dfrac{\sqrt{2}}{\sqrt{18}}$

13. $\dfrac{\sqrt{18}}{\sqrt{32}}$

14. $\dfrac{\sqrt{12}}{\sqrt{75}}$

15. $\dfrac{\sqrt{8x}}{\sqrt{2x}}$

16. $\dfrac{\sqrt{18b}}{\sqrt{2b}}$

17. $\dfrac{\sqrt{63y^3}}{\sqrt{7y}}$

18. $\dfrac{\sqrt{48x^3}}{\sqrt{3x}}$

19. $\dfrac{\sqrt{500a^{10}}}{\sqrt{5a^2}}$

20. $\dfrac{\sqrt{27x^5}}{\sqrt{3x}}$

21. $\dfrac{\sqrt{21a^9}}{\sqrt{7a^3}}$

22. $\dfrac{\sqrt{35t^{13}}}{\sqrt{5t^5}}$

23. $\sqrt{\dfrac{4}{25}}$

24. $\sqrt{\dfrac{9}{49}}$

25. $\sqrt{\dfrac{49}{16}}$

26. $\sqrt{\dfrac{100}{49}}$

27. $-\sqrt{\dfrac{25}{81}}$

28. $-\sqrt{\dfrac{25}{64}}$

29. $\sqrt{\dfrac{2a^5}{50a}}$

30. $\sqrt{\dfrac{7a^5}{28a}}$

31. $\sqrt{\dfrac{6x^7}{32x}}$

32. $\sqrt{\dfrac{4x^3}{50x}}$

33. $\sqrt{\dfrac{21t^9}{28t^3}}$

34. $\sqrt{\dfrac{10t^9}{18t^5}}$

B. Rationalizing Denominators with One Term

Form an equivalent expression by rationalizing each denominator.

35. $\dfrac{1}{\sqrt{3}}$

36. $\dfrac{2}{\sqrt{3}}$

37. $\dfrac{5}{\sqrt{7}}$

38. $\dfrac{3}{\sqrt{11}}$

39. $\dfrac{\sqrt{16}}{\sqrt{27}}$

40. $\dfrac{\sqrt{25}}{\sqrt{8}}$

41. $\dfrac{\sqrt{6}}{\sqrt{5}}$

42. $\dfrac{\sqrt{5}}{\sqrt{7}}$

43. $\dfrac{\sqrt{3}}{\sqrt{50}}$

44. $\dfrac{\sqrt{5}}{\sqrt{18}}$

45. $\dfrac{\sqrt{2a}}{\sqrt{45}}$

46. $\dfrac{\sqrt{3a}}{\sqrt{32}}$

47. $\sqrt{\dfrac{12}{5}}$

48. $\sqrt{\dfrac{8}{3}}$

49. $\sqrt{\dfrac{7}{z}}$

50. $\sqrt{\dfrac{13}{p}}$

51. $\sqrt{\dfrac{a}{200}}$

52. $\sqrt{\dfrac{t}{32}}$

53. $\sqrt{\dfrac{x}{90}}$ **54.** $\sqrt{\dfrac{y}{40}}$ Aha! **55.** $\sqrt{\dfrac{3a}{25}}$

56. $\sqrt{\dfrac{5t}{16}}$ **57.** $\sqrt{\dfrac{5x^3}{12x}}$ **58.** $\sqrt{\dfrac{7t^3}{32t}}$

A. Dividing Radical Expressions: Applications

Depreciation of a Vehicle.   The resale value of certain cars and trucks that originally sold for $18,000 can be estimated using the formula

$$V = \frac{18,500}{\sqrt{t} + 1.0565},$$

where V is the value of the vehicle, in dollars, when it is t years old.

59. Estimate the vehicle's value when it is 3 years old.

60. Estimate the vehicle's value when it is 9 years old.

61. Estimate the vehicle's value when it is 1 year old.

62. Estimate the vehicle's value when it is 4 years old.

Period of a Swinging Pendulum.   The period T of a pendulum is the time that it takes the pendulum to move from one side to the other and back. A formula for the period is

$$T = 2\pi\sqrt{\frac{L}{32}},$$

where T is in seconds and L is the length, in feet. Use 3.14 for π.

63. Find the periods of pendulums of lengths 32 ft and 50 ft.

64. Find the periods of pendulums of lengths 8 ft and 2 ft.

65. The pendulum of a mantle clock is $8/\pi^2$ ft long. How long does it take to swing from one side to the other and back?

66. The pendulum of a grandfather clock is $32/\pi^2$ ft long. How long does it take to swing from one side to the other and back?

67. Ingrid is swinging at the bottom of a 72-ft long bungee cord. How long does it take her to complete one swing back and forth? (*Hint:* $\sqrt{2.25} = 1.5$.)

68. Don is swinging back and forth on an 18-ft long rope swing near the Green River Reservoir. How long does it take him to complete one swing back and forth?

69. Why is it important to know how to multiply radical expressions before learning how to divide them?

70. Describe a method that could be used to rationalize the *numerator* of a radical expression.

Skill Review

Perform the indicated operation and, if possible, simplify.   [4.5]

71. $(4t + 3)(4t - 3)$

72. $(x^3 - 7)^2$

73. $(9n + 1)(9n - 2)$

Factor completely.

74. $3x^2 - 6x + 3$ [5.4]

75. $100 - y^2$ [5.4]

76. $12x^3 - 6x^2 - 12x$ [5.1]

Synthesis

77. When calculating approximations by hand using long division, why is it easier to compute $\sqrt{2}/2$ than $1/\sqrt{2}$?

78. Is it always best to rewrite an expression of the form $\sqrt{a}/\sqrt{b}$ as $\sqrt{a/b}$ before simplifying? Why or why not?

Rationalize each denominator and, if possible, simplify to form an equivalent expression.

79. $\sqrt{\dfrac{7}{1000}}$ **80.** $\sqrt{\dfrac{3}{800}}$

81. $\sqrt{\dfrac{3x^2}{8x^7y^3}}$ **82.** $\sqrt{\dfrac{3x^2y}{a^2x^5}}$

83. $\sqrt{\dfrac{1}{5zw^2}}$ **84.** $\sqrt{\dfrac{2a}{5b^3c^9}}$

Aha! **85.** $\dfrac{2}{\sqrt{\sqrt{5}}}$ **86.** $\dfrac{3}{\sqrt{\sqrt{7}}}$

Simplify. Assume that $0 < x \le y$ and $0 < z \le 1$.

87. $\sqrt{\dfrac{1}{x^2} - \dfrac{2}{xy} + \dfrac{1}{y^2}}$

88. $\sqrt{2 - \dfrac{4}{z^2} + \dfrac{2}{z^4}}$

89. Solve: $\sqrt{\dfrac{2x-3}{8}} = \dfrac{5}{2}$.

90. *Research.* A *Foucault pendulum* is designed to demonstrate the earth's rotation.

a) Find the lengths of several Foucault pendulums, usually found in museums and universities. Calculate the period of each pendulum.

b) Using words, pictures, or a model, explain how a Foucault pendulum demonstrates the earth's rotation.

YOUR TURN ANSWERS: SECTION 8.3

1. 4   **2.** $\dfrac{a}{10}$ **3.** $\dfrac{2}{x^5}$ **4.** $\dfrac{\sqrt{3x}}{3}$ **5.** $\dfrac{\sqrt{15}}{6}$

Quick Quiz: Sections 8.1–8.3

1. Use a calculator to approximate $\sqrt{31}$. Round to three decimal places.   [8.1]

Simplify. Assume that all variables represent nonnegative numbers.

2. $\sqrt{100c^2}$ [8.1]   **3.** $\sqrt{12n^5}$ [8.2]

4. $\sqrt{\dfrac{15a^{11}}{16a}}$ [8.3]

5. Rationalize the denominator: $\dfrac{\sqrt{7}}{\sqrt{2}}$.   [8.3]

Prepare to Move On

Simplify.   [4.3]

1. $9x + 6x$ **2.** $9y - z + 8y$

Multiply.   [4.5], [4.6]

3. $4n(3n^2 + 6n + 1)$

4. $(2x - 3)(2x + 3)$

5. $(5t + 7)(5t - 7)$

| 8.4 | **Radical Expressions with Several Terms** |

A. Adding and Subtracting Radical Expressions   **B.** More with Multiplication
C. Rationalizing Denominators with Two Terms

We now consider addition and subtraction of radical expressions as well as some new types of multiplication and simplification.

A. Adding and Subtracting Radical Expressions

The sum of a rational number and an irrational number, like $5 + \sqrt{2}$, *cannot* be simplified. However, the sum of **like radicals**—that is, radical expressions that have the same radical factor—*can* be simplified.

EXAMPLE 1   Add or subtract, as indicated: **(a)** $3\sqrt{5} + 4\sqrt{5}$; **(b)** $\sqrt{x} - 7\sqrt{x}$.

SOLUTION

a) Recall that to simplify an expression like $3x + 4x$, we use the distributive law:

$$3x + 4x = (3 + 4)x = 7x.$$ The middle step is usually performed mentally.

In this example, x is replaced with $\sqrt{5}$:

$$3\sqrt{5} + 4\sqrt{5} = (3 + 4)\sqrt{5}$$ Using the distributive law to factor out $\sqrt{5}$
$$= 7\sqrt{5}.$$ $3\sqrt{5}$ and $4\sqrt{5}$ are like radicals.

b) $\sqrt{x} - 7\sqrt{x} = 1 \cdot \sqrt{x} - 7\sqrt{x}$ $\sqrt{x} = 1 \cdot \sqrt{x}$

$\qquad\qquad\quad = (1 - 7)\sqrt{x}$ Using the distributive law.

$\qquad\qquad\qquad\qquad\qquad\qquad$ Try to do this mentally.

$\qquad\qquad\quad = -6\sqrt{x}$

1. Add: $\sqrt{10} + 7\sqrt{10}$.

⟵ YOUR TURN

An expression may contain like radicals after individual radicals are themselves simplified.

Study Skills

Studying Together by Phone

Working with a classmate over the telephone can be a very effective way to receive or give help. The fact that you cannot point to figures on paper forces you to verbalize the mathematics, and the act of speaking mathematics will improve your understanding of the material.

2. Simplify: $\sqrt{75} - 2\sqrt{12}$.

EXAMPLE 2   Simplify: **(a)** $4\sqrt{2} - \sqrt{18}$; **(b)** $\sqrt{5} + \sqrt{20} + \sqrt{7}$.

SOLUTION

a) $\left.\begin{array}{l} 4\sqrt{2} - \sqrt{18} = 4\sqrt{2} - \sqrt{9 \cdot 2} \\ \qquad\qquad\quad = 4\sqrt{2} - \sqrt{9}\sqrt{2} \end{array}\right\}$ Simplifying $\sqrt{18}$

$\qquad\qquad\quad = 4\sqrt{2} - 3\sqrt{2}$ We now have like radicals.

$\qquad\qquad\quad = \sqrt{2}$ Using the distributive law mentally:

$\qquad\qquad\qquad\qquad\quad 4\sqrt{2} - 3\sqrt{2} = (4 - 3)\sqrt{2} = 1\sqrt{2} = \sqrt{2}$

b) $\sqrt{5} + \sqrt{20} + \sqrt{7} = \sqrt{5} + \sqrt{4}\sqrt{5} + \sqrt{7}$ Simplifying $\sqrt{20}$

$\qquad\qquad\qquad\quad = \sqrt{5} + 2\sqrt{5} + \sqrt{7}$ We now have two like radicals.

$\qquad\qquad\qquad\quad = 3\sqrt{5} + \sqrt{7}$ Adding like radicals; $3\sqrt{5} + \sqrt{7}$ cannot be simplified.

⟵ YOUR TURN

CAUTION!   It is *not true* that the sum of two square roots is the square root of the sum: $\sqrt{A} + \sqrt{B} \neq \sqrt{A + B}$. For example, $\sqrt{9} + \sqrt{16} \neq \sqrt{9 + 16}$ since $3 + 4 \neq 5$.

↰ **Check Your**
UNDERSTANDING

Determine whether the expressions in each exercise are like radicals. Simplify first, if necessary.

1. $3\sqrt{a}, -\sqrt{a}$

2. $\sqrt{x}, x, 2\sqrt{x}$

3. $\sqrt{4y}, \sqrt{y}$

4. $\sqrt{2a}, a\sqrt{2a}$

5. $\sqrt{200}, \sqrt{40}$

6. $\sqrt{50}, \sqrt{75}$

7. $\sqrt{8}, \sqrt{18}, \sqrt{32}, \sqrt{50}$

B. More with Multiplication

Radical expressions with more than one term are multiplied in much the same way that polynomials with more than one term are multiplied.

EXAMPLE 3   Multiply.

a) $\sqrt{2}(\sqrt{3} + \sqrt{10})$ **b)** $(4 + \sqrt{7})(2 + \sqrt{7})$

c) $(2 - \sqrt{5})(2 + \sqrt{5})$ **d)** $(2 + \sqrt{3})(5 - 4\sqrt{3})$

SOLUTION

a) $\sqrt{2}(\sqrt{3} + \sqrt{10}) = \sqrt{2}\sqrt{3} + \sqrt{2}\sqrt{10}$ Using the distributive law

$\qquad\qquad\qquad\quad = \sqrt{6} + \sqrt{20}$ Using the product rule for square roots

$\qquad\qquad\qquad\quad = \sqrt{6} + 2\sqrt{5}$ Simplifying; $\sqrt{20} = \sqrt{4}\sqrt{5} = 2\sqrt{5}$

b) $(4 + \sqrt{7})(2 + \sqrt{7}) = 4 \cdot 2 + 4 \cdot \sqrt{7} + \sqrt{7} \cdot 2 + \sqrt{7} \cdot \sqrt{7}$ Using FOIL

$\qquad\qquad\qquad\qquad = 8 + 4\sqrt{7} + 2\sqrt{7} + 7$

$\qquad\qquad\qquad\qquad = 15 + 6\sqrt{7}$ Combining like terms

c) Note that $(2 - \sqrt{5})(2 + \sqrt{5})$ is of the form $(A - B)(A + B)$.

$(2 - \sqrt{5})(2 + \sqrt{5}) = 2 \cdot 2 + 2 \cdot \sqrt{5} - \sqrt{5} \cdot 2 - \sqrt{5} \cdot \sqrt{5}$ Using FOIL

$\qquad\qquad\qquad\qquad = 4 + 2\sqrt{5} - 2\sqrt{5} - 5$ The middle terms are opposites.

$\qquad\qquad\qquad\qquad = 4 - 5$ We could have used $(A - B)(A + B) = A^2 - B^2$.

$\qquad\qquad\qquad\qquad = -1$

d) $(2 + \sqrt{3})(5 - 4\sqrt{3}) = 2 \cdot 5 - 2 \cdot 4\sqrt{3} + \sqrt{3} \cdot 5 - \sqrt{3} \cdot 4\sqrt{3}$ Using FOIL

$\qquad\qquad\qquad\qquad = 10 - 8\sqrt{3} + 5\sqrt{3} - 4 \cdot 3$ $2 \cdot 5 = 10; 2 \cdot 4 = 8;$ and $\sqrt{3} \cdot \sqrt{3} = 3$

$\qquad\qquad\qquad\qquad = -2 - 3\sqrt{3}$ Combining like terms

3. Multiply: $\sqrt{15}(\sqrt{5} - \sqrt{3})$.

◀ YOUR TURN

C. Rationalizing Denominators with Two Terms

Note in Example 3(c) that the result has no radicals. This will happen whenever expressions like $\sqrt{a} + \sqrt{b}$ and $\sqrt{a} - \sqrt{b}$ are multiplied:

$$(\sqrt{a} + \sqrt{b})(\sqrt{a} - \sqrt{b}) = (\sqrt{a})^2 - (\sqrt{b})^2 = a - b.$$

Expressions such as $\sqrt{3} - \sqrt{5}$ and $\sqrt{3} + \sqrt{5}$ are said to be **conjugates** of each other. So too are expressions like $2 + \sqrt{7}$ and $2 - \sqrt{7}$. The conjugate of a denominator can be used to rationalize the denominator.

EXAMPLE 4   Rationalize each denominator and, if possible, simplify.

a) $\dfrac{3}{2 + \sqrt{5}}$ **b)** $\dfrac{2}{\sqrt{7} - \sqrt{3}}$

SOLUTION

a) We multiply by a form of 1, using the conjugate of $2 + \sqrt{5}$, which is $2 - \sqrt{5}$, as the numerator and the denominator:

$$\frac{3}{2 + \sqrt{5}} = \frac{3}{2 + \sqrt{5}} \cdot \frac{2 - \sqrt{5}}{2 - \sqrt{5}} \qquad \text{Multiplying by 1}$$

$$= \frac{3(2 - \sqrt{5})}{(2 + \sqrt{5})(2 - \sqrt{5})}$$

$$= \frac{3(2 - \sqrt{5})}{2^2 - (\sqrt{5})^2} \qquad \text{Using } (A + B)(A - B) = A^2 - B^2$$

$$= \frac{3(2 - \sqrt{5})}{-1} \qquad \begin{array}{l}\text{Simplifying the denominator.}\\ \text{See Example 3(c): } 4 - 5 = -1.\end{array}$$

$$= \frac{6 - 3\sqrt{5}}{-1} \qquad \text{Using the distributive law}$$

$$= -6 + 3\sqrt{5}. \qquad \text{Dividing } both \text{ terms in the numerator by } -1$$

b) $\dfrac{2}{\sqrt{7} - \sqrt{3}} = \dfrac{2}{\sqrt{7} - \sqrt{3}} \cdot \dfrac{\sqrt{7} + \sqrt{3}}{\sqrt{7} + \sqrt{3}}$ $\begin{array}{l}\text{Multiplying by 1, using } \sqrt{7} + \sqrt{3},\\ \text{the conjugate of } \sqrt{7} - \sqrt{3}\end{array}$

$$= \frac{2(\sqrt{7} + \sqrt{3})}{(\sqrt{7} - \sqrt{3})(\sqrt{7} + \sqrt{3})}$$

$$= \frac{2(\sqrt{7} + \sqrt{3})}{(\sqrt{7})^2 - (\sqrt{3})^2} \qquad \text{Using } (A - B)(A + B) = A^2 - B^2$$

$$= \frac{2(\sqrt{7} + \sqrt{3})}{7 - 3} \qquad \text{The denominator is free of radicals.}$$

$$= \frac{2(\sqrt{7} + \sqrt{3})}{4} \qquad \begin{array}{l}\text{Since 2 is a common factor of both the}\\ \text{numerator and the denominator, we simplify.}\end{array}$$

$$= \frac{2(\sqrt{7} + \sqrt{3})}{2 \cdot 2} \qquad \begin{array}{l}\text{Factoring and removing a factor equal}\\ \text{to 1: } \dfrac{2}{2} = 1\end{array}$$

$$= \frac{\sqrt{7} + \sqrt{3}}{2}$$

4. Rationalize the denominator and, if possible, simplify:

$$\frac{8}{3 - \sqrt{5}}.$$

◀ YOUR TURN

Technology Connection

Parentheses may be needed to specify the radicand when entering a radical expression on a calculator. For some graphing calculators, we would enter the expression $\sqrt{7 - 2} + 8$ as $\sqrt{\ }(7 - 2) + 8$. This will give us a different result from $\sqrt{7} - 2 + 8$ or $\sqrt{\ }(7 - 2 + 8$.

Use a graphing calculator to approximate each of the following. Round to three decimal places.

1. $\dfrac{-3 + \sqrt{8}}{2}$

2. $\dfrac{\sqrt{2} + \sqrt{5}}{3}$

3. $\sqrt{\sqrt{7} + 5}$

4. Use a calculator to check Examples 4(a) and 4(b) by comparing the values of the original expression and the simplified expression.

8.4 EXERCISE SET

Vocabulary and Reading Check

Choose from the following list the term that best completes each statement. Terms may be used more than once.

conjugates   like radicals

1. The expressions $\sqrt{3}$ and $4\sqrt{3}$ are_____.

2. The expressions $\sqrt{2} + 5$ and $\sqrt{2} - 5$ are
   _____.

3. _____ are used to rationalize denominators with two terms.

4. _____ are necessary in order for a sum of radicals to be simplified.

Concept Reinforcement

In Exercises 5–8, match each item with the appropriate choice from the column on the right.

5. ____ The conjugate of $\sqrt{7} + \sqrt{5}$
6. ____ An example of like radicals
7. ____ The product of $\sqrt{5} - \sqrt{7}$ and its conjugate
8. ____ The conjugate of $\sqrt{7} - \sqrt{5}$

a) $\sqrt{7} + \sqrt{5}$
b) $\sqrt{7} - \sqrt{5}$
c) $3\sqrt{7}$ and $-5\sqrt{7}$
d) -2

A. Adding and Subtracting Radical Expressions

Add or subtract. Simplify by combining like radical terms, if possible.

9. $3\sqrt{10} + 8\sqrt{10}$
10. $5\sqrt{7} + \sqrt{7}$
11. $4\sqrt{2} - \sqrt{2}$
12. $8\sqrt{3} - 5\sqrt{3}$
13. $4\sqrt{t} + 9\sqrt{t}$
14. $6\sqrt{t} + 4\sqrt{t}$
15. $7\sqrt{x} - 8\sqrt{x}$
16. $10\sqrt{a} - 15\sqrt{a}$
17. $5\sqrt{2a} + 3\sqrt{2a}$
18. $5\sqrt{6x} + 2\sqrt{6x}$
19. $9\sqrt{10y} - \sqrt{10y}$
20. $12\sqrt{14y} - \sqrt{14y}$
21. $6\sqrt{7} + 2\sqrt{7} + 4\sqrt{7}$
22. $2\sqrt{5} + 7\sqrt{5} + 5\sqrt{5}$
23. $5\sqrt{2} - 9\sqrt{2} + 8\sqrt{2}$
24. $3\sqrt{6} - 7\sqrt{6} + 2\sqrt{6}$
25. $5\sqrt{3} + \sqrt{8}$
26. $2\sqrt{5} + \sqrt{45}$
27. $\sqrt{x} - \sqrt{16x}$
28. $\sqrt{100a} - \sqrt{a}$
29. $2\sqrt{3} - 4\sqrt{75}$
30. $7\sqrt{50} - 3\sqrt{2}$
31. $6\sqrt{18} + 5\sqrt{8}$
32. $3\sqrt{12} + 2\sqrt{300}$
33. $\sqrt{72} + \sqrt{98}$
34. $\sqrt{45} + \sqrt{80}$
Aha! 35. $9\sqrt{8} + \sqrt{72} - 9\sqrt{8}$
36. $4\sqrt{12} + \sqrt{27} - \sqrt{12}$
37. $5\sqrt{18} - 2\sqrt{32} - \sqrt{50}$
38. $7\sqrt{12} - 2\sqrt{27} + \sqrt{75}$
39. $\sqrt{16a} - 4\sqrt{a} + \sqrt{25a}$
40. $\sqrt{9x} + \sqrt{49x} - 9\sqrt{x}$

B. Multiplying Radical Expressions

Multiply.

41. $\sqrt{3}(\sqrt{2} + \sqrt{7})$
42. $\sqrt{2}(\sqrt{7} + \sqrt{5})$
43. $\sqrt{5}(\sqrt{6} - \sqrt{10})$
44. $\sqrt{6}(\sqrt{15} - \sqrt{11})$
45. $(3 + \sqrt{2})(4 + \sqrt{2})$
46. $(5 + \sqrt{11})(3 + \sqrt{11})$
47. $(\sqrt{7} - 2)(\sqrt{7} - 5)$
48. $(\sqrt{10} + 4)(\sqrt{10} - 7)$
49. $(\sqrt{6} + 5)(\sqrt{6} - 5)$
50. $(2 + \sqrt{3})(2 - \sqrt{3})$
51. $(\sqrt{7} - \sqrt{3})(\sqrt{7} + \sqrt{3})$
52. $(\sqrt{2} + \sqrt{5})(\sqrt{2} - \sqrt{5})$
53. $(2 + 3\sqrt{2})(3 - \sqrt{2})$
54. $(8 - \sqrt{7})(3 + 2\sqrt{7})$
55. $(7 + \sqrt{3})^2$
56. $(2 - \sqrt{5})^2$
57. $(1 - 2\sqrt{3})^2$
58. $(6 + 3\sqrt{5})^2$
59. $(\sqrt{x} - \sqrt{10})^2$
60. $(\sqrt{a} - \sqrt{6})^2$

C. Rationalizing Denominators with Two Terms

Rationalize each denominator and, if possible, simplify to form an equivalent expression.

61. $\dfrac{2}{5 + \sqrt{2}}$
62. $\dfrac{5}{3 - \sqrt{7}}$
63. $\dfrac{2}{1 - \sqrt{3}}$
64. $\dfrac{4}{2 + \sqrt{5}}$
65. $\dfrac{2}{\sqrt{7} + 5}$
66. $\dfrac{6}{\sqrt{10} + 3}$
67. $\dfrac{\sqrt{10}}{\sqrt{10} + 4}$
68. $\dfrac{\sqrt{15}}{\sqrt{15} - 5}$
69. $\dfrac{\sqrt{7}}{\sqrt{7} - \sqrt{3}}$
70. $\dfrac{\sqrt{11}}{\sqrt{11} + \sqrt{7}}$
71. $\dfrac{\sqrt{3}}{\sqrt{5} - \sqrt{3}}$
72. $\dfrac{\sqrt{6}}{\sqrt{7} + \sqrt{6}}$

73. $\dfrac{2}{\sqrt{7} + \sqrt{2}}$

74. $\dfrac{6}{\sqrt{5} - \sqrt{3}}$

75. $\dfrac{\sqrt{6} - \sqrt{x}}{\sqrt{6} + \sqrt{x}}$

76. $\dfrac{\sqrt{10} - \sqrt{x}}{\sqrt{10} + \sqrt{x}}$

77. Why does the product of a pair of conjugates contain no radicals?

78. Describe a method that could be used to rationalize a numerator that is the sum of two radical expressions.

Skill Review

79. In which quadrant or on which axis is the point $(-100, 0)$ located? [3.1]

80. Find the slope of the line containing the points $(6, -7)$ and $(-2, -3)$. [3.5]

81. Find the slope of the line given by $y = 2 - x$. [3.6]

82. Write the slope–intercept equation of the line with slope -1 and y-intercept $(0, -6)$. [3.6]

83. Write the slope–intercept equation of the line with slope 6 that contains the point $(1, 2)$. [3.7]

84. Write the slope–intercept equation of the line containing the points $(2, 4)$ and $(-1, 3)$. [3.7]

Synthesis

85. Monica believes that since $(\sqrt{x})^2 = x$, the square of a sum of radical expressions is not a radical expression. Is she correct? Why or why not?

86. Why must you know how to add and subtract radical expressions before you can rationalize denominators with two terms?

Add or subtract and, if possible, simplify.

87. $\sqrt{\dfrac{25}{x}} + \dfrac{\sqrt{x}}{2x} - \dfrac{5}{\sqrt{2}}$

88. $5\sqrt{\dfrac{1}{2}} + \dfrac{7}{2}\sqrt{18} - 4\sqrt{98}$

89. $\sqrt{8x^6y^3} - x\sqrt{2y^7} - \dfrac{x}{3}\sqrt{18x^2y^9}$

90. $a\sqrt{a^{17}b^9} - b\sqrt{a^{13}b^{11}} + a\sqrt{a^9b^{15}}$

91. $7x\sqrt{12xy^2} - 9y\sqrt{27x^3} + 5\sqrt{300x^3y^2}$

92. For which pairs of nonnegative numbers a and b does $\sqrt{a} + \sqrt{b} = \sqrt{a + b}$?

93. Three students were asked to simplify $\sqrt{10} + \sqrt{50}$. Their answers were $\sqrt{10}(1 + \sqrt{5})$, $\sqrt{10} + 5\sqrt{2}$, and $\sqrt{2}(5 + \sqrt{5})$. Which answer(s), if any, is correct?

YOUR TURN ANSWERS: SECTION 8.4

1. $8\sqrt{10}$ **2.** $\sqrt{3}$ **3.** $5\sqrt{3} - 3\sqrt{5}$ **4.** $6 + 2\sqrt{5}$

Quick Quiz: Sections 8.1–8.4

Add or subtract and, if possible, simplify. [8.4]

1. $7\sqrt{3} - 8\sqrt{3}$

2. $\sqrt{32} + 3\sqrt{2}$

Multiply and, if possible, simplify. Assume that x represents a nonnegative number.

3. $\sqrt{6x^3}\sqrt{15x}$ [8.2]

4. $\sqrt{2}(\sqrt{10} + \sqrt{5})$ [8.4]

5. $(1 + \sqrt{5})(2 - \sqrt{3})$ [8.4]

Prepare to Move On

Solve.

1. $3x + 5 + 2(x - 3) = 4 - 6x$ [2.2]

2. $4(x - 3) - 2 = 5(2x - 1)$ [2.2]

3. $x^2 - 5x = 6$ [5.6]

4. $x^2 + 10 = 7x$ [5.6]

5. The average American household spent \$353 less on entertainment in 2013 than in 2008. The amount spent in 2013 was \$214 more than 80% of the amount spent in 2008. How much was spent each year? [2.5], [7.2]

Data: U.S. Bureau of Labor Statistics

Mid-Chapter Review

Square roots like $\sqrt{7}$ are examples of *radical expressions*. Many radical expressions can be simplified.

Simplifying a perfect-square radicand

$$\sqrt{36x^2} = 6x$$

(We assume $x \geq 0$.)

Factoring and simplifying

$$\sqrt{48a^7} = \sqrt{16} \cdot \sqrt{a^6} \cdot \sqrt{3a} = 4a^3\sqrt{3a}$$

(We assume $a \geq 0$.)

Dividing and simplifying

$$\frac{\sqrt{50t^5}}{\sqrt{2t^{11}}} = \sqrt{\frac{50t^5}{2t^{11}}} = \sqrt{\frac{25}{t^6}} = \frac{5}{t^3}$$

(We assume $t > 0$.)

Combining like radicals

$$\sqrt{100t} - \sqrt{81t} = 10\sqrt{t} - 9\sqrt{t} = \sqrt{t}$$

Multiplying and simplifying

$$\sqrt{3}(\sqrt{6} - 5) = \sqrt{18} - 5\sqrt{3} = 3\sqrt{2} - 5\sqrt{3}$$

Rationalizing the denominator

$$\frac{2}{\sqrt{3}} = \frac{2}{\sqrt{3}} \cdot \frac{\sqrt{3}}{\sqrt{3}} = \frac{2\sqrt{3}}{3}$$

GUIDED SOLUTIONS

1. Multiply and simplify: $\sqrt{6x^3} \cdot \sqrt{12xy}$. [8.2]

Solution

$$\sqrt{6x^3} \cdot \sqrt{12xy} = \sqrt{6 \cdot 12 \cdot \square \cdot y}$$
$$= \sqrt{6 \cdot 6 \cdot \square \cdot x^4 \cdot y}$$
$$= \sqrt{6^2} \cdot \sqrt{x^4} \cdot \sqrt{\square}$$
$$= \square\sqrt{\square}$$

2. Rationalize the denominator: $\sqrt{\dfrac{7}{n}}$. [8.3]

Solution

$$\sqrt{\frac{7}{n}} = \frac{\sqrt{7}}{\sqrt{n}}$$
$$= \frac{\sqrt{7}}{\sqrt{n}} \cdot \frac{\square}{\square}$$
$$= \frac{\sqrt{\square}}{\square}$$

MIXED REVIEW

When simplifying, assume that all variables represent positive numbers.

3. Find the square roots of 121. [8.1]

4. Identify the radicand in the expression $2\sqrt{xy}$. [8.1]

Simplify.

5. $\sqrt{100t^2}$ [8.1]

6. $\sqrt{17x^2}$ [8.2]

7. $\dfrac{\sqrt{20}}{\sqrt{45}}$ [8.3]

8. $\sqrt{15t} + 4\sqrt{15t}$ [8.4]

9. $\sqrt{18} \cdot \sqrt{30}$ [8.2]

10. $\sqrt{6}(\sqrt{10} - \sqrt{33})$ [8.4]

11. $2\sqrt{3} - 5\sqrt{12}$ [8.4]

12. $\sqrt{\dfrac{3a^{10}}{12a^4}}$ [8.3]

13. $(\sqrt{5} + 3)(\sqrt{5} - 3)$ [8.4]

14. $\sqrt{28a^{19}}$ [8.2]

15. $\sqrt{12x^2y^5} \cdot \sqrt{75x^6y^7}$ [8.2]

16. $(\sqrt{5} + 1)(\sqrt{10} + 3)$ [8.4]

17. $\sqrt{500} + \sqrt{125}$ [8.4]

18. $\sqrt{64m^{16}}$ [8.2]

Rationalize each denominator.

19. $\dfrac{x}{\sqrt{8}}$ [8.3]

20. $\dfrac{1}{\sqrt{3} - \sqrt{2}}$ [8.4]

8.5 Radical Equations

A. Solving Radical Equations   **B.** Problem Solving and Applications

An equation in which a variable appears in a radicand is called a **radical equation**. The following are examples of radical equations:

$$\sqrt{2x} - 4 = 7, \qquad 2\sqrt{x+2} = \sqrt{x+10}, \quad \text{and} \quad 3 + \sqrt{27 - 3x} = x.$$

To solve a radical equation, we use a new equation-solving principle: the principle of squaring. Use of this principle does *not* always produce equivalent equations, so checking solutions becomes more important than ever.

A. Solving Radical Equations

An equation with a square root can be rewritten without the radical by using *the principle of squaring*.

THE PRINCIPLE OF SQUARING

If $a = b$, then $a^2 = b^2$.

The principle of squaring does *not* say that if $a^2 = b^2$, then $a = b$. Indeed,

$$(-5)^2 = 5^2 \quad \text{is true, but}$$
$$-5 = 5 \quad \text{is false.}$$

Although the principle of squaring *can* lead us to a solution of a radical equation, it *may* lead us to numbers that are not solutions.

EXAMPLE 1   Solve: **(a)** $\sqrt{x} + 3 = 7$; **(b)** $\sqrt{2x} + 5 = 0$.

SOLUTION

a) Our plan is to isolate the radical term on one side of the equation and then use the principle of squaring:

$$\sqrt{x} + 3 = 7$$
$$\sqrt{x} = 4 \qquad \text{Subtracting 3 to get the radical alone on one side}$$
$$(\sqrt{x})^2 = 4^2 \qquad \text{Squaring both sides (using the principle of squaring)}$$
$$x = 16.$$

Check:
$$\sqrt{x} + 3 = 7$$
$$\begin{array}{c|c} \sqrt{16} + 3 & 7 \\ 4 + 3 & \\ 7 \stackrel{?}{=} 7 & \text{TRUE} \end{array}$$

The solution is 16.

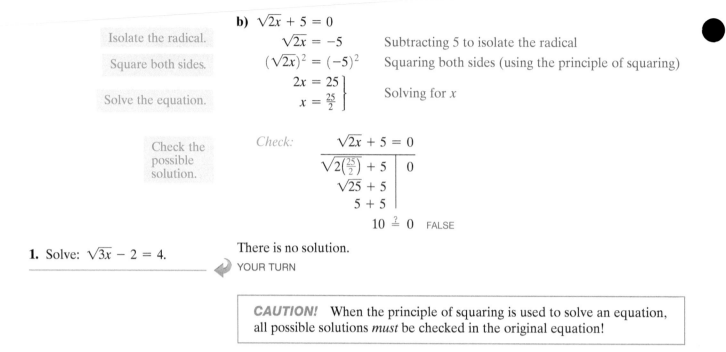

b) $\sqrt{2x} + 5 = 0$

Isolate the radical. $\sqrt{2x} = -5$ Subtracting 5 to isolate the radical

Square both sides. $(\sqrt{2x})^2 = (-5)^2$ Squaring both sides (using the principle of squaring)

Solve the equation. $\left.\begin{array}{l} 2x = 25 \\ x = \frac{25}{2} \end{array}\right\}$ Solving for x

Check the possible solution.

Check: $\sqrt{2x} + 5 = 0$

$$\frac{\sqrt{2\left(\frac{25}{2}\right)} + 5 \,\big|\, 0}{\begin{array}{l} \sqrt{25} + 5 \\ 5 + 5 \\ \hspace{1.5em} 10 \overset{?}{=} 0 \quad \text{FALSE} \end{array}}$$

There is no solution.

1. Solve: $\sqrt{3x} - 2 = 4$.

◀ YOUR TURN

> *CAUTION!* When the principle of squaring is used to solve an equation, all possible solutions *must* be checked in the original equation!

↳ **Check Your UNDERSTANDING**

In each case, determine whether the number is a solution of the equation that follows.

1. 100; $\sqrt{x} = 10$
2. 2; $\sqrt{x} = 4$
3. 3; $\sqrt{2x - 7} = 1$
4. 3; $\sqrt{2x - 7} = -1$
5. 4; $x - 1 = \sqrt{x + 5}$

EXAMPLE 2   Solve: $3\sqrt{x} = \sqrt{x + 32}$.

SOLUTION   We have

$3\sqrt{x} = \sqrt{x + 32}$

$(3\sqrt{x})^2 = (\sqrt{x + 32})^2$ Squaring both sides (using the principle of squaring)

$3^2(\sqrt{x})^2 = x + 32$ Squaring the product on the left; simplifying on the right

$9x = x + 32$ Simplifying on the left

$\left.\begin{array}{l} 8x = 32 \\ x = 4. \end{array}\right\}$ Solving for x

Check: $3\sqrt{x} = \sqrt{x + 32}$

$$\frac{3\sqrt{4} \,\big|\, \sqrt{4 + 32}}{\begin{array}{l} 3 \cdot 2 \,\big|\, \sqrt{36} \\ \hspace{0.5em} 6 \overset{?}{=} 6 \hspace{3em} \text{TRUE} \end{array}}$$

The number 4 checks. The solution is 4.

2. Solve: $2\sqrt{x - 7} = \sqrt{5x}$.

◀ YOUR TURN

We have been using the following strategy.

> **TO SOLVE A RADICAL EQUATION**
> 1. Isolate a radical term.
> 2. Use the principle of squaring (square both sides).
> 3. Solve the new equation.
> 4. Check all possible solutions in the original equation.

In some cases, we apply the principle of zero products after squaring.

EXAMPLE 3 Solve: **(a)** $x - 5 = \sqrt{x + 7}$; **(b)** $3 + \sqrt{27 - 3x} = x$.

SOLUTION

a)

$$x - 5 = \sqrt{x + 7}$$
$$(x - 5)^2 = (\sqrt{x + 7})^2 \qquad \text{Using the principle of squaring}$$
$$x^2 - 10x + 25 = x + 7 \qquad \text{Squaring a binomial on the left side}$$
$$x^2 - 11x + 18 = 0 \qquad \text{Adding } -x - 7 \text{ to both sides}$$
$$(x - 9)(x - 2) = 0 \qquad \text{Factoring}$$
$$x - 9 = 0 \quad or \quad x - 2 = 0 \qquad \text{Using the principle of zero products:}$$
$$\text{If } AB = 0, \text{ then } A = 0 \text{ or } B = 0.$$
$$x = 9 \quad or \quad x = 2$$

Check:

For 9:	For 2:
$x - 5 = \sqrt{x + 7}$	$x - 5 = \sqrt{x + 7}$
$9 - 5 \mid \sqrt{9 + 7}$	$2 - 5 \mid \sqrt{2 + 7}$
$4 \overset{?}{=} 4$ TRUE	$-3 \overset{?}{=} 3$ FALSE

The number 9 checks, but 2 does not. Thus the solution is 9.

b)

$$3 + \sqrt{27 - 3x} = x$$
$$\sqrt{27 - 3x} = x - 3 \qquad \text{Subtracting 3 to isolate the radical}$$
$$(\sqrt{27 - 3x})^2 = (x - 3)^2 \qquad \text{Using the principle of squaring}$$
$$27 - 3x = x^2 - 6x + 9$$
$$0 = x^2 - 3x - 18 \qquad \text{Adding } 3x - 27 \text{ to both sides}$$
$$0 = (x - 6)(x + 3) \qquad \text{Factoring}$$
$$x - 6 = 0 \quad or \quad x + 3 = 0 \qquad \text{Using the principle of zero products}$$
$$x = 6 \quad or \quad x = -3$$

Check:

For 6:	For -3:
$3 + \sqrt{27 - 3x} = x$	$3 + \sqrt{27 - 3x} = x$
$3 + \sqrt{27 - 3 \cdot 6} \mid 6$	$3 + \sqrt{27 - 3 \cdot (-3)} \mid -3$
$3 + \sqrt{9}$	$3 + \sqrt{27 + 9}$
$3 + 3$	$3 + \sqrt{36}$
$6 \overset{?}{=} 6$ TRUE	$3 + 6$
	$9 \overset{?}{=} -3$ FALSE

The number 6 checks, but -3 does not. The solution is 6.

YOUR TURN

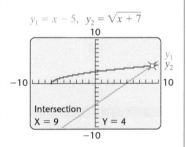
Height h (in feet)	Distance d (in miles)
10	3.9
100	12.2
1000	38.7
5000	86.6

B. Problem Solving and Applications

Many applications translate to radical equations. For example, the table at left lists the distance d, in miles, that one can see to the horizon from a height h, in feet. This distance can be approximated by the formula

$$d = \sqrt{1.5h}.$$

EXAMPLE 4   On DeWayne's first flight in a hot-air balloon, he noticed that he could just see, on the horizon, a landmark that he knew was 30 mi away from the ground under his balloon. How high was his balloon?

SOLUTION

1. **Familiarize.** The formula we need is given above the example: $d = \sqrt{1.5h}$. There is no need to memorize this formula.

2. **Translate.** We substitute 30 for d in the formula:

$$d = \sqrt{1.5h}$$
$$30 = \sqrt{1.5h}. \qquad \text{Substituting}$$

3. **Carry out.** We solve the equation for h:

$$30 = \sqrt{1.5h}$$
$$(30)^2 = (\sqrt{1.5h})^2 \qquad \text{Squaring both sides}$$
$$900 = 1.5h$$
$$600 = h. \qquad \text{Dividing both sides by 1.5}$$

4. **Check.** A height of 600 ft seems a reasonable answer. To check, we let $h = 600$ and calculate d:

$$d = \sqrt{1.5(600)} = \sqrt{900} = 30.$$

5. **State.** DeWayne's balloon was about 600 ft high.

4. From her airplane window, Jorani can just see, on the horizon, a city that is 100 mi away from the ground under her airplane. Use the formula from Example 4 to estimate the altitude of the airplane.

↪ YOUR TURN

CONNECTING 🔗 THE CONCEPTS

We attempt to *simplify expressions* and to *solve equations*. The principle of squaring applies to *equations*, not expressions.

When we simplify an expression, the answer is an equivalent expression.

Example: Add and, if possible, simplify:

$$\sqrt{75x} + \sqrt{12x}.$$

$$\sqrt{75x} + \sqrt{12x} = \sqrt{25 \cdot 3x} + \sqrt{4 \cdot 3x}$$
$$= 5\sqrt{3x} + 2\sqrt{3x} \qquad \text{Factoring and simplifying}$$
$$= 7\sqrt{3x} \qquad \text{Combining like radicals}$$

A simplified, equivalent expression is $7\sqrt{3x}$.

When we solve an equation, the answer is a solution.

Example: Solve: $\sqrt{x + 2} = 6$.

$$\sqrt{x + 2} = 6$$
$$(\sqrt{x + 2})^2 = 6^2 \qquad \text{Squaring both sides}$$
$$\left. \begin{array}{l} x + 2 = 36 \\ x = 34 \end{array} \right\} \quad \text{Solving for } x$$

The solution is 34.

EXERCISES

When simplifying, assume that all variables represent nonnegative numbers. Thus no absolute-value signs are needed in the answer.

 1. Simplify: $\sqrt{25a^2}$.

 2. Solve: $\sqrt{x} + 1 = 7$.

 3. Simplify: $\dfrac{\sqrt{45}}{\sqrt{60}}$.

 4. Solve: $x + 5 = \sqrt{x + 11}$.

 5. Multiply and, if possible, simplify:
    $(\sqrt{3} + 2)(\sqrt{3} - 1)$.

 6. Subtract and, if possible, simplify:
    $6\sqrt{5} - 3\sqrt{20}$.

 7. Multiply and, if possible, simplify:
    $\sqrt{40x^6y^7} \cdot \sqrt{2xy^3}$.

 8. Solve: $x - 1 = 2\sqrt{x - 2}$.

8.5 EXERCISE SET

FOR EXTRA HELP MyMathLab®

↪ Vocabulary and Reading Check

Classify each of the following statements as either true or false.

1. If $a = b$ is a true statement, then $a^2 = b^2$ must be true.

2. If $a^2 = b^2$ is a true statement, then $a = b$ must be true.

3. Every radical equation has at least one solution.

4. To use the principle of squaring, we square both sides of the equation.

A. Solving Radical Equations

Solve.

5. $\sqrt{x} = 6$

6. $\sqrt{x} = 11$

7. $\sqrt{x} - 3 = 9$

8. $\sqrt{x} - 7 = 2$

9. $\sqrt{3x + 1} = 8$

10. $\sqrt{5x - 1} = 7$

11. $2 + \sqrt{3 - y} = 9$

12. $3 + \sqrt{1 - a} = 5$

13. $10 - 2\sqrt{3n} = 0$

14. $4 - 2\sqrt{5n} = 0$

15. $\sqrt{8t + 3} = \sqrt{6t + 7}$

16. $\sqrt{7t - 9} = \sqrt{t + 3}$

Aha! 17. $5\sqrt{y} = -2$

18. $3\sqrt{y} + 1 = 0$

19. $\sqrt{6 - 4t} = \sqrt{2 - 5t}$

20. $\sqrt{11 - 3t} = \sqrt{1 - 5t}$

21. $\sqrt{3x + 1} = x - 3$

22. $x - 7 = \sqrt{x - 5}$

23. $a - 9 = \sqrt{a - 3}$

24. $\sqrt{t + 18} = t - 2$

25. $x + 1 = 6\sqrt{x - 7}$

26. $x - 5 = \sqrt{15 - 3x}$

27. $\sqrt{5x + 21} = x + 3$

28. $\sqrt{22 - x} = x - 2$

29. $t + 4 = 4\sqrt{t + 1}$

30. $1 + 2\sqrt{y - 1} = y$

31. $\sqrt{x^2 + 6} - x + 3 = 0$

32. $\sqrt{x^2 + 7} - x + 2 = 0$

33. $\sqrt{(4x + 5)(x + 4)} = 2x + 5$

34. $\sqrt{(p + 6)(p + 1)} - 2 = p + 1$

35. $\sqrt{8 - 3x} = \sqrt{13 + x}$

36. $\sqrt{3 - 7x} = \sqrt{5 - 2x}$

37. $x = 1 + \sqrt{1 - x}$

38. $x = \sqrt{x + 5} + 7$

39. $2x - 5 = \sqrt{(x - 3)(2x + 1)}$

40. $2x + 1 = \sqrt{(x - 1)(4x + 1)}$

B. Problem Solving and Applications

Speed of a Skidding Car. *The formula $r = 2\sqrt{6L}$ can be used to approximate the speed r, in miles per hour, of a car that has left a skid pattern of length L, in feet, on a dry asphalt road.*

41. How far will a car skid at 36 mph? at 60 mph?

42. How far will a car skid at 50 mph? at 80 mph? Round to the nearest foot.

Temperature and the Speed of Sound. *The formula $s = 21.9\sqrt{5t + 2457}$ can be used to approximate the speed of sound s, in feet per second, at a temperature of t degrees Fahrenheit.*

43. During blasting for avalanche control in Utah's Wasatch Mountains, sound traveled at a rate of 1113 ft/sec. What was the temperature at the time?

44. At a recent concert by the Dave Matthews Band, sound traveled at a rate of 1176 ft/sec. What was the temperature at the time?

Sighting to the Horizon. *At a height of h meters, one can see V kilometers to the horizon, where $V = 3.5\sqrt{h}$.*

45. A scout can see 56 km to the horizon from atop a firetower. What is the altitude of the scout's eyes?

46. Alejandro can see 420 km to the horizon from an airplane window. How high is the airplane?

Speed of Surface Waves. *The speed v, in meters per second, of a wave on the surface of the ocean can be approximated by the formula $v = 3.1\sqrt{d}$, where d is the depth of the water, in meters.*

Data: myweb.dal.ca

47. A wave is traveling 62 m/sec. What is the water depth?

48. A wave is traveling 9.3 m/sec. What is the water depth?

Period of a Swinging Pendulum. The formula $T = 2\pi\sqrt{L/32}$ can be used to find the period T, in seconds, of a pendulum of length L, in feet.

49. A playground swing has a period of 4.4 sec. Find the length of the swing's chain. Use 3.14 for π.

50. The pendulum in Cheri's regulator clock has a period of 2.0 sec. Find the length of the pendulum. Use 3.14 for π.

51. Do you believe that the principle of squaring can be extended to powers other than 2? That is, if $a = b$, does it follow that $a^n = b^n$ for any integer n? Why or why not?

52. Explain in your own words why possible solutions of radical equations must be checked.

Skill Review

Graph.

53. $y = 2x - 1$ [3.6]

54. $2x - 3y = 12$ [3.3]

55. $x = 3$ [3.3]

56. $x = 1 - y$ [3.2]

57. $y < \frac{1}{3}x - 2$ [7.5]

58. $x + y \geq 2$, $x - 2y \leq 2$ [7.6]

Synthesis

59. Mike believes that a radical equation can never have a negative number as a solution. Is he correct? Why or why not?

60. Explain what would have happened in Example 1(a) if we had not isolated the radical before squaring. Could we still have solved the equation? Why or why not?

61. Find a number such that the opposite of three times its square root is -33.

62. Find a number such that 1 less than the square root of twice the number is 7.

Sometimes the principle of squaring must be used more than once in order to solve an equation. Solve Exercises 63–70 by using the principle of squaring as often as necessary. (Hint: Isolate a radical before each use of the principle.)

63. $1 + \sqrt{x} = \sqrt{x + 9}$

64. $5 - \sqrt{x} = \sqrt{x - 5}$

65. $\sqrt{t + 4} = 1 - \sqrt{3t + 1}$

66. $\sqrt{y + 8} - \sqrt{y} = 2$

67. $\sqrt{y + 1} - \sqrt{y - 2} = \sqrt{2y - 5}$

68. $3 + \sqrt{19 - x} = 5 + \sqrt{4 - x}$

69. $2\sqrt{x - 1} - \sqrt{x - 9} = \sqrt{3x - 5}$

70. $x + (2 - x)\sqrt{x} = 0$

71. *Changing Elevations.* A mountain climber pauses to rest and view the horizon. Using the formula $V = 3.5\sqrt{h}$ from Exercises 45 and 46, the climber computes the distance to the horizon and then climbs upward another 100 m. At this higher elevation, the horizon is 20 km farther than before. At what height was the climber when the first computation was made? (*Hint*: Use a system of equations.)

72. Solve $A = \sqrt{1 + \sqrt{a/b}}$ for b.

Graph. Use at least three ordered pairs.

73. $y = \sqrt{x}$

74. $y = \sqrt{x - 4}$

75. $y = \sqrt{x + 2}$

76. $y = \sqrt{x} + 1$

77. *Sound Design.* The following table is one used for time delay for loudspeakers. The time delay, however, can vary in different settings because the speed of sound increases with temperature.

Distance from Sound Source (in feet)	Delay (in seconds)
20	0.0179
40	0.0357
60	0.0536
80	0.0714
100	0.0893

Data: www.gbaudio.co.uk

a) Use the formula given in Exercises 43 and 44 to calculate the time delay for speakers 100 ft from a sound source when the air temperature is 80°F.

b) For what air temperature is the table above accurate?

78. *Safe Driving.* The formula $r = 2\sqrt{5L}$ can be used to find the speed, in miles per hour, that a car was traveling when it leaves skid marks L feet long on a packed gravel road. Police often recommend that drivers allow a space in front of their vehicle of one car length for each 10 miles per hour of speed. Thus a driver traveling 65 mph should leave 6.5 car lengths between his or her vehicle and the car in front of it. Many of today's small cars are approximately 15 ft long. For what speed would the "one car length per 10 mph of speed" rule require a distance equal to that of the skid marks that would be made at that speed?

79. Graph $y = 1 + \sqrt{x}$ and $y = \sqrt{x + 9}$ using the same set of axes. Determine where the graphs intersect in order to estimate a solution of $1 + \sqrt{x} = \sqrt{x + 9}$.

80. Graph $y = x - 7$ and $y = \sqrt{x - 5}$ using the same set of axes. Determine where the graphs intersect in order to estimate a solution of $x - 7 = \sqrt{x - 5}$.

Use a graphing calculator to solve Exercises 81 and 82. Round answers to the nearest hundredth.

81. $\sqrt{x + 3} = 2x - 1$

82. $-\sqrt{x + 3} = 2x - 1$

↪ YOUR TURN ANSWERS: SECTION 8.5
1. 12 **2.** No solution **3.** 1 **4.** About 6700 ft

Quick Quiz: Sections 8.1–8.5

Rationalize each denominator.

1. $\sqrt{\dfrac{2}{x}}$ [8.3]   **2.** $\dfrac{2}{3 + \sqrt{5}}$ [8.4]

3. Find the square roots of 144. [8.1]

4. Simplify: $\sqrt{144}$. [8.1]

5. Solve: $1 + \sqrt{2x + 3} = 6$. [8.5]

Prepare to Move On

Solve. [5.6]

1. $3^2 + 4^2 = x^2$ **2.** $x^2 + (x + 1)^2 = 25$

3. $5^2 + (x - 1)^2 = x^2$

Simplify. [1.8]

4. $|-6 - (-1)|$ **5.** $(2 - 3)^2 + (-2 - 5)^2$

8.6 Applications Using Right Triangles

A. Right Triangles   **B.** Problem Solving   **C.** The Distance Formula

Radicals frequently occur in problem-solving situations in which the Pythagorean theorem is used.

A. Right Triangles

If we know the lengths of two sides of a right triangle, we can use the Pythagorean theorem to find the length of the third side.

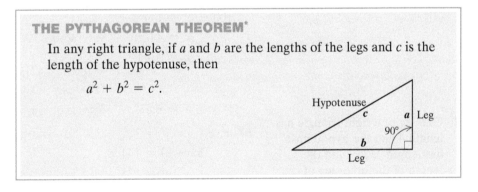

THE PYTHAGOREAN THEOREM*

In any right triangle, if a and b are the lengths of the legs and c is the length of the hypotenuse, then

$$a^2 + b^2 = c^2.$$

*The converse of the Pythagorean theorem also holds. That is, if a, b, and c are the lengths of the sides of a triangle and $a^2 + b^2 = c^2$, then the triangle is a right triangle.

EXAMPLE 1 Find the length of the hypotenuse of the triangle shown. Give an exact answer in radical notation, as well as a decimal approximation to the nearest thousandth.

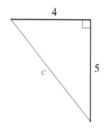

SOLUTION We have

$$a^2 + b^2 = c^2$$
$$4^2 + 5^2 = c^2 \qquad \text{Substituting the lengths of the legs}$$
$$16 + 25 = c^2$$
$$41 = c^2.$$

We now use the fact that if $x^2 = n$, then $x = \sqrt{n}$ or $x = -\sqrt{n}$. In this case, since c is a length, it follows that c is the positive square root of 41:

$$c = \sqrt{41} \qquad \text{This is an exact answer.}$$
$$c \approx 6.403. \qquad \text{Using a calculator for an approximation}$$

1. A right triangle has legs of lengths 3 and 10. Find the length of the hypotenuse of the triangle. Give an exact answer in radical notation as well as a decimal approximation to the nearest thousandth.

YOUR TURN

EXAMPLE 2 Find the length of the indicated leg in each triangle. In each case, give an exact answer as well as a decimal approximation to the nearest thousandth.

a) 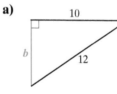   **b)**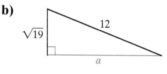

SOLUTION

a) $10^2 + b^2 = 12^2 \qquad \text{Substituting in the Pythagorean equation}$
$100 + b^2 = 144$
$b^2 = 44 \qquad \text{Subtracting 100 from both sides}$
$b = \sqrt{44} \qquad b = \sqrt{44}$ or $b = -\sqrt{44}$. Since the length is positive, only the principal square root is used. The exact answer is $\sqrt{44}$.
$b \approx 6.633 \qquad \text{Approximating } \sqrt{44} \text{ with a calculator}$

b) $a^2 + (\sqrt{19})^2 = 12^2 \qquad \text{Substituting}$
$a^2 + 19 = 144$
$a^2 = 125 \qquad \text{Subtracting 19 from both sides}$
$a = \sqrt{125} \qquad \text{Using only the principal square root. The exact answer is } \sqrt{125}, \text{ or } 5\sqrt{5}.$
$a \approx 11.180 \qquad \text{Using a calculator}$

2. One leg of a right triangle has length 4 and the hypotenuse has length $\sqrt{26}$. Find the length of the other leg of the triangle. Give an exact answer as well as a decimal approximation to the nearest thousandth.

YOUR TURN

B. Problem Solving

The five-step process and the Pythagorean theorem are used in problem solving.

EXAMPLE 3 *Reach of a Ladder.* A ladder is leaning against a house. The bottom of the ladder is 7 ft from the house, and the top of the ladder reaches 25 ft high. How long is the ladder? Give an exact answer and an approximation to the nearest tenth of a foot.

25 ft

7 ft

SOLUTION

1. **Familiarize.** First we make a drawing. In it, there is a right triangle. We label the unknown length l.

2. **Translate.** We use the Pythagorean theorem, substituting 7 for a, 25 for b, and l for c:

$$7^2 + 25^2 = l^2.$$

3. **Carry out.** We solve the equation:

$$7^2 + 25^2 = l^2$$
$$49 + 625 = l^2$$
$$674 = l^2$$
$$\sqrt{674} = l \qquad \text{Using only the principal square root.}$$
$$\phantom{\sqrt{674} = l} \qquad \text{This answer is exact.}$$
$$26.0 \approx l. \qquad \text{Approximating with a calculator}$$

4. **Check.** We check by substituting 7, 25, and $\sqrt{674}$:

$$a^2 + b^2 = c^2$$

$7^2 + 25^2$	$(\sqrt{674})^2$
$49 + 625$	674
$674 \stackrel{?}{=} 674$	TRUE

5. **State.** The ladder is $\sqrt{674}$ ft, or about 26 ft, long.

3. Refer to Example 3. Another ladder is leaning against the house. The bottom of the ladder is 8 ft from the house, and the top of the ladder reaches 22 ft high. How long is the ladder? Give an exact answer and an approximation to the nearest tenth of a foot.

YOUR TURN

EXAMPLE 4 *Emergency Communications.* During disaster relief operations, the Red Cross uses a wire antenna in the shape of an "inverted vee." The length of the wire depends on the frequency being used. For a common frequency, the length of the wire from the top of the pole to the ground should be 42 ft. If a wire antenna is fixed at the top of a 16-ft pole, how far from the base of the pole should the wire be attached to the ground?

Data: www.w0ipl.net

SOLUTION

1. **Familiarize.** We first make a drawing and label the known distances. The wire and the pole form two sides of a right triangle. We let d represent the unknown length, in feet.

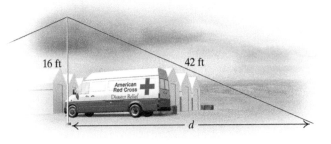
16 ft   42 ft

American
Red Cross
Disaster Relief

d

2. **Translate.** We use the Pythagorean theorem, substituting 16 for a, d for b, and 42 for c:

$$16^2 + d^2 = 42^2.$$

3. **Carry out.** We solve the equation:

$$16^2 + d^2 = 42^2$$
$$256 + d^2 = 1764$$
$$d^2 = 1508$$
$$d = \sqrt{1508} \qquad \text{Using only the principal square root.}$$
The exact answer is $\sqrt{1508}$ ft.
$$d \approx 38.833. \qquad \text{Approximating using a calculator}$$

4. **Check.** To check, we substitute into the Pythagorean equation:

$$a^2 + b^2 = c^2$$

$16^2 + (\sqrt{1508})^2$	42^2
$256 + 1508$	1764
$1764 \stackrel{?}{=} 1764$	TRUE

4. Refer to Example 4. The 42-ft antenna wire is moved from the top of the pole to a tree. If it is attached to the ground 40 ft from the tree, at what height should the antenna be attached to the tree?

5. **State.** For a 42-ft wire antenna, the end of the wire should be attached to the ground about 38.8 ft from the base of the pole.

YOUR TURN

C. The Distance Formula

We can use the Pythagorean theorem to find the distance between two points on a plane.

To find the distance between two points on the number line, we subtract. Depending on the order in which we subtract, the difference may be positive or negative. However, if we take the absolute value of the difference, we always obtain a positive value for the distance.

7 units

$$|4 - (-3)| = |7| = 7;$$
$$|-3 - 4| = |-7| = 7$$

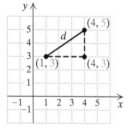

To find the distance between two points on a plane, we use a right triangle and the idea of distance on the number line. Consider the points $(1, 3)$ and $(4, 5)$ and let the distance between the points be d.

The length of the horizontal leg of the triangle is found by subtracting the x-coordinates: $|4 - 1| = 3$. The length of the vertical leg of the triangle is found by subtracting the y-coordinates: $|5 - 3| = 2$. Then, by the Pythagorean theorem, we have

$$d^2 = 2^2 + 3^2 \qquad c^2 = a^2 + b^2$$
$$d^2 = 4 + 9 \qquad \text{Squaring}$$
$$d^2 = 13$$
$$d = \sqrt{13}. \qquad d = -\sqrt{13} \text{ or } d = \sqrt{13}. \text{ We use the principal square root because distance is never negative.}$$

If we follow the same reasoning for two general points on the plane (x_1, y_1) and (x_2, y_2), we can develop a formula for the distance between any two points.

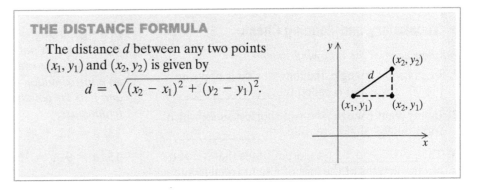

THE DISTANCE FORMULA

The distance d between any two points (x_1, y_1) and (x_2, y_2) is given by

$$d = \sqrt{(x_2 - x_1)^2 + (y_2 - y_1)^2}.$$

Check Your UNDERSTANDING

1. Use the lengths of the sides of the following right triangle to complete the given Pythagorean equation.

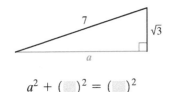

$$a^2 + (\ \ )^2 = (\ \ )^2$$

2. Complete the following formula that gives the distance between the points $(2, 7)$ and $(3, -4)$.

$$d = \sqrt{(\ \ - 2)^2 + (-4 - \ \ )^2}$$

5. Find the distance between $(-2, 5)$ and $(-3, 8)$. Find an exact answer and an approximation to three decimal places.

EXAMPLE 5 Find the distance between $(2, 3)$ and $(8, -1)$. Find an exact answer and an approximation to three decimal places.

SOLUTION Since we will square the differences in the formula, it does not matter which point we regard as (x_1, y_1) and which we regard as (x_2, y_2). We substitute into the distance formula:

$$\begin{aligned}
d &= \sqrt{(x_2 - x_1)^2 + (y_2 - y_1)^2} \\
&= \sqrt{(8 - 2)^2 + (-1 - 3)^2} &&\text{Substituting} \\
&= \sqrt{6^2 + (-4)^2} \\
&= \sqrt{36 + 16} \\
&= \sqrt{52} = 2\sqrt{13} &&\text{This is exact.} \\
&\approx 7.211. &&\text{Using a calculator for an approximation}
\end{aligned}$$

YOUR TURN

ALF
Active Learning Figure

Chapter Resources:
Translating for Success, p. 535;
Decision Making: Connection, p. 536

EXPLORING 🔍 **THE CONCEPT**

Find the distance between each pair of points below using the following steps:

a) Find the horizontal change between the points.
b) Find the vertical change between the points.
c) Use the Pythagorean theorem to find the distance between the points.

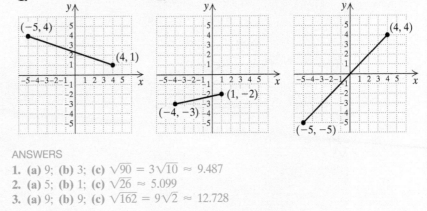

ANSWERS
1. **(a)** 9; **(b)** 3; **(c)** $\sqrt{90} = 3\sqrt{10} \approx 9.487$
2. **(a)** 5; **(b)** 1; **(c)** $\sqrt{26} \approx 5.099$
3. **(a)** 9; **(b)** 9; **(c)** $\sqrt{162} = 9\sqrt{2} \approx 12.728$

8.6 EXERCISE SET

FOR EXTRA HELP MyMathLab®

Vocabulary and Reading Check

Complete each of the following statements.

1. In any right triangle, the longest side is opposite the 90° angle and is called the _____.

2. In any right triangle, the two shortest sides, which form the 90° angle, are called the _____.

3. The _____ theorem states that if a and b are the lengths of the legs of a right triangle and c is the length of the third side, then $a^2 + b^2 = c^2$.

4. The formula $d = \sqrt{(x_2 - x_1)^2 + (y_2 - y_1)^2}$ is used to find the _____ between two points.

A. Right Triangles

Find the length of the third side of each triangle. If an answer is not a whole number, use radical notation to give the exact answer and decimal notation for an approximation to the nearest thousandth.

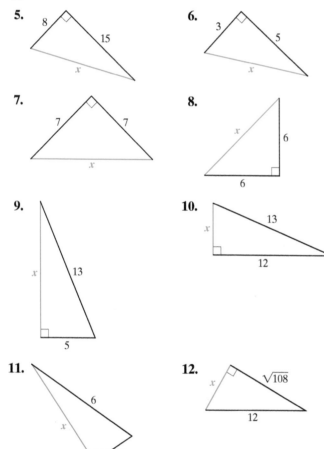

5.

6.

7.

8.

9.

10.

11.

12.

In a right triangle, find the length of the side not given. If an answer is not a whole number, use radical notation to give the exact answer and decimal notation for an approximation to the nearest thousandth. Regard a and b as the lengths of the legs and c as the length of the hypotenuse.

13. $a = 12$, $b = 5$

14. $a = 24$, $b = 10$

15. $a = 9$, $c = 15$

16. $a = 18$, $c = 30$

17. $b = 1$, $c = \sqrt{10}$

18. $b = 1$, $c = \sqrt{2}$

19. $a = 1$, $c = \sqrt{3}$

20. $a = \sqrt{2}$, $b = \sqrt{6}$

21. $c = 10$, $b = 5\sqrt{3}$

22. $a = 5$, $b = 5$

B. Problem Solving

Solve. Don't forget to make drawings. If an answer is not a rational number, use radical notation to give the exact answer and decimal notation for an approximation to the nearest thousandth.

23. *Construction.* Cindi is designing rafters for a house. The rise of the rafter will be 5 ft and the run 12 ft. How long is the rafter length?

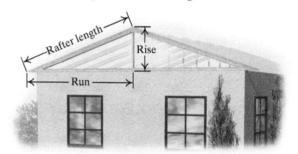

24. *Decorating.* To decorate their boat, Keisha and Casey plan to run a line with flags attached from the top of a 40-ft tall mast to the boat's bow 9 ft from the base of the mast. How long must the line be in order to span that distance?

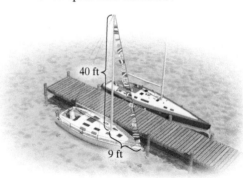

25. *Rollerblading.* Doug and Landon are building a rollerblade jump with a base that is 30 in. long and a ramp that is 33 in. long. How high will the back of the jump be?

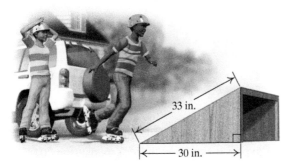

26. *Masonry.* Find the length of a diagonal of a square tile that has sides 4 cm long.

27. *Plumbing.* A new pipe is being cut so that it runs from one corner of a kitchen to the far corner. If the kitchen is 8 ft wide and 12 ft long, how long should the pipe be?

28. *Guy Wires.* How long must a guy wire be in order to reach from the top of a 13-m telephone pole to a point on the ground 9 m from the foot of the pole?

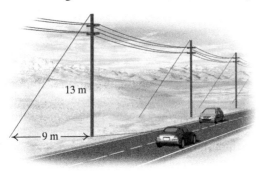

29. *Baseball.* A baseball diamond is a square. The length of each side of the square is 90 ft. How far is it from first base to third base?

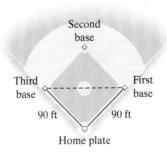

30. *Softball.* A softball diamond is a square. The length of each side of the square is 65 ft. How far is it from home plate to second base?

31. *Lacrosse.* A regulation lacrosse field is 60 yd wide and 110 yd long. Find the length of a diagonal of such a field.

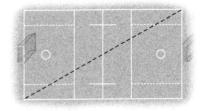

32. *Soccer Fields.* The largest regulation soccer field is 100 yd wide and 130 yd long. Find the length of a diagonal of such a field.

33. *Reptiles.* A snake's cage should be large enough for the snake to stretch out to its full length diagonally. Kristin has a 30-cm by 60-cm rectangular cage. What is the length of the longest snake she should keep in the cage?

34. *Wiring.* JR Electric is installing a security system with a wire that will run diagonally from one corner to the far corner of a 10-ft by 16-ft room. How long will that section of wire need to be?

35. *Home Maintenance.* The hose on a power washer is 32 ft long. If the compressor to which the hose is attached is located at the corner of a 24-ft wide house, how high can the hose reach up the far corner without moving the compressor?

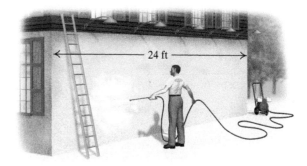

36. *Surveying.* A surveyor has poles located at points *P*, *Q*, and *R*. The distances are as marked in the figure. What is the approximate length of the lake?

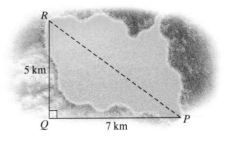

C. The Distance Formula

Find the distance between each pair of points. If an answer is not a whole number, use radical notation to give the exact answer and decimal notation for an approximation to the nearest thousandth.

37. $(2, 3)$ and $(6, 10)$

38. $(1, 0)$ and $(7, 3)$

39. $(0, 3)$ and $(4, 0)$

40. $(-2, -8)$ and $(6, 7)$

41. $(-3, 2)$ and $(-1, 5)$

42. $(2, -7)$ and $(1, -4)$

43. $(-2, 4)$ and $(-8, -4)$

44. $(0, -5)$ and $(-12, 0)$

45. In an *isosceles triangle*, two sides have the same length. Can a right triangle be isosceles? Why or why not?

46. In an *equilateral triangle*, all sides have the same length. Can a right triangle be equilateral? Why or why not?

Skill Review

Perform the indicated operation and, if possible, simplify.

47. $(a^3 - 6a^2 - a) - (a^2 - 3a + 7)$ [4.4]

48. $(t^3 - t + 1) \div (t + 1)$ [4.8]

49. $\dfrac{x^2 - x - 2}{x^3} \div \dfrac{x^2 - 1}{x^2 + x}$ [6.2]

50. $(x^2 + 5x + 6) \cdot \dfrac{x - 3}{x + 3}$ [6.2]

51. $\dfrac{1}{x + 1} - \dfrac{6}{x}$ [6.4]

52. $\dfrac{1}{2x^2 - x - 6} + \dfrac{3}{x^2 - 4}$ [6.4]

Synthesis

53. Should a homeowner use a 28-ft ladder to repair clapboard that is 27 ft above ground level? Why or why not?

54. Can the length of a triangle's hypotenuse ever equal the combined lengths of the two legs? Why or why not?

55. *Holiday Decorations.* From the peak of a 25-ft tall spruce tree in her front yard, Julia plans to run 10 strands of holiday lights to points in her lawn 15 ft from the tree's center. The lights cost $5.99 for 25 ft, $6.99 for 35 ft, and $11.99 for 68 ft. What should Julia purchase in order to make the decorations as economical as possible?

56. *Archaeology.* The base of the Khafre pyramid in Egypt is a square with sides measuring 704 ft. The length of each sloping side of the pyramid is 588 ft. How tall is the pyramid?

57. *Wireless Routers.* Melanie's wireless router has a range of 175 ft. She allows her neighbor Brittany to use her service. If Melanie's router and Brittany's computer are placed at their windows, as shown in the following figure, will Brittany be able to use Melanie's service? Why or why not?

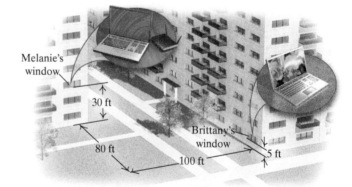

58. *Aviation.* A pilot is instructed to descend from 32,000 ft to 21,000 ft over a horizontal distance of 5 mi. What distance will the plane travel during this descent?

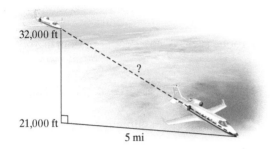

32,000 ft

?

21,000 ft

5 mi

59. The diagonal of a square has a length of $8\sqrt{2}$ ft. Find the length of a side of the square.

60. Find the length of a side of a square that has an area of 7 m².

61. A right triangle has sides with lengths that are consecutive even integers. Find the lengths of the sides.

62. Figure $ABCD$ is a square. Find the length of a diagonal, $\overline{AC}$.

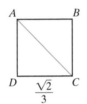

A B

D $\dfrac{\sqrt{2}}{3}$ C

63. Find the length of the diagonal of a cube with sides of length s.

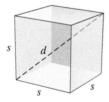

s d

s s

64. The area of square $PQRS$ is 100 ft², and A, B, C, and D are midpoints of the sides on which they lie. Find the area of square $ABCD$.

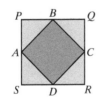

P B Q

A C

S D R

65. Express the height h of an equilateral triangle in terms of the length of a side a.

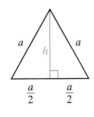

a h a

$\dfrac{a}{2}$ $\dfrac{a}{2}$

66. *Racquetball.* A racquetball court is 20 ft by 20 ft by 40 ft. What is the longest straight-line distance that can be measured in this racquetball court?

67. *Distance Driven.* Two cars leave a service station at the same time. One car travels east at a speed of 50 mph, and the other travels south at a speed of 60 mph. After one half hour, how far apart are they?

68. Solve for x.

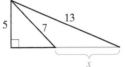

5   7   13

x

69. *Ranching.* If 2 mi of fencing encloses a square plot of land with an area of 160 acres, how large a square, in acres, will 4 mi of fencing enclose?

70. *Construction.* What is the slope of the roof described in Exercise 23?

⤷ **YOUR TURN ANSWERS: SECTION 8.6**
1. $\sqrt{109} \approx 10.440$ **2.** $\sqrt{10} \approx 3.162$
3. $\sqrt{548}$ ft ≈ 23.4 ft
4. About 12.8 ft from the ground
5. $\sqrt{10} \approx 3.162$

Quick Quiz: Sections 8.1–8.6

For Exercises 1–3, assume that all variables represent positive numbers.

1. Simplify: $\sqrt{t^{14}}$.   [8.2]

2. Multiply and simplify: $(\sqrt{10} + 3)^2$.   [8.4]

3. Simplify: $\dfrac{\sqrt{60x^3}}{\sqrt{3x}}$.   [8.3]

4. Solve: $\sqrt{2x + 5} = \sqrt{x - 7}$.   [8.5]

5. Find the length of the hypotenuse of a right triangle with legs of lengths 3 and 5. Give an exact answer and decimal notation for an approximation to the nearest thousandth.   [8.6]

Prepare to Move On

Simplify.   [4.1]

1. $(-2)^5$ **2.** $\left(\dfrac{2}{3}\right)^3$

3. 3^4 **4.** $\left(-\dfrac{1}{4}\right)^2$

5. $(-4)^3$

8.7 Higher Roots and Rational Exponents

A. Higher Roots   **B.** Products and Quotients Involving Higher Roots   **C.** Rational Exponents

In this section, we study *higher* roots, such as cube roots or fourth roots, and exponents that are not integers.

A. Higher Roots

Recall that c is a square root of a if $c^2 = a$. A similar definition exists for *cube roots*.

> **CUBE ROOT**
>
> The number c is the *cube root* of a if $c^3 = a$.

Just as $\sqrt{a}$ represents the square root of a, $\sqrt[3]{a}$ represents the cube root of a. In $\sqrt[3]{a}$, the number 3 is called the **index**, and a is called the **radicand**. The index of a square root $\sqrt{a}$ is 2, although it is rarely written.

EXAMPLE 1   Find the cube root of each number: **(a)** 8; **(b)** -125.

SOLUTION

a) The cube root of 8 is the number whose cube is 8. Since $2^3 = 2 \cdot 2 \cdot 2 = 8$, the cube root of 8 is 2: $\sqrt[3]{8} = 2$.

b) The cube root of -125 is the number whose cube is -125. Since $(-5)^3 = (-5)(-5)(-5) = -125$, the cube root of -125 is -5: $\sqrt[3]{-125} = -5$.

1. Find the cube root of -8.

YOUR TURN

The symbol $\sqrt[n]{a}$ represents the principal nth root of a. For example, $\sqrt[5]{32}$ represents the fifth root of 32. Here the index is 5 and the radicand is 32.

> **nTH ROOT**
>
> For n a natural number greater than 1:
> - The number c is an nth *root* of a if $c^n = a$.
> - If n is odd, there is only one real nth root and $\sqrt[n]{a}$ represents that root.
> - If n is even and a is positive, there are two real nth roots, and $\sqrt[n]{a}$ represents the nonnegative nth root.
> - If n is even and a is negative, $\sqrt[n]{a}$ is not a real number.

EXAMPLE 2   Find each root: **(a)** $\sqrt[4]{16}$; **(b)** $\sqrt[5]{-32}$; **(c)** $\sqrt[4]{-16}$; **(d)** $-\sqrt[3]{64}$.

SOLUTION

a) $\sqrt[4]{16} = 2$ Since $2^4 = 2 \cdot 2 \cdot 2 \cdot 2 = 16$

b) $\sqrt[5]{-32} = -2$ Since $(-2)^5 = (-2)(-2)(-2)(-2)(-2) = -32$

c) $\sqrt[4]{-16}$ is not a real number, because it is an even root of a negative number.

d) $-\sqrt[3]{64} = -(\sqrt[3]{64})$ This is the opposite of $\sqrt[3]{64}$.
        The parentheses are optional.

$\quad\quad\quad = -4 \quad\quad\quad\quad 4^3 = 4 \cdot 4 \cdot 4 = 64$

2. Find $\sqrt[5]{32}$.

YOUR TURN

Some roots arise so frequently that you may want to memorize them.

Square Roots		Cube Roots	Fourth Roots	Fifth Roots
$\sqrt{1} = 1$	$\sqrt{4} = 2$	$\sqrt[3]{1} = 1$	$\sqrt[4]{1} = 1$	$\sqrt[5]{1} = 1$
$\sqrt{9} = 3$	$\sqrt{16} = 4$	$\sqrt[3]{8} = 2$	$\sqrt[4]{16} = 2$	$\sqrt[5]{32} = 2$
$\sqrt{25} = 5$	$\sqrt{36} = 6$	$\sqrt[3]{27} = 3$	$\sqrt[4]{81} = 3$	$\sqrt[5]{243} = 3$
$\sqrt{49} = 7$	$\sqrt{64} = 8$	$\sqrt[3]{64} = 4$	$\sqrt[4]{256} = 4$	
$\sqrt{81} = 9$	$\sqrt{100} = 10$	$\sqrt[3]{125} = 5$	$\sqrt[4]{625} = 5$	
$\sqrt{121} = 11$	$\sqrt{144} = 12$	$\sqrt[3]{216} = 6$		

B. Products and Quotients Involving Higher Roots

The rules for working with products and quotients of square roots extend to products and quotients of nth roots.

THE PRODUCT RULE AND THE QUOTIENT RULE

For any real numbers $\sqrt[n]{A}$ and $\sqrt[n]{B}$,

$$\sqrt[n]{AB} = \sqrt[n]{A}\sqrt[n]{B} \quad \text{and} \quad \sqrt[n]{\frac{A}{B}} = \frac{\sqrt[n]{A}}{\sqrt[n]{B}} \quad (B \neq 0).$$

EXAMPLE 3 Simplify: **(a)** $\sqrt[3]{40}$; **(b)** $\sqrt[3]{\dfrac{125}{27}}$; **(c)** $\sqrt[4]{1250}$; **(d)** $\sqrt[5]{\dfrac{2}{243}}$.

SOLUTION

a) $\sqrt[3]{40} = \sqrt[3]{8 \cdot 5}$ Note that $40 = 2 \cdot 2 \cdot 2 \cdot 5 = 8 \cdot 5$ and 8 is a perfect cube.

$= \sqrt[3]{8} \cdot \sqrt[3]{5}$

$= 2\sqrt[3]{5}$

b) $\sqrt[3]{\dfrac{125}{27}} = \dfrac{\sqrt[3]{125}}{\sqrt[3]{27}}$ $125 = 5 \cdot 5 \cdot 5$ and $27 = 3 \cdot 3 \cdot 3$, so 125 and 27 are perfect cubes.

$= \dfrac{5}{3}$

c) $\sqrt[4]{1250} = \sqrt[4]{625 \cdot 2}$ Note that $1250 = 2 \cdot 5 \cdot 5 \cdot 5 \cdot 5 = 2 \cdot 625$ and 625 is a perfect fourth power.

$= \sqrt[4]{625} \cdot \sqrt[4]{2}$

$= 5\sqrt[4]{2}$

d) $\sqrt[5]{\dfrac{2}{243}} = \dfrac{\sqrt[5]{2}}{\sqrt[5]{243}}$ $243 = 3 \cdot 3 \cdot 3 \cdot 3 \cdot 3$, so 243 is a perfect fifth power.

$= \dfrac{\sqrt[5]{2}}{3}$

Student Notes

Note that

$$2^3\sqrt{5} \quad \text{and} \quad 2\sqrt[3]{5}$$

represent different numbers. Be very careful to place an exponent and an index correctly.

3. Simplify: $\sqrt[4]{\dfrac{81}{16}}$.

YOUR TURN

C. Rational Exponents

Recall that rational numbers can be written as ratios of integers, such as $\frac{1}{3}$, $\frac{5}{2}$, and $\frac{-3}{4}$. These numbers can also be used as exponents. Expressions like $8^{1/3}$, $4^{5/2}$, and $81^{-3/4}$ are defined in such a way that the laws of exponents still hold. For example, if the product rule, $a^m \cdot a^n = a^{m+n}$, is to hold, then

$$a^{1/2} \cdot a^{1/2} = a^{1/2+1/2}$$
$$= a^1 = a.$$

This says that $a^{1/2}$ times itself is a, which means that $a^{1/2}$ is a square root of a. This idea is generalized in the following definition.

THE EXPONENT 1/n

$a^{1/n}$ means $\sqrt[n]{a}$.

$a^{1/2}$ is written $\sqrt{a}$. The index 2 is understood.

If a is negative, then $a^{1/n}$ is a real number only when n is odd.

Thus, $a^{1/2}$ means $\sqrt{a}$, $a^{1/3}$ means $\sqrt[3]{a}$, and so on.

EXAMPLE 4   Simplify: **(a)** $8^{1/3}$; **(b)** $100^{1/2}$; **(c)** $81^{1/4}$; **(d)** $(-243)^{1/5}$.

SOLUTION

a) $8^{1/3} = \sqrt[3]{8} = 2$ Rewriting in radical notation and simplifying

b) $100^{1/2} = \sqrt{100} = 10$ *Check:* $10^2 = 100$

c) $81^{1/4} = \sqrt[4]{81} = 3$ *Check:* $3^4 = 81$

d) $(-243)^{1/5} = \sqrt[5]{-243} = -3$ *Check:* $(-3)^5 = -243$

4. Simplify: $49^{1/2}$.

YOUR TURN

In order to continue multiplying exponents when raising a power to a power, we must have $a^{2/3} = (a^{1/3})^2$ and $a^{2/3} = (a^2)^{1/3}$. This suggests both that $a^{2/3} = (\sqrt[3]{a})^2$ and that $a^{2/3} = \sqrt[3]{a^2}$.

POSITIVE RATIONAL EXPONENTS

For any natural numbers m and n ($n \neq 1$) and any real number a for which $\sqrt[n]{a}$ exists,

$$a^{m/n} \text{ means } (\sqrt[n]{a})^m, \quad \text{or equivalently,} \quad a^{m/n} \text{ means } \sqrt[n]{a^m}.$$

Student Notes

If you are unsure of the rules for when to add or multiply exponents, this would be a good time to review them.

In most cases, it is easiest to simplify using $(\sqrt[n]{a})^m$, because smaller numbers are involved.

EXAMPLE 5   Simplify: **(a)** $27^{2/3}$; **(b)** $8^{5/3}$; **(c)** $81^{3/4}$.

SOLUTION

a) $27^{2/3} = (27^{1/3})^2 = (\sqrt[3]{27})^2 = 3^2 = 9$ Using $a^{m/n} = (\sqrt[n]{a})^m$

b) $8^{5/3} = (8^{1/3})^5 = (\sqrt[3]{8})^5 = 2^5 = 32$

c) $81^{3/4} = (81^{1/4})^3 = (\sqrt[4]{81})^3 = 3^3 = 27$

5. Simplify: $8^{2/3}$.

YOUR TURN

Write an equivalent expression in radical notation. Do not simplify.

1. $64^{1/3}$
2. $64^{1/2}$
3. $16^{3/2}$
4. $16^{3/4}$
5. $100^{-1/2}$

6. Simplify: $81^{-3/4}$.

Negative rational exponents are defined in much the same way that negative integer exponents are.

> **NEGATIVE RATIONAL EXPONENTS**
>
> For any rational number m/n and any nonzero real number a for which $a^{m/n}$ exists,
>
> $$a^{-m/n} = \frac{1}{a^{m/n}}.$$

EXAMPLE 6 Simplify: **(a)** $16^{-1/2}$; **(b)** $8^{-1/3}$; **(c)** $32^{-2/5}$; **(d)** $64^{-3/2}$.

SOLUTION

a) $16^{-1/2} = \dfrac{1}{16^{1/2}} = \dfrac{1}{\sqrt{16}} = \dfrac{1}{4}$

b) $8^{-1/3} = \dfrac{1}{8^{1/3}} = \dfrac{1}{\sqrt[3]{8}} = \dfrac{1}{2}$

c) $32^{-2/5} = \dfrac{1}{32^{2/5}} = \dfrac{1}{(32^{1/5})^2} = \dfrac{1}{(\sqrt[5]{32})^2} = \dfrac{1}{2^2} = \dfrac{1}{4}$

d) $64^{-3/2} = \dfrac{1}{64^{3/2}} = \dfrac{1}{(\sqrt{64})^3} = \dfrac{1}{8^3} = \dfrac{1}{512}$

⟲ YOUR TURN

> **CAUTION!** A negative exponent does not indicate that the expression in which it appears is negative.

A calculator with a key for finding powers can be used to approximate numbers like $\sqrt[5]{8}$. Generally such keys are labeled $\boxed{x^y}$, $\boxed{a^x}$, or ⬤.

We can approximate $\sqrt[5]{8}$ by entering the radicand, 8, pressing the power key, entering the exponent, 0.2 or 1/5, and pressing $\boxed{=}$ or $\boxed{\text{ENTER}}$ to get $\sqrt[5]{8} \approx 1.515716567$. Note that it may be necessary to write 1/5 as (1/5). Consult an owner's manual or your instructor if your calculator works differently.

8.7   EXERCISE SET

FOR EXTRA HELP   MyMathLab®

↳ Vocabulary and Reading Check

Classify each of the following statements as either true or false.

1. $4^{1/2}$ is the same as $\sqrt{4}$, or 2.
2. $a^{m/n}$ is the same as $\sqrt[n]{a^m}$.
3. $a^{m/n}$ is the same as $(\sqrt[n]{a})^m$.
4. $5^{-1/2}$ is a negative number.
5. 3 is the cube root of 9.
6. $\sqrt[4]{-625}$ is not a real number.

A. Higher Roots

Simplify. If an expression does not represent a real number, state this.

7. $\sqrt[3]{-8}$ 8. $\sqrt[3]{-64}$ 9. $\sqrt[3]{-1000}$

10. $\sqrt[3]{-27}$ 11. $\sqrt[3]{125}$ 12. $\sqrt[3]{8}$

13. $-\sqrt[3]{216}$ 14. $-\sqrt[3]{-343}$ 15. $\sqrt[4]{625}$

16. $\sqrt[4]{81}$ 17. $\sqrt[5]{0}$ 18. $\sqrt[5]{1}$

19. $\sqrt[5]{-1}$ 20. $\sqrt[5]{-243}$ 21. $\sqrt[4]{-81}$

22. $\sqrt[4]{-1}$ *Aha!* 23. $\sqrt[4]{10,000}$ 24. $\sqrt[5]{100,000}$

Aha! **25.** $\sqrt[3]{6^3}$ **26.** $\sqrt[4]{2^4}$ **27.** $\sqrt[8]{1}$

28. $\sqrt[6]{64}$ **29.** $\sqrt[7]{a^7}$ **30.** $\sqrt[5]{t^5}$

B. Products and Quotients Involving Higher Roots

Simplify.

31. $\sqrt[3]{54}$ **32.** $\sqrt[3]{24}$ **33.** $\sqrt[4]{48}$

34. $\sqrt[5]{160}$ **35.** $\sqrt[3]{\dfrac{64}{125}}$ **36.** $\sqrt[3]{\dfrac{125}{27}}$

37. $\sqrt[5]{\dfrac{32}{243}}$ **38.** $\sqrt[4]{\dfrac{625}{256}}$ **39.** $\sqrt[3]{\dfrac{7}{8}}$

40. $\sqrt[5]{\dfrac{15}{32}}$ **41.** $\sqrt[4]{\dfrac{14}{81}}$ **42.** $\sqrt[3]{\dfrac{10}{27}}$

C. Rational Exponents

Simplify.

43. $49^{1/2}$ **44.** $36^{1/2}$ **45.** $1000^{1/3}$

46. $27^{1/3}$ **47.** $16^{1/4}$ **48.** $32^{1/5}$

49. $16^{3/4}$ **50.** $8^{4/3}$ **51.** $16^{3/2}$

52. $4^{3/2}$ **53.** $64^{2/3}$ **54.** $32^{2/5}$

55. $1000^{4/3}$ **56.** $16^{5/4}$ **57.** $100^{5/2}$

58. $36^{3/2}$ **59.** $9^{-1/2}$ **60.** $32^{-1/5}$

61. $256^{-1/4}$ **62.** $81^{-1/2}$ **63.** $16^{-3/4}$

64. $81^{-3/4}$ **65.** $81^{-5/4}$ **66.** $32^{-2/5}$

67. $125^{-2/3}$ **68.** $8^{-4/3}$

69. Expressions of the form $a^{m/n}$ can be rewritten as $(\sqrt[n]{a})^m$ or $\sqrt[n]{a^m}$. Which radical expression would you use when simplifying $25^{3/2}$ and why?

70. Explain in your own words why $\sqrt[n]{a}$ is negative when n is odd and a is negative.

Skill Review

71. Solve $2c = \dfrac{a+b}{5}$ for a. [2.3]

72. Determine the degree of the polynomial
$2x^3 - 6x^2 + x^5 - 7$. [4.3]

73. Find the absolute value: $|-0.05|$. [1.4]

74. Write without negative exponents: 6^{-1}. [4.2]

75. Write an inequality with the same meaning as $-6 \geq t$. [1.4]

76. Convert to decimal notation: 4×10^{-3}. [4.2]

Synthesis

77. If $a > b$, does it follow that $a^{1/n} > b^{1/n}$? Why or why not?

78. Under what condition(s) will $a^{-3/5}$ be negative?

▦ *Using a calculator, approximate each of the following to three decimal places.*

79. $8^{4/5}$ **80.** $12^{5/2}$

Simplify.

81. $a^{1/4}a^{3/2}$ **82.** $(x^{2/3})^{7/3}$

83. $m^{-2/3}m^{1/4}m^{3/2}$ **84.** $\dfrac{p^{5/6}}{p^{2/3}}$

Graph.

85. $y = \sqrt[3]{x}$ **86.** $y = \sqrt[4]{x}$

87. *Herpetology.* The daily number of calories c needed by a reptile of weight w pounds can be approximated by $c = 10w^{3/4}$. Find the daily calorie requirement of a green iguana weighing 16 lb.

Data: www.anapsid.org

88. Use a graphing calculator to draw the graphs of $y_1 = x^{2/3}, y_2 = x^1, y_3 = x^{5/4}$, and $y_4 = x^{3/2}$. Use the window $[-1, 17, -1, 32]$ and the SIMULTANEOUS mode. Then determine which curve corresponds to each equation.

↪ YOUR TURN ANSWERS: SECTION 8.7

1. -2 **2.** 2 **3.** $\frac{3}{2}$ **4.** 7 **5.** 4 **6.** $\frac{1}{27}$

Quick Quiz: Sections 8.1–8.7

1. Identify the radicand in $2\sqrt{3x}$. [8.1]

2. Determine whether $\sqrt{12}$ is rational or irrational. [8.1]

3. Find the distance between the points $(1, -3)$ and $(9, 5)$. [8.6]

4. Solve: $\sqrt{2x - 3} = x - 1$. [8.5]

5. Simplify: $100^{1/2}$. [8.7]

Prepare to Move On

Solve. [5.6]

1. $x^2 + 7x + 12 = 0$ **2.** $x^2 - 6x - 7 = 0$

3. $16t^2 - 9 = 0$ **4.** $6n^2 - 5n = 0$

5. $3x^2 - x - 10 = 0$

1. *Coin Mixture.* A collection of nickels and quarters is worth \$9.35. There are 59 coins in all. How many of each type of coin are there?

2. *Diagonal of a Square.* Find the length of a diagonal of a square mural whose sides are 8 ft long.

3. *Shoveling Time.* It takes Ian 55 min to shovel 4 in. of snow from his driveway. It takes Eric 75 min to do the same job. How long would it take if they worked together?

4. *Angles of a Triangle.* The second angle of a triangle is three times as large as the first. The third is 17° less than the sum of the other angles. Find the measures of the angles.

5. *Perimeter.* The perimeter of a rectangular loft is 568 ft. The length is 26 ft greater than the width. Find the length and the width.

Translating for Success

Use after Section 8.6.

Match each application with the most appropriate translation A–O below.

A. $5x + 25y = 9.35,$
  $x + \quad y = 59$

B. $4^2 + x^2 = 8^2$

C. $x(x + 26) = 568$

D. $8 = x \cdot 24$

E. $\dfrac{75}{x} = \dfrac{105}{x + 5}$

F. $\dfrac{75}{x} = \dfrac{55}{x + 5}$

G. $2x + 2(x + 26) = 568$

H. $x + 3x + (x + 3x - 17) = 180$

I. $x + 3x + (3x - 17) = 180$

J. $0.05x + 0.25y = 9.35,$
  $x + \quad y = 59$

K. $8^2 + 8^2 = x^2$

L. $x^2 + (x + 26)^2 = 568$

M. $\dfrac{1}{4} + \dfrac{1}{x} = \dfrac{1}{55}$

N. $\dfrac{1}{55} + \dfrac{1}{75} = \dfrac{1}{x}$

O. $x + 5\% \cdot x = 8568$

Answers on page A-45

An additional, animated version of this activity appears in MyMathLab. *To use MyMathLab, you need a course ID and a student access code. Contact your instructor for more information.*

6. *Bicycle Travel.* One biker travels 75 km in the same time that a biker traveling 5 km/h faster travels 105 km. Find the speed of each biker.

7. *Money Borrowed.* Emma borrows some money at 5% simple interest. After 1 year, \$8568 pays off her loan. How much did she originally borrow?

8. *TV Time.* The average amount of time per day that TV sets in the United States are turned on is 8 hr. What percent of the time are our TV sets on?

Data: Nielsen Media Research

9. *Obstacle Course.* As part of an obstacle course, an 8-ft plank is leaning against a tree. The bottom of the plank is 4 ft from the tree. How high is the top of the plank?

10. *Lengths of a Rectangle.* The area of a rectangle is 568 ft². The length is 26 ft greater than the width. Find the length and the width.

Collaborative Activity   *Lengths and Cycles of a Pendulum*

Focus: Square roots and modeling
Use after: Section 8.1
Time: 25–35 minutes
Group size: 3
Materials: Rulers, watch (to measure seconds), pendulums, calculators

A pendulum is simply a string, a rope, or a chain with a weight of some sort attached at one end. When the unweighted end is held, a pendulum can swing freely from side to side.
   In this activity, each group will develop a mathematical model (formula) that relates a pendulum's length L to the time T that it takes for one complete swing back and forth (one "cycle").

Activity

1. The group should design a pendulum with a length that can be adjusted from 1 ft to 4 ft. One group member should hold the pendulum so that its length is 1 ft. A second group member should lift the weight to one side, keeping the string straight, and then release (do not throw) the weight. The third group member should find the average time, in seconds, for one swing (cycle) back and forth, by timing *five* cycles and dividing by five. Repeat this for each pendulum length listed, so that the following table is completed.

L (in feet)	1	1.5	2	2.5	3	3.5
T (in seconds)						

2. Examine the table your group has created. Can you find one number, a, such that $T \approx aL$ for all pairs of values on the chart?

3. To see if a better model can be found, add a third row to the chart and fill in $\sqrt{L}$ for each value of L listed. Can you find one number, b, such that $T \approx b\sqrt{L}$? Does this appear to be a more accurate model than $T = aL$?

4. Use the model from your answer to part (3) to predict T when L is 4 ft. Then check your prediction by measuring T as you did in part (1) above.

5. A formula relating T to L is
$$T = 2\pi/\sqrt{32} \cdot \sqrt{L}.$$
How does your value of b from part (3) compare with $2\pi/\sqrt{32}$?

Decision Making ℰ Connection   *(Use after Section 8.6.)*

1. *TV Sets.* The size of a TV set is usually given as the diagonal length across the viewing area. An optimal size s (in inches) of a TV depends on the viewing distance d from the TV (in feet) and the source of the TV signal. For cable TV or a TV with an antenna, $s = 4d$. For a DVD or satellite TV, $s = 6d$. For HDTV, $s = 7d$. Shawn has satellite TV and plans to place a TV 9 ft away from his sofa. He is looking at two TVs with screen dimensions as shown below. Which is the better TV for him to purchase?

First TV: 48 in. wide, 27 in. high

Second TV: 40 in. wide, 30 in. high

Data: "TV Buying Guide," by Robert Valdes, on howstuffworks.com

2. *Construction.* The converse of the Pythagorean theorem is also true: If a, b, and c are the lengths of the sides of a triangle and $a^2 + b^2 = c^2$, then the triangle is a right triangle. Each such set of three numbers is called a *Pythagorean triple*. Pythagorean triples provide a handy way of deciding whether an angle is a right angle.
   Gina is building a deck and needs to position an 8-ft piece of lumber so that it forms a right angle with the wall of the house.

a) Find a Pythagorean triple that contains the number 8.

b) Describe how Gina can use a tape measure to ensure that the angle formed is a right angle.

3. *Research.* On the basis of the information in Exercise 1 and your own TV viewing habits, determine whether the TV you most often watch is of optimal size.

4. *Research.* Discover the manner in which a rope with knots was used by ancient Egyptians as a construction tool. Explain how they could use this rope to construct a right angle.

Study Summary

KEY TERMS AND CONCEPTS	EXAMPLES	PRACTICE EXERCISES

SECTION 8.1: *Introduction to Square Roots and Radical Expressions*

The following is an example of a **radical expression**:

$\sqrt{7}$ ←— Radical symbol
←— Radicand

If $c^2 = a$, then c is a **square root** of a. Every positive number has two square roots. The notation $\sqrt{a}$ indicates the **principal**, or nonnegative, square root of a. If a is not a perfect square, then $\sqrt{a}$ is irrational.

The square roots of 25 are -5 and 5.

$$\sqrt{25} = 5$$

$\sqrt{25}$ is rational.

$\sqrt{10}$ is irrational.

1. Find the square roots of 121.

2. Simplify: $\sqrt{9}$.

3. Determine whether $\sqrt{11}$ is rational or irrational.

For any real number A,
$$\sqrt{A^2} = |A|.$$
If we assume that $A \geq 0$, then
$$\sqrt{A^2} = A.$$

Assume that x is any real number.
$$\sqrt{(3x)^2} = |3x| = 3|x|$$
Assume that x is a nonnegative number.
$$\sqrt{(3x)^2} = 3x$$

4. Simplify $\sqrt{25n^2}$. Assume that n is nonnegative.

SECTION 8.2: *Multiplying and Simplifying Radical Expressions*

The Product Rule for Square Roots
For any real numbers $\sqrt{A}$ and $\sqrt{B}$,
$$\sqrt{A} \cdot \sqrt{B} = \sqrt{A \cdot B}.$$

$$\sqrt{2x} \cdot \sqrt{3y} = \sqrt{6xy}$$

5. Multiply: $\sqrt{5n}\sqrt{6y}$.

We can use the product rule in reverse to factor and simplify radical expressions:
$$\sqrt{A \cdot B} = \sqrt{A} \cdot \sqrt{B}.$$

$$\sqrt{75} = \sqrt{25 \cdot 3} = \sqrt{25} \cdot \sqrt{3} = 5\sqrt{3}$$

6. Simplify by factoring: $\sqrt{68}$.

Every even power is a perfect square.

$$\sqrt{x^{12}} = x^6$$

7. Simplify: $\sqrt{m^{16}}$.

After multiplying, we should simplify if possible.

$$\sqrt{6x} \cdot \sqrt{15xy} = \sqrt{6 \cdot 15 \cdot x \cdot xy}$$
$$= \sqrt{2 \cdot 3 \cdot 3 \cdot 5 \cdot x \cdot x \cdot y}$$
$$= \sqrt{3^2} \cdot \sqrt{x^2} \cdot \sqrt{2 \cdot 5 \cdot y}$$
$$= 3x\sqrt{10y} \quad \text{(We assume } x \geq 0.)$$

8. Multiply and, if possible, simplify: $\sqrt{10a}\sqrt{30a^3b}$.

SECTION 8.3: *Quotients Involving Square Roots*

The Quotient Rule for Square Roots
For any real numbers $\sqrt{A}$ and $\sqrt{B}$ with $B \neq 0$,
$$\frac{\sqrt{A}}{\sqrt{B}} = \sqrt{\frac{A}{B}}.$$

$$\frac{\sqrt{18a^9}}{\sqrt{2a^3}} = \sqrt{\frac{18a^9}{2a^3}} = \sqrt{9a^6} = 3a^3 \qquad \text{(We assume } a > 0.)$$

$$\sqrt{\frac{81x^2}{100}} = \frac{\sqrt{81x^2}}{\sqrt{100}} = \frac{9x}{10} \qquad \text{(We assume } x \geq 0.)$$

9. Divide and simplify:
$$\frac{\sqrt{35x}}{\sqrt{5x^3}}.$$

10. Simplify: $\sqrt{\dfrac{x^{30}}{100}}.$

We can **rationalize a denominator** by multiplying by 1.

$$\frac{\sqrt{3}}{\sqrt{5}} = \frac{\sqrt{3}}{\sqrt{5}} \cdot \frac{\sqrt{5}}{\sqrt{5}} = \frac{\sqrt{15}}{5}$$

11. Rationalize the denominator:
$$\sqrt{\frac{x}{3}}.$$

SECTION 8.4: *Radical Expressions with Several Terms*

Like radicals have the same radical factor and can be combined.

$$2\sqrt{3} + 5\sqrt{3} = 7\sqrt{3}$$

12. Combine like radical terms: $\sqrt{2} - 6\sqrt{2}.$

Radical expressions are multiplied in much the same way that polynomials are multiplied.

$$(1 + 5\sqrt{2})(4 - \sqrt{6})$$
$$= 1 \cdot 4 - 1 \cdot \sqrt{6} + 4 \cdot 5\sqrt{2} - 5\sqrt{2} \cdot \sqrt{6}$$
$$= 4 - \sqrt{6} + 20\sqrt{2} - 5\sqrt{2 \cdot 2 \cdot 3}$$
$$= 4 - \sqrt{6} + 20\sqrt{2} - 10\sqrt{3}$$

13. Multiply and simplify:
$$(2 - \sqrt{6})(1 + \sqrt{3}).$$

To rationalize a denominator containing two terms, we use the **conjugate** of the denominator to write a form of 1.

$$\frac{2}{1 - \sqrt{3}} = \frac{2}{1 - \sqrt{3}} \cdot \frac{1 + \sqrt{3}}{1 + \sqrt{3}} = \frac{2(1 + \sqrt{3})}{1 - 3}$$
$$= \frac{2(1 + \sqrt{3})}{-2} = -1 - \sqrt{3}$$

14. Rationalize the denominator:
$$\frac{5}{1 + \sqrt{2}}.$$

SECTION 8.5: *Radical Equations*

The Principle of Squaring
If $a = b$, then $a^2 = b^2$.
Solutions found using the principle of squaring must be checked in the original equation.

Solve: $\sqrt{2x - 3} = 5.$
$$(\sqrt{2x - 3})^2 = 5^2$$
$$2x - 3 = 25$$
$$2x = 28$$
$$x = 14$$

Check: $\dfrac{\sqrt{2x - 3} = 5}{\sqrt{2 \cdot 14 - 3} \mid 5}$
$$\sqrt{25} \mid$$
$$5 \overset{?}{=} 5 \quad \text{TRUE}$$
The solution is 14.

15. Solve:
$$\sqrt{2x + 5} - 3 = 7.$$

SECTION 8.6: *Applications Using Right Triangles*

The Pythagorean Theorem
In any right triangle, the sum of the squares of the lengths of the **legs** is the square of the **hypotenuse**.
$$a^2 + b^2 = c^2$$

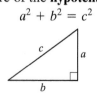

Find the length of the hypotenuse. Give an exact answer as well as an approximation to the nearest thousandth.
$$a^2 + b^2 = c^2$$
$$4^2 + 7^2 = c^2$$
$$16 + 49 = c^2$$
$$65 = c^2$$
$$\sqrt{65} = c \qquad \text{Exact answer}$$
$$8.062 \approx c \qquad \text{Approximation}$$

16. In a right triangle, if $a = 5$ and $c = 8$, find b. Give an exact answer as well as an approximation to the nearest thousandth.

| **The Distance Formula** | Find the distance d between $(-2, 5)$ and $(-3, -1)$. | **17.** Find the distance between $(5, -4)$ and $(-2, 10)$. |

The distance d between any two points (x_1, y_1) and (x_2, y_2) is given by
$$d = \sqrt{(x_2 - x_1)^2 + (y_2 - y_1)^2}.$$

$$d = \sqrt{(-3 - (-2))^2 + (-1 - 5)^2}$$
$$= \sqrt{(-1)^2 + (-6)^2}$$
$$= \sqrt{37} \quad \text{Exact answer}$$
$$\approx 6.083 \quad \text{Approximation}$$

SECTION 8.7: *Higher Roots and Rational Exponents*

The number c is the **cube root** of a if $c^3 = a$.

The number c is the ***n*th root** of a if $c^n = a$.

The nth root of a is written $\sqrt[n]{a}$, and n is called the **index**.

$\sqrt[3]{125} = 5$
$\sqrt[3]{-125} = -5$
$\sqrt[4]{16} = 2$
$\sqrt[4]{-16}$ is not a real number, since any real number raised to an even power is nonnegative.

Simplify.
18. $\sqrt[4]{81}$
19. $\sqrt[3]{-1000}$

$a^{1/n}$ means $\sqrt[n]{a}$.
$a^{m/n}$ means $(\sqrt[n]{a})^m$ or, equivalently, $\sqrt[n]{a^m}$.
$a^{-m/n} = \dfrac{1}{a^{m/n}}$

$64^{1/2} = \sqrt{64} = 8$
$125^{2/3} = (\sqrt[3]{125})^2 = 5^2 = 25$
$8^{-1/3} = \dfrac{1}{8^{1/3}} = \dfrac{1}{2}$

Simplify.
20. $125^{1/3}$
21. $64^{-1/2}$

Review Exercises: Chapter 8

Concept Reinforcement

Classify each of the following statements as either true or false.

1. The sum of two square roots is the square root of the sum of the radicands. [8.4]

2. The product of two square roots is the square root of the product of the radicands. [8.2]

3. The quotient of two cube roots is the cube root of the quotient of the radicands. [8.7]

4. The principal square root of a number is never negative. [8.1]

5. The cube root of a number is never negative. [8.7]

6. The conjugate of $5 - \sqrt{3}$ is $1/(5 - \sqrt{3})$. [8.4]

7. For *any* triangle, if the lengths of two sides are known, the Pythagorean theorem can be used to determine the length of the third side. [8.6]

8. The rules for adding or multiplying integer exponents also apply to rational exponents. [8.7]

9. The distance formula is an application of the Pythagorean theorem. [8.6]

10. The principle of squaring does not always lead to a solution. [8.5]

Find the square roots of each number. [8.1]
11. 16   **12.** 400

Simplify. [8.1]
13. $\sqrt{144}$ **14.** $-\sqrt{169}$

Identify each radicand. [8.1]
15. $3x\sqrt{5x^3y}$ **16.** $a\sqrt{\dfrac{a}{b}}$

Determine whether each square root is rational or irrational. [8.1]
17. $\sqrt{36}$ **18.** $\sqrt{45}$
19. $\sqrt{99}$ **20.** $\sqrt{25}$

Use a calculator to approximate each square root. Round to three decimal places. [8.1]
21. $\sqrt{5}$ **22.** $\sqrt{90}$

Simplify. Assume that x can be any real number. [8.1]
23. $\sqrt{(5x)^2}$ **24.** $\sqrt{(x+2)^2}$

Assume for Exercises 25–56 that all variables represent nonnegative numbers.

Simplify. [8.1]

25. $\sqrt{p^2}$

26. $\sqrt{(3x)^2}$

27. $\sqrt{49n^2}$

28. $\sqrt{(ac)^2}$

Simplify by factoring. [8.2]

29. $\sqrt{48}$

30. $\sqrt{300t^2}$

31. $\sqrt{32p}$

32. $\sqrt{x^{16}}$

33. $\sqrt{12a^{13}}$

34. $\sqrt{36m^{15}}$

Multiply and, if possible, simplify. [8.2]

35. $\sqrt{5}\,\sqrt{11}$

36. $\sqrt{6}\,\sqrt{10}$

37. $\sqrt{3s}\,\sqrt{7t}$

38. $\sqrt{3a}\,\sqrt{8a}$

39. $\sqrt{5x}\,\sqrt{10xy^2}$

40. $\sqrt{20a^3b}\,\sqrt{5a^2b^2}$

Simplify.

41. $\dfrac{\sqrt{35}}{\sqrt{45}}$ [8.3]

42. $\sqrt{\dfrac{49}{64}}$ [8.3]

43. $\sqrt{\dfrac{20}{45}}$ [8.3]

44. $\sqrt{\dfrac{64t}{t^7}}$ [8.3]

45. $10\sqrt{5} + 3\sqrt{5}$ [8.4]

46. $\sqrt{80} - \sqrt{45}$ [8.4]

47. $2\sqrt{x} - \sqrt{25x}$ [8.4]

48. $(2 + \sqrt{3})^2$ [8.4]

49. $(1 + \sqrt{7})(1 - \sqrt{7})$ [8.4]

50. $(1 + 2\sqrt{7})(3 - \sqrt{7})$ [8.4]

Write an equivalent expression by rationalizing each denominator.

51. $\sqrt{\dfrac{1}{5}}$ [8.3]

52. $\dfrac{\sqrt{5}}{\sqrt{8}}$ [8.3]

53. $\sqrt{\dfrac{7}{y}}$ [8.3]

54. $\dfrac{2}{\sqrt{3}}$ [8.3]

55. $\dfrac{4}{2 + \sqrt{3}}$ [8.4]

56. $\dfrac{1 + \sqrt{5}}{2 - \sqrt{5}}$ [8.4]

Solve. [8.5]

57. $\sqrt{x - 5} = 7$

58. $\sqrt{5x + 3} = \sqrt{2x - 1}$

59. $\sqrt{x + 5} = x - 1$

60. $1 + x = \sqrt{1 + 5x}$

61. *Length of a Rope Swing.* The formula $T = 2\pi\sqrt{L/32}$ can be used to find the period T, in seconds, of a pendulum of length L, in feet. A rope swing hanging from a branch over a river has a period of 6.6 sec. Find the length of the rope. Use 3.14 for π. [8.5]

In a right triangle, find the length of the side not given. If the answer is not a whole number, use radical notation to give an exact answer and decimal notation to give an approximation to the nearest thousandth. Assume that a and b are the lengths of the legs and c is the length of the hypotenuse. [8.6]

62. $a = 15$, $c = 25$

63. $a = 1$, $b = \sqrt{2}$

64. Four telephone poles form a square that is 48 ft on each side, with the intersection of Blakely Rd and Rt 7 in the center. A traffic light is to be hung from a wire running diagonally from two of the poles. How long should the wire be? [8.6]

65. Find the distance between the points $(-1, 5)$ and $(-4, 9)$. [8.6]

Simplify. If an expression does not represent a real number, state this.   [8.7]

66. $\sqrt[5]{-32}$

67. $\sqrt[4]{-625}$

68. $\sqrt[3]{\dfrac{8}{27}}$

69. $\sqrt[3]{1000}$

Simplify.   [8.7]

70. $81^{1/2}$

71. $100^{-1/2}$

72. $25^{3/2}$

73. $32^{-4/5}$

Synthesis

 74. When are absolute-value signs necessary for simplifying radical expressions?   [8.1]

75. Why should you simplify each term in a radical expression before attempting to combine like radical terms?   [8.4]

76. Simplify: $\sqrt{\sqrt{\sqrt{256}}}$.   [8.1]

77. Solve: $\sqrt{x^2} = -10$.   [8.1]

78. Use square roots to factor $x^2 - 5$.   [8.4]

Test: Chapter 8   For step-by-step test solutions, access the Chapter Test Prep Videos in MyMathLab.

1. Find the square roots of 49.

Simplify.

2. $\sqrt{16}$

3. $-\sqrt{25}$

4. Identify the radicand in $3t\sqrt{t^2 + 1}$.

Determine whether each square root is rational or irrational.

5. $\sqrt{44}$

6. $\sqrt{49}$

7. Approximate using a calculator. Round to three decimal places: $\sqrt{2}$.

Simplify. Assume $a, y \geq 0$.

8. $\sqrt{a^2}$

9. $\sqrt{64y^2}$

Simplify by factoring. For Exercises 10–27, assume that all variables represent nonnegative numbers.

10. $\sqrt{60}$

11. $\sqrt{27x^6}$

12. $\sqrt{36t^{11}}$

Perform the indicated operation and, if possible, simplify.

13. $\sqrt{5}\sqrt{6}$

14. $\sqrt{5}\sqrt{15}$

15. $\sqrt{7x}\sqrt{2y}$

16. $\sqrt{2t}\sqrt{8t}$

17. $\sqrt{3ab}\sqrt{6ab^3}$

18. $\dfrac{\sqrt{28}}{\sqrt{63}}$

19. $\dfrac{\sqrt{35x}}{\sqrt{80xy^2}}$

20. $\sqrt{\dfrac{144}{a^2}}$

21. $3\sqrt{18} - 5\sqrt{18}$

22. $\sqrt{27} + 2\sqrt{12}$

23. $(4 - \sqrt{5})^2$

24. $(4 - \sqrt{5})(4 + \sqrt{5})$

Rationalize each denominator to form an equivalent expression.

25. $\sqrt{\dfrac{2}{5}}$

26. $\dfrac{2x}{\sqrt{y}}$

27. $\dfrac{10}{4 - \sqrt{5}}$

28. Find the distance between the points $(1, -3)$ and $(4, -1)$.

Solve.

29. $\sqrt{2x} - 3 = 7$

30. $\sqrt{6x + 13} = x + 3$

31. *Guy Wires.*   One wire steadying the Rossis' wind turbine tower stretches from a point 40 ft high on the tower to a point on the ground 25 ft from the base of the tower. How long is the wire?

32. Valerie calculates that she can see 247.49 km to the horizon from an airplane window. How high is the airplane? Use the formula $V = 3.5\sqrt{h}$, where h is the altitude, in meters, and V is the distance to the horizon, in kilometers.

Simplify. If an expression does not represent a real number, state this.

33. $\sqrt[4]{16}$

34. $-\sqrt[6]{1}$

35. $\sqrt[3]{-64}$

36. $\sqrt[4]{-81}$

37. $9^{1/2}$

38. $27^{-1/3}$

39. $100^{3/2}$

40. $16^{-5/4}$

Synthesis

41. Solve: $\sqrt{1 - x} + 1 = \sqrt{6 - x}$.

42. Simplify: $\sqrt{y^{16n}}$.

Cumulative Review: Chapters 1–8

Simplify.

1. $-\frac{1}{2} - \frac{1}{3}$ [1.6]

2. $24 - 2^3 \div 2 \cdot 4 + 3$ [1.8]

3. $(-2x^5)^3$ [4.1]

4. $\left(\dfrac{a}{b}\right)^{-1}$ [4.2]

5. $\dfrac{6a^{-2}}{8a}$ [4.2]

6. $\dfrac{2x^2 + 3x - 2}{x^2 - 4}$ [6.1]

7. $\sqrt[3]{-27}$ [8.7]

8. $\sqrt{121}$ [8.1]

Perform the indicated operation and, if possible, simplify.

9. $(2x^3 - x - 5) - (5x^3 + 3x^2 - x)$ [4.4]

10. $-5x^2(3x^3 - x^2 + 2x)$ [4.5]

11. $(9x^2 + 1)^2$ [4.6]

12. $(x^3 - x^2 + 2) \div (x - 1)$ [4.8]

13. $\dfrac{x^2 - 1}{x - 2} \cdot \dfrac{6x - 12}{x + 1}$ [6.2]

14. $\dfrac{3}{t - 2} + \dfrac{5}{t}$ [6.4]

15. $\sqrt{10} \cdot \sqrt{15}$ [8.2]

16. $\sqrt{\dfrac{4x^2}{81}}$ (Assume $x \geq 0$.) [8.3]

17. $(3 - \sqrt{7})(3 + \sqrt{7})$ [8.4]

Factor completely.

18. $40x^2 - 90$ [5.4]

19. $t^2 - 11t + 10$ [5.2]

20. $6z^3 - 8z^2 + 10z$ [5.1]

21. $6x^2 - 13x + 2$ [5.3]

22. $5a^2b^2 + 10ab + 5$ [5.4]

23. $16 - t^4$ [5.4]

Graph.

24. $y = 3x - 2$ [3.6]

25. $y = \frac{1}{3}x$ [3.2]

26. $3x - 2y = 12$ [3.3]

27. $y = -3$ [3.3]

28. $x - 2y < 2$ [7.5]

29. $x + y \leq 1$,
$x - y \geq 2$ [7.6]

30. Find the slope of the line containing the points $(2, 5)$ and $(-1, 6)$. [3.5]

31. Find the slope of the line given by $y = 5x - 3$. [3.6]

32. Write the slope–intercept equation of the line with slope $-\frac{1}{2}$ and y-intercept $(0, -1)$. [3.6]

Solve.

33. $3x - 2(5x - 1) = 5 - 4x$ [2.2]

34. $t^2 - t = 6$ [5.6]

35. $5 - 2n \geq 1 - (n - 1)$ [2.6]

36. $\sqrt{2x - 5} = 4$ [8.5]

37. $x + y = 5$,
$x - y = -3$ [7.3]

38. $2x - 5 = y$,
$x - y = 7$ [7.2]

39. $\dfrac{1}{2} - \dfrac{1}{2t} = \dfrac{3}{t}$ [6.6]

40. $\sqrt{x + 2} - 2 = x$ [8.5]

41. $6y^2 + 1 = 5y$ [5.6]

42. $\dfrac{1}{x + 1} - \dfrac{9}{4x} = \dfrac{1}{5}$ [6.6]

43. Solve for x: $z = \dfrac{x + y}{2}$. [2.3]

Solve.

44. *Movie Downloads.* It took Ichiro 48 min to download two movies. The second movie took three times as long to download as the first movie. How long did it take to download each movie? [2.5]

45. *Metallurgy.* Alan stocks two alloys that are different purities of gold. The first is three-fourths pure gold and the second is five-twelfths pure gold. How many ounces of each should be combined in order to obtain a 60-oz mixture that is two-thirds pure gold? [7.4]

46. *Construction.* By checking work records, Molly finds that Brady can roof her house in 20 hr. Devin can roof Molly's house in 16 hr. How long would it take if they worked together? [6.7]

47. *Rectangle Dimensions.* The length of a rectangle is 1 cm less than three times the width. The area of the rectangle is 24 cm². Find the dimensions of the rectangle. [5.7]

Synthesis

48. Solve: $2|n| - 5 = 11$. [2.2]

49. Solve: $(x^2 - 9)(x^2 + x - 20) = 0$. [5.6]

Quadratic Equations

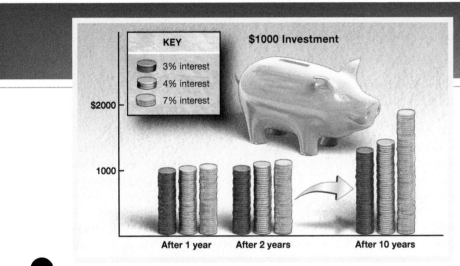

KEY
- 3% interest
- 4% interest
- 7% interest

$1000 Investment

$2000

1000

After 1 year   After 2 years   After 10 years

9.1 Solving Quadratic Equations: The Principle of Square Roots

9.2 Solving Quadratic Equations: Completing the Square

9.3 The Quadratic Formula and Applications

MID-CHAPTER REVIEW

9.4 Formulas

9.5 Complex Numbers as Solutions of Quadratic Equations

CONNECTING THE CONCEPTS

9.6 Graphs of Quadratic Equations

9.7 Functions

CHAPTER RESOURCES

Visualizing for Success
Collaborative Activity
Decision Making: Connection

STUDY SUMMARY

REVIEW EXERCISES
CHAPTER TEST
CUMULATIVE REVIEW/
    FINAL EXAM

A ccording to Bankrate.com, 24% of Americans have no emergency savings at all. Not only does having some savings help avoid borrowing for emergencies, money saved can also grow. The table above shows the value of $1000 after it has been invested for 1 year, 2 years, and 10 years at three different interest rates. In this chapter, we will examine the formula that is used to calculate the values in the table. (*See Exercises 63–66 in Exercise Set 9.3 and Exercise 43 in Exercise Set 9.4.*)

I use math to ensure that I give the best advice to my clients.

Whether figuring taxes or giving advice, Robert Fernandez, a certified public accountant in Fort Worth, Texas, uses math to figure amounts correctly. He also uses math to help his clients understand the financial consequences of decisions.

ALF *Active Learning Figure*   Explore the math using the Active Learning Figure in MyMathLab.

SA *Student Activity*   Do the Student Activity in MyMathLab to see math in action.

Certain quadratic equations can be solved by factoring and using the principle of zero products. In this chapter, we develop methods for solving *any* quadratic equation. These methods are then used in applications and in graphing.

| **9.1** | **Solving Quadratic Equations: The Principle of Square Roots** |

A. The Principle of Square Roots   **B.** Solving Quadratic Equations of the Type $(x + k)^2 = p$

The following are examples of quadratic equations:

$$x^2 - 7x + 9 = 0, \qquad 5t^2 - 4t = 8, \qquad 6y^2 = -9y, \qquad m^2 = 49.$$

One way to solve an equation like $m^2 = 49$ is to subtract 49 from both sides, factor, and then use the principle of zero products:

$$m^2 - 49 = 0$$
$$(m + 7)(m - 7) = 0$$
$$m + 7 = 0 \quad or \quad m - 7 = 0 \qquad \text{Using the principle of zero products:}$$
$$\qquad\qquad\qquad\qquad\qquad\qquad\qquad \text{If } AB = 0, \text{then } A = 0 \text{ or } B = 0.$$
$$m = -7 \quad or \qquad m = 7.$$

Another equation-solving principle, the *principle of square roots*, allows us to solve equations like $m^2 = 49$ without factoring.

A. The Principle of Square Roots

It is possible to solve $m^2 = 49$ just by noting that m must be a square root of 49, namely, -7 or 7.

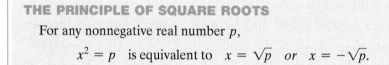

> **THE PRINCIPLE OF SQUARE ROOTS**
>
> For any nonnegative real number p,
>
> $$x^2 = p \quad \text{is equivalent to} \quad x = \sqrt{p} \quad or \quad x = -\sqrt{p}.$$

The notation $x = \pm\sqrt{p}$ is often used to indicate $x = \sqrt{p}$ or $x = -\sqrt{p}$.

EXAMPLE 1   Solve: $x^2 = 16$.

SOLUTION   We use the principle of square roots:

$$x^2 = 16$$
$$x = \sqrt{16} \quad or \quad x = -\sqrt{16} \qquad \text{Using the principle of square roots}$$
$$x = 4 \quad or \quad x = -4. \qquad \text{Simplifying}$$

We check mentally that $4^2 = 16$ and $(-4)^2 = 16$. The solutions are 4 and -4.

1. Solve: $x^2 = 36$.

YOUR TURN

Unlike the principle of zero products, the principle of square roots can be used to solve quadratic equations that have irrational solutions.

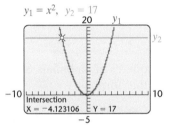
EXAMPLE 2   Solve: **(a)** $x^2 = 17$; **(b)** $5t^2 = 15$; **(c)** $-3x^2 + 7 = 0$.

SOLUTION

a)
$$x^2 = 17$$
$$x = \sqrt{17} \quad or \quad x = -\sqrt{17} \qquad \text{Using the principle of square roots}$$

Check: For $\sqrt{17}$:

$$x^2 = 17$$
$$(\sqrt{17})^2 \;|\; 17$$
$$17 \overset{?}{=} 17 \quad \text{TRUE}$$

For $-\sqrt{17}$:

$$x^2 = 17$$
$$(-\sqrt{17})^2 \;|\; 17$$
$$17 \overset{?}{=} 17 \quad \text{TRUE}$$

The solutions are $\sqrt{17}$ and $-\sqrt{17}$.

b)
$$5t^2 = 15$$
$$t^2 = 3 \qquad \text{Dividing both sides by 5 to isolate } t^2$$
$$t = \sqrt{3} \quad or \quad t = -\sqrt{3} \qquad \text{Using the principle of square roots}$$

We leave the check to the student. The solutions are $\sqrt{3}$ and $-\sqrt{3}$.

c)
$$-3x^2 + 7 = 0$$
$$7 = 3x^2 \qquad \text{Adding } 3x^2 \text{ to both sides}$$
$$\frac{7}{3} = x^2 \qquad \text{Dividing both sides by 3 to isolate } x^2$$
$$x = \sqrt{\frac{7}{3}} \quad or \quad x = -\sqrt{\frac{7}{3}} \qquad \text{Using the principle of square roots}$$

These answers can also be written $\dfrac{\sqrt{21}}{3}$ and $-\dfrac{\sqrt{21}}{3}$ by rationalizing denominators.

Check: For $\sqrt{\frac{7}{3}}$:

$$-3x^2 + 7 = 0$$
$$-3(\sqrt{\tfrac{7}{3}})^2 + 7 \;|\; 0$$
$$-3 \cdot \tfrac{7}{3} + 7$$
$$-7 + 7$$
$$0 \overset{?}{=} 0 \quad \text{TRUE}$$

For $-\sqrt{\frac{7}{3}}$:

$$-3x^2 + 7 = 0$$
$$-3(-\sqrt{\tfrac{7}{3}})^2 + 7 \;|\; 0$$
$$-3 \cdot \tfrac{7}{3} + 7$$
$$-7 + 7$$
$$0 \overset{?}{=} 0 \quad \text{TRUE}$$

The solutions of $-3x^2 + 7 = 0$ are $\sqrt{\frac{7}{3}}$ and $-\sqrt{\frac{7}{3}}$.

YOUR TURN

B. Solving Quadratic Equations of the Type $(x + k)^2 = p$

Equations like $(x - 5)^2 = 9$ or $(t + 2)^2 = 7$ are of the form $(x + k)^2 = p$. The principle of square roots can be used to solve such equations.

EXAMPLE 3   Solve: **(a)** $(x - 5)^2 = 9$; **(b)** $(t + 2)^2 = 7$.

SOLUTION

a)
$$(x - 5)^2 = 9$$
$$x - 5 = 3 \quad or \quad x - 5 = -3 \qquad \text{Using the principle of square roots:}$$

If $A^2 = 9$, then $A = 3$ or $A = -3$. Here $A = x - 5$.

$$x = 8 \quad or \quad x = 2 \qquad \text{Adding 5 to both sides}$$

The solutions are 8 and 2. We leave the check to the student.

b)
$$(t + 2)^2 = 7$$

$$t + 2 = \sqrt{7} \quad \text{or} \quad t + 2 = -\sqrt{7} \qquad \text{Using the principle of square roots}$$

$$t = -2 + \sqrt{7} \quad \text{or} \quad t = -2 - \sqrt{7} \qquad \text{Adding } -2 \text{ to both sides}$$

Check: For $-2 + \sqrt{7}$:   For $-2 - \sqrt{7}$:

$$
\begin{array}{c|c}
(t + 2)^2 = 7 \\
\hline
(-2 + \sqrt{7} + 2)^2 & 7 \\
(\sqrt{7})^2 & \\
7 \overset{?}{=} 7 \quad \text{TRUE}
\end{array}
\qquad
\begin{array}{c|c}
(t + 2)^2 = 7 \\
\hline
(-2 - \sqrt{7} + 2)^2 & 7 \\
(-\sqrt{7})^2 & \\
7 \overset{?}{=} 7 \quad \text{TRUE}
\end{array}
$$

The solutions are $-2 + \sqrt{7}$ and $-2 - \sqrt{7}$, or simply $-2 \pm \sqrt{7}$ (read "-2 plus or minus $\sqrt{7}$").

3. Solve: $(n - 2)^2 = 5$.

⟳ YOUR TURN

In Example 3, the left sides of the equations are squares of binomials. Sometimes factoring can be used to express an equation in that form.

EXAMPLE 4   Solve by factoring and using the principle of square roots.

a) $x^2 + 8x + 16 = 49$ **b)** $x^2 - 6x + 9 = 10$

SOLUTION

a) $x^2 + 8x + 16 = 49$ The left side is a perfect-square trinomial.

$$\qquad (x + 4)^2 = 49 \qquad \text{Factoring}$$

$$\qquad x + 4 = 7 \quad \text{or} \quad x + 4 = -7 \qquad \text{Using the principle of square roots}$$

$$\qquad x = 3 \quad \text{or} \qquad x = -11$$

The solutions are 3 and -11. We leave the check to the student.

Chapter Resource:
Decision Making: Connection,
p. 588

b) $x^2 - 6x + 9 = 10$ The left side is a perfect-square trinomial.

$$\qquad (x - 3)^2 = 10 \qquad \text{Factoring}$$

$$x - 3 = \sqrt{10} \quad \text{or} \quad x - 3 = -\sqrt{10} \qquad \text{Using the principle of square roots}$$

$$x = 3 + \sqrt{10} \quad \text{or} \qquad x = 3 - \sqrt{10}$$

The solutions are $3 + \sqrt{10}$ and $3 - \sqrt{10}$, or simply $3 \pm \sqrt{10}$. We leave the check to the student.

4. Solve by factoring and using the principle of square roots:

$$t^2 + 2t + 1 = 3.$$

⟳ YOUR TURN

9.1   EXERCISE SET

⤵ Vocabulary and Reading Check

Classify each of the following statements as either true or false.

1. To solve $3x^2 - 2 = 5$ using the principle of square roots, we must first isolate x^2.

2. The notation $x = \pm\sqrt{5}$ means $x = -\sqrt{5}$ or $x = \sqrt{5}$.

3. The principle of square roots can be used only to find rational solutions.

4. The principle of square roots can be used to solve equations like $(a + 3)^2 = 8$.

A. The Principle of Square Roots

Solve. Use the principle of square roots.

5. $t^2 = 81$ 6. $n^2 = 64$

7. $x^2 = 1$ 8. $x^2 = 100$

9. $a^2 = 11$ 10. $t^2 = 15$

11. $10x^2 = 40$

12. $5x^2 = 45$

13. $3t^2 = 6$

14. $7t^2 = 21$

15. $4 - 9x^2 = 0$

16. $25 - 4a^2 = 0$

Aha! **17.** $12y^2 + 1 = 1$

18. $4y^2 - 3 = 9$

19. $15x^2 - 25 = 0$

20. $4x^2 - 14 = 0$

B. Solving Quadratic Equations of the Type $(x + k)^2 = p$

Solve.

21. $(x - 1)^2 = 49$

22. $(x - 2)^2 = 25$

23. $(t + 6)^2 = 4$

24. $(t + 9)^2 = 100$

25. $(m + 3)^2 = 6$

26. $(m - 4)^2 = 21$

Aha! **27.** $(a - 7)^2 = 0$

28. $(a + 12)^2 = 81$

29. $(5 - x)^2 = 14$

30. $(7 - x)^2 = 12$

31. $(t + 1)^2 = 1$

32. $(x - 5)^2 = 25$

33. $\left(y - \frac{3}{4}\right)^2 = \frac{17}{16}$

34. $\left(x + \frac{3}{2}\right)^2 = \frac{13}{4}$

35. $x^2 - 10x + 25 = 100$

36. $x^2 - 6x + 9 = 64$

37. $p^2 + 8p + 16 = 1$

38. $y^2 + 14y + 49 = 4$

39. $t^2 - 16t + 64 = 7$

40. $m^2 - 2m + 1 = 5$

41. $x^2 + 12x + 36 = 18$

42. $x^2 + 4x + 4 = 12$

43. Under what conditions is it easier to solve a quadratic equation using the principle of square roots rather than the principle of zero products?

44. Under what conditions is it easier to solve a quadratic equation using the principle of zero products rather than the principle of square roots?

Skill Review

Simplify.

45. $t^{-3} \cdot t^{-7}$ [4.2]

46. $(2m^3)^4$ [4.1]

47. $100^{1/2}$ [8.7]

48. $27^{-1/3}$ [8.7]

49. $\dfrac{68x^8y^3}{6x^6y}$ [4.1]

50. $\dfrac{a^{-7}}{a^{-11}}$ [4.2]

Synthesis

51. Under what conditions does a quadratic equation have only one solution?

52. Is it possible for a quadratic equation with rational coefficients to have $5 + \sqrt{2}$ as a solution, but not $5 - \sqrt{2}$? Why or why not?

Factor the left side of each equation. Then solve.

53. $x^2 - 5x + \frac{25}{4} = \frac{13}{4}$

54. $x^2 + \frac{7}{3}x + \frac{49}{36} = \frac{7}{36}$

55. $t^2 + 3t + \frac{9}{4} = \frac{49}{4}$

56. $m^2 - \frac{3}{2}m + \frac{9}{16} = \frac{17}{16}$

57. $x^2 + 2.5x + 1.5625 = 9.61$

58. $a^2 - 3.8a + 3.61 = 27.04$

Use the graph of
$$y = (x + 3)^2$$
to solve each equation.

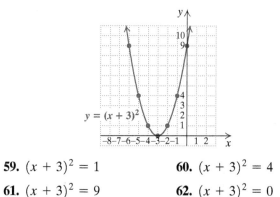

$y = (x + 3)^2$

59. $(x + 3)^2 = 1$

60. $(x + 3)^2 = 4$

61. $(x + 3)^2 = 9$

62. $(x + 3)^2 = 0$

63. *Gravitational Force.* Newton's law of gravitation states that the gravitational force f between objects of mass M and m, at a distance d from each other, is given by
$$f = \frac{kMm}{d^2},$$
where k is a constant. Solve for d.

Prepare to Move On

Multiply. [4.6]

1. $(x - 3)^2$

2. $(x + a)^2$

3. $\left(x + \frac{1}{2}\right)^2$

4. $\left(x - \frac{3}{2}\right)^2$

Factor. [5.4]

5. $x^2 + 2x + 1$

6. $x^2 - 4x + 4$

9.2 Solving Quadratic Equations: Completing the Square

A. Completing the Square   **B.** Solving by Completing the Square

Equations like $x^2 + 8x + 16 = 12$ can be solved using the principle of square roots because the left side is a perfect-square trinomial: $(x + 4)^2 = 12$. We now learn to solve equations like $x^2 - 8x = 2$, in which the left side is not (yet!) a perfect-square trinomial. The new procedure involves *completing the square* and enables us to solve *any* quad-ratic equation.

A. Completing the Square

Recall that

$$(x + 3)^2 = (x + 3)(x + 3)$$
$$= x^2 + 3x + 3x + 9$$
$$= x^2 + 6x + 9 \qquad \text{This is a perfect-square trinomial.}$$

and, in general,

$$(x + a)^2 = x^2 + 2ax + a^2. \qquad \text{This is also a perfect-square trinomial.}$$

For each perfect-square trinomial above, *the last term is the square of half of the coefficient of* x:

$$x^2 + 6x + 9 \quad \Rightarrow \quad \tfrac{1}{2} \cdot 6 = 3 \quad \text{and} \quad 3^2 = 9;$$
$$x^2 + 2ax + a^2 \quad \Rightarrow \quad \tfrac{1}{2} \cdot 2a = a \quad \text{and} \quad a^2 = a^2.$$

Consider $x^2 + 10x = 4$. We would like to add a number to both sides that will make the left side a perfect-square trinomial. Such a number is described above as the square of half of the coefficient of x: $\tfrac{1}{2} \cdot 10 = 5$, and $5^2 = 25$. Thus we add 25 to both sides:

$$x^2 + 10x = 4 \qquad \textit{Think:} \text{ Half of 10 is 5; } 5^2 = 25.$$
$$x^2 + 10x + 25 = 4 + 25 \qquad \text{Adding 25 to both sides}$$
$$(x + 5)^2 = 29. \qquad \text{Factoring the perfect-square trinomial}$$

By adding 25 to $x^2 + 10x$, we have *completed the square*. The resulting equation contains the square of a binomial on one side. Solutions can then be found using the principle of square roots:

$$(x + 5)^2 = 29$$
$$x + 5 = \sqrt{29} \qquad \text{or} \quad x + 5 = -\sqrt{29} \qquad \text{Using the principle of square roots}$$
$$x = -5 + \sqrt{29} \quad \text{or} \qquad x = -5 - \sqrt{29}.$$

The solutions are $-5 \pm \sqrt{29}$.

Technology Connection

One way to check that $-5 - \sqrt{29}$ and $-5 + \sqrt{29}$ are both solutions of $x^2 + 10x = 4$ is to store $-5 - \sqrt{29}$ (or $-5 + \sqrt{29}$) as X in the calculator's memory. We then press (X,T,θ,n) (x²) (+) (1) (0) (X,T,θ,n) (ENTER). The result should be 4.

COMPLETING THE SQUARE

To *complete the square* for an expression like $x^2 + bx$, add half of the coefficient of x, squared. That is, add $(b/2)^2$.

⤷ Check Your
UNDERSTANDING

Fill in the steps to complete the square for each expression.

1. $x^2 + 18x$

Half of 18 is ▢.
$(9)^2$ is ▢.
Add 81: $x^2 + 18x + 81$.
Factor to check:

$$x^2 + 18x + 81 = (\text{▢})^2.$$

2. $x^2 - 20x$

Half of -20 is ▢.
$(-10)^2$ is ▢.
Add 100: $x^2 - 20x + 100$.
Factor to check:

$$x^2 - 20x + 100 = (\text{▢})^2.$$

Following is a visual interpretation of completing the square.

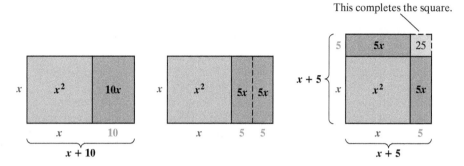

This completes the square.

In each figure above, the sum of the pink area and the purple area is $x^2 + 10x$. However, by splitting the purple area in half, we can "complete" a square by adding the blue area. The blue area is $5 \cdot 5$, or 25 square units.

> **CAUTION!** Completing the square, by itself, does *not* produce an equivalent expression. It will be used with the addition principle to produce equivalent equations.

EXAMPLE 1 For each expression, find the number that completes the square. Check by factoring.

a) $x^2 - 12x$ **b)** $x^2 + 5x$

SOLUTION

Find half of the coefficient of x.

Square that number.

a) To complete the square for $x^2 - 12x$, note that the coefficient of x is -12.

Half of -12 is -6. $\frac{1}{2}(-12) = -6$
$(-6)^2 = 36$.   36 is the number that completes the square.

Thus, $x^2 - 12x + 36$ is the square of $x - 6$. The number 36 completes the square.

Check: $x^2 - 12x + 36 = (x - 6)(x - 6) = (x - 6)^2$.

b) To complete the square for $x^2 + 5x$, we take half of the coefficient of x and square it.

Half of 5 is $\frac{5}{2}$. $\frac{1}{2}(5) = \frac{5}{2}$
$\left(\frac{5}{2}\right)^2 = \frac{25}{4}$. $\left(\frac{5}{2}\right)^2 = \frac{5}{2} \cdot \frac{5}{2} = \frac{25}{4}$

Thus, $x^2 + 5x + \frac{25}{4}$ is the square of $x + \frac{5}{2}$. The number $\frac{25}{4}$ completes the square.

Check: $x^2 + 5x + \frac{25}{4} = \left(x + \frac{5}{2}\right)\left(x + \frac{5}{2}\right) = \left(x + \frac{5}{2}\right)^2$.

1. Find the number that completes the square for $x^2 + 14x$. Check by factoring.

YOUR TURN

> **CAUTION!** In Example 1, we are neither solving an equation nor writing an equivalent expression. Instead, we are learning how to find the number that completes the square. In Examples 2 and 3, we use the number that completes the square, along with the addition principle, to solve equations.

B. Solving by Completing the Square

The concept of completing the square can now be used to solve equations.

EXAMPLE 2 Solve by completing the square.

a) $x^2 + 6x = -8$ **b)** $x^2 - 10x + 14 = 0$

SOLUTION

a) To solve $x^2 + 6x = -8$, we take half of 6 and square it, to get 9. Then we add 9 to both sides of the equation. This makes the left side the square of a binomial:

Complete the square.

$$x^2 + 6x + 9 = -8 + 9 \qquad \text{Adding 9 to both sides to complete the square on the left side}$$

Factor.

$$(x + 3)^2 = 1 \qquad \text{Factoring}$$

$$x + 3 = 1 \quad or \quad x + 3 = -1 \qquad \text{Using the principle of square roots. If } A^2 = 1, \text{ then } A = 1 \text{ or } A = -1.$$

Use the principle of square roots.

$$x = -2 \quad or \qquad x = -4.$$

The solutions are -2 and -4. The check is left to the student.

b) We have

$$x^2 - 10x + 14 = 0$$

$$x^2 - 10x = -14 \qquad \text{Subtracting 14 from both sides to prepare for completing the square}$$

$$x^2 - 10x + 25 = -14 + 25 \qquad \text{Adding 25 to both sides to complete the square: } \left(-\frac{10}{2}\right)^2 = 25$$

$$(x - 5)^2 = 11 \qquad \text{Factoring}$$

$$x - 5 = \sqrt{11} \qquad or \quad x - 5 = -\sqrt{11} \qquad \text{Using the principle of square roots}$$

$$x = 5 + \sqrt{11} \quad or \qquad x = 5 - \sqrt{11}.$$

The solutions are $5 + \sqrt{11}$ and $5 - \sqrt{11}$, or simply $5 \pm \sqrt{11}$. The check is left to the student.

2. Solve by completing the square: $x^2 + 12x = 1$.

YOUR TURN

In order for us to complete the square, the coefficient of x^2 must be 1. When the x^2-coefficient is not 1, we can multiply or divide on both sides to find an equivalent equation with an x^2-coefficient of 1.

EXAMPLE 3 Solve by completing the square.

a) $3x^2 + 24x = 3$ **b)** $2x^2 - 5x + 1 = 0$

SOLUTION

a)
$$3x^2 + 24x = 3$$

$$\left.\begin{array}{l} \frac{1}{3}(3x^2 + 24x) = \frac{1}{3} \cdot 3 \\ x^2 + 8x = 1 \end{array}\right\} \qquad \text{We multiply by } \frac{1}{3} \text{ (or divide by 3) on both sides to ensure an } x^2\text{-coefficient of 1.}$$

$$x^2 + 8x + 16 = 1 + 16 \qquad \text{Adding 16 to both sides to complete the square: } \left(\frac{8}{2}\right)^2 = 16$$

$$(x + 4)^2 = 17 \qquad \text{Factoring}$$

$$x + 4 = \sqrt{17} \qquad or \quad x + 4 = -\sqrt{17} \qquad \text{Using the principle of square roots}$$

$$x = -4 + \sqrt{17} \quad or \qquad x = -4 - \sqrt{17}$$

The solutions are $-4 \pm \sqrt{17}$. The check is left to the student.

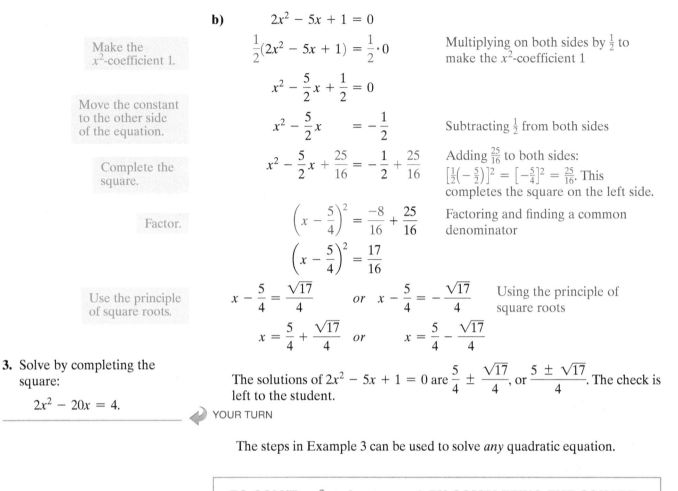

b) $2x^2 - 5x + 1 = 0$

Make the
x^2-coefficient 1.

$\frac{1}{2}(2x^2 - 5x + 1) = \frac{1}{2} \cdot 0$ Multiplying on both sides by $\frac{1}{2}$ to
make the x^2-coefficient 1

$x^2 - \frac{5}{2}x + \frac{1}{2} = 0$

Move the constant
to the other side
of the equation.

$x^2 - \frac{5}{2}x \quad\quad = -\frac{1}{2}$ Subtracting $\frac{1}{2}$ from both sides

Complete the
square.

$x^2 - \frac{5}{2}x + \frac{25}{16} = -\frac{1}{2} + \frac{25}{16}$ Adding $\frac{25}{16}$ to both sides:
$\left[\frac{1}{2}\left(-\frac{5}{2}\right)\right]^2 = \left[-\frac{5}{4}\right]^2 = \frac{25}{16}$. This
completes the square on the left side.

Factor.

$\left(x - \frac{5}{4}\right)^2 = \frac{-8}{16} + \frac{25}{16}$ Factoring and finding a common
denominator

$\left(x - \frac{5}{4}\right)^2 = \frac{17}{16}$

Use the principle
of square roots.

$x - \frac{5}{4} = \frac{\sqrt{17}}{4}$ *or* $x - \frac{5}{4} = -\frac{\sqrt{17}}{4}$ Using the principle of
square roots

$x = \frac{5}{4} + \frac{\sqrt{17}}{4}$ *or* $x = \frac{5}{4} - \frac{\sqrt{17}}{4}$

3. Solve by completing the
square:

$2x^2 - 20x = 4.$

The solutions of $2x^2 - 5x + 1 = 0$ are $\frac{5}{4} \pm \frac{\sqrt{17}}{4}$, or $\frac{5 \pm \sqrt{17}}{4}$. The check is
left to the student.

YOUR TURN

The steps in Example 3 can be used to solve *any* quadratic equation.

TO SOLVE $ax^2 + bx + c = 0$ BY COMPLETING THE SQUARE

1. If $a \neq 1$, multiply both sides of the equation by $1/a$ or divide both
sides by a so that the x^2-coefficient is 1.
2. Once the x^2-coefficient is 1, rewrite the equation in the form

$x^2 + bx = -c,$ or, if step (1) was needed,

$x^2 + \frac{b}{a}x = -\frac{c}{a}.$

3. Take half of the x-coefficient and square it. Add the result to both
sides of the equation. This completes the square.
4. Express the left side as the square of a binomial. (Factor.)
5. Use the principle of square roots and complete the solution.

Chapter Resource:
Collaborative Activity: p. 588

9.2   EXERCISE SET

FOR
EXTRA   MyMathLab®
HELP

Vocabulary and Reading Check

*Choose from the following list the phrase that best
completes each statement.*

a) The number that completes the square
b) Half of the coefficient of x
c) A perfect-square trinomial
d) The principle of square roots

1. For $x^2 + 14x$, 7 is ____.

2. For $x^2 + 14x$, 49 is ____.

3. The expression $x^2 + 14x + 49$ is ____.

4. If $(x + 7)^2 = 5$, then $x + 7 = \sqrt{5}$ *or*
$x + 7 = -\sqrt{5}$ because of ____.

↪ Concept Reinforcement

In Exercises 5–10, match each equation with the equation from the column on the right to which it is equivalent.

5. ____ $x^2 + 6x = 2$ a) $(x + 3)^2 = 10$

6. ____ $x^2 - 6x + 9 = 10$ b) $x^2 - 6x + 9 = 2 + 9$

7. ____ $x^2 + 6x + 9 = 10$ c) $x^2 + 6x + 9 = 2 + 9$

8. ____ $x^2 - 6x = 2$ d) $(x + 4)^2 = 18$

9. ____ $x^2 + 8x + 16 = 18$ e) $(x - 3)^2 = 10$

10. ____ $x^2 - 8x = 2$ f) $x^2 - 8x + 16 = 2 + 16$

A. Completing the Square

Determine the number that will complete the square. Check by factoring.

11. $x^2 + 8x$ 12. $x^2 + 4x$ 13. $x^2 - 2x$

14. $x^2 - 20x$ 15. $x^2 - 3x$ 16. $x^2 - 9x$

17. $t^2 + t$ 18. $a^2 - a$ 19. $x^2 + \frac{3}{2}x$

20. $x^2 + \frac{2}{3}x$ 21. $m^2 - \frac{8}{3}m$ 22. $y^2 + \frac{3}{4}y$

B. Solving by Completing the Square

Solve by completing the square.

23. $x^2 + 7x + 10 = 0$ 24. $x^2 - 3x - 10 = 0$

25. $x^2 - 24x + 21 = 0$ 26. $x^2 - 10x + 20 = 0$

27. $t^2 + 10t + 12 = 0$ 28. $t^2 - 16t + 50 = 0$

29. $2x^2 - 8x = 14$ 30. $3x^2 + 6x = 18$

31. $x^2 + 3x - 3 = 0$ 32. $x^2 - x - 3 = 0$

33. $2t^2 + t - 1 = 0$ 34. $3t^2 + t - 2 = 0$

35. $x^2 - \frac{3}{2}x - 2 = 0$ 36. $x^2 + \frac{5}{2}x - 2 = 0$

37. $2x^2 - 5x - 10 = 0$ 38. $2x^2 - 7x - 10 = 0$

39. $3t^2 + 4t - 5 = 0$ 40. $3t^2 - 8t + 1 = 0$

41. $2x^2 = 5 + 9x$ 42. $3x^2 = 22x - 7$

43. $6x^2 + 11x = 10$ 44. $4x^2 + 12x = 7$

45. How does completing the square allow us to solve equations that we could not otherwise have solved?

46. Explain how the addition principle, the multiplication principle, and the square-root principle were used in this section.

Skill Review

Solve.

47. $2(x - 7) - 3(x - 5) = 5(x + 1)$ [2.2]

48. $3(8 - y) < 7 - y - 3$ [2.6]

49. $x^2 - 6x - 7 = 0$ [5.6]

50. $t = \sqrt{t + 1} + 1$ [8.5]

51. $\dfrac{1}{x} + \dfrac{1}{x + 9} = \dfrac{1}{x - 3}$
[6.6]

52. $x = 3y - 2,$
$2x - y = 7$ [7.2]

Synthesis

53. Ian states that "since solving a quadratic equation by completing the square relies on the principle of square roots, the solutions are always opposites of each other." Is Ian correct? Why or why not?

54. When we are completing the square, what determines if the number being added is a whole number or a fraction?

Find b such that each trinomial is a square.

55. $x^2 + bx + 25$ 56. $x^2 + bx + 81$

57. $x^2 + bx + 45$ 58. $x^2 + bx + 50$

59. $4x^2 + bx + 16$ 60. $x^2 - bx + 48$

Solve each of the following by letting y_1 represent the left side of each equation and y_2 the right side, and graphing y_1 and y_2 on the same set of axes. INTERSECT can then be used to determine the x-coordinate at any point of intersection. Find solutions accurate to two decimal places.

61. $(x + 4)^2 = 13$ 62. $x^2 + 3x - 3 = 0$
(see Exercise 31)

63. $6x^2 + 11x = 10$ 64. $2x^2 = 5 + 9x$
(see Exercise 43)   (see Exercise 41)

Aha! 65. What is the best way to solve $x^2 + 8x = 0$? Why?

↪ **YOUR TURN ANSWERS: SECTION 9.2**
1. $49; x^2 + 14x + 49 = (x + 7)^2$ **2.** $-6 \pm \sqrt{37}$
3. $5 \pm 3\sqrt{3}$

Quick Quiz: Sections 9.1–9.2

1. Determine the number that will complete the square: $x^2 + 20x$.   [9.2]

Solve using the principle of square roots.   [9.1]
2. $3y^2 = 33$ 3. $(7 - x)^2 = 2$

Solve by completing the square.   [9.2]
4. $x^2 + 20x + 15 = 0$ 5. $2n^2 + n - 4 = 0$

Prepare to Move On

Evaluate.   [1.8]
1. $-b$, for $b = 7$ 2. $-b$, for $b = -7$

3. $b^2 - 4ac$, for $a = 3$, $b = -1$, and $c = -1$

Simplify.   [8.4]
4. $\dfrac{2(1 - \sqrt{5})}{2}$ 5. $\dfrac{24 - 3\sqrt{5}}{9}$

9.3 | The Quadratic Formula and Applications

A. The Quadratic Formula   **B.** Problem Solving

When mathematicians use a procedure repeatedly, they often try to find a formula for the procedure. The *quadratic formula* condenses into one calculation the many steps used to solve a quadratic equation by completing the square.

A. The Quadratic Formula

Consider a quadratic equation in *standard form*, $ax^2 + bx + c = 0$, with $a > 0$. Our plan is to solve this equation for x by completing the square. As the steps are performed, compare them with those in Example 3(b) in Section 9.2.

$$ax^2 + bx + c = 0$$

$$\frac{1}{a}(ax^2 + bx + c) = \frac{1}{a} \cdot 0 \qquad \text{Multiplying by } \frac{1}{a} \text{ to make the } x^2\text{-coefficient 1}$$

$$x^2 + \frac{b}{a}x + \frac{c}{a} = 0$$

$$x^2 + \frac{b}{a}x = -\frac{c}{a} \qquad \text{Adding } -\frac{c}{a} \text{ to both sides}$$

$$x^2 + \frac{b}{a}x + \frac{b^2}{4a^2} = -\frac{c}{a} + \frac{b^2}{4a^2} \qquad \text{Adding } \frac{b^2}{4a^2} \text{ to both sides:}$$

$$\left[\frac{1}{2} \cdot \frac{b}{a}\right]^2 = \left[\frac{b}{2a}\right]^2 = \frac{b^2}{4a^2}$$

$$\left(x + \frac{b}{2a}\right)^2 = -\frac{4ac}{4a^2} + \frac{b^2}{4a^2} \qquad \text{Factoring and writing with a common denominator}$$

$$\left(x + \frac{b}{2a}\right)^2 = \frac{b^2 - 4ac}{4a^2}$$

$$x + \frac{b}{2a} = \sqrt{\frac{b^2 - 4ac}{4a^2}} \qquad \text{or} \qquad x + \frac{b}{2a} = -\sqrt{\frac{b^2 - 4ac}{4a^2}} \qquad \text{Using the principle of square roots}$$

$$x + \frac{b}{2a} = \frac{\sqrt{b^2 - 4ac}}{2a} \qquad \text{or} \qquad x + \frac{b}{2a} = -\frac{\sqrt{b^2 - 4ac}}{2a} \qquad \sqrt{4a^2} = 2a \text{ because } a > 0.$$

$$x = -\frac{b}{2a} + \frac{\sqrt{b^2 - 4ac}}{2a} \qquad \text{or} \qquad x = -\frac{b}{2a} - \frac{\sqrt{b^2 - 4ac}}{2a} \qquad \text{Adding } -b/(2a) \text{ to both sides}$$

Thus,

$$x = -\frac{b}{2a} \pm \frac{\sqrt{b^2 - 4ac}}{2a}, \quad \text{or} \quad x = \frac{-b \pm \sqrt{b^2 - 4ac}}{2a}.$$

This last equation is the result we sought. A similar proof would show that this formula also holds when $a < 0$. Unless $b^2 - 4ac$ is 0, the formula produces two different solutions.

> **THE QUADRATIC FORMULA**
>
> The solutions of $ax^2 + bx + c = 0$, $a \neq 0$, are given by
> $$x = \frac{-b \pm \sqrt{b^2 - 4ac}}{2a}.$$

The quadratic formula is so useful that it is worth memorizing.

Student Notes

After identifying which numbers to use as a, b, and c, be careful to substitute exactly for the *letters* in the quadratic formula. Using parentheses, as shown in Example 1, can help in this regard. Also, reading "$-b$" as "the opposite of b" will serve as a reminder of how to substitute correctly.

EXAMPLE 1   Solve using the quadratic formula.

a) $4x^2 + 5x - 6 = 0$ **b)** $t^2 = 4t + 7$ **c)** $x^2 + x = -1$

SOLUTION

a) We identify a, b, and c and substitute into the quadratic formula:

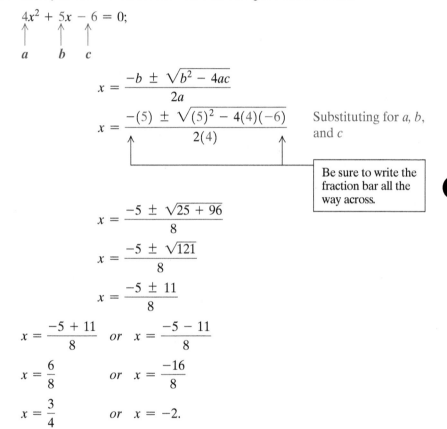

$$4x^2 + 5x - 6 = 0;$$
$$\quad\uparrow\qquad\uparrow\qquad\uparrow$$
$$\quad a\qquad b\qquad c$$

$$x = \frac{-b \pm \sqrt{b^2 - 4ac}}{2a}$$

$$x = \frac{-(5) \pm \sqrt{(5)^2 - 4(4)(-6)}}{2(4)} \qquad \text{Substituting for } a, b, \text{ and } c$$

> Be sure to write the fraction bar all the way across.

$$x = \frac{-5 \pm \sqrt{25 + 96}}{8}$$

$$x = \frac{-5 \pm \sqrt{121}}{8}$$

$$x = \frac{-5 \pm 11}{8}$$

$$x = \frac{-5 + 11}{8} \quad or \quad x = \frac{-5 - 11}{8}$$

$$x = \frac{6}{8} \qquad or \quad x = \frac{-16}{8}$$

$$x = \frac{3}{4} \qquad or \quad x = -2.$$

The solutions are $\frac{3}{4}$ and -2.

b) We rewrite $t^2 = 4t + 7$ in standard form, identify a, b, and c, and solve using the quadratic formula:

$$1t^2 - 4t - 7 = 0; \qquad \text{Subtracting } 4t + 7 \text{ from both sides}$$
$$\uparrow\qquad\uparrow\qquad\uparrow$$
$$a\qquad b\qquad c$$

$$t = \frac{-(-4) \pm \sqrt{(-4)^2 - 4(1)(-7)}}{2 \cdot 1} \qquad \text{Substituting into the quadratic formula}$$

$$t = \frac{4 \pm \sqrt{16 + 28}}{2} = \frac{4 \pm \sqrt{44}}{2}$$

$$t = \frac{4}{2} \pm \frac{\sqrt{44}}{2}$$

$$t = 2 \pm \frac{\sqrt{4}\sqrt{11}}{2}$$

$$t = 2 \pm \frac{2\sqrt{11}}{2}$$ } Simplifying $\sqrt{44}$

$$t = 2 \pm \sqrt{11}.$$ Removing a factor equal to 1: $\frac{2}{2} = 1$

The solutions are $2 + \sqrt{11}$ and $2 - \sqrt{11}$, or $2 \pm \sqrt{11}$.

c) We rewrite $x^2 + x = -1$ in standard form and use the quadratic formula:

$$1x^2 + 1x + 1 = 0;$$ Adding 1 to both sides

$a \quad b \quad c$

$$x = \frac{-(1) \pm \sqrt{1^2 - 4 \cdot 1 \cdot 1}}{2 \cdot 1}$$ Substituting into the quadratic formula

$$x = \frac{-1 \pm \sqrt{1 - 4}}{2}$$

$$x = \frac{-1 \pm \sqrt{-3}}{2}.$$

1. Solve using the quadratic formula:

$$x^2 - 4x - 3 = 0.$$

Since the radicand, -3, is negative, there are no real-number solutions. Later, you may study a number system in which solutions of this equation can be found. For now we simply state, "No real-number solution exists."

YOUR TURN

B. Problem Solving

EXAMPLE 2 *Diagonals in a Polygon.* The number of diagonals d in a polygon that has n sides is given by the formula

$$d = \frac{n^2 - 3n}{2}.$$

If a polygon has 27 diagonals, how many sides does it have?

SOLUTION

1. Familiarize. A sketch can help us to become familiar with the problem. We draw a hexagon (6 sides) and count the diagonals. As the formula predicts, for $n = 6$, there are

$$\frac{6^2 - 3 \cdot 6}{2} = \frac{36 - 18}{2}$$

$$= \frac{18}{2} = 9 \text{ diagonals.}$$

Clearly, the polygon in question must have more than 6 sides. We might suspect that tripling the number of diagonals requires tripling the number of sides. Evaluating the above formula for $n = 18$, you can confirm that this is *not* the case. Rather than continue guessing, we proceed to a translation.

2. Translate. Since the number of diagonals is 27, we substitute 27 for d:

$$27 = \frac{n^2 - 3n}{2}.$$

This gives us a translation.

Technology Connection

To see that no real solutions exist for Example 1(c), we let $y_1 = x^2 + x$ and $y_2 = -1$.

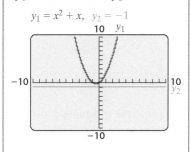

Since the graphs do not intersect, no real-number solution exists.

1. What happens when the INTERSECT feature is used with the graph above?
2. How can the graph of $y = x^2 + x + 1$ be used to provide still another check of Example 1(c)?

Student Notes

Although you can use the quadratic formula to solve any quadratic equation, it is not always the best method. When it is not difficult, factoring is easier than using the quadratic formula.

2. A polygon has 5 diagonals. Use the formula in Example 2 to determine how many sides the polygon has.

3. **Carry out.** We reverse the equation for convenience and solve for n:

$$\frac{n^2 - 3n}{2} = 27$$

$$n^2 - 3n = 54 \qquad \text{Multiplying both sides by 2 to clear fractions}$$
$$n^2 - 3n - 54 = 0 \qquad \text{Subtracting 54 from both sides}$$
$$(n - 9)(n + 6) = 0 \qquad \text{Factoring}$$
$$n - 9 = 0 \quad or \quad n + 6 = 0$$
$$n = 9 \quad or \qquad n = -6.$$

4. **Check.** Since the number of sides cannot be negative, -6 cannot be a solution. We leave it to the student to show by substitution that 9 checks.

5. **State.** The polygon has 9 sides (it is a nonagon).

   YOUR TURN

EXAMPLE 3 *Free-Falling Objects.* The natural arch Pravčická brána in the Czech Republic is 70 ft high. How many seconds will it take a pebble to fall from that height? Round to the nearest hundredth.

SOLUTION

1. **Familiarize.** If we did not know anything about this problem, we might consider looking up a formula in a mathematics or physics book. A formula that fits this situation is

   $$s = 16t^2, \qquad \text{It is useful to remember this formula.}$$

   where s is the distance, in feet, traveled by a body falling freely from rest for t seconds. (This formula does not account for air resistance.) In this problem, the distance s is 70 ft. We want to determine the time t that it takes the pebble to reach the ground.

2. **Translate.** The distance is 70 ft and we need to solve for t. We substitute 70 for s in the formula above:

   $$70 = 16t^2.$$

3. **Carry out.** Because there is no t-term, we can use the principle of square roots to solve:

   $$70 = 16t^2$$
   $$\frac{70}{16} = t^2 \qquad \text{Solving for } t^2$$
   $$\sqrt{\frac{70}{16}} = t \quad or \quad -\sqrt{\frac{70}{16}} = t \qquad \text{Using the principle of square roots}$$
   $$\frac{\sqrt{70}}{4} = t \quad or \quad \frac{-\sqrt{70}}{4} = t$$
   $$2.09 \approx t \quad or \quad -2.09 \approx t. \qquad \text{Using a calculator and rounding to the nearest hundredth}$$

3. The Sipapu Bridge in Natural Bridges National Monument is 220 ft high. Use the formula from Example 3 to find how many seconds that it will take a pebble to fall from that height. Round to the nearest hundredth.

4. **Check.** The number -2.09 cannot be a solution because time cannot be negative in this situation. We substitute 2.09 in the original equation:

   $$s = 16(2.09)^2 = 16(4.3681) = 69.8896.$$

   This is close to 70. Remember that we approximated a solution.

5. **State.** It would take about 2.09 sec for a pebble to reach the ground if dropped from the Pravčická brána.

   YOUR TURN

EXAMPLE 4 *Right Triangles.* The hypotenuse of a right triangle is 6 m long. One leg is 1 m longer than the other. Find the lengths of the legs. Round to the nearest hundredth.

SOLUTION

1. **Familiarize.** We first make a drawing and label it, with s = the length, in meters, of one leg. Then $s + 1$ = the length, in meters, of the other leg.

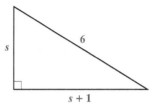

Student Notes

The a, b, and c in the Pythagorean equation represent sides of the triangle. These are not the same as the a, b, and c used in the quadratic formula.

The Pythagorean equation $a^2 + b^2 = c^2$ holds, where $c^2 = 6^2$, or 36. We make and check some guesses.

s	$s + 1$	$s^2 + (s + 1)^2$
3	4	$3^2 + 4^2 = 25$
4	5	$4^2 + 5^2 = 41$

Neither sum is 36.

Although we have not guessed the solution, we expect s to be between 3 and 4.

2. **Translate.** To translate, we use the Pythagorean equation:

$$s^2 + (s + 1)^2 = 6^2.$$

3. **Carry out.** We solve the equation:

$$s^2 + (s + 1)^2 = 6^2$$
$$s^2 + s^2 + 2s + 1 = 36$$
$$2s^2 + 2s - 35 = 0$$

We cannot factor this so we use the quadratic formula: $s = \dfrac{-b \pm \sqrt{b^2 - 4ac}}{2a}$.

$$s = \frac{-2 \pm \sqrt{2^2 - 4 \cdot 2(-35)}}{2 \cdot 2}$$

$$s = \frac{-2 \pm \sqrt{4 + 280}}{4} = \frac{-2 \pm \sqrt{284}}{4}$$

$$s \approx 3.71 \quad or \quad s \approx -4.71.$$

Using a calculator and rounding to the nearest hundredth

4. **Check.** Length cannot be negative, so -4.71 does not check. Note that if the shorter leg is 3.71 m, the other leg is 4.71 m. Then

$$(3.71)^2 + (4.71)^2 = 13.7641 + 22.1841 = 35.9482$$

and since $35.9482 \approx 6^2$, our answer checks. Also, note that the value of s, 3.71, is between 3 and 4, as predicted in step (1).

5. **State.** One leg is about 3.71 m long; the other is about 4.71 m long.

YOUR TURN

⤷ Check Your UNDERSTANDING

For each quadratic equation, determine a, b, and c. Remember to first write in standard form if necessary.

1. $3x^2 + 5x + 2 = 0$
2. $x^2 - 6x + 7 = 0$
3. $2x^2 = 4x + 9$
4. $x^2 = 12x$

4. The hypotenuse of a right triangle is 8 ft long. One leg is 2 ft longer than the other. Find the lengths of the legs. Round to the nearest hundredth.

9.3 EXERCISE SET

FOR EXTRA HELP MyMathLab®

Vocabulary and Reading Check

Classify each of the following statements as either true or false.

1. A quadratic equation $ax^2 + bx + c = 0, a > 0$, is said to be in standard form.

2. The quadratic formula can be derived by completing the square.

3. The quadratic formula can be used to solve any quadratic equation.

4. If $b^2 - 4ac$ is negative, then there is no real-number solution of $ax^2 + bx + c = 0$.

A. The Quadratic Formula

Solve. If no real-number solutions exist, state this.

5. $x^2 - 8x = 20$
6. $x^2 + 3x = 40$

7. $t^2 = 2t - 1$
8. $t^2 = 10t - 25$

9. $3y^2 + 7y + 4 = 0$
10. $3y^2 + 2y - 8 = 0$

11. $4x^2 - 12x = 7$
12. $4x^2 + 4x = 15$

13. $p^2 = 25$
14. $r^2 = 1$

15. $x^2 + 4x - 7 = 0$
16. $x^2 + 2x - 2 = 0$

17. $y^2 - 10y + 19 = 0$
18. $y^2 + 6y - 2 = 0$

Aha! 19. $x^2 - 10x + 25 = 3$
20. $x^2 + 2x + 1 = 3$

21. $3t^2 + 8t + 2 = 0$
22. $3t^2 - 4t - 2 = 0$

23. $2x^2 - 5x = 1$
24. $2x^2 + 2x = 3$

25. $4y^2 + 2y - 3 = 0$
26. $4y^2 - 4y - 3 = 0$

27. $2m^2 - m + 3 = 0$
28. $3p^2 + 2p + 5 = 0$

29. $3x^2 - 5x = 4$
30. $2x^2 + 3x = 1$

31. $2y^2 - 6y = 10$
32. $5m^2 = 3 + 11m$

33. $6t^2 + 26t = 20$
34. $7x^2 + 2 = 6x$

35. $5t^2 - 7t = -4$
36. $15t^2 + 10t = 0$

37. $5y^2 = 60$
38. $4y^2 = 200$

Solve using the quadratic formula. Use a calculator to approximate the solutions to the nearest thousandth.

39. $x^2 + 3x - 2 = 0$
40. $x^2 - 6x - 1 = 0$

41. $y^2 - 5y - 1 = 0$
42. $y^2 + 7y + 3 = 0$

43. $4x^2 + 4x = 1$
44. $4x^2 = 4x + 1$

B. Problem Solving

Solve. If an irrational answer occurs, round to the nearest hundredth.

45. A polygon has 35 diagonals. How many sides does it have? (Use $d = (n^2 - 3n)/2$.)

46. A polygon has 20 diagonals. How many sides does it have? (Use $d = (n^2 - 3n)/2$.)

47. *Free-Fall Time.* At 2723 ft, the Burj Khalifa in Dubai, United Arab Emirates, is the world's tallest man-made structure. How long would it take a marble to fall from the top? (Use $s = 16t^2$ and disregard wind resistance.)

48. *Free-Fall Time.* The skydeck on the 103rd floor of the Willis Tower in Chicago is 1353 ft high. How long would it take a golf ball to fall from the top? (Use $s = 16t^2$ and disregard wind resistance.)

49. *Skateboarding.* On April 6, 2006, skateboarder Danny Way set a free-fall record by dropping 28 ft onto a ramp. For how long was Way airborne? (Use $s = 16t^2$ and disregard wind resistance.)

Data: skateboard.about.com

50. *Cliff Jumping.* While attempting a different jump, skier Fred Syversen set a world record for a jump on skis by falling 351 ft off a cliff. For how long was he airborne? (Use $s = 16t^2$ and disregard wind resistance.)

Data: adventure4everyone.com

51. *Right Triangles.* The hypotenuse of a right triangle is 25 ft long. One leg is 17 ft longer than the other. Find the lengths of the legs.

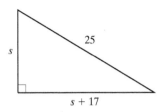

52. *Right Triangles.* The hypotenuse of a right triangle is 26 yd long. One leg is 14 yd longer than the other. Find the lengths of the legs.

53. *Area of a Rectangle.* The length of a rectangle is 4 cm greater than the width. The area is 60 cm^2. Find the length and the width.

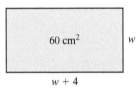

54. *Area of a Rectangle.* The length of a rectangle is 3 m greater than the width. The area is 70 m^2. Find the length and the width.

55. *Plumbing.* A pipe runs diagonally under a rectangular yard that is 6 m longer than it is wide. If the pipe is 30 m long, determine the dimensions of the yard.

56. *Guy Wires.* A 26-ft long guy wire is anchored 10 ft from the base of a telephone pole. How far up the pole does the wire reach?

57. *Right Triangles.* The area of a right triangle is 26 cm^2. One leg is 5 cm shorter than the other. Find the lengths of the legs.

58. *Right Triangles.* The area of a right triangle is 31 m^2. One leg is 2.4 m shorter than the other. Find the lengths of the legs.

59. *Area of a Rectangle.* The length of a rectangle is 5 ft greater than the width. The area is 25 ft^2. Find the length and the width.

60. *Area of a Rectangle.* The length of a rectangle is 3 in. greater than the width. The area is 30 in^2. Find the length and the width.

61. *Area of a Rectangle.* The length of a rectangle is twice the width. The area is 16 m^2. Find the length and the width.

62. *Area of a Rectangle.* The length of a rectangle is twice the width. The area is 20 cm^2. Find the length and the width.

Investments. When interest is paid on interest earned previously as well as on the initial amount invested, we say that interest is compounded. *The following table lists values to which an initial investment of $1000 will grow when interest is compounded annually at several interest rates.*

	After 1 year	After 2 years	After 10 years
$r = 3\%$	$1030	$1060.90	$1343.92
$r = 4\%$	1040	1081.60	1480.24
$r = 7\%$	1070	1144.90	1967.15

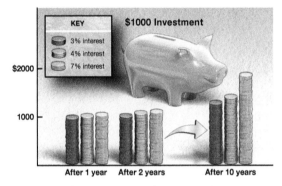

The formula $A = P(1 + r)^t$ is used to find the value A to which P dollars grows when interest is compounded annually for t years at interest rate r. In Exercises 63–66, find the interest rate for the given information.

63. $2500 grows to $2704 in 2 years

64. $5000 grows to $5202 in 2 years

65. $6000 grows to $6615 in 2 years

66. $3125 grows to $3645 in 2 years

67. *Geology.* The Richat Structure in the Sahara Desert is a 700-mi^2 eroded circular dome that is large enough to be visible from space. What is the diameter of the structure?

Data: nasa.gov

68. *Gardening.* Laurie has enough mulch to cover 250 ft² of garden space. How wide is the largest circular flower garden that Laurie can cover with mulch?

69. Under what condition(s) is the quadratic formula *not* the easiest way to solve a quadratic equation?

70. Roy claims to be able to solve any quadratic equation by completing the square. He also claims to be incapable of understanding why the quadratic formula works. Does this strike you as odd? Why or why not?

Skill Review

Perform the indicated operation and simplify.

71. $(3x^4 - x^2 - 7x) - (x^3 + x^2 + 3x)$ [4.4]

72. $(5m - 7n)^2$ [4.7]

73. $(4t + 3)(3t - 5)$ [4.6]

74. $\dfrac{x^2 - 7x}{x^2 - 2x + 1} \cdot \dfrac{7x^2 - 7}{x^2 - 8x + 7}$ [6.2]

75. $\dfrac{3}{x + 1} - \dfrac{x - 2}{x}$ [6.4]

76. $(3x^4 - x^2 - 7x + 2) \div (x - 1)$ [4.8]

Synthesis

77. Sofia can tell from the value of $b^2 - 4ac$ whether or not the solutions of a quadratic equation are rational. How can she do this?

78. Where does the $\pm$ symbol in the quadratic formula come from?

Solve.

79. $5x = -x(x + 6)$

80. $y(3y + 7) = 3y$

81. $3 - x(x - 3) = 4$

82. $x(5x - 7) = 1$

83. $(y + 4)(y + 3) = 15$

84. $t^2 + (t + 2)^2 = 7$

85. $\dfrac{x^2}{x + 3} = \dfrac{11}{x + 3}$

86. $\dfrac{x^2}{x + 5} - \dfrac{3}{x + 5} = 0$

87. $\dfrac{1}{t} + \dfrac{1}{t + 1} = \dfrac{1}{3}$

88. $\dfrac{1}{x} + \dfrac{1}{x + 6} = \dfrac{1}{5}$

89. Find *r* in this figure. Round to the nearest hundredth.

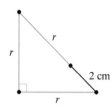

90. *Area of a Square.* Find the area of a square for which the diagonal is 3 units longer than the length of the sides.

91. *Golden Rectangle.* The so-called *golden rectangle* is said to be extremely pleasing visually and was used often by ancient Greek and Roman architects. The length of a golden rectangle is approximately 1.6 times the width. Find the dimensions of a golden rectangle if its area is 9000 m².

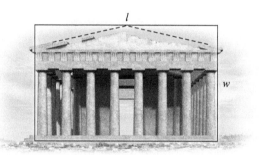

92. *Flagpoles.* A 20-ft flagpole is struck by lightning and, while not completely broken, falls over and touches the ground 10 ft from the bottom of the pole. How high up did the pole break?

93. *Investments.* $4000 is invested at interest rate *r* for 2 years. After 1 year, an additional $2000 is invested, again at interest rate *r*. What is the interest rate if $6510 is in the account at the end of 2 years?

94. *Investments.* Dar needs $5000 in 2 years to pay for dental work. How much does he need to invest now if he can get an interest rate of 3.75%, compounded annually?

95. *Enlarged Strike Zone.* In baseball, a batter's strike zone is a rectangular area about 15 in. wide and 40 in. high. Many batters subconsciously enlarge this area by 40% when fearful that if they don't swing, they will strike out. Assuming that the strike zone is enlarged by an invisible band of uniform width around the actual zone, find the dimensions of the enlarged strike zone.

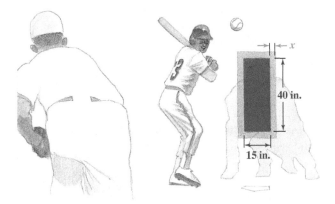

96. Use a graph to approximate to the nearest thousandth the solutions of Exercises 39–44. Compare your answers with those found using a calculator.

97. The Sipapu Bridge in Your Turn Exercise 3 is over three times as high as the Pravčická brána in Example 3. Why does it take less than three times the time for a pebble to fall from the higher bridge?

YOUR TURN ANSWERS: SECTION 9.3
1. $2 \pm \sqrt{7}$ **2.** 5 sides **3.** 3.71 sec
4. 4.57 ft, 6.57 ft

Quick Quiz: Sections 9.1–9.3
Solve.

1. $2n^2 = 20$ [9.1]

2. $3x^2 - 5x + 2 = 0$ [9.3] $\frac{2}{3}, 1$

3. $x^2 - x = 3$ [9.3]

4. $(x - 9)^2 = 8$ [9.1]

5. $(x - 4)(2x + 1) = 0$ [9.1]

Prepare to Move On
Solve. [2.3]

1. $n = \frac{y}{4}$, for y **2.** $t = \frac{c + d}{2}$, for d

3. $a = \frac{4n}{p}$, for p **4.** $y = Ax + B$, for x

5. $y = Ax + Bx$, for x

Mid-Chapter Review

We have now studied four ways of solving quadratic equations. Each method has certain advantages and disadvantages, as outlined in the following chart. Note that although the quadratic formula can be used to solve *any* quadratic equation, other methods are sometimes faster and easier.

Method	Advantages	Disadvantages
The quadratic formula	Can be used to solve *any* quadratic equation.	Can be slower than factoring or the principle of square roots.
Completing the square	Works well on equations of the form $x^2 + bx = -c$, where b is even. Can be used to solve *any* quadratic equation.	Can be complicated when b is odd or a fraction.
The principle of square roots	Fastest way to solve equations of the form $ax^2 = p$, or $(x + k)^2 = p$. Can be used to solve *any* quadratic equation.	Can be slow when completing the square is required.
Factoring	Can be very fast.	Can be used only on certain equations. Many equations are difficult or impossible to solve by factoring.

GUIDED SOLUTIONS

1. Solve using the principle of square roots:

$(x + 3)^2 = 2$. [9.1]

Solution

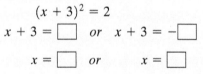

$$(x + 3)^2 = 2$$
$$x + 3 = \square \quad or \quad x + 3 = -\square$$
$$x = \square \quad or \quad x = \square$$

2. Solve using the quadratic formula:

$x^2 - 3x + 1 = 0$. [9.3]

Solution

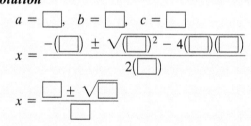

$$a = \square, \quad b = \square, \quad c = \square$$
$$x = \frac{-(\square) \pm \sqrt{(\square)^2 - 4(\square)(\square)}}{2(\square)}$$
$$x = \frac{\square \pm \sqrt{\square}}{\square}$$

MIXED REVIEW

Solve. Examine each problem carefully, and try to solve using the most efficient method. If no real-number solution exists, state this.

3. $(x + 2)(x - 1) = 0$ [9.1]

4. $x^2 = 100$ [9.1]

5. $x^2 + 10x = 7$ [9.3]

6. $2x^2 - 5x - 3 = 0$ [9.1]

7. $x^2 + 2x + 1 = 0$ [9.1]

8. $x^2 + x + 1 = 0$ [9.3]

9. $(x - 2)(x - 3) = 12$ [9.1]

10. $x^2 = 5x$ [9.1]

11. $5x^2 = x + 2$ [9.3]

12. $x^2 + x = 6$ [9.1]

13. $(x - 2)^2 = 5$ [9.1]

14. $x^2 + 8x + 1 = 0$ [9.3]

15. $121x^2 - 1 = 0$ [9.1]

16. $(x + 1)^2 - 3 = 0$ [9.1]

17. $2x^2 - x + 2 = 0$ [9.3]   **18.** $9x^2 = x$ [9.1]

Solve. Round any irrational answers to the nearest hundredth.

19. The "Thumbnail" in South Greenland is a sea cliff 4430 ft high. Use $s = 16t^2$ to estimate how long it would take a stone to fall from the top. Disregard wind resistance. [9.3]

20. The length of a rectangle is 3 ft greater than the width. The area is 20 ft^2. Find the length and the width. [9.3]

9.4 Formulas

A. Solving Formulas

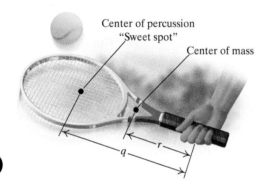

Center of percussion "Sweet spot"

Center of mass

Formulas arise frequently in the natural and social sciences, business, engineering, and health care. We often need to solve such a formula for a variable.

A. Solving Formulas

To solve formulas, we use the same steps that we use to solve equations. Probably the greatest difference is that whereas the solution of an equation is a number, the solution of a formula is generally a variable expression.

EXAMPLE 1 *Sports Engineering.* The "sweet spot," or center of percussion, of a tennis racquet is located q centimeters from the hand, where $q = \dfrac{I}{mr}$, with I the swing weight of the racquet, in kg·cm², m the mass of the racquet, in kilograms, and r the distance from the hand to the center of mass, in centimeters. Solve for m.

Data: www.racquetresearch.com

SOLUTION We have

$$q = \frac{I}{mr}$$

$$mr \cdot q = mr \cdot \frac{I}{mr} \qquad \text{Multiplying both sides by } mr \text{ to clear fractions}$$

$$mrq = I$$

$$m = \frac{I}{rq}. \qquad \text{Dividing both sides by } rq$$

1. Solve $\dfrac{1}{a} = \dfrac{1}{b}$ for a.

YOUR TURN

EXAMPLE 2 *A Work Formula.* The formula $\dfrac{t}{a} + \dfrac{t}{b} = 1$ is used to solve work problems. Solve this formula for t.

SOLUTION We have

$$\frac{t}{a} + \frac{t}{b} = 1$$

$$ab\left(\frac{t}{a} + \frac{t}{b}\right) = ab \cdot 1 \qquad \text{Multiplying by the LCD, } ab, \text{ to clear fractions}$$

$$\left.\begin{array}{c} \dfrac{\cancel{a}bt}{\cancel{a}} + \dfrac{a\cancel{b}t}{\cancel{b}} = ab \\[2mm] bt + at = ab \end{array}\right\} \qquad \begin{array}{l}\text{Multiplying to remove parentheses and} \\ \text{removing factors equal to 1: } \dfrac{a}{a} = 1 \text{ and } \dfrac{b}{b} = 1\end{array}$$

$$(b + a)t = ab \qquad \begin{array}{l}\text{Factoring out } t, \text{ the letter for which we are} \\ \text{solving}\end{array}$$

2. Solve $\dfrac{t}{c} - t = 1$ for t.

$$t = \frac{ab}{b + a}. \qquad \text{Dividing both sides by } b + a$$

YOUR TURN

The answer to Example 2 can be used when the times required to do a job independently (a and b) are known and t represents the time required to complete the task working together.

EXAMPLE 3 *Body Surface Area.* An individual's body surface area A, in square meters, can be estimated using Mosteller's formula

$$A = \frac{\sqrt{hw}}{60},$$

where h is the height, in centimeters, and w is the weight, in kilograms. Solve for w.

SOLUTION

$$A = \frac{\sqrt{hw}}{60}$$

$$60A = \sqrt{hw} \qquad \text{Isolating the radical}$$

$$(60A)^2 = (\sqrt{hw})^2 \qquad \text{Using the principle of powers;}$$
$$\text{squaring both sides}$$

$$3600A^2 = hw \qquad (60A)^2 = 60^2 \cdot A^2 = 3600A^2$$

$$\frac{3600A^2}{h} = w \qquad \text{Solving for } w$$

YOUR TURN

In some cases, the quadratic formula is needed.

EXAMPLE 4 *A Physics Formula.* The distance s from some established point of an object that is steadily accelerating away from that point can be determined using the formula $s = s_0 + v_0 t + \frac{1}{2} at^2$, where s_0 is the object's original distance from the point, v_0 is the object's original velocity, a is the object's acceleration, and t is time. Solve for t.

SOLUTION Note that t appears in two terms, raised to the first power and then to the second power. Thus the equation is "quadratic in t," and the quadratic formula is needed.

$$\frac{1}{2} at^2 + v_0 t + s_0 = s \qquad \text{Rewriting the equation from left to right}$$

$$\underbrace{\frac{1}{2} at^2}_{a} + \underbrace{v_0 t}_{b} + \underbrace{s_0 - s}_{c} = 0 \qquad \text{Subtracting } s \text{ from both sides and identifying the coefficients needed for the quadratic formula}$$

$$t = \frac{-v_0 \pm \sqrt{(v_0)^2 - 4\left(\frac{1}{2} a\right)(s_0 - s)}}{2 \cdot \frac{1}{2} a} \qquad \text{Substituting into the quadratic formula}$$

$$= \frac{-v_0 \pm \sqrt{(v_0)^2 - 2a(s_0 - s)}}{a} \left.\begin{array}{c} \\ \\ \end{array}\right\}$$

$$= \frac{-v_0 \pm \sqrt{(v_0)^2 - 2as_0 + 2as}}{a} \qquad \text{Simplifying}$$

YOUR TURN

---

3. The formula $A = \sqrt{\dfrac{hw}{3131}}$ can be used to estimate the body surface area, in square meters, of an individual with height h, in inches, and weight w, in pounds. Solve for w.

Complete each statement.

1. To clear fractions in
$Q = \dfrac{100m}{c}$, multiply both
sides by ▨.

2. To clear fractions in $\dfrac{x}{y} = \dfrac{z}{x}$,
multiply both sides by
▨.

3. To clear fractions in
$\dfrac{1}{R} = \dfrac{1}{r_1} + \dfrac{1}{r_2}$, multiply both
sides by ▨.

4. Solve $h = \frac{1}{2}at^2 + vt$ for t.

9.4 **EXERCISE SET**

FOR EXTRA HELP MyMathLab®

⤷ Vocabulary and Reading Check

Classify each of the following statements as either true or false.

1. The formula $a = b^2 + c\sqrt{a}$ is already solved for a.

2. In a formula, the letter T and the letter t may represent different quantities.

3. When we solve a formula for a variable, the result is usually a number.

4. We sometimes use the quadratic formula to solve a formula for a letter.

⤷ Concept Reinforcement

Match each formula with the process that would allow one to solve most easily for t.

5. ___ $a = \sqrt{t}$

6. ___ $a_2t^2 + a_1t + a_0 = 0$

7. ___ $v = \dfrac{t}{a}$

8. ___ $a = t^2$

a) Use the principle of square roots.

b) Square both sides.

c) Multiply both sides by a.

d) Use the quadratic formula.

A. Solving Formulas

Solve each formula for the specified variable.

9. $A = \frac{1}{2}bh$, for h
(The area of a triangle)

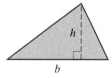

10. $S = 2\pi rh$, for h
(A formula for surface area)

11. $Q = \dfrac{100m}{c}$, for c
(A formula for intelligence quotient)

12. $y = \dfrac{k}{x}$, for x
(A formula for inverse variation)

13. $A = P(1 + rt)$, for t
(An investment formula)

14. $A = \pi r(s + r)$, for s
(The surface area of a cone)

15. $d = c\sqrt{h}$, for h
(A formula for distance to the horizon)

16. $n = c + \sqrt{t}$, for t
(A formula for population)

17. $\dfrac{1}{R} = \dfrac{1}{r_1} + \dfrac{1}{r_2}$, for R
(An electricity formula)

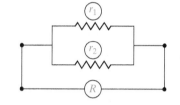

18. $\dfrac{1}{p} + \dfrac{1}{q} = \dfrac{1}{f}$, for f
(An optics formula)

Aha! 19. $ax^2 + bx + c = 0$, for x

20. $(x - d)^2 = k$, for x

21. $Ax + By = C$, for y
(An equation for a line)

22. $T = mg + mf$, for m

23. $S = 2\pi r(r + h)$, for h
(The surface area of a right circular cylinder)

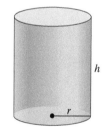

24. $xy - z = wy$, for y

25. $\dfrac{M - g}{t} = r + s$, for t

26. $\dfrac{m}{n} = p - q$, for n

27. $ab = ac + d$, for a

28. $\dfrac{x}{y} = \dfrac{z}{x}$, for x

29. $s = \frac{1}{2}gt^2$, for t
(A physics formula for distance)

30. $E = \frac{1}{2}mv^2$, for v
(A formula for kinetic energy)

31. $\frac{s}{h} = \frac{h}{t}$, for h
(A geometry formula)

32. $v = 8s(t - s)$, for t

33. $d = (r - w)t$, for w

34. $x - y = a(x + y)$, for y

35. $\sqrt{2x - y} + 1 = 6$, for x

36. $\sqrt{n - 2m} - 4 = 3$, for m

37. $\sqrt{2t + s} = \sqrt{s - t}$, for t

38. $\sqrt{m - n} = \sqrt{3t}$, for n

39. $mt^2 + nt - p = 0$, for t
(*Hint:* Use the quadratic formula.)

40. $rs^2 - ts + p = 0$, for s

41. $m + t = \frac{n}{m}$, for m

42. $x - y = \frac{z}{y}$, for y

43. $A = P(1 + r)^2$, for r

44. $y = a(x - h)^2 + k$, for x

45. $n = p - 3\sqrt{t + c}$, for t

46. $T = a + b\sqrt{ct}$, for t

47. Is it easier to solve
$$\frac{1}{25} + \frac{1}{23} = \frac{1}{x} \text{ for } x,$$
or to solve
$$\frac{1}{p} + \frac{1}{q} = \frac{1}{f} \text{ for } f?$$
Explain why.

48. Explain why someone might want to solve
$A = \frac{1}{2}bh$ for h.
(See Exercise 9.)

Skill Review

Factor.

49. $x^2 - 5x - 14$ [5.2]

50. $50y^5 - 5y^4 - 10y^3$ [5.3]

51. $n^2 - 20n + 100$ [5.4]

52. $2x^3 + 3x^2 - 10x - 15$ [5.1]

53. $x^2y^2 - 25$ [5.4]

54. $24x^2y - 30xy^2 + 6xy$ [5.1]

Synthesis

55. As a step in solving a formula for a certain variable, a student takes the reciprocal of both sides of the equation. Is a mistake being made? Why or why not?

56. Describe a situation in which the result of Example 2,
$$t = \frac{ab}{b + a},$$
would be especially useful.

57. *Health Care.*   Young's rule for determining the size of a particular child's medicine dosage c is
$$c = \frac{a}{a + 12} \cdot d,$$
where a is the child's age and d is the typical adult dosage. If a child receives 8 mg of antihistamine when the typical adult receives 24 mg, how old is the child?

Data: Olsen, June Looby, Leon J. Ablon, and Anthony Patrick Giangrasso, *Medical Dosage Calculations*, 6th ed., p. A-31

Solve.

58. $Sr = \frac{rl - a}{r - l}$, for r

59. $fm = \frac{gm - t}{m}$, for m

60. $\frac{n_1}{p_1} + \frac{n_2}{p_2} = \frac{n_2 - n_1}{R}$, for n_2

61. *Economics.*   The formula
$$V = \frac{k}{\sqrt{at + b}} + c$$
can be used to estimate the value of certain items that depreciate over time. Solve for t.

62. *Marine Biology.*   The formula
$$N = \frac{(b + d)f_1 - v}{(b - v)f_2}$$
is used when monitoring the water in fisheries. Solve for v.

63. *Meteorology.* The formula
$$C = \tfrac{5}{9}(F - 32)$$
is used to convert the Fahrenheit temperature F to the Celsius temperature C. At what temperature are the Fahrenheit and Celsius readings the same?

64. *Research.* Choose one of the formulas from Exercises 11, 16, 18, or 30. Find what each variable represents and how the formula is used.

YOUR TURN ANSWERS: SECTION 9.4

1. $a = b$ **2.** $t = \dfrac{c}{1-c}$ **3.** $w = \dfrac{3131A^2}{h}$

4. $t = \dfrac{-v \pm \sqrt{v^2 + 2ah}}{a}$

Quick Quiz: Sections 9.1–9.4

1. Solve by completing the square: $n^2 - 4n = 7$. [9.2]

2. Solve: $x^2 + 3x = 2$. [9.3]

3. The hypotenuse of a right triangle is 3 ft long. One leg is 1 ft longer than the other. Find the lengths of the legs. Round any irrational answers to the nearest hundredth. [9.3]

4. Solve $p = t(2 + ab)$ for a. [9.4]

5. Solve $y = \dfrac{k}{x^2}$ for x. [9.4]

Prepare to Move On
Simplify. [8.2]

1. $\sqrt{48}$ 2. $\sqrt{250}$

3. $\sqrt{20} \cdot \sqrt{30}$ 4. $\sqrt{75} \cdot \sqrt{12}$

9.5   Complex Numbers as Solutions of Quadratic Equations

A. The Complex-Number System   **B.** Solutions of Equations

A. The Complex-Number System

Negative numbers do not have square roots in the real-number system. However, the **complex-number system**, containing the real-number system, is designed so that negative numbers *do* have square roots. To develop the complex-number system, we use the number i to represent the square root of -1.

THE NUMBER i

i is the unique number for which $i^2 = -1$ and $i = \sqrt{-1}$.

The square root of any negative number can be written using i. If $p \geq 0$, then

$$\sqrt{-p} = \sqrt{-1}\sqrt{p} = i\sqrt{p}, \text{ or } \sqrt{p}\,i.$$

EXAMPLE 1   Express in terms of i.

a) $\sqrt{-3}$ b) $\sqrt{-25}$

c) $-\sqrt{-10}$ d) $\sqrt{-24}$

SOLUTION

a) $\sqrt{-3} = \sqrt{-1 \cdot 3} = \sqrt{-1} \cdot \sqrt{3} = i\sqrt{3}$, or $\sqrt{3}\,i$ ← i is *not* under the radical.

b) $\sqrt{-25} = \sqrt{-1 \cdot 25} = \sqrt{-1} \cdot \sqrt{25} = i \cdot 5 = 5i$

c) $-\sqrt{-10} = -\sqrt{-1 \cdot 10} = -\sqrt{-1} \cdot \sqrt{10} = -i\sqrt{10}$, or $-\sqrt{10}i$

d) $\sqrt{-24} = \sqrt{4(-1)6} = \sqrt{4}\sqrt{-1}\sqrt{6} = 2i\sqrt{6}$, or $2\sqrt{6}\,i$

1. Express in terms of i: $\sqrt{-64}$.

YOUR TURN

> ### IMAGINARY NUMBERS
>
> An *imaginary number* is a number that can be written in the form $a + bi$, where a and b are real numbers and $b \neq 0$.

The following are examples of imaginary numbers:

$$3 + 8i, \quad \sqrt{7} - 2i, \quad 4 + \sqrt{6}\,i, \quad \text{and} \quad 4i\,(\text{here } a = 0).$$

Numbers like $4i$ are often called *pure imaginary* because they are of the form $0 + bi$. Imaginary numbers can be thought of as "nonreal." Together, the imaginary numbers and the real numbers form the set of **complex numbers**.*

> ### COMPLEX NUMBERS
>
> A *complex number* is any number that can be written as $a + bi$, where a and b are real numbers. (Note that a and b both can be 0.)

Types of Numbers

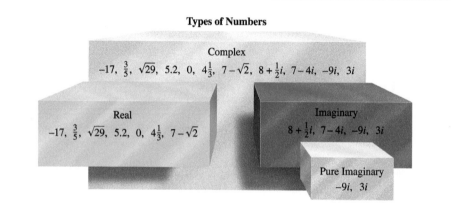

It may help to remember that every real number is a complex number, but not every complex number is real. Numbers like 2 and $\sqrt{3}$, where $b = 0$, are complex and real. Numbers like $7 + 3i$ and $-5i$, where $b \neq 0$, are complex and not real.

B. Solutions of Equations

Not all quadratic equations have real-number solutions. All quadratic equations *do* have complex-number solutions. These solutions are generally written in the form $a + bi$ unless a or b is zero.

EXAMPLE 2 Solve: **(a)** $x^2 = -100$; **(b)** $x^2 + 3x + 4 = 0$; **(c)** $x^2 + 2 = 2x$.

SOLUTION

a) We use the principle of square roots:

$$x^2 = -100$$
$$x = \sqrt{-100} \quad \text{or} \quad x = -\sqrt{-100} \qquad \text{Using the principle of square roots}$$
$$x = \sqrt{-1}\sqrt{100} \quad \text{or} \quad x = -\sqrt{-1}\sqrt{100}$$
$$x = 10i \quad \text{or} \quad x = -10i. \qquad \sqrt{-1} = i \text{ and } \sqrt{100} = 10$$

The solutions are $10i$ and $-10i$.

Student Notes

Do not confuse the a and the b in the expression $a + bi$ with the a and the b used in the quadratic formula.

---

* The names "imaginary" and "complex" should not lead you to believe that these numbers are unimportant or complicated. Imaginary numbers and complex numbers have important applications in engineering and the physical sciences.

Express in terms of i.

1. $\sqrt{-1} \cdot \sqrt{4}$
2. $\sqrt{-1} \cdot \sqrt{5}$
3. $\sqrt{-1 \cdot 16}$
4. $\sqrt{-1 \cdot 7}$
5. $\sqrt{-1 \cdot 9 \cdot 2}$
6. $\sqrt{25(-1)3}$

b) We use the quadratic formula:

$$1x^2 + 3x + 4 = 0 \qquad\qquad a = 1, b = 3, c = 4$$

$$x = \frac{-3 \pm \sqrt{3^2 - 4 \cdot 1 \cdot 4}}{2 \cdot 1} \qquad x = \frac{-b \pm \sqrt{b^2 - 4ac}}{2a}$$

$$x = \frac{-3 \pm \sqrt{-7}}{2} \qquad\qquad \text{Simplifying}$$

$$x = \frac{-3 \pm \sqrt{-1}\sqrt{7}}{2}$$

$$x = \frac{-3 \pm i\sqrt{7}}{2} \qquad\qquad \text{Rewriting } \sqrt{-7} \text{ as } i\sqrt{7}$$

$$x = -\frac{3}{2} \pm \frac{\sqrt{7}}{2}i. \qquad\qquad \text{Writing in the form } a + bi$$

The solutions are $-\dfrac{3}{2} + \dfrac{\sqrt{7}}{2}i$ and $-\dfrac{3}{2} - \dfrac{\sqrt{7}}{2}i$.

c) We have

$$x^2 + 2 = 2x$$

$$1x^2 - 2x + 2 = 0 \qquad \text{Rewriting in standard form; } a = 1, b = -2, c = 2$$

$$x = \frac{-(-2) \pm \sqrt{(-2)^2 - 4 \cdot 1 \cdot 2}}{2 \cdot 1} \qquad x = \frac{-b \pm \sqrt{b^2 - 4ac}}{2a}$$

$$x = \frac{2 \pm \sqrt{-4}}{2} \qquad\qquad \text{Simplifying}$$

$$x = \frac{2 \pm \sqrt{-1}\sqrt{4}}{2}$$

$$x = \frac{2 \pm 2i}{2} \qquad\qquad \sqrt{-1} = i \text{ and } \sqrt{4} = 2$$

$$x = \frac{2}{2} \pm \frac{2}{2}i \qquad\qquad \text{Rewriting in the form } a + bi$$

$$x = 1 \pm i. \qquad\qquad \text{Simplifying}$$

The solutions are $1 + i$ and $1 - i$.

2. Solve: $x^2 + x + 1 = 0$.

↪ YOUR TURN

CONNECTING 🔗 THE CONCEPTS

We have now completed our study of the various types of equations that appear in elementary algebra. Below are examples of each type, along with their solutions.

Linear Equations		*Radical Equations*	
$5x + 3 = 2x + 9$		$\sqrt{2x + 1} - 5 = 2$	
$3x = 6$	Adding $-2x - 3$ to both sides	$\sqrt{2x + 1} = 7$	Adding 5 to both sides
$x = 2$	Dividing both sides by 3	$2x + 1 = 49$	Squaring both sides
The solution is 2.		$2x = 48$	
		$x = 24$	
		The solution is 24.	

(continued)

Rational Equations

$$\frac{5}{2x} + \frac{4}{3x} = 2 \qquad \text{Note that } x \neq 0.$$

$$6x\left(\frac{5}{2x} + \frac{4}{3x}\right) = 6x \cdot 2 \qquad \text{Multiplying both sides by the LCD, } 6x$$

$$15 + 8 = 12x \qquad \text{Simplifying}$$

$$23 = 12x$$

$$\frac{23}{12} = x$$

The solution is $\frac{23}{12}$.

Quadratic Equations

$$2x^2 + 3x = 1$$

$$2x^2 + 3x - 1 = 0 \qquad \text{Subtracting 1 from both sides}$$

$$x = \frac{-3 \pm \sqrt{3^2 - 4(2)(-1)}}{2 \cdot 2} \qquad \begin{array}{l}\text{Using the}\\ \text{quadratic}\\ \text{formula}\end{array}$$

$$x = \frac{-3 \pm \sqrt{17}}{4}$$

The solutions are $\dfrac{-3}{4} + \dfrac{\sqrt{17}}{4}$ and $\dfrac{-3}{4} - \dfrac{\sqrt{17}}{4}$.

You should always check solutions in the original equation in case an error was made in the solution process. For rational equations and radical equations, a check is necessary because the equation-solving principles used may not yield equivalent equations. The checks for the equations above are left to the student.

EXERCISES

Solve.

1. $3x + 8 = 7x + 4$

2. $x^2 = 5x - 6$

3. $3 + \sqrt{x} = 8$

4. $\dfrac{5}{x} + \dfrac{3}{4} = 2$

5. $11 = 4\sqrt{3x + 1} - 5$

6. $n^2 + 2n - 2 = 0$

7. $x + 1 = \dfrac{5}{x - 3}$

8. $4x - 7 = 2(5x - 3)$

9. $t^2 + 100 = 0$

10. $t^2 - 100 = 0$

9.5 EXERCISE SET

FOR EXTRA HELP MyMathLab®

Vocabulary and Reading Check

Classify each of the following statements as either true or false.

1. Every complex number is a real number.

2. Many complex numbers are real numbers.

3. Many complex numbers are imaginary numbers.

4. Every complex number is an imaginary number.

A. The Complex-Number System

Express in terms of i.

5. $\sqrt{-1}$

6. $\sqrt{-4}$

7. $\sqrt{-49}$

8. $\sqrt{-100}$

9. $\sqrt{-5}$

10. $\sqrt{-15}$

11. $\sqrt{-45}$

12. $\sqrt{-18}$

13. $-\sqrt{-50}$

14. $-\sqrt{-12}$

15. $4 + \sqrt{-49}$

16. $7 + \sqrt{-4}$

17. $3 - \sqrt{-9}$

18. $-8 - \sqrt{-36}$

19. $-2 + \sqrt{-75}$

20. $5 - \sqrt{-20}$

B. Solutions of Equations

Solve.

Aha! **21.** $x^2 + 25 = 0$

22. $x^2 + 16 = 0$

23. $x^2 = -28$

24. $x^2 = -48$

25. $t^2 + 4t + 5 = 0$

26. $t^2 - 4t + 6 = 0$

27. $(x - 4)^2 = -9$

28. $(x + 3)^2 = -4$

29. $x^2 + 5 = 2x$

30. $x^2 + 3 = -2x$

31. $t^2 + 7 - 4t = 0$

32. $t^2 + 8 + 4t = 0$

33. $5y^2 + 4y + 1 = 0$

34. $4y^2 + 3y + 2 = 0$

35. $1 + 2m + 3m^2 = 0$

36. $4p^2 + 3 = 6p$

37. Is it possible for a quadratic equation to have one imaginary-number solution and one real-number solution? Why or why not?

38. Under what condition(s) will an equation of the form $x^2 = c$ have imaginary-number solutions?

Skill Review

Simplify. Assume that variables represent nonnegative numbers.

39. $-\sqrt[3]{-1000}$ [8.7]   **40.** $\sqrt{50x^3y^8}$ [8.2]

41. $5\sqrt{8} - 3\sqrt{18}$ [8.4]

42. $(3 + 2\sqrt{5})(3 - 2\sqrt{5})$ [8.4]

43. $\dfrac{\dfrac{1}{x} - \dfrac{3}{4x}}{\dfrac{2}{3x} + \dfrac{5}{x}}$ [6.5]   **44.** $\sqrt{\dfrac{100n^2}{49}}$ [8.3]

Synthesis

45. When the quadratic formula is used to solve an equation, if $b^2 < 4ac$, are the solutions imaginary? Why or why not?

46. Can imaginary-number solutions of a quadratic equation be found using the method of completing the square? Why or why not?

Solve.

47. $(x + 1)^2 + (x + 3)^2 = 0$

48. $(p + 5)^2 + (p + 1)^2 = 0$

49. $\dfrac{2x - 1}{5} - \dfrac{2}{x} = \dfrac{x}{2}$

50. $\dfrac{1}{a - 1} - \dfrac{2}{a - 1} = 3a$

51. Use a graphing calculator to confirm that there are no real-number solutions of Examples 2(a), 2(b), and 2(c).

YOUR TURN ANSWERS: SECTION 9.5

1. $8i$ **2.** $-\dfrac{1}{2} \pm \dfrac{\sqrt{3}}{2}i$

Quick Quiz: Sections 9.1–9.5

1. Express in terms of i: $\sqrt{-36}$.   [9.5]

Solve.

2. $x^2 - 12 = 0$ [9.1]   **3.** $x^2 + 12 = 0$ [9.5]

4. $y^2 - 2y + 1 = 0$ [9.1]   **5.** $p^2 - 4p = 1$ [9.3]

Prepare to Move On

Graph.   [3.2], [3.3], [3.6]

1. $2x - 3y = 10$ **2.** $y = -4x$

3. $x = 2$ **4.** $y = \frac{1}{2}x - 3$

5. $y = -1$

<table>
<tr><td>**9.6**</td><td>**Graphs of Quadratic Equations**</td></tr>
</table>

A. Graphing Equations of the Form $y = ax^2$ **B.** Graphing Equations of the Form $y = ax^2 + bx + c$

Study Skills

Get Some Rest

The final exam is probably your most important math test of the semester. Do yourself a favor and be sure to get a good night's sleep the night before. Being well-rested will help guarantee that you put forth your best work.

In this section, we graph quadratic equations like

$$y = \tfrac{1}{2}x^2, \qquad y = x^2 + 2x - 3, \quad \text{and} \quad y = -5x^2 + 4.$$

Such equations, of the form $y = ax^2 + bx + c$ with $a \neq 0$, have graphs that are cupped either upward or downward. These graphs are symmetric with respect to an **axis of symmetry**, as shown below. When folded along its axis, the graph has two halves that match exactly. These graphs are called **parabolas**.

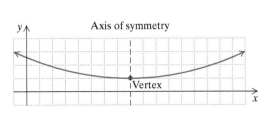

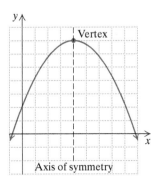

The point at which the graph of a quadratic equation crosses its axis of symmetry is called the **vertex** (plural, vertices). The y-coordinate of the vertex is the graph's largest value of y if the curve opens downward or its smallest value of y if the curve opens upward.

A. Graphing Equations of the Form $y = ax^2$

The simplest parabolas to sketch are given by equations of the form $y = ax^2$.

EXAMPLE 1 Graph: $y = x^2$.

SOLUTION We choose numbers for x and find the corresponding values for y.

If $x = -2$, then $y = (-2)^2 = 4$.   We get the pair $(-2, 4)$.
If $x = -1$, then $y = (-1)^2 = 1$.   We get the pair $(-1, 1)$.
If $x = 0$,   then $y = 0^2 = 0$.   We get the pair $(0, 0)$.
If $x = 1$,   then $y = 1^2 = 1$.   We get the pair $(1, 1)$.
If $x = 2$,   then $y = 2^2 = 4$.   We get the pair $(2, 4)$.

The following table lists these solutions of $y = x^2$. After several ordered pairs are found, we plot them and connect them with a smooth curve.

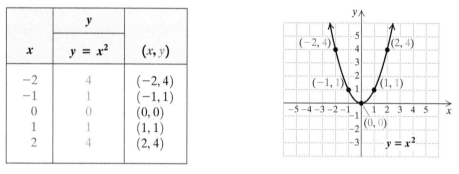

x	y $y = x^2$	(x, y)
-2	4	$(-2, 4)$
-1	1	$(-1, 1)$
0	0	$(0, 0)$
1	1	$(1, 1)$
2	4	$(2, 4)$

1. Graph: $y = 2x^2$.

YOUR TURN

In Example 1, the vertex is $(0, 0)$ and the axis of symmetry is the y-axis. This will be the case for any parabola having an equation of the form $y = ax^2$.

EXAMPLE 2 Graph: $y = -\frac{1}{2}x^2$.

SOLUTION We select numbers for x, find the corresponding y-values, plot the resulting ordered pairs, and connect them with a smooth curve.

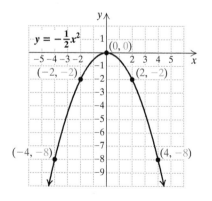

x	y $y = -\frac{1}{2}x^2$	(x, y)
-4	-8	$(-4, -8)$
-2	-2	$(-2, -2)$
0	0	$(0, 0)$
2	-2	$(2, -2)$
4	-8	$(4, -8)$

2. Graph: $y = -\frac{3}{2}x^2$.

YOUR TURN

Student Notes

The graphs of some parabolas do not cross the horizontal axis. For example, if the parabola in Example 3 had the same vertex, but opened downward instead of upward, no x-intercepts would exist.

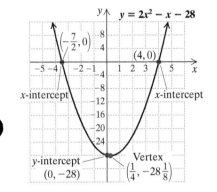

3. Find all y- and x-intercepts of the graph of

$$y = x^2 + 2x - 8.$$

B. Graphing Equations of the Form $y = ax^2 + bx + c$

The points at which a graph crosses the x- and y-axes are called its x- and y-intercepts, respectively. To find the intercepts of a parabola, we use the same approach that we used with lines.

THE INTERCEPTS OF A PARABOLA

To find the y-intercept of the graph of $y = ax^2 + bx + c$, replace x with 0 and solve for y. The result will be c.

To find the x-intercept(s) of the graph of $y = ax^2 + bx + c$, if any exist, replace y with 0 and solve for x. To do this, factor or use the quadratic formula. If no real solution exists, there are no x-intercepts.

EXAMPLE 3   Find all y- and x-intercepts of the graph of $y = 2x^2 - x - 28$.

SOLUTION   To find the y-intercept, we replace x with 0 and solve for y:

$$y = 2 \cdot 0^2 - 0 - 28 \qquad \text{At a } y\text{-intercept, } x = 0.$$
$$y = 0 - 0 - 28 = -28.$$

When x is 0, we have $y = -28$. Thus the y-intercept is $(0, -28)$.

To find the x-intercept(s), we replace y with 0 and solve for x:

$$0 = 2x^2 - x - 28. \qquad \text{At any } x\text{-intercept, } y = 0.$$

The quadratic formula could be used, but factoring is faster:

$$0 = (2x + 7)(x - 4) \qquad \text{Factoring}$$

$$
\begin{aligned}
2x + 7 &= 0 && or && x - 4 = 0 \\
2x &= -7 && or && x = 4 \\
x &= -\tfrac{7}{2} && or && x = 4.
\end{aligned}
$$

The x-intercepts are $(4, 0)$ and $\left(-\tfrac{7}{2}, 0\right)$, and the y-intercept is $(0, -28)$, as shown in the graph at left.

YOUR TURN

Although we were not asked to graph the equation in Example 3, we did so to show that the x-coordinate of the vertex, $\tfrac{1}{4}$, is exactly midway between the x-intercepts. The quadratic formula can be used to show that the x-coordinate of the vertex is $\dfrac{-b}{2a}$. In Exercise 46, you are asked to demonstrate this by averaging the x-coordinates of the x-intercepts.

THE VERTEX OF A PARABOLA

For any parabola given by an equation of the form $y = ax^2 + bx + c$:

1. The x-coordinate of the vertex is $-\dfrac{b}{2a}$.

2. The y-coordinate of the vertex can be found by substituting $-\dfrac{b}{2a}$ for x and solving for y.

EXAMPLE 4   Graph: $y = x^2 + 2x - 3$.

SOLUTION   Our plan is to plot the vertex and some points on either side of the vertex. We will then draw a parabola passing through these points.

To locate the vertex, we use $-b/(2a)$ to find its x-coordinate:

$$x\text{-coordinate of the vertex} = -\frac{b}{2a} = -\frac{2}{2 \cdot (1)}$$
$$= -1.$$

We substitute -1 for x to find the y-coordinate of the vertex:

$$y\text{-coordinate of the vertex} = (-1)^2 + 2(-1) - 3 = 1 - 2 - 3$$
$$= -4.$$

The vertex is $(-1, -4)$. The axis of symmetry is $x = -1$.

Next, we choose some x-values on both sides of the vertex and graph the parabola.

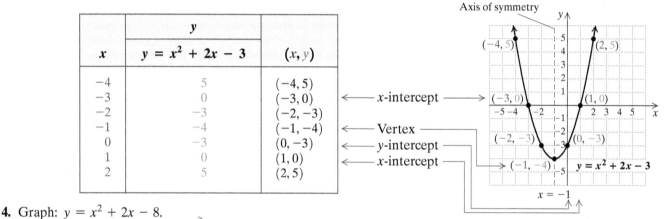

x	y $y = x^2 + 2x - 3$	(x, y)	
-4	5	$(-4, 5)$	
-3	0	$(-3, 0)$	$\longleftarrow$ x-intercept $\longrightarrow$
-2	-3	$(-2, -3)$	
-1	-4	$(-1, -4)$	$\longleftarrow$ Vertex
0	-3	$(0, -3)$	$\longleftarrow$ y-intercept
1	0	$(1, 0)$	$\longleftarrow$ x-intercept
2	5	$(2, 5)$	

4. Graph: $y = x^2 + 2x - 8$.

YOUR TURN

✎ Check Your
UNDERSTANDING

Determine whether the graph of the equation opens upward or downward.

1. $y = 4x^2 - 6x + 7$

2. $y = -2x^2 + 10x + 3$

3. $y = x^2 - 2x + 1$

Find the y-intercept of the graph of the equation.

4. $y = 3x^2 - 6x + 8$

5. $y = x^2 - 5$

Note that in Examples 1, 3, and 4, a is positive and the graph opens upward. In Example 2, a is negative and the graph opens downward. This is true in general: The graph of $y = ax^2 + bx + c$ is cupped upward for $a > 0$ and downward for $a < 0$.

EXAMPLE 5   Graph: $y = -2x^2 + 4x + 1$.

SOLUTION   Since the coefficient of x^2 is negative, we expect the graph to open downward. To locate the vertex, we first find its x-coordinate:

$$x\text{-coordinate of the vertex} = -\frac{b}{2a} = -\frac{4}{2 \cdot (-2)}$$
$$= 1.$$

We substitute 1 for x to find the y-coordinate of the vertex:

$$y\text{-coordinate of the vertex} = -2 \cdot 1^2 + 4 \cdot 1 + 1 = -2 + 4 + 1$$
$$= 3.$$

The vertex is $(1, 3)$. The axis of symmetry is $x = 1$.

We choose some x-values on both sides of the vertex, calculate their corresponding y-values, and graph the parabola.

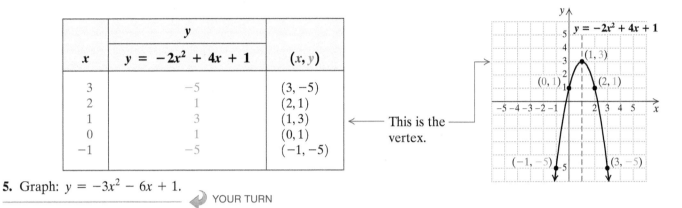

x	y $y = -2x^2 + 4x + 1$	(x, y)
3	-5	$(3, -5)$
2	1	$(2, 1)$
1	3	$(1, 3)$
0	1	$(0, 1)$
-1	-5	$(-1, -5)$

⟵ This is the vertex.

5. Graph: $y = -3x^2 - 6x + 1$.

YOUR TURN

In Examples 1–5, note that any x-value to the left of the vertex is paired with the same y-value as an x-value the same distance to the right of the vertex. Thus, since the vertex for Example 5 is $(1, 3)$ and since the x-values -1 and 3 are both 2 units from 1, we know that -1 and 3 are both paired with the same y-value. This symmetry provides a useful check and allows us to plot *two* points after calculating just *one*.

GUIDELINES FOR GRAPHING QUADRATIC EQUATIONS

1. Graphs of quadratic equations, $y = ax^2 + bx + c$, are parabolas. They are cupped upward for $a > 0$ and downward for $a < 0$.

2. Use the formula $x = -b/(2a)$ to find the x-coordinate of the vertex. After calculating the y-coordinate, plot the vertex and some points on either side of it.

3. Once a point is graphed, a second point with the same y-coordinate can be plotted on the opposite side of the axis of symmetry.

4. Graph the y-intercept and, if requested, any x-intercepts.

Chapter Resource:
Visualizing for Success, p. 587

9.6 EXERCISE SET

FOR EXTRA HELP MyMathLab®

🢒 Vocabulary and Reading Check

Classify each of the following statements as either true or false.

1. The graph of every quadratic equation has two x-intercepts.

2. The sign of the constant a in $y = ax^2 + bx + c$ indicates whether the graph of the equation opens upward or downward.

3. The constant b in $y = ax^2 + bx + c$ gives the x-coordinate of the x-intercepts.

4. The constant c in $y = ax^2 + bx + c$ gives the y-coordinate of the y-intercept.

A. Graphing Equations of the Form $y = ax^2$

Graph each quadratic equation, labeling the vertex and the y-intercept.

5. $y = 2x^2$

6. $y = 3x^2$

7. $y = -2x^2$

8. $y = -1 \cdot x^2$

9. $y = -\frac{1}{3}x^2$

10. $y = \frac{1}{4}x^2$

B. Graphing Equations of the Form $y = ax^2 + bx + c$

Graph each quadratic equation, labeling the vertex and the y-intercept.

11. $y = x^2 - 2$

12. $y = x^2 + 1$

13. $y = x^2 - 2x + 1$

14. $y = x^2 + 4x + 4$

15. $y = x^2 + 3x - 10$

16. $y = x^2 - 2x - 3$

17. $y = -2x^2 + 12x - 13$

18. $y = -3x^2 + 12x - 11$

19. $y = -\frac{1}{2}x^2 + 5$

20. $y = \frac{1}{2}x^2 - 7$

21. $y = x^2 - 3x$

22. $y = -x^2 + 2x$

Graph each equation, labeling the vertex, the y-intercept, and any x-intercepts. If an x-intercept is irrational, use a calculator and round to three decimal places.

23. $y = x^2 + 2x - 8$

24. $y = x^2 + x - 6$

25. $y = 2x^2 - 6x$

26. $y = 2x^2 - 7x$

27. $y = -x^2 - x + 12$

28. $y = -x^2 - 3x + 10$

29. $y = 3x^2 - 6x + 1$

30. $y = 3x^2 + 12x + 11$

31. $y = x^2 + 2x + 3$

32. $y = -x^2 - 2x - 3$

33. $y = 3 - 4x - 2x^2$

34. $y = 1 - 4x - 2x^2$

35. Why is it helpful to know the coordinates of the vertex when graphing a parabola?

36. Suppose that both x-intercepts of a parabola are known. What is the easiest way to find the coordinates of the vertex?

Skill Review

37. Find the LCM of 18 and 30.   [6.3]

38. Simplify $(5.2 \times 10^9)(3.5 \times 10^{12})$. Write the result using scientific notation.   [4.2]

39. Compute: $2 - (1 - 5)^2 \div 2 \cdot 4 - 3$.   [1.8]

40. Find the slope and the y-intercept: $y = \frac{2}{3}x - 9$.   [3.6]

41. Find the slope–intercept equation for the line containing $(-2, 4)$ and having slope $m = -1$.   [3.7]

42. Solve $\dfrac{1}{a} = \dfrac{1}{b} - \dfrac{1}{c}$ for b.   [9.4]

Synthesis

43. Describe a method that could be used to find an equation for a parabola that has x-intercepts $(p, 0)$ and $(q, 0)$.

44. What effect does the size of $|a|$ have on the graph of $y = ax^2 + bx + c$?

45. *Height of a Golf Ball.*   The height H, in feet, of a golf ball with an initial velocity of 96 ft/sec is given by the equation

$$H = -16t^2 + 96t,$$

where t is the number of seconds from launch. Use the graph of this equation, shown below, or any equation-solving technique to answer the following.

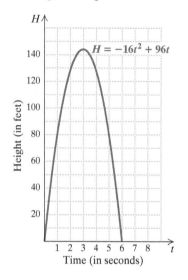

a) How many seconds after launch is the golf ball 128 ft above ground?

b) When does the golf ball reach its maximum height?

Aha! **c)** How many seconds after launch does the golf ball return to the ground?

46. Show that the average of

$$\frac{-b - \sqrt{b^2 - 4ac}}{2a} \quad \text{and} \quad \frac{-b + \sqrt{b^2 - 4ac}}{2a}$$

is $-\dfrac{b}{2a}$.

47. *Stopping Distance.*   In how many feet can a car stop if it is traveling at a speed of r miles per hour? One estimate, developed in Britain, is as follows. The distance d, in feet, is given by

$$d = \underbrace{\text{Thinking distance}}_{\text{(in feet)}} + \underbrace{\text{Stopping distance}}_{\text{(in feet)}}$$

$$d = \quad\;\; r \qquad + \qquad 0.05r^2.$$

a) How many feet would it take to stop a car traveling 25 mph? 40 mph? 55 mph? 65 mph? 75 mph? 100 mph?

b) Graph the equation, assuming $r \geq 0$.

48. On one set of axes, graph $y = x^2$, $y = (x - 3)^2$, and $y = (x + 1)^2$. Describe the effect that h has on the graph of $y = (x - h)^2$.

49. On one set of axes, graph $y = x^2$, $y = x^2 - 5$, and $y = x^2 + 2$. Describe the effect that k has on the graph of $y = x^2 + k$.

50. *Seller's Supply.* As the price of a product increases, the seller is willing to sell, or *supply*, more of the product. Suppose that the supply for a certain product is given by

$$S = p^2 + p + 10,$$

where p is the price in dollars and S is the number supplied, in thousands, at that price. Graph the equation for values of p such that $0 \le p \le 6$.

51. *Consumer's Demand.* As the price of a product increases, consumers purchase, or *demand*, less of the product. Suppose that the demand for a certain product is given by

$$D = (p - 6)^2,$$

where p is the price in dollars and D is the number demanded, in thousands, at that price. Graph the equation for values of p such that $0 \le p \le 6$.

52. *Equilibrium Point.* The price p at which the consumer and the seller agree determines the *equilibrium point*. Find p such that

$$D = S$$

for the demand and supply curves in Exercises 50 and 51. How many units of the product will be sold at that price?

53. Use a graphing calculator to graph $y = x^2 - 5$ and then, using the graph, estimate $\sqrt{5}$ to four decimal places.

54. Explain the graphical significance of the result of Exercise 46.

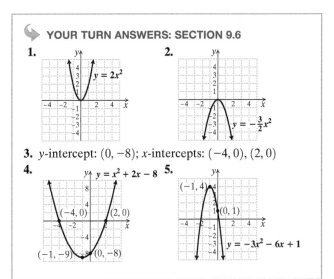

1. [graph labeled $y = 2x^2$]   **2.** [graph labeled $y = -\frac{3}{2}x^2$]

3. y-intercept: $(0, -8)$; x-intercepts: $(-4, 0)$, $(2, 0)$

4. [graph labeled $y = x^2 + 2x - 8$ with points $(-4, 0)$, $(2, 0)$, $(-1, -9)$, $(0, -8)$]   **5.** [graph labeled $y = -3x^2 - 6x + 1$ with points $(-1, 4)$, $(0, 1)$]

Quick Quiz: Sections 9.1–9.6

1. Determine the number that will complete the square: $x^2 + 6x$.   [9.2]

2. Solve $x + \dfrac{1}{x} = y$ for x.   [9.4]

3. Approximate the solutions to the nearest thousandth: $2x^2 + 3x = 4$.   [9.3]

4. Suppose that \$1500 is invested at interest rate r, compounded annually, and that in 2 years, it grows to \$1591.35. What is the interest rate? Use $A = P(1 + r)^2$.   [9.3]

5. Graph $y = x^2 - 4$. Label the vertex, the y-intercept, and any x-intercepts.   [9.6]

Prepare to Move On

Evaluate.   [4.2]

1. $3x^2 - 4x$, for $x = -1$ **2.** $5a^3 + 3a$, for $a = -1$

3. $6 - t^3$, for $t = -2$ **4.** $3t^4 + t^2$, for $t = 10$

5. $(a - 9)^2$, for $a = 8$ **6.** $(a - 9)^2$, for $a = -3$

9.7 Functions

A. Identifying Functions   **B.** Function Notation   **C.** Graphs of Functions
D. Recognizing Graphs of Functions

Functions are enormously important in modern mathematics and science. The more mathematics and science you study, the more you will use functions.

A. Identifying Functions

Functions appear regularly in many contexts. To motivate understanding of functions, consider the following table relating heart rate to average life span.

	Elephant	Horse	Human	Cat	Monkey	Chicken	Hamster
Heart Rate (in beats per minute)	30	44	60	150	190	275	450
Average Life Span (in years)	70	40	70	15	15	15	3

Note that to each heart rate there corresponds *exactly one* average life span. A correspondence of this type is called a **function**.

> **FUNCTION**
>
> A *function* is a correspondence (or rule) that assigns to each member of some set (called the *domain*) exactly one member of another set (called the *range*).

The members of the domain are sometimes called **inputs**, and the members of the range **outputs**. For the information above, we have the following.

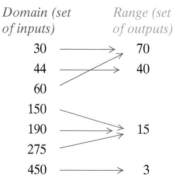

Domain (set of inputs)   *Range (set of outputs)*

The function above can be written as a set of ordered pairs. Note that each input has exactly *one* output, even though some outputs are used more than once:

$$\{(30, 70), \ (44, 40), \ (60, 70), \ (150, 15), \ (190, 15), \ (275, 15), \ (450, 3)\}.$$

Correspondences are often named using letters.

EXAMPLE 1 Determine whether or not each of the following correspondences is a function.

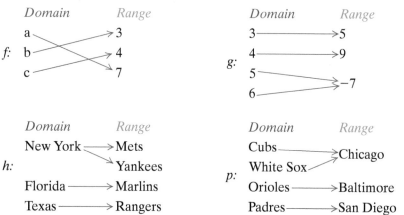

1. Determine whether or not the following correspondence is a function.

Domain   *Range*

Ethiopia → Africa

Kenya

Austria → Europe

Brazil

Argentina → South America

SOLUTION Correspondence *f is* a function because each member of the domain is paired with just one member of the range.

Correspondence *g is* also a function because each member of the domain is paired with just one member of the range.

Correspondence *h is not* a function because one member of the domain, New York, is matched to more than one member of the range.

Correspondence *p is* a function because each member of the domain is paired with just one member of the range.

YOUR TURN

B. Function Notation

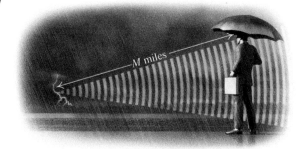

Many functions are described by formulas. Equations like $y = x + 3$ and $y = 4x^2$ are examples of such formulas. Outputs are found by substituting members of the domain for x.

For example, during a thunderstorm, it is possible to calculate how far away lightning is by using the formula $M = \frac{1}{5}t$. Here, M is the distance, in miles, that a storm is from an observer when the sound of thunder arrives t seconds after the lightning appears.

We can make a table of values for this function by substituting values of t and computing M.

For $t = 1$, $M = \frac{1}{5} \cdot 1 = \frac{1}{5}$.   For $t = 2$, $M = \frac{1}{5} \cdot 2 = \frac{2}{5}$.

For $t = 3$, $M = \frac{1}{5} \cdot 3 = \frac{3}{5}$.   For $t = 4$, $M = \frac{1}{5} \cdot 4 = \frac{4}{5}$.

For $t = 5$, $M = \frac{1}{5} \cdot 5 = 1$.   For $t = 10$, $M = \frac{1}{5} \cdot 10 = 2$.

t (in seconds)	1	2	3	4	5	10
M (in miles)	$\frac{1}{5}$	$\frac{2}{5}$	$\frac{3}{5}$	$\frac{4}{5}$	1	2

In the table above, each red input (seconds) corresponds to one blue output (miles). *Function notation* clearly and concisely presents inputs and outputs together. If we name the above function M, the notation $M(t)$, read "M of t," denotes the output that is paired with the input t by the function M:

$$M(2) = \frac{1}{5} \cdot 2 = \frac{2}{5}, \qquad M(3) = \frac{1}{5} \cdot 3 = \frac{3}{5}, \quad \text{and, in general,} \quad M(t) = \frac{1}{5} \cdot t.$$

The notation $M(4) = \frac{4}{5}$ means "when 4 is the input, $\frac{4}{5}$ is the output."

> **CAUTION!** $M(4)$ *does not* mean M times 4 and should not be read or pronounced that way.

Equations for nonvertical lines can be written in function notation. For example, $f(x) = x + 2$, read "f of x equals x plus 2," can be used instead of $y = x + 2$ when we are discussing functions, although both equations describe the same correspondence.

The variable x in $f(x) = x + 2$ is called the **independent variable**. To find a particular function value $f(a)$, we replace every occurrence of x with a.

EXAMPLE 2   For the function given by $f(x) = x + 2$, find each of the following.

a) $f(8)$ **b)** $f(-3)$ **c)** $f(0)$

SOLUTION

a) $f(8) = 8 + 2$, or 10 This function adds 2 to each input.

b) $f(-3) = -3 + 2$, or -1 $f(-3)$ is read "f of -3."

c) $f(0) = 0 + 2$, or 2 $f(0)$ *does not mean* $f \cdot 0$.

2. For the function given by $f(x) = 1 - x$, find $f(7)$.

⤺ YOUR TURN

It is sometimes helpful to think of a function as a machine that gives an output for each input that enters the machine. For example, the function $g(t) = 2t^2 + t$ pairs an input with the sum of the input and twice its square. The following diagram is one way in which the function given by $g(t) = 2t^2 + t$ can be illustrated.

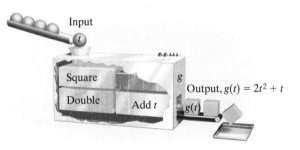

EXAMPLE 3   For the function given by $g(t) = 2t^2 + t - 7$, find each of the following.

a) $g(3)$ **b)** $g(0)$ **c)** $g(-2)$

SOLUTION

a) $g(3) = 2 \cdot 3^2 + 3 - 7$ Using 3 for each occurrence of t

$\quad = 2 \cdot 9 + 3 - 7$

$\quad = 14$

b) $g(0) = 2 \cdot 0^2 + 0 - 7$ Using 0 for each occurrence of t

$\quad = -7$

c) $g(-2) = 2(-2)^2 + (-2) - 7$ Using -2 for each occurrence of t

$\quad = 2 \cdot 4 - 2 - 7$

$\quad = -1$

3. For the function given by $g(t) = t^2 - t$, find $g(-1)$.

⤺ YOUR TURN

Outputs are also called **function values**. In Example 3, $g(-2) = -1$. We can say that the "function value at -2 is -1," or "when x is -2, the value of the function is -1." Most often we simply say "g of -2 is -1."

EXPLORING 🔍 THE CONCEPT

You can think of a function that is defined by an equation as giving a description of what operations to perform with any given input. For example, the function given by $f(x) = 2x - 5$ tells us to double an input and then subtract 5. The result is the output.

Match each of the following descriptions with the equation that performs the same operations.

1. Square an input and then add 3.
2. Triple an input and then subtract 2.
3. Square an input and then multiply by 3.
4. Subtract 3 from an input and then square.

a) $f(x) = (x - 3)^2$
b) $f(x) = 3x - 2$
c) $f(x) = 3x^2$
d) $f(x) = x^2 + 3$

ANSWERS

1. (d)   **2.** (b)   **3.** (c)   **4.** (a)

C. Graphs of Functions

To graph a function, we usually calculate ordered pairs of the form (x, y) or $(x, f(x))$, plot them, and connect the points. The symbols y and $f(x)$ are often used interchangeably when we are working with functions and their graphs.

EXAMPLE 4   Graph: $f(x) = x + 2$.

SOLUTION   A list of some function values is shown in the following table. We plot the points and connect them. The graph is a straight line.

x	$f(x)$
-4	-2
-3	-1
-2	0
-1	1
0	2
1	3
2	4
3	5
4	6

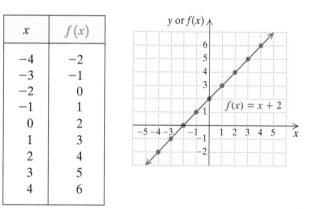

4. Graph: $g(x) = x - 1$.

YOUR TURN

EXAMPLE 5   Graph: $g(x) = 4 - x^2$.

SOLUTION   We calculate some function values and draw the curve. The graph is a parabola.

$$g(0) = 4 - 0^2 = 4 - 0 = 4,$$
$$g(-1) = 4 - (-1)^2 = 4 - 1 = 3,$$
$$g(2) = 4 - (2)^2 = 4 - 4 = 0,$$
$$g(-3) = 4 - (-3)^2 = 4 - 9 = -5$$

x	g(x)
−3	−5
−2	0
−1	3
0	4
1	3
2	0
3	−5

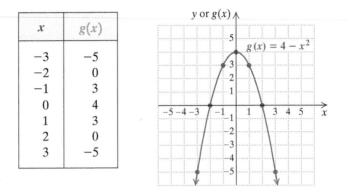

5. Graph: $h(x) = x^2 + 1$.

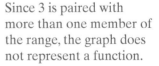 YOUR TURN

EXAMPLE 6   Graph: $h(x) = |x|$.

SOLUTION   A list of some function values is shown in the following table. We plot the points and connect them. The graph is V-shaped and symmetric, rising on either side of the vertical axis.

x	h(x)
−3	3
−2	2
−1	1
0	0
1	1
2	2
3	3

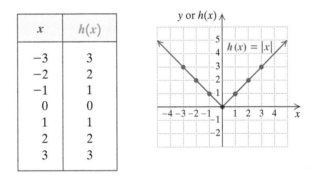

6. Graph: $f(x) = |x + 1|$.

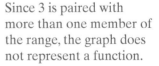 YOUR TURN

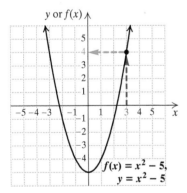

D. Recognizing Graphs of Functions

Consider the function f described by $f(x) = x^2 - 5$. Its graph is shown at left. It is also the graph of the equation $y = x^2 - 5$.

To find a function value, like $f(3)$, from a graph, we locate the input on the horizontal axis, move vertically to the graph of the function, and then horizontally to find the output on the vertical axis, where members of the range are found.

When one member of the domain is paired with two or more different members of the range, the correspondence is not a function. Thus, when a graph contains two or more points with the same first coordinate, it cannot represent a function. Points sharing a common first coordinate are vertically above and below each other, as shown in the following figure.

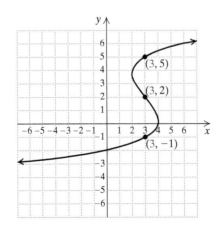

Since 3 is paired with more than one member of the range, the graph does not represent a function.

↳ Check Your
UNDERSTANDING

Find the function values for each value of *x* given.

1.

x	*f*(*x*) = 7 − *x*
−1	
0	
2	
7	

2.

x	*g*(*x*) = *x*² − 1
−2	
−1	
0	
1	
2	

This observation leads to the *vertical-line test*.

> **THE VERTICAL-LINE TEST**
>
> A graph represents a function if it is impossible to draw a vertical line that intersects the graph more than once.

EXAMPLE 7 Determine whether each of the following is the graph of a function.

a)

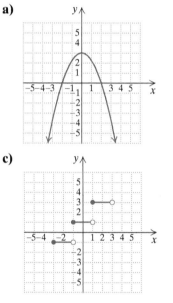

b)

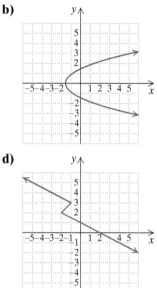

c)

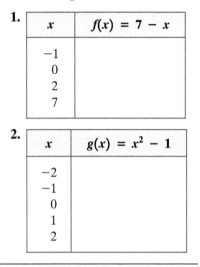

d)

Student Notes

To apply the vertical-line test, you can lay a ruler, a pencil, or an ID card parallel to the *y*-axis at the far left of the graph. As you move the straightedge from left to right, the vertical lines formed cannot cross the graph more than once.

SOLUTION

a) The graph *is* that of a function because it is not possible to draw a vertical line that crosses the graph more than once. This can be confirmed with a ruler or a straightedge.

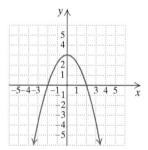

b) The graph *is not* that of a function because it does not pass the vertical-line test. The line *x* = 1 is one of many vertical lines that cross the graph at more than one point.

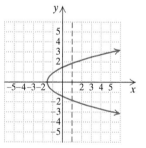

c) The graph *is* that of a function because it is not possible to draw a vertical line that crosses the graph more than once. Note that the open dots indicate the absence of a point.

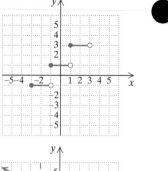

7. Determine whether the following graph is the graph of a function.

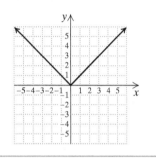

d) The graph *is not* that of a function because it does not pass the vertical-line test. For example, as shown, the line $x = -2$ crosses the graph at more than one point.

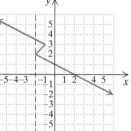

YOUR TURN

9.7 EXERCISE SET

FOR EXTRA HELP   MyMathLab®

Vocabulary and Reading Check

Classify each of the following statements as either true or false.

1. When functions are discussed, the notation $f(3)$ does not mean $f \cdot 3$.

2. If $f(x) = x^2$, then $f(-5) = 25$.

3. In order to pass the vertical-line test, a function must score over 70%.

4. If a graph includes both $(5, 9)$ and $(5, 7)$, it cannot represent a function.

A. Identifying Functions

Determine whether each correspondence is a function.

5. Domain   Range

1 ———→ 1
2 ⟋
3 ———→ 2
5 ———→ 3

6. Domain   Range

3 ———→ −1
4 ⟨⟩ −2
5 ———→ −3

7. Domain   Range

10 ⟍
20 ———→ 0
30 ⟋
40 ⟋

8. Domain   Range

3 ———→ 1
5 ⟋
−3 ———→ −1
−5 ⟋

9. Domain   Range

Texas ⟨ Austin
       Houston
       Dallas

Ohio ⟨ Cleveland
       Toledo
       Cincinnati

10. Domain   Range

Austin ⟍
Houston ———→ Texas
Dallas ⟋

Cleveland ⟍
Toledo ———→ Ohio
Cincinnati ⟋

11. Domain

Pro Football Hall-of-Famers

Les Richter ———→ 48
Bobby Mitchell ———→ 49
Mike Singletary ⟍
Ken Strong ———→ 50
Alex Wojciechowicz ⟋

Range

Primary Jersey Number

12. Domain

Birthday

January 17

January 29

Range

Celebrity

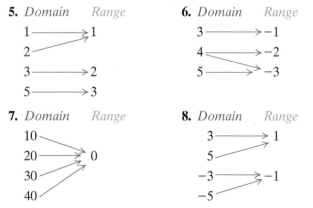

Dolly Parton
Jim Carrey
Michelle Obama
Muhammad Ali
Oprah Winfrey
Tom Selleck

B. Function Notation

Find the indicated outputs.

13. $f(4), f(7)$, and $f(-2)$

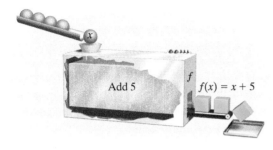

14. $g(1), g(6)$, and $g(13)$

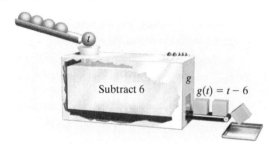

15. $h(p) = -3p$; find $h(-7), h(5)$, and $h(10)$.

16. $f(p) = 10p$; find $f(6), f(-\frac{1}{2})$, and $f(20)$.

17. $g(s) = 3s - 2$; find $g(1), g(-5)$, and $g(2.5)$.

18. $h(t) = 12$; find $h(6), h(-25)$, and $h(14.6)$.

19. $F(x) = 2x^2 + x$; find $F(-1), F(0)$, and $F(2)$.

20. $P(x) = x^3 + x$; find $P(0), P(-1)$, and $P(5)$.

21. $f(t) = (t - 3)^2$; find $f(-2), f(8)$, and $f(\frac{1}{2})$.

22. $g(t) = |t - 3|$; find $g(-2), g(8)$, and $g(\frac{1}{2})$.

23. $h(x) = |2x| - x$; find $h(0), h(-4)$, and $h(4)$.

24. $f(x) = x^4 - x$; find $f(-1), f(2)$, and $f(0)$.

25. *Life Span.* The function given by $l(x) = \dfrac{1700}{x}$ can be used to approximate the life span of an animal with a pulse rate of x beats per minute.

  a) Find the approximate life span of a horse with a pulse rate of 50 beats per minute.

  b) Find the approximate life span of a seal with a pulse rate of 85 beats per minute.

26. *Temperature as a Function of Depth.* The function given by $T(d) = 10d + 20$ gives the temperature, in degrees Celsius, inside the earth as a function of the depth d, in kilometers. Find the temperature at 5 km, 20 km, and 1000 km.

27. *Predicting Heights.* An anthropologist can estimate the height of a male or a female, given the lengths of certain bones. A *humerus* is the bone from the elbow to the shoulder. The height, in centimeters, of a female with a humerus of x centimeters is given by

$$F(x) = 2.75x + 71.48.$$

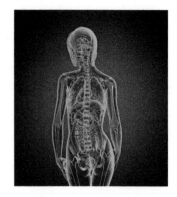

If a humerus is known to be from a female, how tall was the female if the bone is **(a)** 32 cm long? **(b)** 30 cm long?

28. When a humerus (see Exercise 27) is from a male, the function given by $M(x) = 2.89x + 70.64$ is used to find the male's height, in centimeters. If a humerus is known to be from a male, how tall was the male if the bone is **(a)** 30 cm long? **(b)** 35 cm long?

29. *Temperature Conversions.* The function given by $C(F) = \frac{5}{9}(F - 32)$ determines the Celsius temperature that corresponds to F degrees Fahrenheit. Find the Celsius temperature that corresponds to 62°F, 77°F, and 23°F.

30. *Pressure at Sea Depth.* The function given by $P(d) = 1 + (d/33)$ gives the pressure, in *atmospheres* (atm), at a depth of d feet, in the sea. Note that $P(0) = 1$ atm, $P(33) = 2$ atm, and so on. Find the pressure at 20 ft, 30 ft, and 100 ft.

C. Graphs of Functions

Graph each function.

31. $f(x) = 2x - 3$

32. $g(x) = 2x + 5$

33. $g(x) = -x + 4$

34. $f(x) = -\frac{1}{2}x + 2$

35. $f(x) = \frac{1}{2}x + 1$

36. $f(x) = -\frac{3}{4}x - 2$

37. $g(x) = 2|x|$

38. $h(x) = -|x|$

39. $g(x) = x^2$

40. $f(x) = x^2 - 1$

41. $f(x) = x^2 - x - 2$

42. $g(x) = x^2 + 6x + 5$

D. Recognizing Graphs of Functions

Determine whether each graph is that of a function.

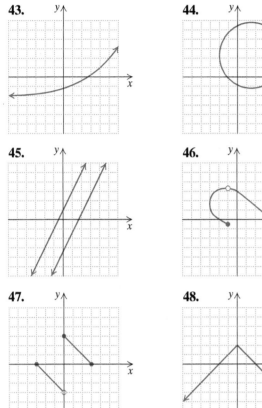

43.

44.

45.

46.

47.

48.

49. Is it possible for a function to have more elements in the range than in the domain? Why or why not?

50. Is it possible for a function to have more elements in the domain than in the range? Why or why not?

Skill Review

Graph on a plane.

51. $y = \frac{4}{5}x$ [3.2]

52. $y = -\frac{1}{2}x + 3$ [3.6]

53. $x - 2y < 4$ [7.5]

54. $y = -3$ [3.3]

55. $y = x^2 - 6x + 1$ [9.6]

56. $3x - y = 6$ [3.3]

Synthesis

57. Explain in your own words why the vertical-line test works.

58. If $f(x) = g(x) + 2$, how do the graphs of f and g compare?

Graph.

59. $g(x) = x^3$

60. $f(x) = 2 + \sqrt{x}$

61. $f(x) = |x| + x$

62. $g(x) = |x| - x$

63. Sketch a graph that is not that of a function.

64. If $f(-1) = -7$ and $f(3) = 8$, find a linear equation for $f(x)$.

65. If $g(0) = -4$, $g(-2) = 0$, and $g(2) = 0$, find a quadratic equation for $g(x)$.

Find the range of each function for the given domain.

66. $f(x) = 5 - 3x$, when the domain is the set of whole numbers less than 4

67. $g(t) = t^2 - t$, when the domain is the set of integers between -4 and 2

68. $f(m) = m^3 + 1$, when the domain is the set of integers between -3 and 3

69. $h(x) = |x| - x$, when the domain is the set of integers between -2 and 20

70. Use a graphing calculator to check your answers to Exercises 59–62 and 64–69.

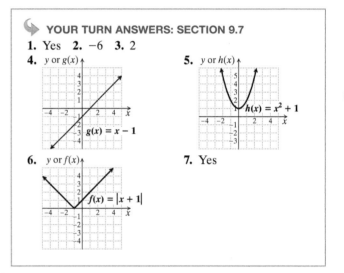

↳ **YOUR TURN ANSWERS: SECTION 9.7**

1. Yes   **2.** -6 **3.** 2

4. y or $g(x)$

$g(x) = x - 1$

5. y or $h(x)$

$h(x) = x^2 + 1$

6. y or $f(x)$

$f(x) = |x + 1|$

7. Yes

Quick Quiz: Sections 9.1–9.7

Solve.

1. $5t^2 + 5 = 0$ [9.5]   **2.** $(x - 4)^2 = 3$ [9.1]

3. $x^2 - 5x - 2 = 0$ [9.3]

4. Graph: $f(x) = 4 - x^2$. [9.7]

5. If $g(x) = |x| - 7$, find $g(-11)$. [9.7]

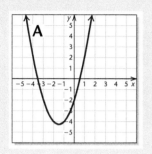

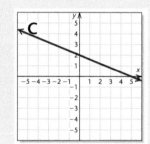

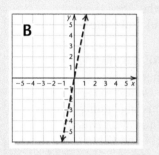

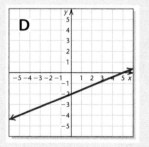

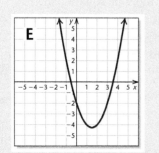

Visualizing for Success

Use after Section 9.6.

Match each equation or inequality with its graph.

1. $y = -4 + 4x - x^2$

2. $y = 5 - x^2$

3. $5x + 2y = -10$

4. $5x + 2y \leq 10$

5. $y < 5x$

6. $y = x^2 - 3x - 2$

7. $2x - 5y = 10$

8. $5x - 2y = 10$

9. $2x + 5y = 10$

10. $y = x^2 + 3x - 2$

Answers on page A-51

An additional, animated version of this activity appears in MyMathLab. To use MyMathLab, you need a course ID and a student access code. Contact your instructor for more information.

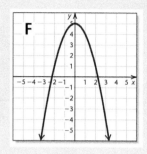

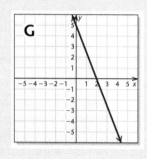

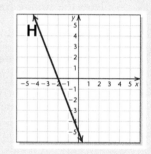

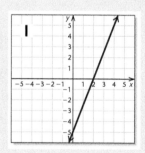

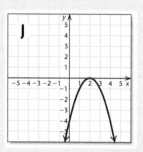

Collaborative Activity   *Using Areas to Complete the Square*

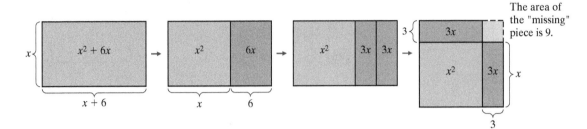

Focus: Visualizing completion of the square
Use after: Section 9.2
Time: 15–25 minutes
Group size: 2
Materials: Rulers and graph paper may be helpful.

To draw a representation of completing the square, we use areas and the fact that the area of any rectangle is given by multiplying the length and the width. For example, the above sequence of figures can be drawn to explain why 9 completes the square for $x^2 + 6x$.

Activity

1. Draw a sequence of four figures, similar to those shown above, to complete the square for $x^2 + 8x$. Group members should take turns, so that each person draws and labels two of the figures.

2. Repeat part (1) to complete the square for $x^2 + 14x$. The person who drew the first drawing in part (1) should take the second turn this time.

3. When we add the area of the missing piece, we increase the original area. For this reason, to complete the square we use the addition principle, adding the "missing" number to both sides to form an equivalent *equation*. Use the work in parts (1) and (2) to solve the equations $x^2 + 8x = 9$ and $x^2 + 14x = 15$.

Decision Making & Connection   (*Use after Section 9.1.*)

Area.   Recall the following area formulas.

Area of a square with side s: $A = s^2$
Area of a circle with radius r: $A = \pi r^2$

1. Kristin purchased a round pizza with a diameter of 16 in. for $17.99. How much did her pizza cost per square inch?

2. Lauren purchased a square pizza for $14.99. If the price per square inch was the same for Lauren's pizza as for Kristin's, what was the length of a side of Lauren's pizza?

3. Tony purchased a circular pizza with a diameter that is half of the diameter of Kristin's pizza. What fraction of the price of Kristin's pizza should Tony expect to pay?

4. Tyler's cake recipe calls for a 9-in. square cake pan. All his cake pans are round. What size round cake pan can he substitute for the 9-in. square pan?

5. *Research.*   Find formulas for the surface area of a cube and of a sphere. Derive a formula giving the radius of a sphere that has the same surface area as a cube with sides of length s.

Study Summary

KEY TERMS AND CONCEPTS	EXAMPLES	PRACTICE EXERCISES

SECTION 9.1: *Solving Quadratic Equations: The Principle of Square Roots*

A **quadratic equation in standard form** is written $ax^2 + bx + c = 0$, with a, b, and c constant and $a \neq 0$.

Some quadratic equations can be solved by factoring and using the principle of zero products.

$$x^2 - 3x - 10 = 0$$
$$(x + 2)(x - 5) = 0 \qquad \text{Factoring}$$
$$x + 2 = 0 \quad or \quad x - 5 = 0 \qquad \text{Using the principle of zero products}$$
$$x = -2 \quad or \qquad x = 5$$

1. Solve:
$$x^2 - x - 6 = 0.$$

Some quadratic equations can be solved using the principle of square roots.

The Principle of Square Roots

For any nonnegative real number p, $x^2 = p$ is equivalent to

$$x = \sqrt{p} \quad or \quad x = -\sqrt{p}.$$

$$x^2 - 8x + 16 = 25$$
$$(x - 4)^2 = 25$$
$$x - 4 = -5 \quad or \quad x - 4 = 5 \qquad \text{Using the principle of square roots}$$
$$x = -1 \quad or \qquad x = 9$$

2. Solve:
$$(x - 5)^2 = 3.$$

SECTION 9.2: *Solving Quadratic Equations: Completing the Square*

Any quadratic equation can be solved by completing the square. To complete the square when solving an equation of the form $x^2 + bx = k$, add half of the coefficient of x, squared, or $(b/2)^2$, to both sides.

$$x^2 + 6x = 1$$
$$x^2 + 6x + \left(\tfrac{6}{2}\right)^2 = 1 + \left(\tfrac{6}{2}\right)^2 \qquad \text{Completing the square}$$
$$x^2 + 6x + 9 = 1 + 9$$
$$(x + 3)^2 = 10 \qquad \text{Factoring}$$
$$x + 3 = \pm\sqrt{10} \qquad \text{Using the principle of square roots}$$
$$x = -3 \pm \sqrt{10}$$

3. Solve by completing the square:
$$x^2 + 4x = 1.$$

SECTION 9.3: *The Quadratic Formula and Applications*

Any quadratic equation can be solved using the quadratic formula.

The Quadratic Formula

The solutions of a quadratic equation $ax^2 + bx + c = 0$ are given by

$$x = \frac{-b \pm \sqrt{b^2 - 4ac}}{2a}.$$

$$3x^2 - 2x - 5 = 0 \qquad a = 3, b = -2, c = -5$$
$$x = \frac{-(-2) \pm \sqrt{(-2)^2 - 4 \cdot 3(-5)}}{2 \cdot 3}$$
$$x = \frac{2 \pm \sqrt{4 + 60}}{6}$$
$$x = \frac{2 \pm \sqrt{64}}{6}$$
$$x = \frac{2 \pm 8}{6}$$
$$x = \frac{10}{6} = \frac{5}{3} \quad or \quad x = \frac{-6}{6} = -1$$

4. Solve using the quadratic formula:
$$x^2 + 2x - 2 = 0.$$

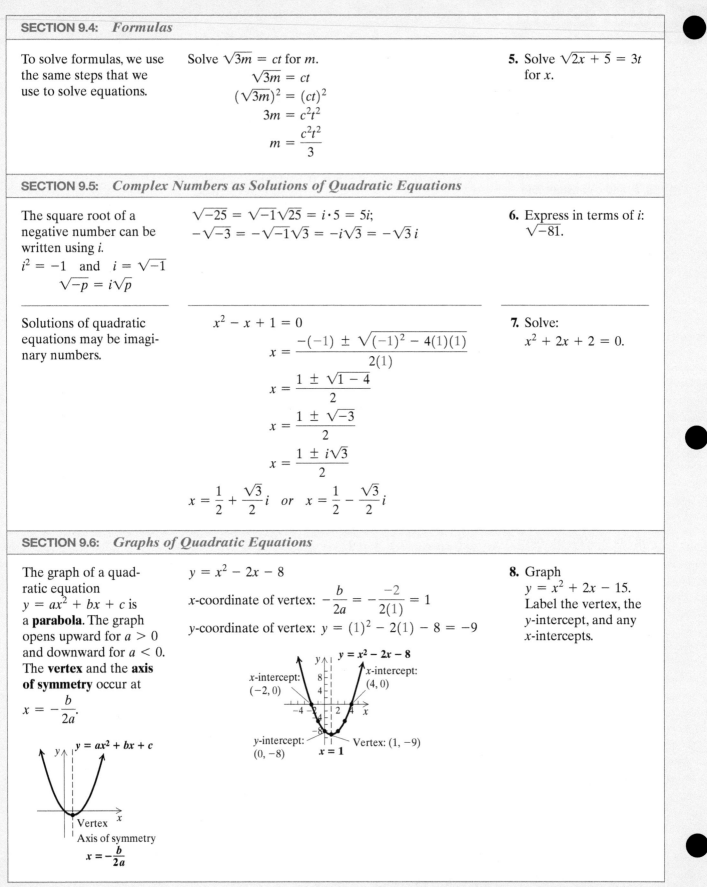

SECTION 9.4: *Formulas*

To solve formulas, we use the same steps that we use to solve equations.

Solve $\sqrt{3m} = ct$ for m.

$$\sqrt{3m} = ct$$
$$(\sqrt{3m})^2 = (ct)^2$$
$$3m = c^2t^2$$
$$m = \frac{c^2t^2}{3}$$

5. Solve $\sqrt{2x + 5} = 3t$ for x.

SECTION 9.5: *Complex Numbers as Solutions of Quadratic Equations*

The square root of a negative number can be written using i.
$i^2 = -1$ and $i = \sqrt{-1}$
$$\sqrt{-p} = i\sqrt{p}$$

$\sqrt{-25} = \sqrt{-1}\sqrt{25} = i \cdot 5 = 5i$;
$-\sqrt{-3} = -\sqrt{-1}\sqrt{3} = -i\sqrt{3} = -\sqrt{3}\,i$

6. Express in terms of i:
$\sqrt{-81}$.

Solutions of quadratic equations may be imaginary numbers.

$$x^2 - x + 1 = 0$$
$$x = \frac{-(-1) \pm \sqrt{(-1)^2 - 4(1)(1)}}{2(1)}$$
$$x = \frac{1 \pm \sqrt{1 - 4}}{2}$$
$$x = \frac{1 \pm \sqrt{-3}}{2}$$
$$x = \frac{1 \pm i\sqrt{3}}{2}$$
$$x = \frac{1}{2} + \frac{\sqrt{3}}{2}i \quad or \quad x = \frac{1}{2} - \frac{\sqrt{3}}{2}i$$

7. Solve:
$x^2 + 2x + 2 = 0$.

SECTION 9.6: *Graphs of Quadratic Equations*

The graph of a quadratic equation
$y = ax^2 + bx + c$ is
a **parabola**. The graph opens upward for $a > 0$ and downward for $a < 0$.
The **vertex** and the **axis of symmetry** occur at
$$x = -\frac{b}{2a}.$$

$$y = x^2 - 2x - 8$$

x-coordinate of vertex: $-\dfrac{b}{2a} = -\dfrac{-2}{2(1)} = 1$

y-coordinate of vertex: $y = (1)^2 - 2(1) - 8 = -9$

x-intercept: $(-2, 0)$
x-intercept: $(4, 0)$
y-intercept: $(0, -8)$
$x = 1$
Vertex: $(1, -9)$
$y = x^2 - 2x - 8$

Vertex
Axis of symmetry
$x = -\dfrac{b}{2a}$
$y = ax^2 + bx + c$

8. Graph
$y = x^2 + 2x - 15$.
Label the vertex, the y-intercept, and any x-intercepts.

SECTION 9.7: *Functions*

A **function** is a correspondence that assigns to each member of the **domain** exactly one member of the **range**. For a function f, the notation $f(a)$ represents the **output** that is paired with the **input** a.

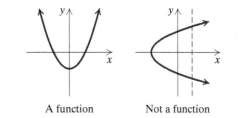

If $f(x) = x^2 - x + 3$, then
$$f(-1) = (-1)^2 - (-1) + 3$$
$$= 1 + 1 + 3$$
$$= 5.$$

9. If $g(t) = 3 - t^2$, find $g(-2)$.

The Vertical-Line Test
A graph represents a function if it is impossible to draw a vertical line that intersects the graph more than once.

A function   Not a function

10. Determine whether the following graph is that of a function.

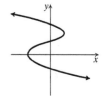

Review Exercises: Chapter 9

Concept Reinforcement

In Exercises 1–8, match each item with the appropriate item from the column on the right.

1. ____ An equation most easily solved by factoring [9.3]

2. ____ An equation most easily solved using the principle of square roots [9.1]

3. ____ An equation most easily solved by the quadratic formula [9.3]

4. ____ A parabola that opens upward [9.6]

5. ____ A parabola that opens downward [9.6]

6. ____ An expression for "the input 2 is paired with the output 5" [9.7]

7. ____ An expression for "the input 5 is paired with the output 2" [9.7]

8. ____ An equation most easily solved by completing the square [9.2]

a) $f(2) = 5$
b) $y = 3x^2 + 4x - 7$
c) $f(5) = 2$
d) $y = -3x^2 + 4x - 7$
e) $3x^2 - 4x - 5 = 0$
f) $(x - 7)^2 = 64$
g) $x^2 + 8x = 3$
h) $x^2 - 7x + 6 = 0$

Solve by completing the square. [9.2]

9. $x^2 - 10x = 1$

10. $2x^2 + 3x - 2 = 0$

Solve.

11. $5x^2 = 30$ [9.1]

12. $5x^2 - 8x + 3 = 0$ [9.2]

13. $x^2 - 2x - 10 = 0$ [9.3]

14. $(x + 3)^2 + 4 = 0$ [9.5]

15. $3y^2 + 5y = 2$ [9.2]

16. $11t^2 = t$ [9.3]

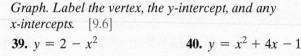

17. $x^2 + 1 = 0$ [9.5]

18. $2y^2 - 6y = 20$ [9.3]

19. $(p + 10)^2 = 12$ [9.1]   **20.** $x^2 + 6x = 9$ [9.3]

21. $x^2 + x + 1 = 0$ [9.5]   **22.** $1 + 4x^2 = 8x$ [9.3]

23. $t^2 + 9 = 6t$ [9.3]   **24.** $40 = 5y^2$ [9.1]

25. $3m = 4 + 5m^2$ [9.5]   **26.** $6x^2 + 11x = 35$ [9.2]

Solve. [9.4]

27. $p = ct - bt$, for t

28. $\sqrt{xy} + z = c$, for y

29. $m + n = \dfrac{p}{n}$, for n

30. $\dfrac{1}{r} + \dfrac{1}{s} = \dfrac{1}{t}$, for t

Approximate the solutions to the nearest thousandth. [9.3]

31. $x^2 + 3 = 5x$ **32.** $4y^2 + 8y + 1 = 0$

33. *Right Triangles.* The hypotenuse of a right tri-angle is 7 m long. One leg is 3 m longer than the other. Find the lengths of the legs. Round to the nearest thousandth. [9.3]

34. *Investments.* $1200 is invested at interest rate r, compounded annually. In 2 years, it grows to $1297.92. What is the interest rate? Use $A = P(1 + r)^2$. [9.3]

35. *Area of a Rectangle.* The width of a rectangle is 3 m less than the length. The area is 108 m². Find the length and the width. [9.3]

36. *Free-Fall Time.* The Troll Wall, in Norway, is 3600 ft high. How long would it take an object to fall from the top? (Use $s = 16t^2$ and disregard air resistance.) [9.3]

Express in terms of i. [9.5]

37. $\sqrt{-9}$ **38.** $-\sqrt{-125}$

Graph. Label the vertex, the y-intercept, and any x-intercepts. [9.6]

39. $y = 2 - x^2$ **40.** $y = x^2 + 4x - 1$

41. $y = 3x^2 - 10x$ **42.** $y = x^2 - 2x + 1$

43. If $f(t) = 3t - 2$, find $f(4), f(-2)$, and $f(1.5)$. [9.7]

44. If $g(x) = |x + 2|$, find $g(-3), g(-1)$, and $g(0)$. [9.7]

45. *Calories.* Each day a moderately active person needs about 15 calories per pound of body weight. The function given by $C(p) = 15p$ approximates the number of calories that are needed to maintain body weight p, in pounds. How many calories are needed in order to maintain a body weight of 130 lb? [9.7]

Graph. [9.7]

46. $g(x) = x - 6$ **47.** $f(x) = x^2 + 2$

48. $h(x) = 3|x|$

Determine whether each graph is that of a function. [9.7]

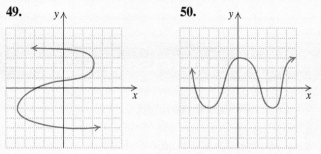

Synthesis

51. Tanner is shown the graphs of a line and a parab-ola, and is told that one is the graph of $y = 3x + 2$ and the other is the graph of $y = 3x^2 + 2$. Explain how he can match each graph with its equation without calculating any points. [9.6]

52. How can $b^2 - 4ac$ be used to determine how many x-intercepts the graph of $y = ax^2 + bx + c$ has? [9.6]

53. Suppose that a graph consisting of three points is not the graph of a function. Where must the points be in relation to each other? [9.7]

54. Two consecutive integers have squares that differ by 63. Find the integers. [9.3]

55. Find b such that the trinomial $x^2 + bx + 49$ is the square of a binomial. [9.2]

56. Solve: $x - 4\sqrt{x} - 5 = 0$. [9.3]

57. A square with sides of length s has the same area as a circle with radius 5 in. Find s. [9.3]

Test: Chapter 9   For step-by-step test solutions, access the Chapter Test Prep Videos in MyMathLab*.

Solve.

1. $8x^2 = 80$

2. $x^2 = -9$

3. $2t^2 - 5t = 0$

4. $(t - 1)^2 = 8$

5. $50 = p^2 - 5p$

6. $3m^2 + 13m = 10$

7. $p^2 - 16p + 64 = 2$

8. $x^2 - 4x = -4$

9. $x^2 - 4x = -5$

10. $m^2 - 4m = 2$

11. $10 = 4x + x^2$

12. $3x^2 - 7x + 1 = 0$

13. Solve by completing the square:
$$x^2 - 4x - 10 = 0.$$

Solve.

14. $3 = n + 2\sqrt{p + 5}$, for p

15. $1 + t = \dfrac{a}{b}$, for b

16. Approximate the solutions to the nearest thousandth:
$$x^2 - 3x - 8 = 0.$$

17. *Area of a Rectangle.* The width of a rectangle is 4 m less than the length. The area is 16.25 m². Find the length and the width.

18. *Diagonals of a Polygon.* A polygon has 44 diagonals. How many sides does it have? (Use $d = (n^2 - 3n)/2$.)

Express in terms of i.

19. $\sqrt{-200}$

20. $-\sqrt{-100}$

Graph. Label the vertex, the y-intercept, and any x-intercepts.

21. $y = -x^2 + x - 5$

22. $y = x^2 + 2x - 15$

23. If $f(x) = \frac{1}{2}x + 1$, find $f(0), f(1)$, and $f(2)$.

24. If $g(t) = -2t^2 + 5t$, find $g(-1), g(0)$, and $g(3)$.

25. *World Records.* The world record for the 10,000-m run has been decreasing steadily since 1940. The function given by $R(t) = 30.18 - 0.06t$ estimates the record, in minutes, t years after 1940. Predict what the record will be in 2016.

Graph.

26. $h(x) = x - 4$

27. $g(x) = x^2 - 4$

Determine whether each graph is that of a function.

28.

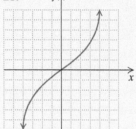

29.

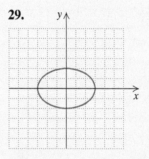

Synthesis

30. Find the area of a square whose diagonal is 5 ft longer than a side.

31. Solve this system for x. Use the substitution method.
$$x - y = 2,$$
$$xy = 4$$

Cumulative Review/Final Exam: Chapters 1–9

1. Evaluate $(1 - x)^2 + 3x$ for $x = -2$.   [1.8]

2. Remove parentheses and simplify:
   $2n - 5 - (5n - 6)$.   [1.8]

Compute and simplify.

3. $2.8 - (-12.2) - 1.63$ [1.6]

4. $-\frac{3}{8} \div \frac{5}{2}$ [1.7]

5. $13 \cdot 6 \div 3 \cdot 2 \div 13$ [1.8]

Simplify. Write the result using scientific notation.

6. $(2.1 \times 10^7)(1.3 \times 10^{-12})$ [4.2]

7. $\dfrac{5.2 \times 10^{-1}}{2.6 \times 10^{-15}}$ [4.2]

Simplify.

8. $x^{-6} \cdot x^2$ [4.2]

9. $\dfrac{y^3}{y^{-4}}$ [4.2]

10. $(3m^4)^2$ [4.1]

Perform the indicated operation and simplify.

11. $(4x^3 + 3x^2 - 5) + (3x^3 - 5x^2 + 4x - 12)$ [4.4]

12. $(6x^2 - 4x + 1) - (-6x^2 - 4x + 7)$ [4.4]

13. $-2y^2(4y^2 - 3y + 1)$ [4.5]

14. $(2t - 3)(3t^2 - 4t + 2)$ [4.5]

15. $\left(t - \frac{1}{4}\right)\left(t + \frac{1}{4}\right)$ [4.6]

16. $(2x + 7)(3x - 8)$ [4.6]

17. $(12x^2y + 10xy^2 - 7) - (3x^2y^2 - 2xy^2 - 5)$ [4.7]

18. $(5p^2 + 2q)^2$ [4.7]

19. $\dfrac{4}{2x - 6} \cdot \dfrac{x - 3}{x + 3}$ [6.2]

20. $\dfrac{3a^4}{a^2 - 1} \div \dfrac{2a^3}{a^2 - 2a + 1}$ [6.2]

21. $\dfrac{3}{3x - 1} + \dfrac{4}{5x}$ [6.4]

22. $\dfrac{2}{x^2 - 16} - \dfrac{x - 3}{x^2 - 9x + 20}$ [6.4]

23. $(x^3 + 7x^2 - 2x + 3) \div (x - 2)$ [4.8]

Factor.

24. $18x^2 - 8$ [5.4]

25. $y^3 - 6y^2 - 5y + 30$ [5.1]

26. $m^2 + 10m + 25$ [5.4]

27. $10x^3 - 11x^2 - 6x$ [5.3]

28. $t^2 - t - 6$ [5.2]

29. $49p^2 + 50p + 1$ [5.3]

30. $4x^2 - 20xy + 25y^2$ [5.4]

31. $3y^4 - 39y^3 + 120y^2$ [5.2]

Simplify.

32. $\dfrac{\dfrac{3}{x} + \dfrac{1}{2x}}{\dfrac{1}{3x} - \dfrac{3}{4x}}$ [6.5]

33. $-\sqrt[3]{-125}$ [8.7]

34. $\sqrt{49t^2}$ [8.1]
    (Assume $t \geq 0$.)

35. $16^{-3/4}$ [8.7]

36. $\sqrt{250x^6y^7}$ [8.2]
    (Assume $x, y \geq 0$.)

37. $\sqrt{\dfrac{16}{9}}$ [8.3]

38. $(1 + 3\sqrt{2})(1 - 3\sqrt{2})$ [8.4]

39. $5\sqrt{12} + 2\sqrt{48}$ [8.4]

40. $\sqrt{32ab}\sqrt{6a^4b^2}$ $(a, b \geq 0)$ [8.2]

41. $\dfrac{\sqrt{72}}{\sqrt{45}}$ [8.3]

Solve. If an equation has no solution, state this.

42. $-5x = 30$ [2.1]

43. $-5x > 30$ [2.6]

44. $3(y - 1) - 2(y + 2) = 0$ [2.2]

45. $x^2 - 8x + 15 = 0$ [5.6]

46. $y - x = 1$,
    $y = 3 - x$ [7.2]

47. $\dfrac{x}{x + 1} = \dfrac{3}{2x} + 1$ [6.6]

48. $4x - 3y = 3$,
    $3x - 2y = 4$ [7.3]

49. $x^2 - 2x = 2$ [9.3]

50. $3 - x = \sqrt{x^2 - 3}$ [8.5]

51. $4 - 9x \leq 18 + 5x$ [2.6]

52. $-\frac{7}{8}x + 7 = \frac{3}{8}x - 3$ [2.2]

53. $0.6x - 1.8 = 1.2x$ [2.2]

54. $x + y = 15,$
$x - y = 15$ [7.3]

55. $x^2 + 2x + 5 = 0$ [9.5]

56. $3y^2 = 30$ [9.1]

57. $(x - 3)^2 = 6$ [9.1]

58. $\dfrac{t}{t + 2} = \dfrac{1}{t}$ [6.6]

59. $12x^2 + x = 20$ [5.6]

Solve each formula for the given letter.

60. $ay = by - ax$, for a [9.4]

61. $\dfrac{1}{t} = \dfrac{1}{m} - \dfrac{1}{n}$, for m [9.4]

62. Approximate the solutions of $4x^2 = 4x + 1$ to the nearest thousandth.   [9.3]

Graph on a plane.

63. $y = \frac{1}{3}x - 5$ [3.6]

64. $2x + 3y = -6$ [3.3]

65. $y = -2$ [3.3]

66. $4x - 3y > 12$ [7.5]

67. $y = x^2 + 2x + 1$ [9.6]

68. $x \geq -3$ [7.5]

69. Graph $y = x^2 + 2x - 5$. Label the vertex, the y-intercept, and any x-intercepts.   [9.6]

70. Graph the following system of inequalities:
$$x + y \leq 6,$$
$$x + y \geq 2,$$
$$x \leq 3,$$
$$x \geq 1. \quad [7.6]$$

71. Find the slope and the y-intercept:
$$-6x + 3y = -24. \quad [3.6]$$

72. Find the slope of the line containing the points $(-5, -6)$ and $(-4, 9)$.   [3.5]

73. Find a point–slope equation for the line containing $(1, -3)$ and having slope $m = -\frac{1}{2}$.   [3.7]

74. For the function f given by
$$f(x) = 2x^2 + 7x - 4,$$
find $f(0), f(-4)$, and $f\left(\frac{1}{2}\right)$.   [9.7]

75. Simplify: $-\sqrt{-1}$.   [9.5]

Solve.

76. *Alternative-Fueled Buses.*   In 2011, approximately 23,400 city buses in the United States ran on alternative fuels. Of these, there were three times as many buses using compressed natural gas or biodiesel as there were hybrid buses. How many hybrid city buses were there in 2011?   [2.5]

Data: usatoday.com; apta.com

77. *Candy Blends.*   The Best Chocolate in Town combined dark-chocolate cheesecake truffles worth \$28.50 per pound with milk-chocolate hazelnut truffles worth \$24.50 per pound to make 20 lb of a custom order worth \$26.90 per pound. How many pounds of each kind of chocolate did they use?   [7.4]

78. *Sighting to the Horizon.*   At a height of h meters, one can see V kilometers to the horizon, where $V = 3.5\sqrt{h}$. Shari can see 35 km to the horizon from the top of a cell-phone tower. What is the height of the tower?   [8.5]

79. *Hemoglobin.*   A normal 10-cc sample of human blood contains 1.2 g of hemoglobin. How much hemoglobin would 16 cc of the same blood contain?   [6.7]

80. The length of a rectangle is 4 ft longer than twice the width. The area of the rectangle is 30 ft^2. Find the dimensions of the rectangle.   [5.7]

81. The length of a rectangle is 7 m more than the width. The length of a diagonal is 13 m. Find the dimensions of the rectangle.   [9.3]

82. The hypotenuse of a right triangle is 40 cm long and the length of a leg is 10 cm. Find the length of the other leg. Give an exact answer and an approximation to three decimal places.   [8.6]

Synthesis

83. Find x.   [8.6]

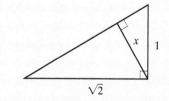

Determine whether each pair of expressions is equivalent.

84. $x^2 - 9$, $(x - 3)(x + 3)$ [4.6]

85. $\dfrac{n^2 + 2n}{2n + 3}$, $\dfrac{n^2}{3}$ [6.1]

86. $(x + 5)^2$, $x^2 + 25$ [4.6]

87. $\sqrt{(t - 3)^2}$, $t - 3$ [8.1]

88. $\sqrt{9x^4}$, $3x^2$ [8.1]

Appendixes

A | Factoring Sums or Differences of Cubes

A. Factoring a Sum or a Difference of Cubes

A. Factoring a Sum or a Difference of Cubes

It is possible to factor both a sum and a difference of two cubes. To see how, consider the following:

$$(A + B)(A^2 - AB + B^2) = A(A^2 - AB + B^2) + B(A^2 - AB + B^2)$$
$$= A^3 - A^2B + AB^2 + A^2B - AB^2 + B^3$$
$$= A^3 + B^3$$

and

$$(A - B)(A^2 + AB + B^2) = A(A^2 + AB + B^2) - B(A^2 + AB + B^2)$$
$$= A^3 + A^2B + AB^2 - A^2B - AB^2 - B^3$$
$$= A^3 - B^3.$$

These equations show how we can factor a sum or a difference of two cubes.

> **TO FACTOR A SUM OR A DIFFERENCE OF CUBES**
>
> $A^3 + B^3 = (A + B)(A^2 - AB + B^2),$
> $A^3 - B^3 = (A - B)(A^2 + AB + B^2)$

Remembering this list of cubes may prove helpful when factoring.

N	0.2	0.1	0	1	2	3	4	5	6
N^3	0.008	0.001	0	1	8	27	64	125	216

EXAMPLE 1   Factor: $x^3 - 8$.

SOLUTION   We have

$$x^3 - 8 = x^3 - 2^3 = (x - 2)(x^2 + x \cdot 2 + 2^2).$$
$$A^3 - B^3 = (A - B)(A^2 + A\ B + B^2)$$

This tells us that $x^3 - 8 = (x - 2)(x^2 + 2x + 4)$. Note that we cannot factor $x^2 + 2x + 4$. (It is not a perfect-square trinomial nor can it be factored by trial and error or grouping.) The check is left to the student.

1. Factor: $y^3 - 125$.

YOUR TURN

597

EXAMPLE 2 Factor: $x^3 + 125$.

SOLUTION We have

$$x^3 + 125 = x^3 + 5^3 = (x + 5)(x^2 - x \cdot 5 + 5^2).$$
$$\uparrow \quad \uparrow \qquad \uparrow \quad \uparrow \quad \uparrow \qquad \uparrow \quad \uparrow \qquad \uparrow$$
$$A^3 + B^3 = (A + B)(A^2 - A\,B + B^2)$$

2. Factor: $p^3 + 64$.

Thus, $x^3 + 125 = (x + 5)(x^2 - 5x + 25)$. The check is left to the student.

YOUR TURN

EXAMPLE 3 Factor: $250 + 128t^3$.

SOLUTION We first look for a common factor:

$$250 + 128t^3 = 2[125 + 64t^3]$$
$$= 2[5^3 + (4t)^3] \qquad \text{This is of the form } A^3 + B^3,$$
$$\text{where } A = 5 \text{ and } B = 4t.$$
$$= 2[(5 + 4t)(25 - 20t + 16t^2)].$$

3. Factor: $m^5 + m^2$.

The check is left to the student. The factorization is $2(5 + 4t)(25 - 20t + 16t^2)$.

YOUR TURN

EXAMPLE 4 Factor: $y^3 - 0.001$.

SOLUTION Since $0.001 = (0.1)^3$, we have a difference of cubes:

$$y^3 - 0.001 = (y - 0.1)(y^2 + 0.1y + 0.01).$$

4. Factor: $c^3 - 0.008$.

The check is left to the student.

YOUR TURN

Difference of cubes: $\quad A^3 - B^3 = (A - B)(A^2 + AB + B^2)$
Sum of cubes: $\qquad\quad A^3 + B^3 = (A + B)(A^2 - AB + B^2)$
Difference of squares: $\;\, A^2 - B^2 = (A + B)(A - B)$
There is no formula for factoring a sum of squares.

A   EXERCISE SET

FOR EXTRA HELP   MyMathLab®

A. Factoring a Sum or a Difference of Cubes

Factor completely.

1. $t^3 + 8$

2. $p^3 + 27$

3. $x^3 + 1$

4. $w^3 - 1$

5. $z^3 - 125$

6. $a^3 + 64$

7. $8a^3 - 1$

8. $27x^3 - 1$

9. $y^3 - 27$

10. $p^3 - 64$

11. $64 + 125x^3$

12. $8 + 27b^3$

13. $125p^3 + 1$

14. $64w^3 + 1$

15. $27m^3 - 64$

16. $8t^3 - 27$

17. $p^3 - q^3$

18. $a^3 + b^3$

19. $x^3 + \frac{1}{8}$

20. $y^3 - \frac{1}{27}$

21. $2y^3 - 128$

22. $3z^3 - 375$

23. $24a^3 + 3$

24. $54x^3 + 2$

25. $rs^3 - 125r$

26. $a^2b^3 + 64a^2$

27. $5x^3 + 40z^3$

28. $2y^3 - 54z^3$

29. $x^3 + 0.008$

30. $y^3 - 0.125$

Synthesis

31. Dino incorrectly believes that
$$a^3 - b^3 = (a - b)(a^2 + b^2).$$
How could you convince him that he is wrong?

32. If $x^3 + c$ is prime, what can you conclude about c? Why?

Factor. Assume that variables in exponents represent natural numbers.

33. $125c^6 + 8d^6$

34. $64x^6 + 8t^6$

35. $3x^{3a} - 24y^{3b}$

36. $\frac{8}{27}x^3 - \frac{1}{64}y^3$

37. $\frac{1}{24}x^3y^3 + \frac{1}{3}z^3$

38. $\frac{1}{16}x^{3a} + \frac{1}{2}y^{6a}z^{9b}$

> **YOUR TURN ANSWERS: APPENDIX A**
> **1.** $(y - 5)(y^2 + 5y + 25)$ $(p + 4)(p^2 - 4p + 16)$
> **2.** $m^2(m + 1)(m^2 - m + 1)$
> **3.** $(c - 0.2)(c^2 + 0.2c + 0.04)$

B | Mean, Median, and Mode

A. Mean **B.** Median **C.** Mode

One way to analyze data is to look for a single representative number, called a **center point**, or **measure of central tendency**. Those most often used are the **mean** (or **average**), the **median**, and the **mode**.

A. Mean

The *mean* of a set of numbers is often called an *average*.

> **MEAN, OR AVERAGE**
> The *mean*, or *average*, of a set of numbers is the sum of the numbers divided by the number of addends.

EXAMPLE 1 Find the mean of the following set of prices for a milkshake:

$$\$2.20, \quad \$2.60, \quad \$3.30, \quad \$4.10, \quad \$5.30.$$

What is the mean price?

SOLUTION We add the numbers and divide by the number of addends, 5:

$$\frac{(2.20 + 2.60 + 3.30 + 4.10 + 5.30)}{5} = \frac{17.50}{5} = 3.50.$$

The mean, or average, price is $3.50.

1. Find the mean:

$$13, 19, 6, 11, 28, 37.$$

YOUR TURN

B. Median

The *median* is useful when we wish to de-emphasize extreme scores. For example, suppose that five workers in a technology company manufactured the following number of components during one day's work:

Sarah:	88
Matt:	92
Pat:	66
Jen:	94
Mark:	91

Let's first list the numbers in order from smallest to largest:

66 88 91 92 94.

↑

Middle number

The middle number—in this case, 91—is the **median**.

> **MEDIAN**
>
> Once a set of data has been arranged from smallest to largest, the *median* of the set of data is the middle number if there is an odd number of data numbers. If there is an even number of data numbers, then there are two middle numbers and the median is the *average* of the two middle numbers.

EXAMPLE 2 Find the median of the following set of household incomes:

$76,000, $58,000, $87,000, $32,500, $64,800, $62,500.

SOLUTION We first rearrange the numbers in order from smallest to largest.

$32,500, $58,000, $62,500, $64,800, $76,000, $87,000

↑

Median

There is an even number of numbers. We look for the middle two, which are $62,500 and $64,800. The median is the average of $62,500 and $64,800:

2. Find the median:

4, 6, 12, 8, 3.

$$\frac{\$62,500 + \$64,800}{2} = \$63,650.$$

YOUR TURN

C. Mode

The last center point we consider is called the *mode*. A number that occurs most often in a set of data is sometimes considered a representative number, or center point.

> **MODE**
>
> The *mode* of a set of data is the number or numbers that occur most often. If each number occurs the same number of times, then there is *no* mode.

EXAMPLE 3 Find the mode of the following data:

23, 24, 27, 18, 19, 27.

3. Find the mode:

1.6, 8.5, 1.6, 3.7, 1.6.

SOLUTION The number that occurs most often is 27. Thus the mode is 27.

YOUR TURN

It is easier to find the mode of a set of data if the data are ordered.

EXAMPLE 4 Find the mode of the following data:

83, 84, 84, 84, 85, 86, 87, 87, 87, 88, 89, 90.

4. Find the mode:

3, 8, 4, 3, 2, 2, 2, 3.

SOLUTION There are two numbers that occur most often, 84 and 87. Thus the modes are 84 and 87.

YOUR TURN

EXAMPLE 5 Find the mode of the following data:

115, 117, 211, 213, 219.

5. Find the mode:

1, 2, 3, 4, 5, 6.

SOLUTION Each number occurs the same number of times. Thus the set of data has *no* mode.

YOUR TURN

B | EXERCISE SET

FOR EXTRA HELP MyMathLab®

A, B, C. Mean, Median, and Mode

For each set of numbers, find the mean (average), the median, and any modes that exist.

1. 13, 21, 18, 13, 20

2. 5, 2, 8, 10, 7, 1, 9

3. 3, 8, 20, 3, 20, 10

4. 19, 19, 8, 16, 8, 7

5. 4.7, 2.3, 4.6, 4.9, 3.8

6. 13.4, 13.4, 12.6, 42.9

7. 234, 228, 234, 228, 234, 278

8. $29.95, $28.79, $30.95, $29.95

9. *Hurricanes.* The following bar graph shows the number of hurricanes that struck the United States by month from 1851 to 2015. What is the average number for the 8 months given? the median? the mode?

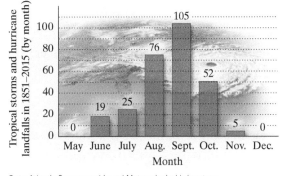

Atlantic Storms and Hurricanes

Data: Atlantic Oceanographic and Meteorological Laboratory

10. *Phone Prices.* A price comparison showed the following online prices for the same used phone in good condition:

$399.99, $329.95, $379.95, $399.99, $288.00.

What was the average price? the median price? the mode?

11. *PBA Scores.* Kelly Kulick rolled scores of 254, 202, 184, 269, 151, 223, 258, 222, and 202 in a recent tour trial for the Professional Bowlers Association. What was her average? her median? her mode?

Data: Professional Bowlers Association

12. *Salmon Prices.* The following prices per pound of Atlantic salmon were found at six fish markets:

$17.99, $15.99, $17.99, $16.99, $17.49, $19.99.

What was the average price per pound? the median price? the mode?

Synthesis

13. *Hank Aaron.* Hank Aaron averaged $34\frac{7}{22}$ home runs per year over a 22-year career. After 21 years, Aaron had averaged $35\frac{10}{21}$ home runs per year. How many home runs did Aaron hit in his final year?

14. *Length of Pregnancy.* Marta was pregnant 270 days, 259 days, and 272 days for her first three pregnancies. In order for Marta's average length of pregnancy to equal the worldwide average of 266 days, how long must her fourth pregnancy last?

15. The ordered set of data 18, 21, 24, a, 36, 37, b has a median of 30 and an average of 32. Find a and b.

16. *Male Height.* Jason's brothers are 174 cm, 180 cm, 179 cm, and 172 cm tall. The average male is 176.5 cm tall. How tall is Jason if he and his brothers have an average height of 176.5 cm?

> ➤ **YOUR TURN ANSWERS: APPENDIX A**
> **1.** 19 **2.** 6 **3.** 1.6 **4.** 2, 3 **5.** No mode

C — Sets

A. Naming Sets **B.** Membership **C.** Subsets **D.** Intersections **E.** Unions

A. Naming Sets

A **set** is a collection of objects. In mathematics the objects, or **elements**, of a set are generally numbers.

To name the set of whole numbers less than 6, we can use *roster notation*, as follows:

$$\{0, 1, 2, 3, 4, 5\}.$$

The set of real numbers x for which x is less than 6 cannot be named by listing all its members because there is an infinite number of them. We name such a set using *set-builder notation*, as follows:

$$\{x \mid x < 6\}.$$

This is read

"The set of all x such that x is less than 6."

The **empty set** contains no elements and is written $\varnothing$. It can also be written $\{\ \}$. Note that $\{0\}$ contains 0 and is thus *not* empty.

B. Membership

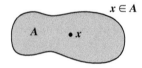

$x \in A$

The symbol $\in$ means *is a member of* or *belongs to*, or *is an element of*. Thus, $x \in A$ means

x is a member of A, or x belongs to A, or x is an element of A.

EXAMPLE 1 Classify each of the following as true or false.
a) $1 \in \{1, 2, 3\}$ **b)** $1 \in \{2, 3\}$
c) $4 \in \{x \mid x \text{ is an even whole number}\}$

SOLUTION
a) Since 1 is listed as a member of the set, $1 \in \{1, 2, 3\}$ is true.
b) Since 1 is *not* a member of $\{2, 3\}$, the statement $1 \in \{2, 3\}$ is false.
c) Since 4 is an even whole number, $4 \in \{x \mid x \text{ is an even whole number}\}$ is true.

1. Classify as true or false.

$5 \in \{x \mid x \text{ is an even whole number}\}$.

YOUR TURN

C. Subsets

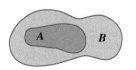

If every element of A is also an element of B, then A is a *subset* of B. This is denoted $A \subseteq B$.

For example, the set of whole numbers is a subset of the set of integers, and the set of rational numbers is a subset of the set of real numbers.

EXAMPLE 2 Classify each of the following as true or false.

a) $\{1, 2\} \subseteq \{1, 2, 3, 4\}$ b) $\{p, q, r, w\} \subseteq \{a, p, r, z\}$

c) $\{x \mid x < 6\} \subseteq \{x \mid x \le 11\}$

SOLUTION

a) Since every element of $\{1, 2\}$ is in the set $\{1, 2, 3, 4\}$, it follows that $\{1, 2\} \subseteq \{1, 2, 3, 4\}$ is true.

b) Since $q \in \{p, q, r, w\}$, but $q \notin \{a, p, r, z\}$, it follows that $\{p, q, r, w\} \subseteq \{a, p, r, z\}$ is false.

2. Classify as true or false:

$\{8, 9, 10\} \subseteq \{8, 9\}$.

c) Since every number that is less than 6 is also less than 11, the statement $\{x \mid x < 6\} \subseteq \{x \mid x \le 11\}$ is true.

YOUR TURN

$A \cap B$

D. Intersections

The *intersection* of sets A and B, denoted $A \cap B$, is the set of members common to both sets.

EXAMPLE 3 Find each intersection.

a) $\{0, 1, 3, 5, 25\} \cap \{2, 3, 4, 5, 6, 7, 9\}$

b) $\{a, p, q, w\} \cap \{m, n, x, y\}$

SOLUTION

a) $\{0, 1, 3, 5, 25\} \cap \{2, 3, 4, 5, 6, 7, 9\} = \{3, 5\}$

3. Find the intersection:

$\{3, 4, 5\} \cap \{3, 4, 5, 6\}$.

b) $\{a, p, q, w\} \cap \{m, n, x, y\} = \varnothing$ These sets have no members in common. Their intersection is the empty set.

YOUR TURN

$A \cup B$ **is shaded.**

E. Unions

Two sets A and B can be combined to form a set that contains the members of both A and B. The new set is called the *union* of A and B, denoted $A \cup B$.

EXAMPLE 4 Find each union.

a) $\{0, 5, 7, 13, 27\} \cup \{0, 2, 3, 4, 5\}$ b) $\{a, c, e, g\} \cup \{b, d, f\}$

SOLUTION

a) $\{0, 5, 7, 13, 27\} \cup \{0, 2, 3, 4, 5\} = \{0, 2, 3, 4, 5, 7, 13, 27\}$

Note that the 0 and the 5 are *not* listed twice in the union.

4. Find the union:

$\{1, 3, 5, 7\} \cup \{2, 3, 5, 8\}$.

b) $\{a, c, e, g\} \cup \{b, d, f\} = \{a, b, c, d, e, f, g\}$

YOUR TURN

C    EXERCISE SET

FOR EXTRA HELP MyMathLab®

A. Naming Sets

Name each set using the roster method.

1. The set of whole numbers 8 through 11

2. The set of whole numbers 83 through 89

3. The set of odd numbers between 40 and 50

4. The set of multiples of 5 between 10 and 40

5. $\{x \mid$ the square of x is $9\}$

6. $\{x \mid x$ is the cube of $\frac{1}{2}\}$

B, C. Membership and Subsets

Classify each of the following statements as either true or false.

7. $5 \in \{x \mid x$ is an odd number$\}$

8. $8 \in \{x \mid x$ is an odd number$\}$

9. Skiing $\in$ The set of all sports

10. Pharmacist $\in$ The set of all professions requiring a college degree

11. $3 \in \{-4, -3, 0, 1\}$

12. $0 \in \{-4, -3, 0, 1\}$

13. $\frac{2}{3} \in \{x \mid x$ is a rational number$\}$

14. $\frac{2}{3} \in \{x \mid x$ is a real number$\}$

15. $\{-1, 0, 1\} \subseteq \{-3, -2, -1, 1, 2, 3\}$

16. The set of vowels $\subseteq$ The set of consonants

17. The set of integers $\subseteq$ The set of rational numbers

18. $\{2, 4, 6\} \subseteq \{1, 2, 3, 4, 5, 6, 7\}$

D. Intersections

Find each intersection.

19. $\{a, b, c, d, e\} \cap \{c, d, e, f, g\}$

20. $\{a, e, i, o, u\} \cap \{q, u, i, c, k\}$

21. $\{1, 2, 3, 4, 6, 12\} \cap \{1, 2, 3, 6, 9, 18\}$

22. $\{1, 2, 3, 4, 6, 12\} \cap \{1, 5, 7, 35\}$

23. $\{2, 4, 6, 8\} \cap \{1, 3, 5, 7\}$

24. $\{a, e, i, o, u\} \cap \{m, n, f, g, h\}$

E. Unions

Find each union.

25. $\{a, e, i, o, u\} \cup \{q, u, i, c, k\}$

26. $\{a, b, c, d, e\} \cup \{c, d, e, f, g\}$

27. $\{1, 2, 3, 4, 6, 12\} \cup \{1, 2, 3, 6, 9, 18\}$

28. $\{1, 2, 3, 4, 6, 12\} \cup \{1, 5, 7, 35\}$

29. $\{2, 4, 6, 8\} \cup \{1, 3, 5, 7\}$

30. $\{a, e, i, o, u\} \cup \{m, n, f, g, h\}$

31. What advantage(s) does set-builder notation have over roster notation?

32. What advantage(s) does roster notation have over set-builder notation?

Synthesis

33. Find the union of the set of integers and the set of whole numbers.

34. Find the intersection of the set of odd integers and the set of even integers.

35. Find the union of the set of rational numbers and the set of irrational numbers.

36. Find the intersection of the set of even integers and the set of positive rational numbers.

37. Find the intersection of the set of rational numbers and the set of irrational numbers.

38. Find the union of the set of negative integers, the set of positive integers, and the set containing 0.

39. For a set A, find each of the following.

   a) $A \cup \varnothing$
   b) $A \cup A$
   c) $A \cap A$
   d) $A \cap \varnothing$

Classify each of the following statements as either true or false.

40. The empty set can be written $\varnothing$, { }, or $\{0\}$.

41. For any set $A, \varnothing \subseteq A$.

42. For any set $A, A \subseteq A$.

43. For any sets A and $B, A \cap B \subseteq A$.

44. A set is *closed* under an operation if, when the operation is performed on its members, the result is in the set. For example, the set of real numbers is closed under the operation of addition since the sum of any two real numbers is a real number.

   a) Is the set of even numbers closed under addition?
   b) Is the set of odd numbers closed under addition?
   c) Is the set $\{0, 1\}$ closed under addition?
   d) Is the set $\{0, 1\}$ closed under multiplication?
   e) Is the set of real numbers closed under multiplication?
   f) Is the set of integers closed under division?

45. Experiment with sets of various types and determine whether the following distributive law for sets is true:

$$A \cap (B \cup C) = (A \cap B) \cup (A \cap C).$$

↘ YOUR TURN ANSWERS: APPENDIX B
1. False   **2.** False   **3.** $\{3, 4, 5\}$ **4.** $\{1, 2, 3, 5, 7, 8\}$

Tables

TABLE 1 Fraction and Decimal Equivalents

Fraction Notation	$\frac{1}{10}$	$\frac{1}{8}$	$\frac{1}{6}$	$\frac{1}{5}$	$\frac{1}{4}$	$\frac{3}{10}$	$\frac{1}{3}$	$\frac{3}{8}$	$\frac{2}{5}$	$\frac{1}{2}$
Decimal Notation	0.1	0.125	$0.16\overline{6}$	0.2	0.25	0.3	$0.333\overline{3}$	0.375	0.4	0.5
Percent Notation	10%	12.5%, or $12\frac{1}{2}\%$	$16.6\overline{6}\%$, or $16\frac{2}{3}\%$	20%	25%	30%	$33.3\overline{3}\%$, or $33\frac{1}{3}\%$	37.5%, or $37\frac{1}{2}\%$	40%	50%
Fraction Notation	$\frac{3}{5}$	$\frac{5}{8}$	$\frac{2}{3}$	$\frac{7}{10}$	$\frac{3}{4}$	$\frac{4}{5}$	$\frac{5}{6}$	$\frac{7}{8}$	$\frac{9}{10}$	$\frac{1}{1}$
Decimal Notation	0.6	0.625	$0.66\overline{6}$	0.7	0.75	0.8	$0.83\overline{3}$	0.875	0.9	1
Percent Notation	60%	62.5%, or $62\frac{1}{2}\%$	$66.6\overline{6}$, or $66\frac{2}{3}\%$	70%	75%	80%	$83.3\overline{3}\%$, or $83\frac{1}{3}\%$	87.5%, or $87\frac{1}{2}\%$	90%	100%

TABLE 2 Squares and Square Roots with Approximations to Three Decimal Places

N	$\sqrt{N}$	N^2	N	$\sqrt{N}$	N^2	N	$\sqrt{N}$	N^2	N	$\sqrt{N}$	N^2
1	1	1	26	5.099	676	51	7.141	2601	76	8.718	5776
2	1.414	4	27	5.196	729	52	7.211	2704	77	8.775	5929
3	1.732	9	28	5.292	784	53	7.280	2809	78	8.832	6084
4	2	16	29	5.385	841	54	7.348	2916	79	8.888	6241
5	2.236	25	30	5.477	900	55	7.416	3025	80	8.944	6400
6	2.449	36	31	5.568	961	56	7.483	3136	81	9	6561
7	2.646	49	32	5.657	1024	57	7.550	3249	82	9.055	6724
8	2.828	64	33	5.745	1089	58	7.616	3364	83	9.110	6889
9	3	81	34	5.831	1156	59	7.681	3481	84	9.165	7056
10	3.162	100	35	5.916	1225	60	7.746	3600	85	9.220	7225
11	3.317	121	36	6	1296	61	7.810	3721	86	9.274	7396
12	3.464	144	37	6.083	1369	62	7.874	3844	87	9.327	7569
13	3.606	169	38	6.164	1444	63	7.937	3969	88	9.381	7744
14	3.742	196	39	6.245	1521	64	8	4096	89	9.434	7921
15	3.873	225	40	6.325	1600	65	8.062	4225	90	9.487	8100
16	4	256	41	6.403	1681	66	8.124	4356	91	9.539	8281
17	4.123	289	42	6.481	1764	67	8.185	4489	92	9.592	8464
18	4.243	324	43	6.557	1849	68	8.246	4624	93	9.644	8649
19	4.359	361	44	6.633	1936	69	8.307	4761	94	9.695	8836
20	4.472	400	45	6.708	2025	70	8.367	4900	95	9.747	9025
21	4.583	441	46	6.782	2116	71	8.426	5041	96	9.798	9216
22	4.690	484	47	6.856	2209	72	8.485	5184	97	9.849	9409
23	4.796	529	48	6.928	2304	73	8.544	5329	98	9.899	9604
24	4.899	576	49	7	2401	74	8.602	5476	99	9.950	9801
25	5	625	50	7.071	2500	75	8.660	5625	100	10	10,000

Photo Credits

Answers

CHAPTER 1

Check Your Understanding, p. 6

1. $100 - x$ **2.** 89

Technology Connection, p. 7

1. 3438   **2.** 47,531

Exercise Set 1.1, pp. 8–11

1. Constant   **2.** Operation   **3.** Evaluate
4. Equation   **5.** Expression   **6.** Equation
7. Equation   **8.** Expression   **9.** Equation
10. Expression   **11.** Equation   **12.** Expression
13. 45   **15.** 8   **17.** 5   **19.** 4   **21.** 5   **23.** 3
25. $24 \, \text{ft}^2$ **27.** $15 \, \text{cm}^2$ **29.** $804 \, \text{ft}^2$ **31.** Let r
represent Ron's age; $r + 5$, or $5 + r$ **33.** $6b$, or $b \cdot 6$
35. $c - 9$ **37.** $6 + q$, or $q + 6$ **39.** $p - t$
41. $y - x$ **43.** $x \div w$, or $\dfrac{x}{w}$
45. Let l represent the length of the box and h the
height; $l + h$, or $h + l$ **47.** $9 \cdot 2m$, or $2m \cdot 9$
49. Let y represent "some number"; $\dfrac{1}{4}y - 13$, or $\dfrac{y}{4} - 13$
51. Let a and b represent the two numbers; $5(a - b)$
53. Let w represent the number of women attending; 64% of w, or $0.64w$ **55.** Let x represent the
unknown number; $73 + x = 201$ **57.** Let x represent
the unknown number; $42x = 2352$ **59.** Let s represent the number of unoccupied squares; $s + 19 = 64$
61. Let w represent the amount of solid waste generated, in millions of tons; 34.5% of $w = 87$, or
$0.345w = 87$ **63.** $f = a + 5$ **65.** $n = m + 2.42$
67. $v = 10,000d$ **69.** (f)   **71.** (d)   **73.** (g)
75. (e)   **77.** ☞   **79.** ☞   **81.** \$450   **83.** 2
85. 6   **87.** $w + 4$ **89.** $l + w + l + w$, or $2l + 2w$
91. $t + 8$ **93.** ☞

Check Your Understanding, p. 16

1. **(a)** Addition and multiplication; **(b)** distributive;
(c) $21 + 7x$ **2.** **(a)** Multiplication; **(b)** associative;
(c) $21x$

Exercise Set 1.2, pp. 16–18

1. Equivalent   **2.** Commutative   **3.** Sum   **4.** Factors
5. Commutative   **6.** Associative   **7.** Distributive
8. Commutative   **9.** Commutative   **10.** Distributive
11. $t + 11$ **13.** $8x + 4$ **15.** $3y + 9x$ **17.** $5(1 + a)$
19. $x \cdot 7$ **21.** ts **23.** $5 + ba$ **25.** $(a + 1)5$
27. $x + (8 + y)$ **29.** $(u + v) + 7$ **31.** $ab + (c + d)$
33. $10(xy)$ **35.** $(2a)b$ **37.** $(3 \cdot 2)(a + b)$
39. $(s + t) + 6$; $(t + 6) + s$ **41.** $17(ab)$; $b(17a)$
43. $(1 + x) + 2 = (x + 1) + 2$ Commutative law
$ = x + (1 + 2)$ Associative law
$ = x + 3$ Simplifying
45. $(m \cdot 3)7 = m(3 \cdot 7)$ Associative law
$ = m \cdot 21$ Simplifying
$ = 21m$ Commutative law
47. $x, xyz, 1$ **49.** $2a, \dfrac{a}{3b}, 5b$ **51.** $4x, 4y$ **53.** $5, n$
55. $3, (x + y)$ **57.** $7, a, b$ **59.** $(a - b), (x - y)$
61. $2x + 30$ **63.** $4 + 4a$ **65.** $90x + 60$
67. $5r + 10 + 15t$ **69.** $2a + 2b$ **71.** $5x + 5y + 10$
73. $2(a + b)$ **75.** $7(1 + y)$ **77.** $2(16x + 1)$
79. $5(x + 2 + 3y)$ **81.** $7(a + 5b)$
83. $11(4x + y + 2z)$ **85.** Commutative law of
addition; distributive law   **87.** Distributive law;
associative law of multiplication   **89.** ☞   **91.** ☞
93. Distributive law; associative law of addition;
commutative law of addition; associative law of addition;
distributive law   **95.** Yes; distributive law   **97.** No;
for example, let $m = 1$. Then $7 \div 3 \cdot 1 = \frac{7}{3}$ and
$1 \cdot 3 \div 7 = \frac{3}{7}$.   **99.** No; for example, let $x = 1$ and
$y = 2$. Then $30 \cdot 2 + 1 \cdot 15 = 60 + 15 = 75$ and
$5[2(1 + 3 \cdot 2)] = 5[2(7)] = 5 \cdot 14 = 70$.   **101.** ☞
103. **(a)** Answers may vary. Aidan: $10(1.5x + 40)$;
Beth: $10 \cdot 40 + 10(1.5)x$; Cody: $15x + 400$; **(b)** all
expressions are equivalent to $15x + 400$.

Quick Quiz: Sections 1.1–1.2, p. 18

1. 9   **2.** $2(m + 3)$ **3.** Let n represent the number;
$\frac{1}{3} \cdot n = 18$ **4.** $3x + 15y + 21$ **5.** $7(2a + t + 1)$

Check Your Understanding, p. 25

1. (b)   **2.** (a)   **3.** (c)   **4.** (c)   **5.** (a)   **6.** (b)

Exercise Set 1.3, pp. 26–28

1. Numerator   **2.** Prime   **3.** Reciprocal   **4.** Add
5. (b)   **6.** (c)   **7.** (d)   **8.** (a)   **9.** Composite
11. Prime   **13.** Composite   **15.** Prime   **17.** Neither
19. $1, 2, 5, 10, 25, 50$ **21.** $1, 2, 3, 6, 7, 14, 21, 42$

23. $3 \cdot 13$ **25.** $2 \cdot 3 \cdot 5$ **27.** $3 \cdot 3 \cdot 3$ **29.** $2 \cdot 3 \cdot 5 \cdot 5$
31. Prime **33.** $2 \cdot 3 \cdot 5 \cdot 7$ **35.** $5 \cdot 23$ **37.** $\frac{3}{5}$
39. $\frac{2}{7}$ **41.** $\frac{1}{4}$ **43.** 4 **45.** $\frac{1}{4}$ **47.** 6 **49.** $\frac{21}{25}$
51. $\frac{60}{41}$ **53.** $\frac{15}{7}$ **55.** $\frac{3}{10}$ **57.** 6 **59.** $\frac{1}{2}$ **61.** $\frac{7}{6}$
63. $\frac{3b}{7a}$ **65.** $\frac{10}{n}$ **67.** $\frac{5}{6}$ **69.** 1 **71.** $\frac{5}{18}$ **73.** $\frac{13}{90}$
75. $\frac{35}{18}$ **77.** 27 **79.** 1 **81.** $\frac{6}{35}$ **83.** 18 **85.**
87. **89.** Row 1: 7, 2, 36, 14, 8, 8; row 2: 9, 18, 2, 10,
12, 21 **91.** $\frac{2}{5}$ **93.** $\frac{5q}{t}$ **95.** $\frac{6}{25}$ **97.** $\frac{5ap}{2cm}$ **99.** $\frac{23r}{18t}$
101. $\frac{28}{45}$ m² **103.** $14\frac{2}{9}$ m **105.** $27\frac{3}{5}$ cm

Quick Quiz: Sections 1.1–1.3, p. 28

1. Let w represent the width of the box; $w - 4$
2. $5(3 + x)$, or $(x + 3)5$ **3.** $2 \cdot 2 \cdot 2 \cdot 5$
4. $\frac{9}{11}$ **5.** $\frac{11}{x}$

Technology Connection, p. 32

1. 2.236067977 **2.** 2.645751311 **3.** 3.605551275
4. 5.196152423

Check Your Understanding, p. 34

1. Z, Q, R **2.** Q, R **3.** I, R **4.** W, Z, Q, R
5. N, W, Z, Q, R **6.** Q, R

Exercise Set 1.4, pp. 34–36

1. Terminating **2.** Integer **3.** Whole number
4. Irrational number **5.** Opposite **6.** Absolute value
7. $-n$ **8.** $|x|$ **9.** $-10 < x$ **10.** $6 \geq y$
11. $-9500; 5000$ **13.** $100; -80$ **15.** $-777.68; 936.42$
17. $8; -5$ **19.** [number line with point at -2, marked $-5 -4 -3 -2 -1\ 0\ 1\ 2\ 3\ 4\ 5$]
21. [number line with point at -4.3, marked $-5 -4 -3 -2 -1\ 0\ 1\ 2\ 3\ 4\ 5$]
23. [number line with point at $\frac{10}{3}$, marked $-5 -4 -3 -2 -1\ 0\ 1\ 2\ 3\ 4\ 5$]
25. [number line with team labels: Lakers, Kings, Nuggets, Pistons, Celtics, Pelicans, Wizards, Bulls, Grizzlies, Spurs; $-7 -6 -5 -4 -3 -2 -1\ 0\ 1\ 2\ 3\ 4\ 5$; $-6.1\ -4.2\ -3.7\ -1.7\ -1.2\ 0.8\ 0.9\ 2.2\ 3.7\ 4.2$] **27.** 0.875
29. -0.75 **31.** $-1.1\overline{6}$ **33.** $0.\overline{6}$ **35.** -0.5
37. $0.1\overline{3}$ **39.** [number line with point at $\sqrt{5}$, marked $-5 -4 -3 -2 -1\ 0\ 1\ 2\ 3\ 4\ 5$]
41. [number line with point at $-\sqrt{22}$, marked $-5 -4 -3 -2 -1\ 0\ 1\ 2\ 3\ 4\ 5$] **43.** $>$ **45.** $<$
47. $<$ **49.** $>$ **51.** $<$ **53.** $<$ **55.** $x < -2$
57. $y \geq 10$ **59.** $-83, -4.7, 0, \frac{5}{9}, 2.\overline{16}, 62$
61. $-83, 0, 62$ **63.** $-83, -4.7, 0, \frac{5}{9}, 2.\overline{16}, \pi, \sqrt{17}, 62$

65. 58 **67.** 12.2 **69.** $\sqrt{2}$ **71.** $\frac{9}{7}$ **73.** 0 **75.** 8
77. **79.** **81.** **83.** $-23, -17, 0, 4$
85. $-\frac{4}{3}, \frac{4}{9}, \frac{4}{8}, \frac{4}{6}, \frac{4}{5}, \frac{4}{3}, \frac{4}{2}$ **87.** $<$ **89.** $=$ **91.** $-2, -1, 0, 1, 2$
93. $\frac{1}{9}$ **95.** $\frac{50}{9}$ **97.** $a < 0$ **99.** $|x| \leq 10$ **101.**

Quick Quiz: Sections 1.1–1.4, p 36

1. 1 **2.** $5(x + y + 3)$ **3.** $\frac{1}{6}$ **4.** 2 **5.** $0 > -0.5$

Mid-Chapter Review: Chapter 1, p. 37

1. $\frac{x - y}{3} = \frac{22 - 10}{3} = \frac{12}{3} = 4$
2. $14x + 7 = 7 \cdot 2x + 7 \cdot 1 = 7(2x + 1)$
3. 15 **4.** 4 **5.** $d - 10$ **6.** Let h represent the
number of hours worked; $8h$ **7.** Let s represent the
number of students originally enrolled; $s - 5 = 27$
8. No **9.** $10x + 7$ **10.** $(3a)b$ **11.** $8x + 32$
12. $6m + 15n + 30$ **13.** $3(6x + 5)$
14. $3(3c + 4d + 1)$ **15.** $2 \cdot 2 \cdot 3 \cdot 7$ **16.** $\frac{3}{7}$ **17.** $\frac{13}{24}$
18. $\frac{44}{45}$ **19.** [number line with point at -2.5, marked $-5 -4 -3 -2 -1\ 0\ 1\ 2\ 3\ 4\ 5$] **20.** -0.15
21. $>$ **22.** $<$ **23.** $9 \leq x$ **24.** 5.6 **25.** 0

Check Your Understanding, p. 41

1. 3 **2.** -3 **3.** 1 **4.** -1

Exercise Set 1.5, pp. 41–43

1. Add; negative **2.** Subtract; positive **3.** Subtract;
negative **4.** Identity **5.** Identity **6.** Like **7.** (f)
8. (d) **9.** (e) **10.** (a) **11.** (b) **12.** (c)
13. -3 **15.** 4 **17.** -7 **19.** -8 **21.** -11
23. -5 **25.** 0 **27.** -41 **29.** 0 **31.** 9 **33.** -36
35. 11 **37.** -43 **39.** 0 **41.** 18 **43.** -16
45. -0.8 **47.** -9.1 **49.** $\frac{3}{5}$ **51.** $\frac{-6}{7}$ **53.** $-\frac{1}{15}$
55. $\frac{2}{9}$ **57.** -3 **59.** 0 **61.** The price dropped 1¢.
63. Her new balance was \$95. **65.** The total gain was
20 yd. **67.** The lake dropped $\frac{7}{10}$ ft. **69.** The elevation of the peak is 13,796 ft. **71.** $17a$ **73.** $9x$
75. $-2m$ **77.** $-10y$ **79.** $-2x + 1$ **81.** $-4m$
83. $-5x - 3.9$ **85.** $12x + 17$ **87.** $7r + 8t + 16$
89. $18n + 16$ **91.** **93.** **95.** \$451.70
97. $-5y$ **99.** $-7m$ **101.** $-7t, -23$
103. 1 under par

Quick Quiz: Sections 1.1–1.5, p. 43

1. $18a + 12c + 3$ **2.** $\frac{7}{5}$ **3.** $\frac{1}{12}$ **4.** 505 **5.** $-2 - 7x$

Check Your Understanding, p. 48

1. $6 + (-8)$ **2.** $-5 + 1$ **3.** $-10 + (-7)$
4. $13 + 4$

Exercise Set 1.6, pp. 48–51

1. Opposites   **2.** Sign   **3.** Opposite   **4.** Difference
5. (d)   **6.** (g)   **7.** (f)   **8.** (h)   **9.** (a)   **10.** (c)
11. (b)   **12.** (e)   **13.** Six minus ten   **15.** Two
minus negative twelve   **17.** The opposite of x minus y
19. Negative three minus the opposite of n **21.** -51
23. $\frac{11}{3}$ **25.** 3.14 **27.** 45 **29.** $\frac{14}{3}$ **31.** -0.101
33. 37 **35.** $-\frac{2}{5}$ **37.** 1 **39.** -15 **41.** -3
43. -6 **45.** -7 **47.** -6 **49.** 0 **51.** -5
53. -10 **55.** 0 **57.** 0 **59.** 8 **61.** -11 **63.** 16
65. -19 **67.** -1 **69.** 17 **71.** 3 **73.** -3
75. -21 **77.** -11 **79.** -8 **81.** -60 **83.** -23
85. -7.3 **87.** 1.1 **89.** -5.5 **91.** -0.928
93. $-\frac{7}{11}$ **95.** $-\frac{4}{5}$ **97.** $-\frac{1}{6}$ **99.** 32 **101.** -62
103. -139 **105.** 0 **107.** $-3y, -8x$ **109.** $9, -5t, -3st$
111. $-3x$ **113.** $-5a + 4$ **115.** $-n - 9$ **117.** $-3x - 6$
119. $-8t - 7$ **121.** $-12x + 3y + 9$ **123.** $8x + 66$
125. -40 **127.** 43 **129.** $3.8 - (-5.2); 9$
131. $114 - (-79); 193$ **133.** 950 m   **135.** $213.8°$F
137. 8.9 points   **139.** ☑   **141.** ☑   **143.** $11{:}00$ P.M.,
on August 14   **145.** False. For example, let $m = -3$ and
$n = -5$. Then $-3 > -5$, but $-3 + (-5) = -8 \not> 0$.
147. True. For example, for $m = 4$ and $n = -4$,
$4 = -(-4)$ and $4 + (-4) = 0$; for $m = -3$ and
$n = 3, -3 = -3$ and $-3 + 3 = 0$.
149. ⎣(-)⎦⎣9⎦⎣–⎦⎣(-)⎦⎣7⎦⎣ENTER⎦

Quick Quiz: Sections 1.1–1.6, p. 51

1. Let p represent the total PC sales; $0.11p = 9.25$
2. $1, 2, 4, 13, 26, 52$ **3.** $2 \cdot 2 \cdot 13$ **4.** -29 **5.** 5

Check Your Understanding, p. 56

1. Negative   **2.** Positive   **3.** Positive
4. Negative   **5.** Negative

Connecting the Concepts, p. 57

1. -10 **2.** 16 **3.** 4 **4.** -6 **5.** -1 **6.** 1
7. -3.77 **8.** -7 **9.** 0 **10.** -92

Exercise Set 1.7, pp. 57–59

1. Positive   **2.** Odd   **3.** Undefined   **4.** Reciprocal
5. Opposite   **6.** Reciprocal   **7.** 1 **8.** 0 **9.** 0
10. 1 **11.** 0 **12.** 1 **13.** 1 **14.** 0 **15.** 1
16. 0 **17.** -40 **19.** -56 **21.** -40 **23.** 72
25. 190 **27.** -132 **29.** -126 **31.** 11.5 **33.** 0
35. $-\frac{2}{7}$ **37.** $\frac{1}{12}$ **39.** -11.13 **41.** $-\frac{5}{12}$ **43.** 252
45. 0 **47.** $\frac{1}{28}$ **49.** 150 **51.** 0 **53.** -720
55. $-30,240$ **57.** -9 **59.** -4 **61.** -7 **63.** 4
65. -9 **67.** 5.1 **69.** $\frac{100}{11}$ **71.** -8 **73.** Undefined
75. -4 **77.** 0 **79.** 0 **81.** $-\frac{8}{3}; \frac{8}{-3}$ **83.** $-\frac{29}{35}; \frac{-29}{35}$

85. $\frac{-7}{3}; \frac{7}{-3}$ **87.** $-\frac{x}{2}; \frac{x}{-2}$ **89.** $-\frac{5}{4}$ **91.** $-\frac{10}{51}$
93. $-\frac{1}{10}$ **95.** $\frac{1}{4.3}$, or $\frac{10}{43}$ **97.** -4 **99.** Does not exist
101. $\frac{21}{20}$ **103.** -1 **105.** 1 **107.** $\frac{3}{11}$ **109.** $-\frac{7}{4}$
111. -12 **113.** -3 **115.** 1 **117.** $\frac{1}{10}$ **119.** $-\frac{7}{6}$
121. Undefined   **123.** $-\frac{17}{30}$ **125.** ☑   **127.** ☑
129. $\dfrac{1}{a + b}$ **131.** $-(a + b)$ **133.** $x = -x$
135. For 2 and 3, the reciprocal of the sum is
$1/(2 + 3)$, or $1/5$. But $1/5 \neq 1/2 + 1/3$.   **137.** $5°$F
139. Positive   **141.** Positive   **143.** Positive
145. Distributive law; law of opposites; multiplicative
property of zero   **147.** 🖥

Quick Quiz: Sections 1.1–1.7, p. 59

1. $11(2x + 1 + 3y)$ **2.**
3. -6.1 **4.** $-x - 8m$ **5.** -2

Check Your Understanding, p. 66

1. (a) 1; (b) 8 **2.** (a) 1; (b) 49 **3.** (a) 100; (b) -100
4. (a) 12; (b) 2 **5.** (a) 1; (b) 16 **6.** (a) 2; (b) 5

Exercise Set 1.8, pp. 66–68

1. (c)   **2.** (b)   **3.** (a)   **4.** (f)   **5.** (e)   **6.** (d)
7. Division   **8.** Subtraction   **9.** Addition
10. Multiplication   **11.** Subtraction
12. Multiplication   **13.** x^6 **15.** $(-5)^3$ **17.** $(3t)^5$
19. $2n^4$ **21.** 16 **23.** 9 **25.** -9 **27.** 64
29. 625 **31.** 7 **33.** -32 **35.** $81t^4$ **37.** $-343x^3$
39. 26 **41.** 51 **43.** -15 **45.** 1 **47.** 298 **49.** 11
51. -36 **53.** 1291 **55.** 152 **57.** 24 **59.** 1
61. -44 **63.** 41 **65.** -10 **67.** -5 **69.** -19
71. -3 **73.** -75 **75.** 9 **77.** 30 **79.** 6 **81.** -17
83. $13x + 33$ **85.** $17x^2 - 17x$ **87.** $21t - r$
89. $-t^3 + 4t$ **91.** $-9x - 1$ **93.** $7n - 8$
95. $-4a + 3b - 7c$ **97.** $-3x^2 - 5x + 1$ **99.** $2x - 7$
101. $-9x + 6$ **103.** $9y - 25z$ **105.** $x^2 + 6$
107. $37a^2 - 23ab + 35b^2$ **109.** $-22t^3 - t^2 + 9t$
111. $2x - 25$ **113.** ☑   **115.** ☑
117. $-6r - 5t + 21$ **119.** $-2x - f$ **121.** ☑
123. True   **125.** False   **127.** 0 **129.** 17
131. $39,000$ **133.** $44x^3$

Quick Quiz: Sections 1.1–1.8, p. 68

1. Let m and n represent the numbers; $\frac{1}{2}(m + n)$
2. $-10 < x$ **3.** $-1.\overline{2}$ **4.** -2.94 **5.** $8x - 9y$

Translating for Success, p. 69

1. H   **2.** E   **3.** K   **4.** B   **5.** O   **6.** L   **7.** M
8. C   **9.** D   **10.** F

Decision Making: Connection, p. 71

1. $-100; 500; -100$; weight-loss diet: Monday, Wednesday; weight-gain diet: Tuesday   **2.** 2649 calories

Study Summary: Chapter 1, pp. 72–75

1. 8   **2.** $4\,\text{ft}^2$ **3.** Let n represent some number; $78 = n - 92$ **4.** $10n + 6$ **5.** $(3a)b$
6. $50m + 90n + 10$ **7.** $13(2x + 1)$ **8.** Composite
9. $2 \cdot 2 \cdot 3 \cdot 7$ **10.** 1   **11.** $\frac{9}{10}$ **12.** $\frac{3}{2}$ **13.** $\frac{9}{20}$
14. $\frac{25}{6}$ **15.** 25   **16.** $0, -15, \frac{30}{3}$ **17.** $-1.\overline{1}$ **18.** $>$
19. 1.5 **20.** -5 **21.** -2.9 **22.** -12 **23.** 15
24. $-7c + 9d - 2$ **25.** 21   **26.** -4 **27.** -100
28. -21 **29.** $a - 2b + 3c$ **30.** $5m + 7n - 15$

Review Exercises: Chapter 1, pp. 75–77

1. True   **2.** True   **3.** False   **4.** True   **5.** False
6. False   **7.** True   **8.** False   **9.** False   **10.** True
11. 24   **12.** -16 **13.** -15 **14.** $y - 7$
15. $xz + 10$, or $10 + xz$ **16.** Let b represent Brandt's speed and w the wind speed; $15(b - w)$
17. Let b represent the number of calories per hour that Katie burns while backpacking; $b = 2 \cdot 237$
18. $c = 200t$ **19.** $t \cdot 3 + 5$ **20.** $2x + (y + z)$
21. $(4x)y, 4(yx), (4y)x$; answers may vary
22. $18x + 30y$ **23.** $40x + 24y + 16$ **24.** $3(7x + 5y)$
25. $11(2a + 9b + 1)$ **26.** $2 \cdot 2 \cdot 2 \cdot 7$ **27.** $\frac{5}{12}$ **28.** $\frac{9}{4}$
29. $\frac{19}{24}$ **30.** $\frac{3}{16}$ **31.** $\frac{3}{5}$ **32.** $\frac{27}{25}$ **33.** $-3600; 1350$
34.

$$\overset{\frac{-1}{3}}{\xleftarrow{\hspace{1cm}}\underset{-5\,-4\,-3\,-2\,-1\ 0\ 1\ 2\ 3\ 4\ 5}{\bullet}\xrightarrow{\hspace{1cm}}}$$

35. $x > -3$ **36.** $<$
37. $-0.\overline{4}$ **38.** 1   **39.** -12 **40.** -10 **41.** $-\frac{7}{12}$
42. -5 **43.** 8   **44.** $-\frac{7}{5}$ **45.** -9.18 **46.** $-\frac{2}{7}$
47. -140 **48.** -7 **49.** -3 **50.** $\frac{9}{4}$ **51.** 48
52. 168   **53.** $\frac{21}{8}$ **54.** 18   **55.** 53   **56.** $\frac{103}{17}$
57. $7a - b$ **58.** $-4x + 5y$ **59.** 7   **60.** $-\frac{1}{7}$
61. $(2x)^4$ **62.** $-125x^3$ **63.** $-3a + 9$ **64.** $3x^4 + 10x$
65. $17n^2 + m^2 + 20mn$ **66.** $5x + 28$
67. ☑ The value of a constant never varies. A variable can represent a variety of numbers.   **68.** ☑ A term is one of the parts of an expression that is separated from the other parts by plus signs. A factor is part of a product.   **69.** ☑ The distributive law is used in factoring algebraic expressions, multiplying algebraic expressions, combining like terms, finding the opposite of a sum, and subtracting algebraic expressions.   **70.** ☑ A negative number raised to an even power is positive; a negative number raised to an odd power is negative.
71. 25,281   **72.** (a) $\frac{3}{11}$; (b) $\frac{10}{11}$ **73.** $-\frac{5}{8}$ **74.** -2.1
75. (i)   **76.** (j)   **77.** (a)   **78.** (h)   **79.** (k)
80. (b)   **81.** (c)   **82.** (e)   **83.** (d)   **84.** (f)
85. (g)

Test: Chapter 1, p. 78

1. [1.1] 4   **2.** [1.1] Let x and y represent the numbers; $xy - 9$ **3.** [1.1] $240\,\text{ft}^2$ **4.** [1.2] $q + 3p$
5. [1.2] $(x \cdot 4) \cdot y$ **6.** [1.1] Let t represent the number of golden lion tamarins living in zoos; $1500 = t + 1050$
7. [1.2] $35 + 7x$ **8.** [1.7] $-5y + 10$
9. [1.2] $11(1 + 4x)$ **10.** [1.2] $7(x + 1 + 7y)$
11. [1.3] $2 \cdot 2 \cdot 3 \cdot 5 \cdot 5$ **12.** [1.3] $\frac{3}{7}$ **13.** [1.4] $<$
14. [1.4] $>$ **15.** [1.4] $\frac{9}{4}$ **16.** [1.4] 3.8   **17.** [1.6] $\frac{2}{3}$
18. [1.7] $-\frac{7}{4}$ **19.** [1.6] 10   **20.** [1.4] $-5 \geq x$
21. [1.6] 7.8   **22.** [1.5] -8 **23.** [1.6] $-\frac{7}{8}$
24. [1.7] -48 **25.** [1.7] $\frac{2}{9}$ **26.** [1.7] $\frac{3}{4}$
27. [1.7] -9.728 **28.** [1.8] 20   **29.** [1.6] 15
30. [1.8] -64 **31.** [1.8] 448   **32.** [1.6] $21a + 22y$
33. [1.8] $16x^4$ **34.** [1.8] $x + 7$
35. [1.8] $9a - 12b - 7$ **36.** [1.8] $-y - 16$ **37.** [1.1] 5
38. [1.8] $9 - (3 - 4) + 5 = 15$ **39.** [1.4] $n \geq 0$
40. [1.8] $4a$ **41.** [1.8] False

CHAPTER 2

Check Your Understanding, p. 85

1. (d)   **2.** (b)   **3.** (c)   **4.** (a)

Exercise Set 2.1, pp. 86–87

1. (c)   **2.** (b)   **3.** (f)   **4.** (a)   **5.** (d)   **6.** (e)
7. No   **9.** Yes   **11.** Yes   **13.** Yes   **15.** 11
17. -25 **19.** -31 **21.** 41   **23.** 19   **25.** -6
27. $\frac{7}{3}$ **29.** $-\frac{1}{10}$ **31.** $\frac{41}{24}$ **33.** $-\frac{1}{20}$ **35.** 9.1
37. -5 **39.** 7   **41.** 12   **43.** -38 **45.** 8
47. -7 **49.** 50   **51.** -0.2 **53.** 8   **55.** 88
57. 20   **59.** -54 **61.** $-\frac{5}{9}$ **63.** 1   **65.** $\frac{9}{2}$
67. -7.6 **69.** -2.5 **71.** -15 **73.** -5 **75.** $-\frac{7}{6}$
77. -128 **79.** $-\frac{1}{2}$ **81.** -15 **83.** 9   **85.** 310.756
87. ☑   **89.** $\frac{1}{3}y - 7$ **90.** $12x + 66$

91. $5(7a + 11c + 1)$ **92.**

$$\overset{-\frac{11}{5}}{\xleftarrow{\hspace{1cm}}\underset{-5\,-4\,-3\,-2\,-1\ 0\ 1\ 2\ 3\ 4\ 5}{\bullet}\xrightarrow{\hspace{1cm}}}$$

93. ☑   **95.** 11.6   **97.** 2   **99.** $-23, 23$
101. 9000   **103.** $2500

Prepare to Move On, p. 87

1. -6 **2.** 2   **3.** $-x + 10$ **4.** $-5x + 28$

Technology Connection, p. 90

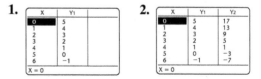

3. 4; not reliable because, depending on the choice of ΔTbl, it is easy to scroll past a solution without realizing it.

Check Your Understanding, p. 93

1. (c) **2.** (a) **3.** (b) **4.** (d) **5.** 12 **6.** 18
7. 3 **8.** 10 **9.** 100 **10.** 100

Exercise Set 2.2, pp. 94–95

1. Addition principle **2.** Multiplication principle
3. Multiplication principle **4.** Distributive law
5. Addition principle **6.** Distributive law
7. (c) **8.** (e) **9.** (a) **10.** (f) **11.** (b) **12.** (d)
13. 8 **15.** 5 **17.** $\frac{10}{3}$ **19.** -7 **21.** -5 **23.** -4
25. 19 **27.** -2.8 **29.** 3 **31.** 15 **33.** -6
35. $-\frac{25}{2}$ **37.** $\frac{9}{8}$ **39.** -3 **41.** -6 **43.** 0 **45.** 10
47. 4 **49.** $\frac{16}{3}$ **51.** $\frac{2}{5}$ **53.** 1 **55.** -4 **57.** $1.\overline{6}$
59. $-\frac{60}{37}$ **61.** 11 **63.** 2 **65.** 0 **67.** 6 **69.** $-\frac{1}{2}$
71. 0 **73.** $\frac{5}{2}$ **75.** $\frac{1}{6}$ **77.** 2 **79.** 8 **81.** $\frac{16}{15}$
83. $-\frac{1}{31}$ **85.** 2 **87.** ☑ **89.** $\frac{7}{18}$ **90.** 1
91. $-0.\overline{1}$ **92.** 16 **93.** ☑ **95.** 500 mi
97. No solution; contradiction **99.** All real numbers
are solutions; identity **101.** No solution; contradiction
103. $\frac{1136}{909}$, or $1.\overline{2497}$ **105.** No solution; contradiction
107. All real numbers are solutions; identity
109. $\frac{2}{3}$ **111.** 0

Quick Quiz: Sections 2.1–2.2, p. 95

1. Yes **2.** -5.4 **3.** 25 **4.** -4 **5.** $\frac{1}{2}$

Prepare to Move On, p. 95

1. -7 **2.** 15 **3.** -15 **4.** -28

Technology Connection, p. 96

1. 14.4

Check Your Understanding, p. 99

1. Yes **2.** Yes **3.** No **4.** No

Exercise Set 2.3, pp. 99–102

1. False **2.** True **3.** Circumference
4. Formula **5.** 309.6 m **7.** 1423 students
9. 8.4734 **11.** 255 mg **13.** $b = \frac{A}{h}$ **15.** $P = \frac{I}{rt}$
17. $m = 65 - H$ **19.** $l = \frac{P - 2w}{2}$, or $l = \frac{P}{2} - w$
21. $\pi = \frac{A}{r^2}$ **23.** $h = \frac{2A}{b}$ **25.** $c^2 = \frac{E}{m}$
27. $d = 2Q - c$ **29.** $q = p + r - 2$ **31.** $r = wf$
33. $T = \frac{550H}{V}$ **35.** $C = \frac{5}{9}(F - 32)$ **37.** $y = 2x - 1$
39. $y = -\frac{2}{5}x + 2$ **41.** $y = \frac{4}{3}x - 2$
43. $y = -\frac{9}{8}x + \frac{1}{2}$ **45.** $y = \frac{3}{5}x - \frac{8}{5}$

47. $x = \frac{z - 13}{2} - y$, or $x = \frac{z - 13 - 2y}{2}$
49. $l = 4(t - 27) + w$ **51.** $t = \frac{A}{a + b}$
53. $h = \frac{2A}{a + b}$ **55.** $L = W - \frac{N(R - r)}{400}$, or
$L = \frac{400W - NR + Nr}{400}$ **57.** ☑ **59.** -10
60. -196 **61.** 0 **62.** -32 **63.** -13 **64.** 65
65. ☑ **67.** 40 years old **69.** 27 in^3
71. $a = \frac{w}{c} \cdot d$ **73.** $c = \frac{d}{a - b}$ **75.** $a = \frac{c}{3 + b + d}$
77. $K = 9.632w + 19.685h - 10.54a + 102.3$

Quick Quiz: Sections 2.1–2.3, p. 102

1. 7 **2.** $\frac{64}{3}$ **3.** -0.025 **4.** -21 **5.** $y = -3x + \frac{1}{2}$

Prepare to Move On, p. 102

1. 0.25 **2.** 1.125 **3.** $0.\overline{6}$ **4.** $0.8\overline{3}$

Mid-Chapter Review: Chapter 2, p. 103

1. $2x + 3 - 3 = 10 - 3$
$\qquad 2x = 7$
$\qquad \frac{1}{2} \cdot 2x = \frac{1}{2} \cdot 7$
$\qquad x = \frac{7}{2}$
2. $6 \cdot \frac{1}{2}(x - 3) = 6 \cdot \frac{1}{3}(x - 4)$
$\qquad 3(x - 3) = 2(x - 4)$
$\qquad 3x - 9 = 2x - 8$
$\qquad 3x - 9 + 9 = 2x - 8 + 9$
$\qquad 3x = 2x + 1$
$\qquad 3x - 2x = 2x + 1 - 2x$
$\qquad x = 1$
3. 1 **4.** 3 **5.** $\frac{5}{3}$ **6.** -8 **7.** 48 **8.** 0.5 **9.** -5
10. $\frac{8}{3}$ **11.** 6 **12.** 0 **13.** $\frac{49}{9}$ **14.** $-\frac{4}{11}$ **15.** $\frac{23}{7}$
16. $A = \frac{E}{w}$ **17.** $y = \frac{C - Ax}{B}$ **18.** $a = \frac{m}{t + p}$
19. $a = \frac{F}{m}$ **20.** $b = vt + f$

Check Your Understanding, p. 107

1. 4 **2.** 20 **3.** 10 **4.** 40 **5.** 0.4

Exercise Set 2.4, pp. 108–112

1. Left **2.** Hundred **3.** Percent **4.** Base
5. Sale **6.** Approximately **7.** (d) **8.** (c) **9.** (e)
10. (b) **11.** (c) **12.** (d) **13.** (f) **14.** (a)
15. (b) **16.** (e) **17.** 0.47 **19.** 0.05 **21.** 0.032
23. 0.1 **25.** 0.0625 **27.** 0.002 **29.** 1.75 **31.** 79%
33. 4.7% **35.** 70% **37.** 0.09% **39.** 106%

41. 60% **43.** 32% **45.** 25% **47.** $46\frac{2}{3}$, or $\frac{140}{3}$
49. 2.5 **51.** 10,000 **53.** 125% **55.** 0.8 **57.** 50%
59. $33.\overline{3}$%, or $33\frac{1}{3}$% **61.** 14.34 million cats
63. 1.912 million cats **65.** 75 credits **67.** 548 at-bats
69. (a) 16%; **(b)** $29 **71.** About 88.6%; about 11.4%
73. $163.20 **75.** 285 women **77.** $19.20 per hour
79. About 462% **81.** About 31.5 lb **83.** About 90%
85. 5 calories **87.** ✏ **89.** $\frac{1}{3}$ **90.** -3 **91.** -12
92. $9x^2$ **93.** ✏ **95.** 18,500 people
97. About 6 ft 7 in. **99.** About 27% **101.** ✏

Quick Quiz: Sections 2.1–2.4, p. 112

1. 0 **2.** $\frac{62}{3}$ **3.** $p = \dfrac{T}{4+m}$ **4.** 0.012 **5.** 25

Prepare to Move On, p. 112

1. Let l represent the length and w represent the width; $2l + 2w$ **2.** $0.05 \cdot 180$ **3.** $10\left(\frac{1}{2}a\right)$
4. Let n represent the number; $3n + 10$
5. Let l represent the board's length and w represent the width; $w = l - 2$ **6.** Let x represent the first number and y represent the second number; $x = 4y$

Check Your Understanding, p. 121

1. $x + 2$; $x + 2$; 32 **2.** $0.06x$; $0.06x$; 36.57

Exercise Set 2.5, pp. 121–127

1. (1) Familiarize. **(2)** Translate. **(3)** Carry out.
(4) Check. **(5)** State. **2.** Carry out. **3.** State.
4. Familiarize. **5.** Translate. **6.** Familiarize.
7. Familiarize. **8.** Check. **9.** 11 **11.** $\frac{11}{2}$
13. 16.4 mi **15.** 260 mi **17.** 1204 and 1205
19. 285 and 287 **21.** 32, 33, 34 **23.** Man: 104 years;
woman: 102 years **25.** *The 13th Warrior*: $98.3 million;
Mars Needs Moms: $111 million **27.** 140 and 141
29. Width: 21 m; length: 25 m **31.** $1\frac{1}{2}$ in. by $3\frac{1}{2}$ in.
33. 30°, 90°, 60° **35.** 70° **37.** Bottom: 144 ft;
middle: 72 ft; top: 24 ft **39.** 12 mph **41.** 1 hr
43. Approximately 15.8% **45.** 12.5% **47.** $1225
49. $280.80 **51.** $320 **53.** $36,000 **55.** $100
57. $4 million **59.** $125,000 **61.** $8\frac{3}{4}$ mi
63. 109.2 mi **65.** 65°, 25° **67.** 140°, 40°
69. Length: 27.9 cm; width: 21.6 cm **71.** $1540
73. 830 points **75.** 2015 **77.** 160 chirps per minute
79. ✏ **81.** $8n + 32t + 4$ **82.** $3(4 + 6x + 7y)$
83. $7x - 18$ **84.** 0 **85.** ✏ **87.** $37 **89.** 20
91. Half-dollars: 5; quarters: 10; dimes: 20; nickels: 60
93. $95.99 **95.** 5 DVDs **97.** 6 mi **99.** ✏
101. Width: 23.31 cm; length: 27.56 cm

Quick Quiz: Sections 2.1–2.5, p. 127

1. $\frac{25}{6}$ **2.** -3 **3.** $v = \dfrac{3}{B}$ **4.** $10.50 **5.** $52\frac{1}{2}$ min

Prepare to Move On, p. 127

1. $<$ **2.** $>$ **3.** $-4 \le x$ **4.** $y < 5$

Check Your Understanding, p. 133

1. Equivalent **2.** Not equivalent **3.** Equivalent
4. Equivalent **5.** Equivalent **6.** Not equivalent

Connecting the Concepts, p. 134

1. 21 **2.** $\{x \mid x \le 21\}$, or $(-\infty, 21]$ **3.** -6
4. $\{x \mid x > -6\}$, or $(-6, \infty)$
5. $\left\{x \mid x \le -\frac{1}{3}\right\}$, or $\left(-\infty, -\frac{1}{3}\right]$ **6.** $-\frac{1}{3}$ **7.** 66
8. $\{n \mid n < 66\}$, or $(-\infty, 66)$ **9.** $\{a \mid a \ge 0\}$, or $[0, \infty)$
10. 0

Exercise Set 2.6, pp. 134–136

1. Solution **2.** Set-builder **3.** Closed **4.** Bracket
5. $\ge$ **6.** $\le$ **7.** $>$ **8.** $\le$ **9.** $\ge$ **10.** $<$
11. $<$ **12.** $>$ **13. (a)** Yes; **(b)** no; **(c)** no
15. (a) Yes; **(b)** no; **(c)** yes **17. (a)** Yes; **(b)** yes; **(c)** yes

19. $y < 2$

21. $x \ge -1$

23. $0 \le t$

25. $-5 \le x < 2$

27. $-4 < x < 0$

29. $\{y \mid y < 6\}$, $(-\infty, 6)$
31. $\{x \mid x \ge -4\}$, $[-4, \infty)$
33. $\{t \mid t > -3\}$, $(-3, \infty)$
35. $\{x \mid x \le -7\}$, $(-\infty, -7]$
37. $\{x \mid x > -4\}$, $(-4, \infty)$ **39.** $\{x \mid x \le 2\}$, $(-\infty, 2]$
41. $\{x \mid x < -1\}$, $(-\infty, -1)$ **43.** $\{x \mid x \ge 0\}$, $[0, \infty)$
45. $\{y \mid y > 3\}$, $(3, \infty)$
47. $\{n \mid n < 17\}$, $(-\infty, 17)$,
49. $\{x \mid x \le -9\}$, $(-\infty, -9]$,
51. $\{t \mid t \le -3\}$, $(-\infty, -3]$,
53. $\{t \mid t > \frac{5}{8}\}$, $\left(\frac{5}{8}, \infty\right)$,
55. $\{x \mid x < 0\}$, $(-\infty, 0)$,
57. $\{t \mid t < 23\}$, $(-\infty, 23)$,
59. $\{x \mid x < 7\}$, $(-\infty, 7)$,

61. $\{t\,|\,t < -3\}$, $(-\infty, -3)$,

63. $\{n\,|\,n \geq -1.5\}$, $[-1.5, \infty)$,

65. $\{y\,|\,y \geq -\frac{1}{10}\}$, $[-\frac{1}{10}, \infty)$,

67. $\{x\,|\,x < -\frac{4}{5}\}$, $(-\infty, -\frac{4}{5})$,

69. $\{x\,|\,x < 6\}$, or $(-\infty, 6)$ **71.** $\{t\,|\,t \leq 7\}$, or $(-\infty, 7]$
73. $\{x\,|\,x > -4\}$, or $(-4, \infty)$
75. $\{y\,|\,y < -\frac{10}{3}\}$, or $\left(-\infty, -\frac{10}{3}\right)$
77. $\{x\,|\,x > -10\}$, or $(-10, \infty)$
79. $\{y\,|\,y < 0\}$, or $(-\infty, 0)$ **81.** $\{x\,|\,x > -4\}$, or $(-4, \infty)$
83. $\{t\,|\,t > 1\}$, or $(1, \infty)$ **85.** $\{x\,|\,x \leq -9\}$, or $(-\infty, -9]$
87. $\{t\,|\,t < 14\}$, or $(-\infty, 14)$
89. $\{y\,|\,y \leq -4\}$, or $(-\infty, -4]$
91. $\{t\,|\,t < -\frac{5}{3}\}$, or $\left(-\infty, -\frac{5}{3}\right)$
93. $\{r\,|\,r > -3\}$, or $(-3, \infty)$
95. $\{x\,|\,x \leq 7\}$, or $(-\infty, 7]$
97. $\{x\,|\,x > -\frac{5}{32}\}$, or $\left(-\frac{5}{32}, \infty\right)$ **99.** ✅
101. $17x - 6$ **102.** $2m - 16n$ **103.** $7x - 8y - 46$
104. $-21x + 32$ **105.** ✅
107. $\{x\,|\,x \text{ is a real number}\}$, or $(-\infty, \infty)$
109. $\{x\,|\,x \leq \frac{5}{6}\}$, or $\left(-\infty, \frac{5}{6}\right]$
111. $\{x\,|\,x \leq -4a\}$, or $(-\infty, -4a]$
113. $\left\{x\,\middle|\,x > \dfrac{y - b}{a}\right\}$, or $\left(\dfrac{y - b}{a}, \infty\right)$
115. $\{x\,|\,x \text{ is a real number}\}$, or $(-\infty, \infty)$
117. **(a)** No; **(b)** There is more than 6 g of fat per serving.

Quick Quiz: Sections 2.1–2.6, p. 136

1. $-\frac{28}{3}$ **2.** $\{x\,|\,x < -\frac{10}{3}\}$, or $\left(-\infty, -\frac{10}{3}\right)$ **3.** $\$15.00$
4. $\$19.34$ **5.** $c = \dfrac{X - 12 + 5d}{5}$

Prepare to Move On, p. 136

1. Let h represent the height of the triangle; $\frac{1}{2} \cdot 3 \cdot h = 5$
2. Let s represent the shortest side of the triangle;
$s + (s + 1) + (s + \frac{1}{2}) = 12$

Check Your Understanding, p. 139

1. No **2.** Yes **3.** Yes **4.** Yes **5.** No

Exercise Set 2.7, pp. 139–143

1. Is at least **2.** Cannot exceed **3.** No less than
4. Minimum **5.** $b \leq a$ **6.** $b < a$ **7.** $a \leq b$
8. $a < b$ **9.** $b \leq a$ **10.** $a \leq b$ **11.** $b < a$
12. $a < b$ **13.** Let n represent the number; $n < 10$
15. Let t represent the temperature; $t \leq -3$
17. Let d represent the number of years of driving
experience; $d \geq 5$ **19.** Let a represent the age of the
altar; $a > 1200$ **21.** Let h represent Bianca's hourly
wage, in dollars; $h \geq 12$ **23.** Let s represent the number of hours of sunshine; $1100 < s < 1600$

25. More than 2.375 hr **27.** At least 2.6
29. Scores greater than or equal to 97 **31.** 8 credits
or more **33.** At least 3 plate appearances
35. Lengths greater than 6 cm **37.** Depths less than
437.5 ft **39.** Blue-book value is greater than or equal
to $\$10{,}625$ **41.** Lengths less than 55 in.
43. Temperatures greater than 37°C **45.** No more
than 3 ft tall **47.** A serving contains at least 16 g of
fat. **49.** 2019 and later **51.** No more than 811 text
messages **53.** 2012 and later **55.** Distances less
than or equal to 107 mi **57.** ✅ **59.** $7 + xy$
60. $3 \cdot 3 \cdot 5 \cdot 5$ **61.** -18 **62.** -5 **63.** ✅
65. Temperatures between -15°C and $-9\frac{4}{9}$°C
67. Lengths less than or equal to 8 cm **69.** At least $\$42$
71. ✅ **73.** 🖥

Quick Quiz: Sections 2.1–2.7, p. 143

1. $\{x\,|\,x \leq 4\}$, or $(-\infty, 4]$ **2.** $\frac{1210}{3}$ **3.** 20%
4. 48 km **5.** 72 or higher

Prepare to Move On, p. 143

1. **2.**
3. 79.5 **4.** 8.5

Translating for Success, p. 144

1. F **2.** I **3.** C **4.** E **5.** D **6.** J **7.** O
8. M **9.** B **10.** L

Decision Making: Connection, p. 145

1. Barry should remain a crew manager.
2. $A = 1.2x$ **3.** 🖥

Study Summary: Chapter 2, pp. 146–148

1. 5 **2.** 4.8 **3.** -1 **4.** $-\frac{1}{10}$ **5.** $c = \dfrac{d}{a - b}$
6. 80 **7.** $47\frac{1}{2}$ mi; $72\frac{1}{2}$ mi **8.** $(-\infty, 0]$
9. $\{x\,|\,x > 7\}$, or $(7, \infty)$ **10.** $\{x\,|\,x \geq -\frac{1}{4}\}$, or $\left[-\frac{1}{4}, \infty\right)$
11. Let d represent the distance Luke runs, in miles;
$d \geq 3$

Review Exercises: Chapter 2, pp. 149–150

1. True **2.** False **3.** True **4.** True **5.** True
6. False **7.** True **8.** True **9.** -25 **10.** 7
11. -65 **12.** 1.11 **13.** $\frac{1}{2}$ **14.** $-\frac{3}{2}$ **15.** -8
16. -4 **17.** $-\frac{1}{3}$ **18.** 4 **19.** 3 **20.** 4
21. 16 **22.** 1 **23.** $-\frac{7}{5}$ **24.** 0 **25.** 4
26. $d = \dfrac{C}{\pi}$ **27.** $B = \dfrac{3V}{h}$ **28.** $y = \frac{5}{2}x - 5$
29. $x = \dfrac{b}{t - a}$ **30.** 0.012 **31.** 44% **32.** 70%

33. 140   **34.** No   **35.** Yes   **36.** No

37.
$$5x - 6 < 2x + 3$$
$-5\,-4\,-3\,-2\,-1\ 0\ 1\ 2\ 3\ 4\ 5$

38.
$$-2 < x \le 5$$
$-5\,-4\,-3\,-2\,-1\ 0\ 1\ 2\ 3\ 4\ 5$

39.
$$t > 0$$
$-5\,-4\,-3\,-2\,-1\ 0\ 1\ 2\ 3\ 4\ 5$

40. $\{t \mid t \ge -\frac{1}{2}\}$, or $\left[-\frac{1}{2}, \infty\right)$ **41.** $\{y \mid y > 3\}$, or $(3, \infty)$
42. $\{y \mid y \le -4\}$, or $(-\infty, -4]$
43. $\{y \mid y > -7\}$, or $(-7, \infty)$
44. $\{x \mid x > -6\}$, or $(-6, \infty)$
45. $\{x \mid x > -\frac{9}{11}\}$, or $\left(-\frac{9}{11}, \infty\right)$
46. $\{t \mid t \le -12\}$, or $(-\infty, -12]$
47. $\{x \mid x \le -8\}$, or $(-\infty, -8]$ **48.** About 933 cats
49. 15 ft, 17 ft   **50.** Chinese students: 194,000; Indian students: 100,000   **51.** About 990,000 students
52. 57, 59   **53.** Width: 11 cm; length: 17 cm   **54.** $160
55. $35°, 85°, 60°$ **56.** $105 or less   **57.** 14 or fewer copies   **58.** ✍ Multiplying both sides of an equation by *any* nonzero number results in an equivalent equation. When we are multiplying on both sides of an inequality, the sign of the number being multiplied by must be considered. If the number is positive, the direction of the inequality symbol remains unchanged; if the number is negative, the direction of the inequality symbol must be reversed to produce an equivalent inequality.   **59.** ✍ The solutions of an equation can usually each be checked. The solutions of an inequality are normally too numerous to check. Checking a few numbers from the solution set found cannot guarantee that the answer is correct, although if any number does not check, the answer found is incorrect.   **60.** About 1 hr 36 min   **61.** Nile: 4160 mi; Amazon: 4225 mi

62. $-23, 23$ **63.** $-20, 20$ **64.** $a = \dfrac{y - 3}{2 - b}$

65. $F = \dfrac{0.3(12w)}{9}$, or $F = 0.4w$

Test: Chapter 2, p. 151

1. [2.1] 9   **2.** [2.1] -3 **3.** [2.1] 49   **4.** [2.1] -12
5. [2.2] 2   **6.** [2.1] -8 **7.** [2.2] $-\frac{23}{67}$ **8.** [2.2] 7
9. [2.2] $-\frac{5}{3}$ **10.** [2.2] $\frac{23}{3}$
11. [2.6] $\{x \mid x > -5\}$, or $(-5, \infty)$
12. [2.6] $\{x \mid x > -13\}$, or $(-13, \infty)$
13. [2.6] $\{y \mid y \le -13\}$, or $(-\infty, -13]$
14. [2.6] $\{n \mid n < -5\}$, or $(-\infty, -5)$
15. [2.6] $\{x \mid x < -7\}$, or $(-\infty, -7)$
16. [2.6] $\{t \mid t \ge -1\}$, or $[-1, \infty)$
17. [2.6] $\{x \mid x \le -1\}$, or $(-\infty, -1]$

18. [2.3] $r = \dfrac{A}{2\pi h}$ **19.** [2.3] $l = 2w - P$

20. [2.4] 2.3   **21.** [2.4] 0.3%   **22.** [2.4] 14.8
23. [2.4] 44%

24. [2.6]
$$y < 4$$
$-10\,-8\,-6\,-4\,-2\ 0\ 2\ 4\ 6\ 8\ 10$

25. [2.6]
$$-2 \le x \le 2$$
$-5\,-4\,-3\,-2\,-1\ 0\ 1\ 2\ 3\ 4\ 5$

26. [2.5] Width: 7 cm; length: 11 cm   **27.** [2.5] 60 mi
28. [2.5] 81 mm, 83 mm, 85 mm   **29.** [2.4] $65
30. [2.7] More than 38 one-way trips per month

31. [2.3] $d = \dfrac{a}{3}$ **32.** [1.4], [2.2] $-15, 15$

33. [2.7] Let $h =$ the number of hours of sun each day; $4 \le h \le 6$ **34.** [2.5] 60 tickets

Cumulative Review: Chapters 1–2, p. 152

1. -12 **2.** $\frac{3}{4}$ **3.** -4.2 **4.** 10   **5.** 134   **6.** $\frac{1}{2}$
7. $2x + 1$ **8.** $-21n + 36$

9.
$$-\tfrac{5}{2}$$
$-5\,-4\,-3\,-2\,-1\ 0\ 1\ 2\ 3\ 4\ 5$ **10.** $2(3x + 2y + 4z)$

11. 16   **12.** 9   **13.** $\frac{13}{18}$ **14.** 1   **15.** $-\frac{7}{2}$

16. $z = \dfrac{x}{4y}$ **17.** $y = \frac{4}{9}x - \frac{1}{9}$ **18.** $n = \dfrac{p}{a + r}$

19. 1.83   **20.** 37.5%

21.
$$-\tfrac{5}{2} \qquad t > -\tfrac{5}{2}$$
$-5\,-4\,-3\,-2\,-1\ 0\ 1\ 2\ 3\ 4\ 5$

22. $\{t \mid t \le -2\}$, or $(-\infty, -2]$
23. $\{t \mid t < 3\}$, or $(-\infty, 3)$
24. $\{x \mid x < 30\}$, or $(-\infty, 30)$
25. $\{n \mid n \ge 2\}$, or $[2, \infty)$ **26.** 48 million   **27.** $14\frac{1}{3}$ m
28. 9 ft, 15 ft   **29.** No more than $52
30. About 54%   **31.** $105°$ **32.** For widths greater than 27 cm   **33.** $4t$ **34.** $-5, 5$ **35.** $1025

CHAPTER 3

Check Your Understanding, p. 156

1. $(-40, -1)$ **2.** $(-30, 1.5)$ **3.** $(0, 0)$ **4.** $(0, -2)$
5. $(10, -1.5)$ **6.** $(40, 0)$

Exercise Set 3.1, pp. 159–163

1. E   **2.** C   **3.** G   **4.** B   **5.** H   **6.** I   **7.** (a)
8. (c)   **9.** (b)   **10.** (d)   **11.** 2 drinks
13. The person weighs more than 140 lb.
15. 12%   **17.** 1985 and 1995   **19.** 1990

21.

23.

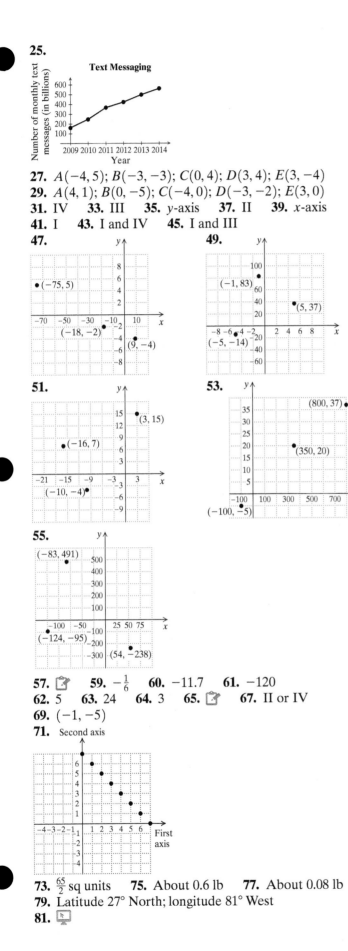

25.

Text Messaging

(graph: Number of monthly text messages (in billions) vs. Year 2009–2014)

27. $A(-4, 5)$; $B(-3, -3)$; $C(0, 4)$; $D(3, 4)$; $E(3, -4)$
29. $A(4, 1)$; $B(0, -5)$; $C(-4, 0)$; $D(-3, -2)$; $E(3, 0)$
31. IV **33.** III **35.** y-axis **37.** II **39.** x-axis
41. I **43.** I and IV **45.** I and III
47. (points: $(-75, 5)$, $(-18, -2)$, $(9, -4)$)
49. (points: $(-1, 83)$, $(5, 37)$, $(-5, -14)$)
51. (points: $(3, 15)$, $(-16, 7)$, $(-10, -4)$)
53. (points: $(800, 37)$, $(350, 20)$, $(-100, -5)$)
55. (points: $(-83, 491)$, $(-124, -95)$, $(54, -238)$)
57. 🖊 **59.** $-\frac{1}{6}$ **60.** -11.7 **61.** -120
62. 5 **63.** 24 **64.** 3 **65.** 🖊 **67.** II or IV
69. $(-1, -5)$
71. (graph with Second axis / First axis)
73. $\frac{65}{2}$ sq units **75.** About 0.6 lb **77.** About 0.08 lb
79. Latitude 27° North; longitude 81° West
81. 🖥

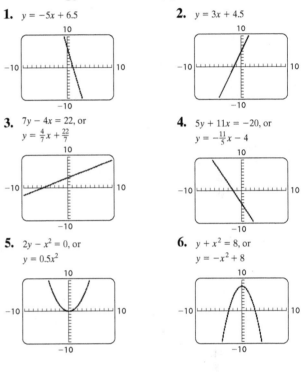

Prepare to Move On, p. 163

1. $y = \frac{2}{5}x$ **2.** $y = -\frac{3}{2}x$ **3.** $y = x - 8$
4. $y = -\frac{2}{5}x + 2$ **5.** $y = \frac{5}{8}x - \frac{1}{8}$

Check Your Understanding, p. 169

1. $9, 5, 3, -1$ **2.** $(-2, 9), (0, 5), (1, 3), (3, -1)$

Technology Connection, p. 169

1. $y = -5x + 6.5$
2. $y = 3x + 4.5$
3. $7y - 4x = 22$, or $y = \frac{4}{7}x + \frac{22}{7}$
4. $5y + 11x = -20$, or $y = -\frac{11}{5}x - 4$
5. $2y - x^2 = 0$, or $y = 0.5x^2$
6. $y + x^2 = 8$, or $y = -x^2 + 8$

Exercise Set 3.2, pp. 170–173

1. False **2.** True **3.** True **4.** True **5.** True
6. False **7.** Yes **9.** No **11.** No
13.

$y = x + 3$		$y = x + 3$	
2	$-1 + 3$	7	$4 + 3$
$2 \overset{?}{=} 2$	True	$7 \overset{?}{=} 7$	True

$(2, 5)$; answers may vary

15.

$y = \frac{1}{2}x + 3$		$y = \frac{1}{2}x + 3$	
5	$\frac{1}{2} \cdot 4 + 3$	2	$\frac{1}{2}(-2) + 3$
	$2 + 3$		$-1 + 3$
$5 \overset{?}{=} 5$	True	$2 \overset{?}{=} 2$	True

$(0, 3)$; answers may vary

17.

$y + 3x = 7$		$y + 3x = 7$	
$1 + 3 \cdot 2$	7	$-5 + 3 \cdot 4$	7
$1 + 6$		$-5 + 12$	
$7 \overset{?}{=} 7$	True	$7 \overset{?}{=} 7$	True

$(1, 4)$; answers may vary

19.

$$\frac{4x - 2y = 10}{4 \cdot 0 - 2(-5) \mid 10}$$
$$\frac{0 + 10}{10 \overset{?}{=} 10 \quad \text{True}}$$

$$\frac{4x - 2y = 10}{4 \cdot 4 - 2 \cdot 3 \mid 10}$$
$$\frac{16 - 6}{10 \overset{?}{=} 10 \quad \text{True}}$$

$(2, -1)$; answers may vary

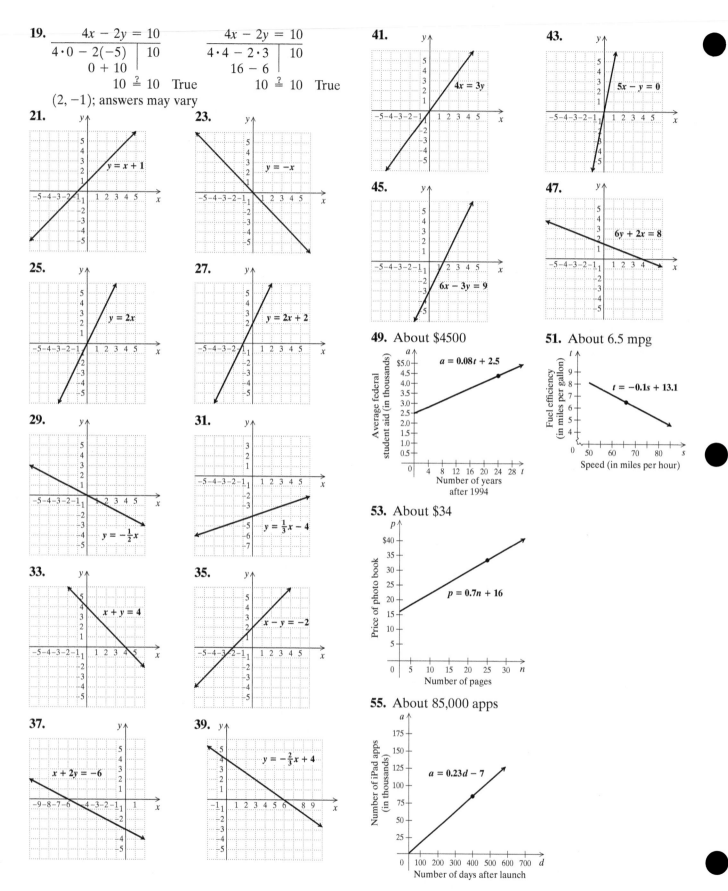

21. $y = x + 1$

23. $y = -x$

25. $y = 2x$

27. $y = 2x + 2$

29. $y = -\frac{1}{2}x$

31. $y = \frac{1}{3}x - 4$

33. $x + y = 4$

35. $x - y = -2$

37. $x + 2y = -6$

39. $y = -\frac{2}{3}x + 4$

41. $4x = 3y$

43. $5x - y = 0$

45. $6x - 3y = 9$

47. $6y + 2x = 8$

49. About $4500

$a = 0.08t + 2.5$

Average federal student aid (in thousands)

Number of years after 1994

51. About 6.5 mpg

$t = -0.1s + 13.1$

Fuel efficiency (in miles per gallon)

Speed (in miles per hour)

53. About $34

$p = 0.7n + 16$

Price of photo book

Number of pages

55. About 85,000 apps

$a = 0.23d - 7$

Number of iPad apps (in thousands)

Number of days after launch

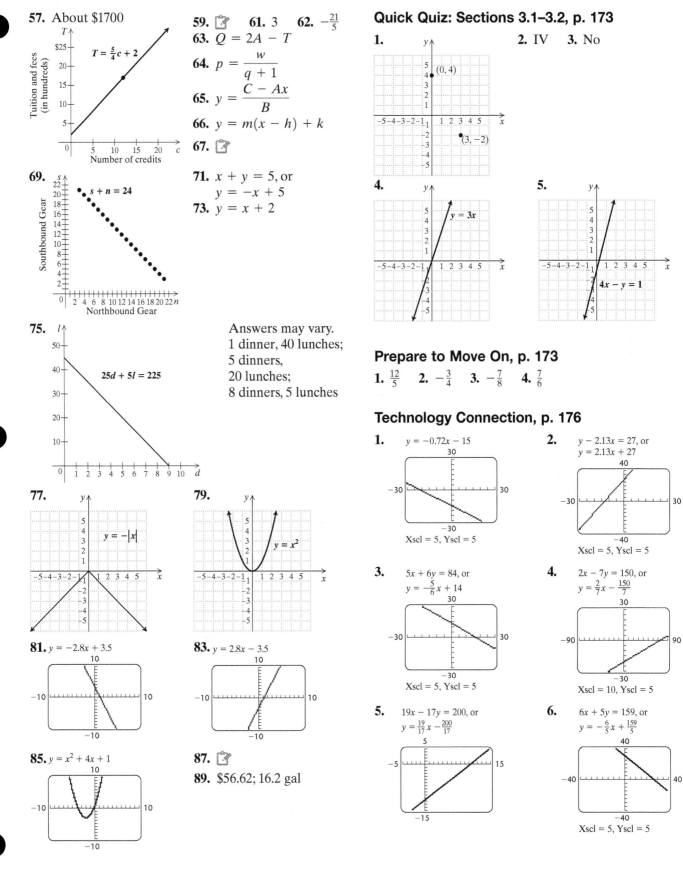

57. About $1700

$T = \frac{5}{4}c + 2$

59. 🖼   **61.** 3   **62.** $-\frac{21}{5}$

63. $Q = 2A - T$

64. $p = \dfrac{w}{q + 1}$

65. $y = \dfrac{C - Ax}{B}$

66. $y = m(x - h) + k$

67. 🖼

69. $s + n = 24$

71. $x + y = 5$, or $y = -x + 5$

73. $y = x + 2$

75. $25d + 5l = 225$

Answers may vary.
1 dinner, 40 lunches;
5 dinners,
20 lunches;
8 dinners, 5 lunches

77. $y = -|x|$

79. $y = x^2$

81. $y = -2.8x + 3.5$

83. $y = 2.8x - 3.5$

85. $y = x^2 + 4x + 1$

87. 🖼

89. $56.62; 16.2 gal

Quick Quiz: Sections 3.1–3.2, p. 173

1. $(0, 4)$ $(3, -2)$ **2.** IV   **3.** No

4. $y = 3x$

5. $4x - y = 1$

Prepare to Move On, p. 173

1. $\frac{12}{5}$ **2.** $-\frac{3}{4}$ **3.** $-\frac{7}{8}$ **4.** $\frac{7}{6}$

Technology Connection, p. 176

1. $y = -0.72x - 15$
Xscl = 5, Yscl = 5

2. $y - 2.13x = 27$, or $y = 2.13x + 27$
Xscl = 5, Yscl = 5

3. $5x + 6y = 84$, or $y = -\frac{5}{6}x + 14$
Xscl = 5, Yscl = 5

4. $2x - 7y = 150$, or $y = \frac{2}{7}x - \frac{150}{7}$
Xscl = 10, Yscl = 5

5. $19x - 17y = 200$, or $y = \frac{19}{17}x - \frac{200}{17}$

6. $6x + 5y = 159$, or $y = -\frac{6}{5}x + \frac{159}{5}$
Xscl = 5, Yscl = 5

Check Your Understanding, p. 178

1.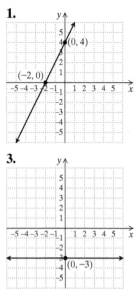

2.

3.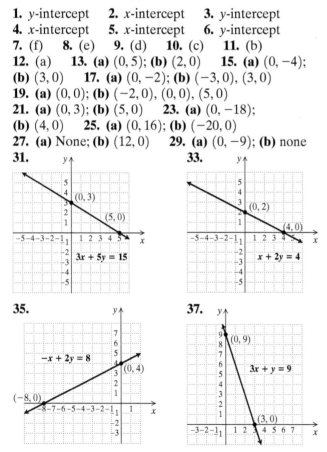

Exercise Set 3.3, pp. 178–181

1. *y*-intercept   **2.** *x*-intercept   **3.** *y*-intercept
4. *x*-intercept   **5.** *x*-intercept   **6.** *y*-intercept
7. (f)   **8.** (e)   **9.** (d)   **10.** (c)   **11.** (b)
12. (a)   **13. (a)** $(0, 5)$; **(b)** $(2, 0)$ **15. (a)** $(0, -4)$;
(b) $(3, 0)$ **17. (a)** $(0, -2)$; **(b)** $(-3, 0)$, $(3, 0)$
19. (a) $(0, 0)$; **(b)** $(-2, 0)$, $(0, 0)$, $(5, 0)$
21. (a) $(0, 3)$; **(b)** $(5, 0)$ **23. (a)** $(0, -18)$;
(b) $(4, 0)$ **25. (a)** $(0, 16)$; **(b)** $(-20, 0)$
27. (a) None; **(b)** $(12, 0)$ **29. (a)** $(0, -9)$; **(b)** none

31.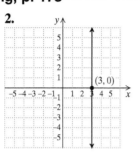

33.

35.

37.

39.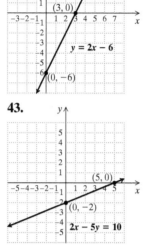

41.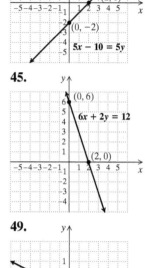

43.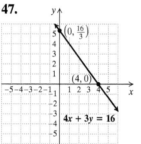

45.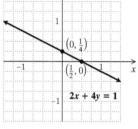

47.

49.

51.

53.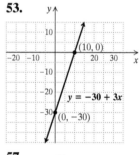

55.

57.

59.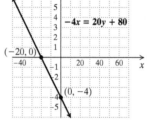

61.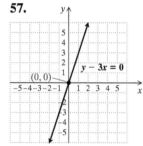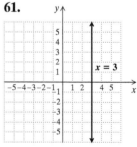

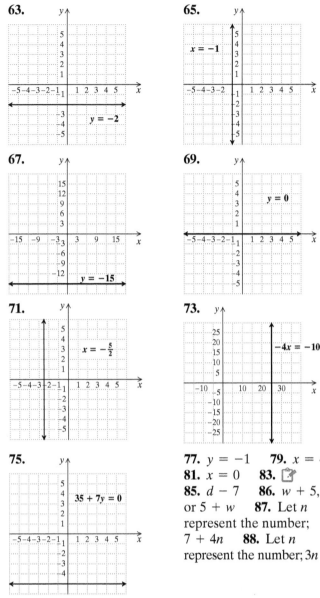

63.

65.

67.

69.

71.

73.

75.

89. Let x and y represent the numbers; $2(x + y)$
90. Let a and b represent the numbers; $\frac{1}{2}(a + b)$
91. ☑   **93.** $y = 0$ **95.** $x = -2$ **97.** $(-3, 4)$
99. $-5x + 3y = 15$, or $y = \frac{5}{3}x + 5$ **101.** -24
103. $\left(\dfrac{C - D}{A}, 0\right)$ **105.** $\left(0, -\frac{80}{7}\right)$, or $(0, -11.\overline{428571})$;
$(40, 0)$ **107.** $(0, -9)$; $(45, 0)$ **109.** $\left(0, \frac{1}{25}\right)$, or
$(0, 0.04)$; $\left(\frac{1}{50}, 0\right)$, or $(0.02, 0)$

Quick Quiz: Sections 3.1–3.3, p. 181

1.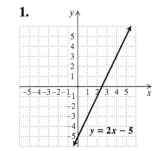

2.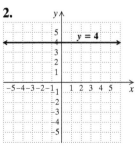

77. $y = -1$ **79.** $x = 4$
81. $x = 0$ **83.** ☑
85. $d - 7$ **86.** $w + 5$,
or $5 + w$ **87.** Let n
represent the number;
$7 + 4n$ **88.** Let n
represent the number; $3n$

3.

4.

5.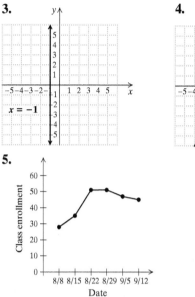

Prepare to Move On, p. 181

1. 40   **2.** 518   **3.** $\frac{1720}{103}$ **4.** 0.6108

Check Your Understanding, p. 183

1. 45 min   **2.** 3 mi   **3.** 15 min/mi, or 4 mi/hr

Exercise Set 3.4, pp. 185–190

1. True   **2.** True   **3.** False   **4.** True
5. Miles per hour, or $\dfrac{\text{miles}}{\text{hour}}$
6. Hours per chapter, or $\dfrac{\text{hours}}{\text{chapter}}$
7. Dollars per mile, or $\dfrac{\text{dollars}}{\text{mile}}$
8. Cups of flour per cake, or $\dfrac{\text{cups of flour}}{\text{cake}}$
9. (a) 30 mpg; **(b)** \$39.33/day; **(c)** 130 mi/day;
(d) 30¢/mi   **11. (a)** 7 mph; **(b)** \$7.50/hr; **(c)** \$1.07/mi
13. (a) \$22/hr; **(b)** 20.6 pages/hr; **(c)** \$1.07/page
15. \$768.6 billion/year   **17. (a)** 14.5 floors/min;
(b) 4.14 sec/floor   **19. (a)** 23.12 ft/min; **(b)** 0.04 min/ft
21.

23.

25.

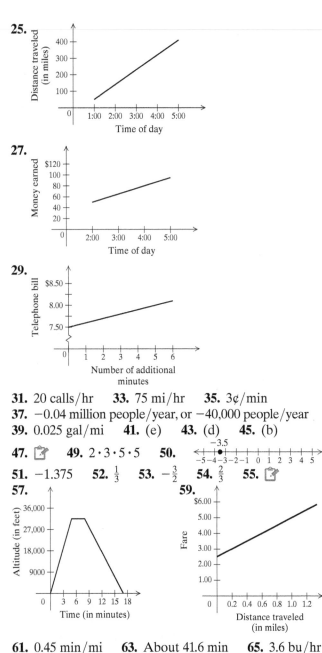

27.

29.

31. 20 calls/hr   **33.** 75 mi/hr   **35.** 3¢/min
37. −0.04 million people/year, or −40,000 people/year
39. 0.025 gal/mi   **41.** (e)   **43.** (d)   **45.** (b)
47. 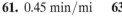   **49.** $2 \cdot 3 \cdot 5 \cdot 5$ **50.**

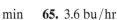

51. −1.375   **52.** $\frac{1}{3}$ **53.** $-\frac{3}{2}$ **54.** $\frac{2}{3}$ **55.**
57.   **59.**

61. 0.45 min/mi   **63.** About 41.6 min   **65.** 3.6 bu/hr

Quick Quiz: Sections 3.1–3.4, p. 190

1. y-axis   **2.** No
3.   **4.**

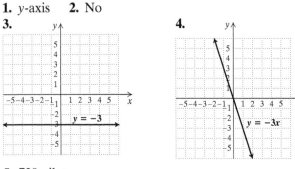

5. 730 pilots per year

Prepare to Move On, p. 190

1. 5   **2.** −6   **3.** −1   **4.** $-\frac{4}{3}$ **5.** $-\frac{4}{3}$ **6.** 1
7. 0   **8.** Undefined

Check Your Understanding, p. 194

1. To the right 3 units   **2.** Down 4 units   **3.** −4
4. 3   **5.** Negative   **6.** $-\frac{4}{3}$

Exercise Set 3.5, pp. 196–201

1. y, x **2.** $\dfrac{\text{change in } y}{\text{change in } x}$ **3.** $\dfrac{\text{rise}}{\text{run}}$ **4.** $\dfrac{y_2 - y_1}{x_2 - x_1}$
5. Positive, negative   **6.** Zero, undefined   **7.** Positive
8. Negative   **9.** Negative   **10.** Positive   **11.** Zero
12. Positive   **13.** Negative   **14.** Zero   **15.** $25/post
17. $-$\$$\frac{1}{2}$ billion/year   **19.** $\frac{1}{2}$ point/$1000 income
21. About −2.1°/min   **23.** $\frac{4}{3}$ **25.** $\frac{1}{3}$ **27.** −1   **29.** 0
31. −2   **33.** Undefined   **35.** $-\frac{1}{3}$ **37.** 5   **39.** $\frac{5}{4}$
41. $-\frac{4}{5}$ **43.** $\frac{2}{3}$ **45.** −1   **47.** $-\frac{1}{2}$ **49.** 0   **51.** 1
53. Undefined   **55.** 0   **57.** Undefined
59. Undefined   **61.** 0   **63.** 14.375%, or $14\frac{3}{8}$ %
65. 35%   **67.** $\frac{29}{98}$, or about 30%   **69.** About 5.1%; yes
71. 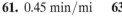   **73.** $12 + 3a$ **74.** $7(2 + 5x)$ **75.** 15
76. $3 \leq x$ **77.** $(5t)^3$ **78.** $16y^4$ **79.**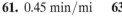
81. 0.364, or 36.4%   **83.** $\{m \mid m \geq \frac{5}{2}\}$ **85.** $\frac{1}{2}$

Quick Quiz: Sections 3.1–3.5, p. 201

1.   **2.**

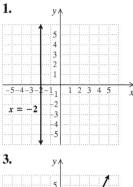

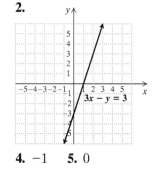

3.   **4.** −1   **5.** 0

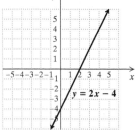

Prepare to Move On, p. 201

1. $y = -\frac{2}{3}x + \frac{7}{3}$ **2.** $y = \frac{3}{4}x - 2$ **3.** $y = \dfrac{c - ax}{b}$

4. $y = \dfrac{ax - c}{b}$

Mid-Chapter Review: Chapter 3, p. 202

1. *y-intercept:* $y - 3 \cdot 0 = 6$
$\qquad y = 6$
The *y*-intercept is $(0, 6)$.
x-intercept: $0 - 3x = 6$
$\qquad -3x = 6$
$\qquad x = -2$
The *x*-intercept is $(-2, 0)$.

2. $m = \dfrac{y_2 - y_1}{x_2 - x_1} = \dfrac{-1 - 5}{3 - 1}$
$\qquad = \dfrac{-6}{2}$
$\qquad = -3$

3. **4.** IV **5.** No

6. **7.**

8. **9.**

10. **11.**

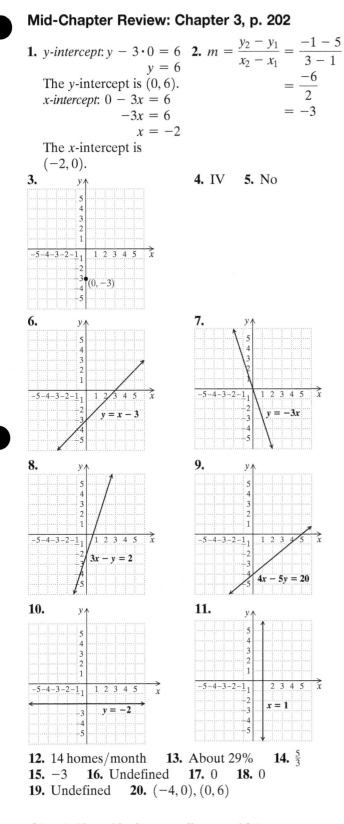

12. 14 homes/month **13.** About 29% **14.** $\frac{5}{3}$
15. -3 **16.** Undefined **17.** 0 **18.** 0
19. Undefined **20.** $(-4, 0), (0, 6)$

Check Your Understanding, p. 207

1. Yes **2.** $\frac{2}{3}$ **3.** $\frac{2}{3}$ **4.** 2 units **5.** 1 **6.** $(0, 1)$
7. Yes

Exercise Set 3.6, pp. 208–210

1. Slope **2.** *y*-intercept **3.** *y*-intercept **4.** Slope
5. *y*-intercept **6.** Slope **7.** (f) **8.** (b) **9.** (d)
10. (c) **11.** (e) **12.** (a)
13. **15.**

17. **19.**

21. **23.**

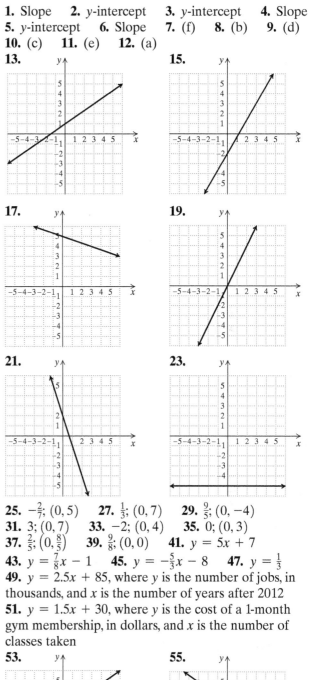

25. $-\frac{2}{7}; (0, 5)$ **27.** $\frac{1}{3}; (0, 7)$ **29.** $\frac{9}{5}; (0, -4)$
31. $3; (0, 7)$ **33.** $-2; (0, 4)$ **35.** $0; (0, 3)$
37. $\frac{2}{5}; \left(0, \frac{8}{5}\right)$ **39.** $\frac{9}{8}; (0, 0)$ **41.** $y = 5x + 7$
43. $y = \frac{7}{8}x - 1$ **45.** $y = -\frac{5}{3}x - 8$ **47.** $y = \frac{1}{3}$
49. $y = 2.5x + 85$, where y is the number of jobs, in
thousands, and x is the number of years after 2012
51. $y = 1.5x + 30$, where y is the cost of a 1-month
gym membership, in dollars, and x is the number of
classes taken
53. **55.**

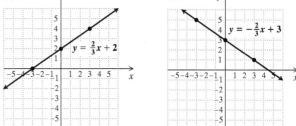

57.

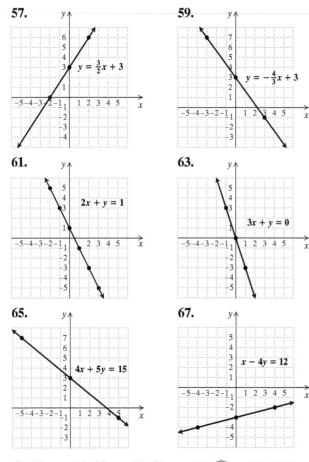

$y = \frac{3}{2}x + 3$

59.

$y = -\frac{4}{3}x + 3$

61.

$2x + y = 1$

63.

$3x + y = 0$

65.

$4x + 5y = 15$

67.

$x - 4y = 12$

69. Yes **71.** No **73.** Yes **75.** ☑ **77.** 12%
78. 250 words **79.** $2400 **80.** Width: 20 ft;
length: 50 ft **81.** ☑ **83.** When $x = 0, y = b$,
so $(0, b)$ is on the line. When $x = 1, y = m + b$, so
$(1, m + b)$ is on the line. Then

$$\text{slope} = \frac{(m + b) - b}{1 - 0} = m.$$

85. $y = \frac{1}{2}x$ **87.** $y = -\frac{4}{5}x + 4$ **89.** Yes **91.** Yes
93. No **95.** $y = -\frac{5}{3}x + 3$ **97.** $y = \frac{5}{2}x + 1$

Quick Quiz: Sections 3.1–3.6, p. 210

1.

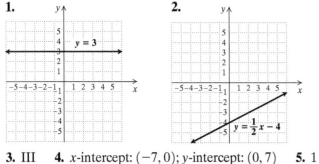

$y = 3$

2.

$y = \frac{1}{2}x - 4$

3. III **4.** x-intercept: $(-7, 0)$; y-intercept: $(0, 7)$ **5.** 1

Prepare to Move On, p. 210

1. $y = m(x - h) + k$ **2.** $y = -2(x + 4) + 9$, or
$y = -2x + 1$ **3.** -7 **4.** 13 **5.** -9

Check Your Understanding, p. 215

1. $m = 2; x_1 = 4; y_1 = 8$ **2.** $m = \frac{1}{5}; x_1 = -1; y_1 = 6$
3. $m = -3; x_1 = 0; y_1 = 7$ **4.** $m = -2; x_1 = -4; y_1 = 3$
or $m = -2; x_1 = 2; y_1 = 0$

Connecting the Concepts, p. 216

1. Slope–intercept form **2.** Standard form
3. None of these **4.** Standard form
5. Point–slope form **6.** None of these
7. $2x - 5y = 10$ **8.** $-2x + y = 7$, or $2x - y = -7$
9. $y = \frac{2}{7}x - \frac{8}{7}$ **10.** $y = -x - 8$

Exercise Set 3.7, pp. 217–220

1. True **2.** True **3.** False **4.** True **5.** (g)
6. (b) **7.** (d) **8.** (h) **9.** (e) **10.** (a)
11. (f) **12.** (c) **13.** $y - 6 = 3(x - 1)$
15. $y - 8 = \frac{3}{5}(x - 2)$ **17.** $y - 1 = -4(x - 3)$
19. $y - (-4) = \frac{3}{2}(x - 5)$ **21.** $y - 6 = -\frac{5}{4}(x - (-2))$
23. $y - (-1) = -2(x - (-4))$
25. $y - 8 = 1(x - (-2))$ **27.** $y = 4x - 7$
29. $y = \frac{7}{4}x - 9$ **31.** $y = -2x + 1$
33. $y = -4x - 9$ **35.** $y = \frac{2}{3}x + \frac{8}{3}$
37. $y = -\frac{5}{6}x + 4$ **39.** $y = -x + 5$
41. $y = \frac{2}{3}x + 3$ **43.** $y = \frac{2}{5}x - 2$
45. $y = 2x + 7$ **47.** $\frac{2}{7}; (8, 9)$ **49.** $-5; (7, -2)$
51. $-\frac{5}{3}; (-2, 4)$ **53.** $\frac{4}{7}; (0, 0)$

55.

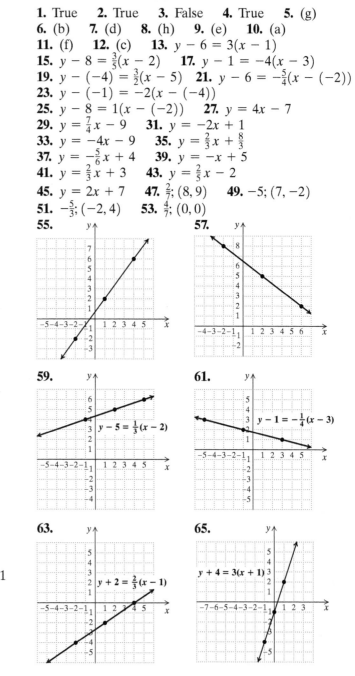

57.

59.

$y - 5 = \frac{1}{3}(x - 2)$

61.

$y - 1 = -\frac{1}{4}(x - 3)$

63.

$y + 2 = \frac{2}{3}(x - 1)$

65.

$y + 4 = 3(x + 1)$

67.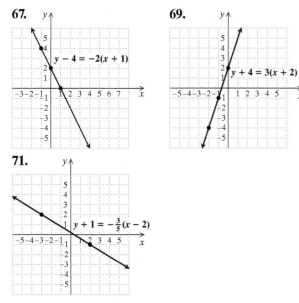

69.

71.

$y - 4 = -2(x + 1)$

$y + 4 = 3(x + 2)$

$y + 1 = -\frac{3}{5}(x - 2)$

3.

4.

5.

$y = -3x + 2$

$3x - 6y = 6$

$y - 1 = 2(x + 1)$

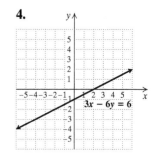
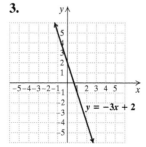
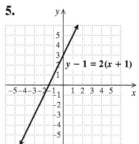

73. (a) $y = -0.875x + 28.75$; **(b)** about 22%;
(c) about 16%   **75. (a)** $y = 0.8x + 31.6$; **(b)** \$58 per
ton; \$66 per ton   **77. (a)** $R = -1.145t + 12.34$;
(b) \$3.18 billion; **(c)** in about 2019   **79. (a)** Let x
represent the number of years after 2000 and y the
percent of the U.S. population living in metropolitan
areas; $y = \frac{17}{100}x + 79$; **(b)** 79.9%; 81.6%   **81.** 📝
83. -64 **84.** 9   **85.** 18   **86.** $\frac{11}{2}$
87. $\{x | x < -\frac{17}{3}\}$, or $\left(-\infty, -\frac{17}{3}\right)$
88. $\{x | x \geq -2\}$, or $[-2, \infty)$ **89.** 📋

91.

$y - 3 = 0(x - 52)$

93. $y = 2x - 9$
95. $y = -\frac{4}{3}x + \frac{23}{3}$
97. $y - 7 = -\frac{2}{3}(x - (-4))$
99. $y = -4x + 7$
101. $y = \frac{10}{3}x + \frac{25}{3}$
103. $(2, 0), (0, 5)$

105. $-\frac{x}{4} + \frac{y}{3} = 1$;

$(-4, 0), (0, 3)$
107. 📝

Quick Quiz: Sections 3.1–3.7, p. 220

1.

$y - 3 = 2x$

2.

$x + 7 = 8$

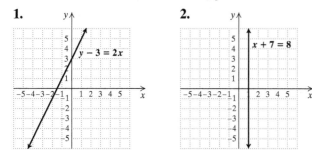

Prepare to Move On, p. 220

1. -125 **2.** 64   **3.** -64 **4.** 8   **5.** -4

Visualizing for Success, p. 221

1. C   **2.** G   **3.** F   **4.** B   **5.** D   **6.** A   **7.** I
8. H   **9.** J   **10.** E

Decision Making: Connection, p. 222

1.

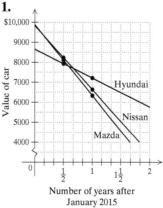

2. Mazda: \$295.83/
month, or \$3550/year;
Hyundai: \$120.83/month,
or \$1450/year; Nissan:
\$266.67/month, or
\$3200/year. The slopes
of the lines illustrate the
depreciation rates.

3. Mazda: $y = -3550x + 9875$; Hyundai:
$y = -1450x + 8650$; Nissan: $y = -3200x + 9825$
4. Mazda: \$2775; Hyundai: \$5750; Nissan: \$3425
5. 📝   **6.** 💻

Study Summary: Chapter 3, pp. 223–224

1.

$(2, -3)$

$(0, -5)$

2. III

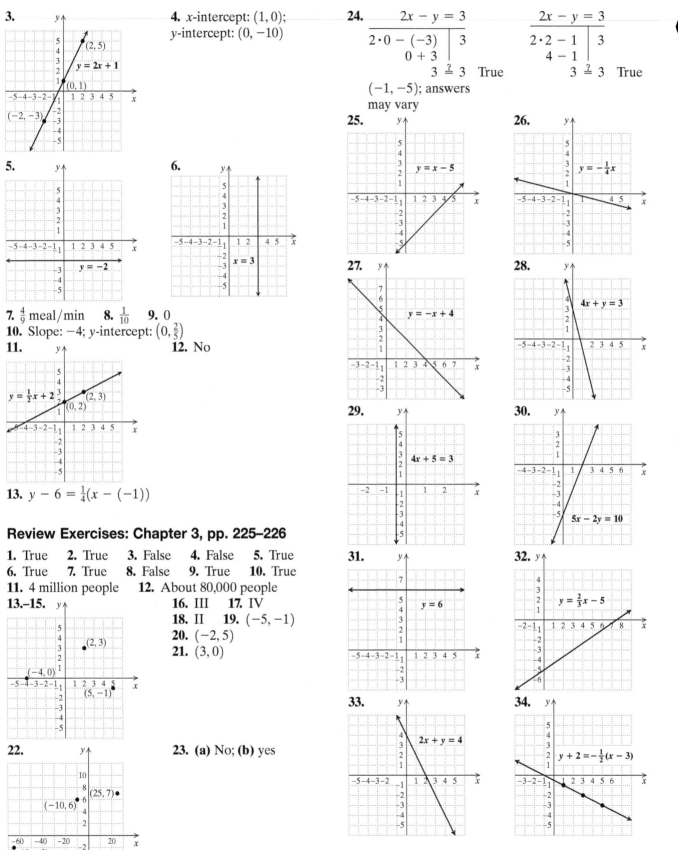

3.

(2, 5), $y = 2x + 1$, (0, 1), (−2, −3)

4. *x*-intercept: $(1, 0)$; *y*-intercept: $(0, -10)$

5. $y = -2$

6. $x = 3$

7. $\frac{4}{9}$ meal/min   **8.** $\frac{1}{10}$ **9.** 0

10. Slope: -4; *y*-intercept: $\left(0, \frac{2}{5}\right)$

11. $y = \frac{1}{2}x + 2$, (0, 2), (2, 3)   **12.** No

13. $y - 6 = \frac{1}{4}(x - (-1))$

Review Exercises: Chapter 3, pp. 225–226

1. True   **2.** True   **3.** False   **4.** False   **5.** True
6. True   **7.** True   **8.** False   **9.** True   **10.** True
11. 4 million people   **12.** About 80,000 people
13.–15.   (2, 3), (−4, 0), (5, −1)

16. III   **17.** IV
18. II   **19.** $(-5, -1)$
20. $(-2, 5)$
21. $(3, 0)$

22. (25, 7), (−10, 6), (−65, −2)

23. **(a)** No; **(b)** yes

24.

$2x - y = 3$	
$2 \cdot 0 - (-3)$	3
$0 + 3$	
	$3 \stackrel{?}{=} 3$ True

$2x - y = 3$	
$2 \cdot 2 - 1$	3
$4 - 1$	
	$3 \stackrel{?}{=} 3$ True

$(-1, -5)$; answers may vary

25. $y = x - 5$

26. $y = -\frac{1}{4}x$

27. $y = -x + 4$

28. $4x + y = 3$

29. $4x + 5 = 3$

30. $5x - 2y = 10$

31. $y = 6$

32. $y = \frac{2}{3}x - 5$

33. $2x + y = 4$

34. $y + 2 = -\frac{1}{2}(x - 3)$

35. 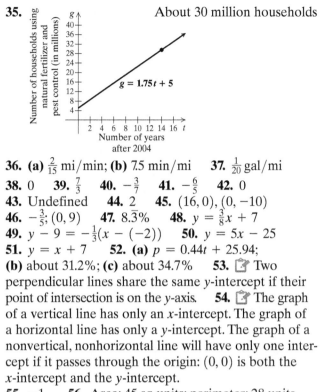 About 30 million households

36. (a) $\frac{2}{15}$ mi/min; **(b)** 7.5 min/mi   **37.** $\frac{1}{20}$ gal/mi
38. 0   **39.** $\frac{7}{3}$ **40.** $-\frac{3}{7}$ **41.** $-\frac{6}{5}$ **42.** 0
43. Undefined   **44.** 2   **45.** $(16, 0), (0, -10)$
46. $-\frac{3}{5}; (0, 9)$ **47.** $8.\overline{3}\%$ **48.** $y = \frac{3}{8}x + 7$
49. $y - 9 = -\frac{1}{3}(x - (-2))$ **50.** $y = 5x - 25$
51. $y = x + 7$ **52. (a)** $p = 0.44t + 25.94$;
(b) about 31.2%; **(c)** about 34.7%   **53.** ☑ Two
perpendicular lines share the same y-intercept if their
point of intersection is on the y-axis.   **54.** ☑ The graph
of a vertical line has only an x-intercept. The graph of
a horizontal line has only a y-intercept. The graph of a
nonvertical, nonhorizontal line will have only one intercept if it passes through the origin: $(0, 0)$ is both the
x-intercept and the y-intercept.
55. -1 **56.** Area: 45 sq units; perimeter: 28 units
57. $(0, 4), (1, 3), (-1, 3)$; answers may vary

Test: Chapter 3, p. 227

1. [3.1] \$4000   **2.** [3.1] 2000 and 2007   **3.** [3.1] III
4. [3.1] II   **5.** [3.1] $(3, 4)$ **6.** [3.1] $(0, -4)$
7. [3.1] $(-5, 2)$
8. [3.2]   **9.** [3.3]

10. [3.3]   **11.** [3.2]

12. [3.2]   **13.** [3.3]

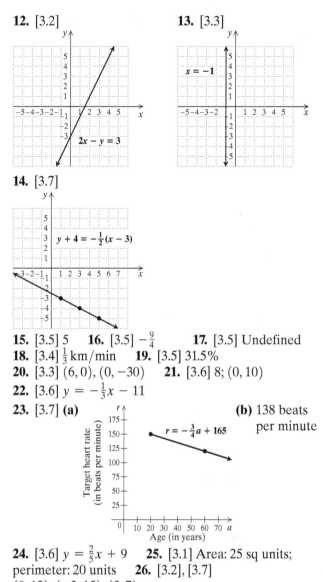

14. [3.7]

15. [3.5] 5   **16.** [3.5] $-\frac{9}{4}$ **17.** [3.5] Undefined
18. [3.4] $\frac{1}{3}$ km/min   **19.** [3.5] 31.5%
20. [3.3] $(6, 0), (0, -30)$ **21.** [3.6] 8; $(0, 10)$
22. [3.6] $y = -\frac{1}{3}x - 11$
23. [3.7] **(a)**   **(b)** 138 beats per minute

24. [3.6] $y = \frac{2}{5}x + 9$ **25.** [3.1] Area: 25 sq units;
perimeter: 20 units   **26.** [3.2], [3.7]
$(0, 12), (-3, 15), (5, 7)$

Cumulative Review: Chapters 1–3, p. 228

1. 7   **2.** $12a - 6b + 18$ **3.** $2(4x - 2y + 1)$
4. $2 \cdot 3^3$ **5.** -0.15 **6.** 0.367 **7.** $\frac{11}{60}$ **8.** 2.6
9. 7.28   **10.** $-\frac{5}{12}$ **11.** -3 **12.** $5x + 11$
13. -27 **14.** 16   **15.** -6 **16.** 2   **17.** $\frac{7}{9}$
18. -17 **19.** $\{x \,|\, x < 16\}$, or $(-\infty, 16)$
20. $h = \dfrac{A - \pi r^2}{2\pi r}$ **21.** IV   **22.** $-1 < x \le 2$

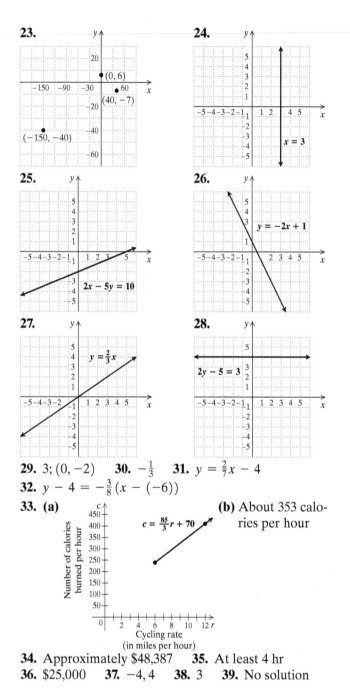

23.

24. $x = 3$

25. $2x - 5y = 10$

26. $y = -2x + 1$

27. $y = \frac{2}{3}x$

28. $2y - 5 = 3$

29. $3; (0, -2)$ **30.** $-\frac{1}{3}$ **31.** $y = \frac{2}{7}x - 4$

32. $y - 4 = -\frac{3}{8}(x - (-6))$

33. (a)

$c = \frac{85}{3}r + 70$

(b) About 353 calories per hour

34. Approximately \$48,387   **35.** At least 4 hr
36. \$25,000   **37.** $-4, 4$ **38.** 3   **39.** No solution

CHAPTER 4

Check Your Understanding, p. 235

1. The Power Rule   **2.** Raising a product to a power
3. The Product Rule   **4.** The Quotient Rule
5. Raising a quotient to a power

Exercise Set 4.1, pp. 235–237

1. (e)   **2.** (f)   **3.** (b)   **4.** (h)   **5.** (g)   **6.** (a)   **7.** (c)
8. (d)   **9.** Base: $2x$; exponent: 5
10. Base: $x + 1$; exponent: 0   **11.** Base: x; exponent: 3
12. Base: y; exponent: 6   **13.** Base: $\frac{4}{y}$; exponent: 7

14. Base: $-5x$; exponent: 4   **15.** d^{13} **17.** a^7 **19.** 6^{15}
21. $(3y)^{12}$ **23.** $5p$ **25.** $(x + 3)^{13}$ **27.** a^5b^9 **29.** r^{10}
31. m^4n^9 **33.** 7^3, or 343   **35.** t^7 **37.** $5a$ **39.** 1
41. $(r + s)^8$ **43.** $\frac{4}{5}d^7$ **45.** $4a^7b^6$ **47.** $x^{12}y^7$ **49.** 1
51. 5   **53.** 2   **55.** -4 **57.** x^{33} **59.** 5^{32} **61.** t^{80}
63. $100x^2$ **65.** $-8a^3$ **67.** $25n^{14}$ **69.** $a^{14}b^7$ **71.** $r^{17}t^{11}$
73. $24x^{19}$ **75.** $\frac{x^3}{125}$ **77.** $\frac{49}{36n^2}$ **79.** $\frac{a^{18}}{b^{48}}$ **81.** $\frac{x^8y^4}{z^{12}}$
83. $\frac{a^{12}}{16b^{20}}$ **85.** $-\frac{125x^{21}y^3}{8z^{12}}$ **87.** 1   **89.** 🖎   **91.** -21
92. $\{x \mid x \le \frac{7}{4}\}$, or $\left(-\infty, \frac{7}{4}\right]$ **93.** -1 **94.** 0
95. $\{n \mid n < \frac{16}{3}\}$, or $\left(-\infty, \frac{16}{3}\right)$ **96.** 1   **97.** 🖎   **99.** 🖎
101. Let $x = 1$; then $3x^2 = 3$, but $(3x)^2 = 9$.
103. Let $t = -1$; then $\frac{t^6}{t^2} = 1$, but $t^3 = -1$.   **105.** y^{6x}
107. x^t **109.** 13   **111.** $<$ **113.** $<$ **115.** $>$
117. 4,000,000; 4,194,304; 194,304
119. 2,000,000,000; 2,147,483,648; 147,483,648
121. 1,536,000 bytes, or approximately 1,500,000 bytes

Prepare to Move On, p. 237

1. -24 **2.** -15 **3.** -11 **4.** 16   **5.** -14 **6.** 4

Check Your Understanding, p. 241

1. (d)   **2.** (c)   **3.** (e)   **4.** Yes   **5.** No   **6.** No

Technology Connection, p. 242

1. 1.71×10^{17} **2.** $5.\overline{370} \times 10^{-15}$ **3.** 3.68×10^{16}

Connecting the Concepts, p. 243

1. x^{14} **2.** $\frac{1}{x^{14}}$ **3.** $\frac{1}{x^{14}}$ **4.** x^{14} **5.** x^{40} **6.** x^{40}
7. c^8 **8.** $\frac{1}{c^8}$ **9.** $\frac{a^{15}}{b^{20}}$ **10.** $\frac{b^{20}}{a^{15}}$

Exercise Set 4.2, pp. 243–246

1. True   **2.** False   **3.** True   **4.** False   **5.** (c)
6. (d)   **7.** (a)   **8.** (b)   **9.** Positive power of 10
10. Negative power of 10   **11.** Negative power of 10
12. Positive power of 10   **13.** Positive power of 10
14. Negative power of 10   **15.** $\frac{1}{x^3} = \frac{1}{8}$ **17.** $\frac{1}{(-2)^6} = \frac{1}{64}$
19. $\frac{1}{t^9}$ **21.** $\frac{8}{x^3}$ **23.** a^8 **25.** $\frac{1}{7}$ **27.** $\frac{3a^8}{b^6}$
29. $\left(\frac{2}{x}\right)^5 = \frac{32}{x^5}$ **31.** $\frac{1}{3x^5z^4}$ **33.** 9^{-2} **35.** y^{-3}
37. t^{-1} **39.** 2^3, or 8   **41.** $\frac{1}{x^{12}}$ **43.** $\frac{1}{t^2}$ **45.** $\frac{10}{a^6b^2}$
47. $\frac{1}{n^{15}}$ **49.** t^{18} **51.** $\frac{1}{m^7n^7}$ **53.** $\frac{9}{x^8}$ **55.** $\frac{25t^6}{r^8}$

57. t^{14} **59.** $\dfrac{1}{y^4}$ **61.** $5y^3$ **63.** $2x^5$ **65.** $\dfrac{-3b^9}{2a^7}$

67. 1   **69.** $3s^2t^4u^4$ **71.** $\dfrac{1}{x^{12}y^{15}}$ **73.** $\dfrac{m^{10}n^6}{9}$ **75.** $\dfrac{b^5c^4}{a^8}$

77. $\dfrac{9}{a^8}$ **79.** $\dfrac{n^{12}}{m^3}$ **81.** $\dfrac{27b^{12}}{8a^6}$ **83.** 1   **85.** $\dfrac{2b^3}{a^4}$

87. $\dfrac{5y^4z^{10}}{4x^{11}}$ **89.** 4920   **91.** 0.00892   **93.** 904,000,000

95. 0.000003497   **97.** 3.6×10^7 **99.** 5.83×10^{-3}
101. 7.8×10^{10} **103.** 1.032×10^{-6} **105.** 6×10^{13}
107. 2.47×10^8 **109.** 3.915×10^{-16} **111.** 2.5×10^{13}
113. 5.0×10^{-6} **115.** 3×10^{-21} **117.** 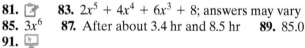   **119.** $-\frac{1}{6}$
120. 7   **121.** $-32a^5$ **122.** 1   **123.** 5   **124.** -15
125.   **127.**   **129.** 2^{-12} **131.** 5   **133.** 5^6
135. $\frac{1}{3} + \frac{1}{4} = \frac{7}{12}$ **137.** 3×10^8 mi
139. 1×10^5 light-years   **141.** 2×10^4 strands

Quick Quiz: Sections 4.1–4.2, p. 246

1. $200n^8m^{11}$ **2.** $\dfrac{1}{6^{11}}$ **3.** $\dfrac{y^3}{8x^6}$ **4.** 1   **5.** 3.007×10^7

Prepare to Move On, p. 246

1. $8x$ **2.** $-3a - 6b$ **3.** $-2x - 7$ **4.** $-4t - r - 5$
5. 1004   **6.** 9

Check Your Understanding, p. 250

1. 4   **2.** 4   **3.** 2   **4.** $4x^2$ and $7x^2$ **5.** -100

Technology Connection, p. 250

1. 1141.0023

Exercise Set 4.3, pp. 251–254

1. (b)   **2.** (f)   **3.** (h)   **4.** (d)   **5.** (g)   **6.** (e)
7. (a)   **8.** (c)   **9.** Yes   **10.** Yes   **11.** No
12. Yes   **13.** Yes   **14.** No   **15.** $8x^3, -11x^2, 6x, 1$
17. $-t^6, -3t^3, 9t, -4$ **19.** Trinomial   **21.** Polynomial
with no special name   **23.** Binomial   **25.** Monomial

27.

Term	Coefficient	Degree of the Term	Degree of the Polynomial
$8x^5$	8	5	
$-\frac{1}{2}x^4$	$-\frac{1}{2}$	4	
$-4x^3$	-4	3	5
$7x^2$	7	2	
6	6	0	

29. Coefficients: 8, 2; degrees: 4, 1
31. Coefficients: 9, -3, 4; degrees: 2, 1, 0
33. Coefficients: 1, -1, 4, -3; degrees: 4, 3, 1, 0
35. **(a)** 1, 3, 4; **(b)** $8t^4$, 8; **(c)** 4   **37.** **(a)** 2, 0, 4; **(b)** $2a^4$, 2;
(c) 4   **39.** **(a)** 0, 2, 1, 5; **(b)** $-x^5$, -1; **(c)** 5
41. $11n^2 + n$ **43.** $4a^4$ **45.** $11b^3 + b^2 - b$
47. $-x^4 - x^3$ **49.** $\frac{1}{15}x^4 + 10$
51. $-1.1a^2 + 5.3a - 7.5$ **53.** $-3; 21$ **55.** $16; 34$
57. $-38; 148$ **59.** $159; 165$ **61.** 1112 ft   **63.** 62.8 cm
65. 153.86 m²   **67.** About 135 ft   **69.** 14; 55 oranges
71. About 2.3 mcg/mL   **73.**

75.

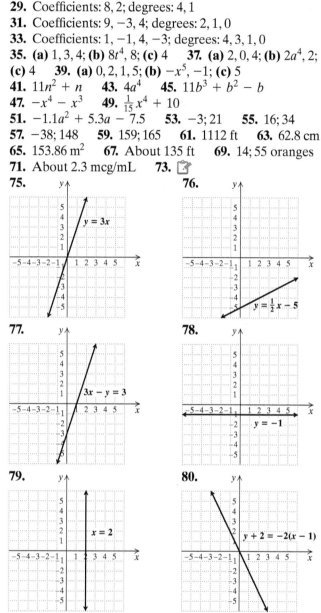

76.

77.   **78.**

79.   **80.**

81.   **83.** $2x^5 + 4x^4 + 6x^3 + 8$; answers may vary
85. $3x^6$ **87.** After about 3.4 hr and 8.5 hr   **89.** 85.0
91.
93.

t	$-t^2 + 10t - 18$
3	3
4	6
5	7
6	6
7	3

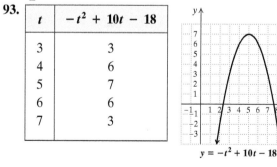

$y = -t^2 + 10t - 18$

Quick Quiz: Sections 4.1–4.3, p. 254

1. $9x^{10}y^8$ **2.** $-16a^4$ **3.** $1, -6, \frac{1}{2}$ **4.** -1290 **5.** 1

Prepare to Move On, p. 254

1. $x - 3$ **2.** $-2x - 6$ **3.** $6a - 3$ **4.** $-t - 1$
5. $-t^4 + 17t$ **6.** $0.4a^2 - a + 11$

Check Your Understanding, p. 258

1. $x^2 - x$ **2.** $x^2 + x$ **3.** $x^2 - x - 1$ **4.** $x^2 - x + 1$
5. $x^2 + x - 1$ **6.** $x^2 + x + 1$

Technology Connection, p. 259

1. In each case, let $y_1 = $ the expression before the addition or subtraction has been performed, $y_2 = $ the simplified sum or difference, and $y_3 = y_2 - y_1$; and note that the graph of y_3 coincides with the x-axis. That is, $y_3 = 0$.

Exercise Set 4.4, pp. 259–262

1. Opposite   **2.** Ascending   **3.** Opposite; sign
4. Like; missing   **5.** x^2 **6.** -6 **7.** $-$ **8.** $+$
9. $4x + 9$ **11.** $-6t + 8$ **13.** $-3x^2 + 6x - 2$
15. $9t^2 + t + 3$ **17.** $8m^3 - 3m - 7$
19. $7 + 13a - a^2 + 7a^3$
21. $2x^6 + 9x^4 + 2x^3 - 4x^2 + 5x$
23. $x^4 + \frac{1}{4}x^3 - \frac{3}{4}x^2 - \frac{5}{6}x + 3$
25. $4.2t^3 + 3.5t^2 - 6.4t - 1.8$
27. $-4x^3 + 4x^2 + 6x$
29. $1.3x^4 + 0.35x^3 + 9.53x^2 + 2x + 0.96$
31. $-(-3t^3 + 4t^2 - 7); 3t^3 - 4t^2 + 7$
33. $-(x^4 - 8x^3 + 6x); -x^4 + 8x^3 - 6x$
35. $-3a^4 + 5a^2 - 1.2$ **37.** $4x^4 - 6x^2 - \frac{3}{4}x + 8$
39. $-2x - 7$ **41.** $-t^2 - 12t + 13$
43. $8a^3 + 8a^2 + a - 10$ **45.** 0
47. $1 + a + 12a^2 - 3a^3$ **49.** $\frac{9}{8}x^3 - \frac{1}{2}x$
51. $0.05t^3 - 0.07t^2 + 0.01t + 1$
53. $2x^2 + 6$ **55.** $-3x^4 - 8x^3 - 7x^2$
57. (a) $5x^2 + 4x$; **(b)** $145; 273$ **59.** $16y + 26$
61. $(r + 11)(r + 9); 9r + 99 + 11r + r^2$
63. $(x + 3)^2; x^2 + 3x + 9 + 3x$ **65.** $m^2 - 40$
67. $\pi r^2 - 49$ **69.** $(x^2 - 12)$ ft^2 **71.** $(z^2 - 36\pi)$ ft^2
73. $\left(144 - \frac{d^2}{4}\pi\right)$ m^2 **75.** ✏   **77.** $y = \frac{1}{3}x + 2$
78. $y = 4x + 24$ **79.** $y = 10$ **80.** $y = -3x + 8$
81. $y = -\frac{4}{7}x - 4$ **82.** $y = -\frac{4}{3}x - \frac{8}{3}$ **83.** ✏
85. $9t^2 - 20t + 11$ **87.** $-6x + 14$
89. $250.591x^3 + 2.812x$ **91.** $20w + 42$ **93.** $2x^2 + 20x$
95. (a) $P = -x^2 + 175x - 5000$; **(b)** 2500; **(c)** 1600

Quick Quiz: Sections 4.1–4.4, p. 262

1. $\dfrac{9y^{10}}{4x^6}$ **2.** $(y + 3)^{12}$ **3.** 4 **4.** $-y^3 - 6y^2 + 2y$
5. $5y^2 + 10$

Prepare to Move On, p. 262

1. $2x^2 - 2x + 6$ **2.** $-15x^2 + 10x + 35$ **3.** x^8
4. y^7 **5.** $2n^3$ **6.** $-6n^{12}$

Mid-Chapter Review: Chapter 4, p. 263

1. $\dfrac{x^{-3}y}{x^{-4}y^7} = x^{-3-(-4)}y^{1-7} = x^1y^{-6} = \dfrac{x}{y^6}$
2. $(x^2 + 7x - 12) - (3x^2 - 6x - 1) =$
$x^2 + 7x - 12 - 3x^2 + 6x + 1 = -2x^2 + 13x - 11$
3. $x^{16}y^{40}$ **4.** 1 **5.** $\frac{1}{4}a^{10}$ **6.** $-\frac{8}{3}b$ **7.** $\dfrac{5}{x^2}$ **8.** $\dfrac{a^2}{bc^3}$
9. $\dfrac{27}{8a^6}$ **10.** $\dfrac{2}{3m^6n^4}$ **11.** 0.00000189 **12.** 2.7×10^{10}
13. 3 **14.** $2a^2 - 5a - 3$ **15.** $3x^2 + 3x + 3$
16. $7x + 7$ **17.** $-6x^2 + 2x - 12$ **18.** $t^9 + 5t^7$
19. $3a^4 - 13a^3 - 13a^2 - 4$ **20.** $-x^5 + 2x^4 - 2x^2 - x$

Check Your Understanding, p. 267

1. (c)   **2.** (d)   **3.** (d)   **4.** (a)   **5.** (c)   **6.** (b)

Technology Connection, p. 268

1. Let $y_1 = (5x^4 - 2x^2 + 3x)(x^2 + 2x)$ and
$y_2 = 5x^6 + 10x^5 - 2x^4 - x^3 + 6x^2$. With the table set in AUTO mode, note that the value in the Y_1- and Y_2-columns match, regardless of how far we scroll up or down.
2. Use TRACE, a table, or a boldly drawn graph to confirm that y_3 is always 0.

Exercise Set 4.5, pp. 268–270

1. (c)   **2.** (d)   **3.** (b)   **4.** (a)   **5.** $21x^5$ **7.** $-x^7$
9. x^8 **11.** $36t^4$ **13.** $-0.12x^9$ **15.** $-\frac{1}{20}x^{12}$
17. $5n^3$ **19.** $72y^{10}$ **21.** $20x^2 + 5x$ **23.** $3a^2 - 27a$
25. $x^5 + x^2$ **27.** $-6n^3 + 24n^2 - 3n$ **29.** $-15t^3 - 30t^2$
31. $4a^9 - 8a^7 - \frac{5}{12}a^4$ **33.** $0.24x^5 - 0.1x^4$
35. $x^2 + 7x + 12$ **37.** $t^2 + 4t - 21$
39. $a^2 - 1.3a + 0.42$ **41.** $x^2 - 9$
43. $28 - 15x + 2x^2$ **45.** $t^2 + \frac{17}{6}t + 2$
47. $\frac{3}{16}a^2 + \frac{5}{4}a - 2$
49.

51.

53.

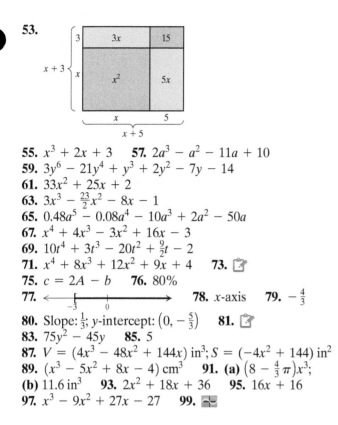

55. $x^3 + 2x + 3$ **57.** $2a^3 - a^2 - 11a + 10$
59. $3y^6 - 21y^4 + y^3 + 2y^2 - 7y - 14$
61. $33x^2 + 25x + 2$
63. $3x^3 - \frac{23}{2}x^2 - 8x - 1$
65. $0.48a^5 - 0.08a^4 - 10a^3 + 2a^2 - 50a$
67. $x^4 + 4x^3 - 3x^2 + 16x - 3$
69. $10t^4 + 3t^3 - 20t^2 + \frac{9}{2}t - 2$
71. $x^4 + 8x^3 + 12x^2 + 9x + 4$ **73.** 📝
75. $c = 2A - b$ **76.** 80%
77. ⟵———┤———┼———⟶   **78.** x-axis   **79.** $-\frac{4}{3}$
                 -3 0
80. Slope: $\frac{1}{3}$; y-intercept: $\left(0, -\frac{5}{3}\right)$ **81.** 📝
83. $75y^2 - 45y$ **85.** 5
87. $V = (4x^3 - 48x^2 + 144x)$ in^3; $S = (-4x^2 + 144)$ in^2
89. $(x^3 - 5x^2 + 8x - 4)$ cm^3 **91. (a)** $\left(8 - \frac{4}{3}\pi\right)x^3$;
(b) 11.6 in^3 **93.** $2x^2 + 18x + 36$ **95.** $16x + 16$
97. $x^3 - 9x^2 + 27x - 27$ **99.** 〰

Quick Quiz: Sections 4.1–4.5, p. 270

1. $-6x^9 + 16x^5 - 24x^4$ **2.** $10x^4 - 15x + 1$
3. $7a^2 - 10a - 2$ **4.** $125x^5y^4$ **5.** Trinomial

Prepare to Move On, p. 270

1. 0.49   **2.** $49x^6$ **3.** $\frac{9}{4}$ **4.** $\frac{3}{2}x$ **5.** $2a$ **6.** $\frac{1}{9}t^8$

Check Your Understanding, p. 275

1. (a) $(A - B)^2 = A^2 - 2AB + B^2$; **(b)** $A = x; B = 2$
2. (a) $(A + B)(A - B) = A^2 - B^2$; **(b)** $A = 4y; B = 7$
3. (a) $(A + B)^2 = A^2 + 2AB + B^2$;
(b) $A = 3a; B = \frac{1}{4}$ ($A = \frac{1}{4}, B = 3a$ is also correct.)
4. (a) $(A - B)^2 = A^2 - 2AB + B^2$;
(b) $A = 2x; B = 3$
5. (a) $(A + B)(A - B) = A^2 - B^2$;
(b) $A = 2p; B = 5$

Exercise Set 4.6, pp. 276–278

1. True   **2.** False   **3.** False   **4.** True
5. $x^3 + 3x^2 + 2x + 6$ **7.** $t^5 + 7t^4 - 2t - 14$
9. $y^2 - y - 6$ **11.** $9x^2 + 21x + 10$
13. $5x^2 + 17x - 12$ **15.** $15 - 13t + 2t^2$
17. $x^4 - 4x^2 - 21$ **19.** $p^2 - \frac{1}{16}$ **21.** $x^2 - 0.09$
23. $100x^4 - 9$ **25.** $1 - 25t^6$ **27.** $t^2 - 4t + 4$
29. $x^2 + 20x + 100$ **31.** $9x^2 + 12x + 4$
33. $1 - 20a + 100a^2$ **35.** $x^6 + 24x^3 + 144$
37. $x^5 + 3x^3 - x^2 - 3$ **39.** $x^2 - 64$

41. $-3n^2 - 19n + 14$ **43.** $x^2 + 6x + 9$
45. $49x^6 - 14x^3 + 1$ **47.** $81a^6 - 1$
49. $x^8 - 0.01$ **51.** $t^2 - \frac{9}{16}$ **53.** $1 - 3t + 5t^2 - 15t^3$
55. $a^2 - \frac{4}{5}a + \frac{4}{25}$ **57.** $t^8 + 6t^4 + 9$
59. $45x^2 - 56x - 45$ **61.** $14n^5 - 7n^3$
63. $a^3 - a^2 - 10a + 12$ **65.** $49 - 42x^4 + 9x^8$
67. $4 - 12x^4 + 9x^8$ **69.** $25t^2 + 60t^3 + 36t^4$
71. $5x^3 + 30x^2 - 10x$ **73.** $q^{10} - 1$
75. $15t^5 - 3t^4 + 3t^3$ **77.** $36x^8 - 36x^5 + 9x^2$
79. $18a^4 + 0.8a^3 + 4.5a + 0.2$ **81.** $\frac{1}{25} - 36x^8$
83. $a^3 + 1$ **85.** $x^2 + 6x + 9$ **87.** $t^2 + 7t + 12$
89. $x^2 + 10x + 21$ **91.** $25t^2 + 20t + 4$
93.

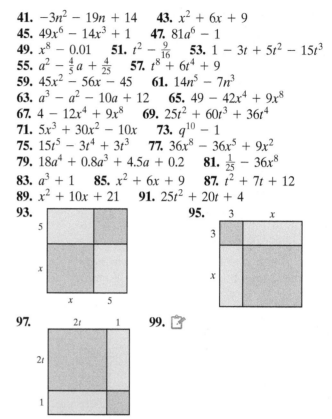

95.

97.

99. 📝

101. Refrigerator: 360 kWh/year; freezer: 720 kWh/year; washing machine: 120 kWh/year
102. For heat pumps costing $6666.67 or less
103. $a = \dfrac{c}{3b}$ **104.** $x = \dfrac{by + c}{a}$ **105.** 📝
107. $16x^4 - 81$ **109.** $81t^4 - 72t^2 + 16$
111. $t^{24} - 4t^{18} + 6t^{12} - 4t^6 + 1$ **113.** 396   **115.** -7
117. $17F + 7(F - 17)$, $F^2 - (F - 17)(F - 7)$; other equivalent expressions are possible.
119. $(y + 1)(y - 1)$, $y(y + 1) - y - 1$; other equivalent expressions are possible.   **121.** $y^2 - 4y + 4$
123. 〰

Quick Quiz: Sections 4.1–4.6, p. 278

1. $-20a^4b^{12}$ **2.** n^{14} **3.** $4x^3 - 13x^2 + 5x - 6$
4. $100x^{10} - 1$ **5.** $9a^2 + 24a + 16$

Prepare to Move On, p. 278

1. 4   **2.** 19   **3.** $12a - 7c$ **4.** $13y - w$

Check Your Understanding, p. 281

1. a, x, y **2.** 5   **3.** $-axy$ **4.** $3ax^2, -axy, 7ax^2$
5. $3ax^2$ and $7ax^2$

Technology Connection, p. 282

1. 36.22   **2.** 22,312

Exercise Set 4.7, pp. 283–286

1. (b) **2.** (d) **3.** (a) **4.** (c) **5.** (a) **6.** (b)
7. (b) **8.** (a) **9.** (c) **10.** (c) **11.** -13
13. -68 **15.** About 735 kg **17.** 73.005 in^2
19. 66.4 m **21.** 3, 2, 2, 0; 3 **23.** 0, 3, 3, 3; 3
25. $2r - 6s$ **27.** $5xy^2 - 2x^2y + x + 3x^2$
29. $9u^2v - 11uv^2 + 11u^2$ **31.** $6a^2c - 7ab^2 + a^2b$
33. $11x^2 - 10xy - y^2$ **35.** $-6a^4 - 8ab + 7ab^2$
37. $-6r^2 - 5rt - t^2$ **39.** $3x^3 - x^2y + xy^2 - 3y^3$
41. $-2y^4x^3 - 3y^3x$ **43.** $-8x + 8y$
45. $12c^2 + 5cd - 2d^2$ **47.** $x^2y^2 + 4xy - 5$
49. $4a^2 - b^2$ **51.** $20r^2t^2 - 23rt + 6$
53. $m^6n^2 + 2m^3n - 48$ **55.** $30x^2 - 28xy + 6y^2$
57. $0.01 - p^2q^2$ **59.** $x^2 + 2xh + h^2$
61. $16a^2 - 40ab + 25b^2$ **63.** $a^2b^2 - c^2d^4$
65. $x^3y^2 + x^2y^3 + 2x^2y^2 + 2xy^3 + 3xy + 3y^2$
67. $a^2 + 2ab + b^2 - c^2$ **69.** $a^2 - b^2 - 2bc - c^2$
71. $x^2 + 2xy + y^2$ **73.** $\frac{1}{2}a^2b^2 - 2$
75. $a^2 + c^2 + ab + 2ac + ad + bc + bd + cd$
77. $m^2 - n^2$
79. We draw a rectangle with dimensions $r + s$ by $u + v$. **81.**

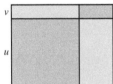
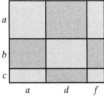

83.   **85.** 6 **86.** 7 **87.** 6 **88.** -21
89. $17a + 3x - 21$ **90.** $5y + 21$ **91.**
93. $4xy - 4y^2$ **95.** $2\pi ab - \pi b^2$
97. $x^3 + 2y^3 + x^2y + xy^2$
99. $2x^2 - 2\pi r^2 + 4xh + 2\pi rh$ **101.** **103.** 40
105. $P + 2Pr + Pr^2$ **107.** $12,960.29

Quick Quiz: Sections 4.1–4.7, p. 286

1. 31 **2.** $4.1x^2 - 2.9x + 6.1$
3. $5x^4 + 15x^3 - x^2 + x + 12$ **4.** $a^4 - 100$ **5.** -21

Prepare to Move On, p. 286

1. $x^2 - 8x - 4$ **2.** $2x^3 - x^2 - x + 4$
3. $-2x + 5$ **4.** $5x^2 + x$

Check Your Understanding, p. 290

1. $x - 2 \overline{) x^2 - 8x + 12}$
2. $x - 1 \overline{) x^3 + 0x^2 + 0x - 1}$
3. $x + 3 \overline{) -3x^3 + 0x^2 + x + 2}$

Exercise Set 4.8, pp. 291–292

1. Divisor **2.** Quotient **3.** Dividend
4. Remainder **5.** $8x^6 - 5x^3$ **7.** $1 - 2u + u^6$

9. $6t^2 - 8t + 2$ **11.** $7x^3 - 6x + \frac{3}{2}$
13. $-4t^2 - 2t + 1$ **15.** $4x - 5 + \frac{1}{2x}$
17. $x + 2x^3y + 3$ **19.** $-3rs - r + 2s$ **21.** $x - 6$
23. $t - 5 + \frac{-45}{t - 5}$ **25.** $2x - 1 + \frac{1}{x + 6}$
27. $t^2 - 3t + 9$ **29.** $a + 5 + \frac{4}{a - 5}$
31. $3x - 5$ **33.** $x - 3 - \frac{3}{5x - 1}$
35. $3a + 1 + \frac{3}{2a + 5}$ **37.** $t^2 - 3t + 1$ **39.** $x^2 + 1$
41. $t^2 - 1 + \frac{3t - 1}{t^2 + 5}$ **43.** $2x^2 + 1 + \frac{-x}{2x^2 - 3}$ **45.**
47.

$y = -\frac{2}{3}x + 4$

48.

$8x = 4y$

49. Slope: 4; y-intercept: $\left(0, \frac{7}{2}\right)$ **50.** $y = \frac{5}{4}x - \frac{9}{2}$
51.   **53.** $5x^{6k} - 16x^{3k} + 14$ **55.** $3t^{2h} + 2t^h - 5$
57. $a + 3 + \frac{5}{5a^2 - 7a - 2}$ **59.** $2x^2 + x - 3$ **61.** 3
63. -1 **65.** (a) $(a + 1)$ cm; (b) $(a^2 + 2a + 1)$ cm^2

Quick Quiz: Sections 4.1–4.8, p. 292

1. $-t^5$ **2.** $-15x^6 - 18x^5 + 3x^4$ **3.** $\frac{1}{9}a^4 + 4a^2 + 36$
4. $8ab^2 - 10ab$ **5.** $4n^3 - 2n^2 + 1$

Prepare to Move On, p. 292

1. $3, x + 7$ **2.** $x - 3, x + 10$ **3.** $x, x + 1, 2x - 5$
4. $3(3x + 5y)$ **5.** $7(2a + c + 1)$

Visualizing for Success, p. 293

1. E, F **2.** B, O **3.** S, K **4.** R, G **5.** D, M
6. J, P **7.** C, L **8.** N, Q **9.** A, H **10.** I, T

Decision Making: Connection, p. 294

1. 80 movies **2.** 32 GB or 64 GB **3.**   **4.**

Study Summary: Chapter 4, pp. 295–297

1. 6 **2.** 1 **3.** x^{16} **4.** 8^7 **5.** y^{15} **6.** $x^{30}y^{10}$
7. $\frac{x^{10}}{7^5}$ **8.** $\frac{1}{10}$ **9.** $\frac{y^3}{x}$ **10.** 9.04×10^{-4} **11.** 690,000
12. $x^2, -10, 5x, -8x^6$ **13.** 1 **14.** 1 **15.** $-8x^6$
16. -8 **17.** 6 **18.** Trinomial **19.** $3x^2 - 4x$ **20.** 4

21. $8x^2 + x$ **22.** $10x^2 - 7x$ **23.** $x^3 - 2x^2 - x + 2$
24. $2x^2 + 11x + 12$ **25.** $25 - 9x^2$
26. $x^2 + 18x + 81$ **27.** $64x^2 - 16x + 1$ **28.** -7
29. 6   **30.** $3cd^2 + 4cd - 7c$ **31.** $-2p^2w$
32. $49x^2y^2 - 14x^3y + x^4$ **33.** $y^3 - 2y + 4$

34. $x - 2 + \dfrac{6}{x + 1}$

Review Exercises: Chapter 4, pp. 298–299

1. True   **2.** True   **3.** True   **4.** False   **5.** False
6. False   **7.** True   **8.** True   **9.** n^{12} **10.** $(7x)^{10}$
11. t^6 **12.** 4^3, or 64   **13.** 1   **14.** $-9c^4d^2$ **15.** $-8x^3y^6$
16. $18x^5$ **17.** a^7b^6 **18.** $\dfrac{4t^{10}}{9s^8}$ **19.** $\dfrac{1}{8^6}$ **20.** a^{-9}

21. $\dfrac{1}{4^2}$, or $\dfrac{1}{16}$ **22.** $\dfrac{2b^9}{a^{13}}$ **23.** $\dfrac{1}{w^{15}}$ **24.** $\dfrac{x^6}{4y^2}$

25. $\dfrac{y^3}{8x^3}$ **26.** 470,000,000   **27.** 1.09×10^{-5}
28. 2.09×10^4 **29.** 5.12×10^{-5}
30. $-4y^5, 7y^2, -3y, -2$ **31.** $7, -\frac{5}{6}, -4, 10$
32. (a) $2, 0, 5$; **(b)** $15t^5, 15$; **(c)** 5
33. (a) $5, 0, 2, 1$; **(b)** $-2x^5, -2$; **(c)** 5   **34.** Trinomial
35. Polynomial with no special name   **36.** Monomial
37. $t - 1$ **38.** $14a^5 - 2a^2 - a - \frac{2}{3}$ **39.** -24 **40.** 16
41. $x^5 + 8x^4 + 6x^3 - 2x - 9$ **42.** $6a^5 - a^3 - 12a^2$
43. $x^5 - 3x^3 - 2x^2 + 8$
44. $\frac{3}{4}x^4 + \frac{1}{4}x^3 - \frac{1}{3}x^2 - \frac{7}{4}x + \frac{3}{8}$
45. $-x^5 + x^4 - 5x^3 - 2x^2 + 2x$
46. (a) $4w + 6$; **(b)** $w^2 + 3w$ **47.** $-30x^5$
48. $49x^2 + 14x + 1$ **49.** $a^2 - 3a - 28$
50. $d^2 - 64$ **51.** $12x^3 - 23x^2 + 13x - 2$
52. $15t^5 - 6t^4 + 12t^3$ **53.** $4a^2 - 81$
54. $x^2 - 1.3x + 0.4$
55. $x^7 + x^5 - 3x^4 + 3x^3 - 2x^2 + 5x - 3$
56. $16y^6 - 40y^3 + 25$ **57.** $2t^4 - 11t^2 - 21$
58. $a^2 + \frac{1}{6}a - \frac{1}{3}$ **59.** $-49 + 4n^2$ **60.** 49
61. Coefficients: $1, -7, 9, -8$; degrees: $6, 2, 2, 0$; 6
62. Coefficients: $1, -1, 1$; degrees: $13, 22, 15$; 22
63. $-4u + 4v - 7$ **64.** $6m^3 + 4m^2n - mn^2$
65. $2a^2 - 16ab$ **66.** $11x^3y^2 - 8x^2y - 6x^2 - 6x + 6$
67. $2x^2 - xy - 15y^2$ **68.** $25a^2b^2 - 10abcd^2 + c^2d^4$
69. $\frac{1}{2}x^2 - \frac{1}{2}y^2$ **70.** $y^4 - \frac{1}{3}y + 4$
71. $3x^2 - 7x + 4 + \dfrac{1}{2x + 3}$ **72.** $t^3 + 2t - 3$

73. 📝 In the expression $5x^3$, the exponent refers only to the x. In the expression $(5x)^3$, the entire expression $5x$ is the base.
74. 📝 It is possible to determine two possibilities for the binomial that was squared by using the equation $(A - B)^2 = A^2 - 2AB + B^2$ in reverse. Since, in $x^2 - 6x + 9, A^2 = x^2$ and $B^2 = 9$, or 3^2, the binomial that was squared was $A - B$, or $x - 3$. If the polynomial is written $9 - 6x + x^2$, then $A^2 = 9$ and $B^2 = x^2$, so the binomial that was squared was $3 - x$. We cannot

determine without further information whether the binomial squared was $x - 3$ or $3 - x$.   **75. (a)** 9; **(b)** 28
76. $64x^{16}$ **77.** $8x^4 + 4x^3 + 5x - 2$
78. $-16x^6 + x^2 - 10x + 25$ **79.** $\frac{94}{13}$
80. 2.28×10^{11} platelets

Test: Chapter 4, p. 300

1. [4.1] x^{13} **2.** [4.1] 3   **3.** [4.1] 1   **4.** [4.1] t^{45}
5. [4.1] $-27y^6$ **6.** [4.1] $-40x^{19}y^4$ **7.** [4.1] $\frac{6}{5}a^5b^3$
8. [4.1] $\dfrac{16p^2}{25q^6}$ **9.** [4.2] $\dfrac{1}{y^7}$ **10.** [4.2] 5^{-6} **11.** [4.2] $\dfrac{1}{t^9}$
12. [4.2] $\dfrac{3y^5}{x^5}$ **13.** [4.2] $\dfrac{b^4}{16a^{12}}$ **14.** [4.2] $\dfrac{c^3}{a^3b^3}$
15. [4.2] 3.06×10^9 **16.** [4.2] 0.00000005
17. [4.2] 1.75×10^{17} **18.** [4.2] 1.296×10^{22}
19. [4.3] Binomial   **20.** [4.3] $3, -1, \frac{1}{9}$
21. [4.3] Degrees of terms: $3, 1, 5, 0$; leading term: $7t^5$; leading coefficient: 7; degree of polynomial: 5
22. [4.3] -7 **23.** [4.3] $\frac{7}{4}y^2 - 4y$
24. [4.3] $4x^3 + 4x^2 + 3$
25. [4.4] $4x^5 + x^4 + 5x^3 - 8x^2 + 2x - 7$
26. [4.4] $5x^4 + 5x^2 + x + 5$
27. [4.4] $-2a^4 + 3a^3 - a - 7$
28. [4.4] $-t^4 + 2.5t^3 - 0.6t^2 - 9$
29. [4.5] $-6x^4 + 6x^3 + 10x^2$
30. [4.6] $x^2 - \frac{2}{3}x + \frac{1}{9}$ **31.** [4.6] $25t^2 - 49$
32. [4.6] $6b^2 + 7b - 5$
33. [4.6] $x^{14} - 4x^8 + 4x^6 - 16$
34. [4.6] $48 + 34y - 5y^2$
35. [4.5] $6x^3 - 7x^2 - 11x - 3$
36. [4.6] $64a^6 + 48a^3 + 9$ **37.** [4.7] 24
38. [4.7] $-4x^3y - x^2y^2 + xy^3 - y^3 + 19$
39. [4.7] $8a^2b^2 + 6ab + 6ab^2 + ab^3 - 4b^3$
40. [4.7] $9x^{10} - y^2$ **41.** [4.8] $4x^2 + 3x - 5$
42. [4.8] $6x^2 - 20x + 26 + \dfrac{-39}{x + 2}$
43. [4.5], [4.6] $V = l(l - 2)(l - 1) = l^3 - 3l^2 + 2l$
44. [4.2] $\frac{1}{2} - \frac{1}{4} = \frac{1}{4}$ **45.** [4.6] About 2.9×10^8 hr

Cumulative Review: Chapters 1–4, p. 301

1. 6   **2.** -8 **3.** $-\frac{7}{45}$ **4.** 6   **5.** $y + 10$ **6.** t^{12}
7. $-4x^5y^3$ **8.** $50a^4b^7$ **9.** $2(5a - 3b + 6)$ **10.** 6
11. II

12.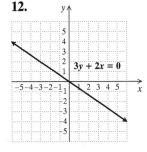

13.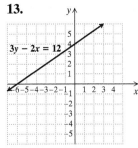

14.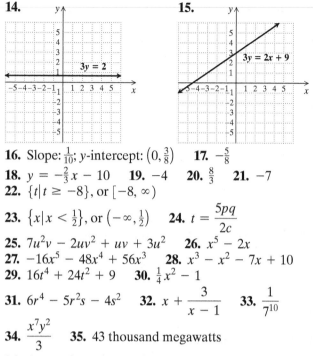

15.

16. Slope: $\frac{1}{10}$; y-intercept: $\left(0, \frac{3}{8}\right)$ **17.** $-\frac{5}{8}$

18. $y = -\frac{2}{3}x - 10$ **19.** -4 **20.** $\frac{8}{3}$ **21.** -7

22. $\{t \mid t \geq -8\}$, or $[-8, \infty)$

23. $\left\{x \mid x < \frac{1}{2}\right\}$, or $\left(-\infty, \frac{1}{2}\right)$ **24.** $t = \dfrac{5pq}{2c}$

25. $7u^2v - 2uv^2 + uv + 3u^2$ **26.** $x^5 - 2x$

27. $-16x^5 - 48x^4 + 56x^3$ **28.** $x^3 - x^2 - 7x + 10$

29. $16t^4 + 24t^2 + 9$ **30.** $\frac{1}{4}x^2 - 1$

31. $6r^4 - 5r^2s - 4s^2$ **32.** $x + \dfrac{3}{x - 1}$ **33.** $\dfrac{1}{7^{10}}$

34. $\dfrac{x^7y^2}{3}$ **35.** 43 thousand megawatts

36. Approximately 3.8 trillion kWh **37.** No more than 4 hr **38.** Approximately \$3.2 billion per year
39. 0 **40.** No solution **41.** $\frac{1}{7} + 1 = \frac{8}{7}$ **42.** $15x^{12}$

CHAPTER 5

Technology Connection, p. 309

1. Correct **2.** Not correct **3.** Not correct
4. Correct **5.** Not correct **6.** Correct

Check Your Understanding, p. 309

1. 8 **2.** $2x^3$ **3.** a^2bc^5 **4.** $10u^2v^3$ **5.** Yes
6. Yes **7.** No **8.** No

Exercise Set 5.1, pp. 310–311

1. False **2.** True **3.** True **4.** False **5.** (h)
6. (f) **7.** (b) **8.** (e) **9.** (c) **10.** (g) **11.** (d)
12. (a) **13.** Answers may vary.
$(14x)(x^2), (7x^2)(2x), (-2)(-7x^3)$ **15.** Answers may vary. $(-15)(a^4), (-5a)(3a^3), (-3a^2)(5a^2)$
17. Answers may vary. $(5t^2)(5t^3), (25t)(t^4), (-5t)(-5t^4)$
19. $8(x + 3)$ **21.** $2(x^2 + x - 4)$ **23.** $t(3t + 1)$
25. $-5y(y + 2)$ **27.** $x^2(x + 6)$ **29.** $8a^2(2a^2 - 3)$
31. $-t^2(6t^4 - 9t^2 + 4)$ **33.** $6x^2(x^6 + 2x^4 - 4x^2 + 5)$
35. $x^2y^2(x^3y^3 + x^2y + xy - 1)$
37. $-5a^2b^2(7ab^2 - 2b + 3a)$ **39.** $(n - 6)(n + 3)$
41. $(x + 3)(x^2 - 7)$ **43.** $(2y - 9)(y^2 + 1)$
45. $(x + 2)(x^2 + 5)$ **47.** $(3n - 2)(3n^2 + 1)$
49. $(t - 5)(4t^2 + 3)$ **51.** $(7x + 5)(x^2 - 3)$
53. $(6a + 7)(a^2 + 1)$ **55.** $(x - 6)(2x^2 - 1)$
57. Not factorable by grouping **59.** $(y + 8)(y^2 - 2)$

61. $(x - 4)(2x^2 - 9)$ **63.** ☑ **65.** $\frac{3}{25}$ **66.** 24
67. $\dfrac{y^4}{2x}$ **68.** $a^7b^7c^3$ **69.** $4x^2 + 5x - 13$
70. $-w^2y + yz + wz$ **71.** ☑ **73.** $(2x^3 + 3)(2x^2 + 3)$
75. $2x(x + 1)(x^2 - 2)$ **77.** $(x - 1)(5x^4 + x^2 + 3)$
79. Answers may vary. $8x^4y^3 - 24x^2y^4 + 16x^3y^4$

Prepare to Move On, p. 311

1. $x^2 + 9x + 14$ **2.** $x^2 - 9x + 14$ **3.** $x^2 - 5x - 14$
4. $x^2 + 5x - 14$ **5.** 1, 2, 3, 4, 5, 6, 10, 12, 15, 20, 30, 60
6. 1, 2, 3, 6, 9, 18

Check Your Understanding, p. 316

1. 1, 30; 2, 15; 3, 10; 5, 6 **2.** 1, 9; 3, 3 **3.** 1, 11
4. 1, 48; 2, 24; 3, 16; 4, 12; 6, 8

Exercise Set 5.2, pp. 317–318

1. False **2.** True **3.** True **4.** True
5. Positive; positive **6.** Negative; negative
7. Negative; positive **8.** Positive; positive
9. Positive **10.** Negative **11.** $(x + 4)(x + 4)$
13. $(x + 1)(x + 10)$ **15.** $(t - 2)(t - 7)$
17. $(b - 4)(b - 1)$ **19.** $(d - 2)(d - 5)$
21. $(x - 5)(x + 3)$ **23.** $(x + 5)(x - 3)$
25. $2(x + 2)(x - 9)$ **27.** $-x(x + 2)(x - 8)$
29. $(y - 5)(y + 9)$ **31.** $(x - 6)(x + 12)$
33. $-5(b - 3)(b + 10)$ **35.** $x^3(x - 2)(x + 1)$
37. Prime **39.** $(t + 4)(t + 8)$ **41.** $(x + 9)(x + 11)$
43. $3x(x - 25)(x + 4)$ **45.** $-4(x + 5)(x + 5)$
47. $(y - 12)(y - 8)$ **49.** $-a^4(a - 6)(a + 15)$
51. $\left(t + \frac{1}{3}\right)\left(t + \frac{1}{3}\right)$ **53.** Prime **55.** $(p - 5q)(p - 2q)$
57. Prime **59.** $(s - 6t)(s + 2t)$
61. $6a^8(a - 2)(a + 7)$ **63.** ☑ **65.** -10
66. 1.13 **67.** $\frac{5}{3}$ **68.** $\frac{5}{2}$ **69.** -11 **70.** 2
71. ☑ **73.** $-5, 5, -23, 23, -49, 49$
75. $(y + 0.2)(y - 0.4)$ **77.** $-\frac{1}{3}a(a - 3)(a + 2)$
79. $(x^m + 4)(x^m + 7)$ **81.** $(a + 1)(x + 2)(x + 1)$
83. $(x + 3)^3$, or $(x^3 + 9x^2 + 27x + 27)$ cubic meters
85. $x^2\left(\frac{3}{4}\pi + 2\right)$, or $\frac{1}{4}x^2(3\pi + 8)$ **87.** $x^2\left(9 - \frac{1}{2}\pi\right)$
89. $(x + 4)(x + 5)$

Quick Quiz: Sections 5.1–5.2, p. 318

1. $6x^2(2x^2 - 5x + 1)$ **2.** $(x - 4)(3x^2 + 5)$
3. $(x - 2)(x - 1)$ **4.** $(a + 4b)(a - 2b)$
5. $3(x + 2)(x + 3)$

Prepare to Move On, p. 318

1. $6x^2 + 17x + 12$ **2.** $6x^2 + x - 12$ **3.** $6x^2 - x - 12$
4. $6x^2 - 17x + 12$ **5.** $5x^2 - 36x + 7$
6. $3x^2 + 13x - 30$

Check Your Understanding, p. 325

1. $x + 5$　　**2.** $5x - 4$　　**3.** $5x + 2$　　**4.** $2x + 9$

Exercise Set 5.3, pp. 326–327

1. (d)　　**2.** (a)　　**3.** (b)　　**4.** (c)　　**5.** $(2x - 1)(x + 4)$
7. $(3x + 1)(x - 6)$　　**9.** $(2t + 1)(2t + 5)$
11. $(5a - 3)(3a - 1)$　　**13.** $(3x + 4)(2x + 3)$
15. $2(3x + 1)(x - 2)$　　**17.** $t(7t + 1)(t + 2)$
19. $(4x - 5)(3x - 2)$　　**21.** $-1(7x + 4)(5x + 2)$, or
$-(7x + 4)(5x + 2)$　　**23.** Prime　　**25.** $(5x + 4)^2$
27. $(20y - 1)(y + 3)$　　**29.** $(7x + 5)(2x + 9)$
31. $-1(x - 3)(2x + 5)$, or $-(x - 3)(2x + 5)$
33. $-3(2x + 1)(x + 5)$　　**35.** $2(a + 1)(5a - 9)$
37. $4(3x - 1)(x + 6)$　　**39.** $(3x + 1)(x + 1)$
41. $(x + 3)(x - 2)$　　**43.** $(4t - 3)(2t - 7)$
45. $(3x + 2)(2x + 5)$　　**47.** $(y + 4)(2y - 1)$
49. $(3a - 4)(2a - 1)$　　**51.** $(16t + 7)(t + 1)$
53. $-1(3x + 1)(3x + 5)$, or $-(3x + 1)(3x + 5)$
55. $10(x^2 + 3x - 7)$　　**57.** $3x(3x - 1)(2x + 3)$
59. $(x + 1)(25x + 64)$　　**61.** $3x(7x + 1)(8x + 1)$
63. $-t^2(2t - 3)(7t + 1)$　　**65.** $2(2y + 9)(8y - 3)$
67. $(2a - b)(a - 2b)$　　**69.** $2(s + t)(4s + 7t)$
71. $3(3x - 4y)^2$　　**73.** $-2(3a - 2b)(4a - 3b)$
75. $x^2(2x + 3)(7x - 1)$　　**77.** $a^6(3a + 4)(3a + 2)$
79.

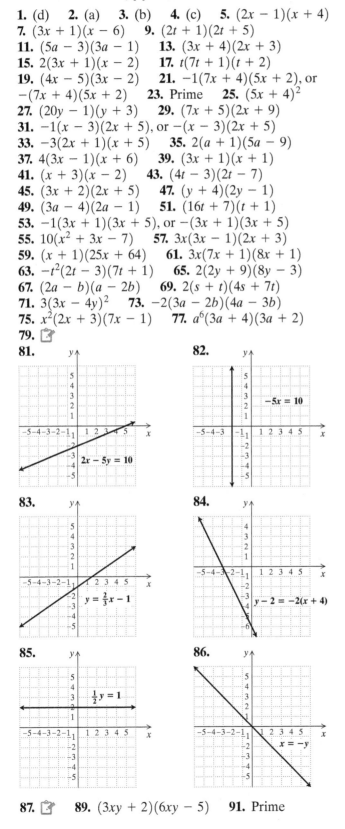

81. $2x - 5y = 10$

82. $-5x = 10$

83. $y = \frac{2}{3}x - 1$

84. $y - 2 = -2(x + 4)$

85. $\frac{1}{2}y = 1$

86. $x = -y$

87. 　　**89.** $(3xy + 2)(6xy - 5)$　　**91.** Prime

93. $(4t^5 - 1)^2$　　**95.** $-1(5x^m - 2)(3x^m - 4)$, or
$-(5x^m - 2)(3x^m - 4)$　　**97.** $(3a^{3n} + 1)(a^{3n} - 1)$
99. $[7(t - 3)^n - 2][(t - 3)^n + 1]$　　**101.** 7 ft

Quick Quiz: Sections 5.1–5.3, p. 327

1. $3ab^3(2a - 3b + 5a^2b^2)$　　**2.** $x(x - 3)(x - 4)$
3. Prime　　**4.** $(2a + 3)(2a + 5)$　　**5.** $(6x + 5)(x - 1)$

Prepare to Move On, p. 327

1. $x^2 - 4x + 4$　　**2.** $x^2 + 4x + 4$　　**3.** $x^2 - 4$
4. $25t^2 - 30t + 9$　　**5.** $16a^2 + 8a + 1$　　**6.** $4n^2 - 49$

Check Your Understanding, p. 332

1. $A = x; B = 5$　　**2.** $A = y; B = 1$
3. $A = 2x; B = 3$　　**4.** $A = 5c; B = 6$
5. $A = x; B = 5$　　**6.** $A = 1; B = 2y$
7. $A = m^2; B = 4$　　**8.** $A = 10n^2; B = 9$

Exercise Set 5.4, pp. 333–334

1. Prime polynomial　　**2.** Difference of squares
3. Difference of squares　　**4.** None of these
5. Perfect-square trinomial　　**6.** Perfect-square trinomial
7. None of these　　**8.** Prime polynomial
9. Difference of squares　　**10.** Perfect-square trinomial
11. Yes　　**13.** No　　**15.** No　　**17.** Yes　　**19.** $(x + 8)^2$
21. $(x - 5)^2$　　**23.** $5(p + 2)^2$　　**25.** $(1 - t)^2$, or $(t - 1)^2$
27. $2(3x + 1)^2$　　**29.** $(7 - 4y)^2$, or $(4y - 7)^2$
31. $-x^3(x - 9)^2$　　**33.** $2n(n + 10)^2$　　**35.** $5(2x + 5)^2$
37. $(7 - 3x)^2$, or $(3x - 7)^2$　　**39.** $(4x + 3)^2$
41. $2(1 + 5x)^2$, or $2(5x + 1)^2$　　**43.** $(3p + 2x)^2$
45. Prime　　**47.** $-1(8m + n)^2$, or $-(8m + n)^2$
49. $-2(4s - 5t)^2$　　**51.** Yes　　**53.** No　　**55.** Yes
57. $(x + 5)(x - 5)$　　**59.** $(p + 3)(p - 3)$
61. $(7 + t)(-7 + t)$, or $(t + 7)(t - 7)$
63. $6(a + 2)(a - 2)$　　**65.** $(7x - 1)^2$
67. $2(10 + t)(10 - t)$　　**69.** $-5(4a + 3)(4a - 3)$
71. $5(t + 4)(t - 4)$　　**73.** $2(2x + 9)(2x - 9)$
75. $x(6 + 7x)(6 - 7x)$　　**77.** Prime
79. $(t^2 + 1)(t + 1)(t - 1)$　　**81.** $-3x(x - 4)^2$
83. $3t(5t + 3)(5t - 3)$　　**85.** $a^6(a - 1)^2$
87. $10(a + b)(a - b)$
89. $(4x^2 + y^2)(2x + y)(2x - y)$
91. $2(3t + 2s)(3t - 2s)$　　**93.** 　　**95.** $2x^3 + x^2 - 2$
96. $2t^2 + 6t - 11$　　**97.** $6x^4 + x^2y - y^2$
98. $5x^3 - 15x^2 + 35x$　　**99.** $7x^2 - x + 3$

100. $x + 4 + \dfrac{10}{x - 5}$　　**101.**

103. $(x^4 + 2^4)(x^2 + 2^2)(x + 2)(x - 2)$, or
$(x^4 + 16)(x^2 + 4)(x + 2)(x - 2)$
105. $2x(3x - \frac{2}{5})(3x + \frac{2}{5})$
107. $[(y - 5)^2 + z^4][(y - 5) + z^2][(y - 5) - z^2]$, or
$(y^2 - 10y + 25 + z^4)(y - 5 + z^2)(y - 5 - z^2)$

109. $-1(x^2 + 1)(x + 3)(x - 3)$, or
$-(x^2 + 1)(x + 3)(x - 3)$ **111.** $(y + 4)^2$
113. $(3p + 5)(3p - 5)^2$
115. $(9 + b^{2k})(3 + b^k)(3 - b^k)$ **117.** $2x^3 - x^2 - 1$
119. $(y + x + 7)(y - x - 1)$ **121.** 16
123. $(x + 1)^2 - x^2 = [(x + 1) + x][(x + 1) - x] =$
$2x + 1 = (x + 1) + x$

Quick Quiz: Sections 5.1–5.4, p. 334

1. $(x + 3)(2x^2 - 1)$ **2.** $(x - 20)(x - 4)$
3. $a(3a - 4)(3a + 2)$ **4.** $(y + 8)^2$
5. $3(z^2 + 1)(z + 1)(z - 1)$

Prepare to Move On, p. 334

1. $8x^6y^{12}$ **2.** $-125x^6y^3$ **3.** $x^3 - 3x^2 + 3x - 1$
4. $p^3 + 3p^2t + 3pt^2 + t^3$

Mid-Chapter Review: Chapter 5, p. 335

1. $12x^3y - 8xy^2 + 24x^2y = 4xy(3x^2 - 2y + 6x)$
2. $3a^3 - 3a^2 - 90a = 3a(a^2 - a - 30) =$
$3a(a - 6)(a + 5)$ **3.** $6x^2(x^3 - 3)$
4. $(x + 2)(x + 8)$ **5.** $(x + 7)(2x - 1)$
6. $(x + 3)(x^2 + 2)$ **7.** $(8n + 3)(8n - 3)$ **8.** Prime
9. $6(p + t)(p - t)$ **10.** $(b - 7)^2$
11. $(3x - 1)(4x + 1)$ **12.** $a(1 - 5a)^2$
13. $10(x^2 + 1)(x + 1)(x - 1)$ **14.** Prime
15. $15(d^2 - 2d + 5)$ **16.** $(3p + 2x)(5p + 2x)$
17. $-2t(t + 2)(t + 3)$ **18.** $10(c + 1)^2$
19. $-1(2x - 5)(x + 1)$ **20.** $2n(m - 5)(m^2 - 3)$

Check Your Understanding, p. 339

1. (a) Yes; (b) 3 **2.** (a) Yes; (b) 2 **3.** (a) No; (b) 4
4. (a) Yes; (b) 3 **5.** (a) No; (b) 3

Exercise Set 5.5, pp. 340–341

1. Common factor **2.** Perfect-square trinomial
3. Grouping **4.** Multiplying **5.** $5(a + 5)(a - 5)$
7. $(y - 7)^2$ **9.** $(3t + 7)(t + 3)$ **11.** $x(x + 9)^2$
13. $(x - 5)^2(x + 5)$ **15.** $3t(3t + 1)(3t - 1)$
17. $3x(3x - 5)(x + 3)$ **19.** Prime
21. $6(y - 5)(y + 8)$ **23.** $-2a^4(a - 2)^2$
25. $5x(x^2 + 4)(x + 2)(x - 2)$ **27.** $(t^2 + 3)(t^2 - 3)$
29. $-x^4(x^2 - 2x + 7)$ **31.** $(p + w)(p - w)$
33. $a(x^2 + y^2)$ **35.** $2\pi r(h + r)$
37. $(a + b)(5a + 3b)$ **39.** $(x + 1)(x + y)$
41. $10a^2(4m^2 + 1)(2m + 1)(2m - 1)$
43. $(a - 2)(a - y)$ **45.** $(3x - 2y)(x + 5y)$
47. $8mn(m^2 - 4mn + 3)$ **49.** $\left(\frac{3}{4} + y\right)\left(\frac{3}{4} - y\right)$
51. $(a - 2b)^2$ **53.** $(4x + 3y)^2$ **55.** Prime
57. $(2x - 3)(5x + 2)$
59. $(a^2b^2 + 4)(ab + 2)(ab - 2)$

61. $4c(4d - c)(5d - c)$ **63.** $(3b - a)(b + 6a)$
65. $-1(xy + 2)(xy + 6)$, or $-(xy + 2)(xy + 6)$
67. $(2t + 5)(4t - 3)$ **69.** $5(pt + 6)(pt - 1)$
71. $b^4(ab - 4)(ab + 8)$ **73.** $x^4(x + 2y)(x - y)$
75. $\left(6a - \frac{5}{4}\right)^2$ **77.** $\left(\frac{1}{9}x - \frac{4}{3}\right)^2$
79. $(1 + 4x^6y^6)(1 + 2x^3y^3)(1 - 2x^3y^3)$
81. $(2ab + 3)^2$ **83.** $z(z + 6)(z^2 - 6)$
85. $(x + 5)(x + 1)(x - 1)$ **87.**
89. Approximately 4.45 billion users
90. $35\frac{5}{6}$ billionaires per year **91.** Mother's Day:
$21.2 billion; Father's Day: $12.7 billion **92.** 5 or
fewer roses **93.**   **95.** $(6 - x + a)(6 - x - a)$
97. $-x(x^2 + 9)(x^2 - 2)$
99. $-1(x^2 + 2)(x + 3)(x - 3)$, or
$-(x^2 + 2)(x + 3)(x - 3)$
101. $(y + 1)(y - 7)(y + 3)$ **103.** $(y + 4 + x)^2$
105. $(a + 3)^2(2a + b + 4)(a - b + 5)$
107. $(7x^2 + 1 + 5x^3)(7x^2 + 1 - 5x^3)$

Quick Quiz: Sections 5.1–5.5, p. 341

1. $10(m + 3)(m - 3)$ **2.** $(a + 3)(a - 2)$
3. $(2x - 1)(6x^2 - 5)$ **4.** $8(mn - 1)^2$
5. $c(3c + 2)(2c - 3)$

Prepare to Move On, p. 341

1. $\frac{9}{8}$ **2.** $-\frac{7}{2}$ **3.** 3 **4.** $-\frac{5}{3}$ **5.** $\frac{1}{4}$ **6.** 11

Check Your Understanding, p. 345

1. $x + 4 = 0; x - 5 = 0$ **2.** $2x - 7 = 0; 3x + 4 = 0$
3. $x = 0; x - 3 = 0$ **4.** $x = 0; x + 7 = 0; x - 9 = 0$
5. $x + 6 = 0; 2x + 1 = 0$

Connecting the Concepts, p. 346

1. Expression **2.** Equation **3.** Expression
4. Equation **5.** $2x^3 + x^2 - 8x$ **6.** $-2x^2 - 11$
7. $-10, 10$ **8.** $6a^2 - 19a + 10$ **9.** $(n - 1)(n - 9)$
10. $2, 8$

Technology Connection, p. 347

1. $-4.65, 0.65$ **2.** $-0.37, 5.37$ **3.** $-8.98, -4.56$
4. No solution **5.** $0, 2.76$

Exercise Set 5.6, pp. 347–349

1. (c) **2.** (a) **3.** (d) **4.** (b) **5.** $-9, -2$
7. $-1, 8$ **9.** $-6, \frac{3}{2}$ **11.** $\frac{1}{7}, \frac{3}{10}$ **13.** $0, 7$ **15.** $\frac{1}{21}, \frac{18}{11}$
17. $-\frac{8}{3}, 0$ **19.** $50, 70$ **21.** $1, 6$ **23.** $-7, 3$
25. $-9, -2$ **27.** $0, 10$ **29.** $-6, 0$ **31.** $-6, 6$
33. $-\frac{7}{2}, \frac{7}{2}$ **35.** -5 **37.** 8 **39.** $0, 2$ **41.** $-\frac{5}{4}, 3$
43. 3 **45.** $0, \frac{4}{3}$ **47.** $-\frac{7}{6}, \frac{7}{6}$ **49.** $-4, -\frac{2}{3}$ **51.** $-3, 1$
53. $-\frac{5}{2}, \frac{4}{3}$ **55.** $-1, 4$ **57.** $-3, 2$ **59.** $(-2, 0), (3, 0)$

61. $(-4, 0), (2, 0)$ **63.** $(-3, 0), (\frac{3}{2}, 0)$ **65.** ☑
67. 65   **68.** $\frac{4}{3}$ **69.** 1.65   **70.** 6.8×10^{-4} **71.** $-0.\overline{6}$
72. 125%   **73.** ☑   **75.** $-7, -\frac{8}{3}, \frac{11}{2}$
77. **(a)** $x^2 - x - 20 = 0$; **(b)** $x^2 - 6x - 7 = 0$;
(c) $4x^2 - 13x + 3 = 0$; **(d)** $6x^2 - 5x + 1 = 0$;
(e) $12x^2 - 17x + 6 = 0$; **(f)** $x^3 - 4x^2 + x + 6 = 0$
79. $-5, 4$ **81.** $-\frac{3}{5}, \frac{3}{5}$ **83.** $-4, 2$ **85.** $-1, 1, 2$
87. **(a)** $2x^2 + 20x - 4 = 0$; **(b)** $x^2 - 3x - 18 = 0$;
(c) $(x + 1)(5x - 5) = 0$; **(d)** $(2x + 8)(2x - 5) = 0$;
(e) $4x^2 + 8x + 36 = 0$; **(f)** $9x^2 - 12x + 24 = 0$
89. ☑   **91.** $2.33, 6.77$ **93.** $-9.15, -4.59$
95. $-3.76, 0$ **97.** -8 and -7, or 7 and 8

Quick Quiz: Sections 5.1–5.6, p. 349

1. $(x - 4)(x + 2)$ **2.** $-2, 4$ **3.** $-10, 10$
4. $(a + 10)(a - 10)$ **5.** $2(a + 10)^2$

Prepare to Move On, p. 349

1. Let n represent the first integer; $[n + (n + 1)]^2$
2. Let n represent the first integer; $n(n + 1)$
3. $140°, 35°, 5°$ **4.** Length: 64 in.; width: 32 in.

Check Your Understanding, p. 354

1. (e)   **2.** (a)   **3.** (c)   **4.** (f)

Exercise Set 5.7, pp. 355–359

1. $x + 1$ **2.** $2w$ **3.** $90°$ **4.** $c^2 = a^2 + b^2$
5. $-2, 3$ **7.** $11, 12$ **9.** -14 and -12; 12 and 14
11. Length: 40 ft; width: 8 ft   **13.** Length: 12 cm;
width: 7 cm   **15.** Foot: 7 ft; height: 12 ft   **17.** Base: 8 ft;
height: 16 ft   **19.** Base: 8 ft; height: 15 ft   **21.** 8 teams
23. 66 handshakes   **25.** 12 players   **27.** 1 min, 3 min
29. 10 knots   **31.** 9 ft   **33.** 300 ft by 400 ft by 500 ft
35. Length: 200 ft; width: 150 ft   **37.** $3, 4, 5$
39. 12 ft, 35 ft, 37 ft   **41.** Dining room: 12 ft by 12 ft;
kitchen: 12 ft by 10 ft   **43.** 20 ft   **45.** 1 sec, 2 sec
47. ☑   **49.** $\varnothing$; contradiction   **50.** 0
51. $\mathbb{R}$; identity   **52.** $\{y \mid y \le -4\}$, or $(-\infty, -4]$
53. $\{x \mid x < -1\}$, or $(-\infty, -1)$ **54.** $y = -\frac{1}{2}x + \frac{7}{2}$
55. ☑   **57.** \$180   **59.** 39 cm   **61.** 4 in., 6 in.
63. 35 ft   **65.** Length: 6 in.; width: $4\frac{1}{4}$ in.
67. 0.8 hr, 6.4 hr   **69.** 🖥

Quick Quiz: Sections 5.1–5.7, p. 359

1. $(x - 5)(3x^2 - 7)$ **2.** $-\frac{1}{5}, \frac{3}{4}$ **3.** $(2x - 3)(5x - 1)$
4. $6(x^4 + 1)(x^2 + 1)(x + 1)(x - 1)$
5. Longer leg: 40 ft; hypotenuse: 41 ft

Prepare to Move On, p. 359

1. $\frac{6}{7}$ **2.** $\frac{45}{44}$ **3.** $\frac{4}{5}$ **4.** 0   **5.** Undefined

Translating for Success, p. 360

1. O   **2.** M   **3.** K   **4.** I   **5.** G   **6.** E   **7.** C
8. A   **9.** H   **10.** B

Decision Making: Connection, p. 362

1. 105 games   **2.** 210 games
3. Even number of teams: $n - 1$ rounds; odd number of
teams: n rounds   **4.** 🖥   **5.** 🖥   **6.** ☑

Study Summary: Chapter 5, pp. 363–365

1. $6x(2x^3 - 3x^2 + 5)$ **2.** $(x - 3)(2x^2 - 1)$
3. $(x - 9)(x + 2)$ **4.** $(3x + 2)(2x - 1)$
5. $(2x - 3)(4x - 5)$ **6.** $(10n + 9)^2$
7. $(12t + 5)(12t - 5)$ **8.** $3x(x - 6)^2$
9. $-10, 3$ **10.** $12, 13$

Review Exercises: Chapter 5, pp. 365–366

1. False   **2.** True   **3.** True   **4.** False   **5.** False
6. True   **7.** True   **8.** False   **9.** Answers may vary.
$(4x)(5x^2), (-2x^2)(-10x), (x^3)(20)$ **10.** Answers may
vary. $(-3x^2)(6x^3), (2x^4)(-9x), (-18x)(x^4)$
11. $6x^3(2x - 3)$ **12.** $(10t + 1)(10t - 1)$
13. $(x + 4)(x - 3)$ **14.** $(x + 7)^2$
15. $3x(2x + 1)^2$ **16.** $(2x + 3)(3x^2 + 1)$
17. $(6a - 5)(a + 1)$ **18.** $(5t - 3)^2$
19. $(9a^2 + 1)(3a + 1)(3a - 1)$
20. $3x(3x - 5)(x + 3)$ **21.** $2(x - 5)(x^2 + 5x + 25)$
22. $(x + 4)(x^3 - 2)$ **23.** $(ab^2 + 8)(ab^2 - 8)$
24. $-4x^4(2x^2 - 8x + 1)$ **25.** $3(2x - 5)^2$ **26.** Prime
27. $-t(t + 6)(t - 7)$ **28.** $(2x + 5)(2x - 5)$
29. $(n + 6)(n - 10)$ **30.** $5(z^2 - 6z + 2)$
31. $(4t + 5)(t + 2)$ **32.** $(2t + 1)(t - 4)$
33. $7x(x + 1)(x + 4)$ **34.** $-6x(x + 5)(x - 5)$
35. $(5 - x)(3 - x)$ **36.** Prime
37. $(xy + 8)(xy - 2)$ **38.** $3(2a + 7b)^2$
39. $(m + 5)(m + t)$ **40.** $(3r + 5s)(2r - 3s)$
41. $-11, 9$ **42.** $-7, 5$ **43.** $-\frac{3}{4}, \frac{3}{4}$ **44.** $\frac{2}{3}, 1$
45. $-2, 3$ **46.** $0, \frac{3}{5}$ **47.** 1   **48.** $-3, 4$ **49.** 5 sec
50. $(-1, 0), (\frac{5}{2}, 0)$ **51.** Height: 14 ft; base: 14 ft
52. 10 holes   **53.** ☑ Answers may vary. Because Celia
did not first factor out the largest common factor, 4, her
factorization will not be "complete" until she removes
a common factor of 2 from each binomial. The answer
should be $4(x - 5)(x + 5)$. Awarding 3 to 7 points
would seem reasonable.   **54.** ☑ The principle of zero
products is used to solve quadratic equations and is not
used to solve linear equations.   **55.** 2.5 cm   **56.** 0, 2
57. Length: 12 cm; width: 6 cm   **58.** $-3, 2, \frac{5}{2}$
59. No real solution

Test: Chapter 5, p. 367

1. [5.2] $(x - 4)(x - 9)$ **2.** [5.4] $(x - 5)^2$
3. [5.1] $2y^2(2y^2 - 4y + 3)$ **4.** [5.1] $(x + 1)(x^2 + 2)$
5. [5.1] $t^5(t^2 - 3)$ **6.** [5.2] $a(a + 4)(a - 1)$
7. [5.3] $2(5x - 6)(x + 4)$ **8.** [5.4] $(2t + 5)(2t - 5)$
9. [5.3] $-3m(2m + 1)(m + 1)$
10. [5.4] $3(w + 5)(w - 5)$ **11.** [5.4] $5(3r + 2)^2$
12. [5.4] $3(x^2 + 4)(x + 2)(x - 2)$
13. [5.4] $(7t + 6)^2$ **14.** [5.1] $(x + 2)(x^3 - 3)$
15. [5.2] Prime **16.** [5.3] $3t(2t + 5)(t - 1)$
17. [5.2] $3(m - 5n)(m + 2n)$ **18.** [5.6] $1, 5$
19. [5.6] $-\frac{3}{2}, 5$ **20.** [5.6] $0, \frac{2}{5}$ **21.** [5.6] $-\frac{1}{5}, \frac{1}{5}$
22. [5.6] $-4, 5$ **23.** [5.6] $(-1, 0), (\frac{8}{3}, 0)$
24. [5.7] Length: 10 m; width: 4 m **25.** [5.7] 10 people
26. [5.7] 5 ft **27.** [5.7] 15 cm by 30 cm
28. [5.2] $(a - 4)(a + 8)$ **29.** [5.6] $-\frac{3}{5}, 0, \frac{2}{5}$

Cumulative Review: Chapters 1–5, p. 368

1. $\frac{1}{2}$ **2.** $\frac{9}{8}$ **3.** 29 **4.** $\dfrac{1}{9x^4y^6}$ **5.** t^{10}

6. $3x^4 + 5x^3 + x - 10$ **7.** $\dfrac{t^8}{4s^2}$ **8.** $-\dfrac{8x^6y^3}{27z^{12}}$ **9.** 8

10. $-x^4$ **11.** $2x^3 - 5x^2 + \frac{1}{2}x - 1$
12. $-4t^{11} + 8t^9 + 20t^8$ **13.** $9x^2 - 30x + 25$
14. $100x^{10} - y^2$ **15.** $(c + 1)(c - 1)$
16. $5(x + y)(1 + 2x)$ **17.** $(2r - t)^2$ **18.** $10(y^2 + 4)$
19. $y(x - 1)(x - 2)$ **20.** $(3x - 2y)(4x + y)$
21. $\frac{1}{12}$ **22.** $-\frac{9}{4}$ **23.** $\{x | x \geq 1\}$, or $[1, \infty)$

24. $-4, 3$ **25.** $-2, 2$ **26.** $0, 4$ **27.** $c = \dfrac{a}{b + d}$

28. 0 **29.** $y = 5x + \frac{5}{3}$
30.

(graph: $4(x + 1) = 8$)

32.

(graph: $y = \frac{3}{2}x - 2$)

31.

(graph: $x + y = 5$)

33.

(graph: $3x + 5y = 10$)

34. 372 min **35.** Bottom of ladder to building: 5 ft; top of ladder to ground: 12 ft **36.** Scores that are 9 and higher

37. (a)

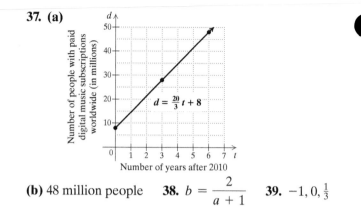

(b) 48 million people **38.** $b = \dfrac{2}{a + 1}$ **39.** $-1, 0, \frac{1}{3}$

CHAPTER 6

Technology Connection, p. 373

1. Correct **2.** Not correct

Check Your Understanding, p. 374

1. $\dfrac{x^2}{y}$ **2.** $\dfrac{t - 6}{t}$ **3.** $\dfrac{x - 2}{x - 1}$ **4.** $\dfrac{y - 7}{y - 4}$

5. $\dfrac{1}{-1}$, or -1

Exercise Set 6.1, pp. 375–376

1. Quotient **2.** Denominator **3.** Factors
4. A factor equal to 1 **5.** (a) **6.** (c) **7.** (d)
8. (b) **9.** 0 **11.** -5 **13.** 5 **15.** $-4, 7$
17. $-6, \frac{1}{2}$ **19.** $\dfrac{5a}{4b^2}$ **21.** $\dfrac{t + 2}{t - 3}$ **23.** $\frac{7}{8}$
25. $\dfrac{a - 3}{a + 1}$ **27.** $-\dfrac{2x^3}{3}$ **29.** $\dfrac{y - 3}{4y}$ **31.** $\dfrac{3(2a + 1)}{7(a + 1)}$
33. $\dfrac{t - 4}{t - 5}$ **35.** $\dfrac{a + 4}{2(a - 4)}$ **37.** $\dfrac{x - 4}{x + 4}$
39. $t - 1$ **41.** $\dfrac{y^2 + 4}{y + 2}$ **43.** $\frac{1}{2}$ **45.** $\dfrac{y}{2y + 1}$
47. $\dfrac{2x - 3}{5x + 2}$ **49.** -1 **51.** -7
53. $-\frac{1}{4}$ **55.** $-\frac{3}{2}$ **57.** -1 **59.** ☝
61. $(x + 5)(3x^2 + 1)$ **62.** $(3x + 1)(x + 5)$
63. $3y^2(6y^2 - 9y + 1)$ **64.** $(5a + 4b)(5a - 4b)$
65. $m(m - 4)^2$ **66.** $5(x - 4)(x - 3)$ **67.** ☝
69. $-(2y + x)$ **71.** $\dfrac{x^3 + 4}{(x^3 + 2)(x^2 + 2)}$
73. $\dfrac{(t - 1)(t - 9)^2}{(t + 1)(t^2 + 9)}$ **75.** $x + 2$
77. $\dfrac{(x - y)^3}{(x + y)^2(x - 5y)}$ **79.** ☝

Prepare to Move On, p. 376

1. $\frac{4}{21}$ **2.** $-\frac{5}{3}$ **3.** $\frac{15}{4}$ **4.** $-\frac{21}{16}$

Check Your Understanding, p. 378

1. $\frac{ac}{bd}$ **2.** $\frac{ad}{bc}$ **3.** $\frac{ac}{b}$ **4.** $\frac{a}{bc}$

Technology Connection, p. 379

1. Let $y_1 = ((x^2 + 3x + 2)/(x^2 + 4))/(5x^2 + 10x)$ and $y_2 = (x + 1)/((x^2 + 4)(5x))$. With the table set in AUTO mode, note that the values in the Y1- and Y2-columns match except for $x = -2$.
2. ERROR messages occur when division by 0 is attempted. Since the simplified expression has no factor of $x + 5$ or $x + 1$ in a denominator, no ERROR message occurs in Y2 for $x = -5$ or -1.

Exercise Set 6.2, pp. 380–382

1. (d)   **2.** (c)   **3.** (a)   **4.** (b)   **5.** (d)   **6.** (c)
7. (a)   **8.** (e)   **9.** (b)   **10.** (f)   **11.** $\frac{3x(x+2)}{8(5x-1)}$
13. $\frac{(a-4)(a+2)}{(a+6)^2}$ **15.** $\frac{(n-4)(n+4)}{(n^2+4)(n^2-4)}$
17. $\frac{6t}{5}$ **19.** $\frac{4}{c^2 d}$ **21.** $\frac{y+4}{4}$ **23.** $\frac{x+2}{x-2}$
25. $\frac{(n-5)(n-1)(n-6)}{(n+6)(n^2+36)}$ **27.** $\frac{7(a+3)}{a(a+4)}$
29. $\frac{y(y-1)}{y+1}$ **31.** $\frac{3v}{v-2}$ **33.** $\frac{t-5}{t+5}$
35. $\frac{9(y-5)}{10(y-1)}$ **37.** 1   **39.** $\frac{(t-2)(t+5)}{(t-5)(t-5)}$
41. $(2x-1)^2$ **43.** $\frac{(7x+5)(3x+4)}{8}$
45. $\frac{2x+5y}{x^2 y}$ **47.** $\frac{9}{2x}$ **49.** $\frac{1}{a^4+3a}$
51. $\frac{x^2}{20}$ **53.** $\frac{a^3}{b^3}$ **55.** $\frac{4(t-3)}{3(t+1)}$ **57.** $4(y-2)$
59. $-\frac{a}{b}$ **61.** $\frac{(n+3)(n+3)}{n-2}$ **63.** $\frac{a-5}{3(a-1)}$
65. $\frac{(2x-1)(2x+1)}{x-5}$ **67.** $\frac{(w-7)(w-8)}{(2w-7)(3w+1)}$
69. $\frac{1}{(c-5)(5c-3)}$ **71.** $\frac{15}{16}$ **73.** $\frac{-x^2-4}{x+2}$
75. $\frac{(x-4y)(x-y)}{(x+y)^3}$ **77.** ☑

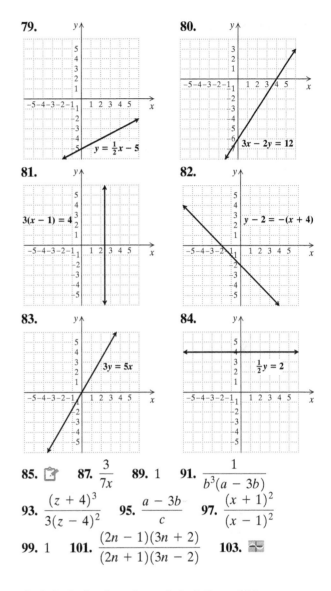

85. ☑   **87.** $\frac{3}{7x}$ **89.** 1   **91.** $\frac{1}{b^3(a-3b)}$
93. $\frac{(z+4)^3}{3(z-4)^2}$ **95.** $\frac{a-3b}{c}$ **97.** $\frac{(x+1)^2}{(x-1)^2}$
99. 1   **101.** $\frac{(2n-1)(3n+2)}{(2n+1)(3n-2)}$ **103.** ▧

Quick Quiz: Sections 6.1–6.2, p. 382

1. $\frac{6(x+3)}{x-4}$ **2.** $\frac{y-3}{2y+1}$ **3.** $\frac{(x-1)^2}{4x}$
4. $\frac{m+3}{6(m+2)}$ **5.** $\frac{(x+4)^2}{(x+5)(x-4)}$

Prepare to Move On, p. 382

1. $\frac{41}{24}$ **2.** $\frac{1}{6}$ **3.** x^2+3 **4.** $-2x^2-4x+1$

Check Your Understanding, p. 387

1. $\frac{a+b}{c}$ **2.** $\frac{a-b}{c}$ **3.** $\frac{x+y}{3}$ **4.** $\frac{5}{n}$
5. $\frac{8}{x+3}$

Exercise Set 6.3, pp. 389–391

1. Numerators; denominator   **2.** Term   **3.** Least common denominator; LCD   **4.** Factor   **5.** $\dfrac{8}{t}$

7. $\dfrac{3x+5}{12}$ **9.** $\dfrac{9}{a+3}$ **11.** $\dfrac{8}{4x-7}$ **13.** $\dfrac{2y+7}{2y}$

15. 6   **17.** $\dfrac{4(x-1)}{x+3}$ **19.** $a+5$ **21.** $y-7$

23. 0   **25.** $\dfrac{1}{x+2}$ **27.** $\dfrac{(3a-7)(a-2)}{(a+6)(a-1)}$

29. $\dfrac{t-4}{t+3}$ **31.** $\dfrac{y+2}{y-4}$ **33.** $-\dfrac{5}{x-4}$, or $\dfrac{5}{4-x}$

35. $-\dfrac{1}{x-1}$, or $\dfrac{1}{1-x}$ **37.** 180   **39.** 72   **41.** 60

43. $18t^5$ **45.** $30a^4b^8$ **47.** $6(y-3)$

49. $(x-5)(x+3)(x-3)$ **51.** $t(t-4)(t+2)^2$

53. $120x^2y^3z^2$ **55.** $(a+1)(a-1)^2$

57. $(2n-1)(n+1)(n+2)$ **59.** $(t+3)(t-3)$

61. $12x^3(x-5)(x-3)(x-1)$

63. $\dfrac{15}{18t^4}, \dfrac{st^2}{18t^4}$ **65.** $\dfrac{21y}{9x^4y^3}, \dfrac{4x^3}{9x^4y^3}$

67. $\dfrac{2x(x+3)}{(x-2)(x+2)(x+3)}, \dfrac{4x(x-2)}{(x-2)(x+2)(x+3)}$

69. ☞   **71.** $-\frac{10}{3}$ **72.** $\{x \mid x > 15\}$, or $(15, \infty)$

73. $-2, 10$ **74.** $\{x \mid x \le -8\}$, or $(-\infty, -8]$ **75.** -1

76. $\frac{8}{3}$ **77.** ☞   **79.** $\dfrac{18x+5}{x-1}$ **81.** $\dfrac{x}{3x+1}$

83. 30 strands   **85.** 60 strands

87. $(2x+5)(2x-5)(3x+4)^4$ **89.** 20 sec

91. 6:15 A.M.   **93.** ☞   **95.** 🖥

Quick Quiz: Sections 6.1–6.3, p. 391

1. $\dfrac{1}{x+7}$ **2.** $x(x-6)(x-1)$ **3.** $\dfrac{1}{(x+3)(x-1)}$

4. $\dfrac{4}{3t^2(t-2)}$ **5.** $\dfrac{8}{x-2}$

Prepare to Move On, p. 391

1. $\frac{-5}{8}, \frac{5}{-8}$ **2.** $\frac{-4}{11}, -\frac{4}{11}$ **3.** $-x+y$, or $y-x$
4. $-3+a$, or $a-3$ **5.** $-2x+7$, or $7-2x$
6. $-a+b$, or $b-a$

Check Your Understanding, p. 396

1. $\dfrac{3(x-5)}{(x+4)(x-5)}$ **2.** $\dfrac{2ab^4}{2a^3b^5}$ **3.** $\dfrac{-1}{y-2}$

Exercise Set 6.4, pp. 396–398

1. LCD   **2.** Missing; denominator   **3.** Numerators; LCD   **4.** Simplify   **5.** $\dfrac{3+5x}{x^2}$ **7.** $-\dfrac{5}{24r}$

9. $\dfrac{3u^2+4v}{u^3v^2}$ **11.** $\dfrac{-2(xy+9)}{3x^2y^3}$ **13.** $\dfrac{7x+1}{24}$

15. $\dfrac{-x-4}{6}$ **17.** $\dfrac{a^2+13a-5}{15a^2}$ **19.** $\dfrac{7z-12}{12z}$

21. $\dfrac{(3c-d)(c+d)}{c^2d^2}$ **23.** $\dfrac{4x^2-13xt+9t^2}{3x^2t^2}$

25. $\dfrac{6x}{(x+2)(x-2)}$ **27.** $\dfrac{(t-3)(t+1)}{(t-1)(t+3)}$

29. $\dfrac{11x+2}{3x(x+1)}$ **31.** $\dfrac{-5t+3}{2t(t-1)}$ **33.** $\dfrac{a^2}{(a-3)(a+3)}$

35. $\dfrac{16}{3(z+4)}$ **37.** $\dfrac{5q-3}{(q-1)^2}$ **39.** $\dfrac{9a}{4(a-5)}$

41. $\dfrac{2y-1}{y(y-1)}$ **43.** $\dfrac{10}{(a-3)(a+2)}$

45. $\dfrac{x-5}{(x+5)(x+3)}$ **47.** $\dfrac{3z^2+19z-20}{(z-2)^2(z+3)}$

49. $\dfrac{-7}{x^2+25x+24}$ **51.** $\dfrac{6x+7}{2x+1}$ **53.** $\dfrac{10-3x}{4-x}$

55. $\dfrac{3x-1}{2}$ **57.** 0   **59.** $y+3$ **61.** 0

63. $\dfrac{p^2+7p+1}{(p-5)(p+5)}$ **65.** $\dfrac{(x+1)(x+3)}{(x-4)(x+4)}$

67. $\dfrac{-a-2}{(a+1)(a-1)}$, or $\dfrac{a+2}{(1+a)(1-a)}$

69. $\dfrac{2(5x+3y)}{(x-y)(x+y)}$ **71.** $\dfrac{2x-3}{2-x}$ **73.** 3   **75.** 0

77. $\dfrac{2}{(x+3)(x+4)}$ **79.** ☞   **81.** 6   **82.** 7

83. 3×10^{14} **84.** $\dfrac{-6a^4}{b}$ **85.** $\dfrac{a^2}{9b^2}$

86. 12   **87.** ☞

89. Perimeter: $\dfrac{2(5x-7)}{(x-5)(x+4)}$; area: $\dfrac{6}{(x-5)(x+4)}$

91. $\dfrac{x}{3x+1}$ **93.** $\dfrac{-29}{(x-3)(x+1)}$

95. $\dfrac{x^4+4x^3-5x^2-126x-441}{(x+2)^2(x+7)^2}$

97. $\dfrac{5(a^2+2ab-b^2)}{(a-b)(3a+b)(3a-b)}$ **99.** $\dfrac{a}{a-b} + \dfrac{3b}{b-a}$; answers may vary.   **101.** ☞, 🖥

Quick Quiz: Sections 6.1–6.4, p. 398

1. $\dfrac{x-8}{4x}$ **2.** $\dfrac{(x-1)(x+6)}{(x-4)(x+2)}$ **3.** $x+2$

4. $\dfrac{t(4t-1)}{(2t+1)(t-1)}$ **5.** $\dfrac{(x-1)^2(2x+1)}{(x-2)^3}$

Prepare to Move On, p. 398

1. $\frac{9}{10}$ **2.** $\frac{16}{27}$ **3.** $\frac{2}{3}$ **4.** $\dfrac{(x-3)(x+2)}{(x-2)(x+3)}$

Mid-Chapter Review: Chapter 6, pp. 399–400

1.
$$\frac{a^2}{a-10} \div \frac{a^2+5a}{a^2-100} = \frac{a^2}{a-10} \cdot \frac{a^2-100}{a^2+5a}$$
$$= \frac{a \cdot a \cdot (a+10) \cdot (a-10)}{(a-10) \cdot a \cdot (a+5)}$$
$$= \frac{a(a-10)}{a(a-10)} \cdot \frac{a(a+10)}{a+5}$$
$$= \frac{a(a+10)}{a+5}$$

2.
$$\frac{2}{x} + \frac{1}{x^2+x} = \frac{2}{x} + \frac{1}{x(x+1)}$$
$$= \frac{2}{x} \cdot \frac{x+1}{x+1} + \frac{1}{x(x+1)}$$
$$= \frac{2x+2}{x(x+1)} + \frac{1}{x(x+1)}$$
$$= \frac{2x+3}{x(x+1)}$$

3. $\dfrac{3x+10}{5x^2}$ **4.** $\dfrac{6}{5x^3}$ **5.** $\dfrac{3x}{10}$ **6.** $\dfrac{3x-10}{5x^2}$

7. $\dfrac{x-3}{15(x-2)}$ **8.** $\frac{1}{3}$ **9.** $\dfrac{x^2-2x-2}{(x-1)(x+2)}$

10. $\dfrac{5x+17}{(x+3)(x+4)}$ **11.** -5 **12.** $\dfrac{5}{x-4}$

13. $\dfrac{(2x+3)(x+3)}{(x+1)^2}$ **14.** $\frac{1}{6}$ **15.** $\dfrac{x(x+4)}{(x-1)^2}$

16. $\dfrac{x+7}{(x-5)(x+1)}$ **17.** $\dfrac{9(u-1)}{16}$ **18.** $(t+5)^2$

19. $\dfrac{a-1}{(a+2)(a-2)^2}$ **20.** $\dfrac{-3x^2+9x-14}{2x}$

Check Your Understanding, p. 402

1. $\dfrac{1}{a} + \dfrac{a}{3}$ **2.** $\dfrac{a^2}{a+3}$ **3.** $\dfrac{1}{a}, \dfrac{a}{3}, \dfrac{a^2}{a+3}$
4. $a, 3, a+3$ **5.** $3a(a+3)$

Exercise Set 6.5, pp. 404–406

1. Complex   **2.** Denominator   **3.** Least common
denominator   **4.** Reciprocal   **5.** (b)   **6.** (a)

7. (a)   **8.** (b)   **9.** 10   **11.** $\frac{5}{11}$ **13.** $\dfrac{5x^2}{4(x^2+4)}$

15. $\dfrac{(x+2)(x-3)}{(x-1)(x+4)}$ **17.** $\dfrac{-10t}{5t-2}$ **19.** $\dfrac{2(2a-5)}{a-7}$

21. $\dfrac{x^2-18}{2(x+3)}$ **23.** $-\frac{1}{5}$ **25.** $\dfrac{1+t^2}{t(1-t)}$ **27.** $\dfrac{x}{x-y}$

29. $\dfrac{c(4c+7)}{3(2c^2-1)}$ **31.** $\dfrac{15(4-a^3)}{14a^2(9+2a)}$ **33.** 1

35. $\dfrac{3a^2+4b^3}{b^3(5-3a^2)}$ **37.** $\dfrac{(t-3)(t+3)}{t^2+4}$

39. $\dfrac{y^2+1}{(y+1)(y-1)}$ **41.** $\dfrac{1}{a(a-h)}$ **43.** $\dfrac{1}{x-y}$

45. $\dfrac{t^2+5t+3}{(t+1)^2}$ **47.** $\dfrac{x^2-2x-1}{x^2-5x-4}$

49. $\dfrac{(a-2)(a-7)}{(a+1)(a-6)}$ **51.** $\dfrac{x+2}{x+3}$ **53.** 📋
55. $(2x-3)(3x^2-2)$ **56.** $4ab(3a+b-2)$
57. $3n(2n+1)(5n-3)$ **58.** $(5a-4b)^2$
59. $(n^2+1)(n+1)(n-1)$ **60.** $(pw+10)(pw-12)$
61. 📋   **63.** 6, 7, 8   **65.** $-3, -\frac{4}{5}, 3$

67. $\dfrac{A}{B} \div \dfrac{C}{D} = \dfrac{\frac{A}{B}}{\frac{C}{D}} = \dfrac{\frac{A}{B}}{\frac{C}{D}} \cdot \dfrac{BD}{BD} = \dfrac{AD}{BC} = \dfrac{A}{B} \cdot \dfrac{D}{C}$

69. $\dfrac{x^2+5x+15}{-x^2+10}$ **71.** 0   **73.** $\dfrac{x-5}{x-3}$

75. $\dfrac{(x-1)(3x-2)}{5x-3}$

Quick Quiz: Sections 6.1–6.5, p. 406

1. $\dfrac{2(2x+1)}{2x-1}$ **2.** $\dfrac{2x(x+3)}{3}$ **3.** $\dfrac{a+3}{a+1}$
4. $\dfrac{-5x^2+21x-2}{(x-3)(x+1)}$ **5.** $\dfrac{5x+8}{6x}$

Prepare to Move On, p. 406

1. -4 **2.** $-\frac{14}{27}$ **3.** 3, 4   **4.** $-15, 2$

Check Your Understanding, p. 409

1. x **2.** $12x$ **3.** $y(y+5)$, or y^2+5y

Connecting the Concepts, pp. 411–412

1. Expression; $\dfrac{19n-2}{5n(2n-1)}$ **2.** Equation; 8

3. Equation; $-\frac{1}{2}$ **4.** Expression; $\dfrac{8(t+1)}{(t-1)(2t-1)}$

5. Expression; $\dfrac{2a}{a-1}$ **6.** Equation; $-10, 10$

Exercise Set 6.6, pp. 412–413

1. False   **2.** True   **3.** True   **4.** True   **5.** $-\frac{2}{5}$
7. $\frac{24}{5}$ **9.** $-6, 6$ **11.** 12   **13.** $\frac{14}{3}$ **15.** $-6, 6$
17. $-4, -1$ **19.** -10 **21.** $-4, -3$ **23.** $\frac{23}{2}$
25. $\frac{5}{2}$ **27.** -1 **29.** No solution   **31.** -10

33. $-\frac{7}{3}$ **35.** $-2, \frac{7}{3}$ **37.** $-3, 13$ **39.** No solution
41. 2 **43.** -8 **45.** No solution **47.** No solution
49. 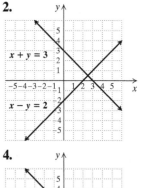 **51.** x-intercept: $(3, 0)$; y-intercept: $(0, -18)$
52. $\frac{1}{9}$ **53.** Slope: -2; y-intercept: $(0, 5)$ **54.** Yes
55. $y = \frac{1}{3}x - 2$ **56.** $y - (-5) = -4(x - 1)$
57. 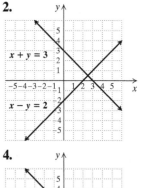 **59.** -2 **61.** 3 **63.** 4 **65.** 4 **67.** -2
69.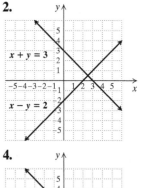

Quick Quiz: Sections 6.1–6.6, p. 413

1. $\dfrac{7(x - 11)}{x + 7}$ **2.** $x(x + 2)(x - 2)(x + 1)$

3. $\dfrac{-2n^2 + 5n + 6}{2n(n + 2)}$ **4.** $\dfrac{x - 2}{x + 2}$ **5.** -7

Prepare to Move On, p. 413

1. 137, 139 **2.** $-8, -6$; 6, 8 **3.** Base: 9 cm;
height: 12 cm **4.** 0.06 cm per day

Check Your Understanding, p. 418

1. x **2.** z **3.** 3

Exercise Set 6.7, pp. 420–425

1. False **2.** True **3.** True **4.** True **5.** True
6. False **7.** $\frac{1}{2}$ cake per hour **8.** $\frac{1}{3}$ cake per hour
9. $\frac{5}{6}$ cake per hour **10.** 1 lawn per hour
11. $\frac{1}{3}$ lawn per hour **12.** $\frac{2}{3}$ lawn per hour **13.** 6 hr
15. $3\frac{3}{7}$ hr **17.** $8\frac{4}{7}$ hr **19.** 21 min
21. $\frac{10}{7}$ min, or $1\frac{3}{7}$ min
23. AMTRAK: 80 km/h; CSX: 66 km/h

	Distance (in km)	Speed (in km/h)	Time (in hours)
CSX	330	$r - 14$	$\dfrac{330}{r - 14}$
AMTRAK	400	r	$\dfrac{400}{r}$

25. Rita: 50 mph; Sean: 65 mph
27. Tau: 5 km/h; Baruti: 9 km/h **29.** 3 hr
31. 20 mph **33.** 10.5 **35.** $\frac{8}{3}$ **37.** 12.6
39. $3\frac{3}{4}$ in. **41.** 20 ft **43.** 15 ft **45.** About 26.7 cm
47. \$65.25 **49.** 702 photos **51.** 52 bulbs
53. $7\frac{1}{2}$ oz **55.** 90 whales **57.**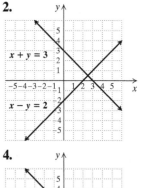
59. $x^3 - x^2 + x - 15$ **60.** $2x^4 + 6x^3 - 7x - 21$
61. $3y^2z + 2yz^2 + 6y^2 - 6yz$ **62.** $2ab^2 + 4b - a$

63. $64n^6 - 9$ **64.** $x^2 + x + \dfrac{7}{x - 1}$ **65.**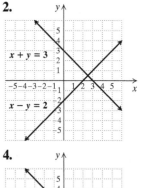

67. $49\frac{1}{2}$ hr

69. Equation 1: $\dfrac{1}{a} \cdot t + \dfrac{1}{b} \cdot t = 1$;

Equation 2: $\left(\dfrac{1}{a} + \dfrac{1}{b}\right)t = 1$;

Equation 3: $\dfrac{t}{a} + \dfrac{t}{b} = 1$;

Equation 4: $\dfrac{1}{a} + \dfrac{1}{b} = \dfrac{1}{t}$

$\dfrac{1}{a} \cdot t + \dfrac{1}{b} \cdot t = 1$	Equation 1
$t\left(\dfrac{1}{a} + \dfrac{1}{b}\right) = 1$	Factoring out t; equation 1 = equation 2
$t \cdot \dfrac{1}{a} + t \cdot \dfrac{1}{b} = 1$	Using the distributive law
$\dfrac{t}{a} + \dfrac{t}{b} = 1$	Multiplying; equation 2 = equation 3
$\dfrac{1}{t} \cdot \left(\dfrac{t}{a} + \dfrac{t}{b}\right) = \dfrac{1}{t} \cdot 1$	Multiplying both sides by $\dfrac{1}{t}$
$\dfrac{1}{t} \cdot \dfrac{t}{a} + \dfrac{1}{t} \cdot \dfrac{t}{b} = \dfrac{1}{t} \cdot 1$	Using the distributive law
$\dfrac{1}{a} + \dfrac{1}{b} = \dfrac{1}{t}$	Multiplying; equation 3 = equation 4

71. $30\frac{22}{31}$ hr **73.** About 14 sec
75. $21\frac{9}{11}$ min after 4:00 **77.** 48 km/h
79. 45 mph **81.** $66\frac{2}{3}$ ft **83.**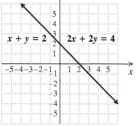

Quick Quiz: Sections 6.1–6.7, p. 425

1. $\dfrac{(y - 3)(y - 2)}{7x(y - 1)}$ **2.** 35

3. $\dfrac{2x}{(x - 2)(x + 1)(x + 2)}$ **4.** -7

5. $\frac{100}{3}$ min, or $33\frac{1}{3}$ min

Prepare to Move On, p. 425

1.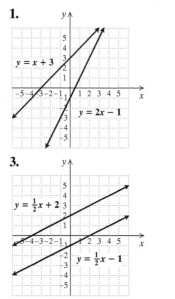

2.

3.

4.

Translating for Success, p. 426

1. K **2.** C **3.** N **4.** D **5.** E **6.** O
7. F **8.** H **9.** B **10.** A

Decision Making: Connection, p. 427

1. (a) 400 messages; **(b)** Yes. He will send more than 400 messages at that rate. **2. (a)** 200 min; **(b)** No. He will talk for fewer than 200 min at that rate. **3.** $36
4. More than 480 messages **5.** Answers will vary.

Study Summary: Chapter 6, pp. 428–431

1. $\dfrac{3(x-1)}{(x-3)}$ **2.** $\dfrac{5(a+1)}{2a}$ **3.** $\dfrac{t^2(t-10)}{(t-2)^2}$

4. $\dfrac{3x-1}{x+5}$ **5.** 1 **6.** $(a+5)(a-5)(a-1)$

7. $\dfrac{(x-3)(2x-1)}{(x-1)(x-2)}$ **8.** $\dfrac{3}{3x+2}$ **9.** 2

10. $\frac{6}{5}$ hr, or $1\frac{1}{5}$ hr **11.** Jogging: 9 mph; walking: 4 mph
12. $\frac{15}{2}$

Review Exercises: Chapter 6, pp. 431–432

1. False **2.** True **3.** False **4.** True **5.** True
6. False **7.** False **8.** False **9.** 0 **10.** $-6, 6$

11. $-6, 5$ **12.** -2 **13.** $\dfrac{x-3}{x+5}$ **14.** $\dfrac{7x+3}{x-3}$

15. $\dfrac{3(y-3)}{2(y+3)}$ **16.** $-5(x+2y)$ **17.** $\dfrac{a-6}{5}$

18. $\dfrac{3(y-2)^2}{4(2y-1)(y-1)}$ **19.** $-32t$ **20.** $\dfrac{2x(x-1)}{x+1}$

21. $\dfrac{(x^2+1)(2x+1)}{(x-2)(x+1)}$ **22.** $\dfrac{(t+4)^2}{t+1}$ **23.** $60a^5b^8$

24. $x^4(x-1)(x+1)$ **25.** $(y-2)(y+2)(y+1)$

26. $\dfrac{15-3x}{x+3}$ **27.** $\dfrac{4}{x-4}$ **28.** $\dfrac{x+5}{2x}$

29. $\dfrac{2a^2-21ab+15b^2}{5a^2b^2}$ **30.** $y-4$

31. $\dfrac{t(t-2)}{(t-1)(t+1)}$ **32.** $d+2$

33. $\dfrac{-x^2+x+26}{(x+1)(x-5)(x+5)}$ **34.** $\dfrac{2(x-2)}{x+2}$

35. $\dfrac{3(7t+2)}{4t(3t+2)}$ **36.** $\dfrac{z}{1-z}$ **37.** $\dfrac{10x}{3x^2+16}$

38. $c-d$ **39.** 4 **40.** $\frac{7}{2}$ **41.** $-6, -1$
42. $-1, 4$ **43.** $5\frac{1}{7}$ hr
44. Jennifer: 70 mph; Elizabeth: 62 mph
45. 55 seals **46.** 6 **47.** 72 radios
48. ☑ The LCM of denominators is used to clear fractions when simplifying a complex rational expression using the method of multiplying by the LCD, and when solving rational equations.

49. ☑ Although multiplying the denominators of the expressions being added results in a common denominator, it is often not the *least* common denominator. Using a common denominator other than the LCD makes the expressions more complicated, requires additional simplifying after the addition has been performed, and leaves more room for error.

50. $\dfrac{5(a+3)^2}{a}$ **51.** $\dfrac{10a}{(a-b)(b-c)}$ **52.** 0
53. 44%

Test: Chapter 6, p. 433

1. [6.1] 0 **2.** [6.1] 1, 2 **3.** [6.1] $\dfrac{3x+7}{x+3}$

4. [6.2] $\dfrac{2t(t+3)}{3(t-1)}$ **5.** [6.2] $\dfrac{(5y+1)(y+1)}{3y(y+2)}$

6. [6.2] $\dfrac{(2a+1)(4a^2+1)}{4a^2(2a-1)}$ **7.** [6.2] $(x+3)(x-3)$

8. [6.3] $(y-3)(y+3)(y+7)$

9. [6.3] $\dfrac{-3x+9}{x^3}$ **10.** [6.3] $\dfrac{-2t+8}{t^2+1}$ **11.** [6.4] 1

12. [6.4] $\dfrac{3x-5}{x-3}$ **13.** [6.4] $\dfrac{11t-8}{t(t-2)}$

14. [6.4] $\dfrac{y^2+3}{(y-1)(y+3)^2}$ **15.** [6.4] $\dfrac{x^2+2x-7}{(x+1)(x-1)^2}$

16. [6.5] $\dfrac{3y+1}{y}$ **17.** [6.5] $x-8$ **18.** [6.6] $\frac{8}{3}$

19. [6.6] $-3, 5$ **20.** [6.7] 12 min **21.** [6.7] $1\frac{1}{4}$ mi
22. [6.7] Ryan: 65 km/h; Alicia: 45 km/h
23. [6.5] a **24.** [6.7] -1

Cumulative Review: Chapters 1–6, p. 434

1. $a+cb$ **2.** -25 **3.** 25 **4.** $-3x+33$
5. 4 **6.** $-10, -1$ **7.** $-7, 7$ **8.** -8 **9.** 1, 4
10. -17 **11.** $-4, \frac{1}{2}$ **12.** $\{x \mid x > 43\}$, or $(43, \infty)$
13. -13 **14.** $b = 3a - c + 9$
15.

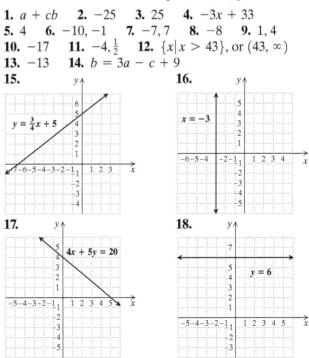

19. -2 **20.** Slope: $\frac{1}{2}$; y-intercept: $(0, -\frac{1}{4})$

21. $y = -\frac{5}{8}x - 4$ **22.** $\dfrac{1}{x^2}$ **23.** $-4a^4b^{14}$

24. $-y^3 - 2y^2 - 2y + 7$ **25.** $2x^5 + x^3 - 6x^2 - x + 3$

26. $36x^2 - 60xy + 25y^2$ **27.** $3n^2 - 13n - 10$

28. $4x^6 - 1$ **29.** $2x(3 - x - 12x^3)$

30. $(4x + 9)(4x - 9)$ **31.** $(t - 4)(t - 6)$

32. $(4x + 3)(2x + 1)$ **33.** $2(3x - 2)(x - 4)$

34. $(5t + 4)^2$ **35.** $(xy - 5)(xy + 4)$

36. $(x + 2)(x^3 - 3)$ **37.** $\dfrac{4}{t + 4}$ **38.** 1

39. $\dfrac{a^2 + 7ab + b^2}{(a + b)(a - b)}$ **40.** $\dfrac{2x + 5}{4 - x}$ **41.** $\dfrac{x}{x - 2}$

42. $15x^3 - 57x^2 + 177x - 529 + \dfrac{1605}{x + 3}$

43. 15 books   **44.** 12 ft   **45.** $\frac{75}{4}$ min, or $18\frac{3}{4}$ min

46. 150 calories   **47.** $-144, 144$ **48.** $-7, 4, 12$

49. 18

CHAPTER 7

Check Your Understanding, p. 438

1. (a)   **2.** (c)   **3.** (b)   **4.** (b)   **5.** (a)   **6.** (c)

Technology Connection, p. 439

1.

Exercise Set 7.1, pp. 440–441

1. Both   **2.** Intersect   **3.** Consistent
4. Dependent   **5.** Yes   **7.** No   **9.** Yes
11. $(3, 1)$ **13.** $(1, 3)$ **15.** $(-3, 5)$ **17.** No solution
19. $(4, -1)$ **21.** $(5, -2)$ **23.** $(-2, 3)$
25. Infinite number of solutions   **27.** $(-6, -2)$
29. Infinite number of solutions   **31.** $(2, -6)$
33. $(-8, -7)$ **35.** $(3, -4)$ **37.** $(2, 6)$
39. 2;   **41.** 5;

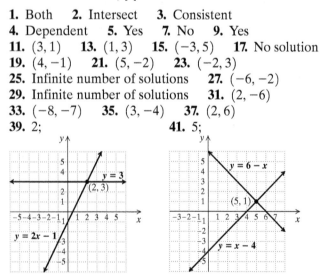

43. 2;   **45.** -4;

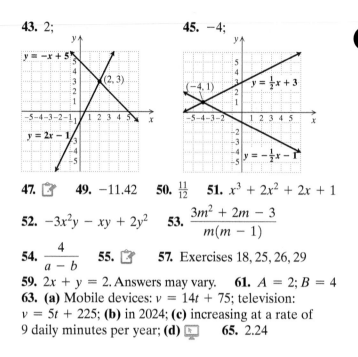

47. 🖊   **49.** -11.42 **50.** $\frac{11}{12}$ **51.** $x^3 + 2x^2 + 2x + 1$

52. $-3x^2y - xy + 2y^2$ **53.** $\dfrac{3m^2 + 2m - 3}{m(m - 1)}$

54. $\dfrac{4}{a - b}$ **55.** 🖊   **57.** Exercises 18, 25, 26, 29

59. $2x + y = 2$. Answers may vary.   **61.** $A = 2$; $B = 4$
63. (a) Mobile devices: $v = 14t + 75$; television:
$v = 5t + 225$; **(b)** in 2024; **(c)** increasing at a rate of
9 daily minutes per year; **(d)** 🖥   **65.** 2.24

Prepare to Move On, p. 441

1. $\frac{13}{5}$ **2.** 13   **3.** $\frac{12}{11}$ **4.** $\frac{7}{10}$ **5.** $y = \frac{3}{4}x - \frac{1}{2}$
6. $x = -\frac{3}{2}y - \frac{1}{2}$

Technology Connection, p. 443

1. 🔲   **2.** Both equations change: The first becomes
$y = 6 - \dfrac{x}{-2}$ and the second becomes $y = 4 - \dfrac{3x}{2}$.

Check Your Understanding, p. 445

1. $(8, 10)$ **2.** $-\frac{1}{2}$ **3.** There is no solution.
4. An infinite number of solutions

Exercise Set 7.2, pp. 446–448

1. False   **2.** True   **3.** True   **4.** True   **5.** $(4, 5)$
7. $(2, 1)$ **9.** $(1, 4)$ **11.** $(-20, 5)$ **13.** $(-3, 2)$
15. No solution   **17.** $(-1, -4)$ **19.** $\left(\frac{17}{3}, \frac{2}{3}\right)$
21. Infinite number of solutions   **23.** Infinite number
of solutions   **25.** $\left(-\frac{13}{22}, \frac{7}{11}\right)$ **27.** $(0, -6)$
29. $(10, 5)$ **31.** No solution   **33.** $(-3, -4)$
35. No solution   **37.** $28, 35$ **39.** $19, 32$ **41.** $8, 10$
43. $55°, 125°$ **45.** $28°, 62°$ **47.** Length: 166 m;
width: 28 m   **49.** Length: 365 mi; width: 275 mi
51. Length: 90 yd; width: 50 yd   **53.** Height: 20 ft;
width: 5 ft   **55.** Print books: 869 million; electronic
books: 82 million   **57.** 🖊   **59.** $-\frac{2}{5}$ **60.** 3.25
61. $2x^3 + x^2 - 6x - 3$ **62.** $a^6 + 2a^3b + b^2$
63. $x + 4 + \dfrac{30}{x - 5}$ **64.** $\dfrac{(x - 1)(x + 2)}{5x(x - 2)}$

65. **67.** $(7, -1)$ **69.** $(4.38, 4.33)$
71. 34 years **73.** $(30, 50, 100)$ **75.**

Quick Quiz: Sections 7.1–7.2, p. 448

1. Yes **2.** $(4, 1)$ **3.** No solution
4. $(-1, -3)$ **5.** Infinite number of solutions

Prepare to Move On, p. 448

1. $2x + y$ **2.** $-11x$ **3.** $-6y - 5$ **4.** $26x$

Check Your Understanding, p. 453

1. y **2.** -2 **3.** -2

Connecting the Concepts, p. 454

1. $(2, 2)$ **2.** $(4, 1)$ **3.** $\left(\frac{5}{3}, \frac{8}{3}\right)$ **4.** $\left(2, \frac{2}{3}\right)$
5. $(-2, -3)$ **6.** No solution

Exercise Set 7.3, pp. 454–456

1. Opposites **2.** Eliminate **3.** Solve
4. Variable **5.** Both **6.** Ordered **7.** $(8, 5)$
9. $(5, 1)$ **11.** $(2, -3)$ **13.** $(-1, 3)$ **15.** $\left(-1, \frac{1}{5}\right)$
17. Infinite number of solutions **19.** $(-2, -1)$
21. $(4, 5)$ **23.** $(3, 10)$ **25.** No solution
27. $\left(\frac{1}{2}, -\frac{1}{2}\right)$ **29.** $(-3, -1)$ **31.** No solution
33. $\left(\frac{1}{2}, 30\right)$ **35.** $(-2, 2)$ **37.** $(2, -1)$ **39.** $\left(\frac{231}{202}, \frac{117}{202}\right)$
41. $727\frac{8}{11}$ mi **43.** $32°, 58°$ **45.** 35,200 pages
47. $37°, 143°$ **49.** Riesling: 340 acres; Chardonnay:
480 acres **51.** Length: 6 ft; width: 3 ft **53.**
55. -8 **56.** $-\frac{1}{3}, \frac{1}{3}$ **57.** $\left\{a \mid a \geq -\frac{1}{9}\right\}$, or $\left[-\frac{1}{9}, \infty\right)$
58. $\frac{1}{3}, 5$ **59.** 1 **60.** $\frac{1}{2}, 1$ **61.** **63.** $(-1, 1)$
65. $(0, 4)$ **67.** $(0, 3)$ **69.** $x = \dfrac{-b - c}{a}; y = 1$
71. Rabbits: 12; pheasants: 23 **73.** Man: 45 years; his
daughter: 10 years

Quick Quiz: Sections 7.1–7.3, p. 456

1. $(2, -1)$ **2.** $\left(\frac{7}{4}, -\frac{1}{4}\right)$ **3.** $\left(\frac{5}{4}, -\frac{17}{16}\right)$ **4.** $\left(\frac{5}{7}, \frac{17}{7}\right)$
5. $(4, 1)$

Prepare to Move On, p. 456

1. 0.122 **2.** 0.005 **3.** 40% **4.** 3.06
5. Let $n =$ the number of liters; $0.12n$
6. Let $x =$ the number of pounds; $0.105x$

Check Your Understanding, p. 461

1. $50 **2.** $7x$ **3.** $4y$ **4.** No; $5.50/lb
5. More peanuts than raisins. The price per pound of
the mixture is closer to the price per pound of the peanuts than it is to the price per pound of the raisins.

Exercise Set 7.4, pp. 462–465

1. (b) **2.** (d) **3.** (c) **4.** (f) **5.** (a) **6.** (e)
7. Two-pointers: 9; three-pointers: 2 **9.** Two-pointers:
207; three-pointers: 56 **11.** 3-credit courses: 19;
4-credit courses: 8 **13.** 5-cent bottles or cans: 336;
10-cent bottles or cans: 94 **15.** 2220 motorcycles
17. 127 adult admissions **19.** 32 texts
21. Wrapped strings: 8; unwrapped strings: 24
23. Decaffeinated: $6\frac{1}{2}$ lb; regular: $1\frac{1}{2}$ lb
25. Cranberry juice cocktail: 12 L; sparkling water: 4 L
27.

Type of Solution	50%-Acid	80%-Acid	68%-Acid Mix
Amount of Solution	x	y	200
Percent Acid	50%	80%	68%
Amount of Acid in Solution	$0.5x$	$0.8y$	136

80 mL of 50%; 120 mL of 80% **29.** 128 L of 80%;
72 L of 30% **31.** 87-octane: 4 gal; 93-octane: 8 gal
33. 1300-word pages: 7; 1850-word pages: 5
35. Foul shots: 28; two-pointers: 36 **37.** Calm
Morning: 25 mL; Relaxed Evening: 75 mL
39. Breadshop Supernatural: 20 lb; New England
Natural Bakers Muesli: 20 lb **41.**

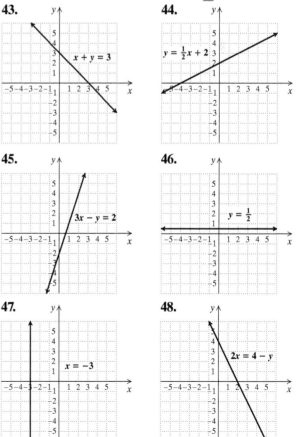

49. 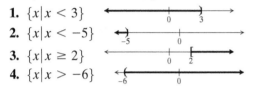 **51.** $19\frac{2}{7}$ lb **53.** 1.8 L **55.** 2.5 gal
57. $13 **59.** 10 workers at $20/hr; 5 workers
at $25/hr **61.** 54 **63.** Tweedledum: 120 lb;
Tweedledee: 121 lb

Quick Quiz: Sections 7.1–7.4, p. 465

1. $(3, -1)$ **2.** $\left(\frac{16}{5}, \frac{3}{5}\right)$ **3.** $\left(2, \frac{1}{2}\right)$ **4.** Length:
15 ft; width: 5 ft **5.** Vegetarian sandwiches: 8;
turkey sandwiches: 22

Prepare to Move On, p. 465

1. $\{x | x < 3\}$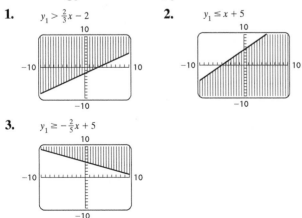
2. $\{x | x < -5\}$
3. $\{x | x \geq 2\}$
4. $\{x | x > -6\}$

Mid-Chapter Review: Chapter 7, p. 466

1.
$$x - y = 3$$
$$x - (2x + 1) = 3$$
$$x - 2x - 1 = 3$$
$$-x - 1 = 3$$
$$-x = 4$$
$$x = -4$$
$$y = 2x + 1$$
$$y = 2(-4) + 1$$
$$y = -7$$
The solution is $(-4, -7)$.

2.
$$8x - 11y = 12$$
$$\underline{3x + 11y = 10}$$
$$11x \qquad = 22$$
$$x = 2$$
$$3x + 11y = 10$$
$$3(2) + 11y = 10$$
$$6 + 11y = 10$$
$$11y = 4$$
$$y = \frac{4}{11}$$
The solution is $\left(2, \frac{4}{11}\right)$.

3. $(8, 5)$ **4.** No solution **5.** $\left(\frac{2}{3}, \frac{17}{3}\right)$ **6.** $\left(-\frac{1}{2}, \frac{13}{4}\right)$
7. $(-9, -5)$ **8.** No solution **9.** $\left(3, -\frac{5}{3}\right)$
10. $\left(\frac{11}{7}, \frac{13}{14}\right)$ **11.** Infinite number of solutions
12. $(-11, -17)$ **13.** $\left(\frac{5}{6}, 2\right)$ **14.** $(0, 0)$ **15.** $(10, 20)$
16. $(12, 8)$ **17.** $64°, 26°$ **18.** 2 months
19. Two-point baskets: 28; three-point baskets: 6
20. 10% pigment: 2 gal; 20% pigment: 6 gal

Technology Connection, p. 467

1. $y_1 > \frac{2}{3}x - 2$

2. $y_1 \leq x + 5$

3. $y_1 \geq -\frac{2}{5}x + 5$

Check Your Understanding, p. 470

1. (a) Dashed; **(b)** yes **2. (a)** Solid; **(b)** yes
3. (a) Dashed; **(b)** no **4. (a)** Dashed; **(b)** yes
5. (a) Solid; **(b)** no **6. (a)** Solid; **(b)** no

Exercise Set 7.5, pp. 470–471

1. True **2.** False **3.** False **4.** True
5. No **7.** Yes

9.
$y \leq x - 1$

11.
$y < x + 4$

13.
$y \geq x - 1$

15.
$y \leq 3x + 2$

17.
$x + y \leq 4$

19.
$x - y > 7$

21.
$y \geq 1 - 2x$

23.
$y + 2x > 0$

25.
$x \geq 4$

27.
$y > -1$

29.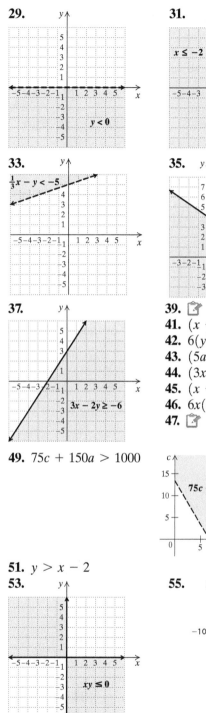

31. $x \le -2$

33. $\frac{1}{3}x - y < -5$

35. $2x + 3y \le 12$

37. $3x - 2y \ge -6$

39.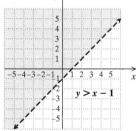

41. $(x - 8)(x + 5)$

42. $6(y^2 + 1)(y + 1)(y - 1)$

43. $(5a + 1)^2$

44. $(3x - 2)(2x + 1)$

45. $(x - 3)(x^2 + 1)$

46. $6x(6x^3 - 7x^2 + 12)$

47.

49. $75c + 150a > 1000$

$75c + 150a > 1000$

51. $y > x - 2$

53. $xy \le 0$

55. $y \le 4.9 - 3x$

Quick Quiz: Sections 7.1–7.5, p. 471

1. $(0, 1)$ **2.** $\left(\frac{4}{11}, \frac{6}{11}\right)$ **3.** $(1, 1)$ **4.** Walk: 8 mi; run: 12 mi   **5.**

$y > x - 1$

Prepare to Move On, p. 471

1. $y = \frac{1}{2}x - \frac{7}{4}$ **2.** Yes   **3.** No

Check Your Understanding, p. 474

1. Yes   **2.** No   **3.** Yes   **4.** Yes   **5.** No
6. No   **7.** No   **8.** Yes

Exercise Set 7.6, pp. 474–475

1. False   **2.** True   **3.** True   **4.** True

5.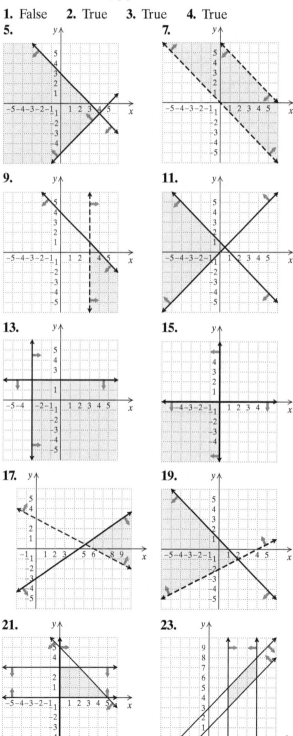

7.

9.

11.

13.

15.

17.

19.

21.

23.

25.

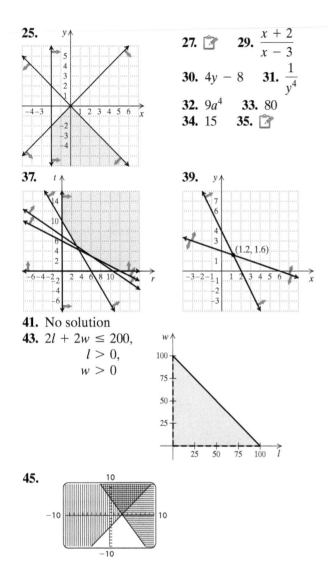

27. 📋 **29.** $\dfrac{x + 2}{x - 3}$

30. $4y - 8$ **31.** $\dfrac{1}{y^4}$

32. $9a^4$ **33.** 80

34. 15 **35.** 📋

37.

39.

$(1.2, 1.6)$

41. No solution

43. $2l + 2w \le 200,$
$\quad l > 0,$
$\quad w > 0$

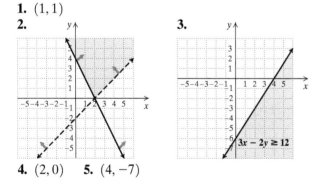

45.

Quick Quiz: Sections 7.1–7.6, p. 475

1. $(1, 1)$

2.

3.

$3x - 2y \ge 12$

4. $(2, 0)$ **5.** $(4, -7)$

Prepare to Move On, p. 475

1. 21 **2.** $\frac{4}{5}$ **3.** $\frac{3}{7}$ **4.** 40 **5.** 16

Check Your Understanding, p. 480

1. Inverse variation **2.** Direct variation
3. Direct variation **4.** Inverse variation
5. Inverse variation **6.** Direct variation

Exercise Set 7.7, pp. 480–482

1. (a) **2.** (f) **3.** (e) **4.** (c) **5.** (b) **6.** (d)
7. Inverse variation **8.** Direct variation **9.** Direct
variation **10.** Inverse variation **11.** $y = 5x$

13. $y = 7x$ **15.** $y = \frac{3}{5}x$ **17.** $y = \frac{2}{3}x$ **19.** $y = \dfrac{120}{x}$

21. $y = \dfrac{1}{x}$ **23.** $y = \dfrac{20}{x}$ **25.** $y = \dfrac{210}{x}$ **27.** \$207

29. 27.75 kWh **31.** 99.86% gold **33.** 2 hr
35. 1.6 ft **37.** 20 people **39.** 📋 **41.** 📋

43. 200 **44.** $c = \dfrac{3a}{b}$ **45.** $-\frac{13}{2}$ **46.** $y = -\frac{2}{5}x + 6$

47. 3.07×10^{-3} **48.** $a^2(a + 1)(a - 1)^2$ **49.** 📋

51. $P = kv^3$ **53.** $S = kv^6$ **55.** $N = k \cdot \dfrac{T}{P}$

57. $P = 8S$ **59.** $A = \pi r^2$

Quick Quiz: Sections 7.1–7.7, p. 482

1. $\left(\frac{23}{11}, -\frac{13}{11}\right)$ **2.** $\left(\frac{13}{7}, \frac{9}{14}\right)$ **3.** 10% peanuts: $7\frac{1}{2}$ lb;
50% peanuts: $2\frac{1}{2}$ lb

4.

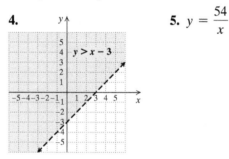

$y > x - 3$

5. $y = \dfrac{54}{x}$

Prepare to Move On, p. 482

1. 100 **2.** 100 **3.** $25x^2$ **4.** $s^6 t^8$ **5.** $x^2 + 2xy + y^2$
6. $4a^2 - 4ab + b^2$

Visualizing for Success, p. 483

1. G **2.** B **3.** F **4.** H **5.** C **6.** D **7.** A
8. J **9.** E **10.** I

Decision Making: Connection, p. 484

1. \$2125/month **2.** \$40/month **3.** Cobblefield
Downs: $c = 940t + 100$, Grassymeadow Park:
$c = 700t + 1180$, where c is the cost, in dollars, after
t months rental; 4.5 months. For the first 4 months,
Cobblefield Downs will be less expensive; for 5 months
and longer, Grassymeadow Park will be less expensive.
4. 💻

Study Summary: Chapter 7, pp. 485–487

1. $(6, 0)$ **2.** $(1, 1)$ **3.** $\left(\frac{17}{16}, -\frac{5}{8}\right)$ **4.** 40%-acid: $\frac{4}{5}$ L;
15%-acid: $\frac{6}{5}$ L

5.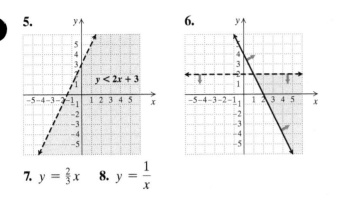

$y < 2x + 3$

6.

7. $y = \frac{2}{3}x$ **8.** $y = \frac{1}{x}$

Review Exercises: Chapter 7, pp. 487–489

1. Ordered   **2.** Infinite   **3.** Parallel   **4.** Parentheses
5. Coefficients   **6.** Below   **7.** Intersection
8. Inversely   **9.** No   **10.** Yes   **11.** $(1, 3)$
12. $(6, -2)$ **13.** No solution   **14.** Infinite number of
solutions   **15.** $(-5, 9)$ **16.** $\left(\frac{10}{3}, \frac{4}{3}\right)$ **17.** No solution
18. $(-3, 9)$ **19.** $(10, 15)$ **20.** $(5, -3)$ **21.** Infinite
number of solutions   **22.** $\left(-\frac{1}{2}, \frac{1}{3}\right)$ **23.** $(-1, -6)$
24. $(3, 2)$ **25.** $\left(4, -\frac{1}{2}\right)$ **26.** Infinite number of
solutions   **27.** $\left(\frac{18}{7}, \frac{4}{7}\right)$ **28.** No solution
29. Length: 25 ft; width: 8 ft   **30.** Two-pointers: 9;
three-pointers: 1   **31.** $250-tablets: 11; $350-tablets: 24
32. Café Rich: $66\frac{2}{3}$ g; Café Light: $133\frac{1}{3}$ g

33.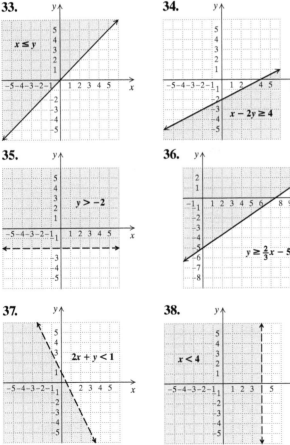

$x \le y$

34.

$x - 2y \ge 4$

35.

$y > -2$

36.

$y \ge \frac{2}{3}x - 5$

37.

$2x + y < 1$

38.

$x < 4$

39.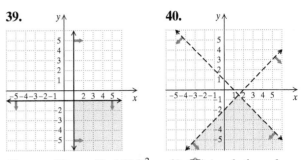

40.

41. $y = 27x$ **42.** 150 ft^2 **43.** 📋 A solution of a
system of two equations is an ordered pair that makes
both equations true. The graph of an equation represents
all ordered pairs that make that equation true. So in
order for an ordered pair to make *both* equations true,
it must be on *both* graphs.   **44.** 📋 The solution sets of
linear inequalities are regions, not lines. Thus the solution
sets can intersect even if the boundary lines do not.
45. $C = 1, D = 3$ **46.** $(2, 1, -2)$ **47.** $(2, 0)$
48. 24   **49.** $1800

Test: Chapter 7, p. 489

1. [7.1] Yes   **2.** [7.1] $(1, 3)$ **3.** [7.1] No solution
4. [7.2] $(-3, 5)$ **5.** [7.2] $\left(\frac{1}{2}, 3\right)$ **6.** [7.3] $(3, 2)$
7. [7.3] $(12, -6)$ **8.** [7.3] $(0, 1)$ **9.** [7.1], [7.2], [7.3]
Infinite number of solutions   **10.** [7.2], [7.3] $\left(-\frac{1}{5}, \frac{14}{5}\right)$
11. [7.2] $23°, 67°$ **12.** [7.4] 40%: 20 L; 25%: 40 L
13. [7.4] $\frac{3}{8}$ mi

14. [7.5]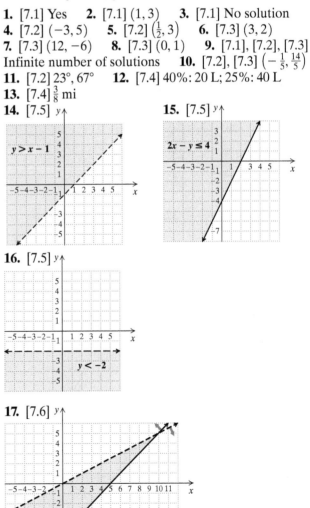

$y > x - 1$

15. [7.5]

$2x - y \le 4$

16. [7.5]

$y < -2$

17. [7.6]

18. [7.6]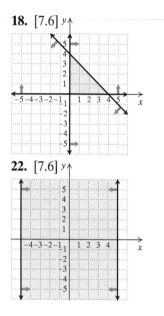

19. [7.7] $y = \dfrac{18}{x}$

20. [7.7] \$1330

21. [7.4] 9 people

22. [7.6]

23. [7.1] $C = -\dfrac{19}{2}, D = \dfrac{14}{3}$

36.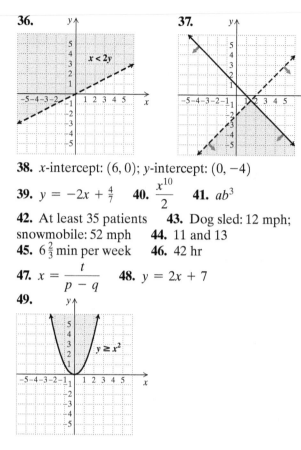

37.

38. x-intercept: $(6, 0)$; y-intercept: $(0, -4)$

39. $y = -2x + \dfrac{4}{7}$ **40.** $\dfrac{x^{10}}{2}$ **41.** ab^3

42. At least 35 patients **43.** Dog sled: 12 mph; snowmobile: 52 mph **44.** 11 and 13

45. $6\frac{2}{3}$ min per week **46.** 42 hr

47. $x = \dfrac{t}{p - q}$ **48.** $y = 2x + 7$

49.

Cumulative Review: Chapters 1–7, p. 490

1. 24 **2.** $13x^2 - y$ **3.** -24 **4.** 2 **5.** $\frac{9}{7}$

6. 2, 3 **7.** $\left\{ y \mid y \geq -\frac{2}{3} \right\}$ **8.** $-10, 10$ **9.** $(1, 2)$

10. -11 **11.** $-\frac{1}{2}, \frac{2}{3}$ **12.** $\left(\frac{9}{11}, \frac{17}{22} \right)$ **13.** $p = \dfrac{3t}{q}$

14. $s = 2A - r$ **15.** $3a^3 + 8a^2 - 14$ **16.** 3

17. $\dfrac{m(m + 2n)}{(m - n)(2m + n)}$ **18.** $-3a^3 + 8a^2 - 12a$

19. $\dfrac{-6p}{p + q}$ **20.** $\dfrac{2x + 4}{x - 2}$ **21.** $25a^4 - b^2$

22. $4n^2 + 20n + 25$ **23.** $8t^5 + 5t^3 + 32t^2 + 20$

24. $\dfrac{(x - 1)(x + 2)}{(x - 2)^2}$ **25.** $2x + 1$

26. $\dfrac{x}{2(x + 1)(x + 2)}$ **27.** $(x + 1)(x^2 + 2)$

28. $(m^2 + 1)(m + 1)(m - 1)$ **29.** $2x(x + 4)(x + 5)$

30. $(x^2 - 3)^2$ **31.** $(2x - 5)(5x - 2)$

32.

33.

34.

35.

CHAPTER 8

Technology Connection, p. 493

1. $\sqrt{8} \approx 2.828427125$; $\sqrt{11} \approx 3.31662479$; $\sqrt{48} \approx 6.92820323$

Technology Connection, p. 494

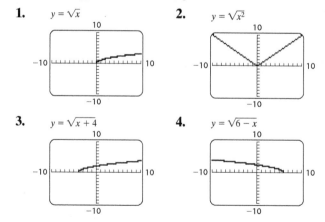

1. $y = \sqrt{x}$

2. $y = \sqrt{x^2}$

3. $y = \sqrt{x + 4}$

4. $y = \sqrt{6 - x}$

Check Your Understanding, p. 495

1. True **2.** False **3.** True **4.** True **5.** False

6. True **7.** False **8.** True

Exercise Set 8.1, pp. 496–498

1. (c) **2.** (e) **3.** (a) **4.** (f) **5.** (b) **6.** (d)
7. True **8.** False **9.** False **10.** True
11. $-10, 10$ **13.** $-6, 6$ **15.** $-1, 1$ **17.** $-12, 12$
19. 10 **21.** -1 **23.** 0 **25.** -11 **27.** 30
29. -12 **31.** $10a$ **33.** $t^3 - 2$ **35.** $\dfrac{7}{x+y}$
37. Rational **39.** Irrational **41.** Irrational
43. Rational **45.** Rational **47.** 2.828 **49.** 3.873
51. 9.110 **53.** $|x|$ **55.** $|10x|$, or $10|x|$
57. $|x+7|$ **59.** $|5-2x|$ **61.** x **63.** $5y$ **65.** $4t$
67. $n+7$ **69.** (a) 15; (b) 14 **71.** 0.870 sec
73. ☐ **75.** $\frac{3}{4}$ **76.** $\frac{20}{3}$ **77.** $\{n \mid n > \frac{19}{8}\}$, or $\left(\frac{19}{8}, \infty\right)$
78. $-\frac{5}{2}, 3$ **79.** $\left(\frac{11}{3}, \frac{1}{3}\right)$ **80.** $-\frac{1}{2}, 1$ **81.** ☐
83. $\frac{1}{10}$ **85.** 7 **87.** 5 **89.** $-6, -5$ **91.** $-4, 4$
93. $-3, 3$ **95.** $\{x \mid x < 10\}$ **97.** $\dfrac{2x^4}{y^3}$ **99.** $\dfrac{20}{m^8}$
101. (a) 1.7; (b) 2.2; (c) 2.6 **103.** 1097.1 ft/sec
105.
$y_1 = \sqrt{x-2};\ y_2 = \sqrt{x+7};$
$y_3 = 5 + \sqrt{x};\ y_4 = -4 + \sqrt{x}$
107. $\{a \mid a < 1\}$

Prepare to Move On, p. 498

1. t^{20} **2.** $50a^{13}$ **3.** $45x^7$ **4.** $300m^9$ **5.** $50y^{15}$

Check Your Understanding, p. 501

1. $2\sqrt{3}$ **2.** $5\sqrt{2}$ **3.** $3\sqrt{7}$ **4.** $4\sqrt{x}$ **5.** $5y$
6. $x\sqrt{a}$ **7.** $2x\sqrt{2x}$ **8.** $5ab\sqrt{3b}$

Exercise Set 8.2, pp. 502–504

1. True **2.** True **3.** False **4.** True **5.** (f)
6. (h) **7.** (d) **8.** (a) **9.** (g) **10.** (c) **11.** (b)
12. (e) **13.** $\sqrt{10}$ **15.** $\sqrt{12}$, or $2\sqrt{3}$
17. $\sqrt{\frac{21}{40}}$, or $\frac{1}{2}\sqrt{\frac{21}{10}}$ **19.** 10 **21.** $\sqrt{75}$, or $5\sqrt{3}$
23. $\sqrt{2x}$ **25.** $\sqrt{14a}$ **27.** $\sqrt{21xy}$ **29.** $\sqrt{6abc}$
31. $2\sqrt{3}$ **33.** $5\sqrt{3}$ **35.** $10\sqrt{5}$ **37.** $4\sqrt{t}$
39. $2\sqrt{5z}$ **41.** $10y$ **43.** $x\sqrt{13}$ **45.** $3b\sqrt{3}$
47. $12x\sqrt{y}$ **49.** $2\sqrt{6}$ **51.** 0 **53.** $2\sqrt{21}$ **55.** a^9
57. x^8 **59.** $r^2\sqrt{r}$ **61.** $t^7\sqrt{t}$ **63.** $2a\sqrt{10a}$
65. $3x^2\sqrt{5x}$ **67.** $10p^{12}\sqrt{2p}$ **69.** $2\sqrt{5}$ **71.** 9
73. $6\sqrt{xy}$ **75.** $17\sqrt{x}$ **77.** $10b\sqrt{5}$ **79.** $12t$
81. $a\sqrt{bc}$ **83.** $x^8\sqrt{x}$ **85.** $7m^3\sqrt{2}$ **87.** $xy^3\sqrt{xy}$
89. $6ab^3\sqrt{2a}$ **91.** 2000 gal/min; 4000 gal/min
93. $65.7\sqrt{318}$ ft/sec; 1172 ft/sec **95.** ☐ **97.** $\frac{3}{4}$
98. $\frac{3}{2}$ **99.** $\dfrac{19x^2y^6}{17}$ **100.** $\dfrac{3x-5}{x+1}$ **101.** 126
102. a **103.** ☐ **105.** 0.1 **107.** 0.25 **109.** $<$

111. $=$ **113.** $>$ **115.** $18(x+1)\sqrt{y(x+1)}$
117. $2x^7\sqrt{5x}$ **119.** $\sqrt{14x^{26}}$ **121.** x^{8n} **123.** $y^{n/2}$

Quick Quiz: Sections 8.1–8.2, p. 504

1. -6 **2.** $|x+5|$ **3.** $7y$ **4.** $3m\sqrt{2}$ **5.** $x+5$

Prepare to Move On, p. 504

1. x^3y^5 **2.** $36n^2$ **3.** $9x^4$ **4.** $\dfrac{21a^2}{16b^2}$ **5.** $\dfrac{2r^3}{7t}$

Check Your Understanding, p. 506

1. $\dfrac{\sqrt{7}}{\sqrt{7}}$ **2.** $\dfrac{\sqrt{x}}{\sqrt{x}}$ **3.** $\dfrac{\sqrt{5}}{\sqrt{5}}$ **4.** $\dfrac{\sqrt{3}}{\sqrt{3}}$ **5.** $\dfrac{\sqrt{b}}{\sqrt{b}}$

Exercise Set 8.3, pp. 507–509

1. True **2.** True **3.** False **4.** False **5.** 10
7. 2 **9.** $\sqrt{11}$ **11.** $\frac{1}{2}$ **13.** $\frac{3}{4}$ **15.** 2 **17.** $3y$
19. $10a^4$ **21.** $a^3\sqrt{3}$ **23.** $\frac{2}{5}$ **25.** $\frac{7}{4}$ **27.** $-\frac{5}{9}$
29. $\dfrac{a^2}{5}$ **31.** $\dfrac{x^3\sqrt{3}}{4}$ **33.** $\dfrac{t^3\sqrt{3}}{2}$ **35.** $\dfrac{\sqrt{3}}{3}$ **37.** $\dfrac{5\sqrt{7}}{7}$
39. $\dfrac{4\sqrt{3}}{9}$ **41.** $\dfrac{\sqrt{30}}{5}$ **43.** $\dfrac{\sqrt{6}}{10}$ **45.** $\dfrac{\sqrt{10a}}{15}$
47. $\dfrac{2\sqrt{15}}{5}$ **49.** $\dfrac{\sqrt{7z}}{z}$ **51.** $\dfrac{\sqrt{2a}}{20}$ **53.** $\dfrac{\sqrt{10x}}{30}$
55. $\dfrac{\sqrt{3a}}{5}$ **57.** $\dfrac{x\sqrt{15}}{6}$ **59.** $9185 **61.** $12,901
63. 6.28 sec; 7.85 sec **65.** 1 sec **67.** 9.42 sec
69. ☐ **71.** $16t^2 - 9$ **72.** $x^6 - 14x^3 + 49$
73. $81n^2 - 9n - 2$ **74.** $3(x-1)^2$
75. $(10+y)(10-y)$ **76.** $6x(2x^2 - x - 2)$
77. ☐ **79.** $\dfrac{\sqrt{70}}{100}$ **81.** $\dfrac{\sqrt{6xy}}{4x^3y^2}$ **83.** $\dfrac{\sqrt{5z}}{5wz}$
85. $\dfrac{2\sqrt{5}\sqrt{5}}{5}$ **87.** $\dfrac{1}{x} - \dfrac{1}{y}$ **89.** $\frac{53}{2}$

Quick Quiz: Sections 8.1–8.3, p. 509

1. 5.568 **2.** $10c$ **3.** $2n^2\sqrt{3n}$ **4.** $\dfrac{a^5\sqrt{15}}{4}$
5. $\dfrac{\sqrt{14}}{2}$

Prepare to Move On, p. 509

1. $15x$ **2.** $17y - z$ **3.** $12n^3 + 24n^2 + 4n$
4. $4x^2 - 9$ **5.** $25t^2 - 49$

Check Your Understanding, p. 510

1. Yes **2.** No **3.** Yes **4.** Yes **5.** No
6. No **7.** Yes

Technology Connection, p. 511

1. -0.086 **2.** 1.217 **3.** 2.765

Exercise Set 8.4, pp. 512–513

1. Like radicals **2.** Conjugates **3.** Conjugates
4. Like radicals **5.** (b) **6.** (c) **7.** (d) **8.** (a)
9. $11\sqrt{10}$ **11.** $3\sqrt{2}$ **13.** $13\sqrt{t}$ **15.** $-\sqrt{x}$
17. $8\sqrt{2a}$ **19.** $8\sqrt{10y}$ **21.** $12\sqrt{7}$ **23.** $4\sqrt{2}$
25. $5\sqrt{3} + 2\sqrt{2}$ **27.** $-3\sqrt{x}$ **29.** $-18\sqrt{3}$ **31.** $28\sqrt{2}$
33. $13\sqrt{2}$ **35.** $6\sqrt{2}$ **37.** $2\sqrt{2}$ **39.** $5\sqrt{a}$
41. $\sqrt{6} + \sqrt{21}$ **43.** $\sqrt{30} - 5\sqrt{2}$ **45.** $14 + 7\sqrt{2}$
47. $17 - 7\sqrt{7}$ **49.** -19 **51.** 4 **53.** $7\sqrt{2}$
55. $52 + 14\sqrt{3}$ **57.** $13 - 4\sqrt{3}$ **59.** $x - 2\sqrt{10x} + 10$
61. $\dfrac{10 - 2\sqrt{2}}{23}$ **63.** $-1 - \sqrt{3}$ **65.** $\dfrac{5 - \sqrt{7}}{9}$
67. $\dfrac{-5 + 2\sqrt{10}}{3}$ **69.** $\dfrac{7 + \sqrt{21}}{4}$ **71.** $\dfrac{\sqrt{15} + 3}{2}$
73. $\dfrac{2\sqrt{7} - 2\sqrt{2}}{5}$ **75.** $\dfrac{6 - 2\sqrt{6x} + x}{6 - x}$ **77.** ☐
79. x-axis **80.** $-\frac{1}{2}$ **81.** -1 **82.** $y = -x - 6$
83. $y = 6x - 4$ **84.** $y = \frac{1}{3}x + \frac{10}{3}$ **85.** ☐
87. $\dfrac{11\sqrt{x} - 5x\sqrt{2}}{2x}$ **89.** $xy\sqrt{2y}\,(2x^2 - y^2 - xy^3)$
91. $37xy\sqrt{3x}$ **93.** All three are correct.

Quick Quiz: Sections 8.1–8.4, p. 513

1. $-\sqrt{3}$ **2.** $7\sqrt{2}$ **3.** $3x^2\sqrt{10}$ **4.** $2\sqrt{5} + \sqrt{10}$
5. $2 - \sqrt{3} + 2\sqrt{5} - \sqrt{15}$

Prepare to Move On, p. 513

1. $\frac{5}{11}$ **2.** $-\frac{3}{2}$ **3.** $-1, 6$ **4.** $2, 5$
5. 2008: \$2835; 2013: \$2482

Mid-Chapter Review: Chapter 8, p. 514

1. $\sqrt{6x^3} \cdot \sqrt{12xy} = \sqrt{6 \cdot 12 \cdot x^4 \cdot y} = \sqrt{6 \cdot 6 \cdot 2 \cdot x^4 \cdot y}$
$= \sqrt{6^2} \cdot \sqrt{x^4} \cdot \sqrt{2y} = 6x^2\sqrt{2y}$
2. $\sqrt{\dfrac{7}{n}} = \dfrac{\sqrt{7}}{\sqrt{n}} = \dfrac{\sqrt{7}}{\sqrt{n}} \cdot \dfrac{\sqrt{n}}{\sqrt{n}} = \dfrac{\sqrt{7n}}{n}$
3. $-11, 11$ **4.** xy **5.** $10t$ **6.** $x\sqrt{17}$ **7.** $\frac{2}{3}$
8. $5\sqrt{15t}$ **9.** $6\sqrt{15}$ **10.** $2\sqrt{15} - 3\sqrt{22}$
11. $-8\sqrt{3}$ **12.** $\dfrac{a^3}{2}$ **13.** -4 **14.** $2a^9\sqrt{7a}$
15. $30x^4y^6$ **16.** $5\sqrt{2} + 3\sqrt{5} + \sqrt{10} + 3$ **17.** $15\sqrt{5}$
18. $8m^8$ **19.** $\dfrac{x\sqrt{2}}{4}$ **20.** $\sqrt{3} + \sqrt{2}$

Check Your Understanding, p. 516

1. Yes **2.** No **3.** No **4.** No **5.** Yes

Technology Connection, p. 517

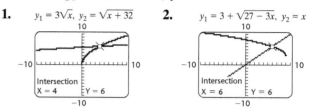

1. $y_1 = 3\sqrt{x},\ y_2 = \sqrt{x + 32}$

2. $y_1 = 3 + \sqrt{27 - 3x},\ y_2 = x$

Connecting the Concepts, p. 518

1. $5a$ **2.** 36 **3.** $\dfrac{\sqrt{3}}{2}$ **4.** -2 **5.** $1 + \sqrt{3}$
6. 0 **7.** $4x^3y^5\sqrt{5x}$ **8.** 3

Exercise Set 8.5, pp. 519–521

1. True **2.** False **3.** False **4.** True **5.** 36
7. 144 **9.** 21 **11.** -46 **13.** $\frac{25}{3}$ **15.** 2
17. No solution **19.** -4 **21.** 8 **23.** 12
25. $11, 23$ **27.** 3 **29.** $0, 8$ **31.** No solution **33.** 5
35. $-\frac{5}{4}$ **37.** 1 **39.** $\frac{7}{2}, 4$ **41.** 54 ft; 150 ft **43.** $25.2°$F
45. 256 m **47.** 400 m **49.** 15.71 ft **51.** ☐
53.

$y = 2x - 1$

54.

$2x - 3y = 12$

55.

$x = 3$

56.

$x = 1 - y$

57.

$y < \frac{1}{3}x - 2$

58.

59. ☐ **61.** 121 **63.** 16 **65.** No solution **67.** 3
69. 10 **71.** $34\frac{569}{784}$ m

73. 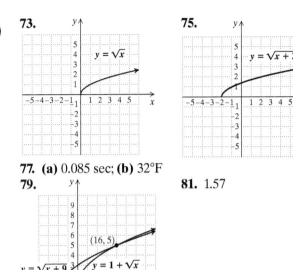   **75.**

77. (a) 0.085 sec; **(b)** 32°F

79.   **81.** 1.57

The solution is 16.

Quick Quiz: Sections 8.1–8.5, p. 521

1. $\dfrac{\sqrt{2x}}{x}$ **2.** $\dfrac{3 - \sqrt{5}}{2}$ **3.** $-12, 12$ **4.** 12   **5.** 11

Prepare to Move On, p. 521

1. $-5, 5$ **2.** $-4, 3$ **3.** 13   **4.** 5   **5.** 50

Check Your Understanding, p. 525

1. $a^2 + \left(\sqrt{3}\right)^2 = (7)^2$

2. $d = \sqrt{(3 - 2)^2 + (-4 - 7)^2}$

Exercise Set 8.6, pp. 526–529

1. Hypotenuse   **2.** Legs   **3.** Pythagorean
4. Distance   **5.** 17   **7.** $\sqrt{98}$, or $7\sqrt{2}$; approximately
9.899   **9.** 12   **11.** $\sqrt{31} \approx 5.568$ **13.** 13   **15.** 12
17. 3   **19.** $\sqrt{2} \approx 1.414$ **21.** 5   **23.** 13 ft
25. $\sqrt{189}$ in. ≈ 13.748 in.   **27.** $\sqrt{208}$ ft ≈ 14.422 ft
29. $\sqrt{16,200}$ ft ≈ 127.279 ft
31. $\sqrt{15,700}$ yd ≈ 125.300 yd
33. $\sqrt{4500}$ cm ≈ 67.082 cm   **35.** $\sqrt{448}$ ft ≈ 21.166 ft
37. $\sqrt{65} \approx 8.062$ **39.** 5   **41.** $\sqrt{13} \approx 3.606$ **43.** 10
45. 🖘   **47.** $a^3 - 7a^2 + 2a - 7$ **48.** $t^2 - t + \dfrac{1}{t + 1}$
49. $\dfrac{(x + 1)(x - 2)}{x^2(x - 1)}$ **50.** $(x + 2)(x - 3)$
51. $\dfrac{-5x - 6}{x(x + 1)}$ **52.** $\dfrac{7x + 11}{(x - 2)(x + 2)(2x + 3)}$ **53.** 🖘
55. 5 68-ft strands for $59.95   **57.** Yes; the distance
between the windows is $\sqrt{17,025}$ ft, or about 130 ft.
59. 8 ft   **61.** 6, 8, 10   **63.** $s\sqrt{3}$ **65.** $\dfrac{\sqrt{3}a}{2}$
67. $\sqrt{1525}$ mi ≈ 39.051 mi   **69.** 640 acres

Quick Quiz: Sections 8.1–8.6, p. 529

1. t^7 **2.** $19 + 6\sqrt{10}$ **3.** $2x\sqrt{5}$ **4.** No solution
5. $\sqrt{34} \approx 5.831$

Prepare to Move On, p. 529

1. -32 **2.** $\frac{8}{27}$ **3.** 81   **4.** $\frac{1}{16}$ **5.** -64

Check Your Understanding, p. 533

1. $\sqrt[3]{64}$ **2.** $\sqrt{64}$ **3.** $\left(\sqrt{16}\right)^3$, or $\sqrt{16^3}$
4. $\left(\sqrt[4]{16}\right)^3$, or $\sqrt[4]{16^3}$ **5.** $\dfrac{1}{\sqrt{100}}$

Exercise Set 8.7, pp. 533–534

1. True   **2.** True   **3.** True   **4.** False   **5.** False
6. True   **7.** -2 **9.** -10 **11.** 5   **13.** -6
15. 5   **17.** 0   **19.** -1 **21.** Not a real number
23. 10   **25.** 6   **27.** 1   **29.** a **31.** $3\sqrt[3]{2}$
33. $2\sqrt[4]{3}$ **35.** $\frac{4}{5}$ **37.** $\frac{2}{3}$ **39.** $\dfrac{\sqrt[3]{7}}{2}$ **41.** $\dfrac{\sqrt[4]{14}}{3}$
43. 7   **45.** 10   **47.** 2   **49.** 8   **51.** 64   **53.** 16
55. 10,000   **57.** 100,000   **59.** $\frac{1}{3}$ **61.** $\frac{1}{4}$ **63.** $\frac{1}{8}$
65. $\frac{1}{243}$ **67.** $\frac{1}{25}$ **69.** 🖘   **71.** $a = 10c - b$ **72.** 5
73. 0.05   **74.** $\frac{1}{6}$ **75.** $t \le -6$ **76.** 0.004   **77.** 🖘
79. 5.278   **81.** $a^{7/4}$ **83.** $m^{13/12}$
85.   **87.** 80 calories

Quick Quiz: Sections 8.1–8.7, p. 534

1. $3x$ **2.** Irrational   **3.** $\sqrt{128} \approx 11.314$ **4.** 2
5. 10

Prepare to Move On, p. 534

1. $-4, -3$ **2.** $-1, 7$ **3.** $-\frac{3}{4}, \frac{3}{4}$ **4.** $0, \frac{5}{6}$ **5.** $-\frac{5}{3}, 2$

Translating for Success, p. 535

1. J   **2.** K   **3.** N   **4.** H   **5.** G   **6.** E   **7.** O
8. D   **9.** B   **10.** C

Decision Making: Connection, p. 536

1. First TV   **2. (a)** 6, 8, 10; 8, 15, 17; **(b)** 🖘   **3.** 🖥
4. 🖥

Study Summary: Chapter 8, pp. 537–539

1. $-11, 11$ **2.** 3 **3.** Irrational   **4.** $5n$ **5.** $\sqrt{30ny}$

6. $2\sqrt{17}$ **7.** m^8 **8.** $10a^2\sqrt{3b}$ **9.** $\dfrac{\sqrt{7}}{x}$ **10.** $\dfrac{x^{15}}{10}$

11. $\dfrac{\sqrt{3x}}{3}$ **12.** $-5\sqrt{2}$ **13.** $2 + 2\sqrt{3} - \sqrt{6} - 3\sqrt{2}$

14. $-5 + 5\sqrt{2}$ **15.** $\frac{95}{2}$ **16.** $b = \sqrt{39}; b \approx 6.245$

17. $\sqrt{245} \approx 15.652$ **18.** 3 **19.** -10 **20.** 5

21. $\frac{1}{8}$

Review Exercises: Chapter 8, pp. 539–541

1. False   **2.** True   **3.** True   **4.** True   **5.** False
6. False   **7.** False   **8.** True   **9.** True   **10.** True
11. $4, -4$ **12.** $20, -20$ **13.** 12 **14.** -13

15. $5x^3y$ **16.** $\dfrac{a}{b}$ **17.** Rational   **18.** Irrational

19. Irrational   **20.** Rational   **21.** 2.236 **22.** 9.487
23. $|5x|$, or $5|x|$ **24.** $|x + 2|$ **25.** p **26.** $3x$
27. $7n$ **28.** ac **29.** $4\sqrt{3}$ **30.** $10t\sqrt{3}$ **31.** $4\sqrt{2p}$
32. x^8 **33.** $2a^6\sqrt{3a}$ **34.** $6m^7\sqrt{m}$ **35.** $\sqrt{55}$
36. $2\sqrt{15}$ **37.** $\sqrt{21st}$ **38.** $2a\sqrt{6}$ **39.** $5xy\sqrt{2}$

40. $10a^2b\sqrt{ab}$ **41.** $\dfrac{\sqrt{7}}{3}$ **42.** $\frac{7}{8}$ **43.** $\frac{2}{3}$ **44.** $\dfrac{8}{t^3}$

45. $13\sqrt{5}$ **46.** $\sqrt{5}$ **47.** $-3\sqrt{x}$ **48.** $7 + 4\sqrt{3}$

49. -6 **50.** $-11 + 5\sqrt{7}$ **51.** $\dfrac{\sqrt{5}}{5}$ **52.** $\dfrac{\sqrt{10}}{4}$

53. $\dfrac{\sqrt{7y}}{y}$ **54.** $\dfrac{2\sqrt{3}}{3}$ **55.** $8 - 4\sqrt{3}$

56. $-7 - 3\sqrt{5}$ **57.** 54 **58.** No solution   **59.** 4
60. $0, 3$ **61.** 35.344 ft   **62.** 20 **63.** $\sqrt{3} \approx 1.732$
64. $48\sqrt{2}$ ft ≈ 67.882 ft   **65.** 5 **66.** -2
67. Not a real number   **68.** $\frac{2}{3}$ **69.** 10 **70.** 9
71. $\frac{1}{10}$ **72.** 125 **73.** $\frac{1}{16}$ **74.** ☞ Absolute-value signs may be necessary when simplifying a radical expression with an even index. For n an even number, if it is possible that A is negative, then $\sqrt[n]{A^n} = |A|$.
75. ☞ Some radical terms that are like terms may not appear to be so until they are in simplified form.
76. 2 **77.** No solution   **78.** $(x + \sqrt{5})(x - \sqrt{5})$

Test: Chapter 8, p. 541

1. [8.1] $7, -7$ **2.** [8.1] 4 **3.** [8.1] -5 **4.** [8.1] $t^2 + 1$
5. [8.1] Irrational   **6.** [8.1] Rational   **7.** [8.1] 1.414
8. [8.1] a **9.** [8.1] $8y$ **10.** [8.2] $2\sqrt{15}$
11. [8.2] $3x^3\sqrt{3}$ **12.** [8.2] $6t^5\sqrt{t}$ **13.** [8.2] $\sqrt{30}$
14. [8.2] $5\sqrt{3}$ **15.** [8.2] $\sqrt{14xy}$ **16.** [8.2] $4t$

17. [8.2] $3ab^2\sqrt{2}$ **18.** [8.3] $\frac{2}{3}$ **19.** [8.3] $\dfrac{\sqrt{7}}{4y}$

20. [8.3] $\dfrac{12}{a}$ **21.** [8.4] $-6\sqrt{2}$ **22.** [8.4] $7\sqrt{3}$

23. [8.4] $21 - 8\sqrt{5}$ **24.** [8.4] 11 **25.** [8.3] $\dfrac{\sqrt{10}}{5}$

26. [8.3] $\dfrac{2x\sqrt{y}}{y}$ **27.** [8.4] $\dfrac{40 + 10\sqrt{5}}{11}$

28. [8.6] $\sqrt{13} \approx 3.606$ **29.** [8.5] 50 **30.** [8.5] $-2, 2$
31. [8.6] $\sqrt{2225}$ ft ≈ 47.170 ft   **32.** [8.5] About 5000 m
33. [8.7] 2 **34.** [8.7] -1 **35.** [8.7] -4
36. [8.7] Not a real number   **37.** [8.7] 3 **38.** [8.7] $\frac{1}{3}$
39. [8.7] 1000 **40.** [8.7] $\frac{1}{32}$ **41.** [8.5] -3 **42.** [8.2] y^{8n}

Cumulative Review: Chapters 1–8, p. 542

1. $-\frac{5}{6}$ **2.** 11 **3.** $-8x^{15}$ **4.** $\dfrac{b}{a}$ **5.** $\dfrac{3}{4a^3}$

6. $\dfrac{2x - 1}{x - 2}$ **7.** -3 **8.** 11 **9.** $-3x^3 - 3x^2 - 5$

10. $-15x^5 + 5x^4 - 10x^3$ **11.** $81x^4 + 18x^2 + 1$

12. $x^2 + \dfrac{2}{x - 1}$ **13.** $6(x - 1)$ **14.** $\dfrac{2(4t - 5)}{t(t - 2)}$

15. $5\sqrt{6}$ **16.** $\dfrac{2x}{9}$ **17.** 2 **18.** $10(2x + 3)(2x - 3)$

19. $(t - 1)(t - 10)$ **20.** $2z(3z^2 - 4z + 5)$
21. $(x - 2)(6x - 1)$ **22.** $5(ab + 1)^2$
23. $(4 + t^2)(2 + t)(2 - t)$

24.

25.

26.

27.

28.

29.

30. $-\frac{1}{3}$ **31.** 5 **32.** $y = -\frac{1}{2}x - 1$ **33.** -1
34. $-2, 3$ **35.** $\{n \mid n \le 3\}$, or $(-\infty, 3]$ **36.** $\frac{21}{2}$
37. $(1, 4)$ **38.** $(-2, -9)$ **39.** 7 **40.** $-2, -1$
41. $\frac{1}{3}, \frac{1}{2}$ **42.** $-5, -\frac{9}{4}$ **43.** $x = 2z - y$
44. 12 min; 36 min   **45.** First type: 45 oz; second type:
15 oz   **46.** $8\frac{8}{9}$ hr   **47.** Length: 8 cm; width: 3 cm
48. $-8, 8$ **49.** $-5, -3, 3, 4$

CHAPTER 9

Technology Connection, p. 545

1.

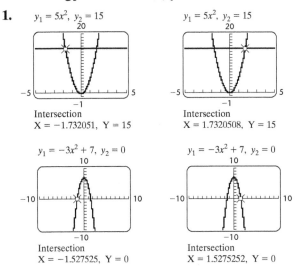

$y_1 = 5x^2, \ y_2 = 15$
Intersection
X = −1.732051, Y = 15

$y_1 = 5x^2, \ y_2 = 15$
Intersection
X = 1.7320508, Y = 15

$y_1 = -3x^2 + 7, \ y_2 = 0$
Intersection
X = −1.527525, Y = 0

$y_1 = -3x^2 + 7, \ y_2 = 0$
Intersection
X = 1.5275252, Y = 0

Check Your Understanding, p. 545

1. $5, -5$ **2.** $5, -5$

Exercise Set 9.1, pp. 546–547

1. True   **2.** True   **3.** False   **4.** True   **5.** $-9, 9$
7. $-1, 1$ **9.** $-\sqrt{11}, \sqrt{11}$ **11.** $-2, 2$ **13.** $-\sqrt{2}, \sqrt{2}$
15. $-\frac{2}{3}, \frac{2}{3}$ **17.** 0 **19.** $-\dfrac{\sqrt{15}}{3}, \dfrac{\sqrt{15}}{3}$ **21.** $-6, 8$
23. $-8, -4$ **25.** $-3 - \sqrt{6}, -3 + \sqrt{6}$, or $-3 \pm \sqrt{6}$
27. 7 **29.** $5 - \sqrt{14}, 5 + \sqrt{14}$, or $5 \pm \sqrt{14}$
31. $-2, 0$ **33.** $\dfrac{3}{4} - \dfrac{\sqrt{17}}{4}, \dfrac{3}{4} + \dfrac{\sqrt{17}}{4}$, or $\dfrac{3}{4} \pm \dfrac{\sqrt{17}}{4}$
35. $-5, 15$ **37.** $-5, -3$ **39.** $8 - \sqrt{7}, 8 + \sqrt{7}$, or
$8 \pm \sqrt{7}$ **41.** $-6 - 3\sqrt{2}, -6 + 3\sqrt{2}$, or $-6 \pm 3\sqrt{2}$
43.   **45.** $\dfrac{1}{t^{10}}$ **46.** $16m^{12}$ **47.** 10 **48.** $\frac{1}{3}$
49. $\dfrac{34x^2 y^2}{3}$ **50.** a^4 **51.** ☑   **53.** $\dfrac{5}{2} - \dfrac{\sqrt{13}}{2}$,
$\dfrac{5}{2} + \dfrac{\sqrt{13}}{2}$, or $\dfrac{5}{2} \pm \dfrac{\sqrt{13}}{2}$ **55.** $-5, 2$ **57.** $-4.35, 1.85$
59. $-4, -2$ **61.** $-6, 0$ **63.** $d = \dfrac{\sqrt{kMmf}}{f}$

Prepare to Move On, p. 547

1. $x^2 - 6x + 9$ **2.** $x^2 + 2ax + a^2$ **3.** $x^2 + x + \frac{1}{4}$
4. $x^2 - 3x + \frac{9}{4}$ **5.** $(x + 1)^2$ **6.** $(x - 2)^2$

Check Your Understanding, p. 549

1. $9; 81; x + 9$ **2.** $-10; 100; x - 10$

Exercise Set 9.2, pp. 551–552

1. (b)   **2.** (a)   **3.** (c)   **4.** (d)   **5.** (c)
6. (e)   **7.** (a)   **8.** (b)   **9.** (d)
10. (f)   **11.** $16; (x + 4)^2 = x^2 + 8x + 16$
13. $1; (x - 1)^2 = x^2 - 2x + 1$
15. $\frac{9}{4}; \left(x - \frac{3}{2}\right)^2 = x^2 - 3x + \frac{9}{4}$
17. $\frac{1}{4}; \left(t + \frac{1}{2}\right)^2 = t^2 + t + \frac{1}{4}$
19. $\frac{9}{16}; \left(x + \frac{3}{4}\right)^2 = x^2 + \frac{3}{2}x + \frac{9}{16}$
21. $\frac{16}{9}; \left(m - \frac{4}{3}\right)^2 = m^2 - \frac{8}{3}m + \frac{16}{9}$
23. $-5, -2$ **25.** $12 \pm \sqrt{123}$ **27.** $-5 \pm \sqrt{13}$
29. $2 \pm \sqrt{11}$ **31.** $-\dfrac{3}{2} \pm \dfrac{\sqrt{21}}{2}$, or $\dfrac{-3 \pm \sqrt{21}}{2}$
33. $-1, \frac{1}{2}$ **35.** $\dfrac{3}{4} \pm \dfrac{\sqrt{41}}{4}$, or $\dfrac{3 \pm \sqrt{41}}{4}$
37. $\dfrac{5}{4} \pm \dfrac{\sqrt{105}}{4}$, or $\dfrac{5 \pm \sqrt{105}}{4}$ **39.** $-\dfrac{2}{3} \pm \dfrac{\sqrt{19}}{3}$,
or $\dfrac{-2 \pm \sqrt{19}}{3}$ **41.** $-\frac{1}{2}, 5$ **43.** $-\frac{5}{2}, \frac{2}{3}$ **45.** ☑
47. $-\frac{2}{3}$ **48.** $\{y \mid y > 10\}$, or $(10, \infty)$ **49.** $-1, 7$
50. 3 **51.** $-3, 9$ **52.** $\left(\frac{23}{5}, \frac{11}{5}\right)$ **53.** ☑
55. $-10, 10$ **57.** $-6\sqrt{5}, 6\sqrt{5}$ **59.** $-16, 16$
61. $-7.61, -0.39$ **63.** $-2.5, 0.67$ **65.** ☑

Quick Quiz: Sections 9.1–9.2, p. 552

1. 100 **2.** $-\sqrt{11}, \sqrt{11}$ **3.** $7 + \sqrt{2}, 7 - \sqrt{2}$
4. $-10 + \sqrt{85}, -10\sqrt{85}$
5. $-\dfrac{1}{4} + \dfrac{\sqrt{33}}{4}, -\dfrac{1}{4} - \dfrac{\sqrt{33}}{4}$

Prepare to Move On, p. 552

1. -7 **2.** 7 **3.** 13 **4.** $1 - \sqrt{5}$ **5.** $\dfrac{8 - \sqrt{5}}{3}$

Technology Connection, p. 555

1. An ERROR message appears.   **2.** The graph has
no x-intercepts, so there is no value of x for which
$x^2 + x + 1 = 0$, or, equivalently, for which $x^2 + x = -1$.

Check Your Understanding, p. 557

1. $a = 3, b = 5, c = 2$ **2.** $a = 1, b = -6, c = 7$
3. $a = 2, b = -4, c = -9$ **4.** $a = 1, b = -12, c = 0$

Exercise Set 9.3, pp. 558–561

1. True   **2.** True   **3.** True   **4.** True   **5.** $-2, 10$
7. 1 **9.** $-\frac{4}{3}, -1$ **11.** $-\frac{1}{2}, \frac{7}{2}$ **13.** $-5, 5$
15. $-2 \pm \sqrt{11}$ **17.** $5 \pm \sqrt{6}$ **19.** $5 \pm \sqrt{3}$
21. $-\dfrac{4}{3} \pm \dfrac{\sqrt{10}}{3}$, or $\dfrac{-4 \pm \sqrt{10}}{3}$ **23.** $\dfrac{5}{4} \pm \dfrac{\sqrt{33}}{4}$, or
$\dfrac{5 \pm \sqrt{33}}{4}$ **25.** $-\dfrac{1}{4} \pm \dfrac{\sqrt{13}}{4}$, or $\dfrac{-1 \pm \sqrt{13}}{4}$

27. No real-number solution   **29.** $\dfrac{5}{6} \pm \dfrac{\sqrt{73}}{6}$, or

$\dfrac{5 \pm \sqrt{73}}{6}$ **31.** $\dfrac{3}{2} \pm \dfrac{\sqrt{29}}{2}$, or $\dfrac{3 \pm \sqrt{29}}{2}$ **33.** $-5, \frac{2}{3}$

35. No real-number solution   **37.** $\pm 2\sqrt{3}$

39. $-3.562, 0.562$ **41.** $-0.193, 5.193$ **43.** $-1.207,$
0.207 **45.** 10 sides   **47.** 13.05 sec   **49.** 1.32 sec

51. 7 ft, 24 ft   **53.** Length: 10 cm; width: 6 cm

55. Length: 24 m; width: 18 m   **57.** 5.13 cm, 10.13 cm

59. Length: 8.09 ft; width: 3.09 ft   **61.** Length: 5.66 m;
width: 2.83 m   **63.** 4%   **65.** 5%   **67.** 29.85 mi

69. ☐   **71.** $3x^4 - x^3 - 2x^2 - 10x$

72. $25m^2 - 70mn + 49n^2$ **73.** $12t^2 - 11t - 15$

74. $\dfrac{7x(x + 1)}{(x - 1)^2}$ **75.** $\dfrac{-x^2 + 4x + 2}{x(x + 1)}$

76. $3x^3 + 3x^2 + 2x - 5 + \dfrac{-3}{x - 1}$ **77.** ☐   **79.** $-11, 0$

81. $\dfrac{3}{2} \pm \dfrac{\sqrt{5}}{2}$, or $\dfrac{3 \pm \sqrt{5}}{2}$ **83.** $-\dfrac{7}{2} \pm \dfrac{\sqrt{61}}{2}$, or

$\dfrac{-7 \pm \sqrt{61}}{2}$ **85.** $\pm \sqrt{11}$ **87.** $\dfrac{5}{2} \pm \dfrac{\sqrt{37}}{2}$, or $\dfrac{5 \pm \sqrt{37}}{2}$

89. 4.83 cm   **91.** Length: 120 m; width: 75 m   **93.** 5%

95. Length: 44.06 in.; width: 19.06 in.   **97.** ☐

Quick Quiz: Sections 9.1–9.3, p. 561

1. $-\sqrt{10}, \sqrt{10}$ **2.** $\frac{2}{3}, 1$ **3.** $\dfrac{1}{2} + \dfrac{\sqrt{13}}{2}, \dfrac{1}{2} - \dfrac{\sqrt{13}}{2}$

4. $9 + 2\sqrt{2}, 9 - 2\sqrt{2}$ **5.** $-\frac{1}{2}, 4$

Prepare to Move On, p. 561

1. $y = 4n$ **2.** $d = 2t - c$ **3.** $p = \dfrac{4n}{a}$

4. $x = \dfrac{y - B}{A}$ **5.** $x = \dfrac{y}{A + B}$

Mid-Chapter Review: Chapter 9, p. 562

1.
$$(x + 3)^2 = 2$$
$$x + 3 = \sqrt{2} \qquad \text{or} \quad x + 3 = -\sqrt{2}$$
$$x = -3 + \sqrt{2} \quad \text{or} \qquad x = -3 - \sqrt{2}$$

2. $a = 1, b = -3, c = 1$
$$x = \dfrac{-(-3) \pm \sqrt{(-3)^2 - 4(1)(1)}}{2(1)}$$
$$x = \dfrac{3 \pm \sqrt{5}}{2}$$

3. $-2, 1$ **4.** $-10, 10$ **5.** $-5 \pm 4\sqrt{2}$ **6.** $-\frac{1}{2}, 3$

7. -1 **8.** No real-number solution   **9.** $-1, 6$

10. $0, 5$ **11.** $\dfrac{1}{10} \pm \dfrac{\sqrt{41}}{10}$, or $\dfrac{1 \pm \sqrt{41}}{10}$ **12.** $-3, 2$

13. $2 \pm \sqrt{5}$ **14.** $-4 \pm \sqrt{15}$ **15.** $-\frac{1}{11}, \frac{1}{11}$

16. $-1 \pm \sqrt{3}$ **17.** No real-number solution

18. $0, \frac{1}{9}$ **19.** 16.64 sec   **20.** Length: 6.22 ft;
width: 3.22 ft

Check Your Understanding, p. 564

1. c **2.** xy **3.** Rr_1r_2

Exercise Set 9.4, pp. 565–567

1. False   **2.** True   **3.** False   **4.** True   **5.** (b)

6. (d)   **7.** (c)   **8.** (a)   **9.** $h = \dfrac{2A}{b}$ **11.** $c = \dfrac{100m}{Q}$

13. $t = \dfrac{A}{Pr} - \dfrac{1}{r}$, or $t = \dfrac{A - P}{Pr}$ **15.** $h = \dfrac{d^2}{c^2}$

17. $R = \dfrac{r_1r_2}{r_1 + r_2}$ **19.** $x = \dfrac{-b \pm \sqrt{b^2 - 4ac}}{2a}$

21. $y = \dfrac{C - Ax}{B}$ **23.** $h = \dfrac{S - 2\pi r^2}{2\pi r}$ **25.** $t = \dfrac{M - g}{r + s}$

27. $a = \dfrac{d}{b - c}$ **29.** $t = \pm\sqrt{\dfrac{2s}{g}}$ **31.** $h = \pm\sqrt{st}$

33. $w = r - \dfrac{d}{t}$, or $w = \dfrac{rt - d}{t}$ **35.** $x = \dfrac{25 + y}{2}$

37. $t = 0$ **39.** $t = \dfrac{-n \pm \sqrt{n^2 + 4mp}}{2m}$

41. $m = \dfrac{-t \pm \sqrt{t^2 + 4n}}{2}$ **43.** $r = -1 \pm \sqrt{\dfrac{A}{P}}$

45. $t = \dfrac{p^2 - 2pn + n^2 - 9c}{9}$ **47.** ☐

49. $(x - 7)(x + 2)$ **50.** $5y^3(2y - 1)(5y + 2)$

51. $(n - 10)^2$ **52.** $(2x + 3)(x^2 - 5)$

53. $(xy + 5)(xy - 5)$ **54.** $6xy(4x - 5y + 1)$

55. ☐   **57.** 6 years old   **59.** $m = \dfrac{g \pm \sqrt{g^2 - 4ft}}{2f}$

61. $t = -\dfrac{b}{a} + \dfrac{k^2}{a(V - c)^2}$ **63.** $-40°$

Quick Quiz: Sections 9.1–9.4, p. 567

1. $2 \pm \sqrt{11}$ **2.** $-\dfrac{3}{2} \pm \dfrac{\sqrt{17}}{2}$ **3.** 1.56 ft, 2.56 ft

4. $a = \dfrac{p - 2t}{tb}$ **5.** $x = \pm\sqrt{\dfrac{k}{y}}$

Prepare to Move On, p. 567

1. $4\sqrt{3}$ **2.** $5\sqrt{10}$ **3.** $10\sqrt{6}$ **4.** 30

Check Your Understanding, p. 569

1. $2i$ **2.** $\sqrt{5}i$ **3.** $4i$ **4.** $\sqrt{7}i$

5. $3\sqrt{2}i$ **6.** $5\sqrt{3}i$

Connecting the Concepts, pp. 569–570

1. 1   **2.** 2, 3   **3.** 25   **4.** 4   **5.** 5   **6.** $-1 \pm \sqrt{3}$

7. $-2, 4$ **8.** $-\frac{1}{6}$ **9.** $-10i, 10i$ **10.** $-10, 10$

Exercise Set 9.5, pp. 570–571

1. False **2.** True **3.** True **4.** False **5.** i
7. $7i$ **9.** $\sqrt{5}\,i$, or $i\sqrt{5}$ **11.** $3\sqrt{5}\,i$, or $3i\sqrt{5}$
13. $-5\sqrt{2}\,i$, or $-5i\sqrt{2}$ **15.** $4 + 7i$ **17.** $3 - 3i$
19. $-2 + 5\sqrt{3}\,i$, or $-2 + 5i\sqrt{3}$ **21.** $\pm 5i$
23. $\pm 2\sqrt{7}\,i$, or $\pm 2i\sqrt{7}$ **25.** $-2 \pm i$ **27.** $4 \pm 3i$
29. $1 \pm 2i$ **31.** $2 \pm \sqrt{3}\,i$, or $2 \pm i\sqrt{3}$ **33.** $-\frac{2}{5} \pm \frac{1}{5}i$

35. $-\dfrac{1}{3} \pm \dfrac{\sqrt{2}}{3}i$ **37.** 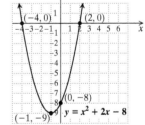 **39.** 10 **40.** $5xy^4\sqrt{2x}$

41. $\sqrt{2}$ **42.** -11 **43.** $\frac{3}{68}$ **44.** $\dfrac{10n}{7}$ **45.**

47. $-2 \pm i$ **49.** $-1 \pm \sqrt{19}i$ **51.**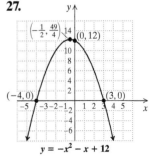

Quick Quiz: Sections 9.1–9.5, p. 571

1. $6i$ **2.** $\pm 2\sqrt{3}$ **3.** $\pm 2\sqrt{3}i$ **4.** 1
5. $2 \pm \sqrt{5}$

Prepare to Move On, p. 571

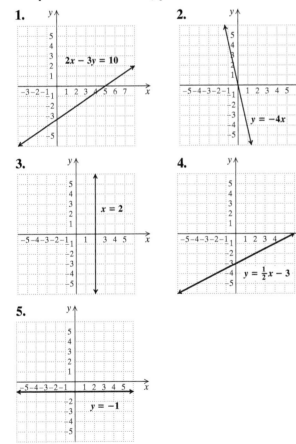

1. $2x - 3y = 10$
2. $y = -4x$
3. $x = 2$
4. $y = \frac{1}{2}x - 3$
5. $y = -1$

Check Your Understanding, p. 574

1. Upward **2.** Downward **3.** Upward
4. $(0, 8)$ **5.** $(0, -5)$

Exercise Set 9.6, pp. 575–577

1. False **2.** True **3.** False **4.** True

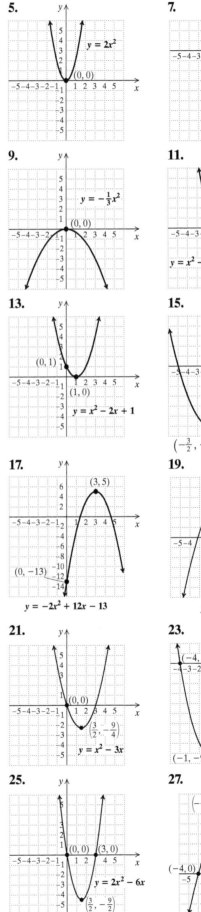

5. $y = 2x^2$; $(0, 0)$
7. $y = -2x^2$; $(0, 0)$
9. $y = -\frac{1}{3}x^2$; $(0, 0)$
11. $y = x^2 - 2$; $(0, -2)$
13. $y = x^2 - 2x + 1$; $(0, 1)$, $(1, 0)$
15. $y = x^2 + 3x - 10$; $(0, -10)$, $\left(-\frac{3}{2}, -\frac{49}{4}\right)$
17. $y = -2x^2 + 12x - 13$; $(3, 5)$, $(0, -13)$
19. $y = -\frac{1}{2}x^2 + 5$; $(0, 5)$
21. $y = x^2 - 3x$; $(0, 0)$, $\left(\frac{3}{2}, -\frac{9}{4}\right)$
23. $y = x^2 + 2x - 8$; $(-4, 0)$, $(2, 0)$, $(0, -8)$, $(-1, -9)$
25. $y = 2x^2 - 6x$; $(0, 0)$, $(3, 0)$, $\left(\frac{3}{2}, -\frac{9}{2}\right)$
27. $y = -x^2 - x + 12$; $\left(-\frac{1}{2}, \frac{49}{4}\right)$, $(0, 12)$, $(-4, 0)$, $(3, 0)$

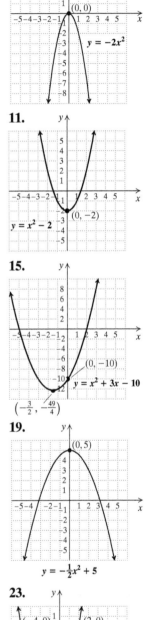

29.

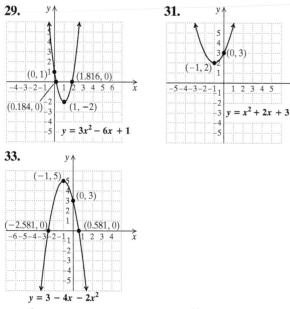

$y = 3x^2 - 6x + 1$

Labeled points: $(0, 1)$, $(1.816, 0)$, $(0.184, 0)$, $(1, -2)$

31.

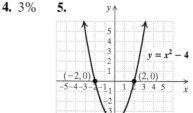

$y = x^2 + 2x + 3$

Labeled points: $(-1, 2)$, $(0, 3)$

33.

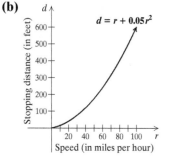

$y = 3 - 4x - 2x^2$

Labeled points: $(-1, 5)$, $(0, 3)$, $(-2.581, 0)$, $(0.581, 0)$

35. 🖉   **37.** 90   **38.** 1.82×10^{22} **39.** -33

40. Slope: $\frac{2}{3}$; y-intercept: $(0, -9)$ **41.** $y = -x + 2$

42. $b = \dfrac{ac}{a + c}$ **43.** 🖉   **45. (a)** 2 sec after launch

and 4 sec after launch; **(b)** 3 sec after launch; **(c)** 6 sec

after launch   **47. (a)** 56.25 ft, 120 ft, 206.25 ft, 276.25 ft,

356.25 ft, 600 ft;

(b)

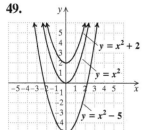

$d = r + 0.05r^2$

Stopping distance (in feet) vs. Speed (in miles per hour)

49.

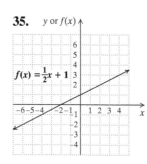

$y = x^2 + 2$, $y = x^2$, $y = x^2 - 5$

We can move the graph of $y = x^2$ up k units if $k \geq 0$ or down $|k|$ units if $k < 0$ to obtain the graph of $y = x^2 + k$.

51.

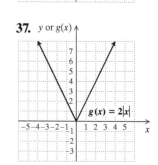

$D = (p - 6)^2$

Number sold (in thousands) vs. Price (in dollars)

53. 🖳 2.2361

Quick Quiz: Sections 9.1–9.6, p. 577

1. 9   **2.** $x = \dfrac{y \pm \sqrt{y^2 - 4}}{2}$ **3.** $-2.351, 0.851$

4. 3%   **5.**

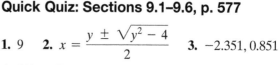

$y = x^2 - 4$

Labeled points: $(-2, 0)$, $(2, 0)$, $(0, -4)$

Prepare to Move On, p. 577

1. 7   **2.** -8 **3.** 14   **4.** 30,100   **5.** 1   **6.** 144

Check Your Understanding, p. 583

1. 8; 7; 5; 0   **2.** 3; 0; -1; 0; 3

Exercise Set 9.7, pp. 584–586

1. True   **2.** True   **3.** False   **4.** True   **5.** Yes

7. Yes   **9.** No   **11.** Yes   **13.** 9; 12; 3

15. 21; -15; -30 **17.** 1; -17; 5.5   **19.** 1; 0; 10

21. 25; 25; $\frac{25}{4}$ **23.** 0; 12; 4

25. (a) 34 years; **(b)** 20 years

27. (a) 159.48 cm; **(b)** 153.98 cm

29. $16.\overline{6}°$C; 25°C; -5°C

31.

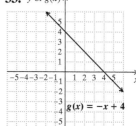

$f(x) = 2x - 3$

33.

$g(x) = -x + 4$

35.

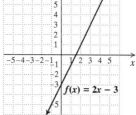

$f(x) = \frac{1}{2}x + 1$

37.

$g(x) = 2|x|$

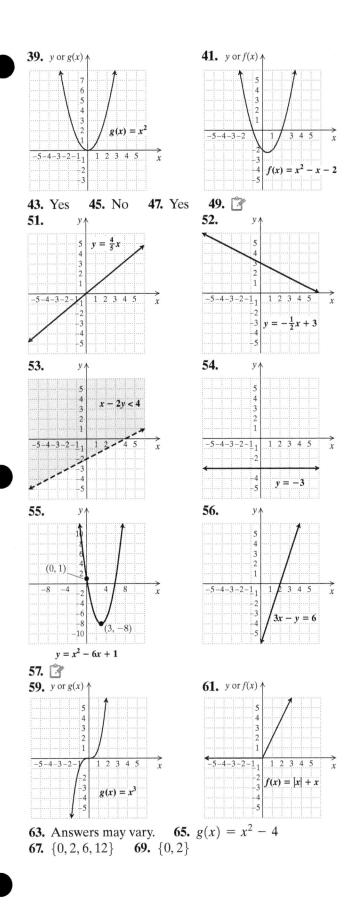

39. y or $g(x)$

$g(x) = x^2$

41. y or $f(x)$

$f(x) = x^2 - x - 2$

43. Yes   **45.** No   **47.** Yes   **49.** 🗒

51.

$y = \frac{4}{5}x$

52.

$y = -\frac{1}{2}x + 3$

53.

$x - 2y < 4$

54.

$y = -3$

55.

$(0, 1)$

$(3, -8)$

$y = x^2 - 6x + 1$

56.

$3x - y = 6$

57. 🗒

59. y or $g(x)$

$g(x) = x^3$

61. y or $f(x)$

$f(x) = |x| + x$

63. Answers may vary.   **65.** $g(x) = x^2 - 4$
67. $\{0, 2, 6, 12\}$ **69.** $\{0, 2\}$

Quick Quiz: Sections 9.1–9.7, p. 586

1. $\pm i$ **2.** $4 \pm \sqrt{3}$ **3.** $\dfrac{5}{2} \pm \dfrac{\sqrt{33}}{2}$

4. y or $f(x)$ **5.** 4

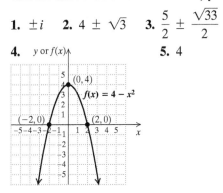

$(0, 4)$

$f(x) = 4 - x^2$

$(-2, 0)$ $(2, 0)$

Visualizing for Success, p. 587

1. J   **2.** F   **3.** H   **4.** G   **5.** B   **6.** E   **7.** D
8. I   **9.** C   **10.** A

Decision Making: Connection, p. 588

1. $8.9¢/\text{in}^2$ **2.** 13 in.   **3.** One-fourth of the price

4. A 10-in. round pan   **5.** $r = \dfrac{s\sqrt{6\pi}}{2\pi}$

Study Summary: Chapter 9, pp. 589–591

1. $-2, 3$ **2.** $5 \pm \sqrt{3}$ **3.** $-2 \pm \sqrt{5}$

4. $-1 \pm \sqrt{3}$ **5.** $x = \dfrac{9t^2 - 5}{2}$ **6.** $9i$

7. $-1 \pm i$

8.

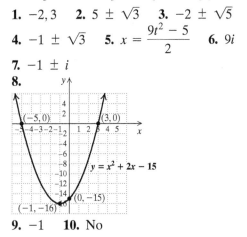

$(-5, 0)$ $(3, 0)$

$y = x^2 + 2x - 15$

$(0, -15)$

$(-1, -16)$

9. -1 **10.** No

Review Exercises: Chapter 9, pp. 591–592

1. (h)   **2.** (f)   **3.** (e)   **4.** (b)   **5.** (d)
6. (a)   **7.** (c)   **8.** (g)   **9.** $5 \pm \sqrt{26}$
10. $-2, \frac{1}{2}$ **11.** $\pm\sqrt{6}$ **12.** $\frac{3}{5}, 1$ **13.** $1 \pm \sqrt{11}$
14. $-3 \pm 2i$ **15.** $-2, \frac{1}{3}$ **16.** $0, \frac{1}{11}$ **17.** $\pm i$
18. $-2, 5$ **19.** $-10 \pm 2\sqrt{3}$ **20.** $-3 \pm 3\sqrt{2}$
21. $-\dfrac{1}{2} \pm \dfrac{\sqrt{3}}{2}i$ **22.** $1 \pm \dfrac{\sqrt{3}}{2}$, or $\dfrac{2 \pm \sqrt{3}}{2}$

23. 3   **24.** $\pm 2\sqrt{2}$ **25.** $\dfrac{3}{10} \pm \dfrac{\sqrt{71}}{10}i$ **26.** $-\frac{7}{2}, \frac{5}{3}$

27. $t = \dfrac{p}{c - b}$ **28.** $y = \dfrac{c^2 - 2cz + z^2}{x}$

29. $n = \dfrac{-m \pm \sqrt{m^2 + 4p}}{2}$ **30.** $t = \dfrac{rs}{r + s}$

31. 0.697, 4.303   **32.** $-1.866, -0.134$

33. 3.217 m, 6.217 m   **34.** 4%   **35.** Length: 12 m; width: 9 m   **36.** 15 sec   **37.** $3i$ **38.** $-5\sqrt{5}\,i$, or $-5i\sqrt{5}$

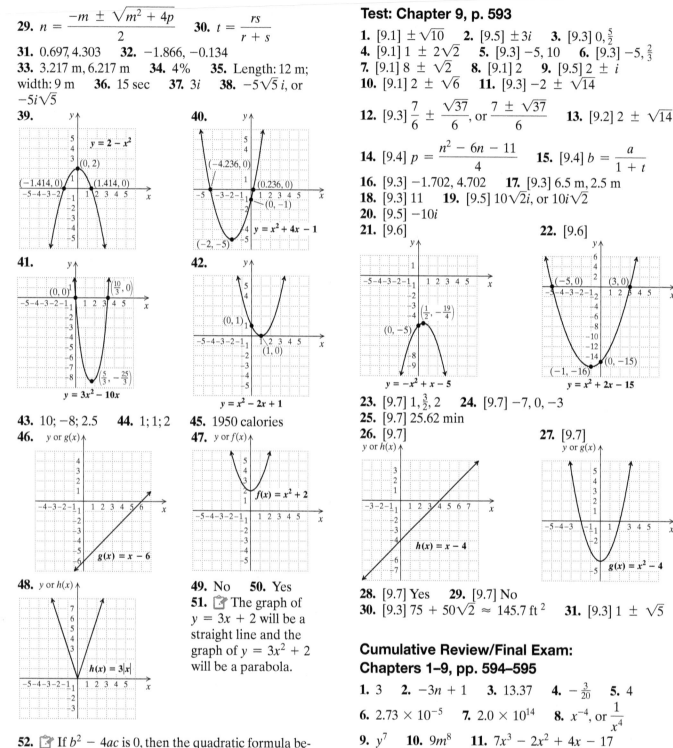

39.
$y = 2 - x^2$
$(0, 2)$
$(-1.414, 0)$ $(1.414, 0)$

40.
$(-4.236, 0)$
$(0.236, 0)$
$(0, -1)$
$(-2, -5)$
$y = x^2 + 4x - 1$

41.
$(0, 0)$ $\left(\frac{10}{3}, 0\right)$
$\left(\frac{5}{3}, -\frac{25}{3}\right)$
$y = 3x^2 - 10x$

42.
$(0, 1)$
$(1, 0)$
$y = x^2 - 2x + 1$

43. $10; -8; 2.5$ **44.** $1; 1; 2$ **45.** 1950 calories

46. y or $g(x)$
$g(x) = x - 6$

47. y or $f(x)$
$f(x) = x^2 + 2$

48. y or $h(x)$
$h(x) = 3|x|$

49. No   **50.** Yes

51. ☞ The graph of $y = 3x + 2$ will be a straight line and the graph of $y = 3x^2 + 2$ will be a parabola.

52. ☞ If $b^2 - 4ac$ is 0, then the quadratic formula becomes $x = -b/(2a)$; thus there is only one x-intercept. If $b^2 - 4ac$ is negative, then there are no real-number solutions and thus no x-intercepts. If $b^2 - 4ac$ is positive, then $x = \dfrac{-b + \sqrt{b^2 - 4ac}}{2a}$ or $x = \dfrac{-b - \sqrt{b^2 - 4ac}}{2a}$ so there must be two x-intercepts.   **53.** ☞ At least two of the points must be above and below each other on the graph.   **54.** -32 and -31; 31 and 32   **55.** ± 14 **56.** 25   **57.** $5\sqrt{\pi}$ in.

Test: Chapter 9, p. 593

1. [9.1] $\pm\sqrt{10}$ **2.** [9.5] $\pm 3i$ **3.** [9.3] $0, \frac{5}{2}$
4. [9.1] $1 \pm 2\sqrt{2}$ **5.** [9.3] $-5, 10$ **6.** [9.3] $-5, \frac{2}{3}$
7. [9.1] $8 \pm \sqrt{2}$ **8.** [9.1] 2   **9.** [9.5] $2 \pm i$
10. [9.1] $2 \pm \sqrt{6}$ **11.** [9.3] $-2 \pm \sqrt{14}$

12. [9.3] $\dfrac{7}{6} \pm \dfrac{\sqrt{37}}{6}$, or $\dfrac{7 \pm \sqrt{37}}{6}$ **13.** [9.2] $2 \pm \sqrt{14}$

14. [9.4] $p = \dfrac{n^2 - 6n - 11}{4}$ **15.** [9.4] $b = \dfrac{a}{1 + t}$

16. [9.3] $-1.702, 4.702$ **17.** [9.3] 6.5 m, 2.5 m
18. [9.3] 11   **19.** [9.5] $10\sqrt{2}i$, or $10i\sqrt{2}$
20. [9.5] $-10i$
21. [9.6]   **22.** [9.6]

23. [9.7] $1, \frac{3}{2}, 2$ **24.** [9.7] $-7, 0, -3$
25. [9.7] 25.62 min
26. [9.7]   **27.** [9.7]

y or $h(x)$
$h(x) = x - 4$

y or $g(x)$
$g(x) = x^2 - 4$

28. [9.7] Yes   **29.** [9.7] No
30. [9.3] $75 + 50\sqrt{2} \approx 145.7$ ft^2 **31.** [9.3] $1 \pm \sqrt{5}$

Cumulative Review/Final Exam: Chapters 1–9, pp. 594–595

1. 3   **2.** $-3n + 1$ **3.** 13.37   **4.** $-\frac{3}{20}$ **5.** 4
6. 2.73×10^{-5} **7.** 2.0×10^{14} **8.** x^{-4}, or $\dfrac{1}{x^4}$
9. y^7 **10.** $9m^8$ **11.** $7x^3 - 2x^2 + 4x - 17$
12. $12x^2 - 6$ **13.** $-8y^4 + 6y^3 - 2y^2$
14. $6t^3 - 17t^2 + 16t - 6$ **15.** $t^2 - \frac{1}{16}$
16. $6x^2 + 5x - 56$ **17.** $12x^2y - 3x^2y^2 + 12xy^2 - 2$
18. $25p^4 + 20p^2q + 4q^2$
19. $\dfrac{2}{x + 3}$ **20.** $\dfrac{3a(a - 1)}{2(a + 1)}$ **21.** $\dfrac{27x - 4}{5x(3x - 1)}$
22. $\dfrac{-x^2 + x + 2}{(x + 4)(x - 4)(x - 5)}$ **23.** $x^2 + 9x + 16 + \dfrac{35}{x - 2}$
24. $2(3x + 2)(3x - 2)$ **25.** $(y - 6)(y^2 - 5)$
26. $(m + 5)^2$ **27.** $x(2x - 3)(5x + 2)$

28. $(t - 3)(t + 2)$ **29.** $(49p + 1)(p + 1)$
30. $(2x - 5y)^2$ **31.** $3y^2(y - 8)(y - 5)$ **32.** $-\frac{42}{5}$
33. 5 **34.** $7t$ **35.** $\frac{1}{8}$ **36.** $5x^3y^3\sqrt{10y}$ **37.** $\frac{4}{3}$
38. -17 **39.** $18\sqrt{3}$ **40.** $8a^2b\sqrt{3ab}$ **41.** $\frac{2\sqrt{10}}{5}$
42. -6 **43.** $\{x | x < -6\}$, or $(-\infty, -6)$ **44.** 7
45. $3, 5$ **46.** $(1, 2)$ **47.** $-\frac{3}{5}$ **48.** $(6, 7)$
49. $1 \pm \sqrt{3}$ **50.** 2 **51.** $\{x | x \geq -1\}$, or $[-1, \infty)$
52. 8 **53.** -3 **54.** $(15, 0)$ **55.** $-1 \pm 2i$
56. $\pm\sqrt{10}$ **57.** $3 \pm \sqrt{6}$ **58.** $-1, 2$
59. $-\frac{4}{3}, \frac{5}{4}$ **60.** $a = \dfrac{by}{x + y}$ **61.** $m = \dfrac{tn}{t + n}$
62. $-0.207, 1.207$

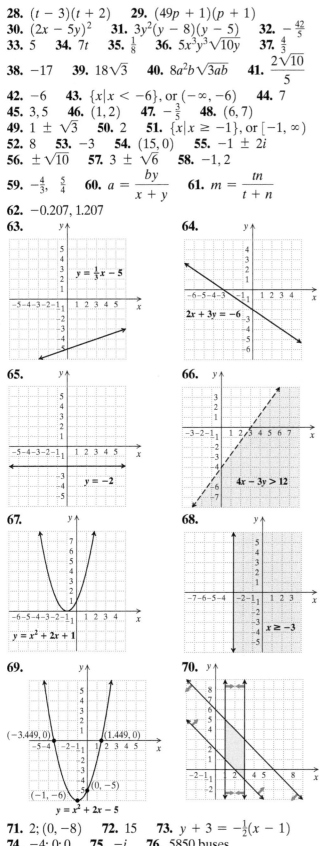

63. $y = \frac{1}{3}x - 5$
64. $2x + 3y = -6$
65. $y = -2$
66. $4x - 3y > 12$
67. $y = x^2 + 2x + 1$
68. $x \geq -3$
69. $(-3.449, 0)$ $(1.449, 0)$ $(0, -5)$ $(-1, -6)$ $y = x^2 + 2x - 5$
70.

71. $2; (0, -8)$ **72.** 15 **73.** $y + 3 = -\frac{1}{2}(x - 1)$
74. $-4; 0; 0$ **75.** $-i$ **76.** 5850 buses
77. Dark-chocolate cheesecake truffles: 12 lb; milk-chocolate hazelnut truffles: 8 lb   **78.** 100 m
79. 1.92 g   **80.** Length: 10 ft; width: 3 ft

81. Length: 12 m; width: 5 m
82. $\sqrt{1500}$ cm ≈ 38.730 cm   **83.** $\dfrac{\sqrt{6}}{3}$ **84.** Yes
85. No   **86.** No   **87.** No   **88.** Yes

APPENDIXES
Exercise Set A, pp. 598–599

1. $(t + 2)(t^2 - 2t + 4)$ **3.** $(x + 1)(x^2 - x + 1)$
5. $(z - 5)(z^2 + 5z + 25)$ **7.** $(2a - 1)(4a^2 + 2a + 1)$
9. $(y - 3)(y^2 + 3y + 9)$
11. $(4 + 5x)(16 - 20x + 25x^2)$
13. $(5p + 1)(25p^2 - 5p + 1)$
15. $(3m - 4)(9m^2 + 12m + 16)$
17. $(p - q)(p^2 + pq + q^2)$ **19.** $\left(x + \frac{1}{2}\right)\left(x^2 - \frac{1}{2}x + \frac{1}{4}\right)$
21. $2(y - 4)(y^2 + 4y + 16)$
23. $3(2a + 1)(4a^2 - 2a + 1)$
25. $r(s - 5)(s^2 + 5s + 25)$
27. $5(x + 2z)(x^2 - 2xz + 4z^2)$
29. $(x + 0.2)(x^2 - 0.2x + 0.04)$
31. 🖽   **33.** $(5c^2 + 2d^2)(25c^4 - 10c^2d^2 + 4d^4)$
35. $3(x^a - 2y^b)(x^{2a} + 2x^ay^b + 4y^{2b})$
37. $\frac{1}{3}\left(\frac{1}{2}xy + z\right)\left(\frac{1}{4}x^2y^2 - \frac{1}{2}xyz + z^2\right)$

Exercise Set B, pp. 601–602

1. Mean: 17; median: 18; mode: 13   **3.** Mean: $10.\overline{6}$;
median: 9; modes: 3, 20   **5.** Mean: 4.06; median: 4.6;
mode: none   **7.** Mean: $239.\overline{3}$; median: 234; mode: 234
9. Average: 35.25; median: 22; mode: 0   **11.** Average:
$218.\overline{3}$; median: 222; mode: 202   **13.** 10 home runs
15. $a = 30, b = 58$

Exercise Set C, pp. 603–604

1. $\{8, 9, 10, 11\}$ **3.** $\{41, 43, 45, 47, 49\}$ **5.** $\{-3, 3\}$
7. True   **9.** True   **11.** False   **13.** True
15. False   **17.** True   **19.** $\{c, d, e\}$ **21.** $\{1, 2, 3, 6\}$
23. $\varnothing$ **25.** $\{a, e, i, o, u, q, c, k\}$
27. $\{1, 2, 3, 4, 6, 9, 12, 18\}$ **29.** $\{1, 2, 3, 4, 5, 6, 7, 8\}$
31. 🖽   **33.** The set of integers   **35.** The set of real
numbers   **37.** $\varnothing$ **39. (a)** A; **(b)** A; **(c)** A; **(d)** $\varnothing$
41. True   **43.** True   **45.** True

Glossary

Absolute value [1.4]   The distance that a number is from 0 on the number line

Additive identity [1.5]   The number 0

Additive inverse [1.6]   A number's opposite; two numbers are additive inverses of each other if when added the result is zero

Algebraic expression [1.1]   An expression consisting of variables and/or numerals, often with operation signs and grouping symbols

Ascending order [4.3, 4.4]   A polynomial in one variable written with the terms arranged according to degree, from least to greatest

Associative law for addition [1.2]   The statement that when three numbers are added, changing the grouping does not change the result

Associative law for multiplication [1.2]   The statement that when three numbers are multiplied, changing the grouping does not change the result

Average [2.7, App A]   Most commonly, the *mean* of a set of numbers, found by adding the numbers and then dividing by the number of addends

Axes (singular, **Axis**) **[3.1]**   Two perpendicular number lines used to identify points in a plane

Axis of symmetry [9.6]   A line that can be drawn through a graph such that the part of the graph on one side of the line is an exact reflection of the part on the opposite side

Bar graph [3.1]   A graphic display of data using bars proportional in length to the numbers represented

Base [1.8]   In exponential notation, the number being raised to a power

Binomial [4.3]   A polynomial with two terms

Center point [App A]   A single representative number used to analyze a set of data; also called *measure of central tendency*

Circle graph [3.1]   A graphic display of data often used to show what percent of the whole each item in a group represents

Circumference [2.3]   The distance around a circle

Closed interval [a, b] [2.6]   The set of all numbers x for which $a \le x \le b$; thus, $[a, b] = \{x | a \le x \le b\}$

Coefficient [2.1, 4.3]   The numerical multiplier of a variable

Common factor [1.2, 5.1]   A factor that appears in every term in an expression

Commutative law for addition [1.2]   The statement that when two numbers are added, changing the order of addition does not affect the answer

Commutative law for multiplication [1.2]   The statement that when two numbers are multiplied, changing the order of multiplication does not affect the answer

Completing the square [9.2]   A method of adding a particular constant to an expression so that the resulting sum is a perfect square

Complex number [9.5]   Any number that can be written in the form $a + bi$, where a and b are real numbers and $i = \sqrt{-1}$

Complex rational expression [6.5]   A rational expression that contains rational expressions within its numerator and/or its denominator

Complex-number system [9.5]   A number system that contains the real-number system and is designed so that negative numbers have defined square roots

Composite number [1.3]   A natural number other than 1 that is not prime

Conjugates [8.4]   Pairs of radical expressions like $\sqrt{a} + \sqrt{b}$ and $\sqrt{a} - \sqrt{b}$

Consecutive even integers [2.5]   Integers that are even and two units apart

Consecutive integers [2.5]   Integers that are one unit apart

Consecutive odd integers [2.5]   Integers that are odd and two units apart

Consistent system of equations [7.1]   A system of equations that has at least one solution

Constant [1.1]   A known number that never changes

Constant of proportionality [7.7]   The constant, k, in an equation of variation; also called *variation constant*

Contradiction [2.2]   An equation that is never true

Coordinates [3.1]   The numbers in an ordered pair

Cube root [8.7]   The number c is the *cube root* of a if $c^3 = a$.

Data [1.1]   A set of numbers used to represent information

Degree of a monomial [4.3, 4.7]   The number of variable factors in the monomial

Degree of a polynomial [4.3, 4.7]   The degree of the term of highest degree in a polynomial

Denominator [1.3]   The number below the fraction bar in a fraction

Dependent equations [7.1]   Equations in a system from which one equation can be removed without changing the solution set

Descending order [4.3, 4.4]   A polynomial in one variable written with the terms arranged according to degree, from greatest to least

Difference [1.6]   The result when two numbers are subtracted

Difference of cubes [App A]   An expression that can be written in the form $A^3 - B^3$

Difference of squares [4.6, 5.4, App A]   An expression that can be written in the form $A^2 - B^2$

Direct variation [7.7]   A situation that can be modeled by a linear function of the form $f(x) = kx$, or $y = kx$, where k is a nonzero constant

Distance formula [8.6]   The formula
$$d = \sqrt{(x_2 - x_1)^2 + (y_2 - y_1)^2},$$ where d is the distance between any two points (x_1, y_1) and (x_2, y_2)

Distributive law [1.2]   The statement that multiplying a factor by the sum of two numbers gives the same result as multiplying the factor by each of the two numbers and then adding

Domain [9.7]   The set of all first coordinates of the ordered pairs in a function

Elements [App C]   The objects of a set

Elimination method [7.3]   An algebraic method that uses the addition principle to solve a system of equations

Empty set [App C]   The set containing no elements, written $\varnothing$ or { }

Equation [1.1]   A number sentence formed by placing an equals sign between two expressions

Equation of direct variation [7.7]   The equation $y = kx$, with k a constant

Equation of inverse variation [7.7]   The equation $y = k/x$, with k a constant

Equivalent equations [2.1]   Equations that have the same solutions

Equivalent expressions [1.2]   Expressions that have the same value for all allowable replacements

Equivalent inequalities [2.6]   Inequalities that have the same solution set

Evaluate [1.1]   To substitute a number for each variable in the expression and calculate the result

Exponent [1.8]   In an expression of the form b^n, the number n is an exponent.

Exponential notation [1.8]   A representation of a number using a base raised to an exponent

Extrapolation [3.7]   The process of estimating a value that goes beyond the given data

Factor [1.2, 5.1]   *Verb:* To write an equivalent expression that is a product; *noun:* part of a product

Factoring [1.2]   The process of rewriting an expression as a product

Factoring by grouping [5.1]   If a polynomial can be split into groups of terms and the groups share a common factor, then the original polynomial can be factored. This method can be tried on any polynomial with four or more terms.

FOIL method [4.6]   To multiply two binomials $A + B$ and $C + D$, multiply the First terms AC, the Outside terms AD, the Inner terms BC, and then the Last terms BD. Then add the results.

Formula [2.3]   An equation using numbers and/or letters to represent a relationship between two or more quantities

Fraction notation [1.3]   A number written using a numerator and a denominator

Function [9.7]   A correspondence between a first set, called the *domain,* and a second set, called the *range,* such that each member of the domain corresponds to *exactly one* member of the range

Grade [3.5]   The ratio of the vertical distance a road rises over the horizontal distance it runs, expressed as a percent

Graph [2.6, 3.1]   A picture or a diagram of the data in a table; a line, a curve, a plane, a collection of points, etc., that represents all the solutions of an equation or an inequality

Grouping method [5.3]   A method for factoring a trinomial of the type $ax^2 + bx + c$, that uses factoring by grouping

Half-open intervals $(a, b]$ and $[a, b)$ [2.6]   An interval that contains one endpoint and not the other; thus, $(a, b] = \{x | a < x \le b\}$ and $[a, b) = \{x | a \le x < b\}$

Horizontal line [3.3, 3.5]   The graph of $y = b$ is a horizontal line, with y-intercept $(0, b)$.

Hypotenuse [5.7, 8.6]   In a right triangle, the side opposite the 90° angle

i [9.5]   The square root of -1; that is, $i = \sqrt{-1}$ and $i^2 = -1$

Identity [2.2]   An equation that is true for all replacements

Identity property of 0 [1.5]   The statement that the sum of 0 and a number is the original number

Identity property of 1 [1.3]   The statement that the product of 1 and a number is the original number

Imaginary number [9.5]   A number that can be written in the form $a + bi$, where a and b are real numbers and $b \ne 0$ and $i = \sqrt{-1}$

Inconsistent system of equations [7.1]   A system of equations for which there is no solution

Independent equations [7.1]   Equations that are not dependent

Index (plural, Indices) [8.7]   In the radical, $\sqrt[n]{a}$, the number n is called the index.

Inequality [1.4, 2.6]   A mathematical sentence using $<, >, \le, \ge,$ or $\ne$

Input [9.7]   An element of the domain of a function

Integers [1.4]   The set of all whole numbers and their opposites: $\{ \ldots, -4, -3, -2, -1, 0, 1, 2, 3, 4, \ldots \}$

Interpolation [3.7]   The process of estimating a value between given values

Intersection of sets A and B [App C]   The set of all elements that are common to both A and B; denoted $A \cap B$

Interval notation [2.6]   The use of a pair of numbers inside parentheses and/or brackets to represent the set of all numbers between and sometimes including those two numbers; see also *open, closed,* and *half-open intervals*

Inverse variation [7.7]   A situation that can be modeled by a rational function of the form $f(x) = k/x$, or $y = k/x$, where k is a nonzero constant

Irrational number [1.4] A real number that cannot be written as the ratio of two integers; when written in decimal notation, an irrational number neither terminates nor repeats

Largest common factor [5.1] The largest common factor of a polynomial is the largest common factor of the coefficients times the largest common factor of the variable(s) in all of the terms.

Leading coefficient [4.3] The coefficient of the term of highest degree in a polynomial

Leading term [4.3] The term of highest degree in a polynomial

Least common denominator (LCD) [6.3] The least common multiple of the denominators of two or more fractions

Least common multiple (LCM) [6.3] The smallest number that is a multiple of two or more numbers

Legs [5.7, 8.6] In a right triangle, the two sides that form the 90° angle

Like radicals [8.4] Radical expressions that have the same indices and radicands

Like terms [1.5, 1.8, 4.3, 4.7] Terms containing the same variable(s) raised to the same power(s); also called *similar terms*

Line graph [3.1] A graph in which quantities are represented as points connected by straight-line segments

Linear equation [2.2, 3.2] In two variables, any equation whose graph is a straight line and can be written in the form $y = mx + b$ or $Ax + By = C$, where x and y are variables and m, b, A, B, and C are constants

Linear inequality [7.5] An inequality whose related equation is a linear equation

Linear regression [3.7] A method for finding an equation for a line that best fits a set of data

Mathematical model [1.1] A mathematical representation of a real-world situation

Mean [2.7, App B] The sum of a set of numbers, divided by the number of addends

Measure of central tendency [App B] A single representative number used to analyze a set of data; also called *center point*

Median [App B] After arranging a set of data from smallest to largest, the middle number if there is an odd number of data numbers, or the average of the two middle numbers if there is an even number of data numbers

Mode [App B] In a set of data, the number or numbers that occur most often; if each number occurs the same number of times, then there is *no* mode

Monomial [4.3] A constant, a variable, or a product of a constant and one or more variables

Motion problem [2.5, 6.7] A problem dealing with distance, rate (or speed), and time

Multiplicative identity [1.3] The number 1

Multiplicative inverses [1.3] *Reciprocals*; two numbers whose product is 1

Multiplicative property of zero [1.7] The statement that the product of 0 and any real number is 0

Natural numbers [1.3] The numbers used for counting: $\{1, 2, 3, 4, 5, \ldots\}$; also called *counting numbers*

Nonlinear equation [3.2] An equation whose graph is not a straight line

nth root [8.7] A number c is called the nth root of a if $c^n = a$.

Numerator [1.3] The number above the fraction bar in a fraction

Open interval (a, b) [2.6] The set of all numbers x for which $a < x < b$; thus, $(a, b) = \{x \mid a < x < b\}$

Opposite of a polynomial [4.4] To find the *opposite* of a polynomial, change the sign of every term; this is the same as multiplying the polynomial by -1

Opposites [1.6] Two expressions whose sum is 0; *additive inverses*

Ordered pair [3.1] A pair of numbers of the form (x, y) for which the order in which the numbers are listed is important

Origin [3.1] The point on a coordinate plane where the two axes intersect

Output [9.7] An element of the range of a function

Parabola [9.6] A graph of a quadratic function

Parallel lines [3.6] Lines in the same plane that never intersect; two lines are parallel if they have the same slope or if both lines are vertical

Percent notation [2.4] The percent symbol % means "per hundred."

Perfect-square trinomial [4.6, 5.4] A trinomial that is the square of a binomial

Perpendicular lines [3.6] Two lines that intersect to form a right angle; two lines are perpendicular if the product of their slopes is -1 or if one line is vertical and the other line is horizontal

Point–slope form [3.7] Any equation of the form $y - y_1 = m(x - x_1)$, where the slope of the line is m and the line passes through (x_1, y_1)

Polynomial [4.3] A monomial or a sum of monomials

Prime factorization [1.3] The factorization of a composite number into a product of prime numbers

Prime number [1.3] A natural number that has exactly two different natural number factors: the number itself and 1

Prime polynomial [5.2] A polynomial that cannot be factored using rational numbers

Principal square root [8.1] The nonnegative square root of a number

Product [1.2] The result when two numbers are multiplied

Proportion [6.7] An equation stating that two ratios are equal

Pythagorean theorem [5.7, 8.6] In any right triangle, if a and b are the lengths of the legs and c is the length of the hypotenuse, then $a^2 + b^2 = c^2$.

Quadrants [3.1] The four regions into which the horizontal axis and the vertical axis divide a plane

Quadratic equation [5.6] An equation equivalent to one of the form $ax^2 + bx + c = 0$, where a, b, and c are constants, with $a \neq 0$

Quadratic formula [9.3]   The formula $x = \dfrac{-b \pm \sqrt{b^2 - 4ac}}{2a}$, which gives the solutions of $ax^2 + bx + c = 0, a \neq 0$

Quotient [1.2]   The result when two numbers are divided

Radical equation [8.5]   An equation in which a variable appears in a radicand

Radical expression [8.1]   Any expression in which a radical sign appears

Radical sign [8.1]   The symbol $\sqrt{}$

Radicand [8.1, 8.7]   The expression under a radical sign

Range [9.7]   The set of all second coordinates of the ordered pairs in a function

Rate [3.4]   A ratio that indicates how two quantities change with respect to each other

Ratio [6.7]   The quotient of two quantities

Rational equation [6.6]   An equation that contains one or more rational expressions

Rational expression [6.1]   A quotient of two polynomials

Rational numbers [1.4]   The set of all numbers a/b, such that a and b are integers and $b \neq 0$

Rationalizing the denominator [8.3]   A procedure for finding an equivalent expression without a radical expression in the denominator

Real numbers [1.4]   The set of all numbers corresponding to points on the number line

Reciprocals [1.3]   Two numbers whose product is 1; *multiplicative inverses*

Repeating decimal [1.4]   A decimal in which a block of digits repeats indefinitely

Revenue [4.8]   The price per item times the quantity of items sold

Right triangle [5.7, 8.6]   A triangle that has a 90° angle

Scientific notation [4.2]   An expression of the type $N \times 10^m$, where N is at least 1 but less than 10 (that is, $1 \leq N < 10$), N is expressed in decimal notation, and m is an integer

Set [1.4, App C]   A collection of objects

Set-builder notation [2.6]   The naming of a set by describing basic characteristics of the elements in the set

Similar terms [1.5, 1.8, 4.3, 4.7]   Terms containing the same variable(s) raised to the same power(s); also called *like terms*

Similar triangles [6.7]   Triangles in which corresponding angles have the same measure and corresponding sides are proportional

Slope [3.5]   The ratio of vertical change to horizontal change for any two points on a line

Slope–intercept equation [3.6]   An equation of the form $y = mx + b$, with slope m and y-intercept $(0, b)$

Solution [2.1, 2.6]   Any replacement or substitution for a variable that makes an equation or inequality true

Solution of a system [7.1]   A solution of a system of two equations makes *both* equations true.

Solution set [2.6, 7.6]   The set of all solutions of an equation, an inequality, or a system of equations or inequalities

Solve [2.1, 2.6, 7.1]   To find all solutions of an equation, an inequality, or a system of equations or inequalities

Speed [3.4]   The speed of an object is found by dividing the distance traveled by the time required to travel that distance.

Square root [8.1]   The number c is a *square root* of a if $c^2 = a$.

Standard form of a linear equation [3.7]   Any equation of the form $Ax + By = C$, where A, B, and C are real numbers and A and B are not both 0

Subset [App C]   If every element of A is also an element of B, then A is a *subset* of B; denoted $A \subseteq B$

Substitute [1.1]   To replace a variable with a number or an expression

Substitution method [7.2]   An algebraic method for solving a system of equations

Sum [1.2]   The result when two numbers are added

Sum of cubes [App A]   An expression that can be written in the form $A^3 + B^3$

System of equations [7.1]   A set of two or more equations, in two or more variables, for which a common solution is sought

System of linear inequalities [7.6]   A set of two or more inequalities that are to be solved simultaneously

Term [1.2, 4.3]   A number, a variable, a product of numbers and/or variables, or a quotient of numbers and/or variables

Terminating decimal [1.4]   A decimal that can be written using a finite number of decimal places

Trinomial [4.3]   A polynomial with three terms

Undefined [1.7]   An expression that has no meaning attached to it

Union of sets A and B [App C]   The collection of elements belonging to A and/or B; denoted $A \cup B$

Value [1.1]   The numerical result after a number has been substituted for a variable in an expression and calculations have been carried out

Variable [1.1]   A letter that represents an unknown number

Variable expression [1.1]   An expression containing a variable

Variation constant [7.7]   The constant k in an equation of direct variation or inverse variation; also called *constant of proportionality*

Vertex (plural, vertices) [9.6]   The "turning point" of the graph of a quadratic equation

Vertical line [3.3, 3.5]   The graph of $x = a$ is a vertical line, with x-intercept $(a, 0)$.

Vertical-line test [9.7]   The statement that if it is possible for a vertical line to cross a graph more than once, then the graph is not the graph of a function

Whole numbers [1.4]   The set of natural numbers and 0: $\{0, 1, 2, 3, 4, 5, \ldots\}$

x-axis [3.1]   The horizontal axis in a coordinate plane

x-intercept [3.3]   A point at which a graph crosses the x-axis

y-axis [3.1]   The vertical axis in a coordinate plane

y-intercept [3.3]   A point at which a graph crosses the y-axis

Index

Note: Page numbers followed by n refer to footnotes.

Abscissa, 156n
Absolute value, 34, 74
  on graphing calculator, 34
  square roots and, 494
Addition. *See also* Sums
  associative law for, 13–14, 72
  commutative law for, 12, 72
  of fractions, 23–24
  of negative numbers, 39
  of polynomials, 254–255, 296
    checking on graphing calculator, 259
    in several variables, 281, 297
  of radical expressions, 509–510, 538
  of rational expressions
    on graphing calculator, 395
    with unlike denominators, 391–394, 429
    when denominators are the same, 383, 428
  of real numbers, 38–41, 74
    with number line, 38–39
    problem solving and, 40–41
    rules for, 39
    without number line, 39–40
    with zero, 39
  words indicating, 4
Addition principle, 81–82, 146
  for inequalities, 130–131, 148
    with multiplication principle, 132–134
    multiplication principle with, 88–89, 132–134
Additive identity, 39, 74
Additive inverses. *See* Opposites
Algebraic expressions, 2
  evaluating, 2–3, 72
  translating to, 3–7, 72
  value of, 2–3, 72
Algebraic fractions. *See* Rational expressions
Apps for graphing calculator, 169
Ascending order, 249
Associative laws
  for addition, 13–14, 72
  for multiplication, 13, 72
at least, 137
at most, 137
Average, 138–139

Axes of graphs, 156, 223
  numbering, 157–159
Axis of symmetry of a parabola, 571, 574, 590

Bar graphs, 154–155
Bases (in exponential notation), 60
Binomials, 247, 296
  dividing polynomials by, 288–290, 297
  multiplication of, with a binomial, FOIL method for, 271–272
  squaring, 273–274, 297
Braces ({ }), grouping symbols, 61
Brackets ([ ]), grouping symbols, 61

Canceling, 23
Cartesian coordinate plane, 156n
Center points, 599
Central tendency measures, 138–139, 599–601
Change, rates of, 181–183
Circles, circumference of, 96
Circumference of a circle, 96
Clearing decimals and fractions, 91–92, 146
Closed interval, 129
Coefficients
  of terms of polynomials, 247–248, 295
    leading, 248, 295
  of variables, 83–84
Collecting like terms, 41, 47, 64, 90–91, 248–249, 296
Combining like terms, 41, 47, 64, 90–91, 248–249, 296
Common factors, 16
  factoring polynomials with terms having, 304–307, 363
Commutative laws
  for addition, 12, 72
  for multiplication, 12, 72
Completing the square, 548–551, 589
  solving quadratic equations by, 550–551
Complex numbers, 568
  solutions of quadratic equations involving, 568–569
Complex rational expressions, 400–404, 429
  multiplying by the LCD, 402–404
  simplifying, by dividing, 400–402
Complex-number system, 567–568

Composite numbers, 20, 72
Conic sections. *See* Circles; Parabolas
Conjugates, 511, 538
Connecting the Concepts, 346
  definitions and properties of exponents, 243
  equations describing a line, 217
  equivalent expressions and equivalent equations and, 411
  operations on real numbers, 57
  solving inequalities versus solving equations, 134
  solving radical equations, 518
  solving systems of equations, 454
  types of equations, 569–570
Consistent systems of equations, 437, 439, 485
Constant(s), 2
  of proportionality (variation constant), 476, 478
Contrast, on graphing calculator, 7
Coordinates, 156–157, 223
Cube(s), sums or differences of, factoring, 597–598
Cube roots, 530, 539
Cursor, on graphing calculator, 7

Data, 6
Decimal(s)
  clearing, 91–93, 146
  repeating, 30, 73
  terminating, 30, 73
Decimal notation, converting between percent notation and, 104–105, 147
Degree(s)
  of a monomial, 247
  of a polynomial, 248, 295
  in several variables, 280
Denominators, 19
  least common, 384–388, 428
  rationalizing, 506–507, 538
    with two terms, 511
  same
    addition of rational expressions with, 383, 428
    subtraction of rational expressions with, 384, 428

Denominators (*continued*)
  unlike
    addition of rational expressions
      with, 391–394, 429
    subtraction of rational expressions
      with, 391–394, 429
Dependent equations, 438, 439, 485
Descartes, René, 156n
Descending order, 249
Differences, 44. *See also* Subtraction
  of cubes, factoring, 597–598
  of squares, 330–331
    factoring, 330–331, 364
    recognizing, 330
Direct variation
  equations of, 476–477, 487
  problem solving with, 477–478
Distributive law, 14–16, 72
  factoring and, 16
  simplification and, 63–64
Division. *See also* Quotient(s)
  of fractions, 21
  by negative one, 84
  of polynomials, 287–290, 297
    by binomials, 288–290, 297
    by monomials, 287–288, 297
  of powers with like bases, 231–232
  of rational expressions, 378–379, 428
    on graphing calculator, 379
  of real numbers, 53–57, 75
    rules for, 54
  simplifying complex rational
    expressions using, 400–402
  of square roots, 504–507, 538
    quotient rule for square roots and,
      504, 538
    rationalizing denominators with
      one term and, 506–507, 538
  using scientific notation, 242–243
  words indicating, 4
  by zero, as undefined, 54, 56, 75
Domains of functions, 578, 591

Elements of a set, 602
Elimination method, for solving systems
  of equations, 449–454, 486
Empty set, 602
Equality, words indicating, 5
Equals symbol (=), 5
Equations
  dependent and independent, 438,
    439, 485
  of direct variation, 476–477, 487
  equivalent, 81
  expressions compared with, 5
  of inverse variation, 478–479, 487
  for a line, writing, 212
  linear. *See* Linear equations
  point–slope, 211–213
  polynomial, solving using graphing
    calculator, 347

Pythagorean, 353
quadratic. *See* Quadratic equations
radical, 515–518, 538
  applications involving, 517–518
  solving, 515–517
  solving using graphing calculator, 517
rational, 407–411, 430
  applications involving, 414–422
  applications using, 430
  solving using graphing calculator,
    410
selecting correct approach to solve,
  85–86
in slope–intercept form, 204–206
solutions of, 80–81, 146
solving, simplifying an expression
  versus, 65
systems of. *See* Systems of equations
translating to, 5–7
in two variables, solutions of, 163–164
writing using point–slope form,
  211–213
Equivalent equations, 81
Equivalent expressions, 81, 341
Equivalent inequalities, 130
Estimations using two points, 214–215
Evaluation
  of algebraic expressions, 2–3, 72
  of formulas, 96
  of polynomials, 249–250, 296
    checking calculations using, 267
    on graphing calculator, 282
    in several variables, 279, 297
Exponent(s), 60, 230–235, 295. *See also*
  Power(s)
  dividing powers with like bases and,
    231–232
  multiplying powers with like bases
    and, 230–231
  negative integers as, 237–240, 295
  raising a power to a power and, 233
  raising a product or quotient to a
    power and, 233–235, 295
  rational, 532–533
    negative, 533
    positive, 532
  zero as, 232–233, 295
Exponential expressions on graphing
  calculator, 61
Exponential notation, 60–61, 75
Expressions
  algebraic, 2
    evaluating, 2–3, 72
    translating to, 3–7, 72
    value of, 2–3, 72
  equations compared with, 5
  equivalent, 81, 341
  exponential, on graphing
    calculator, 61
  fraction. *See* Rational expressions
  indeterminate, 54, 56, 75

radical. *See* Radical expressions
rational. *See* Rational expressions
simplifying, solving an equation
  versus, 65
undefined, 54, 56, 75
variable, 2
Extrapolation, 214

Factor(s), 14
  common, 16
    factoring polynomials with terms
      having, 304–307, 363
  negative exponents and, 239
Factoring, 16, 304–309, 363
  completely, 316, 332
  of differences of squares, 330–331, 364
  distributive law and, 16
  general strategy for, 364
  of monomials, 304
  of polynomials, 336–339, 364
    choosing method for, 336–339, 364
    by grouping, 307–309, 363
    tips for, 307
    when terms have a common factor,
      304–307, 363
  solving quadratic equations by,
    343–346, 365
  sums or differences of cubes, 597–598
  tips for, 332
  of trinomials. *See* Factoring trinomials
Factoring trinomials
  perfect-square, 329–330
  of type $ax^2 + bx + c$, 319–325,
    363–364
    with FOIL, 319–323, 363
    with grouping method, 323–325, 364
    tips for, 322
  of type $x^2 + bx + c$, 311–316, 363
    factoring completely and, 316
    FOIL method in reverse for, 312
    prime polynomials and, 315–316
    when constant term is negative,
      314–315
    when constant term is positive,
      312–313
Factorizations, 16, 19, 304
  checking by multiplication, 309
  checking using graphing calculator, 309
  complete, 332
  prime, 20, 73
First coordinate, 156
FOIL method
  factoring trinomials of type
    $ax^2 + bx + c$ using, 319–323, 363
  in reverse, factoring trinomials of
    type $x^2 + bx + c$ using, 312
Formulas, 96–99, 146
  evaluating, 96
  motion, 118–119
  quadratic, 553–557
    problem solving involving, 555–557

solving for a given variable, 96–99, 563–564, 590
    on graphing calculator, 96
Fraction(s)
    addition of, 23–24
    algebraic. *See* Rational expressions
    clearing, 91–92, 146
    division of, 21
    multiplication of, 20–21
    subtraction of, 23, 25
Fraction expressions. *See* Rational expressions
Fraction notation, 19–26, 73–74
    for one, 21, 73
Function(s), 578–584, 591
    checking solutions using graphing calculator, 580
    defined, 578
    domain of, 578, 591
    graphs of, 581–584
        recognizing, 582–584
    identifying, 578–579
    range of, 578, 591
Function notation, 579–580
Function values, 580

Graph(s), 153–224
    axes of, 156, 223
        numbering, 157–159
    bar, 154–155
    of functions, 581–584
        recognizing, 582–584
    of horizontal lines, 177–178, 223
    of inequalities, 128–129, 148
    intercepts and, 173–178, 223
    line, 155
    of linear equations, 165–169, 223
    ordered pairs on, 156
    origin of, 156, 223
    plotting, 156
    points on, 156
    quadrants of, 157, 223
    of quadratic equations, 571–575, 590
        of form $y = ax^2$, 572
        of form $y = ax^2 + bx + c$, 573–575
    of quadratic functions. *See* Parabolas
    of vertical lines, 177–178, 223
Graphing, 156
    of linear inequalities, 467–469, 486
    point–slope form and, 213–214
    of quadratic equations, guidelines for, 575
    solving systems of equations by, 437–440
    of systems of linear inequalities in two variables, 472–474, 487
Graphing calculator
    absolute value on, 34
    addition of rational expressions using, 395
    apps for, 169
    checking factorizations using, 309

checking function solutions using, 580
checking polynomial addition or subtraction using, 259
checking polynomial multiplication using, 268
checking solutions of quadratic equations using, 548
checking solutions of systems of equations using, 443
contrast on, 7
cursor on, 7
displaying systems of linear inequalities using, 473
division of rational expressions using, 379
evaluating polynomials using, 282
exponential expressions on, 61
finding numbers for which rational expressions are undefined using, 371
graphing expressions containing radical expressions using, 494
graphing linear inequalities using, 467
home screen of, 7
keypad of, 7
linear equations, 169
menus on, 26
negation and subtraction keys on, 46
parentheses to specify radicand using, 511
radicands using, 493
scientific notation and, 242
screen of, 7
simplifying rational expressions using, 373
solving formulas for a variable, 96
solving inequalities using, 130–131, 148
solving polynomial equations using, 347
solving quadratic equations using, 353, 545, 554
solving radical equations using, 517
solving rational equations using, 410
solving systems of equations using, 439
square roots on, 32
subtraction of rational expressions using, 395
TABLE feature of, 90
TRACE key on, 250
viewing intercepts on, 176
viewing window of, 158
    dimensions of, 158
    standard, 158
Greater than or equal to symbol ($\geq$), 33
Greater than symbol ($>$), 33, 74
Grouping method
    factoring polynomials by, 307–309, 363
    factoring trinomials of type $ax^2 + bx + c$ using, 323–325, 364
Grouping symbols, 61

Half-open interval, 129
Home screen of graphing calculator, 7
Horizontal lines
    graphing, 177–178, 223
    slope of, 194–195, 224
Hypotenuse, 353

Identities, 93
    additive, 39, 74
    multiplicative, 22, 73
Identity property of one, 22, 73
Imaginary numbers, 568
Inconsistent systems of equations, 437, 439, 485
Independent equations, 438, 439, 485
Independent variable, 580
Indeterminate expressions, 54, 56, 75
Index, of a root, 530, 539
Inequalities, 33
    addition principle for, 130–131, 148
    equivalent, 130
    graphs of, 128–129, 148
    linear. *See* Linear inequalities
    multiplication principle for, 131–132, 148
    solutions of, 127–128
    solving, 127–134, 148
        using graphing calculator, 130–131, 148
    solving problems with, 137–139, 148
    translating to, 137
Inputs, of domains of functions, 578
Integers, 29, 73
Intercepts
    finding, 174
    graphs and, 173–178, 223
    of a parabola, 573
    viewing on graphing calculator, 176
in terms of, 6
Interpolation, 214
Intersections of sets, 603
Interval notation, 129–130, 148
Inverse(s), additive. *See* Opposites
Inverse variation
    equations of, 478–479, 487
    problem solving with, 479–480
Irrational numbers, 31, 73, 493

Keypad of graphing calculator, 7

Law of opposites, 45
Least common denominator (LCD), 384–388, 428
Least common multiple (LCM), 385
Legs of right triangles, 353
Less than or equal to symbol ($\leq$), 33
Less than symbol ($<$), 33, 74
Like radicals, 509, 538
Like terms
    combining (collecting), 41, 47, 64, 90–91, 248–249, 296
    definition of, 63–64

Line(s)
  horizontal
    graphing, 177–178, 223
    slope of, 194–195, 224
  parallel, slope–intercept form and,
    207–208, 224
  vertical
    graphing, 177–178, 223
    slope of, 195, 224
  writing equations for, 212
Linear equations, 163–169
  applications of, 168–169
  graphing, 165–169, 223
  on graphing calculator, 169
  point–slope form of, 212–217, 224
  slope–intercept form of, 203–208,
    224
    equations in, 204–206
    graphing and, 203–204, 206–207
    parallel lines and, 207–208, 224
  solutions of, 163–164
  systems of. *See* Systems of equations
Linear inequalities
  displaying systems of, using graphing
    calculator, 473
  graphing, 467–469, 486
    using calculator, 467
  in one variable, 469–470
  systems of. *See* Systems of linear
    inequalities in two variables
  in two variables, 467–469, 486
Line graphs, 155

Mathematical models, 6
Mean, 138–139, 599
Measures of central tendency, 138–139,
  599–601
Median, 599–600
Menus on graphing calculator, 26
Mixture problems, 459–461, 486
Mode, 599, 600–601
Models, 6
Monomials, 246, 247, 296
  degree of, 247
  dividing polynomials by, 287–288,
    297
  factoring, 304
  multiplication of, 264
    with polynomials, 264–265
Motion formula, 118–119
Motion problems, 416, 430
Multiplication. *See also* Product(s)
  associative law for, 13, 72
  checking factorizations by, 309
  commutative law for, 12, 72
  of fractions, 20–21
  by LCD, simplifying complex rational
    expressions using, 402–404
  of monomials, 264
    with polynomials, 264–265
  by negative one, 84

  of polynomials
    checking using graphing
      calculator, 268
    with a monomial, 264–265
    with a polynomial, 265–267, 275–276
    in several variables, 281–282, 297
  of powers with like bases, 230–231
  of radical expressions, 498, 510–511, 537
    simplification and, 501
  of rational expressions, 377–378, 428
    when factors are opposites, 394–396
  of real numbers, 51–53, 75
    rules for, 54
  of sums and differences of two terms
    of polynomials, 272–273, 297
  of two binomials, FOIL method for,
    271–272
  using scientific notation, 242–243
  words indicating, 4
Multiplication principle, 83–85, 146
  addition principle with, 88–89, 132–134
  for inequalities, 131–132, 148
    with addition principle, 132–134
Multiplicative identity, 22, 73, 83
Multiplicative inverses, 21, 83
Multiplicative property of zero, 52

Natural numbers, 19
Negation key on graphing calculator, 46
Negative exponents, 237–240, 295
  factors and, 239
  rational, 533
  reciprocals and, 240
Negative numbers
  addition of, 39
  integers, as exponents, 237–240, 295
  product of positive numbers and, 52
  products of two negative numbers
    and, 52–53
Negative one
  division by, 84
  factoring out to reverse subtraction, 308
  multiplying and dividing by, 84
  property of, 53
Notation
  decimal, converting between percent
    notation and, 104–105, 147
  exponential, 60–61, 75
  fraction, 19–26, 73–74
    for 1, 21, 73
  function, 579–580
  interval, 129–130, 148
  percent, converting between decimal
    notation and, 104–105, 147
  roster, 602
  scientific, 240–243, 295
  set-builder, 129, 148, 602
*n*th root, 530, 539
Number(s)
  complex, 567–568
  composite, 20, 72

  imaginary, 568
  irrational, 493
  natural, 19
  negative
    addition of, 39
    integers, as exponents, 237–240, 295
    products of positive numbers
      and, 52
    products of two negative numbers
      and, 52–53
  positive
    addition of, 39
    product of negative numbers and, 52
  prime, 20, 72
  rational, 29–31, 73
  real. *See* Real numbers
  whole, 28, 73
Number line, addition of real numbers
  using, 38–39
Numerators, 19

One
  fraction notation for, 21, 73
  identity property of, 22, 73
  negative
    factoring out to reverse
      subtraction, 308
    multiplying and dividing by, 84
    property of, 53
Open interval, 129
Operation(s), 2
Operation signs, 2
Opposites, 44–46, 74
  factors as, 374–375
    in rational expressions, 374–375,
      394–396
  law of, 45
  of opposites, 44–45
  of polynomials, 255–256
  of a sum, 64–66, 75
Ordered pairs, 156–157, 223
Order of operations, 61–63, 75
  rules for, 61
Ordinate, 156n
Origin of a graph, 156, 223
Outputs, of domains of functions, 578

Parabolas
  axis of symmetry of, 571, 574, 590
  intercepts of, 573
  vertices of, 572, 573–574, 590
Parallel lines, slope–intercept form and,
  207–208, 224
Parentheses (()), as grouping
  symbols, 61
Percent increase/decrease, 120–121
Percent notation, converting
  between decimal notation
  and, 104–105, 147
Percent problems, solving, 105–108
Perfect squares, 493

Perfect-square trinomials, 273–274, 297, 328–330, 364
  factoring, 329–330
  recognizing, 328–329
Pipes (‖), grouping symbols, 61
Plotting, 156
Point(s), estimations and predictions using two points and, 214–215
Point–slope equations, 211–213
Point–slope form, 212–217, 224
  graphing and, 213–214
  writing equations using, 211–213
Polynomial(s), 229–297, 246
  addition of, 254–255, 296
    checking on graphing calculator, 259
    polynomials in several variables and, 281, 297
  applications of, 249–250, 296
  degree of, 248, 295
  division of, 287–290, 297
    by binomials, 288–290, 297
    by monomials, 287–288, 297
  evaluating, 249–250, 296
    checking calculations using, 267
    on graphing calculator, 282
    in several variables, 279, 297
  exponents and. *See* Exponent(s)
  factoring
    by grouping, 307–309, 363
    when terms have a common factor, 304–307, 363
  multiplication of
    checking on graphing calculator, 268
    with a monomial, 264–265
    with a polynomial, 275–276
    polynomials in several variables and, 281–282, 297
  opposites of, 255–256
  prime, factoring, 315–316
  problem solving and, 257–258
  quotients of. *See* Rational expressions
  in several variables, 279–282, 297
    addition of, 281, 297
    degree of, 280
    like terms of, 280
    multiplication of, 281–282, 297
    subtraction of, 281, 297
  subtraction of, 256–257, 296
    checking on graphing calculator, 259
    polynomials in several variables and, 281, 297
Polynomial equations, solving using graphing calculator, 347
Positive numbers
  addition of, 39
  product of negative numbers and, 52
Positive rational exponents, 532
Power(s), 60. *See also* Exponent(s)
  like
    dividing powers with, 231–232
    multiplying powers with, 230–231

raising to a power, 233, 295
  raising a product or quotient to, 233–235, 295
Power rule, 233
Predictions using two points, 214–215
Prime factorization, 20, 73
Prime numbers, 20, 72
Prime polynomials, factoring, 315–316
Principal square root, 492
Principle of square roots, 544–545, 589
Principle of squaring, 515, 538
Principle of zero products, 342–343, 365
Problem solving
  five steps for, 112–121, 147
  tips for, 119
Product(s), 14. *See also* Factorizations; Multiplication
  involving higher roots, 531
  of a negative number and a positive number, 52
  raising to a power, 233–234, 295
  of two negative numbers, 52–53
Product rule, 230–231, 295, 531
  for square roots, 498, 537
Property of −1, 53
Proportion(s), 418
  applications using, 418–420, 430
Proportionality, constant of, 476, 478
Pythagorean equation, 353
Pythagorean theorem, 353–354, 365, 521–522, 524–525, 538

Quadrants, 157, 223
Quadratic equations, 303, 543–595
  applications of, 350–352
  checking solutions using graphing calculator, 548
  completing the square and, 548–551, 589
  of form $y = ax^2$, graphing, 572
  of form $y = ax^2 + bx + c$, graphing, 573–575
  graphing, guidelines for, 575
  graphs of, 571–575, 590
  principle of square roots and, 544–545, 589
  principle of zero products and, 342–343, 365
  solutions involving complex numbers, 568–569
  solving by factoring, 343–346, 365
  solving using graphing calculator, 353, 545, 554
  in standard form, 553, 589
  of type $(x + k)^2 = p$, solving, 545–546
Quadratic formula, 553–557
  problem solving involving, 555–557
Quotient(s). *See also* Division
  involving higher roots, 531
  raising to a power, 234–235, 295
Quotient rule, 231–232, 295, 531
  for square roots, 504, 538

Radical equations, 515–518, 538
  applications involving, 517–518
  solving, 515–517
  using graphing calculator, 517
Radical expressions, 493, 537
  applications involving, 495
  graphing expressions containing, using graphing calculator, 494
  multiplication of, 498, 537
    simplification and, 501
  with several terms, 509–511
    addition of, 509–510, 538
    multiplication of, 510–511
    rationalizing denominators of, 511, 538
    subtraction of, 509–510, 538
  simplification of, 499–501
    factoring and, 499–500
    multiplication and, 501
    square roots of powers and, 500–501
Radicals, like, 509, 538
Radical sign (√), 492
Radicands, 493, 530
  using graphing calculator, 493, 511
Ranges, of functions, 578, 591
Rates, 181–184, 223
  of change, 181–183
  slope and, 190–194
  visualizing, 183–184
Ratio(s), 418
Rational equations, 407–411, 430
  applications using, 414–422, 430
  solving using graphing calculator, 410
Rational exponents, 532–533
  negative, 533
  positive, 532
Rational expressions, 370–375, 428
  addition of
    on graphing calculator, 395
    with unlike denominators, 391–394, 429
    when denominators are the same, 383, 428
  complex, 400–404, 429
    multiplying by the LCD, 402–404
    simplifying by dividing, 400–402
    simplifying using division, 400–402
  conjugates, 511, 538
  division of, 378–379, 428
    on graphing calculator, 379
  factors that are opposites and, 374–375
  least common denominators and, 384–388, 428
  multiplication of, 377–378, 428
    when factors are opposites, 394–396
  restricting replacement values and, 370–371
  simplification of, 371–374
    on graphing calculator, 373

Rational expressions (*continued*)
  subtraction of
    on graphing calculator, 395
    with unlike denominators,
      391–394, 429
    when denominators are the same,
      384, 428
Rational numbers, 29–31, 73
Rationalizing the denominator,
  506–507, 538
  with two terms, 511
Real numbers, 32–33, 73
  addition of, 38–41, 74
    with number line, 38–39
    problem solving and, 40–41
    rules for, 39
    without number line, 39–40
    with zero, 39
  division of, 53–57, 75
    rules for, 54
  multiplication of, 51–53, 75
    rules for, 54
  negative, addition of, 39
  positive, addition of, 39
  subtraction of, 44–48, 74
    problem solving and, 48
Real-number system, 32
Reciprocals, 21
  negative exponents and, 240
Repeating decimals, 30, 73
Right triangles, 353, 521–525, 538
  distance formula and, 524–525, 539
  problem solving involving, 523–524
  Pythagorean theorem and, 521–522,
    524–525, 538
Roots, square, 328
  on graphing calculator, 32
Roster notation, 602

Scientific notation, 240–243, 295
  division using, 242–243
  on graphing calculator, 242
  multiplication using, 242–243
Screen of graphing calculator, 7
Second coordinate, 156
Set(s), 28, 602–603
  elements of, 602
  empty, 93, 602
  of integers, 29
  intersections of, 603
  members of, 602
  naming, 602
  of rational numbers, 30
  of real numbers, 32, 73
  subsets of, 602–603
  unions of, 603
Set-builder notation, 129, 148, 602
Similar terms
  combining, 41
  definition of, 63–64
Similar triangles, 418, 431

Simplification
  of an expression, solving an equation
    versus, 65
  distributive law and, 63–64
  of radical expressions, 499–501
    factoring and, 499–500
    multiplying and, 501
    square roots of powers and, 500–501
  of rational expressions, 371–374
    on graphing calculator, 373
Slope, 190–196, 224
  applications of, 195–196
  graphing lines using y-intercept and,
    203–204
  of horizontal lines, 194–195, 224
  rate and, 190–194
  of vertical lines, 195, 224
Slope–intercept form, 203–208, 224
  equations in, 204–206
  graphing and, 203–204, 206–207
  parallel lines and, 207–208, 224
  perpendicular lines and, 208
Solutions
  of equations, 80–81, 146
    in two variables, 163–164
  of inequalities, 127–128
  linear, 163–164
  of systems of equations, 436–437, 485
Solution sets, 128
Solving formulas for a given variable,
  96–99
  on graphing calculator, 96
Square(s)
  of binomials, 273–274, 297
  completing, 548–551, 589
    solving quadratic equations by,
      550–551
  differences of
    factoring, 330–331, 364
    recognizing, 330
  perfect, 493
  trinomial, 273, 297
Square roots, 328, 492, 537
  absolute value and, 494
  division of, 504–507, 538
    quotient rule for square roots and,
      504, 538
    rationalizing denominators with
      one term and, 506–507, 538
  on graphing calculator, 32
  principal, 492
  principle of, 544–545, 589
  product rule for, 498, 537
  quotient rule for, 504, 538
  simplified form of, 499
Squaring, principle of, 515, 538
Standard form, quadratic equations in,
  553, 589
Study tips
  abbreviations in notes, 530
  aiming for mastery, 257

  answer section use, 88
  asking questions in class, 280
  avoiding interruptions, 138
  checking answers, 414
  continual review of material, 104
  creating your own glossary, 175
  doing the exercises, 22
  doing homework promptly, 377
  emailing questions to your instructor,
    304
  exercise breaks, 370
  familiar topics, 445
  finding all solutions, 407
  finishing a chapter, 350
  getting course information, 7
  helping classmates, 230
  highlighting, 342
  information about topics on final
    exam, 563
  keeping a section ahead, 30
  keeping up your effort, 544
  keeping your focus, 337
  learning by example, 12
  learning multiple methods, 400
  looking ahead in text, 266
  looking for connections, 238
  mastering new topics, 436
  missed classes, 46
  note taking, 190
  organizing your supporting
    work, 312
  organizing your work, 53
  pacing yourself, 98
  planning future courses, 499
  power naps, 476
  professors' errors, 492
  questions that stump you, 392
  reading the instructions, 319
  reading your text, 246
  resources, 85
  reviewing earlier material, 504
  reviewing for exams, 287
  reviewing your mistakes, 214, 578
  rewriting problems in an equivalent
    form, 449
  setting reasonable expectations, 115
  sketching, 521
  sleep, 130, 571
  solving multiple exercises, 203
  study groups, 40
  studying for the final exam,
    515, 548, 553
  studying together by phone, 510
  study partners, 182
  summarizing your notes, 567
  time management, 468
  using a second textbook, 271
  using math in everyday life, 457, 472
  verbalizing your questions, 167
  visualizing the steps, 383
  working with a pencil, 154

writing out missing steps, 328
writing steps when working
     problems, 60
Subsets, 602–603
Substitution method, for solving systems
     of equations, 442–445, 485
Subtraction. *See also* Differences
     factoring out −1 to reverse, 308
     of fractions, 23, 25
     of polynomials, 256–257, 296
          checking on graphing calculator, 259
          in several variables, 281, 297
     of radical expressions, 509–510, 538
     of rational expressions
          on graphing calculator, 395
          with unlike denominators,
               391–394, 429
          when denominators are the same,
               384, 428
     of real numbers, 44–48, 74
          problem solving and, 48
     words indicating, 4
Subtraction key on graphing
     calculator, 46
Sums, 14. *See also* Addition
     of cubes, factoring, 597–598
     opposites of, 64–66, 75
Systems of equations, 436–440
     applications involving, 457–461, 486
     checking solutions using graphing
          calculator, 443
     consistent and inconsistent,
          437, 439, 485
     problems involving, 445–446, 453
     solutions of, 436–437, 485
     solving by graphing, 437–440
     solving using elimination method,
          449–454, 486
     solving using graphing calculator,
          439
     solving using substitution method,
          442–445, 485

Systems of linear inequalities in two
     variables, 472–474, 487
     graphing, 472–474, 487

TABLE feature on graphing
     calculator, 90
Term(s), 14
     like, combining (collecting), 41, 47, 64,
          90–91, 248–249, 296
     of polynomials, 246–247, 295
          coefficient of, 247–248, 295
          degree of, 280
          leading, 248, 295
          like, 280
     radical. *See Radical expressions, with
          several terms*
Terminating decimals, 30, 73
Total-value problems, 457–459, 486
TRACE key on graphing
     calculator, 250
Translation
     to algebraic expressions, 3–7, 72
     to equations, 5–7
Triangles
     right. *See* Right triangles
     similar, 418, 431
Trinomials, 247, 296
     factoring. *See* Factoring trinomials
     perfect-square, 273, 297, 328–330, 364
          factoring, 329–330
          recognizing, 328–329
Trinomial squares, 273, 297

Undefined expressions, 54, 56, 75
     rational, finding numbers for, using
          graphing calculator, 371
Unions of sets, 603

Values
     absolute. *See* Absolute value
     of an algebraic expression, 2–3, 72
     function, 580

Variable(s), 2
     coefficients of, 83–84
     independent, 580
Variable expressions, 2
Variation
     direct
          equations of, 476–477, 487
          problem solving with, 477–478
     inverse
          equations of, 478–479, 487
          problem solving with, 479–480
Variation constant, 476, 478
Vertex(ices), of a parabola, 572,
     573–574, 590
Vertical lines
     graphing, 177–178, 223
     slope of, 195, 224
Vertical-line test, 583–584, 591
Viewing window of graphing
     calculator, 158
     dimensions of, 158
     standard, 158

Whole numbers, 28, 73
Work principle, 415, 430
Work problems, 414–419

x-coordinate of a y-intercept, 173
x-intercept, 173

y-coordinate of an x-intercept, 173
y-intercept, 173
     graphing lines using slope and, 203–204

Zero(s)
     addition of real number and, 39
     division by, as undefined, 54, 56, 75
     as exponent, 232–233
     multiplicative property of, 52
Zero products, principle of, 342–343,
     365

Index of Applications

Agriculture
Apple picking rate, 189
Dairy farming, 465
Gardening, 356, 414–415, 430, 456, 560
Grass seed, 463
Organic gardening, 225
Picking produce, 415
Planting grapes, 455

Astronomy
Composition of the sun, 109
Jupiter's atmosphere, 109
Light-years, 245
Mass of the earth, 240
Milky Way galaxy, 245
Orbit time, 8
Rocket sections, 123

Automotive
Automobile maintenance, 465
Automobile tax, 124
Car depreciation, 222
Commuter parking permit, 458–459
Driving under the influence, 160
Engine size, 172
Fuel economy, 170, 172
Gas mileage, 182, 188, 189, 226, 423
Insurance-covered repairs, 141
Octane ratings, 464, 465
Safe driving, 520
Speed of a skidding car, 491, 503, 519
Stopping distance, 576

Biology
Cricket chirps and temperature, 126
DNA strands, 246
Earth's surface, 104
Ecology, 245, 482
Elephant measurements, 229, 283
Endangered species, 78, 124
Golden lion tamarins, 78
Heart rates of animals, 578
Herpetology, 534
Horticulture, 464
Humpback whale population, 423
Life span of an animal, 578, 585
Marine biology, 566
Monarch butterflies, 124
Predicting heights, 585
Reptiles, 527
Speed of a hummingbird, 183

Veterinary science, 253, 423
Weight of a fish, 102
Wildlife population, 369, 420, 424, 432
Wing aspect ratio, 423
Zoology, 8, 283

Business
Advertising, 124, 426
Billboards, 447
Break-even point, 356
Budget overruns, 110
Call center, 187
Catering, 463
Company's revenue, 292
Copiers, 125, 170, 433, 455
Custom embroidery, 421
Customer satisfaction, 218
Defective radios, 432
Framing, 455
Hairdresser, 187
Jewelry orders, 107
Manufacturing, 351, 599–600
Markup, 111
Monthly business taxes, 87
Online music store, 6–7
Packaging, 27
Printing, 421, 441, 453, 464
Production, 190, 463, 464
Profits and losses, 42
Purchasing tablet computers, 488
Ranching, 529
Restaurant sales, 206
Retail losses due to employee theft
    and error, 301
Retail sales, 197
Returnable bottles, 462
Sales meeting attendance, 360
Sales tax, 360
Selling at an auction, 121
Selling a guitar, 152
Selling a home, 124
Storekeeper requesting change, 126
Ticket sales, 2
Total cost, profit, and revenue, 262

Chemistry
Acid mixtures, 463, 465, 486
Gas volume, 481
Gold, 109, 141
Karat rating, 481
Mass of a hydrogen atom, 240

Metallurgy, 542
Mixing solutions, 464

Construction
Accessible design, 200
Architecture, 357, 419, 422, 475
Blueprints, 150, 419, 422
Bridge construction, 125
Building a deck, 536
Cabin design, 356
Cabinet making, 356
Corner in a building's foundation, 367
Cutting a beam, ribbon, or wire, 149,
    152, 360
Depths of a gutter, 358–359
Designing rafters, 526
Diagonal braces, 356, 366
Dimensions of a porch, 355
Dimensions of a room, 357
Dimensions of lumber, 447
Fencing, 144, 327
Folding sheet metal, 358–359
Grade of a stairway, 201
Hancock Building dimensions, 123
Home restoration, 421
Length of a wire, 540
Masonry, 527
Painting, 286, 420, 466, 486
Paving, 424
Plumbing, 527, 559
Reach of a ladder, 368, 523
Roofing, 117–118, 200, 358, 359, 529, 542
Sanding oak floors, 432
Slope of land, 200
Staining a bookshelf, 430
Two-by-four, 123
Winterizing homes, 202
Wiring, 527

Consumer
Appliances, 391
Banquet costs, 142
Bills, 42, 151, 186
Buying sandwiches, 465
Buying a smartphone, 294
Catering costs, 137–138
Cell-phone budget, 142
Conserving energy and money, 484
Cost, 2, 11, 125, 136, 168–169, 187, 210,
    453, 588
Cost of a fitness club membership, 209

Cost of an overseas internet call, 209
Couponing, 124
Discount, 107–108, 124, 126, 143
eBay purchases, 126
Energy use, 109, 277, 481
Frequent buyer bonus, 143
Furnace repairs, 140
Gasoline purchases, 463
Home maintenance, 424, 527
International calling plans, 455
Mother's Day and Father's Day
    spending, 341
Ordering books, 434
Parking fees, 142, 143
Phone rates, 427
Photography fees, 79
Postage rates, 10
Prepaid calling card, 490
Prices, 143, 150, 171, 465, 599, 601
Purchasing, 79, 115–116
Renting an apartment, 484
Sales and discounts, 145
Spending, 184, 423, 513
Taxi fares, 125, 126, 189, 489
Tipping, 110, 111, 136, 209
Toll charges, 142
Vehicle rental, 125, 182, 185, 455

Economics
Billionaires in China, 341
Consumer's demand, 577
Depreciation, 188, 286, 508, 566
Employment rate, 124
Energy credit, 277
Equilibrium point, 577
Gasoline prices, 42
Income tax deductions, 124
National debt, 185
Natural gas prices, 42, 162
Residential fuel oil prices, 162
Seller's supply, 577
State sales tax, 124
Stock market, 35
Stock prices, 125
Tax bracket, 112
Tax-exempt organizations, 124
Town budget, 124

Education
Answering questions on a quiz, 481
Bachelor's degrees, 160, 226
Chinese and Indian students enrolled in
    U.S. colleges, 150
Class size, 43
College admission tests, 197
College course load, 140
College credits, 462
College enrollment, 99, 107, 181
College graduation, 110
College tuition, 10, 140, 186
Cost of college, 171

Cost per student to visit a museum, 182
Dropout rate, 111
Education debt, 153, 214–215
Final course average, 254
Financial aid, 138–139
First-year students, 107
Foreign student enrollment, 109
Grading, 143, 424
Graduate school, 140
History paper scores, 143
Level of education, 226
Library circulation, 447
Library holdings, 183–184
Math and science instructors, 6
New international students, 150
Online college enrollment, 205–206
Quiz average, 140, 144, 368
Reading assignment, 144
School purchasing a piano, 127
Semester averages, 254
Standard school day, 140
Strong math skills, 108
Student aid, 170
Student enrollment, 37
Student loans and grants, 35, 110
Students enrolled in public schools, 110
Study time, 127
Teachers working in public schools, 110
Test scores, 6, 126
Time spent reading or doing
    homework, 150
Writing an essay, 209

Engineering
Bridge heights, 6
Cell-phone tower, 360
Coal to generate electricity, 301
Coordinates on the globe, 163
Design, 354, 355, 357, 358
Electricity, 481
Furnace output, 99, 102
Grade of a ramp, 200, 201, 226
Grade of a road, 196, 200, 227
Guy wires, 357, 527, 541, 559
Height of an antenna, 524
Height of a telephone pole, 359
Renewable energy sources, 249, 250
Roadway design, 355, 357
Solar capacity, 301
Sound design, 503, 520
Sports engineering, 563
Surveying, 200, 527
Well drilling, 140–141
Width of a quarry, 425
Wind energy, 482
Wiring, 424

Environment
Altitude and temperature, 59, 102
Amount of soil required for a tree, 435,
    477–478

Area of tree canopy, 435, 477–478
Average grade of Longs Peak, 202
Below and above sea level, 29, 35
Coral reefs, 246
Distance from a storm, 99, 579
Elevations, 42–43, 48, 50, 520
Hurricanes, 163, 601
Lake level, 42
Meteorology, 10
National parks, 109
Ozone layer, 161, 163
Paper sent to landfills, 186
Pond depth, 142
Record temperature drop, 171
Recycling, 9
Richat Structure in the Sahara Desert, 559
River lengths, 150
Social Cost of Carbon (SCC), 218
Sorting solid waste, 162
Temperature, 29, 35, 50, 59, 197
Wave height, 303, 356
Wind speed, 303, 356

Finance
Account balance, 42
Banking, 43, 184
Budgeting, 150, 152
Car payments, 140
Coin mixture, 535
Coin value, 464
Compound interest, 101, 286
Credit cards, 40–41, 42, 125
Foreign currency, 10
Household income, 227
Interest rates, 41, 112, 286, 577
Investment, 465, 543, 559, 560, 577, 592
Maximum mortgage payment, 143
Median household income, 600
Money borrowed, 535
Per-capita household debt, 29
Prepaid cards, 466
Retirement account, 124
Savings account, 127, 209, 223
Savings interest, 125

Geometry
Angle measure, 101
Angles of a triangle, 118, 123, 150, 152,
    349, 535
Area of a circular region, 252, 261, 588, 592
Area of a parallelogram, 8–9
Area of a rectangle, 559, 592, 593
Area of a square, 258, 560, 588, 592
Area of a trapezoid, 101
Area of a triangular region, 141, 358
Circumference of a circle, 96, 252
Complementary angles, 125, 446, 447,
    455, 466, 489
Diagonal of a square, 535
Diagonals in a polygon, 555–556, 593
Diameter of a circular region, 559

Dimensions of a box, 359, 367
Dimensions of a leaf, 351–352
Dimensions of a rectangular region, 123, 127, 144, 147, 150, 151, 152, 209, 261, 270, 349, 351, 355, 358, 360, 366, 367, 426, 436, 447, 455, 456, 465, 535, 542, 562, 595
Dimensions of a sail, 355
Dimensions of a state, 447
Dimensions of a triangle, 136, 354, 357, 360, 366, 413, 456
Golden rectangle, 560
Height of a pyramid, 528
Lengths of the sides of a triangle, 127, 136, 151, 354, 359, 567, 595
Maximum dimensions of a postcard, 359
Perimeter of a rectangle, 116–117, 126, 140, 360, 488, 535
Perimeter of a triangle, 122, 140
Right triangles, 357, 557, 559, 592, 595
Sides of a square, 143, 426, 434
Similar triangles, 418, 422, 431, 432
Stacking spheres, 252
Supplementary angles, 125, 445–446, 455
Surface area of a box, 270
Surface area of a cube, 102
Surface area of a right circular cylinder, 98, 279
Surface area of a silo, 283, 286
Volume of a box, 270, 300
Volume of a cube, 270, 292, 318
Volume of a display, 254
Width of a pool sidewalk, 358
Width of the margins in a book, 366

Government
Federal funds rate, 100
Gettysburg Address, 126
President's approval rating, 181

Health/Medicine
Absorption of Ibuprofen, 100
Acid reflux, 359
Aerobic exercise, 227
Basal Metabolic Rate, 71
Blood donors, 299
Blood-alcohol level, 160
Body fat percentage, 111
Body surface area, 564
Body temperature, 141
Calories, 35, 71, 76, 100, 102, 108, 111, 136, 228, 423, 433, 434, 481, 592
Constricted bronchial muscles, 356
Cough syrup, 464
Diabetes, 152
Epinephrine in the bloodstream, 352
Exercise and pulse rate, 155
Fat, 136, 142, 150, 488
Fitness, 120, 196
Heart conditions, 356
Height, 111, 585, 602

Hemoglobin, 595
High cholesterol, 218
Hospital costs, 489
Infant health, 110
Kissing and colds, 111
Length of pregnancy, 601
Life expectancy, 171
Long-term care, 197
Medicine dosage, 102, 566
Nurse practitioner seeing patients, 490
Nutrition, 9, 98, 140, 464
Registered nurses, 209
Sodium content, 111
Treadmill, 200
Vegetarians, 109
Vitamins, 391
Weight gain, 120

Labor
Availability pay, 228
Commission, 144, 189
Compensation package, 489
Cost of self-employment, 110, 145
Cubicle space, 488
Earnings, 154–155, 189, 228
Filing receipts, 434
Hourly rate, 420
Hours worked, 9, 139, 301, 490
Inequality in pay, 111
Job change, 124
Job completion, 481
Jobs for veterinary technicians, 209
Law enforcement, 124, 186
News reporters and correspondents employed in the United States, 197
Overtime pay, 18, 465
Payroll, 426, 465
Proofreading, 185
Raise, 112
Salary, 125, 144, 186, 189, 228
Sharing the workload, 427
Sick leave, 111
Temporary help, 185
Wages, 186, 189, 481
Work time, 8
Working alone, 420, 424
Working together, 414–415, 424, 420, 421, 425, 430, 432, 434, 542

Miscellaneous
African artistry, 390
Ages, 11, 318
Air purifying, 421
Apartment numbers, 122
Archaeology, 357, 528
Aromatherapy, 464
Autoharp strings, 463
Baking, 455
Beard growth, 413
Blending coffee, 459–460, 463, 465
Caged rabbits and pheasants, 456

Cake recipe, 588
Candy blends, 595
Cats adopted through FACE clinic, 149
Celebrity birthdays, 584
Comic-book values, 43
Converting temperature, 567, 585
Cooking, 489
Decorating, 526, 528
Decorating a cake, 420
Elevators, 186, 471
Emergency communications, 523–524
Filling a freighter with oil, 424
Filling a pool, bog, tub, or tank, 421, 424, 426, 481
Flagpoles, 560
Food, 109, 123, 126
Furniture, 355
Gambling, 51
Gold chain, 481
Hands on a clock, 424
House numbers, 490
Household pets, 109, 142
Ice cream cones, 462
Law enforcement, 124
Left-handed females, 111
Length of a rope swing, 540
Light bulbs, 423
Mayan symbols, 68
Mixing nuts, 461, 463, 482
Mowing lawns, 420
Newspapers, 218
Oatmeal servings, 481
Page numbers, 122, 127, 355
Paint colors, 460–461
Photography, 79, 112, 115–116, 423
Pottery, 186
Power outages, 50–51
Preparing appetizers, 425
Pumping water, 421
Quilting, 424
Serving rate, 223
Shoveling time, 535
Shredding, 421
Sighting to the horizon, 517–518, 519, 541, 595
Sizes of envelopes, 141
Sizes of packages, 141
Snow removal, 186
Spam e-mail, 111, 300
Stacking cannonballs, 253
Street addresses, 122
Ticket window line, 489

Physics
Acoustics, 482
Altitude of a launched object, 283–284
Distance of an accelerating object, 564
Falling objects, 503, 556, 558, 561, 562, 592
Gravitational force, 547
Height of a rocket, 358
Lengths and cycles of a pendulum, 536

Lighting, 482
Musical tones, 481
Period of a swinging pendulum, 508, 509, 520
Pressure at sea depth, 585
Speed of sound, 497
Speed of surface waves, 519
Stone tossed down from a cliff, 366
Temperature and speed of sound, 519
Temperature as a function of depth, 585
Water flow from a fire hose, 503
Wavelength of a musical note, 152

Social Sciences
Age, 456
Average minutes per month men talk on cell phones, 368
Charitable organizations, 120–121
Community service, 479–480
Crowd size, 482
Event planning, 96
Fundraising, 147
Handshakes, 356, 367
High-fives, 356
Toasting, 356
Volunteering, 108, 225, 420, 480

Sports/Entertainment
Baseball, 109, 140, 527
Basketball, 1, 35, 50, 116–117, 126, 152, 436, 457–458, 462, 464, 466, 488
Batting average, 110, 432
Battleship® game, 222
Bicycle race, 143, 200
Bicycle tour, 476
Biggest box office flops, 122
Chess, 9, 101
Cliff jumping, 558
Concert tickets, 96, 151
Concerts, 99
Cross country skiing, 421
Designing a tournament, 362
Disneyland admissions, 463
Enlarged strike zone, 561
Games in a sports league, 100, 356
Golfing, 43
Hang time, 497
Hank Aaron, 601
Height of a golf ball, 576
Hiking, 113–114
Hockey, 8, 471
Indy car racing, 122
Jersey numbers of pro football hall-of-famers, 584
Jogging speed, 430
Kayaking, 122, 252
Lacrosse, 447, 527
Literature, 465
Mountaineering, 186
Movie collections, 144

Museum admissions, 463
Music, 140, 463
NASCAR, 109, 122
Obstacle course, 535
Open Championship golfer, 35
Pass completions, 110
Path of the Olympic arrow, 253
PBA scores, 601
Price of a movie ticket, 142
Quilting, 475
Race numbers, 350
Race time, 11, 126
Racquetball, 447, 529
Roller blading, 527
Running, 123, 189, 227, 417, 471
Running records, 142
Sailing, 358
SCAD diving, 252
Scrabble®, 125
Skateboarding, 201, 558
Skiing, 142, 195–196, 201
Skydiving, 252
Sled dog racing, 122
Soccer, 447, 527
Softball, 448, 527
"Sweet spot" of a tennis racquet, 563
Swimming, 111
Team's magic number, 294
Tour de France, 200
Triathlon training, 188–189
TV time, 535
Unicycling, 114
Video viewing, 441
Walking, 183, 421, 430, 471
Women's softball, 9
World running records, 593
Yardage gained and lost in football, 35, 42
Yellowstone Park admissions, 462
Zoo admissions, 463

Statistics/Demographics
Age at marriage, 448
Aging population, 219
Areas of Alaska and Arizona, 111
Birth rate among teenagers in the United States, 197
Home burglaries, 111
Longest marriage, 122
Oldest bride, 122
Population, 144, 188
U.S. Census Bureau, 424
Urban population, 219

Technology
Blogging, 196
Cells in a spreadsheet, 286
Computer games, 218
Cost of computers, 190–192, 203
Downloading a song or television show, 478

E-mail, 122
Flash drives, 423
FLASH ROM on a TI-84 Plus graphing calculator, 237
International messaging, 463
iPad apps, 171
Laser to move data, 245
Movie downloads, 542
Music streaming, 448
Paid subscriptions for digital music services, 368
Printers, 391
Smartphone users worldwide, 341
Technology grant, 35
Text messaging, 160, 419
TV screen dimensions, 536
Value of a color copier, 170
Value of computer software, 171
Video games, 109, 218
Website growth rate, 182
Wireless routers, 528

Transportation
Airplane wing aspect ratio, 423
Alternative-fueled buses, 595
Altitude of an airplane, 518
Average speed, 424
Aviation, 189, 529
Bicycling, 106–107, 151, 172, 426, 535
Boat speed, 422
Boating, 123
Bus schedules, 391
Car speed, 432, 433, 478
Chartering a boat, 481
Commercial pilots, 190
Commuting, 123, 424
Cost of road service, 141
Cycling distance, 426
Driving rate/speed, 118–119, 182, 226, 417, 421
Driving time, 95, 119, 123
Hot-air balloon, 518
Interstate mile markers, 114–115
Local light rail service, 151
Miles driven, 95, 423, 529
Moped speed, 422
Motorcycle's gas consumption, 189
Navigation, 189, 200
Number of drivers, 10
Parking spaces, 355, 495, 496–497
Plane travel, 186
Speed of a bus, 183
Speed of bicyclists, 416–417, 421, 426, 430
Speed of travel, 417, 421
Speeds of a dog sled and snowmobile, 490
Time and speed, 478
Tractor speed, 422
Train speeds, 421
Train travel, 186, 187, 422
Travel to work, 9